Dictionary of Environmental Legal Terms

Dictionary of Environmental Legal Terms

Edited by

C. C. Lee, Ph.D.

Research Program Manager
National Risk Management Research Laboratory
U.S. Environmental Protection Agency
Cincinnati, Ohio 45268

McGraw-Hill

New York San Francisco Washington, D.C. Auckland Bogotá
Caracas Lisbon London Madrid Mexico City Milan
Montreal New Delhi San Juan Singapore
Sydney Tokyo Toronto

Library of Congress Cataloging-in-Publication Data

Lee, C. C.
   Dictionary of environmental legal terms / written and edited by
C. C. Lee.
      p.   cm.
   Includes bibliographical references and index.
   ISBN 0-07-038113-5 (alk. paper)
   1. Environmental law—United States—Dictionaries. I. Title.
KF3775.A68L44 1996
344.73'046'03—dc20
[347.3044603]   96-25123
                                                  CIP

McGraw-Hill

A Division of The McGraw-Hill Companies

*The sponsoring editor for this book was Robert Esposito, the editing
supervisor was Patricia V. Amoroso, and the production supervisor
was Donald F. Schmidt.*

Printed and bound by Quebecor.

This book was written and edited by Dr. C. C. Lee in his private capacity.
No official support or endorsement by the U.S. Environmental Protection
Agency is intended nor should be inferred.

This book is printed on recycled, acid-free paper containing a
minimum of 50% recycled de-inked fiber.

McGraw-Hill books are available at special quantity discounts to use as
premiums and sales promotions, or for use in corporate training programs. For more information, please write to the Director of Special
Sales, McGraw-Hill, 11 West 19th Street, New York, NY 10011. Or contact your local bookstore.

Contents

Preface   vii

Definitions   1

Appendix: Environmental Acronyms   771

References   816

Major Environmental Laws and Environmental Regulations   818

Preface

It has been well recognized that environmental laws and regulations are the driving forces behind environmental protection. However, one of the major frustrations in dealing with environmental problems is locating exact definitions of various regulatory terms. Official regulatory definitions are contained in each of the federal environmental laws and in 40 CFR (Code of Federal Regulations) from Part 1 to Part 1517. Other CFRs, such as 10 CFR and 29 CFR, also contain some environmentally related definitions. The huge volume of CFRs and various environmental laws make information searches extremely difficult. To help resolve this difficulty, this dictionary, a reference tool for environmental problems, was created.

This dictionary systematically compiles countless definitions of environmental terms. It focuses on the definitions of environmental laws and regulations. Other information, such as engineering-related and health-related environmental terms, are also included. Major subjects covered are:

1. Statutory definitions from all major environmental laws such as the Clean Air Act, the Clean Water Act, the Safe Drinking Water Act, the Resource Conservation and Recovery Act, the Toxic Substances Control Act, and the Superfund Act;

2. Regulatory definitions from the entire 40 CFR from Part 1 to Part 1517;

3. Health-related environmental definitions from 29 CFR;

4. Radioactive waste-related environmental definitions from 10 CFR;

5. Common environmental engineering definitions from EPA publications;

6. In addition to the above-mentioned definitions, representative chemical listings from each major environmental law and regulation are provided. These listings allow readers to understand both the breadth and complexity of environmental control activities, and which chemicals are regulated by what regulations. The selected chemical listings are:

   - CAA-regulated compounds listed under the term "hazardous air pollutants."

- CERCLA-regulated compounds listed under the term "extremely hazardous substance."

- CWA-regulated compounds listed under the term "toxic pollutants."

- FIFRA-regulated compounds listed under the term "pesticide chemicals."

- RCRA-regulated compounds listed under the term "hazardous constituent."

- SDWA-regulated compounds listed under the term "drinking water supply."

- TSCA-regulated compounds listed under the term "significant new use notice."

- Radionuclide compounds are provided under the term "radionuclide/radioisotope."

7. Environmental acronyms including most abbreviations of EPA offices (Appendix); and

8. References which provide the sources of information used.

A special feature of this dictionary is the citing of references for each of the terms collected. This allows readers to search for more information, if needed. This dictionary can expeditiously assist users in locating needed definitions at their fingertips. It provides not only exact official (EPA) definitions, but also the definition's origin. This dictionary is intended to be a reference book for those who are involved in the environmental protection of air, water, and land resources. It is an essential tool and will make many environmental jobs much easier.

ABOUT THE AUTHOR

Dr. C. C. Lee is the Medical Waste Research Coordinator and a Research Program Manager at the National Risk Management Research Laboratory of the U.S. Environmental Protection Agency in Cincinnati, Ohio. In addition, he is currently a member of the Policy Review Group to the Center for Clean Technology at the University of California, Los Angeles (UCLA). He is also the Chairman of the Sponsoring Committee to the International Congress on Toxic Combustion By-Product (ICTCB). He initiated the ICTCB and served as the Chairman of the First and Second Congresses, which were held in 1989 and 1991, respectively.

Dr. Lee has more than 20 years of experience in conducting various engineering and research projects, which often involve multi-environmental issues ranging from clean air and clean water control to solid waste disposal. He has been recognized as a worldwide expert in the thermal treatment of medical and hazardous wastes, and lead discussions on medical waste disposal technologies at a meeting conducted by the Congressional Office of Technology Assessment. Also, at the initiation of the U.S. State Department, he served as head of the U.S. delegation to the Conference on "National Focal Points for the Low- and Non-Waste Technology" (sponsored by the United Nations and held in Geneva, Switzerland, on August 28–30, 1978). He has been invited to lecture on various issues regarding solid waste disposal in numerous national and international conferences, and he has published more than 100 papers and reports in various environmental areas.

He received a B.S. from National Taiwan University in 1964, and an M.S. and Ph.D. from North Carolina State University in 1968 and 1972, respectively. Before joining EPA in 1974, he was an Assistant Professor at North Carolina State University.

**Dictionary of
Enviromental
Legal Terms**

************ AAAAA ************

1/R² correction (40CFR60-AA (alt. method 1)): The correction made for the systematic decrease in lidar backscatter signal amplitude with range.

11-AA (40CFR704.25-1) means the chemical substance 11-aminoundecanoic acid, CAS Number 2432-997.

1987 Montreal Protocol (40CFR82.3-y) means the Montreal Protocol, as originally adopted by the parties in 1987.

7Q10 (EPA-94/04): Seven-day, consecutive low flow with a ten-year return frequency; the lowest stream flow for seven consecutive days that would be expected to occur once in ten years.

A (40CFR786.1050-ii) means the symbol of absorbance (optical density) under the definition of UV-VIS absorption spectrum of a solution (see also 40CFR796.3700-ii; 796.3780-ii; 796.3800-ii).

A, as in "cyanide A" (40CFR433.11-b) shall mean amenable to alkaline chlorination.

A/B (40CFR60.471): See afterburner.

A/E services (40CFR33.005; 35.6015): See architectural or engineering services.

A-scale sound level (EPA-94/04): A measurement of sound approximating the sensitivity of the human ear, used to note the intensity or annoyance of sounds.

AALACS (40CFR300-App/A): See ambient aquatic life advisory concentrations.

AAP (40CFR205.151): See acoustical assurance period.

abandoned mine (40CFR434.11-r) means a mine where mining operations have occurred in the past and:

(1) The applicable reclamation bond or financial assurance has been released or forfeited or
(2) If no reclamation bond or other financial assurance has been posted, no mining operations have occurred for five years or more.

abandoned well (40CFR146.3) means a well whose use has been permanently discontinued or which is in a state of disrepair such that it cannot be used for its intended purpose or for observation purposes.

abandoned well (EPA-94/04) means a well whose use has been permanently discontinued or which is in a state of disrepair such that it cannot be used for its intended purpose.

abatement (EPA-94/04): Reducing the degree or intensity of, or eliminating, pollution.

abnormally treated vehicle (40CFR86.085.2) means any diesel light-duty vehicle or diesel light-duty truck that is operated for less than five miles in a 30 day period immediately prior to conducting a particulate emissions test.

aboveground release (40CFR280.12) means any release to the surface of the land or to surface water. This includes, but is not limited to, releases from the aboveground portion of an UST system and aboveground releases associated with overfills and transfer of operations as the regulated substance moves to or from an UST system.

aboveground storage facility (40CFR113.3-a) means a tank or other container, the bottom of which is on a plane not more than 6 inches below the surrounding surface.

aboveground tank (40CFR260.10) means a device meeting the definition of "tank" in 40CFR260.10 and that is situated in such a way that the entire surface area of the tank is completely above the plane of the adjacent surrounding surface and the entire surface area of the tank (including the tank bottom) is able to be visually inspected.

abrasive (29CFR1910.94a) means a solid substance used in an abrasive blasting operation.

abrasive blasting (29CFR1910.94a) means the

forcible application of an abrasive to a surface by pneumatic pressure, hydraulic pressure, or centrifugal force.

abrasive blasting respirator (29CFR1910.94a) means a continuous flow air-line respirator constructed so that it will cover the wearer's head, neck, and shoulders to protect him from rebounding abrasive.

abrasive cutting-off wheels (29CFR1910.94b) means organic-bonded wheels, the thickness of which is not more than one forty-eighth of their diameter for those up to, and including, 20 inches in diameter, and not more than one-sixtieth of their diameter for those larger than 20 inches in diameter, used for a multitude of operations variously known as cutting, cutting off, grooving, slotting, coping, and jointing, and the like. The wheels may be "solid" consisting of organic-bonded abrasive material throughout, "steel centered" consisting of a steel disc with a rim of organic-bonded material molded around the periphery, or of the "inserted tooth" type consisting of a steel disc with organic-bonded abrasive teeth or inserts mechanically secured around the periphery.

absorbance (A) (40CFR796.3700-ii) is defined as the logarithm to the base 10 of the ratio of the initial intensity (I_o) of a beam of radiant energy to the intensity (I) of the same beam after passage through a sample at a fixed wavelength. Thus, $A = \log_{10}(I_o/I)$ (other identical or similar definitions are provided in 40CFR796.3780-ii; 796.3800-ii).

absorbed dose (EPA-94/04): The amount of a chemical that enters the body of an exposed organism.

absorption (EPA-94/04): The uptake of water or dissolved chemicals by a cell or an organism (as tree roots absorb dissolved nutrients in soil).

absorption toxicokinetics (40CFR795.235-b) refers to the bioavailability, i.e., the rate and extent of absorption of the test substance, and metabolism and excretion rates of the test substance after absorption (under TSCA).

academic year (CWA112-33USC1262) means an academic year or its equivalent, as determined by the Administrator.

accelacota (40CFR52.741) means a pharmaceutical coating operation which consists of a horizontally rotating perforated drum in which tablets are placed, a coating is applied by spraying, and the coating is dried by the flow of air across the drum through the perforations.

acceleration test procedure (40CFR205.151-5) means the measurement methodologies specified in Appendix I.

accelerator pump (plunger or diaphragm) (40CFR85.2122(a)(3)(ii)) means a device used to provide a supplemental supply of fuel during increasing throttle opening as required.

acceptable of a batch (40CFR204.51-p) means that the number of non-complying compressors in the batch sample is less than or equal to the acceptance number as determined by the appropriate sampling plan.

acceptable quality level (40CFR204.51-i) means the maximum percentage of failing compressors that, for purposes of sampling inspection, can be considered satisfactory as a process average.

acceptable quality level (AQL) (40CFR205.151-6) means the maximum allowable average percentage of vehicles or exhaust systems that can fail sampling inspection under a Selective Enforcement Audit.

acceptable quality level (AQL) (40CFR86.1002.2001-1) means the maximum percentage of failing engines or vehicles that, for purposes of sampling inspection, can be considered satisfactory as a process average.

acceptable quality level (AQL) (40CFR86.1002.84) means the maximum percentage of failing engines or vehicles that, for purposes of sampling inspection, can be considered satisfactory as a process average.

acceptable quality level (AQL) (40CFR86.602.84-1) means the maximum percentage of failing vehicles that, for purposes of sampling inspection, can be considered satisfactory as a process average (other identical or similar definitions are provided in 40CFR205.51-1).

acceptable quality level (AQL) (40CFR89.502.96)

means the maximum percentage of failing engines that can be considered a satisfactory process average for sampling inspections.

acceptance date (40CFR195.2) means the date on which EPA enters the application into the data system.

acceptance of a batch (40CFR205.51-2) means that the number of noncomplying vehicles in the batch sample is less than or equal to the acceptance number as determined by the appropriate sampling plan.

acceptance of a batch sequence (40CFR204.51-r) means that the number of rejected batches in the sequences is less than or equal to the acceptance number as determined by the appropriate sampling plan.

acceptance of a batch sequence (40CFR205.51-10) means that the number of rejected batches in the sequence is less than or equal to the acceptance number as determined by the appropriate sampling plan.

acceptance of a compressor (40CFR204.51-v) means that the measured noise emissions of the compressor, when measured in accordance with the applicable procedure, conforms to the applicable standard.

acceptance of a vehicle (40CFR205.51-25) means that the measured emissions of the vehicle, when measured in accordance with the applicable procedure, conforms to the applicable standard.

accepted application (40CFR195.2) refers to an application that has been entered into the data system.

access (29CFR1910.20) means the right and opportunity to examine and copy.

accessible (40CFR763.83) when referring to ACM (asbestos containing material) means that the material is subject to disturbance by school building occupants or custodial or maintenance personnel in the course of their normal activities.

accessible environment (40CFR191.12) means:

(1) The atmosphere;
(2) land surfaces;
(3) surface waters;
(4) oceans; and
(5) all of the lithosphere that is beyond the controlled area.

accident (40CFR171.2-1) means an unexpected, undesirable event, caused by the use or presence of a pesticide, that adversely affects man or the environment.

accident site (EPA-94/04): The location of an unexpected occurrence, failure or loss, either at a plant or along a transportation route, resulting in a release of hazardous materials.

accidental occurrence (40CFR264.141-g) means an accident, including continuous or repeated exposure to conditions, which results in bodily injury or property damage neither expected nor intended from the standpoint of the insured (other identical or similar definitions are provided in 40CFR265.141-g).

accidental release (40CFR280.92) means any sudden or nonsudden release of petroleum from an underground storage tank that results in a need for corrective action and/or compensation for bodily injury or property damage neither expected nor intended by the tank owner or operator.

acclimation (40CFR797.1400-1) means the physiological compensation by test organisms to new environmental conditions (e.g., temperature, hardness, pH) (other identical or similar definitions are provided in 40CFR797.1520-1).

acclimation (40CFR797.1600-1) means the physiological or behavioral adaptation of organisms to one or more environmental conditions associated with the test method (e.g., temperature, hardness, pH).

acclimation (40CFR797.1830-1) is the physiological compensation by test organisms to new environmental conditions (e.g., temperature, salinity, pH).

acclimation (40CFR797.2050-1) is the physiological or behavioral adaptation of test animals to environmental conditions and basal diet associated with the test (other identical or similar definitions are provided in 40CFR797.2175-1).

acclimation (40CFR797.2130-i) means the physiological and behavioral adaptation to environmental conditions (e.g., housing and diet) associated with the test procedure (cf. accilimization) (other identical or similar definitions are provided in 40CFR797.2150-i).

acclimatization (EPA-94/04): The physiological and behavioral adjustments of an organism to changes in its environment.

account number (40CFR72.2) means the identification number given by the Administrator to each Allowance Tracking System account pursuant to 40CFR73.31(d) of this chapter.

accreditation (40CFR763.83): See accredited.

accredited asbestos contractor (TSCA202-15USC2642) means a person accredited pursuant to the provisions of section 206 of this title.

accredited or accreditation (40CFR763.83) when referring to a person or laboratory means that such person or laboratory is accredited in accordance with section 206 of Title II of the Act.

accrual date (40CFR14.2-e) means the date of the incident causing the loss or damage or the date on which the loss or damage should have been discovered by the employee through the exercise of reasonable care.

accrued expenditures (40CFR31.3) mean the charges incurred by the grantee during a given period requiring the provision of funds for:
(1) Goods and other tangible property received;
(2) services performed by employees, contractors, subgrantees, subcontractors, and other payees; and
(3) other amounts becoming owed under programs for which no current services or performance is required, such as annuities, insurance claims, and other benefit payments.

accrued income (40CFR31.3) means the sum of:

(1) Earnings during a given period from services performed by the grantee and goods and other tangible property delivered to purchasers, and
(2) amounts becoming owed to the grantee for which no current services or performance is required by the grantee.

accumulated speculatively (40CFR261.1-8): A material is "accumulated speculatively" if it is accumulated before being recycled. A material is not accumulated speculatively, however, if the person accumulating it can show that the material is potentially recyclable and has a feasible means of being recycled; and that--during the calendar year (commencing on January 1)--the amount of material that is recycled, or transferred to a different site for recycling, equals at least 75 percent by weight or volume of the amount of that material accumulated at the beginning of the period. In calculating the percentage of turnover, the 75 percent requirement is to be applied to each material of the same type (e.g., slags from a single smelting process) that is recycled in the same way (i.e., from which the same material is recovered or that is used in the same way). Materials accumulating in units that would be exempt from regulation under 40CFR261.4(c) are not be included in making the calculation. (Materials that are already defined as solid wastes also are not to be included in making the calculation.) Materials are no longer in this category once they are removed from accumulation for recycling, however.

accumulator (40CFR52.741) means the reservoir of a condensing unit receiving the condensate from a surface condenser.

accuracy (40CFR86.082.2) means the difference between a measurement and true value.

ACFM (EPA-94/04): Actual cubic feet per minute.

acid deposition (EPA-94/04): A complex chemical and atmospheric phenomenon that occurs when emissions of sulfur and nitrogen compounds and other substances are transformed by chemical processes in the atmosphere, often far from the original sources, and then deposited on earth in either a wet or dry form. The wet forms, popularly called "acid rain," can fall as rain, snow, or fog. The dry forms are acidic gases or particulates.

acid gas (40CFR60.641) means a gas stream of hydrogen sulfide (H_2S) and carbon dioxide (CO_2) that has been separated from sour natural gas by a sweetening unit.

acid mine drainage (EPA-94/04): Drainage of water from areas that have been mined for coal of other mineral ores. The water has a low pH because of its contact with sulfur bearing material and is harmful to aquatic organisms.

acid mist (40CFR60.81-b) means sulfuric acid mist, as measured by Method 8 of appendix A to this part or an equivalent or alternative method.

acid neutralizing capacity (EPA-94/04): Measure of ability of water or soil to resist changes in pH.

acid or ferruginous mine drainage (40CFR434.11-a) means mine drainage which, before any treatment, either has a pH of less than 6.0 or a total iron concentration equal to or greater than 10 mg/L.

acid rain (EPA-94/04): See acid deposition.

acid rain compliance option (40CFR72.2) means one of the methods of compliance used by an affected unit under the Acid Rain Program as described in a compliance plan submitted and approved in accordance with subpart D of this part, part 74 of this chapter or part 76 of this chapter.

acid rain emissions limitation (40CFR72.2) means:
(1) For the purposes of sulfur dioxide emissions:
   (i) The tonnage equivalent of the allowances authorized to be allocated to an affected unit for use in a calendar year under section 404(a)(1) and (a)(3) of the Act, the basic Phase II allowance allocations authorized to be allocated to an affected unit for use in a calendar year, or the allowances authorized to be allocated to an opt-in source under section 410 of the Act for use in a calendar year;
   (ii) As adjusted:
     (A) By allowances allocated by the Administrator pursuant to section 403, section 405 (a)(2), (a)(3), (b)(2), (c)(4), (d)(3), and (h)(2), and section 406 of the Act;

     (B) By allowances allocated by the Administrator pursuant to subpart D of this part; and thereafter
     (C) By allowance transfers to or from the compliance subaccount for that unit that were recorded or properly submitted for recordation by the allowance transfer deadline as provided in 40CFR73.35 of this chapter, after deductions and other adjustments are made pursuant to 40CFR73.34(c) of this chapter; and
(2) For purposes of nitrogen oxides emissions, the applicable limitation established by regulations promulgated by the Administrator pursuant to section 407 of the Act, as modified by an Acid Rain permit application submitted to the permitting authority, and an Acid Rain permit issued by the permitting authority, in accordance with regulations implementing section 407 of the Act.

acid rain emissions reduction requirement (40CFR72.2) means a requirement under the Acid Rain Program to reduce the emissions of sulfur dioxide or nitrogen oxides from a unit to a specified level or by a specified percentage.

acid rain permit or permit (40CFR72.2) means the legally binding written document, or portion of such document, issued by a permitting authority under this part (following an opportunity for appeal pursuant to part 78 of this chapter or any State administrative appeals procedure), including any permit revisions, specifying the Acid Rain Program requirements applicable to an affected source, to each affected unit at an affected source, and to the owners and operators and the designated representative of the affected source or the affected unit.

acid rain program (40CFR72.2) means the national sulfur dioxide and nitrogen oxides air pollution control and emissions reduction program established in accordance with title IV of the Act, this part, and parts 73, 74, 75, 76, 77, and 78 of this chapter (under CAA).

acid recovery (40CFR420.91-g) means those sulfuric acid pickling operations that include processes for recovering the unreacted acid from spent pickling acid solutions.

acid regeneration (40CFR420.91-h) means those hydrochloric acid pickling operations that include processes for regenerating acid from spent pickling acid solutions.

acidic (EPA-94/04): The condition of water or soil that contains a sufficient amount of acid substances to lower the pH below 7.0.

acoustic descriptor (40CFR211.102-e) means the numeric, symbolic, or narrative information describing a product's acoustic properties as they are determined according to the test methodology that the Agency prescribes.

acoustical assurance period (AAP) (40CFR205.151-7) means a specified period of time or miles driven after sale to the ultimate purchaser during which a newly manufactured vehicle or exhaust system, properly used and maintained, must continue in compliance with the Federal standard.

acquisition cost of an item of purchased equipment (40CFR31.3) means the net invoice unit price of the property including the cost of modifications, attachments, accessories, or auxiliary apparatus necessary to make the property usable for the purpose for which it was acquired. Other charges, such as the cost of installation, transportation, taxes, duty or protective in-transit insurance, shall be included or excluded from the unit acquisition cost in accordance with the grantee's regular accounting practices.

acrylic fiber (40CFR60.601) means a manufactured synthetic fiber in which the fiber-forming substance is any long-chain synthetic polymer composed of at least 85 percent by weight of acrylonitrile units.

act (29CFR1910.2) means the Williams Steiger Occupational Safety and Health Act of 1970 (84 Stat. 1590).

act (Pub. L. 93-523) means the Public Health Service Act, as amended by the Safe Drinking Water Act (other related information is provided in 40CFR141.2; 142.2; 149.101-a).

act (40CFR1508.2) means the National Environmental Policy Act, as amended (42 USC. 4321, et seq.) which is also referred to as NEPA.

act (40CFR17.2-a) means section 504 of Title 5, United States Code, as amended by section 203(a)(1) of the Equal Access to Justice Act, Pub. L. No. 96-481.

act (40CFR192.00) for purposes of subparts A, B, and C of this part, means the Uranium Mill Tailings Radiation Control Act of 1978.

act (40CFR2.309-1) means the Marine Protection, Research, and Sanctuaries Act of 1972, as amended (33USC1401) (other identical or similar definitions are provided in 40CFR220.2-a).

act (40CFR2.311-1) means the Motor Vehicle Information and Cost Savings Act, as amended, 15USC1901 et seq (other identical or similar definitions are provided in 40CFR600.002.85-1).

act (40CFR22.03) means the particular statute authorizing the institution of the proceeding at issue (other identical or similar definitions are provided in 40CFR123.64).

act of God (CWA311-33USC1321) means an act occasioned by an unanticipated grave natural disaster.

act of God (SF101-42USC9601) means an unanticipated grave natural disaster or other natural phenomenon of an exceptional, inevitable, and irresistible character, the effects of which could not have been prevented or avoided by the exercise of due care or foresight.

act or AEA (10CFR20.3) means the Atomic Energy Act of 1954 (68 Stat. 919), including any amendments thereto (other related information is provided in 10CFR30.4; 40.4; 70.4).

act or CAA (42USC7401-7626) means the Clean Air Act of 1970. The Act:
• Is to protect and enhance the quality of the Nation's air resources so as to promote the public health and welfare and the productive capacity of its population.

- Consists of Public Law 159 (July 14, 1955) and subsequent amendments including: CAAA (Clean Air Act Amendments of 1990).

(Other related information is provided in CFR2.301-1; 50.1-b; 51.491; 52.741; 53.1-b; 56.1; 51.100-a; 51.491; 52.741; 57.103-a; 58.1-a; 60.2; 61.02; 63.2; 63.71; 63.101; 63.191; 65.01-a; 66.3-a; 70.2; 72.2; 79.2-a; 80.2-a; 80.2-a; 82.172; 81.1-a; 85.1801-a; 85.1902-a; 85.2102-1; 85.2113-a; 82.285.1502-1; 86.0; 86.402.78; 87.1; 93.101; CAA112.)

act or CERCLA (42USC9601-9657) means the Comprehensive Environmental Response, Compensation, and Liability Act of 1980. The Act:
- Is commonly known as Superfund.
- Provides for liability, compensation, cleanup, and emergency response for hazardous substances released into the environment and the cleanup of inactive hazardous waste disposal sites.
- Created a $1.6 billion Hazardous Substance Response Trust Fund "Superfund" from a special tax on crude oil and commercial chemicals.
- Allowed the EPA to use the money in Superfund to investigate and clean up abandoned or uncontrolled hazardous waste sites. The EPA can either pay for the site cleanup itself or take legal action to force the parties responsible for the contamination to pay for the cleanup.
- Required owners and operators of vessels or facilities handling hazardous wastes to show evidence of financial responsibility. This provision ensures that if a hazardous waste is released, the responsible person can pay the costs of removing the contaminant and restoring damaged natural resources. Persons responsible for the release are liable for all costs incurred from the cleanup and restoration of the environment. Only when a financially responsible defendant cannot be found will the Superfund absorb the costs of removing the released hazardous waste costs.

The amendments of the CERCLA include:
- SARA (Superfund Amendments and Reauthorization Act of 1986).
- EPCR (Emergency Planning and Community Right-to-Know of 1986)
- RGIAQ (Radon Gas and Indoor Air Quality Research of 1986).

(Other related information is provided in 40CFR2.310-1; 35.6015-6; 280.12; 300.5; 300-AA; 302.3; 304.12-a; 355.20; 372.3.)

act or CWA (Pub. L. 92-500) means the Clean Water Act (formerly referred to as the Federal Water Pollution Control Act or Federal Water Pollution Control Act Amendments of 1972), 33USC1251 et seq. The Act:
- Is to restore and maintain the chemical, physical, and biological integrity of the Nation's waters.
- Requires EPA to establish a system of national effluent standards for major water pollutants.
- Requires all municipalities to use secondary sewage treatment by 1988.
- Sets interim goals of making all U.S. waters safe for fishing and swimming.
- Allows point source discharges of pollutants into waterways only with a permit from EPA.
- Requires all industries to use the best practicable technology (BPT) for control of conventional and non-conventional pollutants and to use the best available technology (BAT) that is reasonable or affordable.

CWA has five main elements:
- A permit program.
- A system of minimum national effluent standards for each industry.
- Water quality standards.
- Provisions for special problems such as toxic chemicals and oil spills.
- A construction grant program for publicly owned treatment works (POTWs).

CWA established the national goals to:
- Achieve a level of water quality which provides for the protection and propagation of fish, shellfish, and wildlife and for recreation in and on the water by July 1, 1983.
- Eliminate the discharge of pollutants into United States waters by 1985.

CWA regulated the discharge of 65 categories of priority pollutants [including at least 129 specific chemical substances (see Appendix 1)] by 34 industry:
1. adhesives and sealants.
2. aluminum forming.
3. asbestos manufacturing.
4. auto and other laundries.
5. battery manufacturing.
6. coal mining.

7. coil coating.
8. copper forming.
9. electric and electronic components.
10. electroplating.
11. explosives manufacturing.
12. ferroalloys.
13. foundries.
14. gum and wood chemicals.
15. inorganic chemicals manufacturing.
16. iron and steel manufacturing.
17. leather tanning and finishing.
18. mechanical products manufacturing.
19. nonferrous metals manufacturing.
20. ore mining.
21. organic chemicals manufacturing.
22. pesticides.
23. petroleum refining.
24. pharmaceutical preparations.
25. photographic equipment and supplies.
26. plastic and synthetic materials anufacturing.
27. plastic processing.
28. porcelain enamelling.
29. printing and publishing.
30. pulp and paperboard mills.
31. soap and detergent manufacturing.
32. steam electric power plants.
33. textile mills.
34. timber products processing (cf. existing stationary facility).

(Other related information is provided in 40CFR2.302-1; 35.905; 35.2005-1; 35.1605-1; 51.392; 122.2; 124.2; 125.2; 130.2-a; 131.3-a; 133.101; 233.2; 136.2-a; 230.3-a; 270.2; 406.61-c; 501.2; 503.9.)

act or ERDDAA (42USC1857 et seq.) means the Environmental Research, Development, and Demonstration Authorization Act of 1980: The Act authorized research, development and demonstration activities in the areas of air, water, solid waste, pesticides, toxic substances and radiation.

act or FFDCA (21USC301-392) means the Federal Food, Drug and Cosmetic Act, as amended (other related information is provided in 40CFR2.308-1160.3; 163.2-a; 177.3; 178.3; 179.3; 710.2).

act or FIFRA (7USC136 et seq.) means the Federal Insecticide, Fungicide and Rodenticide Act of 1947, as amended, and its predecessor, 7USC135 et seq.
- In 1910, Congress passed the Insecticide Act to regulate the manufacture of insecticide, Paris green, lead arsenate or fungicide. In 1947, Congress replaced the Insecticide Act with the more comprehensive FIFRA to regulate economic poisons which include not only insecticides and fungicides, but also rodenticides, herbicides, and preparations intended to control other forms or pests which were not subject to the Insecticide Act.
- Major subsequent amendments include:
  - FEPCA (Federal Environmental Pesticide Control Act) of 1972.

(Other related information is provided in 40CFR2.307-1; 152.3-a; 154.3-a; 164.2-a; 165.1-a; 166.3-a; 167.3; 160.3; 152.3-a; 171.2-2; 172.1-a; 177.3; 179.3.)

act or FWPCA (33USC1151, et seq.) means the Federal Water Pollution Control Act of 1972.
- The objective of this FWPCA was to restore and maintain the chemical, physical, and biological integrity of the Nation's waters. The Act was amended by the Clean Water Act of 1977 and is commonly known as the Clean Water Act.
- Major subsequent amendments include:
  - CWA (Clean Water Act) of 1977
  - WQA (Water Quality Act) of 1987
  - Ocean Dumping Ban Act of 1988
  - OPA (Oil Pollution Act) of 1990.

(Other related information is provided in 40CFR21.2-e; 20.2-a; 35.3105-a; 39.105-a; 104.2-a; 108.2-a; 110.1; 113.3-b; 116.3; 121.1-f; 129.2; 220.2; 401.11-a; 403.3-b.)

act or HSWA (42USC6901 et seq.) means the Hazardous and Solid Waste Amendment of 1984. The Act is the 1984 amendment to the Resource Conservation and Recovery Act (RCRA) of 1976. The Act:
- Established strict limits on the land disposal of hazardous waste.
- Established a strict timeline for restricting untreated hazardous waste from land disposal.
- Regulated for the first time more than 100,000 companies that produce only small quantities of hazardous wastes (less than 1,000 kilograms per month). These small quantity generators were exempted from RCRA requirements before the 1984 Amendments.

- Established the following four major policies (OSWER-87):
    - Land disposal restriction policy.
    - Deep-well injection policy.
    - Domestic sewage sludge policy.
    - Waste minimization policy.

act or MPRSA (33USC1401 - 1434) means the Marine Protection, Research, and Sanctuaries Act of 1972, as amended, also known as the Ocean Dumping Act. It outlawed dumping of waste in oceans without an EPA permit and required the EPA to designate sites to be used by permit holders. EPA claimed authority to regulate incineration at sea and had issued research permits until 1987. Because this method was considered extremely controversial, EPA stopped issuing permits since then.

act or MWTA (RCRA subtitle J) means the Medical Waste Tracking Act of 1988. The Act:
- was passed by the Congress in October and signed by the President to make it a law on November 2, 1988. MWTA is Subtitle J of RCRA.
- Authorized EPA to implement a two year demonstration program to track certain medical wastes and to establish requirements for the segregation, handling, and labeling of these wastes from 3 mandated States plus other opt in States. The three States are New Jersey, New York, and Connecticut.

act or NCA ((42USC4901-4918)) means the Noise Control Act of 1972. The Act:
- Gave EPA the authority to set national noise emission standards for: Commercial products (NCA Sec. 6); Aircrafts (NCA Sec. 7); Railroads (NCA Sec. 17); Motor carriers (NCA Sec. 18).
- Required the EPA to assist the Federal Aviation Administration in developing noise regulations for airports and aircrafts.
- Consists of Public Law 95-574 (October 27, 1972) and the amendments by Public Law 95-609 (November 8, 1978).

(Other related information is provided in Pub. L. 92-574, 86 Stat. 1234; 40CFR201.1-a; 202.10-a; 203.1-1; 204.2-1; 205.2-1; 209.3-a; 211.102-b; 2.303-1.)

act or NEPA (42USC4321 et seq.) means the National Environmental Policy Act of 1970 (Public Law 91-190, January 1, 1970). The Act:
- Established the Council on Environmental Quality (CEQ) and required the development of a national policy on the environment.
- Required that Federal agencies include in their decision-making processes appropriate and careful consideration of all environmental effects of proposed actions, analyze potential environmental effects of proposed actions and their alternatives for public understanding and security, avoid or minimize adverse effects of proposed actions, and restore and enhance environmental quality as much as possible.

The purposes of the Act are:
- to declare a national policy which will encourage productive and enjoyable harmony between man and his environment.
- to promote efforts which will prevent or eliminate damage to the environment and biosphere and stimulate the health and welfare of man.
- to enrich the understanding of the ecological systems and natural resources important to the Nation; and to establish a Council on Environmental Quality.

(Other related information is provided in 40CFR51.852; 51.392; 93.101; 51.392.)

act or OSHA (29USC651 et seq.) means the Occupational Safety and Health Act of 1970. The purpose of this Act is to regulate commerce among the several States and with foreign nations and to provide for the general welfare, to assure so far as possible every working man and woman in the Nation safe and healthful working conditions and preserve human resources.

act or PHSA (42USC300f) means the Public Health Service Act. Its major subsequent amendments include: SDWA (Safe Drinking Water Act) of 1974 (Title XIV of PHSA).

act or PPA (Public Law 101-508, November 5, 1990) means the Pollution Prevention Act of 1990. The congress declares it to be the national policy of the United States that pollution should be prevented or reduced at the source whenever feasible; pollution that cannot be prevented should be recycled in an environmentally safe manner, whenever feasible; pollution that cannot be prevented or recycled should be treated in an environmentally safe manner whenever possible; and disposal or other release into the environmental should be employed only as a last resort and should be conducted in an environmental safe manner.

act or RCRA (42USC6901 et seq.) means the Resource Conservation and Recovery Act of 1976 (PL 95-609). The Act amended the Solid Waste Disposal Act. The Act:

- Established a regulatory system to track hazardous substances from the time of generation to disposal. It was designed to prevent new CERCLA sites from ever being created. The congress declares it to be the national policy of the United States that, whenever feasible, the generation of hazardous waste is to be reduced or eliminated as expeditiously as possible. Waste that is nevertheless generated should be treated, stored, or disposed of so as to minimize the present and future threat to human health and the environment (40USC6902);
- Promotes the protection of health and the environment and to conserve valuable material and energy resources (42USC6902); and
- Requires the use of safe and secure procedures in the treatment, transport, storage, and disposal of hazardous wastes to ensure that solid wastes are managed in an environmentally sound manner.

RCRA's major standards include:

- Maximum extraction procedure (EP) toxicity (40CFR261.24).
- Hazardous waste generators (RCRA Sec. 3002; 40CFR262).
- Hazardous waste transporters (RCRA Sec. 3003; 40CFR263).
- Owners and operators of TSD (RCRA Sec. 3004; 40CFR264):
  - Tank system (Subpart J, 40CFR264.190).
  - Surface impoundment (Subpart K, 40CFR264.220).
  - Waste pile (Subpart L, 40CFR264.250).
  - Land treatment (Subpart M, 40CFR264.270).
  - Landfill (Subpart N, 40CFR264.300).
  - Incinerator (Subpart O, 40CFR264.340).
  - Corrective action (Subpart S, 40CFR246.552).

  - Miscellaneous units (Subpart X, 40CFR264.600).

RCRA's major amendments include:

- HSWA (Hazardous and Solid Waste Amendment) of 1984.

(Other related information is provided in 40CFR2.305-1; 124.2; 144.3; 146.3; 248.4-a; 249.04-a; 250.4-a; 252.4-a; 253.4-a; 248.4-a; 249.04-a; 250.4-a; 252.4-a; 253.4; 256.06; 260.10; 270.2.)

act or SARA (40CFR) is the Superfund Amendments and Reauthorization Act of 1986. The Act:

- Operated under the legislative authority of CERCLA and SARA that funds and carries out the EPA solid waste emergency and long-term removal remedial activities. These activities include establishing the National Priorities List, investigating sites for inclusion on the list, determining their priority level on the list, and conducting and/or supervising the ultimately determined cleanup and other remedial actions.
- In addition to certain free-standing provisions of the law, the Act includes amendments to CERCLA, the Solid Waste Disposal Act, and the Internal Revenue Code. Among the free-standing provisions of the law is Title III of SARA, also known as the "Emergency Planning and Community Right-to-Know Act of 1986" and Title IV of SARA, also known as the "Radon Gas and Indoor Air Quality Research Act of 1986." Title V of SARA amending the Internal Revenue Code is also known as the "Superfund Revenue Act of 1986."

(Other related information is provided in 40CFR280.12; 300.5.)

act or SDWA (42USC300f-300j-9) means the Safe Drinking Water Act of 1974. The Act:

- Amended the Public Health Service Act in 1974 under Title XIV, Safety of Public Water Systems and was amended in 1976, 1977, 1979, 1980, 1984 and 1986. In effect, SDWA was also amended by a provision of the Hazardous and Solid Waste Amendments of 1984 (RCRA amendment) because toxic pollutants may enter the drinking water in many ways.
- Requires the EPA to establish a drinking water priority list (DWPL) of contaminants which may have adverse effects on the health of persons

and which are known or anticipated to occur in public water systems and may require regulations under the SDWA. It further requires the EPA to promulgate regulations for 25 new contaminants every three years. The first list was required by January 1, 1988. Because the Congress made no stipulation to the length of the DWPL, EPA may place a contaminant on the list indefinitely.

- The Act set standards for maximum allowable levels (MCL) of certain chemicals and bacteriological pollutants in public drinking water systems; and regulated underground injection systems including deep-well injection. The standards include:

(1) Maximum contaminant level (MCL): MCL means the maximum permissible level of a contaminant in water which is delivered to any user of a public water system. MCL has two control levels:

(A) Primary maximum contaminant levels: The levels of water quality necessary, with an adequate margin of safety, to protect the public health. See 40CFR141.11-141.16 for standards.

(B) Secondary maximum contaminant levels: The levels of water quality necessary, with an adequate margin of safety, to protect the public water systems. See 40CFR143.3 for standards.

(2) Maximum contaminant level goal (MCLG): MCLG means the maximum level of a contaminant in drinking water at which no known or anticipated adverse effect on the health or persons would occur, and which allows an adequate margin of safety. Maximum contaminant level goals are nonenforceable health goals including:

(A) Primary drinking water regulation: The regulation (a) applies to public water systems, (b) specifies contaminants which may have any adverse effect on the health of persons, (c) specifies, for each contaminant, a maximum contaminant level (MCL), (d) contains criteria and procedures to assure a supply of drinking water which dependably complies with MCL (40CFR141; SDWA1401).

(B) Secondary drinking water regulation: The regulation applies to public water systems and which specifies the maximum contaminant levels to protect the welfare.

(Other related information is provided in 40CFR143; SDWA1401.)

act or SWDA (42USC 6901-699li) means the Solid Waste Disposal Act of 1965.

- The Act was to promote the protection of health and environment and to conserve valuable material and energy resources.
- The Act provided research funds and technical assistance for State and local planners to mainly manage the municipal waste.
- The Act consists of Title II of Public Law 89-272 (July 14, 1955) and subsequent amendments including:
  - RCRA (Resource Conservation and Recovery Act) of 1976.
  - HSWA (Hazardous and Solid Waste Act) of 1984.

act or TSCA (15USC2601-2629) means the Toxic Substances Control Act of 1976. The Act:

- Created a screening mechanism for the production of new chemicals and new uses of existing chemicals.
- Prescribed EPA's range of actions to control manufacturing, use, and disposal once information gathered under the first two sections indicates a need for an action; and (4) conferred on EPA extraordinary authority to deal with imminent hazards.
- Banned manufacture and use of polychlorinated biphenyls (PCBs).
- Gives EPA power to require testing of chemical substances that present a risk of injury to health and the environment.
- Regulates the production and distribution of new chemicals and governs the manufacture, processing, distribution, and use of existing chemicals. Among the chemicals controlled by TSCA regulations are PCBs, chloroflurocarbons, and asbestos. In specific cases, there is an interface with RCRA regulations. For example, PCB disposal is generally regulated by TSCA. However, hazardous wastes mixed with PCBs

are covered under RCRA.

- Regulated most heavily in the 1980s were asbestos and PCBs.
- PCBs at concentrations of 500 ppm or greater must be disposed of in an incinerator which complies with TSCA's incineration requirements. See PCB incineration standards in 40CFR761.70.
- The Act consists of Public Law 94-469 (October 11, 1976) and subsequent amendments including:
  - AHERA (Asbestos Hazard Emergency Response Act) of 1986.

(Other related information is provided in 40CFR2.306-1; 704.3; 707.63; 710.2-d; 712.3-p; 716.3; 720.3-b; 723.50; 723.175-1; 723.250-1; 747.115-1; 747.195-1; 747.200-1; 749.68-1; 766.3; 792.3; 763.83; 763.103-a; 763.163; 790.3; 791.3-a.)

actinic solar irradiance in the atmosphere (40CFR796.3800-viii) is related to the sunlight intensity in the atmosphere and is proportional to the average light flux (in the units of photons $cm^{-2}day^{-1}$) that is available to cause photoreaction in the wavelength interval, center at interval, over a 24-hour day at a specific latitude and season date. It is the irradiance which would be measured by a weakly absorbing spherical actinometer exposed to direct solar radiation and sky radiation from all directions (under TSCA).

action level (40CFR141.2) is the concentration of lead or copper in water specified in 40CFR141.8(c) which determines, in some cases, the treatment requirements contained in subpart I of this part that a water system is required to complete.

action level (40CFR763.121) means an airborne concentration of asbestos of 0.1 fiber per cubic centimeter (f/cc) of air calculated as an 8-hour time-weighted average.

action levels (EPA-94/04):
(1) Regulatory levels recommended by EPA for enforcement by FDA and USDA when pesticide residues occur in food or feed commodities for reasons other than the direct application of the pesticide. As opposed to "tolerances" which are established for residues occurring as a direct result of proper usage, action levels are set for inadvertent residues resulting from previous legal use or accidental contamination.
(2) In the Superfund program, the existence of a contaminant concentration in the environment high enough to warrant action or trigger a response under SARA and the National Oil and Hazardous Substances Contingency Plan. The term can be used similarly in other regulatory programs (cf. tolerances).

activated carbon (EPA-94/04): A highly adsorbent form of carbon used to remove odors and toxic substances from liquid or gaseous emissions. In waste treatment it is used to remove dissolved organic matter from waste water. It is also used in motor vehicle evaporative control systems.

activated sludge (EPA-94/04): Sludge that results when primary effluent is mixed with bacteria-laden sludge and then agitated and aerated to promote biological treatment. This speeds breakdown of organic matter in raw sewage undergoing secondary waste treatment.

activation (40CFR300.5) means notification by telephone or other expeditious manner or, when required, the assembly of some or all appropriate members of the RRT (regional response team) or NRT (national response team).

activator (EPA-94/04): A chemical added to a pesticide to increase its activity.

active ingredient (40CFR152.3-b) means any substance (or group of structurally similar substances if specified by the Agency) that will prevent, destroy, repel or mitigate any pest, or that functions as a plant regulator, desiccant, or defoliant within the meaning of FIFRA section 2(a).

active ingredient (40CFR158.153-a) means any substance (or group of structurally similar substances, if specified by the Agency) that will prevent, destroy, repel or mitigate any pest, or that functions as a plant regulator, desiccant, or defoliant within the meaning of FIFRA sec. 2(a).

active ingredient (40CFR455.10-b) means an ingredient of a pesticide which is intended to prevent, destroy, repel, or mitigate any pest.

active ingredient (EPA-94/04): In any pesticide product, the component which kills, or otherwise controls, target pests. Pesticides are regulated primarily on the basis of active ingredients.

active ingredient (FIFRA2-7USC136) means:
(1) in the case of a pesticide other than a plant regulator, defoliant, or desiccant, an ingredient which will prevent, destroy, repel, or mitigate any pest;
(2) in the case of a plant regulator, an ingredient which, through physiological action, will accelerate or retard the rate of growth or rate of maturation or otherwise alter the behavior of ornamental or crop plants or the product thereof;
(3) in the case of a defoliant, an ingredient which will cause the leaves or foliage to drop from a plant; and
(4) in the case of a desiccant, an ingredient which will artificially accelerate the drying of plant tissues.

active institutional control (40CFR191.12) means:
(1) Controlling access to a disposal site by any means other than passive institutional controls;
(2) performing maintenance operations or remedial actions at a site;
(3) controlling or cleaning up releases from a site, or
(4) monitoring parameters related to disposal system performance.

active life (40CFR258.2) means the period of operation beginning with the initial receipt of solid waste and ending at completion of closure activities in accordance with 40CFR258.60 of this part.

active life of a facility (40CFR260.10) means the period from the initial receipt of hazardous waste at the facility until the Regional Administrator receives certification of final closure.

active mine (40CFR61.21-a) means an underground uranium mine which is being ventilated to allow workers to enter the mine for any purpose.

active mining area (40CFR434.11-b) means the area, on and beneath land, used or disturbed in activity related to the extraction, removal, or recovery of coal from its natural deposits. This term excludes coal preparation plants, coal preparation plant associated areas and post-mining areas.

active mining area (40CFR440.132-a) is a place where work or other activity related to the extraction, removal, or recovery of metal ore is being conducted, except, with respect to surface mines, any area of land on or in which grading has been completed to return the earth to desired contour and reclamation work has begun.

active portion (40CFR258.2) means that part of a facility or unit that has received or is receiving wastes and that has not been closed in accordance with 40CFR258.60 of this part.

active portion (40CFR260.10) means that portion of a facility where treatment, storage, or disposal operations are being or have been conducted after the effective date of part 261 of this chapter and which is not a closed portion.

active service (40CFR60.691) means that a drain is receiving refinery wastewater from a process unit that will continuously maintain a water seal.

active sewage sludge unit (40CFR503.21-a) is a sewage sludge unit that has not closed.

active use (40CFR57.103-b) refers to an SO_2 constant control system installed at a smelter before August 7, 1977 and not totally removed from regular service by that date.

active waste disposal site (40CFR61.141) means any disposal site other than an inactive site.

activity (40CFR35.6015-1) means a set of CERCLA-funded tasks that makes up a segment of the sequence of events undertaken in determining, planning, and conducting a response to a release or potential release of a hazardous substance. These include core program, pre-remedial (i.e., preliminary assessments and site inspections), remedial investigation/feasibility studies, remedial design, remedial action, removal, enforcement, and Core Program activities.

activity plans (EPA-94/04): Written procedures in a

school's asbestos-management plan that detail the steps a Local Education Agency (LEA) will follow in performing the initial and additional cleaning, operation and maintenance program tasks; periodic surveillance; and reinspections required by the Asbestos Hazard Emergency Response Act.

activity subject to regulation (40CFR124.41): See regulated activity.

actual 1985 emission rate (CAA402-42USC7651a) for electric utility units, means the annual sulfur dioxide or nitrogen oxide emission rate in pounds per million Btu as reported in the NAPAP Emissions Inventory, Version 2, National Utility Reference File. For nonutility units, the term "actual 1985 emission rate" means the annual sulfur dioxide or nitrogen oxides emission rate in pounds per million Btu as reported in the NAPAP Emission Inventory, Version 2, National Utility Reference File. For nonutility units, the term "actual 1985 emission rate" means the annual sulfur dioxide or nitrogen oxides emission rate in pounds per million Btu as reported in the NAPAP Emission Inventory, Version 2.

actual emissions (40CFR51.165-xii) means:
(A) the actual rate of emissions of a pollutant from an emissions unit as determined in accordance with paragraphs (a)(1)(xii) (B) through (D) of this section.
(B) In general, actual emissions as of a particular date shall equal the average rate, in tons per year, at which the unit actually emitted the pollutant during a two-year period which precedes the particular date and which is representative of normal source operation. The reviewing authority shall allow the use of a different time period upon a determination that it is more representative of normal source operation. Actual emissions shall be calculated using the unit's actual operating hours, production rates, and types of materials processed, stored, or combusted during the selected time period.
(C) The reviewing authority may presume that the source-specific allowable emissions for the unit are equivalent to the actual emissions of the unit.
(D) For any emissions unit (other than an electric utility steam generating unit specified in

paragraph (a)(1)(xii)(E) of this section) which has not begun normal operations on the particular date, actual emissions shall equal the potential to emit of the unit on that date.
(E) For an electric utility steam generating unit (other than a new unit or the replacement of an existing unit) actual emissions of the unit following the physical or operational change shall equal the representative actual annual emissions of the unit, provided the source owner or operator maintains and submits to the reviewing authority, on an annual basis for a period of 5 years from the date the unit resumes regular operation, information demonstrating that the physical or operational change did not result in an emissions increase. A longer period, not to exceed 10 years, may be required by the reviewing authority if it determines such a period to be more representative of normal source post-change operations.

actual emissions (40CFR51.166-21) means:
(i) The actual rate of emissions of a pollutant from an emissions unit, as determined in accordance with paragraphs (b)(21) (ii) through (iv) of this section.
(ii) In general, actual emissions as of a particular date shall equal the average rate, in tons per year, at which the unit actually emitted the pollutant during a two-year period which precedes the particular date and which is representative of normal source operation. The reviewing authority may allow the use of a different time period upon a determination that it is more representative of normal source operation. Actual emissions shall be calculated using the unit's actual operating hours, production rates, and types of materials processed, stored, or combusted during the selected time period.
(iii) The reviewing authority may presume that source-specific allowable emissions for the unit are equivalent to the actual emissions of the unit.
(iv) For any emissions unit (other than an electric utility steam generating unit specified in paragraph (b)(21)(v) of this section) which has not begun normal operations on the particular date, actual emissions shall equal the potential to emit of the unit on that date.

(v) For an electric utility steam generating unit (other than a new unit or the replacement of an existing unit) actual emissions of the unit following the physical or operational change shall equal the representative actual annual emissions of the unit following the physical or operational change, provided the source owner or operator maintains and submits to the reviewing authority, on an annual basis for a period of 5 years from the date the unit resumes regular operation, information demonstrating that the physical or operational change did not result in an emissions increase. A longer period, not to exceed 10 years, may be required by the reviewing authority if it determines such a period to be more representative of normal source post-change operations.

actual emissions (40CFR52.21-21) means:

(i) The actual rate of emissions of a pollutant from an emissions unit, as determined in accordance with paragraphs (b)(21) (ii) through (iv) of this section.

(ii) In general, actual emissions as of a particular date shall equal the average rate, in tons per year, at which the unit actually emitted the pollutant during a two-year period which precedes the particular date and which is representative of normal source operation. The Administrator shall allow the use of a different time period upon a determination that it is more representative of normal source operation. Actual emissions shall be calculated using the unit's actual operating hours, production rates, and types of materials processed, stored, or combusted during the selected time period.

(iii) The Administrator may presume that source-specific allowable emissions for the unit are equivalent to the actual emissions of the unit.

(iv) For any emissions unit (other than an electric utility steam generating unit specified in paragraph (b)(21)(v) of this section) which has not begun normal operations on the particular date, actual emissions shall equal the potential to emit of the unit on that date.

(v) For an electric utility steam generating unit (other than a new unit or the replacement of an existing unit) actual emissions of the unit following the physical or operational change shall equal the representative actual annual emissions

of the unit, provided the source owner or operator maintains and submits to the Administrator on an annual basis for a period of 5 years from the date the unit resumes regular operation, information demonstrating that the physical or operational change did not result in an emissions increase. A longer period, not to exceed 10 years, may be required by the Administrator if he determines such a period to be more representative of normal source post-change operations.

actual emissions (40CFR52.24-13) means:

(i) The actual rate of emissions of a pollutant from an emissions unit, as determined in accordance with paragraphs (f) (ii) through (iv) of this section.

(ii) In general, actual emissions as of a particular date shall equal the average rate, in tons per year, at which the unit actually emitted the pollutant during a two-year period which precedes the particular date and which is representative of normal source operation. The Administrator shall allow the use of a different time period upon a determination that it is more representative of normal source operation. Actual emissions shall be calculated using the unit's actual operating hours, production rates, and types of materials processed, stored, or combusted during the selected time period.

(iii) The Administrator may presume that source-specific allowable emissions for the unit are equivalent to the actual emissions of the unit.

(iv) For any emissions unit (other than an electric utility steam generating unit specified in paragraph (f)(13)(v) of this section) which has not begun normal operations on the particular date, actual emissions shall equal the potential to emit of the unit on that date.

(v) For an electric utility steam generating unit (other than a new unit or the replacement of an existing unit) actual emissions of the unit following the physical or operational change shall equal the representative actual annual emissions of the unit, provided the source owner or operator maintains and submits to the Administrator, on an annual basis for a period of 5 years from the date the unit resumes regular operation, information demonstrating that the physical or operational change did not result in an emissions increase. A longer period, not to exceed 10 years, may be required by the Administrator if he determines such a period to be more representative of normal source post-change operations.

actual emissions (40CFR52.741) means the actual quantity of VOM emissions from an emission source during a particular time period.

actual emissions (40CFR63.2) is defined in subpart D of this part for the purpose of granting a compliance extension for an early reduction of hazardous air pollutants (other identical or similar definitions are provided in 40CFR63.101; 63.191).

actual emissions (40CFR63.51) means the actual rate of emissions of a pollutant, but does not include excess emissions from a malfunction, or startups and shutdowns associated with a malfunction. Actual emissions shall be calculated using the source's actual operating rates, and types of materials processed, stored, or combusted during the selected time period.

actual SO$_2$ emissions rate (40CFR72.2) means the annual average sulfur dioxide emissions rate for the unit (expressed in lb/mmBtu), for the specified calendar year; provided that, if the unit is listed in the NADB, the "1985 actual SO$_2$ emissions rate" for the unit shall be the rate specified by the Administrator in the NADB under the data field "SO2RTE."

acute delayed neurotoxicity (40CFR798.6540) is a prolonged, delayed-onset locomotor ataxia resulting from single administration of the test substance, repeated once if necessary.

acute dermal LD50 (40CFR152.3-c) means a statistically derived estimate of the single dermal dose of a substance that would cause 50 percent mortality to the test population under specified conditions.

acute dermal toxicity (40CFR798.1100-1) is the adverse effects occurring within a short time of dermal application of a single dose of a substance or multiple doses given within 24 hours.

acute exposure (EPA-94/04): A single exposure to a toxic substance which results in severe biological harm or death. Acute exposures are usually characterized as lasting no longer than a day.

acute inhalation LC50 (40CFR152.3-d) means a statistically derived estimate of the concentration of a substance that would cause 50 percent mortality to the test population under specified conditions.

acute inhalation toxicity (40CFR798.1150-1) is the adverse effects caused by a substance following a single uninterrupted exposure by inhalation over a short period of time (24 hours or less) to a substance capable of being inhaled.

acute lethal toxicity (40CFR797.1350-1) means the lethal effect produced on an organism within a short period of time of exposure organism within a short period of time (days) of exposure to a chemical (other identical or similar definitions are provided in 40CFR797.1440-2).

acute oral LD50 (40CFR152.3-e) means a statistically derived estimate of the single oral dose of a substance that would cause 50 percent mortality to the test population under specified conditions (under FIFRA).

acute oral toxicity (40CFR798.1175-1) is the adverse effects occurring within a short time of oral administration of a single dose of a substance or multiple doses given within 24 hours.

acute toxicity (40CFR131.35.d-1) means a deleterious response (e.g., mortality, disorientation, immobilization) to a stimulus observed in 96 hours or less.

acute toxicity (40CFR300-AA) is a measure of toxicological responses that result from a single exposure to a substance or from multiple exposures within a short period of time (typically several days or less). Specific measures of acute toxicity used with the HRS (hazardous ranking system) include lethal dose(50) (LD50) and lethal concentration(50) (LC50), typically measured within a 24-hour to 96-hour period.

acute toxicity (40CFR797.1440-2) is the discernible adverse effects induced in an organism within a short period of time (days) of exposure to a chemical. For aquatic animals this usually refers to continuous exposure to the chemical in water for a period of up to four days. The effects (lethal or sub-lethal) occurring may usually be observed within the period of exposure with aquatic organisms.

acute toxicity (40CFR797.1800-1) is the discernible adverse effects induced in an organism within a short period of time (days) of exposure to a chemical. For aquatic animals this usually refers to continuous exposure to the chemical in water for a period of up to four days. The effects (lethal or sublethal) occurring may usually be observed within the period of exposure with aquatic organisms. In this test guideline, shell deposition is used as the measure of toxicity.

acute toxicity (EPA-94/04): The ability of a substance to cause poisonous effects resulting in severe biological harm or death soon after a single exposure or dose. Also, any severe poisonous effect resulting from a single short-term exposure to a toxic substance (cf. chronic toxicity or toxicity).

acute toxicity test (40CFR797.1400-2) means a method used to determine the concentration of a substance that produces a toxic effect on a specified percentage of test organisms in a short period of time (e.g., 96 hours). In this guideline, death is used as the measure of toxicity.

acutely toxic effects (40CFR721.3): A chemical substance produces acutely toxic effects if it kills within a short time period (usually 14 days):
(1) At least 50 percent of the exposed mammalian test animals following oral administration of a single dose of the test substance at 25 milligrams or less per kilogram of body weight (LD50).
(2) At least 50 percent of the exposed mammalian test animals following dermal administration of a single dose of the test substance at 50 milligrams or less per kilogram of body weight (LD50).
(3) At least 50 percent of the exposed mammalian test animals following administration of the test substance for 8 hours or less by continuous inhalation at a steady concentration in air at 0.5 milligrams or less per liter of air (LC50) (cf. chronic toxicity).

ad valorem tax (40CFR35.905) means a tax based upon the value of real (other identical or similar definitions are provided in 40CFR35.2005-2).

adaptation (40CFR796.3100-i) is the process by which a substance induces the synthesis of any degradative enzymes necessary to catalyze the transformation of that substance.

adaptation (EPA-94/04): Changes in an organism's structure or habit that help it adjust to its surroundings.

add on control (40CFR72.2) means a pollution reduction control technology that operates independent of the combustion process.

add on control device (EPA-94/04): An air pollution control device such as carbon adsorber or incinerator which reduces the pollution in an exhaust gas. The control device usually does not affect the process being controlled and thus is "add-on" technology as opposed to a scheme to control pollution through making some alteration to the basic process.

added ingredients (40CFR407.81-b) shall mean the prepared sauces (prepared from items such as dairy products, starches, sugar, tomato sauce and concentrate, spices, and other related pre-processed ingredients) which are added during the canning and freezing of fruits and vegetables.

additional advance auction (40CFR72.2) means the auction of advance allowances that were offered the previous year for sale in an advance sale.

additions and alterations (40CFR21.2-l) means the act of undertaking construction of any facility.

additive (40CFR79.2-e) means any substance, other than one composed solely of carbon and/or hydrogen, that is intentionally added to a fuel named in the designation (including any added to a motor vehicle's fuel system) and that is not intentionally removed prior to sale or use.

additive (40CFR790.3) means a chemical substance that is intentionally added to another chemical substance to improve its stability or impart some other desirable quality.

additive manufacturer (40CFR79.2-f) means any person who produces or manufactures an additive for use as an additive and/or sells an additive under his own name.

adequate evidence (40CFR32.105-a) means information sufficient to support the reasonable belief that a particular act or emission has occurred.

adequate SO₂ emission limitation (40CFR57.103-c) means a SIP (state implementation plan) emission limitation which was approved or promulgated by EPA as adequate to attain and maintain the NAAQS (national ambient air quality standards program) in the areas affected by the stack emissions without the use of any unauthorized dispersion technique.

adequate storage (40CFR165.1-d) means placing of pesticides in proper containers and in safe areas as per 40CFR165.10 as to minimize the possibility of escape which could result in unreasonable adverse effects on the environment.

adequately wet (EPA-94/04): Asbestos containing material that is sufficiently mixed or penetrated with liquid to prevent the release of particulates.

adequately wetted (40CFR61.141) means sufficiently mixed or coated with water or an aqueous solution to prevent dust emissions.

adhesive (40CFR52.741) means any substance or mixture of substances intended to serve as a joining compound.

adjacent (40CFR230.3-b) means bordering, contiguous, or neighboring Wetlands separated from other waters of the United States by man-made dikes or barriers, natural river berms, beach dunes, and the like are "adjacent wetlands."

adjustable parameter (40CFR89.2) means any device, system, or element of design which is physically capable of being adjusted (including those which are difficult to access) and which, if adjusted, may affect emissions or engine performance during emission testing.

adjusted configuration (40CFR610.11-15) means the test configuration after adjustment of engine calibrations to the retrofit specifications, but excluding retrofit hardware installation.

adjusted loaded vehicle weight (40CFR86.094.2) means the numerical average of vehicle curb weight and GVWR (gross vehicle weight rating).

administering agency (40CFR8.2-a) means any department, agency, and establishment in the Executive Branch of the Government, including any wholly owned Government corporation, which administers a program involving federally assisted construction contracts.

administrative action (40CFR6.601-a) for the sake of this subpart means the issuance by EPA of an NPDES (national pollutant discharge elimination system) permit to discharge as a new source, pursuant to 40CFR124.15.

administrative law judge (40CFR22.03) means an Administrative Law Judge appointed under 5USC3105 (see also Pub. L. 95-251, 92 Stat. 183) (other identical or similar definitions are provided in 40CFR27.2; 57.103-d; 123.64; 164.2-b; 209.3-b).

administrative offset (40CFR13.2-f) means the withholding of money payable by the United States to, or held by the United States for, a person to satisfy a debt the person owes the Government.

administrative order (EPA-94/04): A legal document signed by EPA directing an individual, business, or other entity to take corrective action or refrain from an activity. It describes the violations and actions to be taken, and can be enforced in court. Such orders may be issued, for example, as a result of an administrative complaint whereby the respondent is ordered to pay a penalty for violations of a statute.

administrative order on consent (EPA-94/04): A legal agreement signed by EPA and an individual, business, or other entity through which the violator agrees to pay for correction of violations, take the required corrective or cleanup actions, or refrain from an activity. It describes the actions to be taken, may be subject to a comment period, applies to civil actions, and can be enforced in court.

administrative procedures act (EPA-94/04): A law

that spells out procedures and requirements related to the promulgation of regulations.

administrative record (EPA-94/04): All documents which EPA considered or relied on in selecting the response action at a Superfund site, culminating in the record of decision for remedial action or, an action memorandum for removal actions.

administrative requirements (40CFR31.3) mean those matters common to grants in general, such as financial management, kinds and frequency of reports, and retention of records. These are distinguished from programmatic requirement, which concern matters that can be treated only on a program-by-program or grant-by-grant basis, such as kinds of activities that can be supported by grants under a particular program.

administrator (CAA302) means the Administrator of the U.S. Environmental Protection Agency, or the Administrator's authorized representative (other identical or similar definitions are provided in 40CFR2.201; 7.25; 8.33; 13.2; 15.4; 17.2; 20.2; 22.3; 23.1; 27.2; 51;.100; 52.741; 53.1; 56.1; 57.103; 58.1; 60.2; 61.02; 65.01; 79.2; 80.2; 81.1; 82.3; 82.23; 85.1502; 85.2102; 86.082.2; 86.402.78; 87.1; 104.2-b; 110.1; 117.1; 121.1-c; 122.2; 124.2; 124.41; 125.58-a; 129.2; 136.2-b; 142.2; 144.3; 146.3; 147.2902; 149.101-d; 152.3; 164.2; 165.1; 171.2; 177.3; 178.3; 179.3; 180.1; 191.02; 203.1; 232; 260.10; 204.2; 205.2; 209.3; 211.102; 270.2; 302.3; 304.12; 401.11; 501.2; 600.002.85-2; 710.2-e; 723.250-2; 761.3; 763.121).

administrator and general counsel (40CFR350.1) mean the EPA officers or employees occupying the positions so titled.

administrator or regional administrator (40CFR108.2-e) means the Administrator or a Regional Administrator of the Environmental Protection Agency.

adsorption (EPA-94/04):
(1) Adhesion of molecules of gas, liquid, or dissolved solids to a surface.
(2) An advanced method of treating wastes in which activated carbon removes organic matter from wastewater.

adsorption ratio, K_d (40CFR796.2750-vi) is the amount of test chemical adsorbed by a sediment or soil (i.e., the solid phase) divided by the amount of test chemical in the solution phase, which is in equilibrium with the solid phase, at a fixed solid/solution ratio.

adulterants (EPA-94/04): Chemical impurities or substances that by law do not belong in a food, or pesticide.

adulterated (EPA-94/04) means:
(1) Any pesticide whose strength or purity falls below the quality stated on its label; or
(2) A food, feed, or product that contains illegal pesticide residues.

adulterated (FIFRA2-7USC136) applies to any pesticide if:
(1) its strength or purity falls below the professed standard of quality as expressed on its labeling under which it is sold;
(2) any substance has been substituted wholly or in part for the pesticide; or
(3) any valuable constituent of the pesticide has been wholly or in part abstracted.

advance allowance (40CFR72.2) means an allowance that may be used for purposes of compliance with a unit's acid rain sulfur dioxide emissions limitation requirements beginning no earlier than seven years following the year in which the allowance is first offered for sale.

advance auction (40CFR72.2) means an auction of advance allowances.

advance sale (40CFR72.2) means a sale of advance allowances.

advanced air emission control devices (40CFR426.11-c) shall mean air pollution control equipment, such as electrostatic precipitators and high energy scrubbers, that are used to treat an air discharge which has been treated initially by equipment including knockout chambers and low energy scrubbers.

advanced treatment (EPA-94/04): A level of wastewater treatment more stringent than secondary treatment; requires an 85-percent reduction in conventional pollutant concentration or a significant reduction in nonconventional pollutants.

advanced wastewater treatment (EPA-94/04): Any treatment of sewage that goes beyond the secondary or biological water treatment stage and includes the removal of nutrients such as phosphorus and nitrogen and a high percentage of suspended solids (cf. primary treatment or secondary treatment).

adversary adjudication (40CFR17.2-c) means an adjudication required by statute to be held pursuant to 5USC554 in which the position of the United States is represented by counsel or otherwise, but excludes an adjudication for the purpose of granting or renewing a license.

adverse environmental effect (CAA112-42USC7412) means any significant and widespread adverse effect, which may reasonably be anticipated, to wildlife, aquatic life, or other natural resources, including adverse impacts on populations of endangered or threatened species or significant degradation of environmental quality over broad areas.

adverse impact on visibility (40CFR51.301-a) means, for purposes of section 307, visibility impairment which interferes with the management, protection, preservation, or enjoyment of the visitor's visual experience of the Federal Class I area. This determination must be made on a case-by-case basis taking into account the geographic extent, intensity, duration, frequency and time of visibility impairments, and how these factors correlate with
(1) times of visitor use of the Federal Class I area, and
(2) the frequency and timing of natural conditions that reduce visibility. This term does not include effects on integral vistas.

adverse impact on visibility (40CFR52.21-29) means visibility impairment which interferes with the management, protection, preservation or enjoyment of the visitor's visual experience of the Federal Class I area. This determination must be made on a case-by-case basis taking into account the geographic extent, intensity, duration, frequency and time of visibility impairment, and how these factors correlate with:

(1) times of visitor use of the Federal Class I area, and
(2) the frequency and timing of natural conditions that reduce visibility.

adverse weather (40CFR112.2) means the weather conditions that make it difficult for response equipment and personnel to cleanup or remove spilled oil, and that will be considered when identifying response systems and equipment in a response plan for the applicable operating environment. Factors to consider include significant wave height as specified in Appendix E to this part, as appropriate, ice conditions, temperatures, weather-related visibility, and currents within the area in which the systems or equipment are intended to function.

advertised engine displacement (40CFR205.151-8) means the rounded of volumetric engine capacity used for marketing purposes by the motorcycle manufacturer.

advisory (EPA-94/04): A non-regulatory document that communicates risk information to those who may have to make risk management decisions.

AECD (40CFR86.082.2; 600.002.85): See auxiliary emission control device.

aerated lagoon (EPA-94/04): A holding and/or treatment pond that speeds up the natural process of biological decomposition of organic waste by stimulating the growth and activity of bacteria that degrade organic waste.

aeration (EPA-94/04): A process which promotes biological degradation of organic matter in water. The process may be passive (as when waste is exposed to air), or active (as when a mixing or bubbling device introduces the air).

aeration tank (EPA-94/04): A chamber used to inject air into water.

aerobic (EPA-94/04): Life or processes that require, or are not destroyed by, the presence of oxygen (cf. anaerobic).

aerobic digestion (40CFR503.31-a) is the biochemical

decomposition of organic matter in sewage sludge into carbon dioxide and water by microorganisms in the presence of air.

aerobic treatment (EPA-94/04): Process by which microbes decompose complex organic compounds in the presence of oxygen and use the liberated energy for reproduction and growth. (Such processes include extended aeration, trickling filtration, and rotating biological contactors.)

aerodynamic diameter (40CFR798.1150-2) applies to the size of particles of aerosols. It is the diameter of a sphere of unit density which behaves aerodynamically as the particle of the test substance. It is used to compare particles of different size and densities and to predict where in the respiratory tract such particles may be deposited. This term is used in contrast to measured or geometric diameter which is representative of actual diameters which in themselves cannot be related to deposition within the respiratory tract (other identical or similar definitions are provided in 40CFR798.2450-2).

aerodynamic diameter (40CFR798.4350-2) applies to the behavioral size of particles of aerosols. It is the diameter of a sphere of unit density which behaves aerodynamically like the particles of the test substance. It is used to compare particles of different sizes, shapes, and densities and to predict where in the respiratory tract such particles may be deposited. This term is used in contrast to optical, measured or geometric diameters which are representations of actual diameters which in themselves cannot be related to deposition within the respiratory tract..

aerosol (EPA-94/04): A suspension of liquid or solid particles in a gas.

aerosol propellant (40CFR762.3) means a liquefied or compressed gas in a container where the purpose of the liquefied or compressed gas is to expel from the container liquid or solid material different from the aerosol propellant.

affected (40CFR35.4010) means subject to an actual or potential health, economic or environmental threat arising from a release or a threatened release at a facility listed on the National Priorities List (NPL) or proposed for listing under the National Oil and

Hazardous Substances Pollution Contingency Plan (NCP) where a response action under CERCLA (Comprehensive Environmental Response, Compensation, and Liability Act) has begun. Examples of affected parties include individuals who live in areas adjacent to NPL facilities, who depend on water sources endangered by releases of hazardous substances at the facility, or whose economic interests are directly threatened or harmed.

affected business (40CFR2.201-d) means, with reference to an item of business information, a business which has asserted (and not waived or withdrawn) a business confidentiality claim covering the information, or a business which could be expected to make such a claim if it were aware that disclosure of the information to the public was proposed.

affected facility (40CFR52.1161-5) means any employment facility at which 50 or more persons are employees or any educational facility at which 250 or more persons are students and employees.

affected facility (40CFR60.2) means, with reference to a stationary source, any apparatus to which a standard is applicable.

affected Federal land manager (40CFR51.852) means the Federal agency or the Federal official charged with direct responsibility for management of an area designated as Class I under the Act (42USC7472) that is located within 100 km of the proposed Federal action (other identical or similar definitions are provided in 40CFR93.152).

affected public (EPA-94/04): The people who live and/or work near a hazardous waste site.

affected source (40CFR51.491) means any stationary, area, or mobile source of a criteria pollutant(s) to which an EIP (economic incentive program) applies. This term applies to sources explicitly included at the start of a program, as well as sources that voluntarily enter (i.e., opt into) the program.

affected source (40CFR63.2) for the purposes of this part, means the stationary source, the group of stationary sources, or the portion of a stationary source that is regulated by a relevant standard or other requirement established pursuant to section 112 of the Act (Clean Air Act). Each relevant standard will define the "affected source" "for the purposes of that standard. The term "affected source," as used in this part, is separate and distinct from any other use of that term in EPA regulations such as those implementing title IV of the Act. Sources regulated under part 60 or part 61 of this chapter are not affected sources for the purposes of part 63 (other identical or similar definitions are provided in 40CFR63.101; 63.191).

affected source (40CFR70.2) shall have the meaning given to it in the regulations promulgated under title IV of the Act.

affected source (40CFR72.2) means a source that includes one or more affected units.

affected source (CAA402-42USC7651a) means a source that includes one or more affected units.

affected source (CAA501-42USC7661) shall have the meaning given such term in title IV.

affected States (40CFR70.2) are all States:
(1) Whose air quality may be affected and that are contiguous to the State in which a part 70 permit, permit modification or permit renewal is being proposed; or
(2) That are within 50 miles of the permitted source.

affected unit (40CFR70.2) shall have the meaning given to it in the regulations promulgated under title IV of the Act.

affected unit (40CFR72.2) means a unit that is subject to any Acid Rain emissions reduction requirement or Acid Rain emissions limitation under 40CFR72.6 or part 74 of this chapter (under the Clean Air Act).

affected unit (CAA402-42USC7651a) means a unit that is subject to emission reduction requirements or limitations under this title.

affected units (40CFR52.145) means the steam-generating unit(s) at the Navajo Generating Station, all of which are subject to the emission limitation in paragraph (d)(2) of this section, that has accumulated at least 365 boiler operating days since the passage of the date defined in paragraph (d)(6) of this section applicable to it.

affecting (40CFR1508.3) means will or may have an effect on.

affiliate (40CFR32.105-b): Persons are affiliates of each another if, directly or indirectly, either one controls or has the power to control the other, or a third person controls or has the power to control both. Indicia of control include, but are not limited to: interlocking management or ownership, identity of interest among family members, shared facilities and equipment, common use of employees, or a business entity organized following suspension or debarment of a person which has the same or similar management, ownership, or principal employees as the suspended, debarred, ineligible, or voluntarily excluded person.

affiliate (40CFR72.2) shall have the meaning set forth in section 2(a)(11) of the Public Utility Holding Company Act of 1935, 15 U.S.C. 79b(a)(11), as of November 15, 1990.

affiliated entity (40CFR66.3-b) means a person who directly, or indirectly through one or more intermediaries, controls, is controlled by, or is under common control with the owner or operator of a source.

afterburner (40CFR52.741) means a control device in which materials in gaseous effluent are combusted.

afterburner (A/B) (40CFR60.471) means an exhaust gas incinerator used to control emissions of particulate matter.

afterburner (EPA-94/04): In incinerator technology, a burner located so that the combustion gases are made to pass through its flame in order to remove smoke and odors. It may be attached to or be separated from the incinerator proper.

aftermarket part (40CFR85.2113-b) means any part offered for sale for installation in or on a motor vehicle after such vehicle has left the vehicle manufacturers production line.

aftermarket part manufacturer (40CFR85.2113-c) means:

(1) A manufacturer of an aftermarket part or,

(2) A party that markets aftermarket parts under its own brand name, or,

(3) A rebuilder of original equipment or aftermarket parts, or

(4) A party that licenses others to sell its parts.

age tank (EPA-94/04): A tank used to store a chemical solution of known concentration for feed to a chemical feeder. Also called a day tank.

aged catalytic converter (40CFR85.2122(a)(15)(ii)-(F)) means a converter that has been installed on a vehicle or engine stand and operated through a cycle specifically designed to chemically age, including exposure to representative lead concentrations, and mechanically stress the catalytic converter in a manner representative of in-use vehicle or engine conditions.

agency (40CFR13.2) means the United States Environmental Protection Agency (other identical or similar definitions are provided in 40CFR 13.2; 15.4; 22.03; 32.105; 34.105; 51.301; 104.2-c; 142.2; 152.3; 163.2; 164.2; 165.1; 166.3; 177.3; 178.3; 179.3; 180.1; 204.2; 205.2; 209.3; 211.102; 761.3; 763.163; 791.3-b).

agency head (40CFR32.105-x) means the Administrator of the Environmental Protection Agency.

agency trial staff (40CFR124.78-1) means those Agency employees, whether temporary or permanent, who have been designated by the Agency under 40CFR124.77 or 124.116 as available to investigate, litigate, and present the evidence, arguments, and position of the Agency in the evidentiary hearing or nonadversary panel hearing. Any EPA employee, consultant, or contractor who is called as a witness by EPA trial staff, or who assisted in the formulation of the draft permit which is the subject of the hearing, shall be designated as a member of the Agency trial staff.

agency trial staff (40CFR57.809-1) means those Agency employees, whether temporary or permanent, who have been designated by the Agency as available to investigate, litigate, and present the evidence arguments and position of the Agency in the evidentiary hearing or non-adversary panel hearing. Appearance as a witness does not necessarily require a person to be designated as a member of the Agency trial staff.

agent orange (EPA-94/04): A toxic herbicide and defoliant used in the Vietnam conflict, containing 2,4,5-trichlorophenoxyacetic acid (2,4,5-T) and 2-4 dichlorophenoxyacetic acid (2,4-D) with trace amounts of dioxin.

aggregate costs (40CFR35.9010) means the total cost of all research, surveys, studies, modeling, and other technical work completed by a Management Conference during a fiscal year to develop a Comprehensive Conservation and Management Plan for the estuary.

aggregate facility (40CFR60.691) means an individual drain system together with ancillary downstream sewer lines and oil-water separators, down to and including the secondary oil-water separator, as applicable.

agreement State (10CFR30.4) means any state with which the Atomic Energy Commission or the Nuclear Regulatory Commission has entered into an effective agreement under subsection 274b of the Act. Non-Agreement State means any other State (under AEA).

agreement State (10CFR40.4) means any State with which the Atomic Energy Commission or the Nuclear Regulatory Commission has entered into an effective agreement under subsection 274b of the Atomic Energy Act of 1954, as amended.

agreement State (10CFR70.4) as designated in part 150 of this chapter means any State with which the Commission has entered into an effective agreement under subsection 274b of the Act. Non-agreement State means any other State.

agreement State (40CFR191.02-f) means any State with which the Commission or the Atomic Energy Commission has entered into an effective agreement under subsection 274b of the Atomic Energy Act of 1954, as amended (68 Stat. 919).

agreement State (40CFR61.101-a) means a State with which the Atomic Energy Commission or the Nuclear Regulatory Commission has entered into an effective agreement under subsection 274(b) of the Atomic Energy Act of 1954, as amended.

agricultural commodity (40CFR171.2-5) means any plant, or part thereof, or animal, or animal product, produced by a person (including farmers, ranchers, vineyardists, plant propagators, Christmas tree growers, aquaculturists, floriculturists, orchardists, foresters, or other comparable persons) primarily for sale, consumption, propagation, or other use by man or animals.

agricultural employer (40CFR170.3) means any person who hires or contracts for the services of workers, for any type of compensation, to perform activities related to the production of agricultural plants, or any person who is an owner of or is responsible for the management or condition of an agricultural establishment that uses such workers.

agricultural establishment (40CFR170.3) means any farm, forest, nursery, or greenhouse.

agricultural land (40CFR503.11-) is land on which a food crop, a feed crop, or a fiber crop is grown. This includes range land and land used as pasture.

agricultural plant (40CFR170.3) means any plant grown or maintained for commercial or research purposes and includes, but is not limited to, food, feed, and fiber plants; trees; turfgrass; flowers, shrubs; ornamentals; and seedlings.

agricultural pollution (EPA-94/04): Farming wastes, including runoff and leaching of pesticides and fertilizers; erosion and dust from plowing; improper disposal of animal manure and carcasses; crop residues, and debris.

agricultural solid waste (40CFR243.101-b) means the solid waste that is generated by the rearing of animals, and the producing and harvesting of crops or trees (other identical or similar definitions are provided in 40CFR246.101-a).

agro-ecosystem (EPA-94/04): Land used for crops, pasture, and livestock; the adjacent uncultivated land that supports other vegetation and wildlife; and the associated atmosphere, the underlying soils, groundwater, and drainage networks.

agronomic rate (40CFR503.11-b) is the whole sludge application rate (dry weight basis) designed:
(1) To provide the amount of nitrogen needed by the food crop, feed crop, fiber crop, cover crop, or vegetation grown on the land; and
(2) To minimize the amount of nitrogen in the sewage sludge that passes below the root zone of the crop or vegetation grown on the land to the ground water.

AHERA designated person (ADP) (EPA-94/04): A person designated by a Local Education Agency to ensure that the AHERA requirements for asbestos management and abatement are properly implemented.

air assisted airless spray (40CFR52.741) means a spray coating method which combines compressed air with hydraulic pressure to atomize the coating material into finer droplets than is achieved with pure airless spray. Lower hydraulic pressure is used than with airless spray.

air binding (EPA-94/04): Situation where air enters the filter media and harms both the filtration and backwash processes.

air bleed to intake manifold retrofit (40CFR52.2039-1) means a system or device (such as a modification to the engine's carburetor or positive crankcase ventilation system) that results in engine operation at an increased air-fuel ratio so as to achieve reductions in exhaust emissions of hydrocarbons and carbon monoxide from 1967 and earlier light-duty vehicles of at least 21 percent and 58 percent, respectively.

air bleed to intake manifold retrofit (40CFR52.2490-1) means a system or device (such as modification to the engine's carburetor) that results in engine operation at an increased air-fuel ratio so as to achieve reduction in exhaust emissions of hydrocarbon and carbon monoxide from 1967 and earlier light-duty vehicles of at least 21 and 58 percent, respectively, and from 1973 and earlier medium-duty vehicles of at least 15 and 30 percent, respectively.

air changes per hour (ACH) (EPA-94/04): The movement of a volume of air in a given period of time; if a house has one air change per hour, it means that all of the air in the house will be replaced in a one-hour period.

air cleaner filter element (40CFR85.2122(a)(16)(ii)-(A)) means a device to remove particulates from the primary air that enters the air induction system of the engine.

air contaminant (40CFR52.741) means any solid, liquid, or gaseous matter, any odor, or any form of energy, that is capable of being released into the atmosphere from an emission source (cf. air pollutant).

air contaminant (40CFR52.274.d.2-vii) means nitrogen dioxide, particulate matter, and/or sulfur dioxide and particulate matter combined.

air contaminant (EPA-94/04): Any particulate matter, gas, or combination thereof, other than water vapor (cf. air pollutant).

air curtain (EPA-94/04): A method of containing oil spills. Air bubbling through a perforated pipe causes an upward water flow that slows the spread of oil. It can also be used to stop fish from entering polluted water.

air dried coatings (40CFR52.741) means any coatings that dry by use of air or forced air at temperatures up to 363.15 K (194 F).

air emissions (40CFR129.2-o) means the release or discharge of a toxic pollutant by an owner or operator into the ambient air either (1) by means of a stack or (2) as a fugitive dust, mist or vapor as a result inherent to the manufacturing or formulating process.

air erosion (40CFR763.83) means the passage of air over friable ACBM (asbestos containing building material) which may result in the release of asbestos fibers.

air gap (EPA-94/04): Open vertical gap or empty space that separates drinking water supply to be protected from another water system in a treatment plant or other location. The open gap protects the drinking water from contamination by backflow or backsiphonage.

air mass (EPA-94/04): A large volume of air with certain meteorological or polluted characteristics, e.g., a heat inversion or smogginess while in one location. The characteristics can change as the air mass moves away.

air monitoring (EPA-94/04): See monitoring.

air oxidation reactor (40CFR60.611) means any device or process vessel in which one or more organic reactants are combined with air, or a combination of air and oxygen, to produce one or more organic compounds. Ammoxidation and oxychlorination reactions are included in this definition.

air oxidation reactor (40CFR63.101) means a device or vessel in which air, or a combination of air and oxygen, is used as an oxygen source in combination with one or more organic reactants to produce one or more organic compounds. Air oxidation reactor includes the product separator and any associated vacuum pump or steam jet.

air oxidation reactor (40CFR63.111) means a device or vessel in which air, or a combination of air and oxygen, is used as an oxygen source in combination with one or more organic reactants to produce one or more organic compounds. Air oxidation reactor includes the product separator and any associated vacuum pump or steam jet.

air oxidation reactor recovery train (40CFR60.611) means an individual recovery system receiving the vent stream from at least one air oxidation reactor, along with all air oxidation reactors feeding vent streams into this system.

air oxidation unit process (40CFR60.611) means a unit process, including ammoxidation and oxychlorination unit process, that uses air, or a combination of air and oxygen, as an oxygen source in combination with one or more organic reactants to produce one or more organic compounds.

air padding (EPA-94/04): Pumping dry air into a

container to assist with the withdrawal of liquid or to force a liquified gas such as chlorine out of the container.

air plenum (EPA-94/04): Any space used to convey air in a building, furnace, or structure. The space above a suspended ceiling is often used as an air plenum.

air pollutant (CAA302-42USC7602) means any air pollutant agent or combination of such agents, including any physical, chemical, biological, radioactive (including source material, special nuclear material, and byproduct material) substance or matter which is emitted into or otherwise enters the ambient air.

air pollutant (EPA-94/04): Any substance in air that could, in high enough concentration, harm man, other animals, vegetation, or material. Pollutants may include almost any natural or artificial composition of airborne matter capable of being airborne. They may be in the form of solid particles, liquid droplets, gases, or in combination thereof. Generally, they fall into two main groups:
(1) those emitted directly from identifiable sources and
(2) those produced in the air by interaction between two or more primary pollutants, or by reaction with normal atmospheric constituents, with or without photoactivation.
Exclusive of pollen, fog, and dust, which are of natural origin, about 100 contaminants have been identified and fall into the following categories: solids, sulfur compounds, volatile organic chemicals, nitrogen compounds, oxygen compounds, halogen compounds, radioactive compounds, and odors (under CAA).

air pollution (40CFR52.741) means the presence in the atmosphere of one or more air contaminants in sufficient quantities and of such characteristics and duration as to be injurious to human, plant, or animal life, to health, or to property, or to unreasonably interfere with the enjoyment of life or property.

air pollution (EPA-94/04): The presence of contaminant or pollutant substances in the air that do not disperse properly and interfere with human health or welfare, or produce other harmful environmental effects.

air pollution control agency (40CFR15.4) means any agency which is defined in action 302(b) or section 302(c) of the CAA.

air pollution control agency (CAA302-42USC7602) means any of the following:
(1) A single State agency designated by the Governor of that State as the official State air pollution control agency for purposes of this Act;
(2) An agency established by two or more States and having substantial powers or duties pertaining to the prevention and control of air pollution;
(3) A city, county, or other local government health authority, or, in the case of any city, county, or other local government in which there is an agency other than the health authority charged with responsibility for enforcing ordinances or laws relating to the prevention and control of air pollution, such other agency; or
(4) An agency of two or more municipalities located in the same State or in different States and having substantial powers or duties pertaining to the prevention and control of air pollution (cf. interstate air pollution control agency).

air pollution control device (40CFR503.41-a) is one or more processes used to treat the exit gas from a sewage sludge incinerator stack.

air pollution control device (EPA-94/04): Mechanism or equipment that cleans emissions generated by an incinerator by removing pollutants that would otherwise be released to the atmosphere.

air pollution control equipment (40CFR52.741) means any equipment or facility of a type intended to eliminate, prevent, reduce or control the emission of specified air contaminants to the atmosphere.

air pollution emergency (40CFR51-AL-1.0): This regulation is designed to prevent the excessive buildup of air pollutants during air pollution episodes, thereby preventing the occurrence of an emergency due to the effects of these pollutants on the health of persons.

air pollution episode (EPA-94/04): A period of abnormally high concentration of air pollutants, often due to low winds and temperature inversion, that can cause illness and death (cf. episode).

air quality control region (EPA-94/04): Federally designated area that is required to meet and maintain federal ambient air quality standards. May include nearby locations in the same state or nearby states that share common air pollution problems.

air quality criteria (EPA-94/04): The levels of pollution and lengths of exposure above which adverse health and welfare effects may occur.

air quality restricted operation of a spray tower (40CFRd417.151-e) shall mean an operation utilizing formulations (e.g., those with high non-ionic content) which require a very high rate of wet scrubbing to maintain desirable quality of stack gases, and thus generate much greater quantities of waste water than can be recycled to process.

air quality standards (EPA-94/04): The level of pollutants prescribed by regulations that may not be exceeded during a given time in a defined area.

air stripping (EPA-94/04): A treatment system that removes volatile organic compounds (VOCs) from contaminated ground water or surface water by forcing an airstream through the water and causing the compounds to evaporate.

air stripping operation (40CFR264.1031) is a desorption operation employed to transfer one or more volatile components from a liquid mixture into a gas (air) either with or without the application of heat to the liquid. Packed towers, spray towers, and bubble-cap, sieve, or valve-type plate towers are among the process configurations used for contacting the air and a liquid.

air suspension coater/dryer (40CFR52.741) means a pharmaceutical coating operation which consists of vertical chambers in which tablets or particles are placed, and a coating is applied and then dried while the tablets or particles are kept in a fluidized state by the passage of air upward through the chambers.

air to fuel ratio (40CFR60-AA (method 28A-2.2)) means the ratio of the mass of dry combustion air introduced into the firebox, to the mass of dry fuel consumed (grams of dry air per gram of dry wood burned).

air toxics (EPA-94/04): Any air pollutant for which a national ambient air quality standard does not exist (i.e., excluding ozone, carbon monoxide, PM-10, sulfur dioxide, nitrogen oxide) that may reasonably be anticipated to cause cancer, developmental effects, reproductive dysfunctions, neurological disorders, heritable gene mutations, or other serious or irreversible chronic or acute health effects in humans.

airborne particulates (EPA-94/04): Total suspended particulate matter found in the atmosphere as solid particles or liquid droplets. Chemical composition of particulates varies widely, depending on location and time of year. Airborne particulates include: windblown dust, emissions from industrial processes, smoke from the burning of wood and coal, and motor vehicle or non-road engine exhausts.

airborne radioactive material (10CFR20.3) means any radioactive material dispersed in the air in the form of dusts, fumes, mists, vapors, or gases.

airborne release (EPA-94/04): Release of any chemical into the air.

aircraft (40CFR87.1) means any airplane for which a U.S. standard airworthiness certificate or equivalent foreign airworthiness certificate is issued.

aircraft engine (40CFR87.1) means a propulsion engine which is installed in or which is manufactured for installation in an aircraft.

aircraft gas turbine engine (40CFR87.1) means a turboprop, turbofan, or turbojet aircraft engine (under CAA).

airless spray (40CFR52.741) means a spray coating method in which the coating is atomized by forcing it through a small opening at high pressure. The coating liquid is not mixed with air before exiting from the nozzle.

airport (40CFR257.3.8-1) means public-use airport open to the public without prior permission and without restrictions within the physical capabilities of available facilities.

alachlor (EPA-94/04): A herbicide, marketed under the trade name Lasso, used mainly to control weeds in corn and soybean fields.

alar (EPA-94/04): Trade name for daminozide, a pesticide that makes apples redder, firmer, and less likely to drop off trees before growers are ready to pick them. It is also used to a lesser extent on peanuts, tart cherries, concord grapes, and other fruits.

ALA (EPA-94/04): Delta-aminolevulinic acid.

Alaskan north slope (40CFR60.591) means the approximately 69,000 square mile area extending from the Brooks Range to the Arctic Ocean (other identical or similar definitions are provided in 40CFR60.631).

alcohol abuse (40CFR7.25) means any misuse of alcohol which demonstrably interferes with a person's health, interpersonal relations or working ability.

aldicarb (EPA-94/04): An insecticide sold under the trade name Temik. It is made from ethyl isocyanate.

aldrin/dieldrin (40CFR129.4-a):
(1) Aldrin means the compound aldrin as identified by the chemical name, 1,2,3,4,10,10-hexachloro-1,4,4a,5,8,8a-hexahydro-1,4-endo-5, 8-exo-dimethanonaphthalene;
(2) Dieldrin means the compound dieldrin as identified by the chemical name 1,2,3,4,10,10-hexachloro-6,7-epoxy-1,4,4a,5,6,7,8,8a-octahydro-1,4-endo-5,8-exo-dimethanonaphth-alene.

aldrin/dieldrin formulator (40CFR129.100-2) means a person who produces, prepares or processes a formulated product comprising a mixture of either aldrin or dieldrin and inert materials or other diluents, into a product intended for application in any use registered under the Federal Insecticide, Fungicide and Rodenticide Act, as amended (7USC135, et seq.).

aldrin/dieldrin manufacturer (40CFR129.100-1) means a manufacturer, excluding any source which is

exclusively an aldrin/dieldrin formulator, who produces, prepares or processes technical aldrin or dieldrin or who uses aldrin or dieldrin as a material in the production, preparation or processing of another synthetic organic substance.

alert (10CFR30.4) means events may occur, are in progress, or have occurred that could lead to a release of radioactive material but that the release is not expected to require a response by offsite response organizations to protect persons offsite (other identical or similar definitions are provided in 10CFR40.4).

alert (10CFR70.4) means events may occur, are in progress, or have occurred that could lead to a release of radioactive material[s] but that the release is not expected to require a response by offsite response organizations to protect persons offsite.

algae (EPA-94/04): Simple rootless plants that grow in sunlit waters in proportion to the amount of available nutrients. They can affect water quality adversely by lowering the dissolved oxygen in the water. They are food for fish and small aquatic animals.

algal blooms (EPA-94/04): Sudden spurts of algal growth, which can affect water quality adversely and indicate potentially hazardous changes in local water chemistry.

algicidal (40CFR797.1050-1) means having the property of killineans the compound dieldrin as identified by the chemical name 1,2,3,4,10,10-hexachloro-6,7-epoxy-1,4,4a,5,6,7,8,8a-octahydro-1,4-endo-5,8-exo-dimethanonaphth-alene.

aldrin/dieldrin formulator (40CFR129.100-2) means a person who produces, prepares or processes a formulated product comprising a mixture of either aldrin or dieldrin and inert materials or other diluents, into a product intended for application in any use registered under the Federal Insecticide, Fungicide and Rodenticide Act, as amended (7USC135, et seq.).

aldrin/dieldrin manufacturer (40CFR129.100-1) means a manufacturer, excluding any source which isly an aldrin/dieldrin formulator, who produces, prepares or processes technical aldrin or dieldrin or who uses aldrin or dieldrin as a material in the production, preparation or processing of another synthetic organic substance.

alert (10CFR30.4) means events may occur, are in progress, or have occurred that could lead to a release of radioactive material but that the release is not expected to require a response by offsite response organizations to protect persons offsite (other identical or similar definitions are provided in 10CFR40.4).

alert (10CFR70.4) means events may occur, are in progress, or have occurred that could lead to a release of radioactive material[s] but that the release is not expected to require a response by offsite response organizations to protect persons offsite.

algae (EPA-94/04): Simple rootless plants that grow in sunlit waters in proportion to the amount of available nutrients. They can affect water quality adversely by lowering the dissolved oxygen in the water. They are food for fish and small aquatic animals.

algal blooms (EPA-94/04): Sudden spurts of algal growth, which can affect water quality adversely and indicate potentially hazardous changes in local water chemistry.

algicidal (40CFR797.1050-1) means having the property of killing algae.

algicide (EPA-94/04): Substance or chemical used specifically to kill or control algae.

algistatic (40CFR797.1050-2) means having the property of inhibiting algal growth.

ALJ: Administrative law judge (EPA-94/04)

alkaline (EPA-94/04): The condition of water or soil which contains a sufficient amount of alkali substance to raise the pH above 7.0.

alkaline cleaning (40CFR471.02-c) uses a solution (bath), usually detergent, to remove lard, oil, and other such compounds from a metal surface. Alkaline cleaning is usually followed by a water rinse.

The rinse may consist of single or multiple stage rinsing. For the purposes of this part, an alkaline cleaning operation is defined as a bath followed by a rinse, regardless of the number of rinse stages. Each alkaline cleaning bath and rinse combination is entitled to a discharge allowance.

alkaline cleaning bath (40CFR468.02-a) shall mean a bath consisting of an alkaline cleaning solution through which a workpiece is processed.

alkaline cleaning rinse (40CFR468.02-b) shall mean a rinse following an alkaline cleaning bath through which a workpiece is processed. A rinse consisting of a series of rinse tanks is considered as a single rinse (under CWA).

alkaline cleaning rinse for forged parts (40CFR468.02-s) shall mean a rinse following an alkaline cleaning bath through which a forged part is processed. A rinse consisting of a series of rinse tanks is considered as a single rinse.

alkalinity (EPA-94/04): The capacity of water to neutralize acids.

all electric melter (40CFR60.291) means a glass melting furnace in which all the heat required for melting is provided by electric current from electrodes submerged in the molten glass, although some fossil fuel may be charged to the furnace as raw material only.

allegation (40CFR717.3-a) means a statement, made without formal proof or regard for evidence, that a chemical substances or mixture has caused significant adverse reaction to health or the environment (under TSCA).

allergic contact dermatitis (40CFR798.4100-1): See skin sensitization.

alley collection (40CFR243.101-a) means the collection of solid waste from containers placed adjacent to or in an alley.

allocate or allocation (40CFR72.2) means the initial crediting of an allowance by the Administrator to an Allowance Tracking System unit account or general account.

allotment (40CFR35.105) means an amount representing a State's share of funds requested in the President's budget or appropriated by Congress for an environmental program, as EPA determines after considering any factors indicated by this regulation. The allotment is not an entitlement but rather the objective basis for determining the range for a State's planning target.

allotment management plan (FLPMA10343USC-1702) means a document prepared in consultation with the lessees or permittees involved, which applies to livestock operations on the public lands or on lands within National Forests in the eleven contiguous Western States and which:
(1) prescribes the manner and extent to, which livestock operations will be conducted in order to meet the multiple-use, sustained yield, economic and other needs and objectives as determined for the lands by the Secretary concerned; and
(2) describes the type, location, ownership, and general specifications for the range improvements to be installed and maintained on the lands to meet the livestock grazing and other objectives of land management; and
(3) contains such other provisions relating to livestock grazing and other objectives found by the Secretary concerned to be consistent with the provisions of this Act and other applicable law.

allowable 1985 emissions rate (CAA402-42USC7651a) means a federally enforceable emissions limitation for sulfur dioxide or oxides of nitrogen, applicable to the unit in 1985 or the limitation applicable in such other subsequent year as determined by the Administrator if such a limitation for 1985 does not exist. Where the emissions limitation for a unit is not expressed in pounds of emissions per million Btu, or the averaging period of that emissions limitation is not expressed on an annual basis, the Administrator shall calculate the annual equivalent of that emissions limitation in pounds per million Btu to establish the allowable 1985 emissions rate.

allowable costs (40CFR30.200) means those project costs that are eligible, reasonable, necessary, and allocable to the project; permitted by the appropriate

Federal cost principles, and approved by EPA in the assistance agreement.

allowable costs (40CFR35.6015-2) means those project costs that are: eligible, reasonable, necessary, and allocable to the project; permitted by the appropriate Federal cost principles; and approved by EPA in the cooperative agreement and/or Superfund State Contract.

allowable emissions (40CFR51.165-xi) means the emissions rate of a stationary source calculated using the maximum rated capacity of the source (unless the source is subject to federally enforceable limits which restrict the operating rate, or hours of operation, or both) and the most stringent of the following:
(A) The applicable standards set forth in 40CFR part 60 or 61;
(B) Any applicable State Implementation Plan emissions limitation including those with a future compliance date; or
(C) The emissions rate specified as a federally enforceable permit condition, including those with a future compliance date.

allowable emissions (40CFR51.166-16) means the emissions rate of a stationary source calculated using the maximum rated capacity of the source (unless the source is subject to federally enforceable limits which restrict the operating rate, or hours of operation, or both) and the most stringent of the following:
(i) The applicable standards as set forth in 40CFR parts 60 and 61;
(ii) The applicable State Implementation Plan emissions limitation, including those with a future compliance date; or
(iii) The emissions rate specified as a federally enforceable permit condition.

allowable emissions (40CFR51.491) means the emissions of a pollutant from an affected source determined by taking into account the most stringent of all applicable SIP emissions limits and the level of emissions consistent with source compliance with all Federal requirements related to attainment and maintenance of the NAAQS and the production rate associated with the maximum rated capacity and hours of operation (unless the source is subject to federally enforceable limits which restrict the operating rate, or hours of operation, or both).

allowable emissions (40CFR51-AS-11) means the emissions rate of a stationary source calculated using the maximum rated capacity of the source (unless the source is subject to federally enforceable limits which restrict the operating rate, or hours of operation, or both) and the most stringent of the following:

(A) The applicable standards set forth in 40CFR60 or 61;

(B) Any applicable State Implementation Plan emissions limitation including those with a future compliance date; or

(C) The emissions rate specified as a federally enforceable permit condition, including those with a future compliance date.

allowable emissions (40CFR52.21-16) means the emissions rate of a stationary source calculated using the maximum rated capacity of the source (unless the source is subject to federally enforceable limits which restrict the operating rate, or hours of operation, or both) and the most stringent of the following:

(i) The applicable standards as set forth in 40CFR parts 60 and 61;

(ii) The applicable State Implemenation Plan emissions limitation, including those with a future compliance date; or

(iii) The emissions rate specified as a federally enforceable permit condition, including those with a future compliance date.

allowable emissions (40CFR52.24-11) means the emissions rate of a stationary source calculated using the maximum rated capacity of the source (unless the source is subject to federally enforceable limits which restrict the operating rate, or hours of operation, or both) and the most stringent of the following:

(i) The applicable standards set forth in 40CFR parts 60 and 61;

(ii) Any applicable State Implementation Plan emissions limitation, including those with a future compliance date; or

(iii) The emissions rate specified as a federally enforceable permit condition, including those with a future compliance date.

allowable emissions (40CFR52.741) means the quantity of VOM emissions during a particular time period from a stationary source calculated using the maximum rated capacity of the source (unless restricted by federally enforceable limitations on operating rate, hours of operation, or both) and the most stringent of:

(A) The applicable standards in 40CFR60 and 61;

(B) The applicable implementation plan; or

(C) A federally enforceable permit.

allowable pressure change (40CFR60-AA (method 27-2.8)): The allowable amount of decrease in pressure during the static pressure test, within the time period t, as specified in the appropriate regulation, in mm H_2O.

allowable SO_2 emissions rate (40CFR72.2) means the most stringent federally enforceable emissions limitation for sulfur dioxide (in lb/mmBtu (pounds per million Btu)) applicable to the unit or combustion source for the specified calendar year, or for such subsequent year as determined by the Administrator where such a limitation does not exist for the specified year; provided that, if a Phase I or Phase II unit is listed in the NADB (national atmospheric data bank), the "1985 allowable SO_2 emissions rate" for the Phase I or Phase II unit shall be the rate specified by the Administrator in the NADB under the data field "1985 annualized boiler SO_2 emission limit."

allowable vacuum change (40CFR60-AA (method 27-2.9)): The allowable amount of decrease in vacuum during the static vacuum test, within the time period t, as specified in the appropriate regulation, in mm H_2O.

allowance (40CFR35.2005-3) means an amount based on a percentage of the projects allowable building cost, computed in accordance with Appendix B.

allowance (40CFR72.2) means an authorization by the Administrator under the Acid Rain Program to emit up to one ton of sulfur dioxide during or after a specified calendar year.

allowance (CAA402-42USC7651a) means an authorization, allocated to an affected unit by the

Administrator under this title, to emit, during or after a specified calendar year, one ton of sulfur dioxide.

allowance deduction, or deduct (40CFR72.2) when referring to allowances, means the permanent withdrawal of allowances by the Administrator from an Allowance Tracking System compliance subaccount, or future year subaccount, to account for the number of tons of SO_2 emissions from an affected unit for the calendar year, for tonnage emissions estimates calculated for periods of missing data as provided in part 75 of this chapter, or for any other allowance surrender obligations of the Acid Rain Program.

allowance reserve (40CFR72.2) means any bank of allowances established by the Administrator in the Allowance Tracking System pursuant to sections 404(a)(2) (Phase I extension reserve), 404(g) (energy conservation and renewable energy reserve), or 416(b) (special allowance reserve) of the Act, and implemented in accordance with part 73, subpart B of this chapter.

allowance Tracking or ATS (40CFR72.2) means the Acid Rain Program system by which the Administrator allocates, records, deducts, and tracks allowances.

allowance Tracking System account (40CFR72.2) means an account in the Allowance Tracking System established by the Administrator for purposes of allocating, holding, transferring, and using allowances.

allowance transfer deadline (40CFR72.2) means midnight of January 30 or, if January 30 is not a business day, midnight of the first business day thereafter and is the deadline by which allowances may be submitted for recordation in an affected unit's compliance subaccount for the purposes of meeting the unit's Acid Rain emissions limitation requirements for sulfur dioxide for the previous calendar year.

allowances held or hold allowances (40CFR72.2) means the allowances recorded by the Administrator, or submitted to the Administrator for recordation in accordance with 40CFR73.50 of this chapter, in an Allowance Tracking System account.

alluvial (EPA-94/04): Relating to mud and/or sand deposited by flowing water.

alluvial valley floors (SMCRA701-30USC1291) means the unconsolidated stream laid deposits holding streams where water availability is sufficient for sub-irrigation or flood irrigation agricultural activities but does not include upland areas which are generally overlain by a thin veneer of colluvial deposits composed chiefly of debris from sheet erosion, deposits by unconcentrated runoff or slope wash, together with talus, other mass movement accumulation and windblown deposits.

altered discharge (40CFR125.58-b) means any discharge other than a current discharge or improved discharge, as defined in this regulation.

alternate method (EPA-94/04): Any method of sampling and analyzing for an air pollutant that is not a reference or equivalent method but that has been demonstrated in specific cases:
- to EPA's satisfaction
- to produce results adequate for compliance monitoring.

alternative contemporaneous annual emission limitation (40CFR76.2) means the maximum allowable NO_x emission rate (on a lb/mmBtu (pounds per million Btu), annual average basis) assigned to an individual unit in a NO_x emissions averaging plan pursuant to 40CFR76.10.

alternative courses of action (ESA3-16USC1531) means all alternatives and thus is not limited to original project objectives and agency jurisdiction.

alternative effluent limitations (40CFR125.71-a) means all effluent limitations or standards of performance for the control of the thermal component of any discharge which are established under section 316(a) and this subpart.

alternative emission limitation (40CFR63.2) means conditions established pursuant to sections 112(i)(5) or 112(i)(6) of the Act by the Administrator or by a State with an approved permit program.

alternative emission standard (40CFR63.2) means an alternative means of emission limitation that, after notice and opportunity for public comment, has been demonstrated by an owner or operator to the Administrator's satisfaction to achieve a reduction in emissions of any air pollutant at least equivalent to the reduction in emissions of such pollutant achieved under a relevant design, equipment, work practice, or operational emission standard, or combination thereof, established under this part pursuant to section 112(h) of the Act.

alternative fuels (EPA-94/04): Substitutes for traditional liquid, oil-derived motor vehicle fuels like gasoline and diesel. Includes methanol, ethanol, compressed natural gas, and others.

alternative method (40CFR60.2) means any method of sampling and analyzing for an air pollutant which is not a reference or equivalent method but which has been demonstrated to the Administrator's satisfaction to, in specific cases, produce results adequate for his determination of compliance (other identical or similar definitions are provided in 40CFR61.02).

alternative method of compliance (CAA402-42USC7651a) means a method of compliance in accordance with one or more of the following authorities:
(A) a substitution plan submitted and approved in accordance with subsections 404 (b) and (c);
(B) a Phase I extension plan approved by the Administrator under section 404(d), using qualifying Phase I technology as determined by the Administrator in accordance with that section; or
(C) repowering with a qualifying clean coal technology under section 409.

alternative monitoring system (40CFR72.2) means a system or a component of a system designed to provide direct or indirect data of mass emissions per time period, pollutant concentrations, or volumetric flow, that is demonstrated to the Administrator as having the same precision, reliability, accessibility, and timeliness as the data provided by a certified CEMS or certified CEMS component in accordance with part 75 of this chapter.

alternative remedial contract strategy contractors (EPA-94/04): Government contractors who provide project management and technical services to support remedial response activities at National Priorities List sites.

alternative technology (40CFR35.2005-4) means proven wastewater treatment processes and techniques which provide for the reclaiming and reuse of water, productively recycle wastewater constituents or otherwise eliminate the discharge of pollutants, or recover energy. Specifically, alternative technology includes land application of effluent and sludge; aquifer recharge; aquaculture; direct reuse (non-potable); horticulture; revegetation of disturbed land; containment ponds; sludge composting and drying prior to land application; self-sustaining incineration; and methane recovery.

alternative technology (40CFR76.2) means a control technology for reducing NO_x emissions that is outside the scope of the definition of low NO_x burner technology. Alternative technology does not include overfire air as applied to wall-fired boilers or separated overfire air as applied to tangentially fired boilers.

alternative technology (EPA-94/04): Approach that aims to use resources efficiently or to substitute resources in order to do minimum damage to the environment. This approach permits a large degree of personal user control over the technology.

alternative test method (40CFR63.2) means any method of sampling and analyzing for an air pollutant that is not a test method in this chapter and that has been demonstrated to the Administrator's satisfaction, using Method 301 in Appendix A of this part, to produce results adequate for the Administrator's determination that it may be used in place of a test method specified in this part.

alternative to conventional treatment works for a small community (40CFR35.2005-5) for purposes of 40CFR35.2020 and 35.2032, alternative technology used by treatment works in small communities include alternative technologies defined in paragraph (4), as well as, individual and onsite systems; small diameter gravity, pressure or vacuum sewers conveying treated or partially treated wastewater. These systems can also include small diameter gravity sewers carrying raw wastewater to cluster systems.

alternative water supplies (40CFR300.5), as defined by section 101(34) of CERCLA, includes, but is not limited to, drinking water and household water supplies.

alternative water supplies (SF101-42USC9601) includes, but is not limited to, drinking water and household water supplies.

altitude performance adjustments (40CFR86.1602) are adjustments or modifications made to vehicle, engine, or emission control functions in order to improve emission control performance at altitudes other than those for which the vehicles were designed.

aluminum basis material (40CFR465.02-g) means aluminum, aluminum alloys and aluminum coated steels which are processed in coil coating.

aluminum casting (40CFR464.02-a) means the remelting of aluminum or an aluminum alloy to form a cast intermediate or final product by pouring or forcing the molten metal into a mold, except for ingots, pigs, or other cast shapes related to nonferrous (primary and secondary) metals manufacturing (40CFR421) and aluminum forming (40CFR467). Processing operations following the cooling of castings not covered under aluminum forming, except for grinding scrubber operations which are covered here, are covered under the electroplating and metal finishing point source categories (40CFR413 and 433).

aluminum equivalent (40CFR60.191) means an amount of aluminum which can be produced from a Mg of anodes produced by an anode bake plant as determined by 40CFR60.195(g).

aluminum forming (40CFR467.02-a) is a set of manufacturing operations in which aluminum and aluminum alloys are made into semifinished products by hot or cold working.

ambient air (40CFR50.1-e) means that portion of the atmosphere, external to buildings, to which the general public has access.

ambient air (40CFR57.103-f) shall have the meaning given by 40CFR50.1(e), as that definition appears upon promulgation of this subpart, or as hereafter amended.

ambient air (EPA-94/04): Any unconfined portion of the atmosphere: open air, surrounding air.

ambient air quality (40CFR57.103-g) refers only to concentrations of sulfur dioxide in the ambient air, unless otherwise specified.

ambient air quality standards (40CFR52.741) means those standards designed to protect the public health and welfare codified in 40CFR50 and promulgated from time to time by the USEPA pursuant to authority contained in section 108 of the Clean Air Act, 42USC7401 et seq., as amended from time to time.

ambient air quality standards (EPA-94/04): See criteria pollutants and national ambient air quality standards.

ambient air quality violation (40CFR52.1475-ii) means any single ambient concentration of sulfur dioxide that exceeds any National Ambient Air Quality Standard for sulfur dioxide at any point in a designated liability area, as specified in paragraph (e)(8) of this section.

ambient aquatic life advisory concentrations (AALACS) (40CFR300-AA) means EPA's advisory concentration limit for acute or chronic toxicity to aquatic organisms as established under section 304(a)(1) of the Clean Water Act, as amended (cf. ambient water criterion).

ambient temperature (EPA-94/04): Temperature of the surrounding air or other medium.

ambient water criterion (40CFR129.2-g) means that concentration of a toxic pollutant in a navigable water that, based upon available data, will not result in adverse impact on important aquatic life, or on consumers of such aquatic life, after exposure of that aquatic life for periods of time exceeding 96 hours and continuing at least through one reproductive cycle; and will not result in a significant risk of adverse health effects in a large human population based on available information such as mammalian laboratory toxicity data, epidemiological studies of human occupational exposures, or human exposure data, or any other relevant data.

ambient water criterion for aldrin/dieldrin in navigable waters (40CFR129.100-3) is 0.003 ug/L.

ambient water criterion for benzidine in navigable waters (40CFR129.104-3) is 0.1 ug/L.

ambient water criterion for DDT in navigable waters (40CFR129.101-3) is 0.001 ug/L.

ambient water criterion for endrin in navigable waters (40CFR129.102-3) is 0.004 ug/L.

ambient water criterion for PCBs in navigable waters (40CFR129.105-4) is 0.001 ug/L.

ambient water criterion for toxaphene in navigable waters (40CFR129.103-3) is 0.005 ug/L.

ambient water quality criteria (AWQC) (40CFR300-AA) means EPA's maximum acute or chronic toxicity concentrations for protection of aquatic life and its uses as established under section 304(a)(1) of the Clean Water Act, as amended.

amendment review (40CFR152.403-e) means review of any application requiring Agency approval to amend the registration of a currently registered product, or for which an application is pending Agency decision, not entailing a major change to the use pattern of an active ingredient.

ammonia-N (or ammonia-nitrogen) (40CFR420.02-c) means the value obtained by manual distillation (at pH 9.5) followed by the Nesslerization method specified in 40CFR136.3.

ammonium sulfate dryer (40CFR60.421) means a unit or vessel into which ammonium sulfate is charged for the purpose of reducing the moisture content of the product using a heated gas stream. The unit includes foundations, superstructure, material charger systems, exhaust systems, and integral control systems and instrumentation.

ammonium sulfate feed material streams (40CFR60.421) means the sulfuric acid feed stream to the reactor/crystallizer for synthetic and coke oven byproduct ammonium sulfate manufacturing plants; and means the total or combined feed streams (the oximation ammonium sulfate stream and the rearrangement reaction ammonium sulfate stream) to the crystallizer stage, prior to any recycle streams.

ammonium sulfate manufacturing plant (40CFR60.421) means any plant which produces ammonium sulfate.

amount of pesticide or active ingredient (40CFR169.1-a) means the weight or volume of the pesticide or active ingredient used in producing a pesticide expressed as weight for solid or semi-solid products and as weight or volume of liquid products.

amount of pesticide product (40CFR167.3) means quantity, expressed in weight or volume of the product, and is to be reported in pounds for solid or semi-solid pesticides and active ingredients or gallons for liquid pesticides and active ingredients, or number of individual retail units for devices.

amperometric titration (EPA-94/04): A way of measuring concentrations of certain substances in water using an electric current that flows during a chemical reaction.

amplitude (40CFR798.6850-3) is the voltage excursion recorded during the process of recording the compound nerve action potential. It is an indirect measure of the number of axons firing.

anaerobic (EPA-94/04): A life or process that occurs in, or is not destroyed by, the absence of oxygen.

anaerobic decomposition (EPA-94/04): Reduction of the net energy level and change in chemical composition of organic matter caused by microorganisms in an oxygenfree environment.

anaerobic digestion (40CFR503.31-b) is the biochemical decomposition of organic matter in sewage sludge into methane gas and carbon dioxide by microorganisms in the absence of air.

analysis using exposure or medical records (29CFR1910.20) means any compilation of data or any statistical study based at least in part on information collected from individual employee exposure or medical records of information collected from health insurance claims records, provided that either the analysis has been reported to the employer or no further work is currently being done by the person responsible for preparing the analysis (under OSHA).

analytical sensitivity (40CFR763-AA-1) means the airborne asbestos concentration represented by each fiber counted under the electron microscope. It is determined by the air volume collected and the proportion of the filter examined. This method requires that the analytical sensitivity be no greater than 0.005 structures/cm^3.

analyzer (40CFR53.1-i): See automated method.

analyzer calibration error (40CFR60-AA (method 6C-3.4)) means the difference between the gas concentration exhibited by the gas analyzer and the known concentration of the calibration gas when the calibration gas is introduced directly to the analyzer (other identical or similar definitions are provided in 40CFR60-AA (method 7E-3.2)).

ancillary equipment (40CFR260.10) means any device including, but not limited to, such devices as piping, fittings, flanges, valves, and pumps, that is used to distribute, meter, or control the flow of hazardous waste from its point of generation to a storage or treatment tank(s) between hazardous waste storage and treatment tanks to a point of disposal onsite, or to a point of shipment for disposal off-site.

ancillary equipment (40CFR280.12) means any devices including, but not limited to, such devices as piping, fittings, flanges, valves, and pumps used to distribute, meter, or control the flow of regulated substances to and from an UST (underground storage tank).

ancillary equipment (40CFR63.321) means the equipment used with a dry cleaning machine in a dry cleaning system including, but not limited to, emission control devices, pumps, filters, muck cookers, stills, solvent tanks, solvent containers, water separators, exhaust dampers, diverter valves, interconnecting piping, hoses, and ducts (under CAA).

ancillary operation (40CFR467.02-b) is a manufacturing operation that has a large flow, discharges significant amounts of pollutants, and may not be present at every plant in a subcategory, but when present is an integral part of the aluminum forming process.

ancillary operation (40CFR467.11-b) shall mean any operation not previously included in the core, performed on-site, following or preceding the rolling operation. The ancillary operations shall include continuous rod casting, continuous sheet casting, solution heat treatment, cleaning or etching.

ancillary operation (40CFR467.21-b) shall mean any operation not previously included in the core, performed on-site, following or preceding the rolling operation. The ancillary operations shall include direct chill casting, solution heat treatment, cleaning or etching, and degassing.

ancillary operation (40CFR467.31-c) shall mean any operation not previously included in the core, performed on-site, following or preceding the extrusion operation. The ancillary operations shall include direct chill casting, press or solution heat treatment, cleaning or etching, degassing, and extrusion press hydraulic fluid leakage (under the Clean Water Act).

ancillary operation (40CFR467.41-b) shall mean any operation not previously included in the core, performed on-site, following or preceding the forging operation. The ancillary operations shall include forging air pollution scrubbers, solution heat treatment, and cleaning or etching.

ancillary operation (40CFR467.51-b) shall mean any operation not previously included in the core, performed on-site, following or preceding the drawing operation. The ancillary operation shall include continuous rod casting, solution heat treatment, and cleaning or etching.

ancillary operation (40CFR467.61-b) shall mean any operation not previously included in the core, performed on-site, following or preceding the drawing operation. The ancillary operations shall include continuous rod casting, solution heat treatment and cleaning or etching.

ancillary operation (40CFR468.02-c) shall mean any operation associated with a primary forming operation. These ancillary operations include surface and heat treatment, hydrotesting, sawing, and surface coating.

ancillary operations (40CFR461.2-c) means all of the operations specific to battery manufacturing and not included specifically within anode or cathode manufacture (ancillary operations are primarily associated with battery assembly and chemical production of anode or cathode active materials) (under CWA).

anemometer (29CFR1910.66) means an instrument for measuring wind velocity.

angle of projection (40CFR60-AB-2.9) means the angle that contains all of the radiation projected from the lamp assembly of the analyzer at a level of greater than 2.5 percent of the peak illuminance.

angle of view (40CFR60-AB-2.8) means the angle that contains all of the radiation detected by the photodetector assembly of the analyzer at a level greater than 2.5 percent of the peak detector response.

anhydrous product (40CFR417.11-b) shall mean the theoretical product that would result if all water were removed from the actual product (other identical or similar definitions are provided in 40CFR417.21-b; 417.31-b; 417.41-b; 417.51-b; 417.61-b; 417.71-b; 417.81-b; 417.91-b; 417.101-b; 417.111-b; 417.121-b; 417.131-b; 417.141-b; 417.151-b; 417.161-b; 417.171-b; 417.181-b; 417.191-b).

animal (FIFRA2-7USC136) means all vertebrate and invertebrate species including but not limited to man and other mammals, birds, fish, and the shellfish.

animal feed (40CFR257.3.5-1) means any crop grown for consumption by animals, such as pasture crops, forage, and grain.

animal feeding operation (40CFR122.23-1) means a lot or facility (other than an aquatic animal production facility) where the following conditions are met:
(i) Animals (other than aquatic animals) have been,

are, or will be stabled or confined and fed or maintained for a total of 45 days or more in any 12-month period, and
(ii) Crops, vegetation forage growth, or post-harvest residues are not sustained in the normal growing season over any portion of the lot or facility.

animal studies (EPA-94/04): Investigations using animals as surrogates for humans with the expectation that the results are pertinent to humans (under FIFRA).

animals (40CFR116.3) means appropriately sensitive animals which carry out respiration by means of a lung structure permitting gaseous exchange between air and the circulatory system.

annealing with oil (40CFR468.02-d) shall mean the use of oil to quench a workpiece as it passes from an annealing furnace.

annealing with water (40CFR468.02-e) shall mean the use of a water spray or bath, of which water is the major constituent, to quench a workpiece as it passes from an annealing furnace.

annual (40CFR704.3) means the corporate fiscal year.

annual average (40CFR407.61-x) shall mean the maximum allowable discharge of BOD5 or TSS as calculated by multiplying the total mass (kkg or 1000 lb) of each raw commodity processed for the entire processing season or calendar year by the applicable annual average limitation (other identical or similar definitions are provided in 40CFR407.71-v).

annual average (40CFR407.81-n) shall mean the maximum allowable discharge of BOD5 or TSS, as calculated by multiplying the total mass (kkg or 1000 lb) of each final product produced for the entire processing season or calendar year by the applicable annual average limitation.

annual capacity factor (40CFR60.41b) means the ratio between the actual heat input to a steam generating unit from the fuels listed in 40CFR60.42b(a) or 40CFR60.43b(a), or 40CFR60.44b(a), as applicable, during a calendar year and the potential heat input to the steam generating unit had it been operated for 8,760 hours during a calendar at the maximum steady state design heat input capacity. In the case of steam generating units that are rented or leased, the actual heat input shall be determined based on the combined heat input from all operations of the affected facility in a calendar year.

annual capacity factor (40CFR60.41c) means the ratio between the actual heat input to a steam generating unit from an individual fuel or combination of fuels during a period of 12 consecutive calendar months and the potential heat input to the steam generating unit from all fuels had the steam generating unit been operated for 8,760 hours during that 12-month period at the maximum design heat input capacity. In the case of steam generating units that are rented or leased, the actual heat input shall be determined based on the combined heat input from all operations of the affected facility during a period of 12 consecutive calendar months.

annual coke production (40CFR61.131) means the coke produced in the batteries connected to the coke byproduct recovery plant over a 12-month period. The first 12-month period concludes on the first December 31 that comes at least 12 months after the effective date or after the date of initial startup if initial startup is after the effective date.

annual committed effective dose (40CFR191.12) means the committed effective dose resulting from one-year intake of radionuclides released plus the annual effective dose caused by direct radiation from facilities or activities subject to subparts B and C of this part.

annual document log (40CFR761.3) means the detailed information maintained at the facility on the PCB waste handling at the facility.

annual pollutant loading rate (40CFR503.11-c) is the maximum amount of a pollutant that can be applied to a unit area of land during a 365 day period (under CWA).

annual precipitation and annual evaporation (40CFR440.132-b) are the mean annual precipitation and mean annual lake evaporation, respectively, as established by the U.S. Department of Commerce, Environmental Science Services Administration, Environmental Data Services, or equivalent regional rainfall and evaporation data.

annual report (40CFR761.3) means the written document submitted each year by each disposer and commercial storer of PCB waste to the appropriate EPA Regional Administrator. The annual report is a brief summary of the information included in the annual document log.

annual research period (40CFR85.402) means the time period from August 1 of a previous calendar year to July 31 of the given calendar year, e.g., the 1981 annual research period would be the time period from August 1, 10 to July 31, 11.

annual whole sludge application rate (40CFR503.11-d) is the maximum amount of sewage sludge (dry weight basis) that can be applied to a unit area of land during a 365 day period.

annual work plan (40CFR35.9010) means the plan, developed by the Management Conference each year, which documents projects to be undertaken during the upcoming year. The Annual Work Plan is developed within budgetary targets provided by EPA.

anode bake plant (40CFR60.191) means a facility which produces carbon anodes for use in a primary aluminum reduction plant.

ANSI S3.19-1974 (40CFR211.203-c) means a revision of the ANSI Z24.22-1957 measurement procedure using one-third octave band stimuli presented under diffuse (reverberant) acoustic field conditions.

ANSI Z24.22-1957 (40CFR211.203-b) means a measurement procedure published by the American National Standards Institute (ANSI) for obtaining hearing protector attenuation values at nine of the one-third octave band center frequencies by using pure tone stimuli presented to ten different test subjects under anechoic conditions.

antagonism (EPA-94/04): Interference or inhibition of the effect of one chemical by the action of another.

antarctic "ozone hole" (EPA-94/04): Refers to the seasonal depletion of ozone in a large area over Antarctica.

anthracite (40CFR60.41a-23) means coal that is classified as anthracite according to the American Society of Testing and Materials' (ASTM) Standard Specification for Classification of Coals by Rank D388-77 (incorporated by reference, 40CFR60.17) (under CAA).

anti-degradation clause (EPA-94/04): Part of federal air quality and water quality requirements prohibiting deterioration where pollution levels are above the legal limit.

antimony (40CFR415.661-e) shall mean the total antimony present in the process wastewater stream exiting the wastewater treatment system.

apparent plateau (40CFR797.1560-5): See steady-state.

appliance (40CFR82.152-a) means any device which contains and uses a class I or class II substance as a refrigerant and which is used for household or commercial purposes, including any air conditioner, refrigerator, chiller, or freezer.

appliance (CAA601-42USC7671) means any device which contains and uses a class I or class II substance as a refrigerant and which is used for household or commercial purposes, including any air conditioner, refrigerator, chiller, or freezer.

applicable implementation plan (40CFR51.392) defined in section 302(q) of the CAA and means the portion (or portions) of the implementation plan, or most recent revision thereof, which has been approved under section 110, or promulgated under section 110(c), or promulgated or approved pursuant to regulations promulgated under section 301(d) and which implements the relevant requirements of the CAA (other identical or similar definitions are provided in 40CFR93.101).

applicable implementation plan or applicable plan (40CFR52.138-1) means the portion (or portions) of the implementation plan, or most recent revision thereof, which has been approved under section 110 of the Clean Air Act, 42USC7410, or promulgated under section 110(c) of the CAA, 42USC7410(c).

applicable implementation plan or applicable SIP (40CFR51.852) means the portion (or portions) of the SIP or most recent revision thereof, which has been approved under section 110 of the Act, or promulgated under section 110(c) of the Act (Federal implementation plan), or promulgated or approved pursuant to regulations promulgated under section 301(d) of the Act and which implements the relevant requirements of the Act (other identical or similar definitions are provided in 40CFR93.152).

applicable legal requirements (40CFR66.3) means any of the following:
(1) In the case of any major source, any emission limitation, emission standard, or compliance schedule under any EPA-approved State implementation plan (regardless of whether the source is subject to a Federal or State consent decree);
(2) In the case of any source, an emission limitation, emission standard, standard of performance, or other requirement (including, but not limited to, work practice standards) established under section 111 or 112 of the Act;
(3) In the case of a source that is subject to a federal or federally approved state judicial consent decree or EPA approved extension, order, or suspension, any interim emission control requirement or schedule of compliance under that consent decree, extension, order or suspensions;
(4) In the case of a nonferrous smelter which has received a primary nonferrous smelter order issued or approved by EPA under section 119 of the Act, any interim emission control requirement (including a requirement relating to the use of supplemental or intermittent controls) or schedule of compliance under that order.

applicable marine water quality criteria (40CFR227.31) means the criteria given for marine waters in the EPA publication "Quality Criteria for Water" as published in 1976 and amended by subsequent supplements or additions.

applicable or appropriate requirements (EPA-94/04): Any state or federal statute that pertains to protection of human life and the environment in addressing specific conditions or use of a particular cleanup technology at a Superfund site.

applicable plan (40CFR60.21-d) means the plan, or most recent revision thereof, which has been approved under 40CFR60.27(b) or promulgated under 40CFR60.27(d).

applicable requirement (40CFR70.2) means all of the following as they apply to emissions units in a part 70 source (including requirements that have been promulgated or approved by EPA through rulemaking at the time of issuance but have future-effective compliance dates):

1. Any standard or other requirement provided for in the applicable implementation plan approved or promulgated by EPA through rulemaking under title I of the Act that implements the relevant requirements of the Act, including any revisions to that plan promulgated in part 52 of this chapter;
2. Any term or condition of any preconstruction permits issued pursuant to regulations approved or promulgated through rulemaking under title I, including parts C or D, of the Act;
3. Any standard or other requirement under section 111 of the Act, including section 111(d);
4. Any standard or other requirement under section 112 of the Act, including any requirement concerning accident prevention under section 112(r)(7) of the Act;
5. Any standard or other requirement of the acid rain program under title IV of the Act or the regulations promulgated thereunder;
6. Any requirements established pursuant to section 504(b) or section 114(a)(3) of the Act;
7. Any standard or other requirement governing solid waste incineration, under section 129 of the Act;
8. Any standard or other requirement for consumer and commercial products, under section 183(e) of the Act;
9. Any standard or other requirement for tank vessels under section 183(f) of the Act;
10. Any standard or other requirement of the program to control air pollution from outer continental shelf sources, under section 328 of the Act;
11. Any standard or other requirement of the

regulations promulgated to protect stratospheric ozone under title VI of the Act, unless the Administrator has determined that such requirements need not be contained in a title V permit; and
12. Any national ambient air quality standard or increment or visibility requirement under part C of title I of the Act, but only as it would apply to temporary sources permitted pursuant to section 504(e) of the Act.

applicable requirements (40CFR300.5) means those cleanup standards, standards of control, and other substantive requirements, criteria, or limitations promulgated under federal environmental or state environmental or facility siting laws that specifically address a hazardous substance, pollutant, contaminant, remedial action, location, or other circumstance found at a CERCLA site. Only those state standards that are identified by a state in a timely manner and that are more stringent than federal requirements may be applicable.

applicable standard (40CFR130.10.a.4):
(1) For the purposes of listing waters under 40CFR130.10-(d)(2), means a numeric criterion for a priority pollutant promulgated as part of a state water quality standard. Where a state numeric criterion for a priority pollutant is not promulgated as part of a state water quality standard,
(2) for the purposes of listing waters, means the state narrative water quality criterion to control a priority pollutant (e.g., no toxics in toxic amounts) interpreted on a chemical-by-chemical basis by applying a proposed state criterion, an explicit state policy or regulation, or an EPA national water quality criterion, supplement with other relevant information.

applicable standard (40CFR21.2-c) means any requirement, not subject to an exception under 40CFR21.6 relating to the quality of water containing or potentially containing pollutants, if such requirement is imposed by:
(1) The Act;
(2) EPA regulations promulgated thereunder or permits issued by EPA or a State thereunder;
(3) Regulations by any other Federal Agency promulgated thereunder;

(4) Any State standard or requirement as applicable under section 510 of the Act;

(5) Any requirements necessary to comply with an areawide management plan approved pursuant to section 208(b) of the Act;

(6) Any requirements necessary to comply with a facilities plan developed under section 201 of the Act (cf. 35CFR, subpart E);

(7) Any State regulations or laws controlling the disposal of aqueous pollutants that may affect groundwater.

applicable standards and limitations (40CFR122.2) means all State, interstate, and federal standards and limitations to which a "discharge" a "sewage sludge use or disposal practice," or a related activity is subject under the CWA, including "effluent limitations," water quality standards, standards of performance, toxic effluent standards or prohibitions, "best management practices," pretreatment standards and "standards for sewage sludge use or disposal" under sections 301, 302, 303, 304, 306, 307, 308, 403, and 405 of CWA.

applicable standards and limitations (40CFR124.2) means all State, interstate, and federal standards and limitations to which a "discharge," a "sludge use or disposal practice" or a related activity is subject under the CWA, including "standards for sewage sludge use or disposal," "effluent limitations," water quality standards, standards of performance, toxic effluent standards or prohibitions, "best management practices," and pretreatment standards under sections 301, 302, 303, 304, 306, 307, 308, 403, and 405 of CWA.

applicable underground injection control program with respect to a State (SDWA1422-42USC300h.l) means the program (or most recent amendment thereof):

(1) which has been adopted by the State and which has been approved under subsection (b), or

(2) which has been prescribed by the Administrator under subsection (c).

applicable water quality standards (40CFR110.1) means State water quality standards adopted by the State pursuant to section 303 of the Act or promulgated by EPA pursuant to that section (FWPCA).

applicant (40CFR2.307-2) means any person who has submitted to EPA (or to a predecessor agency with responsibility for administering the Act) a registration statement or application for registration under the Act (FIFRA) of a pesticide or of an establishment (other identical or similar definitions are provided in 40CFR6.501-e; 6.601-b; 7.25; 8.2-d; 15.4; 20.2-c; 30.200; 35.4010; 53.1-k; 125.58-c; 152.3-h; 164.2-e; 172.1-b; 195.2).

application (40CFR2.309-3) means an application for a permit (other identical or similar definitions are provided in 40CFR35.4010; 82.172; 85.2113-k; 122.2; 124.2; 124.41; 125.58-d; 144.3; 146.3; 232.2; 195.2; 270.2; 160.3).

application questionnaire (40CFR125.58-e) means EPA's "Applicant Questionnaire for Modification of Secondary Treatment Requirements." Individual questionnaires for small applicants and for large applicants are published as Appendix A and Appendix B to this subpart, respectively.

applicator (40CFR52.741) means a device used in a coating line to apply coating.

applied coating solids (40CFR60.391) means the volume of dried or cured coating solids which is deposited and remains on the surface of the automobile or light-duty truck body.

applied coating solids (40CFR60.451) means the coating solids that adhere to the surface of the large appliance part being coated.

apply sewage sludge or sewage sludge applied to the land (40CFR503.9-a) means land application of sewage sludge.

approach angle (40CFR86.084.2) means the smallest angle in a plan side view of an automobile, formed by the level surface on which the automobile is standing and a line tangent to the front tire static loaded radius arc and touching the underside of the automobile forward of the front tire.

appropriate (40CFR157.21-a) when used with respect to child-resistant packaging, means that the packaging is chemically compatible with the pesticide contained therein.

appropriate act and regulations (40CFR124.41) means the Clean Air Act and applicable regulations promulgated under it.

appropriate Act and regulations (40CFR124.2) means the Clean Water Act (CWA); the Solid Waste Disposal Act, as amended by the Resource Conservation Recovery Act (RCRA); or Safe Drinking Water Act (SDWA), whichever is applicable; and applicable regulations promulgated under those statutes. In the case of an "approved State program" appropriate Act and regulations includes program requirements.

appropriate Act and regulations (40CFR144.3) means the Solid Waste Disposal Act, as amended by the Resource Conservation and Recovery Act (RCRA); or Safe Drinking Water Act (SDWA), whichever is applicable; and applicable regulations promulgated under those statutes.

appropriate program official (40CFR6.701) means the official at each decision level within ORD to whom the Assistant Administrator has delegated responsibility for carrying out the environmental review process.

appropriate sensitive benthic marine organisms (40CFR227.27-c) means at least one species each representing filter-feeding, deposit-feeding, and burrowing species chosen from among the most sensitive species accepted by EPA as being reliable test organisms to determine the anticipated impact on the site; provided, however, that until sufficient species are adequately tested and documented, interim guidance on appropriate organisms available for use will be provided by the Administrator, Regional Administrator, or the District Engineer, as the case may be.

appropriate sensitive marine organisms (40CFR227.27-d) means at least one species each representative of phytoplankton or zooplankton, crustacean or mollusk, and fish species chosen from among the most sensitive species documented in the scientific literature or accepted by EPA as being reliable test organisms to determine the anticipated impact of the wastes on the ecosystem at the disposal site. Bioassays, except on phytoplankton or zooplankton, shall be run for a minimum of 96 hours under temperature, salinity, and dissolved oxygen conditions representing the extremes of environmental stress at the disposal site. Bioassays on phytoplankton or zooplankton may be run for shorter periods of time as appropriate for the organisms tested at the discretion of EPA, or EPA and the Corps of Engineers, as the case may be (under the Marine Protection, Research, and Sanctuaries Act).

appropriate treatment of the recycle water (40CFR440.132-c.) in subpart J, section 440.104 includes, but is not limited to pH adjustment settling and pH adjustment, settling, and mixed media filtration.

approval authority (40CFR403.3-c) means the Director in an NPDES State with an approved State pretreatment program and the appropriate Regional Administrator in a non NPDES State or NPDES State without an approved State pretreatment program.

approval of the facilities plan (40CFR6.501-h) means approval of the facilities plan for a proposed wastewater treatment works pursuant to 40CFR35, subpart E or I.

approved clean coal technology demonstration project (40CFR76.2) means a project using funds appropriated under the Department of Energy's "Clean Coal Technology Demonstration Program," up to a total amount of $2,500,000,000 for commercial demonstration of clean coal technology, or similar projects funded through appropriations for the Environmental Protection Agency. The Federal contribution for a qualifying project shall be at least 20 percent of the total cost of the demonstration project.

approved equipment testing organization (40CFR82.152-b) means any organization which has applied for and received approval from the Administrator pursuant to 40CFR 82.160.

Approved Independent Standards Testing Organization (40CFR82.32-a) means any organization which has applied for and received approval from the Administrator pursuant to 40CFR82.38.

approved measure (40CFR57.103-h) refers to one contained in an NSO which is in effect.

approved permit program (40CFR60.2) means a State permit program approved by the Administrator as meeting the requirements of part 70 of this chapter or a Federal permit program established in this chapter pursuant to title V of the Act (42 U.S.C. 7661) (other identical or similar definitions are provided in 40CFR61.02).

approved permit program (40CFR63.2) means a State permit program approved by the Administrator as meeting the requirements of part 70 of this chapter or a Federal permit program established in this chapter pursuant to title V of the Act (42 U.S.C. 7661) (other identical or similar definitions are provided in 40CFR63.101; 63.191).

approved POTW pretreatment program or program or POTW pretreatment program (40CFR403.3-d) means a program administered by a POTW that meets the criteria established in this regulation (40CFR403.8 and 403.9) and which has been approved by a Regional Administrator or State Director in accordance with 40CFR403.11 of this regulation.

approved program (40CFR124.41) means a State implementation plan providing for issuance of PSD permits which has been approved by EPA under the Clean Air Act and 40CFR51. An "approved State" is one administering an "approved program." "State Director" as used in 40CFR124.4 means the person(s) responsible for issuing PSD permits under an approved program, or that person's delegated representative.

approved program (40CFR232.2) means a State program which has been approved by the Regional Administrator under part 233 of this chapter or which is deemed approved under section 404(h)(3), 33 U.S.C. 1344(h)(3).

approved program or approved State (40CFR122.2) means a State or interstate program which has been approved or authorized by EPA under 40CFR123 (under FWPCA).

approved program or approved State (40CFR270.2) means a State which has been approved or authorized by EPA under 30CFR271.

approved refrigerant recycling equipment (40CFR82.32-b) means equipment certified by the Administrator or an organization approved under 40CFR82.38 as meeting either one of the standards in 40CFR82.36. Such equipment extracts and recycles refrigerant or extracts refrigerant for recycling on-site or reclamation off-site.

approved section 120 program (40CFR66.3-d) means a State program to assess and collect section 120 penalties that has been approved by the Administrator.

approved State (40CFR122.2; 233.3; 270.2): See approved program.

approved State primacy program (40CFR142.2) consists of those program elements listed in 40CFR142.11(a) that were submitted with the initial State application for primary enforcement authority and approved by the EPA Administrator and all State program revisions thereafter that were approved by the EPA Administrator.

approved State program (40CFR501.2) means a State program which has received EPA approval under this part.

approved State program (40CFR144.3) means a UIC program administered by the State or Indian Tribe that has been approved by EPA according to SDWA sections 1422 and/or 1425.

approximate original contour (SMCRA701-30USC1291) means that surface configuration achieved by backfilling and grading of the mined area so that the reclaimed area, including any terracing or access roads, closely resembles the general surface configuration of the land prior to mining and blends into and complements the drainage pattern of the surrounding terrain, with all highwalls and spoil piles eliminated; water impoundments may be permitted where the regulatory authority determines that they are in compliance with section 1265(b)(8) of this title.

apricots (40CFR407.61-b) shall include the

processing of apricots into the following product styles: Canned and frozen, pitted and unpitted, peeled and unpeeled, whole, halves, slices, nectar, and concentrate.

aquaculture project (40CFR122.25-1) means a defined managed water area which uses discharges of pollutants into that designated area for the maintenance or production of harvestable freshwater, estuarine, or marine plants or animals.

aquatic animals (40CFR116.3) means appropriately sensitive wholly aquatic animals which carry out respiration by means of a gill structure permitting gaseous exchange between the water and the circulatory system.

aquatic ecosystem (40CFR230.3-c) See aquatic environment.

aquatic environment and aquatic ecosystem (40CFR230.3-c) mean waters of the United States, including wetlands, that serve as habitat for interrelated and interacting communities and populations of plants and animals.

aquatic flora (40CFR116.3) means plant life associated with the aquatic eco-system including, but not limited to, algae and higher plants (under FWPCA).

aqueous (EPA-94/04): Something made up of, similar to, or containing water; watery.

aquifer (40CFR144.3) means a geological formation, group of formations, or part of a formation that is capable of yielding a significant amount of water to a well or spring (other identical or similar definitions are provided in 40CFR146.3; 147.2902; 149.2-a).

aquifer (40CFR149.101-k) means the Edwards Underground Reservoir.

aquifer (40CFR191.12) means an underground geological formation, group of formations, or part of a formation that is capable of yielding a significant amount of water to a well or spring.

aquifer (40CFR257.3.4-1) means a geologic formation, group of formations, or portion of a formation capable of yielding usable quantities of ground water to wells or springs.

aquifer (40CFR258.2) means a geological formation, group of formations, or portion of a formation capable of yielding significant quantities of ground water to wells or springs.

aquifer (40CFR260.10) means a geologic formation, group of formations, or part of a formation capable of yielding a significant amount of ground water to wells or springs.

aquifer (40CFR270.2) means a geological formation, group of formations, or part of a formation that is capable of yielding a significant amount of water to a well or spring.

aquifer (40CFR503.21-b) is a geologic formation, group of geologic formations, or a portion of a geologic formation capable of yielding ground water to wells or springs.

aquifer (EPA-94/04): An underground geological formation, or group of formations, containing usable amounts of groundwater that can supply wells and springs.

arbitrator (40CFR304.12-c) means the person appointed in accordance with 40CFR304.22 of this part and governed by the provisions of this part (under CERCLA).

architectural coatings (EPA-94/04): Coverings such as paint and roof tar that are used on exteriors of buildings.

architectural or engineering (A/E) services (40CFR33.005) means consultation, investigations, reports, or services for design-type projects withn the scope of the practice of architecttre or professional enneering as defined by the laws of the State or territory in which the recipient is located (other identical or similar definitions are provided in 40CFR35.6015-3).

architectural or engineering services (40CFR35.2005-6) means consultation, investigations, reports, or services for design-type projects within the scope of the practice of architecture or professional engineering as defined by the laws of the State or territory in which the grantee is located.

area (40CFR61.251-a) means the vertical projection of the pile upon the earth's surface.

area coated (40CFR466.02-d) means the area of basis material covered by each coating of enamel.

area committee (CWA311-33USC1321) means an Area Committee established under subsection (j).

area contingency plan (CWA311-33USC1321) means an Area Contingency Plan prepared under subsection (j).

area of concern (CWA118-33USC1268) means a geographic area located within the great lakes, in which beneficial uses are impaired and which has been officially designated as such under Annex 2 of the Great lakes Water Quality Agreement.

area of review (40CFR144.3) means the area surrounding an injection well described according to the criteria set forth in 40CFR146.06 or in the case of an area permit, the project area plus a circumscribing area the width of which is either 1/4 of a mile or a number calculated according to the criteria set forth in 40CFR146.06 (other identical or similar definitions are provided in 40CFR146.3).

area of review (40CFR147.3001) means the area surrounding an injection well or project area described according to the criteria set forth in 40CFR147.3009 of this subpart.

area of review (EPA-94/04): In the UIC program, the area surrounding an injection well that is reviewed during the permitting process to determine if flow between aquifers will be induced by the injection operation.

area processed (40CFR465.02-d) means the area actually exposed to process solutions. Usually this includes both sides of the metal strip.

area processed (40CFR466.02-c) means the total basis material area exposed to processing solutions.

area source (40CFR51.100-l) means any small

residential, governmental, institutional, commercial, or industrial fuel combustion operations; onsite solid waste disposal facility, motor vehicles, aircraft vessels, or other transportation facilities or other miscellaneous sources identified through inventory techniques similar to those described in the AEROS Manual series, Vol. II AEROS Users Manual, EPA450-2-76-029 December 1976.

area source (40CFR51.491): Stationary and nonroad sources that are small and/or too numerous to be individually included in a stationary source emissions inventory.

area source (40CFR63.2) means any stationary source of hazardous air pollutants that is not a major source as defined in this part.

area source (40CFR63.321) means any perchloroethylene dry cleaning facility that meets the conditions of 40CFR63.320(h).

area source (CAA112-42USC7412) means any stationary source of hazardous air pollutants that is not a major source. For purposes of this section, the term "area source" shall not include motor vehicles or nonroad vehicles subject to regulation under title II.

area source (EPA-94/04): Any small source of non-natural air pollution that is released over a relatively small area but which cannot be classified as a point source. Such sources may include vehicles and other small engines, small businesses and household activities.

areas of critical environmental concern (FLPMA103-43USC1702) means areas within the public lands where special management attention is required (when such areas are developed or used or where no development is required) to protect and prevent irreparable damage to important historic, cultural, or scenic values, fish and wildlife resources or other natural systems or processes, or to protect life and safety from natural hazards.

areawide agency (40CFR130.2-l) means an agency designated under section 208 of the Act, which has responsibilities for WQM planning within a specified area of a State.

areawide agency (40CFR21.2-q) means an areawide management agency designated under section 208(c)(1) of the Act.

areawide air quality modeling analysis (40CFR51.852) means an assessment on a scale that includes the entire nonattainment or maintenance area which uses an air quality dispersion model to determine the effects of emissions on air quality (other identical or similar definitions are provided in 40CFR93.152).

argon oxygen decarburization vessel (AOD vessel) (40CFR60.271a) means any closed-bottom, refractory-lined converter vessel with submerged tuyeres through which gaseous mixtures containing argon and oxygen or nitrogen may be blown into molten steel for further refining.

aromatic (EPA-94/04): A type of hydrocarbon, such as benzene or toluene, added to gasoline in order to increase octane. Some aromatics are toxic.

aromatic content (40CFR80.2-z) is the aromatic hydrocarbon content in volume percent as determined by ASTM standard test method D 1319-88, entitled "Standard Test Method for Hydrocarbon Types in Liquid Petroleum Products by Fluorescent Indicator Adsorption." ASTM test method D 1319-88 is incorporated by reference. This incorporation by reference was approved by the Director of the Federal Register in accordance with 5 U.S.C. 552(a) and 1CFR part 51. A copy may be obtained from the American Society for Testing and Materials, 1916 Race Street, Philadelphia, PA 19103. A copy may be inspected at the Air Docket section (A-130), room M-1500, U.S. Environmental Protection Agency, Docket No. A-86-03, 401 M Street SW., Washington, DC 20460 or at the Office of the Federal Register, 1100 L Street NW., room 8401 Washington, DC 20005.

arrest (40CFR303.11-a): Restraint of an arrestee's liberty or the equivalent through the service of judicial process compelling such a person to respond to a criminal accusation.

arsenic (40CFR415.671-c) shall mean the total arsenic present in the process wastewater stream exiting the wastewater treatment system.

arsenic containing glass type (40CFR61.161) means any glass that is distinguished from other glass solely by the weight percent of arsenic added as a raw material and by the weight percent of arsenic in the glass produced. Any two or more glasses that have the same weight percent of arsenic in the raw materials as well as in the glass produced shall be considered to belong to one arsenic-containing glass type, without regard to the recipe used or any other characteristics of the glass or the method of production.

arsenic kitchen (40CFR61.181) means a baffled brick chamber where inorganic arsenic vapors are cooled, condensed. and removed in a solid form.

arsenicals (EPA-94/04): Pesticides containing arsenic.

artesian (aquifer or well) (EPA-94/04): Water held under pressure in porous rock or soil confined by impermeable geologic formations.

article (40CFR372.3) means a manufactured item:
(1) Which is formed to a specific shape or design during manufacture;
(2) which has end use function(s) dependent in whole or in part upon its shape or design during end use; and
(3) which does not release a toxic chemical under normal conditions or processing or use of that item at the facility or establishments.

article (40CFR704.3) means a manufactured item:
(1) which is formed to a specific shape or design during manufacture,
(2) which has end use function(s) dependent in whole or in part upon its shape or design during end use, and
(3) which has either no change of chemical compositions during its end use or only those changes of composition which have no commercial purpose separate from that of the article, and that result from a chemical reaction that occurs upon end use of other chemical substances, mixtures, or articles; except that fluids and particles are not considered articles regardless of shape or design.

article (40CFR710.2-f) is a manufactured item:
(1) Which is formed to a specific shape or design

during manufacture,
(2) which has end use function(s) dependent in whole or in part upon its shape or design during end use, and
(3) which has either no change of chemical composition during its end use or only those changes of composition which have no commercial purpose separate from that of the article and that may occur as described in 40CFR710.4(d)(5); except that fluids and particles are not considered articles regardless of shape or design.

article (40CFR720.3-c) is a manufactured item:
(1) Which is formed to a specific shape or design during manufacture,
(2) which has end use function(s) dependent in whole or in part upon its shape or design during end use, and
(3) which has either no change of chemical composition during its end use or only those changes of composition which have no commercial purpose separate from that of the article and that may occur as described in 40CFR720.36(g)(5); except that fluids and particles are not considered articles regardless of shape or design.
(Other identical or similar definitions are provided in 40CFR723.50-2; 747.115-1; 747.195-1; 747.200-1.)

article (40CFR723.175-2) is a manufactured item:
(i) which is formed to a specific shape or design during manufacture,
(ii) which has end use function(s) dependent in whole or in part upon its shape or design during end use, and
(iii) which has either no change of chemical composition during its end use or only those changes of composition which have no commercial purpose separate from that of the article and that may occur as described in 40CFR710.2 of this chapter except that fluids and particles are not considered articles regardless of shape or design.

articles (40CFR63.321) mean clothing, garments, textiles, fabrics, leather goods, and the like, that are dry cleaned.

artificially or substantially greater emissions (40CFR63.51) means abnormally high emissions such as could be caused by equipment malfunctions, accidents, unusually high production or operating rates compared to historical rates, or other unusual circumstances.

as applied (40CFR52.741) means the exact formulation of a coating during application on or impregnation into a substrate.

as expeditiously as practicable (CAA169A-42USC7491) means as expeditiously as practicable but in no event later than five years after the date of approval of a plan revision under this section (or the date or promulgation of such a plan revision in the case of action by the Administrator under section 110(c) for purposes of this section).

as expeditiously as practicable considering technological feasibility (40CFR192.31-k) means as quickly as possible considering: the physical characteristics of the tailings and the site; the limits of available technology; the need for consistency with mandatory requirements of other regulatory programs; and factors beyond the control of the licensee. The phrase permits consideration of the cost of compliance only to the extent specifically provided for by use of the term "available technology."

as fired (40CFR72.2) means the taking of a fuel sample just prior to its introduction into the unit for combustion.

asbestiform (40CFR763-AA-2) means a specific type of mineral fibrosity in which the fibers and fibrils possess high tensile strength and flexibility (cf. asbestos).

asbestos (40CFR61.141) means the asbestiform varieties of serpentinite (chrysotile), riebeckite (crocidolite), cummingtonite-grunerite, anthophyllite, and actinolite-tremolite.

asbestos (40CFR763.163) means the asbestiform varieties of: chrysotile (serpentine); crocidolite (riebeckite); amosite (cummingtonite-grunerite); tremolite; anthophyllite; and actinolite.

asbestos (40CFR763.63-a) means the asbestiform

varieties of: chrysotile (serpentine); crocidolite (riebeckite); amosite (cummingtonite-grunerite); anthophyllite; tremolite; and actinolite (other identical or similar definitions are provided in 40CFR763.83; 763.103-b; 763.121).

asbestos (EPA-94/04): A mineral fiber that can pollute air or water and cause cancer or asbestosis when inhaled. EPA has banned or severely restricted its use in manufacturing and construction.

asbestos (TSCA-AIA1) means:
(A) chrysotile, amosite, or crocidolite, or
(B) in fibrous form, tremolite, anthophyllite; or actinolite.

asbestos (TSCA202-15USC2642) means the asbestiform varieties of:
(A) chrysotile (serpentine);
(B) crocidolite (riebeckite);
(C) amosite (cummingtonite-grunerite);
(D) anthophyllite;
(E) tremolite; or
(F) actinolite.

asbestos abatement (EPA-94/04): Procedures to control fiber release from asbestos-containing materials in a building or to remove them entirely, including removal, encapsulation, repair, enclosure, encasement, and operations and maintenance programs.

asbestos abatement project (40CFR763.121) means any activity involving the removal, enclosure, or encapsulation of friable asbestos material.

asbestos containing building material (ACBM) (40CFR763.83) means surfacing ACM, thermal system insulation ACM, or miscellaneous ACM that is found in or on interior structural members or other parts of a school building.

asbestos containing material (ACM) (40CFR763.83) when referring to school buildings means any material or product which contains more than 1 percent asbestos.

asbestos containing material (TSCA-AIA1) means any material containing more than 1 percent asbestos by weight.

asbestos containing material (TSCA202-15USC2642) means any material which contains more than 1 percent asbestos by weight (other identical or similar definitions are provided in 40CFR763.103-c).

asbestos containing product (40CFR763.163) means any product to which asbestos is deliberately added in any concentration or which contains more than 1.0 percent asbestos by weight or area.

asbestos containing waste materials (40CFR61.141) means mill tailings or any waste that contains commercial asbestos and is generated by a source subject to the provisions of this subpart. This term includes filters from control devices, friable asbestos waste material, and bags or other similar packaging contaminated with commercial asbestos. As applied to demolition and renovation operations, this term also includes regulated asbestos-containing material waste and materials contaminated with asbestos including disposable equipment and clothing (under CAA).

asbestos containing waste materials (ACWM) (EPA-94/04): Mill tailings or any waste that contains commercial asbestos and is generated by a source covered by the Clean Air Act Asbestos NESHAPS (national standards for hazardous air pollutants).

asbestos debris (40CFR763.83) means pieces of ACBM that can be identified by color, texture, or composition, or means dust, if the dust is determined by an accredited inspector to be ACM.

asbestos mill (40CFR61.141) means any facility engaged in converting, or in any intermediate step in converting, asbestos ore into commercial asbestos. Outside storage of asbestos material is not considered a part of the asbestos mill.

asbestos mixture (40CFR763.63-b) means a mixture which contains bulk asbestos or another asbestos mixture as an intentional component. An asbestos mixture may be either amorphous or a sheet, cloth fabric, or another structure. This term does not include mixtures which contain asbestos as a contaminant or impurity.

asbestos program manager (EPA-94/04): A building owner or designated representative who supervises all aspects of the facility asbestos management and control program.

asbestos tailings (40CFR61.141) means any solid waste that contains asbestos and is a product of asbestos mining or milling operations.

asbestos waste from control devices (40CFR61.141) means any waste material that contains asbestos and is collected by a pollution control device.

asbestosis (EPA-94/04): A disease associated with inhalation of asbestos fibers. The disease makes breathing progressively more difficult and can be fatal.

ash (EPA-94/04): The mineral content of a product remaining after complete combustion.

ASME (40CFR60.51a) means the American Society of Mechanical Engineers.

aspect ratio (40CFR763-AA-3) means a ratio of the length to the width of a particle. Minimum aspect ratio as defined by this method is equal to or greater than 5:1.

asphalt (40CFR52.741) means the dark-brown to black cementitious material (solid, semisolid, or liquid in consistency) of which the main constituents are bitumens which occur naturally or as a residue of petroleum refining.

asphalt processing (40CFR60.471) means the storage and blowing of asphalt.

asphalt processing plant (40CFR60.471) means a plant which blows asphalt for use in the manufacture of asphalt products.

asphalt roofing plant (40CFR60.471) means a plant which produces asphalt roofing products (shingles, roll roofing, siding, or saturated felt).

asphalt storage tank (40CFR60.471) means any tank used to store asphalt at asphalt roofing plants, petroleum refineries, and asphalt processing plants. Storage tanks containing cutback asphalts (asphalts diluted with solvents to reduce viscosity for low temperature applications) and emulsified asphalts

(asphalts dispersed in water with an emulsifying agent) are not subject to this regulation.

assay (EPA-94/04): A test for a particular chemical or effect.

assessment (EPA-94/04): In the asbestos-in-schools program, the evaluation of the physical condition and potential for damage of all friable asbestos containing materials and thermal insulation systems.

assets (40CFR144.61-d) means all existing and all probable future economic benefits obtained or controlled by a particular entity (other identical or similar definitions are provided in 40CFR264.141-f; 265.141-f).

assimilation (EPA-94/04): The ability of a body of water to purify itself of pollutants.

assimilative capacity (EPA-94/04): The capacity of a natural body of water to receive wastewaters or toxic materials without deleterious effects and without damage to aquatic life or humans who consume the water.

assistance agreement (40CFR30.200) means the legal instrument EPA uses to transfer money, property, services, or anything of value to a recipient to accomplish a public purpose. It is either a grant or a cooperative agreement and will specify: budget and project periods; the Federal share of eligible project costs; description of the work to be accomplished; and any special conditions.

Assistant Administrator (40CFR15.4) means the Assistant Administrator for Enforcement and compliance Monitoring, United States Environmental Protection Agency, or his or her designee.

Assistant Administrator (40CFR178.3) means the Agency's Assistant Administrator for Pesticides and Toxic Substances, or any officer or employee of the Agency's Office of Pesticides and Toxic Substances to whom the Assistant Administrator delegates the authority to perform functions under this part.

Assistant Administrator (40CFR179.3) means the Agency's Assistant Administrator for Pesticides and Toxic Substances, or any officer or employee of

OPTS (Office of Pesticides and Toxic Substances) to whom the Assistant Administrator has delegated the authority to perform functions under this part.

Assistant Administrator (40CFR723.50-3) means the EPA Assistant Administrator for Pesticides and Toxic Substances or any employee designated by the Assistant Administrator to carry out the Assistant Administrator's functions under this section.

Assistant Administrator for Air and Radiation (40CFR57.103-i) means the Assistant Administrator for Air and Radiation of the U.S. Environmental Protection Agency.

Assistant Attorney General (40CFR12.103) means the Assistant Attorney General, Civil Rights Division, U.S. Department of Justice.

Assistant Attorney General (40CFR7.25) is the head of the Civil Rights Division, U.S. Department of Justice.

association (40CFR304.12-d) means the organization offering arbitration services selected by EPA to conduct arbitrations pursuant to this part.

association of boards of certification (EPA-94/04): An international organization representing boards which certify the operators of waterworks and waste water facilities.

at retail (40CFR60.531) means the sale by a commercial owner of a wood heater to the ultimate purchaser.

at-the-source (40CFR421.31-c) means at or before the commingling of delacquering scrubber liquor blowdown with other process or nonprocess wastewaters.

atomic energy (10CFR70.4) means all forms of energy released in the course of nuclear fission or nuclear transformation.

atomic weapon (10CFR70.4) means any device utilizing atomic energy, exclusive of the means for transporting or propelling the device (where such means is a separable and divisible part of the device), the principal purpose of which is for use as, or for development of, a weapon, a weapon prototype, or a weapon test device.

atomization (40CFR471.02-d) is the process in which a stream of water or gas impinges upon a molten metal stream, breaking it into droplets which solidify as powder particles.

attainment and maintenance of national standards: See 40CFR51.110.

attainment area (40CFR51.491) means any area of the country designated or redesignated by the EPA at 40CFR part 81 in accordance with section 107(d) as having attained the relevant NAAQS for a given criteria pollutant. An area can be an attainment area for some pollutants and a nonattainment area for other pollutants.

attainment area (EPA-94/04): An area considered to have air quality as good as or better than the national ambient air quality standards as defined in the Clean Air Act. An area may be an attainment area for one pollutant and a non-attainment area for others.

attainment demonstration (40CFR51.491) means the requirement in section 182(b)(1)(A) of the Act to demonstrate that the specific annual emissions reductions included in a SIP are sufficient to attain the primary NAAQS by the date applicable to the area.

attenuation (EPA-94/04): The process by which a compound is reduced in concentration over time, through absorption, adsorption, degradation, dilution, and/or transformation.

attractant (EPA-94/04): A chemical or agent that lures insects or other pests by stimulating their sense of smell.

attrition (EPA-94/04): Wearing or grinding down of a substance by friction. Dust from such processes contributes to air pollution.

auction subaccount (40CFR72.2) means a subaccount in the Special Allowance Reserve, as specified in section 416(b) of the Act, which contains allowances to be sold at auction in the amount of

150,000 per year from calendar year 1995 through 1999, inclusive, and 200,000 per year for each year begnning in calendar year 2000, subject to the adjustments noted in the regulations in part 73, subpart E of this chapter.

authority head (40CFR27.2) means the Administrator of a judicial officer.

authorized account representative (40CFR72.2) means a responsible natural person who is authorized, in accordance with part 73 of this chapter, to transfer and otherwise dispose of allowances held in an Allowance Tracking System general account; or, in the case of a unit account, the designated representative of the owners and operators of the affected unit.

authorized person (40CFR763.121) means any person authorized by the employer and required by work duties to be present in regulated ares (under TSCA).

authorized representative (40CFR260.10) means the person responsible for the overall operation of a facility or an operational unit (i.e., part of a facility), e.g., the plant manager, superintendent or person of equivalent responsibility.

automated data acquisition and handling system (40CFR72.2) means that component of the CEMS, COMS, or other emissions monitoring system approved by the Administrator for use in the Acid Rain Program, designed to interpret and convert individual output signals from pollutant concentration monitors, flow monitors, diluent gas monitors, opacity monitors, and other component parts of the monitoring system to produce a continuous record of the measured parameters in the measurement units required by part 75 of this chapter.

automated method or analyzer (40CFR53.1-i) means a method for measuring concentrations of an ambient air pollutant in which sample collection, analysis, and measurement are performed automatically.

automated monitoring and recording system (40CFR63.111) means any means of measuring values of monitored parameters and creating a hard copy or computer record of the measured values that does not require manual reading of monitoring instruments and manual transcription of data values. Automated monitoring and recording systems include, but are not limited to, computerized systems and strip charts.

automatic temperature compensator (40CFR60.431) means a device that continuously senses the temperature of fluid flowing through a metering device and automatically adjusts the registration of the measured volume to the corrected equivalent volume at a base temperature.

automobile (40CFR52.741) means a motor vehicle capable of carrying no more than 12 passengers (other identical or similar definitions are provided in 40CFR60.391).

automobile (40CFR600.002.85-4) means:
(i) Any four-wheel vehicle propelled by a combustion engine using onboard fuel or by an electric motor drawing current from rechargeable storage batteries or other portable energy storage devices (rechargeable using energy from a source off the vehicle such as residential electric service),
(ii) Which is manufactured primarily for use on public streets, roads, or highways (except any vehicle operated on a rail or rails),
(iii) Which is rated at not more than 8,500 pounds gross vehicle weight, which has a curb weight of not more than 6,000 pounds, and which has a basic vehicle frontal area of not more than 45 square feet, or
(iv) Is a type of vehicle which the Secretary determines is substantially used for the same purposes.

automobile (40CFR610.11-2) means any four-wheeled vehicle propelled by fuel which is manufactured primarily for use on public streets, roads, and highways (except any vehicle operated exclusively on a rail or rails), and which is rated at 6,000 lbs. gross vehicle weight or less.

automobile and light-duty truck body (40CFR60.391) means the exterior surface of an automobile or light-duty truck including hoods, fenders, cargo boxes, doors, and grill opening panels.

automobile or light-duty truck assembly plant (40CFR52.741) means a facility where parts are assembled or finished for eventual inclusion into a finished automobile or light-duty truck ready for sale to vehicle dealers, but not including customizers, body shops, and other repainters.

automobile or light-duty truck refinishing (40CFR52.741) means the repainting of used automobiles and light-duty trucks.

auxiliary aids (40CFR12.103) means services or devices that enable persons with impaired sensory, manual, or speaking skills to have an equal opportunity to participate in, and enjoy the benefits of, programs or activities conducted by the agency. For example, auxiliary aids useful for persons with impaired vision include readers, Brailled materials, audio recordings, and other similar services and devices. Auxiliary aids useful for persons with impaired hearing include telephone handset amplifiers, telephones compatible with hearing aids, telecommunication devices for deaf persons (TDD's), interpreters, notetakers, written materials, and other similar services and devices.

auxiliary emission control device (AECD) (40CFR600.002.85-29) means an element of design as defined in part 86.

auxiliary emission control device (AECD) (40CFR86.082.2) means any element of design which senses temperature, vehicle speed, engine RPM, transmission gear, manifold vacuum, or any other parameter for the purpose of activating, modulating, delaying, or deactivating the operation of any part of the emission control system.

auxiliary emission control device (AECD) (40CFR89.2) means any element of design that senses temperature, vehicle speed, engine RPM, transmission gear, or any other parameter for the purpose of activating, modulating, delaying, or deactivating the operation of any part of the emission control system.

auxiliary fuel (40CFR503.41-b) is fuel used to augment the fuel value of sewage sludge. This includes, but is not limited to, natural gas, fuel oil, coal, gas generated during anaerobic digestion of sewage sludge, and municipal solid waste (not to exceed 30 percent of the dry weight of sewage sludge and auxiliary fuel together). Hazardous wastes are not auxiliary fuel.

availability session (EPA-94/04): Informal meeting at a public location where interested citizens can talk with EPA and state officials on a one-to-one basis.

available chlorine (EPA-94/04): A measure of the amount of chlorine available in chlorinated lime, hypochlorite compounds, and other materials used as a source of chlorine when compared with that of liquid or gaseous chlorine.

available information (40CFR63.51) means, for purposes of conducting a MACT floor finding and identifying control technology options for emission units subject to the provisions of this subpart, information contained in the following information sources as of the section 112(j) deadline:
(1) A relevant proposed regulation, including all supporting information;
(2) Background information documents for a draft or proposed regulation;
(3) Any regulation, information or guidance collected by the Administrator establishing a MACT floor finding and/or MACT determination;
(4) Data and information available from the Control Technology Center developed pursuant to section 112(l)(3) of the Act;
(5) Data and information contained in the Aerometric Informational Retrieval System (AIRS) including information in the MACT database;
(6) Any additional information that can be expeditiously provided by the Administrator;
(7) Any information provided by applicants in an application for a permit, permit modification, administrative amendment, or Notice of MACT Approval pursuant to the requirements of this subpart; and
(8) Any additional relevant information provided by the applicant.

available purchase power (40CFR60.41a-18) means the lesser of the following:
(a) The sum of available system capacity in all neighboring companies.

(b) The sum of the rated capacities of the power interconnection devices between the principal company and all neighboring companies, minus the sum of the electric power load on these interconnections.

(c) The rated capacity of the power transmission lines between the power interconnection devices and the electric generating units (the unit in the principal company that has the malfunctioning flue gas desulfurization system and the unit(s) in the neighboring company supplying replacement electrical power) less the electric power load on these transmission lines.

available system capacity (40CFR60.41a-16) means the capacity determined by subtracting the system load and the system emergency reserves from the net system capacity.

available technology (40CFR192.31-m) means technologies and methods for emplacing a permanent radon barrier on uranium mill tailings piles or impoundments. This term shall not be construed to include extraordinary measures or techniques that would impose costs that are grossly excessive as measured by practice within the industry or one that is reasonably analogous (such as, by way of illustration only, unreasonable overtime, staffing or transportation requirements, etc., considering normal practice in the industry; laser fusion, of soils, etc.), provided there is reasonable progress toward emplacement of a permanent radon barrier. To determine grossly excessive costs, the relevant baseline against which cost increases shall be compared is the cost estimate for tailings impoundment closure contained in the licensee's tailings closure plan, but costs beyond such estimates shall not automatically be considered grossly excessive.

average concentration (40CFR423.11-k) as it relates to chlorine discharge means the average of analyses made over a single period of chlorine release which does not exceed two hours.

average concentration (40CFR63.111), as used in the wastewater provisions, means the flow-weighted annual average concentration, as determined according to the procedures specified in 40CFR63.144(b) of this subpart.

average flow rate (40CFR63.111), as used in the wastewater provisions, means the annual average flow rate, as determined according to the procedures specified in 40CFR63.144(e) of this subpart.

average fuel economy (40CFR2.311-2) has the meaning given it in section 501(4) of the Act, 15USC2001(4).

average fuel economy (40CFR600.002.85-14) means the unique fuel economy value as computed under 40CFR600.510 for a specific class of automobiles produced by a manufacturer that is subject to average fuel economy standards.

average monthly discharge limitation (40CFR122.2) means the highest allowable average of "daily discharges" over a calendar month, calculated as the sum of all "daily discharges" measured during a calendar month divided by the number of "daily discharges" measured during that month.

average of daily values for 30 consecutive days (40CFR435.11-b) shall be the average of the daily values obtained during any 30 consecutive day period.

average of daily values for thirty consecutive days (40CFR407.61-y; 407.71-w; 407.81-o): See maximum for any one day.

average process water usage flow rate of a cleaning water process in liters per day (40CFR463.21-a) is equal to the volume of process water (liters) used per year by a process divided by the numbers of days per year the process operates. The "average process water usage flow rate" for a plant with more than one plastics molding and forming process that uses cleaning water is the sum of the "average process water usage flow rate" for the cleaning processes (other identical or similar definitions are provided in 40CFR463.31-a).

average process water usage flow rate of a contact cooling and heating water process in liters per day (40CFR463.11-a) is equal to the volume of process water (liters) used per year by a process divided by the number of days per year the process operates. The "average process water usage flow rate" for a plant with more than one plastics molding and forming process that uses contact cooling and heating water is the sum of the "average process usage flow rates" for the contact cooling and heating water processes.

average weekly discharge limitation (40CFR122.2) means the highest allowable average of "daily discharges" over a calendar week, calculated as the sum of all "daily discharges" measured during a calendar week divided by the number of "daily discharges" measured during that week (under FWPCA).

averaging compliance records (40CFR80.126-a) shall include the calculations used to determine compliance with relevant standards on average, for each averaging period and for each quantity of gasoline for which standards must be achieved separately.

averaging for clean-fuel vehicles (40CFR88.202.94) means the sale of clean-fuel vehicles that meet more stringent standards than required, which allows the manufacturer to sell fewer clean-fuel vehicles than would otherwise be required.

averaging for heavy-duty engines (40CFR86.090.2) means the exchange of NO_x and particulate emission credits among engine families within a given manufacturer's product line.

averaging for nonroad engines (40CFR89.202.96) means the exchange of emission credits among engine families within a given manufacturer's product line.

averaging set (40CFR86.090.2) means a subcategory of heavy-duty engines within which engine families can average and trade emission credits with one other.

award (40CFR35.4010) means the technical assistance grant agreement signed by both EPA and the recipient.

award (40CFR72.2) means the conditional set-aside by the Administrator, based on the submission of an early ranking application pursuant to subpart D of this part, of an allowance from the Phase I extension reserve, for possible future allocation to a Phase I extension applicant's Allowance Tracking System unit account.

award official (40CFR7.25) means the EPA official with the authority to approve and execute assistance agreements and to take other assistance related actions authorized by this part and by other EPA regulations or delegation of authority (other identical or similar definitions are provided in 40CFR30.200; 35.4010; 35.6015-4).

awarding agency (40CFR31.3) means:
(1) with respect to a grant, the Federal agency, and
(2) with respect to a subgrant, the party that awarded the subgrant.

axenic (40CFR797.1160-1) means a culture of Lemna fronds free from other organisms.

axle clearance (40CFR86.084.2) means the vertical distance from the level surface on which an automobile is standing to the lowest point on the axle differential of the automobile.

axle ratio (40CFR600.002.85-28) means the number of times the input shaft to the differential (or equivalent) turns for each turn of the drive wheels.

axle ratio (40CFR86.602.84-2) means all ratios within plus or minus 3% of the axle ratio specified in the configuration in the test order.

azimuth angle (40CFR60-AA (alt. method 1)) means the angle in the horizontal plane that designates where the laser beam is pointed. It is measured from an arbitrary fixed reference line in that plane.

*********** BBBBB ***********

baby foods (40CFR407.81-c) shall mean the processing of canned fresh fruits and vegetables, meats, eggs, fruit juices, cereal, formulated entrees, desserts, and snacks using fresh, preprocessed, or any combination of these and other food ingredients necessary for the production of infant foods (cf. food).

back pressure (EPA-94/04): A pressure that can cause water to backflow into the water supply when a user's water system is at a higher pressure than the public system.

backflow/back siphonage (EPA-94/04): A reverse flow condition created by a difference in water pressures that causes water to flow back into the distribution pipes of a drinking water supply from any source other than an intended one.

background conditions (40CFR131.35.d-2) means the biological, chemical, and physical conditions of a water body, upstream from the point or non-point source discharge under consideration. Background sampling location in an enforcement action will be upstream from the point of discharge, but not upstream from other inflows. If several discharges to any water body exist, and an enforcement action is being taken for possible violations to the standards, background sampling will be undertaken immediately upstream from each discharge.

background level (EPA-94/04): In air pollution control, the concentration of air pollutants in a definite area during a fixed period of time prior to the starting up or on the stoppage of a source of emission under control. In toxic substances monitoring, the average presence in the environment, originally referring to naturally occurring phenomena.

background soil pH (40CFR257.3.5-2) means the pH of the soil prior to the addition of substances that alter the hydrogen ion concentration.

backscatter (40CFR60-AA (alt. method 1)) means the scattering of laser light in a direction opposite to that of the incident laser beam due to reflection from particulates along the beam's atmospheric path which may include a smoke plume.

backscatter signal (40CFR60-AA (alt. method 1)) means the general term for the lidar return signal which results from laser light being backscattered by atmospheric and smoke plume particulates.

backwashing (EPA-94/04): Reversing the flow of water back through the filter media to remove the entrapped solids.

bacteria (EPA-94/04): (Singular: bacterium) Microscopic living organisms that can aid in pollution control by metabolizing organic matter in sewage, oil spills or other pollutants. However, bacteria in soil, water or air can also cause human, animal and plant health problems.

baffle (EPA-94/04): A flat board or plate, deflector, guide or similar device constructed or placed in flowing water or slurry systems to cause more uniform flow velocities, to absorb energy, and to divert, guide, or agitate liquids.

baffle chamber (EPA-94/04): In incinerator design, a chamber designed to promote the settling of fly ash and coarse particulate matter by changing the direction and/or reducing the velocity of the gases produced by the combustion of the refuse or sludge.

bagging operation (40CFR60.671) means the mechanical process by which bags are filled with nonmetallic minerals.

baghouse filter (EPA-94/04): Large fabric bag, usually made of glass fibers, used to eliminate intermediate and large (greater than 20 microns in diameter) particles. This device operates like the bag of an electric vacuum cleaner, passing the air and smaller particles while entrapping the larger ones (cf. air pollution control device).

bailer (EPA-94/04): A long pipe with a valve at the lower end, used to remove slurry from the bottom or side of a well as it is being drilled.

bake oven (40CFR60.311) means a device which uses heat to dry or cure coatings (other identical or similar definitions are provided in 40CFR60.391).

baked coatings (40CFR52.741) means any coating which is cured or dried in an oven where the oven air temperature exceeds 90°C (194°F).

balanced, indigenous community (40CFR125.71-c) is synonymous with the term "balanced, indigenous population" in the Act and means a biotic community typically characterized by diversity, the capacity to sustain itself through cyclic seasonal changes, presence of necessary food chain species and by a lack of domination by pollution tolerant species. Such a community may include historically non-native species introduced in connection with a program of wildlife management and species whose presence or abundance results from substantial, irreversible environmental modifications. Normally, however, such a community will not include species whose presence or abundance is attributable to the introduction of pollutants that will be eliminated by compliance by all sources with section 301(b)(2) of the Act; and may not include species whose presence or abundance is attributable to alternative effluent limitations imposed pursuant to section 316(a).

balanced, indigenous population (40CFR125.58-f) means an ecological community which:
(1) Exhibits characteristics similar to those of nearby, healthy communities existing under comparable but unpolluted environmental conditions; or
(2) May reasonably be expected to become re-established in the polluted water body segment from adjacent waters if sources of pollution were removed.

baler (40CFR246.101-b) means a machine used to compress solid wastes, primary materials, or recoverable materials, with or without binding, to a density or from which will support handling and transportation as a material unit rather than requiring a disposable or reuseable container. This specifically excludes briquetters and stationary compaction equipment which are used to compact materials into disposable or reuseable containers.

baling (EPA-94/04): Compacting solid waste into blocks to reduce volume and simplify handling (cf. baler).

ballast (40CFR419.11-c) shall mean the flow of waters, from a ship, that is treated along with refinery wastewaters in the main treatment system.

ballistic separator (EPA-94/04): A machine that sorts organic from inorganic matter for composting.

band application (EPA-94/04): The spreading of chemicals over, or next to, each row of plants in a field.

bank (40CFR39.105-b) means the Federal Financing Bank established pursuant to the Federal Financing Bank Act of 1973 (12USC2281 et seq.).

banking (40CFR86.090.2) means the retention of heavy-duty engine NO_x and particulate emission credits, by the manufacturer generating the emission credits, for use in future model year certification programs as permitted by regulation.

banking (40CFR88.202.94) means the retention of credits, by the manufacturer generating the emissions credits, for use in future model-year certification as permitted by regulation.

banking (40CFR89.202.96) means the retention of nonroad engine emission credits by the manufacturer generating the emission credits for use in future model year averaging or trading as permitted by these regulations.

banking (EPA-94/04): A system for recording qualified air emission reductions for later use in bubble, offset, or netting transactions (cf. emissions trading).

bar, billet and bloom (40CFR420.91-l) means those acid pickling operations that pickle bar, billet or bloom products.

bar screen (EPA-94/04): In wastewater treatment, a device used to remove large solids.

barometric condensing operations (40CFR409.11-b) shall mean those operations or processes directly associated with or related to the concentration and crystallization of sugar solutions.

barrel (CWA311-33USC1321) means 42 United States gallons at 60 degrees Fahrenheit (other identical or similar definitions are provided in 40CFR113.3-c; SF101-42USC9601).

barrel finishing (40CFR471.02-mm): See tumbling.

barrier (40CFR191.12) means any material or structure that prevents or substantially delays movement of water or radionuclides toward the accessible environment. For example, a barrier may be a geologic structure, a canister, a waste form with physical and chemical characteristics that significantly decrease the mobility of radionuclides, or a material placed over and around waste, provided that the material or structure substantially delays movement of water or radionuclides.

barrier coating(s) (EPA-94/04): A layer of a material that obstructs or prevents passage of something through a surface that is to be protected, e.g. grout, caulk, or various sealing compounds; sometimes used with polyurethane membranes to prevent corrosion or oxidation of metal surfaces, chemical impacts on various materials, or, for example, to prevent radon infiltration through walls, cracks, or joints in a house.

basal application (EPA-94/04): In pesticides, the application of a chemical on plant stems or tree trunks just above the soil line.

basal diet (40CFR797.2050-8) means the food or diet as it is prepared or received from the supplier, without the addition of any carrier, diluent, or test substance.

basal diet (40CFR797.2130-iv) means the untreated form of the diet, such as the diet obtained from a commercial (other identical or similar definitions are provided in 40CFR797.2150-iv).

base date (40CFR52.1161-8) means the date set forth in paragraph (d) of this section as of which the base number of single-passenger commuter vehicles at a particular employment facility or educational institution must be determined.

base date (40CFR52.2297-11) means the date which is used as a reference for determination of compliance with this regulation. The base date for all facilities shall be November 1, 1977, with the exception listed in paragraph (h) of this section (under CAA).

base date period (40CFR52.2297-12) means the thirty day period immediately preceding the base date; "compliance date period" means the thirty day period immediately preceding the compliance date. In situations where the averaging periods are not

appropriate, approval of an alternate period may be requested from the Regional Administrator (under CAA).

base film (40CFR60.711-1) means the substrate that is coated to produce magnetic tape.

base flood (40CFR503.9-b) is a flood that has a one percent chance of occurring in any given year (i.e., a flood with a magnitude equalled once in 100 years).

base flood (40CFR6-AA-a) means that flood which has a one percent chance of occurrence in any given year (also known as a 100-year flood). This term is used in the National Flood Insurance Program (NFIP) to indicate the minimum level of flooding to be used by a community in its floodplain management regulations (under Executive Order 11988 entitled "Flooding Plain Management" dated May 24, 1977).

base floodplain (40CFR6-AA-b) means the land area covered by a 100-year flood (one percent chance floodplain). Also see definition of floodplain (under Executive Order 11988 entitled "Flooding Plain Management" dated May 24, 1977).

base gasoline (CAA241-42USC7581) means gasoline which meets the following specifications:

    API gravity: 57.8
    Sulfur, ppm: 317
    Color: Purple
    Benzene, vol. %: 1.35
    Reid vapor pressure: 8.7
    Drivability: 1195
    Antiknock index: 87.3
Distillation, D-86 F
    IBP: 92
    10%: 126
    50%: 219
    90%: 327
    EP: 414
Hydrocarbon Type, Vol. % FIA:
    Aromatics: 30.9
    Olefins: 8.2
    Saturates: 60.9
The Administrator shall modify the definitions of NMOG (nonmethane organic gas), base gasoline, and the methods for making reactivity adjustments, to conform to the definitions and method used in

California under the Low-Emission Vehicle and Clean Fuel Regulations of the California Air Resources Board, so long as the California definitions are, in the aggregate, at least as protective of public health and welfare as the definitions in this section.

base level (40CFR600.002.85-23) means a unique combination of basic engine inertia weight class and transmission class.

base load (40CFR60.331-j) means the load level at which a gas turbine is normally operated (under CAA).

base pair mutagens (40CFR798.5265-2) are agents which cause a base change in the DNA. In a reversion assay, this change may occur at the site of the original mutation or at a second site in the chromosome.

base pair mutagens (40CFR798.5300-2) are agents which cause a base change in the DNA.

base temperature (40CFR60.431) means an arbitrary reference temperature for determining liquid densities or adjusting the measured volume of a liquid quantity.

base vehicle (40CFR600.002.85-40) means the lowest priced version of each body style that makes up a car line.

based flood (40CFR257.3.1-1) means a flood that has a 1 percent or greater chance of recurring in any year or a flood of a magnitude equalled or exceeded over in 100 years on the average over a significantly long period.

baseline (40CFR72.2) means the annual average quantity of fossil fuel consumed by a unit, measured in millions of British Thermal Units (expressed in mmBtu) for calendar years 1985 through 1987; provided that in the event that a unit is listed in the NADB, the baseline will be calculated for each unit-generator pair that includes the unit, and the unit's baseline will be the sum of such unit-generator baselines. The unit-generator baseline will be as provided in the NADB under the data field "BASE8587", as adjusted by the outage hours listed

in the NADB under the data field "OUTAGEHR" in accordance with the following equation:

$$\text{Baseline} = \text{BASE8587} \times \{26280/(26280 - \text{OUTAGEHR})\} \times \{36/(36 - \text{months not on line})\} \times 10^6$$

"Months not on line" is the number of months during January 1985 through December 1987 prior to the commencement of firing for units that commenced firing in that period, i.e., the number of months, in that period, prior to the on-line month listed under the data field "BLRMNONL" and the on-line year listed in the data field "BLRYRONL" in the NADB (under CAA).

baseline (CAA402-42USC7651a) means the annual quantity of fossil fuel consumed by an affected unit, measured in millions of British Thermal Units (mmBtu), calculated as follows:

(A) For each utility unit that was in commercial operation prior to January 1, 1985, the baseline shall be the annual average quantity of mmBtu's consumed in fuel during calendar years 1985, 1986, and 1987, as recorded by the Department of Energy pursuant to Form 767. For any utility unit for which such form was not filed, the baseline shall be the level specified for such unit in the 1985 National Acid Precipitation Assessment Program (NAPAP) Emissions Inventory, Version 2, National Utility Reference File (NURF) or in a corrected data base as established by the Administrator pursuant to paragraph (3). For non-utility units, the baseline is the NAPAP Emissions Inventory, Version 2. The Administrator, in the Administrator's sole discretion, may exclude periods during which a unit is shutdown for a continuous period of four calendar months or longer, and make appropriate adjustments under this paragraph. Upon petition of the owner or operator of any unit, the Administrator may make appropriate baseline adjustments for accidents that caused prolonged outages.

(B) For any other nonutility unit that is not included in the NAPAP Emissions Inventory, Version 2, or a corrected data base as established by the Administrator pursuant to paragraph (3), the baseline shall be the annual average quantity, in mmBtu consumed in fuel by that unit, as calculated pursuant to a method which the administrator shall prescribe by regulation to be promulgated not later than eighteen months after enactment of the Clean Air Act Amendments of 1990.

(C) The Administrator shall, upon application or on his own motion, by December 31, 1991, supplement data needed in support of this title and correct any factual errors in data from which affected Phase II units' baselines or actual 1985 emission rates have been calculated. Corrected data shall be used for purposes of issuing allowances under the title. Such corrections shall not be subject to judicial review, nor shall the failure of the Administrator to correct an alleged factual error in such reports be subject to judicial review.

baseline area (40CFR51.166-15) means:

(i) Any intrastate area (and every part thereof) designated as attainment or unclassifiable under section 107(d)(1)(D) or (E) of the Act in which the major source or major modification establishing the minor source baseline date would construct or would have an air quality impact equal to or greater than 1 ug/m^3 (annual average) of the pollutant for which the minor source baseline date is established.

(ii) Area redesignations under section 107(d)(1) (D) or (E) of the Act cannot intersect or be smaller than the area of impact of any major stationary source or major modification which:

(a) Establishes a minor source baseline date; or

(b) Is subject to 40CFR52.21 or under regulations approved pursuant to 40CFR51.166, and would be constructed in the same state as the state proposing the redesignation.

(iii) Any baseline area established originally for the TSP increments shall remain in effect and shall apply for purposes of determining the amount of available PM-10 increments, except that such baseline area shall not remain in effect if the permit authority rescinds the corresponding minor source baseline date in accordance with paragraph (b)(14)(iv) of this section.

baseline area (40CFR52.21-15) means:

(i) Any intrastate area (and every part thereof) designated as attainment or unclassifiable under section 107(d)(1) (D) or (E) of the Act in which

the major source or major modification establishing the minor source baseline date would construct or would have an air quality impact equal to or greater than 1 ug/m^3 (annual average) of the pollutant for which the minor source baseline date is established.

(ii) Area redesignations under section 107(d)(1) (D) or (E) of the Act cannot intersect or be smaller than the area of impact of any major stationary source or major modification which:

(a) Establishes a minor source baseline date; or

(b) Is subject to 40CFR52.21 and would be constructed in the same state as the state proposing the redesignation.

(iii) Any baseline area established originally for the TSP increments shall remain in effect and shall apply for purposes of determining the amount of available PM-10 increments, except that such baseline area shall not remain in effect if the Administrator rescinds the corresponding minor source baseline date in accordance with paragraph (b)(14)(iv) of this section (under CAA).

baseline concentration (40CFR51.166-13) means that:

(i) ambient concentration level which exists in the baseline area at the time of the applicable minor source baseline date. A baseline concentration is determined for each pollutant for which a minor source baseline date is established and shall include:

(a) The actual emissions representative of sources in existence on the applicable minor source baseline date, except as provided in paragraph (b)(13)(ii) of this section;

(b) The allowable emissions of major stationary sources which commenced construction before the major source baseline date, but were not in operation by the applicable minor source baseline date.

(ii) The following will not be included in the baseline concentration and will affect the applicable maximum allowable increase(s):

(a) Actual emissions from any major stationary source on which construction commenced after the major source baseline date; and

(b) Actual emissions increases and decreases at

any stationary source occurring after the minor source baseline date.

baseline concentration (40CFR52.21-13) means:

(i) That ambient concentration level which exists in the baseline area at the time of the applicable minor source baseline date. A baseline concentration is determined for each pollutant for which a baseline date is established and shall include:

    (a) The actual emissions representative of sources in existence on the applicable minor source baseline date, except as provided in paragraph (b)(13)(ii) of this section;

    (b) The allowable emissions of major stationary sources which commenced construction before the major source baseline date but were not in operation by the applicable minor source baseline date.

(ii) The following will not be included in the baseline concentration and will affect the applicable maximum allowable increase(s):

    (a) Actual emissions from any major stationary source on which construction commenced after the major source baseline date; and

    (b) Actual emissions increases and decreases at any stationary source occurring after the minor source baseline date.

baseline concentration (CAA169-42USC7479) means, with respect to a pollutant, the ambient concentration levels which exist at the time of the first application for a permit in an area subject to this part, based on air quality data available in the Environmental Protection Agency or a State air pollution control agency and on such monitoring data as the permit applicant is required to submit. Such ambient concentration levels shall take into account all projected emissions in, or which may affect, such area from any major emitting facility on which construction commenced prior to January 6, 1975, but which has not begun operation by the date of the baseline air quality concentration determination. Emissions of sulfur oxides and particulate matter from any major emitting facility on which construction commenced after January 6, 1975, shall not be included in the baseline and shall be counted against the maximum allowable increases in pollutant concentrations established under this part.

baseline configuration (40CFR610.11-14) means the unretrofitted test configuration, tuned in accordance with the automobile manufacturers specifications (under CAA).

baseline consumption allowances (40CFR82.3-b) means the consumption allowances apportioned under 40CFR82.6.

baseline date (40CFR51.166-14): See major source baseline date.

baseline gasoline (CAA211.k-42USC7545):

(i) Summertime--The term baseline gasoline means in the case of gasoline sold during the high ozone period (as defined by the Administrator) a gasoline which meets the following specifications:

BASELINE GASOLINE FUEL PROPERTIES

API Gravity	57.4
Sulfur, ppm	339
Benzene, %	1.53
RVP, psi	8.7
Octane, R+M/2	87.3
IBP, F	91
10%, F	128
50%, F	218
90%, F	330
End Point, F	415
Aromatics, %	32.0
Olefins, %	9.2
Saturates, %	58.8

(ii) Wintertime--Te administrator shall establish the specifications of baseline gasoline for gasoline sold at times other than the high ozone period (as defined by the Administrator). Such specifications shall be the specifications of 1990 industry average gasoline sold during such period.

baseline or trend assessment survey (40CFR228.2-b) means the planned sampling or measurement of parameters at set stations or in set areas in and near disposal sites for a period of time sufficient to provide synoptic data for determining water quality, benthic, or biological conditions as a result of ocean disposal operations. The minimum requirements of rush surveys are given in 40CFR228.13.

baseline production allowances (40CFR82.3-c) means

the production allowances apportioned under 40CFR82.5.

baseline vehicles (CAA211.k-42USC7545) mean representative model year 1990 vehicles.

baseline year (CAA601-42USC7671) means:
(A) the calendar year 1986, in the case of any class I substance listed in Group I or II under section 602(a),
(B) the calendar year 1989, in the case of any class I substance listed in Group III, IV, or V under section 602(a), and
(C) a representative calendar year selected by the Administrator, in the case of:
  (i) any substance added to the list of class I substances after the publication of the initial list under section 602(a), and
  (ii) any class II substance.

basic engine (40CFR600.002.85-21) means a unique combination of manufacturer, engine displacement, number of cylinders, fuel system (as distinguished by number of carburetor barrels or use of fuel injection), catalyst usage, and other engine and emission control system characteristics specified by the Administrator. For electric vehicles, basic engine means a unique combination of manufacturer and electric traction motor, motor controller, battery configuration, electrical charging system, energy storage device, and other components as specified by the Administrator.

basic engine (40CFR86.082.2) means unique combination of manufacturer, engine displacement, number of cylinders, fuel system (as distinguished by number of carburetor barrels or use of fuel injection), catalyst usage, and other engine and emission control system characteristics specified by the Administrator.

basic oxygen furnace steelmaking (40CFR420.41-a) means the production of steel from molten iron, steel scrap, fluxes, and various combinations thereof, in refractory lined fuel-fired furnaces by adding oxygen.

basic oxygen process furnace (40CFR60.141-a) means any furnace with a refractory lining in which molten steel is produced by charging scrap metal, molten iron, and flux materials or alloy additions into

a vessel and by introducing a high volume of oxygen-rich gas. Open hearth, blast, and reverberatory furnaces are not included in this definition (other identical or similar definitions are provided in 40CFR60.141a).

basic phase II allowance allocations (CAA402-42USC7651a) means:
(A) For calendar years 2000 through 2009 inclusive, allocations of allowances made by the Administrator pursuant to section 403 and subsections (b)(1), (3), and (4); (c)(1), (2), (3), and (5); (d)(1), (2), (4), and (5); (e); (f); (g)(1), (2), (3), (4), and (5); (h)(1); (i) and (j) of section 405.
(B) For each calendar year beginning in 2010, allocations of allowances made by the Administrator pursuant to section 403 and subsections (b)(1), (3), and (4); (c)(1), (2), (3), and (5); (d)(1), (2), (4) and (5); (e); (f); (g)(1), (2), (3), (4), and (5); (h)(1) and (3); (i) and (j) of section 405.

basic phase II allowance allocations (40CFR72.2) means:
(1) For calendar years 2000 through 2009 inclusive, allocations of allowances made by the Administrator pursuant to section 403 and section 405 (b)(1), (3), and (4); (c)(1), (2), (3), and (5); (d)(1), (2), (4), and (5); (e); (f); (g)(1), (2), (3), (4), and (5); (h)(1); (i); and (j).
(2) For each calendar year beginning in 2010, allocations of allowances made by the Administrator pursuant to section 403 and section 405 (b)(1), (3), and (4); (c)(1), (2), (3), and (5); (d)(1), (2), (4), and (5); (e); (f); (g)(1), (2), (3), (4), and (5); (h)(1) and (3); (i); and (j).

basic vehicle frontal area (40CFR86.082.2) means the area enclosed by the geometric projection of the basic vehicle along the longitudinal axis, which includes tires but excludes mirrors and air deflectors, onto a plane perpendicular to the longitudinal axis of the vehicle.

basis material (40CFR465.02-c) means the coiled strip which is processed.

basis material (40CFR466.02-b) means the metal part or base onto which porcelain enamel is applied.

BAT (40CFR467.02-u) means the best available technology economically achievable under section 304(b)(2)(B) of the Act.

batch (40CFR160.3) means a specific quantity or lot of a test or control substance that has been characterized according to 40CFR160.105(a).

batch (40CFR169.1-b) means a quantity of a pesticide product or active ingredient used in producing a pesticide made in one operation or lot or if made in a continuous or semi-continuous process or cycle, the quantity produced during an interval of time to be specified by the producer.

batch (40CFR204.51-j) means the collection of compressors of the same category or configuration, as designated by the Administrator in a test request, from which a batch sample is to be randomly drawn and inspected to determine conformance with the acceptability criteria.

batch (40CFR205.51-3) means the collection of vehicles of the same category, configuration or subgroup thereof as designated by the Administrator in a test request, from which a batch sample is to be drawn, and inspected to determine conformance with the acceptability criteria.

batch (40CFR420.81-g) means those descaling operations in which the products are processed in discrete batches.

batch (40CFR420.91-e) means those pickling operations which process steel products such as coiled wire, rods, and tubes in discrete batches or bundles.

batch (40CFR792.3) means a specific quantity or lot of a test or control substance that has been characterized according to 40CFR792.105(a).

batch distillation operation (40CFR60.661) means a noncontinuous distillation operation in which a discrete quantity or batch of liquid feed is charged into a distillation unit and distilled at one time. After the initial charging of the liquid feed, no additional liquid is added during the distillation operation.

batch MWC (40CFR60.51a) means an MWC (municipal waste combustor) unit designed such that it cannot combust MSW (municipal solid waste) continuously 24 hours per day because the design does not allow waste to be fed to the unit or ash to be removed while combustion is occurring.

batch of reformulated gasoline (40CFR80.2-gg) means a quantity of reformulated gasoline which is homogeneous with regard to those properties which are specified for reformulated gasoline certification.

batch operation (40CFR60.701) means any noncontinuous reactor process that is not characterized by steady-state conditions and in which reactants are not added and products are not removed simultaneously.

batch operation (40CFR63.101) means a noncontinuous operation in which a discrete quantity or batch of feed is charged into a chemical manufacturing process unit and distilled or reacted at one time. Batch operation includes noncontinuous operations in which the equipment is fed intermittently or discontinuously. Addition of raw material and withdrawal of product do not occur simultaneously in a batch operation. After each batch operation, the equipment is generally emptied before a fresh batch is started.

batch operation (40CFR63.111) means a noncontinuous operation in which a discrete quantity or batch of feed is charged into a chemical manufacturing process unit and distilled or reacted at one time. Batch operation includes noncontinuous operations in which the equipment is fed intermittently or discontinuously. Addition of raw material and withdrawal of product do not occur simultaneously in a batch operation. After each batch operation, the equipment is generally emptied before a fresh batch is started.

batch process (40CFR63.161) means a process in which the equipment is fed intermittently or discontinuously. Processing then occurs in this equipment after which the equipment is generally emptied. Examples of industries that use batch processes include pharmaceutical production and pesticide production.

batch product process equipment train

(40CFR63.161) means the collection of equipment (e.g., connectors, reactors, valves, pumps, etc.) configured to produce a specific product or intermediate by a batch process.

batch, rod and wire (40CFR420.81-d) means those descaling operations that remove surface scale from rod and wire products in batch processes.

batch sample (40CFR204.51-k) means the collection of compressors that are drawn from a batch.

batch sample (40CFR205.51-5) means the collection of vehicles of the same category, configuration or subgroup thereof which are drawn from a batch and from which test samples are drawn.

batch sample size (40CFR204.51-l) means the number of compressors of the same category or configuration which is randomly drawn from the batch sample and which will receive emissions tests (under the Noise Control Act).

batch sample size (40CFR205.51-6) means the number of vehicles of the same category or configuration in a batch sample.

batch, sheet and plate (40CFR420.81-c) means those descaling operations that remove surface scale from sheet and plate products in batch processes.

batch size (40CFR204.51-n) means the number as designated by the Administrator in the test request, of compressors of the same category or configuration in a batch.

batch size (40CFR205.51-4) means the number as designated by the Administrator in the test request of vehicles of the same category or configuration.

batt insulation (40CFR248.4-b): See blanket insulation.

battery (40CFR461.2-a) means a modular electric power source where part or all of the fuel is contained within the unit and electric power is generated directly from a chemical reaction rather than indirectly through a heat cycle engine. In this regulation there is no differentiation between a single cell and a battery.

battery configuration (40CFR600.002.85-49) means the electrochemical type, voltage, capacity (in Watt-hours at the c/3 rate), and physical characteristics of the battery used as the tractive energy storage device.

battery manufacturing operations (40CFR461.2-b) means all of the specific processes used to produce a battery including the manufacture of anodes and cathode and associated ancillary operations. These manufacturing operations are excluded from regulation under any other point source category.

bauxite (40CFR421.11-b) shall mean ore containing alumina monohydrate or alumina trihydrate which serves as the principal raw material for the production of alumina by the Bayer process or by the combination process.

BCT (40CFR430.223) means the best conventional pollutant control technology, under section 304(b)(4) of the Act (other identical or similar definitions are provided in 40CFR467.02-v).

bead (40CFR60.541) means rubber-covered strands of wire, wound into a circular form, which ensure a seal between a tire and the rim of the wheel onto which the tire is mounted.

bead cementing operation (40CFR60.541) means the system that is used to apply cement to the bead rubber before or after it is wound into its final circular form. A bead cementing operation consists of a cement application station, such as a dip tank, spray booth and nozzles, cement trough and roller or swab applicator, and all other equipment necessary to apply cement to would beads or bead rubber and to allow evaporation of solvent from cemented beads.

bed load (EPA-94/04): Sediment particles resting on or near the channel bottom that are pushed or rolled along by the flow of water.

beehive cokemaking (40CFR420.11-a) means those operations in which coal is heated with the admission of air in controlled amounts for the purpose of producing coke. There are no byproduct recovery operations associated with beehive cokemaking operations.

Beer Lambert law (40CFR796.3700-iii) states that

the absorbance of a solution of a given chemical species, at a fixed wavelength, is proportional to the thickness of the solution (1), or the light pathlength, and the concentration of the absorbing species (C) (other identical or similar definitions are provided in 40CFR796.3780-iii; 796.3800-iii).

beets (40CFR407.71-b) shall include the processing of beets into the following product styles: Canned and peeled, whole, sliced, diced, French style, sections, irregular, and other cuts but not dehydrated beets.

begin actual construction (40CFR51.165-xv) means in general, initiation of physical on-site construction activities on an emissions unit which are of a permanent nature. Such activities include, but are not limited to, installation of building supports and foundations, laying of underground pipework, and construction of permanent storage structures. With respect to a change in method of operating this term refers to those on-site activities other than preparatory activities which mark the initiation of the change.

begin actual construction (40CFR51.166-11) means, in general, initiation of physical on-site construction activities on an emissions unit which are of a permanent nature. Such activities include, but are not limited to, installation of building supports and foundations, laying of underground pipework, and construction of permanent storage structures. With respect to a change in method of operation this term refers to those on-site activities, other than preparatory activities, which mark the initiation of the change.

begin actual construction (40CFR51-AS-17) means, in general, initiation of physical on-site construction activities on an emissions unit which are of a permanent nature. Such activities include, but are not limited to, installation of building supports and foundations, laying of underground pipework, and construction of permanent storage structures. With respect to a change in method of operating this term refers to those on-site activities other than preparatory activities which mark the initiation of the change.

begin actual construction (40CFR52.21-11) means, in

general, initiation of physical on-site construction activities on an emissions unit which are of a permanent nature. Such activities include, but are not limited to, installation of building supports and foundations, laying underground pipework and construction of permanent storage structures. With respect to a change in method of operations, this term refers to those on-site activites other than preparatory activities which mark the initiation of the change.

begin actual construction (40CFR52.24-17) means, in general, initiation of physical on-site construction activities on an emissions unit which are of a permanent nature. Such activities include, but are not limited to, installation of building supports and foundations, laying of underground pipework, and construction of permanent storage structures. With respect to a change in method of operations, this term refers to those on-site activities other than preparatory activities which mark the initiation of the change.

belowground release (40CFR280.12) means any release to the subsurface of the land and to ground water. This includes, but is not limited to, releases from the belowground portions of an underground storage tank system and belowground releases associated with overfills and transfer operations as the regulated substances move to or from an underground storage tank.

belowground storage facility (40CFR113.3-d) means a tank or other container located other than as defined as aboveground.

belt conveyor (40CFR60.671) means a conveying device that transports material from one location to another by means of an endless belt that is carried on a series of idlers and routed around a pulley at each end.

BEN (EPA-94/04): EPA's computer model for analyzing a violator's economic gain from not complying with the law.

bench scale tests (EPA-94/04): Laboratory testing of potential cleanup technologies (cf. treatability study).

beneath the surface of the ground (40CFR280.12)

means beneath the ground surface or otherwise covered with earthen materials.

beneficial organism (40CFR166.3-c) means any pollinating insect, or any pest predator, parasite, pathogen or other biological control agent which functions naturally or as part of an integrated pest management program to control another pest.

beneficiation (40CFR60.401-g) means the process of washing the rock to remove impurities or to separate size fractions.

beneficiation area (40CFR440.141-1) means the area of land used to stockpile ore immediately before the beneficiation process, the area of land used for the beneficiation process, the area of land used to stockpile the tailings immediately after the beneficiation process, and the area of land from the stockpiled tailings to the treatment system (e.g., holding pond or settling pond, and the area of the treatment system).

beneficiation process (40CFR440.141-2) means the dressing or processing of gold bearing ores for the purpose of:
(i)   Regulating the size of, or recovering, the ore or product,
(ii) Removing unwanted constituents from the ore, and
(iii) Improving the quality, purity, or assay grade of a desired product.

benefit (40CFR27.2) means, in the context of "statement," anything of value, including but not limited to any advantage, preference, privilege, license, permit, favorable decision, ruling status, or loan guarantee.

benzene (or priority pollutant No. 4) (40CFR420.02-m) means the value obtained by the standard method Number 602 specified in 44 FR 69464, 69570 (December 3, 1979).

benzene concentration (40CFR61.341) means the fraction by weight of benzene in a waste as determined in accordance with the procedures specified in 40CFR1.355 of this subpart.

benzene storage tank (40CFR61.131) means any tank, reservoir, or container used to collect or store refined benzene.

benzidine (40CFR129.4-e) means the compound benzidine and its salts as identified by the chemical name 4,4-diaminobiphenyl.

benzidine based dye applicator (40CFR129.104-2) means an owner or operator who uses benzidine based dyes in the dyeing of textiles, leather or paper (under CWA).

benzidine manufacturer (40CFR129.104-1) means a manufacturer who produces benzidine or who produces benzidine as an intermediate product in the manufacture of dyes commonly used for textile, leather and paper dyeing.

benzo(a)pyrene (or priority pollutant No. 73) (40CFR420.02-n) means the value obtained by the standard method Number 610 specified in 44 FR 69464, 69570 (December 3, 1979).

beryllium (40CFR61.31-a) means the element beryllium. Where weights or concentrations are specified, such weights or concentrations apply to beryllium only, excluding the weight or concentration of any associated elements.

beryllium (EPA-94/04): An airborne metal hazardous to human health when inhaled. It is discharged by machine shops, ceramic and propellant plants, and foundries.

beryllium alloy (40CFR61.31-j) means any metal to which beryllium has been added in order to increase its beryllium content and which contains more than 0.1 percent beryllium by weight.

beryllium containing waste (40CFR61.31-g) means material contaminated with beryllium and/or beryllium compounds used or generated during any process or operation performed by a source subject to this subpart.

beryllium copper alloy (40CFR468.02-y) shall mean any copper alloy that is alloyed to contain 0.10 percent or greater beryllium.

beryllium ore (40CFR61.31-c) means any naturally

occurring material mined or gathered for its beryllium content.

beryllium propellant (40CFR61.41-b) means any propellant incorporating beryllium.

best available control (CAA183.e-42USC7511b) means the degree of emissions reduction that the Administrator determines, on the basis of technological and economic feasibility, health, environmental, and energy impacts, is achievable through the application of the most effective equipment, measures, processes, methods, systems or techniques, including chemical reformulation, product or feedstock substitution, repackaging, and directions for use, consumption, storage, or disposal.

best available control measures (BACM) (EPA-94/04): A term used to refer to the most effective measures (according to EPA guidance) for controlling small or dispersed particulates from sources such as roadway dust, soot and ash from woodstoves and open burning of rush, timber, grasslands, or trash.

best available control technology (40CFR51.166-12) means an emissions limitation (including a visible emissions standard) based on the maximum degree of reduction for each pollutant subject to regulation under the Act which would be emitted from any proposed major stationary source or major modification which the reviewing authority, on a case-by-case basis, taking into account energy, environmental, and economic impacts and other costs, determines is achievable for such source or modification through application of production processes or available methods, systems, and techniques, including fuel cleaning or treatment or innovative fuel combination techniques for control of such pollutant. In no event shall application of best available control technology result in emissions of any pollutant which would exceed the emissions allowed by any applicable standard under 40CFR parts 60 and 61. If the reviewing authority determines that technological or economic limitations on the application of measurement methodology to a particular emissions unit would make the imposition of an emissions standard infeasible, a design, equipment, work practice, operational standard or combination thereof, may be prescribed instead to satisfy the requirement for the application of best available control technology. Such standard shall, to the degree possible, set forth the emissions reduction achievable by implementation of such design, equipment, work practice or operation, and shall provide for compliance by means which achieve equivalent results.

best available control technology (40CFR52.21-12) means an emissions limitation (including a visible emission standard) based on the maximum degree of reduction for each pollutant subject to regulation under Act which would be emitted from any proposed major stationary source or major modification which the Administrator, on a case-by-case basis, taking into account energy, environmental, and economic impacts and other costs, determines is achievable for such source or modification through application of production processes or available methods, systems, and techniques, including fuel cleaning or treatment or innovative fuel combustion techniques for control of such pollutant. In no event shall application of best available control technology result in emissions of any pollutant which would exceed the emissions allowed by any applicable standard under 40CFR parts 60 and 61. If the Administrator determines that technological or economic limitations on the application of measurement methodology to a particular emissions unit would make the imposition of an emissions standard infeasible, a design, equipment, work practice, operational standard, or combination thereof, may be prescribed instead to satisfy the requirement for the application of best available control technology. Such standard shall, to the degree possible, set forth the emissions reduction achievable by implementation of such design, equipment, work practice or operation, and shall provide for compliance by means which achieve equivalent results.

best available control technology (BACT) (CAA169-42USC7479) means an emission limitation based on the maximum degree of reduction of each pollutant subject to regulation under this Act emitted from or which results from any major emitting facility, which the permitting authority, on a case-by-case basis, taking into account energy, environmental, and economic impacts and other costs, determines is achievable for such facility through application of production processes and available methods, systems, and techniques, including fuel cleaning or treatment or innovative fuel combustion techniques for control of each such pollutant. In no event shall application of "best available control technology" result in emissions of any pollutants which will exceed the emissions allowed by any applicable standard established pursuant to section 111 or 112 of this Act.

best available control technology (BACT) (EPA-94/04): For any specific source, the necessary technology that would produce the greatest reduction of each pollutant regulated by the Clean Air Act, taking into account energy, environmental, economic and other costs.

best available retrofit technology (BART) (40CFR51.301-c) means an emission limitation based on the degree of reduction achievable through the application of the best system of continuous emission reduction for each pollutant which is emitted by an existing stationary facility. The emission limitation must be established, on a case-by-case basis, taking into consideration the technology available, the costs of compliance, the energy and nonair quality environmental impacts of compliance, any pollution control equipment in use or in existence at the source, the remaining useful life of the source, and the degree of improvement in visibility which may reasonably be anticipated to result from the use of such technology.

best available technology or BAT (40CFR141.2) means the best technology, treatment techniques, or other means which the Administrator finds, after examination for efficacy under field conditions and not solely under laboratory conditions, are available (taking cost into consideration). For the purposes of setting MCLs for synthetic organic chemicals, any BAT must be at least as effective as granular activated carbon.

best conventional pollutant control technology (BCT) (40CFR430.223; 467.02-v) is defined under section 304(b)(4) of the Act.

best demonstrated available technology (BDAT) (EPA-94/04): As identified by EPA, the most effective commercially available means of treating specific types of hazardous waste. The BDATs may change with advances in treatment technologies.

best management practice (BMP) (40CFR130.2-m) means methods, measures or practices selected by an agency to meet its nonpoint source control needs. BMPs include but are not limited to structural and nonstructural controls and operation and maintenance procedures. BMPs can be applied before, during and after pollution-producing activities to reduce or eliminate the introduction of pollutants into receiving waters.

best management practice (BMP) (EPA-94/04): Methods that have been determined to be the most effective, practical means of preventing or reducing pollution from nonpoint sources.

best management practices (BMPs) (40CFR122.2) means schedules of activities, prohibitions of practices, maintenance procedures, and other management practices to prevent or reduce the pollution of "waters of the United States." BMPs also include treatment requirements, operating procedures, and practices to control plant site runoff, spillage or leaks, sludge or waste disposal, or drainage from raw material storage. BMPs means "best management practices." Class I sludge management facility means any POTW identified under 40CFR403.8(a) as being required to have an approved pretreatment program (including such POTWs located in a State that has elected to assume local program responsibilities pursuant to 40CFR403.10(e)) and any other treatment works treating domestic sewage classified as a Class I sludge management facility by the Regional Administrator, or, in the case of approved State programs, the Regional Administrator, in conjunction with the State Director, because of the potential for its sludge use or disposal practices to adversely affect public health and the environment.

best management practices (BMPs) (40CFR232.2) means schedules of activities, prohibitions of practices, maintenance procedures, and other management practices to prevent or reduce the pollution of waters of the United States from discharges of dredged or fill material. BMPs include methods, measures, practices, or design and performance standards which facilitate compliance with the section 404(b)(1) Guidelines (40CFR part 230), effluent limitations or prohibitions under section 307(a), and applicable water quality standards.

best practicable control technology currently available (BPT) (40CFR430.222; 467.02-t) is defined under section 304(b)(1) of the Act.

best practicable waste treatment technology (40CFR35.2005-7) means the cost-effective technology that can treat wastewater, combined sewer overflows and nonexcessive infiltration and inflow in publicly owned or individual wastewater treatment works, to meet the applicable provisions of:
(i) 40CFR133--secondary treatment of wastewater;
(ii) 40CFR125, subpart G--marine discharge waivers;
(iii) 40CFR122.44(d)--more stringent water quality standards and State standards; or
(iv) 41 FR 610 (February 11, 1976)--Alternative Waste Management Techniques for Best Practicable Waste Treatment (treatment and discharge, land application techniques and utilization practices, and reuse).

beverage (40CFR244.101-a) means carbonated natural or mineral waters; soda water and similar carbonated soft drinks; and beer or other carbonated malt drinks in liquid form and intended for human onsumption.

beverage can (40CFR60.491-1) means any two-piece steel or aluminum container in which soft drinks or beer, including malt liquor, are packaged. The definition does not include containers in which fruit or vegetable juices are packaged.

beverage container (40CFR244.101-b) means an airtight container containing a beverage under pressure of carbonation. Cups and other open receptacles are specifically excluded from this definition.

BIA (40CFR147.2902) means the "Bureau of Indian Affairs," United States Department of Interior.

bias (40CFR72.2) means systematic error, resulting in measurements that will be either consistently low or high relative to the reference value (under the Clean Air Act).

bicycle (40CFR52.1162-1) means a two-wheel nonmotor-powered vehicle.

bicycle (40CFR52.2494-2) means a nonmotor-powered, 2-wheeled vehicle.

bicycle parking facility (40CFR52.1162-6) means any facility for the temporary storage of bicycles which allows the frame and both wheels of the bicycle to be locked so as to minimize the risk of theft and vandalism.

bicycle lane (40CFR52.2494-3) means a route for the exclusive use of bicycles, whether constructed especially for that purpose or converted from an existing traffic lane.

BID (EPA-94/04) means buoyancy induced dispersion.

bike lane (40CFR52.1162-3) means a street lane restricted to bicycles and so designated by means of painted lanes, pavement coloring or other appropriate markings. A "peak hour" bike lane means a bike lane effective only during times of heaviest auto commuter traffic.

bike path (40CFR52.1162-2) means a route for the exclusive use of bicycles separated by grade or other physical barrier from motor traffic.

bike route (40CFR52.1162-4) means a route in which bicycles share road space with motorized vehicles.

bikeway (40CFR52.1162-5) means bike paths, bike lanes and bike routes.

bimetal (EPA-94/04): Beverage containers with steel bodies and aluminum tops; handled differently from pure aluminum in recycling.

binders (40CFR52.741) means organic materials and resins which do not contain VOM's.

binding commitment (40CFR35.3105-b) means a legal obligation by the State to a local recipient that defines the terms for assistance under the SRF (State Water Pollution Control Revolving Fund).

bioaccumulants (EPA-94/04): Substances that increase in concentration in living organisms as they take in contaminated air, water, or food because the substances are very slowly metabolized or excreted.

bioaccumulation (bioretention) (40CFR798.7100) is the uptake and, at least temporary, storage of a chemical by an exposed animal. The chemical can be retained in its original form and/or as modified by enzymatic and non-enzymatic reactions in the body.

bioassay (EPA-94/04): Study of living organisms to measure the effect of a substance, factor, or condition by comparing before-and-after exposure or other data.

bioavailability (40CFR795.223-1) refers to the rate and extent to which the administered compound is absorbed, i.e., reaches the systemic circulation (other identical or similar definitions are provided in 40CFR795.230).

bioavailability (40CFR795.228-1) refers to the rate and relative amount of administered test substance which reaches the systemic circulation (other identical or similar definitions are provided in 40CFR795.231-1).

bioavailability (40CFR795.232-1) refers to the relative amount of administered test substance which reaches the systemic circulation and the rate at which this process occurs.

biochemical oxygen demand (BOD) (EPA-94/04): A measure of the amount of oxygen consumed in the biological processes that break down organic matter in water. The greater the BOD, the greater the degree of pollution.

biochemical specific locus mutation (40CFR798.5195-1) is a genetic change resulting from a DNA lesion causing alterations in proteins that can be detected by electrophoretic methods.

bioconcentration (40CFR797.1520-2) is the net accumulation of a substance directly from water into and onto aquatic organisms.

bioconcentration (40CFR797.1560-1) is the increase in concentration of test material in or on test organisms (or specified tissues thereof) relative to the concentration of test material in the ambient water (under TSCA).

bioconcentration (40CFR797.1830-2) is the net accumulation of a chemical directly from water into and onto aquatic organisms.

bioconcentration (EPA-94/04): The accumulation of a chemical in tissues of an organism (such as a fish) to levels greater than in the surrounding medium in which the organism lives.

bioconcentration factor (BCF) (40CFR300-AA) means the measure of the tendency for a substance to accumulate in the tissue of an aquatic organism. BCF is determined by the extent of partitioning of a substance, at equilibrium, between the tissue of an aquatic organism and water. As the ratio of concentration of a substance in the organism divided by the concentration in water, higher BCF values reflect a tendency for substances to accumulate in the tissue of aquatic organisms (unitless).

bioconcentration factor (BCF) (40CFR797.1520-3) is the quotient of the concentration of a test substance in aquatic organisms at or over a discrete time period of exposure divided by the concentration in the test water at or during the same time period.

bioconcentration factor (BCF) (40CFR797.1560-2) is the ratio of the test substance concentration in the test fish (C_f) to the concentration in the test water (C_w) at steady-state.

bioconcentration factor (BCF) (40CFR797.1830-3) is the quotient of the concentration of a test chemical in tissues of aquatic organisms at or over a discrete time period of exposure divided by the concentration of test chemical in the test water at or during the same time period.

biodegradable (EPA-94/04): Capable of decomposing rapidly under natural conditions.

biodegradation (40CFR300-AA) means the chemical reaction of a substance induced by enzymatic activity of microorganisms.

biodiversity (EPA-94/04): Refers to the variety and variability among living organisms and the ecological complexes in which they occur. Diversity can be defined as the number of different items and their relative frequencies. For biological diversity, these items are organized at many levels, ranging from complete ecosystems to the biochemical structures that are the molecular basis of heredity. Thus, the term encompasses different ecosystem, species, and genes.

biological additives (40CFR300.5) **means** microbiological cultures, enzymes, or nutrient additives that are deliberately introduced into an oil discharge for the specific purpose of encouraging biodegradation to mitigate the effects of the discharge.

biological control (EPA-94/04): In pest control, the use of animals and organisms that eat or otherwise kill or out-compete pests.

biological control agent (40CFR152.3-i) means any living organism applied to or introduced into the environment that is intended to function as a pesticide against another organism declared to be a pest by the Administrator.

biological magnification (EPA-94/04): Refers to the process whereby certain substances such as pesticides or heavy metals move up the food chain, work their way into rivers or lakes, and are eaten by aquatic organisms such as fish, which in turn are eaten by large birds, animals or humans. The substances become concentrated in tissues or internal organs as they move up the chain (cf. bioaccumulative).

biological monitoring (40CFR401.11-m) shall be defined in accordance with section 502 of the Act unless the context otherwise requires (under the Clean Water Act).

biological monitoring (CWA502-33USC1362) shall mean the determination of the effects on aquatic life, including accumulation of pollutants in tissue, in receiving waters due to the discharge of pollutants:
(A) by techniques and procedures, including sampling of organisms representative of appropriate levels of the food chain appropriate to the volume and the physical, chemical, and biological characteristics of the effluent, and
(B) at appropriate frequencies and locations.

biological oxidation (EPA-94/04): Decomposition of complex organic materials by microorganisms. Occurs in self-purification of water bodies and in activated sludge wastewater treatment.

biological oxygen demand (BOD) (EPA-94/04): An indirect measure of the concentration of biologically degradable material present in organic wastes. It usually reflects the amount of oxygen consumed in five days by biological processes breaking down organic waste.

biological treatment (EPA-94/04): A treatment technology that uses bacteria to consume organic waste.

biologicals (40CFR259.10.b) means preparations made from living organisms and their products, including vaccines, cultures, etc., intended for use in diagnosing, immunizing or treating humans or animals or in research pertaining thereto.

biologicals (EPA-94/04): Vaccines, cultures and other preparations made from living organisms and their products, intended for use in diagnosing, immunizing, or treating humans or animals, or in related research.

biomass (EPA-94/04): All of the living material in a given area; often refers to vegetation.

biome (EPA-94/04): Entire community of living organisms in a single major ecological area (cf. biotic community).

biomonitoring (EPA-94/04):
(1) The use of living organisms to test the suitability of effluents for discharge into receiving waters and to test the quality of such waters downstream from the discharge.
(2) Analysis of blood, urine, tissues, etc., to measure chemical exposure in humans.

biopolymer (40CFR723.250-3) means a polymer directly produced by living or once-living cells or cellular components.

bioremediation (EPA-94/04): Use of living organisms

to clean up oil spills or remove other pollutants from soil, water, or wastewater; use of organisms such as non-harmful insects to remove agricultural pests or counteract diseases of trees, plants, and garden soil (cf. remedial action).

bioretention (40CFR798.7100): See bioaccumulation.

biosphere (EPA-94/04): The portion of Earth and its atmosphere that can support life.

biostabilizer (EPA-94/04): A machine that converts solid waste into compost by grinding and aeration.

biota (EPA-94/04): The animal and plant life of a given region.

biotechnology (EPA-94/04): Techniques that use living organisms or parts of organisms to produce a variety of products (from medicines to industrial enzymes) to improve plants or animals or to develop microorganisms to remove toxics from bodies of water, or act as pesticides.

biotic community (EPA-94/04): A naturally occurring assemblage of plants and animals that live in the same environment and are mutually sustaining and interdependent (cf. biome).

biotransformation (EPA-94/04): Conversion of a substance into other compounds by organisms; includes biodegradation.

bird hazard (40CFR257.3.8-2) means an increase in the likelihood of bird/aircraft collisions that may cause damage to the aircraft or injury to its occupants.

bitterns (40CFR415.161-c) shall mean the saturated brine solution remaining after precipitation of sodium chloride in the solar evaporation process.

bituminous coal (40CFR60.251-b) means solid fossil fuel classified as bituminous coal by ASTM Designation D38877 (incorporated by reference, see 40CFR60.17).

bituminous coatings (40CFR52.741) means black or brownish coating materials which are soluble in carbon disulfide, which consist mainly of hydrocarbons, and which are obtained from natural deposits or as residues from the distillation of crude oils or of low grades of coal.

biweekly (40CFR63.321) means any 14-day period of time.

black liquor oxidation system (40CFR60.281-g) means the vessels used to oxidize, with air or oxygen, the black liquor, and associated storage tank(s).

black liquor solids (40CFR60.281-k) means the dry weight of the solids which enter the recovery furnace in the black liquor.

blackwater (EPA-94/04): Water that contains animal, human, or food waste.

blanket insulation (40CFR248.4-b) means relatively flat and flexible insulation in coherent sheet form, furnished in units of substantial area. Batt insulation is included in this term.

blast cleaning barrel (29CFR1910.94a) means a complete enclosure which rotates on an axis, or which has an internal moving tread to tumble the parts, in order to expose various surfaces of the parts to the action of an automatic blast spray.

blast cleaning room (29CFR1910.94a) means a complete enclosure in which blasting operations are performed and where the operator works inside of the room to operate the blasting nozzle and direct the flow of the abrasive material.

blast furnace (40CFR60.131-d) means any furnace used to recover metal from slag.

blast furnace (40CFR60.181-e) means any reduction furnace to which sinter is charged and which forms separate layers of molten slag and lead bullion.

blasting cabinet (29CFR1910.94a) means an enclosure where the operator stands outside and operates the blasting nozzle through an opening or openings in the enclosure.

bleached papers (40CFR250.4-b) means paper made of pulp that has been treated with bleaching agents (under RCRA).

blend fertilizer (40CFR418.71-c) shall mean a mixture of dry, straight and mixed fertilizer materials.

blood products (40CFR259.10.b) means any product derived from human blood, including but not limited to blood plasma, platelets, red or white blood corpuscles, and other derived licensed products, such as interferon, etc.

blood products (EPA-94/04): Any product derived from human blood, including but not limited to blood plasma, platelets, red or white corpuscles, and derived licensed products such as interferon.

bloom (EPA-94/04): A proliferation of algae and/or higher aquatic plants in a body of water; often related to pollution, especially when pollutants accelerate growth.

blowdown (40CFR401.11-p) means the minimum discharge of recirculating water for the purpose of discharging materials contained in the water, the further buildup of which would cause concentration in amounts exceeding limits established by best engineering practice (other identical or similar definitions are provided in 40CFR423.11-j).

blowing (40CFR61.171) means the injection of air or oxygen-enriched air into a molten converter bath.

blowing still (40CFR60.471) means the equipment in which air is blown through asphalt flux to change the softening point and penetration rate.

blowing tap (40CFR60.261-i) means any tap in which an evolution of gas forces or projects jets of flame or metal sparks beyond the ladle, runner, or collection hood.

BMPs (40CFR122.2): Best management practices.

board (CAA212-42USC7546) means the Low-Emission Vehicle Certification Board.

board insulation (40CFR248.4-c) means semirigid insulation preformed into rectangular units having a degree of suppleness, particularly related to their geometrical dimensions.

board of arbitrators or board (40CFR305.12) means a panel of one or more persons selected in accordance with section 112(b)(4)(A) of CERCLA and governed by the provisions in 40CFR305 (other identical or similar definitions are provided in 40CFR306.12).

BOD (40CFR133.101-d): The five day measure of the pollutant parameter biochemical oxygen demand (BOD).

BOD5 (EPA-94/04): The amount of dissolved oxygen consumed in five days by biological processes breaking down organic matter.

BOD5 input (40CFR405.11-b) shall mean the biochemical oxygen demand of the materials entered into process. It can be calculated by multiplying the fats, proteins and carbohydrates by factors of 0.890, 1.031 and 0.691 respectively. Organic acids (e.g., lactic acids) should be included as carbohydrates. Composition of input materials may be based on either direct analyses or generally accepted published values (other identical or similar definitions are provided in 40CFR405.21-b; 405.31-b; 405.41-b; 405.51-b; 405.61-b; 405.71-b; 405.81-b; 405.91-b; 405.101-b; 405.111-b; 405.121-b).

BOD7 (40CFR417.151-g) shall mean the biochemical oxygen demand as determined by incubation at 20 degrees C for a period of 7 days using an acclimated seed. Agitation employing a magnetic stirrer set at 200 to 500 rpm may be used (other identical or similar definitions are provided in 40CFR417.161-f; 417.171-d; 417.181-d).

bodily injury (40CFR280.92) shall have the meaning given to this term by applicable state law; however, this term shall not include those liabilities which, consistent with standard insurance industry practices, are excluded from coverage in liability insurance policies for bodily injury.

bodily injury and property damage (40CFR264.141-g) shall have the meanings given these terms by applicable State law. However, these terms do not include those liabilities which, consistent with standard industry practices, are excluded from coverage in liability policies for bodily injury and property damage. The Agency intends the meanings of other terms used in the liability insurance requirements to be consistent with their common meanings within the insurance industry. The definitions given below of several of the terms are intended to assist in the understanding of these regulations and are not intended to limit their meanings in a way that conflicts with general insurance industry usage (other identical or similar definitions are provided in 40CFR265.141-g).

body fluids (40CFR259.10.b) means liquid emanating or derived from humans and limited to blood: cerebrospinal, synovial, pleural, peritoneal and pericardial fluids; and semen and vaginal secretions.

body style (40CFR600.002.85-33) means a level of commonality in vehicle construction as defined by number of doors and roof treatment (e.g., sedan, convertible, fastback, hatchback) and number of seats (i.e., front, second, or third seat) requiring seat belts pursuant to National Highway Traffic Safety Administration safety regulations. Station wagons and light trucks are identified as car lines.

body style (40CFR86.082.2) means level of commonality in vehicle construction as defined by number of doors and roof treatment (e.g., sedan, convertible, fastback, hatchback).

body type (40CFR86.082.2) means name denoting a group of vehicles that are either in the same car line or in different car lines provided the only reason the vehicles qualify to be considered in different car lines is that they are produced by a separate division of a single manufacturer.

bog (EPA-94/04): A type of wetland that accumulates appreciable peat deposits. Bogs depend primarily on precipitation for their water source, and are usually acidic and rich in plant residue with a conspicuous mat of living green moss.

boiler (40CFR260.10) means an enclosed device using controlled flame combustion and having the following characteristics:
(1) (i) The unit must have physical provisions for recovering and exporting thermal energy in the form of steam, heated fluids, or heated gases; and
(ii) The unit's combustion chamber and primary energy recovery sections(s) must be of integral design. To be of integral design, the combustion chamber and the primary energy recovery section(s) (such as waterwalls and superheaters) must be physically formed into one manufactured or assembled unit. A unit in which the combustion chamber and the primary energy recovery section(s) are joined only by ducts or connections carrying flue gas is not integrally designed; however, secondary energy recovery equipment (such as economizers or air preheaters) need not be physically formed into the same unit as the combustion chamber and the primary energy recovery section. The following units are not precluded from being boilers solely because they are not of integral design: process heaters (units that transfer energy directly to a process stream), and fluidized bed combustion units;
(iii) While in operation, the unit must maintain a thermal energy recovery efficiency of at least 60 percent, calculated in terms of the recovered energy compared with the thermal value of the fuel; and
(iv) The unit must export and utilize at least 75 percent of the recovered energy, calculated on an annual basis. In this calculation, no credit shall be given for recovered heat used internally in the same unit. (Examples of internal use are the preheating of fuel or combustion air, and the driving of induced or forced draft fans or feedwater pumps); or
(2) The unit is one which the Regional Administrator has determined, on a case-by-case basis, to be a boiler, after considering the standards in 40CFR260.32.

boiler (40CFR60.531) means a solid fuel burning appliance used primarily for heating spaces, other than the space where the appliance is located, by the distribution through pipes of a gas or fluid heated in the appliance. The appliance must be tested and listed as a boiler under accepted American or Canadian safety testing codes. A manufacturer may request an exemption in writing from the Administrator by stating why the testing and listing requirement is not practicable and by demonstrating that his appliance is otherwise a boiler.

boiler (40CFR60.561) means any enclosed combustion device that extracts useful energy in the form of steam (other identical or similar definitions are provided in 40CFR60.611; 60.661).

boiler (40CFR60.701) means any enclosed combustion device that extracts useful energy in the form of steam and is not an incinerator (other identical or similar definitions are provided in 40CFR63.111).

boiler (40CFR72.2) means an enclosed fossil or other fuel-fired combustion device used to produce heat and to transfer heat to recirculating water, steam, or any other medium.

boiler operating day (40CFR52.145) for each of the boiler units at the Navajo Generating Station is defined as a 24-hour calendar day (the period of time between 12:01 a.m. and 12:00 midnight in Page, Arizona) during which coal is combusted in that unit for the entire 24 hours.

boiler operating day (40CFR60.41a-28) means a 24-hour period during which fossil fuel is combusted in a steam generating unit for the entire 24 hours.

boiler operating day (40CFR60-AG) means a 24-hour period during which any fossil fuel is combusted in either the Unit 1 or Unit 2 steam generating unit and during which the provisions of 40CFR60.43(e) are applicable.

bond paper (40CFR250.4-c) means a generic category of paper used in a variety of end use applications such as forms (cf. "form bond"), offset printing, copy paper, stationery, etc. In the paper industry, the term was originally very specific but is now very general.

bond release (40CFR434.11-d) means the time at which the appropriate regulatory authority returns a reclamation or performance bond based upon its determination that reclamation work (including, in the case of underground mines, mine sealing and abandonment procedures) has been satisfactorily completed.

book paper (40CFR250.4-d) means a generic category of papers produced in a variety of forms,

weights, and finishes for use in books and other graphic arts applications, and related grades such as tablet, envelope, and converting papers.

boom (EPA-94/04):
1. A floating device used to contain oil on a body of water.
2. A piece of equipment used to apply pesticides from a tractor or truck (cf. sonic boom).

borosilicate recipe (40CFR60.291) means glass product composition of the following approximate ranges of weight proportions: 60 to 80 percent silicon dioxide, 4 to 10 percent total R_2O (e.g., Na_2O and K_2O), 5 to 35 percent boric oxides, and 0 to 13 percent other oxides.

borrower (40CFR15.4) means any recipient of a loan as defined below (cf. loan definition).

Boston intrastate region (40CFR52.1128-2) means the Metropolitan Boston Intrastate Air Quality Control Region, as defined in 40CFR81.19 of this part.

Boston proper (40CFR52.1128-5) means that portion of the city of Boston, Massachusetts, contained within the following boundaries: The Charles River and Boston Inner Harbor on the northwest, north, and northeast, the Inner Harbor, Fort Point Channel, Fitzgerald Expressway, and the Massachusetts Avenue Expressway access branch on the east and southeast, and Massachusetts Avenue on the west. Where a street or roadway forms a boundary, the entire right-of-way of the street is within the Boston proper area as here defined.

botanical pesticide (EPA-94/04): A pesticide whose active ingredient is a plant-produced chemical such as nicotine or strychnine. Also called a plant-derived pesticide.

bottle bill (EPA-94/04): Proposed or enacted legislation which requires a returnable deposit on beer or soda containers and provides for retail store or other redemption. Such legislation is designed to discourage use of throwaway containers.

bottom ash (40CFR240.101-b) means the solid material that remains on a hearth or falls off the grate after thermal processing is complete (cf. fly ash).

bottom ash (40CFR423.11-f) means the ash that drops out of the furnace gas stream in the furnace and in the economizer sections. Economizer ash is included when it is collected with bottom ash.

bottom ash (EPA-94/04): The non-airborne combustion residue from burning pulverized coal in a boiler; the material which falls to the bottom of the boiler and is removed mechanically; a concentration of the non-combustible materials, which may include toxics.

bottom blown furnace (40CFR60.141a) means any BOPF (basic oxygen process furnace) in which oxygen and other combustion gases are introduced to the bath of molten iron through tuyeres in the bottom of the vessel or through tuyeres in the bottom and sides of the vessel.

bottom land hardwoods (EPA-94/04): Forested freshwater wetlands adjacent to rivers in the southeastern United States, especially valuable for wildlife breeding, nesting and habitat.

bottoms receiver (40CFR264.1031) means a container or tank used to receive and collect the heavier bottoms fractions of the distillation feed stream that remain in the liquid phase.

bottoms receiver (40CFR63.101) means a tank that collects distillation bottoms before the stream is sent for storage or for further downstream processing (other identical or similar definitions are provided in 40CFR63.161; 63.191).

bounce (40CFR85.2122(a)(5)(ii)(B)) means unscheduled point contact opening(s) after initial closure and before scheduled reopening.

BPT (40CFR467.02-t) means the best practicable control technology currently available (BPT) under section 304(b)(1) the Act.

brackish (EPA-94/04) means mixed fresh and salt water.

brass or bronze (40CFR60.131-a) means any metal

alloy containing copper as its predominant constituent, and lesser amounts of zinc, tin, lead, or other metals.

breakdown voltage (40CFR85.2122(a)(6)(ii)(C)) means the voltage level at which the capacitor fails.

breaker point (40CFR85.2122(a)(5)(ii)(A)) means mechanical switch operated by the distributor cam to establish and interrupt the primry ignition coil current.

breakover angle (40CFR86.084.2) means the supplement of the largest angle, in the plan side view of an automobile, that can be formed by two lines tangent to the front and rear static loaded radii arcs and intersecting at a point on the underside the automobile.

breakpoint chlorination (EPA-94/04): Addition of chlorine to water until the chlorine demand has been satisfied.

breakthrough (EPA-94/04): A crack or break in a filter bed that allows the passage of floc or particulate matter through a filter; will cause an increase in filter effluent turbidity.

brine mud (EPA-94/04): Waste material, often associated with well-drilling or mining, composed of mineral salts or other inorganic compounds.

broadcast application (EPA-94/04): The spreading of pesticides over an entire area.

broccoli (40CFR407.71-c) shall include the processing of broccoli into the following product styles: Frozen, chopped, spears, and miscellaneous cuts.

brood stock (40CFR797.1300-1) means the animals which are cultured to produce test organisms through reproduction (other identical or similar definitions are provided in 40CFR797.1330-1).

brown papers (40CFR250.4-e) means papers usually made from unbleached kraft pulp and used for bags, sacks, wrapping paper, and so forth.

brown stock washer system (40CFR60.281-e) means

brown stock washers and associated knotters, vacuum pumps, and filtrate tanks used to wash the pulp following the digestion system. Diffusion washers are excluded from this definition.

brownfield coke oven battery (40CFR63.301) means a new coke oven battery that replaces an existing coke oven battery or batteries with no increase in the design capacity of the coke plant as of November 15, 1990 (including capacity qualifying under 40CFR63.304(b)(6)), and the capacity of any coke oven battery subject to a construction permit on November 15, 1990, which commenced operation before October 27, 1993.

brush or wipe coating (40CFR52.741) means a manual method of applying a coating using a brush, cloth, or similar object.

BTX (40CFR61.131) means benzene-toluene-xylene.

BTX storage tank (40CFR61.131) means any tank, reservoir, or container used to collect or store benzene-toluene-xylene or other light-oil fractions (under CAA).

bubble (EPA-94/04): A system under which existing emissions sources can propose alternate means to comply with a set of emissions limitations; under the bubble concept, sources can control more than required at one emission point where control costs are relatively low in return for a comparable relaxation of controls at a second emission point where costs are higher.

bubble policy (EPA-94/04): See emissions trading.

bubbling fluidized bed combustor (40CFR60.51a) means a fluidized bed combustor in which the majority of the bed material remains in a fluidized state in the primary combustion zone.

bucket elevator (40CFR60.381) means a conveying device for metallic minerals consisting of a head and foot assembly that supports and drives an endless single or double strand chain or belt to which buckets are attached.

bucket elevator (40CFR60.671) means a conveying device of nonmetallic minerals consisting of a head

and foot assembly which supports and drives an endless single or double strand chain or belt to which buckets are attached.

budget (40CFR35.4010) means the financial plan for the spending of all Federal and matching funds (including in-kind contributions) for a technical assistance grant project as proposed by the applicant, and negotiated with and approved by the Award Official.

budget period (40CFR30.200) means the length of time EPA specifies in an assistance agreement during which the recipient may expend or obligate Federal Funds.

budget period (40CFR35.4010) means the length of time specified in a grant agreement during which the recipient may spend or obligate Federal funds. The budget period may not exceed 3 years. A technical assistance grant project period may be comprised of several budget periods.

budget period (40CFR35.6015-5) means the length of time EPA specifies in a cooperative agreement during which the recipient may expend or obligate Federal funds.

buffer (EPA-94/04): A solution or liquid whose chemical makeup neutralizes acids or bases without a great change in pH.

buffer strips (EPA-94/04): Strips of grass or other erosion-resisting vegetation between or below cultivated strips or fields.

building (40CFR35.2005-8) means the erection, acquisition, alteration, remodeling, improvement or extension of treatment works.

building (40CFR60.671) means any frame structure with a roof.

building completion (40CFR35.2005-9) means the date when all but minor components of a project have been built, all equipment is operational and the project is capable of functioning as designed (under CWA).

building cooling load (EPA-94/04): The hourly

amount of heat that must be removed from a building to maintain indoor comfort.

building insulation (40CFR248.4-d) means a material, primarily designed to resist heat flow, which is installed between the conditioned volume of a building and adjacent unconditioned volumes or the outside. This term includes but is not limited to insulation products such as blanket, board, spray-in-place, and loose-fill that are used as ceiling, floor, foundation, and wall insulation.

building, structure, facility, or installation (40CFR51.165-ii) means all of the pollutant-emitting activities which belong to the same industrial grouping, are located on one or more contiguous or adjacent properties, and are under the control of the same person (or persons under common control) except the activities of any vessel. Pollutant-emitting activities shall be considered as part of the same industrial grouping if they belong to the same Major Group (i.e., which have the same two-digit code) as described in the Standard Industrial Classification Manual, 1972, as amended by the 1977 Supplement (U.S. Government Printing Office stock numbers 4101-0065 and 003-005-00176-0, respectively).

building, structure, facility, or installation (40CFR51.166-6) means all of the pollutant-emitting activities which belong to the same industrial grouping, are located on one or more contiguous or adjacent properties, and are under the control of the same person (or persons under common control) except the activities of any vessel. Pollutant-emitting activities shall be considered as part of the same industrial grouping if they belong to the same Major Group (i.e., which have the same two-digit code) as described in the Standard Industrial Classification Manual, 1972, as amended by the 1977 Supplement (U.S. Government Printing Office stock numbers 4101-0066 and 003-005-00176-0, respectively).

building, structure, facility or installation (40CFR51-AS-2) means all of the pollutant-emitting activities which belong to the same industrial grouping, are located on one or more contiguous or adjacent properties, and are under the control of the same person (or persons under common control) except the activities of any vessel. Pollutant-emitting activities shall be considered as part of the same

industrial grouping if they belong to the same "Major Group" (i.e., which have the same two digit code) as described in the Standard Industrial Classification Manual 1972, as amended by the 1977 Supplement (U.S. Government Printing Office stock numbers 4101-0066 and 003-005-00176-0, respectively).

building, structure, facility, or installation (40CFR52.21-6) means all of the pollutant-emitting activities which belong to the same industrial grouping are located on one or more contiguous or adjacent properties, and are under the control of the same person (or persons under common control) except the activities of any vessel. Pollutant-emitting activities shall be considered as part of the same industrial grouping if they belong to the same "Major Group" (i.e., which have the same first two digit code) as described in the Standard Industrial Classification Manual, 1972, as amended by the 1977 Supplement (U.S. Government Printing Office stock numbers 4101-0066 and 003-005-00176-0, respectively).

building, structure, facility, or installation (40CFR52.24-2) means all of the pollutant-emitting activities which belong to the same industrial grouping are located on one or more contiguous or adjacent properties, and are under the control of the same person (or persons under common control) except the activities of any vessel. Pollutant-emitting activities shall be considered as part of the same industrial grouping if they belong to the same "Major Group" (i.e., which have the same two-digit code) as described in the following document, Standard Industrial Classification Manual, 1972, as amended by the 1977 Supplement (U.S. Government Printing Office stock numbers 4101-0066 and 003-005-00176-0, respectively).

building, structure, or facility (40CFR51.301-d) means all of the pollutant-emitting activities which belong to the same industrial grouping are located on one or more contiguous or adjacent properties, and are under the control of the same person (or persons under common control). Pollutant-emitting activities must be considered as part of the same industrial grouping if they belong to the same "Major Group" (i.e., which have the same two-digit code) as described in the Standard Industrial Classification Manual, 1972 as amended by the 1977 Supplement

(U.S. Government Printing Office stock numbers 4101-0066 and 003-005-00176-0 respectively) (cf. installation).

bulk asbestos (40CFR763.63) means any quantity of asbestos fiber of any type or grade, or combination of types or grades, that is mined or milled with the purpose of obtaining asbestos. This term does not include asbestos that is produced or processed as a contaminant or an impurity.

bulk container (40CFR246.101-c) means a large container that can either be pulled or lifted mechanically onto a service vehicle or emptied mechanically into a service vehicle.

bulk gasoline plant (40CFR52.741) means a gasoline storage and distribution facility with an average throughput of 76,000 liter (20,000 gal) or less on a 30-day rolling average that distributes gasoline to gasoline dispensing facilities.

bulk gasoline plant (40CFR60.111b) means any gasoline distribution facility that has a gasoline throughput less than or equal to 75,700 liters per day. Gasoline throughput shall be the maximum calculated design throughput as may be limited by compliance with an enforceable condition under Federal requirement or Federal, State or local law, and discoverable by the Administrator and any other person.

bulk gasoline terminal (40CFR60.501) means any gasoline facility which receives gasoline by pipeline, ship or barge, and has a gasoline throughput greater than 75,700 liters per day. Gasoline throughput shall be the maximum calculated design throughput as may be limited by compliance with an enforceable condition under Federal, State or local law and discoverable by the Administrator and any other person.

bulk resin (40CFR61.61-i) means a resin which is produced by a polymerization process in which no water is used.

bulk sample (EPA-94/04): A small portion (usually thumbnail size) of a suspect asbestos containing building material collected by an asbestos inspector for laboratory analysis to determine asbestos content.

bulk sewage sludge (40CFR503.11-e) is sewage sludge that is not sold or given away in a bag or other container for application to the land.

bulk terminal (40CFR61.301) means any facility which receives liquid product containing benzene by pipelines, marine vessels, tank trucks, or railcars, and loads the product for further distribution into tank trucks, railcars, or marine vessels.

bulky waste (40CFR243.101-c) means large items of solid waste such as household appliances, furniture, large auto parts, trees, branches, stumps, and other oversize wastes whose large size precludes or complicates their handling by normal solid wastes collection, processing, or disposal methods.

bulky waste (EPA-94/04): Large items of waste materials, such as appliances, furniture, large auto parts, trees, stumps.

bundle (40CFR763-AA-4) means a structure composed of three or more fibers in a parallel arrangement with each fiber closer than one fiber diameter.

bureau (FLPMA103-43USC1702) means the Bureau of Land Management.

burial ground (graveyard) (EPA-94/04): A disposal site for radioactive waste materials that uses earth or water as a shield.

burial operation (40CFR257.3.6-5): See trenching operation.

burn rate (40CFR60-AA (method 28-2.1)) means the rate at which test fuel is consumed in a wood heater. Measured in kilogram of wood (dry basis) per hour (kg/hr) (other identical or similar definitions are provided in 40CFR60-AA (method 28A-2.1)).

burning agents (40CFR300.5) means those additives that, through physical or chemical means, improve the combustibility of the materials to which they are applied.

burnishing (40CFR468.02-u): See tumbling.

burnishing (40CFR471.02-e) is a surface finishing

process in which minute surface irregularities are displaced rather than removed.

bus (40CFR86.093.2) means a heavy heavy-duty diesel-powered passenger-carrying vehicle with a load capacity of fifteen or more passengers that is centrally fueled, and all urban buses. This definition only applies in the context of 40CFR86.093-11 and 86.093-35.

bus/carpool lane (40CFR52.2050-2) means a lane on a street or highway, which lane is open only to buses (or buses and carpools), whether constructed especially for that purpose or converted from existing lanes.

bus/carpool lane (40CFR52.2294-1) means a lane on a street or highway open only to buses (or buses and carpools), whether constructed specially for that purpose or converted from existing lanes (under CAA).

business (40CFR2.201-b) means any person engaged in a business, trade, employment, calling or profession, whether or not all or any part of the net earnings derived from such engagement by such person inure (or may lawfully inure) to the benefit of any private shareholder or individual.

business confidentiality claim or claim (40CFR2.201-h) means a claim or allegation that business information is entitled to confidential treatment for reasons of business confidentiality, or request for a determination that such information is entitled to such treatment.

business confidentiality or confidential business information (40CFR350.1) includes the concept of trade secrecy and other related legal concepts which give (or may give) a business the right to preserve the confidentiality of business information and to limit its use or disclosure by others in order that the business may obtain or retain business advantages it derives from its right in the information. The definition is meant to encompass any concept which authorizes a Federal agency to withhold business information under 5USC552(b)(4), as well as any concept which requires EPA to withhold information from the public for the benefit of a business under 18USC1905.

business information or information (40CFR2.201-c) means any information which pertains to the interests of any business, which was developed or acquired by that business, and (except where the context otherwise requires) which is possessed by EPA in recorded form (cf. business confidentiality claim or claim).

business machine (40CFR60.721) means a device that uses electronic or mechanical methods to process information, perform calculations, print or copy information, or convert sound into electrical impulses for transmission, such as:
(1) Products classified as typewriters under SIC Code 3572;
(2) Products classified as calculating and accounting machines under SIC Code 3574;
(3) Products classified as calculating and accounting machines under SIC Code 3574;
(4) Products classified as telephone and telegraph equipment under SIC Code 3661;
(5) Products classified as office machines, not elsewhere classified, under SIC Code 3579; and
(6) Photocopy machines, a subcategory of products classified as photographic equipment under SIC Code 3861.

butadiene furfural cotrimer (R-11) (40CFR63.191) means the product of reaction of butadiene with excess furfural in a liquid phase reactor. R-11 is usually used as an insect repellant and as a delousing agent for cows in the dairy industry.

by compound (40CFR60.611) means by individual stream components, not carbon equivalents (other identical or similar definitions are provided in 40CFR60.661; 60.701).

by compound (40CFR63.111) means by individual stream components, not carbon equivalents.

bypass (40CFR122.41.m-i) means the intentional diversion of waste streams from any portion of a treatment facility.

bypass (40CFR403.17-1) means the intentional diversion of wastestreams from any portion of an Industrial User's treatment facility.

bypass (40CFR60.61-b) means any system that

prevents all or a portion of the kiln or clinker cooler exhaust gases from entering the main control device and ducts the gases through a separate control device. This does not include emergency systems designed to duct exhaust gases directly to the atmosphere in the event of a malfunction of any control device controlling kiln or clinker cooler emissions.

bypass/bleeder stack (40CFR63.301) means a stack, duct, or offtake system that is opened to the atmosphere and used to relieve excess pressure by venting raw coke oven gas from the collecting main to the atmosphere from a byproduct coke oven battery, usually during emergency conditions (cf. emergency vent stream).

bypass stack (40CFR60.61-c) means the stack that vents exhaust gases to the atmosphere from the bypass control device.

bypass stack (40CFR72.2) means any duct, stack, or conduit through which emissions from an affected unit may or do pass to the atmosphere, which either augments or substitutes for the principal stack exhaust system or ductwork during any portion of the unit's operation.

bypass the control device (40CFR61.161) means to operate the glass melting furnace without operating the control device to which that furnace's emissions are directed routinely.

byproduct (40CFR261.1-3) is a material that is not one of the primary products of a production process and is not solely or separately produced by the production process. Examples are process residues such as slags or distillation column bottoms. The term does not include a co-product that is produced for the general public's use and is ordinarily used in the form it is produced by the process (under RCRA).

byproduct (40CFR63.101) means a chemical that is produced coincidentally during the production of another chemical.

byproduct (40CFR704.3) means a chemical substance produced without a separate commercial intent during the manufacture, processing, use, or disposal of another chemical substance(s) or mixture(s).

byproduct (40CFR710.2-g) means a chemical substance produced without separate commercial intent during the manufacture or processing of another chemical substance(s) or mixture(s) (other identical or similar definitions are provided in 40CFR723.175-3).

byproduct (40CFR712.3-a) means any chemical substance or mixture produced without a separate commercial intent during the manufacture, processing, use, or disposal of another chemical substance or mixture.

byproduct (40CFR716.3) means a chemical substance produced without a separate commercial intent during the manufacture, processing, use, or disposal of another chemical substance(s) or mixture(s).

byproduct (40CFR720.3-d) means a chemical substance produced without a separate commercial intent during the manufacture, processing, use, or disposal of another chemical substance or mixture (other identical or similar definitions are provided in 40CFR723.50-2; 747.200-1).

byproduct (40CFR761.3) means a chemical substance produced without separate commercial intent during the manufacturing or processing of another chemical substance(s) or mixture(s).

byproduct (40CFR791.3-c) refers a chemical substance produced without a separate commercial intent during the manufacture, processing, use or disposal of another chemical substance or mixture.

byproduct (EPA-94/04): Material, other than the principal product, generated as a consequence of an industrial process.

byproduct coke oven (40CFR52.1881-i) gas means the gas produced during the production of metallurgical coke in slot-type, byproduct coke batteries.

byproduct coke oven battery (40CFR63.301) means a source consisting of a group of ovens connected by common walls, where coal undergoes destructive distillation under positive pressure to produce coke and coke oven gas, from which byproducts are recovered. Coke oven batteries in operation as of April 1, 1992, are identified in appendix A to this subpart.

byproduct cokemaking (40CFR420.11-b) means those cokemaking operations in which coal is heated in the absence of air to produce coke. In this process, byproducts may be recovered from the gases and liquids driven from the coal during cokemaking (under CWA).

byproduct material (10CFR20.3) means any radioactive material (except special nuclear material) yielded in or made radioactive by exposure to the radiation incident to the process of producing or utilizing special nuclear material (other identical or similar definitions are provided in 10CFR30.4).

byproduct material (10CFR40.4) means the tailings or wastes produced by the extraction or concentration of uranium or thorium from any ore processed primarily for its source material content, including discrete surface wastes resulting from uranium solution extraction processes. Underground ore bodies depleted by such solution extraction operations do not constitute "byproduct material" within this definition. With the exception of "byproduct material" as defined in section 1le of the Act, other terms defined in section 11 of the Act shall have the same meaning when used in the regulations in this part.

byproduct material (40CFR710.2-c) shall have the meaning contained in the Atomic Energy Act of 1954, 42USC2014 et seq., and the regulations issued thereunder (other identical or similar definitions are provided in 40CFR720.3).

byproduct/waste (40CFR60.41b) means any liquid or gaseous substance produced at chemical manufacturing plants or petroleum refineries (except natural gas, distillate oil, or residual oil) and combusted in a steam generating unit for heat recovery or for disposal. Gaseous substances with carbon dioxide levels greater than 50 percent or carbon monoxide levels greater than 10 percent are not byproduct/waste for the purposes of this subpart.

************ CCCCC ***********

CAA: See act or CAA.

cab over axle or cab over engine (40CFR205.51-7) means the cab which contains the operator-passenger compartment is directly above the engine and front axle and the entire cab can be tilted forward to permit access to the engine compartment.

cab over engine (40CFR205.51-7): See cab over axle.

cadmium (40CFR415.451) shall mean the total cadmium present in the process wastewater stream exiting the wastewater treatment system (other identical or similar definitions are provided in 40CFR415.631).

cadmium (40CFR415.641-c) shall mean the total cadmium present in the process wastewater stream exiting the wastewater treatment system.

cadmium (Cd) (EPA-94/04): A heavy metal element that accumulates in the environment.

calcine (40CFR60.161-d) means the solid materials produced by a roaster.

calciner (40CFR60.401-d) means a unit in which the moisture and organic matter of phosphate rock is reduced within a combustion chamber.

calciner or nodulizing kiln (40CFR61.121-b) means a unit in which phosphate rock is heated to high temperatures to remove organic material and/or to convert it to a nodular form. For the purpose of this subpart, calciners and nodulizing kilns are considered to be similar units.

calcium carbide (40CFR60.261-p) means material containing 70 to 85 percent calcium carbide by weight.

calcium silicon (40CFR60.261-v) means that alloy as defined by ASTM Designation A495-76 (incorporated by reference, see 40CFR60.17).

calcium sulfate storage pile runoff (40CFR418.11-f) shall mean the calcium sulfate transport water runoff from or through the calcium sulfate pile, and the precipitation which falls directly on the storage pile and which may be collected in a seepage ditch at the base of the outer slopes of the storage pile, provided such seepage ditch is protected from the incursion of surface runoff from areas outside of the outer perimeter of the seepage ditch.

calculated level (40CFR82.3-d) means the weighted amount of a controlled substance determined by multiplying the amount (in kilograms) of the controlled substance by that substance's ozone depletion potential (ODP) weight listed in appendix A or appendix B to this subpart.

calendar quarter (10CFR20.3) means not less than 12 consecutive weeks nor more than 14 consecutive weeks. The first calendar quarter of each year shall begin in January and subsequent calendar quarters shall be such that no day is included in more than one calendar quarter or omitted from inclusion within a calendar quarter. No licensee shall change the method observed by him of determining calendar quarters except at the beginning of a calendar year.

calibrating gas (40CFR86.082.2) means a gas of known concentration which is used to establish the response curve of an analyzer.

calibration (40CFR86.082.2) means the set of specifications, including tolerances, unique to particular design, version, or application of a component or components assembly capable of functionally describing its operation over its working range (other identical or similar definitions are provided in 40CFR600.002.85-31).

calibration blank (40CFR136-AC-13) means a volume of deionized, distilled water acidified with HNO_3 and HCl.

calibration curve (40CFR60-AA (method 6C-3.10)) means a graph or other systematic method of establishing the relationship between the analyzer response and the actual gas concentration introduced to the analyzer.

calibration drift (40CFR60-AA (method 25A-2.5)) means the difference in the measurement system response to a mid-level calibration gas before and after a stated period of operation during which no unscheduled maintenance, repair or adjustment took place.

calibration drift (40CFR60-AA (method 6C-3.7)) means the difference in the measurement system output reading from the initial calibration response at a mid-range calibration value after a stated period of operation during which no unscheduled maintenance, repair, or adjustment took place (other identical or similar definitions are provided in 40CFR60-AA (method 7E-3.2)).

calibration drift (40CFR60-AB-2.14) means the difference in the CEMS (continuous emission monitoring system) output readings from the upscale calibration value after a stated period of normal continuous operation during which no unscheduled maintenance, repair, or adjustment took place (under CAA).

calibration drift (CD) (40CFR60-AF-2.5) means the difference in the CEMS (continuous emission monitoring system) output reading from a reference value after a period of operation during which no unscheduled maintenance, repair or adjustment took place. The reference value may be supplied by a cylinder gas, gas cell, or optical filter and need not be certified.

calibration error (40CFR60-AA (method 25A-2.7)) means the difference between the gas concentration indicated by the measurement system and the known concentration of the calibration gas.

calibration error (40CFR60-AB-2.12) means the difference between the opacity values indicated by the CEMS (continuous emission monitoring system) and the known values of a series of calibration attenuators (filters or screens).

calibration error (40CFR72.2) means the difference between:
(1) The response of gaseous monitor to a

calibration gas and the known concentration of the calibration gas;

(2) The response of a flow monitor to a reference signal and the known value of the reference signal; or

(3) The response of a continuous opacity monitoring system to an attenuation filter and the known value of the filter after a stated period of operation during which no unscheduled maintenance, repair, or adjustment took place.

calibration gas (40CFR60-AA (method 21-2.3)) means the VOC compound used to adjust the instrument meter reading to a known value. The calibration gas is usually the reference compound at a concentration approximately equal to the leak definition concentration.

calibration gas (40CFR60-AA (method 25A-2.3)) means a known concentration of a gas in an appropriate diluent gas.

calibration gas (40CFR60-AA (method 6C-3.3)) means a known concentration of a gas in an appropriate diluent gas (other identical or similar definitions are provided in 40CFR60-AA (method 7E-3.2)).

calibration gas (40CFR72.2) means:

(1) A standard reference material;

(2) A NIST Traceable Reference Material (NTRM);

(3) A Protocol 1 gas; or

(4) Zero ambient air material.

calibration of equipment (40CFR171.2-6) means measurement of dispersal or output of application equipment and adjustment of such equipment to control the rate of dispersal, and droplet or particle size of a pesticide dispersed by the equipment.

calibration precision (40CFR60-AA (method 21-2.6)) means the degree of agreement between measurements of the same known value, expressed as the relative percentage of the average difference between the meter readings and the known concentration to the known concentration.

calibration standards (40CFR136-AC-10) means a series of known standard solutions used by the analyst for calibration of the instrument (i.e., preparation of the analytical curve).

California only certificate (40CFR86.902.93) is a Certificate of Conformity issued by EPA which only signifies compliance with the emission standards established by California.

can (40CFR465.02-h) means a container formed from sheet metal and consisting of a body and two ends or a body and a top.

can (40CFR52.741) means any metal container, with or without a top, cover, spout or handles, into which solid or liquid materials are packaged.

can be centrally fueled (40CFR88.302.94) means the sum of those vehicles that are centrally fueled and those vehicles that are capable of being centrally fueled:

(1) Capable of being centrally fueled means a fleet, or that part of a fleet, consisting of vehicles that could be refueled 100 percent of the time at a location that is owned, operated, or controlled by the covered fleet operator, or is under contract with the covered fleet operator. The fact that one or more vehicles in a fleet is/are not capable of being centrally fueled does not exempt an entire fleet from the program.

(2) Centrally fueled means a fleet, or that part of a fleet, consisting of vehicles that are fueled 100 percent of the time at a location that is owned, operated, or controlled by the covered fleet operator, or is under contract with the covered fleet operator. Any vehicle that is under normal operations garaged at home at night but that is, in fact, centrally fueled 100 percent of the time shall be considered to be centrally fueled for the purpose of this definition. The fact that one or more vehicles in a fleet is/are not centrally fueled does not exempt an entire fleet from the program. The fact that a vehicle is not centrally fueled does not mean it could not be centrally fueled in accordance with the definition of "capable of being centrally fueled."

(3) Location means any building, structure, facility, or installation which is owned or operated by a person, or is under the control of a person; is located on one or more contiguous properties

and contains or could contain a fueling pump or pumps for the use of the vehicles owned or controlled by that person.

can coating (40CFR52.741) means any coating applied on a single walled container that is manufactured from metal sheets thinner than 29 gauge (0.0141 in.).

can coating facility (40CFR52.741) means a facility that includes one or more can coating line(s).

can coating line (40CFR52.741) means a coating line in which any protective, decorative, or functional coating is applied onto the surface of cans or can components.

cancellation (EPA-94/04): Refers to section 6 (b) of the Federal Insecticide, Fungicide and Rodenticide Act (FIFRA) which authorizes cancellation of a pesticide registration if unreasonable adverse effects to the environment and public health develop when a product is used according to widespread and commonly recognized practice, or if its labeling or other material required to be submitted does not comply with FIFRA provisions.

candidate method (40CFR53.1-g) means a method of sampling and analyzing the ambient air for an air pollutant for which an application for a reference method determination or an equivalent method determination is submitted in accordance with 40CFR53.4, or a method tested at the initiative of the Administrator in accordance with 40CFR53.7 (under CAA).

caneberries (40CFR407.61-c) shall include the processing of the following berries: Canned and frozen blackberries, blueberries, boysenberries, currants, gooseberries, loganberries, ollalieberries, raspberries, and any other similar cane or bushberry but not strawberries or cranberries.

canmaking (40CFR465.02-i) means the manufacturing process or processes used to manufacture a can from a basic metal.

canned meat processor (40CFR432.91-b) shall mean an operation which prepares and cans meats (such as stew, sandwich spreads, or similar products) alone or in combination with other finished products at rates greater than 2730 kg (6000 lb.) per day.

canned onions (40CFR407.71-l) shall mean the processing of onions into the following product styles: Canned, frozen, and fried (canned), peeled, whole, sliced, and any other piece size but not including frozen, battered onion rings or dehydrated onions.

cap (EPA-94/04): A layer of clay, or other impermeable material installed over the top of a closed landfill to prevent entry of rainwater and minimize leachate.

capable of transportation of property on a street or highway (40CFR205.51-12) means that the vehicle:
(i)   Is self propelled and is capable of transporting any material or fixed apparatus, or is capable of drawing a trailer or semitrailer;
(ii) Is capable of maintaining a cruising speed of at least 25 mph over level, paved surface;
(iii) Is equipped or can readily be equipped with features customarily associated with practical street or highway use, such features including but not being limited to: A reverse gear and a differential, fifth wheel, cargo platform or cargo enclosure; and
(iv) Does not exhibit features which render its use on a street or highway impractical, or highly unlikely, such features including, but not being limited to, tracked road means, an inordinate size or features ordinarily associated with combat or tactical vehicles.

capacitance (40CFR85.2122(a)(6)(ii)(A)) means the property of a device which permits storage of electrically-separated charges when differences in electrical potential exist between the conductors and measured as the ratio of stored charge to the difference in electrical potential between conductors.

capacitor (40CFR761.3) means a device for accumulating and holding a charge of electricity and consisting of conducting surfaces separated by a dielectric. Types of capacitors are as follows:
(1) Small capacitor means a capacitor which contains less than 1.36 kg (3 lbs.) of dielectric fluid. The following assumptions may be used if the actual weight of the dielectric fluid is unknown. A capacitor whose total volume is less

than 1,639 cubic centimeters (100 cubic inches) may be considered to contain less than 1.36 kgs (3 lbs.) of dielectric fluid and a capacitor whose total volume is more than 3,278 cubic centimeters (200 cubic inches) must be considered to contain more than 1.36 kg (3 lbs.) of dielectric fluid. A capacitor whose volume is between 1,639 and 3,278 cubic centimeters may be considered to contain less then 1.36 kg (3 lbs.) of dielectric fluid if the total weight of the capacitor is less than 4.08 kg (9 lbs.).

(2) Large high voltage capacitor means a capacitor which contains 1.36 kg (3 lbs.) or more of dielectric fluid and which operates at 2,000 volts (a.c. or d.c.) or above.

(3) Large low voltage capacitor means a capacitor which contains 1.36 kg (3 lbs.) or more of dielectric fluid and which operates below 2,000 volts (a.c. or d.c.).

capacitor/condenser (40CFR85.2122(a)(6)(ii)(D)) means a device for the storage of electrical energy consisting of no oppositely charged conducting plates separated by a dielectric and which resists the flow of direct current.

capacity (40CFR60.671) means the cumulative rated capacity of all initial crushers that are part of the plant.

capacity assurance plan (EPA-94/04): A statewide plan which supports a state's ability to manage the hazardous waste generated within its boundaries over a twenty year period.

capacity factor (40CFR51.100-aa) means the ratio of the average load on a machine or equipment for the period of time considered to the capacity rating of the machine or equipment.

capacity factor (40CFR72.2) means the ratio of a unit's actual annual electric output (expressed in Mwe hr) and the unit's nameplate capacity times 8,760 hours.

capacity factor (CAA402-42USC7651a) means the ratio between the actual electric output from a unit and the potential electric output from that unit.

capillary action (EPA-94/04): Movement of water

through very small spaces due to molecular forces called capillary forces.

capillary fringe (EPA-94/04): The porous material just above the water table which may hold water by capillarity (a property of surface tension that draws water upwards) in the smaller void spaces.

capital expenditure (40CFR60.2) means an expenditure for a physical or operational change to an existing facility which exceeds the product of the applicable "annual asset guideline repair allowance percentage" specified in the latest edition of Internal Revenue Service (IRS) Publication 534 and the existing facility's basis, as defined by section 1012 of the Internal Revenue Code. However, the total expenditure for a physical or operational change to an existing facility must not be reduced by any "excluded additions" as defined in IRS Publication 534, as would be done for tax purposes.

capital expenditure (40CFR60.481) means, in addition to the definition in 40CFR60.2, an expenditure for a physical or operational change to an existing facility that: (a) Exceeds P, the product of the facility's replacement cost, R, and an adjusted annual asset guideline repair allowance, A, as reflected by the following equation: $P = RXA$, where:

(1) The adjusted annual asset guideline repair allowance, A, is the product of the percent of the replacement cost, Y, and the applicable basic annual asset guideline repair allowance, B, as reflected by the following equation: $A = YX(B/100)$;

(2) The percent Y is determined from the following equation: $Y = 1.0 - 0.575 \log X$, where X is 1982 minus the year of construction; and

(3) The applicable basic annual asset guideline repair allowance, B, is selected from the table in this definition, consistent with the applicable subpart.

(See the Table next to this definition for determining applicable for B.)

capital expenditure (40CFR60.561) means, in addition to the definition in 40CFR60.2, an expenditure for a physical or operational change to an existing facility that exceeds P, the product of the facility's replacement cost, R, and an adjusted annual asset guideline repair allowance, A, as reflected by the following equation: P = RXA, where:

(a) The adjusted annual asset guideline repair allowance, A, is the product of the percent of the replacement cost, Y, and the applicable basic annual asset guideline repair allowance, B, as reflected by the following equation: A = YX(B/100);

(b) The percent Y is determined from the following equation: Y = 1.0 - 0.57 log X, where X is 1986 minus the year of construction; and

(c) The applicable basic annual asset guideline repair allowance, B, is equal to 12.5.

capital expenditure (40CFR61.02) means an expenditure for a physical or operational change to a stationary source which exceeds the product of the applicable "annual asset guideline repair allowance percentage" specified in the latest edition of the Internal Revenue Service (IRS) Publication 534 and the stationary source's basis, as defined by section 1012 of the Internal Revenue Code. However, the total expenditure for a physical or operational change to a stationary source must not be reduced by any "excluded additions" as defined for stationary sources constructed after December 31, 1981, in IRS Publication 534, as would be done for tax purposes. In addition, "annual asset guideline repair allowance" may be used even though it is excluded for tax purposes in IRS Publication 534.

capitalization grant (40CFR35.3105-c) means the assistance agreement by which the EPA obligates and awards funds allotted to a State for purposes of capitalizing that State's revolving fund.

caprolactam byproduct ammonium sulfate manufacturing plant (40CFR60.421) means any plant which produces ammonium sulfate as a byproduct from process streams generated during caprolactam manufacture.

captafol[R] (40CFR63.191) means the fungicide captafol ([cis-N(1,1,2,2-tetrachloroethyl)-thio]-4-cylcohexene-1,2-dicarboximide). The category includes any production process units that store, react, or otherwise process 1,3-butadiene in the production of Captafol.

captan[R] (40CFR63.191) means the fungicide captan[R]. The production process typically includes, but is not limited to, the reaction of tetrahydrophthalimide and perchloromethyl mercaptan with caustic.

capture (40CFR52.741) means the containment or recovery of emissions from a process for direction into a duct which may be exhausted through a stack or sent to a control device. The overall abatement of emissions from a process with an add-on control device is a function both of the capture efficiency and of the control device.

capture device (40CFR52.741) means a hood, enclosed room floor sweep or other means of collecting solvent or other pollutants into a duct. The pollutant can then be directed to a pollution control device such as an afterburner or carbon adsorber. Sometimes the term is used loosely to include the control device.

capture efficiency (40CFR52.741) means the fraction of all VOM generated by a process that are directed to an abatement or recovery device.

capture efficiency (EPA-94/04): The fraction of organic vapors generated by a process that are directed to an abatement or recovery device.

capture system (40CFR52.741) means all equipment (including, but not limited to, hoods, ducts, fans, ovens, dryers, etc.) used to contain, collect and transport an air pollutant to a control device (under CAA).

capture system (40CFR60.261-m) means the equipment (including hoods, ducts, fans, dampers, etc.) used to capture or transport particulate matter generated by an affected electric submerged arc furnace to the control device.

capture system (40CFR60.271-d) means the equipment (including ducts, hoods, fans, dampers, etc.) used to capture or transport particulate matter generated by an EAF to the air pollution control device.

capture system (40CFR60.271a) means the equipment (including ducts, hoods, fans, dampers, etc.) used to capture or transport particulate matter generated by an electric arc furnace or AOD vessel to the air pollution control device.

capture system (40CFR60.301-i) means the equipment such as sheds, hoods, ducts, fans, dampers, etc., used to collect particulate matter generated by an affected facility at a grain elevator.

capture system (40CFR60.381) means the equipment used to capture and transport particulate matter generated by one or more affected facilities to a control device.

capture system (40CFR60.671) means the equipment (including enclosures, hoods, ducts, fans, dampers, etc.) used to capture and transport particulate matter generated by one or more process operations to a control device.

capture system (40CFR60.711-2) means any device or combination of devices that contains or collects an airborne pollutant and directs it into a duct.

car coupling sound (40CFR201.1-b) means a sound which is heard and identified by the observer as that of car coupling impact, and that causes a sound level meter indicator (FAST) to register an increase of at least ten decibels above the level observed immediately before hearing the sound.

car line (40CFR86.082.2) means a name denoting a group of vehicles within a make or car division which has a degree of commonality in construction (e.g., body, chassis). Car line does not consider any level of decor or opulence and is not generally distinguished by characteristics as roof line, number of doors, seats, or windows, except for station wagons or light-duty trucks. Station wagons and light-duty trucks are considered to be different car lines than passenger cars (other identical or similar definitions are provided in 40CFR600.002.85-20).

car seal (40CFR60.701) means a seal that is placed on a device that is used to change the position of a valve (e.g., from opened to closed) in such a way that the position of the valve cannot be changed without breaking the seal (other identical or similar definitions are provided in 40CFR61.341).

car seal (40CFR63.111) means a seal that is placed on a device that is used to change the position of a valve (e.g., from opened to closed) in such a way that the position of the valve cannot be changed without breaking the seal.

car sealed (40CFR60.561) means, for purposes of these standards, a seal that is placed on the device used to change the position of a valve (e.g., from opened to closed) such that the position of the valve cannot be changed without breaking the seal and requiring the replacement of the old seal once broken with a new seal.

car sealed (40CFR61.301) means having a seal that is placed on the device used to change the position of a valve (e.g. from open to closed) such that the position of the valve cannot be changed without breaking the seal and requiring the replacement of the old seal, once broken, with a new seal.

carbon (40CFR420.71-k): See carbon hot forming operation.

carbon absorber (EPA-94/04): An add-on control device that uses activated carbon to absorb volatile organic compounds from a gas stream. (The VOCs are later recovered from the carbon.)

carbon adsorber (40CFR63.321) means a bed of activated carbon into which an air-perchloroethylene gas-vapor stream is routed and which adsorbs the perchloroethylene on the carbon.

carbon adsorption (EPA-94/04): A treatment system that removes contaminants from ground water or surface water by forcing it through tanks containing activated carbon treated to attract the contaminants.

carbon hot forming operation (or carbon) (40CFR420.71-k) means those hot forming operations which produce a majority, on a tonnage basis, of carbon steel products.

carbon monoxide (CO) (EPA-94/04): A colorless, odorless, poisonous gas produced by incomplete fossil fuel combustion.

carbon monoxide national ambient air quality standard (CO NAAQS) (40CFR52.138-2) means the standards for carbon monoxide promulgated by the

Administrator under section 109, 42 U.S.C. 7409, of the Clean Air Act and found in 40CFR50.8.

carbon regeneration unit (40CFR260.10) means any enclosed thermal treatment device used to regenerate spent activated carbon.

carbon steel (40CFR420.71-j) means those steel products other than specialty steel products.

carboxyhemoglobin (EPA-94/04): Hemoglobin in which the iron is bound to carbon monoxide (CO) instead of oxygen.

carcinogen (EPA-94/04): Any substance that can cause or aggravate cancer.

carcinogenic effect (CAA112-42USC7412) shall have the meaning provided by the Administrator under Guidelines for Carcinogenic Risk Assessment as of the date of enactment. Any revisions in the existing Guidelines shall be subject to notice and opportunity for comment.

carpool (40CFR52.2043-1) means two or more persons utilizing the same vehicle.

carpool (40CFR52.2050-1) means a vehicle containing three or more persons.

carpool (40CFR52.2296) means two or more persons utilizing the same vehicle.

carpool (40CFR52.2297-9) means a private motor vehicle occupied by two or more persons traveling together.

carpool matching (40CFR52.2043-2) means assembling lists of commuters sharing similar travel needs and providing a mechanism by which persons on such lists may be put in contact with each other for the purpose of forming carpools.

carpool matching (40CFR52.2492-1) means assembling lists of commuters whose daily travel plans indicate they might carpool with each other and making such lists available to such commuters to aid them in forming carpools.

carrier (40CFR160.3) means any material, including but not limited to, feed, water, soil, and nutrient media, with which the test substance is combined for administration to a test system (other identical or similar definitions are provided in 40CFR792.3).

carrier (40CFR201.1-c) means a common carrier by railroad, or partly by railroad and partly by water, within the continental United States, subject to the Interstate Commerce Act, as amended, excluding street, suburban, and inter-urban electric railways unless operated as a part of a general railroad system of transportation.

carrier (40CFR797.1400-3) means a solvent used to dissolve a test substance prior to delivery to the test chamber.

carrier (40CFR797.1520-4) is a solvent used to dissolve a test substance prior to delivery of the test substance to the test chamber.

carrier (40CFR797.1600-2) is a solvent or other agent used to dissolve or improve the solubility of the test substance in dilution water.

carrier (40CFR80.2-t) means any distributor who transports or stores or causes the transportation or storage of gasoline or diesel fuel without taking title to or otherwise having any ownership of the gasoline or diesel fuel, and without altering either the quality or quantity of the gasoline or diesel fuel.

carrier (EPA-94/04): The inert liquid or solid material added to an active ingredient in a pesticide.

carrier of contaminant (40CFR230.3-d) means dredged or fill material that contains contaminants.

carrots (40CFR407.71-d) shall include the processing of carrots into the following product styles: Canned and frozen, peeled, whole, sliced, diced, nuggets, crinkle cut, julienne, shoestrings, chunks, chips and other irregular cuts, and juices but not dehydrated carrots.

carrying capacity (EPA-94/04): 1. In recreation management, the amount of use a recreation area can sustain without loss of quality. 2. In wildlife management, the maximum number of animals an area can support during a given period.

carrying case (40CFR211.203-d) means the container used to store reusable hearing protectors.

carryout collection (40CFR243.101-e) means collection of solid waste from a storage area proximate to the dwelling unit(s) or establishment.

cartridge filter (40CFR60.621) means a discrete filter unit containing both filter paper and activated carbon that traps and removes contaminants from petroleum solvent, together with the piping and ductwork used in the installation of this device.

CAS Number (40CFR704.3) means Chemical Abstracts Service Registry Number.

CAS Number (40CFR721.3) means Chemical Abstracts Service Registry Number assigned to a chemical substance on the Inventory.

CAS registration number (EPA-94/04): A number assigned by the Chemical Abstracts Service to identify a chemical.

case examiner (40CFR15.4) means an EPA official familiar with pollution control issues who is designated by the Assistant Administrator to conduct a listing or removal proceeding and to determine whether a facility will be placed on the List of Violating Facilities or removed from such list. The Case Examiner may not be:
(1) The Listing Official;
(2) the Recommending Person or anyone subordinate to the Recommending Person; or
(3) closely involved in the underlying enforcement action.

cash contribution (40CFR35.4010) means actual non-Federal dollars, or Federal dollars if expressly authorized by statute to do so, that a recipient spends for goods and services and real or personal property used to satisfy the matching funds requirement.

cash contributions (40CFR31.3) means the grantee's cash outlay, including the outlay of money contributed to the grantee or subgrantee by other public agencies and institutions, and private organizations and individuals. When authorized by Federal legislation, Federal funds received from other assistance agreements may be considered as grantee or subgrantee cash contributions.

cash draw (40CFR35.3105-d) means the transfer of cash under a letter of credit (LOC) from the Federal Treasury into the State's SRF.

casing (40CFR146.3) means a pipe or tubing of appropriate material, of varying diameter and weight, lowered into a borehole during or after drilling in order to support the sides of the hole and thus prevent the walls from caving, to prevent loss of drilling mud into porous ground, or to prevent water, gas, or other fluid from entering or leaving the hole.

casing (40CFR147.2902) means a pipe or tubing of varying diameter and weight, lowered into a borehole during or after drilling in order to support the sides of the hole and, thus, prevent the walls from caving, to prevent loss of drilling mud into porous ground, or to prevent water, gas, or other fluid from entering the hole.

cask (EPA-94/04): A thick-walled container (usually lead) used to transport radioactive material. Also called a coffin.

cast iron (40CFR464.31-b) means an iron containing carbon in excess of the solubility in the austentite that exists in the alloy at the eutectic temperature. Cast iron also is defined here to include any iron-carbon alloys containing 1.2 percent or more carbon by weight.

casting (40CFR471.02-f) is pouring molten metal into a mold to produce an object of desired shape.

catalyst (40CFR60.471) means a substance which, when added to asphalt flux in a blowing still, alters the penetrating-softening point relationship or increases the rate of oxidation of the flux.

catalyst (EPA-94/04): A substance that changes the speed or yield of a chemical reaction without being consumed or chemically changed by the chemical reaction.

catalytic converter (40CFR85.2122(a)(15)(ii)(A)) means a device installed in the exhaust system of an internal combustion engine that utilizes catalytic action to oxidize hydrocarbon (HC and carbon monoxide (CO)) emissions to carbon dioxide (CO_2) and water (H_2O).

catalytic converter (EPA-94/04): An air pollution abatement device that removes pollutants from motor vehicle exhaust, either by oxidizing them into carbon dioxide and water or reducing them to nitrogen and oxygen.

catalytic incinerator (EPA-94/04): A control device that oxidizes volatile organic compounds (VOCs) by using a catalyst to promote the combustion process. Catalytic incinerators require lower temperatures than conventional thermal incinerators, thus saving fuel and other costs.

catastrophic collapse (40CFR146.3) means the sudden and utter failure of overlying "strata" caused by removal of underlying materials (under the Safe Drinking Water Act).

catch basin (40CFR60.691) means an open basin which serves as a single collection point for stormwater runoff received directly from refinery surfaces and for refinery wastewater from process drains.

categorical exclusion (40CFR1508.4) means a category of actions which do not individually or cumulatively have a significant effect on the human environment and which have been found to have no such effect in procedures adopted by a Federal agency in implementation of these regulations (40CFR1507.3) and for which, therefore, neither an environmental assessment nor an environmental impact statement is required. An agency may decide in its procedures or otherwise to prepare environmental assessments for the reasons stated in section 150 even though it is not required to do so. Any procedures under this section shall provide for extraordinary circumstances in which a normally excluded action may have a significant environmental effect.

categorical exclusion (EPA-94/04): A class of actions which either individually or cumulatively would not have a significant effect on the human environment and therefore would not require preparation of an environmental assessment or environmental impact statement under the National Environmental Policy Act (NEPA).

categorical pretreatment standard (EPA-94/04): A technology-based effluent limitation for an industrial facility discharging into a municipal sewer system. Analogous in stringency to Best Availability Technology (BAT) for direct dischargers.

category (40CFR204.51-e) means a group of compressor configurations which are identical in all aspects with respect to the parameters listed in paragraph (c)(l)(i) of 40CFR204.55.2.

category (40CFR205.151-9) means a group of vehicle configurations which are identical in all material aspects with respect to the parameters listed in 40CFR205.157.2 of this subpart.

category (40CFR205.165-1) means a group of exhaust systems which are identical in all material aspects with respect to the parameters listed in 40CFR205.168 of this subpart.

category (40CFR205.51-8) means a group of vehicle configurations which are identical in all material aspects with respect to the parameters listed in 40CFR205.55.2.

category (40CFR211.203-e) means a group of hearing protectors which are identical in all aspects to the parameters listed in 40CFR211.210-2(c).

category I nonfriable asbestos containing material (ACM) (40CFR61.141) means asbestos-containing packings, gaskets, resilient floor covering, and asphalt roofing products containing more than 1 percent asbestos as determined using the method specified in appendix A, subpart F, 40CFR763, section 1, Polarized Light Microscopy.

category II nonfriable ACM (40CFR61.141) means any material, excluding Category I nonfriable ACM, containing more than 1 percent asbestos as determined using the methods specified in appendix A, subpart F, 40CFR763, section 1, Polarized Light Microscopy that, when dry, cannot be crumbled, pulverized, or reduced to powder by hand pressure.

category of chemical substances (40CFR723.50-4;

723.175-4; 723.250-4) has the same meaning as in section 26(c)(2) of the Act (Toxic Substances Control Act).

category of chemical substances (TSCA26-15USC2625) means a group of chemical substances the members of which are similar in molecular structure, in physical, chemical, or biological properties, in use, or in mode of entrance into the human body or into the environment, or the members of which are in some other way suitable for classification as such for purposes of this Act, except that such term does not mean a group of chemical substances which are grouped together solely on the basis of their being new chemical substances.

category of mixtures (TSCA26-15USC2625) means a group of mixtures the members of which are similar in molecular structure, in physical, chemical, or biological properties, in use, or in the mode of entrance into the human body or into the environment, or the members of which are in some other way suitable for classification as such for purposes of this Act.

cathode ray tubes (40CFR469.31-a) means electronic devices in which electrons focus through a vacuum to generate a controlled image on a luminescent surface. This definition does not include receiving and transmitting tubes.

cathodic protection (40CFR280.12) is a technique to prevent corrosion of a metal surface by making that surface the cathode of an electrochemical cell. For example, a tank system can be cathodically protected through the application of either galvanic anodes or impressed current.

cathodic protection (EPA-94/04): A technique to prevent corrosion of a metal surface by making it the cathode of an electrochemical cell.

cathodic protection tester (40CFR280.12) means a person who can demonstrate an understanding of the principles and measurements of all common types of cathodic protection systems as applied to buried or submerged metal piping and tank systems. At a minimum, such persons must have education and experience in soil resistivity, stray current, structure-to-soil potential, and component electrical isolation measurements of buried metal piping and tank systems.

cation exchange capacity (40CFR257.3.5-3) means the sum of exchangeable cations a soil can absorb expressed in milliequivalents per 100 grams of soil as determined by sampling the soil to the depth of cultivation or solid waste placement, whichever is greater, and analyzing by the summation method for distinctly acid soils or the sodium acetate method for neutral, calcareous or saline soils ("Methods of Soil Analysis, Agronomy Monograph No. 9." C. A. Black, ed., American Society of Agronomy, Madison, Wisconsin. pp 891-901, 1965).

cation exchange capacity (CEC) (40CFR796.2700-i) is the sum total of exchangeable cations that a soil can adsorb. The CEC is expressed in milliequivalents of negative charge per 100 grams (meq/100 g) or milliequivalents of negative charge per gram (meq/g) of soil.

cation exchange capacity (CEC) (40CFR796.2750-i) is the sum total of exchangeable cations that a sediment or soil can adsorb. The CEC is expressed in milliequivalents of negative charge per 100 grams (meq/100 g) or milliequivalents of negative charge per gram (meq/g) of soil or sediment.

cationic polymer (40CFR723.250-5) means a polymer that contains a net positively charged atom(s) or associated groups of atoms covalently linked to its polymer molecule.

cause (40CFR52.138-3) means resulting in a violation of the CO NAAQS (national ambient air quality standard) in an area which previously did not have ambient CO concentrations above the CO NAAQS.

cause or contribute to a new violation (40CFR51.852) means a Federal action that:

(1) Causes a new violation of a national ambient air quality standard (NAAQS) at a location in a nonattainment or maintenance area which would otherwise not be in violation of the standard during the future period in question if the Federal action were not taken; or

(2) Contributes, in conjunction with other reasonably foreseeable actions, to a new violation of a NAAQS at a location in a

nonattainment or maintenance area in a manner that would increase the frequency or severity of the new violation.
(Other identical or similar definitions are provided in 40CFR93.152.)

cause or contribute to a new violation for a project (40CFR51.392) means:
(1) To cause or contribute to a new violation of a standard in the area substantially affected by the project or over a region which would otherwise not be in violation of the standard during the future period in question, if the project were not implemented; or
(2) To contribute to a new violation in a manner that would increase the frequency or severity of a new violation of a standard in such area.
(Other identical or similar definitions are provided in 40CFR93.101).

caused by (40CFR51.852), as used in the terms "direct emissions" and "indirect emissions," means emissions that would not otherwise occur in the absence of the Federal action (other identical or similar definitions are provided in 40CFR93.152) (under CAA).

cavitation (EPA-94/04): The formation and collapse of gas pockets or bubbles on the blade of an impeller or the gate of a valve; collapse of these pockets or bubbles drives water with such force that it can cause pitting of the gate or valve surface.

CBOD5 (40CFR133.101-e) means the five day measure of the pollutant parameter carbonaceous biochemical oxygen demand (CBOD5).

ceiling insulation (40CFR248.4-e) means a material, primarily designed to resist heat flow, which is installed between the conditioned area of a building and an unconditioned attic as well as common ceiling floor assemblies between separately conditioned units in multi-unit structures. Where the conditioned area of a building extends to the roof, ceiling insulation includes such a material used between the underside and upperside of the roof.

cell (40CFR241.101-a) means compacted solid wastes that are enclosed by natural soil or cover material in a land disposal site.

cell burner boiler (40CFR76.2) means a wall-fired boiler that utilizes two or three circular burners combined into a single vertically oriented assembly that results in a compact, intense flame. Any low NO_x retrofit of a cell burner boiler that reuses the existing cell burner, close-coupled wall opening configuration would not change the designation of the unit as a cell burner boiler.

cell room (40CFR61.51-k) means a structure(s) housing one or more mercury electrolytic chlor-alkali cells.

cells (EPA-94/04): 1. In solid waste disposal, holes where waste is dumped, compacted, and covered with layers of dirt on a daily basis. 2. The smallest structural part of living matter capable of functioning as an independent unit.

cellular polyisocyanurate insulation (40CFR248.4-f) means insulation produced principally by the polymerization of polymeric polyisocyanates, usually in the presence of polyhydroxl compounds with the addition of catalysts, cell stabilizers, and blowing agents.

cellular polystyrene insulation (40CFR248.4-g) means an organic foam composed principally of polymerized styrene resin processed to form a homogenous rigid mass of cells.

cellular polyurethane insulation (40CFR248.4-h) means insulation composed principally of the catalyzed reaction product of polyisocyanurates and polyhydroxl compounds, processed usually with a blowing agent to form a rigid foam having a predominantly closed cell structure.

cellulose (40CFR248.4-i) means vegetable fiber such as paper, wood, and cane.

cellulose fiber fiberboard (40CFR248.4-j) means insulation composed principally of cellulose fibers usually derived from paper, paperboard stock, cane, or wood, with or without binders.

cellulose fiber loose-fill (40CFR248.4-k) means a basic material of recycled wood-based cellulosic fiber made from selected paper, paperboard stock, or ground wood stock, excluding contaminated materials which may reasonably be expected to be retained in the finished product, with suitable chemicals introduced to provide properties such as flame resistance, processing and handling characteristics. The basic cellulosic material may be processed into a form suitable for installation by pneumatic or pouring methods.

cementing (40CFR146.3) means the operation whereby a cement slurry is pumped into a drilled hole and/or forced behind the casing (other identical or similar definitions are provided in 40CFR147.2902).

cementing (40CFR147.2902) means the operation whereby a cement slurry is pumped into a drilled hole and/or forced behind the casing.

cementitious (EPA-94/04): Densely packed and nonfibrous friable materials.

CEMS precision or precision (40CFR72.2) as applied to the monitoring requirements of part 75 of this chapter, means the closeness of a measurement to the actual measured value expressed as the uncertainty associated with repeated measurements of the same sample or of different samples from the same process (e.g., the random error associated with simultaneous measurements of a process made by more than one instrument). A measurement technique is determined to have increasing "precision" as the variation among the repeated measurements decreases.

central business district (40CFR52.2294-2) is defined for each of the major cities in the affected areas as follows:
(i)   For the City of Houston, in Harris County, that area bounded on the northwest by Interstate 45, on the southwest by Interstate 45, on the northeast by Franklin Street, and on the southeast by Crawford Street.
(ii) For the City of Dallas, in Dallas County, that area bounded on the west by Interstate 35, on the south by Interstate 20, on the east by Central Expressway (U.S. 75) and on the north by Woodall Rogers Freeway right of way.
(iii) For the City of Fort Worth, in Tarrant County, that area bounded on the west by Henderson Street, on the south by Interstate 20, on the east

by Interstate 35 West, and on the north by Belknap Street.
(iv) For the City of San Antonio, in Bexar County, that area bounded on the west and northwest by U.S. 81, on the south and southeast by Alamo Street from U.S. 81 to Victoria Street, and by Victoria Street to the Southern and Pacific Railroad tracks, on the east by the Southern and Pacific Railroad tracks, and on the northeast by Jones Avenue to U.S. 81.

central collection point (40CFR259.10.b) means a location where a generator consolidates regulated medical waste brought together from original generation points prior to its transport off-site or its treatment-site (e.g., incineration).

central collection point (EPA-94/04): Location where a generator of regulated medical waste consolidates wastes originally generated at various locations in this facility. The wastes are gathered together for treatment on-site or for transportation elsewhere for treatment and/or disposal. This term could also apply to community hazardous waste collections, industrial and other waste management systems.

central core area (40CFR52.732-1) means that area within the City of Chicago bounded by and including Wacker Drive on the North and West, Michigan Avenue on the East, and Harrison Street on the South.

centrally fueled bus (40CFR86.093.2) means a bus that is refueled at least 75 percent of the time at one refueling facility that is owned, operated, or controlled by the bus operator.

centrifugal collector (EPA-94/04): A mechanical system using centrifugal force to remove aerosols from a gas stream or to de-water sludge.

centroidal area (40CFR72.2) means a representational concentric area that is geometrically similar to the stack or duct cross section, and is not greater than 1 percent of the stack or duct cross-sectional area.

CEQ Regulations (40CFR6.101-b) means the regulations issued by the Council on Environmental Quality (CEQ) on November 29, 1978 (see 43 FR

55978), which implement Executive Order 11991. The CEQ Regulations will often be referred to throughout this regulation by reference to 40CFR1500 et al.

ceramic plant (40CFR61.31-e) means a manufacturing plant producing ceramic items.

CERCLA: See act or CERCLA.

CERCLIS (40CFR300.5) is the abbreviation of the CERCLA Information System, EPA's comprehensive data base and management system that inventories and tracks releases addressed or needing to be addressed by the Superfund program. CERCLIS contains the official inventory of CERCLA sites and supports EPA's site planning and tracking functions. Sites that EPA decides do not warrant moving further in the site evaluation process are given a "No Further Response Action Planned" (NFRAP) designation in CERCLIS. This means that no additional federal steps under CERCLA will be taken at the site unless future information so warrants. Sites are not removed from the data base after completion of evaluations in order to document that these evaluations took place and to preclude the possibility that they be needlessly repeated. Inclusion of a specific site or area in the CERCLIS data base does not represent a determination of any party's liability, nor does it represent a finding that any response action is necessary. Sites that are deleted from the NPL (national priority list) are not designated NFRAP sites. Deleted sites are listed in a separate category in the CERCLIS data base (cf. record of decision).

cereal (40CFR406.81-b) shall mean breakfast cereal (other identical or similar definitions are provided in 40CFR406.91-b).

ceremonial and religious water use (40CFR131.35.d-3) means activities involving traditional Native American spiritual practices which involve, among other things, primary (direct) contact with water (under FWPCA).

certificate holder (40CFR85.1502-4) is the entity in whose name the certificate of conformity for a class of motor vehicles or motor vehicle engines has been issued.

certificate of conformity (40CFR85.1502-3) is the document issued by the Administrator under section 206(a) of the Act.

certificate of conformity (40CFR89.602.96): The document issued by the Administrator under section 213 and section 206(a) of the Act.

certificate of representation (40CFR72.2) means the completed and signed submission required by 40CFR72.20, for certifying the appointment of a designated representative for an affected source or a group of identified affected sources authorized to represent the owners and operators of such source(s) and of the affected units at such source(s) with regard to matters under the Acid Rain Program.

certification (40CFR163.2-g) means a certification by the Director that a pesticide chemical is useful for the purpose for which a tolerance or exemption is sought under the act.

certification (40CFR171.2) means the recognition by a certifying agency that a person is competent and thus authorized to use or supervise the use of restricted use pesticides.

certification (40CFR26.102-j) means the official notification by the institution to the supporting department or agency, in accordance with the requirements of this policy, that a research project or activity involving human subjects has been reviewed and approved by an IRB (institutional review board) in accordance with an approved assurance.

certification (40CFR260.10) means a statement of professional opinion based upon knowledge and belief.

certification (40CFR761.3) means a written statement regarding a specific fact or representation that contains the following language:
"Under civil and criminal penalties of law for the making or submission of false or fraudulent statements or representations (18 U.S.C. 1001 and 15 U.S.C. 2615), I certify that the information contained in or accompanying this document is true, accurate, and complete. As to the identified section(s) of this document for which I cannot personally verify truth and

accuracy, I certify as the company official having supervisory responsibility for the persons who, acting under my direct instructions, made the verification that this information is true, accurate, and complete."

certification (40CFR89.2) means, with respect to new nonroad engines, obtaining a certificate of conformity for an engine family complying with the nonroad engine emission standards and requirements specified in this part.

certification or audit test (40CFR60-AA (method 28-2.2)) means a series of at least four test runs conducted for certification or audit purposes that meets the burn rate specifications in section 5.

certification request (40CFR86.902.93) means a manufacturer's request for certification evidence by the submission of an application for certification, ESI (engine system information) data sheet, or ICI (independent commercial importer) carryover data sheet.

certification short test (40CFR86.096.2) means the test, for gasoline-fueled Otto-cycle light-duty vehicles and light-duty trucks, performed in accordance with the procedures contained in 40CFR part 86 subpart O.

certification vehicle (40CFR600.002.85-15) means a vehicle which is selected under 40CFR86.084-24(b)(1) and used to determine compliance under 40CFR86.084-30 for issuance of an original certificate of conformity.

certification vehicle emission margin (40CFR85.2113-j) for a certified engine family means the difference between the EPA emission standards and the average FTP (federal test procedure) emission test results of that engine family's emission-data vehicles at the projected applicable useful life mileage point (i.e., useful life mileage for light-duty vehicles is 50,000 miles and for light-duty trucks is 120,000 miles for 1985 and later model years or 50,000 miles for 1984 and earlier model years).

certified aftermarket part (40CFR85.2113-e) means any aftermarket part which has been certified pursuant to this subpart.

certified applicator (40CFR171.2-8) means any individual who is certified to use or supervise the use of any restricted use pesticides covered by his certification.

certified applicator, etc. (FIFRA2-7USC136):
(1) Certified applicator means any individual who is certified under section 4 as authorized to use or supervise the use of any pesticide which is classified for restricted use. Any applicator who holds or applies registered pesticides, or uses dilutions of registered pesticides consistent with section 2(ee) of this Act, only to provide a service of controlling pests without delivering any unapplied pesticide to any person so served is not deemed to be a seller or distributor of pesticides under this Act.
(2) Private applicator means a certified applicator who uses or supervises the use of any pesticide which is classified for restricted use for purposes of producing any agricultural commodity on property owned or rented by him or his employer or (if applied without compensation other than trading of personal services between producers of agricultural commodities) on the property of another person.
(3) Commercial applicator means an applicator (whether or not the applicator is a private applicator with respect to some uses) who uses or supervises the use of any pesticide which is classified for restricted use for any purpose or on any property other than as provided by paragraph (2).
(4) "Under The Direct Supervision of A Certified Applicator"--Unless otherwise prescribed by its labeling, a pesticide shall be considered to be applied under the direct supervision of a certified applicator if it is applied by a competent person acting under the instructions and control of a certified applicator who is available if and when needed, even though such certified applicator is not physically present at the time and place the pesticide is applied.

certified equipment or retrofit/rebuild equipment (40CFR85.1402) means equipment certified in accordance with the certification regulations contained in this subpart.

certified observer (40CFR63.301) means a visual

emission observer, certified under (if applicable) Method 303 and Method 9 (if applicable) and employed by the Administrator, which includes a delegated enforcement agency or its designated agent. For the purpose of notifying an owner or operator of the results obtained by a certified observer, the person does not have to be certified.

certified part (40CFR85.2102-3) means a part certified in accordance with the aftermarket part certification regulations contained in this subpart.

certified refrigerant recovery or recycling equipment (40CFR82.152-g) means equipment certified by an approved equipment testing organization to meet the standards in 40CFR 82.158 (b) or (d), equipment certified pursuant to 40CFR 82.36(a), or equipment manufactured before November 15, 1993, that meets the standards in 40CFR 82.158 (c), (e), or (g).

certifying agency (40CFR121.1-e) means the person or agency designated by the Governor of a State, by statute, or by other governmental act, to certify compliance with applicable water quality standards. If an interstate agency has sole authority to so certify for the area within its jurisdiction, such interstate agency shall be the certifying agency. Where a State agency and an interstate agency have concurrent authority to certify, the State agency shall be the certifying agency. Where water quality standards have been promulgated by the Administrator pursuant to section 10(c)(2) of the Act, or where no State or interstate agency has authority to certify, the Administrator shall be the certifying agency.

certifying official (40CFR72.2), for purposes of part 73 of this chapter, means:
(1) For a corporation, a president, secretary, treasurer, or vice-president of the corporation in charge of a principal business function, or any other person who performs similar policy or decision-making functions for the corporation;
(2) For partnership or sole proprietorship, a general partner or the proprietor, respectively; and
(3) For a local government entity or State, federal, or other public agency, either a principal executive officer or ranking elected official (cf. certified applicator).

cetane index or calculated cetane index (40CFR80.2-w) is a number representing the ignition properties of diesel fuel oils from API gravity and mid-boiling point as determined by ASTM standard method D 976-80, entitled "Standard Methods for Calculated Cetane Index of Distillate Fuels." ASTM test method D 976-80 is incorporated by reference. This incorporation by reference was approved by the Director of the Federal Register in accordance with 5 U.S.C. 552(a) and 1CFR part 51. A copy may be obtained from the American Society for Testing and Materials, 1916 Race Street, Philadelphia, PA 19103. A copy may be inspected at the Air Docket section (A-130), Room M-1500, U.S. Environmental Protection Agency, Docket No. A-86-03, 401 M Street SW., Washington, DC 20460 or at the Office of the Federal Register, 1100 L Street NW., Room 8401, Washington, DC 20005.

challenge exposure (40CFR798.4100-4) is an experimental exposure of a previously treated subject to a test substance following an induction period, to determine whether the subject will react in a hypersensitive manner.

change order (40CFR35.6015-7) means a written order issued by a recipient, or its designated agent, to its contractor authorizing an addition to, deletion from, or revision of, a contract, usually initiated at the contractor's request.

channelization (EPA-94/04): Straightening and deepening streams so water will move faster, a marsh-drainage tactic that can interfere with waste assimilation capacity, disturb fish and wildlife habitats, and aggravate flooding.

characteristic (EPA-94/04): Any one of the four categories used in defining hazardous waste: ignitability, corrosivity, reactivity, and toxicity.

charge (40CFR60.271-e) means the addition of iron and steel scrap or other materials into the top of an electric arc furnace.

charge (40CFR60.271a) means the addition of iron and steel scrap or other materials into the top of an electric arc furnace or the addition of molten steel or other materials into the top of an AOD vessel.

charge chrome (40CFR60.261-r) means that alloy

containing 52 to 70 percent by weight chromium, 5 to 8 percent by weight carbon, and 3 to 6 percent by weight silicon.

charge or charging period (40CFR63.301) means, for a byproduct coke oven battery, the period of time that commences when coal begins to flow into an oven through a topside port and ends when the last charging port is recapped. For a nonrecovery coke oven battery, charge or charging period means the period of time that commences when coal begins to flow into an oven and ends when the push side door is replaced.

charging (40CFR61.171) means the addition of a molten or solid material to a copper converter.

charging period (40CFR60.271-f) means the time period commencing at the moment an EAF starts to open and ending either three minutes after the EAF roof is returned to its closed position or six minutes after commencement of opening of the roof, whichever is longer.

chemical (40CFR790.3) means a chemical substance or mixture.

chemical agents (40CFR300.5) means those elements, compounds, or mixtures that coagulate, disperse, dissolve, emulsify, foam, neutralize, precipitate, reduce, solubilize, oxidize, concentrate, congeal, entrap, fix, make the pollutant mass more rigid or viscous, or otherwise facilitate the mitigation of deleterious effects or the removal of the pollutant from the water.

chemical composition (40CFR79.2-h) means the name and percentage by weight of each compound in an additive and the name and percentage by weight of each element in an additive (chemical agents).

chemical manufacturing plant (40CFR61.341) means any facility engaged in the production of chemicals by chemical, thermal, physical, or biological processes for use as a product, co-product, byproduct, or intermediate including but not limited to industrial organic chemicals, organic pesticide products, pharmaceutical preparations, paint and allied products, fertilizers, and agricultural chemicals.

Examples of chemical manufacturing plants include facilities at which process units are operated to produce one or more of the following chemicals: benzenesulfonic acid, benzene, chlorobenzene, cumene, cyclohexane, ethylene, ethylbenzene, hydroquinone, linear alklylbenzene, nitrobenzene, resorcinol, sulfolane, or styrene.

chemical manufacturing plants (40CFR60.41b) means industrial plants which are classified by the Department of Commerce under Standard Industrial Classification (SIC) Code 28.

chemical manufacturing process unit (40CFR63.101) means the equipment assembled and connected by pipes or ducts to process raw materials and to manufacture an intended product. For the purpose of this subpart, chemical manufacturing process unit includes air oxidation reactors and their associated product separators and recovery devices; reactors and their associated product separators and recovery devices; distillation units and their associated distillate receivers and recovery devices; associated unit operations (as defined in this section); and any feed, intermediate and product storage vessels, product transfer racks, and connected ducts and piping. A chemical manufacturing process unit includes pumps, compressors, agitators, pressure relief devices, sampling connection systems, open-ended valves or lines, valves, connectors, instrumentation systems, and control devices or systems. A chemical manufacturing process unit is identified by its primary product.

chemical metal cleaning waste (40CFR423.11-c) means any wastewater resulting from the cleaning of any metal process equipment with chemical compounds, including, but not limited to, boiler tube cleaning.

chemical name (40CFR721.3) means the scientific designation of a chemical substance in accordance with the nomenclature system developed by the International Union of Pure and Applied Chemistry or the Chemical Abstracts Service's rules of nomenclature, or a name which will clearly identify a chemical substance for the purpose of conducting a hazard evaluation.

chemical oxygen demand (COD) (EPA-94/04): A

measure of the oxygen required to oxidize all compounds, both organic and inorganic, in water.

chemical protective clothing (40CFR721.3) means items of clothing that provide a protective barrier to prevent dermal contact with chemical substances of concern. Examples can include, but are not limited to: full body protective clothing, boots, coveralls, gloves, jackets, and pants.

chemical structure (40CFR79.2-i) means the molecular structure of a compound in an additive.

chemical substance (40CFR2.306-2) has the meaning given it in section 3(2) of the Act, 15USC2602(2).

chemical substance (40CFR710.2-h) means any organic or inorganic substance of a particular molecular identity, including any combination of such substances occurring in whole or in part as a result of a chemical reaction or occurring in nature, and any chemical element or uncombined radical; except that "chemical substance" does not include:
(1) Any mixture,
(2) Any pesticide when manufactured, processed, or distributed in commerce for use as a pesticide,
(3) Tobacco or any tobacco product, but not including any derivative products,
(4) Any source material, special nuclear material, or byproduct material,
(5) Any pistol, firearm, revolver, shells, and cartridges, and
(6) Any food, food additive, drug, cosmetic, or device, when manufactured, processed, or distributed in commerce for use as a food, food additive, drug, cosmetic, or device (under TSCA).

chemical substance (40CFR720.3-e) means any organic or inorganic substance of a particular molecular identity, including any combination of such substances occurring in whole or in part as a result of a chemical reaction or occurring in nature, and any chemical element or uncombined radical; except that "chemical substance" does not include:
(1) Any mixture,
(2) Any pesticide when manufactured, processed, or distributed in commerce for use as a pesticide,
(3) Tobacco or any tobacco product, but not including any derivative products,

(4) Any source material, special nuclear material, or byproduct material,
(5) Any pistol, firearm, revolver, shells, and cartridges, and
(6) Any food, food additive, drug, cosmetic, or device, when manufactured, processed, or distributed in commerce for use as a food, food additive, drug, cosmetic, or device.
(Other identical or similar definitions are provided in 40CFR723.250-6; 747.115-1; 747.195-1; 747.200-1.)

chemical substance (40CFR761.3):
(1) Except as provided in paragraph (2) of this definition, means any organic or inorganic substance of a particular molecular identity, including: Any combination of such substances occurring in whole or part as a result of a chemical reaction or occurring in nature, and any element or uncombined radical.
(2) Such term does not include: Any mixture; any pesticide (as defined in the Federal Insecticide, Fungicide, and Rodenticide Act) when manufactured, processed, or distributed in commerce for use as a pesticide; tobacco or any tobacco product; any source material, special nuclear material, or byproduct material (as such terms are defined in the Atomic Energy Act of 1954 and regulations issued under such Act); any article the sale of which is subject to the tax imposed by section 4181 of the Internal Revenue Code of 1954 (determined without regard to any exemptions from such tax provided by section 4182 or section 4221 or any provisions of such Code); and any food, food additive, drug, cosmetic, or device (as such terms are defined in section 201 of the Federal Food, Drug, and Cosmetic Act) when manufactured, processed, or distributed in commerce for use as a food, food additive, drug, cosmetic, or device.

chemical substance (TSCA3-15USC2602):
(A) Except as provided in subparagraph (B), the term "chemical substance" means any organic or inorganic substance of a particular molecular identity, including:
(i) any combination of such substances occurring in whole or in part as a result of a chemical reaction or occurring in nature, and

(ii) any element or uncombined radical.

(B) Such term does not include:

(i) any mixture,

(ii) any pesticide (as defined in the Federal Insecticide, Fungicide, and Rodenticide Act) when manufactured, processed, or distributed in commerce for use as a pesticide,

(iii) tobacco or any tobacco product,

(iv) any source material, special nuclear material, or byproduct material (as such terms are defined in the Atomic Energy Act of 1954 and regulations issued under such Act),

(v) any article the sale of which is subject to the tax imposed by section 4181 of the Internal Revenue Code of 1954 (determined without regard to any exemptions from such tax provided by section 4182 or 4221 or any other provision of such Code), and

(vi) any food, food additive, drug, cosmetic, or device (as such terms are defined in section 201 of the Federal Food, Drug, and Cosmetic Act) when manufactured, processed, or distributed in commerce for use as a food, food additive, drug, cosmetic, or device.

(Other identical or similar definitions are provided in 40CFR723.175-5; 723.175-d; 763.163.)

chemical treatment (EPA-94/04): Any one of a variety of technologies that use chemicals or a variety of chemical processes to treat waste.

chemical waste landfill (40CFR761.3) means a landfill at which protection against risk of injury to health or the environment from migration of PCBs to land, water, or the atmosphere is provided from PCBs and PCB Items deposited therein by locating, engineering, and operating the landfill as specified in 40CFR761.75.

chemigation (40CFR170.3) means the application of pesticides through irrigation systems.

chemnet (EPA-94/04): Mutual aid network of chemical shippers and contractors that assigns a contracted emergency response company to provide technical support if a representative of the firm whose chemicals are involved in an incident is not readily available.

chemosterilant (EPA-94/04): A chemical that controls pests by preventing reproduction.

chemterc (EPA-94/04): The industry-sponsored Chemical Transportation Emergency Center; provides information and/or emergency assistance to emergency responders.

cherries, brined (40CFR407.61-f) shall include the processing of all varieties of cherries into the following brined product styles: Canned, bottled and bulk, sweet and sour, pitted and unpitted, bleached, sweetened, colored and flavored, whole, halved and chopped.

cherries, sour (40CFR407.61-e) shall include the processing of all sour varieties of cherries into the following products styles: Frozen and canned, pitted and unpitted, whole, halves, juice and concentrate.

cherries, sweet (40CFR407.61-d) shall include the processing of all sweet varieties of cherries into the following products styles: Frozen and canned, pitted and unpitted, whole, halves, juice and concentrate.

Chief Executive Officer of the tribe (40CFR350.1) means the person who is recognized by the Bureau of Indian Affairs as the chief elected administrative officer of the tribe (other identical or similar definitions are provided in 40CFR355.20; 370.2; 372.3).

chief facility operator (40CFR60.51a) means the person in direct charge and control of the operation of an MWC and who is responsible for daily on-site supervision, technical direction, management, and overall performance of the facility.

chief financial officer, in the case of local government owners and operators (40CFR280.92), means the individual with the overall authority and responsibility for the collection, disbursement, and use of funds by the local government.

child-resistant packaging (40CFR157.21-b) means packaging that is designed and constructed to be significantly difficult for children under 5 years of age to open or obtain a toxic or harmful amount of the substance contained therein within a reasonable time, and that is not difficult for normal adults to use properly.

chilled water loop (40CFR749.68-2) means any closed cooling water system that transfers heat from air handling units or refrigeration equipment to a refrigeration machine or chiller.

chilling effect (EPA-94/04): The lowering of the Earth's temperature because of increased particles in the air blocking the sun's rays (cf. greenhouse effect).

chips, corn (40CFR407.81-e) shall mean the processing of fried corn, made by soaking, rinsing, milling and extruding into a fryer without toasting. In terms of finished corn chips, 1 kg (lb) of finished product is equivalent to 0.9 kg (lb) of raw material.

chips, potato (40CFR407.81-d) shall mean the processing of fried chips, made from fresh or stored white potatoes, all varieties. In terms of finished potato chips, 1 kg (lb) of finished product is equivalent to 4 kg (lb) of raw material.

chips, tortilla (40CFR407.81-f) shall mean the processing of fried corn, made by soaking, rinsing, milling, rolling into sheets, toasting and frying. In terms of finished tortilla chips, 1 kg (lb) of finished product is equivalent to 0.9 kg (lb) of raw material.

chisel plowing (EPA-94/04): Preparing croplands by using a special implement that avoids complete inversion of the soil as in with conventional plowing. Chisel plowing can leave a protective cover or crop residues on the soil surface to help prevent erosion and improve filtration.

chlorinated hydrocarbons (EPA-94/04): These include a class of persistent, broad-spectrum insecticides that linger in the environment and accumulate in the food chain. Among them are DDT, aldrin, dieldrin, heptachlor, chlordane, lindane, endrin, mirex, hexachloride, and toxaphene. Other examples include TCE (trichloroethylene), used as an industrial solvent.

chlorinated paraffins (40CFR63.191) means dry chlorinated paraffins, which are mainly straight-chain, saturated hydrocarbons. The category includes, but is not limited to, production of chlorinated paraffins by passing gaseous chlorine into a paraffin hydrocarbon or by chlorination by using solvents, such as carbon tetrachloride, under reflux.

chlorinated solvent (EPA-94/04): An organic solvent containing chlorine atoms, e.g., methylene chloride and 1,1,1-trichloromethane, used in aerosol spray containers and in highway paint.

chlorinated terphenyl (40CFR704.45-1) means a chemical substance, CAS No. 61788-336, comprised of chlorinated ortho-, meta-, and paraterphenyl.

chlorination (EPA-94/04): The application of chlorine to drinking water, sewage, or industrial waste to disinfect or to oxidize undesirable compounds.

chlorinator (EPA-94/04): A device that adds chlorine, in gas or liquid form, to water or sewage to kill infectious bacteria.

chlorine (40CFR415.661-d) shall mean the total residual chlorine present in the process wastewater stream exiting the wastewater treatment system.

chlorine-contact chamber (EPA-94/04): That part of a water treatment plant where effluent is disinfected by chlorine.

chlorofluorocarbons (CFCs) (EPA-94/04): A family of inert, nontoxic, and easily liquified chemicals used in refrigeration, air conditioning, packaging, insulation, or as solvents and aerosol propellants. Because CFCs are not destroyed in the lower atmosphere they drift into the upper atmosphere where their chlorine components destroy ozone.

chlorophenoxy (EPA-94/04): A class of herbicides that may be found in domestic water supplies and cause adverse health effects.

chlorosis (EPA-94/04): Discoloration of normally green plant parts caused by disease, lack of nutrients, or various air pollutants.

chlorothalonil (40CFR63.191) means the agricultural fungicide, bactericide and nematocide Chlorothalonil (Daconil). The category includes any process units utilized to dissolve tetrachlorophthalic acid chloride in an organic solvent, typically carbon tetrachloride, with the subsequent addition of ammonia.

choke pull-off (40CFR85.2122(a)(1)(ii)(F)): See vacuum break.

cholinesterase (EPA-94/04): An enzyme found in animals that regulates nerve impulses. Cholinesterase inhibition is associated with a variety of acute symptoms such as nausea, vomiting, blurred vision, stomach cramps, and rapid heart rate.

chromatid type aberrations (40CFR798.5375-2) are damage expressed as breakage of single chromatids or breakage and/or reunion between chromatids (other identical or similar definitions are provided in 40CFR798.5385-2).

chrome pigments (40CFR415.341-b) means chrome yellow, chrome orange, molybdate chrome orange, anhydrous and hydrous chromium oxide, chrome green, and zinc yellow.

chrome tan (40CFR425.02-f) means the process of converting hide into leather using a form of chromium.

chromium (40CFR415.661-c) shall mean the total chromium present in the process wastewater stream exiting the wastewater treatment system (under CWA).

chromium (40CFR420.02-g) means total chromium and is determined by the method specified in 40CFR136.3.

chromium (40CFR428.101-b) shall mean total chromium.

chromium (EPA-94/04): See heavy metals.

chromium VI (40CFR420.02-h): See hexavalent chromium.

chromosome mutations (40CFR798.5955-1) are chromosomal changes resulting from breakage and reunion of chromosomes. Chromosomal mutations are also produced through nondisjunction of chromosomes during cell division.

chromosome type aberrations (40CFR798.5375-1) are changes which result from damage expressed in both sister chromatids at the same time (other identical or similar definitions are provided in 40CFR798.5385-1).

chronic effect (EPA-94/04): An adverse effect on a human or animal in which symptoms recur frequently or develop slowly over a long period of time.

chronic toxicity (40CFR131.35.d-4) means the lowest concentration of a constituent causing observable effects (i.e., considering lethality, growth, reduced reproduction, etc.) over a relatively long period of time, usually a 28-day test period for small fish test species.

chronic toxicity (40CFR300-AA) means the measure of toxicological responses that result from repeated exposure to a substance over an extended period of time (typically 3 months or longer). Such responses may persist beyond the exposure or may not appear until much later in time than the exposure. HRS measures of chronic toxicity include Reference Dose (RfD) values.

chronic toxicity (EPA-94/04): The capacity of a substance to cause long-term poisonous human health effects (cf. acute toxicity).

chronic toxicity test (40CFR797.1330-2) means a method used to determine the concentration of a substance in water that produces an adverse effect on a test organism over an extended period of time. In this test guideline, mortality and reproduction (and optionally, growth) are the criteria of toxicity (under TSCA).

chronic toxicity test (40CFR797.1950-1) means a method used to determine the concentration of a substance that produces an adverse effect from prolonged exposure of an organism to that substance. In this test, mortality, number of young per female and growth are used as measures of chronic toxicity.

circle of influence (EPA-94/04): The circular outer edge of a depression produced in the water table by the pumping of water from a well (cf. cone of influence or cone of depression).

circulating fluidized bed combustor (40CFR60.51a) means a fluidized bed combustor in which the majority of the fluidized bed material is carried out of the primary combustion zone and is transported back to the primary zone through a recirculation loop.

cistern (EPA-94/04): Small tank or storage facility used to store water for a home or farm; often used to store rain water.

city fuel economy (40CFR600.002.85-11) means the fuel economy determined by operating a vehicle (or vehicles) over the driving schedule in the Federal emission test procedure.

city fuel economy test (40CFR610.11-6): See federal test procedure.

civil judgement (40CFR32.105-d) means the disposition of a civil action by any court of competent jurisdiction, whether entered by verdict, decision, settlement, stipulation, or otherwise creating a civil liability for the wrongful acts complained of; or a final determination of liability under the Program Fraud Civil Remedies Act of 1988 (31USC3801-12).

cladding or metal cladding (40CFR471.02-g) is the art of producing a composite metal containing two or more layers that have been metallurgically bonded together by roll bonding (co-rolling), solder application (or brazing), or explosion bonding (under CWA).

claim (40CFR14.2-b) means a demand for payment by an employee or his/her representative for the value or the repair cost of an item of personal property damaged, lost or destroyed as an incident to government service.

claim (40CFR2.201-h): See business confidentiality claim.

claim (40CFR211.203-f) means an assertion made by a manufacturer regarding the effectiveness of his product.

claim (40CFR27.2) means any request, demand, or submission:
(a) Made to the Authority for property, services, or money (including money representing grants, loans, insurance, or benefit);
(b) Made to a recipient of property, services, or money from the Authority or to a party to a contract with the Authority:
  (1) For property or services if the United States:
    (i) Provided such property or services;
    (ii) Provided any portion of the funds for the purchase of such property or services; or
    (iii) Will reimburse such recipient or party for the purchase of such property or services; or
  (2) For the payment of money (including money representing grants, loans, insurance, or benefits) if the United States:
    (i) Provided any portion of the money requested or demanded: or
    (ii) Will reimburse such recipient or party for any portion of the money paid on such request or demand; or
(c) Made to the Authority which has the effect of decreasing an obligation to pay or account for property, services, or money.

claim (40CFR300.5), as defined by section 101(4) of CERCLA, means a demand in writing for a sum certain.

claim (40CFR304.12-e) means the amount sought by EPA as recovery of response costs incurred and to be incurred by the United States at a facility, which does not exceed $500,000, excluding interest.

claim (40CFR35.6015-8) means a demand or written assertion by a contractor seeking, as a matter of right, changes in contract duration costs, or other provisions, which originally have been rejected by the recipient.

claim (OPA1001) means a request, made in writing for a sum certain, for compensation for damages or removal costs resulting from an incident.

claim (SF101-42USC9601) means a demand in writing for a sum certain.

claimant (40CFR350.1) means a person submitting a claim of trade secrecy to EPA in connection with a chemical otherwise required to be disclosed in a report or other filing made under Title III.

claimant (OPA1001) means any person or government who presents a claim for compensation under this title.

claimant (SF101-42USC9601) means any person who presents a claim for compensation under this Act.

clarification (EPA-94/04): Clearing action that occurs during wastewater treatment when solids settle out. This is often aided by centrifugal action and chemically induced coagulation in wastewater.

clarifier (EPA-94/04): A tank in which solids settle to the bottom and are subsequently removed as sludge.

class (40CFR205.151-10) means a group of vehicles which are identical in all material aspects with respect to the parameters listed in 40CFR205.155 of this subpart.

class (40CFR86.402.78): See 40CFR86.419-78 (engine displacement, motorcycle classes).

class I (40CFR82.3-e) refers to the controlled substances listed in appendix A to this subpart.

class I area (EPA-94/04): Under the Clean Air Act, a Class I area is one in which visibility is protected more stringently than under the national ambient air quality standards; includes national parks, wilderness area, monuments and other areas of special national and cultural significance.

class I or class II (40CFR82.172) means the specific ozone-depleting compounds described in section 602 of the Act.

class I sludge management facility (40CFR122.2) means any POTW (public owned treatment works) identified under 40CFR403.8(a) as being required to have an approved pretreatment program (including such POTWs located in a State that has elected to assume local program responsibilities pursuant to 40CFR403.10(e)) and any other treatment works treating domestic sewage classified as a Class I sludge management facility by the Regional Administrator, or, in the case of approved State programs, the

Regional Administrator in conjunction with the State Director, because of the potential for its sludge use or disposal practices to adversely affect public health and the environment.

class I sludge management facility (40CFR501.2122.2) means any POTW identified under 40CFR403.8(a) as being required to have an approved pretreatment program (including such POTWs located in a State that has elected to assume local program responsibilities pursuant to 40CFR403.10(e)) and any other treatment works treating domestic sewage classified as a Class I sludge management facility by the Regional Administrator in conjunction with the State Program Director because of the potential for its sludge use or disposal practices to adversely affect public health or the environment.

class I sludge management facility (40CFR503.9-c) is any publicly owned treatment works (POTW), as defined in 40CFR501.2, required to have an approved pretreatment program under 40CFR403.8(a) (including any POTW located in a State that has elected to assume local program responsibilities pursuant to 40CFR403.10(e)) and any treatment works treating domestic sewage, as defined in 40CFR122.2, classified as a Class I sludge management facility by the EPA Regional Administrator, or, in the case of approved State programs, the Regional Administrator in conjunction with the State Director, because of the potential for its sewage sludge use or disposal practice to affect public health and the environment adversely.

class I substance (40CFR82.104-a) means any substance designated as class I in 40CFR part 82, appendix A to subpart A, including chlorofluorocarbons, halons, carbon tetrachloride and methyl chloroform and any other substance so designated by the Agency at a later date.

class I substance (40CFR82.82-a) means any substance designated as class I by EPA pursuant to 42 U.S.C. 7671(a), including but not limited to chlorofluorocarbons, halons, carbon tetrachloride and methyl chloroform.

class I substance (CAA601-42USC7671) means each of the substances listed as provided in section 602(a).

class II (40CFR82.3-f) refers to the controlled substances listed in appendix B to this subpart.

class II substance (40CFR82.104-b) means any substance designated as class II in 40CFR part 82, appendix A to subpart A, including hydrochlorofluorocarbons and any other substance so designated by the Agency at a later date.

class II substance (40CFR82.82-b) means any substance designated as class II by EPA pursuant to 42 U.S.C. 7671(a), including but not limited to hydrochlorofluorocarbons.

class II substance (CAA601-42USC7671) means each of the substances listed as provided in section 602(b).

class II wells (40CFR147.2902) means wells which inject fluids:
(a) Which are brought to the surface in connection with conventional oil or natural gas production and may be commingled with waste waters from gas plants which are an integral part of production operations, unless those waters would be classified as a hazardous waste at the time of injection;
(b) For enhanced recovery of oil or natural gas; and
(c) For storage of hydrocarbons which are liquid at standard temperature and pressure.

class T3 (40CFR87.1) means all aircraft gas turbine engines of the JT3D model family.

class T8 (40CFR87.1) means all aircraft gas turbine engines of the JT8D model family.

class TF (40CFR87.1) means all turbofan or turbojet aircraft engines except engines of Class T3, T8, and TSS.

class TP (40CFR87.1) means all aircraft turboprop erines.

class TSS (40CFR87.1) means all aircraft gas turbine engines employed for propulsion of aircraft designed to operate at supersonic flight speeds.

classification of railroads (40CFR201.1-d) means the division of railroad industry operating companies by the Interstate Commerce Commission into three categories. As of 1, Class I railroads must have annual revenues of $50 million or greater, Class II railroads must have annual revenues of between $10 and $50 million and Class III railroads must have less than $10 million in annual revenues.

classified information (40CFR11.4-a) means official information which has been assigned a security classification category in the interest of the national defense or foreign relations of the United States.

classified material (40CFR11.4-b) means any document, apparatus, model, film, recording, or any other physical object from which classified information can be derived by study, analysis, observation, or use of the material involved.

classified waste (40CFR246.101-d) means waste material that has been given security classification in accordance with 50USC401 and Executive Order 11652.

claus sulfur recovery plant (40CFR60.101-i) means a process unit which recovers sulfur from hydrogen sulfide by a vapor-phase catalytic reaction of sulfur dioxide and hydrogen sulfide.

clay mineral analysis (40CFR796.2750) is the estimation or determination of the kinds of clay-size minerals and the amount present in a sediment or soil.

clay soil (EPA-94/04): Soil material containing more than 40 percent clay, less than 45 percent sand, and less than 40 percent silt.

clean air (29CFR1910.94a) means air of such purity that it will not cause harm or discomfort to an individual if it is inhaled for extended periods of time.

Clean Air Act: See act or CAA.

clean air standards (40CFR15.4) means any enforceable rules, regulations, guidelines, standards, limitations, orders, controls, prohibitions, or other requirements which are contained in, issued under, or otherwise adopted pursuant to the CAA or Executive Order 11738, an applicable implementation plan as described in section 110(d) of the CAA, and approved implementation procedure or plan under section 111(c) or section 111(d), respectively, of the CAA or an approved implementation procedure under section 112(d) of the CAA.

clean alternative fuel (CAA241-42USC7581) means any fuel (including methanol, ethanol, or other alcohols (including any mixture thereof containing 85 percent or more by volume of such alcohol with gasoline or other fuels), reformulated gasoline, diesel, natural gas, liquefied petroleum gas, and hydrogen) or power source (including electricity) used in a clean-fuel vehicle that complies with the standards and requirements applicable to such vehicle under this title when using such fuel or power source. In the case of any flexible fuel vehicle or dual fuel vehicle, the term "clean alternative fuel" means only a fuel with respect to which such vehicle was certified as a clean-fuel vehicle meeting the standards applicable to clean-fuel vehicles under section 243(d)(2) when operating on clean alternative fuel (or any CARB standards which replaces such standards pursuant to section 243(e)).

clean area (40CFR763-AA-5) means a controlled environment which is maintained and monitored to assure a low probability of asbestos contamination to materials in that space. Clean areas used in this method have HEPA filtered air under positive pressure and are capable of sustained operation with an open laboratory blank which on subsequent analysis has an average of less than 18 structures/mm^2 in an area of 0.057 mm^2 (nominally 10 200-mesh grid openings) and a maximum of 53 structures/mm^2 for any single preparation for that same area.

clean coal technology (40CFR51.165-xxiii) means any technology, including technologies applied at the precombustion, combustion, or post combustion stage, at a new or existing facility which will achieve significant reductions in air emissions of sulfur dioxide or oxides of nitrogen associated with the utilization of coal in the generation of electricity, or process steam which was not in widespread use as of November 15, 1990.

clean coal technology (40CFR51.166-33) means any

technology, including technologies applied at the precombustion, combustion, or post combustion stage, at a new or existing facility which will achieve significant reductions in air emissions of sulfur dioxide or oxides of nitrogen associated with the utilization of coal in the generation of electricity, or process steam which was not in widespread use as of November 15, 1990.

clean coal technology (40CFR52.21-34) means any technology, including technologies applied at the precombustion, combustion, or post combustion stage, at a new or existing facility which will achieve significant reductions in air emissions of sulfur dioxide or oxides of nitrogen associated with the utilization of coal in the generation of electricity, or process steam which was not in widespread use as of November 15, 1990.

clean coal technology (40CFR52.24-22) means any technology, including technologies applied at the precombustion, combustion, or post combustion stage, at a new or existing facility which will achieve significant reductions in air emissions of sulfur dioxide or oxides of nitrogen associated with the utilization of coal in the generation of electricity, or process steam which was not in widespread use as of November 15, 1990.

clean coal technology (EPA-94/04): Any technology not in widespread use prior to the Clean Air Act amendments of 1990. This Act will achieve significant reductions in pollutants associated with the burning of coal.

clean coal technology demonstration project (40CFR51.165-xxiv) means a project using funds appropriated under the heading "Department of Energy-Clean Coal Technology," up to a total amount of $2,500,000,000 for commercial demonstration of clean coal technology, or similar projects funded through appropriations for the Environmental Protection Agency. The Federal contribution for a qualifying project shall be at least 20 percent of the total cost of the demonstration project.

clean coal technology demonstration project (40CFR51.166-34) means a project using funds appropriated under the heading "Department of

Energy-Clean Coal Technology," up to a total amount of $2,500,000,000 for commercial demonstration of clean coal technology, or similar projects funded through appropriations for the Environmental Protection Agency. The Federal contribution for a qualifying project shall be at least 20 percent of the total cost of the demonstration project.

clean coal technology demonstration project (40CFR52.21-35) means a project using funds appropriated under the heading "Department of Energy-Clean Coal Technology," up to a total amount of $2,500,000,000 for commercial demonstration of clean coal technology, or similar projects funded through appropriations for the Environmental Protection Agency. The Federal contribution for a qualifying project shall be at least 20 percent of the total cost of the demonstration project.

clean coal technology demonstration project (40CFR52.24-23) means a project using funds appropriated under the heading "Department of Energy-Clean Coal Technology," up to a total amount of $2,500,000,000 for commercial demonstration of clean coal technology, or similar projects funded through appropriations for the Environmental Protection Agency. The Federal contribution for a qualifying project shall be at least 20 percent of the total cost of the demonstration project.

clean coal technology demonstration project (40CFR60.2) means a project using funds appropriated under the heading "Department of Energy-Clean Coal Technology," up to a total amount of $2,500,000,000 for commercial demonstration of clean coal technology, or similar projects funded through appropriations for the Environmental Protection Agency.

clean fuel vehicle (CAA241-42USC7581) means a vehicle in a class or category of vehicles which has been certified to meet for any model year the clean-fuel vehicle standards applicable under this part for that model year to clean-fuel vehicles in that class or category.

clean fuels (EPA-94/04): Blends or substitutes for gasoline fuels, including compressed natural gas, methanol, ethanol, liquified petroleum gas, and others.

clean room (40CFR763.121) means an uncontaminated room having facilities for the storage of employees' street clothing and uncontaminated materials and equipment.

clean water standards (40CFR15.4) means any enforceable limitation, control, condition, prohibition, standard, or other requirement which is established pursuant to the CWA or contained in a permit issued to a discharger by the United States Environmental Protection Agency, or by a State under an approved program, as authorized by section 402 of the CWA, or by a local government to ensure compliance with pretreatment regulations as required by section 307 of the Clean Water Act.

Clean Water Act: See act or CWA.

cleaning or etching (40CFR467.02-i) is a chemical solution bath and a rinse or series of rinses designed to produce a desired surface finish on the workpiece. This term includes air pollution control scrubbers which are sometimes used to control fumes from chemical solution baths. Conversion coating and anodizing when performed as an integral part of the aluminum forming operations are considered cleaning or etching operations. When conversion coating or anodizing are covered here they are not subject to regulation under the provisions of 40CFR433, Metal Finishing.

cleaning water (40CFR463.2-d) is process water used to clean the surface of an intermediate or final plastic product or to clean the surfaces of equipment used in plastics molding and forming that contact an intermediate or final plastic product. It includes water used in both the detergent wash and rinse cycles of a cleaning process.

cleanup (EPA-94/04): Actions taken to deal with a release or threat of release of a hazardous substance that could affect humans and/or the environment. The term "cleanup" is sometimes used interchangeably with the terms remedial action, removal action, response action, or corrective action (under CERCLA).

clear coating (40CFR52.741) means coatings that lack color and opacity or are transparent using the undercoat as a reflectant base or undertone color.

clear cut (EPA-94/04): Harvesting all the trees in one area at one time, a practice that can encourage fast rainfall or snowmelt runoff, erosion, sedimentation of streams and lakes, flooding, and destroys vital habitat.

clear topcoat (40CFR52.741) means the final coating which contains binders, but not opaque pigments, and is specifically formulated to form a transparent or translucent solid protective film.

clear well (EPA-94/04): A reservoir for storing filtered water of sufficient quantity to prevent the need to vary the filtration rate with variations in demand. Also used to provide chlorine contact time for disinfection.

clearance (40CFR797.1560-3): See depuration.

CLEANS (EPA-94/04): Clinical laboratory for evaluation and assessment of toxic substances.

cloning (EPA-94/04): In biotechnology, obtaining a group of genetically identical cells from a single cell; making identical copies of a gene.

closed cooling water system (40CFR749.68-3) means any configuration of equipment in which any heat is transferred by circulating water that is contained within the equipment and not discharged to the air; chilled water loops are included.

closed course competition event (40CFR205.151-11) means any organized competition event covering an enclosed, repeated or confined route intended for easy viewing of the entire route by all spectators. Such events include short track, dirt track, drag race, speedway, hillclimb, ice race, and the Bonneville Speed Trials.

closed loop recycling (EPA-94/04): Reclaiming or reusing wastewater for non-potable purposes in an enclosed process.

closed loop system (40CFR63.161) means an enclosed system that returns process fluid to the process and is not vented to the atmosphere except through a closed-vent system.

closed portion (40CFR260.10) means that portion of a facility which an owner or operator has closed in accordance with the approved facility closure plan and all applicable closure requirements (cf. active portion or inactive portion).

closed purge system (40CFR63.161) means a system or combination of system and portable containers, to capture purged liquids. Containers must be covered or closed when not being filled or emptied.

closed vent system (40CFR264.1031) means a system that is not open to the atmosphere and that is composed of piping, connections, and, if necessary, flow-inducing devices that transport gas or vapor from a piece or pieces of equipment to a control device.

closed vent system (40CFR52.741) means a system that is not open to the atmosphere and is composed of piping, connections, and, if necessary, flow inducing devices that transport gas or vapor from an emission source to a control device.

closed vent system (40CFR60.481) means a system that is not open to the atmosphere and that is composed of piping, connections, and, if necessary, flow inducing devices that transport gas or vapor from a piece or pieces of equipment to a control device (other identical or similar definitions are provided in 40CFR60.561).

closed vent system (40CFR60.691) means a system that is not open to the atmosphere and is composed of piping, connections, and, if necessary, flow inducing devices that transport gas or vapor from an emission source to a control device.

closed vent system (40CFR61.241) means a system that is not open to atmosphere and that is composed of piping, connections, and, if necessary, flow-inducing devices that transport gas or vapor from a piece or pieces of equipment to a control device.

closed vent system (40CFR61.341) means a system that is snot open to the atmosphere and is composed of piping, ductwork, connections, and, if necessary, flow inducing devices that transport gas or vapor from an emission source to a control device.

closed vent system (40CFR63.111) means a system that is not open to the atmosphere and is composed of piping, ductwork, connections, and, if necessary, flow inducing devices that transport gas or vapor from an emission point to a control device or back into the process.

closed vent system (40CFR63.161) means a system that is not open to the atmosphere and that is composed of hard-piping, ductwork, connections and, if necessary, flow-inducing devices that transport gas or vapor from a piece or pieces of equipment to a control device or back into a process.

closeout (40CFR35.6015-9) means the final EPA or recipient actions taken to assure satisfactory completion of project work and to fulfill administrative requirements, including financial settlement submission of acceptable required final reports, and resolution of any outstanding issues under the cooperative agreement and/or Superfund State Contract.

closing rpm (40CFR205.151-12) means the engine speed in Figure 2 of Appendix I.

closure (40CFR270.2) means the act of securing a Hazardous Waste Management facility pursuant to the requirements of 40CFR264.

closure (EPA-94/04): The procedure a landfill operator must follow when a landfill reaches its legal capacity for solid waste: ceasing acceptance of solid waste and placing a cap on the landfill site.

closure period (40CFR192.31-h) means the period of time beginning with the cessation, with respect to a waste impoundment, of uranium ore processing operations and ending with completion of requirements specified under a closure plan.

closure plan (40CFR192.31-i) means the plan required under 40CFR264.112 of this chapter.

closure plan (40CFR264.141-a) means the plan for closure prepared in accordance with the requirements of 40CFR264.112.

closure plan (40CFR265.141-a) means the plan for closure prepared in accordance with the requirements of 40CFR265.112.

cluster (40CFR763-AA-6) means a structure with fibers in a random arrangement such that all fibers are intermixed and no single fiber is isolated from the group. Groupings must have more than two intersections.

CN,A (40CFR413.02-a) shall mean cyanide amenable to chlorination as defined by 40CFR136.

CN,T (40CFR413.02-b) shall mean cyanide, total.

CO (40CFR58.1-g; CAA302-42USC7602) means carbon monoxide.

co-fire (EPA-94/04): Burning of two fuels in the same combustion unit, e.g., coal and natural gas, or oil and coal.

co-permittee (40CFR122.26-1) means a permittee to a NPDES permit that is only responsible for permit conditions relating to the discharge for which it is operator.

co-product (40CFR63.101) means a chemical that is produced during the production of another chemical.

coagulation (40CFR141.2) means a process using coagulant chemicals and mixing by which colloidal and suspended materials are destabilized and agglomerated into flocs.

coagulation (EPA-94/04): Clumping of particles in wastewater to settle out impurities, often induced by chemicals such as lime, alum, and iron salts.

coal (40CFR60.251-c) means all solid fossil fuels classified as anthracite, bituminous, subbituminous, or lignite by ASTM (American Society and Testing and Materials) Designation D38877 (incorporated by reference, see 40CFR60.17).

coal (40CFR60.41-f) means all solid fuels classified as anthracite, bituminous, subbituminous, or lignite by the American Society and Testing and Materials, Designation D388-77 (incorporated by reference, see 40CFR60.17).

coal (40CFR60.41b) means all solid fuels classified as anthracite, bituminous, subbituminous, or lignite by the American Society of Testing and Materials in ASTM D388-77, Standard Specification for Classification of Coals by Rank (IBR, see 40CFR60.17), coal refuse, and petroleum coke. Coal-derived synthetic fuels, including but not limited to solvent refined coal, gasified coal, coal-oil mixtures and coal-water mixtures, are also included in this definition for the purposes of this subpart.

coal (40CFR60.41c) means all solid fuels classified as anthracite, bituminous, subbituminous, or lignite by the American Society for Testing and Materials in ASTM D388-77, "Standard Specification for Classification of Coals by Rank" (incorporated by reference, 40CFR60.17); coal refuse; and petroleum coke. Synthetic fuels derived from coal for the purpose of creating useful heat, including but not limited to solvent-refined coal, gasified coal, coal-oil mixtures, and coal-water mixtures, are included in this definition for the purposes of this subpart.

coal (40CFR72.2) means all solid fuels classified as anthracite, bituminous, subbituminous, or lignite by the American Society for Testing and Materials Designation ASTM D388-92 "Standard Classification of Coals by Rank" (as incorporated by reference in 40CFR72.13).

coal bunker (40CFR60-AG) means a single or group of coal trailers, hoppers, silos or other containers that:
(1) are physically attached to the affected facility; and
(2) provide coal to the coal pulverizers.

coal cleaning technology (EPA-94/04): Precombustion process by which coal is physically or chemically treated to remove some of its sulfur so as to reduce sulfur dioxide emissions.

coal derived fuel (40CFR72.2) means any fuel, whether in a solid, liquid, or gaseous state, produced by the mechanical, thermal, or chemical processing of coal (e.g., pulverized coal, coal refuse, liquified or gasified coal, washed coal, chemically cleaned coal, coal-oil mixtures, and coke).

coal fired (40CFR72.2) means the combustion of fuel consisting of coal or any coal-derived fuel (except a coal-derived gaseous fuel with a sulfur content no greater than natural gas), alone or in combination with any other fuel, where:
(1) For purposes of the requirements of part 75 of this chapter, a unit is "coal-fired" independent of the percentage of coal or coal-derived fuel consumed in any calendar year (expressed in mmBtu); and
(2) For all other purposes under the Acid Rain Program (including for calculating allowance allocations pursuant to part 73 of this chapter and applicability of the requirements of section 407 of the Act), a unit is "coal-fired" if it uses coal or coal-derived fuel as its primary fuel (expressed in mmBtu); provided that, if the unit is listed in the NADB, the primary fuel is the fuel listed in the NADB under the data field "PRIMFUEL."

coal fired utility unit (40CFR76.2) means a utility unit in which the combustion of coal (or any coal-derived fuel) on a Btu basis exceeds 50.0 percent of its annual heat input, for Phase I units in calendar year 1990 and, for Phase II units in the calendar year 1995. For the purposes of this part, this definition shall apply notwithstanding the definition at 40CFR72.2 of this chapter.

coal gasification (EPA-94/04): Conversion of coal to a gaseous product by one of several available technologies.

coal laboratory (SMCRA701-30USC1291), as used in subchapter VIII of this chapter, means a university coal research laboratory established and operated pursuant to a designation made under section 1311 of this title.

coal only heater (40CFR60.531) means an enclosed, coal-burning appliance capable of space heating, or domestic water heating, which has all of the following characteristics:
(a) An opening for emptying ash that is located near the bottom or the side of the appliance,
(b) A system that admits air primarily up and through the fuel bed,
(c) A grate or other similar device for shaking or disturbing the fuel bed or power-driven mechanical stoker,

(d) Installation instructions that state that the use of wood in the stove, except for coal ignition purposes, is prohibited by law, and

(e) The model is listed by a nationally recognized safety-testing laboratory for use of coal only, except for coal ignition purposes.

coal pile runoff (40CFR423.11-m) means the rainfall runoff from or through any coal storage pile.

coal preparation plant (40CFR434.11-e) means a facility where coal is subjected to cleaning, concentrating, or other processing or preparation in order to separate coal from its impurities and then is loaded for transit to a consuming facility.

coal preparation plant (40CFR60.251-a) means any facility (excluding underground mining operations) which prepares coal by one or more of the following processes: breaking, crushing, screening, wet or dry cleaning, and thermal drying.

coal preparation plant associated areas (40CFR434.11-f) means the coal preparation plant yards, immediate access roads, coal refuse piles and coal storage piles and facilities.

coal preparation plant water circuit (40CFR434.11-g) means all pipes, channels, basins, tanks, and all other structures and equipment that convey, contain, treat, or process any water that is used is coal preparation processes within a coal preparation plant.

coal processing and conveying equipment (40CFR60.251-g) means any machinery used to reduce the size of coal or to separate coal from refuse, and the equipment used to convey coal to or remove coal and refuse from the machinery. This includes, but is not limited to, breakers, crushers, screens, and conveyor belts.

coal/RDF mixed fuel fired combustor (40CFR60.51a) means a combustor that fires coal and RDF simultaneously.

coal refuse (40CFR60.41-c) means waste-products of coal mining, cleaning, and coal preparation operations (e.g. culm, gob, etc.) containing coal, matrix material, clay, and other organic and inorganic material.

coal refuse (40CFR60.41a-6) means waste products of coal mining, physical coal cleaning, and coal preparation operations (e.g. culm, gob, etc.) containing coal, matrix material, clay, and other organic and inorganic material.

coal refuse (40CFR60.41b) means any byproduct of coal mining or coal cleaning operations with an ash content greater than 50 percent, by weight, and a heating value less than 13,900 kJ/kg (6,000 Btu/lb) on a dry basis.

coal refuse (40CFR60.41c) means any byproduct of coal mining or coal cleaning operations with an ash content greater than 50 percent (by weight) and a heating value less than 13,900 kilojoules per kilogram (kJ/kg) (6,000 Btu per pound (Btu/lb) on a dry basis).

coal refuse disposal pile (40CFR434.11-p) means any coal refuse deposited on the earth and intended as permanent disposal or long-term storage (greater than 180 days) of such material, but does not include coal refuse deposited within the active mining area or coal refuse never removed from the active mining area.

coal remining operation (CWA301.p-33USC1311) means a coal mining operation which begins after the date of the enactment of this subsection at a site on which coal mining was conducted before the effective date of the Surface Mining Control and Reclamation Act of 1977.

coal storage system (40CFR60.251-h) means any facility used to store coal except for open storage piles.

coarse papers (40CFR250.4-f) means papers used for industrial purposes, as distinguished from those used for cultural or sanitary purposes.

Coast Guard District Response Group (CWA311-33USC1321) means a Coast Guard District Response Group established under subsection (j).

coastal (40CFR435.41-e) shall mean:
(1) any body of water landward of the territorial seas as defined in 40CFR125.1(gg),1 or
(2) any wetlands adjacent to such waters.

coastal energy activity (CZMA304-16USC1453) means any of the following activities if, and to the extent that:

(A) the conduct, support, or facilitation of such activity requires and involves the siting, construction, expansion, or operation of any equipment or facility; and

(B) any technical requirement exists which, in the determination of the Secretary, necessitates that the siting, construction, expansion, or operation of such equipment or facility be carried out in, or in close proximity to, the coastal zone of any coastal state:

  (i) Any outer Continental Shelf energy activity.

  (ii) Any transportation, conversion, treatment, transfer, or storage of liquefied natural gas.

  (iii) Any transportation, transfer, or storage of oil, natural gas, or coal (including, but not limited to, by means of any deepwater port, as defined in section 1502(10) of title 33). For purposes of this paragraph, the siting, construction, expansion, or operation of any equipment or facility shall be "in close proximity to" the coastal zone of any coastal state if such siting, construction, expansion, or operation has, or is likely to have, a significant effect on such coastal zone.

coastal resource of national significance (CZMA304-16USC1453) means any coastal wetland, beach, dune, barrier island, reef, estuary, or fish and wildlife habitat, if any such area is determined by a coastal state to be of substantial biological or natural storm protective value.

coastal state (CZMA304-16USC1453) means a state of the United States in, or bordering on, the Atlantic, Pacific, or Arctic Ocean, the Gulf of Mexico, Long Island Sound, or one or more of the Great Lakes. For the purposes of this chapter, the term also includes Puerto Rico, the Virgin Islands, Guam, the Commonwealth of the Northern Mariana Islands, and the Trust Territories of the Pacific Islands, and American Samoa.

coastal waters (40CFR300.5), for the purposes of classifying the size of discharges, means the waters of the coastal zone except for the Great Lakes and specified ports and harbors on inland rivers.

coastal waters (CZMA304-16USC1453) means:

(A) in the Great Lakes area, the waters within the territorial jurisdiction of the United States consisting of the Great Lakes, their connecting waters, harbors. roadsteads, and estuary-type areas such as bays, shallows, and marshes and

(B) in other areas, those waters, adjacent to the shorelines, which contain a measurable quantity or percentage of sea water, including, but not limited to, sounds, bays, lagoons, bayous, ponds, and estuaries.

coastal zone (40CFR300.5), as defined for the purpose of the NCP, means all United States waters subject to the tide, United States waters of the Great Lakes, specified ports and harbors on the inland rivers, waters of the contiguous zone, other waters of the high seas subject to the NCP, and the land surface or land substrata, ground waters, and ambient air proximal to those waters. The term coastal zone delineates an area of federal responsibility for response action. Precise boundaries are determined by EPA/USCG agreements and identified in Federal regional contingency plans (under CERCLA).

coastal zone (CZMA304-16USC1453) means the coastal waters (including the lands therein and thereunder) and the adjacent shorelands (including the waters therein and thereunder), strongly influenced by each other and in proximity to the shorelines of the several coastal states, and includes islands, transitional and intertidal areas, salt marshes, wetlands, and beaches. The zone extends, in Great Lakes waters, to the international boundary between the United States and Canada and, in other areas, seaward to the outer limit of the United States territorial sea. The zone extends inland from the shorelines only to the extent necessary to control shorelands, the uses of which have a direct and significant impact on the coastal waters. Excluded from the coastal zone are lands the use of which is by law subject solely to the discretion of or which is held in trust by the Federal Government, its officers or agents.

coastal zone (EPA-94/04): Lands and waters adjacent to the coast that exert an influence on the uses of the sea and its ecology, or whose uses and ecology are affected by the sea.

coating (40CFR52.741) means a material applied onto or impregnated into a substrate for protective, decorative, or functional purposes. Such materials include, but are not limited to, paints, varnishes, sealers, adhesives, thinners, diluents, and inks.

coating (40CFR60.461) means any organic material that is applied to the surface of metal coil.

coating application station (40CFR60.451) means that portion of the large appliance surface coating operation where a prime coat or a top coat is applied to large appliance parts or products (e.g., dip tank, spray booth, or flow coating unit).

coating application station (40CFR60.461) means that portion of the metal coil surface coating operation where the coating is applied to the surface of the metal coil. Included as part of the coating application station is the flashoff area between the coating application station and the curing oven.

coating applicator (40CFR52.741) means equipment used to apply a coating.

coating applicator (40CFR60.441) means an apparatus used to apply a surface coating to a continuous web.

coating applicator (40CFR60.711-3) means any apparatus used to apply a coating to a continuous base film.

coating applicator (40CFR60.741) means any apparatus used to apply a coating to a continuous substrate.

coating blow (40CFR60.471) means the process in which air is blown through hot asphalt flux to produce coating asphalt. The coating blow starts when the air is turned on and stops when the air is turned off.

coating line (40CFR52.741) means an operation consisting of a series of one or more coating applicators and any associated flash-off areas, drying areas, and ovens wherein a surface coating is applied, dried, or cured. (It is not necessary for an operation to have an oven, or flash-off area, or drying area to be included in this definition.)

coating line (40CFR60.441) means any number or combination of adhesive, release, or precoat coating applicators, flashoff areas, and ovens which coat a continuous web, located between a web unwind station and a web rewind station, to produce pressure sensitive tape and label materials (under CAA).

coating mix preparation equipment (40CFR60.711-4) means all mills, mixers, holding tanks, polishing tanks, and other equipment used in the preparation of the magnetic coating formulation but does not include those mills that do not emit VOC (colatile organic compound) because they are closed, sealed, and operated under pressure.

coating mix preparation equipment (40CFR60.741) means all mixing vessels in which solvent and other materials are blended to prepare polymeric coatings.

coating operation (40CFR60.711-5) means any coating applicator, flashoff area, and drying oven located between a base film unwind station and a base film rewind station that coat a continuous base film to produce magnetic tape.

coating operation (40CFR60.721) means the use of a spray booth for the application of a single type of coating (e.g., prime coat); the use of the same spray booth for the application of another type of coating (e.g., texture coat) constitutes a separate coating operation for which compliance determinations are performed separately.

coating operation (40CFR60.741) means any coating applicator(s), flashoff area(s), and drying oven(s) located between a substrate unwind station and a rewind station that coats a continuous web to produce a substrate with a polymeric coating. Should the coating process not employ a rewind station, the end of the coating operation is after the last drying oven in the process.

coating operations (40CFR466.02-e) means all of the operations associated with preparation and application of the vitreous coating. Usually this includes ballmilling, slip transport, application of slip to the workpieces, cleaning and recovery of faulty parts, and firing (fusing) of the enamel coat (under CWA).

coating plant (40CFR52.741) means any plant that contains one or more coating line(s).

coating solids applied (40CFR60.441) means the solids content of the coated adhesive, release, or precoat as measured by Reference Method 24.

coating solids applied (40CFR60.721) means the coating solids that adhere to the surface of the plastic business machine part being coated.

cobalt (40CFR415.651-c) shall mean the total cobalt present in the process wastewater stream exiting the wastewater treatment system.

COD (40CFR427.61-b; 427.71-d) shall mean COD added to the process waste water.

coefficient of haze (COH) (EPA-94/04): A measurement of visibility interference in the atmosphere.

cofired combustor (40CFR60.51a) means a unit combusting MSW or RDF with a non-MSW fuel and subject to a Federally enforceable permit limiting the unit to combusting a fuel feed stream, 30 percent or less of the weight of which is comprised, in aggregate, of MSW or RDF as measured on a 24-hour daily basis. A unit combusting a fuel feed stream, more than 30 percent of the weight of which is comprised, in aggregate, of MSW or RDF shall be considered an MWC unit and not a cofired combustor. Cofired combustors which fire less than 30 percent segregated medical waste and no other municipal solid waste are not covered by this subpart (under CAA).

cogeneration steam generating unit (40CFR60.41c) means a steam generating unit that simultaneously produces both electrical (or mechanical) and thermal energy from the same primary energy source.

cogeneration unit (40CFR72.2) means a unit that has equipment used to produce electric energy and forms of useful thermal energy (such as heat or steam) for industrial, commercial, heating or cooling purposes, through the sequential use of energy.

COH (40CFR52.274.d.2-ii) means coefficient of haze.

coil (40CFR465.02-a) means a strip of basis material rolled into a roll for handling.

coil (40CFR52.741) means any flat metal sheet or strip that is rolled or wound in concentric rings.

coil (40CFR85.2122(a)(9)(ii)(A)) means a device used to provide high voltage in an inductive ignition system.

coil coating (40CFR465.02-b) means the process of converting basis material strip into coated stock. Usually cleaning, conversion coating, and painting are performed on the basis material. This regulation covers processes which perform any two or more of the three operations.

coil coating (40CFR52.741) means any coating applied on any flat metal sheet or strip that comes in rolls or coils.

coil coating facility (40CFR52.741) means a facility that includes one or more coil coating line(s).

coil coating line (40CFR52.741) means a coating line in which any protective, decorative or functional coating is applied onto the surface of flat metal sheets, strips, rolls, or coils for industrial or commercial use.

coin-operated dry cleaning machine (40CFR63.321) means a dry cleaning machine that is operated by the customer (that is, the customer places articles into the machine, turns the machine on, and removes articles from the machine).

coke burn-off (40CFR60.101-h) means the coke removed from the surface of the fluid catalytic cracking unit catalyst by combustion in the catalyst regenerator. The rate of coke burn-off is calculated by the formula specified in 40CFR60.106.

coke byproduct recovery plant (40CFR61.131) means any plant designed and operated for the separation and recovery of coal tar derivatives (byproducts) evolved from coal during the coking process of a coke oven battery.

coke byproduct recovery plant (40CFR61.341) means any facility designed and operated for the separation and recovery of coal tar derivatives (byproducts) evolved from coal during the coking process of a coke oven battery.

coke oven (EPA-94/04): An industrial process which converts coal into coke, one of the basic materials used in blast furnaces for the conversion of iron ore into iron.

coke oven battery (40CFR63.301) means either a byproduct or nonrecovery coke oven battery.

coke oven byproduct ammonium sulfate manufacturing plant (40CFR60.421) means any plant which produces ammonium sulfate by reacting sulfuric acid with ammonia recovered as a byproduct from the manufacture of coke.

coke oven door (40CFR63.301) means each end enclosure on the pusher side and the coking side of an oven. The chuck, or leveler-bar, door is part of the pusher side door. A coke oven door includes the entire area on the vertical face of a coke oven between the bench and the top of the battery between two adjacent buckstays.

cold cleaning (40CFR52.741) means the process of cleaning and removing soils from surfaces by spraying, brushing, flushing, or immersion while maintaining the organic solvent below its boiling point. Wipe cleaning is not included in this definition (under CAA).

cold idle coke oven battery (40CFR63.301) means an existing coke oven battery that has been shut down, but is not dismantled.

cold rolling (40CFR468.02-f) shall mean the process of rolling a workpiece below the recrystallization temperature of the copper or copper alloy.

cold temperature CO (EPA-94/04): A standard for automobile carbon monoxide (CO) emissions to be met at a low temperature (i.e. 20 degrees Fahrenheit). Conventional automobile catalytic convertors are less efficient upon start-up at low temperatures.

cold worked pipe and tube (40CFR420.101-f) means those cold forming operations that process unheated

tube products using either water or oil solutions for cooling and lubrication.

coliform index (EPA-94/04): A rating of the purity of water based on a count of fecal bacteria.

coliform organism (EPA-94/04): Microorganisms found in the intestinal tract of humans and animals. Their presence in water indicates fecal pollution and potentially adverse contamination by pathogens.

collecting main (40CFR63.301) means any apparatus that is connected to one or more offtake systems and that provides a passage for conveying gases under positive pressure from the byproduct coke oven battery to the byproduct recovery system.

collecting main repair (40CFR63.301) means any measure to stop a collecting main leak on a long-term basis. A repair measure in general is intended to restore the integrity of the collecting main by returning the main to approximately its design specifications or its condition before the leak occurred. A repair measure may include, but is not limited to, replacing a section of the collecting main or welding the source of the leak.

collection (40CFR243.101-e) means the act of removing solid waste (or materials which have been separated for the purpose of recycling) from a central storage point (other identical or similar definitions are provided in 40CFR246.101-e).

collection frequency (40CFR243.101-f) means the number of times collection is provided in a given period of time.

collector sewer (40CFR35.2005-10) means the common lateral sewers, within a Publicly owned treatment system, which are primarily installed to receive wastewaters directly from facilities which convey wastewater from individual systems, or from private property, and which include service Y connections designed for connection with those facilities including:
(i) Crossover sewers connecting more than one property on one side of a major street, road, or highway to a lateral sewer on the other side when more cost effective than parallel sewers; and

(ii) Except as provided in paragraph (b)(10)(iii) of this section, pumping units and pressurized lines serving individual structures or groups of structures when such units are cost effective and are owned and maintained by the grantee.

(iii) This definition excludes other facilities which convey wastewater from individual structures, from private property to the public lateral sewer, or its equivalent and also excludes facilities associated with alternatives to conventional treatment works in small communities.

collector sewers (EPA-94/04): Pipes used to collect and carry wastewater from individual sources to an interceptor sewer that will carry it to a treatment facility.

colloidal dispersion (40CFR796.1840-i) is a mixture resembling a true solution but containing one or more substances that are finely divided but large enough to prevent passage through a semipermeable membrane. It consists of particles which are larger than molecules, which settle out very slowly with time, which scatter a beam of light, and which are too small for resolution with an ordinary light microscope.

colloids (EPA-94/04): Very small, finely divided solids (that do not dissolve) that remain dispersed in a liquid for a long time due to their small size and electrical charge.

colony (40CFR797.1160-2) means an aggregate of mother and daughter fronds attached to each other.

color coat (40CFR60.721) means the coat applied to a part that affects the color and gloss of the part, not including the prime coat or texture coat. This definition includes fog coating, but does not include conductive sensitizers or electromagnetic interference/radio frequency interference shielding coatings.

colorimetric detector tube (40CFR63.321) means a glass tube (sealed prior to use) containing material impregnated with a chemical that is sensitive to perchloroethylene and is designed to measure the concentration of perchloroethylene in air.

column dryer (40CFR60.301-m) means any equipment used to reduce the moisture content of grain in which the grain flows from the top to the bottom in one or more continuous packed columns between two perforated metal sheets.

combination (40CFR420.101-b) means those cold rolling operations which include recirculation of rolling solutions at one or more mill stands, and once through use of rolling solutions at the remaining stand or stands.

combination acid pickling (40CFR420.91-c) means those operations in which steel products are immersed in solutions of more than one acid to chemically remove scale and oxides, and those rinsing steps associated with such immersions (under the Clean Water Act).

combination heavy-duty vehicle (40CFR88.302.93) means a vehicle with a GVWR greater than 8,500 pounds (3,900 kilograms) which is comprised of a truck-tractor and one or more pieces of trailered equipment. The truck-tractor is a self-propelled motor vehicle built on one chassis which encompasses the engine, passenger compartment, and a means of coupling to a cargo carrying trailer(s). The truck-tractor itself is not designed to carry cargo.

combined cycle gas turbine (40CFR60.331-d) means any stationary gas turbine which recovers heat from the gas turbine exhaust gases to heat water or generate steam.

combined cycle gas turbine (40CFR60.41a-8) means a stationary turbine combustion system where heat from the turbine exhaust gases is recovered by a steam generating unit.

combined cycle system (40CFR60.41b) means a system in which a separate source such as a gas turbine, internal combustion engine, kiln, etc., provides exhaust gas to a heat recovery steam generating unit.

combined cycle system (40CFR60.41c) means a system in which a separate source (such as a stationary gas turbine, internal combustion engine, or kiln) provides exhaust gas to a steam generating unit (under CAA).

combined fuel economy (40CFR600.002.85-13) means:
(i) the fuel economy value determined for a vehicle (or vehicles) by harmonically averaging the city and highway fuel economy values, weighted 0.55 and 0.45 respectively, for gasoline-fueled and diesel vehicles.
(ii) For electric vehicles, the term means the equivalent petroleum-based fuel economy value as determined by the calculation procedure promulgated by the Secretary of Energy.

combined metals (40CFR421.261) mean the total of gold, platinum and palladium.

combined sewer (40CFR35.2005-11) means a sewer that is designed as a sanitary sewer and a storm sewer.

combined sewer (40CFR35.905) means a sewer intended to serve as a sanitary sewer and a storm sewer, or as an industrial sewer and a storm sewer.

combined sewer overflows (EPA-94/04): Discharge of a mixture of storm water and domestic waste when the flow capacity of a sewer system is exceeded during rainstorms.

combined sewers (EPA-94/04): A sewer system that carries both sewage and storm-water runoff. Normally, its entire flow goes to a waste treatment plant, but during a heavy storm, the volume of water may be so great as to cause overflows of untreated mixtures of storm water and sewage into receiving waters. Storm-water runoff may also carry toxic chemicals from industrial areas or streets into the sewer system.

combustibles (40CFR240.101-c) means materials that can be ignited at a specific temperature in the presence of air to release heat energy.

combustion (EPA-94/04):
(1) Burning, or rapid oxidation, accompanied by release of energy in the form of heat and light. A basic cause of air pollution.
(2) Refers to controlled burning of waste, in which heat chemically alters organic compounds, converting into stable inorganics such as carbon dioxide and water.

combustion chamber (EPA-94/04): The actual compartment where waste is burned in an incinerator.

combustion device (40CFR60.701) means an individual unit of equipment, such as an incinerator, flare, boiler, or process heater, used for combustion of a vent stream discharged from the process vent (under CAA).

combustion device (40CFR63.111) means an individual unit of equipment, such as a flare, incinerator, process heater, or boiler, used for the combustion of organic hazardous air pollutant vapors.

combustion product (EPA-94/04): Substance produced during the burning or oxidation of a material.

combustion source (40CFR72.2) means a stationary fossil fuel fired boiler, turbine, or internal combustion engine that has submitted or intends to submit an opt-in permit application under 40CFR74.14 of this chapter to enter the Opt-in Program.

comfort cooling towers (40CFR749.68-4) means cooling towers that are dedicated exclusively to and are an integral part of heating, ventilation, and air conditioning or refrigeration systems.

command post (EPA-94/04): Facility located at a safe distance upwind from an accident site, where the on-scene coordinator, responders, and technical representatives make response decisions, deploy manpower and equipment, maintain liaison with news media, and handle communications.

commence (40CFR52.2486-5) means to undertake a continuous program of on-site construction or modification.

commence as applied to construction of a major stationary source or major modification (40CFR51.166-9) means that the owner or operator has all necessary preconstruction approvals or permits and either has:
(i) Begun, or caused to begin, a continuous program of actual on-site construction of the

source, to be completed within a reasonable time; or

(ii) Entered into binding agreements or contractual obligations, which cannot be cancelled or modified without substantial loss to the owner or operator, to undertake a program of actual construction of the source to be completed within a reasonable time.

commence as applied to construction of a major stationary source or major modification (40CFR52.21-9) means that the owner or operator has all necessary preconstruction approvals or permits and either has:

(i) Begun, or caused to begin, a continuous program of actual on-site construction of the source, to be completed within a reasonable time; or

(ii) Entered into binding agreements or contractual obligations, which cannot be cancelled or modified without substantial loss to the owner or operator, to undertake a program of actual construction of the source to be completed within a reasonable time.

commence as applied to construction of a major stationary source or major modification (40CFR51.165-xvi) means that the owner or operator has all necessary preconstruction approvals or permits and either has:

(A) Begun, or caused to begin, a continuous program of actual on-site construction of the source, to be completed within a reasonable time; or

(B) Entered into binding agreements or contractual obligations, which cannot be canceled or modified without substantial loss to the owner or operator, to undertake a program of actual construction of the source to be completed within a reasonable time.

commence as applied to construction of a major stationary source or major modification (40CFR52.24-15) means that the owner or operator has all necessary preconstruction approvals or permits and either has:

(i) Begun, or caused to begin, a continuous program of actual on-site construction of the source, to be completed within a reasonable time; or

(ii) Entered into binding agreements or contractual obligations, which cannot be cancelled or modified without substantial loss to the owner or operator, to undertake a program of actual construction of the source to be completed within a reasonable time.

commence commercial operation (40CFR72.2) means to have begun to generate electricity for sale, including the sale of test generation.

commence construction (40CFR52.1135-1) means to engage in a continuous program of on-site construction including site clearance, grading, dredging, or land filling specifically designed for a parking facility in preparation for the fabrication, erection, or installation of the building components of the facility. For the purpose of this paragraph, interruptions resulting from acts of God, strikes, litigation, or other matters beyond the control of the owner shall be disregarded in determining whether a construction or modification program is continuous (cf. commence modification).

commence construction (40CFR72.2) means that an owner or operator has either undertaken a continuous program of construction or has entered into a contractual obligation to undertake and complete, within a reasonable time, a continuous program of construction.

commence modification (40CFR52.1135-2) means to engage in a continuous program of on-site modification including site clearance, grading, dredging, or land filling in preparation for a specific modification of the parking facility (cf. commence construction).

commence operation (40CFR72.2) means to have begun any mechanical, chemical, or electronic process, including start-up of an emissions control technology or emissions monitor or of a unit's combustion chamber.

commenced (40CFR52.01-b) means that an owner or operator has undertaken a continuous program of construction or modification.

commenced (40CFR60.2) means, with respect to the definition of "new source" in section 111(a)(2) of the

Act, that an owner or operator has undertaken a continuous program of construction or modification or that an owner or operator has entered into a contractual obligation to undertake and complete, within a reasonable time, a continuous program of construction or modification.

commenced (40CFR60.666) means that an owner or operator has undertaken a continuous program of component replacement or that an owner or operator has entered into a contractual obligation to undertake and complete, within a reasonable time, a continuous program of component replacement.

commenced (40CFR61.02) means, with respect to the definition of the "new source" in section 111(a)(2) of the new Act, that an owner or operator has undertaken a continuous program of construction or modification or that an owner or operator has entered into a contractual obligation to undertake and complete, within a reasonable time, a continuous program of construction or modification.

commenced (40CFR63.2) means, with respect to construction or reconstruction of a stationary source, that an owner or operator has undertaken a continuous program of construction or reconstruction or that an owner or operator has entered into a contractual obligation to undertake and complete, within a reasonable time, a continuous program of construction or reconstruction (other identical or similar definitions are provided in 40CFR63.101; 63.191).

commenced (CAA169-42USC7479) as applied to construction of a major emitting facility means that the owner or operator has obtained all necessary preconstruction approvals or permits required by Federal, State, or local air pollution emissions and air quality laws or regulations and either has (i) begun, or caused to begin, a continuous program of physical on-site construction of the facility or (ii) entered into binding agreements or contractual obligations, which cannot be canceled or modified without substantial loss to the owner or operator, to undertake a program of construction of the facility to be completed within a reasonable time.

commenced (CAA402-42USC7651a) as applied to construction of any new electric utility unit means that an owner or operator has undertaken a continuous program of construction or that an owner or operator has entered into a contractual obligation to undertake and complete, within a reasonable time, a continuous program of construction.

commenced commercial operation (CAA402-42USC7651a) means to have begun to generate electricity for sale.

commencement of construction (10CFR30.4) means any clearing of land, excavation, or other substantial action that would adversely affect the natural environment of a site but does not include changes desirable for the temporary use of the land for public recreational uses, necessary borings to determine site characteristics or other preconstruction monitoring to establish background information related to the suitability of a site or to the protection of environmental values (other identical or similar definitions are provided in 40CFR40.4; 70.4).

comment period (EPA-94/04): Time provided for the public to review and comment on a proposed EPA action or rulemaking after publication in the Federal Register.

commerce (40CFR205.2-17) means trade, traffic, commerce, or transportation: (i) Between a place in a State and any place outside thereof, or (ii) Which affects trade, traffic, commerce, or transportation described in paragraph (a)(17)(i) of this section.

commerce (40CFR710.2-i) means trade, traffic, transportation, or other commerce:
(1) Between a place in a State and any place outside of such State, or
(2) which affects trade, traffic, transportation, or commerce described in paragraph (i)(1) of this section.

commerce (40CFR720.3-f) means trade, traffic, transportation, or other commerce:
(1) between a place in a State and any place in a State and any place outside of such State, or
(2) which affects trade, traffic, transportation, or commerce between a place in a State and any place outside of such State.
(Other identical or similar definitions are provided in 40CFR747.115-1; 747.195-1; 747.200-1.)

commerce (40CFR761.3) means trade, traffic, transportation, or other commerce:
(1) Between a place in a State and any place outside of such State, or
(2) Which affects trade, traffic, transportation, or commerce described in paragraph (1) of this definition.

commerce (CAA216-42USC7550) means:
(A) commerce between any place in any State and any place outside thereof; and
(B) commerce wholly within the District of Columbia.

commerce (NCA3-42USC4902) means trade, traffic, commerce, or transportation (A) between a place in a State and any place outside thereof, or (B) which affects trade, traffic, commerce, or transportation described in subparagraph (A).

commerce (OSHA3-29USC652) means trade, traffic, commerce, transportation, or communication among the several States, or between a State and any place outside thereof, or within the District of Columbia, or a possession of the United States (other than the Trust Territory of the Pacific Islands), or between points in the same State but through a point outside thereof (other identical or similar definitions are provided in 29CFR1910.2).

commerce (SMCRA701-30USC1291) means trade, traffic, commerce, transportation, transmission, or communication among the several States, or between a State and any other place outside thereof, or between points in the same State which directly or indirectly affect interstate commerce.

commerce (TSCA3-15USC2602) means trade, traffic, transportation, or other commerce:
(A) between a place in a State and any place outside of such State, of
(B) which affects trade, traffic, transportation, or commerce described in clause (A).
(Other identical or similar definitions are provided in 40CFR762.3-d; 763.163.)

commercial activity (ESA3-16USC1531) means all activities of industry and trade, including but not limited to the buying or selling of commodities and activities conducted for the purpose of facilitating such buying and selling: Provided, however, that it does not include exhibition of commodities by museum or similar cultural or historical organizations.

commercial aircraft engine (40CFR87.1) means any aircraft engine used or intended for use by an air carrier (including those engaged in intrastate air transportation) or a commercial operator (including those engaged in intrastate air transportation) as these terms are defined in the Federal Aviation Act and the Federal Aviation Regulations.

commercial aircraft gas turbine engine (40CFR87.1) means a turboprop, turbofan, or turbojet commercial aircraft engine.

commercial applicator (40CFR171.2-9) means a certified applicator (whether or not he is a private applicator with respect to some uses) who uses or supervises the use of any pesticide which is classified for restricted use for any purpose or on any property other than as provided by the definition of private applicator.

commercial arsenic (40CFR61.161) means any form of arsenic that is produced by extraction from any arsenic-containing substance and is intended for sale or for intentional use in a manufacturing process. Arsenic that is a naturally occurring trace constituent of another substance is not considered "commercial arsenic."

commercial asbestos (40CFR61.141) means any material containing asbestos that is extracted from ore and has value because of its asbestos content (under CAA).

commercial establishment (40CFR246.101-f) means stores, offices, restaurants, warehouses and other non-manufacturing activities.

commercial hexane (40CFR799.2155-1), for purposes of this section, is a product obtained from crude oil, natural gas liquids, or petroleum refinery processing in accordance with the American Society for Testing and Materials Designation D 1836-83 (ASTM D 1836), consists primarily of six-carbon alkanes or cycloalkanes, and contains at least 40 liquid volume percent n-hexane (CAS No. 110-543) and at least 5 liquid volume percent methylcyclopentane (MCP; CAS No. 96-377). ASTM D 1836, formally entitled "Standard Specification for Commercial Hexanes," is published in 1986 Annual Book of ASTM Standards: Petroleum Products and Lubricants, ASTM D 1836-83, pp. 966-967, 1986, is incorporated by reference, and is available for public inspection at the Office of the Federal Register, Room 8301, 1100 L Street NW. Washington DC.

commercial hexane test substance (40CFR799.2155-2) for purposes of this section, is a product which conforms to the specifications of ASTM D 1836 and contains no more than 40 liquid volume percent n-hexane and no less than 10 liquid volume percent MCP (methylcyclopentane).

commercial item descriptions (40CFR248.4) are a series of simplified item descriptions under the Federal specifications-and-standards program used in the acquisition of commercial off-the-shelf and commercial type products.

commercial owner (40CFR60.531) means any person who owns or controls a wood heater in the course of the manufacture, importation, distribution, or sale of the wood heater.

commercial paper (40CFR763.163) means an asbestos-containing product which is made of paper intended for use as general insulation paper or muffler paper. Major applications of commercial papers are insulation against fire, heat transfer, and corrosion in circumstances that require a thin, but durable, barrier.

commercial parking facility (also called facility) (40CFR52.1135-5) means any lot, garage, building or structure, or combination or portion thereof, on or in which motor vehicles are temporarily parked for a fee, excluding:
(i)   a parking facility, the use of which is limited exclusively to residents (and guests of residents) of a residential building or group of buildings under common control, and
(ii) parking on public streets.

commercial parking space (40CFR52.1135-3) means a space used for parking a vehicle in a commercial parking facility.

commercial pesticide handling establishment (40CFR170.3) means any establishment, other than an agricultural establishment, that:
(1)   Employs any person, including a self-employed person, to apply on an agricultural establishment, pesticides used in the production of agricultural plants.
(2)   Employs any person, including a self-employed person, to perform on an agricultural establishment, tasks as a crop advisor.

commercial property (40CFR201.1-e) means any property that is normally accessible to the public and that is used for any of the purposes described in the following standard land use codes (reference Standard Land Use Coding Manual, U.S. DOT/FHWA(Department of Transportation/Federal Highway Administration) reprinted March 1977): 53-59, Retail Trade; 61-64, Finance, Insurance, Real Estate, Personal, Business and Repair Services; 652-659, Legal and other professional services; 671, 672, and 673 Governmental Services: 692 and 699, Welfare, Charitable and Other Miscellaneous Services; 712 and 719, Nature exhibitions and other Cultural Activities; 721, 723, and 729, Entertainment, Public and other Public Assembly; and 74-79, Recreational, Resort, Park and other Cultural Activities.

commercial refrigeration (40CFR82.152-d) means, for the purposes of 40CFR 82.156(i), the refrigeration appliances utilized in the retail food and cold storage warehouse sectors. Retail food includes the refrigeration equipment found in supermarkets, convenience stores, restaurants and other food service establishments. Cold storage includes the equipment used to store meat, produce, dairy products, and other perishable goods. All of the equipment contains large refrigerant charges, typically over 75 pounds.

commercial solid waste (40CFR243.101-g) means all types of solid wastes generated by stores, offices, restaurants, warehouses, and other non-manufacturing activities, excluding residential and industrial wastes.

commercial solid waste (40CFR245.101-a) means all types of solid waste generated by stores, offices, restaurants, warehouses, and other such nonmanufacturing activities, and non-processing waste generated at industrial facilities such as office and packing wastes.

commercial solid waste (40CFR246.101-g) means all types of solid wastes generated by stores, offices, restaurants, warehouses and other non-manufacturing activities, and non-processing wastes such as office and packing wastes generated at industrial facilities.

commercial solid waste (40CFR258.2) means all types of solid waste generated by stores, offices, restaurants, warehouses, and other nonmanufacturing activities, excluding residential and industrial wastes.

commercial storer of PCB waste (40CFR761.3) means the owner or operator of each facility which is subject to the PCB storage facility standards of 40CFR761.65, and who engages in storage activities involving PCB waste generated by others, or PCB waste that was removed while servicing the equipment owned by others and brokered for disposal. The receipt of a fee or any other form of compensation for storage services is not necessary to qualify as a commercial storer of PCB waste. It is sufficient under this definition that the facility stores PCB waste generated by others or the facility removed the PCB waste while servicing equipment owned by others. A generator who stores only the generator's own waste is subject to the storage requirements of 40CFR761.65, but is not required to seek approval as a commercial storer. If a facility's storage of PCB waste at no time exceeds 500 liquid gallons of PCBs, the owner or operator is not required to seek approval as a commercial storer of PCB waste.

commercial use (40CFR721.3) means the use of a chemical substance or any mixture containing the chemical substance in a commercial enterprise providing saleable goods or a service to consumers (e.g., a commercial dry cleaning establishment or painting contractor).

commercial use request (40CFR2.100-e) refers to a request from or on behalf of one who seeks information for a use or purpose that furthers the commercial, trade or profit interests of the requestor or the person on whose behalf the request is made.

In determining whether a requestor properly belongs in this category, EPA must determine the use to which a requestor will put the documents requested. Moreover, where EPA has reasonable cause to doubt the use to which a requestor will put the records sought, or where that use is not clear from the request itself, EPA may seek additional clarification before assigning the request to a specific category.

commercial vessels (CWA312-33USC1322) means those vessels used in the business of transporting property for compensation or hire, or in transporting property in the business of the owner, lessee, or operator of the vessel.

commercial waste (EPA-94/04): All solid waste emanating from business establishments such as stores, markets, office buildings, restaurants, shopping centers, and theaters.

commercial waste management facility (EPA-94/04): A treatment, storage, disposal, or transfer facility which accepts waste from a variety of sources, as compared to a private facility which normally manages a limited waste stream generated by its own operations.

commingled recyclables (EPA-94/04): Mixed recyclables that are collected together.

comminuter (EPA-94/04): A machine that shreds or pulverizes solids to make waste treatment easier.

comminution (EPA-94/04): Mechanical shredding or pulverizing of waste. Used in both solid waste management and wastewater treatment.

commission (10CFR20.3) means the Nuclear Regulatory Commission and its duly authorized representatives (other identical or similar definitions are provided in 10CFR30.4; 40.4; 70.4).

commission (40CFR191.02-c) means the Nuclear Regulatory Commission.

commission (40CFR350.1) means the emergency response commission for the State in which the facility is located except where the facility is located in Indian Country, in which case, commission means the emergency response commission for the tribe under whose jurisdiction the facility is located. In the absence of an emergency response commission, the Governor and the chief executive officer, respectively, shall be the commission. Where there is a cooperative agreement between a State and a Tribe, the commission shall be the entity identified in the agreement.

commission (40CFR355.20) means the emergency response commission for the State in which the facility is located except where the facility is located in Indian Country, in which case, commission means the emergency response commission for the tribe under whose jurisdiction the facility is located. In the absence of an emergency response commission, the Governor and the chief executive officer, respectively, shall be the commission. Where there is a cooperative agreement between a State and a Tribe, the commission shall be the entity identified in the agreement (other identical or similar definitions are provided in 40CFR370.2).

commission (OSHA3-29USC652) means the Occupational Safety and Health Review Commission established under this Act.

commission finishing (40CFR410.01-d) shall mean the finishing of textile materials, 50 percent or more of which are owned by others, in mills that are 51 percent or more independent (i.e., only a minority ownership by company(ies) with greige or integrated operations); the mills must process 20 percent or more of their commissioned production through batch, noncontinuous processing operations with 50 percent or more of their commissioned orders processed in 5000 yard or smaller lots.

commission scouring (40CFR410.11-c) shall mean the scouring of wool, 50 percent or more of which is owned by others, in mills that are 51 percent or more independent (i.e., only a minority ownership by company(ies) with greige or integrated operations); the mills must process 20 percent or more of their commissioned production through batch, noncontinuous processing operations.

commitment of Federal financial assistance (40CFR149.101-h) means a written agreement entered into by a department, agency, or instrumentality of the Federal Government to provide financial assistance as defined in paragraph (g) of this section. Renewal of a commitment which the issuing agency determines has lapsed shall not constitute a new commitment unless the Regional Administrator determines that the project's impact on the aquifer has not been previously reviewed under section 1424(e). The determination of a Federal agency that a certain written agreement constitutes a commitment shall be conclusive with respect to the existence of such a commitment.

committee (40CFR164.2-f) means a group of qualified scientists designated by the National Academy of Sciences according to agreement under the Act to submit an independent report to the Administrative Law Judge on questions of scientific fact referred from a hearing under subpart B of this part.

committee (NCA15-42USC4914) means the Low-Noise-Emission Product Advisory Committee.

committee (OSHA3-29USC652) means the National Advisory Committee on Occupational Safety and Health established under this Act.

committee or local emergency planning committee (40CFR355.20) means the local emergency planning committee appointed by the emergency response commission (other identical or similar definitions are provided in 40CFR370.2).

common carrier by motor vehicle (40CFR202.10-b) means any person who holds himself out to the general public to engage in the transportation by motor vehicle in interstate or foreign commerce of passengers or property or any class or classes thereof for compensation, whether over regular or irregular routes.

common emission control device (40CFR60.711-6) means a control device controlling emissions from the coating operation as well as from another emission source within the plant.

common emission control device (40CFR60.741) means a device controlling emissions from an affected coating operation as well as from any other emission source.

common exposure route (40CFR171.2-12) means a likely way (oral, dermal, respiratory) by which a pesticide may reach and/or enter an organism.

common name (40CFR721.3) means any designation or identification such as code name, code number, trade name, brand name, or generic chemical name used to identify a chemical substance other than by its chemical name.

common stack (40CFR72.2) means the exhaust of emissions from two or more units through a single flue.

commonwealth (40CFR52.1128-8) means the Commonwealth of Massachusetts.

community (EPA-94/04): In ecology, a group of interacting populations in time and space. Sometimes, a particular subgrouping may be specified, such as the fish community in a lake or the soil arthropod community in a forest.

community relations (40CFR300.5) means EPA's program to inform and encourage public participation in the Superfund process and to respond to community concerns. The term "public" includes citizens directly affected by the site, other interested citizens or parties, organized groups, elected officials, and potentially responsible parties.

community relations (EPA-94/04): The EPA effort to establish two-way communication with the public to create understanding of EPA programs and related actions, to assure public input into decision-making processes related to affected communities, and to make certain that the Agency is aware of and responsive to public concerns. Specific community relations activities are required in relation to Superfund remedial actions.

community relations coordinator (40CFR300.5) means lead agency staff who work with the OSC/RPM (on scene coordinator/remedial project manager) to involve and inform the public about the Superfund process and response actions in accordance with the interactive community relations requirements set forth in the NCP (national contingency plan).

community relations plan (CRP) (40CFR35.6015-10) means a management and planning tool outlining the specific community relations activities to be undertaken during the course of a response. It is designed to provide for two-way communication between the affected community and the agencies responsible for conducting a response action, and to assure public input into the decision-making process related to the affected communities.

community water system (40CFR141.2) means a public water system which serves at least 15 service connections used by year-round residents or regularly serves at least 25 year-round residents.

community water system (EPA-94/04): A public water system which serves at least 15 service connections used by year-round residents or regularly serves at least 25 year-round residents.

commuter (40CFR52.1161-6) means both an "employee" and a "student."

commuter (40CFR52.2296-3) means an employee who travels regularly to a place of employment.

commuter (40CFR52.2297-3) means an employee or a student who travels regularly to and from a facility.

compaction (EPA-94/04): Reduction of the bulk of solid waste by rolling and tamping.

compactor collection vehicle (40CFR243.101-h) means a vehicle with an enclosed body containing mechanical devices that convey solid waste into the main compartment of the body and compress it into a smaller volume of greater density.

compartment (40CFR60-AA (method 27-2.3)) means a liquid-tight division of a delivery tank (under CAA).

compartmentalized vehicle (40CFR246.101-i) means a collection vehicle which has two or more compartments for placement of solid wastes or recyclable materials. The compartments may be within the main truck body or on the outside of that body as in the form of metal racks.

compatibility (40CFR171.2-10) means that property of a pesticide which permits its use with other chemicals without undesirable results being caused by the combination.

compatible (40CFR280.12) means the ability of two or more substances to maintain their respective physical and chemical properties upon contact with one another for the design life of the tank system under conditions likely to be encountered in the UST.

compensating unit (40CFR72.2) means an affected unit that is not otherwise subject to Acid Rain emissions limitation or Acid Rain emissions reduction requirements during Phase I and that is designated as a Phase I unit in a reduced utilization plan under 40CFR72.43; provided that an opt-in source shall not be a compensating unit.

competent (40CFR171.2-11) means properly qualified to perform functions associated with pesticide application, the degree of capability required being directly related to the nature of the activity and the associated responsibility.

competent person (40CFR763.121) means one who is capable of identifying existing asbestos hazards in the workplace and who has the authority to take prompt corrective measures to eliminate them. The duties of the competent person include at least the following: Establishing the negative-pressure enclosure, ensuring its integrity, and controlling entry to and exit from the enclosure; supervising any employee exposure to monitoring required by this subpart, ensuring that all employees working within such an enclosure wear the appropriate personal protective equipment, are trained in the use of appropriate methods of exposure control, and the use of hygiene facilities and decontamination procedures specified in this subpart; and ensuring that engineering controls in use are in proper operating condition and are functioning properly (under TSCA).

competition motorcycle (40CFR205.151-3) means any motorcycle designed and marketed solely for use in closed course competition events.

complainant (40CFR209.3-e) means the Agency acting through any person authorized by the Administrator to issue a complaint to alleged violators of the Act. The complainant shall not be the judicial officer or the Administrator.

complainant (40CFR22.03) means any person authorized to issue a complaint on behalf of the Agency to persons alleged to be in violation of the Act. The complainant shall not be the Judicial Officer, Regional Judicial Officer, or any other person who will participate or advise in the decision.

complaint (40CFR22.03) means a written communication, alleging one or more violations of specific provisions of the Act, or regulations or a permit promulgated thereunder, issued by the complainant to a person under 40CFR22.13 and 22.14.

complaint (40CFR27.2) means the administrative complaint served by the reviewing official on the defendant under 40CFR27.2.

complete (40CFR51.166-22) means, in reference to an application for a permit, that the application contains all the information necessary for processing the application. Designating an application complete for purposes of permit processing does not preclude the reviewing authority from requesting or accepting any additional information.

complete (40CFR52.21-22) means, in reference to an application for a permit, that the application contains all of the information necessary for processing the application.

complete complaint (40CFR12.103) means a written statement that contains the complainant's name and address and describes the agency's alleged discriminatory action in sufficient detail to inform the agency of the nature and date of the alleged violation of section 504. It shall be signed by the complainant or by someone authorized to do so on his or her behalf. Complaints filed on behalf of classes or third parties shall describe or identify (by name, if possible) the alleged victims of discrimination.

complete destruction of pesticides (40CFR165.1-e) means alteration by physical or chemical processes to inorganic forms.

complete treatment (EPA-94/04): A method of

treating water that consists of the addition of coagulant chemicals, flash mixing, coagulation-flocculation, sedimentation, and filtration. Also called conventional filtration.

complete waste treatment system (40CFR35.2005-12) consists of all the treatment works necessary to meet the requirements of title III of the Act, involving:
(i) The transport of wastewater from individual homes or buildings to a plant or facility where treatment of the wastewater is accomplished;
(ii) the treatment of the wastewater to remove pollutants; and
(iii) the ultimate disposal, including recycling or reuse, of the treated wastewater and residues which result from the treatment Process.

complete waste treatment system (40CFR35.905) consists of all the treatment works necessary to meet the requirements of title III of the Act, involved in:
(a) The transport of wastewaters from individual homes or buildings to a plant or facility where treatment of the wastewater is accomplished;
(b) The treatment of the wastewaters to remove pollutants; and
(c) The ultimate disposal, including recycling or reuse, of the treated wastewaters and residues which result from the treatment Process. One complete waste treatment system would, normally, include one treatment plant or facility, but also includes two or more connected or integrated treatment plants or facilities.

completely closed drain system (40CFR60.691) means an individual drain system that is not open to the atmosphere and is equipped and operated with a closed vent system and control device complying with the requirements of 40CFR60.692-5 (under CAA).

completely destroy (40CFR82.3-g) means to cause the expiration of a controlled substance at a destruction efficiency of 98 percent or greater, using one of the destruction technologies approved by the parties.

complex (40CFR112.2) means a facility possessing a combination of transportation-related and non-transportation-related components that is subject to the jurisdiction of more than one Federal agency under section 311(j) of the Clean Water Act.

collection (40CFR243.101-e) means the act of removing solid waste (or materials which have been separated for the purpose of recycling) from a central storage point (other identical or similar definitions are provided in 40CFR246.101-e).

complex manufacturing operation (40CFR410.41-b) shall mean "simple" unit processes (desizing, fiber preparation and dyeing) plus any additional manufacturing operations such as printing, water proofing, or applying stain resistance or other functional fabric finishes (cf. simple manufacturing operation) (other identical or similar definitions are provided in 40CFR410.51-b).

complex manufacturing operation (40CFR410.61-c) shall mean "simple" unit processes (fiber preparation, dyeing and carpet backing) plus any additional manufacturing operations such as printing or dyeing and printing (cf. simple manufacturing operation).

complex slaughterhouse (40CFR432.21-c) shall mean a slaughterhouse that accomplishes extensive byproduct processing, usually at least three of such operations as rendering, paunch and viscera handling, blood processing, hide processing, or hair processing.

compliance (40CFR15.4) means compliance with clean air standards or clean water standards. For the purpose of these regulations, compliance also shall mean compliance with a schedule or plan ordered or approved by a court of competent jurisdiction, the United States Environmental Protection Agency, or an air or water pollution control agency, in accordance with the requirements of the CAA or the CWA and regulations issued pursuant thereto.

compliance agency (40CFR8.2-e) means the agency designated by the Director on a geographical, industry, or other basis to conduct compliance reviews and to undertake such other responsibilities in connection with the administration of the order as the Director may determine to be appropriate. In the absence of such a designation the Compliance Agency will be determined as follows:
(1) In the case of a prime contractor not involved in construction work, the Compliance Agency will

be the agency whose contracts with the prime contractor have the largest aggregate dollar value;

(2) In the case of a subcontractor not involved in construction work, the Compliance Agency will be the Compliance Agency of the prime contractor with which the subcontractor has the largest aggregate value of subcontracts or purchase orders for the performance of work under contracts;

(3) In the case of a prime contractor or subcontractor involved in construction work, the Compliance Agency for such construction project will be the agency providing the largest dollar value for the construction projects; and

(4) In the case of a contractor who is both a prime contractor and subcontractor the Compliance Agency will be determined as if such contractor is a prime contractor only.

compliance certification (40CFR72.2) means a submission to the Administrator or permitting authority, as appropriate, that is required by this part, by parts 73, 74, 75, 76, 77, or 78 of this chapter, to report an affected source or an affected unit's compliance or non-compliance with a provision of the Acid Rain Program and that is signed and verified by the designated representative in accordance with subparts B and I of this part and the Acid Rain Program regulations generally.

compliance coal (EPA-94/04): Any coal that emits less than 1.2 pounds of sulfur dioxide per million Btu when burned. Also known as low sulfur coal.

compliance coating (EPA-94/04): A coating whose volatile organic compound content does not exceed that allowed by regulation.

compliance cycle (40CFR141.2) means the nine-year calendar year cycle during which public water systems must monitor. Each compliance cycle consists of three three-year compliance periods. The first calendar year cycle begins January 1, 1993 and ends December 31, 2001; the second begins January 1, 2002 and ends December 31, 2010; the third begins January 1, 2011 and ends December 31, 2019.

compliance cycle (EPA-94/04): The nine-year calendar year cycle, beginning January 1, 1993, during which public water systems must monitor. Each cycle consists of three three-year compliance periods.

compliance date (40CFR63.161) means the dates specified in 40CFR63.100(k) or 40CFR63.100(l)(3) of subpart F of this part for process units subject to subpart F of this part; the dates specified in 40CFR63.190(e) of subpart I of this part for process units subject to subpart I of this part. For sources subject to other subparts in 40CFR part 63 that reference this subpart, compliance date will be defined in those subparts. However, the compliance date for 40CFR63.170 shall be no later than 3 years after the effective date of those subparts unless otherwise specified in such other subparts.

compliance date (40CFR63.2) means the date by which an affected source is required to be in compliance with a relevant standard, limitation, prohibition, or any federally enforceable requirement established by the Administrator (or a State with an approved permit program) pursuant to section 112 of the Act (other identical or similar definitions are provided in 40CFR63.101; 63.191).

compliance level (40CFR86.1002.2001-2) means an emission level determined during a Production Compliance Audit pursuant to subpart L of this part.

compliance level (40CFR86.1002.84) means an emission level determined during a Production Compliance Audit Pursuant to subpart L of this part.

compliance level (40CFR86.1102.87) means the deteriorated pollutant emissions level at the 60th percentile point for a population of heavy-duty engines or heavy-duty vehicles subject to production, Compliance Audit testing pursuant to the requirements of this subpart. A compliance level can only be determined for a pollutant for which an upper limit has been established in this subpart (under CAA).

compliance monitoring (EPA-94/04): Collection and evaluation of data, including selfmonitoring reports, and verification to show whether pollutant concentrations and loads contained in permitted discharges are in compliance with the limits and conditions specified in the permit.

compliance period (40CFR141.2) means a three-year calendar year period within a compliance cycle. Each compliance cycle has three three-year compliance periods. Within the first compliance cycle, the first compliance period runs from January 1, 1993 to December 31, 1995; the second from January 1, 1996 to December 31, 1998; the third from January 1, 1999 to December 31, 2001.

compliance plan (40CFR63.2) means a plan that contains all of the following:
(1) A description of the compliance status of the affected source with respect to all applicable requirements established under this part;
(2) A description as follows:
    (i) For applicable requirements for which the source is in compliance, a statement that the source will continue to comply with such requirements;
    (ii) For applicable requirements that the source is required to comply with by a future date, a statement that the source will meet such requirements on a timely basis;
    (iii) For applicable requirements for which the source is not in compliance, a narrative description of how the source will achieve compliance with such requirements on a timely basis;
(3) A compliance schedule, as defined in this section; and
(4) A schedule for the submission of certified progress reports no less frequently than every 6 months for affected sources required to have a schedule of compliance to remedy a violation.

compliance plan (40CFR72.2) for the purposes of the Acid Rain Program, means the document submitted for an affected source in accordance with subpart C of this part or subpart E of part 74 of this chapter, or part 76 of this chapter, specifying the method(s) including one or more Acid Rain compliance options as provided under subpart D of this part or subpart E of part 74 of this chapter, or part 76 of this chapter by which each affected unit at the source will meet the applicable Acid Rain emissions limitation and Acid Rain emissions reduction requirements.

compliance plan (CAA402-42USC7651a) means, for purposes of the requirements of this title, either (A) a statement that the source will comply with all applicable requirements under this title, or (B) where applicable, a schedule and description of the method or methods for compliance and certification by the owner or operator that the source is in compliance with the requirements of this title.

compliance schedule (40CFR51.100-p) means the date or dates by which a source or category of sources is required to comply with specific emission limitations contained in an implementation plan and with any increments of progress toward such compliance.

compliance schedule (40CFR60.21-g) means a legally enforceable schedule specifying a date or dates by which a source or category of sources must comply with specific emission standards contained in a plan or with any increments of progress to achieve such compliance.

compliance schedule (40CFR61.02) means the date or dates by which a source or category of sources is required to comply with the standards of this part and with any steps toward such compliance which are set forth in a waiver of compliance under 40CFR61.11.

compliance schedule (40CFR63.2) means:
(1) In the case of an affected source that is in compliance with all applicable requirements established under this part, a statement that the source will continue to comply with such requirements; or
(2) In the case of an affected source that is required to comply with applicable requirements by a future date, a statement that the source will meet such requirements on a timely basis and, if required by an applicable requirement, a detailed schedule of the dates by which each step toward compliance will be reached; or
(3) In the case of an affected source not in compliance with all applicable requirements established under this part, a schedule of remedial measures, including an enforceable sequence of actions or operations with milestones and a schedule for the submission of certified progress reports, where applicable, leading to compliance with a relevant standard,

limitation, prohibition, or any federally enforceable requirement established pursuant to section 112 of the Act for which the affected source is not in compliance. This compliance schedule shall resemble and be at least as stringent as that contained in any judicial consent decree or administrative order to which the source is subject. Any such schedule of compliance shall be supplemental to, and shall not sanction noncompliance with, the applicable requirements on which it is based.

compliance schedule (EPA-94/04): A negotiated agreement between a pollution source and a government agency that specifies dates and procedures by which a source will reduce emissions and, thereby, comply with a regulation.

compliance subaccount (40CFR72.2) means the subaccount in an affected unit's Allowance Tracking System account, established pursuant to 40CFR73.31 (a) or (b) of this chapter, in which are held, from the date that allowances for the current calendar year are recorded under 40CFR73.34(a) until December 31, allowances available for use by the unit in the current calendar year and, after December 31 until the date that deductions are made under 40CFR73.35(b), allowances available for use by the unit in the preceding calendar year, for the purpose of meeting the unit's Acid Rain emissions limitation for sulfur dioxide.

compliance use date (40CFR72.2) means the first calendar year for which an allowance may be used for purposes of meeting a unit's Acid Rain emissions limitation for sulfur dioxide.

complying with the Protocol (40CFR82.3-h), when referring to a foreign state not party to the 1987 Montreal Protocol, the London Amendments, or the Copenhagen Amendments, means that the non-Party has been determined as complying with the Protocol, as indicated in appendix C to this subpart, by a meeting of the parties as noted in the records of the directorate of the United Nations Secretariat.

component (40CFR260.10) means either the tank or ancillary equipment of a tank system.

component (40CFR270.2) means any constituent part of a unit or any group of constituent parts of a unit which are assembled to perform a specific function (e.g., a pump seal, pump, kiln liner, kiln thermocouple).

component (40CFR35.936.13-ii) means any article, material, or supply directly incorporated in construction material.

component (40CFR52.741) means, with respect to synthetic organic chemical and polymer manufacturing equipment, and petroleum refining and related industries, any piece of equipment which has the potential to leak VOM including, but not limited to, pump seals, compressor seals, seal oil degassing vents, pipeline valves, pressure relief devices, process drains, and open ended pipes. This definition excludes valves which are not externally regulated, flanges, and equipment in heavy liquid service. For purposes of paragraph (i) of this section, this definition also excludes bleed ports of gear pumps in polymer service.

component (40CFR60.541) means a piece of tread, combined tread/sidewall, or separate sidewall rubber, or other rubber strip that is combined into the sidewall of a finished tire.

composite NO_x standard (40CFR86.088.2), for a manufacturer which elects to average light-duty trucks subject to the NO_x standard of 40CFR86.088.9(a)(iii)(A) together with those subject to the NO_x standard of 40CFR86.088.9(a)(iii)(B) in the light-duty truck NO_x averaging program, means that standard calculated according to the equation (set forth in 40CFR86.088.2) and rounded to the nearest one-tenth gram per mile.

composite particulate standard (40CFR86.085.2) for a manufacturer which elects to average diesel light-duty vehicles and diesel light-duty trucks together in either the particulate averaging program, means that standard calculated according to the equation (set forth in 40CFR86.085.2 under this definition) and rounded to the nearest one-hundredth (0.01) of a gram per mile.

composite particulate standard (40CFR86.087.2) for a manufacturer which elects to average diesel light-duty vehicles and diesel light-duty trucks with a loaded vehicle weight equal to or less than 3,750 lbs (LDDT1s) together in the particulate averaging program, means that standard calculated according to the equation (set forth in 40CFR86.087.2 under this definition) and rounded to the nearest one-hundredth (0.01) gram per mile (LDDT means light duty diesel truck).

composite particulate standard (40CFR86.090.2) for a manufacturer which elects to average light-duty vehicles and light-duty trucks together in either the petroleum-fueled or methanol-fueled light-duty particulate averaging program, means that standards calculated using the following equation and rounded to the nearest one-hundredth (0.01) of a gram per mile:

- $(PROD_{LDV}) \times [(STD_{LDV}) + (PROD_{LDT})] \times (STD_{LDT})/[(PROD_{LDV}) + (PROD_{LDT})]$
  = Manufacturer composite particulate standard

Where:

- $PROD_{LDV}$ represents the manufacturer's total petroleum-fueled diesel or methanol-fueled diesel light-duty vehicle production for those engine families being included in the appropriate average for a given model year.
- STD_{LDV} represents the light-duty vehicle particulate standard.
- $PROD_{LDT}$ represents the manufacturer's total petroleum-fueled diesel or methanol-fueled diesel light-duty truck production for those engine families being included in the appropriate average for a given model year.
- STD_{LDT} represents the light-duty truck particulate standard.

composite sample (40CFR471.02-qq) is a sample composed of no less than eight grab samples taken over the compositing period.

composite sample (EPA-94/04): A series of water samples taken over a given period of time and weighted by flow rate.

compost (EPA-94/04): The relatively stable humus material that is produced from a composting process in which bacteria in soil mixed with garbage and degradable trash break down the mixture into organic fertilizer.

composting (EPA-94/04): The controlled biological decomposition of organic material in the presence of air to form a humus-like material. Controlled methods of composting include mechanical mixing and aerating, ventilating the materials by dropping them through a vertical series of aerated chambers, or placing the compost in piles out in the open air and mixing it or turning it periodically.

Comprehensive Environmental Response, Compensation, and Liability Act: See act or CERCLA.

compressed natural gas (CNG) (EPA-94/04): An alternative fuel for motor vehicles; considered one of cleanest because of low hydrocarbon emissions and its vapors are relatively non-ozone producing. However, it does emit a significant quantity of nitrogen oxides.

compressor (40CFR204.51-a): See portable air compressor.

compressor configuration (40CFR204.51-d) means the basic classification unit of a manufacturers product line and is comprised of compressor lines, models or series which are identical in all material respects with regard to the parameters listed in 40CFR204.55.3.

computer aided carpool matching (40CFR52.2492-2) means a carpool matching system in which the work of assembling lists of commuters with similar daily travel patterns is done by computer (under the Clean Air Act).

computer paper (40CFR250.4-g) means a type of paper used in manifold business forms produced in rolls and/or fan folded. It is used with computers and word processors to print out data, information, letters, advertising, etc. It is commonly called computer printout.

computer printout (40CFR250.4-g): See computer paper.

computer program (40CFR66.3-e) means the computer program used to calculate noncompliance penalties under section 120 of the Clean Air Act. This computer program appears as Appendix C to these regulations.

concentrated animal feeding operation (40CFR122.23-3) means an "animal feeding operation" which meets the criteria in Appendix B of this part, or which the Director designates under paragraph (c) of this section.

concentrated aquatic animal production facility (40CFR122.24) means a hatchery, fish farm, or other facility which meets the criteria in Appendix C of this part, or which the Director designates under paragraph (c) of this section.

concentration (40CFR798.4350-5) refers to an exposure level. Exposure is expressed as weight or volume of test substance per volume of air (mg/L), or as parts per million (ppm).

concentration of a solution (40CFR796.1840-iii) is the amount of solute in a given amount of solvent and can be expressed as a weight/weight or weight/volume relationship. The conversion from a weight relationship to one of volume incorporates density as a factor. For dilute aqueous solutions, the density of the solvent is approximately equal to the density of the solutions; thus, concentrations in mg/L are approximately equal to $10^{-3}g/10^3g$ or parts per million (ppm); ones in ug/L are approximately equal to $10^{-6}g/10^3g$ or parts per billion (ppb). In addition, concentration can be expressed in terms of molarity, normality, molality, and mole fraction. For example, to convert from weight/volume to molarity one incorporates molecular mass as a factor.

concentration of a solution (40CFR796.1860-i) is the amount of solute in a given amount of solvent or solution and can be expressed as a weight/weight or weight/volume relationship. The conversion from a weight relationship to one of volume incorporates density as a factor. For dilute aqueous solutions, the density of the solvent is approximately equal to the density of the solution; thus, concentrations in mg/L are approximately equal to $10^{-3}g/10^3g$ or parts per million (ppm); ones in ug/L are approximately equal to $10^{-6}g/10^3g$ or parts per billion (ppb). In addition, concentration can be expressed in terms of molarity, normality, molality, and mole fraction. For example, to convert from weight/volume to molarity one incorporates molecular mass as a factor.

concentration vs. time study (40CFR796.1840-ii)

results in a graph which plots the measured concentration of a given compound in a solution as a function of elapsed time. Usually, it provides a more reliable determination of equilibrium water solubility of hydrophobic compounds than can be obtained by single measurements of separate samples.

concrete curing compounds (40CFR52.741) means any coating applied to freshly poured concrete to retard the evaporation of water.

concurrent (40CFR60.711-7) means construction of a control device is commenced or completed within the period beginning 6 months prior to the date construction of affected coating mix preparation equipment commences and ending 2 years after the date construction of affected coating mix preparation equipment is completed.

concurrent (40CFR60.741) means the period of time in which construction of an emission control device serving an affected facility is commenced or completed, beginning 6 months prior to the date that construction of the affected facility commences and ending 2 years after the date that construction of the affected facility is completed.

condensate (40CFR52.741) means volatile organic liquid separated from its associated gases, which condenses due to changes in the temperature or pressure and remains liquid at standard conditions.

condensate (40CFR60.111-f) means hydrocarbon liquid separated from natural gas which condenses due to changes in the temperature and/or pressure and remains liquid at standard conditions.

condensate (40CFR60.111a-e) means hydrocarbon liquid separated from natural gas which condenses due to changes in the temperature or pressure, or both, and remains liquid at standard conditions.

condensate (40CFR60.111b-b) means hydrocarbon liquid separated from natural gas that condenses due to changes in the temperature or pressure, or both, and remains liquid at standard conditions (under CAA).

condensate stripper system (40CFR60.281-o) means

a column, and associated condensers, used to strip, with air or steam, TRS compounds from condensate streams from various processes within a kraft pulp mill.

condenser (40CFR264.1031) means a heat-transfer device that reduces a thermodynamic fluid from its vapor phase to its liquid phase.

condenser stack gases (40CFR61.51-d) mean the gaseous effluent evolved from the stack of processes utilizing heat to extract mercury metal from mercury ore.

conditional registration (EPA-94/04): Under special circumstances, the Federal Insecticide, Fungicide, and Rodenticide Act (FIFRA) permits registration of pesticide products that is "conditional" upon the submission of additional data. These special circumstances include a finding by the EPA Administrator that a new product or use of an existing pesticide will not significantly increase the risk of unreasonable adverse effects. A product containing a new (previously unregistered) active ingredient may be conditionally registered only if the Administrator finds that such conditional registration is in the public interest, that a reasonable time for conducting the additional studies has not elapsed, and the use of the pesticide for the period of conditional registration will not present an unreasonable risk.

conditionally exempt generators (CE) (EPA-94/04): Persons or enterprises which produce less than 220 pounds of hazardous waste per month. Exempt from most regulation, they are required merely to determine whether their waste is hazardous, notify appropriate state or local agencies, and ship it by permitted facility for proper disposal.

conditioned (40CFR248.4-m) means heated and/or mechanically cooled.

conditioning (40CFR797.1400-4) means the exposure of construction materials, test chambers, and testing apparatus to dilution water or to test solutions prior to the start of a test in order to minimize the sorption of the test substance onto the test facilities or the leaching of substances from the test facilities into the dilution water or test solution.

conditioning (40CFR797.1600-3) is the exposure of construction materials, test chambers, and testing apparatus to dilution water or to the test solution prior to the start of the test in order to minimize the sorption of test substance onto the test facilities or the leaching of substances from test facilities into the dilution water or the test solution.

conditioning period (40CFR60-AB-2.6) means a period of time (168 hours minimum) during which the CEMS is operated without any unscheduled maintenance, repair, or adjustment prior to initiation of the operational test period.

conductance (EPA-94/04): A rapid method of estimating the dissolved-solids content of a water supply by determining the capacity of a water sample to carry an electrical current.

conduction velocity (40CFR798.6850-2) is the speed at which the compound nerve action potential traverses a nerve.

conductive sensitizer (40CFR60.721) means a coating applied to a plastic substrate to render it conductive for purposes of electrostatic application of subsequent prime, color, texture, or touch-up coats (under CAA).

cone of depression (EPA-94/04): A depression in the water table that develops around a pumped well.

cone of influence (40CFR146.61) means that area around the well within which increased injection zone pressures caused by injection into the hazardous waste injection well would be sufficient to drive fluids into a underground source of drinking water.

cone of influence (EPA-94/04): The depression, roughly conical in shape, produced in the water table by the pumping of water from a well.

confidence limits (40CFR797.1350-2) are the limits within which, at some specified level of probability, the true value of a result lies (other identical or similar definitions are provided in 40CFR797.1440-3).

confidential (40CFR11.4-f): See security classification category.

confidential business information (40CFR154.3-d) means trade secrets or confidential commercial or financial information under FIFRA section 10(b) or 5USC552(b)(3) or (4).

confidential business information (40CFR155.23) means trade secrets or confidential commercial or financial information under FIFRA sec. 10(b) or 5 U.S.C. 552(b) (3) or (4).

confidential business information (40CFR350.1): See business confidentiality.

configuration (40CFR205.151-13) means the basic classification unit of a manufacturers product line and is comprised of all vehicle designs, models or series which are identical in all material aspects with respect to the parameters listed in 40CFR205.157.3 of this subpart.

configuration (40CFR205.51-9) means the basic classification unit of a manufacturers product line and is comprised of all vehicle designs, models or series which are identical in material aspects with respect to the parameters listed in 40CFR205.55.3.

configuration (40CFR610.11-13) means the mechanical arrangement, calibration and condition of a test automobile, with particular respect to carburetion, ignition timing, and emission control systems.

configuration (40CFR86.082.2) means a subclassification of an engine-system combination on the basis of engine code, inertia weight class, transmission type and gear ratios, final drive ratio, and other parameters which may be designated by the Administrator.

configuration (40CFR86.1002.2001-3) means a subclassification, if any, of a heavy-duty engine family for which a separate projected sales figure is listed in the manufacturer's Application for Certification and which can be described on the basis of emission control system, governed speed, injector size, engine calibration and other parameters which may be designated by the Administrator, or for light-duty trucks a subclassification of a light-duty truck engine family/emission control system combination on the basis of engine code, inertia weight class, transmission type and gear ratios, axle ratio, and other parameters which may be designated by the Administrator and/or a subclassification of a light-duty truck evaporative/refueling emission family/emission control system.

configuration (40CFR86.1002.84) means a subclassification, if any, of a heavy-duty engine family for which a separate projected sales figure is listed in the manufacturers Application for Certification and which can be described on the basis of emission control system, governed speed, injector size, engine calibration, and other parameters which may be designated by the Administrator, or a subclassification of a light-duty truck engine family/emission control system combination on the basis of engine code, inertia weight class, transmission type and gear ratios, axle ratio, and other parameters which may be designated by the Administrator.

configuration (40CFR86.1102.87) means a subdivision, if any, of a heavy-duty engine family for which a separate projected sales figure is listed in the manufacturers Application for Certification and which can be described on the basis of emission control system, governed speed, injector size, engine calibration, or other parameters which may be designated by the Administrator, or a subclassification of light-duty truck engine family emission control system combination on the basis of engine code, inertia weight class, transmission type and gear ratios, rear axle ratio, or other parameters which may be designated by the Administrator (under CAA).

configuration (40CFR86.602.84-3) means a subclassification of an engine-system combination on the basis of engine code, inertia weight class, transmission type and gear ratios, axle ratio, and other parameters which may be designated by the Administrator.

configuration (40CFR86.602.98-(b)(3)(i)), when used for LDV exhaust emissions testing, means a subclassification of an engine-system combination on the basis of engine code, inertia weight class, transmission type and gear ratios, axle ratio, and other parameters which may be designated by the Administrator.

configuration (40CFR86.602.98-(b)(3)(ii)), when used for LDV refueling emissions testing, means a subclassification of an evaporative/refueling emission family on the basis of evaporative and refueling control system and other parameters which may be designated by the Administrator.

configuration (40CFR89.502.96) means any subclassification of an engine family which can be described on the basis of gross power, emission control system, governed speed, injector size, engine calibration, and other parameters as designated by the Administrator.

confined aquifer (40CFR260.10) means an aquifer bounded above and below by impermeable beds or by beds of distinctly lower permeability than that of the aquifer itself; an aquifer containing confined ground water.

confined aquifer (EPA-94/04): An aquifer in which ground water is confined under pressure which is significantly greater than atmospheric pressure (under RCRA).

confining bed (40CFR146.3) means a body of impermeable or distinctly less permeable material stratigraphically adjacent to one or more aquifers (other identical or similar definitions are provided in 40CFR147.2902).

confining zone (40CFR146.3) means a geological formation, group of formations, or part of a formation that is capable of limiting fluid movement above an injection zone.

confining zone (40CFR147.2902) means a geologic formation, group of formations, or part of a formation that is capable of limiting fluid movement above an injection zone.

confluent growth (40CFR141.2) means a continuous bacterial growth covering the entire filtration area of a membrane filter, or a portion thereof, in which bacterial colonies are not discrete.

confluent growth (EPA-94/04): A continuous bacterial growth covering the entire filtration area of a membrane filter, or a portion thereof, in which bacterial colonies are not discrete.

congener (40CFR766.3) means any one particular member of a class of chemical substances. A specific congener is denoted by unique chemical structure, for example 2,3,7,8-tetrachlorodibenzofuran.

connected piping (40CFR280.12) means all underground piping including valves, elbows, joints, flanges, and flexible connectors attached to a tank system through which regulated substances flow. For the purpose of determining how much piping is connected to any individual UST system, the piping that joins two UST systems should be allocated equally between them.

connector (40CFR264.1031) means flanged, screwed, welded, or other joined fittings used to connect two pipelines or a pipeline and a piece of equipment. For the purposes of reporting and recordkeeping, connector means flanged fittings that are not covered by insulation or other materials that prevent location of the fittings.

connector (40CFR60.481) means flanged, screwed, welded, or other joined fittings used to connect two pipe lines or a pipe line and a piece of process equipment.

connector (40CFR61.241) means flanged, screwed, welded, or other joined fittings used to connect two pipe lines or a pipe line and a piece of equipment. For the purpose of reporting and recordkeeping, connector means flanged fittings that are not covered by insulation or other materials that prevent location of the fittings.

connector (40CFR63.161) means flanged, screwed, or other joined fittings used to connect two pipe lines or a pipe line and a piece of equipment. A common connector is a flange. Joined fittings welded completely around the circumference of the interface are not considered connectors for the purpose of this regulation. For the purpose of reporting and recordkeeping, connector means joined fittings that are not inaccessible, glass, or glass-lined as described in 40CFR63.174(h) of this subpart.

consecutive charges (40CFR63.301) means charges observed successively, excluding any charge during which the observer's view of the charging system or topside ports is obscured.

consent agreement (40CFR22.03) means any written document, signed by the parties, containing stipulations or conclusions of fact or law and a proposed penalty or proposed revocation or suspension acceptable to both complainant and respondent.

consent decree (EPA-94/04): A legal document, approved by a judge, that formalizes an agreement reached between EPA and potentially responsible parties (PRPs) through which PRPs will conduct all or part of a cleanup action at a Superfund site; cease or correct actions or processes that are polluting the environment; or otherwise comply with EPA initiated regulatory enforcement actions to resolve the contamination at the Superfund site involved. The consent decree describes the actions PRPs will take and may be subject to a public comment period.

conservation (EPA-94/04): Preserving and renewing, when possible, human and natural resources. The use, protection, and improvement of natural resources according to principles that will assure their highest economic or social benefits.

conservation and management (MMPA3-16USC1362) means the collection and application of biological information for the purposes of increasing and maintaining the number of animals within species and populations of marine mammals at their optimum sustainable population. Such terms include the entire scope of activities that constitute a modern scientific resource program, including but not limited to, research, census, law enforcement, and habitat acquisition and improvement. Also included within these terms, when and where appropriate, is the periodic or total protection of species or populations as well as regulated taking.

conservation verification protocol (40CFR72.2) means a methodology developed by the Administrator for calculating the kilowatt hour savings from energy conservation measures and improved unit efficiency measures for the purposes of title IV of the Act.

conserve, conserving, and conservation (ESA3-16USC1531) mean to use and the use of all methods and procedures which are necessary to bring any endangered species or threatened species to the point at which the measures provided pursuant to this chapter are no longer necessary. Such methods and procedures include, but are not limited to, all activities associated with scientific resources management such as research, census, law enforcement, habitat acquisition and maintenance, propagation, live trapping, and transplantation, and, in the extraordinary case where population pressures within a given ecosystem cannot be otherwise relieved, may include regulated taking.

consignee (40CFR262.51) means the ultimate treatment, storage or disposal facility in a receiving country to which the hazardous waste will be sent.

consistent removal (40CFR403.7(b)(1)) shall mean the average of the lowest 50 percent of the removal measured according to paragraph (b)(2) of this section. All sample data obtained for the measured pollutant during the time period prescribed in paragraph (b)(2) of this section must be reported and used in computing Consistent Removal. If a substance is measurable in the influent but not in the effluent, the effluent level may be assumed to be the limit of measurement, and those data may be used by POTW at its discretion and subject to approval by the Approval Authority. If the substance is not measurable in the influent, the date may not be used. Where the number of samples with concentrations equal to or above the limit of measurement is between 8 and 12, the average of the lowest 6 removals shall be used. If there are less than 8 samples with concentrations equal to or above the limit of measurement, the Approval Authority may approve alternate means for demonstrating consistent removal.

consolidate (40CFR29.12-2) means that a State may meet statutory and regulatory requirements by combining two or more plans into one document and that the State can select the format, submission date, and planning period for the consolidated plan.

consolidated assistance (40CFR30.200) means an assistance agreement awarded under more than one EPA program authority or funded together with one or more other Federal agencies. Applicants for consolidated assistance submit only one application.

consolidated PMN (40CFR700.43): See consolidated

premanufacture notice (PMN) or consolidated PMN (under TSCA).

consolidated premanufacture notice or consolidated PMN (40CFR700.43) means any PMN submitted to EPA that covers more than ore chemical substance (each being assigned a separate PMN number by EPA) as a result of a prenotice agreement with EPA (see FR 21734).

consortium (40CFR790.3) means an association of manufacturers and/or Processors who have made an agreement to jointly sponsor testing.

constant controls, control technology, and continuous emission reduction technology (40CFR57.103-j) mean systems which limit the quantity, rate, or concentration, excluding the use of dilution, and emissions of air pollutants on a continuous basis.

construction (CAA402-42USC7651a) means fabrication, erection, or installation of an affected unit (other identical or similar definitions are provided in 40CFR21.2-k; 33.005; 35.905; 35.2005-13; 35.6015-11; 40.115-1; 52.01-c; 51.165-xviii; 51.166-8; 51-AS-14; 52.21-8; 52.24-14; 52.2486-3; 60.2; 61.02; 63.321; 63.2; 63.101; 63.191; 72.2; 124.41; 125.91; 129.2-l; 249.04-b; CAA169-42USC7479; CWA212-33USC1292; CWA306-33USC1316; RCRA1004-42USC6903).

construction and demolition waste (40CFR243.101-i) means the waste building materials, packaging, and rubble resulting from construction, remodeling, repair, and demolition operations on pavements, houses, commercial buildings, and other structures (other identical or similar definitions are provided in 40CFR246.101-h).

construction and demolition waste (EPA-94/04): Waste building materials, dredging materials, tree stumps, and rubble resulting from construction, remodeling, repair, and demolition of homes, commercial buildings and other structures and pavements. May contain lead, asbestos, or other hazardous substances.

construction ban (EPA-94/04): If, under the Clean Air Act, EPA disapproves an area's planning requirements for correcting nonattainment, the Agency can ban the construction or modification of any major stationery source of the pollutant for which the area is in non-attainment.

construction work (40CFR8.2-f) means the construction, rehabilitation, alteration, conversion, extension, demolition or repair of buildings, highways, or other changes or improvements to real property, including facilities providing utility services. The term also includes the supervision, inspection, and other on-site functions incidental to the actual construction (cf. modificaiton).

consultation with the Regional Administrator (40CFR124.62(a)(2)) (40CFR124.2) means review by the Regional Administrator following evaluation by a panel of the technical merits of all 301(k) applications approved by the Director. The panel (to be appointed by the Director of the Office of Water Enforcement and Permits) will consist of Headquarters, Regional, and State personnel familiar with the industrial category in question.

consumer (40CFR244.101-c) means any person who purchases a beverage in a beverage container for final use or consumption.

consumer (40CFR721.3) means a private individual who uses a chemical substance or any product containing the chemical substance in or around a permanent or temporary household or residence, during recreation, or for any personal use or enjoyment.

consumer (40CFR82.104-d) means a commercial or non-commercial purchaser of a product or container that has been introduced into interstate commerce (under CAA).

consumer or commercial product (CAA183.e-42USC7511b) means any substance, product (including paints, coatings, and solvents), or article (including any container or packaging) held by any person, the use, consumption, storage, disposal, destruction, or decomposition of which may result in the release of volatile organic compounds. The term does not include fuels or fuel additives regulated under section 211, or motor vehicles, non-road vehicles, and non-road engines as defined under section 216.

consumer price index or CPI (40CFR72.2) means, for purposes of the Acid Rain Program, the U.S. Department of Labor, Bureau of Labor Statistics unadjusted Consumer Price Index for All Urban Consumers for the U.S. city average, for All Items on the latest reference base, or if such index is no longer published, such other index as the Administrator in his or her discretion determines meets the requirements of the Clean Air Act Amendments of 1990:

(1) CPI (1990) means the CPI for all urban consumers for the month of August 1989. The "CPI (1990)" is 124.6 (with 1982-1984=100). Beginning in the month for which a new reference base is established, "CPI (1990)" will be the CPI value for August 1989 on the new reference base.

(2) CPI (year) means the CPI for all urban consumers for the month of August of the previous year.

consumer product (40CFR302.3) shall have the meaning stated in 15USC2052.

consumer product (40CFR721.3) means a chemical substance that is directly, or as part of a mixture, sold or made available to consumers for their use in or around a permanent or temporary household or residence, in or around a school, or in recreation.

consumption (40CFR82.3-i) means the production plus imports minus exports of a controlled substance (other than transhipments, or used controlled substances).

consumption allowances (40CFR82.3-j) means the privileges granted by this subpart to produce and import class I controlled substances; however, consumption allowances may be used to produce class I controlled substances only in conjunction with production allowances. A person's consumption allowances are the total of the allowances obtained under 40CFR82.6 and 40CFR82.7 and 82.10, as may be modified under 40CFR 82.12 (transfer of allowances).

consumptive use (40CFR280.12) with respect to heating oil means consumed on the premises.

consumptive use (EPA-94/04): Water removed from available supplies without return to a water resource system (uses such as manufacturing, agriculture, and food preparation).

contact cooling and heating water (40CFR463.2-b) is process water that contacts the raw materials or plastic product for the purpose of heat transfer during the plastics molding and forming process.

contact cooling water (40CFR467.02-c) is any wastewater which contacts the aluminum workpiece or the raw materials used in forming aluminum.

contact cooling water (40CFR471.02-h) is any water which contacts the metal workpiece or the raw materials used in forming metals for the purpose of removing heat from the metal.

contact material (40CFR60.101-p) means any substance formulated to remove metals, sulfur, nitrogen, or any other contaminant from petroleum derivatives.

contact pesticide (EPA-94/04): A chemical that kills pests when it touches them, instead of by ingestion. Also, soil that contains the minute skeletons of certain algae that scratch and dehydrate waxy-coated insects.

contact resistance (40CFR85.2122(a)(5)(ii)(D)) means the opposition to the flow of current between the mounting bracket and the insulated terminal.

container (40CFR165.1-f) means any package, can, bottle, bag, barrel, drum, tank, or other containing-device (excluding sprag applicator tanks) used to enclose a pesticide or pesticide-related waste.

container (40CFR259.10.b) means any portable device in which a regulated medical waste is stored, transported, disposed or otherwise handled. The term container as used in this part does not include items in the Table or Regulated Medical Waste at 40CFR259.30(a) of this part.

container (40CFR260.10) means any portable device in which a material is stored, transported, treated, disposed of, or otherwise handled.

container (40CFR266.111) means any portable

device in which hazardous waste is transported, stored, treated, or otherwise handled, and includes transport vehicles that are containers themselves (i.e., tank trucks, tanker-trailers, and rail tank cars), and containers placed on or in a transport vehicle.

container (40CFR61.341) means any portable waste management unit in which a material is stored, transported, treated, or otherwise handled. Examples of containers are drums, barrels, tank trucks, barges, dumpsters, tank cars, dump trucks, and ships.

container (40CFR63.111), as used in the wastewater provisions, means any portable waste management unit that has a capacity greater than or equal to 0.1 m^3 in which a material is stored, transported, treated, or otherwise handled. Examples of containers are drums, barrels, tank trucks, barges, dumpsters, tank cars, dump trucks, and ships.

container (40CFR749.68-5) means bag, barrel, bottle, box, can, cylinder, drum, or the like that.

container (40CFR82.104-e) means the immediate vessel in which a controlled substance is stored or transported.

container containing (40CFR82.104-e) means a container that physically holds a controlled substance within its structure that is intended to be transferred to another container, vessel or piece of equipment in order to realize its intended use.

container glass (40CFR60.291) means glass made of soda-lime recipe, clear or colored, which is pressed and/or blown into bottles, jars, ampoules, and other products listed in Standard Industrial Classification 3221 (SIC 3221).

containment building (40CFR260.10) means a hazardous waste management unit that is used to store or treat hazardous waste under the provisions of subpart DD of part 264 or 265 of this chapter (under RCRA).

contaminant (EPA-94/04): Any physical, chemical, biological, or radiological substance or matter that has an adverse affect on air, water, or soil.

contaminant (SDWA1401-42USC300f) means any physical, chemical, biological, or radiological substance or matter in water (other identical or similar definitions are provided in 40CFR2.304-2; 141.2; 142.2; 143.2-b; 144.3; 146.3; 147.2902; 149.101-b; 230.3-e; 300.5; 310.11-h).

contaminate (40CFR257.3.4-2) means to introduce a substance that would cause:
(i) The concentration of that substance in the ground water to exceed the maximum contaminant level specified in Appendix I, or
(ii) An increase in the concentration of that substance in the ground water where the existing concentration of that substance exceeds the maximum contaminant level specified in Appendix I.

contaminate an aquifer (40CFR503.21-c) means to introduce a substance that causes the maximum contaminant level for nitrate in 40CFR141.11 to be exceeded in ground water or that causes the existing concentration of nitrate in ground water to increase when the existing concentration of nitrate in the ground water exceeds the maximum contaminant level for nitrate in 40CFR141.11.

contaminated nonprocess waste water (40CFR422.51-c) shall mean any water including precipitation runoff, which during manufacturing or processing, comes into incidental contact with any raw material, intermediate product, finished product, byproduct or waste product by means of:
(1) Precipitation runoff,
(2) accidental spills,
(3) accidental leaks caused by the failure of process equipment and which are repaired or the discharge of pollutants therefrom contained or terminated within the shortest reasonable time which shall not exceed 24 hours after discovery or when discovery should reasonably have been made, whichever is earliest, and
(4) discharges from safety showers and related personal safety equipment, and from equipment washings for the purpose of safe entry, inspection and maintenance; have been taken to prevent, reduce, eliminate and control to the maximum extent feasible such contact and provided further that all reasonable measures have been taken that will mitigate the effects of such contact once it has occurred.

contaminated nonprocess wastewater (40CFR415.91-f) shall mean any water which, during manufacturing or processing, comes into incidental contact with any raw material, intermediate product, finished product, byproduct or waste product by means of:
(1) rainfall runoff;
(2) accidental spills;
(3) accidental leaks caused by the failure of process equipment, which are repaired within the shortest reasonable time not to exceed 24 hours after discovery; and
(4) discharges from safety showers and related personal safety equipment: Provided, that all reasonable measures have been taken:
  (i) to prevent, reduce and control such contact to the maximum extent feasible; and
  (ii) to mitigate the effects of such contact once it has occurred.
(Other identical or similar definitions are provided in 40CFR415.241-e; 415.311-d; 415.331-e; 415.381-d; 415.401-d; 415.411-d; 415.431-d; 415.441-d; 415.531-e; 415.551-d; 415.601-d; 415.631-d.)

contaminated nonprocess wastewater (40CFR418.11-c) shall mean any water including precipitation runoff which, during manufacturing or processing, comes into incidental contact with any raw material, intermediate product, finished product, byproduct or waste product by means of:
(1) Precipitation runoff;
(2) accidental spills;
(3) accidental leaks caused by the failure of process equipment and which are repaired or the discharge of pollutants therefrom contained or terminated within the shortest reasonable time which shall not exceed 24 hours after discovery or when discovery should reasonably have been made, whichever is earliest; and
(4) discharges from safety showers and related personal safety equipment, and from equipment washings for the purpose of safe entry, inspection and maintenance; provided that all reasonable measures have been taken to prevent, reduce, eliminate and control to the maximum extent feasible such contact and provided further that all reasonable measures have been taken that will mitigate the effects of such contact once it has occurred.

contaminated nonprocess wastewater (40CFR422.41-

c) shall mean any water including precipitation runoff, which during manufacturing or processing comes into incidental contact with any raw material, intermediate product finished product, byproduct or waste product by means of:
(1) Precipitation runoff,
(2) accidental spills,
(3) accidental leaks caused by the failure of process equipment and which are repaired or the discharge of pollutants therefrom contained or terminated within the shortest reasonable time which shall not exceed 24 hours after discovery or when discovery should reasonably have been made, whichever is earliest, and
(4) discharges from safety showers and related personal safety equipment, and from equipment washings for the purpose of entry, inspection and maintenance. Provided that all reasonable measures to reduce, eliminate and control to the maximum extent feasible such contact and provided further that all reasonable measures have been taken that will mitigate the effects of such contact once it has occurred.

contaminated runoff (40CFR419.11-g) shall mean runoff which comes into contact with any raw material, intermediate product, finished product, byproduct or waste product located on petroleum refinery property.

contamination (EPA-94/04): Introduction into water, air and soil of microorganisms, chemicals, toxic substances, wastes, or wastewater in a concentration that makes the medium unfit for its next intended use. Also applies to surfaces of objects and buildings, and various household and agricultural use products.

contiguous zone (40CFR401.11-m) shall be defined in accordance with section 502 of the Act unless the context otherwise requires.

contiguous zone (CWA311-33USC1321 means the entire zone established or to be established by the United States under article 24 of the Convention on the Territorial Sea and the Contiguous Zone (other identical or similar definitions are provided in 40CFR110.1; 122.2; 116.3; 117.1; 122.2; 300.5; CWA502-33USC1362).

contingency plan (40CFR260.10) means a document

setting out an organized, planned, and coordinated course of action to be followed in case of a fire, explosion, or release of hazardous waste or hazardous waste constituents which could threaten human health or the environment.

contingency plan (EPA-94/04): A document setting out an organized, planned, and coordinated course of action to be followed in case of a fire, explosion, or other accident that releases toxic chemicals, hazardous waste, or radioactive materials that threaten human health or the environment (cf. national oil and hazardous substances contingency plan).

continuation award (40CFR30.200) means an assistance agreement after the initial award, for a project which has more than one budget period in its approved project period, or annual awards, after the first award, to State, Interstate, or local agencies for continuing environmental programs (cf. section 30.306).

continuation award (40CFR35.105) means any assistance award after the first award to a State, interstate, or local agency for a continuing environmental program.

continuing environmental programs (40CFR35.105) means those pollution control programs which will not be completed within a definable time period.

continuous (40CFR420.111-b) means those alkaline cleaning operations which process steel products other than in discrete batches or bundles (under CWA).

continuous (40CFR420.81-f) means those descaling operations that remove surface scale from the sheet or wire products in continuous processes.

continuous (40CFR420.91-f) means those pickling operations which process steel products other than in discrete batches or bundles.

continuous casting (40CFR467.02-d) is the production of sheet, rod, or other long shapes by solidifying the metal while it is being poured through an open-ended mold using little or no contact cooling water. Continuous casting of rod and sheet generates

spent lubricants and rod casting also generates contact cooling water.

continuous casting (40CFR471.02-i) is the production of sheet, rod, or other long shapes by solidifying the metal while it is being poured through an open-ended mold.

continuous discharge (40CFR122.2) means a "discharge" which occurs without interruption throughout the operating hours of the facility, except for infrequent shutdowns for maintenance, process changes, or other similar activities.

continuous discharge (EPA-94/04): A routine release to the environment that occurs without interruption, except for infrequent shutdowns for maintenance, process changes, etc.

continuous disposal (40CFR61.251-b) means a method of tailings management and disposal in which tailings are dewatered by mechanical methods immediately after generation. The dried tailings are then placed in trenches or other disposal areas and immediately covered to limit emissions consistent with applicable Federal standards.

continuous emission monitoring system (40CFR60-AF-2.1) means the total equipment required for the determination of a gas concentration or emission rate.

continuous emission monitoring system (CEMS) (40CFR60-AB-2.1) means the total equipment required for the determination of opacity. The system consists of the following major subsystems:

(1) Sample interface: That portion of CEMS that protects the analyzer from the effects of the stack effluent and aids in keeping the optical surfaces clean.

(2) Analyzer: That portion of the CEMS that senses the pollutant and generates an output that is a function of the opacity.

(3) Data recorder: That portion of the CEMS that provides a permanent record of the analyzer output in terms of opacity. The data recorder may include automatic data-reduction capabilities.

continuous emission monitoring system (CEMS)

(40CFR63.2) means the total equipment that may be required to meet the data acquisition and availability requirements of this part, used to sample, condition (if applicable), analyze, and provide a record of emissions (e.g., CO, CO_2, O_2, NO_2, and SO_2 monitors) (other identical or similar definitions are provided in 40CFR63.101).

continuous emission monitoring system (CEMS) (CAA402-42USC7651a) means the equipment as required by section 412, used to sample, analyze, measure, and provide on a continuous basis a permanent record of emissions and flow (expressed in pounds per million British thermal units (lbs/mmBtu), pounds per hour (lbs/hr) or such other form as the Administrator may prescribe by regulations under section 412).

continuous emission monitoring system or CEMS (40CFR60.51a) means a monitoring system for continuously measuring the emissions of a pollutant from an affected facility.

continuous emission monitoring system or CEMS (40CFR72.2) means the equipment required by part 75 of this chapter used to sample, analyze, measure, and provide, by readings taken at least once every 15 minutes, a permanent record of emissions, expressed in pounds per hour (lb/hr) for sulfur dioxide and in pounds per million British thermal units (lb/mmBtu) for nitrogen oxides. The following systems are component parts included in a continuous emission monitoring system:
(1) Sulfur dioxide pollutant concentration monitor;
(2) Flow monitor;
(3) Nitrogen oxides pollutant concentration monitors;
(4) Diluent gas monitor (oxygen or carbon dioxide);
(5) A continuous moisture monitor when such monitoring is required by part 75 of this chapter; and
(6) A data acquisition and handling system.

continuous emission reduction technology (40CFR57.103-j): See constant controls.

continuous emissions (40CFR60.561) means any gas stream containing VOC that is generated essentially continuously when the process line or any piece of equipment in the process line is operating.

continuous monitoring system (40CFR60.2) means the total equipment, required under the emission monitoring sections in applicable subparts, used to sample and condition (if applicable), to analyze, and to provide a permanent record of emissions or process parameters.

continuous monitoring system (CMS) (40CFR63.2) is a comprehensive term that may include, but is not limited to, continuous emission monitoring systems, continuous opacity monitoring systems, continuous parameter monitoring systems, or other manual or automatic monitoring that is used for demonstrating compliance with an applicable regulation on a continuous basis as defined by the regulation (under the Clean Air Act).

continuous opacity monitoring system (COMS) (40CFR63.2) means a continuous monitoring system that measures the opacity of emissions (under the Clean Air Act).

continuous opacity monitoring system or COMS (40CFR72.2) means the equipment required by part 75 of this chapter to sample, measure, analyze, and provide, with readings taken at least once every 6 minutes, a permanent record of opacity or transmittance. The following systems are component parts included in a continuous opacity monitoring system:
(1) Opacity monitor; and
(2) A data acquisition and handling system.

continuous operations (40CFR471.02-tt) means that the industrial user introduces regulated wastewaters to the POTW throughout the operating hours of the facility, except for infrequent shutdowns for maintenance, process changes, or other similar activities.

continuous parameter monitoring system (40CFR63.2) means the total equipment that may be required to meet the data acquisition and availability requirements of this part, used to sample, condition (if applicable), analyze, and provide a record of process or control system parameters (other identical or similar definitions are provided in 40CFR63.101).

continuous process (40CFR52.741) means, with respect to polystyrene resin, a method of manufacture in which the styrene raw material is delivered on a continuous basis to the reactor in which the styrene is polymerized to polystyrene.

continuous process (40CFR60.561) means a polymerization process in which reactants are introduced in a continuous manner and products are removed either continuously or intermittently at regular intervals so that the process can be operated and polymers produced essentially continuously.

continuous record (40CFR63.111) means documentation, either in hard copy or computer readable form, of data values measured at least once every 15 minutes and recorded at the frequency specified in 40CFR63.152(f) of this subpart.

continuous recorder (40CFR60.611) means a data recording device recording an instantaneous data value at least once every 15 minutes (other identical or similar definitions are provided in 40CFR60.661; 60.701; 264.1031).

continuous recorder (40CFR63.111) means a data recording device that either records an instantaneous data value at least once every 15 minutes or records 15-minute or more frequent block average values.

continuous release (40CFR302.8) is a release that occurs without interruption or abatement or that is routine, anticipated, and intermittent and incidental to normal operations or treatment processes.

continuous sample (EPA-94/04): A flow of water from a particular place in a plant to the location where samples are collected for testing; may be used to obtain grab or composite samples.

continuous seal (40CFR63.111) means a seal that forms a continuous closure that completely covers the space between the wall of the storage vessel and the edge of the floating roof. A continuous seal may be a vapor-mounted, liquid-mounted, or metallic shoe seal.

continuous vapor processing system (40CFR60.501) means a vapor processing system that treats total organic compound vapors collected from gasoline tank trucks on a demand basis without intermediate accumulation in a vapor holder.

continuous vapor processing system (40CFR63.111) means a vapor processing system that treats total organic compound vapors collected from tank trucks or railcars on a demand basis without intermediate accumulation in a vapor holder.

contour plowing (EPA-94/04): Soil tilling method that follows the shape of the land to discourage erosion.

contour strip farming (EPA-94/04): A kind of contour farming in which row crops are planted in strips, between alternating strips of close-growing, erosion resistant forage crops.

contract (40CFR15.4) means any contract or other agreement made with an Executive Branch agency for the procurement of goods, materials, or services (including construction), and includes any subcontract made thereunder.

contract (40CFR31.3) means (except as used in the definitions for grant and subgrant in this section and except where qualified by Federal) a procurement contract under a grant or subgrant, and means a procurement subcontract under a contract.

contract (40CFR35.4010) means a written agreement between the recipient and another party (other than a public agency) for services or supplies necessary to complete the TAG project. Contracts include contracts and subcontracts for personal and professional services or supplies necessary to complete the TAG (technical assistance grant) project, and agreements with consultants, and purchase orders.

contract (40CFR35.6015-12) means a written agreement between an EPA recipient and another party (other than another public agency) or between the recipient's contractor and the contractor's first tier subcontractor.

contract (40CFR8.2-g) means any Government contract or any federally assisted construction contract.

contract carrier by motor vehicle (40CFR202.10) means any person who engages in transportation by motor vehicle of passengers or property in interstate or foreign commerce for compensation (other than transportation referred to in Paragraph (b) of this section) under continuing contracts with one person or a limited number of persons either:

(1) for the furnishing of transportation services through the assignment of motor vehicles for a continuing period of time to the exclusive use of each person served or

(2) or the furnishing of transportation services designed to meet the distinct need of each individual customer.

contract laboratory program (CLP) (40CFR300-AA) means analytical program developed for CERCLA waste site samples to fill the need for legally defensible analytical results supported by a high level of quality assurance and documentation.

contract labs (EPA-94/04): Laboratories under contract to EPA, which analyze samples taken from waste, soil, air, and water or carry out research projects.

contract or other approved (40CFR112.2) means:

(1) A written contractual agreement with an oil spill removal organization(s) that identifies and ensures the availability of the necessary personnel and equipment within appropriate response times; and/or

(2) A written certification by the owner or operator that the necessary personnel and equipment resources, owned or operated by the facility owner or operator, are available to respond to a discharge within appropriate response times; and/or

(3) Active membership in a local or regional oil spill removal organization(s) that has identified and ensures adequate access through such membership to necessary personnel and equipment to respond to a discharge within appropriate response times in the specified geographic areas; and/or

(4) Other specific arrangements approved by the Regional Administrator upon request of the owner or operator.

contract required detection limit (CRDL) (40CFR300-AA) means the term equivalent to contract-required quantitation limit, but used primarily for inorganic substances.

contract required quantitation limit (CRQL) (40CFR300-AA) means the substance-specific level that a CLP laboratory must be able to routinely and reliably detect in specific sample matrices. It is not the lowest detectable level achievable, but rather the level that a CLP laboratory should reasonably quantify. The CRQL may or may not be equal to the quantitation limit of a given substance in a given sample. For HRS purposes, the term CRQL refers to both the contract-required quantitation limit and the contract-required detection limit.

contract specifications (40CFR249.04-c) means the set of specifications prepared for an individual construction project, which contains design, performance, and material requirements for that project.

contractor (40CFR15.4) means any person with whom an Executive Branch agency has entered into, extended, or renewed a contract as defined above, and includes subcontractors or any person holding a subcontract.

contractor (40CFR30.200) means any party to whom a recipient awards a subagreement (other identical or similar definitions are provided in 40CFR33.005).

contractor (40CFR35.4010) means any party (i.e., technical advisor) to whom a recipient awards a subagreement.

contractor (40CFR35.6015-13) means any party to whom a recipient awards a contract.

contractor (40CFR35.936.1-c) means a party to whom a subagreement is awarded.

contractor (40CFR8.2-h) means, unless otherwise indicated, a prime contractor or subcontractor.

contractual relationship (SF101-42USC9601):

(A) For the purpose of section 107(b)(3), includes, but is not limited to, land contracts, deeds or other instruments transferring title or possession, unless the real property on which the facility concerned is located was acquired by the defendant after the disposal or placement of the hazardous substance on, in, or at the facility, and one or more of the circumstances described in clause (i), (ii), or (iii) is also established by the defendant by a preponderance of the evidence:

(i) At the time the defendant acquired the facility the defendant did not know and had no reason to know that any hazardous substance which is the subject of the release or threatened release was disposed of on, in, or at the facility.

(ii) The defendant is a government entity which acquired the facility by escheat, or through any other involuntary transfer or acquisition, or through the exercise of eminent domain authority by purchase or condemnation.

(iii) The defendant acquired the facility by inheritance or bequest. In addition to establishing the foregoing, the defendant must establish that he has satisfied the requirements of section 107(b)(3)(a) and (b).

(B) To establish that the defendant had no reason to know, as provided in clause (i) of subparagraph (A) of this paragraph, the defendant must have undertaken, at the time of acquisition, all appropriate inquiry into the previous ownership and uses of the property consistent with good commercial or customary practice in an effort to minimize liability. For purposes of the preceding sentence the court shall take into account any specialized knowledge or experience on the part of the defendant, the relationship of the purchase price to the value of the property if uncontaminated, commonly known or reasonably ascertainable information about the property, the obviousness of the presence or likely presence of contamination at the property, and the ability to detect such contamination by appropriate inspection.

(C) Nothing in this paragraph or in section 107(b)(3) shall diminish the liability of any previous owner or operator of such facility who would otherwise be liable under this Act. Notwithstanding this paragraph, if the defendant obtained actual knowledge of the release or threatened release of a hazardous substance at such facility when the defendant owned the real property and then subsequently transferred ownership of the property to another person without disclosing such knowledge, such

defendant shall be treated as liable under section 107(a)(1) and no defense under section 107(b)(3) shall be available to such defendant.

(D) Nothing in this paragraph shall affect the liability under this Act of a defendant who, by any act or omission, caused or contributed to the release or threatened release of a hazardous substance which is the subject of the action relating to the facility.

contribute (40CFR52.138-4) means resulting in measurably higher average 8-hour ambient CO concentrations over the NAAQS or an increased number of violations of the NAAQS in an area which currently experiences CO levels above the standard.

control (40CFR192.01-c) means any remedial action intended to stabilize, inhibit future misuse of, or reduce emissions of effluents from residual radioactive materials.

control (40CFR192.31-c) means any action to stabilize, inhibit future misuse of, or reduce emissions or effluents from uranium byproduct materials.

control (40CFR66.3-f) (including the terms "controlling," "controlled by," and "under common control with") means the power to direct or cause the direction of the management and policies of a person or organization, whether by the ownership of stock, voting rights, by contract, or otherwise.

control (40CFR704.3; 720.3-y): See possession.

control (40CFR797.1600-4) is an exposure of test organisms to dilution water only or dilution water containing the test solvent or carrier (no toxic agent is intentionally or inadvertently added).

control (40CFR88.302.94) means:

(1) When it is used to join all entities under common management, means any one or a combination of the following:

(i) A third person or firm has equity ownership of 51 percent or more in each of two or more firms;

(ii) Two or more firms have common corporate officers, in whole or in substantial part, who are responsible for the day-to-day operation of the companies.

   (iii) One firm leases, operates, supervises, or in 51 percent or greater part owns equipment and/or facilities used by another person or firm, or has equity ownership of 51 percent or more of another firm.

(2) When it is used to refer to the management of vehicles, means a person has the authority to decide who can operate a particular vehicle, and the purposes for which the vehicle can be operated.

(3) When it is used to refer to the management of people, means a person has the authority to direct the activities of another person or employee in a precise situation, such as at the workplace. Conversion configuration means any combination of vehicle/engine conversion hardware and a base vehicle of a specific engine family.

control area (40CFR80.2-pp) means a geographic area in which only oxygenated gasoline under the oxygenated gasoline program may be sold or dispensed, with boundaries determined by section 211(m) of the Act.

control authority (40CFR403.12-a) as it is used in this section refers to:

(1) The POTW if the POTW's Submission for its pretreatment program (40CFR403.3(t)(1)) has been approved in accordance with the requirements of 40CFR403.11; or

(2) the Approval Authority if the Submission has not been approved.

control authority (40CFR413.02-g) is defined as the POTW if it has an approved pretreatment program; in the absence of such a program, the NPDES State if it has an approved pretreatment program or EPA if the State does not have an approved program (other identical or similar definitions are provided in 40CFR466.02-g; 471.02-ss).

control device (40CFR264.1031) means an enclosed combustion device, vapor recovery system, or flare. Any device the primary function of which is the recovery of capture of solvents or other organics for use, reuse, or sale (e.g., a primary condenser on a solvent recovery unit) is not a control device (under RCRA).

control device (40CFR52.741) means equipment (such as an afterburner or adsorber) used to remove or prevent the emission of air pollutants from a contaminated exhaust stream.

control device (40CFR60.261-l) means the air pollution control equipment used to remove particulate matter generated by an electric submerged arc furnace from an effluent gas stream.

control device (40CFR60.271-c) means the air pollution control equipment used to remove particulate matter generated by an EAF(s) from the effluent gas stream.

control device (40CFR60.271a) means the air pollution control equipment used to remove particulate matter from the effluent gas stream generated by an electric arc furnace or AOD vessel.

control device (40CFR60.381) means the air pollution control equipment used to reduce particulate matter emissions released to the atmosphere from one or more affected facilities at a metallic mineral processing plant.

control device (40CFR60.481) means an enclosed combustion device, vapor recovery system, or flare (other identical or similar definitions are provided in 40CFR60.561).

control device (40CFR60.671) means the air pollution control equipment used to reduce particulate matter emissions released to the atmosphere from one or more process operations at a nonmetallic mineral processing plant.

control device (40CFR60.691) means an enclosed combustion device, vapor recovery system or flare (under CAA).

control device (40CFR60.711-8) means any apparatus that reduces the quantity of a pollutant emitted to the air (other identical or similar definitions are provided in 40CFR60.741).

control device (40CFR61.171) means the air pollution control equipment used to collect particulate matter emissions (other identical or similar definitions are provided in 40CFR61.181).

control device (40CFR61.241) means an enclosed combustion device, vapor recovery system, or flare.

control device (40CFR61.301) means all equipment used for recovering or oxidizing benzene vapors displaced from the affected facility.

control device (40CFR61.341) means an enclosed combustion device, vapor recovery system, or flare.

control device (40CFR63.101) means any equipment used for recovering or oxidizing organic hazardous air pollutant vapors. Such equipment includes, but is not limited to, absorbers, carbon adsorbers, condensers, incinerators, flares, boilers, and process heaters. For process vents (as defined in this section), recovery devices are not considered control devices.

control device (40CFR63.111) means any equipment used for recovering or oxidizing organic hazardous air pollutant vapors. Such equipment includes, but is not limited to, absorbers, carbon adsorbers, condensers, incinerators, flares, boilers, and process heaters. For process vents, recovery devices are not considered control devices and for a steam stripper, a primary condenser is not considered a control device.

control device (40CFR63.161) means any equipment used for recovering or oxidizing organic hazardous air pollutant vapors. Such equipment includes, but is not limited to, absorbers, carbon adsorbers, condensers, flares, boilers, and process heaters (under CAA).

control device efficiency (40CFR52.741) means the ratio of pollution prevented by a control device and the pollution introduced to the control device, expressed as a percentage.

control device shutdown (40CFR264.1031) means the cessation of operation of a control device for any purpose.

control efficiency (40CFR503.41-c) is the mass of a pollutant in the sewage sludge fed to an incinerator minus the mass of that pollutant in the exit gas from the incinerator stack divided by the mass of the pollutant in the sewage sludge fed to the incinerator (cf. destruction and removal efficiency).

control period (40CFR80.2-qq) means the period during which oxygenated gasoline must be sold or dispensed in any control area, pursuant to section 211(m)(2) of the Act.

control period (40CFR82.3-k) means the period from January 1, 1992 through December 31, 1992, and each twelve-month period from January 1 through December 31, thereafter.

control strategy (40CFR51.100-n) means a combination of measures designated to achieve the aggregate reduction of emissions necessary for attainment and maintenance of national standards including, but not limited to, measures such as:
(1) Emission limitations.
(2) Federal or State emission charges or taxes or other economic incentives or disincentives.
(3) Closing or relocation of residential, commercial, or industrial facilities.
(4) Changes in schedules or methods of operation of commercial or industrial facilities or transportation systems, including, but not limited to, short-term changes made in accordance with standby plans.
(5) Periodic inspection and testing of motor vehicle emission control systems, at such time as the Administrator determines that such programs are feasible and practicable.
(6) Emission control measures applicable to in-use motor vehicles, including, but not limited to, measures such as mandatory maintenance, installation of emission control devices, and conversion to gaseous fuels.
(7) Any transportation control measure including those transportation measures listed in section 108(f) of the Clean Air Act as amended.
(8) Any variation of, or alternative to any measure delineated herein.
(9) Control or prohibition of a fuel or fuel additive used in motor vehicles, if such control or prohibition is necessary to achieve a national primary or secondary air quality standard and is approved by the Administrator under section 211(c)(4)(C) of the Act.

control strategy implementation plan revision (40CFR51.392) is the applicable implementation plan which contains specific strategies for controlling the emissions of and reducing ambient levels of pollutants in order to satisfy CAA requirements for demonstrations of reasonable further progress and attainment (CAA sections 182(b)(1), 182(c)(2)(A), 182(c)(2)(B), 187(a)(7), 189(a)(1)(B), and 189(b)(1)(A); and sections 192(a) and 192(b), for nitrogen dioxide) (other identical or similar definitions are provided in 40CFR93.101) (under CAA).

control strategy period with respect to particulate (40CFR93.101) matter less than 10 microns in diameter (PM10), carbon monoxide (CO), nitrogen dioxide (NO_2), and/or ozone precursors (volatile organic compounds and oxides of nitrogen), means that period of time after EPA approves control strategy implementation plan revisions containing strategies for controlling PM10, NO_2, CO, and/or ozone, as appropriate. This period ends when a State submits and EPA approves a request under section 107(d) of the CAA for redesignation to an attainment area.

control substance (40CFR160.3) means any chemical substance or mixture or any other material other than a test substance that is administered to the test system in the course of a study for the purpose of establishing a basis for comparison with the test substance for known chemical or biological measurements (other identical or similar definitions are provided in 40CFR792.3).

control technique guidelines (CTG) (EPA-94/04): EPA documents issued to assist state and local pollution control authorities to achieve and maintain air quality standards for certain sources through reasonably available control technologies. CTGs can be quite specific, e.g., one was written to control organic emissions from solvent metal cleaning, known as degreasing.

control technology (40CFR57.103-j): See constant controls.

control technology (40CFR63.51) means measures, processes, methods, systems, or techniques to limit the emission of hazardous air pollutants including, but not limited to, measures which:
(1) Reduce the quantity, or eliminate emissions, of such pollutants through process changes, substitution of materials or other modifications;

(2) Enclose systems or processes to eliminate emissions;
(3) Collect, capture, or treat such pollutants when released from a process, stack, storage or fugitive emissions point;
(4) Are design, equipment, work practice, or operational standards (including requirements for operator training or certification) as provided in 42 USC 7412(h); or
(5) Are a combination of paragraphs (1) through (4) of this definition.

control unit (40CFR72.2) means a unit employing a qualifying Phase I technology in accordance with a Phase I extension plan under 40CFR72.42 (under CAA).

controlled area (40CFR191.12) means:
(1) A surface location, to be identified by passive institutional controls, that encompasses no more than 100 square kilometers and extends horizontally no more than five kilometers in any direction from the outer boundary of the original location of the radioactive wastes in a disposal system; and
(2) the subsurface underlying such a surface location.

controlled product (40CFR82.3-L) means:
(1) a product that contains a controlled substance listed as a Class I, Group I or II substance in appendix A to this subpart and that belongs to one or more of the following six categories of products:
(i) Automobile and truck air conditioning units (whether incorporated in vehicles or not);
(ii) Domestic and commercial refrigeration and air-conditioning/heat pump equipment (whether containing controlled substances as a refrigerant and/or in insulating material of the product), e.g. Refrigerators, Freezers, Dehumidifiers, Water coolers, Ice machines, Air-conditioning and heat pump units;
(iii) Aerosol products, except medical aerosols;
(iv) Portable fire extinguishers;
(v) Insulation boards, panels and pipe covers;
(vi) Pre-polymers.
(2) Controlled products include, but are not limited

to, those products listed in appendix D of this subpart.

controlled reaction (EPA-94/04): A chemical reaction under temperature and pressure conditions maintained within safe limits to produce a desired product or process.

controlled substance (40CFR32.605-1) means a controlled substance in schedules I through V of the Controlled Substances Act (21USC812), and as further defined by regulation at 21CFR1308.11 through 1308.15.

controlled substance (40CFR82.3) means any substance listed in Appendix A to this part, whether existing alone or in a mixture, but excluding any such substance or mixture that is in a manufactured product other than a container used for the transportation or storage of the substance or mixture. Any amount of a listed substance which is not part of a use system containing the substance is a controlled substance. If a listed substance or mixture must first be transferred from a bulk container to another container, vessel, or piece of equipment in order to realize its intended use, the listed substance or mixture is a controlled substance. Controlled substances are divided into five groups, Group I, Group II, Group III, Group IV, and Group V, as set forth in Appendix A to this part (under CAA).

controlled substance (40CFR82.3-m) means any substance listed in appendix A or appendix B to this subpart, whether existing alone or in a mixture, but excluding any such substance or mixture that is in a manufactured product other than a container used for the transportation or storage of the substance or mixture. Thus, any amount of a listed substance in appendix A or appendix B to this subpart that is not part of a use system containing the substance is a controlled substance. If a listed substance or mixture must first be transferred from a bulk container to another container, vessel, or piece of equipment in order to realize its intended use, the listed substance or mixture is a "controlled substance." The inadvertent or coincidental creation of insignificant quantities of a listed substance in appendix A or appendix B to this subpart; during a chemical manufacturing process, resulting from unreacted feedstock, from the listed substance's use as a process agent present as a trace quantity in the chemical substance being manufactured, or as an unintended byproduct of research and development applications, is not deemed a controlled substance. Controlled substances are divided into two classes, Class I in appendix A to this subpart, and Class II listed in appendix B to this subpart. Class I substances are further divided into seven groups, Group I, Group II, Group III, Group IV, Group V, Group VI, and Group VII, as set forth in appendix A to this subpart.

controlled substance (40CFR82.82-c) means a class I or class II ozone-depleting substance (other identical or similar definitions are provided in 40CFR82.104-f).

controlled surface mine drainage (40CFR434.11-q) means any surface mine drainage that is pumped or siphoned from the active mining area.

controlled vehicles (40CFR51-AN) means light-duty vehicles sold nationally (except in California) in the 1968 model-year and later and light-duty vehicles sold in California in the 1966 model-year and later (under CAA).

controlling interest (40CFR280.92) means direct ownership of at least 50 percent of the voting stock of another entity.

conveniently available service facility and spare parts for small volume manufacturers (40CFR86.092.2-6) means that the vehicle manufacturer has a qualified service facility at or near the authorized point of sale or delivery of its vehicles and maintains an inventory of all emission-related spare parts or has made arrangements for the part manufacturers to supply the parts by expedited shipment (e.g., utilizing overnight express delivery service, UPS, etc.) (under CAA).

convention (ESA3-16USC1531) means the Convention on International Trade in Endangered Species of Wild Fauna and Flora, signed on March 3, 1973, and the appendices thereto.

conventional filtration (treatment) (EPA-94/04): See complete treatment.

conventional filtration treatment (40CFR141.2) means a series of processes including coagulation, flocculation, sedimentation, and filtration resulting in substantial particulate removal.

conventional gasoline (40CFR80.2-ff) means any gasoline which has not been certified under 40CFR80.40.

conventional gasoline (CAA211.k-42USC7545) means any gasoline which does not meet specifications set by a certification under this subsection (cf. reformulated gasoline).

conventional mine (40CFR146.3) means an open pit or underground excavation for the production of minerals.

conventional pollutants (40CFR401.16[aaa]): The following comprise the list of conventional pollutants designated pursuant to section 304.a.4 of the CWA:
(1) Biochemical oxygen demand (BOD);
(2) Total suspended solids (nonfilterable) (TSS);
(3) pH;
(4) Fecal coliform bacteria; and
(5) Oil and grease.

conventional pollutants (EPA-94/04): Statutorily listed pollutants understood well by scientists. These may be in the form of organic waste, sediment, acid, bacteria, viruses, nutrients, oil and grease, or heat.

conventional systems (EPA-94/04): Systems that have been traditionally used to collect municipal wastewater in gravity sewers and convey it to a central primary or secondary treatment plant prior to discharge to surface waters.

conventional technology (40CFR35.2005-14) means wastewater treatment processes and techniques involving the treatment of wastewater at a centralized treatment plant by means of biological or physical/chemical unit processes followed by direct point source discharge to surface waters.

conventional technology (40CFR60.41b) means wet flue gas desulfurization (FGD) technology, dry FGD technology, atmospheric fluidized bed combustion technology, and oil hydrodesulfurization technology (under CAA).

conventional technology (40CFR60.41c) means wet flue gas desulfurization technology, dry flue gas desulfurization technology, atmospheric fluidized bed combustion technology, and oil hydrodesulfurization technology.

conventional tilling (EPA-94/04): Tillage operations considered standard for a specific location and crop and that tend to bury the crop residues; usually considered as a base for determining the cost effectiveness of control practices.

convergence distance (40CFR60-AA(alt. method 1)): The distance from the lidar to the point of overlap of the lidar receiver's field-of-view and the laser beam (under CAA).

conversion efficiency (40CFR85.2122(a)(15)(ii)(B)) means the measure of the catalytic converter's ability to oxidize HC/CO to CO_2/H_2O under fully warmed-up conditions stated as a percentage calculated by the formula: (inlet conc. - outlet conc.)/(inlet conc.) x 100.

converter (40CFR60.181-h) means any vessel to which lead concentrate or bullion is charged and refined.

converter arsenic charging rate (40CFR61.171) means the hourly rate at which arsenic is charged to the copper converters in the copper converter department based on the arsenic content of the copper matte and of any lead matte that is charged to the copper converters.

conveyance loss (EPA-94/04): Water loss in pipes, channels, conduits, ditches by leakage or evaporation.

conveying system (40CFR60.671) means a device for transporting materials from one piece of equipment or location to another location within a plant. Conveying systems include but are not limited to the following: Feeders, belt conveyors, bucket elevators and pneumatic systems.

conveyor belt transfer point (40CFR60.381) means a point in the conveying operation where the metallic mineral or metallic mineral concentrate is transferred to or from a conveyor belt except where the metallic mineral is being transferred to a stockpile.

conveyorized degreasing (40CFR52.741) means the continuous process of cleaning and removing soils from surfaces utilizing either cold or vaporized solvents.

conviction (40CFR303.11-b) means a judgment of guilt entered in U.S. District Court, upon a verdict rendered by the court or petit jury or by a plea of guilty, including a plea of nolo contendere (under CERCLA).

conviction (40CFR32.105-e) means a judgment of conviction of a criminal offense by any court of competent jurisdiction, whether entered upon a verdict or a plea, including a plea of nolo contendere.

conviction (40CFR32.605-2) means a finding of guilt (including a plea of nolo contendere) or imposition of sentence, or both, by any judicial body charged with the responsibility to determine violations of the Federal or State criminal drug statutes.

cookstove (40CFR60.531) means a wood-fired appliance that is designed primarily for cooking food and that has the following characteristics:
(a) An oven, with a volume of 0.028 cubic meters (1 cubic foot) or greater, and an oven rack,
(b) A device for measuring oven temperatures,
(c) A flame path that is routed around the oven,
(d) A shaker grate,
(e) An ash pan,
(f) An ash clean-out door below the oven, and
(g) The absence of a fan or heat channels to dissipate heat from the appliance.

cooling electricity use (EPA-94/04): Amount of electricity used to meet the building cooling load (cf. building cooling load).

cooling system (40CFR749.68-7) means any cooling tower or closed cooling water system.

cooling tower (40CFR749.68-6) means an open water recirculating device that uses fans or natural draft to draw or force ambient air through the device to cool warm water by direct contact.

cooling tower (EPA-94/04): A structure that helps remove heat from water used as a coolant; e.g., in electric power generating plants.

cooperating agency (40CFR1508.5) means any Federal agency other than a lead agency which has jurisdiction by law or special expertise with respect to any environmental impact involved in a proposal (or a reasonable alternative) for legislation or other major Federal action significantly affecting the quality of the human environment. The selection and responsibilities of a cooperating agency are described in 40CFR1501.6. A State or local agency of similar qualifications or, when the effects are on a reservation, an Indian Tribe, may by agreement with the lead agency become a cooperating agency (under NEPA).

cooperative agreement (40CFR30.200) means an assistance agreement in which substantial EPA involvement is anticipated during the performance of the project (does not include fellowships).

cooperative agreement (40CFR300.5) is a legal instrument EPA uses to transfer money, property, services, or anything of value to a recipient to accomplish a public purpose in which substantial EPA involvement is anticipated during the performance of the project.

cooperative agreement (40CFR35.6015-14) means a legal instrument EPA uses to transfer money, property, services, or anything of value to a recipient to accomplish a public purpose in which substantial EPA involvement is anticipated during the performance of the project.

cooperative agreement (EPA-94/04): An assistance agreement whereby EPA transfers money, property, services or anything of value to a state for the accomplishment of CERCLA-authorized activities or tasks.

cooperator (40CFR172.1-c) means any person who grants permission to a permittee or a permittee's designated participant for the use of an experimental use pesticide at an application site owned or controlled by the cooperator.

Copenhagen Amendments (40CFR82.3-n) means the Montreal Protocol on Substances That Deplete the Ozone Layer, as amended at the Fourth Meeting of the parties to the Montreal Protocol in Copenhagen in 1992.

copolymer (40CFR60.561) means a polymer that has two different repeat units in its chain.

copper (40CFR415.361-c) shall mean the total copper present in the process wastewater stream exiting the wastewater treatment system (other identical or similar definitions are provided in 40CFR415.471-d; 415.651-d).

copper (40CFR420.02-i) means total copper and is determined by the method specified in 40CFR136.3.

copper casting (40CFR464.02-b) means the remelting of copper or a copper alloy to form a cast intermediate or final product by pouring or forcing the molten metal into a mold, except for ingots, pigs, or other cast shapes related to nonferrous (primary and secondary) metals manufacturing (40CFR421). Also excluded are casting of beryllium alloys in which beryllium is present at 0.1 or greater percent by weight and precious metals alloys In which the precious metal is present at 30 or greater percent by weight. Except for grinding scrubber operations which are covered here, processing operations following the cooling of castings are covered under the electroplating and metal finishing point source categories (40CFR413 and 433).

copper converter (40CFR60.161-g) means any vessel to which copper matte is charged and oxidized to copper.

copper converter (40CFR61.171) means any vessel in which copper matte is charged and is oxidized to copper.

copper converter department (40CFR61.171) means all copper converters at a primary copper smelter.

copper matte (40CFR61.171) means any molten solution of copper and iron sulfides produced by smelting copper sulfide ore concentrates or calcines.

coproduct (40CFR704.3) means a chemical substance produced for a commercial purpose during the manufacture, processing, use or disposal of another chemical substance or mixture.

coproduct (40CFR716.3) means a chemical substance produced for a commercial purpose during

the manufacture, processing, use, or disposal of another chemical substance(s) or mixture(s).

copy of study (40CFR716.3) means the written presentation of the purpose and methodology of a study and its results.

copy paper (40CFR250.4-uu): See xerographic paper.

core (EPA-94/04): The uranium-containing heart of a nuclear reactor, where energy is released.

core of the drawing with emulsions or soaps subcategory (40CFR467.61) shall include drawing using emulsions or soaps, stationary casting, artificial aging, annealing, degreasing, sawing, and swaging.

core of the drawing with neat oils subcategory (40CFR467.51-a) shall include drawing using neat oils, stationary casting, artificial aging, annealing, degreasing, sawing, and swaging (other identical or similar definitions are provided in 40CFR467.61-a).

core of the extrusion subcategory (40CFR467.31-a) shall include extrusion die cleaning, dummy block cooling, stationary casting, artificial aging, annealing, degreasing, and sawing.

core of the forging subcategory (40CFR467.41-a) shall include forging, artificial aging, annealing, degreasing, and sawing.

core of the rolling with emulsions subcategory (40CFR467.21-a) shall include rolling using emulsions, roll grinding, stationary casting, homogenizing, artificial aging, annealing, and sawing (cf. core of the forging subcategory).

core of the rolling with neat oils subcategory (40CFR467.11-a) shall include rolling using neat oils, roll grinding, sawing, annealing, stationary casting, homogenizing artificial aging, degreasing, and stamping.

core program cooperative agreement (40CFR35.6015-15) means a cooperative agreement that provides funds to a State or Federally-recognized Indian Tribe to conduct CERCLA implementation activities that are not assignable to specific sites, but are intended to develop and maintain a States ability to participate in the CERCLA response program.

core program cooperative agreement (EPA-94/04): An assistance agreement whereby EPA supports states or tribal governments with funds to help defray the cost of nonitem-specific administrative and training activities.

corn (40CFR406.11-b) shall mean the shelled corn delivered to a plant before processing (other identical or similar definitions are provided in 40CFR406.21-b).

corn, canned (40CFR407.71-e) shall mean the processing of corn into the following product styles: Canned, yellow and white, whole kernel, cream style, and on-the-cob.

corn, frozen (40CFR407.71-f) shall mean the processing of corn into the following product styles: Frozen, yellow and white, whole kernel and whole cob.

corps (40CFR233.2) means the U.S. Army Corps of Engineers.

corrective action management unit or CAMU (40CFR260.10) means an area within a facility that is designated by the Regional Administrator under part 264 subpart S, for the purpose of implementing corrective action requirements under 40CFR264.101 and RCRA section 3008(h). A CAMU shall only be used for the management of remediation wastes pursuant to implementing such corrective action requirements at the facility.

corrective Action Management Unit or CAMU (40CFR270.2) means an area within a facility that is designated by the Regional Administrator under part 264 subpart S, for the purpose of implementing corrective action requirements under 40CFR264.101 and RCRA section 3008(h). A CAMU shall only be used for the management of remediation wastes pursuant to implementing such corrective action requirements at the facility.

corresponding onshore area (CAA328-42USC7627) means, with respect to any OCS source, the onshore attainment or nonattainment area that is closest to the source, unless the Administrator determines that another area with more stringent requirements with respect to the control and abatement of air pollution may reasonably be expected to be affected by such emissions. Such determination shall be based on the potential for air pollutants from the OCS source to reach the other onshore area and the potential of such air pollutants to affect the efforts of the other onshore area to attain or maintain any Federal or State ambient air quality standard or to comply with the provisions of part C of title I.

corresponding onshore area (COA) (40CFR55.2) means, with respect to any existing or proposed OCS source located within 25 miles of a state's seaward boundary, the onshore area that is geographically closest to the source or another onshore area that the Administrator designates as the COA, pursuant to 40CFR55.5 of this part.

corrosion (EPA-94/04): The dissolution and wearing away of metal caused by a chemical reaction such as between water and the pipes, chemicals touching a metal surface, or contact between two metals (cf. erosion).

corrosion expert (40CFR260.10) means a person who, by reason of his knowledge of the physical sciences and the principles of engineering and mathematics, acquired by a professional education and related practical experience, is qualified to engage in the practice of corrosion control on buried or submerged metal piping systems and metal tanks. Such a person must be certified as being qualified by the National Association of Corrosion Engineers (NACE) or be a registered professional engineer who has certification or licensing that includes education and experience in corrosion control on buried or submerged metal piping systems and metal tanks.

corrosion expert (40CFR280.12) means a person who, by reason of thorough knowledge of the physical sciences and the principles of engineering and mathematics acquired by a professional education and related practical experience, is qualified to engage in the practice of corrosion control on buried or submerged metal piping systems and metal tanks. Such a person must be accredited or certified as being qualified by the National

Association of Corrosion Engineers or be a registered professional engineer who has certification or licensing that includes education and experience in corrosion control of buried or submerged metal piping systems and metal tanks.

corrosion inhibitor (40CFR141.2) means a substance capable of reducing the corrosivity of water toward metal plumbing materials, especially lead and copper, by forming a protective film on the interior surface of those materials.

corrosive (EPA-94/04): A chemical agent that reacts with the surface of a material causing it to deteriorate or wear away.

corrugated box (40CFR246.101-k) means a container for goods which is composed of an inner fluting of material (corrugating medium) and one or two outer liners of material (linerboard).

corrugated boxes (40CFR250.4-h) means boxes made of corrugated paperboard, which, in turn, is made from a fluted corrugating medium pasted to two flat sheets of paperboard (linerboard) multiple layers may be used.

corrugated container waste (40CFR246.101-j) means discarded corrugated boxes.

corrugated paper (40CFR763.163) means an asbestos-containing product made of corrugated paper, which is often cemented to a flat backing, may be laminated with foils or other materials, and has a corrugated surface. Major applications of asbestos corrugated paper include: thermal insulation for pipe coverings; block insulation; panel insulation in elevators; insulation in appliances; and insulation in low-pressure steam, hot water, and process lines.

corrugating medium furnish subdivision mills (40CFR430.51) are mills where only recycled corrugating medium is used in the production of paperboard (cf. noncorrugating medium furnish subdivision mills).

cosmetic (40CFR710.2-a) shall have the meaning contained in the Federal Food, Drug, and Cosmetic Act (FFDCA), 21USC321 et seq., and the regulations issued under such Act. The meaning of the term, cosmetic, defined under the Act was provided below:

Cosmetic (21USC321(a)(1)(i)) means:
(1) articles intended to be rubbed, poured, sprinkled, or sprayed on, introduced into, or otherwise applied to the human body or any part thereof for cleansing, beautifying, promoting attractiveness, or altering the appearance, and
(2) articles intended for use as a component of any such articles; except that such term shall not include soap.

(Other identical or similar definitions are provided in 40CFR720.3-a.)

cost analysis (40CFR33.005) means the review and evaluation of each element of subagreement cost to determine reasonableness, allocability and allowability.

cost analysis (40CFR35.6015-16) means the review and evaluation of each element of contract cost to determine reasonableness, allocability, and allocability.

cost/benefit analysis (EPA-94/04): A quantitative evaluation of the costs which would be incurred versus the overall benefits to society of a proposed action such as the establishment of an acceptable dose of a toxic chemical.

cost effective alternative (EPA-94/04): An alternative control or corrective method identified after analysis as being the best available in terms of reliability, performance, and cost. Although costs are one important consideration, regulatory and compliance analysis does not require EPA to choose the least expensive alternative. For example, when selecting or approving a method for cleaning up a Superfund site the Agency balances costs with the long-term effectiveness of the methods proposed and the potential danger posed by the site.

cost of production of a car line (40CFR600.502.81-2) shall mean the aggregate of the products of:
(i) The average U.S. dealer wholesale price for such car line as computed from each official dealer price list effective during the course of a model year, and
(ii) The number of automobiles within the car line

produced during the part of the model year that the price list was in effect.

cost recovery (EPA-94/04): A legal process by which potentially responsible parties who contributed to contamination at a Superfund site can be required to reimburse the Trust Fund for money spent during any cleanup actions by the federal government.

cost share (40CFR35.6015-17) means the portion of allowable project costs that a recipient contributes toward completing its project (i.e., non-Federal share, matching share).

cost sharing (40CFR30.200) means the portion of allowable project costs that a recipient contributes toward completing its project (i.e., non-Federal share, matching share).

cost sharing (EPA-94/04): A publicly financed program through which society, as a beneficiary of environmental protection, shares part of the cost of pollution control with those who must actually install the controls. In Superfund, the government may pay part of the cost of a cleanup action with those responsible for the pollution paying the major share.

cost sharing or matching (40CFR31.3) means the value of the third party in-kind contributions and the portion of the costs of federally assisted project or program not borne by the Federal Government.

cost-type contract (40CFR31.3) means a contract or subcontract under a grant in which the contractor or subcontractor is paid on the basis of the costs it incurs, with or without a fee.

cotton fiber content papers (40CFR250.4-i) means paper that contains a minimum of 25 percent and up to 100 percent cellulose fibers derived from lint cotton, cotton linters, and cotton or linen cloth cuttings. It is also known as rag content paper or rag paper. It is used for stationery, currency, ledgers, wedding invitations, maps, and other specialty papers.

cotton fiber furnish subdivision mills (40CFR430.181-b) are those mills where significant quantities of cotton fibers (equal to or greater than 4 percent of the total product) are used in the production of fine papers (cf. wood fiber furnish subdivision mills).

council (40CFR1508.6) means the Council on Environmental Quality established by Title II of the Act.

council (40CFR1517.2) shall mean the Council on Environmental Quality established under Title II of the National Environmental Policy Act of 1969 (42USC4321 through 4347).

council (SDWA1401-42USC300f) means the National Drinking Water Advisory Council established under section 1446.

council or tribal council (40CFR131.35.d-5) means the Colville Business Council of the Colville Confederated Tribes.

cover (40CFR503.21-d) is soil or other material used to cover sewage sludge placed on an active sewage sludge unit.

cover (40CFR60.711-9) means, with respect to coating mix preparation equipment, a device that lies over the equipment opening to prevent VOC from escaping and that meets the requirements found in 40CFR60.712(c)(1)-(5).

cover (40CFR60.741) means, with respect to coating mix preparation equipment, a device that fits over the equipment opening to prevent emissions of volatile organic compounds (VOC) from escaping.

cover (40CFR61.341) means a device or system which is placed on or over a waste placed in a waste management unit so that the entire waste surface area is enclosed and sealed to minimize air emissions. A cover may have openings necessary for operation, inspection, and maintenance of the waste management unit such as access hatches, sampling ports, and gauge wells provided that each opening is closed and sealed when not in use. Example of covers include a fixed roof installed on a tank, a lid installed on a container, and an air-supported enclosure installed over a waste management unit.

cover (40CFR63.111), as used in the wastewater provisions, means a device or system which is placed on or over a waste management unit containing wastewater or residuals so that the entire surface area is enclosed and sealed to minimize air emissions.

A cover may have openings necessary for operation, inspection, and maintenance of the waste management unit such as access hatches, sampling ports, and gauge wells provided that each opening is closed and sealed when not in use. Examples of covers include a fixed roof installed on a wastewater tank, a lid installed on a container, and an air-supported enclosure installed over a waste management unit.

cover crop (40CFR503.9-d) is a small grain crop, such as oats, wheat, or barley, not grown for harvest.

cover crop (EPA-94/04): A crop that provides temporary protection for delicate seedlings and/or provides a cover canopy for seasonal soil protection and improvement between normal crop production periods.

cover material (40CFR241.101-b) means soil or other suitable material that is used to cover compacted solid wastes in a land disposal site.

cover material (EPA-94/04): Soil used to cover compacted solid waste in a sanitary landfill.

cover paper (40CFR250.4-j): See cover stock.

cover stock or cover paper (40CFR250.4-j) means a heavyweight paper commonly used for covers, books, brochures, pamphlets, and the like.

coverage period (40CFR704.203) means a time-span which is 1 day less than 2 years, as identified in subpart D, and is the time-span which a person uses to determine his/her reporting year. Subject manufacturing or processing activities may or may not have occurred during the coverage period.

covered area (40CFR80.2-hh) means each of the geographic areas specified in 40CFR80.70 in which only reformulated gasoline may be sold or dispensed to ultimate consumers.

covered area (CAA211.k-42USC7545) means the 9 ozone nonattainment areas having a 1980 population in excess of 250,000 and having the highest ozone design value during the period 1987 through 1989 shall be "covered areas" for purposes of this subsection. Effective one year after the reclassification of any ozone nonattainment area as a Severe ozone nonattainment area under section 181(b), such Severe area shall also be a "covered area" for purposes of this subsection.

covered federal action (40CFR34.105-b) means any of the following Federal actions:
(1) The awarding of any Federal contract;
(2) The making of any Federal grant;
(3) The making of any Federal loan;
(4) The entering into of any cooperative agreement; and,
(5) The extension, continuation, renewal, amendment, or modification of any Federal contract, grant, loan, or cooperative agreement.
Covered Federal action does not include receiving from an agency a commitment providing for the United States to insure or guarantee a loan. Loan guarantees and loan insurance are addressed independently within this part.

covered fleet (CAA241-42USC7581) means 10 or more motor vehicles which are owned or operated by a single person. In determining the number of vehicles owned or operated by a single person for purposes of this paragraph, all motor vehicles owned or operated, leased or otherwise controlled by such person, by any person who controls such person, by any person controlled by such person, and by any person under common control with such person shall be treated as owned by such person. The term "covered fleet" shall not include motor vehicles held for lease or rental to the general public, motor vehicles held for sale by motor vehicle dealers (including demonstration vehicles), motor vehicles used for motor vehicle manufacturer product evaluations or tests, law enforcement and other emergency vehicles, or nonroad vehicles (including farm and construction vehicles).

covered fleet operator (40CFR88.302.94) means a person who operates a fleet of at least ten covered fleet vehicles (as defined in section 241(6) of the Act) and that fleet is operated in a single covered area (even if the covered fleet vehicles are garaged outside of it). For purposes of this definition, the vehicle types described in the definition of covered fleet (section 241(5) of the Act) as exempt from the program will not be counted toward the ten-vehicle criterion.

covered fleet vehicle (CAA241-42USC7581) means only a motor vehicle which is:

(i) in a vehicle class for which standards are applicable under this part; and

(ii) in a covered fleet which is centrally fueled (or capable of being centrally fueled).

No vehicle which under normal operations is garaged at a personal residence at night shall be considered to be a vehicle which is capable of being centrally fueled within the meaning of this paragraph.

covered States (40CFR259.10.b) means those States that are participating in the demonstration medical waste tracking program and includes: Connecticut, New Jersey, New York, Rhode Island, and Puerto Rico. Any other State is a Non-Covered State.

Cr(+6) (40CFR415.171-d) shall mean hexavalent chromium.

Cr(T) (40CFR415.171-c) shall mean total chromium.

cradle to grave or manifest system (EPA-94/04): A procedure in which hazardous materials are identified and followed as they are produced, treated, transported, and disposed of by a series of permanent, linkable, descriptive documents (e.g., manifests). Commonly referred to as the cradle to grave system.

cranberries (40CFR407.61-g) shall mean the processing of cranberries into the following product styles: Canned, bottled, and frozen, whole, sauce, jelly, juice and concentrate.

crankcase emissions (40CFR86.082.2) means airborne substances emitted to the atmosphere from any portion of the engine crankcase ventilation or lubrication (other identical or similar definitions are provided in 40CFR86.402.78).

credit trading records (40CFR80.126-b) shall include worksheets and EPA reports showing actual and complying totals for oxygen and benzene; credit calculation worksheets; contracts; letter agreements; and invoices and other documentation evidencing the transfer of credits.

creditor agency (40CFR13.2-g) means the Federal agency to which the debt is owed.

criminal drug statute (40CFR32.605-3) means a Federal or non-Federal criminal statute involving the manufacture, distribution, dispensing, use, or possession of any controlled substance.

criteria (40CFR131.3-b) are elements of State water quality standards, expressed as constituent concentrations, levels, or narrative statements, representing a quality of water that supports a particular use. When criteria are met, water quality will generally protect the designated use.

criteria (40CFR220.2-g) means the criteria set forth in part 227 of this Subchapter H.

criteria (40CFR256.06) means the "Criteria for Classification of Solid Waste Disposal Facilities," 40CFR257, promulgated under section 4004(a) of the Act.

criteria (EPA-94/04): Descriptive factors taken into account by EPA in setting standards for various pollutants. These factors are used to determine limits on allowable concentration levels, and to limit the number of violations per year. When issued by EPA, the criteria provide guidance to the states on how to establish their standards.

criteria pollutant or standard (40CFR51.852) means any pollutant for which there is established a NAAQS (national ambient air quality standards) at 40CFR part 50 (EPA has established the following six criteria pollutants):

(1) carbon monoxide (40CFR50.8);

(2) lead (40CFR50.12);

(3) nitrogen dioxide (40CFR50.11)

(4) ozone (40CFR50.9);

(5) sulfur dioxide (40CFR50.5)

(6) particulate matter (40CFR50.6).

(Other identical or similar definitions are provided in 40CFR93.152.)

criteria pollutants (EPA-94/04): The 1970 amendments to the Clean Air Act required EPA to set National Ambient Air Quality Standards for certain pollutants known to be hazardous to human health. EPA has identified and set standards to protect human health and welfare for six pollutants: ozone, carbon monoxide, total suspended particulates, sulfur dioxide, lead, and nitrogen oxide.

The term, "criteria pollutants" derives from the requirement that EPA must describe the characteristics and potential health and welfare effects of these pollutants. It is on the basis of these criteria that standards are set or revised.

critical aquifer protection area (SDWA1427-42USC300h.6) means either of the following:
(1) All or part of an area located within an area for which an application or designation as a sole or principal source aquifer pursuant to section 1424(e) has been submitted and approved by the Administrator not later than 24 months after the enactment of the Safe Drinking Water Act Amendments of 1986 and which satisfies the criteria established by the Administrator under subsection (d).
(2) All or part of an area which is within an aquifer designated as a sole source aquifer as of the enactment of the Safe Drinking Water Act Amendments of 1986 and for which an areawide ground water quality protection plan has been approved under section 208 of the Clean Water Act prior to such enactment.

critical emission-related components (40CFR86.088.2) are those components which are designed primarily for emission control, or whose failure may result in a significant increase in emissions accompanied by no significant impairment (or perhaps even an improvement) in performance, driveability, and/or fuel economy as determined by the Administrator.

critical emission-related maintenance (40CFR86.088.2) means that maintenance to be performed on critical emission-related components.

critical habitat (ESA3-16USC1531):
(A) For a threatened or endangered species means:
(i) the specific areas within the geographical area occupied by the species at the time it is listed in accordance with the provisions of section 1533 of this title, on which are found those physical or biological features (I) essential to the conservation of the species and (II) which may require special management considerations or protection and
(ii) specific areas outside the geographical area

occupied by the species at the time it listed in accordance with the provisions of section 1533 of this title, upon a determination by the Secretary that such areas are essential for the conservation of the species.
(B) Critical habitat may be established for those species now listed as threatened or endangered species for which no critical habitat has heretofore been established as set forth in subparagraph (A) of this paragraph.
(C) Except in those circumstances determined by the Secretary, critical habitat shall not include the entire geographical area which can be occupied by the threatened or endangered species.

critical organ (40CFR191.02-q) means the most exposed human organ or tissue exclusive of the integumentary system (skin) and the cornea.

critical pollutant (40CFR58-AG-f) means the pollutant with the highest subindex during the reporting period.

crop advisor (40CFR170.3) means any person who is assessing pest numbers or damage, pesticide distribution, or the status or requirements of agricultural plants. The term does not include any person who is performing hand labor tasks.

crop consumptive use (EPA-94/04): The amount of water transpired during plant growth plus what evaporated from the soil surface and foliage in the crop area.

crop rotation (EPA-94/04): Planting a succession of different crops on the same land area as opposed to planting the same crop time after time.

crops for direct human consumption (40CFR257.3.6-1) means crops that are consumed by humans without processing to minimize pathogens prior to distribution to the consumer.

cross connection (EPA-94/04): Any actual ore potential connection between a drinking water system and an unapproved water supply or other source of contamination.

cross recovery furnace (40CFR60.281-j) means a furnace used to recover chemicals consisting primarily of sodium and sulfur compounds by burning black liquor which on a quarterly basis contains more than 7 weight percent of the total pulp solids from the neutral sulfite semichemical process and has a green liquor sulfidity of more than 28 percent.

cross section (40CFR796.3800-iv) is defined as the proportionality constant in the Beer-Lambert law (see 40CFR796.3800-iv for more details).

crude intermediate plastic material (40CFR463.2-g) is plastic material formulated in an on-site polymerization process.

crude oil (40CFR52.741) means a naturally occurring mixture which consists of hydrocarbons and sulfur, nitrogen, or oxygen derivatives of hydrocarbons and which is a liquid at standard conditions.

crude oil gathering (40CFR52.741) means the transportation of crude oil or condensate after custody transfer between a production facility and a reception point.

crusher (40CFR60.381) means a machine used to crush any metallic mineral and includes feeders or conveyors located immediately below the crushing surfaces. Crushers include, but are not limited to, the following types: jaw, gyratory, cone, and hammermill.

crusher (40CFR60.671) means a machine used to crush any nonmetallic minerals, and includes, but is not limited to, the following types: jaw, gyratory, cone, roll, rod mill, hammermill, and impactor.

Cr,VI (40CFR413.02-c) shall mean hexavalent chromium.

CT or CTcalc (40CFR141.2) is the product of "residual disinfectant concentration." (C) in mg/l determined before or at the first customer, and the corresponding "disinfectant contact time" (T) in minutes, i.e., "C" x "T." If a public water system applies disinfectants at more than one point prior to the first customer, it must determine the CT of each disinfectant sequence before or at the first customer to determine the total percent inactivation or total inactivation ratio. In determining the total inactivation ratio, the public water system must determine the residual disinfectant concentration of each disinfection sequence and corresponding contact time before any subsequent disinfection application point(s). "CT99.9" is the CT value required for 99.9 percent (3-log) inactivation of Giardia lamblia cysts. CT99.9 for a variety of disinfectants and conditions appear in Tables 1.1-1.6, 2.1, and 3.1 of 40CFR141.74(b)(3).

- (CTcalc)/(CT99.9) is the inactivation ratio.
- The sum of the inactivation ratios, or total inactivation ratio shown as Σ = (CTcalc)/(CT99.9) is calculated by adding together the inactivation ratio for each disinfection sequence.
- A total inactivation ratio equal to or greater than 1.0 is assumed to provide a 3-log inactivation of Giardia lamblia cysts.

CTG (CAA302-42USC7602) means a Control Technique Guideline published by the Administrator under section 108.

cubic feet or cubic meters of production in subpart A (40CFR429.11-b) means the cubic feet or cubic meters of logs from which bark is removed.

cubic feet per minute (CFM) (EPA-94/04): A measure of the volume of a substance flowing through air within a fixed period of time. With regard to indoor air, refers to the amount of air, in cubic feet, that is exchanged with indoor air in a minute's time, i.e., the air exchange rate.

cullet (40CFR426.21-b) shall mean any broken glass generated in the manufacturing process (other identical or similar definitions are provided in 40CFR426.31-b; 426.41-b; 426.101-c).

cullet (40CFR61.161) means waste glass recycled to a glass melting furnace.

cullet (EPA-94/04): Crushed glass.

cullet water (40CFR426.11-b) shall mean that water which is exclusively and directly applied to molten glass in order to solidify the glass.

cultural eutrophication (EPA-94/04): Increasing rate at which water bodies "die" by pollution from human activities.

cultures and stocks (EPA-94/04): Infectious agents and associated biologicals including: cultures from medical and pathological laboratories; cultures and stocks of infectious agents from research and industrial laboratories; waste from the production of biologicals; discarded live and attenuated vaccines; and culture dishes and devices used to transfer, inoculate, and mix cultures.

cumulative exposure (EPA-94/04): The summation of exposures of an organism to a chemical over a period of time.

cumulative impact (40CFR1508.5) is the impact on the environment which results from the incremental impact of the action when added to other past, present, and reasonably foreseeable future actions regardless of what agency (Federal or non-Federal) or person undertakes such other actions. Cumulative impacts can result from individually minor but collectively significant actions taking place over a period of time.

cumulative pollutant loading rate (40CFR503.11-f) is the maximum amount of an inorganic pollutant that can be applied to an area of land (under the Clean Water Act).

cumulative toxicity (40CFR795.260-4) is the adverse effects of repeated doses occurring as a result of prolonged action on, or increased concentration of, the administered substance or its metabolites in susceptible tissue.

cumulative toxicity (40CFR798.2250-4) is the adverse effects of repeated doses occurring as a result of prolonged action on, or increased concentration of the administered test substance or its metabolites in susceptible tissues.

cumulative toxicity (40CFR798.2450-7) is the adverse effects of repeated doses occurring as a result of prolonged action on, or increased concentration of the administered substance or its metabolites in susceptible tissues (other identical or similar definitions are provided in 40CFR798.2650-4; 798.2675-4).

cumulative working level months (CWLM) (EPA-94/04): The sum of lifetime exposure to radon working levels expressed in total working level months.

curb collection (40CFR243.101-i) means collection of solid waste placed adjacent to a street (under RCRA).

curb idle (40CFR86.082.2), for manual transmission code heavy-duty engines, means the manufacturer's recommended engine speed with the transmission in neutral or with the clutch disengaged. For automatic transmission code heavy-duty engines, curb idle means the manufacturer's recommended engine speed with the automatic transmission in gear and the output shaft stalled.

curb idle (40CFR86.084.2):
(1) For manual transmission code light-duty trucks, the engine speed with the transmission in neutral or with the clutch disengaged and with the air conditioning system, if present, turned off. For automatic transmission code light-duty trucks, curb-idle means the engine speed with the automatic transmission in the Park position (or Neutral position if there is no Park position), and with the air conditioning system, if present, turned off.
(2) For manual transmission code heavy-duty engines, the manufacturer's recommended engine speed with the clutch disengaged. For automatic transmission code heavy-duty engines, curb idle means the manufacturer's recommended engine speed with the automatic transmission in gear and the output shaft stalled. (Measured idle speed may be used in lieu of curb-idle speed for the emission tests when the difference between measured idle speed and curb idle speed is sufficient to cause a void test under either 40CFR86.1341 or 40CFR86.884-7 but not sufficient to permit adjustment in accordance with 40CFR86.085-25).

curb mass (40CFR86.402.78) means the actual or manufacturer's estimated mass of the vehicle with fluids at nominal capacity and with all equipment specified by the Administrator.

curb stop (EPA-94/04): A water service shutoff valve located in a water service pipe near the curb and between the water main and the building.

curbside collection (EPA-94/04): Method of collecting recyclable materials at homes, community districts or businesses.

curie (10CFR30.4) means that amount of radioactive material which disintegrates at the rate of 37 billion atoms per second.

curie (Ci) (40CFR190.02-g) means that quantity of radioactive material producing 37 billion nuclear transformations per second. (One millicurie (mCi = 0.001 Ci)).

curie (Ci) (40CFR192.01-f) means the amount of radioactive material that produces 37 billion nuclear transformation per second. One picocurie (pCi) = 10^{-12} Ci.

curie (Ci) (40CFR300-AA) means a measure used to quantify the amount of radioactivity. One curie equals 37 billion nuclear transformations per second, and one picocurie (pCi) equals 10^{-12} Ci.

curing oven (40CFR60.451) means a device that uses heat to dry or cure the coating(s) applied to large appliance parts or products.

curing oven (40CFR60.461) means the device that uses heat or radiation to dry or cure the coating applied to the metal coil.

current assets (40CFR144.61-d) means cash or other assets or resources commonly identified as those which are reasonably expected to be realized in cash or sold or consumed during the normal operating cycle of the business (other identical or similar definitions are provided in 40CFR264.141-f; 265.141-f).

current closure cost estimate (40CFR264.141-b) means the most recent of the estimates prepared in accordance with 40CFR264.142 (a), (b), and (c) (under RCRA).

current closure cost estimate (40CFR265.141-b) means the most recent of the estimates prepared in accordance with 40CFR265.142 (a), (b), and (c).

current discharge (40CFR125.58-g) means the volume, composition, and location of an applicant's discharge as of anytime between December 27, 1977, and December 29, 1982, as designated by the applicant.

current liabilities (40CFR144.61-d) means obligations whose liquidation is reasonably expected to require the use of existing resources properly classifiable as current assets or the creation of other current liabilities (other identical or similar definitions are provided in 40CFR264.141-f; 265.141-f).

current plugging and abandonment cost estimate (40CFR264.141-f) means the most recent of the estimates prepared in accordance with 40CFR144.62(a), (b), and (c) of this title (other identical or similar definitions are provided in 40CFR265.141-f) (under RCRA).

current plugging cost estimate (40CFR144.61-b) means the most recent of the estimates prepared in accordance with 40CFR144.62(a), (b) and (c) (under SDWA).

current post closure cost estimate (40CFR264.141-c) means the most recent of the estimates prepared in accordance with 40CFR264.144 (a), (b), and (c) (other identical or similar definitions are provided in 40CFR 265.141-c).

current production (40CFR167.3) means amount of planned production in the calendar year in which the pesticides report is submitted, including new products not previously sold or distributed.

current year subaccount (40CFR72.2) means the subaccount in an Allowance Tracking System general account, established pursuant to 40CFR73.31(c) of this chapter, in which are held allowances that may be transferred to a unit's compliance subaccount for use by the unit for the purpose of meeting its Acid Rain sulfur dioxide emissions limitation.

currently valid certificate of conformity (40CFR89.602.96): A certificate of conformity for which the current date is within the effective period as specified on the certificate of conformity, and which has not been withdrawn, superseded, voided, suspended, revoked, or otherwise rendered invalid.

curtail (40CFR61.181) means to cease operations to

the extent technically feasible to reduce emissions (under CAA).

custody transfer (40CFR52.741) means the transfer of produced petroleum and/or condensate after processing and/or treating in the producing operations, from storage tanks or automatic transfer facilities to pipelines or any other forms of transportation.

custody transfer (40CFR60.111-g) means the transfer of produced petroleum and/or condensate, after processing and/or treating in the producing operations, from storage tanks or automatic transfer facilities to pipelines or any other forms of transportation (other identical or similar definitions are provided in 40CFR60.111a-k; 60.111b-c).

custom blender (40CFR167.3) means any establishment which provides the service of mixing pesticides to customer's specifications, usually a pesticide(s)-fertilizer(s), pesticide-pesticide, or pesticide-animal feed mixture, when:
(1) The blend is prepared to the order of the customer and is not held in inventory by the blender;
(2) the blend is to be used on the customer's property (including leased or rented property);
(3) the pesticide(s) used in the blend bears end-use labeling directions which do not prohibit use of the product in such a blend;
(4) the blend is prepared from registered pesticides;
(5) the blend is delivered to the end-user along with a copy of the end-use labeling of each pesticide used in the blend and a statement specifying the composition of mixture; and
(6) no other pesticide production activity is performed at the establishment.

custom molded device (40CFR211.203-g) means a hearing protective device that is made to conform to a specific ear canal. This is usually accomplished by using moldable compound to obtain impression of the ear and ear canal. The compound is subsequently permanently hardened to retain this shape (under the Noise Control Act).

customer (40CFR704.3) means any person to whom a manufacturer, importer, or processor directly distributes any quantity of a chemical substance,

mixture, mixture containing the substance or mixture, or article containing the substance or mixture, whether or not a sale is involved (under TSCA).

customer (40CFR72.2) means a purchaser of electricity not for purposes of transmission or resale (under CAA).

customer (40CFR721.3) means any person to whom a manufacturer, importer, or processor distributes any quantity of a chemical substance, or of a mixture containing the chemical substance, whether or not a sale is involved.

customs territory of the United States (40CFR372.3) means customs territory of the United States means the 50 States, Puerto Rice, and the District of Columbia (other identical or similar definitions are provided in 40CFR720.3-g).

customs territory of the United States (40CFR763.163) means the 50 States, Puerto Rico, and the District of Columbia.

cutie pie (EPA-94/04): An instrument used to measure radiation levels.

cutout or by-pass or similar devices (40CFR202.10-d) means devices which vary the exhaust system gas flow so as to discharge the exhaust gas and acoustic energy to the atmosphere without passing through the entire length of the exhaust system, including all exhaust system sound attenuation components.

cutting (40CFR61.141) means to penetrate with a sharp-edged instrument and includes sawing, but does not include shearing, slicing, or punching (under CAA).

CWA: See act or CWA.

cyanide (40CFR420.02-d) means total cyanide and is determined by the method specified in 40CFR136.3 (under CWA).

cyanide A (40CFR415.421-c) means those cyanides amenable to chlorination and is determined by the methods specified in 40CFR136.3.

cyanide A (40CFR415.91-c) shall mean those

cyanides amenable to chlorination and is determined by the methods specified in 40CFR136.3.

cyanide destruction unit (40CFR439.1-b) shall mean a treatment system designed specifically to remove cyanide.

cyclone boiler (40CFR76.2) means a boiler with one or more water-cooled horizontal cylindrical chambers in which coal combustion takes place. The horizontal cylindrical chamber(s) is (are) attached to the bottom of the furnace. One or more cylindrical chambers are arranged either on one furnace wall or on two opposed furnace walls. Gaseous combustion products exiting from the chamber(s) turn 90 degrees to go up through the boiler while coal ash exits the bottom of the boiler as a molten slag.

cyclone collector (EPA-94/04): A device that uses centrifugal force to pull large particles from polluted air.

cyclonic flow (40CFR60.251-d) means a spiraling movement of exhaust gases within a duct or stack.

************ DDDDD ************

dacthal(TM) (40CFR63.191) means the pre-emergent herbicide dacthal(TM), also known as DCPA, DAC, and dimethyl ester 2,3,5,6-tetrachloroterephthalic acid. The category includes, but is not limited to, chlorination processes and the following production process units: photochlorination reactors, thermal chlorination reactors, and condensers (under FIFRA).

DAFGDS (40CFR60-AG) means the dual alkali flue gas desulfurization system for the Newton Unit 1 steam generating unit.

daily cover (40CFR241.101-c) means cover material that is spread and compacted on the top and side slopes of compacted solid waste at least at the end of each operating day in order to control vectors, fire, moisture, and erosion and to assure an aesthetic appearance.

daily discharge (40CFR122.2) means the "discharge of a pollutant" measured during a calendar day or any 24-hour period that reasonably represents the calendar day for purposes of sampling. For pollutants with limitations expressed in units of mass, the "daily discharge" is calculated as the total mass of the pollutant discharged over the day. For pollutants with limitations expressed in other units of measurement, the "daily discharge" is calculated as the average measurement of the pollutant over the day.

daily maximum limitation (40CFR429.21-j), for the subcategories for which numerical limitations are given, is a value that should not be exceeded by any one effluent measurement. The 30-day limitation is a value that should not be exceeded by the average of daily measurements taken during any 30-day period.

daily values (40CFR435.11-c) as applied to produced water effluent limitations and NSPS shall refer to the daily measurements used to assess compliance with the maximum for any one day.

daily weighted average VOM content (40CFR52.741) means the average VOM content of two or more coatings as applied on a coating line during any day, taking into account the fraction of total coating volume that each coating represents, as calculated with the equation (other identical or similar definitions are provided in 40CFR52.741).

damage (OPA1001) means damages specified in section 1002(b) of this Act, and includes the cost of assessing these damages.

damage (SF101-42USC9601) means damages for injury or loss of natural resources as set forth in section 107(a) or 111(b) of this Act.

damaged friable miscellaneous ACM (40CFR763.83) means friable miscellaneous ACM which has deteriorated or sustained physical injury such that the internal structure (cohesion) of the material is inadequate or, if applicable, which has delaminated such that its bond to the substrate (adhesion) is inadequate or which for any other reason lacks fiber cohesion or adhesion qualities. Such damage or deterioration may be illustrated by the separation of ACM into layers; separation of ACM from the substrate; flaking, blistering, or crumbling of the ACM surface; water damage; significant or repeated water stains, scrapes, gouges, mars or other signs of physical injury on the ACM. Asbestos debris originating from the ACBM (asbestos containing building material) in question may also indicate damage.

damaged friable surfacing ACM (40CFR763.83) means friable surfacing ACM which has deteriorated or sustained physical injury such that the internal structure (cohesion) of the material is inadequate or which has delaminated such that its bond to the substrate (adhesion) is inadequate, or which, for any other reason, lacks fiber cohesion or adhesion qualities. Such damage or deterioration may be illustrated by the separation of ACM into layers; separation of ACM from the substrate; flaking, blistering, or crumbling of the ACM surface; water damage; significant or repeated water stains, scrapes, gouges, mars or other signs of physical injury on the ACM. Asbestos debris originating from the ACBM in question may also indicate damage.

damaged or significantly damaged thermal system insulation ACM (40CFR763.83) means thermal system insulation ACM on pipes, boilers, tanks, ducts, and other thermal system insulation equipment where the insulation has lost its structural integrity, or its covering, in whole or in part, is crushed, water-stained, gouged, punctured, missing, or not intact such that it is not able to contain fibers. Damage may be further illustrated by occasional punctures, gouges or other signs of physical injury to ACM; occasional water damage on the protective coverings/jackets; or exposed ACM ends or joints. Asbestos debris originating from the ACBM in question may also indicate damage.

data call in (EPA-94/04): A part of the Office of Pesticide Programs (OPP) process of developing key required test data, especially on the long-term, chronic effects of existing pesticides, in advance of scheduled Registration Standard reviews. Data Call-In from manufacturers is an adjunct of the Registration Standards program intended to expedite re-registration.

data fleet (40CFR610.11-17) means a fleet of automobiles tested at zero device-miles in baseline configuration, the retrofitted configuration and in some cases the adjusted configuration, in order to determine the changes in fuel economy and exhaust emissions due to the retrofitted configuration, and where applicable the changes due to the adjusted configuration, as compared to the fuel economy and exhaust emissions of the baseline configuration (under CAA).

data gap (40CFR152.83-a) means the absence of any valid study or studies in the Agency files which would satisfy a specific data requirement for a particular pesticide product.

data submitters list (40CFR152.83-b) means the current Agency list, entitled Pesticide Data Submitters by Chemical, of persons who have submitted data to the Agency.

date of completion (40CFR310.11-a) means the date when all field work has been completed and all deliverables (e.g., lab results, technical expert reports) have been received by the local government.

day (40CFR52.741) means the consecutive 24 hours beginning at 12:00 a.m. (midnight) local time.

day (40CFR60.51) means 24 hours.

day tank (EPA-94/04): See age tank.

day night sound level (40CFR201.1-g) means the 24-hour time of day weighted equivalent sound level, in decibels, for any continuous 4-hour period, obtained after addition of ten decibels to sound levels produced in the hours from 10 p.m. to 7 a.m. (2200-000). It is abbreviated as L_{dn}.

days (40CFR85.1801-b) shall mean calendar days.

dB(A) (40CFR201.1-f) is an abbreviation meaning a-weighted sound level in decibels, reference: 20 micropascals.

dB(A) (40CFR202.10-e) means the standard

abbreviation for A-weighted sound level in decibels (other identical or similar definitions are provided in 40CFR204.2-12; 205.2-9).

DDT (40CFR129.4-b) means the compounds DDT, DDD, and DDE as identified by the chemical names: (DDT)-1,1,1-trichloro-2,2-bis(p-chlorophenyl)ethane and some o,p-isomers; (DDD) or (TDE)-1,1-dichloro-2,2-bis(p-chlorophenyl) ethane and some o,p-isomers; (DDE)-1,1-dichloro-2,2-bis(p-chlorophenyl) ethylene.

DDT (EPA-94/04): The first chlorinated hydrocarbon insecticide chemical name: Dichloro-Diphenyl-Trichloroethane). It has a half-life of 15 years and can collect in fatty tissues of certain animals. EPA banned registration and interstate sale of DDT for virtually all but emergency uses in the United States in 1972 because of its persistence in the environment and accumulation in the food chain (under TSCA).

DDT formulator (40CFR129.101-2) means a person who produces, prepares or processes a formulated product comprising a mixture of DDT and inert materials or other diluents into a product intended for application in any use registered under the Federal Insecticide, Fungicide and Rodenticide Act, as amended (7USC135, et seq.).

DDT manufacturer (40CFR129.101-1) means a manufacturer, excluding any source which is exclusively a DDT formulator, who produces, prepares or processes technical DDT, or who uses DDT as a material in the production, preparation or processing of another synthetic organic substance (under FWPCA).

dead end (EPA-94/04): The end of a water main which is not connected to other parts of the distribution system.

dealer (40CFR244.101-d) means any person who engages in the sale of beverages in beverage containers to a consumer.

dealer (40CFR600.002.85-18) means a person who resides or is located in the United States, any territory of the United States, or the District of Columbia and who is engaged in the sale or

distribution of new automobiles to the ultimate purchaser.

dealer (CAA216-42USC7550) means any person who is engaged in the sale or the distribution of new motor vehicles or new motor vehicle engines to the ultimate purchaser.

dealer demonstration vehicle (40CFR88.302.94) means any vehicle that is operated by a motor vehicle dealer (as defined in section 216(4) of the Act) solely for the purpose of promoting motor vehicle sales, either on the sales lot or through other marketing or sales promotions, or for permitting potential purchasers to drive the vehicle for pre-purchase or pre-lease evaluation.

dealership (40CFR171.2) means any site owned or operated by a restricted use pesticide retail dealer where any restricted use pesticide is made available for use, or where the dealer offers to make available for use any such pesticide.

death (40CFR795.120) means the lack of reaction of a test organism to gentle prodding (other identical or similar definitions are provided in 40CFR797.1930-1; 797.1950-2; 797.1970-1).

death (40CFR797.1400-5) means the lack of opercular movement by a test fish.

debarment (40CFR32.105-f) means an action taken by a debarring official in accordance with these regulations to exclude a person from participating in covered transactions. A person so excluded is "debarred."

debarring official (40CFR32.105-g) means an official authorized to impose debarment. The debarring official is either: (1) The agency head, or (2) An official designated by the agency head. (3) The Director, Grants Administration Division, is the authorized debarring official.

debris (40CFR268.2-g) means solid material exceeding a 60 mm particle size that is intended for disposal and that is: A manufactured object; or plant or animal matter; or natural geologic material. However, the following materials are not debris: Any material for which a specific treatment standard is provided in subpart D, part 268, namely lead acid batteries, cadmium batteries, and radioactive lead solids; process residuals such as smelter slag and residues from the treatment of waste, wastewater, sludges, or air emission residues; and Intact containers of hazardous waste that are not ruptured and that retain at least 75% of their original volume. A mixture of debris that has not been treated to the standards provided by 40CFR 268.45 and other material is subject to regulation as debris if the mixture is comprised primarily of debris, by volume, based on visual inspection.

debris (40CFR429.11-i) means woody material such as bark, twigs, branches, heartwood or sapwood that will not pass through a 2.54 cm (1.0 in) diameter round opening and is present in the discharge from a wet storage facility.

debt (40CFR13.2-a) means an amount owed to the United States from sources which include loans insured or guaranteed by the United States and all other amounts due the United States from fees, grants, contracts, leases, rent, royalties, services, sales of real or personal property, overpayments, fines, penalties, damages, interest, forfeiture (except those arising under the Uniform Code of Military Justice), and all other similar sources. As used in this regulation, the terms debt and claim are synonymous (under CAA).

debtor (40CFR13.2-c) means an individual, organization, association, corporation, or a State or local government indebted to the United States or a person or entity with legal responsibility for assuming debtor's obligation.

decant (EPA-94/04): To draw off the upper layer of liquid after the heavier material (a solid or another liquid) has settled.

decay (29CFR1910.25) is disintegration of wood substance due to action of wood-destroying fungi. It is also known as dote and rot.

decay product (40CFR300-AA) means an isotope formed by the radioactive decay of some other isotope. This newly formed isotope possesses physical and chemical properties that are different from those of its parent isotope, and may also be radioactive.

decay products (EPA-94/04): Degraded radioactive materials, often referred to as "daughters" or "progeny;" radon decay products of most concern from a public health standpoint are polonium-214 and polonium-218.

dechlorination (EPA-94/04): Removal of chlorine from a substance by chemically replacing it with hydrogen or hydroxide ions in order to detoxify a substances.

decibel (40CFR201.1-h) means the unit measure of sound level, abbreviated as dB.

decision (40CFR82.172) means any final determination made by the Agency under section 612 of the Act on the acceptability or unacceptability of a substitute for a class I or II compound.

decisional body (40CFR124.78-2) means any Agency employee who is or may reasonably be expected to be involved in the decisional process of the proceeding including the Administrator, Judicial Officer, Presiding Officer, the Regional Administrator (if he or she does not designate himself or herself as a member of the Agency trial staff), and any of their staff participating in the decisional process. In the case of a nonadversary panel hearing, the decisional body shall also include the panel members, whether or not permanently employed by the Agency.

decisional body (40CFR57.809-2) means any Agency employee who is or may reasonably be expected to be involved in the decisional process of the proceeding including the Administrator, Judicial Officer, Presiding Officer, the Regional Administrator (if he does not designate himself as a member of the Agency trial staff), and any of their staff participating in the decisional process. In the case of a non-adversary panel hearing, the decisional body shall also include the panel members whether or not permanently employed by the Agency.

decisional body (40CFR72.2) means any EPA employee who is or may reasonably be expected to act in a decision-making role in a proceeding under part 78 of this chapter, including the Administrator, a member of the Environmental Appeals Board, and a Presiding Officer, and any staff of any such person who are participating in the decisional process (under CAA).

deck drainage (40CFR435.11-d) shall refer to any waste resulting from deck washings, spillage, rainwater, and runoff from gutters and drains including drip pans and work areas within facilities subject to this subpart.

declared value of imported components (40CFR600.502.81-1) shall be:
(i) The value at which components are declared by the importer to the U.S. Customs Service at the date of entry into the customs territory of the United States, or
(ii) With respect to imports into Canada, the declared value of such components as if they were declared as imports into the United States at the date of entry into Canada, or
(iii) With respect to imports into Mexico (when 40CFR600.511-80(b)(3) applies), the declared value of such components as if they were declared as imports into the United States at the date of entry into Mexico.

decommission (10CFR30.4) means to remove (as a facility) safely from service and reduce residual radioactivity to a level that permits release of the property for unrestricted use and termination of license (other identical or similar definitions are provided in 10CFR40.4; 70.4).

decomposition (40CFR60.561) means, for the purposes of these standards, an event in a polymerization reactor that advances to the point where the polymerization reaction becomes uncontrollable, the polymer begins to break down (decompose), and it becomes necessary to relieve the reactor instantaneously in order to avoid catastrophic equipment damage or serious adverse personnel safety consequences.

decomposition (EPA-94/04): The breakdown of matter by bacteria and fungi, changing the chemical makeup and physical appearance of materials.

decomposition emissions (40CFR60.561) refers to those emissions released from a polymer production process as the result of a decomposition or during attempts to prevent a decomposition.

decontamination (40CFR259.10.b) means the process of reducing or eliminating the presence of harmful substances, such as infectious agents, so as to reduce the likelihood of disease transmission from those substances.

decontamination (EPA-94/04): Removal of harmful substances such as noxious chemicals, harmful bacteria or other organisms, or radioactive material from exposed individuals, rooms and furnishings in buildings, or the exterior environment.

decontamination area (40CFR763.121) means an enclosed are adjacent and connected to the regulated area and consisting of an equipment room, shower area, and clean room, which is used for the decontamination of workers, materials, and equipment contaminated with asbestos.

decontamination/detoxification (40CFR165.1-g) means processes which will convert pesticides into nontoxic compounds.

deep well injection (EPA-94/04): Deposition of raw or treated, filtered hazardous waste by pumping it into deep wells, where it is contained in the pores of permeable subsurface rock.

deepwater port (40CFR110.1) means an offshore facility as defined in section (3)(10) of the Deepwater Port Act of 1974 (33USC1502(10)).

deepwater port (OPA1001-33USC1501-1524) means a facility licensed under the Deepwater Port of Act of 1974.

defeat device (40CFR86.082.2) means an AECD that reduces the effectiveness of the emission control system under conditions which may reasonably be expected to be encountered in normal urban vehicle operation and use, unless:
(1) such conditions are substantially included in the Federal emission test procedure,
(2) the need for the AECD is justified in terms of protecting the vehicle against damage or accident, or
(3) the AECD does not go beyond the requirements of engine starting.

defeat device (40CFR86.094.2) means an auxiliary

emission control device (AECD) that reduces the effectiveness of the emission control system under conditions which may reasonably be expected to be encountered in normal vehicle operation and use, unless:
(1) Such conditions are substantially included in the Federal emission test procedure;
(2) The need for the AECD is justified in terms of protecting the vehicle against damage or accident; or
(3) The AECD does not go beyond the requirements of engine starting.

defendant (40CFR27.2) means any person alleged in a complaint under 40CFR27.2 to be liable for a civil penalty or assessment under 40CFR27.2.

deficiency (40CFR52.30-3) means the failure to perform a required activity as defined in paragraph (a)(2) of this section.

deflocculating agent (EPA-94/04): A material added to a suspension to prevent settling.

defluoridation (EPA-94/04): The removal of excess fluoride in drinking water to prevent the staining of teeth.

defoliant (EPA-94/04): An herbicide that removes leaves from trees and growing plants.

defoliant (FIFRA2-7USC136) means any substance or mixture of substances intended for causing the leaves or foliage to drop from a plant, with or without causing abscission.

degasification (EPA-94/04): A water treatment that removes dissolved gases from he water.

degassing (40CFR467.02-e) is the removal of dissolved hydrogen from the molten aluminum prior to casting. Chemicals are added and gases are bubbled through the molten aluminum. Sometimes a wet scrubber is used to remove excess chlorine gas (under CWA).

degradation products (40CFR165.1-h) means those chemicals resulting from partial decomposition or chemical breakdown of pesticides (under the Federal Insecticide, Fungicide, and Rodenticide Act).

degreaser (40CFR52.741) means any equipment or system used in solvent cleaning.

degreasing (40CFR471.02-j) is the removal of oils and greases from the surface of the metal workpiece. This process can be accomplished with detergents as in alkaline cleaning or by the use of solvents.

dehydrated onions and garlic (40CFR407.71-g) shall mean the processing of dehydrated onions and garlic into the following product styles: Air, vacuum, and freeze dried, all varieties, diced, strips, and other piece sizes ranging from large slices to powder but not including green onions, chives, or leeks.

dehydrated vegetables (40CFR407.71-h) shall mean the processing of dehydrated vegetable in the following product styles: Air, vacuum and freeze dried, blanched and unblanched, peeled and unpeeled, beets, bell peppers, cabbage, carrots, celery, chili pepper, horseradish, turnips, parsnips, parsley, asparagus, tomatoes, green beans, corn, spinach, green onion tops, chives, leeks, whole, diced, and any other piece size ranging from sliced to powder.

delayed compliance order (40CFR65.01-e) shall mean an order issued by a State or by the Administrator to a stationary source which postpones the date by which the source is required to comply with any requirement contained in the applicable State implementation plan.

delayed compliance order (CAA302-42USC7602) means an order issued by the State or by the Administrator to an existing stationary source, postponing the date required under an applicable implementation plan for compliance by such source with any requirement of such plan.

delegated agency (40CFR55.2) means any agency that has been delegated authority to implement and enforce requirements of this part by the Administrator, pursuant to 40CFR55.11 of this part. It can refer to a state agency, a local agency, or an Indian tribe, depending on the delegation status of the program.

delegated state (EPA-94/04): A state (or other governmental entity such as a tribal government) that has received authority to administer an environmental regulatory program in lieu of a federal counterpart. As used in connection with NPDES, UIC, and PWS programs, the term does not connote any transfer of federal authority to a state.

delinquent debt (40CFR13.2-b) means any debt which has not been paid by the date specified by the Government for payment or which has not been satisfied in accordance with a repayment agreement.

delist (EPA-94/04): Use of the petition process to have a facility's toxic designation rescinded.

delivery tank (40CFR60-AA (method 27-2.2)) means any container, including associated pipes and fittings, that is attached to or forms a part of any truck, trailer, or railcar used for the transport of gasoline (under CAA).

delivery tank vapor collection equipment (40CFR60-AA (method 27-2.4)) means any piping, hoses, and devices on the delivery tank used to collect and route gasoline vapors either from the tank to a bulk terminal vapor control system or from a bulk plant or service station into the tank.

delivery vessel (40CFR52.2285-5) means tank trucks and tank trailers used for the delivery of gasoline (other identical or similar definitions are provided in 40CFR52.2286-5).

delivery vessel (40CFR52.741) means any tank truck or trailer equipped with a storage tank that is used for the transport of gasoline to a stationary storage tank at a gasoline dispensing facility, bulk gasoline plant, or bulk gasoline terminal.

demand side measure (40CFR72.2) means a measure:
(1) To improve the efficiency of consumption of electricity from a utility by customers of the utility; or
(2) To reduce the amount of consumption of electricity from a utility by customers of the utility without increasing the use by the customer of fuel other than: Biomass (i.e., combustible energy-producing materials from biological sources, which include wood, plant residues, biological wastes, landfill gas, energy

crops, and eligible components of municipal solid waste), solar, geothermal, or wind resources; or industrial waste gases where the party making the submission involved certifies that there is no net increase in sulfur dioxide emissions from the use of such gases. "Demand-side measure" includes the measures listed in part 73, Appendix A, section 1 of this chapter.

demand-side waste management (EPA-94/04): Prices whereby consumers use purchasing decisions to communicate to product manufacturers that they prefer environmentally sound products packaged with the least amount of waste, made from recycled or recyclable materials, and containing no hazardous substances.

demineralization (EPA-94/04): A treatment process that removes dissolved minerals from water.

demolition (40CFR61.141) means the wrecking or taking out of any load-supporting structural member of a facility together with any related handling operations or the intentional burning of any facility.

demolition (40CFR763.121) means the wrecking or taking out of any load-supporting structural member and any related razing, removing, or stripping of asbestos products.

demonstration (RCRA1004-42USC6903) means the initial exhibition of a new technology process or practice or a significantly new combination or use of technologies, processes or practices, subsequent to the development stage, for the purpose of proving technological feasibility and cost effectiveness.

demonstration period (40CFR76.2) means a period of time not less than 15 months, approved under 40CFR76.10, for demonstrating that the affected unit cannot meet the applicable emission limitation under 40CFR76.5, 76.6, or 76.7 and establishing the minimum NO_x emission rate that the unit can achieve during long-term load dispatch operation.

denitrification (EPA-94/04): The anaerobic biological reduction of nitrate to nitrogen gas.

density (40CFR60.431) means the mass of a unit volume of liquid, expressed as grams per cubic centimeter, kilograms per liter, or pounds per gallon, at a specified temperature.

density (40CFR796.1840-iv) is the mass of a unit volume of a material. It is a function of temperature, hence the temperature at which it is measured should be specified. For a solid, it is the density of the impermeable portion rather than the bulk density. For solids and liquids, suitable units of measurement are g/cm^3. The density of a solution is the mass of a unit volume of the solution and suitable units of measurement are g/cm^3 (other identical or similar definitions are provided in 40CFR796.1860-ii).

density (EPA-94/04): A measure of how heavy a solid, liquid, or gas is for its size.

density of microorganisms (40CFR503.31-c) is the number of microorganisms per unit mass of total solids (dry weight) in the sewage sludge (under CWA).

dentist (10CFR30.4) means an individual licensed by a State or Territory of the United States, the District of Columbia, or the Commonwealth of Puerto Rico to practice dentistry.

denuder (40CFR61.51-g) means a horizontal or vertical container which is part of a mercury chlor-alkali cell and in which water and alkali metal amalgam are converted to alkali metal hydroxide, mercury, and hydrogen gas in a short-circuited, electrolytic reaction.

deny or restrict the use of any defined area for specification (40CFR231.2-c) is to deny or restrict the use of any area for the present or future discharge of any dredged or fill material.

department (40CFR191.02-d) means the Department of Energy.

department (FLPMA103-43USC1702) means a unit of the executive branch of the Federal Government which is headed by a member of the President's Cabinet and the term "agency" means a unit of the executive branch of the Federal Government which is not under the jurisdiction of a head of a department.

department, agency and instrumentality of the United States (40CFR82.82-d) refers to any executive department, military department, or independent establishment within the meaning of 5 U.S.C. 101, 102, and 104(1), respectively, any wholly owned Government corporation, the United States Postal Service and Postal Rate Commission, and all parts of and establishments within the legislative and judicial branches of the United States.

department or agency head (40CFR26.102-a) means the head of any federal department or agency and any other officer or employee of any department or agency to whom authority has been delegated.

departure angle (40CFR86.084.2) means the smallest angle, in a plan side view of an automobile, formed by the level surface on which the automobile is standing and a line tangent to the rear tire static loaded radius arc and touching the underside of the automobile rearward of the rear tire. This definition applies beginning with the 14 model year.

depleted uranium (10CFR40.4) means the source material uranium in which the isotope uranium-235 is less than 0.711 weight percent of the total uranium present. Depleted uranium does not include special nuclear material.

depletion curve (EPA-94/04): In hydraulics, a graphical representation of water depletion from storage-stream channels, surface soil, and groundwater. A depletion curve can be drawn for base flow, direct runoff, or total flow.

depletion or depleted (MMPA3-16USC1362) means any case in which:
(A) the Secretary, after consultation with the Marine Mammal Commission and the Committee of Scientific Advisors on Marine Mammals established under subchapter III of this chapter, determines that a species or population stock is below its optimum sustainable population;
(B) a State, to which authority for the conservation and management of a species or population stock is transferred under section 1379 of this title, determines that such species or stock is below its optimum sustainable population; or
(C) a species or population stock is listed as an endangered species or a threatened species

under the Endangered Species Act of 1973 [16USC1531 et seq.].

deposit (40CFR244.101-e) means the sum paid to the dealer by the consumer when beverages are purchased in returnable beverage containers, and which is refunded when the beverage container is returned.

depository site (40CFR192.01-e) means a disposal site (other than a processing site) selected under section 104(b) or 105(b) of the Act.

depreciation (40CFR14.2-f) is the reduction in value of an item caused by the elapse of time between the date of acquisition and the date of loss or damage (under CAA).

depressurization (EPA-94/04): A condition that occurs when the air pressure inside a structure is lower that the air pressure outside. Depressurization can occur when household appliances such as fireplaces or furnaces, that consume or exhaust house air, are not supplied with enough makeup air. Radon may be drawn into a house more rapidly under depressurized conditions.

depuration (40CFR797.1520-5) is the elimination of a test substance from test organism.

depuration (40CFR797.1830-4) is the elimination of a test chemical from test organism.

depuration or clearance or elimination (40CFR797.1560-3) is the process of losing test material from the test organisms.

depuration phase (40CFR797.1520-6) is the portion of a bioconcentration test after the uptake phase during which the organisms are in flowing water to which no test substance is added.

depuration phase (40CFR797.1830-5) is the portion of a bioconcentration test after the uptake phase during which the organisms are in flowing water to which no test chemical is added.

depuration rate constant (K_2) (40CFR797.1560-4) is the mathematically determined value that is used to define the depuration of test material from

previously exposed test animals when placed in untreated dilution water, usually reported in units per hour.

dermal corrosion (40CFR798.4470-2) is the production of irreversible tissue damage in the skin following the application of the test substance.

dermal exposure (EPA-94/04): Contact between a chemical and the skin.

dermal irritation (40CFR798.4470-1) is the production of reversible inflammatory changes in the skin following the application of a test substance (under TSCA).

dermal toxicity (EPA-94/04): The ability of a pesticide or toxic chemical to poison people or animals by contact with the skin (cf. contact pesticide).

DES (EPA-94/04): A synthetic estrogen, diethylstilbestrol, is used as a growth stimulant in food animals. Residues in meat are thought to be carcinogenic.

desalination (desalinization) (EPA-94/04):
(1) Removing salts from ocean or brackish water by using various technologies.
(2) Removal of salts from soil by artificial means, usually leaching.

desalinization (40CFR35.1605-7): Any mechanical procedure or process where some or all of the salt is removed from lake water and the freshwater portion is returned to the lake.

desiccant (EPA-94/04): A chemical agent that absorbs moisture; some desiccants are capable of drying out plants or insects, causing death.

desiccant (FIFRA2-7USC136) means any substance or mixture of substances intended for artificially accelerating the drying of plant tissue.

design capacity (40CFR240.101-d) means the weight of solid waste of a specified gross calorific value that a thermal processing facility is designed to process in 24 hours of continuous operation; usually expressed in tons per day.

design capacity (40CFR63.301) means the original design capacity of a coke oven battery, expressed in megagrams per year of furnace coke.

design capacity (EPA-94/04): The average daily flow that a treatment plant or other facility is designed to accommodate.

design concept (40CFR51.392) means the type of facility identified by the project, e.g., freeway, expressway, arterial highway, grade-separated highway, reserved right-of-way rail transit, mixed-traffic rail transit, exclusive busway, etc (other identical or similar definitions are provided in 40CFR93.101).

design scope (40CFR51.392) means the design aspects which will affect the proposed facility's impact on regional emissions, usually as they relate to vehicle or person carrying capacity and control, e.g., number of lanes or tracks to be constructed or added, length of project, signalization, access control including approximate number and location of interchanges, preferential treatment for high-occupancy vehicles, etc (other identical or similar definitions are provided in 40CFR93.101).

design value (EPA-94/04): The monitored reading used by EPA to determine an area's air quality status, e.g., for ozone, the fourth highest reading measured over the most recent three years is the design value, for carbon monoxide, the second highest non-overlapping 8-hour concentration for one year is the design value.

designated facility (40CFR260.10) means a hazardous waste treatment, storage, or disposal facility which:
(1) has received a permit (or interim status) in accordance with the requirements of parts 270 and 124 of this chapter,
(2) has received a permit (or interim status) from a State authorized in accordance with part 271 of this chapter, or
(3) is regulated under 40CFR261.6(c)(2) or subpart F of part 266 of this chapter, and
(4) that has been designated on the manifest by the generator pursuant to 40CFR262.20.
If a waste is destined to a facility in an authorized State which has not yet obtained authorization to regulate that particular waste as hazardous, then the designated facility must be a facility allowed by the receiving State to accept such waste.

designated facility (40CFR60.21-b) means any existing facility (see 40CFR60.2(aa)) which emits a designated pollutant and which would be subject to a standard of performance for that pollutant if the existing facility were an affected facility (see also 40CFR60.2).

designated facility (40CFR761.3) means the off-site disposer or commercial storer of PCB waste designated on the manifest as the facility that will receive a manifested shipment of PCB waste.

designated liability area (40CFR52.1475-iv) means the geographic area within which emissions from a source may significantly affect the ambient air quality.

designated management agency (DMA) (40CFR130.2-n) means an agency identified by a WQM plan and designated by the Governor to implement specific control recommendations.

designated pollutant (40CFR60.21-a) means any air pollutant, emissions of which are subject to a standard of performance for new stationary sources but for which air quality criteria have not been issued, and which is not included on a list published under section 108(a) or section 112(b)(1)(A) of the Act.

designated pollutant (EPA-94/04): An air pollutant which is neither a criteria nor hazardous pollutant, as described in the Clean Air Act, but for which new source performance standards exist. The Clean Air Act does require states to control these pollutants, which include acid mist, total reduced sulfur (TRS), and fluorides.

designated project area (40CFR122.25-2) means the portions of the waters of the United States within which the permittee or permit applicant plans to confine the cultivated species, using a method or plan or operation (including, but not limited to, physical confinement) which, on the basis of reliable scientific evidence, is expected to ensure that specific individual organisms comprising an aquaculture crop will enjoy increased growth attributable to the discharge of pollutants, and be harvested within a defined geographic area.

designated representative (29CFR1910.20) means any individual or organization to whom an employee gives written authorization to exercise a right of access. For the purposes of access to employee exposure records and analyses using exposure or medical records, a recognized or certified collective bargaining agent shall be treated automatically as a designated representative without regard to written employee authorization.

designated representative (40CFR70.2) shall have the meaning given to it in section 402(26) of the Act and the regulations promulgated thereunder.

designated representative (40CFR72.2) means a responsible natural person authorized by the owners and operators of an affected source and of all affected units at the source or by the owners and operators of a combustion source or process source, as evidenced by a certificate of representation submitted in accordance with subpart B of this part, to represent and legally bind each owner and operator, as a matter of federal law, in matters pertaining to the Acid Rain Program. Whenever the term "responsible official" is used in part 70 of this chapter, in any other regulations implementing title V of the Act, or in a State operating permit program, it shall be deemed to refer to the "designated representative" with regard to all matters under the Acid Rain Program.

designated representative (CAA402-42USC7651a) means a responsible person or official authorized by the owner or operator of a unit to represent the owner or operator in matters pertaining to the holding, transfer, or disposition of allowances allocated to a unit, and the submission of and compliance with permits, permit applications, and compliance plans for the unit.

designated state agency (40CFR172.21-2) means the State agency designated by State law or other authority to be responsible for registering pesticides to meet special local needs.

designated uses (40CFR131.3-f) are those uses specified in water quality standards for each water body or segment whether or not they are being attained.

designated uses (EPA-94/04): Those water uses identified in state water quality standards that must be achieved and maintained as required under the Clean Water Act. Uses can include cold water fisheries, public water supply, irrigation, etc.

designated volatility attainment area (40CFR80.2-dd) means an area not designated as being in nonattainment with the National Ambient Air Quality Standard for ozone pursuant to rulemaking under section 107(d)(4)(A)(ii) of the Clean Air Act.

designated volatility nonattainment area (40CFR80.2-cc) means any area designated as being in nonattainment with the National Ambient Air Quality Standard for ozone pursuant to rulemaking under section 107(d)(4)(A)(ii) of the Clean Air Act.

designation records (40CFR80.126-c) shall include laboratory analysis reports that identify whether gasoline meets the requirements for a given designation; operational and accounting reports of product storage; and product transfer documents.

designer bugs (EPA-94/04): Popular term for microbes developed through biotechnology that can degrade specific toxic chemicals at their source in toxic waste dumps or in ground water.

desizing facilities (40CFR410.41-c), for NSPS (40CFR410.45), shall mean those facilities that desize more than 50 percent of their total production. These facilities may also perform other processing such as fiber preparation, scouring, mercerizing, functional finishing, bleaching, dyeing and printing.

desorption (40CFR63.321) means regeneration of a carbon adsorber by removal of the perchloroethylene adsorbed on the carbon.

desorption efficiency of a particular compound applied to a sorbent and subsequently extracted with a solvent (40CFR796.1950-i) is the weight of the compound which can be recovered from the sorbent divided by the weight of the compound originally sorbed.

destination facility (40CFR259.10.b) means the disposal facility, the incineration facility, or the facility that both treats and destroys regulated medical waste, to which a consignment of such is intended to be shipped, specified in Box 8 of the Medical Waste Tracking Form.

destination facility (EPA-94/04): The facility to which regulated medical waste is shipped for treatment and destruction, incineration, and/or disposal.

destratification (EPA-94/04): Vertical mixing within a lake or reservoir to totally or partially eliminate separate layers of temperature, plant, or animal life.

destroyed medical waste (EPA-94/04): Regulated medical waste that has been ruined, torn apart, or mutilated through thermal treatment, melting, shredding, grinding, tearing, or breaking, so that it is no longer generally recognized as medical waste, but has not yet been treated (excludes compacted regulated medical waste) (under the Medical Waste Tracking Act).

destroyed regulated medical waste (40CFR259.10.b) means regulated medical waste that is no longer generally recognizable as medical waste because the waste has been ruined, torn apart, or mutilated (it does not mean compaction) through:
(1) Processes such as thermal treatment or melting, during which treatment and destruction could occur; or
(2) Processes such as shredding, grinding, tearing, or breaking, during which only destruction would take place.

destruction (40CFR82.3-o) means the expiration of a controlled substance to the destruction efficiency actually achieved, unless considered completely destroyed as defined in this section. Such destruction does not result in a commercially useful end product and uses one of the following controlled processes approved by the parties to the protocol:
(1) Liquid injection incineration;
(2) Reactor cracking;
(3) Gaseous/fume oxidation;
(4) Rotary kiln incineration; or
(5) Cement kiln.

destruction and removal efficiency (DRE) (EPA-94/04): A percentage that represents the number of molecules of a compound removed or destroyed in an incinerator relative to the number of molecules entered the system (e.g., a DRE of 99.99 percent means that 9,999 molecules are destroyed for every 10,000 that enter; 99.99 percent is known as "four nines"; for some pollutants, the RCRA removal requirement may be a stringent as "six nines").

destruction facility (40CFR259.10.b) means a facility that destroys regulated medical waste by ruining or mutilating it, or tearing it apart.

destruction facility (EPA-94/04): A facility that destroys regulated medical waste by mashing or mutilating it.

destruction or adverse modification (40CFR257.3.2-2) means a direct or indirect alteration of critical habitat which appreciably diminishes the likelihood of the survival and recovery of threatened or endangered species using that habitat.

desulfurization (40CFR72.2) refers to various procedures whereby sulfur is removed from petroleum during or apart from the refining process. "Desulfurization" does not include such processes as dilution or blending of low sulfur content diesel fuel with high sulfur content diesel fuel from a diesel refinery not eligible under 40CFR part 73, subpart G.

desulfurization (EPA-94/04): Removal of sulfur from fossil fuels to reduce pollution (see also air pollution control device).

detectable leak rate (EPA-94/04): The smallest leak (from a storage tank), expressed in terms of gallons- or liters-per-hour, that a test can reliably discern with a certain probability of detection or false alarm.

detection criterion (EPA-94/04): A predetermined rule to ascertain whether a tank is leaking or not. Most volumetric tests use a threshold value as the detection criterion (cf. volumetric tank tests).

detection limit (40CFR136.2-f) means the minimum concentration of an analyte (substance) that can be measured and reported with 99% confidence that the analyte concentration is greater than zero as determined by the procedure set forth at Appendix B of this part.

detection limit (DL) (40CFR300-AA) means the lowest amount that can be distinguished from the normal random "noise" of an analytical instrument or method. For HRS purposes, the detection limit used is the method detection limit (MDL) or, for real-time field instruments, the detection limit of the instrument as used in the field.

detention time (EPA-94/04):
1. The theoretical calculated time required for a small amount of water to pass through a tank at a given rate of flow.
2. The actual time that a small amount of water is in a settling basin, flocculating basin, or rapid-mix chamber.
3. In storage reservoirs, the length of time water will be held before being used.

detergent (EPA-94/04): Synthetic washing agent that helps to remove dirt and oil. Some contain compounds which kill useful bacteria and encourage algae growth when they are in wastewater that reaches receiving waters.

detoxification (40CFR165.1-g): See decontamination.

development effects (EPA-94/04): Adverse effects such as altered growth, structural abnormality, functional deficiency, or death observed in a developing organism.

development facility (40CFR435.11-e) shall mean any fixed or mobile structure subject to this subpart that is engaged in the drilling of productive wells.

developmental toxicity (40CFR798.4350-1) is the property of a chemical that causes in utero death, structural or functional abnormalities or growth retardation during the period of development (other identical or similar definitions are provided in 40CFR798.4500).

developmental toxicity (40CFR798.4420) is the capability of an agent to induce in utero death, structural or functional abnormalities or growth retardation after contact with the pregnant animal (under TSCA).

developmental toxicity (40CFR798.4900-1) is the property of a chemical that causes in utero death, structural or functional abnormalities or growth retardation during the period of development.

device (40CFR167.3) means any device or class of devices as defined by the Act and determined by the Administrator pursuant to section 25(c) to be subject to the provisions of section 7 of the Act.

device (40CFR169.1-c) means any device or class of device as defined by the Act and determined by the Administrator to be subject to the provisions of the Act.

device (40CFR610.11-1): See retrofit device.

device (40CFR710.2-a) shall have the meanings contained in the Federal Food, Drug, and Cosmetic Act, 21USC321 et seq., and the regulations issued under such Act. The meaning of the term, device, defined under the Act was provided below:

Device (21USC321(a)(1)(h)) (except when used in paragraph (n) of this section and in sections 301(i), 403(f), 502(c), and 602(c) [21 USCS 331(i), 343(f), 352(c), 362(c)]) means an instrument, apparatus, implement, machine, contrivance, implant, in vitro reagent, or other similar or related article, including any component, part, or accessory, which is
(1) recognized in the official National Formulary, or the United States Pharmacopeia, or any supplement to them,
(2) intended for use in the diagnosis of disease or other conditions, or in the cure, mitigation, treatment, or prevention of disease, in man or other animals, or
(3) intended to affect the structure or any function of the body of man or other animals, and which does not achieve any of its principal intended purposes through chemical action within or on the body of man or other animals and which is not dependent upon being metabolized for the achievement of any of its principal intended purposes.

(Other identical or similar definitions are provided in 40CFR720.3-a.)

device (FIFRA2-7USC136) means any instrument or

contrivance (other than a firearm) which is intended for trapping, destroying, repelling, or mitigating any pest or any other form of plant or animal life (other than man and other than bacteria, virus, or other microorganism on or in living man or other living animals); but not including equipment used for the application of pesticides when sold separately therefrom.

device integrity (40CFR610.11-9) means the durability of a device and effect of its malfunction on vehicle safety or other parts of the vehicle system (under CWA).

device/measurement device (40CFR195.2) means a unit, component, or system designed to measure radon gas or radon decay products.

dewater (EPA-94/04):
1. Remove or separate a portion of the water in a sludge or slurry to dry the sludge so it can be handled and disposed.
2. Remove or drain the water from a tank or trench.

dewatered (40CFR61.251-c) means to remove the water from recently produced tailings by mechanical or evaporative methods such that the water content of the tailings does not exceed 30 percent by weight.

diagnostic feasibility study (40CFR35.1605-8): A two-part study to determine a lake's current condition and to develop possible methods for lake restoration and protection:
(a) The diagnostic portion of the study includes gathering information and data to determine the limnological, morphological, demographic, socio-economic, and other pertinent characteristics of the lake and its watershed. This information will provide recipients an understanding of the quality of the lake, specifying the location and loading characteristics of significant sources polluting the lake.
(b) The feasibility portion of the study includes:
(1) Analyzing the diagnostic information to define methods and procedures for controlling the sources of pollution;
(2) determining the most energy and cost efficient procedures to improve the quality of the lake for maximum public benefit;

(3) developing a technical plan and milestone schedule for implementing pollution control measures and in-lake restoration procedures; and
(4) if necessary, conducting pilot scale evaluations.

diakinesis and metaphase I (40CFR798.5460-2) are stages of meiotic prophase scored cytologically for the presence of multivalent chromosome association characteristic of translocation carriers.

diaphragm (40CFR85.2122(a)(3)(ii)): See accelerator pump.

diaphragm displacement (40CFR85.2122(a)(1)(ii)-(A)) means the distance through which the center of the diaphragm moves when activated. In the case of a non-modulated stem, diaphragm displacement corresponds to stem displacement.

diatomaceous earth (diatomite) (EPA-94/04): A chalk-like material (fossilized diatoms) used to filter out solid waste in wastewater treatment plants, also used as an active ingredient in some powdered pesticides.

diatomaceous earth filtration (40CFR141.2) means a process resulting in substantial particulate removal in which:
(1) a precoat cake of diatomaceous earth filter media is deposited on a support membrane (septum), and
(2) while the water is filtered by passing through the cake on the septum, additional filter media known as body feed is continuously added to the feed water to maintain the permeability of the filter cake.

diazinon (EPA-94/04): An insecticide. In 1986, EPA banned its use on open areas such as sod farms and golf courses because it posed a danger to migratory birds. The ban did not apply to agricultural, home lawn or commercial establishment uses.

dibenzo-p-dioxin or dioxin (40CFR766.3) means any of a family of compounds which has as a nucleus a triple-ring structure consisting of two benzene rings connected through a pair of oxygen atoms (under TSCA).

dibenzofuran (40CFR766.3) means any of a family of compounds which has as a nucleus a triple-ring structure consisting of two benzene rings connected through a pair of bridges between the benzene rings. The bridges are a carbon-carbon bridge and a carbon-oxygen-carbon bridge at both substitution positions.

dibenzofurans (EPA-94/04): A group of highly toxic organic compounds.

dicofol (EPA-94/04): A pesticide used on citrus fruits.

dielectric material (40CFR280.12) means a material that does not conduct direct electrical current. Dielectric coatings are used to electrically isolate UST systems from the surround soils. Dielectric bushings are used to electrically isolate portions of the UST system (e.g., tank from piping).

dielectric strength (40CFR85.2122(a)(7)(ii)(B)) means the ability of the material of the cap and/or rotor to resist the flow of electric current.

dielectric strength (40CFR85.2122(a)(8)(ii)(E)) means the ability of the spark plugs ceramic insulator material to resist electrical breakdown.

dielectric strength (40CFR85.2122(a)(9)(ii)(C)) means the ability of the material of the coil to resist electrical breakdown.

diesel (40CFR86.090.2) means type of engine with operating characteristics significantly similar to the theoretical Diesel combustion cycle. The non-use of a throttle during normal operation is indicative of a diesel engine.

diesel fuel (40CFR72.2) means a low sulfur fuel oil of grades 1-D or 2-D, as defined by the American Society for Testing and Materials ASTM D975-91, "Standard Specification for Diesel Fuel Oils" (as incorporated by reference in 40CFR72.13).

diesel fuel (40CFR80.2-x) means any fuel sold in any State and suitable for use in diesel motor vehicles and diesel motor vehicle engines, and which is commonly or commercially known or sold as diesel fuel.

diesel oil (40CFR435.11-f) shall refer to the grade of distillate fuel oil, as specified in the American Society for Testing and Materials Standard Specification D975-81, that is typically used as the continuous phase in conventional oil-based drilling fluids. This incorporation by reference was approved by the Director of the Federal Register in accordance with 5 U.S.C. 552(a) and 1CFR part 51. Copies may be obtained from the American Society for Testing and Materials, 1916 Race Street, Philadelphia, PA 19103. Copies may be inspected at the Office of the Federal Register, 800 North Capitol Street, NW., suite 700, Washington, DC.

diesel reciprocating engine unit (40CFR72.2) means an internal combustion engine that combusts only diesel fuel and that thereby generates electricity through the operation of pistons, rather than by heating steam or water.

dietary LC50 (40CFR152.161-a) means a statistically derived estimate of the concentration of a test substance in the diet that would cause 50 percent mortality to the test population under specified conditions.

diffused air (EPA-94/04): A type of aeration that forces oxygen into sewage by pumping air through perforated pipes inside a holding tank.

diffusion (EPA-94/04): The movement of suspended or dissolved particles from a more concentrated to a less concentrated area. The process tends to distribute the particles more uniformly.

digester (EPA-94/04): In wastewater treatment, a closed tank; in solid-waste conversion, a unit in which bacterial action is induced and accelerated in order to break down organic matter and establish the proper carbon to nitrogen ratio (cf. biological treatment).

digester system (40CFR60.281-d) means each continuous digester or each batch digester used for the cooking of wood in white liquor, and associated flash tank(s), below tank(s), chip steamer(s), and condenser(s).

digestion (EPA-94/04): The biochemical decomposition of organic matter, resulting in partial gasification, liquefaction, and mineralization of pollutants.

dike (40CFR260.10) means an embankment or ridge of either natural or man-made materials used to prevent the movement of liquids, sludges, solids, or other materials.

dike (EPA-94/04): A low wall that can act as a barrier to prevent a spill from spreading.

diluent (40CFR165.1-i) means the material added to a pesticide by the user or manufacturer to reduce the concentration of active ingredient in the mixture.

diluent (EPA-94/04): Any liquid or solid material used to dilute or carry an active ingredient.

diluent gas (40CFR60-AF-2.2) means a major gaseous constituent in a gaseous pollutant mixture. For combustion sources, CO_2 and O_2 are the major gaseous constituents of interest.

diluent gas (40CFR72.2) means a major gaseous constituent in a gaseous pollutant mixture, which in the case of emissions from fossil fuel-fired units are carbon dioxide and oxygen.

diluent gas monitor (40CFR72.2) means that component of the continuous emission monitoring system that measures the diluent gas concentration in a unit's flue gas.

dilution ratio (EPA-94/04): The relationship between the volume of water in a stream and the volume of incoming water. It affects the ability of the stream to assimilate waste.

dilution water (40CFR797.1520-7) is the water to which the test substance is added and in which the organisms undergo exposure.

dilution water (40CFR797.1600-5) is the water used to produce the flow-through conditions of the test to which the test substance is added and to which the test species is exposed.

dilution weight (40CFR300-AA) means a parameter in the HRS (hazard ranking system) surface water migration pathway that reduces the point value assigned to targets as the flow or depth of the relevant surface water body increases [unitless] (under CERCLA).

dilution zone (40CFR131.35.d-8): See mixing zone.

dimitic (EPA-94/04): Lakes and reservoirs that freeze over and normally go through two stratification and two mixing cycles a year.

dinocap (EPA-94/04): A fungicide used primarily by apple growers to control summer diseases. EPA proposed restrictions on its use in 1986 when laboratory tests found it caused birth defects in rabbits.

dinoseb (EPA-94/04): A herbicide that is also used as a fungicide and insecticide. It was banned by EPA in 1986 because it posed the risk of birth defects and sterility.

dioxin (40CFR766.3): See dibenzo-p-dioxin or dioxin.

dioxin (EPA-94/04): Any of a family of compounds known chemically as dibenzo-p-dioxins. Concern about them arises from their potential toxicity and contaminants in commercial products. Tests on laboratory animals indicate that it is one of the more toxic man-made compounds.

dioxin/furan (40CFR60.51a) means total tetra-through octachlorinated dibenzo-p-dioxins and dibenzofurans.

dip coating (40CFR52.741) means a method of applying coatings in which the part is submerged in a tank filled with the coating.

dip coating (40CFR60.311) means a method of applying coatings in which the part is submerged in a tank filled with the coatings.

direct application (40CFR420.101-c) means those cold rolling operations which include once-through use of rolling solutions at all mill stands.

direct chill casting (40CFR467.02-f) is the pouring of molten aluminum into a water-cooled mold. Contact cooling water is sprayed onto the aluminum as it is dropped into the mold, and the aluminum ingot falls into a water bath at the end of the casting process (under CWA).

direct chill casting (40CFR471.02-k) is the pouring of molten nonferrous metal into a water-cooled mold. Contact cooling water is sprayed onto the metal as it is dropped into the mold, and the metal ingot falls into a water bath at the end of the casting process (under CWA).

direct discharge (40CFR122.2) means the "discharge of a pollutant."

direct discharger (EPA-94/04): A municipal or industrial facility which introduces pollution through a defined conveyance or system such as outlet pipes; a point source.

direct emissions (40CFR51.852) means those emissions of a criteria pollutant or its precursors that are caused or initiated by the Federal action and occur at the same time and place as the action (other identical or similar definitions are provided in 40CFR93.152).

direct filtration (40CFR141.2) means a series of processes including coagulation and filtration but excluding sedimentation resulting in substantial particulate removal.

direct filtration (EPA-94/04): A method of treating water which consists of the addition of coagulant chemicals, flash mixing, coagulation, minimal flocculation, and filtration. Sedimentation is not used.

direct photolysis (40CFR796.3700-xii) is defined as the direct absorption of light by a chemical followed by a reaction which transforms the parent chemical into one or more products (other identical or similar definitions are provided in 40CFR796.3780-xiii; 796.3800-xii).

direct public utility ownership (40CFR72.2) means direct ownership of equipment and facilities by one or more corporations, the principal business of which is sale of electricity to the public at retail. Percentage ownership of such equipment and facilities shall be measured on the basis of book value.

direct runoff (EPA-94/04): Water that flows over the ground surface or through the ground directly into streams, rivers, and lakes.

direct sale subaccount (40CFR72.2) means a subaccount in the Special Allowance Reserve, as specified in section 416(b) of the Act, which contains Phase II allowances to be sold in the amount of 25,000 per year, from calendar year 1993 to 1999, inclusive, and of 50,000 per year for each year beginning in calendar year 2000, subject to the adjustments noted in the regulations at part 73, subpart E of this chapter.

direct shell evacuation control system (DEC system) (40CFR60.271a) means a system that maintains a negative pressure within the electric arc furnace above the slag or metal and ducts emissions to the control device.

direct shell evacuation system (40CFR60.271-n) means any system that maintains a negative pressure within the EAF above the slag or metal and ducts these emissions to the control device.

direct training (40CFR5.2) means all technical and managerial training conducted directly by EPA for personnel of State and local governmental agencies, other Federal agencies, private industries, universities, and other non-EPA agencies and organizations.

direct transfer equipment (40CFR266.111) means any device (including, but not limited to, such devices as piping, fittings, flanges, valves, and pumps) that is used to distribute, meter, or control the flow of hazardous waste between a container (i.e., transport vehicle) and a boiler of industrial furnace.

directionally-sound strategies (40CFR51.491) are strategies for which adequate procedures to quantify emissions reductions or specify a program baseline are not defined as part of the EIP.

director (40CFR122.2) means the Regional Administrator or the State Director, as the context requires, or an authorized representative. When there is no "approved State program," and there is an EPA administered program, "Director" means the Regional Administrator. When there is an approved State program, "Director" normally means the State

Director. In some circumstances, however, EPA retains the authority to take certain actions even when there is an approved State program. (For example, when EPA has issued an NPDES permit prior to the approval of a State program, EPA may retain jurisdiction over that permit after program approval, see 40CFR123.1.) In such cases, the term "Director" means the Regional Administrator and not the State Director (other identical or similar definitions are provided in 40CFR124.2; 124.2; 124.41; 136.2-d; 124.2; 124.2-b; 144.3; 146.3; 163.2-b; 403.3-e; 270.2; 720.3-d; 723.250-6; 1517.2).

director (40CFR8.2-i) means the Director, Office of Federal Contract Compliance, U.S. Department of Labor, or any person to whom he delegates authority under the regulations of the Secretary of Labor.

director (40CFR8.33-c) means the Director of the Office of Civil Rights and Urban Affairs.

director (OSHA3-29USC652) means the director of the National Institute for Occupational Safety and Health.

director of an approved state (40CFR258.2) means the chief administrative officer of a State agency responsible for implementing the State municipal solid waste permit program or other system of prior approval that is deemed to be adequate by EPA under regulations published pursuant to sections 2002 and 4005 of RCRA.

director of the implementing agency (40CFR280.92) means the EPA Regional Administrator, or, in the case of a state with a program approved under section 9004, the Director of the designated state or local agency responsible for carrying out an approved UST (underground storage tank) program.

director of the office of pollution prevention and toxics (40CFR721.3) means the Director of the EPA Office of Pollution Prevention and Toxics or any EPA employee delegated by the Office Director to carry out the Office Director's functions under this part (other identical or similar definitions are provided in 40CFR723.50-5; 723.175-6).

disadvantaged business concern (CAA1001-42USC7601) means:

(i) a concern which is at least 51 percent owned by one or more socially and economically disadvantaged individuals or, in the case of a publicly traded company, at least 51 percent of the stock of which is owned by one or more socially and economically disadvantaged individuals; and

(ii) the management and daily business operations of which are controlled by such individuals.

disbursement (40CFR35.3105-e) means the transfer of cash from an SRF (State Water Pollution Control Revolving Fund) to an assistance recipient.

disc wheels (29CFR1910.94b) means all power-driven rotatable discs faced with abrasive materials, artificial or natural, and used for grinding or polishing on the side of the assembled disc.

discarded material (40CFR261.2) means a discarded material is any material which is:

(i) Abandoned, as explained in paragraph (b) of this section; or

(ii) Recycled, as explained in paragraph (c) of this section; or

(iii) Considered inherently waste-like, as explained in paragraph (d) of this section.

discharge (40CFR109.2-b) includes, but is not limited to, any spilling, leaking, pumping, pouring, emitting, emptying, or dumping (other identical or similar definitions are provided in 40CFR113.3-e).

discharge (40CFR110.1), when used in relation to section 311 of the Act, includes, but is not limited to, any spilling, leaking, pumping, pouring, emitting, emptying, or dumping, but excludes:

(A) discharges in compliance with a permit under section 402 of the Act,

(B) discharges resulting from circumstances identified and reviewed and made a part of the public record with respect to a permit issued or modified under section 402 of the Act, and subject to a condition in such permit, and

(C) continuous or anticipated intermittent discharges from a point source, identified in a permit or permit application under section 402 of the Act, that are caused by events occurring within the scope of relevant operating or treatment systems.

discharge (40CFR112.2) includes, but is not limited to, any spilling, leaking, pumping, pouring, emitting, emptying or dumping. For purposes of this part, the term discharge shall not include any discharge of oil which is authorized by a permit issued pursuant to section 13 of the River and Harbor Act of 1899 (30 Stat. 1121, 33 U.S.C. 407), or sections 402 or 405 of the FWPCA Amendments of 1972 (86 Stat. 816 et seq., 33 U.S.C. 1251 et seq.).

discharge (40CFR116.3) includes, but is not limited to, any spilling, leaking, pumping, pouring, emitting, emptying or dumping, but excludes:
(A) discharges in compliance with a permit under section 402 of this Act,
(B) discharges resulting from circumstances identified and reviewed and made a part of the public record with respect to a permit issued or modified under section 402 of this Act, and subject to a condition in such permit, and
(C) continuous or anticipated intermittent discharges from a point source, identified in a permit or permit application under section 402 of this Act, which are caused by events occurring within the scope of relevant operating or treatment systems.

discharge (40CFR122.2) when used without qualification means the discharge of a pollutant.

discharge (40CFR140.1-b) includes, but is not limited to, any spilling, leaking, pumping, pouring, emitting, emptying, or dumping.

discharge (40CFR240.101-e) means water-borne pollutants released to a receiving stream directly or indirectly or to a sewerage system.

discharge (40CFR300.5), as defined by section 311(a)(2) the of CWA, includes, but is not limited to, any spilling, leaking, pumping, pouring, emitting, emptying, or dumping of oil, but excludes discharges in compliance with a permit under section 402 of the CWA, discharges resulting from circumstances identified and reviewed and made a part of the public record with respect to a permit issued or modified under section 402 of the CWA, and subject to a condition in such permit, or continuous or anticipated intermittent discharges from a point source, identified in a permit or permit application

under section 402 of the CWA, that are caused by events occurring within the scope of relevant operating or treatment systems. For purposes of the NCP, discharge also means threat of discharge.

discharge (40CFR403.3-g): See indirect discharge or discharge.

discharge (CWA311-33USC1321) includes, but is not limited to, any spilling, leaking, pumping, pouring, emitting, emptying or dumping, but excludes:
(A) discharges in compliance with a permit under section 402 of this Act,
(B) discharges resulting from circumstances identified and reviewed and made a part of the public record with respect to a permit issued or modified under section 402 of this Act, and subject to a condition in such permit, and
(C) continuous or anticipated intermittent discharges from a point source, identified in a permit or permit application under section 402 of this Act, which are caused by events occurring within the scope of relevant operating or treatment systems.

discharge (CWA312-33USC1322) includes, but is not limited to, any spilling, leaking, pumping, pouring, emitting, emptying or dumping.

discharge (CWA502-33USC1362) when used without qualification includes a discharge of a pollutant, and a discharge of pollutants.

discharge (EPA-94/04): Flow of surface water in a stream or canal or the outflow of ground water from a flowing artesian well, ditch, or spring. Can also apply to discharge of liquid effluent from a facility or of chemical emissions into the air through designated venting mechanisms.

discharge (OPA1001) means any emission (other than natural seepage), intentional or unintentional, and includes, but is not limited to, spilling, leaking, pumping, pouring, emitting, emptying, or dumping.

discharge allowance (40CFR461.2-e) means the amount of pollutant (mg per kg of production unit) that a plant will be permitted to discharge. For this category the allowances are specific to battery manufacturing operations.

discharge in connection with activities under the Outer Continental Shelf Lands Act or the Deepwater Port Act of 1974, or which may affect natural resources belonging to, appertaining to, or under the exclusive management authority of the United States (including resources under the Fishery Conservation and Management Act of 1976) (40CFR116.3) means:

(1) A discharge into any waters beyond the contiguous zone from any vessel or onshore or offshore facility, which vessel or facility is subject to or is engaged in activities under the Outer Continental Shelf Lands Act or the Deepwater Port Act of 1974, and

(2) any discharge into any waters beyond the contiguous zone which contain, cover, or support any natural resource belonging to, appertaining to, or under the exclusive management authority of the United States (including resources under the Fishery Conservation and Management Act of 1976) (under FWPCA).

discharge monitoring report (DMR) (40CFR122.2) means the EPA uniform national form, including any subsequent additions, revisions, or modifications for the reporting of self-monitoring results by permittees. DMRs must be used by "approved States" as well as by EPA. EPA will supply DMRs to any approved State upon request. The EPA national forms may be modified to substitute the State Agency name, address, logo, and other similar information, as appropriate, in place of EPA's.

discharge of a pollutant (40CFR122.2) means:

(a) Any addition of any "pollutant" or combination of pollutants to "waters of the United States" from any "point source," or

(b) Any addition of any pollutant or combination of pollutants to the waters of the "contiguous zone" or the ocean from any point source other than a vessel or other floating craft which is being used as a means of transportation. This definition includes additions of pollutants into waters of the United States from: surface runoff which is collected or channelled by man; discharges through pipes, sewers, or other conveyances owned by a State, municipality, or other person which do not lead to a treatment works; and discharges through pipes, sewers, or

other conveyances, leading into privately owned treatment works. This term does not include an addition of pollutants by any "indirect discharger."

discharge of a pollutant and discharge of pollutants (CWA502-33USC1362) each means:

(A) any addition of any pollutant to navigable waters from any point source,

(B) any addition of any pollutant to the waters of the contiguous zone or the ocean from any point source other than a vessel or other floating craft.

discharge of dredged material (40CFR232.2):

(1) Except as provided below in paragraph (2), the term discharge of dredged material means any addition of dredged material into, including any redeposit of dredged material within, the waters of the United States. The term includes, but is not limited to, the following:

(i) The addition of dredged material to a specified discharge site located in waters of the Untied States;

(ii) The runoff or overflow, associated with a dredging operation, from a contained land or water disposal area; and

(iii) Any addition, including any redeposit, of dredged material, including excavated material, into waters of the United States which is incidental to any activity, including mechanized landclearing, ditching, channelization, or other excavation.

(2) The term discharge of dredged material does not include the following:

(i) Discharges of pollutants into waters of the United States resulting from the onshore subsequent processing of dredged material that is extracted for any commercial use (other than fill). These discharges are subject to section 402 of the Clean Water Act even though the extraction and deposit of such material may require a permit from the Corps or applicable state.

(ii) Activities that involve only the cutting or removing of vegetation above the ground (e.g., mowing, rotary cutting, and chainsawing) where the activity neither substantially disturbs the root system nor involves mechanized pushing, dragging, or other similar activities that redeposit excavated soil material.

(3) Section 404 authorization is not required for the following:

    (i) Any incidental addition, including redeposit, of dredged material associated with any activity that does not have or would not have the effect of destroying or degrading an area of waters of the U.S. as defined in paragraphs (4) and (5) of this definition; however, this exception does not apply to any person preparing to undertake mechanized landclearing, ditching, channelization and other excavation activity in a water of the United States, which would result in a redeposit of dredged material, unless the person demonstrates to the satisfaction of the Corps, or EPA as appropriate, prior to commencing the activity involving the discharge, that the activity would not have the effect of destroying or degrading any area of waters of the United States, as defined in paragraphs (4) and (5) of this definition. The person proposing to undertake mechanized landclearing, ditching, channelization or other excavation activity bears the burden of demonstrating that such activity would not destroy or degrade any area of waters of the United States.

    (ii) Incidental movement of dredged material occurring during normal dredging operations, defined as dredging for navigation in navigable waters of the United States, as that term is defined in 33CFR part 329, with proper authorization from the Congress or the Corps pursuant to 33CFR part 322; however, this exception is not applicable to dredging activities in wetlands, as that term is defined at 40CFR232.2(r) of this Chapter.

    (iii) Those discharges of dredged material associated with ditching, channelization or other excavation activities in waters of the United States, including wetlands, for which Section 404 authorization was not previously required, as determined by the Corps district in which the activity occurs or would occur, provided that prior to August 25, 1993, the excavation activity

commenced or was under contract to commence work and that the activity will be completed no later that August 25, 1994. This provision does not apply to discharges associated with mechanized landclearing. For those excavation activities that occur on an ongoing basis (either continuously or periodically), e.g., mining operations, the Corps retains the authority to grant, on a case-by-case basis, an extension of this 12-month grandfather provision provided that the discharger has submitted to the Corps within the 12-month period an individual permit application seeking Section 404 authorization for such excavation activity. In no event can the grandfather period under this paragraph extend beyond August 25, 1996.

    (iv) Certain discharges, such as those associated with normal farming, silviculture, and ranching activities, are not prohibited by or otherwise subject to regulation under Section 404. See 40CFR232.3 for discharges that do not require permits.

(4) For purposes of this section, an activity associated with a discharge of dredged material destroys an area of waters of the United States if it alters the area in such a way that it would no longer be a water of the United States. Note: Unauthorized discharges into waters of the United States do not eliminate Clean Water Act jurisdiction, even where such unauthorized discharges have the effect of destroying waters of the United States.

(5) For purposes of this section, an activity associated with a discharge of dredged material degrades an area of waters of the United States if it has more than a de minimis (i.e., inconsequential) effect on the area by causing an identifiable individual or cumulative adverse effect on any aquatic function.

discharge of dredged material (40CFR257.3.3) can be found in the Clean Water Act, as amended, 33USC1251 et seq., and implementing regulations, specifically 33CFR323 (42FR37122, July 19, 1977).

discharge of fill material (40CFR232.2):
(1) The term discharge of fill material means the

addition of fill material into waters of the United States. The term generally includes, without limitation, the following activities: Placement of fill that is necessary for the construction of any structure in a water of the United States; the building of any structure or impoundment requiring rock, sand, dirt, or other material for its construction; site-development fills for recreational, industrial, commercial, residential, and other uses; causeways or road fills; dams and dikes; artificial islands; property protection and/or reclamation devices such as riprap, groins, seawalls, breakwaters, and revetments; beach nourishment; levees; fill for structures such as sewage treatment facilities, intake and outfall pipes associated with power plants and subaqueous utility lines; and artificial reefs.

(2) In addition, placement of pilings in waters of the United States constitutes a discharge of fill material and requires a Section 404 permit when such placement has or would have the effect of a discharge of fill material. Examples of such activities that have the effect of a discharge of fill material include, but are not limited to, the following: Projects where the pilings are so closely spaced that sedimentation rates would be increased; projects in which the pilings themselves effectively would replace the bottom of a waterbody; projects involving the placement of pilings that would reduce the reach or impair the flow or circulation of waters of the United States; and projects involving the placement of pilings which would result in the adverse alteration or elimination of aquatic functions:

(i) Placement of pilings in waters of the United States that does not have or would not have the effect of a discharge of fill material shall not require a Section 404 permit. Placement of pilings for linear projects, such as bridges, elevated walkways, and powerline structures, generally does not have the effect of a discharge of fill material. Furthermore, placement of pilings in waters of the United States for piers, wharves, and an individual house on stilts generally does not have the effect of a discharge of fill material. All pilings, however, placed in the navigable waters of the United States, as that term is defined in 33CFR part 329,

require authorization under section 10 of the Rivers and Harbors Act of 1899 (see 33CFR part 322).

(ii) [Reserved]

discharge or hazardous waste discharge (40CFR260.10) means the accidental or intentional spilling, leaking, pumping, pouring, emitting, emptying, or dumping of hazardous waste into or on any land or water.

discharge point (40CFR230.3-h) means the point within the disposal site at which the dredged or fill material is released.

discharge of pollutants (40CFR401.11-h) means:
(1) The addition of any pollutant to navigable waters from any point source and
(2) any addition of any pollutant to the waters of the contiguous zone or the ocean from any point source, other than from a vessel or other floating craft. The term "discharge" includes either the discharge of a single pollutant or the discharge of multiple pollutants.

discrete wet scrubbing device (40CFR464.31-i) means a discrete wet scrubbing device is a distinct, stand-alone device that removes particulates and fumes from a contaminated gas stream by bringing the gas stream into contact with a scrubber liquor, usually water, and from which there is a wastewater discharge. Examples of discrete wet scrubbing devices are: spray towers and chambers, venturi scrubbers (fixed and variable), wet caps, packed bed scrubbers, quenchers, and orifice scrubbers. Semi-wet scrubbing devices where water is added and totally evaporates prior to dry air pollution control are not considered to be discrete wet scrubbing devices. Ancillary scrubber operations such as fan washes and backwashes are not considered to be discrete wet scrubber devices. These ancillary operations are covered by the mass limitations of the associated scrubber. Aftercoolers are not considered to be discrete wet scrubbing devices, and water dischargers from aftercooling are not regulated as a process wastewater in this category.

discretionary economic incentive program (40CFR51.491) means any EIP submitted to the EPA as an implementation plan revision for purposes other than to comply with the statutory requirements of sections 182(g)(3), 182(g)(5), 187(d)(3), or 187(g) of the Act.

disease vector (40CFR257.3.6-2) means rodents, flies, and mosquitoes capable of transmitting disease to humans.

disinfectant (40CFR141.2) means any oxidant, including but not limited to chlorine, chlorine dioxide, chloramines, and ozone added to water in any part of the treatment or distribution process, that is intended to kill or inactivate pathogenic microorganisms.

disinfectant (EPA-94/04): A chemical or physical process that kills pathogenic organisms in water. Chlorine is often used to disinfect sewage treatment effluent, water supplies, wells, and swimming pools.

disinfectant contact time (T in CT calculations) (40CFR141.2) means the time in minutes that it takes for water to move from the point of disinfectant application or the previous point of disinfectant residual measurement to a point before or at the point where residual disinfectant concentration ("C") is measured. Where only one "C" is measured, "T" is the time in minutes that it takes for water to move from the point of disinfectant application to a point before or at where residual disinfectant concentration ("C") is measured. Where more than one "C" is measured, "T" is:

(a) for the first measurement of "C," the time in minutes that it takes for water to move from the first or only point of disinfectant application to a point before or at the point where the first "C" is measured and

(b) for subsequent measurements of "C," the time in minutes that it takes for water to move from the previous "C" measurement point to the "C" measurement point for which the particular "T" is being calculated. Disinfectant contact time in pipelines must be calculated based on "plug flow" by dividing the internal volume of the pipe by the maximum hourly flow rate through that pipe. Disinfectant contact time within mixing basins and storage reservoirs must be determined by tracer studies or an equivalent demonstration.

disinfectant time (EPA-94/04): The time it takes

water to move from the point of disinfectant application (or the previous point of residual disinfectant measurement) to a point before or at the point where the residual disinfectant is measured. In pipelines, the time is calculated by dividing the internal volume of the pipe by the maximum hourly flow rate; within mixing basins and storage reservoirs it is determined by tracer studies of an equivalent demonstration.

disinfection (40CFR141.2) means a process which inactivates pathogenic organisms in water by chemical oxidants or equivalent agents.

disinfection byproduct (EPA-94/04): A compound formed by the reaction of a disinfectant such as chlorine with organic material in the water supply (under SDWA).

dispatch (40CFR72.2) means the assignment within a dispatch system of generating levels to specific units and generators to effect the reliable and economical supply of electricity, as customer demand rises or falls, and includes:

(1) The operation of high-voltage lines, substations, and related equipment; and

(2) The scheduling of generation for the purpose of supplying electricity to other utilities over interconnecting transmission lines.

dispatch system (40CFR72.2) means either:

(1) A specified unit and generator or specified group of units, and portions of units, and generators that are interconnected and centrally dispatched, provided that the requirements of 40CFR72.33 are met; or

(2) In the event the requirements specified in paragraph (1) of this definition are not met, the unit and generator or group of units and generators that make up one utility system.

dispensed fuel temperature (40CFR86.098.2) means the temperature (deg.F or deg.C may be used) of the fuel being dispensed into the tank of the test vehicle during a refueling test.

dispenser (40CFR211.203-h) means the permanent (intended to be refilled) or disposable (discarded when empty) container designed to hold more than one complete set of hearing protector(s) for the express purpose of display to promote sale or display to promote use or both.

dispersant (EPA-94/04): A chemical agent used to break up concentrations of organic material such as spilled oil.

dispersants (40CFR300.5) means those chemical agents that emulsify, disperse, or solubilize oil into the water column or promote the surface spreading of oil slicks to facilitate dispersal of the oil into the water column.

dispersion factor (40CFR503.41-d) is the ratio of the increase in the ground level ambient air concentration for a pollutant at or beyond the property line of the site where the sewage sludge incinerator is located to the mass emission rate for the pollutant from the incinerator stack.

dispersion resin (40CFR61.61-g) means a resin manufactured in such a way as to form fluid dispersions when dispersed in a plasticizer or plasticizer/diluent mixtures.

dispersion technique (40CFR51.100-hh) means:
(1) Any technique which attempts to affect the concentration of a pollutant in the ambient air by:
  (i)   Using that portion of a stack which exceeds good engineering practice stack height;
  (ii) Varying the rate of emission of a pollutant according to atmospheric conditions or ambient concentrations of that pollutant; or
  (iii) Increasing final exhaust gas plume rise by manipulating source process parameters, exhaust gas parameters, stack parameters, or combining exhaust gases from several existing stacks into one stack; or other selective handling of exhaust gas streams so as to increase the exhaust gas plume rise.
(2) The preceding sentence does not include:
  (i)   The reheating of a gas stream, following use of a pollution control system, for the purpose of returning the gas to the temperature at which it was originally discharged from the facility generating the gas stream;
  (ii) The merging of exhaust gas streams where:

(A) The source owner or operator demonstrates that the facility was originally designed and constructed with such merged gas streams;
(B) After July 8, 1985 such merging is part of a change in operation at the facility that includes the installation of pollution controls and is accompanied by a net reduction in the allowable emissions of a pollutant. This exclusion from the definition of dispersion techniques shall apply only to the emission limitation for the pollutant affected by such change in operation; or
(C) Before July 8, 1985, such merging was part of a change in operation at the facility that included the installation of emissions control equipment or was carried out for sound economic or engineering reasons. Where there was an increase in the emission limitation or, in the event that no emission limitation was in existence prior to the merging, an increase in the quantity of pollutants actually emitted prior to the merging, the reviewing agency shall presume that merging was significantly motivated by an intent to gain emissions credit for greater dispersion. Absent a demonstration by the source owner or operator that merging was not significantly motivated by such intent, the reviewing agency shall deny credit for the effects of such merging in calculating the allowable emissions for the source;
  (iii) Smoke management in agricultural or silvicultural prescribed burning programs;
  (iv) Episodic restrictions on residential woodburning and open burning; or
  (v)   Techniques under 40CFR51.100(hh)(1)(iii) which increase final exhaust gas plume rise where the resulting allowable emissions of sulfur dioxide from the facility do not exceed 5,000 tons per year.

displacement (40CFR264.18(a)(2)(ii)) means the relative movement of any two sides of a fault measured in any direction.

displacement (40CFR503.21-e) is the relative movement of any two sides of a fault measured in any direction.

displacement and displacement class (40CFR86.402.78): See 40CFR86.419.

disposable device (40CFR211.203-i) means a hearing protective device that is intended to be discarded after one period of use.

disposable pay (40CFR13.2-h) means that part of current basic pay, special pay, incentive pay, retired pay, retainer pay, or in the case of an employee not entitled to basic pay, other authorized pay remaining after the deduction of any amount described in 5CFR581.105 (b) through (f). These deductions include but are not limited to: Social security withholdings; Federal, State and local tax withholdings; health insurance premiums; retirement contributions; and life insurance premiums.

disposables (EPA-94/04): Consumer products, other items, and packaging used once or a few times and discarded.

disposal (40CFR191.02-l) means permanent isolation of spent nuclear fuel or radioactive waste from the accessible environment with no intent of recovery, wether or not such isolation permits the recovery of such fuel or waste. For example, disposal of waste in a mined geologic repository occurs when all of the shafts to the repository are backfilled and sealed.

disposal (40CFR245.101-b) means the collection, storage, treatment, utilization, processing, or final disposal of solid waste.

disposal (40CFR257.2) means the discharge, deposit, injection, dumping, spilling, leaking, or placing of any solid waste or hazardous waste into or on any land or water so that such solid waste or hazardous waste or any constituent thereof may enter the environment or be emitted into the air or discharged into any waters, including ground waters.

disposal (40CFR260.10) means the discharge, deposit, injection, dumping, spilling, leaking, or placing of any solid waste or hazardous waste into or on any land or water so that such solid waste or

hazardous waste or any constituent thereof may enter the environment or be emitted into the air or discharged into any waters, including ground waters.

disposal (40CFR270.2) means the discharge, deposit, injection, dumping, spilling, leaking, or placing of any hazardous waste into or on any land or water so that such hazardous waste or any constituent thereof may enter the environment or be emitted into the air or discharged into any waters, including ground water.

disposal (40CFR373.4-d) means the discharge, deposit, injection, dumping, spilling, leaking or placing of any hazardous substance into or on any land or water so that such hazardous substance or any constituent thereof may enter the environment or be emitted into the air or discharged into any waters, including groundwater.

disposal (40CFR761.3) means intentionally or accidentally to discard, throw away, or otherwise complete or terminate the useful life of PCBs and PCB Items. Disposal includes spills, leaks, and other uncontrolled discharges of PCBs as well as actions related to containing, transporting, destroying, degrading, decontaminating, or confining PCBs and PCB Items.

disposal (40CFR82.152-e) means the process leading to and including:
(1) The discharge, deposit, dumping or placing of any discarded appliance into or on any land or water;
(2) The disassembly of any appliance for discharge, deposit, dumping or placing of its discarded component parts into or on any land or water; or
(3) The disassembly of any appliance for reuse of its component parts.

disposal (EPA-94/04): Final placement or destruction of toxic, radioactive, or other wastes; surplus or banned pesticides or other chemicals; polluted soils; and drums containing hazardous materials from removal actions or accidental releases. Disposal may be accomplished through use of approved secure landfills, surface impoundments, land farming, deep-well injection, ocean dumping, or incineration.

disposal (RCRA1004-42USC6903) means the

discharge, deposit, injection, dumping, spilling, leaking, or placing of any solid waste or hazardous waste into or on any land or water so that such solid waste to hazardous waste or any constituent thereof may enter the environment or be emitted into the air or discharged into any waters, including ground waters.

disposal area (40CFR192.31-f) means the region within the perimeter of an impoundment or pile containing uranium by product materials to which the post-closure requirements of 40CFR192.32(b)(1) of this subpart apply.

disposal facility (40CFR260.10) means a facility or part of a facility at which hazardous waste is intentionally placed into or on any land or water, and at which waste will remain after closure.

disposal facility (40CFR270.2) means a facility or part of a facility at which hazardous waste is intentionally placed into or on the land or water, and at which hazardous waste will remain after closure.

disposal, hazardous waste, and treatment (SF101-42USC9601) shall have the meaning provided in section 1004 of the Solid Waste Disposal Act.

disposal site (40CFR192.01-d) means the region within the smallest perimeter of residual radioactive material (excluding cover materials) following completion of control activities.

disposal site (40CFR192.31-e) means a site selected pursuant to section 83 of the Act.

disposal site (40CFR228.2-a) means an interim for finally approved and precise geographical area within which ocean dumping of wastes is permitted under conditions specified in permits issued under sections 102 and 103 of the Act. Such sites are identified by boundaries established by:
(1) coordinates of latitude and longitude for each corner, or by
(2) coordinates of latitude and longitude for the center point and a radius in nautical miles from that point.
Boundary coordinates shall be identified as precisely as is warranted by the accuracy with which the site can be located with existing navigational aids or by

the implantation of transponders, buoys or other means of marking the site.

disposal site (40CFR230.3-i) means that portion of the "waters of the United States" where specific disposal activities are permitted and consist of a bottom surface area and any overlying volume of water. In the case of wetlands on which surface water is not present, the disposal site consists of the wetland surface area.

disposal site designation study (40CFR228.2-d) means the collection, analysis and interpretation of all available pertinent data and information on a proposed disposal site prior to use, including but not limited to, that from baseline surveys, special purpose surveys of other Federal agencies, public data archives, and social and economic studies and records of areas which would be affected by use of the proposed site.

disposal site evaluation study (40CFR228.2-c) means the collection, analysis, and interpretation of all pertinent information available concerning an existing disposal site, including but not limited to, data and information from trend assessment surveys, monitoring surveys, special purpose surveys of other Federal agencies, public data archives, and social and economic studies and records of affected areas.

disposal system (40CFR191.12) means any combination of engineered and natural barriers that isolate spent nuclear fuel or radioactive waste after disposal.

disposal well (40CFR146.3) means a well used for the disposal of waste into a subsurface (other identical or similar definitions are provided in 40CFR147.2902).

disposer of PCB waste (40CFR761.3), as the term is used in subparts J and K of this part, means any person who owns or operates a facility approved by EPA for the disposal of PCB waste which is regulated for disposal under the requirements of subpart D of this part.

dispute (40CFR791.3-d) refers to a present controversy between parties subject to a test rule over the amount or method of reimbursement for the cost of developing health and environmental data on the test chemical.

dissociation (40CFR796.1370) is the reversible splitting into two or more chemical species which may be ionic. The process is indicated generally by the $RX = R^+ + X^-$ and the concentration equilibrium constant governing the reaction is $K = ([R^+][X^-])/[RX]$. For example, in the particular case where R is hydrogen (the substance is and acid), the constant is $K_a = [H^+] \cdot [X^-]/[HX]$ or $pK_a = pH - log([X^-]/[HX])$.

dissolved (40CFR136-AC-1) means those elements which will pass through a 0.45 μm membrane filter.

dissolved oxygen (DO) (EPA-94/04): The oxygen freely available in water, vital to fish and other aquatic life and for the prevention of odors. DO levels are considered a most important indicator of a water body's ability to support desirable aquatic life. Secondary and advanced waste treatment are generally designed to ensure adequate DO in waste-receiving waters.

dissolved solids (EPA-94/04): Disintegrated organic and inorganic material in water. Excessive amounts make water unfit to drink or use in industrial processes.

distance piece (40CFR60.481) means an open or enclosed casing through which the piston rod travels, separating the compressor cylinder from the crankcase.

distance weight (40CFR300-AA) means a parameter in the HRS air migration, ground water migration, and soil exposure pathways that reduces the point value assigned to targets as their distance increases from the site [unitless].

distillate oil (40CFR60.41b, see also 40CFR60.41c) means fuel oils that contain 0.05 weight percent nitrogen or less and comply with the specifications for fuel oils numbers 1 and 2, as defined by the American Society of Testing and Materials in ASTM D396-78, Standard Specifications for Fuel Oils (incorporated by reference, see 40CFR60.17).

distillate receiver (40CFR264.1031) means a

container or tank used to receive and collect liquid material (condensed) from the overhead condenser of a distillation unit and from which the condensed liquid is pumped to larger storage tanks or other process units.

distillate receiver (40CFR63.101) means overhead receivers, overhead accumulators, reflux drums, and condenser(s) including ejector-condenser(s) associated with a distillation unit.

distillate receiver (40CFR63.111) means overhead receivers, overhead accumulators, reflux drums, and condenser(s) including ejector-condenser(s) associated with a distillation unit.

distillation (EPA-94/04): The act of purifying liquids through boiling, so that the steam condenses to a pure liquid and the pollutants remain in a concentrated residue.

distillation operation (40CFR264.1031) means an operation, either batch or continuous, separating one or more feed stream(s) into two or more exit streams, each exit stream having component concentrations different from those in the feed stream(s). The separation is achieved by the redistribution of the components between the liquid and vapor phase as they approach equilibrium within the distillation unit.

distillation operation (40CFR60.661) means an operation separating one or more feed stream(s) into two or more exit stream(s), each exit stream having component concentrations different from those in the feed stream(s). The separation is achieved by the redistribution of the components between the liquid and vapor phase as they approach equilibrium within the distillation unit.

distillation unit (40CFR60.661) means a device or vessel in which distillation operations occur, including all associated internals (such as trays or packing) and accessories (such as reboiler, condenser, vacuum pump, steam jet, etc.) plus any associated recovery system.

distillation unit (40CFR63.101) means a device or vessel in which one or more feed streams are separated into two or more exit streams, each exit stream having component concentrations different from those in the feed stream(s). The separation is achieved by the redistribution of the components between the liquid and the vapor phases by vaporization and condensation as they approach equilibrium within the distillation unit. Distillation unit includes the distillate receiver, reboiler, and any associated vacuum pump or steam jet.

distillation unit (40CFR63.111) means a device or vessel in which one or more feed streams are separated into two or more exit streams, each exit stream having component concentrations different from those in the feed stream(s). The separation is achieved by the redistribution of the components between the liquid and the vapor phases by vaporization and condensation as they approach equilibrium within the distillation unit. Distillation unit includes the distillate receiver, reboiler, and any associated vacuum pump or steam jet.

distribute in commerce (40CFR205.2-18) means sell in, offer for sale in, or introduce or deliver for introduction into, commerce.

distribute in commerce (40CFR720.3-i) means to sell in commerce, to introduce or deliver for introduction into commerce, or to hold after introduction into commerce.

distribute in commerce (40CFR763.163) has the same meaning as in section 3 of the Act, but the term does not include actions taken with respect to an asbestos-containing product (to sell, resale, deliver, or hold) in connection with the end use of the product by persons who are users (persons who use the product for its intended purpose after it is manufactured or processed). The term also does not include distribution by manufacturers, importers, and processors, and other persons solely for purposes of disposal of an asbestos-containing product.

distribute in commerce (NCA3-42USC4902) means sell in, offer for sale in, or introduce or deliver for introduction into, commerce.

distribute in commerce (TSCA3-15USC2602): See distribute in commerce and distribution in commerce.

distribute in commerce and distribution in commerce

(40CFR710.2-j) when used to describe an action taken with respect to a chemical substance or mixture or article containing a substance or mixture, mean to sell or the sale of, the substance, mixture, or article in commerce; to introduce or deliver for introduction into commerce, or the introduction or delivery for introduction into commerce of, the substance, mixture, or article; or to hold, or the holding of, the substance, mixture, or article after its introduction into commerce (under the Toxic Substances Control Act).

distribute in commerce and distribution in commerce (40CFR761.3) when used to describe an action taken with respect to a chemical substance, mixture, or article containing a substance or mixture means to sell, or the sale of, the substance, mixture, or article in commerce; to introduce or deliver for introduction into commerce, or the introduction or delivery for introduction into commerce of the substance, mixture, or article; or to hold or the holding of, the substance, mixture, or article after its introduction into commerce.

distribute in commerce and distribution in commerce (TSCA3-15USC2602) when used to describe an action taken with respect to a chemical substance or mixture or article containing a substance or mixture mean to sell, or the sale of, the substance, mixture, or article in commerce; to introduce or deliver for introduction into commerce, or the introduction or delivery for introduction into commerce of, the substance, mixture, or article after its introduction into commerce (other identical or similar definitions are provided in 40CFR710.2-j; 723.175-5; 762.3-d).

distribute in commerce solely for export (40CFR747.115-4; 747.195-4): See process or distribute in commerce solely for export.

distribute or sell, distributed or sold, or distribution or sale (40CFR152.3-j) means the acts of distributing, selling, offering for sale, holding for sale, shipping, holding for shipment, delivering for shipment, or receiving and (having so received) delivering or offering to deliver, or releasing for shipment to any person in any State.

distributed or sold (40CFR152.3-j): See distribute or sell.

distribution coefficient (Kd) (40CFR300-AA) means a measure of the extent of partitioning of a substance between geologic materials (for example, soil, sediment, rock) and water (also called partition coefficient). The distribution coefficient is used in the HRS in evaluating the mobility of a substance for the ground water migration pathway [ml/g].

distribution in commerce (TSCA3-15USC2602): See distribute in commerce and distribution in commerce.

distribution law which applies only to individual molecular species in solution: See 40CFR796.1550.

distribution or sale (40CFR152.3-j): See distribute or sell.

distributor (40CFR244.101-f) means any person who engages in the sale of beverages, in beverage containers, to a dealer, including any manufacturer who engages in such sale.

distributor (40CFR749.68-8) means any person who distributes in commerce water treatment chemicals for use in cooling systems.

distributor (40CFR80.2-l) means any person who transports or stores or causes the transportation or storage of gasoline or diesel fuel at any point between any gasoline or diesel fuel refinery or importer's facility and any retail outlet or wholesale purchaser-consumer's facility.

distributor (40CFR82.104-g) means a person to whom a product is delivered or sold for purposes of subsequent resale, delivery or export (under the Clean Air Act).

distributor (40CFR85.2122(a)(11)(ii)(A)) means a device for directing the secondary current from the induction coil to the park plugs at the proper intervals and in the proper firing order (under the Clean Air Act).

distributor firing angle (40CFR85.2122(a)(11)(ii)(B)) means the angular relationship of breaker point opening from one opening to the next in the firing sequence.

district court (FIFRA2-7USC136) means a United States district court, the District Court of Guam, the District Court of the Virgin Islands, and the highest court of American Samoa.

district court of the United States (MMPA3-16USC1362) includes the District Court of Guam, District Court of the Virgin Islands, District Court of Puerto Rico, District Court of the Canal Zone, and, in the case of American Samoa and the Trust Territory of the Pacific Islands, the District Court of the United States for the District of Hawaii.

diurnal breathing losses (40CFR86.082.2) means evaporative emissions as a result of the daily range in temperature.

diurnal breathing losses (40CFR86.096.2) means diurnal emissions.

diurnal emissions (40CFR86.096.2) means evaporative emissions resulting from the daily cycling of ambient temperatures.

diversion (EPA-94/04):
1. Use of part of a stream flow as a water supply.
2. A channel with a supporting ridge on the lower side constructed across a slope to divert water at a non-erosive velocity to sites where it can be used or disposed of.

diversion rate (EPA-94/04): The percentage of waste materials diverted from traditional disposal such as landfilling or incineration to be recycled, composted, or re-used.

diverter valve (40CFR63.321) means a flow control device that prevents room air from passing through a refrigerated condenser when the door of the dry cleaning machine is open.

DMA (40CFR130.2-n): Designated management agency.

DMR (40CFR122.2): Discharge monitoring report.

DNA hybridization (EPA-94/04): Use of a segment of DNA, called a DNA probe, to identify its complementary DNA; used to detect specific genes.

doilies (40CFR250.4-k) means paper place mats used

on food service trays in hospitals and other institutions.

domestic (40CFR704.3) means within the geographical boundaries of the 50 United States, including the District of Columbia, the Commonwealth of Puerto Rico, the Virgin Islands, Guam, American Samoa, the Northern Mariana Islands, and any other territory or possession of the United States.

domestic construction material (40CFR35.936.13-iii) means an unmanufactured construction material which has been mined or produced in the United States, or a manufactured construction material which has been manufactured in the United States if the cost of its components which are mined, produced, or manufactured in the United States exceeds 50 percent of the cost of all its components.

domestic or other non-distribution system plumbing problem (40CFR141.2) means a coliform contamination problem in a public water system with more than one service connection that is limited to the specific service connection from which the coliform-positive sample was taken.

domestic septage (40CFR503.9-f) is either liquid or solid material removed from a septic tank, cesspool, portable toilet, Type III marine sanitation device, or similar treatment works that receives only domestic sewage. Domestic septage does not include liquid or solid material removed from a septic tank, cesspool, or similar treatment works that receives either commercial wastewater or industrial wastewater and does not include grease removed from a grease trap at a restaurant.

domestic sewage (40CFR261.4-(a)(1)(ii)) means untreated sanitary wastes that pass through a sewer system.

domestic sewage (40CFR503.9-g) is waste and wastewater from humans or household operations that is discharged to or otherwise enters a treatment works.

domestic waste (40CFR435.11-g) shall refer to materials discharged from sinks, showers, laundries, safety showers, eye-wash stations, hand-wash stations, fish cleaning stations, and galleys located within facilities subject to this subpart.

dominant lethal mutation (40CFR798.5450) is one occurring in a germ cell which does not cause dysfunction of the gamete, but which is lethal to the fertilized egg or developing embryo.

dosage/dose (EPA-94/04): The actual quantity of a chemical administered to an organism or to which it is exposed.

dose (40CFR795.260-2) is the amount of test substance administered. Dose is expressed as weight of test substance (g, mg) per unit weight of test animal (e.g., mg/kg), or as weight of test substance per unit weight of food or drinking water.

dose (40CFR798.1100-2) is the amount of test substance applied. Dose is expressed as weight of test substance (g, mg) per unit weight of test animal (e.g., mg/kg).

dose (40CFR798.1175-2) is the amount of test substance administered. Dose is expressed as weight of test substance (g, mg) per unit weight of test animal (e.g., mg/kg) (other identical or similar definitions are provided in 40CFR798.4900-2).

dose (40CFR798.2250-2) in a dermal test, is the amount of test substance applied to the skin (applied daily in subchronic tests). Dose is expressed as weight of the substance (g, mg) per unit weight of test animal (e.g., mg/kg).

dose (40CFR798.2450-5) refers to an exposure level. Exposure is expressed as weight or volume of test substance per volume of air (mg/L), or as parts per million (ppm).

dose (40CFR798.2650-2) is the amount of test substance administered. Dose is expressed as weight of test substance (g, mg) per unit weight of test animal (e.g., mg/kg), or as weight of test substance per unit weight of food or drinking water (other identical or similar definitions are provided in 40CFR798.2675-2).

dose equivalent (40CFR141.2) means the product of the absorbed dose from ionizing radiation and such factors as account for differences in biological effectiveness due to the type of radiation and its distribution in the body as specified by the International Commission on Radiological Units and Measurements (ICRU).

dose equivalent (40CFR190.02-h) means the product of absorbed dose and appropriate factors to account for differences in biological effectiveness due to the quality of radiation and its partial distribution in the body. The unit of dose equivalent is the "rem." (One millirem (mrem)=0.001 rem.)

dose equivalent (40CFR191.12) means the product of absorbed dose and appropriate factors to account for differences in biological effectiveness due to the quality of radiation and its spatial distribution in the body; the unit of dose equivalent is the "rem" ("sievert" in SI units).

dose equivalent (EPA-94/04): The product of the absorbed does from ionizing radiation and such factors as account for biological differences due to the type of radiation and its distribution in the body as specified by the International Commission on Radiological Units and Measurements.

dose response (40CFR798.1100-3) is the relationship between the dose and the proportion of a population sample showing a defined effect (other identical or similar definitions are provided in 40CFR798.1175-3).

dose response (40CFR798.1150-5) is the relationship between the dose (or concentration) and the proportion of a population sample showing a defined effect.

dose response (EPA-94/04): How a biological organism's response to a toxic substance quantitatively shifts as its overall exposure to the substance changes (e.g., a small dose of carbon monoxide may cause drowsiness; a large dose can be fatal).

dose response assessment (EPA-94/04): Estimating the potency of a chemical.

dose response relationship (EPA-94/04): The quantitative relationship between the amount of exposure to a substance and the extent of toxic injury or disease produced.

dosimetry processor (10CFR20.3) means an individual or an organization that processes and evaluates personnel monitoring equipment in order to determine the radiation dose delivered to the equipment.

DOT (40CFR51.392) means the United States Department of Transportation (other identical or similar definitions are provided in 40CFR93.101).

DOT reportable quantity (EPA-94/04): The quantity of a substance specified in U.S. Department of Transportation regulation that triggers labelling, packaging and other requirements related to shipping such substances.

double block and bleed system (40CFR60.481) means two block valves connected in series with a bleed valve or line that can vent the line between the two block valves (other identical or similar definitions are provided in 40CFR61.241; 264.1031).

double block and bleed system (40CFR63.161) means two block valves connected in series with a bleed valve or line that can vent the line between the two block valves.

double wash/rinse (40CFR761.123) means a minimum requirement to cleanse solid surfaces (both impervious and nonimpervious) two times with an appropriate solvent or other material in which PCB's are at least 5 percent soluble (by weight). A volume of PCB-free fluid sufficient to cover the contaminated surface completely must be used in each wash/rinse. The wash/rinse requirement does not mean the mere spreading of solvent or other fluid over the surface, nor does the requirement mean a once-over wipe with a soaked cloth. Precautions must be taken to contain any runoff to dispose properly of wastes generated during the cleaning.

downgradient (EPA-94/04): The direction that groundwater flows; similar to "downstream" for surface water.

downtown transit service (40CFR52.2493-6) means

regularly scheduled transit operation directly to a central business district.

draft (EPA-94/04):
1. The act of drawing or removing water from a tank or reservoir.
2. The water which is drawn or removed.

draft acid rain permit or draft permit (40CFR72.2) means the version of the Acid Rain permit, or the Acid Rain portion of an operating permit, that a permitting authority offers for public comment.

draft permit (40CFR122.2) means a document prepared under 40CFR124.6 indicating the Director's tentative decision to issue or deny, modify, revoke and reissue, terminate, or reissue a "permit." A notice of intent to terminate a permit, and a notice of intent to deny a permit, as discussed in 40CFR124.5, are types of "draft permits." A denial of a request for modification, revocation and reissuance, or termination, as discussed in 40CFR124.5, is not a "draft permit." A "proposed permit" is not a "draft permit."

draft permit (40CFR124.2) means a document prepared under 40CFR124.6 indicating the Director's tentative decision to issue or deny, modify, revoke and reissue, terminate, or reissue a "permit." A notice of intent to terminate a permit and a notice of intent to deny a permit as discussed in 40CFR124.5, are types of "draft permits." A denial of a request for modification, revocation and reissuance or termination, as discussed in 40CFR124.5, is not a "draft permit." A "proposal permit" is not a "draft permit."

draft permit (40CFR124.41) shall have the meaning set forth in 40CFR124.2.

draft permit (40CFR144.3) means a document prepared under 40CFR124.6 indicating the Director's tentative decision to issue or deny, modify, revoke and reissue, terminate, or reissue a "permit." A notice of intent to terminate a permit, and a notice of intent to deny a permit, as discussed in 40CFR124.5 are types of "draft permits." A denial of a request for modification, revocation and reissuance, or termination, as discussed in 40CFR124.5 is not a "draft permit."

draft permit (40CFR270.2) means a document prepared under 124.6 indicating the Director's tentative decision to issue or deny, modify, revoke and reissue, terminate, or reissue a permit. A notice of intent to terminate a permit, and a notice of intent to deny a permit, as discussed in 124.5, are types of draft permits. A denial of a request for modification, revocation and reissuance, or termination, as discussed in 124.5 is not a "draft permit." A proposed permit is not a draft permit.

draft permit (40CFR70.2) means the version of a permit for which the permitting authority offers public participation under 40CFR70.7(h) or affected State review under 40CFR70.8 of this part (under CWA).

draft permit (EPA-94/04): A preliminary permit drafted and published by EPA; subject to public review and comment before final action on the application.

drainage (EPA-94/04): Improving the productivity of some agricultural land by removing excess water from the soil by such means as ditches or subsurface drainage tiles.

drainage basin (EPA-94/04): The area of land that drains water, sediment, and dissolved materials to a common outlet at some point along a stream channel.

drainage water (40CFR440.141-3) means incidental surface waters from diverse sources such as rainfall, snow melt or permafrost melt.

drainage well (EPA-94/04): A well drilled to carry excess water off agricultural fields. Because they act as a funnel from the surface to the ground water below, drainage wells can contribute to ground water contamination.

drawdown (EPA-94/04):
1. The drop in the water table or level of water in the ground when water is being pumped from a well.
2. The amount of water used from a tank or reservoir.
3. The drop in the water level of a tank or reservoir.

drawing (40CFR467.02-g) is the process of pulling metal through a die or succession of dies to reduce a die or succession of dies to reduce the metal's diameter or alter its shape. There are two aluminum forming subcategories based on the drawing process. In the drawing with neat oils subcategory, the drawing with neat oils subcategory, the drawing process uses a pure or neat oil as a lubricant. In the drawing with emulsions or soaps subcategory, the drawing process uses an emulsion or soap solution as a lubricant.

drawing (40CFR468.02-g) shall mean pulling the workpiece through a die or succession of dies to reduce the diameter or alter its shape.

drawing (40CFR471.02-l) is the process of pulling a metal through a die or succession of dies to reduce the metal's diameter or alter its cross-sectional shape.

dredge (40CFR440.141-4) means a self-contained combination of an elevating excavator (e.g., bucket line dredge), the beneficiation or gold-concentrating plant, and a tailings disposal plant, all mounted on a floating barge.

dredged material (40CFR232.2) means material that is excavated or dredged from waters of the United States.

dredged material permit (40CFR220.2-h) means a permit issued by the Corps of Engineers under section 103 of the Act (see 33CFR209.120) and any Federal projects reviewed under section 103(e) of the Act (see 33CFR209.145).

dredging (EPA-94/04): Removal of mud from the bottom of water bodies. This can disturb the ecosystem and causes silting that kills aquatic life. Dredging of contaminated muds can expose biota to heavy metals and other toxics. Dredging activities may be subject to regulation under section 404 of the Clean Water Act.

dried fruit (40CFR407.61-h) shall mean the processing of various fruits into the following products styles: Air, vacuum, and freeze dried, pitted and unpitted, blanched and unbalanced, whole, halves, slices and other similar styles of apples, apricots, figs, peaches, pears, prunes, canned extracted prune juice and pulp from rehydrated and cooked dehydrated prunes but not including dates or raisins.

drill cuttings (40CFR435.11-h) shall refer to the particles generated by drilling into subsurface geologic formations and carried to the surface with the drilling fluid.

drilling and production facility (40CFR60.111-h) means all drilling and servicing equipment, wells, flow lines, separators, equipment, gathering lines, and auxiliary nontransportation-related equipment used in the production of petroleum but does not include natural gasoline plants.

drilling fluid (40CFR435.11-i) shall refer to the circulating fluid (mud) used in the rotary drilling of wells to clean and condition the hole and to counterbalance formation pressure. A water-based drilling fluid is the conventional drilling mud in which water is the continuous phase and the suspending medium for solids, whether or not oil is present. An oil-based drilling fluid has diesel oil, mineral oil, or some other oil as its continuous phase with water as the dispersed phase.

drilling mud (40CFR144.3) means a heavy suspension used in drilling an "injection well," introduced down the drill pipe and through the drill bit.

drinking water equivalent level (EPA-94/04): Protective level of exposure related to potentially non-carcinogenic effects of chemicals that are also known to cause cancer.

drinking water supply (40CFR300.5), as defined by section 101(7) of CERCLA means any raw or finished water source that is or may be used by a public water system (as defined in the Safe Drinking Water Act) or as drinking water by one or more individuals (see below for the listing of potential adverse health chemicals regulated under the Safe Drinking Water Act. The chemicals provided herewith in both alphabetical and numerical order were excerpted from 40CFR141.32, 1995 edition).

	potential adverse health compounds
number	(alphabetical order)
23	Acrylamide

24	Alachlor
25	Aldicarb
27	Aldicarb sulfone
26	Aldicarb sulfoxide
53	Antimony (Sb)
15	Asbestos (asbestiform minerals)
28	Atrazine
16	Barium (Ba)
5	Benzene
58	Benzo[a]pyrene
54	Beryllium (Be)
17	Badmium (Cd)
29	Carbofuran
2	Carbon tetrachloride
30	Chlordane
18	Chromium
12	Coliform, fecal
11	Coliform, total
14	Copper
55	Cyanide
36	D(2,4-)
59	Dalapon
61	Di(2-ethylhexyl) adipate
62	Di(2-ethylhexyl) phthalate
31	Dibromochloropropane (DBCP)
32	Dichlorobenzene(o-)
7	Dichlorobenzene(p-)
3	Dichloroethane(1,2-)
6	Dichloroethylene(1,1-)
33	Dichloroethylene(cis-1,2-)
34	Dichloroethylene(trans-1,2-)
60	Dichloromethane
35	Dichloropropane(1,2-)
63	Dinoseb
64	Diphenyl hydrazine(1,2-)
65	Endothall
66	Endrin
37	Epichlorohydrin
38	Ethylbenzene
39	Ethylene dibromide (EDB)
9	Fluoride
67	Glyphosate
41	Heptachlor epoxide
40	Heptachlor
68	Hexachlorobenzene
69	Hexachlorocyclopentadiene
13	Lead
42	Lindane
19	Mercury
43	Methoxychlor

10	Microbiological contaminants
44	Monochlorobenzene (MCB)
56*	Nickel (Ni), (* = reserved)
20	Nitrate (as N)
21	Nitrite
70	Oxamyl
46	Pentachlorophenol
71	Picloram
45	Polychlorinated biphenyls (PCBs)
22	Selenium (Se)
72	Simazine
47	Styrene (monomer)
75	Tetrachlorodibenzo-p-dioxin(2,3,7,8-)
48	Tetrachloroethylene
57	Thallium (Tl)
49	Toluene
50	Toxaphene
51	TP(2,4,5-)
73	Trichlorobenzene(1,2,4-)
8	Trichloroethane(1,1,1-)
74	Trichloroethane(1,1,2-)
1	Trichloroethylene
4	Vinyl chloride
52	Xylenes

number (order)	potential adverse health compounds
1	Trichloroethylene
2	Carbon tetrachloride
3	Dichloroethane(1,2-)
4	Vinyl chloride
5	Benzene
6	Dichloroethylene(1,1-)
7	Dichlorobenzene(p-)
8	Trichloroethane(1,1,1-)
9	Fluoride
10	Microbiological contaminants
11	Coliform, total
12	Coliform, fecal
13	Lead
14	Copper
15	Asbestos (asbestiform minerals)
16	Barium (Ba)
17	Badmium (Cd)
18	Chromium
19	Mercury
20	Nitrate (as N)
21	Nitrite
22	Selenium (Se)
23	Acrylamide
24	Alachlor

25	Aldicarb
26	Aldicarb sulfoxide
27	Aldicarb sulfone
28	Atrazine
29	Carbofuran
30	Chlordane
31	Dibromochloropropane (DBCP)
32	Dichlorobenzene(o-)
33	Dichloroethylene(cis-1,2-)
34	Dichloroethylene(trans-1,2-)
35	Dichloropropane(1,2-)
36	D(2,4-)
37	Epichlorohydrin
38	Ethylbenzene
39	Ethylene dibromide (EDB)
40	Heptachlor
41	Heptachlor epoxide
42	Lindane
43	Methoxychlor
44	Monochlorobenzene (MCB)
45	Polychlorinated biphenyls (PCBs)
46	Pentachlorophenol
47	Styrene (monomer)
48	Tetrachloroethylene
49	Toluene
50	Toxaphene
51	TP(2,4,5-)
52	Xylenes
53	Antimony (Sb)
54	Beryllium (Be)
55	Cyanide
56*	Nickel (Ni), (* = reserved)
57	Thallium (Tl)
58	Benzo[a]pyrene
59	Dalapon
60	Dichloromethane
61	Di(2-ethylhexyl) adipate
62	Di(2-ethylhexyl) phthalate
63	Dinoseb
64	Diphenyl hydrazine(1,2-)
65	Endothall
66	Endrin
67	Glyphosate
68	Hexachlorobenzene
69	Hexachlorocyclopentadiene
70	Oxamyl
71	Picloram
72	Simazine
73	Trichlorobenzene(1,2,4-)
74	Trichloroethane(1,1,2-)
75	Tetrachlorodibenzo-p-dioxin(2,3,7,8-)

drinking water supply (SF101-42USC9601) means any raw or finished water source that is or may be used by a public water system (as defined in the Safe Drinking Water Act) or as drinking water by one or more individuals.

drip pad (40CFR260.10) is an engineered structure consisting of a curbed, free-draining base, constructed of non-earthen materials and designed to convey preservative kick-back or drippage from treated wood, precipitation, and surface water run-on to an associated collection system at wood preserving plants.

drive system (40CFR600.002.85-50) is determined by the number and location of drive axles (e.g., front wheel drive, rear wheel drive, four wheel drive) and any other feature of the drive system if the Administrator determines that such other features may result in a fuel economy difference (under the Motor Vehicle Information and Cost Savings Act).

drive train configuration (40CFR86.082.2) means unique combination of engine code, transmission configuration, and axile ratio.

drop off (EPA-94/04): Recyclable materials collection method in which individuals bring them to a designated collection site.

dross reverberatory furnace (40CFR60.181-f) means any furnace used for the removal or refining of impurities from lead bullion (under the Clean Air Act).

drug (40CFR710.2-a) shall have the meanings contained in the Federal Food, Drug, and Cosmetic Act, 21USC321 et seq., and the regulations issued under such Act. The meaning of the term, drug, defined under the Act was provided below:
(1) Drug (21USC321(a)(1)(g)) means
   (A) articles recognized in the official United States Pharmacopoeia, official Homoeopathic Pharmacopoeia of the United States, or official National Formulary, or any supplement to any of them; and
   (B) articles intended for use in the diagnosis,

cure, mitigation, treatment, or prevention of disease in man or other animals; and

(C) articles (other than food) intended to affect the structure or any function of the body of man or other animals; and

(D) articles intended for use as a component of any article specified in clause (A), (B), or (C); but does not include devices or their components, parts, or accessories.

(2) Counterfeit drug (21USC321(a)(1)(g)) means a drug which, or the container or labeling of which, without authorization, bears the trademark, trade name, or other identifying mark, imprint, or device, or any likeness thereof, of a drug manufacturer, processor, packer, or distributor other than the person or persons who in fact manufactured, processed, packed, or distributed such drug and which thereby falsely purports or is represented to be the product of, or to have been packed or distributed by, such other drug manufacturer, processor, packer, or distributor (other identical or similar definitions are provided in 40CFR720.3-a).

drug abuse (40CFR7.25) means:

(a) The use of any drug or substance listed by the Department of Justice in 21CFR1308.11, under authority of the Controlled Substances Act, 21USC801, as a controlled substance unavailable for prescription because:

(1) The drug or substance has a high potential for abuse,

(2) The drug or other substance has no currently accepted medical use in treatment in the United States,

(3) There is a lack of accepted safety for use of the drug or other substance under medical supervision.

(b) The misuse of any drug or substance listed by the Department of Justice in 21CFR1308.12-1308.15 under authority of the Controlled Substances Act as a controlled substance available for prescription.

drug free workplace (40CFR32.605-4) means a site for the performance of work done in connection with a specific grant at which employees of the grantee are prohibited from engaging in the unlawful manufacture, distribution, dispensing, possession, or use of a controlled substance.

drum (40CFR52.741) means any cylindrical metal shipping container of 13- to 110-gallon capacity.

dry beans (40CFR407.71-i) shall mean the production of canned pinto, kidney, navy, great northern, red, pink or related type, with and without formulated sauces, meats and gravies.

dry bottom (40CFR76.2) means the boiler has a furnace bottom temperature below the ash melting point and the bottom ash is removed as a solid.

dry cleaning (40CFR63.321) means the process of cleaning articles using perchloroethylene.

dry cleaning cycle (40CFR63.321) means the washing and drying of articles in a dry-to-dry machine or transfer machine system.

dry cleaning facility (40CFR63.321) means an establishment with one or more dry cleaning systems.

dry cleaning machine (40CFR63.321) means a dry-to-dry machine or each machine of a transfer machine system.

dry cleaning machine drum (40CFR63.321) means the perforated container inside the dry cleaning machine that holds the articles during dry cleaning.

dry cleaning operation (40CFR52.1088) means that process by which an organic solvent is used in the commercial cleaning of garments and other fabric materials (other identical or similar definitions are provided in 40CFR52.1107; 52.2440-1).

dry cleaning system (40CFR63.321) means a dry-to-dry machine and its ancillary equipment or a transfer machine system and its ancillary equipment.

dry flue gas desulfurization technology (40CFR60.41b) means a sulfur dioxide control system that is located downstream of the steam generating unit and removes sulfur oxides from the combustion gases of the steam generating unit by contacting the combustion gases with an alkaline slurry or solution and forming a dry powder material. This definition includes devices where the dry powder material is subsequently converted to another form. Alkaline slurries or solutions used in dry flue gas desulfurization technology include but are not limited to lime and sodium.

dry flue gas desulfurization technology (40CFR60.41c) means a sulfur dioxide (SO$_2$) control system that is located between the steam generating unit and the exhaust vent or stack, and that removes sulfur oxides from the combustion gases of the steam generating unit by contacting the combustion gases with an alkaline slurry or solution and forming a dry powder material. This definition includes devices where the dry powder material is subsequently converted to another form. Alkaline reagents used in dry flue gas desulfurization systems include, but are not limited to, lime and sodium compounds.

dry lot (40CFR412.21-f) shall mean a confinement facility for growing ducks in confinement with a dry litter floor cover and no access to swimming areas.

dry solid form (40CFR721.3): See powder.

dry to dry machine (40CFR63.321) means a one-machine dry cleaning operation in which washing and drying are performed in the same machine.

dry weight basis (40CFR503.9-h) means calculated on the basis of having been dried at 105 degrees Celsius until reaching a constant mass (i.e., essentially 100 percent solids content).

dryer (40CFR60.161-b) means any facility in which a copper sulfide ore concentrate charge is heated in the presence of air to eliminate a portion of the moisture from the charge, provided less than 5 percent of the sulfur contained in the charge is eliminated in the facility.

dryer (40CFR60.401-c) means a unit in which the moisture content of phosphate rock is reduced by contact with a heated gas stream.

dryer (40CFR60.621) means a machine used to remove petroleum solvent from articles of clothing or other textile or leather goods, after washing and removing of excess petroleum solvent, together with the piping and ductwork used in the installation of this device.

dryer (40CFR63.321) means a machine used to remove perchloroethylene from articles by tumbling them in a heated air stream (cf. reclaimer).

drying area (40CFR60.541) means the area where VOC from applied cement or green tire sprays is allowed to evaporate.

drying oven (40CFR60.711-10) means a chamber in which heat is used to bake, cure, polymerize, or dry a surface coating.

drying oven (40CFR60.741) means a chamber within which heat is used to dry a surface coating; drying may be the only process or one of multiple processes performed in the chamber.

dual fuel reciprocating engine unit (40CFR72.2) means an internal combustion engine that combusts any combination of natural gas and diesel fuel and that thereby generates electricity through the operation of pistons, rather than by heating steam or water.

duct burner (40CFR60.41b) means a device that combusts fuel and that is placed in the exhaust duct from another source, such as a stationary gas turbine, internal combustion engine, kiln, etc., to allow the firing of additional fuel before the exhaust gases enter a heat recovery steam generating unit (under CAA).

duct burner (40CFR60.41c) means a device that combusts fuel and that is placed in the exhaust duct from another source (such as a stationary gas turbine, internal combustion engine, kiln, etc.) to allow the firing of additional fuel to heat the exhaust gases before the exhaust gases enter a steam generating unit.

duct work (40CFR63.111) means a conveyance system such as those commonly used for heating and ventilation systems. It is often made of sheet metal and often has sections connected by screws or crimping. Hard-piping is not ductwork.

duct work (40CFR63.161) means a conveyance system such as those commonly used for heating and ventilation systems. It is often made of sheet metal and often has sections connected by screws or crimping. Hard-piping is not ductwork.

ductile iron (40CFR464.31-c) means a cast iron that has been treated while molten with a master alloy containing an element such as magnesium or cerium to induce the formation of free graphite as nodules or spherules, which imparts a measurable degree of ductility to the cast metal.

dump (EPA-94/04): A site used to dispose of solid waste without environmental controls.

dumping (40CFR220.2-e) means a disposition of material: Provided, that it does not mean a disposition of any effluent from any outfall structure to the extent that such disposition is regulated under the provisions of the FWPCA, under the provisions of section 13 of the River and Harbor Act of 1899, as amended (33USC407), or under the provisions of the Atomic Energy Act of 1954, as amended (42USC2011), nor does it mean a routine discharge of effluent incidental to the propulsion of, or operation of motor-driven equipment on, vessels: Provided further, that it does not mean the construction of any fixed structure or artificial island nor the intentional placement of any device in ocean waters or on or in the submerged land beneath such waters, for a purpose other than disposal, when such construction or such placement is otherwise regulated by Federal or State law or occurs pursuant to an authorized Federal or State program; and provided further, that it does not include the deposit of oyster shells, or other materials when such deposit is made for the purpose of developing, maintaining, or harvesting fisheries resources and is otherwise regulated by Federal or State law or occurs pursuant to an authorized Federal or State program.

duplication (40CFR2.100-k) refers to the process of making a copy of document necessary to respond to an FOIA request. Such copies can take the form of paper copy, microform, audiovisual materials, or machine readable documentation (e.g., magnetic tape or disk), among others. The copy provided must be in form that is reasonably usable by requesters.

duplicator paper (40CFR250.4-l) means writing papers used for masters or copy sheets in the aniline ink or hectograph process of reproduction (commonly called spirit machines).

durability fleet (40CFR610.11-18) means a fleet of automobiles operated for mileage accumulation used to assess deterioration effects associated with the retrofit device.

durability useful life (40CFR86.094.2) means the highest useful life mileage out of the set of all useful life mileages that apply to a given vehicle. The durability useful life determines the duration of service accumulation on a durability data vehicle. The determination of durability useful life shall reflect any alternative useful life mileages approved by the Administrator under 40CFR86.094-21(f). The determination of durability useful life shall exclude any standard and related useful life mileage for which the manufacturer has obtained a waiver of emission data submission requirements under 40CFR86.094-23(c).

dust collector (29CFR1910.94a) means a device or combination of devices for separating dust from the air handled by an exhaust ventilation system (under OSHA).

dust handling equipment (40CFR60.261-k) means any equipment used to handle particulate matter collected by the air pollution control device (and located at or near such device) serving any electric submerged arc furnace subject to this subpart (under CAA).

dust handling equipment (40CFR60.271-b) means any equipment used to handle particulate matter collected by the control device and located at or near the control device for an EAF subject to this subpart.

dust handling system (40CFR60.271a) means equipment used to handle particulate matter collected by the control device for an electric arc furnace or AOD vessel subject to this subpart. For the purposes of this subpart, the dust-handling system shall consist of the control device dust hoppers, the dust-conveying equipment, any central dust storage equipment, the dust-treating equipment (e.g., pug mill, pelletizer), dust transfer equipment (from storage to truck), and any secondary control devices used with the dust transfer equipment (under CAA).

dustfall jar (EPA-94/04): An open container used to

collect large particles from the air for measurement and analysis.

dwell angle (40CFR85.2122(a)(11)(ii)(C)) means the number of degrees of distributor mechanical rotation during which the breaker points are capable of conducting current.

dwell angle (40CFR85.2122(a)(5)(ii)(C)) means the number of degrees of distributor mechanical rotation during which the breaker points are conducting current.

dye penetrant testing (40CFR471.02-m) is a nondestructive method for finding discontinuities that are open to the surface of the metal. A dye is applied to the surface of metal and the excess is rinsed off. Dye that penetrates surface discontinuities will not be rinsed away thus marking these discontinuities.

dynamometer idle for automatic transmission code heavy-duty engines (40CFR86.082.2) means the manufacturer's recommended engine speed without a transmission that simulates the recommended engine speed with a transmission and with the transmission in neutral.

dystrophic lakes (EPA-94/04): Acidic, shallow bodies of water that contain much humus and/or other organic matter; contain many plants but few fish (cf. eutrophic lakes).

*********** EEEEE ***********

ear insert device (40CFR211.203-j) means a hearing protective device that is designed to be inserted into the ear canal, and to be held in place principally by virtue of its fit inside the ear canal.

ear muff device (40CFR211.203-k) means a hearing protective device that consists of two acoustic enclosures which fit over the ears and which are held in place by a spring-like headband to which the enclosures are attached.

early entry (40CFR170.3) means entry by a worker into a treated area on the agricultural establishment after a pesticide application is complete, but before any restricted-entry interval for the pesticide has expired.

early life stage toxicity test (40CFR797.1600-6) is a test to determine the minimum concentration of a substance which produces a statistically significant observable effect on hatching, survival, development and/or growth of a fish species continuously exposed during the period of their early development (under TSCA).

EC-X (40CFR797.1075-1) means the experimentally ny derived chemical concentration n that is calculated to effect X percent of the test criterion.

EC50 (40CFR141.43) means that experimentally derived concentration of test substance in dilution water that is calculated to affect 50 percent of a test population during continuous exposure over a specified period of time. In this guideline, the effect measured is immobilization (other identical or similar definitions are provided in 40CFR797.1300-2; 797.1330-3).

EC50 (40CFR797.1800-2) is that experimentally derived concentration of a chemical in water that is calculated to induce shell deposition 50 percent less than that of the controls in a test batch of organisms during continuous exposure within a particular exposure period which should be stated.

EC50 (40CFR797.1830-6) is that experimentally derived concentration of a chemical in water that is calculated to induce shell deposition 50 percent less than that of the controls in a test batch of organisms during continuous exposure within a particular period of exposure (which should be stated).

ecological impact (EPA-94/04): The effect that a man-made or natural activity has on living organisms and their non-living (abiotic) environment.

ecological indicator (EPA-94/04): A characteristic of

the environment that, when measured, quantifies magnitude of stress, habitat characteristics, degree of exposure to a stressor, or ecological response to exposure. The term is a collective term for response, exposure, habitat, and stressor indicators.

ecological risk assessment (EPA-94/04): The application of a formal framework, analytical process, or model to estimate the effects of human actions(s) on a natural resource and to interpret the significance of those effects in light of the uncertainties identified in each component of the assessment process. Such analysis includes initial hazard identification, exposure and dose response assessments, and risk characterization (cf. risk assessment).

ecology (EPA-94/04): The relationship of living things to one another and their environment, or the study of such relationships.

economic incentive program (EIP) (40CFR51.491) means a program which may include State established emission fees or a system of marketable permits, or a system of State fees on sale or manufacture of products the use of which contributes to O_3 formation, or any combination of the foregoing or other similar measures, as well as incentives and requirements to reduce vehicle emissions and vehicle miles traveled in the area, including any of the transportation control measures identified in section 108(f). Such programs may be directed toward stationary, area, and/or mobile sources, to achieve emissions reductions milestones, to attain and maintain ambient air quality standards, and/or to provide more flexible, lower-cost approaches to meeting environmental goals. Such programs are categorized into the following three categories: Emission-limiting, market-response, and directionally-sound strategies.

economic poison (40CFR163.2-e) shall have the same meaning as it has under the Federal Insecticide, Fungicide, and Rodenticide Act (7USC135-135k) and the regulations issued thereunder.

economic poisons (EPA-94/04): Chemicals used to control pests and to defoliate cash crops such as cotton.

economizer (40CFR76.2) means the lowest temperature heat exchange section of a utility boiler where boiler feed water is heated by the flue gas (under CAA).

ecosphere (EPA-94/04): The "bio-bubble" that contains life on earth, in surface waters, and in the air (cf. biosphere).

ecosystem (EPA-94/04): The interacting system of a biological community and its non-living environmental surroundings.

ecosystem structure (EPA-94/04): Attributes related to instantaneous physical state of an ecosystem; examples include species population density, species richness or evenness, and standing crop biomass.

ecotone (EPA-94/04): A habitat created by the juxtaposition of distinctly different habitats; an edge habitat; or an ecological zone or boundary where two or more ecosystems meet.

EC_x (40CFR797.1050-3) means the experimentally derived chemical concentration that is calculated to effect X percent of the test criterion (other identical or similar definitions are provided in 40CFR797.1060-1; 797.1160-3; 797.2750-1; 797.2800-1; 797.2850-1).

ED (40CFR763-AA-7) means electron diffraction.

ED10 (10 percent effective dose) (40CFR300-AA) means an estimated dose associated with a 10 percent increase in response over control groups. For HRS purposes, the response considered is cancer [milligrams toxicant per kilogram body weight per day (mg/kg-day)].

educational facility (40CFR52.2297-5) means any single location of an educational nature of college level or of vocational training above the secondary level with 1,000 or more commuters.

educational institution (40CFR2.100-g) refers to a preschool, a public or private elementary or secondary school, an institution of graduate higher education, an institution of undergraduate higher education, an institution or professional education, and an institution of vocational education, which operates a program or programs of scholarly research.

educational institution (40CFR52.1161-2) means any person or entity which has 250 or more employees and students at any time during the academic year at an educational facility offering secondary level or higher training including vocational training located in the Boston Intrastate Region.

educational institution (40CFR52.2297-8) means any person or entity controlling an education facility.

EDXA (40CFR763-AA-8) means energy dispersive X-ray analysis.

effective corrosion inhibitor residual (40CFR141.2), for the purpose of subpart I of this part only, means a concentration sufficient to form a passivating film on the interior walls of a pipe.

effective date (40CFR61.02) is the date of promulgation in the Federal Register of an applicable standard or other regulation under this part.

effective date (40CFR63.2) means:
(1) With regard to an emission standard established under this part, the date of promulgation in the Federal Register of such standard; or
(2) With regard to an alternative emission limitation or equivalent emission limitation determined by the Administrator (or a State with an approved permit program), the date that the alternative emission limitation or equivalent emission limitation becomes effective according to the provisions of this part. The effective date of a permit program established under title V of the Act (42 U.S.C. 7661) is determined according to the regulations in this chapter establishing such programs.
(Other identical or similar definitions are provided in 40CFR63.101; 63.191.)

effective date of a UIC program (40CFR146.3) means the date that a State UIC program is approved or established by the Administrator.

effective date of an NSO (40CFR57.103-k) means the effective date listed in the Federal Register

publication of EPA's issuance or approval of an NSO.

effective dose (40CFR191.12) means the sum over specified tissues of the products of the dose equivalent received following an exposure of, or an intake of radionuclides into, specified tissues of the body, multiplied by appropriate weighting factors. This allows the various tissue-specific health risks to be summed into an overall health risk. The method used to calculate effective dose is described in Appendix B of this part.

effective dose equivalent (10CFR30.4) means the sum of the products of the dose equivalent to the organ or tissue and the weighting factors applicable to each of the body organs or tissues that are irradiated. Weighting actors are: 0.25 or gonads, 0.15 for breast, 0.12 for red bone marrow, 0.12 for lungs, 0.03 for thyroid, 0.03 for bone surface, and 0.06 for each of the other five organs receiving the highest dose equivalent (other identical or similar definitions are provided in 10CFR70.4).

effective dose equivalent (40CFR61.101-b) means the sum of the products of absorbed dose and appropriate factors to account for differences in biological effectiveness due to the quality of radiation and its distribution in the body of reference man. The unit of the effective dose equivalent is the rem. For purposes of this subpart doses caused by radon-222 and its decay products formed after the radon is released from the facility are not included. The method for calculating effective dose equivalent and the definition of reference man are outlined in the International Commission on Radiological Protection's Publication No. 26.

effective dose equivalent (40CFR61.21-b) means the sum of the products of absorbed dose and appropriate factors to account for differences in biological effectiveness due to the quality of radiation and its distribution in the body of reference man. The unit of the effective dose equivalent is the rem. The method for calculating effective dose equivalent and the definition of reference man are outlined in the International Commission on Radiological Protection's Publication No. 26.

effective dose equivalent (40CFR61.91-a) means the

sum of the products of absorbed dose and appropriate factors to account for differences in biological effectiveness due to the quality of radiation and its distribution in the body of reference man. The unit of the effective dose equivalent is the rem. For purposes of this subpart, doses caused by radon-222 and its respective decay products formed after the radon is released from the facility are not included. The method for calculating effective dose equivalent and the definition of reference man are outlined in the International Commission on Radiological Protection's Publication No. 26.

effective kilogram (10CFR40.4) means
(1) for the source material uranium in which the uranium isotope uranium-235 is greater than 0.005 (0.5 weight percent) of the total uranium present: 10,000 kilograms, and
(2) for any other source material: 20.000 kilograms.

effective kilograms of special nuclear material (10CFR70.4) means:
(1) For plutonium and uranium-233 their weight in kilograms;
(2) For uranium with an enrichment in the isotope U-235 of 0.01 (1%) and above, its element weight in kilograms multiplied by the square of its enrichment expressed as a decimal weight fraction; and
(3) For uranium with an enrichment in the isotope U-235 below 0.01 (1%), by its element weight in kilograms multiplied by 0.0001.

effects (40CFR1508.8) include:
(a) Direct effects, which are caused by the action and occur at the same time and place.
(b) Indirect effects, which are caused by the action and are later in time or farther removed in distance, but are still reasonably foreseeable. Indirect effects may include growth inducing effects and other effects related to induced changes in the pattern of land use, population density or growth rate, and related effects on air and water and other natural systems, including ecosystems. Effects and impacts as used in these regulations are synonymous. Effects includes ecological (such as the effects on natural resources and on the components, structures, and functioning of affected ecosystems), aesthetic, historic, cultural, economic, social, or health, whether direct, indirect, or cumulative. Effects may also include those resulting from actions which may have both beneficial and detrimental effects, even if on balance the agency believes that the effect will be beneficial (under NEPA).

efficiency (40CFR60.331-h) means the gas turbine manufacturer's rated heat rate at peak load in terms of heat input per unit of power output based on the lower heating value of the fuel.

efficiency (40CFR85.2122(a)(16)(ii)(C)) means the ability of the air cleaner or the unit under test to remove contaminant.

effluent (40CFR232.2) means dredged material or fill material, including return flow from confined sites (under CWA).

effluent (EPA-94/04): Wastewater-treated or untreated-that flows out of a treatment plant, sewer, or industrial outfall. Generally refers to wastes discharged into surface waters.

effluent concentrations consistently achievable through proper operation and maintenance (40CFR133.101-f): For a given pollutant parameter, means:
(1) the 95th percentile value for the 30-day average effluent quality achieved by a treatment works in a period of at least two years excluding values attributable to upsets, bypasses, operational errors, or other unusual conditions, and
(2) a 7-day average value equal to 1.5 times the value derived under paragraph (f)(1) of this section.

effluent data (40CFR2.302-2) means:
(i) with reference to any source of discharge of any pollutant (as that term is defined in section 502(6) of the Act, 33USC1362(6)):
(A) Information necessary to determine the identity, amount, frequency, concentration, temperature, or other characteristics (to the extent related to water quality) of any pollutant which has been discharged by the source (or of any pollutant resulting from any discharge from the source), or any combination of the foregoing;

(B) Information necessary to determine the identity, amount, frequency, concentration, temperature, or other characteristics (to the extent related to water quality) of the pollutants which, under an applicable standard or limitation, the source was authorized to discharge (including, to the extent necessary for such purpose, a description of the manner or rate of operation of the source); and

(C) A general description of the location and/or nature of the source to the extent necessary to identify the source and to distinguish it from other sources (including, to the extent necessary or such purposes, a description of the device, installation, or operation constituting the source).

(ii) Notwithstanding paragraph (a)(2)(i) of this section, the following information shall be considered to be effluent data only to the extent necessary to allow EPA to disclose publicly that a source is (or is not) in compliance with an applicable standard or limitation, or to allow EPA to demonstrate the feasibility, practicability, or attainability (or lack thereof) of an existing or proposed standard or limitation:

(A) Information concerning research, or the results of research, on an product, method, device, or installation (or any component thereof) which was produced, developed, installed, and used only for research purposes; and

(B) Information concerning any product, method, device, or installation (or any component thereof) designed and intended to be marketed or used commercially but not yet so marketed or used.

effluent guidelines (EPA-94/04): Technical EPA documents which set effluent limitations for given industries and pollutants.

effluent limitation (40CFR108.2-b) means any effluent limitation which is established as a condition of a permit issued or proposed to be issued by a State or by the Environmental Protection Agency pursuant to section 402 of the Act; any toxic or pretreatment effluent standard established under section 307 of the Act; any standard of performance established under section 306 of the Act; and any

effluent limitation established under section 302, section 316, or section 318 of the Act.

effluent limitation (40CFR122.2) means any restriction imposed by the Director on quantities, discharge rates, and concentrations of "pollutants" which are "discharged" from "point sources" into "waters of the United States," the waters of the "contiguous zone," or the ocean.

effluent limitation (40CFR401.11-i) means any restriction established by the Administrator on quantities, rates, and concentrations of chemical, physical, biological and other constituents which are discharged from point sources, other than new sources, into navigable waters, the waters of the contiguous zone or the ocean.

effluent limitation (CWA502-33USC1362) means any restriction established by a State or the Administrator on quantities, rates, and concentrations of chemical, physical, biological, and other constituents which are discharged from point sources into navigable waters, the waters of the contiguous zone, or the ocean, including schedules of compliance.

effluent limitation (EPA-94/04): Restrictions established by a State or EPA on quantities, rates, and concentrations in wastewater discharges (under CWA).

effluent limitations guidelines (40CFR122.2) means a regulation published by the Administrator under section 304(b) of CWA to adopt or revise "effluent limitations."

effluent limitations guidelines (40CFR401.11-j) means any effluent limitations guidelines issued by the Administrator pursuant to section 304(b) of the Act.

effluent standard (40CFR104.2-g) means any effluent standard or limitation, which may include a prohibition of any discharge, established or proposed to be established for any toxic pollutant under section 307(a) of the Act.

effluent standard (40CFR129.2-c) means, for purposes of section 307, the equivalent of "effluent limitation" as that term is defined in section 502(11) of the Act with the exception that it does not include a schedule of compliance.

effluent standard (EPA-94/04): See effluent limitation.

eggs cracked (40CFR797.2130-ii) are eggs determined to have cracked shells when inspected with a candling lamp. Fine cracks cannot be detected without using a candling lamp and if undetected will bias data by adversely affecting embryo development. Values are expressed as a percentage of eggs laid by all hens during the test (other identical or similar definitions are provided in 40CFR797.2150-ii).

eggs laid (40CFR797.2130-i) refers to the total egg production during the test, which normally includes 10 weeks of laving. Values are expressed as numbers of eggs per pen per season (or test) (other identical or similar definitions are provided in 40CFR797.2150-i).

eggs set (40CFR797.2130-iii) are all eggs placed under incubation, i.e., total eggs minus cracked eggs and those selected for analysis of eggshell thickness. The number of eggs set, itself, is an artificial number, but it is essential for the statistical analysis of other development parameters (other identical or similar definitions are provided in 40CFR797.2150-iii).

eggshell thickness (40CFR797.2130-viii) is the thickness of the shell and the membrane of the egg at several points around the girth after the egg has been opened, washed out, and the shell and membrane dried for at least 4 hours at room temperature. Values are expressed as the average thickness of the several measured points in millimeters (other identical or similar definitions are provided in 40CFR797.2150-viii).

eight 8-hour time weighted average (40CFR704.102-5) means the cumulative exposure for an 8-hour work shift computed (by an equation in this part, for complete definition, see 40CFR704.102-5).

ejector (EPA-94/04): A device used to disperse a chemical solution into water being treated.

electric arc furnace (EAF) (40CFR60.271-a) means a furnace that produces molten steel and heats the charge materials with electric arcs from carbon electrodes. Furnaces that continuously feed direct-reduced iron ore pellets as the primary source of iron are not affected facilities within the scope of this definition.

electric arc furnace steelmaking (40CFR420.41-c) means the production of steel principally from steel scrap and fluxes in refractory lined furnaces by passing an electric current through the scrap or steel bath.

electric furnace (40CFR60.131-c) means any furnace which uses electricity to produce over 50 percent of the heat required in the production of refined brass or bronze.

electric smelting furnace (40CFR60.181-g) means any furnace in which the heat necessary for smelting of the lead sulfide ore concentrate charge is generated by passing an electric current through a portion of the molten mass in the furnace (under CAA).

electric submerged arc furnace (40CFR60.261-a) means any furnace wherein electrical energy is converted to heat energy by transmission of current between electrodes partially submerged in the furnace charge.

electric traction motor (40CFR600.002.85-45) means an electrically powered motor which provides tractive energy to the wheels of a vehicle.

electric utility combined cycle gas turbine (40CFR60.41a-21) means any combined cycle gas turbine used for electric generation that is constructed for the purpose of supplying more than one-third of its potential electric output capacity and more than 25 MW electrical output to any utility power distribution system for sale. Any steam distribution system that is constructed for the purpose of providing steam to a steam electric generator that would produce electrical power for sale is also considered in determining the electrical energy output capacity of the affected facility.

electric utility company (40CFR60.41a-10) means the largest interconnected organization, business, or governmental entity that generates electric power for sale (e.g., a holding company with operating subsidiary companies).

electric utility stationary gas turbine (40CFR60.331-q) means any stationary gas turbine constructed for the purpose of supplying more than one-third of its potential electric output capacity to any utility power distribution system for sale.

electric utility steam generating unit (40CFR51.165-xx) means any steam electric generating unit that is constructed for the purpose of supplying more than one-third of its potential electric output capacity and more than 25 MW electrical output to any utility power distribution system for sale. Any steam supplied to a steam distribution system for the purpose of providing steam to a steam-electric generator that would produce electrical energy for sale is also considered in determining the electrical energy output capacity of the affected facility.

electric utility steam generating unit (40CFR51.166-30) means any steam electric generating unit that is constructed for the purpose of supplying more than one-third of its potential electric output capacity and more than 25 MW electrical output to any utility power distribution system for sale. Any steam supplied to a steam distribution system for the purpose of providing steam to a steam-electric generator that would produce electrical energy for sale is also considered in determining the electrical energy output capacity of the affected facility.

electric utility steam generating unit (40CFR52.21-31) means any steam electric generating unit that is constructed for the purpose of supplying more than one-third of its potential electric output capacity and more than 25 MW electrical output to any utility power distribution system for sale. Any steam supplied to a steam distribution system for the purpose of providing steam to a steam-electric generator that would produce electrical energy for sale is also considered in determining the electrical energy output capacity of the affected facility.

electric utility steam generating unit (40CFR52.24-19) means any steam electric generating unit that is constructed for the purpose of supplying more than one-third of its potential electric output capacity and more than 25 MW electrical output to any utility

power distribution system for sale. Any steam supplied to a steam distribution system for the purpose of providing steam to a steam-electric generator that would produce electrical energy for sale is also considered in determining the electrical energy output capacity of the affected facility.

electric utility steam generating unit (40CFR60.2) means any steam electric generating unit that is constructed for the purpose of supplying more than one-third of its potential electric output capacity and more than 25 MW electrical output to any utility power distribution system for sale. Any steam supplied to a steam distribution system for the purpose of providing steam to a steam-electric generator that would produce electrical energy for sale is also considered in determining the electrical energy output capacity of the affected facility (under CAA).

electric utility steam generating unit (40CFR60.41a-2) means any steam electric generating unit that is constructed for the purpose of supplying more than one-third of its potential electric output capacity and more than 25 MW electrical output to any utility power distribution system for sale. Any steam supplied to a steam distribution system for the purpose of providing steam to a steam-electric generator that would produce electrical energy for sale is also considered in determining the electrical energy output capacity of the affected facility (under CAA).

electric utility steam generating unit (CAA112-42USC7412) means any fossil fuel fired combustion unit of more than 25 megawatts that serves a generator that produces electricity for sale. A unit that cogenerates steam and electricity and supplies more than one-third of its potential electric output capacity and more than 25 megawatts electrical output to any utility power distribution system for sale shall be considered an electric utility steam generating unit.

electrical capacitor manufacturer (40CFR129.105-2) means a manufacturer who produces or assembles electrical capacitors in which PCB or PCB-containing compounds are part of the dielectric.

electrical charging system (40CFR600.002.85-48)

means a device to convert 60 Hz alternating electric current, as commonly available in residential electric service in the United States, to a proper form for recharging the energy storage device.

electrical equipment (40CFR280.12) means underground equipment that contains dielectric fluid that is necessary for the operation of equipment such as transformers and buried electrical cables.

electrical transformer manufacturer (40CFR129.105-3) means a manufacturer who produces or assembles electrical transformers in which PCB or PCB-containing compounds are part of the dielectric.

electrically-heated choke (40CFR85.2122(a)(2)(iii)-(C)) means a device which contains a means for applying heat to the thermostatic coil by electrical current.

electrocoating (40CFR471.02-o) is the electrodeposition of a metallic or nonmetallic coating onto the surface of a workpiece.

electrodeposition (EDP) (40CFR60.311) means a method of applying coatings in which the part is submerged in a tank filled with the coatings and in which an electrical potential is used to enhance deposition of the coatings on the part.

electrodeposition (EDP) (40CFR60.391) means a method of applying a prime coat by which the automobile or light-duty truck body is submerged in a tank filled with coating material and an electrical field is used to effect the deposition of the coating material on the body.

electrodeposition (EDP) (40CFR60.451) means a method of coating application in which the large appliance part or product is submerged in a tank filled with coating material suspended in water and an electrical potential is used to enhance deposition of the material on the part or product.

electrodialysis (EPA-94/04): A process that uses electrical current applied to permeable membranes to remove minerals from water. Often used to desalinize salty or brackish water.

electroless plating (40CFR413.71-b) shall mean the deposition of conductive material from an autocatalytic plating solution without application of electrical current.

electromagnetic interference/radio frequency interference (EMI/RFI) shielding coating (40CFR60.721) means a conductive coating that is applied to a plastic substrate to attenuate EMI/RFI signals.

electronic crystals (40CFR469.22-b) means crystals or crystalline material which because of their unique structural and electronic properties are used in electronic devices. Examples of these crystals are crystals comprised of quartz, ceramic, silicon, gallium arsenide, and indium arsenide.

electroplating process wastewater (40CFR413.02-d) shall mean process wastewater generated in operations which are subject to regulation under any of subparts A through H of this part.

electrostatic bell or disc spray (40CFR52.741) means an electrostatic spray coating method in which a rapidly-spinning bell- or disc-shaped applicator is used to create a fine mist and apply the coating with high transfer efficiency.

electrostatic precipitator (ESP) (40CFR60.471) means an air pollution control device in which solid or liquid particulates in a gas stream are charged as they pass through an electric field and precipitated on a collection surface.

electrostatic precipitator (ESP) (EPA-94/04): A device that removes particles from a gas stream (smoke) after combustion occurs. The ESP imparts an electrical charge to the particles, causing them to adhere to metal plates inside the precipitator. Rapping on the plates causes the particles to fall into a hopper for disposal.

electrostatic spray (40CFR52.741) means a spray coating method in which opposite electrical charges are applied to the substrate and the coating. The coating is attracted to the object due to the electrostatic potential between them.

electrostatic spray application (40CFR60.311) means a spray application method that uses an electrical potential to increase the transfer efficiency of the coatings.

electrostatic spray application (40CFR60.391) means a spray application method that uses an electrical potential to increase the transfer efficiency of the coating solids. Electrostatic spray application can be used for prime coat, guide coat, or topcoat operations.

element of design (40CFR86.094.2) means any control system (i.e., computer software, electronic control system, emission control system, computer logic), and/or control system calibrations, and/or the results of systems interaction, and/or hardware items on a motor vehicle or motor vehicle engine.

elemental phosphorus plant or plant (40CFR61.121-a) means any facility that processes phosphate rock to produce elemental phosphorus. A plant includes all buildings, structures, operations, calciners and nodulizing kilns on one contiguous site.

elementary neutralization unit (40CFR260.10) means a device which:
(1) Is used for neutralizing wastes that are hazardous wastes only because they exhibit the corrosivity characteristic defined in 261.22 of this chapter, or they are listed in subpart D of part 261 of the chapter only for this reason; and,
(2) Meets the definition of tank, tank system, container, transport vehicle, or vessel in 40CFR260.10 of this chapter.

elementary neutralization unit (40CFR270.2) means a device which:
(a) Is used for neutralizing wastes only because they exhibit the corrosivity characteristic defined in 261.22 of this chapter, or are listed in subpart D of part 261 of this chapter only for this reason; and
(b) Meets the definition of tank, tank system, container, transport vehicle, or vessel in 40CFR260.10 of this chapter.

elevation angle (40CFR60-AA(alt. method 1)) means the angle of inclination of the laser beam referenced to the horizontal plane.

eleven contiguous Western States (FLPMA103-

43USC1702) means the States of Arizona, California, Colorado, Idaho, Montana, Nevada, New Mexico, Oregon, Utah, Washington, and Wyoming.

eligible coastal state (CZMA306a-16USC1455a) means a coastal state that for any fiscal year for which a grant is applied for under this section:
(A) has a management program approved under section 1455 of this title; and
(B) in the judgment of the Secretary, is making satisfactory progress in activities designed to result in significant improvement in achieving the coastal management objectives specified in section 1452(2)(A) through (I) of this title.

eligible costs (EPA-94/04): The construction costs for waste-water treatment works upon which EPA grants are based.

eligible Indian Tribe (40CFR35.105) means for purposes of the Clean Water Act, any federally recognized Indian Tribe that meets the requirements set forth at 40CFR130.6(d).

eligible individual (CAA1101-29USC1662e) means an individual who:
(A) is an eligible dislocated worker, as that term is defined in section 301(a), and
(B) has been terminated or laid off, or has received a notice of termination or lay off, as a consequence of compliance with the Clean Air Act.

elimination (40FR797.1560): See depuration.

ELWK (equivalent live weight killed) (40CFR432.11-e) shall mean the total weight of the total number of animals slaughtered at locations other than the slaughterhouse or packinghouse, which animals provide hides, blood, viscera or renderable materials for processing at that slaughterhouse, in addition to those derived from animals slaughtered on site (other identical or similar definitions are provided in 40CFR432.21-e; 432.31-e; 432.41-e).

EMAP data (EPA-94/04): Environmental monitoring data collected under the auspices of the Environmental Monitoring and Assessment Program. All EMAP data share the common attribute of being of known quality, having been collected in the context of explicit data quality objectives (DQOs) and a consistent quality assurance program.

embryo (40CFR797.2750-2) means the young sporophytic plant before the start of germination.

embryo cup (40CFR797.1600-7) is a small glass jar or similar container with a screened bottom in which the embryos of some species (i.e., minnow) are placed during the incubation period and which is normally oscillated to ensure a flow of water through the cup.

emergency (40CFR51.852) means a situation where extremely quick action on the part of the Federal agencies involved is needed and where the timing of such Federal activities makes it impractical to meet the requirements of this subpart, such as natural disasters like hurricanes or earthquakes, civil disturbances such as terrorist acts, and military mobilizations (other identical or similar definitions are provided in 40CFR93.152).

emergency (chemical) (EPA-94/04): A situation created by an accidental release or spill of hazardous chemicals that poses a threat to the safety of workers, residents, the environment, or property.

emergency action plan (29CFR1910.35) means a plan for a workplace, or parts thereof, describing what procedures the employer and employees must take to ensure employee safety from fire or other emergencies.

emergency condition (40CFR166.3-d) means an urgent, non-routine situation that requires the use of a pesticide(s) and shall be deemed to exist when:
(1) No effective pesticides are available under the Act that have labeled uses registered for control of the pest under the conditions of the emergency; and
(2) No economically or environmentally feasible alternative practices which provide adequate control are available; and
(3) The situation:
    (i) Involves the introduction or dissemination of a pest new to or not theretofore known to be widely prevalent or distributed within or throughout the United States and its territories; or

(ii) Will present significant risks to human health; or
(iii) Will present significant risks to threatened or endangered species, beneficial organisms, or the environment; or
(iv) Will cause significant economic loss due to:
    (A) An outbreak or an expected outbreak of a pest; or
    (B) A change in plant growth or development caused by unusual environmental conditions where such change can be rectified by the use of a pesticide(s).

emergency condition (40CFR60.41a-20) means that period of time when:
(a) The electric generation output of an affected facility with a malfunctioning flue gas desulfurization system cannot be reduced or electrical output must be increased because:
    (1) All available system capacity in the principal company interconnected with the affected facility is being operated, and
    (2) All available purchase power interconnected with the affected facility is being obtained, or
(b) The electric generation demand is being shifted as quickly as possible from an affected facility with a malfunctioning flue gas desulfurization system to one or more electrical generating units held in reserve by the principal company or by a neighboring company, or
(c) An affected facility with a malfunctioning flue gas desulfurization system becomes the only available unit to maintain a part or all of the principal company's system emergency reserves and the unit is operated in spinning reserve at the lowest practical electric generation load consistent with not causing significant physical damage to the unit. If the unit is operated at a higher load to meet load demand, an emergency condition would not exist unless the conditions under (a) of this definition apply (under the Clean Air Act).

emergency episode (EPA-94/04): See air pollution episode.

emergency escape route (29CFR1910.35) means the route that employees are directed to follow in the event they are required to evacuate the workplace or seek a designated refuge area.

emergency fuel (40CFR60.331-r) is a fuel fired by a gas turbine only during circumstances, such as natural gas supply curtailment or breakdown of delivery system, that make it impossible to fire natural gas in the gas turbine.

emergency gas turbine (40CFR60.331-e) means any stationary gas turbine which operates as a mechanical or electrical power source only when the primary power source for a facility has been rendered inoperable by an emergency situation.

emergency permit (40CFR144.3) means a UIC "permit" issued in accordance with 40CFR144.34.

emergency planning and community right-to-know act of 1986 (40CFR310.11-b) means Title III-- Emergency Planning and Community Right-To-Know of the Superfund Amendments and Reauthorization Act of 1986 (EPCRA) (Pub. L. 99-499, 42 U.S.C. 960).

emergency project (40CFR763.121) means a project involving the removal, enclosure, or encapsulation of friable asbestos-containing material that was not planned but results from a sudden unexpected event (under TSCA).

emergency renovation operation (40CFR61.141) means a renovation operation that was not planned but results from a sudden, unexpected event that, if not immediately attended to, presents a safety or public hazard, is necessary to protect equipment from damage, or is necessary to avoid imposing an unreasonable financial burden. This term includes operations necessitated by nonroutine failures of equipment.

emergency response values (EPA-94/04): Concentrations of chemicals, published by various groups, defining acceptable levels for short-term exposures in emergencies.

emergency situation for continuing use of a PCB transformer (40CFR761.3) exists when:
(1) Neither a non-PCB Transformer nor a PCB-Contaminated transformer is currently in storage for reuse or readily available (i.e., available within 24 hours) for installation.
(2) Immediate replacement is necessary to continue service to power users.

emergency vehicle (40CFR88.302.94) means any vehicle that is legally authorized by a governmental authority to exceed the speed limit to transport people and equipment to and from situations in which speed is required to save lives or property, such as a rescue vehicle, fire truck, or ambulance (under CAA).

emergency vent stream (40CFR60.561) means, for the purposes of these standards, an intermittent emission that results from a decomposition, attempts to prevent decompositions, power failure, equipment failure, or other unexpected cause that requires immediate venting of gases from process equipment in order to avoid safety hazards or equipment damage. This includes intermittent vents that occur from process equipment where normal operating parameters (e.g., pressure or temperature) are exceeded such that the process equipment cannot be returned to normal operating conditions using the design features of the system and venting must occur to avoid equipment failure or adverse safety personnel consequences and to minimize adverse effects of the runaway reaction. This does not include intermittent vents that are designed into the process to maintain normal operating conditions of process vessels including those vents that regulate normal process vessel pressure.

emerging technology (40CFR60.41b) means any sulfur dioxide control system that is not defined as a conventional technology under this section, and for which the owner or operator of the facility has applied to the Administrator and received approval to operate as an emerging technology under 40CFR60.49b(a)(4) (other identical or similar definitions are provided in 40CFR60.41c).

emergnecy permit (40CFR270.2) means a RCRA permit issued in accordance with 40CFR270.61.

emission (40CFR240.101-f) means gas-borne pollutants released to the atmosphere.

emission (EPA-94/04): Pollution discharged into the

atmosphere from smokestacks, other vents, and surface areas of commercial or industrial facilities; from residential chimneys; and from motor vehicle, locomotive, or aircraft exhausts.

emission cap (EPA-94/04): A limit designed to prevent projected growth in emissions from existing and future stationary sources from eroding any mandated reduction. Generally, such provisions require any emission growth from facilities under the restrictions be offset by equivalent reductions at other facilities under the same cap (cf. emissions trading).

emission control device (40CFR60.581) means any solvent recovery or solvent destruction device used to control volatile organic compounds (VOC) emissions from flexible vinyl and urethane rotogravure printing lines.

emission control system (40CFR60.581) means the combination of an emission control device and a vapor capture system for the purpose of reducing VOC emissions from flexible vinyl and urethane rotogravure printing lines.

emission control system (40CFR89.2) means any device, system, or element of design which controls or reduces the emission of substances from an engine.

emission credits (40CFR86.090.2) mean the amount of emission reductions or exceedances, by a heavy-duty engine family, below or above the emission standard, respectively. Emission credits below the standard are considered as "positive credits," while emission credits above the standard are considered as "negative credits." In addition, "projected credits" refer to emission credits based on the projected U.S. production volume of the engine family. "Reserved credits" are emission credits generated within a model year waiting to be reported to EPA at the end of the model year. "Actual credits" refer to emission credits based on actual U.S. production volumes as contained in the end-of-year reports submitted to EPA. Some or all of these credits may be revoked if EPA review of the end of year reports or any subsequent audit actions uncover problems or errors.

emission credits (40CFR89.202.96) represent the amount of emission reduction or exceedance, by a nonroad engine family, below or above the emission standard, respectively. Emission reductions below the standard are considered as "positive credits," while emission exceedances above the standard are considered as "negative credits." In addition, "projected credits" refer to emission credits based on the projected applicable production/sales volume of the engine family. "Reserved credits" are emission credits generated within a model year waiting to be reported to EPA at the end of the model year. "Actual credits" refer to emission credits based on actual applicable production/sales volume as contained in the end-of-year reports submitted to EPA. Some or all of these credits may be revoked if EPA review of the end-of-year reports or any subsequent audit action(s) uncovers problems or errors.

emission critical parameters (40CFR85.2113-g) means those critical parameters and tolerances which, if equivalent from one part to another, will not cause the vehicle to exceed applicable emission standards with such parts installed.

emission data (40CFR2.301-2) means:
(i) with reference to any source of emission of any substance into the air:
   (A) Information necessary to determine the identity, amount, frequency, concentration, or other characteristics (to the extent related to air quality) of any emission which has been emitted by the source (or of any pollutant resulting from any emission by the source), or any combination of the foregoing;
   (B) Information necessary to determine the identity, amount, frequency, concentration, or other characteristics (to the extent related to air quality) of the emissions which, under an applicable standard or limitation, the source was authorized to emit (including, to the extent necessary for such purposes, a description of the manner or rate of operation of the source); and
   (C) A general description of the location and/or nature of the source to the extent necessary to identify the source and to distinguish it from other sources (including, to the extent necessary for such purposes,

a description of the device, installation, or operation constituting the source).

(ii) Notwithstanding paragraph (a)(2)(i) of this section, the following information shall be considered to be emission data only to the extent necessary to allow EPA to disclose publicly that a source is (or is not) in compliance with an applicable standard or imitation, or to allow EPA to demonstrate the feasibility, practicability, or attainability (or lack thereof) of an existing or proposed standard or limitation:

(A) Information concerning research, or the results of research, on any project, method, device or installation (or any component thereof which was produced, developed, installed, and used only for research purposes; and

(B) Information concerning any product, method, device, or installation (or any component thereof) designed and intended to be marketed or used commercially but not yet so marketed or used.

emission factor (EPA-94/04): The relationship between the amount of pollution produced and the amount of raw material processed. For example, an emission factor for a blast furnace making iron would be the number of pounds of particulates per ton of raw materials.

emission frequency (40CFR60-AA (method 22-3.1)) means percentage of time that emissions are visible during the observation period.

emission guideline (40CFR60.21-e) means a guideline set forth in subpart C of this part, or in a final guideline document published under 40CFR60.22(a), which reflects the degree of emission reduction achievable through the application of the best system of emission reduction which (taking into account the cost of such reduction) the Administrator has determined has been adequately demonstrated for designated facilities.

emission inventory (EPA-94/04): A listing, by source, of the amount of air pollutants discharged into the atmosphere of a community; used to establish emission standards.

emission limitation and emission standard
(40CFR51.100-z) mean a requirement established by a State, local government, or the Administrator which limits the quantity, rate, or concentration of emissions of air pollutants on a continuous basis, including any requirements which limit the level of opacity, prescribe equipment, set fuel specifications, or prescribe operation or maintenance procedures for a source to assure continuous emission reduction (under CAA).

emission limitation and emission standard (CAA302-42USC7602) mean a requirement established by the State or the Administrator which limits the quantity, rate, or concentration of emissions of air pollutants on a continuous basis, including any requirement relating to the operation or maintenance of a source to assure continuous emission reduction.

emission limiting strategies (40CFR51.491) are strategies that directly specify limits on total mass emissions, emission-related parameters (e.g., emission rates per unit of production, product content limits), or levels of emissions reductions relative to a program baseline that are required to be met by affected sources, while providing flexibility to sources to reduce the cost of meeting program requirements.

emission measurement system (40CFR87.1) means all of the equipment necessary to transport and measure the level of emissions. This includes the sample system and the instrumentation system.

emission offset interpretative ruling: See appendix S to 40CFR51.

emission performance warranty (40CFR85.2102-4) means that warranty given pursuant to this subpart and section 20(b) of the Act.

emission point (40CFR63.101) means an individual process vent, storage vessel, transfer rack, wastewater stream, or equipment leak.

emission point (40CFR63.51) means any part or activity of a major source that emits or has the potential to emit, under current operational design, any hazardous air pollutant.

emission related defect (40CFR85.1902-b) shall mean a defect in design, materials, or workmanship in a device, system, or assembly described in the approved Application for Certification (required by 40CFR86.077.22 and like provisions of part 85 and part 86 of Title 40 of the Code of Federal Regulations) which affects any parameter or specification enumerated in Appendix VIII.

emission related maintenance (40CFR86.084.2) means that maintenance which does substantially affect emissions or which is likely to affect the deterioration of the vehicle or engine with respect to emissions, even if the maintenance is performed at some time other than that which is recommended. This definition applies beginning with the 14 model year.

emission related maintenance (40CFR86.088.2) means that maintenance which does substantially affect emissions or which is likely to affect the emissions deterioration of the vehicle or engine during normal in-use operation, even if the maintenance is performed at some time other than that which is recommended.

emission related parts (40CFR85.1402) means those parts installed for the specific purpose of controlling emissions or those components, systems, or elements of design which must function properly to assure continued emission compliance.

emission related parts (40CFR85.2102-14) means those parts installed for the specific purpose of controlling emissions or those components, systems, or elements of design which must function properly to assure continued vehicle emission compliance.

emission source and source (40CFR52.741) mean any facility from which VOM (volatile organic material) is emitted or capable of being emitted into the atmosphere.

emission standard (40CFR60.21-f) means a legally enforceable regulation setting forth an allowable rate of emissions into the atmosphere, or prescribing equipment specifications for control of air pollution emissions.

emission standard (40CFR63.2) means a national standard, limitation, prohibition, or other regulation promulgated in a subpart of this part pursuant to

sections 112(d), 112(h), or 112(f) of the Act (other identical or similar definitions are provided in 40CFR63.101).

emission standard (CAA302; 51.100-z): See emission limitation.

emission standard (EPA-94/04): The maximum amount of air polluting discharge legally allowed from a single source, mobile or stationary (cf. emission limitation).

emission standard or limitation under this Act (CAA304-42USC7604) means:
(1) a schedule or timetable of compliance, emission limitation, standard of performance or emission standard,
(2) a control or prohibition respecting a motor vehicle fuel or fuel additive, which is in effect under this Act (including a requirement applicable by reason of section 118) or under an applicable implementation plan, or
(3) any condition or requirement of a permit under part C of title I (relating to significant deterioration of air quality) or part D of title I (relating to nonattainment), any condition or requirement of section 113(d) (relating to certain enforcement orders), section 119 (relating to primary nonferrous smelter orders), any condition or requirement under an applicable implementation plan relating to transportation control measures, air quality maintenance plans, vehicle inspection and maintenance programs or vapor recovery requirements, section 221 (e) and (f) (relating to fuels and fuel additives), section 169A (relating to visibility protection), any condition or requirement under part B of title I (relating to ozone protection), or any requirement under section 111 or 112 (without regard to whether such requirement is expressed as an emission standard or otherwise). [2] which is in effect under this Act (including a requirement applicable by reason of section 118) or under an applicable implementation plan. [1] So in original public law. Period probably should be a comma. [2] So in original. Subsection (a) enacted without paragraph (2).

emission time (40CFR60-AA (method 22-3.2))

means accumulated amount of time that emissions are visible during the observation period.

emission unit (40CFR63.51) means any building, structure, facility, or installation. This could include an emission point or collection of emission points, within a major source, which the permitting authority determines is the appropriate entity for making a MACT determination under section 112(j), i.e., any of the following:
(1) An emission point that can be individually controlled.
(2) The smallest grouping of emission points, that, when collected together, can be commonly controlled by a single control device or work practice.
(3) Any grouping of emission points, that, when collected together, can be commonly controlled by a single control device or work practice.
(4) A grouping of emission points that are functionally related. Equipment is functionally related if the operation or action for which the equipment was specifically designed could not occur without being connected with or without relying on the operation of another piece of equipment.
(5) The entire geographical entity comprising a major source in a source category subject to a MACT determination under section 112(j) (under CAA).

emission warranty (40CFR85.2113-f) means those warranties given by vehicle manufacturers pursuant to section 207 of the Act.

emissions (40CFR72.2) means air pollutants exhausted from a unit or source into the atmosphere, as measured, recorded, and reported to the Administrator by the designated representative and as determined by the Administrator, in accordance with the emissions monitoring requirements of part 75 of this chapter.

emissions allowable under the permit (40CFR70.2) means a federally enforceable permit term or condition determined at issuance to be required by an applicable requirement that establishes an emissions limit (including a work practice standard) or a federally enforceable emissions cap that the source has assumed to avoid an applicable requirement to which the source would otherwise be subject.

emissions averaging (40CFR63.2) is a way to comply with the emission limitations specified in a relevant standard, whereby an affected source, if allowed under a subpart of this part, may create emission credits by reducing emissions from specific points to a level below that required by the relevant standard, and those credits are used to offset emissions from points that are not controlled to the level required by the relevant standard (other identical or similar definitions are provided in 40CFR63.101).

emissions budgets (40CFR51.852) are those portions of the applicable SIP's (state implementation plan) projected emissions inventories that describe the levels of emissions (mobile, stationary, area, etc.) that provide for meeting reasonable further progress milestones, attainment, and/or maintenance for any criteria pollutant or its precursors (other identical or similar definitions are provided in 40CFR93.152).

emissions offsets (40CFR51.852), for purposes of 40CFR51.858, are emissions reductions which are quantifiable, consistent with the applicable SIP (state implementation plan) attainment and reasonable further progress demonstrations, surplus to reductions required by, and credited to, other applicable SIP provisions, enforceable at both the State and Federal levels, and permanent within the time frame specified by the program (other identical or similar definitions are provided in 40CFR93.152).

emissions that a Federal agency has a continuing program responsibility for (40CFR51.852) means emissions that are specifically caused by an agency carrying out its authorities, and does not include emissions that occur due to subsequent activities, unless such activities are required by the Federal agency. Where an agency, in performing its normal program responsibilities, takes actions itself or imposes conditions that result in air pollutant emissions by a non-Federal entity taking subsequent actions, such emissions are covered by the meaning of a continuing program responsibility (other identical or similar definitions are provided in 40CFR93.152).

emissions trading (EPA-94/04): The creation of

surplus emission reductions at certain stacks, vents, or similar emissions sources and the use of this surplus to meet or redefine pollution requirements applicable to other emission sources. This allows one source to increase emissions when another sources reduces them, maintaining an overall constant emission level. Facilities that reduce emissions substantially may "bank" their "credits" or sell them to other industries.

emissions unit (40CFR51.165-vii) means any part of a stationary source which emits or would have the potential to emit any pollutant subject to regulation under the the Act.

emissions unit (40CFR51.166-7) means any part of a stationary source which emits or would have the potential to emit any pollutant subject to regulation under the Act.

emissions unit (40CFR51-AS-7) means any part of a stationary source which emits or would have the potential to emit any pollutant subject to regulation under the Act.

emissions unit (40CFR52.21-7) means any part of a stationary source which emits or would have the potential to emit any pollutant subject to regulation under the Act.

emissions unit (40CFR52.24-7) means any part of a stationary source which emits or would have the potential to emit any pollutant subject to regulation under the Act.

emissions unit (40CFR70.2) means any part or activity of a stationary source that emits or has the potential to emit any regulated air pollutant or any pollutant listed under section 112(b) of the Act. This term is not meant to alter or affect the definition of the term "unit" for purposes of title IV of the Act.

employee (40CFR3.102-a) means any officer or employee of the Environmental Protection Agency, Public Health Service commissioned officers assigned to EPA employees detailed to EPA from other federal agencies and employees detailed or assigned to EPA under the Intergovernmental Personnel Act. The term does not include special Government employees (other identical or similar definitions are

provided in 40CFR13.2-i; 14.2-c; 32.605-5; 52.1161-3; 52.2297-1; 311.2; OSHA3-29USC652).

employee exposure (40CFR763.121) means that exposure to airborne asbestos would occur if the employee were not using respiratory protective equipment.

employee exposure record (29CFR1910.20) means a record containing any of the following kinds of information:

(i) Environmental (workplace) monitoring or measuring of a toxic substance or harmful physical agent, including personal, area, grab, wipe, or other form of sampling as well as related collection and analytical methodologies, calculations, and other background data relevant to interpretation of the results obtained:

(ii) Biological monitoring results which directly assess the absorption of a toxic substance or harmful physical agent by body systems (e.g., the level of a chemical in the blood, urine, breath, hair, fingernails, etc.) but not including results which assess the biological effect of a substance or agent or which assess an employee's use of alcohol or drugs:

(iii) Material safety data sheets indicating that the material may pose a hazard to human health; or

(iv) In the absence of the above, a chemical inventory or any other record which reveals where and when used and the identity (e.g., chemical, common, or trade name) of a toxic substance or harmful physical agent.

employee salary offset (40CFR13.2-k) means the administrative collection of a debt by deductions at one or more officially established pay intervals from the current pay account of an employee without the employee's consent.

employer (40CFR52.1161-1) means any person or entity which employs 50 or more employees at any time during a calendar year at an employment facility located in the Boston Intrastate Region (other identical or similar definitions are provided in 40CFR; 52.2297-7; 721.3; 763.121; OSHA3-29USC652; 29CFR1910.2).

employment facility (40CFR52.2297-4) means any single location of a business nature with 250 or more employees working, at a minimum, the same six core hours.

emulsions (40CFR467.02-h) are stable dispersions of two immiscible liquids. In the aluminum forming category this is usually an oil and water mixture.

emulsions (40CFR471.02-n) are stable dispersions of two immiscible liquids. In the Nonferrous Metals Forming and Metal Powders Point Source category, this is usually an oil and water mixture.

enamel (40CFR52.741) means a coating that cures by chemical cross- linking of its base resin. Enamels can be distinguished from lacquers because enamels are not readily resoluble in their original solvent.

encapsulate (40CFR165.1-j) means to seal a pesticide, and its container if appropriate, in an impervious container made of plastic, glass, or other suitable material which will not be chemically degraded by the contents. This container then should be sealed within a durable container made from steel, plastic, concrete, or other suitable material of sufficient thickness and strength to resist physical damage during and subsequent to burial or storage.

encapsulation (40CFR763.83) means the treatment of ACBM with a material that surrounds or embeds asbestos fibers in an adhesive matrix to prevent the release of fibers, as the encapsulant creates a membrane over the surface (bridging encapsulant) or penetrates the material and binds its components together (penetrating encapsulant).

encapsulation (EPA-94/04): The treatment of asbestos-containing material with a liquid that covers the surface with a protective coating or embeds fibers in an adhesive matrix to prevent their release into the air.

enclose (40CFR52.741) means to cover any VOL surface that is exposed to the atmosphere.

enclosed process (40CFR704.25-2) means a process that is designed and operated so that there is no intentional release of any substance present in the process. A process with fugitive, inadvertent, or emergency pressure relief releases remains an enclosed process so long as measures are taken to

prevent worker exposure to an environmental contamination from the releases (other identical or similar definitions are provided in 40CFR704.104-2).

enclosed process (40CFR704.3) means a manufacturing or processing operation that is designed and operated so that there is no intentional release into the environment of any substance present in the operation. An operation with fugitive, inadvertent, or emergency pressure relief releases remains an enclosed process so long as measures are taken to prevent worker exposure to and environmental contamination from the releases.

enclosed storage area (40CFR60.381) means any area covered by a roof under which metallic minerals are stored prior to further processing or loading.

enclosed truck or railcar loading station (40CFR60.671) means that portion of a nonmetallic mineral processing plant where nonmetallic minerals are loaded by an enclosed conveying system into enclosed trucks or railcars.

enclosure (40CFR60.541) means a structure that surrounds a VOC (cement, solvent, or spray) application area and drying area, and that captures and contains evaporated VOC and vents it to a control device. Enclosures may have permanent and temporary openings.

enclosure (40CFR763.83) means an airtight, impermeable, permanent barrier around ACBM (asbestos containing building material) to prevent the release of asbestos fibers into the air.

enclosure (40FR60.441): See hood.

enclosure (EPA-94/04): Putting an airtight, impermeable, permanent barrier around asbestos-containing materials to prevent the release of asbestos fibers into the air.

end box (40CFR61.51-i) means a container(s) located on one or both ends of a mercury chlor-alkali electrolyzer which serves as a connection between the electrolyzer and denuder for rich and stripped amalgam.

end box ventilation system (40CFR61.51-j) means a

ventilation system which collects mercury emissions from the end boxes, the mercury pump sumps, and their water collection systems.

end finisher (40CFR60.561) means a polymerization reaction vessel operated under very low pressures, typically at pressures of 2 torr or less, in order to produce high viscosity poly(ethylene terephthalate). An end finisher is preceded in a high viscosity poly(ethylene terephthalate) process line by one or more polymerization vessels operated under less severe vacuums, typically between 5 and 10 torr. A high viscosity poly(ethylene terephthalate) process line may have one or more end finishers.

end sealing compound coat (40CFR52.741) means a compound applied to can ends which functions as a gasket when the end is assembled onto the can (under CAA).

end use (40CFR82.172) means processes or classes of specific applications within major industrial sectors where a substitute is used to replace an ozone-depleting substance.

end use product (40CFR152.3-k) means a pesticide product whose labeling:
(1) Includes directions for use of the product (as distributed or sold, or after combination by the user with other substances) for controlling pests or defoliating, desiccating, or regulating the growth of plants, and
(2) Does not state that the product may be used to manufacture or formulate other pesticide products.

end use product (40CFR158.153-b) means a pesticide product whose labeling:
(1) Includes directions for use of the product (as distributed or sold, or after combination by the user with other substances) for controlling pests or defoliating, desiccating or regulating growth of plants, and
(2) Does not state that the product may be used to manufacture or formulate other pesticide products.

endangered or threatened species (40CFR257.3.2-1) means any species listed as such pursuant to section 4 of the Endangered Species Act.

endangered species (EPA-94/04): Animals, birds, fish, plants, or other living organisms threatened with extinction by man-made or natural changes in their environment. Requirements for declaring a species endangered are contained in the Endangered Species Act.

endangered species (ESA3-16USC1531) means any species which is in danger of extinction throughout all or a significant portion of its range other than a species of the Class Insecta determined by the Secretary to constitute a pest whose protection under the provisions of this chapter would present an overwhelming and overriding risk to man.

endangerment assessment (EPA-94/04): A study to determine the nature and extent of contamination at a site on the National Priorities List and the risks posed to public health or the environment. EPA or the state conduct the study when a legal action is to be taken to direct potentially responsible parties to clean up a site or pay for it. An endangerment assessment supplements a remedial investigation.

endrin (40CFR129.4-c) means the compound endrin as identified by the chemical name 1,2,3,4,10,10-hexachloro-6,7-epoxy-1,4,4a,5,6,7,8,8a-octahydro-1,4-endo-5,8-endodimethanonaphthalene.

endrin (40CFR704.102-1) means the pesticide 2,7:3,6-Dimethanonaphth[2,3-b]oxirene,3,4,5,6,9,9-hexachloro-1a,2,2a,3,6,6a,7,7a-octahydro-, (1aalpha, 2beta, 2abeta, 3alpha, 6alpha, 6abeta, 7beta, 7aalpha)-, CAS Number 72-208.

endrin (EPA-94/04): a pesticide toxic to freshwater and marine aquatic life that produces adverse health effects in domestic water supplies.

endrin formulator (40CFR129.101-2) means a person who produces, prepares or processes a formulated product comprising a mixture of endrin and inert materials or other diluents into a product intended for application in any use registered under the Federal Insecticide, Fungicide and Rodenticide Act, as amended (7USC135, et seq.).

endrin manufacturer (40CFR129.101-1) means a manufacturer, excluding any source which is exclusively an endrin formulator, who produces, prepares or processes technical endrin or who uses endrin as a material in the production, preparation or processing of another synthetic organic substance.

energy average level (40CFR201.1-i) means a quantity calculated by taking ten times the common logarithm of the arithmetic average of the antilogs of one-tenth of each of the levels being averaged. The levels may be of any consistent type, e.g., maximum sound levels, sound exposure levels, and day-night sound levels.

energy facilities (CZMA304-16USC1453) means any equipment or facility which is or will be used primarily:
(A) in the exploration for, or the development, production, conversion, storage, transfer, processing, or transportation of, any energy resource; or
(B) for the manufacture, production, or assembly of equipment, machinery, products, or devices which are involved in any activity described in subparagraph (A). The term includes, but is not limited to:
   i.   electric generating plants;
   ii.   petroleum refineries and associated facilities;
   iii.   gasification plants;
   iv.   facilities used for the transportation, conversion, treatment, transfer, or storage of liquefied natural gas;
   v.   uranium enrichment or nuclear fuel processing facilities;
   vi.   oil and gas facilities, including platforms, assembly plants, storage depots, tank farms, crew and supply bases, and refining complexes;
   vii.   facilities including deepwater ports, for the transfer of petroleum;
   viii.   pipelines and transmission facilities; and
   ix.   terminals which are associated with any of the foregoing.

energy recovery (EPA-94/04): Obtaining energy from waste through a variety of processes.

energy storage device (40CFR600.002.85-46) means a rechargeable means of storing tractive energy on board a vehicle such as storage batteries or a flywheel.

energy summation of levels (40CFR201.1-j) means a quantity calculated by taking ten times the common logarithm of the sum of the antilogs of one-tenth of each of the levels being summed. The levels may be of any consistent type, e.g., day-night sound level or equivalent sound level (under the Noise Control Act).

enforceable requirements (EPA-94/04): Conditions or limitations in permits issued under the Clean Water Act, section 402 or 404 that, if violated, could result in the issuance of a compliance order or initiation of a civil or criminal action under federal or applicable state laws. If a permit has not been issued, the term includes any requirement which, in the Regional Administrator's judgement, would be included in the permit when issued. Where no permit applies, the term includes any requirement which the RA determines is necessary for the best practical waste treatment technology to meet applicable criteria.

enforceable requirements of the Act (40CFR35.2005-15) means those conditions or limitations of section 402 or 404 permits which, if violated, could result in the issuance of a compliance order or initiation of a civil or criminal action under section 30 of the Act or applicable State laws. If a permit has not been issued, the term shall include any requirement which, in the Regional Administrator's judgment, would be included in the permit when issued. Where no permit applies, the term shall include any requirement which the Regional Administrator determines is necessary for the best practicable waste treatment technology to meet applicable criteria.

enforceable requirements of the Act (40CFR35.905) are those conditions or limitations of section 402 or 404 permits which, if violated, could result in the issuance of a compliance order or initiation of a civil or criminal action under section 30 of the Act. If a permit has not been issued, the term shall include any requirement which, in the Regional Administrator's judgment, would be included in the permit when issued. Where no permit applies, the term shall include any requirement which the Regional Administrator determines is necessary to meet applicable criteria for best practicable waste treatment technology (BPWTT).

enforcement (EPA-94/04): EPA, state, or local legal actions to obtain compliance with environmental laws, rules, regulations, or agreements and/or obtain penalties or criminal sanctions for violations. Enforcement procedures may vary, depending on the requirements of different environmental laws and related implementing regulations. Under CERCLA, for example, EPA will seek to require potentially responsible parties to clean up a Superfund site, or pay for the cleanup, whereas under the Clean Air Act the agency may invoke sanctions against cities failing to meet ambient air quality standards that could prevent certain types of construction or federal funding. In other situations, if investigations by EPA and state agencies uncover willful violations, criminal trials and penalties are sought.

enforcement decision document (EDD) (EPA-94/04): A document that provides an explanation to the public of EPA's selection of the cleanup alternative at enforcement sites on the National Priorities List. Similar to a Record of Decision.

engine (40CFR89.2), as used in this part, refers to nonroad engine.

engine code (40CFR600.002.85-25) means, for gasoline-fueled and diesel vehicles, a unique combination, within an engine-system combination (as defined in part 86 of this chapter), of displacement, carburetor (or fuel injection) calibration, distributor calibration, choke calibration, auxiliary emission control devices, and other engine and emission control system components specified by the Administrator. For electric vehicles, engine code means a unique combination of manufacturer, electric traction motor, motor configuration, motor controller, and energy storage device.

engine code (40CFR86.082.2) means unique combination, within an engine system combination, of displacement, carburetor (or fuel injection) calibration, choke calibration, distributor calibration, auxiliary emission control devices, and other engine and mission control system components specified by the Administrator.

engine configuration (40CFR85.1402) means the set of components, tolerances, specifications, design parameters, and calibrations related to the emissions performance of the engine and specific to a subset of an engine family having a unique combination of displacement, fuel injection calibration, auxiliary emission control devices and emission control system components.

engine configuration (40CFR85.2113-i): See vehicle or engine configuration.

engine displacement (40CFR205.151-14) means volumetric engine capacity as defined in 40CFR205.153.

engine displacement-system combination (40CFR86.402.78) means an engine family-displacement-emission control system combination.

engine family (40CFR85.2113-h) means the basic classification unit of a vehicles product line for a single model year used for the purpose of emission-data vehicle or engine selection and as determined in accordance with 40CFR86.078-24.

engine family (40CFR86.082.2) means the basic classification unit of a manufacturer's product line used for the purpose of test fleet selection and determined in accordance with section 6.02-24.

engine family (40CFR86.402.78) means the basic classification unit of manufacturers product line used for the purpose of test fleet selection and determined in accordance with 40CFR86.420.

engine family group (40CFR86.082.2) means a combination of engine families for the purpose of determining a minimum deterioration factor under the Alternative Durability program.

engine lubricating oils (40CFR252.4-b) means petroleum-based oils used for reducing friction in engine parts.

engine manufacturer (40CFR89.2) means any person engaged in the manufacturing or assembling of new nonroad engines or importing such engines for resale, or who acts for and is under the control of any such person in connection with the distribution of such engines. Engine manufacturer does not include any dealer with respect to new nonroad engines received by such person in commerce.

engine model (40CFR87.1) means all commercial aircraft turbine engines which are of the same general series, displacement, and design characteristics and are usually approved under the same type certificate.

engine rebuild (40CFR85.1402) means an activity, occurring over one or more maintenance events, involving:
(1) Disassembly of the engine including the removal of the cylinder head(s); and
(2) The replacement or reconditioning of more than one major cylinder component in more than half of the cylinders.

engine replacement (40CFR85.1402) means the removal of an engine from the coach followed by the installation of another engine.

engine system combination (40CFR86.082.2) means an engine family-exhaust emission control system combination.

engine system combination (40CFR86.902.93) as defined in 40CFR86.082-2, means an engine family-exhaust emission control system combination.

engine used in a locomotive (40CFR89.2) means either an engine placed in the locomotive to move other equipment, freight, or passenger traffic, or an engine mounted on the locomotive to provide auxiliary power.

engine warm-up cycle (40CFR86.094.2) means sufficient vehicle operation such that the coolant temperature has risen by at least 40°F from engine starting and reaches a minimum temperature of 160°F.

enhanced inspection and maintenance (EPA-94/04): An improved automobile inspection and maintenance program-aimed at reducing automobile emissions-that contains, at a minimum, more vehicle types and model years, tighter inspection, and better management practices. It may also include annual computerized or centralized inspections, under-the-hood inspection for signs of tampering with pollution control equipment, and increased repair waiver cost.

enhanced review (40CFR63.51) means a review process containing all administrative steps needed to ensure that the terms and conditions resulting from the review process can be incorporated into the title V permit by an administrative amendment (under CAA).

enrichment (EPA-94/04): The addition of nutrients (e.g., nitrogen, phosphorus, carbon compounds) from sewage effluent or agricultural runoff to surface water, greatly increases the growth potential for algae and other aquatic plants.

entrain (EPA-94/04): To trap bubbles in water either mechanically through turbulence or chemically through a reaction.

entry loss (29CFR1910.94b): The loss in static pressure caused by air flowing into a duct or hood. It is usually expressed in inches of water gauge.

envelopes (40CFR250.4-m) means brown, manila, padded, or other mailing envelopes not included with "stationery."

environment (40CFR300.5), as defined by section 101(8) of CERCLA, means the navigable waters, the waters of the contiguous zone, and the ocean waters of which the natural resources are under the exclusive management authority of the United States under the Magnuson Fishery Conservation and Management Act; and any other surface water, ground water, drinking water supply, land surface and subsurface strata, or ambient air within the United States or under the jurisdiction of the United States (other identical or similar definitions are provided in 40CFR302.3).

environment (40CFR6.1003) means the natural and physical environment and excludes social, economic and other environments.

environment (EPA-94/04): The sum of all external conditions affecting the life, development and survival of an organism.

environment (FIFRA2-7USC136) includes water, air, and land and all plants and man and other animals living therein, and interrelationship which exists among these (other identical or similar definitions are provided in 40CFR171.2-13; 355.20; 370.2).

environment (SF101-42USC9601) means:

(A) the navigable waters, the waters of the contiguous zone, and the ocean waters of which the natural resources are under the exclusive management authority of the United States under the Fishery Conservation and Management Act of 1976, and

(B) any other surface water, ground water, drinking water supply, land surface or subsurface strata, or ambient air within the United States or under the jurisdiction of the United States.

environment (TSCA3-15USC2602) includes water, air, and land and the interrelationship which exists among and between water, air, and land and all living things (other identical or similar definitions are provided in 40CFR723.50-6; 723.175-5; 723.250-2).

Environmental Appeals Board (40CFR124.2) shall mean the Board within the Agency described in 40CFR1.25(e) of this title. The Administrator delegates authority to the Environmental Appeals Board to issue final decisions in RCRA, PSD, UIC, or NPDES permit appeals filed under this subpart, including informal appeals of denials of requests for modification, revocation and reissuance, or termination of permits under section 124.5(b) than to the Environmental Appeals Board, will not be considered. This delegation does not preclude the Environmental Appeals Board from referring an appeal or a motion under this subpart to the Administrator when the Environmental Appeals Board, in its discretion, deems it appropriate to do so. When an appeal or motion is referred to the Administrator by the Environmental Appeals Board, all parties shall be so notified and the rules in this subpart referring to the Environmental Appeals Board shall be interpreted as referring to the Administrator.

Environmental Appeals Board (40CFR72.2) means the three-member board established pursuant to 40CFR1.25(e) of this chapter and authorized to hear appeals pursuant to part 78 of this chapter.

Environmental Appeals Board (40CFR85.1807-6) means the Board within the Agency described in 40CFR1.25 of this title. The Administrator delegates authority to the Environmental Appeals Board to issue final decisions in appeals filed under this part.

An appeal directed to the Administrator, rather than to the Environmental Appeals Board, will not be considered. This delegation of authority to the Environmental Appeals Board does not preclude the Environmental Appeals Board from referring an appeal or a motion filed under this part to the Administrator for decision when the Environmental Appeals Board, in its discretion, deems it appropriate to do so. When an appeal or motion is referred to the Administrator, all parties shall be so notified and the rules in this part referring to the Environmental Appeals Board shall be interpreted as referring to the Administrator (other identical or similar definitions are provided in 40CFR27.2; 22.03; 66.3-g; 86.614.84-5; 86.1014.84; 86.1115.87; 124.72; 164.2-g; 209.3-k).

environmental assessment (40CFR1508.9):

(a) means a concise public document for which a Federal agency is responsible that serves to:

    (1) Briefly provide sufficient evidence and analysis for determining whether to prepare an environmental impact statement or a finding of no significant impact.

    (2) Aid an agency's compliance with the Act when no environmental impact statement is necessary.

    (3) Facilitate preparation of a statement when one is necessary.

(b) Shall include brief discussions of the need for the proposal, of alternatives as required by 40CFR102(2)(E), of the environmental impacts of the proposed action and alternatives, and a listing of agencies and persons consulted.

environmental assessment (EPA-94/04): An environmental analysis prepared pursuant to the National Environmental Policy Act to determine whether a federal action would significantly affect the environment and thus require a more detailed environmental impact statement.

environmental audit (EPA-94/04): An independent assessment of the current status of a party's compliance with applicable environmental requirements or of a party's environmental compliance policies, practices, and controls.

environmental document (40CFR1508.10) includes the documents specified in 40CFR1508.9

(environmental assessment), 40CFR1508.11 (environmental impact statement), 40CFR1508.13 (finding of no significant impact), and 40CFR1508.22 (notice of intent).

environmental education and environmental education and training (40CFR47.105-a) mean educational activities and training activities involving elementary, secondary, and postsecondary students, as such terms are defined in the State in which they reside, and environmental education personnel, but does not include technical training activities directed toward environmental management professionals or activities primarily directed toward the support of noneducational research and development.

environmental equity (EPA-94/04): Equal protection from environmental hazards of individuals, groups or communities regardless of race, ethnicity, or economic status.

environmental exposure (EPA-94/04): Human exposure to pollutants originating from facility emissions. Threshold levels are not necessarily surpassed, but low level chronic pollutant exposure is one of the most common forms of environmental exposure (cf. threshold level).

environmental impact statement (EIS) (40CFR1508.11) means a detailed written statement as required by 40CFR102(2)(C) of the Act (NEPA) (see 40CFR6.201 for the format of EIS preparation).

environmental impact statement (EPA-94/04): A document required of federal agencies by the National Environmental Policy Act for major projects or legislative proposals significantly affecting the environment. A tool for decision making, it describes the positive and negative effects of the undertaking and cites alternative actions.

environmental indicator (EPA-94/04): A measurement, statistic or value that provides a proximate gauge or evidence of the effects of environmental management programs or of the state or condition of the environment.

environmental information document (40CFR6.101-d) means any written analysis prepared by an applicant, grantee or contractor describing the environmental impacts of a proposed action. This document will be of sufficient scope to enable the responsible official to prepare an environmental assessment as described in the remaining subparts of this regulation.

environmental justice (EPA-94/04): The fair treatment of all races, cultures, incomes, and educational levels with respect to the development, implementation, and enforcement of environmental laws, regulations, and policies. Fair treatment implies that no population of people should be forced to shoulder a disproportionate share of the negative environmental impacts of pollution or environmental hazards due to a lack of political or economic strength levels.

environmental noise (NCA3-42USC4902) means the intensity, duration, and the character of sounds from all sources (other identical or similar definitions are provided in 40CFR205.2-21).

Environmental Protection Agency (EPA) (40CFR122.2) means the United States Environmental Protection Agency (other identical or similar definitions are provided in 40CFR144.3; 146.3; 270.2; 401.11-c).

Environmental Research, Development, and Demonstration Authorization Act of 1980, 42USC1857 et seq: See act or ERDDAA.

environmental response team (EPA-94/04): EPA experts located in Edison, N.J., and Cincinnati, OH, who can provide around-the-clock technical assistance to EPA regional offices and states during all types of hazardous waste site emergencies and spills of hazardous substances.

environmental review (40CFR6.101-c) means the process whereby an evaluation is undertaken by EPA to determine whether a proposed Agency action may have a significant impact on the preparation on the EIS (environmental impact statement).

environmental transformation product (40CFR723.50-7) means any chemical substance resulting from the action of environmental processes on a parent compound that changes the molecular identity of the parent compound.

environmentally related measurements (40CFR30.200) means any data collection activity or investigation involving the assessment of chemical, physical, or biological factors in the environment which affect human health or the quality of life. The following are examples of environmentally related measurements:

(a) A determination of pollutant concentrations from sources or in the ambient environment, including studies of pollutant transport and fate;

(b) a determination of the effects of pollutants on human health and on the environment;

(c) a determination of the risk/benefit of pollutants in the environment;

(d) a determination of the quality of environmental data used in economic studies; and

(e) a determination of the environmental impact of cultural and natural processes.

environmentally transformed A chemical substance (40CFR721.3) is "environmentally transformed" when its chemical structure changes as a result of the action of environmental processes on it (under TSCA).

EPA (40CFR2.100) means the United States Environmental Protection Agency (40CFR2.100, other similar definitions are provided in 40CFR7.25; 16.2; 17.2; 122.2; 124.2; 144.3; 146.3; 147.2902; 160.3; 233.3; 270.2; 300.5; 372.3; 503.9-i; 704.3; 707.63-a; 710.2-k; 712.3-b; 716.3; 720.3-j; 723.50-2; 723.250-6; 749.68-9; 763.63-d; 790.3; 792.3).

EPA acknowledgement of consent (40CFR262.51) means the cable sent to EPA from the U.S. Embassy in a receiving country that acknowledges the written consent of the receiving country to accept the hazardous waste and describes the terms and conditions of the receiving country's consent to the shipment.

EPA and the Agency (40CFR57.103-l) means the Administrator of the U.S. Environmental Protection Agency, or the Administrator's authorized representative.

EPA assistance (40CFR7.25) means any grant or cooperative agreement, loan, contract (other than a procurement contract or a contract of insurance or guaranty), or any other arrangement by which EPA provides or otherwise makes available assistance in the form of:

(1) Funds;

(2) Services of personnel; or

(3) Real or personal property or any interest in or use of such property including:

  (i) Transfers or leases of such property for less than fair market value or for reduced consideration; and

  (ii) Proceeds from a subsequent transfer or lease of such property if EPA's share of its fair market value is not returned to EPA.

EPA Claims Officer (40CFR14.2-a) is the Agency official delegated the responsibility by the Administrator to carry out the provisions of the Act.

EPA conditional method (40CFR63.51) means any method of sampling and analyzing for air pollutants that has been validated by the Administrator but that has not been published as an EPA Reference Method.

EPA enforcement officer (40CFR89.2) means any officer or employee of the Environmental Protection Agency so designated in writing by the Administrator (or by his or her designee) (other identical or similar definitions are provided in 40CFR86.082.2; 86.402.78).

EPA hazardous waste number (40CFR260.10) means the number assigned by EPA to each hazardous waste listed in part 261, subpart D, of this chapter and to each characteristic identified in part 261, subpart C, of this chapter (under the Resource Conservation Recovery Act).

EPA identification number (40CFR260.10) means the number assigned by EPA to each generator, transporter, and treatment, storage, or disposal facility.

EPA identification number (40CFR761.3) means the 12-digit number assigned to a facility by EPA upon notification of PCB waste activity under 40CFR761.205.

EPA legal office (40CFR2.201-n) means the EPA General Counsel and any EPA office over which the General Counsel exercises supervisory authority, including the various Offices of Regional Counsel. (see also paragraph (m) of this section).

EPA office (40CFR2.201-m) means any organizational element of EPA at any level or location. (The terms EPA office and EPA legal office are used in this subpart for the sake of brevity and ease of reference. When this subpart requires that an action be taken by an EPA office or by an EPA legal office, it is the responsibility of the officer or employee in charge of that office to take the action or ensure that it is taken).

EPA or the Administrator (40CFR70.2) means the Administrator of the EPA or his designee.

EPA record or record (40CFR2.100-b) means any document, writing, photograph, sound or magnetic recording, drawing, or other similar thing by which information has been preserved, from which the information can be retrieved and copied, and over which EPA has possession or control. It may include copies of the records of other Federal agencies (see 40CFR2.111(d)). The term includes informal writings (such as drafts and the like), and also includes information preserved in a form which must be translated or deciphered by machine in order to be intelligible to humans. The term includes documents and the like which were created or acquired by EPA, its predecessors, its officers, and its employees by use of Government funds or in the course of transacting official business. However, the term does not include materials which are the personal records of an EPA officer or employee. Nor does the term include materials published by non-Federal organizations which are readily available to the public, such as books, journals, and periodicals available through reference libraries, even if such materials are in EPA's possession.

EPA reference method (40CFR63.51) means any method of sampling and analyzing for an air pollutant as described in appendix A of part 60 of this chapter, appendix B of part 61 of this chapter, or appendix A of part 63.

EPA region (40CFR260.10) means the states and territories found in any one of the following ten regions:
Region I

Maine, Vermont, New Hampshire, Massachusetts, Connecticut, and Rhode Island.
Region II
New York, New Jersey, Commonwealth of Puerto Rico, and the U.S. Virgin Islands.
Region III
Pennsylvania, Delaware, Maryland, West Virginia, Virginia, and the District of Columbia.
Region IV
Kentucky, Tennessee, North Carolina, Mississippi, Alabama, Georgia, South Carolina, and Florida.
Region V
Minnesota, Wisconsin, Illinois, Michigan, Indiana and Ohio.
Region VI
New Mexico, Oklahoma, Arkansas, Louisiana, and Texas.
Region VII
Nebraska, Kansas, Missouri, and Iowa.
Region VIII
Montana, Wyoming, North Dakota, South Dakota, Utah, and Colorado.
Region IX
California, Nevada, Arizona, Hawaii, Guam, American Samoa, Commonwealth of the Northern Mariana Islands.
Region X
Washington, Oregon, Idaho, and Alaska.

EPA trial staff (40CFR72.2) means an employee of EPA, whether temporary or permanent, who has been designated by the Administrator to investigate, litigate, and present evidence, arguments, and positions of EPA in any evidentiary hearing under part 78 of this chapter. Any EPA or permitting authority employee, consultant, or contractor who is called as a witness in the evidentiary hearing by EPA trial staff shall be deemed to be "EPA trial staff."

ephippium (40CFR797.1300-3) means a resting egg which develops under the carapace in response to stress conditions in daphnids (other identical or similar definitions are provided in 40CFR797.1330-4).

epidemiology (EPA-94/04): Study of the distribution of disease, or other health-related states and events in human populations, as related to age, sex, occupation, ethnic, and economic status in order to identify and alleviate health problems and promote better health.

epilimnion (EPA-94/04): The upper layer of water in a thermally stratified lake or reservoir. This layer consists of the warmest water.

episode (EPA-94/04): An air pollution incident in a given area caused by a concentration of atmospheric pollutants under meteorological conditions that may result in a significant increase in illnesses or deaths. May also describe water pollution events or hazardous material spills.

episode criteria (40CFR51-AL-1.1) means conditions justifying the proclamation of an air pollution alert, air pollution warning, or air pollution emergency shall be deemed to exist whenever the Director determines that the accumulation of air pollutants in any place is attaining or has attained levels which could, if such levels are sustained or exceeded, lead to a substantial threat to the health of persons. In making this determination, the Director will be guided by the following criteria:
(a) Air pollution forecast;
(b) Alert;
(c) Warning;
(d) Emergency; and
(e) Termination.

equal opportunity clause (40CFR8.2-j) means the contract provisions set forth in section 4 (a) or (b), as appropriate.

equilibrium (EPA-94/04): In relation to radiation, the state at which the radioactivity of consecutive elements within a radioactive series is neither increasing nor decreasing.

equipment (40CFR31.3) means tangible, nonexpendable, personal property having a useful life of more than one year and an acquisition cost of $5,000 or more per unit. A grantee may use its own definition of equipment provided that such definition would at least include all equipment defined above (other identical or similar definitions are provided in 40CFR35.6015-18; 60.481; 60.591; 60.631; 61.131; 61.241; 63.161; 264.1031).

equipment leak (40CFR63.101) means emissions of organic hazardous air pollutants from a pump, compressor, agitator, pressure relief device, sampling connection system, open-ended valve or line, valve, surge control vessel, bottoms receiver, or instrumentation system in organic hazardous air pollutant service as defined in 40CFR63.161 of subpart H of this part.

equipment leaks (40CFR63.51) means leaks from pumps, compressors, pressure relief devices, sampling connection systems, open-ended valves or lines, valves, connectors, agitators, accumulator vessels, and instrumentation systems in hazardous air pollutant service.

equipment room (change room) (40CFR763.121) means a contaminated room located within the decontamination area that is supplied with impermeable base or containers for the disposal of contaminated protective clothing and equipment.

equivalence data (40CFR790.3) means chemical data or biological test data intended to show that two substances or mixtures are equivalent.

equivalent (40CFR790.3) means that a chemical substance or mixture is able to represent or substitute for another in a test or series of tests, and that the data from one substance can be used to make scientific and regulatory decisions concerning the other substance.

equivalent diameter (40CFR60.711-11) means four times the area of an opening divided by its perimeter (other identical or similar definitions are provided in 40CFR60.741).

equivalent diameter (40CFR72.2) means a value, calculated using the equation in paragraph 2.1 of Method 1 in part 60, Appendix A of this chapter, and used to determine the upstream and downstream distances for locating CEMS or CEMS components in flues or stacks with rectangular cross sections.

equivalent emission limitation (40CFR63.2) means the maximum achievable control technology emission limitation (MACT emission limitation) for hazardous air pollutants that the Administrator (or a State with an approved permit program) determines on a case-by-case basis, pursuant to section 112(g) or section 112(j) of the Act, to be equivalent to the emission standard that would apply to an affected source if such standard had been promulgated by the Administrator under this part pursuant to section 112(d) or section 112(h) of the Act (other identical or similar definitions are provided in 40CFR63.101; 63.191).

equivalent emission limitation (40CFR63.51) means an emission limitation, established under section 112(j) of the Act, which is at least as stringent as the MACT standard that EPA would have promulgated under section 112(d) or section 112(h) of the Act (under CAA).

equivalent live weight killed (40FR432.11-e; 432.21-e; 432.31-e; 432.41-e): See ELWK.

equivalent method (40CFR260.10) means any testing or analytical method approved by the Administrator under 40CFR260.20 and 40CFR260.21.

equivalent method (40CFR50.1-g) means a method of sampling and analyzing the ambient air for an air pollutant that has been designated as an equivalent method in accordance with part 53 of this chapter; it does not include a method for which an equivalent method designated has been cancelled in accordance with 40CFR53.11 or 40CFR53.16 of this chapter.

equivalent method (40CFR53.1-f) means a method of sampling and analyzing the ambient air for an air pollutant that has been designated as an equivalent method in accordance with this part; it does not include a method for which an equivalent method designation has been cancelled in accordance with 40CFR53.11 or 40CFR53.16.

equivalent method (40CFR60.2) means any method of sampling and analyzing for an air pollutant which has been demonstrated to the Administrator's satisfaction to have a consistent and quantitatively known relationship to the reference method, under specified conditions.

equivalent method (EPA-94/04): Any method of sampling and analyzing for air pollution which has been demonstrated to the EPA Administrator's satisfaction to be, under specific conditions, an acceptable alternative to normally used reference methods.

equivalent P_2O_5 feed (40CFR60.201-c) means the quantity of phosphorus, expressed as phosphorous pentoxide, fed to the process (other identical or similar definitions are provided in 40CFR60.211-c).

equivalent P_2O_5 feed (40CFR60.221-c) means the quantity of phosphorus, expressed as phosphorus pentoxide, fed to the process (other identical or similar definitions are provided in 40CFR60.231-d).

equivalent P_2O_5 stored (40CFR60.241-c) means the quantity of phosphorus, expressed as phosphorus pentoxide, being cured or stored in the affected facility.

equivalent petroleum-based fuel economy value (40CFR600.502.81-3) means a number which represents the average number of miles traveled by an electric vehicle per gallon of gasoline.

equivalent projects (40CFR35.3105-f) means those section 212 wastewater treatment projects constructed in whole or in part before October 1, 1994, with funds "directly made available by" the capitalization grant. These projects must comply with the requirements of section 602(b)(6) of the Act.

equivalent sound level (40CFR201.1-k) means the level, in decibels, of the mean-square A-weighted sound pressure during a stated time period, with reference to the square of the standard reference sound pressure of 20 micropascals. It is the level of the sound exposure divided by the time period and is abbreviated as L_{eq}.

equivalent test weight (40CFR86.094.2) means the weight, within an inertia weight class, which is used in the dynamometer testing of a vehicle and which is based on its loaded vehicle weight or adjusted loaded vehicle weight in accordance with the provisions of subparts A and B of this part.

erosion (EPA-94/04): The wearing away of land surface by wind or water, intensified by land-clearing practices related to farming, residential or industrial development, road building, or logging (cf. corrosion).

established federal standard (29CFR1910.2) means any operative standard established by any agency of the United States and in effect on April 28, 1971, or contained in any Act of Congress in force on the date of enactment of the Williams-Steiger Occupational Safety and Health Act.

established federal standard (OSHA3-29USC652) means any operative occupational safety and health standard established by an agency of the United States and presently in effect, or contained in any Act of Congress in force on the date of enactment of this.

establishment (40CFR167.3) means any site where a pesticidal product, active ingredient, or device is produced, regardless of whether such site is independently owned or operated, and regardless of whether such site is domestic and producing a pesticidal product for export only, or whether the site is foreign and producing any pesticidal product for import into the United States.

establishment (40CFR372.3) means an economic unit, generally at a single physical location, where business is conducted or where services or industrial operations are performed.

establishment (FIFRA2-7USC136) means any place where a pesticide or device or active ingredient used in producing a pesticide is produced, or held, for distribution or sale.

estuarine sanctuary (CZMA304-16USC1453-90) means a research area which may include any part or all of an estuary and any island, transitional area, and upland in, adjoining, or adjacent to such estuary, and which constitutes to the extent feasible a natural unit, set aside to provide scientists and students the opportunity to examine over a period of time the ecological relationships within the area.

estuary (CZMA304-16USC1453) means that part of a river or stream or other body of water having unimpaired connection with the open sea, where the sea water is measurably diluted with fresh water derived from land drainage. The term includes estuary-type areas of the Great Lakes.

estuary (EPA-94/04): Regions of interaction between rivers and near-shore ocean waters, where tidal action and river flow mix fresh and salt water. Such areas include bays, mouths of rivers, salt marshes, and lagoons. These brackish water ecosystems shelter and feed marine life, birds, and wildlife (cf. wetlands).

etching (40CFR467.02-i): See cleaning.

ethanol (EPA-94/04): An alternative automotive fuel derived from grain and corn; usually blended with gasoline to form gasohol.

ethanol blender (40CFR80.2-v) means any person who owns, leases, operates, controls, or supervises an ethanol blending plant.

ethanol blending plant (40CFR80.2-u) means any refinery at which gasoline is produced solely through the addition of ethanol to gasoline, and at which the quality or quantity of gasoline is not altered in any other manner.

ethnic foods (40CFR407.81-g) shall mean the production of canned and frozen Chinese and Mexican specialties utilizing fresh and pre-processed bean sprouts, bamboo shoots, water chestnuts, celery, cactus, tomatoes, and other similar vegetables necessary for the production of the various characteristic product styles.

ethylene dibromide (EDB) (EPA-94/04): A chemical used as an agricultural fumigant and in certain industrial processes. Extremely toxic and found to be a carcinogen in laboratory animals, EDB has been banned for most agricultural uses in the United States.

ethylene dichloride plant (40CFR61.61-a) includes any plant which produces ethylene dichloride by reaction of oxygen and hydrogen chloride with ethylene.

ethylene dichloride purification (40CFR61.61-o) includes any part of the process of ethylene dichloride purification following ethylene dichloride formation, but excludes crude, intermediate, and final ethylene dichloride storage tanks.

ethylene process or ethylene process unit

(40CFR63.101) means a chemical manufacturing process unit in which ethylene and/or propylene are produced by separation from petroleum refining process streams or by subjecting hydrocarbons to high temperatures in the presence of steam. The ethylene process unit includes the separation of ethylene and/or propylene from associated streams such as a C4 product, pyrolysis gasoline, and pyrolysis fuel oil. The ethylene process does not include the manufacture of SOCMI chemicals such as the production of butadiene from the C4 stream and aromatics from pyrolysis gasoline.

ethylidene norbornene (40CFR63.191) means the diene with CAS number 16219-753. Ethylidene norbornene is used in the production of ethylene-propylene rubber products (under the Clean Air Act).

eutrophic lake (40CFR35.1605-5): A lake that exhibits any of the following characteristics:
(A) Excessive biomass accumulations of primary producers;
(B) rapid organic and/or inorganic sedimentation and shallowing; or
(C) seasonal and/or diurnal dissolved oxygen deficiencies that may cause obnoxious odors, fish kills, or a shift in the composition of aquatic fauna to less desirable forms (under the Clean Water Act).

eutrophic lakes (EPA-94/04): Shallow, murky bodies of water with concentrations of plant nutrients causing excessive production of algae (cf. dystrophic lakes).

eutrophication (EPA-94/04): The slow aging process during which a lake, estuary, or bay evolves into a bog or marsh and eventually disappears. During the later stages of eutrophication the water body is choked by abundant plant life due to higher levels of nutritive compounds such as nitrogen and phosphorus. Human activities can accelerate the process.

evaluation program or program (40CFR610.11-21) means the sequence of analyses and tests prescribed by the Administrator as described in 40CFR610.13 in order to evaluate the performance of a retrofit device.

evaporation ponds (EPA-94/04): Areas where sewage sludge is dumped and dried.

evaporative emission code (40CFR86.082.2) means a unique combination, in an evaporative emission family evaporative emission control system combination, of purge system calibrations, fuel tank and carburetor bowl vent calibrations and other fuel system and evaporative emission control system components and calibrations specified by the Administrator.

evaporative emissions (40CFR86.082.2) means hydrocarbons emitted into the atmosphere from a motor vehicle, other than exhaust and crankcase emissions.

evaporative/refueling emission control system (40CFR86.098.2) means a unique combination within an evaporative/refueling family of canister adsorptive material, purge system configuration, purge strategy, and other parameters determined by the Administrator to affect evaporative and refueling emission control system durability or deterioration factors.

evaporative/refueling emission family (40CFR86.098.2) means the basic classification unit of a manufacturers' product line used for the purpose of evaporative and refueling emissions test fleet selection and determined in accordance with 40CFR86.098-24.

evaporative vehicle configuration (40CFR86.082.2) means a unique combination of basic engine, engine code, body type, and evaporative emission code.

evapotranspiration (EPA-94/04): The loss of water from the soil both by evaporation and by transpiration from the plants growing in the soil.

ex parte communication (40CFR124.78-3) means any communication, written or oral, relating to the merits of the proceeding between the decisional body and an interested person outside the Agency or the Agency trial staff which was not originally filed or stated in the administrative record or in the hearing. Ex parte communications do not include:
(i) Communications between Agency employees other than between the Agency trial staff and

the members of the decisional body;

(ii) Discussions between the decisional body and either:
   (A) Interested persons outside the Agency, or
   (B) The Agency trial staff, if all parties have received prior written notice of the proposed communications and have been given the opportunity to be present and participate therein.

ex parte communication (40CFR304.12-f) means any communication, written or oral, relating to the merits of the arbitral proceeding, between the Arbitrator and any interested person, which was not originally filed or stated in the administrative record of the proceeding. Such communication is not "ex parte communication" if all parties to the proceeding have received prior written notice of the proposed communication and have been given the opportunity to be present and to participate therein (under CERCLA).

ex parte communication (40CFR57.809-3) means any communication, written or oral, relating to the merits of the proceeding between the decisional body and an interested person outside the Agency or the Agency trial staff which was not originally filed or stated in the administrative record or in the hearing. Ex parte communications do not include:

(i) Communications between Agency employees other than between the Agency trial staff and the member of the decisional body;

(ii) Discussions between the decisional body and either:
   (A) Interested persons outside the Agency, or
   (B) The Agency trial staff if all parties have received prior written notice of such proposed communications and have been given the opportunity to be present and participate therein.

ex parte communication (40CFR72.2) means any communication, written or oral, relating to the merits of an adjudicatory proceeding under part 78 of this chapter, that was not originally included or stated in the administrative record, in a pleading, or in an evidentiary hearing or oral argument under part 78 of this chapter, between the decisional body and any interested person outside EPA or any EPA trial staff. Ex parte communication shall not include:

(1) Communication between EPA employees other than between EPA trial staff and a member of the decisional body; or

(2) Communication between the decisional body and interested persons outside the Agency, or EPA trial staff, where all parties to the proceeding have received prior written notice of the proposed communication and are given an opportunity to be present and to participate therein.

excavation zone (40CFR280.12) means the volume containing the tank system and backfill material bounded by the ground surface, walls, and floor of the pit and trenches into which the UST system is placed at the time of installation.

exceedance (EPA-94/04): Violation of the pollutant levels permitted by environmental protection standards.

excess ammonia-liquor storage tank (40CFR61.131) means any tank, reservoir, or container used to collect or store a flushing liquor solution prior to ammonia or phenol recovery.

excess emissions (40CFR51.100-bb) means emissions of an air pollutant in excess of an emission standard.

excess emissions (40CFR57.304): For the purposes of this subpart, any emissions greater than those permitted by the NSO provisions established under 40CFR57.302 (performance level of interim constant controls) or 40CFR 57.303 (plantwide emission limitation) of this subpart shall constitute excess emissions. Emission of any gas stream identified under 40CFR57.301 (a), (b), (c), (d) or (e) of this subpart that is not treated by a sulfur dioxide constant control system shall also constitute an excess emission under this subpart.

excess emissions (40CFR72.2) means:

(1) Any tonnage of sulfur dioxide emitted by an affected unit during a calendar year that exceeds the Acid Rain emissions limitation for sulfur dioxide for the unit; and

(2) Any tonnage of nitrogen oxide emitted by an affected unit during a calendar year that exceeds the annual tonnage equivalent of the Acid Rain emissions limitation for nitrogen oxides

applicable to the affected unit taking into account the unit's heat input for the year.

excess emissions and continuous monitoring system performance report (40CFR63.2) is a report that must be submitted periodically by an affected source in order to provide data on its compliance with relevant emission limits, operating parameters, and the performance of its continuous parameter monitoring systems.

excess emissions and monitoring systems performance report (40CFR60.2) is a report that must be submitted periodically by a source in order to provide data on its compliance with stated emission limits and operating parameters, and on the performance of its monitoring systems.

excess pesticides (40CFR165.1-p): See pesticide.

excess property (40CFR35.6015-19) means any property under the control of a Federal agency that is not required for immediate or foreseeable needs and thus is a candidate for disposal.

excessive concentration (40CFR51.100-kk) is defined for the purpose of determining good engineering practice stack height under 40CFR51.100(ii)(3) and means:

(1) For sources seeking credit for stack height exceeding that established under 40CFR51.100(ii)(2) a maximum ground-level concentration due to emissions from a stack due in whole or part to downwash, wakes, and eddy effects produced by nearby structures or nearby terrain features which individually is at least 40 percent in excess of the maximum concentration experienced in the absence of such downwash, wakes, or eddy effects and which contributes to a total concentration due to emissions from all sources that is greater than an ambient air quality standard. For sources subject to the prevention of significant deterioration program (40CFR 51.166 and 52.21), an excessive concentration alternatively means a maximum ground-level concentration due to emissions from a stack due in whole or part to downwash, wakes, or eddy effects produced by nearby structures or nearby terrain features which individually is at least 40 percent in excess of the maximum concentration experienced in the absence of such downwash, wakes, or eddy effects and greater than a prevention of significant deterioration increment. The allowable emission rate to be used in making demonstrations under this part shall be prescribed by the new source performance standard that is applicable to the source category unless the owner or operator demonstrates that this emission rate is infeasible. Where such demonstrations are approved by the authority administering the State implementation plan, an alternative emission rate shall be established in consultation with the source owner or operator.

(2) For sources seeking credit after October 11, 1983, for increases in existing stack heights up to the heights established under 40CFR51.100(ii)(2), either:

    (i) a maximum ground-level concentration due in whole or part to downwash, wakes or eddy effects as provided in paragraph (kk)(1) of this section, except that the emission rate specified by any applicable State implementation plan (or, in the absence of such a limit, the actual emission rate) shall be used, or

    (ii) the actual presence of a local nuisance caused by the existing stack, as determined by the authority administering the State implementation plan; and

(3) For sources seeking credit after January 12, 1979 for a stack height determined under 40CFR51.100(ii)(2) where the authority administering the State implementation plan requires the use of a field study or fluid model to verify GEP stack height, for sources seeking stack height credit after November 9, 1984 based on the aerodynamic influence of cooling towers, and for sources seeking stack height credit after December 31, 1970 based on the aerodynamic influence of structures not adequately represented by the equations in 40CFR51.100(ii)(2), a maximum ground-level concentration due in whole or part to downwash, wakes or eddy effects that is at least 40 percent in excess of the maximum concentration experienced in the absence of such downwash, wakes, or eddy effects.

excessive infiltration/inflow (40CFR35.2005-16)

means the quantities of infiltration/inflow which can be economically eliminated from a sewer system as determined in a cost effectiveness analysis that compares the costs for correcting the infiltration/inflow conditions to the total costs for transportation and treatment of the infiltration/inflow (see also 40CFR35.2005(b) (28) and (29) and 35.2120).

excessive infiltration/inflow (40CFR35.905) means the quantities of infiltration/inflow which can be economically eliminated from a sewerage system by rehabilitation, as determined in a cost-effectiveness analysis that compares the costs for correcting the infiltration/inflow conditions to the total costs for transportation and treatment of the infiltration/inflow, subject to the provisions in section 35.927.

excessive release (40CFR52.741) means a discharge of more than 295 g (0.65 lbs) of mercaptans and/or hydrogen sulfide into the atmosphere in any 5-minute period.

excluded manufacturing process (40CFR761.3) means a manufacturing process in which quantities of PCBs, as determined in accordance with the definition of inadvertently generated PCBs, calculated as defined, and from which releases to products, air, and water meet the requirements of paragraphs (1) through (5) of this definition, or the importation of products containing PCBs as unintentional impurities, which products meet the requirements of paragraphs (1) and (2) of this definition:

(1) The concentration of inadvertently generated PCBs in products leaving any manufacturing site or imported into the United States must have an annual average of less than 25 ppm, with a 50 ppm maximum.

(2) The concentration of inadvertently generated PCBs in the components of detergent bars leaving the manufacturing site or imported into the United States must be less than 5 ppm.

(3) The release of inadvertently generated PCBs at the point at which emissions are vented to ambient air must be less than 10 ppm.

(4) The amount of inadvertently generated PCBs added to water discharged from a manufacturing site must be less than 100 micrograms per

resolvable gas chromatographic peak per liter of water discharged.

(5) Disposal of any other process wastes above concentrations of 50 ppm PCB must be in accordance with subpart D of this part (under TSCA).

excluded PCB products (40CFR761.3) means PCB materials which appear at concentrations less than 50 ppm, including but not limited to:

(1) Non-Aroclor inadvertently generated PCBs as a byproduct or impurity resulting from a chemical manufacturing process.

(2) Products contaminated with Aroclor or other PCB materials from historic PCB uses (investment casting waxes are one example).

(3) Recycled fluids and/or equipment contaminated during use involving the products described in paragraphs (1) and (2) of this definition (heat transfer and hydraulic fluids and equipment and other electrical equipment components and fluids are examples).

(4) Used oils, provided that in the cases of paragraphs (1) through (4) of this definition:

 (i) The products or source of the products containing < 50 ppm concentration PCBs were legally manufactured, processed, distributed in commerce, or used before October 1, 1984.

 (ii) The products or source of the products containing < 50 ppm concentrations PCBs were legally manufactured, processed, distributed in commerce, or used, i.e., pursuant to authority granted by EPA regulation, by exemption petition, by settlement agreement, or pursuant to other Agency-approved programs;

 (iii) The resulting PCB concentration (i.e. below 50 ppm) is not a result of dilution, or leaks and spills of PCBs in concentrations over 50 ppm.

exclusion (EPA-94/04): In the asbestos program, one of several situations that permit a Local Education Agency (LEA) to delete one or more of the items required by the Asbestos Hazard Emergency Response Act (AHERA), e.g., records of previous asbestos sample collection and analysis may be used by the accredited inspector in lieu of AHERA bulk sampling.

exclusionary ordinance (EPA-94/04): Zoning that excludes classes of persons or businesses from a particular neighborhood or area.

exclusive bus lane (40CFR52.2493-5) means a lane on a street or highway for the exclusive use of buses, whether constructed especially for that purpose or converted from an existing lane.

exclusive economic zone (OPA1001) means the zone established by Presidential Proclamation Numbered 5030, dated March 10, 1983, including the ocean waters of the areas referred to as "eastern special areas" in Article 3(1) of the Agreement between the United States of America and the Union of Soviet Socialist Republics on the Maritime Boundary, signed June 1, 1990.

exclusive use study (40CFR152.83-c) means a study that meets each of the following requirements:
(1) The study pertains to a new active ingredient (new chemical) or new combination of active ingredients (new combination) first registered after September 30, 1978;
(2) The study was submitted in support of, or as a condition of approval of, the application resulting in the first registration of product containing such new chemical or new combination (first registration), or an application to amend such registration to add a new use; and
(3) The study was not submitted to satisfy a data requirement imposed under FIFRA section 3(c)(2)(B); provided that, a study is an exclusive use study only during the 10-year period following the date of the first registration.

executive order (40CFR8.33-d) means Executive Order 11246, 30 FR 12319, as amended.

exempt solvent (EPA-94/04): Specific organic compounds not subject to requirements of regulation because are deemed by EPA to be of negligible photochemical reactivity.

exempted aquifer (40CFR144.3) means an "aquifer" or its portion that meets the criteria in the definition of "underground source of drinking water" but which has been exempted according to the procedures in 40CFR144.7.

exempted aquifer (40CFR146.3) means an aquifer or its portion that meets the criteria in the definition of "underground source of drinking water" but which has been exempted according to the procedures in 40CFR144.8(b).

exempted aquifer (EPA-94/04): Underground bodies of water defined in the Underground Injection Control program as aquifers that are potential sources of drinking water though not being used as such, and thus exempted from regulations barring underground injection activities.

exemption (40CFR790.3) means an exemption from a testing requirement of a test rule promulgated under section 4 of the Act and part of this chapter (under TSCA).

exemption (40CFR89.902.96) means exemption from the prohibitions of 40CFR89.1006.

exemption (EPA-94/04): A state with primacy may relieve a public water system from a requirement respecting an MCL, treatment technique, or both by granting an exemption if the system cannot comply due to compelling economic or other factors; the system was in operation on the effective date of the requirement or MCL; and the exemption will not create an unreasonable public health risk (cf. variance).

exemption application (40CFR700.43) means any application submitted to EPA under section 5(h)(2) of the Act.

exemption category (40CFR723.175-7) means a category of chemical substances for which a person(s) has applied for or been granted an exemption under section 5(h)(4) of the Act (15USC2604).

exemption holder (40CFR791.3-e) refers to a manufacturer or processor, subject to a test rule, that has received an exemption under sections 4(c)(1) or 4(c)(2) of TSCA from the requirement to conduct a test and submit data.

exemption notice (40CFR700.43) means any notice submitted to EPA under section 40CFR723.15 of this chapter.

exhaust damper (40CFR63.321) means a flow control device that prevents the air-perchloroethylene gas-vapor stream from exiting the dry cleaning machine into a carbon adsorber before room air is drawn into the dry cleaning machine.

exhaust emissions (40CFR86.082.2) means substances emitted to the atmosphere from any opening downstream from the exhaust port of a motor vehicle (other identical or similar definitions are provided in 40CFR86.402.78).

exhaust emissions (40CFR87.1) means substances emitted to the atmosphere from the exhaust discharge nozzle of an aircraft or aircraft engine.

exhaust gas (40CFR61.61-x) means any offgas (the constituents of which may consist of any fluids, either as a liquid and/or gas) discharged directly or ultimately to the atmosphere that was initially contained in or was in direct contact with the equipment for which exhaust gas limits are prescribed in 40CFR61.62(a) and (b); 40CFR61.63(a); 40CFR61.64(a)(1), (b), (c), and (d); 40CFR61.65(b)(1)(ii), (b)(2), (b)(3), (b)(5), (b)(6)(ii), (b)(7), and (b)(9)(ii); and 61.65(d). A leak as defined in paragraph (w) of this section is not an exhaust gas. Equipment which contains exhaust gas is subject to 40CFR61.65(b)(8), whether or not that equipment contains 10 percent by volume vinyl chloride.

exhaust gas recirculation (EGR)-air bleed (40CFR52.2491-1) means a system or device (such as modification of the engine's carburetor or positive crankcase ventilation system) that results in engine operation at an increased airfuel ratio so as to achieve reductions in exhaust emissions of hydrocarbons and carbon monoxide of 25 percent and 40 percent, respectively, from light-duty vehicles of model years 1968-1970.

exhaust header pipe (40CFR205.165-2) means any tube of constant diameter which conducts exhaust gas from an engine exhaust port to other exhaust system components which provide noise attenuation. Tubes with cross connections or internal baffling are not considered to be exhaust header pipes.

exhaust system (29CFR1910.94b) means a system consisting of branch pipes connected to hoods or enclosures, one or more header pipes, an exhaust fan, means for separating solid contaminants from the air flowing in the system, and a discharge stack to outside.

exhaust system (40CFR202.10-f) means the system comprised of a combination of components which provides for enclosed flow of exhaust gas from engine parts to the atmosphere.

exhaust system (40CFR205.151-15) means the combination of components which provides for the enclosed flow of exhaust gas from the engine exhaust port to the atmosphere. Exhaust system further means any constituent components of the combination which conduct exhaust gases and which are sold as separate products. Exhaust system does not mean any of the constituent components of the combination, alone, which do not conduct exhaust gases, such as brackets and other mounting hardware.

exhaust system (40CFR205.51-13) means the system comprised of a combination of components which provides for enclosed flow of exhaust gas from engine exhaust port to the atmosphere.

exhaust ventilation system (29CFR1910.94a) means a system for removing contaminated air from a space, comprising two or more of the following elements:
(a) enclosure or hood,
(b) duct work,
(c) dust collecting equipment,
(d) exhauster, and
(e) discharge stack.

exhauster (40CFR61.131) means a fan located between the inlet gas flange and outlet gas flange of the coke oven gas line that provides motive power for coke oven gases.

existing (40CFR63.321) means commenced construction or reconstruction before December 9, 1991.

existing class II wells (40CFR147.2902) means wells that were authorized by BIA and constructed and completed before the effective date of this program.

existing component (40CFR260.10): See existing tank system.

existing control device (40CFR60.561) means, for the purposes of these standards, an air pollution control device that has been in operation on or before September 30, 1987, or that has been in operation between September 30, 1987, and January 10, 1989, on those continuous or intermittent emissions from a process section that is marked by an "--" in Table 1 of this subpart.

existing control device is reconstructed (40CFR60.561) means, for the purposes of these standards, the capital expenditure of at least 50 percent of the replacement cost of the existing control device.

existing control device is replaced (40CFR60.561) means, for the purposes of these standards, the replacement of an existing control device with another control device.

existing facility (40CFR260.10): See existing hazardous waste management (HWM) facility (under RCRA).

existing facility (40CFR60.2) means, with reference to a stationary source, any apparatus of the type for which a standard is promulgated in this part, and the construction or modification of which was commenced before the date of proposal of that standard; or any apparatus which could be altered in such a way as to be of that type.

existing hazardous waste management (HWM) facility or existing facility (40CFR260.10) means a facility which was in operation or for which construction commenced on or before November 19, 1980. A facility has commenced construction if:
(1) The owner or operator has obtained the Federal, State and local approvals or permits necessary to begin physical construction; and either (2)(i) A continuous on-site, physical construction program has begun; or
(ii) The owner or operator has entered into contractual obligations--which cannot be cancelled or modified without substantial loss-- for physical construction of the facility to be completed within a reasonable time.

existing hazardous waste management (HWM) facility or existing facility (40CFR270.2) means a facility which was in operation or for which construction commenced on or before November 19, 1980. A facility has commenced construction if:
(a) The owner or operator has obtained the Federal, State and local approvals or permits necessary to begin physical construction; and either
(b) (1) A continuous on-site, physical construction program has begun; or
(2) The owner or operator has entered into contractual obligations which cannot be cancelled or modified without substantial loss for physical construction of the facility to be completed within a reasonable time.

existing HWM facility (40FR270.2): See existing hazardous waste management facility.

existing impoundment (40CFR61.251-d) means any uranium mill tailings impoundment which is licensed to accept additional tailings and is in existence as of December 15, 1989.

existing indirect dischargers (40CFR420.31-c) means only those two iron blast furnace operations with discharges to publicly owned treatment works prior to May 27, 1982.

existing injection well (40CFR144.3) means an "injection well" other than a "new injection well" (other identical or similar definitions are provided in 40CFR146.3).

existing major source (40CFR63.51) means a major source, construction or reconstruction of which is commenced before EPA proposed a standard, applicable to the major source, under section 112 (d) or (h), or if no proposal was published, then on or before the section 112(j) deadline.

existing MSWLF unit (40CFR258.2) means any municipal solid waste landfill unit that is receiving solid waste as of the appropriate dates specified in 40CFR258.1(e). Waste placement in existing units must be consistent with past operating practices or modified practices to ensure good management.

existing OCS source (CAA328-42USC7627) means

any OCS (outer continental shelf) source other than a new OCS source.

existing portion (40CFR192.31-j) means that land surface area of an existing surface impoundment on which significant quantities of uranium byproduct materials have been placed prior to promulgation of this standard.

existing portion (40CFR260.10) means that land surface area of an existing waste management unit, included in the original part A permit application, on which wastes have been placed prior to the issuance of a permit.

existing solid waste incineration unit (CAA129.g-42USC7429) means a solid waste unit which is not a new or modified solid waste incineration unit.

existing source (40CFR122.29-3) means any source which is not a new source or a new discharger (other identical or similar definitions are provided in 40CFR129.2-i).

existing source (40CFR61.02) means any stationary source which is not a new source.

existing source (40CFR63.2) means any affected source that is not a new source (other identical or similar definitions are provided in 40CFR63.101; 63.191).

existing source (40CFR63.51) means any source as defined in 40CFR63.72, the construction or reconstruction of which commenced prior to proposal of an applicable section 112(d) standard.

existing source (CAA111-42USC7411) means any stationary source other than a new source (other identical or similar definitions are provided in CAA112-42USC7412).

existing source or existing OCS source (40CFR55.2) shall have the meaning given in the applicable requirements incorporated into 40CFR55.13 and 55.14 of this part, except that for two years following the date of promulgation of this part the definition given in 40CFR55.3 of this part shall apply for the purpose of determining the required date of compliance with this part.

existing stationary facility (40CFR51.301-e) means any of the following stationary sources of air pollutants, including any reconstructed source, which was not in operation prior to August 7, 1962, and was in existence on August 7, 1977, and has the potential to emit 250 tons per year or more of any air pollutant. In determining potential to emit, fugitive emissions, to the extent quantifiable, must be counted:
1. Fossil-fuel fired steam electric plants of more than 250 million British thermal units per hour heat input,
2. Coal cleaning plants (thermal dryers),
3. Kraft pulp mills,
4. Portland cement plants,
5. Primary zinc smelters,
6. Iron and steel mill plants,
7. Primary aluminum ore reduction plants,
8. Primary copper smelters,
9. Municipal incinerators capable of charging more than 250 tons of refuse per day,
10. Hydrofluoric, sulfuric, and nitric acid plants,
11. Petroleum refineries,
12. Lime plants,
13. Phosphate rock processing plants,
14. Coke oven batteries,
15. Sulfur recovery plants,
16. Carbon black plants (furnace process.,
17. Primary lead smelters,
18. Fuel conversion plants,
19. Sintering plants,
20. Secondary metal production facilities,
21. Chemical process plants,
22. Fossil-fuel boilers of more than 250 million British thermal units per hour heat input,
23. Petroleum storage and transfer facilities with a capacity exceeding 300,000 barrels,
24. Taconite ore processing facilities,
25. Glass fiber processing plants, and
26. Charcoal production facilities.

existing tank system (40CFR280.12) means a tank system used to contain an accumulation of regulated substances or for which installation has commenced on or before December 22, 1988. Installation is considered to have commenced if:
(a) The owner or operator has obtained all federal, state, and local approvals or permits necessary to begin physical construction of the site or installation of the tank system; and if,

(b) (1) Either a continuous on-site physical construction or installation program has begun; or

(2) The owner or operator has entered into contractual obligations--which cannot be cancelled or modified without substantial loss--for physical construction at the site or installation of the tank system to be completed within a reasonable time.

existing tank system or existing component (40CFR260.10) means a tank system or component that is used for the storage or treatment of hazardous waste and that is in operation or for which installation has commenced on or prior to July 14, 1986. Installation will be considered to have commenced if the owner or operator has obtained all Federal, State, and local approvals or permits necessary to begin physical construction of the site or installation of the tank system and if either

(1) a continuous on-site physical construction or installation program has begun, or

(2) the owner or operator has entered into contractual obligations--which cannot be canceled or modified without substantial loss-- for physical construction of the site or installation of the tank system to be completed within a reasonable time.

existing unit (40CFR72.2) means a unit (including a unit subject to section 111 of the Act) that commenced commercial operation before November 15, 1990 and that on or after November 15, 1990 served a generator with nameplate capacity of greater than 25 MWe. "Existing unit" does not include simple combustion turbines or any unit that on or after November 15, 1990 served only generators with a nameplate capacity of 25 MWe or less. Any "existing unit" that is modified, reconstructed, or repowered after November 15, 1990 shall continue to be an "existing unit."

existing unit (CAA402-42USC7651a) means a unit (including units subject to section 111) that commenced commercial operation before the date of enactment of the Clean Air Act Amendments of 1990. Any unit that commenced commercial operation before the date of enactment of the Clean Air Act Amendments of 1990 which is modified, reconstructed, or repowered after the date of

enactment of the Clean Air Act Amendments of 1990 shall continue to be an existing unit for the purposes of this title. For the purposes of this title, existing units shall not include simple combustion turbines, or units which serve a generator with a nameplate capacity of 25 MWe or less.

existing uses (40CFR131.3-e) are those uses actually attained in the water body on or after November 28, 1975, whether or not they are included in the water quality standards (cf. designated uses).

existing vapor processing system (40CFR60.501) means a vapor processing system [capable of achieving emissions to the atmosphere no greater than 80 milligrams of total organic compounds per liter of gasoline loaded], the construction or refurbishment of which was commenced before December 7, 1980, and which was not constructed or refurbished after that date.

existing vessel (40CFR140.1-f) refers to any vessel on which construction was initiated before January 30, 1975.

existing vessel (CWA312-33USC1322) includes every description of watercraft or other artificial contrivance used, or capable of being used, as a means of transportation on the navigable waters, the construction of which is initiated before promulgation of standards and regulations under this section.

existing well (40CFR146.61) means a Class I well which was authorized prior to August 25, 1988 by an approved State program, or an EPA-administered program or a well which has become a Class I well as a result of a change in the definition of the injected waste which would render the waste hazardous under 40CFR261.3 of this part.

exotic species (EPA-94/04): A species that is not indigenous to a region.

expandable polystyrene (40CFR60.561) means a polystyrene bead to which a blowing agent has been added using either an in-situ suspension process or a post-impregnation suspension process.

expedited hearing (40CFR164.2-h) means a hearing commenced as the result of the issuance of a notice of intention to suspend or the suspension of a registration of a pesticide by an emergency order, and is limited to a consideration as to whether a pesticide presents an imminent hazard which justifies such suspension.

expendable personal property (40CFR30.200) means all tangible personal property other than nonexpendable personal property.

expenditure report (40CFR31.3) means:
(1) For nonconstruction grants, the SF-269 "Financial Status Report" (or other equivalent report);
(2) for construction grants, the SF-271 "Outlay Report and Request for Reimbursement" (or other equivalent report).

experimental animals (40CFR172.1-d) means individual animals or groups of animals, regardless of species, intended for use and used solely for research purposes and does not include animals intended to be used for any food purposes.

experimental furnace (40CFR60.291) means a glass melting furnace with the sole purpose of operating to evaluate glass melting processes, technologies, or glass products. An experimental furnace does not produce glass that is sold (except for further research and development purposes) or that is used as a raw material for nonexperimental furnaces.

experimental process line (40CFR60.561) means a polymer or copolymer manufacturing process line with the sole purpose of operating to evaluate polymer manufacturing processes, technologies, or products. An experimental process line does not produce a polymer or resin that is sold or that is used as a raw material for nonexperimental process lines.

experimental start date (40CFR160.3) means the first date the test substance is applied to the test system (other identical or similar definitions are provided in 40CFR792.3).

experimental technology (40CFR146.3) means a technology which has not been proven feasible under the conditions in which it is being tested (under SDWA).

experimental termination date (40CFR160.3) means the last date on which data are collected directly from the study (other identical or similar definitions are provided in 40CFR792.3).

experimental use permit (EPA-94/04): Obtained by manufacturers for testing new pesticides or uses of thereof whenever they conduct experimental field studies to support registration on 10 acres or more on land or one acre or more of water.

experimental use permit review (40CFR152.403-f) means review of an application for a permit pursuant to section 5 of FIFRA to apply a limited quantity of a pesticide in order to accumulate information necessary to register the pesticide. The application may be for a new chemical or for a new use of an old chemical. The fee applies to such experimental uses of a single unregistered active ingredient (no limit on the number of other active ingredients, in a tank mix, already registered for the crops involved) and no more than three crops. This fee does not apply to experimental use permits required for small-scale field testing of microbial pest control agents (40CFR172.3).

exploratory facility (40CFR435.11-j) shall mean any fixed or mobile structure subject to this subpart that is engaged in the drilling of wells to determine the nature of potential hydrocarbon reservoirs.

exploratory source or exploratory OCS source (40CFR55.2) means any OCS (outer continental shelf) source that is a temporary operation conducted for the sole purpose of gathering information. This includes an operation conducted during the exploratory phase to determine the characteristics of the reservoir and formation and may involve the extraction of oil and gas.

explosive gas (40CFR257.3.8-3) means methane (CH_4).

explosive limits (EPA-94/04): The amounts of vapor in the air that form explosive mixtures; limits are expressed as lower and upper limits and give the range of vapor concentrations in air that will explode if an ignition source is present.

export (40CFR82.104-h) means the transport of

virgin, used, or recycled class I or class II substances or products manufactured or containing class I or class II substances from inside the United States or its territories to persons outside the United States or its territories, excluding United States military bases and ships for on-board use.

export (40CFR82.3) means the transport of virgin, used or recycled controlled substances from inside the United states or its territories to persons outside the United States or its territories, excluding United States military bases and ships for on-board use (under CAA).

export (40CFR82.3-p) means the transport of virgin or used controlled substances from inside the United States or its territories to persons outside the United States or its territories, excluding United States military bases and ships for on-board use.

export exemption (40CFR204.2-4) means an exemption from the prohibitions of section 10(a) (1), (2), (3), and (4) of the Act, granted by statute under section 10(b)(2) of the Act for the purpose of exporting regulated (other identical or similar definitions are provided in 40CFR205.2-4).

export exemption (40CFR211.102-f) means an exemption from the prohibitions of section 10(a)(3) and (4) of the Act; this type of exemption is granted by statute under section 10(b)(2) of the Act for the purpose of exporting regulated products.

export exemption (40CFR85.1702-1) means an exemption granted by statute under section 203(b)(3) of the Act for the purpose of exporting new motor vehicles or new motor vehicle engines.

export exemption (40CFR89.902.96) means an exemption granted under 40CFR89.1004(b) for the purpose of exporting new nonroad engines.

exporter (40CFR82.3-q) means the person who contracts to sell controlled substances for export or transfers controlled substances to his affiliate in another country (other identical or similar definitions are provided in 40CFR82.104-i; 707.63-b).

exposure (EPA-94/04): The amount of radiation or pollutant present in a given environment that represents a potential health threat to living organisms.

exposure assessment (EPA-94/04): Identifying the pathways by which toxicants may reach individuals, estimating how much of a chemical an individual is likely to be exposed to, and estimating the number likely to be exposed.

exposure assessment (RCRA9003): An assessment to determine the extent of exposure of, or potential for exposure of, individuals to chemical substances based on such factors as the nature and extent of contamination and the existence of or potential for pathways of human exposure (including ground or surface water contamination, air emissions, and food chain contamination), the size of community within the likely pathways of exposure, and the comparison of expected human exposure levels to the short-term and long-term health effects associated with identified contaminants and any available recommended exposure or tolerance limits for such contaminants. Such assessment shall not delay corrective action to abate immediate hazards or reduce exposure (cf. risk assessment).

exposure indicator (EPA-94/04): A characteristic of the environment measured to provide evidence of the occurrence or magnitude of a response indicator's exposure to a chemical or biological stress.

exposure level (EPA-94/04): The amount (concentration) of a chemical at the absorptive surfaces of an organism.

exposure or exposed (29CFR1910.20) means that an employee is subjected to a toxic substance or harmful physical agent in the course of employment through any route of entry (inhalation, ingestion, skin contact or absorption, etc.), and includes past exposure and potential (e.g., accidental or possible) exposure, but does not include situations where the employer can demonstrate that the toxic substance or harmful physical agent is not used, handled, stored, generated, or present in the workplace in any manner different from typical non-occupational situations.

exposure period (40CFR797.2050-4) is the 5-day period during which test birds are offered a diet containing the test substance.

extent of chlorination (40CFR704.45-2) means the percent by weight of chlorine for each isomer (ortho, meta, and para).

exterior base coat (40CFR52.741) means a coating applied to the exterior of a can body, or flat sheet to provide protection to the metal or to provide background for any lithographic or printing operation.

exterior base coating operation (40CFR60.491-2) means the system on each beverage can surface coating line used to apply a coating to the exterior of a two-piece beverage can body. The exterior base coat provides corrosion resistance and a background for lithography or printing operations. The exterior base coat operation consists of the coating application station, flashoff area, and curing oven. The exterior base coat may be pigmented or clear (unpigmented).

exterior end coat (40CFR52.741) means a coating applied to the exterior end of a can to provide protection to the metal.

external floating roof (40CFR61.341) means a pontoon-type or double-deck type cover with certain rim sealing mechanisms that rests on the liquid surface in a waste management unit with no fixed roof.

external floating roof (40CFR63.111) means a pontoon-type or double-deck-type cover that rests on the liquid surface in a storage vessel or waste management unit with no fixed roof.

external floating roof (40CFR52.741) means a cover over an open top storage tank consisting of a double deck or pontoon single deck which rests upon and is supported by the volatile organic liquid being contained and is equipped with a closure seal or seals to close the space between the roof edge and tank shell.

extraction plant (40CFR61.31-b) means a facility chemically processing beryllium ore to beryllium metal, alloy, or oxide, or performing any of the intermediate steps in these processes.

extraction procedure (EP toxic) (EPA-94/04): Determining toxicity by a procedure which simulates leaching; if a certain concentration of a toxic substance can be leached from a waste, that waste is considered hazardous, i.e., "EP Toxic."

extraction site (40CFR230.3-k) means the place from which the dredged or fill material proposed for discharge is to be removed.

extractor column (40CFR796.1720-iii) is used to extract the solute from the aqueous solution produced by the generator column.

extractor column (40CFR796.1860-vi) is used to extract the solute from the saturated solutions produced by the generator column. After extraction onto chromatographic support, the solute is eluted with solvent/water mixture and subsequently analyzed by high pressure liquid chromatography (HPLC). A detailed description of the preparation of the extractor column is given in paragraph (b)(1)(i) of this section.

extreme environmental conditions (40CFR52.741) means exposure to any or all of the following: ambient weather conditions; temperatures consistently above 95°C (203°F); detergents; abrasive and scouring agents; solvents; or corrosive atmospheres.

extreme performance coating (40CFR52.741) means any coating which during intended use is exposed to extreme environmental conditions (under the Clean Air Act).

extremely hazardous substance (40CFR355.20) means a substance listed in the Appendices A and B of this part (see below for the listing of extremely hazardous substances regulated under Title III-- Emergency Planning and Community Right-To-Know of the Superfund Amendments and Reauthorization Act of 1986. The substances provided herewith in both alphabetical and CAS order were excerpted from Appendix A to 40CFR355.20, 1995 edition).

CAS #	chemical name (alphabetical order)
75-865	Acetone Cyanohydrin
1752-303	Acetone Thiosemicarbazide
107-028	Acrolein
79-061	Acrylamide
107-131	Acrylonitrile

814-686	Acrylyl Chloride	26419-738	Carbamic Acid, Methyl-, 0-(((2,4-Dimethyl-1, 3-Dithiolan-2-yl) Methylene)Amino)-
111-693	Adiponitrile		
116-063	Aldicarb		
309-002	Aldrin	1563-662	Carbofuran
107-186	Allyl Alcohol	75-150	Carbon Disulfide
107-119	Allylamine	786-196	Carbophenothion
20859-738	Aluminum Phosphide	57-749	Chlordane
54-626	Aminopterin	470-906	Chlorfenvinfos
78-535	Amiton	7782-505	Chlorine
3734-972	Amiton Oxalate	24934-916	Chlormephos
7664-417	Ammonia	999-815	Chlormequat Chloride
300-629	Amphetamine	79-118	Chloroacetic Acid
62-533	Aniline	107-073	Chloroethanol
88-051	Aniline, 2,4,6-Trimethyl-	627-112	Chloroethyl Chloroformate
7783-702	Antimony Pentafluoride	67-663	Chloroform
1397-940	Antimycin A	542-881	Chloromethyl ether
86-884	ANTU	107-302	Chloromethyl methyl ether
1303-282	Arsenic pentoxide	3691-358	Chlorophacinone
1327-533	Arsenous oxide	1982-474	Chloroxuron
7784-341	Arsenous trichloride	21923-239	Chlorthiophos
7784-421	Arsine	10025-737	Chromic Chloride
2642-719	Azinphos-Ethyl	62207-765	Cobalt, ((2,2'-(1,2-Ethanediylbis (Nitrilomethylidyn))Bis(6-Fluorophenolato)) (2-)-N,N',O,O')-,
86-500	Azinphos-Methyl		
98-873	Benzal Chloride		
98-168	Benzenamine, 3-(Trifluoromethyl)-	10210-681	Cobalt Carbonyl
100-141	Benzene, 1-(Chloromethyl)-4-Nitro-	64-868	Colchicine
98-055	Benzenearsonic Acid	56-724	Coumaphos
3615-212	Benzimidazole, 4,5-Dichloro-2-(Trifluoromethyl)-	5836-293	Coumatetralyl
		95-487	Cresol, o-
98-077	Benzotrichloride	535-897	Crimidine
100-447	Benzyl Chloride	4170-303	Crotonaldehyde
140-294	Benzyl Cyanide	123-739	Crotonaldehyde, (E)-
15271-417	Bicyclo[221] Heptane-2-Carbonitrile, 5-Chloro-6-(((((Methylamino) Carbonyl)Oxy) Imino)-, (1s-(1-alpha, 2-beta, 4-alpha, 5-alpha, 6E))-	506-683	Cyanogen Bromide
		506-785	Cyanogen Iodide
		2636-262	Cyanophos
		675-149	Cyanuric Fluoride
534-076	Bis(Chloromethyl) Ketone	66-819	Cycloheximide
4044-659	Bitoscanate	108-918	Cyclohexylamine
10294-345	Boron Trichloride	17702-419	Decaborane(14)
7637-072	Boron Trifluoride	8065-483	Demeton
353-424	Boron Trifluoride Compound With Methyl Ether (1:1)	919-868	Demeton-S-Methyl
		10311-849	Dialifor
28772-567	Bromadiolone	19287-457	Diborane
7726-956	Bromine	111-444	Dichloroethyl ether
1306-190	Cadmium Oxide	149-746	Dichloromethylphe-nylsilane
2223-930	Cadmium Stearate	62-737	Dichlorvos
7778-441	Calcium arsenate	141-662	Dicrotophos
8001-352	Camphechlor	1464-535	Diepoxybutane
56-257	Cantharidin	814-493	Diethyl Chlorophospate
51-832	Carbachol Chloride	1642-542	Diethylcarbamazine Citrate

71-636	Digitoxin	23422-539	Formetanate Hydrochloride
2238-075	Diglycidyl Ether	2540-821	Formothion
20830-755	Digoxin	17702-577	Formparanate
115-264	Dimefox	21548-323	Fosthietan
60-515	Dimethoate	3878-191	Fuberidazole
2524-030	Dimethyl Phosphor-ochloridothioate	110-009	Furan
77-781	Dimethyl sulfate	13450-903	Gallium Trichloride
75-785	Dimethyldichlorosilane	77-474	Hexachlorocyclope-ntadiene
57-147	Dimethylhydrazine	4835-114	Hexamethylenediamine, N,N'-
99-989	Dimethyl-p-Phenylenediamine		Dibutyl-
644-644	Dimetilan	302-012	Hydrazine
534-521	Dinitrocresol	74-908	Hydrocyanic Acid
88-857	Dinoseb	7647-010	Hydrogen chloride (gas only)
1420-071	Dinoterb	7664-393	Hydrogen Fluoride
78-342	Dioxathion	7722-841	Hydrogen Peroxide (Conc >52%)
82-666	Diphacinone	7783-075	Hydrogen Selenide
152-169	Diphosphoramide, Octamethyl-	7783-064	Hydrogen Sulfide
298-044	Disulfoton	123-319	Hydroquinone
514-738	Dithiazanine Iodide	13463-406	Iron, Pentacarbonyl-
541-537	Dithiobiuret	297-789	Isobenzan
316-427	Emetine, Dihydrochloride	78-820	Isobutyronitrile
115-297	Endosulfan	102-363	Isocyanic Acid, 3, 4-Dichlorophenyl
2778-043	Endothion		Ester
72-208	Endrin	465-736	Isodrin
106-898	Epichlorohydrin	55-914	Isofluorphate
2104-645	EPN	4098-719	Isophorone Diisocyanate
50-146	Ergocalciferol	108-236	Isopropyl Chloroformate
379-793	Ergotamine Tartrate	119-380	Isoproplymethylpy-razolyl
1622-328	Ethanesulfonyl Chloride, 2-Chloro-		Dimethylcarbamate
10140-871	Ethanol, 12-Dichloro-, Acetate	78-977	Lactonitrile
563-122	Ethion	21609-905	Leptophos
13194-484	Ethoprophos	541-253	Lewisite
538-078	Ethylbis(2-Chloroethyl)Amine	58-899	Lindane
371-620	Ethylene Fluorohydrin	7580-678	Lithium Hydride
75-218	Ethylene oxide	109-773	Malononitrile
107-153	Ethylenediamine	12108-133	Manganese, Tricarbonyl
151-564	Ethyleneimine		Methylcyclopentadieny
542-905	Ethylthiocyanate	51-752	Mechlorethamine
22224-926	Fenamiphos	950-107	Mephosfolan
122-145	Fenitrothion	1600-277	Mercuric Acetate
115-902	Fensulfothion	7487-947	Mercuric Chloride
4301-502	Fluenetil	21908-532	Mercuric Oxide
7782-414	Fluorine	10476-956	Methacrolein Diacetate
640-197	Fluoroacetamide	760-930	Methacrylic Anhydride
144-490	Fluoroacetic Acid	126-987	Methacrylonitrile
359-068	Fluoroacetyl Chloride	920-467	Methacryloyl Chloride
51-218	Fluorouracil	30674-807	Methacryloyloxyethyl Isocyanate
944-229	Fonofos	10265-926	Methamidophos
50-000	Formaldehyde	558-258	Methanesulfonyl Fluoride
107-164	Formaldehyde Cyanohydrin	950-378	Methidathion

2032-657	Methiocarb	4418-660	Phenol, 2,2'-Thiobis(4-Chloro-6-
16752-775	Methomyl		Methyl)-
151-382	Methoxyethylmercuric Acetate	64-006	Phenol, 3-(1-Methylethyl)-,
80-637	Methyl 2-Chloroacrylate		Methylcarbamate
74-839	Methyl Bromide	58-366	Phenoxarsine, 10,10'-Oxydi-
79-221	Methyl Chloroformate	696-286	Phenyl Dichloroarsine
60-344	Methyl Hydrazine	59-881	Phenylhydrazine Hydrochloride
624-839	Methyl Isocyanate	62-384	Phenylmercury Acetate
556-616	Methyl Isothiocyanate	2097-190	Phenylsilatrane
74-931	Methyl Mercaptan	103-855	Phenylthiourea
3735-237	Methyl Phenkapton	298-022	Phorate
676-971	Methyl Phosphonic Dichloride	4104-147	Phosacetim
556-649	Methyl Thiocyanate	947-024	Phosfolan
78-944	Methyl Vinyl Ketone	75-445	Phosgene
502-396	Methylmercuric Dicyanamide	732-116	Phosmet
75-796	Methyltrichlorosilane	13171-216	Phosphamidon
1129-415	Metolcarb	7803-512	Phosphine
7786-347	Mevinphos	2703-131	Phosphonothioic Acid, Methyl-, O-
315-184	Mexacarbate		Ethyl O-(4-(Methylthio) Phenyl)
50-077	Mitomycin C		Ester
6923-224	Monocrotophos	50782-699	Phosphonothioic Acid, Methyl-, S-(2-
2763-944	Muscimol		(Bis(1-Methylethyl)Amino) Ethyl O-
505-602	Mustard Gas		Ethyl Ester
13463-393	Nickel carbonyl	2665-307	Phosphonothioic Acid, Methyl-, O-
54-115	Nicotine		(4-Nitrophenyl) O-Phenyl Ester
65-305	Nicotine sulfate	3254-635	Phosphoric Acid, Dimethyl 4-
7697-372	Nitric Acid		(Methylthio) Phenyl Ester
10102-439	Nitric Oxide	2587-908	Phosphorothioic Acid, O,O-
98-953	Nitrobenzene		Dimethyl-S-(2-Methylthio) Ethyl
1122-607	Nitrocyclohexane		Ester
10102-440	Nitrogen Dioxide	7723-140	Phosphorus
62-759	Nitrosodimethylamine	10025-873	Phosphorus Oxychloride
991-424	Norbormide	10026-138	Phosphorus Pentachloride
0	Organorhodium Complex (PMN-82-	1314-563	Phosphorus Pentoxide
	147)	7719-122	Phosphorus Trichloride
630-604	Ouabain	57-476	Physostigmine
23135-220	Oxamyl	57-647	Physostigmine, Salicylate (1:1)
78-717	Oxetane, 3,3-Bis (Chloromethyl)-	124-878	Picrotoxin
2497-076	Oxydisulfoton	110-894	Piperidine
10028-156	Ozone	23505-411	Pirimifos-Ethyl
1910-425	Paraquat dichloride	10124-502	Potassium arsenite
2074-502	Paraquat Methosulfate	151-508	Potassium Cyanide
56-382	Parathion	506-616	Potassium Silver Cyanide
298-000	Parathion-Methyl	2631-370	Promecarb
12002-038	Paris Green	106-967	Propargyl Bromide
19624-227	Pentaborane	57-578	Propiolactone, Beta-
2570-265	Pentadecylamine	107-120	Propionitrile
79-210	Peracetic Acid	542-767	Propionitrile, 3-Chloro-
594-423	Perchloromethylmercaptan	70-699	Propiophenone, 4-Amino-
108-952	Phenol	109-615	Propyl Chloroformate

CAS #	Chemical Name
75-569	Propylene Oxide
75-558	Propyleneimine
2275-185	Prothoate
129-000	Pyrene
140-761	Pyridine, 2-Methyl-5-Vinyl-
504-245	Pyridine, 4-Amino-
1124-330	Pyridine, 4-Nitro-, 1-Oxide
53558-251	Pyriminil
14167-181	Salcomine
107-448	Sarin
7783-008	Selenious Acid
7791-233	Selenium Oxychloride
563-417	Semicarbazide Hydrochloride
3037-727	Silane, (4-Aminobutyl) Diethoxymethyl-
7631-892	Sodium Arsenate
7784-465	Sodium arsenite
26628-228	Sodium Azide (Na(N3))
124-652	Sodium Cacodylate
143-339	Sodium Cyanide (Na(CN))
62-748	Sodium Fluoroacetate
13410-010	Sodium Selenate
10102-188	Sodium Selenite
10102-202	Sodium Tellurite
900-958	Stannane, Acetoxytriphenyl-
57-249	Strychnine
60-413	Strychnine sulfate
3689-245	Sulfotep
3569-571	Sulfoxide, 3-Chloropropyl Octyl
7446-095	Sulfur Dioxide
7783-600	Sulfur Tetrafluoride
7446-119	Sulfur Trioxide
7664-939	Sulfuric Acid
77-816	Tabun
13494-809	Tellurium
7783-804	Tellurium Hexafluoride
107-493	TEPP
13071-799	Terbufos
78-002	Tetraethyllead
597-648	Tetraethyltin
75-741	Tetramethyllead
509-148	Tetranitromethane
10031-591	Thallium Sulfate
6533-739	Thallous Carbonate
7791-120	Thallous Chloride
2757-188	Thallous Malonate
7446-186	Thallous Sulfate
2231-574	Thiocarbazide
39196-184	Thiofanox
297-972	Thionazin
108-985	Thiophenol
79-196	Thiosemicarbazide
5344-821	Thiourea, (2-Chlorophenyl)-
614-788	Thiourea, (2-Methylphenyl)-
7550-450	Titanium Tetrachloride
584-849	Toluene 2,4-Diisocyanate
91-087	Toluene 2,6-Diisocyanate
110-576	Trans-1,4-Dichlorobutene
1031-476	Triamiphos
24017-478	Triazofos
76-028	Trichloroacety Chloride
115-219	Trichloroethylsilane
327-980	Trichloronate
98-135	Trichlorophenylsilane
1558-254	Trichloro (Chloromethyl) Silane
27137-855	Trichloro (Dichlorophenyl) Silane
998-301	Triethoxysilane
75-774	Trimethylchlorosilane
824-113	Trimethylolpropane Phosphite
1066-451	Trimethyltin Chloride
639-587	Triphenyltin Chloride
555-771	Tris(2-Chloroethyl) Amine
2001-958	Valinomycin
1314-621	Vanadium Pentoxide
108-054	Vinyl Acetate Monomer
81-812	Warfarin
129-066	Warfarin sodium
28347-139	Xylylene Dichloride
58270-089	Zinc, Dichloro(4,4-Dimethyl-5((((Methylamino) Carbonyl)Oxy) Imino) Pentanenitrile)-, (T-4)-
1314-847	Zinc Phosphide

CAS # (order)	chemical name
0	Organorhodium Complex (PMN-82-147)
50-000	Formaldehyde
50-077	Mitomycin C
50-146	Ergocalciferol
51-218	Fluorouracil
51-752	Mechlorethamine
51-832	Carbachol Chloride
54-115	Nicotine
54-626	Aminopterin
55-914	Isofluorphate
56-257	Cantharidin
56-382	Parathion
56-724	Coumaphos
57-147	Dimethylhydrazine
57-249	Strychnine

57-476	Physostigmine	79-196	Thiosemicarbazide
57-578	Propiolactone, Beta-	79-210	Peracetic Acid
57-647	Physostigmine, Salicylate (1:1)	79-221	Methyl Chloroformate
57-749	Chlordane	80-637	Methyl 2-Chloroacrylate
58-366	Phenoxarsine, 10,10'-Oxydi-	81-812	Warfarin
58-899	Lindane	82-666	Diphacinone
59-881	Phenylhydrazine Hydrochloride	86-500	Azinphos-Methyl
60-344	Methyl Hydrazine	86-884	ANTU
60-413	Strychnine sulfate	88-051	Aniline, 2,4,6-Trimethyl-
60-515	Dimethoate	88-857	Dinoseb
62-384	Phenylmercury Acetate	91-087	Toluene 2,6-Diisocyanate
62-533	Aniline	95-487	Cresol, o-
62-737	Dichlorvos	98-055	Benzenearsonic Acid
62-748	Sodium Fluoroacetate	98-077	Benzotrichloride
62-759	Nitrosodimethylamine	98-135	Trichlorophenylsilane
64-006	Phenol, 3-(1-Methylethyl)-, Methylcarbamate	98-168	Benzenamine, 3-(Trifluoromethyl)-
64-868	Colchicine	98-873	Benzal Chloride
65-305	Nicotine sulfate	98-953	Nitrobenzene
66-819	Cycloheximide	99-989	Dimethyl-p-Phenylenediamine
67-663	Chloroform	100-141	Benzene, 1-(Chloromethyl)-4-Nitro-
70-699	Propiophenone, 4-Amino-	100-447	Benzyl Chloride
71-636	Digitoxin	102-363	Isocyanic Acid, 3, 4-Dichlorophenyl Ester
72-208	Endrin	103-855	Phenylthiourea
74-839	Methyl Bromide	106-898	Epichlorohydrin
74-908	Hydrocyanic Acid	106-967	Propargyl Bromide
74-931	Methyl Mercaptan	107-028	Acrolein
75-150	Carbon Disulfide	107-073	Chloroethanol
75-218	Ethylene oxide	107-119	Allylamine
75-445	Phosgene	107-120	Propionitrile
75-558	Propyleneimine	107-131	Acrylonitrile
75-569	Propylene Oxide	107-153	Ethylenediamine
75-741	Tetramethyllead	107-164	Formaldehyde Cyanohydrin
75-774	Trimethylchlorosilane	107-186	Allyl Alcohol
75-785	Dimethyldichlorosilane	107-302	Chloromethyl methyl ether
75-796	Methyltrichlorosilane	107-448	Sarin
75-865	Acetone Cyanohydrin	107-493	TEPP
76-028	Trichloroacety Chloride	108-054	Vinyl Acetate Monomer
77-474	Hexachlorocyclope-ntadiene	108-236	Isopropyl Chloroformate
77-781	Dimethyl sulfate	108-918	Cyclohexylamine
77-816	Tabun	108-952	Phenol
78-002	Tetraethyllead	108-985	Thiophenol
78-342	Dioxathion	109-615	Propyl Chloroformate
78-535	Amiton	109-773	Malononitrile
78-717	Oxetane, 3,3-Bis (Chloromethyl)-	110-009	Furan
78-820	Isobutyronitrile	110-576	Trans-1,4-Dichlorobutene
78-944	Methyl Vinyl Ketone	110-894	Piperidine
78-977	Lactonitrile	111-444	Dichloroethyl ether
79-061	Acrylamide	111-693	Adiponitrile
79-118	Chloroacetic Acid	115-219	Trichloroethylsilane

115-264	Dimefox	534-076	Bis(Chloromethyl) Ketone
115-297	Endosulfan	534-521	Dinitrocresol
115-902	Fensulfothion	535-897	Crimidine
116-063	Aldicarb	538-078	Ethylbis(2-Chloroethyl)Amine
119-380	Isoproplymethylpy-razolyl Dimethylcarbamate	541-253	Lewisite
		541-537	Dithiobiuret
122-145	Fenitrothion	542-767	Propionitrile, 3-Chloro-
123-319	Hydroquinone	542-881	Chloromethyl ether
123-739	Crotonaldehyde, (E)-	542-905	Ethylthiocyanate
124-652	Sodium Cacodylate	555-771	Tris(2-Chloroethyl) Amine
124-878	Picrotoxin	556-616	Methyl Isothiocyanate
126-987	Methacrylonitrile	556-649	Methyl Thiocyanate
129-000	Pyrene	558-258	Methanesulfonyl Fluoride
129-066	Warfarin sodium	563-122	Ethion
140-294	Benzyl Cyanide	563-417	Semicarbazide Hydrochloride
140-761	Pyridine, 2-Methyl-5-Vinyl-	584-849	Toluene 2,4-Diisocyanate
141-662	Dicrotophos	594-423	Perchloromethylmercaptan
143-339	Sodium Cyanide (Na(CN))	597-648	Tetraethyltin
144-490	Fluoroacetic Acid	614-788	Thiourea, (2-Methylphenyl)-
149-746	Dichloromethylphe-nylsilane	624-839	Methyl Isocyanate
151-382	Methoxyethylmercuric Acetate	627-112	Chloroethyl Chloroformate
151-508	Potassium Cyanide	630-604	Ouabain
151-564	Ethyleneimine	639-587	Triphenyltin Chloride
152-169	Diphosphoramide, Octamethyl-	640-197	Fluoroacetamide
297-789	Isobenzan	644-644	Dimetilan
297-972	Thionazin	675-149	Cyanuric Fluoride
298-000	Parathion-Methyl	676-971	Methyl Phosphonic Dichloride
298-022	Phorate	696-286	Phenyl Dichloroarsine
298-044	Disulfoton	732-116	Phosmet
300-629	Amphetamine	760-930	Methacrylic Anhydride
302-012	Hydrazine	786-196	Carbophenothion
309-002	Aldrin	814-493	Diethyl Chlorophospate
315-184	Mexacarbate	814-686	Acrylyl Chloride
316-427	Emetine, Dihydrochloride	824-113	Trimethylolpropane Phosphite
327-980	Trichloronate	900-958	Stannane, Acetoxytriphenyl-
353-424	Boron Trifluoride Compound With Methyl Ether (1:1)	919-868	Demeton-S-Methyl
		920-467	Methacryloyl Chloride
359-068	Fluoroacetyl Chloride	944-229	Fonofos
371-620	Ethylene Fluorohydrin	947-024	Phosfolan
379-793	Ergotamine Tartrate	950-107	Mephosfolan
465-736	Isodrin	950-378	Methidathion
470-906	Chlorfenvinfos	991-424	Norbormide
502-396	Methylmercuric Dicyanamide	998-301	Triethoxysilane
504-245	Pyridine, 4-Amino-	999-815	Chlormequat Chloride
505-602	Mustard Gas	1031-476	Triamiphos
506-616	Potassium Silver Cyanide	1066-451	Trimethyltin Chloride
506-683	Cyanogen Bromide	1122-607	Nitrocyclohexane
506-785	Cyanogen Iodide	1124-330	Pyridine, 4-Nitro-, 1-Oxide
509-148	Tetranitromethane	1129-415	Metolcarb
514-738	Dithiazanine Iodide	1303-282	Arsenic pentoxide

1306-190	Cadmium Oxide	3689-245	Sulfotep
1314-563	Phosphorus Pentoxide	3691-358	Chlorophacinone
1314-621	Vanadium Pentoxide	3734-972	Amiton Oxalate
1314-847	Zinc Phosphide	3735-237	Methyl Phenkapton
1327-533	Arsenous oxide	3878-191	Fuberidazole
1397-940	Antimycin A	4044-659	Bitoscanate
1420-071	Dinoterb	4098-719	Isophorone Diisocyanate
1464-535	Diepoxybutane	4104-147	Phosacetim
1558-254	Trichloro (Chloromethyl) Silane	4170-303	Crotonaldehyde
1563-662	Carbofuran	4301-502	Fluenetil
1600-277	Mercuric Acetate	4418-660	Phenol, 2,2'-Thiobis(4-Chloro-6-
1622-328	Ethanesulfonyl Chloride, 2-Chloro-		Methyl)-
1642-542	Diethylcarbamazine Citrate	4835-114	Hexamethylenediamine, N,N'-
1752-303	Acetone Thiosemicarbazide		Dibutyl-
1910-425	Paraquat dichloride	5344-821	Thiourea, (2-Chlorophenyl)-
1982-474	Chloroxuron	5836-293	Coumatetralyl
2001-958	Valinomycin	6533-739	Thallous Carbonate
2032-657	Methiocarb	6923-224	Monocrotophos
2074-502	Paraquat Methosulfate	7446-095	Sulfur Dioxide
2097-190	Phenylsilatrane	7446-119	Sulfur Trioxide
2104-645	EPN	7446-186	Thallous Sulfate
2223-930	Cadmium Stearate	7487-947	Mercuric Chloride
2231-574	Thiocarbazide	7550-450	Titanium Tetrachloride
2238-075	Diglycidyl Ether	7580-678	Lithium Hydride
2275-185	Prothoate	7631-892	Sodium Arsenate
2497-076	Oxydisulfoton	7637-072	Boron Trifluoride
2524-030	Dimethyl Phosphor-ochloridothioate	7647-010	Hydrogen chloride (gas only)
2540-821	Formothion	7664-393	Hydrogen Fluoride
2570-265	Pentadecylamine	7664-417	Ammonia
2587-908	Phosphorothioic Acid, O,O-	7664-939	Sulfuric Acid
	Dimethyl-S-(2-Methylthio) Ethyl	7697-372	Nitric Acid
	Ester	7719-122	Phosphorus Trichloride
2631-370	Promecarb	7722-841	Hydrogen Peroxide (Conc >52%)
2636-262	Cyanophos	7723-140	Phosphorus
2642-719	Azinphos-Ethyl	7726-956	Bromine
2665-307	Phosphonothioic Acid, Methyl-, O-	7778-441	Calcium arsenate
	(4-Nitrophenyl) O-Phenyl Ester	7782-414	Fluorine
2703-131	Phosphonothioic Acid, Methyl-, O-	7782-505	Chlorine
	Ethyl O-(4-(Methylthio) Phenyl)	7783-008	Selenious Acid
	Ester	7783-064	Hydrogen Sulfide
2757-188	Thallous Malonate	7783-075	Hydrogen Selenide
2763-944	Muscimol	7783-600	Sulfur Tetrafluoride
2778-043	Endothion	7783-702	Antimony Pentafluoride
3037-727	Silane, (4-Aminobutyl)	7783-804	Tellurium Hexafluoride
	Diethoxymethyl-	7784-341	Arsenous trichloride
3254-635	Phosphoric Acid, Dimethyl 4-	7784-421	Arsine
	(Methylthio) Phenyl Ester	7784-465	Sodium arsenite
3569-571	Sulfoxide, 3-Chloropropyl Octyl	7786-347	Mevinphos
3615-212	Benzimidazole, 4,5-Dichloro-2-	7791-120	Thallous Chloride
	(Trifluoromethyl)-	7791-233	Selenium Oxychloride

7803-512	Phosphine
8001-352	Camphechlor
8065-483	Demeton
10025-737	Chromic Chloride
10025-873	Phosphorus Oxychloride
10026-138	Phosphorus Pentachloride
10028-156	Ozone
10031-591	Thallium Sulfate
10102-188	Sodium Selenite
10102-202	Sodium Tellurite
10102-439	Nitric Oxide
10102-440	Nitrogen Dioxide
10124-502	Potassium arsenite
10140-871	Ethanol, 12-Dichloro-, Acetate
10210-681	Cobalt Carbonyl
10265-926	Methamidophos
10294-345	Boron Trichloride
10311-849	Dialifor
10476-956	Methacrolein Diacetate
12002-038	Paris Green
12108-133	Manganese, Tricarbonyl Methylcyclopentadieny
13071-799	Terbufos
13171-216	Phosphamidon
13194-484	Ethoprophos
13410-010	Sodium Selenate
13450-903	Gallium Trichloride
13463-393	Nickel carbonyl
13463-406	Iron, Pentacarbonyl-
13494-809	Tellurium
14167-181	Salcomine
15271-417	Bicyclo[221] Heptane-2-Carbonitrile, 5-Chloro-6-((((Methylamino) Carbonyl)Oxy) Imino)-, (1s-(1-alpha, 2-beta, 4-alpha, 5-alpha, 6E))-
16752-775	Methomyl
17702-419	Decaborane(14)
17702-577	Formparanate
19287-457	Diborane
19624-227	Pentaborane
20830-755	Digoxin
20859-738	Aluminum Phosphide
21548-323	Fosthietan
21609-905	Leptophos
21908-532	Mercuric Oxide
21923-239	Chlorthiophos
22224-926	Fenamiphos
23135-220	Oxamyl
23422-539	Formetanate Hydrochloride
23505-411	Pirimifos-Ethyl

24017-478	Triazofos
24934-916	Chlormephos
26419-738	Carbamic Acid, Methyl-, 0-(((2,4-Dimethyl-1, 3-Dithiolan-2-yl) Methylene)Amino)-
26628-228	Sodium Azide (Na(N3))
27137-855	Trichloro (Dichlorophenyl) Silane
28347-139	Xylylene Dichloride
28772-567	Bromadiolone
30674-807	Methacryloyloxyethyl Isocyanate
39196-184	Thiofanox
50782-699	Phosphonothioic Acid, Methyl-, S-(2-(Bis(1-Methylethyl)Amino) Ethyl O-Ethyl Ester
53558-251	Pyriminil
58270-089	Zinc, Dichloro(4,4-Dimethyl-5(((((Methylamino) Carbonyl)Oxy) Imino) Pentanenitrile)-, (T-4)-
62207-765	Cobalt, ((2,2'-(1,2-Ethanediylbis (Nitrilomethylidyn))Bis(6-Fluorophenolato)) (2-)-N,N',O,O')-,

extremely hazardous substance (40CFR370.2) means a substance listed in the Appendices to 40CFR355, Emergency Planning and Notification.

extremely hazardous substance (SF329) means a substance on the list described in section 302(a)(2).

extremely hazardous substances (EPA-94/04): Any of 406 chemicals identified by EPA as toxic, and listed under SARA Title III. The list is subject to periodic revision.

extrusion (40CFR467.02-j) is the application of pressure to a billet of aluminum, forcing the aluminum to flow through a die orifice. The extrusion subcategory is based on the extrusion process.

extrusion (40CFR468.02-h) shall mean the application of pressure to a copper workpiece, forcing the copper to flow through a the orifice (under CWA).

extrusion (40CFR471.02-p) is the application of pressure to a billet of metal, forcing the metal to flow through a die orifice.

extrusion die cleaning (40CFR467.31-b) shall mean

the process by which the steel dies used in extrusion of aluminum are cleaned. The term includes a dip into a concentrated caustic bath to dissolve the aluminum followed by a water rinse. It also includes the use of a wet scrubber with the die cleaning operation.

extrusion heat treatment (40CFR468.02-i) shall mean the spray application of water to a workpiece immediately following extrusions for the purpose of heat treatment.

eye corrosion (40CFR798.4500-2) means the production of irreversible tissue damage in the eye following application of a test substance to the anterior surface of the eye.

eye irritation (40CFR798.4500-1) means the production of reversible changes in the eye following the application of a test substance to the anterior surface of the eye.

*********** FFFFF ***********

FAA (NCA7-49USC4906-87) means Administrator of the Federal Aviation Administration.

fabric coating (40CFR52.741) means any coating applied on textile fabric. Fabric coating includes the application of coatings by impregnation.

fabric coating facility (40CFR52.741) means a facility that includes one or more fabric coating lines (under CAA).

fabric coating line (40CFR52.741) means a coating line in which any protective, decorative, or functional coating or reinforcing material is applied on or impregnated into a textile fabric.

fabric filter (EPA-94/04): A cloth device that catches dust particles from industrial emissions.

fabricating (40CFR61.141) means any processing (e.g., cutting, sawing, drilling) of a manufactured product that contains commercial asbestos, with the exception of processing at temporary sites (field fabricating) for the construction or restoration of facilities. In the case of friction products, fabricating includes bonding, debonding, grinding, sawing, drilling, or other similar operations performed as part of fabricating.

facial tissue (40CFR250.4-n) means a class of soft absorbent papers in the sanitary tissue group.

facilities eligible for treatment equivalent to secondary treatment (40CFR133.101-g): Facilities eligible for treatment equivalent to secondary treatment treatment works shall be eligible for consideration for effluent limitations described for treatment equivalent to secondary treatment (40CFR133.105), if:
(1) The BOD(5) and SS effluent concentrations consistently achievable through proper operation and maintenance (40CFR133.101(f)) of the treatment works exceed the minimum level of the effluent quality set forth in 40CFR133.102(a) and 133.102(b),
(2) A trickling filter or waste stabilization pond is used as the principal process, and
(3) The treatment works provide significant biological treatment of municipal wastewater.

facilities or equipment (40CFR122.29-5) means buildings, structures, process or production equipment or machinery which form a permanent part of the new source and which will be used in its operation, if these facilities or equipment are of such value as to represent a substantial commitment to construct. It excludes facilities or equipment used in connection with feasibility, engineering, and design studies regarding the source or water pollution treatment for the source.

facilities plans (EPA-94/04): Plans and studies related to the construction of treatment works necessary to comply with the Clean Water Act or RCRA. A facilities plan investigates needs and provides information on the cost effectiveness of alternatives, a recommended plan, an environmental assessment of the recommendations, and descriptions of the treatment works, costs, and a completion schedule.

facility (40CFR12.103) means all or any portion of buildings, structures, equipment, roads, walks, parking lots, rolling stock or other conveyances, or other real or personal property (other identical or similar definitions are provided in 40CFR2.310-3; 7.25; 8.2-k; 15.4; 20.2-f; 21.2-j; 51.166; 52.21; 52.24-2; 52.2297-6; 61.91-b; 61.101; 61.141; 61.191-a; 61.341; 72.2; 82.3-r; 240.101-g; 245.101-c; 256.06; 257.2; 258.2; 259.10.a; 260.10; 300.5; 302.3; 350.1; 355.20; 370.2; 372.3; 721.3; RCRA9003; SF101-42USC9601; SF329).

facility and practice, criteria for classification of solid waste disposal (40CFR257.3): Solid waste disposal facilities or practices which violate any of the following criteria pose a reasonable probability of adverse effects on health and the environment:
(1) Floodplains;
(2) Endangered species;
(3) Surface water;
(4) Ground water;
(5) Application to land used for the production of food-chain crops (interim final);
(6) Disease;
(7) Air; and
(8) Safety.

facility component (40CFR61.141) means any part of a facility including equipment.

facility emergency coordinator (EPA-94/04): Representative of a facility covered by environmental law (e.g, a chemical plant) who participates in the emergency reporting process with the Local Emergency Planning Committee (LEPC).

facility mailing list (40CFR270.2) means the mailing list for a facility maintained by EPA in accordance with 40CFR124.10(c)(viii).

facility or activity (40CFR122.2) means any NPDES "point source" or any other facility or activity (including land or appurtenances thereto) that is subject to regulation under the NPDES program.

facility or activity (40CFR124.2) means any "HWM facility," UIC "injection well," NPDES "point source" or "treatment works treating domestic sewage" or State 404 dredge or fill activity, or any other facility or activity (including land or appurtenances thereto)

that is subject to regulation under the RCRA, UIC, NPDES, or 404 programs.

facility or activity (40CFR124.41) means a "major PSD stationary source" or "major PSD."

facility or activity (40CFR144.3) means any UIC "injection well," or an other facility or activity that is subject to regulation under the UIC program.

facility or activity (40CFR144.70-c) means any "underground injection well" or any other facility or activity that is subject to regulation under the Underground Injection Control Program.

facility or activity (40CFR146.3) means any "HWM facility," UIC "injection well," NPDES "point source," or State 404 dredge and fill activity, or any other facility or activity (including land or appurtenances thereto) that is subject to regulation under the RCRA, UIC, NPDES, or 404 programs.

facility or activity (40CFR270.2) means any HWM facility or any other facility or activity (including land or appurtenances thereto) that is subject to regulation under the RCRA program.

facility personnel (40CFR260.10): See personnel.

facility structures (40CFR257.3.8) means any buildings and sheds or utility or drainage lines on the facility.

factors beyond the control of the licensee (40CFR192.31-o) means factors proximately causing delay in meeting the schedule in the applicable license for timely emplacement of the permanent radon barrier notwithstanding the good faith efforts of the licensee to achieve compliance. These factors may include, but are not limited to, physical conditions at the site; inclement weather or climatic conditions; an act of God; an act of war; a judicial or administrative order or decision, or change to the statutory, regulatory, or other legal requirements applicable to the licensee's facility that would preclude or delay the performance of activities required for compliance; labor disturbances; any modifications, cessation or delay ordered by state, Federal or local agencies; delays beyond the time reasonably required in obtaining necessary governmental permits, licenses, approvals or consent for activities described in the tailings closure plan (radon) proposed by the licensee that result from agency failure to take final action after the licensee has made a good faith, timely effort to submit legally sufficient applications, responses to requests (including relevant data requested by the agencies), or other information, including approval of the tailings closure plan by NRC or the affected Agreement State; and an act or omission of any third party over whom the licensee has no control.

facultative bacteria (EPA-94/04): Bacteria that can live under aerobic or anaerobic conditions.

failing compressor (40CFR204.51-u) means that the measured noise emissions of the compressor, when measured in accordance with the applicable procedure, exceeds the applicable standard.

failing exhaust system (40CFR205.165-3) means that, when installed on any Federally regulated motorcycle for which it is designed and marketed, that motorcycle and exhaust system exceed the applicable standards.

failing vehicle (40CFR205.151-16) means a vehicle whose noise level is in excess of the applicable standard.

failing vehicle (40CFR205.51-24) means that the measured emissions of the vehicle, when measured in accordance with the applicable procedure, exceeds the applicable standard.

fair market value (40CFR35.6015-20) means the amount at which property would change hands between a willing buyer and a willing seller, neither being under any compulsion to buy or sell and both having reasonable knowledge of the relevant facts. Fair market value is the price in cash, or its equivalent, for which the property would have been sold on the open market.

fall time (40CFR53.23-e): The time interval between initial response and 95 percent of final response after a step decrease in input concentration.

family emission limit (FEL) (40CFR86.090.2) means an emission level declared by the manufacturer which serves in lieu of an emission standard for certification purposes in any of the averaging, trading, or banking programs. FELs must be expressed to the same number of decimal places as the applicable emission standard. The FEL for an engine family using NO_x or particulate NCPs must equal the value of the current NO_x or particulate emission standard.

family emission limit (FEL) (40CFR89.2) means an emission level that is declared by the manufacturer to serve in lieu of an emission standard for certification purposes and for the averaging, banking, and trading program. A FEL must be expressed to the same number of decimal places as the applicable emission standard.

family NO_x emission limit (40CFR86.088.2) means the NO_x emission level to which an engine family is certified in the light-duty truck NO_x averaging program, expressed to one-tenth of a gram per mile accuracy.

far region (40CFR60-AA (alt. method 1)) means the region of the atmosphere's path along the lidar line-of-sight beyond or behind the plume being measured.

farm (40CFR170.3) means any operation, other than a nursery or forest, engaged in the outdoor production of agricultural plants.

farm (40CFR280.12) includes fish hatcheries, rangeland and nurseries with growing operations.

farm tank (40CFR280.12) is a tank located on a tract of land devoted to the production of crops or raising animals, including fish, and associated residences and improvements. A farm tank must be located on the farm property.

fast meter response (40CFR201.1-l) means that the fast response of the sound level meter shall be used. The fast dynamic response shall comply with the meter dynamic characteristics in paragraph 5.3 of the American National Standard Specification for Sound Level Meters, ANSI S1.4-1971. This publication is available from the American National Standards Institute, Inc., 1430 Broadway, New York, New York 1001.

fast meter response (40CFR202.10-g) means that the

fast dynamic response of the sound level meter shall be used. The fast dynamic response shall comply with the meter dynamic characteristics in paragraph 5.3 of the American National Standard Specification for Sound Level Meters, ANSI SI.4-1971. This publication is available from the American National Standards Institute, Inc., 1430 Broadway, New York, New York 10018 (other identical or similar definitions are provided in 40CFR205.2-11).

fast turnaround operation of a spray drying tower (40CFR417.151-f) shall mean operation involving more than 6 changes of formulation in a 30 consecutive day period that are of such degree and type (e.g., high phosphate to no phosphate) as to require cleaning of the tower to maintain minimal product quality.

fast turnaround operation of automated fill lines (40CFR417.161-e) shall mean an operation involving more than 8 changes of formulation in a 30 consecutive day period that are of such degree and type as to require thorough purging and washing of the fill line to maintain minimal product quality.

fault (40CFR146.3) means a surface or zone of rock fracture along which there has been displacement (other identical or similar definitions are provided in 40CFR147.2902).

fault (40CFR264.18-(a)(2)(i)) means a fracture along which rocks on one side have been displaced with respect to those on the other side.

fault (40CFR503.21-f) is a fracture or zone of fractures in any materials along which strata on one side are displaced with respect to strata on the other side.

FDA (40CFR160.3; 792.3) means the U.S. Food and Drug Administration.

feasibility study (EPA-94/04):
1. Analysis of the practicability of a proposal; e.g., a description and analysis of potential cleanup alternatives for a site such as one on the National Priorities List. The feasibility study usually recommends selection of a cost-effective alternative. It usually starts as soon as the remedial investigation is underway; together,

they are commonly referred to as the RI/FS (remedial investigation/feasibility study).
2. A small-scale investigation of a problem to ascertain whether a proposed research approach is likely to provide useful data.

feasibility study (FS) (40CFR300.5) means a study undertaken by the lead agency to develop and evaluate options for remedial action. The FS emphasizes data analysis and is generally performed concurrently and in an interactive fashion with the remedial investigation (RI), using data gathered during the RI. The RI data are used to define the objectives of the response action, to develop remedial action alternatives, and to undertake an initial screening and detailed analysis of the alternatives. The term also refers to a report that describes the results of the study.

fecal coliform bacteria (40CFR140.1-g) are those organisms associated with the intestines of warm-blooded animals that are commonly used to indicate the presence of fecal material and the potential presence of organisms capable of causing human disease.

fecal coliform bacteria (EPA-94/04): Bacteria found in the intestinal tracts of mammals. Their presence in water or sludge is an indicator of pollution and possible contamination by pathogens.

federal act (40CFR109.2-f) means the Federal Water Pollution Control Act, as amended, 33USC1151, et seq.

federal action (40CFR51.852) means any activity engaged in by a department, agency, or instrumentality of the Federal government, or any activity that a department, agency or instrumentality of the Federal government supports in any way, provides financial assistance for, licenses, permits, or approves, other than activities related to transportation plans, programs, and projects developed, funded, or approved under title 23 U.S.C. or the Federal Transit Act (49 U.S.C. 1601 et seq.). Where the Federal action is a permit, license, or other approval for some aspect of a non-Federal undertaking, the relevant activity is the part, portion, or phase or the non-Federal undertaking that requires the Federal permit, license, or approval (other identical or similar definitions are provided in 40CFR93.152).

federal agency (40CFR142.2) means any department, agency, or instrumentality of the United States (other identical or similar definitions are provided in 40CFR51.852; 93.151; 205.2-20; 244.101-g; 248.4-n; 249.04-d; 250.4-o; 252.4-c; 253.4; 260.10; 1508.12; ESA3-16USC1531;NCA3-42USC4902;RCRA1004-42USC6903; SDWA1401-42USC300f; SDWA1447-42USCj.6).

federal agency or agency (40CFR32.605-6) means any United States executive department, military department, government corporation, government controlled corporation, any other establishment in the executive branch (including the Executive Office of the President), or any independent regulatory agency.

federal agency or agency of the United States (40CFR47.105-b) means any department, agency or other instrumentality of the Federal Government, any independent agency or establishment of the Federal Government including any Government corporation.

federal certificate (40CFR86.902.93) is a Certificate of Conformity issued by EPA which signifies compliance with emission standards in 40CFR part 86, subpart A.

federal class I area (40CFR51.301-f) means any Federal land that is classified or reclassified "Class I."

federal contract (40CFR34.105) means an acquisition contract awarded by an agency, including those subject to the Federal Acquisition Regulation (FAR), and any other acquisition contract for real or personal property or services not subject to the FAR.

federal cooperative agreement (40CFR34.105-d) means a cooperative agreement entered into by an agency.

federal delayed compliance order (40CFR65.01-f) shall mean a delayed compliance order issued by the Administrator under section 113(d) (1), (3), (4) or (5) of the Act.

federal emission test procedure (40CFR600.002.85-7) refers to the dynamometer driving schedule, dynamometer procedure, and sampling and analytical procedures described in part 86 for the respective model year, which are used to derive city fuel economy data for gasoline-fueled or diesel vehicles (under the Motor Vehicle Information and Cost Savings Act).

federal facility (40CFR35.4010) means a facility that is owned or operated by a department, agency, or instrumentality of the United States (other identical or similar definitions are provided in 40CFR61.101-d; 243.101-k; 244.101-h; 246.101-l).

federal financial assistance (40CFR149.101-g) means any financial benefits provided directly as aid to a project by a department, agency, or instrumentality of the Federal government in any form including contracts, grants, and loan guarantees. Actions or programs carried out by the Federal government itself such as dredging performed by the Army Corp of Engineers do not involve Federal financial assistance. Actions performed for the Federal government by contractors, such as construction of roads on Federal lands by a contractor under the supervision of the Bureau of Land Management, should be distinguished from contracts entered into specifically for the purpose of providing financial assistance, and will not be considered programs or actions receiving Federal financial assistance. Federal financial assistance is limited to benefits earmarked for a specific program or action and directly award to the program or action. Indirect assistance, e.g., in the form of a loan to a developer by a lending institution which in turn receives Federal assistance not specifically related to the project in question is not Federal financial assistance under section 1424(e).

federal Food, Drug, and Cosmetic Act: See act or FFDCA.

federal government (40CFR203.1-2) includes the legislative, executive, and judicial branches of the Government of the United States, and the government of the District of Columbia.

federal government (NCA15-42USC4914) includes the legislative, executive, and judicial branches of the Government of the United States, and the government of the District of Columbia.

federal grant (40CFR34.105-e) means an award of financial assistance in the form of money, or property in lieu of money, by the Federal Government or a direct appropriation made by law to any person. The term does not include technical assistance which provides service instead of money, or other assistance in the form of revenue sharing, loans, loan guarantees, loan insurance, interest subsidies, insurance, or direct United States cash assistance to an individual.

federal highway fuel economy test procedure (40CFR600.002.85-8) refers to the dynamometer driving schedule, dynamometer procedure, and sampling and analytical procedures described in subpart B of this part and which are used to derive highway fuel economy data for gasoline-fueled or diesel vehicles.

federal implementation plan (CAA302-42USC7602) means a plan (or portion thereof) promulgated by the Administrator to fill all or a portion of a gap or otherwise correct all or a portion of an inadequacy in a State implementation plan, and which includes enforceable emission limitations or other control measures, means or techniques (including economic incentives, such as marketable permits or auctions of emissions allowances), and provides for attainment of the relevant national ambient air quality standard.

federal implementation plan (EPA-94/04): Under current law, a federally implemented plan to achieve attainment of air quality standards, used when a state is unable to develop an adequate plan.

federal Indian reservation (40CFR122.2) means all land within the limits of any Indian reservation under the jurisdiction of the United States Government, notwithstanding the issuance of any patent, and including rights-of-way running through the reservation (other identical or similar definitions are provided in 40CFR35.105; 124.2; 131.3-k; 232.2) (under FWPCA).

federal Indian reservation (40CFR501.2) means all land within the limits of any Indian reservation under the jurisdiction of the United States Government, notwithstanding the issuance of any patent, and including rights-of-way running through the reservation.

Federal Insecticide, Fungicide, and Rodenticide Act: See act or FIFRA.

federal land (SMCRA701-30USC1291) means any land, including mineral interests, owned by the United States without regard to how the United States acquired ownership of the land and without regard to the agency having responsibility for management thereof, except Indian lands: Provided that for the purposes of this chapter lands or mineral interests east of the one hundredth meridian west longitude owned by the United States and entrusted to or managed by the Tennessee Valley Authority shall not be subject to sections 1304 (surface owner protection) and 1305 (federal lessee protection) of this title.

Federal Land Manager (40CFR51.166-24) means, with respect to any lands in the United States, the Secretary of the department with authority over such lands (other identical or similar definitions are provided in 40CFR51.301-g; 52.21-24; 124.41; CAA302-42USC7602).

federal lands program (SMCRA701-30USC1291) means a program established by the Secretary pursuant to section 1273 of this title to regulate surface coal mining and reclamation operations on Federal lands.

federal loan (40CFR34.105-f) means a loan made by an agency. The term does not include loan guarantee or loan insurance.

federal motor vehicle control program (EPA-94/04): All federal actions aimed at controlling pollution from motor vehicles by such efforts as establishing and enforcing tailpipe and evaporative emission standards for new vehicles, testing methods development, and guidance to states operating inspection and maintenance programs.

federal on-scene coordinator (CWA311-33USC1321) means a Federal On-Scene Coordinator designated in the National Contingency Plan.

federal program (SMCRA701-30USC1291) means a program established by the Secretary pursuant to section 1254 of this title to regulate surface coal mining and reclamation operations on lands within a

State in accordance with the requirements of this chapter.

federal register document (40CFR23.1-a) means a document intended for publication in the Federal Register and bearing in its heading an identification code including the letters FRL.

federal standards (40CFR205.165-5) means, for the purpose of this subpart, the standards specified in 40CFR205.152(a)(1), (2) and (3).

Federal, State and local approvals or permits necessary to begin physical construction (40CFR260.10) means permits and approvals required under Federal, State or local hazardous waste control statutes, regulations or ordinances (other identical or similar definitions are provided in 40CFR270.2).

federal test procedure or city fuel economy test (40CFR610.11-6) means the test procedures specified in 40CFR86, except as those procedures are modified in these protocols.

federal test procedure (40CFR85.1502): See FTP.

Federal Water Pollution Control Act: See act or FWPCA.

federally assisted construction contract (40CFR8.2-l) means any agreement or modification thereof between any applicant and any person for construction work which is paid for in whole or in part with funds obtained from the Agency or borrowed on the credit of the Agency pursuant to any Federal program involving a grant, contract, loan, insurance, or guarantee, or undertaken pursuant to any Federal program involving such grant, contract, loan, insurance, or guarantee, or any application or modification thereof approved by the Agency for a grant, contract, loan, insurance, or guarantee under which the applicant itself participates in the construction work.

federally enforceable (40CFR51-AS-12) means all limitations and conditions which are enforceable by the Administrator, including those requirements developed pursuant to 40CFR60 and 61, requirements within any applicable State implementation plan, and any permit requirements established pursuant to 40CFR52.21 or under regulations approved pursuant to this section, 40CFR51 subpart I, including operating permits issued under an EPA-approved program that is incorporated into the State implementation plan and expressly requires adherence to any permit issued under such program.

federally enforceable (40CFR51.165-xiv) means all limitations and conditions which are enforceable by the Administrator, including those requirements developed pursuant to 40CFR parts 60 and 61, requirements within any applicable State implementation plan, any permit requirements established pursuant to 40CFR52.21 or under regulations approved pursuant to 40CFR part 51, subpart I, including operating permits issued under an EPA-approved program that is incorporated into the State implementation plan and expressly requires adherence to any permit issued under such program.

federally enforceable (40CFR51.166) means all limitations and conditions which are enforceable by the Administrator, including those requirements developed pursuant to 40CFR60 and 61, requirements within any applicable State implementation plan, any permit requirements established pursuant to 40CFR52.21 or under regulations approved pursuant to 40CFR51, subpart I, including operating permits issued under an EPA-approved program that is incorporated into the State implementation plan and expressly requires adherence to any permit issued under such program (other identical or similar definitions are provided in 40CFR52.21; 52.24-12).

federally enforceable (40CFR51.166-17) means all limitations and conditions which are enforceable by the Administrator, including those requirements developed pursuant to 40CFR parts 60 and 61, requirements within any applicable State implementation plan, any permit requirements established pursuant to 40CFR52.21 or under regulations approved pursuant to 40CFR part 51, subpart I, including operating permits issued under an EPA-approved program that is incorporated into the State implementation plan and expressly requires adherence to any permit issued under such program (under CAA).

federally enforceable (40CFR51.301-h) means all limitations and conditions which are enforceable by the Administrator under the Clean Air Act including those requirements developed pursuant to parts 60 and 61 of this title, requirements within any applicable State Implementation Plan, and any permit requirements established pursuant to 40CFR52.21 of this chapter or under regulations approved pursuant to 40CFR51, 52, or 60 of this title.

federally enforceable (40CFR52.21-17) means all limitations and conditions which are enforceable by the Administrator, including those requirements developed pursuant to 40CFR parts 60 and 61, requirements within any applicable State implementation plan, any permit requirements established pursuant to 40CFR52.21 or under regulations approved pursuant to 40CFR part 51, subpart I, including operating permits issued under an EPA-approved program that is incorporated into the State implementation plan and expressly requires adherence to any permit issued under such program (under CAA).

federally enforceable (40CFR52.24-12) means all limitations and conditions which are enforceable by the Administrator, including those requirements developed pursuant to 40CFR parts 60 and 61, requirements within any applicable State implementation plan, any permit requirements established pursuant to 40CFR52.21 or under regulations approved pursuant to 40CFR part 51, subpart I, including operating permits issued under an EPA-approved program that is incorporated into the State implementation plan and expressly requires adherence to any permit issued under such program (under CAA).

federally enforceable (40CFR52.741) means all limitations and conditions which are enforceable by the Administrator including those requirements developed pursuant to 40CFR60 and 61; requirements within any applicable implementation plan; and any permit requirements established pursuant to 40CFR52.21 or under regulations approved pursuant to 40CFR51 subpart I and 40CFR51.166.

federally enforceable (40CFR60.41b) means all limitations and conditions that are enforceable by the Administrator, including those requirements of 40CFR60 and 61, requirements within any applicable State Implementation Plan, and any permit requirements established under 40CFR52.21 or under 40CFR51.18 and 40CFR51.24 (other identical or similar definitions are provided in 40CFR60.41c).

federally enforceable (40CFR63.2) means all limitations and conditions that are enforceable by the Administrator and citizens under the Act or that are enforceable under other statutes administered by the Administrator. Examples of federally enforceable limitations and conditions include, but are not limited to:

(1) Emission standards, alternative emission standards, alternative emission limitations, and equivalent emission limitations established pursuant to section 112 of the Act as amended in 1990;

(2) New source performance standards established pursuant to section 111 of the Act, and emission standards established pursuant to section 112 of the Act before it was amended in 1990;

(3) All terms and conditions in a title V permit, including any provisions that limit a source's potential to emit, unless expressly designated as not federally enforceable;

(4) Limitations and conditions that are part of an approved State Implementation Plan (SIP) or a Federal Implementation Plan (FIP);

(5) Limitations and conditions that are part of a Federal construction permit issued under 40CFR52.21 or any construction permit issued under regulations approved by the EPA in accordance with 40CFR part 51;

(6) Limitations and conditions that are part of an operating permit issued pursuant to a program approved by the EPA into a SIP as meeting the EPA's minimum criteria for Federal enforceability, including adequate notice and opportunity for EPA and public comment prior to issuance of the final permit and practicable enforceability;

(7) Limitations and conditions in a State rule or program that has been approved by the EPA under subpart E of this part for the purposes of implementing and enforcing section 112; and

(8) Individual consent agreements that the EPA has legal authority to create.

(Other identical or similar definitions are provided in 40CFR63.101; 63.191.)

federally enforceable (40CFR60.51a) means all limitations and conditions that are enforceable by the Administrator including the requirements of 40CFR60 and 61, requirements within any applicable State implementation plan, and any permit requirements established under 40CFR52.21 or under 40CFR51.18 and 40CFR51.24 (under the Clean Air Act).

federally permitted release (SF101-42USC9601) means:

(A) discharges in compliance with a permit under section 402 of the Federal Water Pollution Control Act,

(B) discharges resulting from circumstances identified and reviewed and made part of the public record with respect to a permit issued or modified under section 402 of the Federal Water Pollution Control Act and subject to a condition of such permit,

(C) continuous or anticipated intermittent discharges from a point source, identified in a permit or permit application under section 402 of the Federal Water Pollution Control Act, which are caused by events occurring within the scope of relevant operating or treatment systems,

(D) discharges in compliance with a legally enforceable permit under section 404 of the Federal Water Pollution Control Act,

(E) releases in compliance with a legallyfinal permit issued pursuant to section 3005 (a) through (d) of the Solid Waste Disposal Act from a hazardous waste treatment, storage, or disposal facility when such permit specifically identifies the hazardous substances and makes such substances subject to a standard of practice, control procedure or bioassay limitation or condition, or other control on the hazardous substances in such release,

(F) any release in compliance with a legally enforceable permit issued under section 102 or section 103 of the Marine Protection, Research, and Sanctuaries Act of 1972, (G) any injection of fluids authorized under Federal underground injection control programs or State programs submitted for Federal approval (and not disapproved by the Administrator of the Environmental Protection Agency) pursuant to Part C of the Safe Drinking Water Act,

(H) any emission into the air subject to a permit or control regulation under section 111, section 112, title I part C, title I part D, or State implementation plans submitted in accordance with section 110 of the Clean Air Act (and not disapproved by the Administrator of the Environmental Protection Agency), including any schedule or waiver granted, promulgated, or approved under these sections,

(I) any injection of fluids or other materials authorized under applicable State law:

   (i) for the purpose of stimulating or treating wells for the production of crude oil, natural gas, or water,

   (ii) for the purpose of secondary, tertiary, or other enhanced recovery of crude oil or natural gas, or

   (iii) which are brought to the surface in conjunction with the production of crude oil or natural gas and which are reinjected,

(J) the introduction of any pollutant into a publicly owned treatment works when such pollutant is specified in and in compliance with applicable pretreatment standards of section 307 (b) or (c) of the Clean Water Act and enforceable requirements in a pretreatment program submitted by a State or municipality for Federal approval under section 402 of such Act, and

(K) any release of source, special nuclear, or byproduct material, as those terms are defined in the Atomic Energy Act of 1954, in compliance with a legally enforceable license, permit, regulation, or order issued pursuant to the Atomic Energy Act of 1954.

federally recognized Indian tribal government (40CFR31.3) means the governing body or a governmental agency of any Indian tribe, band, nation, or other organized group or community (including any Native village as defined in section 3 of the Alaska Native Claims Settlement Act, 85 Stat 688) certified by the Secretary of the Interior as eligible for the special programs and services provided by him through the Bureau of Indian Affairs.

federally registered (40CFR162.151-a) means currently registered under sec. 3 of the Act, after having been initially registered under the Federal Insecticide, Fungicide, and Rodenticide Act (FIFRA) of 1947 (Pub. L. 86-139; 73 Stat. 286; June 25, 1947) by the Secretary of Agriculture or under FIFRA by the Administrator.

federally regulated motorcycle (40CFR205.165-4) means, for the purpose of this subpart, any motorcycle subject to the noise standards of subpart D of this part.

feed crops (40CFR503.9-j) are crops produced primarily for consumption by animals.

feed gas (40CFR85.2122(a)(15)(ii)(E)) means the chemical composition of the exhaust gas measured at the converter inlet.

feedlot (40CFR412.11-b) shall mean a new concentrated confined animal or poultry growing operation for meat, milk or egg production, or stabling, in pens or houses wherein the animals or poultry are fed at the place of confinement and crop or forage growth or production is not sustained in the area of confinement.

feedlot (40CFR412.21-b) shall mean a concentrated, confined animal or poultry growing operation for meat, milk or egg production, or stabling, in pens or houses wherein the animals or poultry are fed at the place of confinement and crop or forage production or growth is not sustained in the area of confinement.

feedlot (EPA-94/04): A confined area for the controlled feeding of animals. Tends to concentrate large amounts of animal waste that cannot be absorbed by the soil and, hence, may be carried to nearby streams or lakes by rainfall runoff.

feedstock (40CFR419.11-d) shall mean the crude oil and natural gas liquids fed to the topping units (under CWA).

fen (EPA-94/04): A type of wetland that accumulates peat deposits. Fens are less acidic than bogs, deriving most of their water from groundwater rich in calcium and magnesium (cf. wetlands).

ferrochrome silicon (40CFR60.261-t) means that

alloy as defined by ASTM Designation A482-76 (incorporated by reference, see 40CFR60.17).

ferromanganese blast furnace (40CFR420.31-a) means those blast furnaces which produce molten iron containing more than fifty percent manganese (under CWA).

ferromanganese silicon (40CFR60.261-y) means that alloy containing 63 to 66 percent by weight manganese, 28 to 32 percent by weight silicon, and a maximum of 0.08 percent by weight carbon.

ferrosilicon (40CFR60.261-w) means that alloy as defined by ASTM Designation A100-69 (Reapproved 1974) (incorporated by reference, 40CFR60.17) grades A, B, C, D, and E, which contains 50 or more percent by weight silicon.

ferrous casting (40CFR464.02-c) means the remelting of ferrous metals to form a cast intermediate or finished product by pouring the molten metal into a mold. Except for grinding scrubber operations which are covered here, processing operations following the cooling of castings are covered under the electroplating and metal finishing point source categories (40CFR413 and 433).

FFDCA: See act or FFDCA.

FHWA (40CFR51.392; 93.101) means the Federal Highway Administration of DOT.

FHWA/FTA project (40CFR51.392), for the purpose of this subpart, is any highway or transit project which is proposed to receive funding assistance and approval through the Federal-Aid Highway program or the Federal mass transit program, or requires Federal Highway Administration (FHWA) or Federal Transit Administration (FTA) approval for some aspect of the project, such as connection to an interstate highway or deviation from applicable design standards on the interstate system (other identical or similar definitions are provided in 40CFR93.101).

fiber (40CFR410.21-a) shall mean the dry wool and other fibers as received at the wool finishing mill for processing into wool and blended products.

fiber (40CFR763.121) means a particulate form of asbestos, 5 micrometers or longer, with a length-to-diameter ratio of at least 3 to 1.

fiber (40CFR763-AA-9) means a structure greater than or equal to 0.5 μm in length with an aspect ratio (length to width) of 5:1 or greater and having substantially parallel sides.

fiber crops (40CFR503.9-k) are crops such as flax and cotton.

fiber length and diameter distributions: See method B in section 40CFR796.1520.

fiber or fiberboard boxes (40CFR250.4-p) means boxes made from containerboard, either solid fiber or corrugated paperboard (general term) or boxes made from solid paperboard of the same material throughout (specific term).

fiber release episode (40CFR763.83) means any uncontrolled or unintentional disturbance of ACBM resulting in visible emission.

fiberglass insulation (40CFR248.4-o) means insulation which is composed principally of glass fibers, with or without binders.

field gas (40CFR60.631) means feedstock gas entering the natural gas processing plant (under CAA).

field testing (40CFR35.2005-17) means practical and generally small-scale testing of innovative or alternative technologies directed to verifying performance and/or refining design parameters not sufficiently tested to resolve technical uncertainties which prevent the funding of a promising improvement in innovative or alternative treatment technology.

FIFRA: See act or FIFRA.

FIFRA pesticide ingredient (EPA-94/04): An ingredient of a pesticide that must be registered with EPA under the Federal Insecticide, Fungicide, and Rodenticide Act. Products making pesticide claims must register under FIFRA and may be subject to labeling and use requirements.

fifteen working day hold period (40CFR89.602.96): The period of time between a request for final admission and the automatic granting of final admission (unless EPA intervenes) for a nonconforming nonroad engine conditionally imported pursuant to 40CFR89.605-96 or 40CFR89.609-96. Day one of the hold period is the first working day (see definition below) after the Manufacturers Operations Division of EPA receives a complete and valid application for final admission.

fifty (50) percent cutpoint (40CFR53.43-b): The particle size for which the sampling effectiveness of the sampler is 50 percent.

file (40CFR72.2) means to send or transmit a document, information, or correspondence to the official custody of the person specified to take possession in accordance with the applicable regulation. Compliance with any "filing" deadline shall be determined by the date that person receives the document, information, or correspondence (under CAA).

filing form (40CFR86.902.93) means the MVECP Fee Filing Form to be sent with payment of the MVECP fee.

fill (40CFR60.111b-d) means the introduction of VOL into a storage vessel but not necessarily to complete capacity.

fill material (40CFR232.2) means any "pollutant" which replaces portions of the "waters of the United States" with dry land or which changes the bottom elevation of a water body for any purpose .

fill or filling (40CFR63.111) means the introduction of organic hazardous air pollutant into a storage vessel or the introduction of a wastewater stream or residual into a waste management unit, but not necessarily to complete capacity.

filling (EPA-94/04): Depositing dirt, mud or other materials into aquatic areas to create more dry land, usually for agricultural or commercial development purposes, often with ruinous ecological consequences.

filter (40CFR63.321) means a porous device through which perchloroethylene is passed to remove contaminants in suspension. Examples include, but are not limited to, lint filter (button trap), cartridge filter, tubular filter, regenerative filter, prefilter, polishing filter, and spin disc filter.

filter background level (40CFR763-AA-13) means the concentration of structures per square millimeter of filter that is considered indistinguishable from the concentration measured on a blank (filters through which no air has been drawn). For this method the filter background level is defined as 70 structures/mm^2.

filter strip (EPA-94/04): Strip or area of vegetation used for removing sediment, organic matter, and other pollutants from runoff and waste water.

filtration (40CFR141.2) means a process for removing particulate matter from water by passage through porous media.

filtration (EPA-94/04): A treatment process, under the control of qualified operators, for removing solid (particulate) matter from water by means of porous media such as sand or a man-made filter; often used to remove particles that containing pathogens.

final approval (40CFR281.12) means the approval received by a state program that meets the requirements in 40CFR281.11(b).

final authorization (40CFR270.2) means approval by EPA of a State program which has met the requirements of section 3006(b) of RCRA and the applicable requirements of part 271, subpart A.

final closure (40CFR260.10) means the closure of all hazardous waste management units at the facility in accordance with all applicable closure requirements so that hazardous waste management activities under parts 264 and 265 of this Chapter are no longer conducted at the facility unless subject to the provisions in 40CFR262.34.

final cover (40CFR241.101-d) means cover material that serves the same functions as daily cover but, in addition, may be permanently exposed on the surface.

final cover (40CFR503.21-g) is the last layer of soil or other material placed on a sewage sludge unit at closure.

final order (40CFR22.03) means:
(a) an order issued by the Administrator after an appeal of an initial decision, accelerated decision, decision to dismiss, or default order, disposing of a matter in controversy between the parties, or
(b) an initial decision which becomes a final order under 40CFR22.2(c).

final permit (40CFR70.2) means the version of a part 70 permit issued by the permitting authority that has completed all review procedures required by 40CFR70.7 and 70.8 of this part.

final printed labeling (40CFR152.3-l) means the label or labeling of the product when distributed or sold. Final printed labeling does not include the Package of the product, unless the labeling is an integral part of the package.

final product (40CFR700.43) means a new chemical substance (as new chemical substance is defined in 40CFR720.3 of this chapter) that is manufactured by a person for distribution in commerce, or for use by the person other than as an intermediate (under TSCA).

final repair coat (40CFR52.741) means the repainting of any topcoat which is damaged during vehicle assembly.

financial assurance for closure (EPA-94/04): Documentation or proof that an owner or operator of a facility such as a landfill or other waste repository is capable of paying the projected costs of closing the facility and monitoring it afterwards as provided in RCRA regulations.

financial reporting year (40CFR280.92) means the latest consecutive twelve-month period for which any of the following reports used to support a financial test is prepared:
(1) a 10-K report submitted to the SEC;
(2) an annual report of tangible net worth submitted to Dun and Bradstreet; or
(3) annual reports submitted to the Energy Information Administration or the Rural Electrification Administration. "Financial

reporting year" may thus comprise a fiscal or a calendar year period.

finding of no significant impact (40CFR1508.13) means a document by a Federal agency briefly presenting the reasons why an action, not otherwise excluded (40CFR1508.4), will not have a significant effect on the human environment and for which an environmental impact statement therefore will not be prepared. It shall include the environmental assessment or a summary of it and shall note any other environmental documents related to it (40CFR1501.7(a)(5)). If the assessment is included, the finding need not repeat any of the discussion in the assessment but may incorporate it by reference.

finding of no significant impact (EPA-94/04): A document prepared by a federal agency showing why a proposed action would not have a significant impact on the environment and thus would not require preparation of an Environmental Impact Statement. An FNSI is based on the results of an environmental assessment.

finish coat operation (40CFR60.461) means the coating application station, curing oven, and quench station used to apply and dry or cure the final coating(s) on the surface of the metal coil. Where only a single coating is applied to the metal coil, that coating is considered a finish coat.

finished product (40CFR432.51-c) shall mean the final manufactured product as fresh meat cuts, hams, bacon or other smoked meats, sausage, luncheon meats, stew, canned meats or related products.

finished product (40CFR432.61-c) shall mean the final manufactured product as fresh meat cuts including, but not limited to, steaks, roasts, chops, or boneless meats.

finished product (40CFR432.71-c) shall mean the final manufactured product as fresh meat cuts including steaks, roasts, chops or boneless meat, bacon or other smoked meats (except hams) such as sausage, bologna or other luncheon meats, or related products (except canned meats).

finished products (40CFR432.81-c) shall mean the final manufactured product as fresh meat cuts including steaks, roasts, chops or boneless meat, smoked or cured hams, bacon or other smoked meats, sausage, bologna or other luncheon meats (except canned meats).

finished products (40CFR432.91-c) shall mean the final manufactured product as fresh meat cuts including steaks, roasts, chops or boneless meat, hams, bacon or other smoked meats, sausage, bologna or other luncheon meats, stews, sandwich spreads or other canned meats.

finished water (EPA-94/04): Water that has passed through a water treatment plant; all the treatment processes are completed or "finished." The water is ready to be delivered to consumers.

finishing water (40CFR463.2-e) is processed water used to remove waste plastic material generated during a finishing process or to lubricate a plastic product during a finishing process. It includes water used to machine or assemble intermediate or final plastic products.

fire-fighting turbine (40CFR60.331-k) means any stationary gas turbine that is used solely to pump water for extinguishing fires.

firebox (40CFR52.741) means the chamber or compartment of a boiler or furnace in which materials are burned, but not the combustion chamber of afterburner of an incinerator.

firebox (40CFR60-AA (method 28-2.3)) means the chamber in the wood heater in which the test fuel charge is placed and combusted.

firm or company (40CFR717.3-b) means any person that is subject to this part, as defined in 40CFR717.5.

first attempt at repair (40CFR63.111) means to take action for the purpose of stopping or reducing leakage of organic material to the atmosphere.

first attempt at repair (40CFR264.1031) means to take rapid action for the purpose of stopping or reducing leakage of organic material to the atmosphere using best practices.

first attempt at repair (40CFR60.481) means to take

rapid action for the purpose of stopping or reducing leakage of organic material to atmosphere using best practices (other identical or similar definitions are provided in 40CFR61.241).

first attempt at repair (40CFR63.161) means to take action for the purpose of stopping or reducing leakage of organic material to the atmosphere.

first draw (EPA-94/04): The water that comes out when a tap is first opened, likely to have the highest level of lead contamination from plumbing materials.

first draw sample (40CFR141.2) means a one-liter sample of tap water, collected in accordance with 40CFR141.86(b)(2), that has been standing in plumbing pipes at least 6 hours and is collected without flushing the tap.

first federal official (40CFR300.5) means the first federal representative of a participating agency of the National Response Team to arrive at the scene of a discharge or a release. This official coordinates activities under the NCP and may initiate, in consultation with the OSC, any necessary actions until the arrival of the predesignated OSC. A state with primary jurisdiction over a site covered by a cooperative agreement will act in the stead of the first federal official for any incident at the site.

first food use (40CFR166.3-e) refers to the use of a pesticide on a food or in a manner which otherwise would be expected to result in residues in a food, if no permanent tolerance, exemption from the requirement of a tolerance, or food additive regulation for residues of the pesticide on any food has been established for the pesticide under section 408 (d) or (e) or 409 of the Federal Food, Drug, and Cosmetic Act.

first order half-life ($t_{1/2}$) (40CFR796.3780-vii) of a chemical is defined as the time required for the concentration of the chemical to be reduced to one-half its initial value.

first order reaction (40CFR796.3700-v) is defined as a reaction in which the rate of disappearance of a chemical is directly proportional to the concentration of the chemical and is not a function of the concentration of any other chemical present in the reaction mixture (other identical or similar definitions are provided in 40CFR796.3780-v; 796.3800-v).

fish and wildlife and sensitive environments (40CFR112.2) means areas that may be identified by either their legal designation or by evaluations of Area Committees (for planning) or members of the Federal On-Scene Coordinator's spill response structure (during responses). These areas may include wetlands, National and State parks, critical habitats for endangered/threatened species, wilderness and natural resource areas, marine sanctuaries and estuarine reserves, conservation areas, preserves, wildlife areas, wildlife refuges, wild and scenic rivers, recreational areas, national forests, Federal and State lands that are research national areas, heritage program areas, land trust areas, and historical and archeological sites and parks. These areas may also include unique habitats such as: aquaculture sites and agricultural surface water intakes, bird nesting areas, critical biological resource areas, designated migratory routes, and designated seasonal habitats.

fish or wildlife (ESA3-16USC1531) means any member of the animal kingdom, including without limitation any mammal, fish, bird (including any migratory, nonmigratory, or endangered bird for which protection is also afforded by treaty or other international agreement), amphibian, reptile, mollusk, crustacean, arthropod or other invertebrate, and includes any part, product, egg, or offspring thereof, or the dead body or parts thereof.

five (5)-year, 6-hour precipitation event (40CFR440.141-5) means the maximum 6-hour precipitation event with a probable recurrence interval of once in 5 years as established by the U.S. Department of Commerce, National Oceanic and Atmospheric Administration, National Weather Service, or equivalent regional or rainfall probability information.

five year State/EPA conference agreement (40CFR35.9010) means the agreement negotiated among the States represented in a Management Conference and the EPA shortly after the Management Conference is convened. The agreement identifies milestones to be achieved during the term of the Management Conference.

fix sample (EPA-94/04): A sample is "fixed" in the field by adding chemicals that prevent water quality indicators of interest in the sample from changing before laboratory measurements are made.

fixed capital cost (40CFR51.301-i) means the capital needed to provide all of the depreciable components (under CAA).

fixed capital cost (40CFR60.15-c) means the capital needed to provide all the depreciable components (under CAA).

fixed capital cost (40CFR63.2) means the capital needed to provide all the depreciable components of an existing source (other identical or similar definitions are provided in 40CFR63.101).

fixed capital cost of the new components (40CFR60.666) as used in 40CFR60.15, includes the fixed capital cost of all depreciable components which are or will be replaced pursuant to all continuous programs of component replacement which are commenced within any 2-year period following December 30, 1983.

fixed plant (40CFR60.671) means any nonmetallic mineral processing plant at which the processing equipment specified in 40CFR60.670(a) is attached by a cable, chain, turnbuckle, bolt or other means (except electrical connections) to any anchor, slab, or structure including bedrock.

fixed roof (40CFR60.691) means a cover that is mounted to a tank or chamber in a stationary manner and which does not move with fluctuations in wastewater levels.

fixed roof (40CFR61.341) means a cover that is mounted on a waste management unit in a stationary manner and that does not move with fluctuations in liquid level.

fixed roof (40CFR63.111) means a cover that is mounted on a waste management unit or storage vessel in a stationary manner and that does not move with fluctuations in liquid level.

fixed roof tank (40CFR52.741) means a cylindrical shell with a permanently affixed roof.

fixed source (40CFR247.101-a) means, for the purpose of these guidelines, a stationary facility that converts fossil fuel into energy, such as steam, hot water, electricity, etc.

flame zone (40CFR60.561) means that portion of the combustion chamber in a boiler occupied by the flame envelope.

flame zone (40CFR60.611) means the portion of the combustion chamber in a boiler occupied by the flame envelope (other identical or similar definitions are provided in 40CFR60.661; 60.701; 264.1031).

flame zone (40CFR63.111) means the portion of the combustion chamber in a boiler occupied by the flame envelope.

flare (EPA-94/04): A control device that burns hazardous materials to prevent their release into the environment; may operate continuously or intermittently, usually on top of a stack.

flashoff area (40CFR60.311) means the portion of a surface coating operation between the coating application area and bake oven.

flashoff area (40CFR60.391) means the structure on automobile and light-duty truck assembly lines between the coating application system (dip tank or spray booth) and the bake oven.

flashoff area (40CFR60.441) means the portion of a coating line after the coating applicator and usually before the oven entrance.

flashoff area (40CFR60.451) means the portion of a surface coating line between the coating application station and the curing oven.

flashoff area (40CFR60.711-12) means the portion of a coating operation between the coating applicator and the drying oven where solvent begins to evaporate from the coated base film.

flashoff area (40CFR60.741) means the portion of a coating operation between the coating applicator and the drying oven where VOC (volatile organic compound) begins to evaporate from the coated substrate.

flashover (40CFR85.2122(a)(7)(ii)(A)) means the discharge of ignition voltage across the surface of the distributor cap and/or rotor rather than at the spark plug gap.

flashover (40CFR85.2122(a)(8)(ii)(F)) means the discharge of ignition voltage at any point other than at the spark plug gap.

flashover (40CFR85.2122(a)(9)(ii)(B)) means the discharge of ignition voltage across the coil (under CAA).

flat glass (40CFR60.291) means glass made of soda-lime recipe and produced into continuous flat sheets and other products listed in SIC 3211.

flat mill (40CFR420.71-d) means those steel hot forming operations that reduce heated slabs to plates, strip and sheet, or skelp.

flexible fuel vehicle (or engine) (40CFR86.090.2) means any motor vehicle (or motor vehicle engine) engineered and designed to be operated on a petroleum fuel, a methanol fuel, or any mixture of the two. Methanol-fueled vehicles that are only marginally functional when using gasoline (e.g., the engine has a drop in power output of more than 80 percent) are not flexible fuel vehicles (under the Clean Air Act).

flexible operation unit (40CFR63.101) means a chemical manufacturing process unit that manufactures different chemical products periodically by alternating raw materials or operating conditions. These units are also referred to as campaign plants or blocked operations.

flexible vinyl and urethane products (40CFR60.581) mean those products, except for resilient floor coverings (977 Standard Industry Code 3996) and flexible packaging, that are more than 50 micrometers (0.002 inches) thick, and that consist of or contain a vinyl or urethane sheet or a vinyl or urethane coated web.

flexographic printing (40CFR52.741) means the application of words, designs, and pictures to a substrate by means of a roll printing technique in which the pattern to be applied is raised above the printing roll and the image carrier is made of elastomeric materials.

flexographic printing line (40CFR52.741) means a printing line in which each roll printer uses a roll with raised areas for applying an image such as words, designs, or pictures to a substrate. The image carrier on the roll is made of rubber or other elastomeric material.

floating roof (40CFR52.741) means a roof on a stationary tank, reservoir, or other container which moves vertically upon change in volume of the stored material.

floating roof (40CFR60.111-j) means a storage vessel cover consisting of a double deck, pontoon single deck, internal floating cover or covered floating roof, which rests upon and is supported by the petroleum liquid being contained, and is equipped with a closure seal or seals to close the space between the roof edge and tank wall.

floating roof (40CFR60.691) means a pontoon-type or double-deck type cover that rests on the liquid surface.

floating roof (40CFR61.341) means a cover with certain rim sealing mechanisms consisting of a double deck, pontoon single deck, internal floating cover or covered floating roof, which rests upon and is supported by the liquid being contained, and is equipped with a closure seal or seals to close the space between the roof edge and unit wall (under CAA).

floating roof (40CFR63.111) means a cover consisting of a double deck, pontoon single deck, internal floating cover or covered floating roof, which rests upon and is supported by the liquid being contained, and is equipped with a closure seal or seals to close the space between the roof edge and waste management unit or storage vessel wall.

floc (EPA-94/04): A clump of solids formed in sewage by biological or chemical action.

flocculation (40CFR141.2) means a process to enhance agglomeration or collection of smaller floc particles into larger, more easily settleable particles through gentle stirring by hydraulic or mechanical means.

flocculation (EPA-94/04): Process by which clumps of solids in water or sewage aggregate through biological or chemical action so they can be separated from water or sewage.

flood or flooding (40CFR6-AA-c) means a general and temporary condition of partial or complete inundation of normally dry land areas from the overflow of inland and/or tidal waters, and/or runoff of surface waters from any source, or flooding from any other source.

floodplain (40CFR257.3.1-2) means the lowland and relatively flat areas adjoining inland and coastal waters, including flood-prone areas of offshore islands which are inundated by the base flood (under RCRA).

floodplain (40CFR6-AA-d) means the lowland and relatively flat areas adjoining inland and coastal waters and other floodprone areas such as offshore islands, including at a minimum, that area subject to a one percent or greater chance of flooding in any given year. The base floodplain shall be used to designate the 100-year floodplain (one percent chance floodplain). The critical action floodplain is defined as the 500-year floodplain (0.2 percent chance floodplain).

floodplain (EPA-94/04): The flat or nearly flat land along a river or stream or in a tidal area that is covered by water during a flood.

floodproofing (40CFR6-AA-e) means modification of individual structures and facilities, their sites, and their contents to protect against structural failure, to keep water out or to reduce effects of water entry.

floor insulation (40CFR248.4-p) means a material, primarily designed to resist heat flow, which is installed between the first level conditioned area of a building and an unconditioned basement, a crawl space, or the outside beneath it. Where the first level conditioned area of a building is on a ground level concrete slab, floor insulation includes such a material installed around the perimeter of or on the slab. In the case of mobile homes, floor insulation

also means skirting to enclose the space between the building and the ground.

floor sweep (EPA-94/04): Capture of heavier-than-air gases that collect at floor level.

flooring felt (40CFR763.163) means an asbestos-containing product which is made of paper felt intended for use as an underlayer for floor coverings, or to be bonded to the underside of vinyl sheet flooring.

flow channels (40CFR60.291) means appendages used for conditioning and distributing molten glass to forming apparatuses and are a permanently separate source of emissions such that no mixing of emissions occurs with emissions from the melter cooling system prior to their being vented to the atmosphere.

flow coating (40CFR60.311) means a method of applying coatings in which the part is carried through a chamber containing numerous nozzles which direct unatomized streams of coatings from many different angles onto the surface of the part.

flow indicator (40CFR264.1031) means a device that indicates whether gas flow is present in a vent stream.

flow indicator (40CFR60.611) means a device which indicates whether gas flow is present in a vent stream (other identical or similar definitions are provided in 40CFR60.661).

flow indicator (40CFR60.701) means a device which indicates whether gas flow is present in a line (other identical or similar definitions are provided in 40CFR63.111).

flow indicator (40CFR61.341) means a device which indicates whether gas flow is present in a line or vent system.

flow meter accuracy (40CFR72.2) means the closeness of the measurement made by a flow meter to the reference value of the fuel flow being measured, expressed as the difference between the measurement and the reference value.

flow monitor (40CFR72.2) means a component of

the continuous emission monitoring system that measures the volumetric flow of exhaust gas.

flow proportional composite sample (40CFR471.02-rr) is composed of grab samples collected continuously or discretely in proportion to the total flow at time of collection or to the total flow since collection of the previous grab sample. The grab volume or frequency of grab collection may be varied in proportion to flow.

flow rate (40CFR146.3) means the volume per time unit given to the flow of gases or other fluid substance which emerges from an orifice, pump, turbine or passes along a conduit or channel.

flow rate (EPA-94/04): The rate, expressed in gallons or liters-per-hour, at which a fluid escapes from a hole or fissure in a tank. Such measurements are also made of liquid waste, effluent, and surface water movement.

flow rate stability (40CFR53.43-d): Freedom from variation in the operating flow rate of the sampler under typical sampling conditions.

flow through (40CFR795.120) means a continuous or an intermittent passage of test solution or dilution water through a test chamber or a holding or acclimation tank, with no recycling.

flow through (40CFR797.1300-4) means a continuous or an intermittent passage of test solution or dilution water through a test chamber or culture tank with no recycling (other identical or similar definitions are provided in 40CFR797.1330-5).

flow through (40CFR797.1400-6) means a continuous or an intermittent passage of test solution or dilution water through a test chamber, or a holding or acclimation tank with no recycling (other identical or similar definitions are provided in 40CFR797.1930-2; 797.1950-3).

flow through (40CFR797.1600-8) refers to the continuous or very frequent passage of fresh test solution through a test chamber with no recycling.

flow through (40CFR797.1970-2) means a continuous passage of test solution or dilution water

through a test chamber, holding or acclimation tank with no recycling.

flow through process tank (40CFR280.12) is a tank that forms an integral part of a production process through which there is a steady, variable, recurring, or intermittent flow of materials during the operation of the process. Flow-through process tanks do not include tanks used for the storage of materials prior to their introduction into the production process or for the storage of finished products or byproducts from the production process.

flow through test (40CFR797.1350-7) is a toxicity test in which water is renewed contstantly in the test chambers, the test chemical being transported with the water used to renew the test medium (other identical or similar definitions are provided in 40CFR797.1440-7).

flowmeter (EPA-94/04): A gauge indicating the velocity of wastewater moving through a treatment plant or of any liquid moving through various industrial processes.

flue (40CFR72.2) means a conduit or duct through which gases or other matter are exhausted to the atmosphere.

flue gas (40CFR76.2) means the combustion products arising from the combustion of fossil fuel in a utility boiler.

flue gas (EPA-94/04): The air coming out of a chimney after combustion in the burner it is venting. It can include nitrogen oxides, carbon oxides, water vapor, sulfur oxides, particles and many chemical pollutants.

flue gas desulfurization (40CFR52.1881-ii) means any pollution control process which treats stationary source combustion flue gas to remove sulfur oxides (under CAA).

flue gas desulfurization (EPA-94/04): A technology that employs a sorbent, usually lime or limestone, to remove sulfur dioxide from the gases produced by burning fossil fuels. Flue gas desulfurization is current state-of-the art technology for major SO_2 emitters, like power plants.

fluid (40CFR144.3) means any material or substance which flows or moves whether in a semisolid, liquid, sludge, gas, or any other form or state.

fluid (40CFR146.3) means material or substance which flows or moves whether in a semisolid, liquid, sludge, gas, or any other form or state (other identical or similar definitions are provided in 40CFR147.2902).

fluid catalytic cracking unit (40CFR60.101-m) means a refinery process unit in which petroleum derivatives are continuously charged; hydrocarbon molecules in the presence of a catalyst suspended in a fluidized bed are fractured into smaller molecules, or react with a contact material suspended in a fluidized bed to improve feedstock quality for additional processing; and the catalyst or contact material is continuously regenerated by burning off coke and other deposits. The unit includes the riser, reactor, regenerator, air blowers, spent catalyst or contact material stripper, catalyst or contact material recovery equipment, and regenerator equipment for controlling air pollutant emissions and for heat recovery.

fluid catalytic cracking unit catalyst regenerator (40CFR60.101-n) means one or more regenerators (multiple regenerators) which comprise that portion of the fluid catalytic cracking unit in which coke burn-off and catalyst or contact material regeneration occurs, and includes the regenerator combustion air blower(s).

fluidized (EPA-94/04): A mass of solid particles that is made to flow like a liquid by injection of water or gas is said to have been fluidized. In water treatment, a bed of filter media is fluidized by backwashing water through the filter.

fluidized bed combustion technology (40CFR60.41b) means combustion of fuel in a bed or series of beds (including but not limited to bubbling bed units and circulating bed units) of limestone aggregate (or other sorbent materials) in which these materials are forced upward by the flow of combustion air and the gaseous products of combustion.

fluidized bed combustion technology (40CFR60.41c) means a device wherein fuel is distributed onto a bed

(or series of beds) of limestone aggregate (or other sorbent materials) for combustion; and these materials are forced upward in the device by the flow of combustion air and the gaseous products of combustion. Fluidized bed combustion technology includes, but is not limited to, bubbling bed units and circulating bed units.

fluidized bed incinerator (40CFR503.41-e) is an enclosed device in which organic matter and inorganic matter in sewage sludge are combusted in a bed of particles suspended in the combustion chamber gas.

fluidized bed incinerator (EPA-94/04): An incinerator that uses a bed of hot sand or other granular material to transfer heat directly to waste. Used mainly for destroying municipal sludge.

flume (EPA-94/04): A natural or man-made channel that diverts water.

fluorescent light ballast (40CFR761.3) means a device that electrically controls fluorescent light fixtures and that includes a capacitor containing 0.1 kg or less of dielectric.

fluoridation (EPA-94/04): The addition of a chemical to increase the concentration of fluoride ions drinking water to reduce the incidence of tooth decay in children.

fluorides (EPA-94/04): Gaseous, solid, or dissolved compounds containing fluorine that result from industrial processes. Excessive amounts in food can lead to fluorosis.

fluorocarbons (FCs) (EPA-94/04): Any of a number of organic compounds analogous to hydrocarbons in which one or more hydrogen atoms are replaced by fluorine. Once used in the United States as a propellant for domestic aerosols, they are now found mainly in coolants and some industrial processes. FCs containing chlorine are called chlorofluorocarbons (CFCs). They are believed to be modifying the ozone layer in the stratosphere, thereby allowing more harmful solar radiation to reach the Earth's surface.

flush (EPA-94/04):
1. To open a cold-water tap to clear out all the

water which may have been sitting for a long time in the pipes. In new homes, to flush a system means to send large volumes of water gushing through the unused pipes to remove loose particles of solder and flux.

2. To force large amounts of water through liquid to clean out piping or tubing, storage or process tanks.

flushing liquor circulation tank (40CFR61.131) means any vessel that functions to store or contain flushing liquor that is separated from the tar in the tar decanter and is recirculated as the cooled liquor to the gas collection system.

flux (EPA-94/04): A flowing or flow.

fly ash (40CFR240.101-h) means suspended particles, charred paper, dust, soot, and other partially oxidized matter carried in the products of combustion (cf. bottom ash).

fly ash (40CFR249.04-e) means the component of coal which results from the combustion of coal, and is the finely divided mineral residue which is typically collected from boiler stack gases by electrostatic precipitator or mechanical collection devices.

fly ash (40CFR423.11-e) means the ash that is carried out of the furnace by the gas stream and collected by mechanical precipitators, electrostatic precipitators, and/or fabric filters. Economizer ash is included when it is collected with fly ash.

fly ash (EPA-94/04): Non-combustible residual particles expelled by flue gas.

foam in place insulation (40CFR248.4-q) foam is rigid cellular foam produced by catalyzed chemical reactions that hardens at the site of the work. The term includes spray-applied and injected applications such as spray-in-place foam and pour-in-place.

fog coating (also known as mist coating and uniforming) (40CFR60.721) means a thin coating applied to plastic parts that have molded-in color or texture or both to improve color uniformity.

fogging (EPA-94/04): Applying a pesticide by rapidly heating the liquid chemical so that it forms very fine droplets that resemble smoke or fog. Used to destroy mosquitoes, black flies, and similar pests.

folding boxboard (40CFR250.4-q) means a paperboard suitable for the manufacture of folding cartons.

food (40CFR166.3-f) means any article used for food or drink for man or animals.

food (40CFR710.2-a) shall have the meanings contained in the Federal Food, Drug, and Cosmetic Act, 21USC321 et seq., and the regulations issued under such Act. The meaning of the term, food, defined under the Act was provided below:

Food (21USC321(a)(1)(f)) means:
(1) articles used for food or drink for man or other animals,
(2) chewing gum, and
(3) articles used for components of any such article (21USC321(a)(1)(f)).

In addition, the term food includes poultry and poultry products, as defined in the Poultry Products Inspection Act, 21USC453 et seq.; meats and meat food products, as defined in the Federal Meat Inspection Act, 21USC60 et seq.; and eggs and egg products, as defined in the Egg Products Inspection Act, 21USC1033 et seq. (other identical or similar definitions are provided in 40CFR720.3-a).

food (TSCA3-15USC2602) as used in clause (vi) of this subparagraph includes poultry and poultry products (as defined in sections 4(e) and 4(f) of the Poultry Products Inspection Act), meat and meat food products (as defined in section 1(j) of the Federal Meat Inspection Act), and eggs and egg products (as defined in section 4 of the Egg Products Inspection Act).

food additive (40CFR177.3) means any substance the intended use of which results or may reasonably be expected to result, directly or indirectly, in its becoming a component of or otherwise affecting the characteristics of any food (including any such substance intended for use in producing, manufacturing, packing, processing, preparing, treating, packaging, transporting, or holding food), except that such term does not include:
(1) A pesticide chemical in or on a raw agricultural commodity.

(2) A pesticide chemical to the extent that it is intended for use or is used in the production, storage, or transportation of any raw agricultural commodity.

(3) A color additive.

(4) Any substance used in accordance with a sanction or approval granted prior to September 6, 1958, pursuant to the FFDCA, the Poultry Products Inspection Act, or the Federal Meat Inspection Act.

(5) A new animal drug.

(6) A substance that is generally recognized, among experts qualified by scientific training and experience to evaluate its safety, as having been adequately shown through scientific procedures (or, in the case of a substance used in food prior to January, 1, 1958, through either scientific procedures or experience based on common use in food) to be safe under the conditions of its intended use.

food additive (40CFR710.2-a) shall have the meaning contained in the Federal Food, Drug, and Cosmetic Act (FFDCA), 21USC321 et seq., and the regulations issued under such Act. The meaning of the term, food additive, defined under the Act was provided below:

Food additive (21USC321(a)(1)(s)) means any substance the intended use of which results or may reasonably be expected to result, directly or indirectly, in its becoming a component or otherwise affecting the characteristics of any food (including any substance intended for use in producing, manufacturing, packing, processing, preparing, treating, packaging, transporting, or holding food; and including any source of radiation intended for any such use), if such substance is not generally recognized, among experts qualified by scientific training and experience to evaluate its safety, as having been adequately shown through scientific procedures (or, in the case of a substance used in food prior to January 1, 1958, through either scientific procedures or experience based on common use in food) to be safe under the conditions of its intended use; except that such term does not include:

(1) a pesticide chemical in or on a raw agricultural commodity; or

(2) a pesticide chemical to the extent that it is intended for use or is used in the production, storage, or transportation of any raw agricultural commodity; or

(3) a color additive; or

(4) any substance used in accordance with a sanction or approval granted prior to the enactment of this paragraph pursuant to this Act [enacted Sept. 6, 1958], the Poultry Products Inspection Act (21 U.S.C. 451 and the following) or the Meat Inspection Act of March 4, 1907 (34 Stat. 1260), as amended and extended (21 U.S.C. 71 and the following); or

(5) a new animal drug.

(Other identical or similar definitions are provided in 40CFR720.3-a.)

food additive regulation (40CFR177.3) means a regulation issued pursuant to FFDCA section 409 that states the conditions under which a food additive may be safely used. A food additive regulation under this part ordinarily establishes a tolerance for pesticide residues in or on a particular processed food or a group of such foods. It may also specify:

(1) The particular food or classes of food in or on which a food additive may be used.

(2) The maximum quantity of the food additive which may be used in or on such food.

(3) The manner in which the food additive may be added to or used in or on such food.

(4) Directions or other labeling or packaging requirements for the food additive (under FIFRA).

Food and Drug Administration Action Level (FDAAL) (40CFR300-AA): Under section 408 of the Federal Food, Drug and Cosmetic Act, as amended, concentration of a poisonous or deleterious substance in human food or animal feed at or above which FDA will take legal action to remove adulterated products from the market. Only FDAALs established for fish and shellfish apply in the HRS.

food chain (EPA-94/04): A sequence of organisms, each of which uses the next, lower member of the sequence as a food source.

food chain crops (40CFR257.3.5-4) means tobacco, crops grown for human consumption, and animal feed for animals whose products are consumed by humans.

food chain crops (40CFR260.10) means tobacco, crops grown for human consumption, and crops grown for feed for animals whose products are consumed by humans.

food crops (40CFR503.9-l) are crops consumed by humans. These include, but are not limited to, fruits, vegetables, and tobacco.

food waste (40CFR243.101-l) means the organic residues generated by the handling, storage, sale, preparation, cooking, and serving of foods; commonly called garbage (other identical or similar definitions are provided in 40CFR246.101-m).

force account work (40CFR30.200) means the use of the recipients own employees or equipment for construction, construction-related activities (including A and E services), or for repair or improvement to a facility.

forced outage (40CFR72.2) means the removal of a unit from service due to an unplanned component failure or other unplanned condition that requires such removal immediately or within 7 days from the onset of the unplanned component failure or condition. For purposes of 40CFR72.43, 72.91, and 72.92, "forced outage" also includes a partial reduction in the heat input or electrical output due to an unplanned component failure or other unplanned condition that requires such reduction immediately or within 7 days from the onset of the unplanned component failure or condition (under CAA).

forecast period (40CFR51.392) with respect to a transportation plan is the period covered by the transportation plan pursuant to 23CFR part 450 (other identical or similar definitions are provided in 40CFR93.101).

foreign awards (40CFR30.200) means an EPA award of assistance when all or part of the project is performed in a foreign country by:
(a) a U.S. recipient,
(b) a foreign recipient, or
(c) an international organization.

foreign commerce (ESA3-16USC1531) includes, among other things, any transaction:
(A) between persons within one foreign country;
(B) between persons in two or more foreign countries;
(C) between a person within the United States and a person in a foreign country; or
(D) between persons within the United States, where the fish and wildlife in question are moving in any country or countries outside the United States.

foreign offshore unit (OPA1001) means a facility which is located, in whole or in part, in the territorial sea or on the continental shelf of a foreign country and which is or was used for one or more of the following purposes: exploring for, drilling for, producing, storing, handling, transferring, processing, or transporting oil produced from the seabed beneath the foreign country's territorial sea or form the foreign country's continental shelf.

foreign state (40CFR82.3-s) means an entity which is recognized as a sovereign nation or country other than the United States of America.

foreign state not party to or non-party (40CFR82.3-t) means a foreign state that has not deposited instruments of ratification, acceptance, or other form of approval with the Directorate of the United Nations Secretariat, evidencing the foreign state's ratification of the provisions of the 1987 Montreal Protocol, the London Amendments, or of the Copenhagen Amendments, as specified.

forest (40CFR170.3) means any operation engaged in the outdoor production of any agricultural plant to produce wood fiber or timber products (under FIFRA).

forest (40CFR171.2-14) means a concentration of trees and related vegetation in non-urban areas sparsely inhabited by and infrequently used by humans; characterized by natural terrain and drainage patterns.

forest (40CFR503.11-g) is a tract of land thick with trees and underbrush.

forging (40CFR467.02-k) is the exertion of pressure

on dies or rolls surrounding heated aluminum stock, forcing the stock to change shape and in the case where dies are used to take the shape of the die. The forging subcategory is based on the forging process (under CWA).

forging (40CFR471.02-q) is deforming metal, usually hot, with compressive force into desired shapes, with or without dies. Where dies are used, the metal is forced to take the shape of the die.

form bond (40CFR250.4-r) means a lightweight commodity paper designed primarily for business forms including computer printout and carbonless paper forms (cf. manifold business forms).

formal amendment (40CFR30.200) means a written modification of an assistance agreement signed by both the authorized representative of the recipient and the award official.

formal hearing (40CFR124.2-c) means any evidentiary hearing under subpart E or any panel hearing under subpart F but does not mean a public hearing conducted under 40CFR124.12.

formaldehyde (EPA-94/04): A colorless, pungent, and irritating gas, CH_2O, used chiefly as a disinfectant and preservative and in synthesizing other compounds like resins.

formation (40CFR144.3) means a body of consolidated or unconsolidated rock characterized by a degree of lithologic homogeneity which is prevailingly, but not necessarily, tabular and is mappable on the earth's surface or traceable in the subsurface.

formation (40CFR146.3) means a body of rock characterized by a degree of lithologic homogeneity which is prevailingly, but not necessarily, tabular and is mappable on the earth's surface or traceable in the subsurface (other identical or similar definitions are provided in 40CFR147.2902).

formation fluid (40CFR144.3) means "fluid" present in a "formation" under natural conditions as opposed to introduced fluids, such as "drilling mud" (other identical or similar definitions are provided in 40CFR146.3).

former employee (40CFR3.102-c) means a former Environmental Protection Agency employee, or a former special Government employee.

forming (40CFR471.02-b) is a set of manufacturing operations in which metals and alloys are made into semifinished products by hot or cold working.

formula quantity (10CFR70.4) means strategic special nuclear material in any combination in a quantity of 5000 grams or more computed by the formula, grams=(grams contained U^{235}) + 2.5(grams U^{235} + grams plutonium).

formulation (40CFR158.153-c) means:
(1) The process of mixing, blending, or dilution of one or more active ingredients with one or more other active or inert ingredients, without an intended chemical reaction, to obtain manufacturing use product or an end use product, or
(2) The repackaging of any registered product.

formulation (EPA-94/04): The substances comprising all active and inert ingredients in a pesticide.

formulator (40CFR82.172) means any person engaged in the preparation or formulation of a substitute, after chemical manufacture of the substitute or its components, for distribution or use in commerce.

forward mutation (40CFR798.5140) is a gene mutation from the wild (parent) type to the mutant condition (cf. reverse mutation) (other identical or similar definitions are provided in 40CFR798.5250) (under TSCA).

forward mutation assay (40CFR798.5300-1) detects a gene mutation from the parental type to the mutant form which gives rise to a change in an enzymatic or functional protein.

fossil fuel (40CFR52.1881-iii) means natural gas, refinery fuel gas, coke oven gas, petroleum, coal and any form of solid, liquid, or gaseous fuel derived from such materials.

fossil fuel (40CFR60.161-i) means natural gas, petroleum, coal, and any form of solid, liquid, or gaseous fuel derived from such materials for the purpose of creating useful heat.

fossil fuel (40CFR60.41-b) means natural gas, petroleum, coal, and any form of solid, liquid, or gaseous fuel derived from such materials for the purpose of creating useful heat.

fossil fuel (40CFR60.41a-3) means natural gas, petroleum, coal, and any form of solid, liquid, or gaseous fuel derived from such materials for the purpose of creating useful heat.

fossil fuel (40CFR72.2) means natural gas, petroleum, coal, or any form of solid, liquid, or gaseous fuel derived from such material.

fossil fuel (EPA-94/04): Fuel derived from ancient organic remains, e.g., peat, coal, crude oil, and natural gas.

fossil fuel and wood residue-fired steam generating unit (40CFR60.41-d) means a furnace or boiler used in the process of burning fossil fuel and wood residue for the purpose of producing steam by heat transfer (under CAA).

fossil fuel-fired (40CFR72.2) means the combustion of fossil fuel or any derivative of fossil fuel, alone or in combination with any other fuel, independent of the percentage of fossil fuel consumed in any calendar year (expressed in mmBtu).

fossil fuel-fired steam generating unit (40CFR52.1881-iv) means a furnace or boiler used in the process of burning fossil fuel for the purpose of producing steam by heat transfer.

fossil fuel-fired steam generator (40CFR51.100-ee) means a furnace or boiler used in the process of burning fossil fuel for the primary purpose of producing steam by heat transfer.

fossil-fuel fired steam generating unit (40CFR60.41-a) means a furnace or boiler used in the process of burning fossil fuel for the purpose of producing steam by heat transfer.

foundation insulation (40CFR248.4-r) means a material, primarily designed to resist heat flow, which is installed in foundation walls between conditioned volumes and unconditioned volumes and the outside or surrounding earth, at the perimeters of concrete slab-on-grade foundations, and at common foundation wall assemblies between conditioned basement volumes.

foundry (40CFR61.31-f) means a facility engaged in the melting or casting of beryllium metal or alloy.

foundry coke (40CFR61.131) means coke that is produced from raw materials with less than 26 percent volatile material by weight and that is subject to a coking period of 24 hours or more. Percent volatile material of the raw materials (by weight) is the weighted average percent volatile material of all raw materials (by weight) charged to the coke oven per coking cycle.

foundry coke byproduct recovery plant (40CFR61.131) means a coke byproduct recovery plant connected to coke batteries whose annual coke production is at least 75 percent foundry coke.

foundry coke producer (40CFR63.301) means a coke producer that is not and was not on January 1, 1992, owned or operated by an integrated steel producer and had on January 1, 1992, an annual design capacity of less than 1.25 million megagrams per year (not including any capacity satisfying the requirements of 40CFR63.300(d)(2) or 40CFR63.304(b)(6)).

fountain solution (40CFR52.741) means the solution which is applied to the image plate to maintain hydrophilic properties of the non-image areas.

four hour block average or 4-hour block average (40CFR60.51a) means the average of all hourly emission rates when the affected facility is operating and combusting MSW measured over 4-hour periods of time from 12 midnight to 4 a.m., 4 a.m. to 8 a.m., 8 a.m. to 12 noon, 12 noon to 4 p.m., 4 p.m. to 8 p.m., and 8 p.m. to 12 midnight.

four wheel drive general utility vehicle (40CFR600.002.85-42) means a four-wheel-drive, general purpose automobile capable of off-highway operation that has a wheelbase not more than 110 inches and that has a body shape similar to a 1977

Jeep CJ-5 or CJ-7, or the 1977 Toyota Land Cruiser, as defined by the Secretary of Transportation at 49CFR 553.4.

fourteen day (14-day) old survivors (40CFR797.2130-vii) are birds that survive for weeks following hatch. Values are expressed both as a percentage of hatched eggs and as the number per pen per season (test) (other identical or similar definitions are provided in 40CFR797.2150-vii).

fractionation operation (40CFR264.1031) means a distillation operation or method used to separate a mixture of several volatile components of different boiling points in successive stages, each stage removing from the mixture some proportion of one of the components.

frameshift mutagens (40CFR798.5265-3) are agents which cause the addition or deletion of single or multiple base pairs in the DNA molecule (other identical or similar definitions are provided in 40CFR798.5300-3).

free available chlorine (40CFR423.11-l) shall mean the value obtained using the amperometric titration method for free available chlorine described in "Standard Methods for the Examination of Water and Wastewater," page 112 (13th edition).

free liquids (40CFR260.10) means liquids which readily separate from the solid portion of a waste under ambient temperature and pressure.

free moisture (40CFR240.101-i) means liquid that will drain freely by gravity from solid materials (other identical or similar definitions are provided in 40CFR241.101-e).

free product (40FR280.12) refers to a regulated substance that is present as a non-aqueous phase liquid (e.g., liquid not dissolved in water) (under RCRA).

free stall barn (40CFR412.11-i) shall mean specialized facilities wherein producing cows are permitted free movement between resting and feeding areas.

freeboard (40CFR260.10) means the vertical distance between the top of a tank or surface impoundment dike, and the surface of the waste contained therein.

freeboard (EPA-94/04):
1. Vertical distance from the normal water surface to the top of the confining wall.
2. The vertical distance from the sand surface to the underside of a trough in a sand filter.

freeze (40CFR52.1135-6) means to maintain at all times after October 15, 1973, the total quantity of commercial parking spaces available for use at the same amounts as were available for use prior to said date; provided, that such quantity may be increased by spaces the construction of which commenced prior to October 15, 1973, or as specifically permitted by paragraphs (n), (p) and (q) of this section; provided further that such additional spaces do not result in an increase of more than 10 percent in the total commercial parking spaces available for use on October 15, 1973, in any municipality within the freeze area or at Logan International Airport ("Logan Airport"). For purposes of the last clause of the previous sentence, the 10 percent limit shall apply to each municipality and Logan Airport separately (under CAA).

freeze area (40CFR52.1128-4) means that portion of the Boston Intrastate Region enclosed within the following boundaries. The City of Cambridge; that portion of the City of Boston from the Charles River and the Boston Inner Harbor on north and northeast of pier 4 on NOrthern Avenue; by the east side of pier 4 to B Street, B Street extension of B Street to B Street, B Street, Dorchester Avenue, and the Preble Street to Old Colony Avenue, then east to the water, then by the water's edge around Columbia Point on various courses generally easterly, southerly, and westerly to the center of the bridge on Morrissey Boulevard, on the east and southeast; then due west to Freeport Street, Freeport Street, Dorchester Avenue, Southeast Expressway, Southampton Street, Reading Street, Island Street, Chadwick Street, Carlow Street, Albany Street, Hunneman Street, Madison Street, Windsor Street, Cabot Street, Ruggles Street, Parker Street, Ward Street, Huntington Avenue, Brookline-Boston municipal boundary, Mountford Street to the Boston University Bridge on the southwest and west; and the Logan International Airport. Where a street or roadway forms a boundary the entire right-of-way of the street is within the freeze area as defined.

fresh feed (40CFR60.101-o) means any petroleum derivative feedstock stream charged directly into the riser or reactor of a fluid catalytic cracking unit except for petroleum derivatives recycled within the fluid catalytic cracking unit, fractionator, or gas recovery unit.

fresh granular triple superphosphate (40CFR60.241-d) means granular triple superphosphate produced no more than 10 days prior to the date of the performance test.

fresh water (40CFR147.2902) means underground source of drinking water.

fresh water (EPA-94/04): Water that generally contains less than 1,000 milligrams per liter of dissolved solids.

freshwater lake (40CFR35.1605-2): Any inland pond, reservoir, impoundment, or other similar body of water that has recreational value, that exhibits no oceanic and tidal influences, and that has a total dissolved solids concentration of less than 1 percent (cf. eutrophic lake).

friable (40CFR763.83) when referring to material in a school building means that the material, when dry, may be crumbled, pulverized, or reduced to powder by hand pressure, and includes previously nonfriable material after such previously nonfriable material becomes damaged to the extent that when dry it may be crumbled, pulverized, or reduced to powder by hand pressure.

friable (EPA-94/04): Capable of being crumbled, pulverized, or reduced to powder by hand pressure.

friable asbestos (EPA-94/04): Any material containing more than one percent asbestos, and that can be crumbled or reduced to powder by hand pressure. (May include previously non-friable material which becomes broken or damaged by mechanical force.)

friable asbestos containing material (TSCA202-15USC2642) means any asbestos-containing material

applied on ceiling, walls, structures members, piping, duct work, or any other part of a building which when dry may be crumbled, pulverized, or reduced to powder by hand pressure. The term includes non-friable asbestos-containing material after such previously non-friable material becomes damaged to the extent that when dry it may be crumbled, pulverized, or reduced to powder by hand pressure.

friable asbestos material (40CFR61.141) means any material containing more than 1 percent asbestos as determined using the method specified in appendix A, subpart F, 40CFR763, section 1, Polarized Light Microscopy, that, when dry, cannot be crumbled, pulverized, or reduced to powder by hand pressure. If the asbestos content is less than 10 percent as determined by a method other than point counting by polarized light microscopy (PLM), verify the asbestos content by point counting using PLM.

friable asbestos material (40CFR763.121) means any material containing more than 1 percent asbestos by weight which, when dry, may be crumbled, pulverized or reduced to powder by hand pressure.

friable material (40CFR763.103-d) means any material applied onto ceilings, walls, structural members, piping, ductwork, or any other part of the building structure which, when dry, may be crumbled, pulverized, or reduced to powder by hand pressure.

frond (40CFR797.1160-4) means a single Lemna leaf-like structure.

frond mortality (40CFR797.1160-5) means dead fronds which may be identified by a total discoloration (yellow, white, black or clear) of the entire frond.

FTP (40CFR85.1502): The federal test procedure at part 86.

fuel (40CFR600.002.85-9) means:
(i) gasoline and diesel fuel for gasoline-or diesel-powered automobiles or
(ii) electrical energy for electrically powered automobiles.

fuel (40CFR79.2-c) means any material which is

capable of releasing energy or power by combustion or other chemical or physical reaction.

fuel combustion emission source (40CFR52.741) means any furnace, boiler, or similar equipment used for the primary purpose of producing heat or power by indirect heat transfer.

fuel economy (40CFR2.311-3) has the meaning given it in section 501(6) of the Act, 15USC2001(6).

fuel economy (40CFR600.002.85-10) means:
(i) the average number of miles traveled by an automobile or group of automobiles per gallon of gasoline or diesel fuel consumed as computed in 40CFR600.113 or 40CFR600.207 or
(ii) the equivalent petroleum-based fuel economy for an electrically powered automobile as determined by the Secretary of Energy.

fuel economy (40CFR610.11-3) means the average number of miles traveled by an automobile per gallon of gasoline (or equivalent amount of other fuel) consumed, as determined by the Administrator in accordance with procedures established under subparts D or F.

fuel economy basic engine (40CFR86.902.93) means a unique combination of manufacturer, engine displacement, number of cylinders, fuel system, catalyst usage, and other characteristics specified by the Administrator.

fuel economy data (40CFR2.311-4) means any measurement or calculation of fuel economy for any model type and average fuel economy of a manufacturer under section 503(d) of the Act, 15USC2003(d).

fuel economy data vehicle (40CFR600.002.85-16) means a vehicle used for the purpose of determining fuel economy which is not a certification vehicle.

fuel economy standard (EPA-94/04): The Corporate Average Fuel Economy Standard (CAFE) effective in 1978. It enhanced the national fuel conservation effort imposing a miles-per-gallon floor for motor vehicles.

fuel efficiency (EPA-94/04): The proportion of the energy released on combustion of a fuel that is converted into useful energy.

fuel evaporative emissions (40CFR86.082.2) means vaporized fuel emitted into the atmosphere from the fuel system of a motor vehicle (under the Clean Air Act).

fuel gas (40CFR60.101-d) means any gas which is generated at a petroleum refinery and which is combusted. Fuel gas also includes natural gas when the natural gas is combined and combusted in any proportion with a gas generated at a refinery. Fuel gas does not include gases generated by catalytic cracking unit catalyst regenerators and fluid coking burners.

fuel gas combustion device (40CFR60.101-g) means any equipment, such as process heaters, boilers and flares used to combust fuel gas, except facilities in which gases are combusted to produce sulfur or sulfuric acid.

fuel gas system (40CFR52.741) means a system for collection of refinery fuel gas including, but not limited to, piping for collecting tail gas from various process units, mixing drums and controls, and distribution piping.

fuel manufacturer (40CFR79.2-d) means any person who, for sale or introduction into commerce, produces or manufactures a fuel or causes or directs the alteration of the chemical composition of, or the mixture of chemical compounds in, a bulk fuel by adding to it an additive.

fuel oil (40CFR72.2) means any petroleum-based fuel (including diesel fuel or petroleum derivatives such as oil tar) as defined by the American Society for Testing and Materials in ASTM D396-90a, "Standard Specification for Fuel Oils" (incorporated by reference in 40CFR72.13), and any recycled or blended petroleum products or petroleum byproducts used as a fuel whether in a liquid, solid or gaseous state; provided that for purposes of the monitoring requirements of part 75 of this chapter, "fuel oil" shall be limited to the petroleum-based fuels for which applicable ASTM methods are specified in Appendices D, E, or F of part 75 of this chapter.

fuel pretreatment (40CFR60.41b) means a process that removes a portion of the sulfur in a fuel before combustion of the fuel in a steam generating unit (other identical or similar definitions are provided in 40CFR60.41c).

fuel supply agreement (40CFR72.2) means a legally binding agreement between a new IPP or a firm associated with a new IPP and a fuel supplier that establishes the terms and conditions under which the fuel supplier commits to provide fuel to be delivered to the new IPP.

fuel switching (EPA-94/04):
1. A precombustion process whereby a low-sulfur coal is used in place of a higher sulfur coal in a power plant to reduce sulfur dioxide emissions.
2. Illegally using leaded gasoline in a car designed to use only unleaded.

fuel system (40CFR86.082.2) means the combination of fuel tank(s), fuel pump, fuel lines, and carburetor or fuel injection components, and includes all fuel system vents and fuel evaporative emission control system components.

fuel system (40CFR86.402.78) means the combination of fuel tank, fuel pump, fuel lines, oil injection metering system, and carburetor or fuel injection components, and includes all fuel system vents.

fuel venting emissions (40CFR87.1) means raw fuel, exclusive of hydrocarbons in the exhaust emissions, discharged from aircraft gas turbine engines during all normal ground and tight operations.

fugitive dust, mist or vapor (40CFR129.2-p) means dust, mist or vapor containing a toxic pollutant regulated under this part which is emitted from any source other than through a stack.

fugitive emission (40CFR60.301-h) means the particulate matter which is not collected by a capture system and is released directly into the atmosphere from an affected facility at a grain elevator (under CAA).

fugitive emission (40CFR60.671) means particulate matter that is not collected by a capture system and

is released to the atmosphere at the point of generation.

fugitive emissions (40CFR51.165-ix) means those emissions which could not reasonably pass through a stack, chimney, vent or other functionally equivalent opening.

fugitive emissions (40CFR51.166-20) means those emissions which could not reasonably pass through a stack, chimney, vent, or other functionally equivalent opening.

fugitive emissions (40CFR51.301-j) means those emissions which could not reasonably pass through a stack, chimney, vent, or other functionally equivalent opening (other identical or similar definitions are provided in 40CFR51-AS-9).

fugitive emissions (40CFR52.21-20) means those emissions which could not reasonably pass through a stack, chimney, vent, or other functionally equivalent opening.

fugitive emissions (40CFR52.24-9) means those emissions which could not reasonably pass through a stack, chimney, vent, or other functionally equivalent opening.

fugitive emissions (40CFR57.103-m) means any air pollutants emitted to the atmosphere other than from a stack.

fugitive emissions (40CFR60-AA (method 22-3.3)) means pollutant generated by an affected facility which is not collected by a capture system and is released to the atmosphere.

fugitive emissions (40CFR63.2) means those emissions from a stationary source that could not reasonably pass through a stack, chimney, vent, or other functionally equivalent opening. Under section 112 of the Act, all fugitive emissions are to be considered in determining whether a stationary source is a major source.

fugitive emissions (40CFR70.2) are those emissions which could not reasonably pass through a stack, chimney, vent, or other functionally equivalent opening.

fugitive emissions (EPA-94/04): Emissions not caught by a capture system.

fugitive emissions equipment (40CFR60.561) means each pump, compressor, pressure relief device, sampling connection system, open-ended valve or line, valve, and flange or other connector in VOC service and any devices or systems required by subpart VV of this part.

fugitive source (40CFR61.141) means any source of emissions not controlled by an air pollution control device.

fugitive volatile organic compounds (40CFR60.441) means any volatile organic compounds which are emitted from the coating applicator and flashoff areas and are not emitted in the oven.

full capacity (40CFR60.41b) means operation of the steam generating unit at 90 percent or more of the maximum steady-state design heat input capacity.

full time employee (40CFR372.3) means 2,000 hours per year of full time equivalent employment. A facility would calculate the number of full time employees by totaling the hours worked during the calendar year by all employees, including contract employees, and dividing that total by 2,000 hours.

full time fellow (40CFR46.120): An individual enrolled in an academic educational program directly related to pollution abatement and control, and taking a minimum of 30 credit hours or an academic workload otherwise defined by the institution as a full time curriculum for a school year. The fellow need not be pursuing a degree.

fume (EPA-94/04): Tiny particles trapped in vapor in a gas stream.

fume scrubber (40CFR420.121-d) means wet air pollution control devices used to remove and clean fumes originating from hot coating operations.

fume scrubber (40CFR420.91-d) means those pollution control devices used to remove and clean fumes originating in pickling operations.

fume suppression system (40CFR60.141a) means the equipment comprising any system used to inhibit the generation of emissions from steelmaking facilities with an inert gas, flame, or steam blanket applied to the surface of molten iron or steel.

fumigant (40CFR170.3) means any pesticide product that is a vapor or gas, or forms a vapor or gas on application, and whose method of pesticidal action is through the gaseous state.

fumigant (EPA-94/04): A pesticide vaporized to kill pests. Used in buildings and greenhouses.

functional equivalent (EPA-94/04): Term used to describe EPA's decision-making process and its relationship to the environmental review conducted under the National Environmental Policy Act (NEPA). A review is considered functionally equivalent when it addresses the substantive components of a NEPA review.

functional space (40CFR763.83) means a room, group of rooms, or homogeneous area (including crawl spaces or the space between a dropped ceiling and the floor or roof deck above), such as classroom(s), a cafeteria, gymnasium, hallway(s), designated by a person accredited to prepare management plans, design abatement projects, or conduct response actions.

functionally equivalent component (40CFR270.2) means a component which performs the same function or measurement and which meets or exceeds the performance specifications of another component.

fund (CZMA304-16USC1453) means the Coastal Energy Impact Fund established by section 1456a(h) of this title.

fund (OPA1001) means the Oil Spill Liability Trust Fund, established by section 9509 of the Internal Revenue code of 1986 (26USC9509).

fund (SMCRA701-30USC1291) means the Abandoned Mine Reclamation Fund established pursuant to section 1231 of this title.

fund or trust fund (40CFR300.5) means the Hazardous Substance Superfund established by section 9507 of the Internal Revenue Code of 1986 (under CERCLA).

fund or trust fund (SF101-42USC9601) means the Hazardous Substance Response Fund established by section 221 of this Act or, in the case of a hazardous waste disposal facility for which liability has been transferred under section 107(k) of this Act, the Post-closure Liability Fund established by section 232 of this Act.

funds directly made available by capitalization grants (40CFR35.3105-g) means funds equaling the amount of the grant.

fungi (EPA-94/04): (Singular: Fungus) Molds, mildews, yeasts, mushrooms, and puffballs, a group organisms lacking in chlorophyll (i.e., are not photosynthetic) and which are usually non-mobile, filamentous, and multicellular. Some grow in soil, others attach themselves to decaying trees and other plants whence they obtain nutrients. Some are pathogens, others stabilize sewage and digest composted waste.

fungicide (EPA-94/04): Pesticides which are used to control, deter, or destroy fungi.

fungistat (EPA-94/04): A chemical that keeps fungi from growing.

fungus (FIFRA2-7USC136) means any non-chlorophyll-bearing thallophyte (that is, any non-chlorophyll-bearing plant of a lower order than mosses and liverworts), as for example, rust, smut, mildew, mold, yeast, and bacteria, except those on or in living man or other animals and those on or in processed food, beverage, or pharmaceuticals.

furnace (40CFR240.101-j) means the chambers of the combustion train where drying, ignition, and combustion of waste material and evolved gases occur.

furnace (40CFR60.531) means a solid fuel burning appliance that is designed to be located outside the ordinary living areas and that warms spaces other than the space where the appliance is located, by the distribution of air heated in the appliance through ducts. The appliance must be tested and listed as a furnace under accepted American or Canadian safety testing codes unless exempted from this provision by the Administrator. A manufacturer may request an exemption in writing from the Administrator by stating why the testing and listing requirement is not practicable and by demonstrating that his appliance is otherwise a furnace.

furnace charge (40CFR60.261-b) means any material introduced into the electric submerged arc furnace, and may consist of, but is not limited to, ores, slag, carbonaceous material, and limestone.

furnace coke (40CFR61.131) means coke produced in byproduct ovens that is not foundry coke.

furnace coke byproduct recovery plant (40CFR61.131) means a coke byproduct recovery plant that is not a foundry coke byproduct recovery plant.

furnace cycle (40CFR60.261-g) means the time period from completion of a furnace product tap to the completion of the next consecutive product tap (under CAA).

furnace power input (40CFR60.261-j) means the resistive electrical power consumption of an electric submerged arc furnace as measured in kilowatts (under CAA).

furnace pull (40CFR426.81-b) shall mean that amount of glass drawn from the glass furnace or furnaces (other identical or similar definitions are provided in 40CFR426.101-b; 426.111-b; 426.121-b) (under CWA).

furrow irrigation (EPA-94/04): Irrigation method in which water travels through the field by means of small channels between each row or groups of rows.

future liability (EPA-94/04): Refers to potentially responsible parties' obligations to pay for additional response activities beyond those specified in the Record of Decision or Consent Decree.

future year subaccount (40CFR72.2) means a subaccount in an Allowance Tracking System account, established by the Administrator pursuant to 40CFR73.31 of this chapter, in which allowances are held for one of the 30 years following the later of 1995 or a current calendar year following 1995.

FWPCA: See act or FWPCA.

************ GGGGG ************

G1 (generation 1) (40CFR797.1950-4) means those mysids which are used to begin the test, also referred to as adults.

G2 (generation 2) (40CFR797.1950-4) are the young produced by G1.

galvanized basis material (40CFR465.02-f) means zinc coated steel, galvalum, brass and other copper base strip which is processed in coil coating (under CWA).

galvanizing (40CFR420.121-a) means coating steel products with zinc by the hot dip process including the immersion of the steel product in a molten bath of zinc metal, and the related operations preceding and subsequent to the immersion phase.

game fish (EPA-94/04): Species like trout, salmon, or bass, caught for sport. Many of them show more sensitivity to environmental change than "rough" fish (under rough).

gap location (40CFR85.2122(a)(8)(ii)(D)) means the position of the electrode gap in the combustion chamber.

gap spacing (40CFR85.2122(a)(8)(ii)(C)) means the distance between the center electrode and the ground electrode where the high voltage ignition arc is discharged.

garbage (EPA-94/04): Animal and vegetable waste resulting from the handling, storage, sale, preparation, cooking, and serving of foods.

garrison facility (40CFR60.331-o) means any permanent military installation.

gas chromatograph/mass spectrometer (EPA-94/04): Highly sophisticated instrument that identifies the molecular composition and concentrations of various chemicals in water and soil samples.

gas fired (40CFR72.2) means the combustion of: Natural gas, or a coal-derived gaseous fuel with a sulfur content no greater than natural gas, for at least 90 percent of the average annual heat input during the previous three calendar years and for at least 85 percent of the annual heat input in each of those calendar years; and any fuel other than coal or any other coal-derived fuel for the remaining heat input, if any; provided that for purposes of the monitoring exceptions of part 75 of this chapter, the supplemental fuel used in addition to natural gas, if any, shall be limited to gaseous fuels containing no more sulfur than natural gas or, as specified in the monitoring exceptions, fuel oil.

gas/gas method (40CFR52.741) means either of two methods for determining capture which rely only on gas phase measurements. The first method requires construction of a temporary total enclosure (TTE) to ensure that all would-be fugitive emissions are measured. The second method uses the building or room which houses the facility as an enclosure. The second method requires that all other VOM sources within the room be shut down while the test is performed, but all fans and blowers within the room must be operated according to normal procedures.

gas phase process (40CFR60.561) means a polymerization process in which the polymerization reaction is carried out in the gas phase; i.e., the monomer(s) are gases in a fluidized bed of catalyst particles and granular polymer.

gas service (40CFR52.741) means that the component contains process fluid that is in the gaseous state at operating conditions.

gas tight (40CFR60.691) means operated with no detectable emissions.

gas turbine model (40CFR60.331-p) means a group of gas turbines having the same nominal air flow, combuster inlet pressure, combuster inlet temperature, firing temperature, turbine inlet temperature and turbine inlet pressure.

gas well (40CFR435.61-d) shall mean any well which produces natural gas in a ratio to the petroleum liquids produced greater than 15,000 cubic feet of gas per 1 barrel (42 gallons) of petroleum liquids.

gasification (EPA-94/04): Conversion of solid material such as coal into a gas for use as a fuel.

gasohol (EPA-94/04): Mixture of gasoline and ethanol derived from fermented agricultural products containing at least nine percent ethanol. Gasohol emissions contain less carbon monoxide than those from gasoline.

gasoline (40CFR52.1101) means any petroleum distillate having a Reid vapor pressure of 4 pounds or greater (other identical or similar definitions are provided in 40CFR52.1102; 52.2042; 52.2438; 52.2439).

gasoline (40CFR52.2285-1) means any petroleum distillate having a Reid vapor pressure of 4 pounds or greater which is produced for use as a motor fuel and is commonly called gasoline (other identical or similar definitions are provided in 40CFR52.2286-1).

gasoline (40CFR52.741) means any petroleum distillate or petroleum distillate/alcohol blend having a Reid vapor pressure of 27.6 kPa or greater which is used as a fuel for internal combustion engines (under CAA).

gasoline (40CFR60.501) means any petroleum distillate or petroleum distillate/alcohol blend having a Reid vapor pressure of 27.6 kilopascals or greater which is used as a fuel for internal combustion engines.

gasoline (40CFR60-AA (method 27-2.1)) means any petroleum distillate or petroleum distillate/alcohol blend having a Reid vapor pressure of 27.6 kilopascals or greater which is used as a fuel for internal combustion engines.

gasoline (40CFR80.2) means any fuel sold in any State (see note 1) for use in motor vehicles and motor vehicle engines, and commonly or commercially known or sold as gasoline.

gasoline (40CFR80.2-c) means any fuel sold in any State{1} for use in motor vehicles and motor vehicle engines, and commonly or commercially known or sold as gasoline. {1}State means a State, the District of Columbia, the Commonwealth of Puerto Rico, the Virgin Islands, Guam, and American Samoa.

gasoline blending stock or component (40CFR80.2) means any liquid compound which is blended with other liquid compounds or with lead additives to produce gasoline.

gasoline blending stock or component (40CFR80.2-s) means any liquid compound which is blended with other liquid compounds or with lead additives to produce gasoline.

gasoline dispensing facility (40CFR52.741) means any site where gasoline is transferred from a stationary storage tank to a motor vehicle gasoline tank used to provide fuel to the engine of that motor vehicle.

gasoline service stations (40CFR60.111b-e) means any site where gasoline is dispensed to motor vehicle fuel tanks from stationary storage tanks.

gasoline tank truck (40CFR60.501) means a delivery tank truck used at bulk gasoline terminals which is loading gasoline or which has loaded gasoline on the immediately previous load.

gasoline volatility (EPA-94/04): The property of gasoline whereby it evaporates into a vapor. Gasoline vapor is a volatile organic compound.

gathering lines (40CFR280.12) means any pipeline, equipment, facility, or building used in the transportation of oil or gas during oil or gas production or gathering operations.

gear oils (40CFR252.4-d) means petroleum-based oils used for lubricating machinery gears.

general account (40CFR72.2) means an Allowance Tracking System account that is not a unit account.

general counsel (40CFR15.4) means the General

Counsel of the U.S. Environmental Protection Agency, or his or her designee.

general counsel (40CFR23.1-c) means the General Counsel of EPA or any official exercising authority delegated by the General Counsel.

general environment (40CFR190.02-c) means the total terrestrial, atmospheric and aquatic environments outside sites upon which any operation which is part of a nuclear fuel cycle is conducted.

general environment (40CFR191.02-o) means the total terrestrial, atmospheric, and aquatic environments outside sites within which any activity, operation, or process associated with the management and storage of spent nuclear fuel or radioactive waste is conducted.

general permit (40CFR122.2) means an NPDES "permit" issued under 40CFR122.28 authorizing a category of discharges under the CWA within a geographical area.

general permit (40CFR232.2) means a permit authorizing a category of discharges of dredged or fill material under the Act. General permits are permits for categories of discharge with are similar in nature, will cause only minimal adverse environmental effects when performed separately, and will have only minimal cumulative adverse effect on the environment.

general permit (40CFR70.2) means a part 70 permit that meets the requirements of 40CFR70.6(d).

general permit (EPA-94/04): A permit applicable to a class or category of dischargers.

General permit (NPDES and 404) (40CFR124.2) means an NPDES or 404 "permit" authorizing a category of discharges or activities under the CWA within a geographical area. For NPDES, a general permit means a permit issued under 40CFR122.28. For 404, a general permit means a permit issued under 40CFR233.37.

general processing (40CFR410.31-a) shall mean the internal subdivision of the low water use processing subcategory for facilities described in 40CFR410.30

that do not qualify under the water jet weaving subdivision.

general purpose unit of local government (40CFR310.11-c) means the governing body of a county, parish, municipality, city, town, township, Federally recognized Indian tribe or similar governing body.

general reporting facility (EPA-94/04): A facility having one or more hazardous chemicals above the 10,000 pound threshold for planning quantities. Such facilities must file MSDS and emergency inventory information with the SERC and LEPC and local fire departments.

generation (40CFR243.101-m) means the act or process of producing solid waste (other identical or similar definitions are provided in 40CFR246.101-n) (under RCRA).

generation 1 (40CFR797.1950-4): See G1.

generation 2 (40CFR797.1950-4): See G2.

generator (40CFR144.3) means any person, by site location, whose act or process produces hazardous waste identified or listed in 40CFR261 (other identical or similar definitions are provided in 40CFR146.3; 259.10.a; 260.10; 270.2).

generator (40CFR72.2) means a device that produces electricity and was or would have been required to be reported as a generating unit pursuant to the United States Department of Energy Form 860 (1990 edition).

generator (CAA402-42USC7651a) means a device that produces electricity and which is reported as a generating unit pursuant to Department of Energy Form 860.

generator (EPA-94/04):
1. A facility or mobile source that emits pollutants into the air or releases hazardous waste into water or soil.
2. Any person, by site, whose act or process produces regulated medical waste or whose act first causes such waste to become subject to regulation. In a case where more than one

person (e.g., doctors with separate medical practices) is located in the same building, each business entity is a separate generator.

generator column (40CFR796.1720-ii) is used to partition the test substance between the octanol and water phases.

generator column (40CFR796.1860-v) is used to produce or generate saturated solutions of a solute in a solvent. The column (see Figure 1 under paragraph (b)(l)(i)(A) of this section) is packed with a solid support coated with the solute, i.e., the organic compound whose solubility is to be determined. When water (the solvent) is pumped through the column, saturated solutions of the solute are generated. Preparation of the generator column is described under paragraph (b)(l)(i) of this section.

generator of PCB waste (40CFR761.3) means any person whose act or process produces PCBs that are regulated for disposal under subpart D of this part, or whose act first causes PCBs or PCB Items to become subject to the disposal requirements of subpart D of this part, or who has physical control over the PCBs when a decision is made that the use of the PCBs has been terminated and therefore is subject to the disposal requirements of subpart D of this part. Unless another provision of this part specifically requires a site-specific meaning, "generator of PCB waste" includes all of the sites of PCB waste generation owned or operated by the person who generates PCB waste.

generator output capacity (40CFR72.2) means the full-load continuous rating of a generator under specific conditions as designed by the manufacturer (under CAA).

genetic engineering (EPA-94/04): A process of inserting new genetic information into existing cells in order to modify any organism for the purpose of changing one of its characteristics.

geographic information system (GIS) (EPA-94/04): A computer system designed for storing, manipulating, analyzing, and displaying data in a geographic context.

geological log (EPA-94/04): A detailed description of all underground features (depth, thickness, type of formations) discovered during the drilling of a well.

geometric mean (40CFR131.35.d-6) means the nth root of a production of n factors.

geometric mean diameter or median diameter (40CFR798.1150-3) is the calculated aerodynamic diameter which divides the particles of an aerosol in half based on the weight of the particles. Fifty percent of the particles by weight will be larger than the median diameter and 50 percent of the particles will be smaller than the median diameter. The median diameter and its geometric standard deviation are used to statistically describe the particle size distribution of any aerosol based on the weight and size of the particles (other identical or similar definitions are provided in 40CFR798.4350-3).

geometric mean diameter or the median diameter (40CFR798.2450-3) is the calculated aerodynamic diameter which divides the particles of an aerosol in half based on the weight of the particles. Fifty percent of the particles by weight will be larger than the median diameter and 50 percent of the particles will be smaller than the median diameter. The median diameter describes the particle size distribution of any aerosol based on the weight and size of the particles.

geophysical log (EPA-94/04): A record of the structure and composition of the earth encountered when drilling a well or similar type of test hold or boring.

GEP stack height: (40FR51.100-ii) See good engineering practice stack height.

germ line (40CFR798.5195-2) is comprised of the cells in the gonads of higher eukaryotes, which are the carriers of the genetic information for the species.

germ line (40CFR798.5200-2) is the cells in the gonads of higher eukaryotes which are the carriers of the genetic information for the species (under TSCA).

germicide (EPA-94/04): Any compound that kills disease-causing microorganisms.

germination (40CFR797.2750-3) means the resumption of active growth by an embryo. The primary root should attain a length of 5 mm for the seed to be counted as having germinated.

germination (40CFR797.2800-2) means the resumption of active growth by an embryo.

giardia lamblia (EPA-94/04): Protozoan in the feces of man and animals that can cause severe gastrointestinal diseases when it contaminates drinking water.

gigawatt year (40CFR190.02-j) refers to the quantity of electrical energy produced at the busbar of a generating station. A gigawatt is equal to one billion watts. A gigawatt-year is equivalent to the amount of energy output represented by an average electric power level of one gigawatt sustained for one year.

glass fiber reinforced polyisocyanurate/polyurethane foam (40CFR248.4-s) means cellular poly-isocyanurate or cellular polyurethane insulation made with glass fibers within the foam core.

glass melting furnace (40CFR60.291) means a unit comprising a refractory vessel in which raw materials are charged, melted at high temperature, refined, and conditioned to produce molten glass. The unit includes foundations, superstructure and retaining walls, raw material charger systems, heat exchangers, melter cooling system, exhaust system, refractory brick work, fuel supply and electrical boosting equipment, integral control systems and instrumentation, and appendages for conditioning and distributing molten glass to forming apparatuses. The forming apparatuses, including the float bath used in flat glass manufacturing and flow channels in wool fiberglass and textile fiberglass manufacturing, are not considered part of the glass melting furnace.

glass melting furnace (40CFR61.161) means a unit comprising a refractory vessel in which raw materials are charged, melted at high temperature, refined, and conditioned to produce molten glass. The unit includes foundations, superstructure and retaining walls, raw material charger systems heat exchangers, melter cooling system, exhaust system, refractory brick work, fuel supply and electrical boosting equipment, integral control systems and

instrumentation, and appendages for conditioning and distributing molten glass to forming apparatuses. The forming apparatuses, including the float bath used in flat glass manufacturing, are not considered part of the glass melting furnace.

glass produced (40CFR60.291) means the weight of the glass pulled from the glass melting furnace.

glass produced (40CFR61.161) means the glass pulled from the glass melting furnace.

glass pull rate (40CFR60.681) means the mass of molten glass utilized in the manufacture of wool fiberglass insulation at a single manufacturing line in a specified time period.

global commons (40CFR6.1003) is that area (land, air, water) outside the jurisdiction of any nation.

glove bag (40CFR61.141) means a sealed compartment with attached inner gloves used for the handling of asbestos-containing materials. Properly installed and used, glove bags provide a small work area enclosure typically used for small-scale asbestos stripping operations. Information on glove-bag installation, equipment and supplies, and work practices is contained in the Occupational Safety and Health Administration's (OSHA's) final rule on occupational exposure to asbestos (appendix G to 29CFR1926.58).

glovebag (EPA-94/04): A polyethylene or polyvinyl chloride bag-like enclosure affixed around an asbestos-containing source (most often thermal system insulation) permitting the material to be removed while minimizing release of airborne fibers in the surrounding atmosphere.

good engineering practice (GEP) stack height (40CFR51.100-ii) means the greater of:
(1) 65 meters, measured from the ground-level elevation at the base of the stack:
(2) (i) For stacks in existence on January 12, 1979, and for which the owner or operator had obtained all applicable permits or approvals required under 40CFR51 and 52. H(g) = 2.5H, provided the owner or operator produces evidence that this equation was actually relied on in

establishing an emission limitation:

(ii) For all other stacks. H(g) = H + 1.5L, where H(g) = good engineering practice stack height, measured from the ground-level elevation at the base of the stack. H = height of nearby structure(s) measured from the ground-level elevation at the base of the stack. L = lesser dimension, height or projected width, of nearby structure(s) provided that the EPA, State or local control agency may require the use of a field study or fluid model to verify GEP stack height for the source; or

(3) The height demonstrated by a fluid model or a field study approved by the EPA State or local control agency, which ensures that the emissions from a stack do not result in excessive concentrations of any air pollutant as a result of atmospheric downwash, wakes, or eddy effects created by the source itself, nearby structures or nearby terrain features.

gooseneck (EPA-94/04): A portion of a service connection between the distribution system water main and a meter. Sometimes called a pigtail.

government (40CFR27.2) means the United States Government.

government (40CFR31.3) means a State or local government or a federally recognized Indian tribal government.

government (40CFR8.2-m) means the Government of the United States of America.

government agency (10CFR20.3) means any executive department, commission, independent establishment, corporation, wholly or partly owned by the United States of America which is an instrumentality of the United States, or any board, bureau, division, service, office, officer, authority, administration, or other establishment in the executive branch of the Government (other identical or similar definitions are provided in 10CFR30.4; 40.4; 70.4).

government contract (40CFR8.2-n) means any agreement or modification thereof between any contracting agency and any person for the furnishing of supplies or services or for the use of real or personal property, including lease arrangements. The term services, as used in this definition includes, but is not limited to, the following services: Utility, construction, transportation, research, insurance, and fund depository. The term government contract does not include:

(1) agreements in which the parties stand in the relationship of employer and employee, and

(2) federally assisted construction contracts.

governor (40CFR15.4) means the governor or principal executive officer of a state.

governor (40CFR52.1128-7) means the Governor of the Commonwealth or the head of such executive office of the Commonwealth as the Governor shall designate as responsible for carrying out specific provisions of this subpart.

grab sample (40CFR471.02-pp) is a single sample which is collected at a time and place most representative of total discharge.

grab sample (EPA-94/04): A single sample collected at a particular time and place that represents the composition of the water only at that time and place.

grade of resin (40CFR61.61-f) means the subdivision of resin classification which describes it as a unique resin, i.e., the most exact description of a resin with no further subdivision.

grain (40CFR60.301-a) means corn, wheat, sorghum, rice, rye, oats, barley, and soybeans.

grain elevator (40CFR60.301-b) means any plant or installation at which grain is unloaded, handled, cleaned, dried, stored, or loaded.

grain handling operations (40CFR60.301-l) include bucket elevators or legs (excluding legs used to unload barges or ships), scale hoppers and surge bins (garners), turn heads, scalpers, cleaners, trippers, and the headhouse and other such structures.

grain loading (EPA-94/04): The rate at which particles are emitted from a pollution source. Measurement is made by the number of grains per cubic foot of gas emitted.

grain loading station (40CFR60.301-k) means that portion of a grain elevator where the grain is transferred from the elevator to a truck, railcar, barge, or ship.

grain storage elevator (40CFR60.301-f) means any grain elevator located at any wheat flour mill, wet corn mill, dry corn mill (human consumption), rice mill, or soybean oil extraction plant which has a permanent grain storage capacity of 35,200 m^3 (ca. 1 million bushels) (ca. = capacity) (under CAA).

grain terminal elevator (40CFR60.301-c) means any grain elevator which has a permanent storage capacity of more than 88,100 m^3 (ca. 2.5 million U.S. bushels), except those located at animal food manufacturers, pet food manufacturers, cereal manufacturers, breweries, and livestock feedlots (ca. = capacity).

grain unloading station (40CFR60.301-j) means that portion of a grain elevator where the grain is transferred from a truck, railcar, barge, or ship to a receiving hopper.

grant (40CFR6.101-e) as used in this part means an award of funds or other assistance by a written grant agreement or cooperative agreement under 40CFR Chapter I, subpart B (other identical or similar definitions are provided in 40CFR15.4; 31.3; 32.605-7).

grant agreement (40CFR30.200) means an assistance agreement that does not substantially involve EPA in the project and where the recipient has the authority and capability to complete all elements of the program (does not include fellowships) (other identical or similar definitions are provided in 40CFR35.936.1-a; 35.4010).

grantee (40CFR6.501-f) means any individual, agency, or entity which has been awarded wastewater treatment construction grant assistance under 40CFR35, subpart E or I (other identical or similar definitions are provided in 40CFR15.4; 31.3; 32.605-8; 35.936.1-d; SDWA1445-42USC300j.4).

grantor (40CFR144.70-a) means the owner or operator who enters into this Agreement and any successors or assigns of the Grantor.

granular activated carbon (GAC) treatment (EPA-94/04): A filtering system often used in small water systems and individual homes to remove organics. GAC can be highly effective in removing elevated levels of radon from water (cf. air pollution contrl device).

granular diammonium phosphate plant (40CFR60.221-a) means any plant manufacturing granular diammonium phosphate by reacting phosphoric acid with ammonia.

granular triple superphosphate storage facility (40CFR60.241-a) means any facility curing or storing granular triple superphosphate.

grape juice canning (40CFR407.61-i) shall mean the processing of grape juice into the following products and product styles: Canned and frozen, fresh and stored, natural grape juice for the manufacture of juices, drinks, concentrates jams, jellies, and other related finished products but not wine or other spirits. In terms of raw material processed 1000 kg (1000 lb) of grapes are equivalent to 834 liters (100 gallons) of grape juice.

grape pressing (40CFR407.61-j) shall mean the washing and subsequent handling including pressing, heating, and filtration of natural juice from all varieties of grapes for the purpose of manufacturing juice, drink, concentrate, and jelly but not wine or other spirits. In terms of raw material processed 1000 kg (1000 lb) of grapes are equivalent to 834 liters (100 gallons) of grape juice.

grassed waterway (EPA-94/04): Natural or constructed watercourse or outlet that is shaped or graded and established in suitable vegetation for the disposal of runoff water without erosion.

gravity separation methods (40CFR440.141-6) means the treatment of mineral particles which exploits differences between their specific gravities. The separation is usually performed by means of sluices, jigs, classifiers, spirals, hydrocyclones, or shaking tables.

gravur (40CFR240.101-k) means the materials that fall from the solid waste fuel bed through the grate openings.

gravure cylinder (40CFR60.431) means a printing cylinder with an intaglio image consisting of minute cells or indentations specially engraved or etched into the cylinder's surface to hold ink when continuously revolved through a fountain of ink.

gravure cylinder (40CFR60.581) means a plated cylinder with a printing image consisting of minute cells or indentations, specifically engraved or etched into the cylinder's surface to hold ink when continuously revolved through a fountain of ink.

gray iron (40CFR464.31-d): A cast iron that gives a gray fracture due to the presence of flake graphite (under CWA).

gray water (EPA-94/04): Domestic wastewater composed of wash water from kitchen, bathroom, and laundry sinks, tubs, and washers.

graywater (CWA312-33USC1322) means galley, bath, and shower water.

grazing permit and lease (FLPMA103-43USC1702-90) means any document authorizing use of public lands or lands in National Forests in the eleven contiguous western States for the purpose of grazing domestic livestock.

Great Lakes (CWA118-33USC1268) means Lake Ontario, Lake Erie, Lake Huron (including Lake St. Clair), Lake Michigan, and Lake Superior, and the connecting channels (Saint Mary's River, Saint Clair River, Detroit River, Niagara River, and Saint Lawrence River to the Canadian Boarder).

Great Lakes states (CWA118-33USC1268) means the States of Illinois, Indiana, Michigan, Minnesota, New York, Ohio, Pennsylvania, and Wisconsin.

Great Lakes system (CWA118-33USC1268) means all the streams, rivers, lakes, and other bodies of water within the drainage basin of the Great Lakes.

Great Lakes water quality agreement (CWA118-33USC1268) means the bilateral agreement, between the United States and Canada, which was signed in 1978 and amended by the Protocol of 1987.

green liquor sulfidity (40CFR60.281-l) means the sulfidity of the liquor which leaves the smelt dissolving tank.

green tire (40CFR60.541) means an assembled, uncured tire.

green tire spraying operation (40CFR60.541) means the system used to apply a mold release agent and lubricant to the inside and/or outside of green tires to facilitate the curing process and to prevent rubber from sticking to the curing press. A green tire spraying operation consists of a booth where spraying is performed, the spray application station, and related equipment, such as the lubricant supply system.

greenfield coke oven battery (40CFR63.301) means a coke oven battery for which construction is commenced at a plant site (where no coke oven batteries previously existed) after December 4, 1992.

greenhouse (40CFR170.3) means any operation engaged in the production of agricultural plants inside any structure or space that is enclosed with nonporous covering and that is of sufficient size to permit worker entry. This term includes, but is not limited to, polyhouses, mushroom houses, rhubarb houses, and similar structures. It does not include such structures as malls, atriums, conservatories, arboretums, or office buildings where agricultural plants are present primarily for aesthetic or climatic modification.

greenhouse effect (EPA-94/04): The warming of the Earth's atmosphere attributed to a buildup of carbon dioxide or other gases; some scientists think that this buildup allows the sun's rays to heat the Earth, while infrared radiation makes the atmosphere opaque to a counterbalancing loss of heat.

grid (40CFR763-AA-10) means an open structure for mounting on the sample to aid in its examination in the TEM. The term is used here to denote a 200-mesh copper lattice approximately 3 mm in diameter (under TSCA).

grid casting facility (40CFR60.371-a) means the facility which includes all lead melting pots and machines used for casting the grid used in battery manufacturing.

grinder (40CFR60.401-e) means a unit which is used to pulverize dry phosphate rock to the final product size used in the manufacture of phosphate fertilizer and does not include crushing devices used in mining.

grinder pump (EPA-94/04): A mechanical device that shreds solids and raises sewage to a higher elevation through pressure sewers.

grinding (40CFR471.02-r) is the process of removing stock from a workpiece by the use of a tool consisting of abrasive grains held by a rigid or semi-rigid grinder. Grinding includes surface finishing, sanding, and slicing.

grinding (40CFR61.141) means to reduce to powder or small fragments and includes mechanical chipping or drilling.

grinding mill (40CFR60.671) means a machine used for the wet or dry fine crushing of any nonmetallic mineral. Grinding mills include, but are not limited to, the following types: hammer, roller, rod, pebble and ball, and fluid energy. The grinding mill includes the air conveying system, air separator, or air classifier, where such systems are used.

grinding wheels (29CFR1910.94b) means all power-driven rotatable grinding or abrasive wheels, except disc wheels as defined in this standards, consisting of abrasive particles held together by artificial or natural bonds and used for peripheral grinding.

gross alpha/beta particle activity (EPA-94/04): The total radioactivity due to emission of alpha or beta particles. Used as the screening measurement for radioactivity generally due to naturally occurring radionuclides. Activity is commonly measured in picocuries.

gross alpha particle activity (40CFR141.2) means the total radioactivity due to alpha particle emission as inferred from measurements on a dry sample.

gross beta particle activity (40CFR141.2) means the total radioactivity due to beta particle emission as inferred from measurements on a dry sample.

gross calorific value (40CFR240.101-l) means heat liberated when waste is burned completely and the products of combustion are cooled to the initial temperature of the waste. Usually expressed in British thermal units per pound (under the Resource Conservation Recovery Act).

gross cane (40CFR409.41-b) shall mean that amount of crop material as harvested, including field trash and other extraneous material (other identical or similar definitions are provided in 40CFR409.61-b; 409.81-b).

gross combination weight rating (GCWR) (40CFR202.10-i) means the value specified by the manufacturer as the loaded weight of a combination vehicle (other identical or similar definitions are provided in 40CFR205.51-14).

gross power (40CFR89.2) means the power measured at the crankshaft or its equivalent, the engine being equipped only with the standard accessories (such as oil pumps, coolant pumps, and so forth) necessary for its operation on the test bed. Alternators must be used, if necessary, to run the engine. Fans, air conditioners, and other accessories may be used at the discretion of the manufacturer, but no power adjustments for these accessories may be made.

gross production of fiberboard products (40CFR429.11-d) means the air dry weight of hardboard or insulation board following formation of the mat and prior to trimming and finishing operations.

gross ton (OPA1001) has the meaning given that term by the Secretary under part J of title 46, United States Code.

gross vehicle weight (40CFR52.741) means the manufacturer's gross weight rating for the individual vehicle (other identical or similar definitions are provided in 40CFR86.082.2).

gross vehicle weight rating (40CFR600.002.85-37) means the manufacturer's gross weight rating for the individual vehicle.

gross vehicle weight rating (GVWR) (40CFR202.10-h) means the value specified by the manufacturer as the loaded weight of a single vehicle (other identical or similar definitions are provided in 40CFR205.51-15).

gross vehicle weight rating (GVWR) (40CFR52.741) means the value specified by the manufacturer as the maximum design loaded weight of a single vehicle.

gross vehicle weight rating (GVWR) (40CFR86.082.2) means the value specified by the manufacturer as the maximum design loaded weight of a single vehicle.

Grotthus-Draper law (40CFR796.3700-ix) means the first law of photochemistry, states that only light which is absorbed can be effective in reducing a chemical transformation (other identical or similar definitions are provided in 40CFR796.3780-x; 796.3800-ix).

ground cover (EPA-94/04): Plants grown to keep soil from eroding.

ground phosphate rock handling and storage system (40CFR60.401-f) means a system which is used for the conveyance and storage of ground phosphate rock from grinders at phosphate rock plants.

ground water (40CFR144.3) means water below the land surface in a zone of saturation (other identical or similar definitions are provided in 40CFR146.3; 147.2902; 191.12; 257.3.4-3; 258.2; 260.10; 270.2).

ground water (40CFR300.5), as defined by section 101(12) of CERCLA, means water in a saturated zone or stratum beneath the surface of land or water.

ground water (40CFR503.9-m) is water below the land surface in the saturated zone (under CWA).

ground water (EPA-94/04): The supply of fresh water found beneath the Earth's surface, usually in aquifers, which supply wells and springs. Because ground water is a major source of drinking water, there is growing concern over contamination from leaching agricultural or industrial pollutants or leaking underground storage tanks (under the Safe Drinking Water Act).

ground water (SF101-42USC9601) means water in a saturated zone or stratum beneath the surface of land or water.

ground water discharge (EPA-94/04): Ground water entering near coastal waters which has been contaminated by landfill leachate, deep well injection of hazardous wastes, septic tanks, etc (under SDWA).

ground water under the direct influence (UDI) of surface water (EPA-94/04): Any water beneath the surface if the ground with:
1. significant occurrence of inserts or other microorganisms, algae, or large-diameter pathogens;
2. significant and relatively rapid shifts in water characteristics such as turbidity, temperature, conductivity, or pH which closely correlate to climatological or surface water conditions. Direct influence must be determined for individual sources in accordance with criteria established by the state.

ground water under the direct influence of surface water (40CFR141.2) means any water beneath the surface of the ground with:
(1) significant occurrence of insects or other macroorganisms, algae, or large-diameter pathogens such as Giardia lamblia, or
(2) significant and relatively rapid shifts in water characteristics such as turbidity, temperature, conductivity, or pH which closely correlate to climatological or surface water conditions. Direct influence must be determined for individual sources in accordance with criteria established by the State. The State determination of direct influence may be based on site-specific measurements of water quality and/or documentation of well construction characteristics and geology with field evaluation.

groundwater (40CFR241.101-f) means water present in the saturated zone of an aquifer (under RCRA).

groundwater infiltration (40CFR440.132-d) in 40CFR440.131 means that water which enters the treatment facility as a result of the interception of natural springs, aquifers, or run-off which percolates into the round and seeps into the treatment facility's tailings pond or wastewater holding facility and that cannot be diverted by ditching or grouting the tailings pond or wastewater holding facility (under the Clean Water Act).

group 1 boiler (40CFR76.2) means a tangentially fired boiler or a dry bottom wall-fired boiler (other than a unit applying cell burner technology) (under CAA).

group 2 boiler (40CFR76.2) means a wet bottom wall-fired boiler, a cyclone boiler, a boiler applying cell burner technology, a vertically fired boiler, an arch-fired boiler, or any other type of utility boiler (such as a fluidized bed or stoker boiler) that is not a Group 1 boiler (under CAA).

group 1 process vent (40CFR63.111) means a process vent for which the flow rate is greater than or equal to 0.005 standard cubic meter per minute, the total organic HAP concentration is greater than or equal to 50 parts per million by volume, and the total resource effectiveness index value, calculated according to 40CFR63.115 of this subpart, is less than or equal to 1.0 (under CAA).

group 2 process vent (40CFR63.111) means a process vent for which the flow rate is less than 0.005 standard cubic meter per minute, the total organic HAP concentration is less than 50 parts per million by volume or the total resource effectiveness index value, calculated according to 40CFR63.115 of this subpart, is greater than 1.0 (under the Clean Air Act).

group 1 storage vessel (40CFR63.111) means a storage vessel that meets the criteria for design storage capacity and stored-liquid maximum true vapor pressure specified in table 5 of this subpart for storage vessels at existing sources, and in table 6 of this subpart for storage vessels at new sources (under CAA).

group 2 storage vessel (40CFR63.111) means a storage vessel that does not meet the definition of a Group 1 storage vessel.

group 1 transfer rack (40CFR63.111) means a transfer rack that annually loads greater than or equal to 0.65 million liter of liquid products that contain organic hazardous air pollutants with a rack

weighted average vapor pressure greater than or equal to 10.3 kilopascals (under CAA).

group 2 transfer rack (40CFR63.111) means a transfer rack that does not meet the definition of Group 1 transfer rack (under CAA).

group 1 wastewater stream (40CFR63.111) means a process wastewater stream from a process unit at an existing or new source with a total volatile organic hazardous air pollutant average concentration greater than or equal to 10,000 parts per million by weight of compounds listed in table 9 of this subpart at any flowrate; or a process wastewater stream from a process unit at an existing or new source that has an average flow rate greater than or equal to 10 liters per minute and a total volatile organic hazardous air pollutant average concentration greater than or equal to 1,000 parts per million by weight. A process wastewater stream from a process unit at a new source that has an average flow rate greater than or equal to 0.02 liter per minute and an average concentration of 10 parts per million by weight or greater of any one of the compounds listed in table 8 of this subpart is also considered a Group 1 wastewater stream. Average flow rate and total volatile organic hazardous air pollutant average concentration are determined for the point of generation of each process wastewater stream (under CAA).

group 2 wastewater stream (40CFR63.111) means any process wastewater stream that does not meet the definition of a Group 1 wastewater stream (under the Clean Air Act).

growth (40CFR797.1050-4) means a relative measure of the viability of an algal population based on the number and/or weight of algal cells per volume of nutrient medium or test solution in a specified period of time.

growth rate (40CFR797.1060-2) means an increase in biomass or cell numbers of algae per unit time (under TSCA).

growth rate (40CFR797.1075-2) means the rate at which the algal population grows, estimated by the increase in cell numbers or biomass over a specified period of time.

guaranteed loan program (40CFR39.105-c) means the program established pursuant to Pub. L. 4-55 which amended the Act by adding section 213 (under FWPCA).

guarantor (OPA1001) means any person, other than the responsible party, who provides evidence of financial responsibility for a responsible party under this Act.

guarantor (SF101-42USC9601) means any person, other than that owner or operator, which provides evidence of financial responsibility for an owner or operator under this Act.

guidance for controlling asbestos-containing material in buildings (TSCA202-15USC2642) means the Environmental Protection Agency document with such title as in effect on March 31, 1986.

guide coat operation (40CFR60.391) means the guide coat spray booth, flash-off area and bake oven(s) which are used to apply and dry or cure a surface coating between the prime coat and topcoat operation on the components of automobile and light-duty truck bodies.

guide specification (40CFR249.04-f) means a general specification--often referred to as a design standard or design guideline--which is a model standard and is suggested or required for use in the design of all of the construction projects of an agency (under RCRA).

guidelines (40CFR766.3) means the Midwest Research Institute (MRI) publication Guidelines for the Determination of Polyhalogenated Dioxins and Dibenzofurans in Commercial Products, EPA contract No. 68-02-3938; MRI Project No. 8201-A(41), 1985.

gully erosion (EPA-94/04): Severe erosion in which trenches are cut to a depth greater than 30 centimeters (a foot). Generally, ditches deep enough to cross with farm equipment are considered gullies.

GVW (40CFR51.731): See gross vehicle weight.

GVWR (40CFR86.082.2): See gross vehicle weight rating.

*********** HHHHH ***********

habitat (EPA-94/04): The place where a population (e.g., human, animal, plant, microorganism) lives and its surroundings, both living and non-living. Habitat Indicator: A physical attribute of the environment measured to characterize conditions necessary to support an organism, population, or community in the absence of pollutants, e.g., salinity of esturine waters or substrate type in streams or lakes.

hair pulp (40CFR425.02-d) means the removal of hair by chemical dissolution.

hair save (40CFR425.02-e) means the physical or mechanical removal of hair which has not been chemically dissolved, and either selling the hair as a byproduct or disposing of it as a solid waste.

half life (40CFR300-AA) means length of time required for an initial concentration of a substance to be halved as a result of loss through decay. The HRS considers five decay processes: biodegradation, hydrolysis, photolysis, radioactive decay, and volatilization.

half life (EPA-94/04):
1. The time required for a pollutant to lose half its affect on the environment. For example, the biochemical half-life of DDT in the environment is 15 years and of radium 1,580 years.
2. The time required for half of the atoms of a radioactive element to undergo self-transmutation or decay.
3. The time required for the elimination of one half a total dose from the body.

half life ($t_{1/2}$) of a chemical (40CFR796.3700-vi) is defined as the time required for the concentration of the chemical being tested to be reduced to one-half its initial value (other identical or similar definitions are provided in 40CFR796.3800-vi).

halocarbon (CAA111-42USC7411) means the chemical compounds $CFCl_3$ and CF_2Cl_2 and such other halogenated compounds as the Administrator determines may reasonably be anticipated to contribute to reductions in the concentration of ozone in the stratosphere.

halogen (40CFR141.2) means one of the chemical elements chlorine, bromine or iodine.

halogen acid furnaces (HAFs) (40CFR260.10): See industrial furnace.

halogenated organic compounds or HOCs (40CFR268.2-a) means those compounds having a carbon-halogen bond which are listed under appendix III to this part.

halogenated vent stream (40CFR60.611) means any vent stream determined to have a total concentration (by volume) of compounds containing halogens of 20 ppmv (by compound) or greater (other identical or similar definitions are provided in 40CFR60.661; 60.701).

halogenated vent stream or halogenated stream (40CFR63.111) means a vent stream from a process vent or transfer operation determined to have a mass emission rate of halogen atoms contained in organic compounds of 0.45 kilograms per hour or greater determined by the procedures presented in 40CFR63.115(d)(2)(v) of this subpart.

halogens and hydrogen halides (40CFR63.111) means hydrogen chloride (HCl), chlorine (Cl_2), hydrogen bromide (HBr), bromine (Br_2), and hydrogen fluoride (HF).

halon (EPA-94/04): Bromine-containing compounds with long atmospheric lifetimes whose breakdown in the stratosphere causes depletion of ozone. Halons are used in fire-fighting.

ham processor (40CFR432.81-b) shall mean an operation which manufactures hams alone or in combination with other finished products at rates greater than 2730 kg (6000 lb) per day (under CWA).

hammermill (EPA-94/04): A high-speed machine that uses hammers and cutters to crush, grind, chip, or shred solid waste.

hand glass melting furnace (40CFR60.291) means a glass melting furnace where the molten glass is removed from the furnace by a glassworker using a blowpipe or a pontil.

hand labor (40CFR170.3) means any agricultural activity performed by hand or with hand tools that causes a worker to have substantial contact with surfaces (such as plants, plant parts, or soil) that may contain pesticide residues. These activities include, but are not limited to, harvesting, detasseling, thinning, weeding, topping, planting, sucker removal, pruning, disbudding, roguing, and packing produce into containers in the field. Hand labor does not include operating, moving, or repairing irrigation or watering equipment or performing the tasks of crop advisors. Handler means any person, including a self-employed person:

(1) Who is employed for any type of compensation by an agricultural establishment or commercial pesticide handling establishment to which subpart C of this part applies and who is:

i. Mixing, loading, transferring, or applying pesticides.

ii. Disposing of pesticides or pesticide containers.

iii. Handling opened containers of pesticides.

iv. Acting as a flagger.

v. Cleaning, adjusting, handling, or repairing the parts of mixing, loading, or application equipment that may contain pesticide residues.

vi. Assisting with the application of pesticides.

vii. Entering a greenhouse or other enclosed area after the application and before the inhalation exposure level listed in the labeling has been reached or one of the ventilation criteria established by this part (40CFR170.110(c)(3)) or in the labeling has been met:

(A) To operate ventilation equipment.

(B) To adjust or remove coverings used in fumigation.

(C) To monitor air levels.

viii. Entering a treated area outdoors after application of any soil fumigant to adjust or remove soil coverings such as tarpaulins.

ix. Performing tasks as a crop advisor:
   (A) During any pesticide application.
   (B) Before the inhalation exposure level listed in the labeling has been reached or one of the ventilation criteria established by this part (40CFR170.110(c)(3)) or in the labeling has been met.
   (C) During any restricted-entry interval.

(2) The term does not include any person who is only handling pesticide containers that have been emptied or cleaned according to pesticide product labeling instructions or, in the absence of such instructions, have been subjected to triple-rinsing or its equivalent.

handicapped person (40CFR7.25):
(a) handicapped person means any person who:
   (1) has a physical or mental impairment which substantially limits one or more major life activities,
   (2) has a record of such an impairment, or
   (3) is regarded as having such an impairment. For purposes of employment, the term handicapped person does not include any person who is an alcoholic or drug abuser whose current use of alcohol or drugs prevents such individual from performing the duties of the job in question or whose employment, by reason of such current drug or alcohol abuse, would constitute a direct threat to property or the safety of others.
(b) As used in this paragraph, the phrase:
   (1) Physical or mental impairment means:
      (i) any physiological disorder or condition, cosmetic disfigurement, or anatomical loss affecting one or more of the following body systems: Neurological; musculoskeletal; special sense organs; respiratory, including speech organs; cardiovasular; reproductive; digestive; genito-urinary; hemic and lymphatic; skin; and endocrine; and
      (ii) any mental or psychological disorder, such as mental retardation, organic brain syndrome, emotional or mental illness, and specific learning disabilities.
   (2) Major life activities means functions such as caring for one's self, performing manual tasks, walking, seeing, hearing, speaking, breathing, learning, and working.
   (3) Has a record of such an impairment means has a history of, or has been misclassified as having, a mental or physical impairment that substantially limits one or more major life activities.
   (4) Is regarded as having an impairment means:
      (i) Has a physical or mental impairment that does not substantially limit major life activities but that is treated by a recipient as constituting such a limitation;
      (ii) Has a physical or mental impairment that substantially limits major life activities only as a result of the attitudes of others toward such impairment; or
      (iii) Has none of the impairments defined above but is treated by a recipient as having such an impairment.

handler employer (40CFR170.3) means any person who is self-employed as a handler or who employs any handler, for any type of compensation (under FIFRA).

hang up (40CFR86.082.2) refers to the process of hydrocarbon molecules being adsorbed, condensed, or by any other method removed from the sample flow prior to reaching the instrument detector. It also refers to any subsequent desorption of the molecules into the sample flow when they are assumed to be absent.

hard piping (40CFR63.111) means tubing that is manufactured and properly installed using good engineering judgement and standards, such as ANSI B31-3.

hard piping (40CFR63.161) means pipe or tubing that is manufactured and properly installed using good engineering judgement and standards, such as ANSI B31-3.

hard water (EPA-94/04): Alkaline water containing dissolved salts that interfere with some industrial processes and prevent soap from sudsing (cf. soft water).

hardboard (40CFR429.11-e) means a panel manufactured from interfelted ligno-cellulosic fibers consolidated under heat and pressure to a density of 0.5 g/cu cm (31 lb/cu ft) or greater (under the Clean Water Act).

hardness (40CFR797.1600-9) is the total concentration of the calcium and magnesium ions in water expressed as calcium carbonate (mg $CaCO_3$/liter).

hardness (water) (EPA-94/04): Characteristic of water caused by presence of various salts. Hard water may interfere with some industrial processes and prevent soap from lathering.

harmful physical agent (29CFR1910.20): See toxic substance.

has a record of such an impairment (40CFR7.25): See handicapped person.

hatch (40CFR797.2050-7) means eggs or young birds that are the same age and that are derived from the same adult breeding population, where the adults are of the same strain and stock.

hatch (40CFR797.2175-5): Eggs or birds that are the same age and that are derived from the same adult breeding population, where the adults are of the same strain and stock.

hatchability (40CFR797.2130-vi) means embryos that mature, pip the shell, and liberate themselves from the eggs on day 23 or 24 of incubation. Values are expressed as percentage of viable embryos (fertile eggs).

hatchability (40CFR797.2150-vi) means embryos that mature, pip the shell, and liberate themselves from their eggs on day 25, 26, or 27 of incubation. Values are expressed as a percentage of viable embryos (fertile eggs).

hatchback (40CFR600.002.85-34) means a passenger automobile where the conventional luggage compartment, i.e., trunk, is replaced by a cargo area which is open to the passenger compartment and accessed vertically by a rear door which encompasses the rear window.

hauler (EPA-94/04): Garbage collection company that offers complete refuse removal service; many also will also collect recyclables.

hazard (40CFR171.2-15) means a probability that a given pesticide will have an adverse effect on man or the environment in a given situation, the relative likelihood of danger or ill effect being dependent on a number of interrelated factors present at any given time.

hazard category (40CFR370.2) means any of the following:
(1) "Immediate (acute) health hazard," including "highly toxic," "toxic," "irritant," "sensitizer," "corrosive," (as defined under 40CFR910.1200 of Title 29 of the Code of Federal Regulations) and other hazardous chemicals that cause an adverse effect to a target organ and which effect usually occurs rapidly as a result of short term exposure and is of short duration;
(2) "Delayed (chronic) health hazard," including "carcinogens" (as defined under 40CFR910.1200 of Title 29 of the Code of Federal Regulations) and other hazardous chemicals that cause an adverse effect to a target organ and which effect generally occurs as a result of long term exposure and is of long duration;
(3) "Fire hazard," including "flammable," combustible liquid," "pyrophoric," and "oxidizer" (as defined under 40CFR910.1200 of Title 29 of the Code of Federal Regulations);
(4) "Sudden release of pressure," including "explosive" and "compressed gas" (as defined under 40CFR910.1200 of Title 29 of the Code of Federal Regulations); and
(5) "Reactive," including "unstable reactive," "organic peroxide," and "water reactive" (as defined under 40CFR910.1200 of Title 29 of the Code of Federal Regulations).

hazard communication standard (EPA-94/04): An OSHA regulation that requires chemical manufacturers, suppliers, and importers to assess the hazards of the chemicals that they make, supply, or import, and to inform employers, customers, and workers of these hazards through MSDS sheets.

hazard contents, ordinary (40CFR1910.35): Ordinary hazard contents shall be classified as those which are liable to burn with moderate rapidity and to give off a considerable volume of smoke but from which neither poisonous fumes nor explosions are to be feared in case of fire.

hazard evaluation (EPA-94/04): A component of risk evaluation that involves gathering and evaluating data on the types of health injury or disease that may be produced by a chemical and on the conditions of exposure under which such health effects are produced.

hazard identification (EPA-94/04): Determining if a chemical can cause adverse health effects in humans and what those affects might be.

hazard ranking system (HRS) (40CFR300.5) means the method used by EPA to evaluate the relative potential of hazardous substance releases to cause health or safety problems, or ecological or environmental damage.

hazardous air pollutant (40CFR63.2) means any air pollutant listed in or pursuant to section 112(b) of the Act (other identical or similar definitions are provided in 40CFR63.101; 63.191).

hazardous air pollutant (HAP) (40CFR63.51) means any air pollutant listed pursuant to section 112(b) of the Act.

hazardous air pollutants (EPA-94/04): Air pollutants which are not covered by ambient air quality standards but which, as defined in the Clean Air Act, may reasonably be expected to cause or contribute to irreversible illness or death. Such pollutants include asbestos, beryllium, mercury, benzene, coke oven emissions, radionuclides, and vinyl chloride (under CAA).

hazardous air pollutants (HAPs) (CAA112-42USC7412) means any air pollutant listed pursuant to subsection (b) of the Clean Air Act Amendments (see below for the listing of 189 hazardous air pollutants regulated under the Clean Air Act. The pollutants provided herewith in both alphabetical and CAS order were excerpted from CAA112).

CAS #	chemical name (alphabetical order)
75-070	Acetaldehyde
60-355	Acetamide
75-058	Acetonitrile
98-862	Acetophenone
53-963	2-Acetylaminofluorene
107-028	Acrolein
79-061	Acrylamide
79-107	Acrylic acid
107-131	Acrylonitrile
107-051	Allyl chloride
92-671	4-Aminobiphenyl
62-533	Aniline
90-040	o-Anisidine
1332-214	Asbestos
71-432	Benzene (including benzene from gasoline)
92-875	Benzidine
98-077	Benzotrichloride
100-447	Benzyl chloride
92-524	Biphenyl
117-817	Bis(2-ethylhexyl)phthalate (DEHP)
542-881	Bis(chloromethyl)ether
75-252	Bromoform
106-990	1,3-Butadiene
156-627	Calcium cyanamide
105-602	Caprolactam
133-062	Captan
63-252	Carbaryl
75-150	Carbon disulfide
56-235	Carbon tetrachloride
463-581	Carbonyl sulfide
120-809	Catechol
133-904	Chloramben
57-749	Chlordane
7782-505	Chlorine
79-118	Chloroacetic acid
532-274	2-Chloroacetophenone
108-907	Chlorobenzene
510-156	Chlorobenzilate
67-663	Chloroform
107-302	Chloromethyl methyl ether
126-998	Chloroprene
1319-773	Cresols/Cresylic acid (isomers and mixture)
95-487	o-Cresol
108-394	m-Cresol
106-445	p-Cresol
98-828	Cumene
94-757	2,4-D, salts and esters
3547-044	DDE
334-883	Diazomethane
132-649	Dibenzofurans

96-128	1,2-Dibromo-3-chloropropane
84-742	Dibutylphthalate
106-467	1,4-Dichlorobenzene(p)
91-941	3,3-Dichlorobenzidene
111-444	Dichloroethyl ether (Bis(2-chloroethyl)ether)
542-756	1,3-Dichloropropene
62-737	Dichlorvos
111-422	Diethanolamine
121-697	N,N-Diethyl aniline (N,N-Dimethylaniline)
64-675	Diethyl sulfate
119-904	3,3-Dimethoxybenzidine
60-117	Dimethyl aminoazobenzene
119-937	3,3'-Dimethyl benzidine
79-447	Dimethyl carbamoyl chloride
68-122	Dimethyl formamide
57-147	1,1-Dimethyl hydrazine
131-113	Dimethyl phthalate
77-781	Dimethyl sulfate
534-521	4,6-Dinitro-o-cresol, and salts
51-285	2,4-Dinitrophenol
121-142	2,4-Dinitrotoluene
123-911	1,4-Dioxane (1,4-Diethyleneoxide)
122-667	1,2-Diphenylhydrazine
106-898	Epichlorohydrin (1-Chloro-2,3-epoxypropane)
106-887	1,2-Epoxybutane
140-885	Ethyl acrylate
100-414	Ethyl benzene
51-796	Ethyl carbamate (Urethane)
75-003	Ethyl chloride (Chloroethane)
106-934	Ethylene dibromide (Dibromoethane)
107-062	Ethylene dichloride (1,2-Dichloroethane
107-211	Ethylene glycol
151-564	Ethylene imine (Aziridine)
75-218	Ethylene oxide
96-457	Ethylene thiourea
75-343	Ethylidene dichloride (1,1-Dichloroethane)
50-000	Formaldehyde
76-448	Heptachlor
118-741	Hexachlorobenzene
87-683	Hexachlorobutadiene
77-474	Hexachlorocyclopentadiene
67-721	Hexachloroethane
822-060	Hexamethylene-1,6-diisocyanate
680-319	Hexamethylphosphoramide
110-543	Hexane
302-012	Hydrazine
7647-010	Hydrochloric acid
7664-393	Hydrogen fluoride (Hydrofluoric acid)
7783-064	Hydrogen sulfide
123-319	Hydroquinone
78-591	Isophorone
58-899	Lindane (all isomers)
108-316	Maleic anhydride
67-561	Methanol
72-435	Methoxychlor
74-839	Methyl bromide (Bromomethane)
74-873	Methyl chloride (Chloromethane)
71-556	Methyl chloroform (1,1,1-Trichloroethane)
78-933	Methyl ethyl ketone (2-Butanone)
60-344	Methyl hydrazine
74-884	Methyl iodide (Iodomethane)
108-101	Methyl isobutyl ketone (Hexone)
624-839	Methyl isocyanate
80-626	Methyl methacrylate
1634-044	Methyl tert butyl ether
101-144	4,4-Methylene bis(2-chloroaniline)
75-092	Methylene chloride (Dichloromethane)
101-688	Methylene diphenyl diisocyanate (MDI)
101-779	4,4'-Methylenedianiline
91-203	Naphthalene
98-953	Nitrobenzene
92-933	4-Nitrobiphenyl
100-027	4-Nitrophenol
79-469	2-Nitropropane
684-935	N-Nitroso-N-methylurea
62-759	N-Nitrosodimethylamine
59-892	N-Nitrosomorpholine
56-382	Parathion
82-688	Pentachloronitrobenzene (Quintobenzene)
87-865	Pentachlorophenol
108-952	Phenol
106-503	p-Phenylenediamine
75-445	Phosgene
7803-512	Phosphine
7723-140	Phosphorus
85-449	Phthalic anhydride
1336-363	Polychlorinated biphenyls (Aroclors)
1120-714	1,3-Propane sultone
57-578	beta-Propiolactone

123-386	Propionaldehyde
114-261	Propoxur (Baygon)
78-875	Propylene dichloride (1,2-Dichloropropane)
75-569	Propylene oxide
75-558	1,2-Propylenimine (2-Methyl aziridine)
91-225	Quinoline
106-514	Quinone
100-425	Styrene
96-093	Styrene oxide
1746-016	2,3,7,8-Tetrachlorodibenzo-p-dioxin
79-345	1,1,2,2-Tetrachloroethane
127-184	Tetrachloroethylene (Perchloroethylene)
7550-450	Titanium tetrachloride
108-883	Toluene
95-807	2,4-Toluene diamine
584-849	2,4-Toluene diisocyanate
95-534	o-Toluidine
8001-352	Toxaphene (chlorinated camphene)
120-821	1,2,4-Trichlorobenzene
79-005	1,1,2-Trichloroethane
79-016	Trichloroethylene
95-954	2,4,5-Trichlorophenol
88-062	2,4,6-Trichlorophenol
121-448	Triethylamine
1582-098	Trifluralin
540-841	2,2,4-Trimethylpentane
108-054	Vinyl acetate
593-602	Vinyl bromide
75-014	Vinyl chloride
75-354	Vinylidene chloride (1,1-Dichloroethylene)
1330-207	Xylenes (isomers and mixture)
95-476	o-Xylenes
108-383	m-Xylenes
106-423	p-Xylenes
(N/A)	Antimony Compounds
(N/A)	Arsenic Compounds (inorganic including arsine)
(N/A)	Beryllium Compounds
(N/A)	Cadmium Compounds
(N/A)	Chromium Compounds
(N/A)	Cobalt Compounds
(N/A)	Coke Oven Emissions
(N/A)	Cyanide Compounds[1]
(N/A)	Glycol ethers[2]
(N/A)	Lead Compounds
(N/A)	Manganese Compounds
(N/A)	Mercury Compounds
(N/A)	Fine mineral fibers[3]
(N/A)	Nickel Compounds
(N/A)	Polycylic Organic Matter[4]
(N/A)	Radionuclides (including radon)[5]
(N/A)	Selenium Compounds

CAS # (order)	chemical name
50-000	Formaldehyde
51-285	2,4-Dinitrophenol
51-796	Ethyl carbamate (Urethane)
53-963	2-Acetylaminofluorene
56-235	Carbon tetrachloride
56-382	Parathion
57-147	1,1-Dimethyl hydrazine
57-578	beta-Propiolactone
57-749	Chlordane
58-899	Lindane (all isomers)
59-892	N-Nitrosomorpholine
60-117	Dimethyl aminoazobenzene
60-344	Methyl hydrazine
60-355	Acetamide
62-533	Aniline
62-737	Dichlorvos
62-759	N-Nitrosodimethylamine
63-252	Carbaryl
64-675	Diethyl sulfate
67-561	Methanol
67-663	Chloroform
67-721	Hexachloroethane
68-122	Dimethyl formamide
71-432	Benzene (including benzene from gasoline)
71-556	Methyl chloroform (1,1,1-Trichloroethane)
72-435	Methoxychlor
74-839	Methyl bromide (Bromomethane)
74-873	Methyl chloride (Chloromethane)
74-884	Methyl iodide (Iodomethane)
75-003	Ethyl chloride (Chloroethane)
75-014	Vinyl chloride
75-058	Acetonitrile
75-070	Acetaldehyde
75-092	Methylene chloride (Dichloromethane)
75-150	Carbon disulfide
75-218	Ethylene oxide
75-252	Bromoform
75-343	Ethylidene dichloride (1,1-Dichloroethane)

75-354	Vinylidene chloride (1,1-Dichloroethylene)	100-027	4-Nitrophenol
		100-414	Ethyl benzene
75-445	Phosgene	100-425	Styrene
75-558	1,2-Propylenimine (2-Methyl aziridine)	100-447	Benzyl chloride
		101-144	4,4-Methylene bis(2-chloroaniline)
75-569	Propylene oxide	101-688	Methylene diphenyl diisocyanate (MDI)
76-448	Heptachlor		
77-474	Hexachlorocyclopentadiene	101-779	4,4'-Methylenedianiline
77-781	Dimethyl sulfate	105-602	Caprolactam
78-591	Isophorone	106-423	p-Xylenes
78-875	Propylene dichloride (1,2-Dichloropropane)	106-445	p-Cresol
		106-467	1,4-Dichlorobenzene(p)
78-933	Methyl ethyl ketone (2-Butanone)	106-503	p-Phenylenediamine
79-005	1,1,2-Trichloroethane	106-514	Quinone
79-016	Trichloroethylene	106-887	1,2-Epoxybutane
79-061	Acrylamide	106-898	Epichlorohydrin (1-Chloro-2,3-epoxypropane)
79-107	Acrylic acid		
79-118	Chloroacetic acid	106-934	Ethylene dibromide (Dibromoethane)
79-345	1,1,2,2-Tetrachloroethane		
79-447	Dimethyl carbamoyl chloride	106-990	1,3-Butadiene
79-469	2-Nitropropane	107-028	Acrolein
80-626	Methyl methacrylate	107-051	Allyl chloride
82-688	Pentachloronitrobenzene (Quintobenzene)	107-062	Ethylene dichloride (1,2-Dichloroethane
84-742	Dibutylphthalate	107-131	Acrylonitrile
85-449	Phthalic anhydride	107-211	Ethylene glycol
87-683	Hexachlorobutadiene	107-302	Chloromethyl methyl ether
87-865	Pentachlorophenol	108-054	Vinyl acetate
88-062	2,4,6-Trichlorophenol	108-101	Methyl isobutyl ketone (Hexone)
90-040	o-Anisidine	108-316	Maleic anhydride
91-203	Naphthalene	108-383	m-Xylenes
91-225	Quinoline	108-394	m-Cresol
91-941	3,3-Dichlorobenzidene	108-883	Toluene
92-524	Biphenyl	108-907	Chlorobenzene
92-671	4-Aminobiphenyl	108-952	Phenol
92-875	Benzidine	110-543	Hexane
92-933	4-Nitrobiphenyl	111-422	Diethanolamine
94-757	2,4-D, salts and esters	111-444	Dichloroethyl ether (Bis(2-chloroethyl)ether)
95-476	o-Xylenes		
95-487	o-Cresol	114-261	Propoxur (Baygon)
95-534	o-Toluidine	117-817	Bis(2-ethylhexyl)phthalate (DEHP)
95-807	2,4-Toluene diamine	118-741	Hexachlorobenzene
95-954	2,4,5-Trichlorophenol	119-904	3,3-Dimethoxybenzidine
96-093	Styrene oxide	119-937	3,3'-Dimethyl benzidine
96-128	1,2-Dibromo-3-chloropropane	120-809	Catechol
96-457	Ethylene thiourea	120-821	1,2,4-Trichlorobenzene
98-077	Benzotrichloride	121-142	2,4-Dinitrotoluene
98-828	Cumene	121-448	Triethylamine
98-862	Acetophenone	121-697	N,N-Diethyl aniline (N,N-Dimethylaniline)
98-953	Nitrobenzene		

122-667	1,2-Diphenylhydrazine
123-319	Hydroquinone
123-386	Propionaldehyde
123-911	1,4-Dioxane (1,4-Diethyleneoxide)
126-998	Chloroprene
127-184	Tetrachloroethylene (Perchloroethylene)
131-113	Dimethyl phthalate
132-649	Dibenzofurans
133-062	Captan
133-904	Chloramben
140-885	Ethyl acrylate
151-564	Ethylene imine (Aziridine)
156-627	Calcium cyanamide
302-012	Hydrazine
334-883	Diazomethane
463-581	Carbonyl sulfide
510-156	Chlorobenzilate
532-274	2-Chloroacetophenone
534-521	4,6-Dinitro-o-cresol, and salts
540-841	2,2,4-Trimethylpentane
542-756	1,3-Dichloropropene
542-881	Bis(chloromethyl)ether
584-849	2,4-Toluene diisocyanate
593-602	Vinyl bromide
624-839	Methyl isocyanate
680-319	Hexamethylphosphoramide
684-935	N-Nitroso-N-methylurea
822-060	Hexamethylene-1,6-diisocyanate
1120-714	1,3-Propane sultone
1319-773	Cresols/Cresylic acid (isomers and mixture)
1330-207	Xylenes (isomers and mixture)
1332-214	Asbestos
1336-363	Polychlorinated biphenyls (Aroclors)
1582-098	Trifluralin
1634-044	Methyl tert butyl ether
1746-016	2,3,7,8-Tetrachlorodibenzo-p-dioxin
3547-044	DDE
7550-450	Titanium tetrachloride
7647-010	Hydrochloric acid
7664-393	Hydrogen fluoride (Hydrofluoric acid)
7723-140	Phosphorus
7782-505	Chlorine
7783-064	Hydrogen sulfide
7803-512	Phosphine
8001-352	Toxaphene (chlorinated camphene)

NOTE: For all listings above which contain the word "compounds" and for glycol ethers, the following

applies: Unless otherwise specified, these listings are defined as including any unique chemical substance that contains the named chemical (i.e., antimony, arsenic, etc.) as part of that chemical's infrastructure.

[1] X'CN where X = H' or any other group where a formal dissociation may occur. For example KCN or Ca(CN)$_2$.

[2] includes mono- and di-ethers of ethylene glycol, diethylene glycol, and triethylene glycol R-(OCH2CH2)$_n$-OR'.
Where n = 1, 2, or 3. R = alkyl or aryl groups. R' = R, H, or groups which, when removed, yield glycol ethers with the structure: R-(OCH2CH)(n)-OH. Polymers are excluded from the glycol category.

[3] Includes mineral fiber emissions from facilities manufacturing or processing glass, rock, or slag fibers (or other mineral derived fibers) of average diameter 1 micrometer or less.

[4] Includes organic compounds with more than one benzene ring, and which have a boiling point greater than or equal to 100° C.

[5] A type of atom which spontaneously undergoes radioactive decay.

(N/A): Not available.

hazardous chemical (40CFR355.20) means any hazardous chemical as defined under 29CFR1910.1200(c) of Title 29 of the Code of Federal Regulations, except that such term does not include the following substances:

(1) Any food, food additive, color additive, drug, or cosmetic regulated by the Food and Drug Administration.

(2) Any substance present as a solid in any manufactured item to the extent exposure to the substance does not occur under normal conditions of use.

(3) Any substance to the extent it is used for personal, family, or household purposes, or is present in the same form and concentration as a product packaged for distribution and use by the general public.

(4) Any substance to the extent it is used in a research laboratory or a hospital or other medical facility under the direct supervision of a technically qualified individual.

(5) Any substance to the extent it is used in routine agricultural operations or is a fertilizer held for sale by a retailer to the ultimate customer.

(Other identical or similar definitions are provided in 40CFR370.2.)

hazardous chemical (EPA-94/04): An EPA designation for any hazardous material requiring an MSDS under OSHA's Hazard Communication Standard. Such substances are capable of producing fires and explosions or adverse health effects like cancer and dermatitis. Hazardous chemicals are distinct from hazardous waste (cf. hazardous waste under RCRA).

hazardous chemical (SF329) has the meaning given such term by section 311(e).

hazardous constituent or constituents (40CFR268.2-b) means those constituents listed in appendix VIII to part 261 of this Chapter (see below for the listing of hazardous constituents regulated under the Resource Conservation and Recovery Act. The constituents provided herewith in both alphabetical and CAS order were excerpted from 40CFR261 Appendix VIII, 1995 edition).

CAS #	chemical name (alphabetical order)
30558-431	A2213
75-058	Acetonitrile
98-862	Acetophenone
53-963	2-Acetylaminefluarone
75-365	Acetyl chloride
591-082	1-Acetyl-2-thiourea
107-028	Acrolein
79-061	Acrylamide
107-131	Acrylonitrile
1402-682	Aflatoxins
116-063	Aldicarb
1646-884	Aldicarb sulfone
309-002	Aldrin
107-186	Allyl alcohol
107-186	Allyl chloride
20859-738	Aluminum phosphide
92-671	4-Aminobiphenyl
2763-964	5-(Aminomethyl)-3-isoxazolol
504-245	4-Aminopyridine
61-825	Amitrole
7803-556	Ammonium vanadate
62-533	Aniline
7440-360	Antimony
(N/A)	Antimony compounds, NOS{1}
140-578	Aramite
7440-382	Arsenic
(N/A)	Arsenic compounds, NOS{1}
7778-394	Arsenic acid
1303-282	Arsenic pentoxide
1327-533	Arsenic trioxide
492-808	Auramine
115-026	Azaserine
101-279	Barban
7440-393	Barium
(N/A)	Barium compounds, NOS{1}
542-621	Barium cyanide
22781-233	Bendiocarb
22961-826	Bendiocarb phenol
17804-352	Benomyl
225-514	Benz[c]acridine
56-553	Benz[a]anthracene
98-873	Benzal chloride
71-432	Benzene
98-055	Benzenearsonic acid
92-875	Benzidine
205-992	Benzo[b]fluoranthene
205-823	Benzo[j]fluoranthene
207-089	Benzo(k)fluoranthene
50-328	Benzo[a]pyrene
106-514	p-Benzoquinone
98-077	Benzotrichloride
100-447	Benzyl chloride
7440-417	Beryllium powder
(N/A)	Beryllium compounds, NOS{1}
120-547	Bis (pentamethylene)-thiuram tetrasulfide
598-312	Bromoacetone
75-252	Bromoform
101-553	4-Bromophenyl phenyl ether
357-573	Brucine
2008-415	Butylate
85-687	Butyl benzyl phthalate
75-605	Cacodylic acid
7440-439	Cadmium
(N/A)	Cadmium compounds, NOS{1}
13765-190	Calcium chromate
592-018	Calcium cyanide
63-252	Carbaryl
10605-217	Carbendazim
1563-662	Carbofuran
1563-388	Carbofuran phenol
75-150	Carbon disulfide
353-504	Carbon oxyfluoride
56-235	Carbon tetrachloride
55285-148	Carbosulfan
75-876	Chloral

305-033	Chlorambucil
57-749	Chlordane
(N/A)	Chlordane (alpha and gamma isomers)
(N/A)	Chlorinated benzenes, NOS{1}
(N/A)	Chlorinated ethane, NOS{1}
(N/A)	Chlorinated fluorocarbons, NOS{1}
(N/A)	Chlorinated naphthalene, NOS{1}
(N/A)	Chlorinated phenol, NOS{1}
494-031	Chlornaphazin
107-200	Chloroacetaldehyde
(N/A)	Chloroalkyl ethers, NOS{1}
106-478	p-Chloroaniline
108-907	Chlorobenzene
510-156	Chlorobenzilate
59-507	p-Chloro-m-cresol
110-758	2-Chloroethyl vinyl ether
67-663	Chloroform
107-302	Chloromethyl methyl ether
91-587	beta-Chloronaphthalene
95-578	o-Chlorophenol
5344-821	1-(o-Chlorophenyl)thiourea
126-998	Chloroprene
542-767	3-Chloropropionitrile
7440-473	Chromium
(N/A)	Chromium compounds, NOS{1}
218-019	Chrysene
6358-538	Citrus red No 2
8007-452	Coal tar creosote
544-923	Copper cyanide
137-291	Copper dimethyldithiocarbamate
(N/A)	Creosote
1319-773	Cresol (Cresylic acid)
4170-303	Crotonaldehyde
64-006	m-Cumenyl methylcarbamate
(N/A)	Cyanides (soluble salts and complexes) NOS{1}
460-195	Cyanogen
506-683	Cyanogen bromide
506-774	Cyanogen chloride
14901-087	Cycasin
1134-232	Cycloate
131-895	2-Cyclohexyl-4,6-dinitrophenol
50-180	Cyclophosphamide
94-757	2,4-D
(N/A)	2,4-D, salts, esters
20830-813	Daunomycin
533-744	Dazomet
72-548	DDD
72-559	DDE
50-293	DDT
2303-164	Diallate
226-368	Dibenz[a,h]acridine
224-420	Dibenz[a,j]acridine
53-703	Dibenz[a,h]anthracene
194-592	7H-Dibenzo[c,g]carbazole
192-654	Dibenzo[a,e]pyrene
189-640	Dibenzo[a,h]pyrene
189-559	Dibenzo[a,i]pyrene
96-128	1,2-Dibromo-3-chloropropane
84-742	Dibutyl phthalate
95-501	o-Dichlorobenzene
541-731	m-Dichlorobenzene
106-467	p-Dichlorobenzene
25321-226	Dichlorobenzene, NOS{1}
91-941	3,3'-Dichlorobenzidine
764-410	1,4-Dichloro-2-butene
75-718	Dichlorodifluoromethane
25323-302	Dichloroethylene, NOS{1}
75-354	1,1-Dichloroethylene
156-605	1,2-Dichloroethylene
111-444	Dichloroethyl ether
108-601	Dichloroisopropyl ether
111-911	Dichloromethoxy ethane
542-881	Dichloromethyl ether
120-832	2,4-Dichlorophenol
87-650	2,6-Dichlorophenol
696-286	Dichlorophenylarsine
26638-197	Dichloropropane, NOS{1}
26545-733	Dichloropropanol, NOS{1}
26952-238	Dichloropropene, NOS{1}
542-756	1,3-Dichloropropene
60-571	Dieldrin
1464-535	1,2:3,4-Diepoxybutane
692-422	Diethylarsine
5952-261	Diethylene glycol, dicarbamate
123-911	1,4-Diethyleneoxide
117-817	Diethylhexyl phthalate
1615-801	N,N'-Diethylhydrazine
3288-582	O,O-Diethyl S-methyl dithiophosphate
311-455	Diethyl-p-nitrophenyl phosphate
84-662	Diethyl phthalate
297-972	O,O-Diethyl O-pyrazinyl phosphoro-thioate
56-531	Diethylstilbesterol
94-586	Dihydrosafrole
55-914	Diisopropylfluorophosphate (DFP)
60-515	Dimethoate
119-904	3,3'-Dimethoxybenzidine

60-117	p-Dimethylaminoazobenzene
57-976	7,12-Dimethylbenz[a]anthracene
119-937	3,3'-Dimethylbenzidine
79-447	Dimethylcarbamoyl chloride
57-147	1,1-Dimethylhydrazine
540-738	1,2-Dimethylhydrazine
122-098	alpha,alpha-Dimethylphenethylamine
105-679	2,4-Dimethylphenol
131-113	Dimethyl phthalate
77-781	Dimethyl sulfate
644-644	Dimetilan
25154-545	Dinitrobenzene, NOS{1}
534-521	4,6-Dinitro-o-cresol
(N/A)	4,6-Dinitro-o-cresol salts
51-285	2,4-Dinitrophenol
121-142	2,4-Dinitrotoluene
606-202	2,6-Dinitrotoluene
88-857	Dinoseb
117-840	Di-n-octyl phthalate
122-394	Diphenylamine
122-667	1,2-Diphenylhydrazine
621-647	Di-n-propylnitrosamine
97-778	Disulfiram
298-044	Disulfoton
541-537	Dithiobiuret
115-297	Endosulfan
145-733	Endothall
72-208	Endrin
(N/A)	Endrin metabolites
106-898	Epichlorohydrin
51-434	Epinephrine
759-944	EPTC
51-796	Ethyl carbamate (urethane)
107-120	Ethyl cyanide
111-546	Ethylenebisdithiocarbamic acid
(N/A)	Ethylenebisdithiocarbamic acid, salts and esters
106-934	Ethylene dibromide
107-062	Ethylene dichloride
110-805	Ethylene glycol monoethyl ether
151-564	Ethyleneimine
75-218	Ethylene oxide
96-457	Ethylenethiourea
75-343	Ethylidene dichloride
97-632	Ethyl methacrylate
62-500	Ethyl methanesulfonate
14324-551	Ethyl Ziram
52-857	Famphur
14484-641	Ferbam
206-440	Fluoranthene
7782-414	Fluorine
640-197	Fluoroacetamide
62-748	Fluoroacetic acid, sodium salt
50-000	Formaldehyde
23422-539	Formetanate hydrochloride
64-186	Formic acid
17702-577	Formparanate
765-344	Glycidylaldehyde
(N/A)	Halomethanes, NOS{1}
76-448	Heptachlor
1024-573	Heptachlor epoxide
(N/A)	Heptachlor epoxide (alpha, beta, and gamma isomers)
(N/A)	Heptachlorodibenzofurans
(N/A)	Heptachlorodibenzo-p-dioxins
118-741	Hexachlorobenzene
87-683	Hexachlorobutadiene
77-474	Hexachlorocyclopentadiene
(N/A)	Hexachlorodibenzo-p-dioxins
(N/A)	Hexachlorodibenzofurans
67-721	Hexachloroethane
70-304	Hexachlorophene
1888-717	Hexachloropropene
757-584	Hexaethyl tetraphosphate
302-012	Hydrazine
74-908	Hydrogen cyanide
7664-393	Hydrogen fluoride
7783-064	Hydrogen sulfide
193-395	Indeno[1,2,3-cd]pyrene
55406-536	3-Iodo-2-propynyl n-butylcarbamate
78-831	Isobutyl alcohol
465-736	Isodrin
119-380	Isolan
120-581	Isosafrole
143-500	Kepone
303-341	Lasiocarpine
7439-921	Lead
(N/A)	Lead compounds, NOS1
301-042	Lead acetate
7446-277	Lead phosphate
1335-326	Lead subacetate
58-899	Lindane
108-316	Maleic anhydride
123-331	Maleic hydrazide
109-773	Malononitrile
15339-363	Manganese dimethyldithiocarbamate
148-823	Melphalan
7439-976	Mercury
(N/A)	Mercury compounds, NOS1
628-864	Mercury fulminate

137-428	Metam Sodium		55-630	Nitroglycerin
126-987	Methacrylonitrile		100-027	p-Nitrophenol
91-805	Methapyrilene		79-469	2-Nitropropane
2032-657	Methiocarb		35576-911	Nitrosamines, NOS{1}
16752-775	Methomyl		924-163	N-Nitrosodi-n-butylamine
72-435	Methoxychlor		1116-547	N-Nitrosodiethanolamine
74-839	Methyl bromide		55-185	N-Nitrosodiethylamine
74-873	Methyl chloride		62-759	N-Nitrosodimethylamine
79-221	Methyl chlorocarbonate		759-739	N-Nitroso-N-ethylurea
71-556	Methyl chloroform		10595-956	N-Nitrosomethylethylamine
56-495	3-Methylcholanthrene		684-935	N-Nitroso-N-methylurea
101-144	4,4'-Methylenebis(2-chloroaniline)		615-532	N-Nitroso-N-methylurethane
74-953	Methylene bromide		4549-400	N-Nitrosomethylvinylamine
75-092	Methylene chloride		59-892	N-Nitrosomorpholine
78-933	Methyl ethyl ketone (MEK)		16543-558	N-Nitrosonornicotine
1338-234	Methyl ethyl ketone peroxide		100-754	N-Nitrosopiperidine
60-344	Methyl hydrazine		930-552	N-Nitrosopyrrolidine
74-884	Methyl iodide		13256-229	N-Nitrososarcosine
624-839	Methyl isocyanate		99-558	5-Nitro-o-toluidine
75-865	2-Methyllactonitrile		152-169	Octamethylpyrophosphoramide
80-626	Methyl methacrylate		20816-120	Osmium tetroxide
66-273	Methyl methanesulfonate		23135-220	Oxamyl
298-000	Methyl parathion		123-637	Paraldehyde
56-042	Methylthiouracil		56-382	Parathion
1129-415	Metolcarb		1114-712	Pebulate
315-184	Mexacarbate		608-935	Pentachlorobenzene
50-077	Mitomycin C		(N/A)	Pentachlorodibenzo-p-dioxins
70-257	MNNG		(N/A)	Pentachlorodibenzofurans
2212-671	Molinate		76-017	Pentachloroethane
505-602	Mustard gas		82-688	Pentachloronitrobenzene (PCNB)
91-203	Naphthalene		87-865	Pentachlorophenol
130-154	1,4-Naphthoquinone		62-442	Phenacetin
134-327	alpha-Naphthylamine		108-952	Phenol
91-598	beta-Naphthylamine		25265-763	Phenylenediamine
86-884	alpha-Naphthylthiourea		62-384	Phenylmercury acetate
7440-020	Nickel		103-855	Phenylthiourea
(N/A)	Nickel compounds, NOS{1}		75-445	Phosgene
13463-393	Nickel carbonyl		7803-512	Phosphine
557-197	Nickel cyanide		298-022	Phorate
54-115	Nicotine		(N/A)	Phthalic acid esters, NOS{1}
(N/A)	Nicotine salts		85-449	Phthalic anhydride
10102-439	Nitric oxide		57-476	Physostigmine
100-016	p-Nitroaniline		57-647	Physostigmine salicylate
98-953	Nitrobenzene		109-068	2-Picoline
10102-440	Nitrogen dioxide		(N/A)	Polychlorinated biphenyls, NOS{1}
51-752	Nitrogen mustard		151-508	Potassium cyanide
(N/A)	Nitrogen mustard, hydrochloride salt		128-030	Potassium dimethyldithiocarbamate
126-852	Nitrogen mustard N-oxide		51026-289	Potassium hyroxymethyl-n-methyl-dithiocarbamate
(N/A)	Nitrogen mustard, N-oxide, hydro-chloride salt		(N/A)	Potassium n-methyldithiocarbamate

7778-736	Potassium pentachlorophenate
506-616	Potassium silver cyanide
2631-370	Promecarb
23950-585	Pronamide
1120-714	1,3-Propane sultone
122-429	Propham
114-261	Propoxur
107-108	n-Propylamine
107-197	Propargyl alcohol
78-875	Propylene dichloride
75-558	1,2-Propylenimine
51-525	Propylthiouracil
52888-809	Prosulfocarb
110-861	Pyridine
50-555	Reserpine
108-463	Resorcinol
81-072	Saccharin
(N/A)	Saccharin salts
94-597	Safrole
7782-492	Selenium
(N/A)	Selenium compounds, NOS{1}
7783-008	Selenium dioxide
7488-564	Selenium sulfide
144-343	Selenium, tetrakis (dimethyl-dithiocarbamate
630-104	Selenourea
7440-224	Silver
(N/A)	Silver compounds, NOS{1}
506-649	Silver cyanide
93-721	Silvex (2,4,5-TP)
143-339	Sodium cyanide
136-301	Sodium dibutyldithiocarbamate
148-185	Sodium diethyldithiocarbamate
128-041	Sodium dimethyldithiocarbamate
131-522	Sodium pentachlorophenate
18883-664	Streptozotocin
57-249	Strychnine
(N/A)	Strychnine salts
95-067	Sulfallate
1746-016	TCDD
1634-022	Tetrabutylthiuram disulfide
97-745	Tetrabutylthiuram monosulfide
95-943	1,2,4,5-Tetrachlorobenzene
(N/A)	Tetrachlorodibenzo-p-dioxins
(N/A)	Tetrachlorodibenzofurans
25322-207	Tetrachloroethane, NOS{1}
630-206	1,1,1,2-Tetrachloroethane
79-345	1,1,2,2-Tetrachloroethane
127-184	Tetrachloroethylene
58-902	2,3,4,6-Tetrachlorophenol

53535-276	2,3,4,6-tetrachlorophenol, potassium salt
25567-559	2,3,4,6-tetrachlorophenol, sodium salt
3689-245	Tetraethyldithiopyrophosphate
78-002	Tetraethyl lead
107-493	Tetraethyl pyrophosphate
509-148	Tetranitromethane
7440-280	Thallium
(N/A)	Thallium compounds, NOS{1}
1314-325	Thallic oxide
563-688	Thallium(I) acetate
6533-739	Thallium(I) carbonate
7791-120	Thallium(I) chloride
10102-451	Thallium(I) nitrate
12039-520	Thallium selenite
7446-186	Thallium(I) sulfate
62-555	Thioacetamide
39196-184	Thiofanox
74-931	Thiomethanol
108-985	Thiophenol
79-196	Thiosemicarbazide
62-566	Thiourea
59669-260	Thiodicarb
23564-058	Thiophanate-methyl
137-268	Thiram
26419-738	Tirpate
108-883	Toluene
25376-458	Toluenediamine
95-807	Toluene-2,4-diamine
823-405	Toluene-2,6-diamine
496-720	Toluene-3,4-diamine
26471-625	Toluene diisocyanate
95-534	o-Toluidine
636-215	o-Toluidine hydrochloride
106-490	p-Toluidine
8001-352	Toxaphene
2303-175	Triallate
120-821	1,2,4-Trichlorobenzene
79-005	1,1,2-Trichloroethane
79-016	Trichloroethylene
75-707	Trichloromethanethiol
75-694	Trichloromonofluoromethane
95-954	2,4,5-Trichlorophenol
88-062	2,4,6-Trichlorophenol
93-765	2,4,5-T
25735-299	Trichloropropane, NOS{1}
96-184	1,2,3-Trichloropropane
121-448	Triethylamine
126-681	O,O,O-Triethyl phosphorothioate
99-354	1,3,5-Trinitrobenzene

CAS #	chemical name	CAS #	chemical name
52-244	Tris(1-aziridinyl)phosphine sulfide	58-899	Lindane
126-727	Tris(2,3-dibromopropyl) phosphate	58-902	2,3,4,6-Tetrachlorophenol
72-571	Trypan blue	59-507	p-Chloro-m-cresol
66-751	Uracil mustard	59-892	N-Nitrosomorpholine
1314-621	Vanadium pentoxide	60-117	p-Dimethylaminoazobenzene
1929-777	Vernolate	60-344	Methyl hydrazine
75-014	Vinyl chloride	60-515	Dimethoate
81-812	Warfarin	60-571	Dieldrin
81-812	Warfarin	61-825	Amitrole
(N/A)	Warfarin salts, when present at concentrations less than 03%	62-384	Phenylmercury acetate
		62-442	Phenacetin
(N/A)	Warfarin salts, when present at concentrations greater than 03%	62-500	Ethyl methanesulfonate
		62-533	Aniline
557-211	Zinc cyanide	62-555	Thioacetamide
1314-847	Zinc phosphide	62-566	Thiourea
1314-847	Zinc phosphide	62-748	Fluoroacetic acid, sodium salt
137-304	Ziram	62-759	N-Nitrosodimethylamine
CAS #		63-252	Carbaryl
(order)	chemical name	64-006	m-Cumenyl methylcarbamate
50-000	Formaldehyde	64-186	Formic acid
50-077	Mitomycin C	66-273	Methyl methanesulfonate
50-180	Cyclophosphamide	66-751	Uracil mustard
50-293	DDT	67-663	Chloroform
50-328	Benzo[a]pyrene	67-721	Hexachloroethane
50-555	Reserpine	70-257	MNNG
51-285	2,4-Dinitrophenol	70-304	Hexachlorophene
51-434	Epinephrine	71-432	Benzene
51-525	Propylthiouracil	71-556	Methyl chloroform
51-752	Nitrogen mustard	72-208	Endrin
51-796	Ethyl carbamate (urethane)	72-435	Methoxychlor
52-244	Tris(1-aziridinyl)phosphine sulfide	72-548	DDD
52-857	Famphur	72-559	DDE
53-703	Dibenz[a,h]anthracene	72-571	Trypan blue
53-963	2-Acetylaminefluarone	74-839	Methyl bromide
54-115	Nicotine	74-873	Methyl chloride
55-185	N-Nitrosodiethylamine	74-884	Methyl iodide
55-630	Nitroglycerin	74-908	Hydrogen cyanide
55-914	Diisopropylfluorophosphate (DFP)	74-931	Thiomethanol
56-042	Methylthiouracil	74-953	Methylene bromide
56-235	Carbon tetrachloride	75-014	Vinyl chloride
56-382	Parathion	75-058	Acetonitrile
56-495	3-Methylcholanthrene	75-092	Methylene chloride
56-531	Diethylstilbesterol	75-150	Carbon disulfide
56-553	Benz[a]anthracene	75-218	Ethylene oxide
57-147	1,1-Dimethylhydrazine	75-252	Bromoform
57-249	Strychnine	75-343	Ethylidene dichloride
57-476	Physostigmine	75-354	1,1-Dichloroethylene
57-647	Physostigmine salicylate	75-365	Acetyl chloride
57-749	Chlordane	75-445	Phosgene
57-976	7,12-Dimethylbenz[a]anthracene	75-558	1,2-Propylenimine

75-605	Cacodylic acid	95-501	o-Dichlorobenzene
75-694	Trichloromonofluoromethane	95-534	o-Toluidine
75-707	Trichloromethanethiol	95-578	o-Chlorophenol
75-718	Dichlorodifluoromethane	95-807	Toluene-2,4-diamine
75-865	2-Methyllactonitrile	95-943	1,2,4,5-Tetrachlorobenzene
75-876	Chloral	95-954	2,4,5-Trichlorophenol
76-017	Pentachloroethane	96-128	1,2-Dibromo-3-chloropropane
76-448	Heptachlor	96-184	1,2,3-Trichloropropane
77-474	Hexachlorocyclopentadiene	96-457	Ethylenethiourea
77-781	Dimethyl sulfate	97-632	Ethyl methacrylate
78-002	Tetraethyl lead	97-745	Tetrabutylthiuram monosulfide
78-831	Isobutyl alcohol	97-778	Disulfiram
78-875	Propylene dichloride	98-055	Benzenearsonic acid
78-933	Methyl ethyl ketone (MEK)	98-077	Benzotrichloride
79-005	1,1,2-Trichloroethane	98-862	Acetophenone
79-016	Trichloroethylene	98-873	Benzal chloride
79-061	Acrylamide	98-953	Nitrobenzene
79-196	Thiosemicarbazide	99-354	1,3,5-Trinitrobenzene
79-221	Methyl chlorocarbonate	99-558	5-Nitro-o-toluidine
79-345	1,1,2,2-Tetrachloroethane	100-016	p-Nitroaniline
79-447	Dimethylcarbamoyl chloride	100-027	p-Nitrophenol
79-469	2-Nitropropane	100-447	Benzyl chloride
80-626	Methyl methacrylate	100-754	N-Nitrosopiperidine
81-072	Saccharin	101-144	4,4'-Methylenebis(2-chloroaniline)
81-812	Warfarin	101-279	Barban
81-812	Warfarin	101-553	4-Bromophenyl phenyl ether
82-688	Pentachloronitrobenzene (PCNB)	103-855	Phenylthiourea
84-662	Diethyl phthalate	105-679	2,4-Dimethylphenol
84-742	Dibutyl phthalate	106-467	p-Dichlorobenzene
85-449	Phthalic anhydride	106-478	p-Chloroaniline
85-687	Butyl benzyl phthalate	106-490	p-Toluidine
86-884	alpha-Naphthylthiourea	106-514	p-Benzoquinone
87-650	2,6-Dichlorophenol	106-898	Epichlorohydrin
87-683	Hexachlorobutadiene	106-934	Ethylene dibromide
87-865	Pentachlorophenol	107-028	Acrolein
88-062	2,4,6-Trichlorophenol	107-062	Ethylene dichloride
88-857	Dinoseb	107-108	n-Propylamine
91-203	Naphthalene	107-120	Ethyl cyanide
91-587	beta-Chloronaphthalene	107-131	Acrylonitrile
91-598	beta-Naphthylamine	107-186	Allyl alcohol
91-805	Methapyrilene	107-186	Allyl chloride
91-941	3,3'-Dichlorobenzidine	107-197	Propargyl alcohol
92-671	4-Aminobiphenyl	107-200	Chloroacetaldehyde
92-875	Benzidine	107-302	Chloromethyl methyl ether
93-721	Silvex (2,4,5-TP)	107-493	Tetraethyl pyrophosphate
93-765	2,4,5-T	108-316	Maleic anhydride
94-586	Dihydrosafrole	108-463	Resorcinol
94-597	Safrole	108-601	Dichloroisopropyl ether
94-757	2,4-D	108-883	Toluene
95-067	Sulfallate	108-907	Chlorobenzene

108-952	Phenol	137-304	Ziram
108-985	Thiophenol	137-428	Metam Sodium
109-068	2-Picoline	140-578	Aramite
109-773	Malononitrile	143-339	Sodium cyanide
110-758	2-Chloroethyl vinyl ether	143-500	Kepone
110-805	Ethylene glycol monoethyl ether	144-343	Selenium, tetrakis (dimethyl-
110-861	Pyridine		dithiocarbamate
111-444	Dichloroethyl ether	145-733	Endothall
111-546	Ethylenebisdithiocarbamic acid	148-185	Sodium diethyldithiocarbamate
111-911	Dichloromethoxy ethane	148-823	Melphalan
114-261	Propoxur	151-508	Potassium cyanide
115-026	Azaserine	151-564	Ethyleneimine
115-297	Endosulfan	152-169	Octamethylpyrophosphoramide
116-063	Aldicarb	156-605	1,2-Dichloroethylene
117-817	Diethylhexyl phthalate	189-559	Dibenzo[a,i]pyrene
117-840	Di-n-octyl phthalate	189-640	Dibenzo[a,h]pyrene
118-741	Hexachlorobenzene	192-654	Dibenzo[a,e]pyrene
119-380	Isolan	193-395	Indeno[1,2,3-cd]pyrene
119-904	3,3'-Dimethoxybenzidine	194-592	7H-Dibenzo[c,g]carbazole
119-937	3,3'-Dimethylbenzidine	205-823	Benzo[j]fluoranthene
120-547	Bis (pentamethylene)-thiuram	205-992	Benzo[b]fluoranthene
	tetrasulfide	206-440	Fluoranthene
120-581	Isosafrole	207-089	Benzo(k)fluoranthene
120-821	1,2,4-Trichlorobenzene	218-019	Chrysene
120-832	2,4-Dichlorophenol	224-420	Dibenz[a,j]acridine
121-142	2,4-Dinitrotoluene	225-514	Benz[c]acridine
121-448	Triethylamine	226-368	Dibenz[a,h]acridine
122-098	alpha,alpha-Dimethylphenethylamine	297-972	O,O-Diethyl O-pyrazinyl phosphoro-
122-394	Diphenylamine		thioate
122-429	Propham	298-000	Methyl parathion
122-667	1,2-Diphenylhydrazine	298-022	Phorate
123-331	Maleic hydrazide	298-044	Disulfoton
123-637	Paraldehyde	301-042	Lead acetate
123-911	1,4-Diethyleneoxide	302-012	Hydrazine
126-681	O,O,O-Triethyl phosphorothioate	303-341	Lasiocarpine
126-727	Tris(2,3-dibromopropyl) phosphate	305-033	Chlorambucil
126-852	Nitrogen mustard N-oxide	309-002	Aldrin
126-987	Methacrylonitrile	311-455	Diethyl-p-nitrophenyl phosphate
126-998	Chloroprene	315-184	Mexacarbate
127-184	Tetrachloroethylene	353-504	Carbon oxyfluoride
128-030	Potassium dimethyldithiocarbamate	357-573	Brucine
128-041	Sodium dimethyldithiocarbamate	460-195	Cyanogen
130-154	1,4-Naphthoquinone	465-736	Isodrin
131-113	Dimethyl phthalate	492-808	Auramine
131-522	Sodium pentachlorophenate	494-031	Chlornaphazin
131-895	2-Cyclohexyl-4,6-dinitrophenol	496-720	Toluene-3,4-diamine
134-327	alpha-Naphthylamine	504-245	4-Aminopyridine
136-301	Sodium dibutyldithiocarbamate	505-602	Mustard gas
137-268	Thiram	506-616	Potassium silver cyanide
137-291	Copper dimethyldithiocarbamate	506-649	Silver cyanide

506-683	Cyanogen bromide	1314-621	Vanadium pentoxide
506-774	Cyanogen chloride	1314-847	Zinc phosphide
509-148	Tetranitromethane	1314-847	Zinc phosphide
510-156	Chlorobenzilate	1319-773	Cresol (Cresylic acid)
533-744	Dazomet	1327-533	Arsenic trioxide
534-521	4,6-Dinitro-o-cresol	1335-326	Lead subacetate
540-738	1,2-Dimethylhydrazine	1338-234	Methyl ethyl ketone peroxide
541-537	Dithiobiuret	1402-682	Aflatoxins
541-731	m-Dichlorobenzene	1464-535	1,2:3,4-Diepoxybutane
542-621	Barium cyanide	1563-388	Carbofuran phenol
542-756	1,3-Dichloropropene	1563-662	Carbofuran
542-767	3-Chloropropionitrile	1615-801	N,N'-Diethylhydrazine
542-881	Dichloromethyl ether	1634-022	Tetrabutylthiuram disulfide
544-923	Copper cyanide	1646-884	Aldicarb sulfone
557-197	Nickel cyanide	1746-016	TCDD
557-211	Zinc cyanide	1888-717	Hexachloropropene
563-688	Thallium(I) acetate	1929-777	Vernolate
591-082	1-Acetyl-2-thiourea	2008-415	Butylate
592-018	Calcium cyanide	2032-657	Methiocarb
598-312	Bromoacetone	2212-671	Molinate
606-202	2,6-Dinitrotoluene	2303-164	Diallate
608-935	Pentachlorobenzene	2303-175	Triallate
615-532	N-Nitroso-N-methylurethane	2631-370	Promecarb
621-647	Di-n-propylnitrosamine	2763-964	5-(Aminomethyl)-3-isoxazolol
624-839	Methyl isocyanate	3288-582	O,O-Diethyl S-methyl
628-864	Mercury fulminate		dithiophosphate
630-104	Selenourea	35576-911	Nitrosamines, NOS{1}
630-206	1,1,1,2-Tetrachloroethane	3689-245	Tetraethyldithiopyrophosphate
636-215	o-Toluidine hydrochloride	4170-303	Crotonaldehyde
640-197	Fluoroacetamide	4549-400	N-Nitrosomethylvinylamine
644-644	Dimetilan	5344-821	1-(o-Chlorophenyl)thiourea
684-935	N-Nitroso-N-methylurea	5952-261	Diethylene glycol, dicarbamate
692-422	Diethylarsine	6358-538	Citrus red No 2
696-286	Dichlorophenylarsine	6533-739	Thallium(I) carbonate
757-584	Hexaethyl tetraphosphate	7439-921	Lead
759-739	N-Nitroso-N-ethylurea	7439-976	Mercury
759-944	EPTC	7440-020	Nickel
764-410	1,4-Dichloro-2-butene	7440-224	Silver
765-344	Glycidylaldehyde	7440-280	Thallium
823-405	Toluene-2,6-diamine	7440-360	Antimony
924-163	N-Nitrosodi-n-butylamine	7440-382	Arsenic
930-552	N-Nitrosopyrrolidine	7440-393	Barium
1024-573	Heptachlor epoxide	7440-417	Beryllium powder
1114-712	Pebulate	7440-439	Cadmium
1116-547	N-Nitrosodiethanolamine	7440-473	Chromium
1120-714	1,3-Propane sultone	7446-186	Thallium(I) sulfate
1129-415	Metolcarb	7446-277	Lead phosphate
1134-232	Cycloate	7488-564	Selenium sulfide
1303-282	Arsenic pentoxide	7664-393	Hydrogen fluoride
1314-325	Thallic oxide	7778-394	Arsenic acid

7778-736	Potassium pentachlorophenate	30558-431	A2213
7782-414	Fluorine	39196-184	Thiofanox
7782-492	Selenium	51026-289	Potassium hyroxymethyl-n-methyl-
7783-008	Selenium dioxide		dithiocarbamate
7783-064	Hydrogen sulfide	52888-809	Prosulfocarb
7791-120	Thallium(I) chloride	53535-276	2,3,4,6-tetrachlorophenol, potassium
7803-512	Phosphine		salt
7803-556	Ammonium vanadate	55285-148	Carbosulfan
8001-352	Toxaphene	55406-536	3-Iodo-2-propynyl n-butylcarbamate
8007-452	Coal tar creosote	59669-260	Thiodicarb
10102-439	Nitric oxide	(N/A)	2,4-D, salts, esters
10102-440	Nitrogen dioxide	(N/A)	4,6-Dinitro-o-cresol salts
10102-451	Thallium(I) nitrate	(N/A)	Antimony compounds, NOS{1}
10595-956	N-Nitrosomethylethylamine	(N/A)	Arsenic compounds, NOS{1}
10605-217	Carbendazim	(N/A)	Barium compounds, NOS{1}
12039-520	Thallium selenite	(N/A)	Beryllium compounds, NOS{1}
13256-229	N-Nitrososarcosine	(N/A)	Cadmium compounds, NOS{1}
13463-393	Nickel carbonyl	(N/A)	Chlordane (alpha and gamma
13765-190	Calcium chromate		isomers)
14324-551	Ethyl Ziram	(N/A)	Chlorinated benzenes, NOS{1}
14484-641	Ferbam	(N/A)	Chlorinated ethane, NOS{1}
14901-087	Cycasin	(N/A)	Chlorinated fluorocarbons, NOS{1}
15339-363	Manganese dimethyldithiocarbamate	(N/A)	Chlorinated naphthalene, NOS{1}
16543-558	N-Nitrosonornicotine	(N/A)	Chlorinated phenol, NOS{1}
16752-775	Methomyl	(N/A)	Chloroalkyl ethers, NOS{1}
17702-577	Formparanate	(N/A)	Chromium compounds, NOS{1}
17804-352	Benomyl	(N/A)	Creosote
18883-664	Streptozotocin	(N/A)	Cyanides (soluble salts and
20816-120	Osmium tetroxide		complexes) NOS{1}
20830-813	Daunomycin	(N/A)	Endrin metabolites
20859-738	Aluminum phosphide	(N/A)	Ethylenebisdithiocarbamic acid, salts
22781-233	Bendiocarb		and esters
22961-826	Bendiocarb phenol	(N/A)	Halomethanes, NOS{1}
23135-220	Oxamyl	(N/A)	Heptachlor epoxide (alpha, beta, and
23422-539	Formetanate hydrochloride		gamma isomers)
23564-058	Thiophanate-methyl	(N/A)	Heptachlorodibenzo-p-dioxins
23950-585	Pronamide	(N/A)	Heptachlorodibenzofurans
25154-545	Dinitrobenzene, NOS{1}	(N/A)	Hexachlorodibenzo-p-dioxins
25265-763	Phenylenediamine	(N/A)	Hexachlorodibenzofurans
25321-226	Dichlorobenzene, NOS{1}	(N/A)	Lead compounds, NOS1
25322-207	Tetrachloroethane, NOS{1}	(N/A)	Mercury compounds, NOS1
25323-302	Dichloroethylene, NOS{1}	(N/A)	Nickel compounds, NOS{1}
25376-458	Toluenediamine	(N/A)	Nicotine salts
25567-559	2,3,4,6-tetrachlorophenol, sodium salt	(N/A)	Nitrogen mustard, hydrochloride salt
25735-299	Trichloropropane, NOS{1}	(N/A)	Nitrogen mustard, N-oxide, hydro-
26419-738	Tirpate		chloride salt
26471-625	Toluene diisocyanate	(N/A)	Pentachlorodibenzo-p-dioxins
26545-733	Dichloropropanol, NOS{1}	(N/A)	Pentachlorodibenzofurans
26638-197	Dichloropropane, NOS{1}	(N/A)	Phthalic acid esters, NOS{1}
26952-238	Dichloropropene, NOS{1}	(N/A)	Polychlorinated biphenyls, NOS{1}

(N/A)   Potassium n-methyldithiocarbamate
(N/A)   Saccharin salts
(N/A)   Selenium compounds, NOS{1}
(N/A)   Silver compounds, NOS{1}
(N/A)   Strychnine salts
(N/A)   Tetrachlorodibenzo-p-dioxins
(N/A)   Tetrachlorodibenzofurans
(N/A)   Thallium compounds, NOS{1}
(N/A)   Warfarin salts, when present at
        concentrations less than 03%
(N/A)   Warfarin salts, when present at
        concentrations greater than 03%

{1} The abbreviation NOS (not otherwise specified) signifies those members of the general class not specifically listed by name in this appendix.
(N/A): Not available.

hazardous debris (40CFR268.2-h) means debris that contains a hazardous waste listed in subpart D of part 261 of this chapter, or that exhibits a characteristic of hazardous waste identified in subpart C of part 261 of this chapter.

hazardous ranking system (EPA-94/04): The principle screening tool used by EPA to evaluate risks to public health and the environment associated with abandoned or uncontrolled hazardous waste sites. The HRS calculates a score based on the potential of hazardous substances spreading from the site through the air, surface water, or ground water, and on other factors such as density and proximity of human population. This score is the primary factor in deciding if the site should be on the National Priorities List and, if so, what ranking it should have compared to other sites on the list.

hazardous substance (40CFR122.2) means any substance designated under 40CFR116 pursuant to section 311 of CWA.

hazardous substance (40CFR2.310-4) has the meaning given it in section 101(14) of the Act (CERCLA), 42USC9601(14).

hazardous substance (40CFR300.5), as defined by section 101(14) of CERCLA, means: Any substance designated pursuant to section 311(b)(2)(A) of the CWA; any element, compound, mixture, solution, or substance designated pursuant to section 102 of CERCLA; any hazardous waste having the characteristics identified under or listed pursuant to section 3001 of the Solid Waste Disposal Act (but not including any waste the regulation of which under the Solid Waste Disposal Act has been suspended by Act of Congress); any toxic pollutant listed under section 307(a) of the CWA; any hazardous air pollutant listed under section 112 of the Clean Air Act; and any imminently hazardous chemical substance or mixture with respect to which the Administrator has taken action pursuant to section 7 of the Toxic Substances Control Act. The term does not include petroleum, including crude oil or any fraction thereof, which is not otherwise specifically listed or designated as a hazardous substance in the first sentence of this paragraph, and the term does not include natural gas, natural gas liquids, liquefied natural gas or synthetic gas usable for fuel (or mixtures of natural gas and such synthetic gas).

hazardous substance (40CFR300-AA) means CERCLA hazardous substances, pollutants, and contaminants as defined in CERCLA sections 101(14) and 101(33), except where otherwise specifically noted in the HRS.

hazardous substance (40CFR302.3) means any substance designated pursuant to 40CFR302.

hazardous substance (40CFR310.11-d), as defined by section 101(14) of CERCLA, means:
(1) Any substance designated pursuant to section 311(b)(2)(A) of the Federal Water Pollution Control Act,
(2) Any element, compound, mixture, solution, or substance designated pursuant to section 102 of CERCLA,
(3) Any hazardous waste having the characteristics identified under or listed pursuant to section 3001 of the Solid Waste Disposal Act (but not including any waste the regulation of which under the Solid Waste Disposal Act has been suspended by Act of Congress),
(4) Any toxic pollutant listed under section 307(a) of the Federal Water Pollution Control Act,
(5) Any hazardous air pollutant listed under section 112 of the Clean Air Act, and
(6) An imminently hazardous chemical substance or mixture with respect to which the Administrator has taken action pursuant to section 7 of the

Toxic Substances Control Act. The term does not include petroleum, including crude oil or any fraction thereof that is not otherwise specifically listed or designated as a hazardous substance under paragraphs (d)(1) through (d)(6) of this paragraph, and the term does not include natural gas, natural gas liquids, liquefied natural gas, or synthetic gas usable for fuel (or mixtures or natural gas and such synthetic gas).

hazardous substance (CWA309) means:
(A) any substance designated pursuant to section 311(b)(2)(A) of this Act,
(B) any element, compound, mixture, solution, or substance designated pursuant to section 102 of the Comprehensive Environmental Response, Compensation, and Liability Act of 1980,
(C) any hazardous waste having the characteristics identified under or listed pursuant to section 3001 of the Solid Waste Disposal Act (but not including any waste the regulation of which under the Solid Waste Disposal Act has been suspended by Act of Congress),
(D) any toxic pollutant listed under section 307(a) of this Act, and
(E) any imminently hazardous chemical substance or mixture with respect to which the Administrator has taken action pursuant to section 7 of the Toxic Substances Control Act.

hazardous substance (CWA311-33USC1321) means any substance designated pursuant to subsection (b)(2) of this section.

hazardous substance (EPA-94/04):
1. Any material that poses a threat to human health and/or the environment. Typical hazardous substances are toxic, corrosive, ignitable, explosive, or chemically reactive.
2. Any substance designated by EPA to be reported if a designated quantity of the substance is spilled in the waters of the United States or if otherwise released into the environment.

hazardous substance (SF101-42USC9601) means:
(A) any substance designated pursuant to section 311(b)(2)(A) of the Federal Water Pollution Control Act,
(B) any element, compound, mixture, solution, or

substance designated pursuant to section 102 of this Act,
(C) any hazardous waste having the characteristics identified under or listed pursuant to section 3001 of the Solid Waste Disposal Act (but not including any waste the regulation of which under the Solid Waste Disposal Act has been suspended by Act of Congress),
(D) any toxic pollutant listed under section 307(a) of the Federal Water Pollution Control Act,
(E) any hazardous air pollutant listed under section 112 of the Clean Air Act, and
(F) any imminently hazardous chemical substance or mixture which respect to which the Administrator has taken action pursuant to section 7 of the Toxic Substances Control Act.
The term does not include petroleum, including crude oil or any fraction thereof which is not otherwise specifically listed or designated as a hazardous substance under subparagraphs (A) through (F) of this paragraph, and the term does not include natural gas, natural gas liquids, liquified natural gas, or synthetic gas usable for fuel (or mixtures of natural gas and such synthetic gas).

hazardous substance UST system (40CFR280.12) means an underground storage tank system that contains a hazardous substance defined in section 101(14) of the Comprehensive Environmental Response, Compensation and Liability Act of 1980 (but not including any substance regulated as a hazardous waste under subtitle C) or any mixture of such substances and petroleum, and which is not a petroleum UST system.

hazardous substances (40CFR373.4-a) means hazardous substances means that group of substances defined as hazardous under CERCLA 101(14), and that appear at 40CFR302.4.

hazardous waste (40CFR144.3) means a hazardous waste as defined in 40CFR261.3 (other identical or similar definitions are provided in 40CFR146.3; 270.2).

hazardous waste (40CFR2.305-3) has the meaning given it in section 1004(5) of the Act (RCRA), 42USC6903(5).

hazardous waste (40CFR240.101-m) means any

waste or combination of wastes which pose a substantial present or potential hazard to human health or living organisms because such wastes are nondegradable or persistent in nature or because they can be biologically magnified, or because they can be lethal, or because they may otherwise cause or tend to cause detrimental cumulative effects (other identical or similar definitions are provided in 40CFR241.101-g).

hazardous waste (40CFR243.101-n) means a waste or combination of wastes of a solid, liquid, contained gaseous, or semisolid form which may cause, or contribute to, an increase in mortality or an increase in serious irreversible, or incapacitating reversible illness, taking into account the toxicity of such waste, its persistence and degradability in nature, its potential for accumulation or concentration in tissue, and other factors that may otherwise cause or contribute to adverse acute or chronic effects on the health of persons or other organisms (under the Resource Conservation Recovery Act).

hazardous waste (40CFR302.3) shall have the meaning provided in 40CFR261.3.

hazardous waste (EPA-94/04): Byproducts of society that can pose a substantial or potential hazard to human health or the environment when improperly managed. Possesses at least one of four characteristics (ignitability, corrosivity, reactivity, or toxicity), or appears on special EPA lists (cf. hazardous waste characteristics).

hazardous waste (RCRA1004-42USC6903) means a solid waste, or combination of solid wastes, which because of its quantity, concentration, or physical, chemical, or infectious characteristics may:
(A) cause, or significantly contribute to an increase in mortality or an increase in serious irreversible, or incapacitating reversible, illness; or
(B) pose a substantial present or potential hazard to human health or the environment when improperly treated, stored, transported, or disposed of, or otherwise managed.

hazardous waste characteristics (40CFR261.20[aaa]): See the following:
- Characteristics of corrosivity: See subpart C in 40CFR261.22.
- Characteristics of ignitability: See subpart C in 40CFR261.21.
- Characteristics of reactivity: See subpart C in 40CFR261.23.
- Characteristics of toxicity: See subpart C in 40CFR261.24.

hazardous waste constituent (40CFR260.10) means a constituent that caused the Administrator to list the hazardous waste in part 261, subpart D, of this chapter, or a constituent listed in Table 1 of 40CFR261.24 of this chapter (cf. hazardous constituent in Appendix VIII to 40CFR261).

hazardous waste discharge (40CFR260.10): See discharge.

hazardous waste generation (RCRA1004-42USC6903) means the act or process of producing hazardous waste.

hazardous waste landfill (EPA-94/04): An excavated or engineered site where hazardous waste is deposited and covered.

hazardous waste listing criteria: See subpart B in 40CFR261.11.

hazardous waste management (40CFR260.10): See management.

hazardous waste management (RCRA1004-42USC6903) means the systematic control of the collection, source separation, storage, transportation, processing, treatment, recovery, and disposal of hazardous wastes.

hazardous waste management facility (HWM facility) (40CFR144.3) means all contiguous land, and structures, other appurtenances, and improvements on the land used for treating, storing, or disposing of hazardous waste. A facility may consist of several treatment, storage, or disposal operational units (for example, one or more landfills, surface impoundments, or combination of them) (other identical or similar definitions are provided in 40CFR146.3; 270.2).

hazardous waste management unit (40CFR260.10) is a contiguous area of land on or in which hazardous waste is placed, or the largest area in which there is significant likelihood of mixing hazardous waste constituents in the same area. Examples of hazardous waste management units include a surface impoundment, a waste pile, a land treatment area, a landfill cell, an incinerator, a tank and its associated piping and underlying containment system and a container storage area. A container alone does not constitute a unit; the unit includes containers and the land or pad upon which they are placed.

hazardous waste management unit shutdown (40CFR264.1031) means a work practice or operational procedure that stops operation of a hazardous waste management unit or part of a hazardous waste management unit. An unscheduled work practice or operational procedure that stops operation of a hazardous waste management unit or part of a hazardous waste management unit for less than 24 hours is not a hazardous waste management unit shutdown. The use of spare equipment and technically feasible bypassing of equipment without stopping operation are not hazardous waste management unit shutdowns.

hazardous waste: See 40CFR261.3

hazardous wastestream (40CFR300-AA) means material containing CERCLA hazardous substances (as defined in CERCLA section 101(14)) that was deposited, stored, disposed, or place in, or that otherwise migrated to, a source.

hazards analysis (EPA-94/04): Procedures used to:
(1) identify potential sources of release of hazardous materials from fixed facilities or transportation accidents;
(2) determine the vulnerability of a geographical area to a release of hazardous materials; and
(3) compare hazards to determine which present greater or lesser risks to a community.

hazards identification (EPA-94/04): Providing information on which facilities have extremely hazardous substances, what those chemicals are, how much there is at each facility, how the chemicals are stored, and whether they are used at high temperatures.

HDD or 2,3,7,8-HDD (40CFR766.3) means any of the dibenzo-p-dioxins totally chlorinated or totally brominated at the following positions on the molecular structure: 2,3,7,8; 1,2,3,7,8; 1,2,3,4,7,8; 1,2,3,6,7,8; 1,2,3,7,8,9; and 1,2,3,4,7,8,9).

HDF or 2,3,7,8-HDF (40CFR766.3) means any of the dibenzofurans totally chlorinated or totally brominated at the following positions on the molecular structure: 2,3,7,8; 1,2,3,7,8; 2,3,4,7,8; 1,2,3,4,7,8; 1,2,3,6,7,8; 1,2,3,7,8,9; 2,3,4,6,7,8; 1,2,3,4,6,7,8; and 1,2,3,4,7,8,9.

headband (40CFR211.203-l) means the component of hearing protective device which applies force to, and holds in place on the head, the component which is intended to acoustically seal the ear canal.

health advisory level (EPA-94/04): A non-regulatory health-based reference level of chemical traces (usually in ppm) in drinking water at which there are no adverse health risks when ingested over various periods of time. Such levels are established for one day, 10 days, long term and life-time exposure periods. They contain a large margin of safety.

health and safety data (40CFR2.306-3) means:
(i) the information described in Paragraphs (a)(3)(i) (A), (B), and (C) of this section with respect to any chemical substance or mixture offered for commercial distribution (including for test marketing purposes and for use in research and development), any chemical substance included on the inventory of chemical substances under section of the Act (15USC2607), or any chemical substance or mixture for which testing is required under section 4 of the Act (15USC2603) or for which notification is required under section 5 of the Act (15USC2604).
(A) Any study of any effect of a chemical substance or mixture on health, on the environment, or on both, including underlying data and epidemiological studies; studies of occupational exposure to a chemical substance or mixture; and toxicological, clinical, and ecological studies of a chemical substance or mixture;
(B) Any test performed under the Act; and
(C) Any data reported to, or otherwise obtained by, EPA from a study described

in paragraph (a)(3)(i)(A) of this section or a test described in paragraph (a)(3)(i)(B) of this section.

(ii) Notwithstanding paragraph (a)(3)(i) of this section, no information shall be considered to be health and safety data if disclosure of the information would:

(A) In the case of a chemical substance or mixture, disclose processes used in the manufacturing or processing the chemical substance or mixture or,

(B) In the case of a mixture, disclose the portion of the mixture comprised by any of the chemical substances in the mixture (under TSCA).

health and safety plan (40CFR35.6015-21) means a plan that specifies the procedures that are sufficient to protect on-site personnel and surrounding communities from the physical, chemical, and/or biological hazards of the site. The health and safety plan outlines:
i. Site hazards;
ii. work areas and site control procedures;
iii. air surveillance procedures;
iv. levels of protection;
v. decontamination and site emergency plans;
vi. arrangements for weather-related problems; and
vii. responsibilities for implementing the health and safety plan.

health and safety study (TSCA3-15USC2602) means any study of any effect of a chemical substance or mixture on health or the environment or on both, including underlying data and epidemiological studies, studies of occupational exposure to a chemical substance or mixture, toxicological, clinical, and ecological studies of a chemical substance or mixture, and any test performed pursuant to this Act.

health and safety study or study (40CFR82.172) means any study of any effect of a substitute or its components on health and safety, or the environment or both, including underlying data and epidemiological studies, studies of occupational, ambient, and consumer exposure to a substitute, toxicological, clinical, and ecological, or other studies of a substitute and its components, and any other pertinent test. Chemical identity is always part of a health and safety study. Information which arises as a result of a formal, disciplined study is included in the definition. Also included is information relating to the effects of a substitute or its components on health or the environment. Any available data that bear on the effects of a substitute or its components on health or the environment would be included. Examples include:

(1) Long- and short-term tests of mutagenicity, carcinogenicity, or teratogenicity; data on behavioral disorders; dermatoxicity; pharmacological effects; mammalian absorption, distribution, metabolism, and excretion; cumulative, additive, and synergistic effects; acute, subchronic, and chronic effects; and structure/activity analyses;

(2) Tests for ecological or other environmental effects on invertebrates, fish, or other animals, and plants, including: Acute toxicity tests, chronic toxicity tests, critical life stage tests, behavioral tests, algal growth tests, seed germination tests, microbial function tests, bioconcentration or bioaccumulation tests, and model ecosystem (microcosm) studies;

(3) Assessments of human and environmental exposure, including workplace exposure, and effects of a particular substitute on the environment, including surveys, tests, and studies of: Biological, photochemical, and chemical degradation; air, water and soil transport; biomagnification and bioconcentration; and chemical and physical properties, e.g., atmospheric lifetime, boiling point, vapor pressure, evaporation rates from soil and water, octanol/water partition coefficient, and water solubility;

(4) Monitoring data, when they have been aggregated and analyzed to measure the exposure of humans or the environment to a substitute; and

(5) Any assessments of risk to health or the environment resulting from the manufacture, processing, distribution in commerce, use, or disposal of the substitute or its components.

(Other identical or similar definitions are provided in 40CFR716.3; 720.3-k; 723.50-2.)

health assessment (EPA-94/04): An evaluation of available data on existing or potential risks to human health posed by a Superfund site. The Agency for Toxic Substances and Disease Registry (ATSDR) of the Department of Health and Human Services (DHHS) is required to perform such an assessment at every site on the National Priorities List.

health professional (40CFR1910.20) means a physician, occupational health nurse, industrial hygienist, toxicologist, or epidemiologist, providing medical or other occupational health services to exposed employees.

hearing (40CFR8.33-e) means a hearing conducted as specified in this subpart to enable the Agency to decide whether to impose sanctions on a respondent for violations of the Executive Order and rules, regulations, and orders thereunder (other identical or similar definitions are provided in 40CFR22.03; 123.64; 164.2-i).

hearing clerk (40CFR22.03) means the Hearing Clerk, A-110, U.S. Environmental Protection Agency, 401 M St. SW., Washington, DC 20460 (other identical or similar definitions are provided in 40CFR27.2; 72.2; 104.2-d; 124.72; 164.2-j; 85.1807-1; 86.614.84-1; 86.1014.84-1; 86.1115.87-1; 209.3-f).

hearing examiner (40CFR8.33-f, other similar definitions are provided in 40CFR8.2-o) means hearing examiner appointed by the Assistant Administrator for Enforcement and General Counsel.

Hearing Officer (40CFR142.202-a) means an Environmental Protection Agency employee who has been delegated by the Administrator the authority to preside over a public hearing held pursuant to section 1414(g)(2) of the Safe Drinking Water Act, 42 U.S.C. 300g-3(g)(2).

hearing protective device (40CFR211.203-m) means any device or material, capable of being worn on the head or in the ear canal, that is sold wholly or in part on the basis of its ability to reduce the level of sound entering the ear. This includes devices of which hearing protection may not be the primary function, but which are nonetheless sold partially as providing hearing protection to the user. This term is used interchangeably with the terms, "hearing protector" and "device."

heat cycle (40CFR60.271a) means the period

beginning when scrap is charged to an empty EAF and ending when the EAF tap is completed or beginning when molten steel is charged to an empty AOD vessel and ending when the AOD vessel tap is completed.

heat exchange system (40CFR63.101) means any cooling tower system or once-through cooling water system (e.g., river or pond water). A heat exchange system can include more than one heat exchanger and can include an entire recirculating or once-through cooling system.

heat input (40CFR52.01-g) means the total gross calorific value (where gross calorific value is measured by ASTM Method D2015-66, D240-64, or D1826-64) of all fuels burned.

heat input (40CFR52.1881-v) means the total gross calorific value (where gross calorific value is measured by ASTM Method D2015-66, D240-64, or D1826-64) of all fossil and non-fossil fuels burned. Where two or more fossil fuel-fired steam generating units are vented to the same stack the heat input shall be the aggregate of all units vented to the stack.

heat input (40CFR60.41b) means heat derived from combustion of fuel in a steam generating unit and does not include the heat input from preheated combustion air, recirculated flue gases, or exhaust gases from other sources, such as gas turbines, internal combustion engines, kilns, etc.

heat input (40CFR60.41c) means heat derived from combustion of fuel in a steam generating unit and does not include the heat derived from preheated combustion air, recirculated flue gases, or exhaust gases from other sources (such as stationary gas turbines, internal combustion engines, and kilns).

heat input (40CFR72.2) means the product (expressed in mmBtu/time) of the gross calorific value of the fuel (expressed in Btu/lb) and the fuel feed rate into the combustion device (expressed in mass of fuel/time) and does not include the heat derived from preheated combustion air, recirculated flue gases, or exhaust from other sources (under CAA).

heat island effect (EPA-94/04): A "dome" of elevated

temperatures over an urban area caused by structural and pavement heat fluxes, and pollutant emissions.

heat rating (40CFR85.2122(a)(8)(ii)(B)) means that measurement of engine indicated mean effective pressure (IMEP) value obtained on the engine at a point when the supercharge pressure is 25.4 mm (one inch) Hg below the preignition point of the spark plug, as rated according to SAE J549A Recommended Practice.

heat release rate (40CFR60.41b) means the steam generating unit design heat input capacity (in MW or Btu/hour) divided by the furnace volume (in cubic meters or cubic feet); the furnace volume is that volume bounded by the front furnace wall where the burner is located, the furnace side waterwall, and extending to the level just below or in front of the first row of convection pass tubes.

heat time (40CFR60.271-l) means the period commencing when scrap is charged to an empty EAF and terminating when the EAF tap is completed.

heat transfer medium (40CFR60.41b) means any material which is used to transfer heat from one point to another point.

heat transfer medium (40CFR60.41c) means any material that is used to transfer heat from one point to another point.

heat treatment (40CFR467.02-l) is the application of heat of specified temperature and duration to change the physical properties of the metal (other identical or similar definitions are provided in 40CFR471.02-s).

heat treatment (40CFR468.02-j) shall mean the application or removal of heat to a workpiece to change the physical properties of the metal.

heated airless spray (40CFR52.741) means an airless spray coating method in which the coating is heated just prior to application.

heating coil (40CFR63.321) means the device used to heat the air stream circulated from the dry cleaning machine drum, after perchloroethylene has been condensed from the air stream and before the stream reenters the dry cleaning machine drum.

heating oil (40CFR280.12) means petroleum that is No. 1, No. 2, No. 4--light, No. 4--heavy, No. 5--light, No. 5--heavy, and No. 6 technical grades of fuel oil; other residual fuel oils (including Navy Special Fuel Oil and Bunker C); and other fuels when used as substitutes for one of these fuel oils. Heating oil is typically used in the operation of heating equipment, boilers, or furnaces.

heatset (40CFR52.741) means a class of web-offset lithography which requires a heated dryer to solidify the printing inks.

heatset web offset lithographic printing line (40CFR52.741) means a lithographic printing line in which a blanket cylinder is used to transfer ink from a plate cylinder to a substrate continuously fed from a roll or an extension process and an oven is used to solidify the printing inks.

heavy duty engine (40CFR86.082.2) means any engine which the engine manufacturer could reasonably expect to be used for motive power in a heavy-duty vehicle.

heavy duty vehicle (40CFR86.082.2) means any motor vehicle rated at more than 8,500 pounds GVWR or that has a vehicle curb weight of more than 6,000 pounds or that has a basic vehicle frontal area in excess of 45 square feet.

heavy duty vehicle (CAA202-42USC7521) means a truck, bus, or other vehicle manufactured primarily for use on the public streets, roads, and highways (not including any vehicle operated exclusively on a rail or rails) which has a gross vehicle weight (as determined under regulations promulgated by the Administrator) in excess of six thousand pounds. Such term includes any such vehicle which has special features enabling off-street or off-highway operation and use (cf. light duty vehicles and engines).

heavy light-duty truck (40CFR86.094.2) means any light-duty truck rated greater than 6000 lbs GVWR.

heavy light-duty truck (40CFR88.101.94) means any light-duty truck rated greater than 6000 lbs. GVWR.

heavy liquid (40CFR52.741) means liquid with a true vapor pressure of less than 0.3 kPa (0.04 psi) at 294.3°K (70°F) established in a standard reference text or as determined by ASTM method D2879-86 (incorporated by reference as specified in 40CFR52.742); or which has 0.1 Reid Vapor Pressure as determined by ASTM method D323-82 (incorporated by reference as specified in 40CFR52.742); or which when distilled requires a temperature of 421.95°K (300°F) or greater to recover 10 percent of the liquid as determined by ASTM method D86-82 (incorporated by reference as specified in 40CFR52.742).

heavy metal (40CFR191.12) means all uranium, plutonium, or thorium placed into a nuclear reactor (under AEA).

heavy metals (40CFR165.1-k) means metallic elements of higher atomic weights, including but not limited to arsenic, cadmium, copper, lead, mercury, manganese, zinc, chromium, tin, thallium, and selenium.

heavy metals (EPA-94/04): Metallic elements with high atomic weights, e.g., mercury, chromium, cadmium, arsenic, and lead; can damage living things at low concentrations and tend to accumulate in the food chain. Heptachlor: An insecticide that was banned on some food products in 1975 and all of them 1978. It was allowed for use in seed treatment until 1983. More recently it was found in milk and other dairy products in Arkansas and Missouri where dairy cattle were illegally fed treated seed.

heavy off-highway vehicle products (40CFR52.741) means, for the purpose of paragraph (e) of this section, heavy construction, mining, farming, or material handling equipment; heavy industrial engines; diesel-electric locomotives and associated power generation equipment; and the components of such equipment or engines.

heavy off-highway vehicle products coating facility (40CFR52.741) means a facility that includes one or more heavy off-highway vehicle products coating line(s).

heavy off-highway vehicle products coating line (40CFR52.741) means a coating line in which any

protective, decorative, or functional coating is applied onto the surface of heavy off-highway vehicle products.

heavy passenger cars (40CFR86.084.2) means, for the 1984 model year only, a passenger car or passenger car derivative capable of seating 12 passengers or less, rated at 6,000 pounds GVW or more and having an equivalent test weight of 5,000 pounds or more.

Henry's law constant (40CFR300-AA) means the measure of the volatility of a substance in a dilute solution of water at equilibrium. It is the ratio of the vapor pressure exerted by a substance in the gas phase over a dilute aqueous solution of that substance to its concentration in the solution at a given temperature. For HRS purposes, use the value reported at or near 25° C. [atmosphere-cubic meters per mole (atm-m^3/mol)].

HEPA (40CFR763.83): See high efficiency particulate air.

HEPA filter: (40FR763.121) See high efficiency particulate air filter.

herbicide (EPA-94/04): A chemical pesticide designed to control or destroy plants, weeds, or grasses.

herbivore (EPA-94/04): An animal that feeds on plants.

heritable translocation (40CFR798.5460-1) is one in which distal segments of nonhomologous chromosomes are involved in a reciprocal exchange (under TSCA).

heritable translocations (40CFR798.5955-3) are reciprocal translocations transmitted from parent to the succeeding progeny.

heteroallelic diploids (40CFR798.5575-2) are diploid strains of yeast carrying two different, inactive alleles of the same gene locus causing a nutritional requirement.

heterotrophic organisms (EPA-94/04): Species that are dependent on organic matter for food.

HEX-BCH (40CFR704.102-2) means the chemical substance 1,2,3,4,7,7-hexachloronorbornadiene, CAS Number 3389-717.

hexavalent chromium (40CFR749.68-10) means the oxidation state of chromium with an oxidation number of +6; a coordination number of 4 and tetrahedral geometry.

hexavalent chromium (or chromium VI) (40CFR420.02-h) means the value obtained by the method specified in 40CFR136.3.

hexavalent chromium-based water treatment chemicals (40CFR749.68-11) means any chemical containing hexavalent chromium which can be used to treat water, either alone or in combination with other chemicals, where the mixture can be used to treat water.

HFPO (40CFR704.104-1) means the chemical substance hexafluoropropylene oxide, CAS Number 428-591. [Listed in TSCA Inventory as oxirane, trifluoro(trifluoromethyl)-.]

hide (40CFR425.02-b) means any animal pelt or skin as received by a tannery as raw material to be processed.

high altitude (40CFR86.082.2) means any elevation over 1,219 meters (4,000 feet).

high altitude conditions (40CFR86.082.2) means a test altitude of 1,620 meters (5,315 feet), plus or minus 100 meters (328 feet), or equivalent observed barometric test conditions of 83.3 plus or minus 1 kilopascals.

high altitude reference point (40CFR86.082.2) means an elevation of 1,620 meters (5,315 feet) plus or minus 100 meters (32 feet), or equivalent observed barometric test conditions of 83.3 kPa (24.2 inches Hg), plus or minus 1 kPa (0.30 Hg).

high carbon ferrochrome (40CFR60.261-q) means that alloy as defined by ASTM Designation A101-73 (incorporated by reference, see 40CFR60.17) grades HC1 through HC6.

high concentration PCBs (40CFR761.123) means

PCBs that contain 500 ppm or greater PCBs, or those materials which EPA requires to be assumed to contain 500 ppm or greater PCBs in the absence of testing.

high contact industrial surface (40CFR761.123) means a surface in an industrial setting which is repeatedly touched, often for relatively long periods of time. Manned machinery and control panels are examples of high-contact industrial surfaces. High-contact industrial surfaces are generally of impervious solid material. Examples of low-contact industrial surfaces include ceilings, walls, floors, roofs, roadways and side walks in the industrial area, utility poles, unmanned machinery, concrete pads beneath electrical equipment, curbing, exterior structural building components, indoor vaults, and pipes (under TSCA).

high contact residential commercial surface (40CFR761.123) means a surface in a residential-commercial area which is repeatedly touched, often for relatively long periods of time. Doors, wall areas below 6 feet in height, uncovered flooring, windowsills, fencing, banisters, stairs, automobiles, and children's play areas such as outdoor patios and sidewalks are examples of high-contact residential-commercial surfaces. Examples of low-contact residential-commercial surfaces include interior ceilings, interior wall areas above 6 feet in height, roofs, asphalt roadways, concrete roadways, wooden utility poles, unmanned machinery, concrete pads beneath electrical equipment, curbing, exterior structural building components, (e.g., aluminum/vinyl siding, cinder block, asphalt tiles), and pipes.

high density polyethylene (HDPE) (40CFR60.561) means a thermoplastic polymer or copolymer comprised of at least 50 percent ethylene by weight and having a density of greater than 0.940 g/cm^3.

high density polyethylene (EPA-94/04): A material used to make plastic bottles and other products that produces toxic fumes when burned.

high dose (40CFR795.232-5) shall not exceed the lower explosive limit (LEL) and ideally should induce minimal toxicity (cf. low dose).

high efficiency particulate air (HEPA)

(40CFR763.83) refers to a filtering system capable of trapping and retaining at least 99.97 percent of all monodispersed particles 0.3 μm in diameter or larger.

high efficiency particulate air (HEPA) filter (40CFR763.121) means a filter capable of trapping and retaining at least 99.97 percent of all monodispersed particles of 0.3 micrometer in diameter or larger.

high grade paper (40CFR246.101-o) means letterhead, dry copy papers, miscellaneous business forms, stationery, typing paper, tablet sheets, and computer printout paper and cards, commonly sold as "white ledger," "computer printout" and "tab card" grade by the wastepaper industry (under the Resource Conservation Recovery Act).

high hazard contents (29CFR1910.35) shall be classified as those which are liable to burn with extreme rapidity or from which poisonous fumes or explosions are to be feared in the event of fire.

high heat release rate (40CFR60.41b) means a heat release rate greater than 730,000 J/sec-m^3 (70,000 Btu/hour-ft^3).

high level nuclear waste facility (EPA-94/04): Plant designed to handle disposal of used nuclear fuel, high level radioactive waste, and plutonium waste.

high level of volatile impurities (40CFR60.161-l) means a total smelter charge containing more than 0.2 weight percent arsenic, 0.1 weight percent antimony, 4.5 weight percent lead or 5.5 weight percent zinc, on a dry basis.

high level radioactive waste (HLRW) (40CFR191.02-h) means high-level radioactive waste as defined in the Nuclear Waste Policy Act of 1982 (Pub. L. 97-425).

high level radioactive waste (HLRW) (40CFR227.30) means the aqueous waste resulting from the operation of the first cycle solvent extraction system, or equivalent, and the concentrated waste from subsequent extraction cycles, or equivalent, in a facility for reprocessing irradiated reactor fuels or irradiated fuel from nuclear power reactors.

high level radioactive waste (HLW) (EPA-94/04): Waste generated in core fuel of a nuclear reactor, found at nuclear reactors or by nuclear fuel reprocessing; is a serious threat to anyone who comes near the waste without shielding (cf. low level radioactive waste.)

high line jumpers (EPA-94/04): Pipes or hoses connected to fire hydrants and laid on top of the ground to provide emergency water service for an isolated portion of a distribution system.

high pressure appliance (40CFR82.152-f) means an appliance that uses a refrigerant with a boiling point between -50 and 10 degrees Centigrade at atmospheric pressure (29.9 inches of mercury). This definition includes but is not limited to appliances using refrigerants -12, -22, -114, -500, or -502 (under CAA).

high pressure process (40CFR60.561) means the conventional production process for the manufacture of low density polyethylene in which a reaction pressure of about 15,000 psig or greater is used.

high processing packinghouse (40CFR432.41-c) shall mean a packinghouse which processes both animals slaughtered at the site and additional carcasses from outside sources (cf. low processing packinghouse).

high risk community (EPA-94/04): A community located within the vicinity of numerous sites or facilities or other potential sources of environmental exposure/health hazards which may result in high levels of exposure to contaminants or pollutants. In determining risk or potential risk, factors such as total weight of toxic contaminants, toxicity, routes of exposure, and other factors may be used.

high risk pollutant (40CFR63.51) means a hazardous air pollutant listed in Table 1 of 40CFR63.74.

high temperature aluminum coating (40CFR52.741) means a coating that is certified to withstand a temperature of 537.8°C (1000°F) for 24 hours (under CAA).

high terrain (40CFR51.166-25) means any area having an elevation 900 feet or more above the base of the stack of a source.

high terrain (40CFR52.21-25) means any area having an elevation 900 feet or more above the base of the stack of a source.

high to low dose extrapolation (EPA-94/04): The process of prediction of low exposure risk to humans from the measured high exposure-high risk data involving rodents.

high velocity air filter (HVAF) (40CFR60.471) means an air pollution control filtration device for the removal of sticky, oily, or liquid aerosol particulate matter from exhaust gas streams.

high viscosity poly(ethylene terephthalate) (40CFR60.561) means poly(ethylene terephthalate) that has an intrinsic viscosity of 0.9 or higher and is used in such applications as tire cord and seat belts.

highway (40CFR202.10-j) means the streets, roads, and public ways in any State (other identical or similar definitions are provided in 40CFR205.2-10).

highway fuel economy (40CFR600.002.85-12) means the fuel economy determined by operating a vehicle (or vehicles) over the driving schedule in the Federal highway fuel economy test procedure.

highway fuel economy test (40CFR610.11-7) means the test procedure described in 40CFR600.111(b).

highway project (40CFR51.392) is an undertaking to implement or modify a highway facility or highway-related program. Such an undertaking consists of all required phases necessary for implementation. For analytical purposes, it must be defined sufficiently to:
(1) Connect logical termini and be of sufficient length to address environmental matters on a broad scope;
(2) Have independent utility or significance, i.e., be usable and be a reasonable expenditure even if no additional transportation improvements in the area are made; and
(3) Not restrict consideration of alternatives for other reasonably foreseeable transportation improvements.
(Other identical or similar definitions are provided in 40CFR93.101.)

holder (FLPMA103-43USC1702-90) means any state or local governmental entity, individual, partnership, corporation, association, or other business entity receiving or using a right-of-way under subchapter V of this chapter.

holding of a copper converter (40CFR61.171) means suspending blowing operations while maintaining in a heated state the molten bath in the copper converter.

holding pond (EPA-94/04): A pond or reservoir, usually made of earth, built to store polluted runoff.

holocene (40CFR264.18-(a)(2)(iii)) means the most recent epoch of the quaternary period, extending from the end of the Pleistocene to the present.

holocene time (40CFR503.21-h) is the most recent epoch of the Quaternary period, extending from the end of the Pleistocene epoch to the present.

homeowner water system (EPA-94/04): Any water system which supplies piped water to a single residence.

homogeneous area (40CFR763.83) means an area of surfacing material, thermal system insulation material, or miscellaneous material that is uniform in color and texture.

homogeneous area (EPA-94/04): In accordance with Asbestos Hazard and Emergency Response Act (AHERA) definitions, an area of surfacing materials, thermal surface insulation, or miscellaneous material that is uniform in color and texture.

homolog (40CFR766.3) means a group of isomers that have the same degree of halogenation. For example, the homologous class of tetrachlorodibenzo-p-dioxins consists of all dibenzo-p-dioxins containing four chlorine atoms. When the homologous classes discussed in this part are referred to, the following abbreviations for the prefix denoting the number of halogens are used: tetra-, T (4 atoms); penta-, Pe (5 atoms); hexa-, Hx (6 atoms); hepta-, Hp (7 atoms).

hood (40CFR52.741) means a partial enclosure or canopy for capturing and exhausting, by means of a draft, the organic vapors or other fumes rising from a coating process or other source.

hood capture efficiency (40CFR52.741) means the emissions from a process which are captured by the hood and directed into a control device, expressed as a percentage of all emissions.

hood capture efficiency (EPA-94/04): Ratio of the emissions captured by a hood and directed into a control or disposal device, expressed as a percent of all emissions.

hood or enclosure (40CFR60.441) means any device used to capture fugitive volatile organic compounds.

hoods and enclosures (29CFR1910.94b) means the partial or complete enclosure around the wheel or disc through which air enters an exhaust system during operation.

horizon year (40CFR51.392) is a year for which the transportation plan describes the envisioned transportation system according to 40CFR51.404 (other identical or similar definitions are provided in 40CFR93.101).

horizontal double-spindle disc grinder (29CFR1910.94b) means a grinding machine carrying two power-driven, rotatable, coaxial, horizontal spindles upon the inside ends of which are mounted abrasive disc wheels used for grinding two surfaces simultaneously.

horizontal single-spindle disc grinder (29CFR1910.94b) means a grinding machine carrying an abrasive disc wheel upon one or both ends of a power-driven, rotatable single horizontal spindle.

hosiery products (40CFR410.51-c), for NSPS (40CFR410.55), shall mean the internal subdivision of the knit fabric finishing subcategory for facilities that are engaged primarily in dyeing or finishing hosiery of any type.

host (40CFR171.2-16) means any plant or animal on or in which another lives for nourishment, development, or protection.

host (EPA-94/04):
1. In genetics, the organism, typically a bacterium, into which a gene from another organism is transplanted.

2. In medicine, an animal infected or parasitized by another organism.

hot forming (40CFR420.71-a) means hot form whose steel operations in which solidified, heated steel is shaped by rolls.

hot metal transfer station (40CFR60.141a) means the facility where molten iron is emptied from the railroad torpedo car or hot metal car to the shop ladle. This includes the transfer of molten iron from the torpedo car or hot metal car to a mixer (or other intermediate vessel) and from a mixer (or other intermediate vessel) to the ladle. This facility is also known as the reladling station or ladle transfer station.

hot mix asphalt facility (40CFR60.91) means any facility, as described in 40CFR60.90, used to manufacture hot mix asphalt by heating and drying aggregate and mixing with asphalt cements.

hot pressing (40CFR471.02-t) is forming a powder metallurgy compact at a temperature high enough to effect concurrent sintering.

hot soak emissions (40CFR86.096.2) means evaporative emissions after termination of engine operation.

hot soak losses (40CFR86.082.2) means evaporative emissions after termination of engine operation (under CAA).

hot soak losses (40CFR86.096.2) means hot soak emissions.

hot spot analysis (40CFR51.392) is an estimation of likely future localized CO and PM10 pollutant concentrations and a comparison of those concentrations to the national ambient air quality standards. Pollutant concentrations to be estimated should be based on the total emissions burden which may result from the implementation of a single, specific project, summed together with future background concentrations (which can be estimated using the ratio of future to current traffic multiplied by the ratio of future to current emission factors) expected in the area. The total concentration must be estimated and analyzed at appropriate receptor locations in the area substantially affected by the project. Hot-spot analysis assesses impacts on a scale smaller than the entire nonattainment or maintenance area, including, for example, congested roadway intersections and highways or transit terminals, and uses an air quality dispersion model to determine the effects of emissions on air quality (other identical or similar definitions are provided in 40CFR93.101).

hot spot analysis (40CFR93.101) is an estimation of likely future localized CO and PM10 pollutant concentrations and a comparison of those concentrations to the national ambient air quality standards. Pollutant concentrations to be estimated should be based on the total emissions burden which may result from the implementation of a single, specific project, summed together with future background concentrations (which can be estimated using the ratio of future to current traffic multiplied by the ratio of future to current emission factors) expected in the area. The total concentration must be estimated and analyzed at appropriate receptor locations in the area substantially affected by the project. Hot-spot analysis assesses impacts on a scale smaller than the entire nonattainment or maintenance area, including, for example, congested roadway intersections and highways or transit terminals, and uses an air quality dispersion model to determine the effects of emissions on air quality (under CAA).

hot strip and sheet mill (40CFR420.71-h) means those steel hot forming operations that produce flat hot-rolled products other than plates.

hot water seal (40CFR467.02-m) is a heated water bath (heated to approximately 180°F) used to seal the surface coating on formed aluminum which has been anodized and coated. In establishing an effluent allowance for this operation, the hot water seal shall be classified as a cleaning or etching rinse.

hot well (40CFR264.1031) means a container for collecting condensate as in a steam condenser serving a vacuum-jet or steam-jet ejector.

hot well (40CFR52.741) means the reservoir of a condensing unit receiving the condensate from a barometric condenser.

hour (40CFR52.741) means a block period of 60 minutes (e.g., 1 a.m. to 2 a.m.).

hourly average (40CFR503.41-f) is the arithmetic mean of all measurements, taken during an hour. At least two measurements must be taken during the hour.

housed lot (40CFR412.11-g) shall mean totally roofed buildings which may be open or completely enclosed on the sides wherein animals or poultry are housed over solid concrete or dirt floors, slotted (partially open) floors over pits or manure collection areas in pens, stalls or cages, with or without bedding materials and mechanical ventilation. For the purposes hereof, the term "housed lot" is synonymous with the terms "slotted floor" buildings (swine, beef), "barn" (dairy cattle) or "stable" (horses), "houses" (turkeys, chickens), which are terms widely used in the industry.

household waste (40CFR258.2) means any solid waste (including garbage, trash, and sanitary waste in septic tanks) derived from households (including single and multiple residences, hotels and motels, bunkhouses, ranger stations, crew quarters, campgrounds, picnic grounds, and day-use recreation areas).

household waste (40CFR261.4-(b)(1)) means any material (including garbage, trash and sanitary wastes in septic tanks) derived from households (including single and multiple residences, hotels and motels, bunkhouses, ranger stations, crew quarters, campgrounds, picnic grounds and day-use recreation areas).

household waste (domestic waste) (EPA-94/04): Solid waste, composed of garbage and rubbish, which normally originated in a private home or apartment house. Domestic waste may contain a significant amount of toxic or hazardous waste.

HRGC (40CFR766.3) means high resolution gas chromatography.

HRMS (40CFR766.3) means high resolution mass spectrometry.

HRS factor (40CFR300-AA) means primary rating

elements internal to the HRS (hazard ranking system).

HRS factor category (40CFR300-AA) means a set of HRS factors (that is, likelihood of release [or exposure], waste characteristics, targets) (under CERCLA).

HRS migration pathways (40CFR300-AA) means HRS ground water, surface water, and air migration pathways.

HRS pathway (40CFR300-AA) means a set of HRS factor categories combined to produce a score to measure relative risks posed by a site in one of four environmental pathways (that is, ground water, surface water, soil, and air).

HRS site score (40CFR300-AA) means composite of the four HRS pathway scores.

human environment (40CFR1508.14) shall be interpreted comprehensively to include the natural and physical environment and the relationship of people with that environment. This means that economic or social effects are not intended by themselves to require preparation of an environmental impact statement. When an environmental impact statement is prepared and economic or social and natural or physical environmental effects are interrelated, then the environmental impact statement will discuss all of these effects on the human environment (cf. effects in 40CFR1508.8).

human equivalent dose (EPA-94/04): A dose which, when administered to humans, produces an effect equal to that produced by a dose in animals.

human exposure evaluation (EPA-94/04): Describing the nature and size of the population exposed to a substance and the magnitude and duration of their exposure. The evaluation could concern past, current, or anticipated exposures.

human health risk (EPA-94/04): The likelihood that a given exposure or series of exposures may have or will damage the health of individuals.

human subject (40CFR26.102-f) means a living individual about whom an investigator (whether professional or student) conducting research obtains:
(1) Data through intervention or interaction with the individual, or
(2) Identifiable private information.
Intervention includes both physical procedures by which data are gathered (for example, venipuncture) and manipulations of the subject or the subject's environment that are performed for research purposes. Interaction includes communication or interpersonal contact between investigator and subject. "Private information" includes information about behavior that occurs in a context in which an individual can reasonably expect that no observation or recording is taking place, and information which has been provided for specific purposes by an individual and which the individual can reasonably expect will not be made public (for example, a medical record). Private information must be individually identifiable (i.e., the identity of the subject is or may readily be ascertained by the investigator or associated with the information) in order for obtaining the information to constitute research involving human subjects.

humane (MMPA3-16USC1362) in the context of the taking of a marine mammal means that method of taking which involves the least possible degree of pain and suffering practicable to the mammal involved.

HWM facility (40CFR144.3) means Hazardous Waste Management facility (other identical or similar definitions are provided in 40CFR146.3; 270.2).

hybrid generation facility (40CFR72.2) means a plant that generates electrical energy derived from a combination of qualified renewable energy (wind, solar, biomass, or geothermal) and one or more other energy resources.

hydraulic barking (40CFR429.11-a) means a wood processing operation that removes bark from wood by the use of water under a pressure of 6.8 atm (100 psia) or greater.

hydraulic fluids (40CFR252.4-e) means petroleum-based hydraulic fluids.

hydraulic gradient (EPA-94/04): In general, the

direction of groundwater flow due to changes in the depth of the water table.

hydraulic lift tank (40CFR280.12) means a tank holding hydraulic fluid for a closed-loop mechanical system that uses compressed air or hydraulic fluid to operate lifts, elevators, and other similar devices (under RCRA).

hydrocarbon (40CFR60.111-e) means any organic compound consisting predominantly of carbon and hydrogen.

hydrocarbons (HC) (EPA-94/04): Chemical compounds that consist entirely of carbon and hydrogen.

hydrochloric acid pickling (40CFR420.91-b) means those operations in which steel products are immersed in hydrochloric acid solutions to chemically remove oxides and scale, and those rinsing operations associated with such immersions.

hydrogen gas stream (40CFR61.51-h) means a hydrogen stream formed in the chlor-alkali cell denuder.

hydrogen sulfide (HS) (EPA-94/04): Gas emitted during organic decomposition. Also a byproduct of oil refining and burning. Smells like rotten eggs and, in heavy concentration, can kill or cause illness.

hydrogeologic cycle (EPA-94/04): The natural process recycling water from the atmosphere down to (and through) the earth and back to the atmosphere again.

hydrogeology (EPA-94/04): The geology of ground water, with particular emphasis on the chemistry and movement of water.

hydrologic cycle (EPA-94/04): Movement or exchange of water between the atmosphere and the earth.

hydrology (EPA-94/04): The science dealing with the properties, distribution, and circulation of water.

hydrolysis (40CFR300-AA) means chemical reaction of a substance with water.

hydrophilic (EPA-94/04): Having a strong affinity for water.

hydrophobic (EPA-94/04): Having a strong aversion for water.

hydropneumatic (EPA-94/04): A water system, usually small, in which a water pump is automatically controlled by the air pressure in a compressed air tank.

hydrotesting (40CFR471.02-u) is the testing of piping or tubing by filling with water and pressurizing to test for integrity.

hypalon$^{(TM)}$ (chlorosulfonated polyethylene) (40CFR63.191) means a synthetic rubber produced by reacting polyethylene with chloric and sulfur dioxide, transforming the thermoplastic polyethylene into a vulcanized elastomer. The reaction is conducted in a solvent (carbon tetrachloride) reaction medium.

hypocotyl (40CFR797.2750-4) means that portion of the axis of an embryo or seedling situated between the cotyledons (seed leaves) and the radicle.

hypolimnion (EPA-94/04): Bottom waters of a thermally stratified lake. The hypolimnion of a eutrophic lake is usually low or lacking in oxygen.

*********** IIIII ***********

ice fog (40CFR60.331-f) means an atmospheric suspension of highly reflective ice crystals.

identification code or EPA I.D. number (EPA-94/04): The unique code assigned to each generator, transporter, and treatment, storage, or disposal facility by regulating agencies to facilitate identification and tracking of chemicals or hazardous waste.

identification number (40CFR89.2) means a specification (for example, model number/serial number combination) which allows a particular nonroad engine to be distinguished from other similar engines.

identifying characteristics (TSCA-AIA1) means a description of asbestos or asbestos-containing material, including:
(A) the mineral or chemical constituents (or both) of the asbestos or material by weight or volume (or both),
(B) the types or classes of the product in which the asbestos or material is contained,
(C) the designs, patterns, or textures of the product in which the asbestos or material is contained, and
(D) the means by which the product in which the asbestos or containing asbestos or asbestos-containing material.

identity (40CFR721.3) means any chemical or common name used to identify a chemical substance or a mixture containing that substance.

idle (40CFR201.1-m) means that condition where all engines capable of providing motive power to the locomotive are set at the lowest operating throttle position; and where all auxiliary non-motive power engines are not operating.

idle adjustments (40CFR51-AN) means a series of adjustments which include idle revolutions per minute, idle air/fuel ratio, and basic timing.

idle emission test (40CFR51-AN) means a sampling procedure for exhaust emissions which requires operation of the engine in the idle mode only. At a minimum, the idle test must consist of the following procedures carried out on a fully warmed-up engine: A verification that the idle revolutions per minute is within manufacturer's specified limits and a measurement of the exhaust carbon monoxide and/or hydrocarbon concentrations during the period of time from 15 to 25 seconds after the engine either was used to move the car or was run at 2,000 to 2,500 r/min with no load for 2 or 3 seconds.

ignitable (EPA-94/04): Capable of burning or causing a fire.

illicit discharge (40CFR122.26-2) means any discharge to a municipal separate storm sewer that is not composed entirely of storm water except discharges pursuant to a NPDES permit (other than the NPDES permit for discharges from the municipal separate storm sewer) and discharges resulting from fire fighting activities.

Imhoff cone (EPA-94/04): A clear cone-shaped container used to measure the volume of settleable solids in a specific volume of water.

immediate family (40CFR170.3) includes only spouse, children, stepchildren, foster children, parents, stepparents, foster parents, brothers, and sisters.

immediate use a chemical substance (40CFR721.3) is for the "immediate use" of a person if it is under the control of, and used only by, the person who transferred it from a labeled container and will only be used by that person within the work shift in which it is transferred from the labeled container.

immediately dangerous to life and health (IDLH) (EPA-94/04): The maximum level to which a healthy individual can be exposed to a chemical for 30 minutes and escape without suffering irreversible health effects or impairing symptoms. Used as a "level of concern." (See level of concern and risk assessment.)

imminent danger to the health and safety of the public (SMCRA701-30USC1291-90) means the existence of any condition or practice, or any violation of a permit or other requirement of this chapter in a surface coal mining and reclamation operation, which condition, practice, or violation could reasonably be expected to cause substantial physical harm to persons outside the permit area before such condition, practice, or violation can be abated. A reasonable expectation of death or serious injury before abatement exists if a rational person, subjected to the same conditions or practices giving rise to the peril, would not expose himself or herself to the danger during the time necessary for abatement.

imminent hazard (40CFR165.1-l) means a situation which exists when the continued use of a pesticide during the time required for cancellation proceedings would be likely to result in unreasonable adverse effects on the environment or will involve unreasonable hazard to the survival of a species declared endangered by the Secretary of the Interior under Pub. L. 91-135.

imminent hazard (40CFR165.1-l) means a situation which exists when the continued use of a pesticide during the time required for cancellation proceedings would be likely to result in unreasonable adverse effects on the environment or will involve unreasonable hazard to the survival of a species declared endangered by the Secretary of the Interior under Pub. L. 91-135.

imminent hazard (FIFRA2-7USC136) means a situation which exists when the continued use of a pesticide during the time required for cancellation proceedings would be likely to result in unreasonable adverse hazard to the survival of a species declared endangered by the Secretary of the Interior under Pub.L. 91-135.

immobilization (40CFR797.1300-5) means the lack of movement by the test organisms except for minor activity of the appendages.

immobilization (40CFR797.1330-6) means the lack of movement by daphnids except for minor activity of the appendages.

impairment of visibility (CAA169A): See visibility impairment.

impermeable (EPA-94/04): Not easily penetrated, The property of a material or soil that does not allow, or allows only with great difficulty, the movement or passage of water.

impervious chemical protective clothing (40CFR721.3) is "impervious" to a chemical substance if the substance causes no chemical or mechanical degradation, permeation, or penetration of the chemical protective clothing under the conditions of, and the duration of, exposure.

impervious solid surfaces (40CFR761.123) means solid surfaces which are nonporous and thus unlikely to absorb spilled PCBs within the short period of time required for cleanup of spills under this policy. Impervious solid surfaces include, but are not limited to, metals, glass, aluminum siding, and enameled or laminated surfaces.

implementation (40CFR249.04-g) means putting a plan into practice by carrying out planned activities, or ensuring that these activities are carried out (under RCRA).

implementation (40CFR256.06) means putting the plan into practice by carrying out planned activities, including compliance and enforcement activities, or ensuring such activities are carried out.

implementation (RCRA1004-42USC6903): For purposes of Federal financial assistance (other than rural communities assistance), the term "implementation" does not include the acquisition, leasing, construction, or modification of facilities or equipment or the acquisition, leasing, or improvement of land.

implementing agency (40CFR191.12) as used in this subpart, means the Commission for spent nuclear fuel or high-level or transuranic wastes to be disposed of in facilities licensed by the Commission in accordance with the Energy Reorganization Act of 1974 and the Nuclear Waste Policy Act of 1982, and it means the Department for all other radioactive wastes covered by this part.

implementing agency (40CFR191.12) means:
(1) The Commission for facilities licensed by the Commission;
(2) The Agency for those implementation responsibilities for the Waste Isolation Pilot Plant, under this part, given to the Agency by the Waste Isolation Pilot Plant Land Withdrawal Act (Pub. L. 102-579, 106 Stat. 4777) which, for the purposes of this part, are:
  (i) Determinations by the Agency that the Waste Isolation Pilot Plant is in compliance with subpart A of this part;
  (ii) Issuance of criteria for the certifications of compliance with subparts B and C of this part of the Waste Isolation Pilot Plant's compliance with subparts B and C of this part;
  (iii) Certifications of compliance with subparts

B and C of this part of the Waste Isolation Pilot Plant's compliance with subparts B and C of this part;

(iv) If the initial certification is made, periodic recertification of the Waste Isolation Pilot Plant's continued compliance with subparts B and C of this part;

(v) Review and comment on performance assessment reports of the Waste Isolation Pilot Plant; and

(vi) Concurrence by the Agency with the Department's determination under 40CFR191.02(i) that certain wastes do not need the degree of isolation required by subparts B and C of this part; and

(3) The Department of Energy for any other disposal facility and all other implementation responsibilities for the Waste Isolation Pilot Plant, under this part, not given to the Agency (under AEA).

implementing agency (40CFR280.12) means EPA, or, in the case of a state with a program approved under section 9004 (or pursuant to a memorandum of agreement with EPA), the designated state or local agency responsible for carrying out an approved UST program.

import (40CFR763.163) means to bring into the customs territory of the United States, except for:

(1) Shipment through the customs territory of the United States for export without any use, processing, or disposal within the customs territory of the United States; or

(2) entering the customs territory of the United States as a component of a product during normal personal or business activities involving use of the product.

import (40CFR82.3-u) means to land on, bring into, or introduce into, or attempt to land on, bring into, or introduce into any place subject to the jurisdiction of the United States whether or not such landing, bringing, or introduction constitutes an importation within the meaning of the customs laws of the United States, with the following exemptions:

(1) Off-loading used or excess controlled substances or controlled products from a ship during servicing,

(2) Bringing controlled substances into the U.S.

from Mexico where the controlled substance had been admitted into Mexico in bond and was of U.S. origin, and

(3) Bringing a controlled product into the U.S. when transported in a consignment of personal or household effects or in a similar non-commercial situation normally exempted from U.S. Customs attention.

(Other identical or similar definitions are provided in 40CFR82.104-j; 704.3; 704.43-2; 716.3; 372.3; CAA601-42USC7671; ESA3-16USC1531.)

import for commercial purposes (40CFR704.3) means to import with the purpose of obtaining an immediate or eventual commercial advantage for the importer, and includes the importation of any amount of a chemical substance or mixture. If a chemical substance or mixture containing impurities is imported for commercial purposes, then those impurities also are imported for commercial purposes.

import for commercial purposes (40CFR716.3) means to import with the purpose of obtaining an immediate or eventual commercial advance for the importer, and includes the importation of any amount of a chemical substance or mixture. If a chemical substance or mixture containing impurities is imported for commercial purposes, then those impurities are also imported for commercial purposes.

import in bulk form (40CFR704.3) means to import a chemical substance (other than as part of a mixture or article) in any quantity, in cans, bottles, drums, barrels, packages, tanks, bags, or other containers, if the chemical substance is intended to be removed from the container and the substance has an end use or commercial purpose separate from the container.

import in bulk form (40CFR712.3-c) means to import a chemical substance (other than as part of a mixture or article) in any quantity, in cans, bottles, drums, barrels, packages, tanks, bags, or other containers used for purposes of transportation or containment, if the chemical substance has an end use or commercial purpose separate from the container.

importer (40CFR720.3-l) means any person who

imports a chemical substance as part of a mixture or article, into the customs territory of the United States. Importer includes the person primarily liable for the payment of any duties of the merchandise or an authorized agent acting on his or her behalf. The term also includes, as appropriate:

(1) The consignee.

(2) The importer of record.

(3) The actual owner if an actual owner's declaration and superseding bond has been filed in accordance with 19CFR141.20;

(4) The transferee, if the right to draw merchandise in a bonded warehouse has been transferred in accordance with subpart C of 19CFR part 144 (see principal importer).

(Other identical or similar definitions are provided in 40CFR723.50-2; 723.250-6; 747.115-1; 747.195; 747.200-1; 763.63-e.)

importer (40CFR763.163) means anyone who imports a chemical substance, including a chemical substance as part of a mixture or article, into the customs territory of the United States. Importer includes the person primarily liable for the payment of any duties on the merchandise or an authorized agent acting on his or her behalf. The term includes as appropriate:

(1) The consignee.

(2) The importer of record.

(3) The actual owner if an actual owner's declaration and superseding bond has been filed in accordance with 19CFR 141.20.

(4) The transferee, if the right to withdraw merchandise in a bonded warehouse has been transferred in accordance with subpart C of 19CFR part 144 (under the Toxic Substances Control Act).

importer (40CFR82.3) means the importer of record listed on U.S. Customs Service Form 7501 for imported controlled substances (other identical or similar definitions are provided in 40CFR80.2-r; 82.3-v; 82.172; 82.104-k) (under the Clean Air Act).

impoundment (40CFR260.10): See surface impoundment or impoundment.

impoundment (EPA-94/04): A body of water or sludge confined by a dam, dike, floodgate, or other barrier.

impregnation (40CFR471.02-v) is the process of filling pores of a formed powder part, usually with a liquid such as a lubricant, or mixing particles of a nonmetallic substance in a matrix of metal powder.

improved discharge (40CFR125.58-h) means the volume, composition and location of an applicant's discharge following:

(1) Construction of planned outfall improvements, including, without limitation, outfall relocation, outfall repair, or diffuser modification; or

(2) Construction of planned treatment system improvements to treatment levels or discharge characteristics; or

(3) Implementation of a planned program to improve operation and maintenance of an existing treatment system or to eliminate or control the introduction of pollutants into the applicant's treatment works.

impulsive noise (40CFR211.203-n) means an acoustic vent characterized by very short rise time and duration.

impurity (40CFR158.153-d) means any substance (or group of structurally similar substances if specified by the Agency) in a pesticide product other than an active ingredient or an inert ingredient, including unreacted starting materials, side reaction products, contaminants, and degradation products.

impurity (40CFR63.101) means a substance that is produced coincidentally with the primary product, or is present in a raw material. An impurity does not serve a useful purpose in the production or use of the primary product and is not isolated.

impurity (40CFR704.3) means a chemical substance which is unintentionally present with another chemical substance (other identical or similar definitions are provided in 40CFR710.2-m; 712.3-e; 716.3; 720.3-m; 723.50-2; 723.175-3; 723.250-6; 747.115-1; 747.195-1; 747.200-1; 790.3; 791.3-f).

impurity (40CFR761.3) means a chemical substance which is unintentionally present with another chemical substance.

impurity (40CFR79.2-j) means any chemical element present in an additive that is not included in the chemical formula or identified in the breakdown by element in the chemical composition of such additive.

impurity associated with an active ingredient (40CFR158.153-e) means:

(1) Any impurity present in the technical grade of active ingredient; and

(2) Any impurity which forms in the pesticide product through reactions between the active ingredient and any other component of the product or packaging of the product.

in benzene service (40CFR61.111) means that a piece of equipment either contains or contacts a fluid (liquid or gas) that is at least 10 percent benzene by weight as determined according to the provisions of 40CFR61.245(d). The provisions of 40CFR61.245(d) also specify how to determine that a piece of equipment is not in benzene service.

in benzene service (40CFR61.131) means a piece of equipment, other than an exhauster, that either contains or contacts a fluid (liquid or gas) at least 1 percent benzene by weight as determined by the provisions of 40CFR61.137(b). The provisions of 40CFR61.137(b) also specify how to determine that a piece of equipment is not in benzene service.

in existence (40CFR51.301-k) means that the owner or operator has obtained all necessary preconstruction approvals or permits required by Federal, State, or local air pollution emissions and air quality laws or regulations and either has:

(1) begun, or caused to begin, a continuous program of physical on-site construction of the facility or

(2) entered into binding agreements or contractual obligations, which cannot be cancelled or modified without substantial loss to the owner or operator, to undertake a program of construction of the facility to be completed in a reasonable time.

in food/medical service (40CFR63.161) means that a piece of equipment in organic hazardous air pollutant service contacts a process stream used to manufacture a Food and Drug Administration regulated product where leakage of a barrier fluid into the process stream would cause any of the following:

(1) A dilution of product quality so that the product would not meet written specifications,

(2) An exothermic reaction which is a safety hazard,

(3) The intended reaction to be slowed down or stopped, or

(4) An undesired side reaction to occur.

in gas/vapor service (40CFR264.1031) means that the piece of equipment contains or contacts a hazardous waste stream that is in the gaseous state at operating conditions.

in gas/vapor service (40CFR60.481) means that the piece of equipment contains process fluid that is in the gaseous state at operating conditions.

in gas/vapor service (40CFR61.241) means that a piece of equipment contains process fluid that is in the gaseous state at operating conditions.

in gas/vapor service (40CFR63.161) means that a piece of equipment in organic hazardous air pollutant service contains a gas or vapor at operating conditions.

in heavy liquid service (40CFR60.481) means that the piece of equipment is not in gas/vapor service or in light liquid service (cf. in light liquid service) (other identical or similar definitions are provided in 40CFR264.1031).

in heavy liquid service (40CFR63.161) means that a piece of equipment in organic hazardous air pollutant service is not in gas/vapor service or in light liquid service.

in hydrogen service (40CFR60.591) means that a compressor contains a process fluid that meets the conditions specified in 40CFR60.593(b) (under CAA).

in kind contribution (40CFR30.200) means the value of a non-cash contribution to meet a recipient's cost sharing requirements. An in-kind contribution may consist of charges for real property and equipment or the value of goods and services directly benefiting the EPA funded project.

in kind contribution (40CFR35.4010) means the value of non-cash contribution used to meet a recipient's matching funds requirement in accordance with 40CFR30.30(b). An in-kind contribution may consist of charges for equipment or the value of goods and services necessary to and directly benefiting the EPA-funded project.

in kind contribution (40CFR35.6015-22) means the value of a non-cash contribution (generally from third parties) to meet a recipient's cost sharing requirements in a cooperative agreement only. An in-kind contribution may consist of charges for real property and equipment or the value of goods and services directly benefiting the CERCLA funded project.

in light liquid service (40CFR264.1031) means that the piece of equipment contains or contacts a waste stream where the vapor pressure of one or more of the components in the stream is greater than 0.3 kilopascals (kPa) at 20°C, the total concentration of the pure components having a vapor pressure greater than 0.3 kPa at 20°C is equal to or greater than 20 percent by weight, and the fluid is a liquid at operating conditions.

in light liquid service (40CFR60.481) means that the piece of equipment contains a liquid that meets the conditions specified in 60.485(e) (cf. in heavy liquid service) (other identical or similar definitions are provided in 40CFR60.631).

in light liquid service (40CFR60.591) means that the piece of equipment contains a liquid that meets the conditions specified in 40CFR60.593(c).

in light liquid service (40CFR63.161) means that a piece of equipment in organic hazardous air pollutant service contains a liquid that meets the following conditions:
(1) The vapor pressure of one or more of the organic compounds is greater than 0.3 kilopascals at 20°C,
(2) The total concentration of the pure organic compounds constituents having a vapor pressure greater than 0.3 kilopascals at 20°C is equal to or greater than 20 percent by weight of the total process stream, and
(3) The fluid is a liquid at operating conditions. (Note: Vapor pressures may be determined by the methods described in 40CFR60.485(e)(1).)

in line filtration (EPA-94/04): Pre-treatment method in which chemical coagulants directly to the filter inlet pipe. The chemicals are mixed by the flowing water; commonly used in pressure filtration installations. Eliminates need for flocculation and sedimentation.

in liquid service (40CFR61.241) means that a piece of equipment is not in gas/vapor service.

in liquid service (40CFR63.161) means that a piece of equipment in organic hazardous air pollutant service is not in gas/vapor service.

in operation (40CFR260.10) refers to a facility which is treating, storing, or disposing of hazardous waste.

in operation (40CFR270.2) means a facility which is treating, storing, or disposing of hazardous waste (under RCRA).

in operation (40CFR51.301-m) means engaged in activity related to the primary design function of the source.

in or near commercial buildings (40CFR761.3) means within the interior of, on the roof of, attached to the exterior wall of, in the parking area serving, or within 30 meters of a non-industrial non-substation building. Commercial buildings are typically accessible to both members of the general public and employees, and include:
(1) Public assembly properties,
(2) educational properties,
(3) institutional properties,
(4) residential properties,
(5) stores,
(6) office buildings, and
(7) transportation centers (e.g., airport terminal buildings, subway stations, bus stations, or train stations).

in organic hazardous air pollutant or in organic HAP service (40CFR63.161) means that a piece of equipment either contains or contacts a fluid (liquid or gas) that is at least 5 percent by weight of total organic HAP's as determined according to the provisions of 40CFR63.180(d) of this subpart. The provisions of 40CFR63.180(d) of this subpart also specify how to determine that a piece of equipment is not in organic HAP (hazardous air pollutants) service.

in organic hazardous air pollutant service or in organic HAP service (40CFR63.191) means that a piece of equipment either contains or contacts a fluid (liquid or gas) that is at least 5 percent by weight of the designated organic HAP's listed in 40CFR63.190(b) of this subpart.

in poor condition (40CFR61.141) means the binding of the material is losing its integrity as indicated by peeling, cracking, or crumbling of the material.

in process control technology (40CFR467.02-n) is the conservation of chemicals and water throughout the production operations to reduce the amount of wastewater to be discharged (other identical or similar definitions are provided in 40CFR471.02-w).

in process tank (40CFR52.741) means a container used for mixing, blending, heating, reacting, holding, crystallizing, evaporating or cleaning operations in the manufacture of pharmaceuticals.

in situ leach methods (40CFR440.132-e) means the processes involving the purposeful introduction of suitable leaching solutions into a uranium ore body to dissolve the valuable minerals in place and the purposeful leaching of uranium ore in static or semistatic condition either by gravity through an open pile, or by flooding confined ore pile. It does not include the natural dissolution of uranium by ground waters, the incidental leaching of uranium by mine drainage, nor the rehabilitation of aquifers and the monitoring of these aquifers.

in situ sampling systems (40CFR60.481) means nonextractive samplers or in-line samplers (other identical or similar definitions are provided in 40CFR61.241; 63.161; 264.1031).

in situ stripping (EPA-94/04): Treatment system that remove or "strips" volatile organic compounds from contaminated ground or surface water by forcing an airstream through the water and causing the compounds to evaporate.

in situ suspension process (40CFR60.561) means a manufacturing process in which styrene, blowing agent, and other raw materials are added together within a reactor for the production of expandable polystyrene.

in the hands of the manufacturer (40CFR86.602.84-8) means that vehicles are still in the possession of the manufacturer and have not had their bills of lading transferred to another person for the purpose of transporting.

in use aircraft gas turbine engine (40CFR87.1) means an aircraft gas turbine engine which is in service.

in use compliance period for purposes of in-use testing (40CFR85.1402) means a period of 150,000 miles.

in vacuum service (40CFR52.741) means, for the purpose of paragraph (i) of this section, equipment which is operating at an internal pressure that is at least 5 kPa (0.73 psia) below ambient pressure.

in vacuum service (40CFR60.481) means that equipment is operating at an internal pressure which is at least 5 kilopascals (kPa) below ambient pressure (other identical or similar definitions are provided in 40CFR61.241; 264.1031).

in vacuum service (40CFR63.161) means that equipment is operating at an internal pressure which is at least 5 kilopascals below ambient pressure (under CAA).

in VHAP service (40CFR61.241) means that a piece of equipment either contains or contacts a fluid (liquid or gas) that is at least 10 percent by weight a volatile hazardous air pollutant (VHAP) as determined according to the provisions of 40CFR61.245(d). The provisions of 40CFR61.245(d) also specify how to determine that a piece of equipment is not in VHAP service.

in vinyl chloride service (40CFR61.61-l) means that a piece of equipment either contains or contacts a liquid that is a least 10 percent vinyl chloride by weight or a gas that is at least 10 percent by volume vinyl chloride as determined according to the provisions of 40CFR61.67(h). The provisions of 40CFR61.67(h) also specify how to determine that a piece of equipment is not in vinyl chloride service. For the purposes of this subpart, this definition must be used in place of the definition of "in VHAP service" in subpart V of this part.

in VOC service (40CFR60.481) means that the piece of equipment contains or contacts a process fluid that is at least 10 percent VOC by weight. (The provisions of 40CFR60.485(d) specify how to determine that a piece of equipment is not in VOC service).

in VOC service (40CFR61.241) means, for the purposes of this subpart, that:
(a) the piece of equipment contains or contacts a process fluid that is at least 10 percent VOC by weight (see 40CFR60.2 for the definition of volatile organic compound or VOC and 40CFR60.485(d) to determine whether a piece of equipment is not in VOC service) and
(b) the piece of equipment is not in heavy liquid service as defined in 40CFR60.481 (under CAA).

in volatile organic compound or in VOC service (40CFR63.161) means, for the purposes of this subpart, that:
(1) The piece of equipment contains or contacts a process fluid that is at least 10 percent VOC by weight (see 40CFR60.2 for the definition of VOC, and 40CFR60.485(d) to determine whether a piece of equipment is not in VOC service); and
(2) The piece of equipment is not in heavy liquid service as defined in 40CFR60.481.

in wet gas service (40CFR60.631) means that a piece of equipment contains or contacts the field gas before the extraction step in the process (under CAA).

inability (40CFR169.1-d) means the incapacity of any person to maintain, furnish or permit access to any records under this Act and regulations, where such incapacity arises out of causes beyond the control and without the fault or negligence of such person. Such causes may include, but are not restricted to acts of God or of the public enemy, fires, floods, epidemics, quarantine restrictions, strikes, and unusually severe weather, but in every case, the

failure must be beyond the control and without the fault or negligence of said person.

inactive facility (40CFR256.06) means a facility which no longer receives solid waste.

inactive portion (40CFR260.10) means that portion of a facility which is not operated after the effective date of part 261 of this chapter (cf. active portion or closed portion).

inactive stack (40CFR61.201-a) means a stack to which no further routine additions of phosphogypsum will be made and which is no longer used for water management associated with the production of phosphogypsum. If a stack has not been used for either purpose for two years it is presumed to be inactive.

inactive waste disposal site (40CFR61.141) means any disposal site or portion of it where additional asbestos-containing waste material will not be deposited within the past year.

incident (OPA1001) means any occurrence or series of occurrences having the same origin, involving one or more vessels, facilities, or any combustion thereof, resulting in the discharge or substantial threat of discharge of oil.

incident command post (EPA-94/04): A facility located at a safe distance from an emergency site, where the incident commander, key staff, and technical representatives can make decisions and deploy emergency manpower and equipment.

incident command system (ICS) (EPA-94/04): The organizational arrangement wherein one person, normally the Fire Chief of the impacted district, is in charge of an integrated, comprehensive emergency response organization and the emergency incident site, backed by an Emergency Operations Center staff with resources, information, and advice.

incineration (40CFR240.101-n) means the controlled process which combustible solid, liquid, or gaseous wastes are burned and changed into noncombustible gases.

incineration (40CFR503.41-g) is the combustion of

organic matter and inorganic matter in sewage sludge by high temperatures in an enclosed device (under RCRA).

incineration (EPA-94/04): A treatment technology involving destruction of waste by controlled burning at high temperatures, e.g., burning sludge to remove the water and reduce the remaining residues to a safe, non-burnable ash that can be disposed of safely on land, in some waters, or in underground locations.

incineration at sea (EPA-94/04): Disposal of waste by burning at sea on specially designed incinerator ships.

incineration vessel (SF101-42USC9601) means any vessel which carries hazardous substances for the purpose of incineration of such substances, so long as such substances or residues of such substances are on board.

incinerator (40CFR240.101-o) means a facility consisting of one or more furnaces in which wastes are burned.

incinerator (40CFR260.10) means any enclosed device that:
(1) Uses controlled flame combustion that neither meets the criteria for classification as a boiler, sludge dryer, or carbon regeneration unit, nor is listed as an industrial furnace; or
(2) Meets the definition of infrared incinerator or plasma arc incinerator.

incinerator (40CFR52.741) means a combustion apparatus in which refuse is burned.

incinerator (40CFR60.51) means any furnace used in the process of burning solid waste for the purpose of reducing the volume of the waste by removing combustible matter.

incinerator (40CFR60.561) means an enclosed combustion device that is used for destroying VOC (under CAA).

incinerator (40CFR60.611) means any enclosed combustion device that is used for destroying organic compounds and does not extract energy in the form of steam or process heat (other identical or similar definitions are provided in 40CFR60.661).

incinerator (40CFR60.701) means an enclosed combustion device that is used for destroying organic compounds. If there is energy recovery, the energy recovery section and the combustion chambers are not of integral design. That is, the energy recovery section and the combustion section are not physically formed into one manufactured or assembled unit but are joined by ducts or connections carrying flue gas.

incinerator (40CFR61.301) means any enclosed combustion device that is used for destroying organic compounds and that does not extract energy in the form of steam or process heat. These devices do not rely on the heating value of the waste gas to sustain efficient combustion. Auxiliary fuel is burned in the device and the heat from the fuel flame heats the waste gas to combustion temperature. Temperature is controlled by controlling combustion air or fuel (under CAA).

incinerator (40CFR61.31-h) means any furnace used in the process of burning waste for the primary purpose of reducing the volume of the waste by removing combustible matter.

incinerator (40CFR63.111) means an enclosed combustion device that is used for destroying organic compounds. Auxiliary fuel may be used to heat waste gas to combustion temperatures. Any energy recovery section present is not physically formed into one manufactured or assembled unit with the combustion section; rather, the energy recovery section is a separate section following the combustion section and the two are joined by ducts or connections carrying flue gas. The above energy recovery section limitation does not apply to an energy recovery section used solely to permit the incoming vent stream or combustion air.

incinerator (40CFR761.3) means an engineered device using controlled flame combustion to thermally degrade PCBs and PCB items. Examples of devices used for incineration include rotary kilns, liquid injection incinerators, cement kilns, and high temperature boilers.

incinerator (EPA-94/04): A furnace for burning waste under controlled conditions.

incipient LC50 (40CFR797.1400-7) means that test

substance concentration, calculated from experimentally derived mortality data, that is lethal to 50 percent of a test population when exposure to the test substance is continued until the mean increase in mortality does not exceed 10 percent in any concentration over a 24-hour period.

incompatible waste (40CFR260.10) means a hazardous waste which is unsuitable for:
(1) Placement in a particular device or facility because it may cause corrosion or decay of containment materials (e.g., container inner liners or tank walls); or
(2) Commingling with another waste or material under uncontrolled conditions because the commingling might produce heat or pressure, fire or explosion, violent reaction, toxic dusts, mists, fumes, or gases, or flammable fumes or gases.

incompatible waste (EPA-94/04): A waste unsuitable for mixing with another waste or material because it may react to form a hazard.

incomplete data area (40CFR51.392) means any ozone nonattainment area which EPA has classified, in 40CFR part 81, as an incomplete data area (other identical or similar definitions are provided in 40CFR93.101).

incomplete gasoline-fueled heavy-duty vehicle (40CFR86.085.2) means any gasoline-fueled heavy-duty vehicle which does not have the primary load-carrying device, or passenger compartment, or engine compartment or fuel system attached (under the Clean Air Act).

incomplete truck (40CFR86.082.2) means any truck which does not have the primary load carrying device or container attached.

incorporated into the soil (40CFR257.3.5-5) means the injection of solid waste beneath the surface of the soil or the mixing of solid waste with the surface soil (other identical or similar definitions are provided in 40CFR257.3.6-3).

incorporated place (40CFR122.26-3) means the District of Columbia, or a city, town, township, or village that is incorporated under the laws of the State in which it is located (under the Federal Water Pollution Control Act).

increase the frequency or severity (40CFR51.392) means to cause a location or region to exceed a standard more often or to cause a violation at a greater concentration than previously existed and/or would otherwise exist during the future period in question, if the project were not implemented (other identical or similar definitions are provided in 40CFR93.101).

increase the frequency or severity of any existing violation of any standard in any area (40CFR51.852) means to cause a nonattainment area exceed a standard more often or to cause a violation at a greater concentration than previously existed and/or would otherwise exist during the future period in question, if the project were not implemented (other identical or similar definitions are provided in 40CFR93.152) (under CAA).

increments of progress (40CFR51.100-q) means steps toward compliance which will be taken by a specific source, including:
(1) Date of submittal of the source's final control plan to the appropriate air pollution control agency;
(2) Date by which contracts for emission control systems or process modifications will be awarded; or date by which orders will be issued for the purchase of component parts to accomplish emission control or process modification;
(3) Date of initiation of on-site construction or installation of emission control equipment or process change;
(4) Date by which on-site construction or installation of emission control equipment or process modification is to be completed; and
(5) Date by which final compliance is to be achieved.

increments of progress (40CFR60.21-h) means steps to achieve compliance which must be taken by an owner or operator of a designated facility, including:
(1) Submittal of a final control plan for the designated facility to the appropriate air pollution control agency;
(2) Awarding of contracts for emission control

systems or for process modifications, or issuance of orders for the purchase of component parts to accomplish emission control or process modification;

(3) Initiation of on-site construction or installation of emission control equipment or process change;

(4) Completion of on-site construction or installation of emission control equipment or process change; and

(5) Final compliance.

independent auditor (40CFR72.2) means a professional engineer who is not an employee or agent of the source being audited (under CAA).

independent commercial importer (ICI) (40CFR85.1502-6) is an importer who is not an original equipment manufacturer (OEM) or does not have a contractual agreement with an OEM to act as its authorized representative for the distribution of motor vehicles or motor vehicle engines in the U.S. market.

independent commercial importer (ICI) (40CFR89.602.96): An importer who is not an original engine manufacturer (OEM), but is the entity in whose name a certificate of conformity for a class of nonroad engines has been issued (under CAA).

independent laboratory (40CFR610.11-20) means a test facility operated independently of any motor vehicle, motor vehicle engine, or retrofit device manufacturer capable of performing retrofit device evaluation tests. Additionally, the laboratory shall have no financial interests in the outcome of these tests other than a fee charged for each test performed.

independent power production facility (IPP) (40CFR72.2) means a source that:

(1) Is nonrecourse project financed, as defined by the Secretary of Energy at 10CFR part 715;

(2) Is used for the generation of electricity, eighty percent or more of which is sold at wholesale; and

(3) Is a new unit required to hold allowances under Title IV of the Clean Air Act but only if direct public utility ownership of the equipment

comprising the facility does not exceed 50 percent.

independent printed circuit board manufacturer (40CFR433.11-d) shall mean a facility which manufacturers printed circuit boards principally for sale to other companies.

independently audited (40CFR144.61-d) refers to an audit performed by an independent certified public accountant in accordance with generally accepted auditing standards (other identical or similar definitions are provided in 40CFR264.141-f; 265.141-f).

index mark (40CFR85.2122(a)(2)(iii)(H)): See index or index mark.

index or index mark (40CFR85.2122(a)(2)(iii)(H)) means a mark on a choke thermostat housing, located in a fixed relationship to the thermostatic coil tang position to aid in assembly and service adjustment of the choke.

Indian country (40CFR350.1) means Indian country as defined in 18USC1151. That section defines Indian country as:

(a) All land within the limits of any Indian reservation under the jurisdiction of the United States government, notwithstanding the issuance of any patent, and including rights-of-way running through the reservation;

(b) All dependent Indian communities within the borders of the United States whether within the original or subsequently acquired territory thereof, and whether within or without the limits of a State; and

(c) All Indian allotments, the Indian titles to which have not been extinguished, including rights-of-way running through the same.

(Other identical or similar definitions are provided in 40CFR355.20; 370.2; 372.3.)

Indian governing body (40CFR51.166-28) means the governing body of any tribe, band, or group of Indians subject to the jurisdiction of the United States and recognized by the United States as possessing power of self-government (other identical or similar definitions are defined in 40CFR51.491; 52.21-28; 58.1-o; 124.41).

Indian lands (40CFR144.3) means Indian country as defined in 18 U.S.C. 1151. That section defines Indian country as:
(a) All land within the limits of any Indian reservation under the jurisdiction of the United States government, notwithstanding the issuance of any patent, and, including rights-of-way running through the reservation;
(b) All dependent Indian communities within the borders of the United States whether within the original or subsequently acquired territory thereof, and whether within or without the limits of a State; and
(c) All Indian allotments, the Indian titles to which have not been extinguished, including rights-of-way running through the same.

Indian lands (SMCRA701-30USC1291) means all lands, including mineral interests, within the exterior boundaries of any Federal Indian reservation, notwithstanding the issuance of any patent, and including rights-of-way, and all lands including mineral interests held in trust for or supervised by an Indian tribe.

Indian lands or Indian country (40CFR258.2) means:
(1) All land within the limits of any Indian reservation under the jurisdiction of the United States Government, notwithstanding the issuance of any patent, and including rights-of-way running throughout the reservation;
(2) All dependent Indian communities within the borders of the United States whether within the original or subsequently acquired territory thereof, and whether within or without the limits of the State; and
(3) All Indian allotments, the Indian titles to which have not been extinguished, including rights of way running through the same (under the Resource Conservation Recovery Act).

Indian reservation (40CFR51.166-27) means any federally recognized reservation established by Treaty, Agreement, Executive Order, or Act of Congress.

Indian reservation (40CFR52.21-27) means any federally recognized reservation established by Treaty, Agreement, executive order, or act of Congress.

Indian reservation (40CFR58.1-n) means any federally recognized reservation established by Treaty, Agreement, Executive Order, or Act of Congress.

Indian tribe (40CFR35.105) means, within the context of the Public Water System Supervision and Underground Water Source Protection grants, any Indian Tribe having a Federally recognized governing body carrying out substantial governmental duties and powers over a defined area (other identical or similar definitions are provided in 40CFR35.105; 35.1605; 35.6015-23; 122.2; 124.2; 130.2-b; 131.3-l; 142.2; 144.3; 146.3; 232.2; 258.2; 300.5; 350.1; 355.20; 370.2; 372.3; 501.2; CAA302-42USC7602; OPA1001; SDWA1401-42USC300f; SMCRA701-30USC1291; SF101-42USC9601).

Indian tribe and tribal organization (40CFR34.105-g) have the meaning provided in section 4 of the Indian Self-Determination and Education Assistance Act (25USC450B). Alaskan Natives are included under the definitions of Indian tribes in that Act.

indicator (EPA-94/04): 1. In biology, an organism, species, or community whose characteristics show the presence of specific environmental conditions. 2. In chemistry, a substance that shows a visible change, usually of color, at a desired point in a chemical reaction. 3. A device that indicates the result of a measurement, e.g., a pressure gauge or a moveable scale.

indictment (40CFR32.105-h) means indictment for a criminal offense. An information or other filing by competent authority charging a criminal offense shall be given the same effect as an indictment.

indirect ammonia recovery system (40CFR420.11-f) means those systems which recover ammonium hydroxide as a byproduct from coke oven gases and waste ammonia liquors.

indirect discharge (EPA-94/04): Introduction of pollutants from a non-domestic source into a publicly owned waste-treatment system. Indirect dischargers can be commercial or industrial facilities whose wastes enter local sewers.

indirect discharge or discharge (40CFR403.3-g)

means the introduction of pollutants into a POTW from any non-domestic source regulated under section 307(b), (c) or (d) of the Act.

indirect discharger (40CFR122.2) means a nondomestic discharger introducing "pollutants" to a "publicly owned treatment works" (cf. new discharger).

indirect emissions (40CFR51.852) means those emissions of a criteria pollutant or its precursors that:
(1) Are caused by the Federal action, but may occur later in time and/or may be farther removed in distance from the action itself but are still reasonably foreseeable; and
(2) The Federal agency can practicably control and will maintain control over due to a continuing program responsibility of the Federal agency.
(Other identical or similar definitions are provided in 40CFR93.152.)

indirect heat transfer (40CFR52.741) means transfer of heat in such a way that the source of heat does not come into direct contact with process materials.

indirect source (EPA-94/04): Any facility or building, property, road or parking facility that attracts motor vehicle traffic and, indirectly, causes pollution.

individual (10CFR20.3) means any human being.

individual (40CFR1516.2) means a citizen of the United States or an alien lawfully admitted for permanent residence.

individual (40CFR27.2; 32.605-9) means a natural person.

individual (40CFR303.11-c) means a natural person, not a corporation or other legal entity nor an association of persons.

individual drain system (40CFR60.691) means all process drains connected to the first common downstream junction box. The term includes all such drains and common junction box, together with their associated sewer lines and other junction boxes, down to the receiving oil-water separator.

individual drain system (40CFR61.341) means the system used to convey waste from a process unit, product storage tank, or waste management unit to a waste management unit. The term includes all process drains and common junction boxes, together with their associated sewer lines and other junction boxes, down to the receiving waste management unit (under CAA).

individual drain system (40CFR63.111) means the system used to convey wastewater streams from a process unit, product storage tank, feed storage tank, or waste management unit to a waste management unit. The term includes all process drains and junction boxes, together with their associated sewer lines and other junction boxes, manholes, sumps, and lift stations, down to the receiving waste management unit. A segregated stormwater sewer system, which is a drain and collection system designed and operated for the sole purpose of collecting rainfall-runoff at a facility, and which is segregated from all other individual drain systems, is excluded from this definition.

individual generation site (40CFR260.10) means the contiguous site at or on which one or more hazardous wastes are generated. An individual generation site, such as a large manufacturing plant, may have one or more sources of hazardous waste but is considered a single or individual generation site if the site or property is contiguous.

individual, maintain, record, system of records, and routine use (40CFR16.2-a) shall have the meaning given to it by 5USC552(a)(2), (a)(3), (a)(4), (a)(5), and (a)(7), respectively.

individual proficiency/RMP exam (40CFR195.2) means the exam which evaluates individuals who provide radon measurement services in a residential environment.

individual systems (40CFR35.2005-18) means privately owned alternative wastewater treatment works (including dual waterless/gray water systems) serving one or more principal residences, or small commercial establishments. Normally these are onsite systems with localized treatment and disposal of wastewater, but may be systems utilizing small diameter gravity, pressure or vacuum sewers conveying treated or partially treated wastewater.

These system can also include small diameter gravity sewers carrying raw wastewater to cluster systems.

individual with handicaps (40CFR12.103) means any person who has a physical or mental impairment that substantially limits one or more major life activities has a record of such an impairment, or is regarded as having such an impairment, or is regarded as having such an impairment. As used in this definition, the phrase:
(1) Physical or mental impairment includes:
  (i) Any physiological disorder or condition, cosmetic disfigurement, or anatomical loss affecting one or more of the following body systems: Neurological; musculoskeletal; special sense organs; respiratory, including speech organs; cardiovascular; reproductive, digestive, genitourinary; hemic and lymphatic; skin, and endocrine; or
  (ii) Any mental or psychological disorder, such as mental retardation, organic brain syndrome, emotional or mental illness, and specific learning disabilities. The term physical or mental impairment includes, but is not limited to, such diseases and conditions as orthopedic, visual, speech, and hearing impairments, cerebral palsy, epilepsy, muscular dystrophy, multiple sclerosis, cancer, heart disease, diabetes, mental retardation, emotional illness, and drug addiction and alcoholism.
(2) Major life activities includes functions such as caring for one's self, performing manual tasks, walking, seeing, hearing, speaking, breathing, learning, and working.
(3) Has a record of such an impairment means has a history of, or has been misclassified as having, a mental or physical impairment that substantially limits one or more major life activities.
(4) Is regarded as having an impairment means:
  (i) Has a physical or mental impairment that does not substantially limit major life activities but is treated by the agency as constituting such a limitation;
  (ii) Has a physical or mental impairment that substantially limits major life activities only as a result of the attitudes of others toward such impairment; or

(iii) Has none of the impairments defined in subparagraph (1) of this definition but is treated by the agency as having such an impairment.

indoor air (EPA-94/04): The breathing air inside a habitable structure or conveyance.

indoor air pollution (EPA-94/04): Chemical, physical, or biological contaminants in indoor air.

indoor climate (EPA-94/04): Temperature, humidity, lighting, and noise levels in a habitable structure or conveyance. Indoor climate can affect indoor air pollution.

induction exposure (40CFR798.4100-3) is an experimental exposure of a subject to a test substance with the intention of inducing a hypersensitive state.

induction period (40CFR798.4100-2) is a period of at least 1 week following a sensitization exposure during which a hypersensitive state is developed (under TSCA).

industrial building (40CFR761.3) means a building directly used in manufacturing or technically productive enterprises. Industrial buildings are not generally or typically accessible to other than workers. Industrial buildings include buildings used directly in the production of power, the manufacture of products, the mining of raw materials, and the storage of textiles, petroleum products, wood and paper products, chemicals, plastics, and metals.

industrial cooling tower (40CFR749.68-12) means any cooling tower used to remove heat from industrial processes, chemical reactions, or plants producing electrical power.

industrial cost recovery (40CFR35.905) means:
(a) The grantee's recovery from the industrial users of a treatment works of the grant amount allocable to the treatment of waste from such users under section 204(b) of the Act and this subpart.
(b) The grantee's recovery from the commercial users of an individual system of the grant amount allocable to the treatment of waste from

such users under section 201(h) of the Act and this subpart.

industrial cost recovery period (40CFR35.905) means that period during which the grant amount allocable to the treatment of wastes from industrial users is recovered from the industrial users of such works.

industrial furnace (40CFR260.10) means any of the following enclosed devices that are integral components of manufacturing processes and that use thermal treatment to accomplish recovery of materials or energy:

1. Cement kilns
2. Lime kilns
3. Aggregate kilns
4. Phosphate kilns
5. Coke ovens
6. Blast furnaces
7. Smelting, melting and refining furnaces (including pyrometallurgical devices such as cupolas, reverberator furnaces, sintering machine, roasters, and foundry furnaces)
8. Titanium dioxide chloride process oxidation reactors
9. Methane reforming furnaces
10. Pulping liquor recovery furnaces
11. Combustion devices used in the recovery of sulfur values from spent sulfuric acid
12. Halogen acid furnaces (HAFs) for the production of acid from halogenated hazardous waste generated by chemical production facilities where the furnace is located on the site of a chemical production facility, the acid product has a halogen acid content of at least 3%, the acid product is used in a manufacturing process, and, except for hazardous waste burned as fuel, hazardous waste fed to the furnace has a minimum halogen content of 20% as-generated.
13. Such other devices as the Administrator may, after notice and comment, add to this list on the basis of one or more of the following factors:
    (i) The design and use of the device primarily to accomplish recovery of material products;
    (ii) The use of the device to burn or reduce raw materials to make a material product;
    (iii) The use of the device to burn or reduce secondary materials as effective substitutes for raw materials, in processes using raw

materials as principal feedstocks;
    (iv) The use of the device to burn or reduce secondary materials as ingredients in an industrial process to make a material product;
    (v) The use of the device in common industrial practice to produce a material product; and
    (vi) Other factors, as appropriate (cf. incinerator).

industrial pollution prevention (EPA-94/04): Combination of industrial source reduction and toxic chemical use substitution.

industrial process refrigeration (40CFR82.152-g) means, for the purposes of 40CFR 82.156(i), complex customized appliances used in the chemical, pharmaceutical, petrochemical and manufacturing industries. This sector also includes industrial ice machines, appliances used directly in the generation of electricity, and ice rinks.

industrial solid waste (40CFR243.101-o) means the solid waste generated by industrial processes and manufacturing (other identical or similar definitions are provided in 40CFR246.101-p) (under the Resource Conservation Recovery Act).

industrial solid waste (40CFR258.2) means solid waste generated by manufacturing or industrial processes that is not a hazardous waste regulated under subtitle C of RCRA. Such waste may include, but is not limited to, waste resulting from the following manufacturing processes: Electric power generation; fertilizer--agricultural chemicals; food and related products--byproducts; inorganic chemicals; iron and steel manufacturing; leather and leather products; nonferrous metals manufacturing-- foundries; organic chemicals; plastics and resins manufacturing; pulp and paper industry; rubber and miscellaneous plastic products; stone, glass, clay, and concrete products; textile manufacturing; transportation equipment; and water treatment. This term does not include mining waste or oil and gas waste.

industrial source (40CFR125.58-i) means any source of nondomestic pollutants regulated under section 307 (b) or (c) of the Clean Water Act which discharges into a POTW.

industrial source (CAA402-42USC7651a): Industrial unit means a unit that does not serve a generator that produces electricity, a "nonutility unit" as defined in this section, or a process source as defined in section 410(e).

industrial source reduction (EPA-94/04): Practices that reduce the amount of any hazardous substance, pollutant, or contaminant entering any waste stream or otherwise released into the environment; also reduces the threat to public health and the environment associated with such releases. Term includes equipment or technology modifications, substitution of raw materials, and improvements in housekeeping, maintenance, training or inventory control.

industrial user (40CFR35.2005-19) means any nongovernmental, nonresidential user of a publicly owned treatment works which is identified in the Standard Industrial Classification Manual, 1972, Office of Management and Budget, as amended and supplemented, under one of the following divisions: Division A. Agriculture, Forestry, and Fishing; Division B. Mining; Division D. Manufacturing; Division E. Transportation, Communications, Electric, Gas, and Sanitary Services; Division I. Services.

industrial user (40CFR35.905) means:

(a) Any nongovernmental, nonresident user of a publicly owned treatment works which discharges more than the equivalent of 25,000 gallons per day (gpd) of sanitary wastes and which is identified in the Standard Industrial Classification manual, 1972, Office of Management and Budget, as amended and supplemented under one of the following divisions:

- Division A. Agriculture, Forestry, and Fishing.
- Division B. Mining.
- Division D. Manufacturing.
- Division E. Transportation, Communications, Electric, Gas, and Sanitary Services.
- Division I. Services.

(1) In determining the amount of a user's discharge for purposes of industrial cost recovery, the grantee may exclude domestic wastes or discharges from sanitary conveniences.

(2) After applying the sanitary waste exclusion in paragraph (b)(1) of this section (if the grantee chooses to do so), dischargers in the above divisions that have a volume exceeding 25,000 gpd or the weight of biochemical oxygen demand (BOD) or suspended solids (SS) equivalent to that weight found in 25,000 gpd of sanitary waste are considered industrial users. Sanitary wastes, for purposes of this calculation of equivalency, are the wastes discharged from residential users. The grantee, with the Regional Administrator's approval, shall define the strength of the residential discharges in terms of parameters including, as a minimum, BOD and SS per volume of flow.

(b) Any nongovernmental user of a publicly owned treatment works which discharges wastewater to the treatment works which contains toxic pollutants or poisonous solids, liquids, or gases in sufficient quantity either singly or by interaction with other wastes, to contaminate the sludge of any municipal systems, or to injure or to interfere with any sewage treatment process, or which constitutes a hazard to humans or animals, creates a public nuisance, or creates any hazard in or has an adverse effect on the waters receiving any discharge from the treatment works.

(c) All commercial users of an individual system constructed with grant assistance under section 201(h) of the Act and this subpart. (See 40CFR35.918(a)(3).)

industrial user (40CFR401.11-m) shall be defined in accordance with section 502 of the Act unless the context otherwise requires.

industrial user (CWA502-33USC1362) means those industries identified in the Standard Industrial Classification Manual, Bureau of the Budget, 1967, as amended and supplemented, under the category "Division D--Manufacturing" and such other classes of significant waste products as, by regulation, the Administrator deems appropriate.

industrial user or user (40CFR403.3-h) means a

source of Indirect Discharge (under the Clean Water Act).

industrial waste (EPA-94/04): Unwanted materials from an industrial operation; may be liquid, sludge, solid, or hazardous waste.

industrial wastewater (40CFR503.9-n) is wastewater generated in a commercial or industrial process.

industrial wipers (40CFR250.4-s) means paper towels especially made for industrial cleaning and wiping.

ineligible (40CFR32.105-i), Excluded from participation in Federal nonprocurement programs pursuant to a determination of ineligibility under statutory, means the executive order, or regulatory authority, other than Executive Order 12549 and its agency implementing regulations; for example, excluded pursuant to the Davis-Bacon Act and its implementing regulations, the equal employment opportunity acts and executive orders, or the environmental protection acts and executive orders. A person is ineligible where the determination of ineligibility to participate in more than one covered transaction.

inert ingredient (40CFR152.3-m) means any substance (or group of structurally similar substances if designated by the Agency), other than an active ingredient, which is intentionally included in a pesticide product.

inert ingredient (40CFR158.153-f) means any substance (or group of structurally similar substances if designated by the Agency), other than an active ingredient, which is intentionally included in pesticide product.

inert ingredient (EPA-94/04): Pesticide components such as solvents, carriers, dispersants, and surfactants that are not active against target pests. Not all inert ingredients are innocuous.

inert ingredients (FIFRA2-7USC136) means an ingredient which is not active.

inertia weight class (40CFR86.082.2) means the class, which is a group of test weights, into which a vehicle is grouped based on its loaded vehicle weight in accordance with the provisions of part 86 (other identical or similar definitions are provided in 40CFR600.002.85-26).

inertial separator (EPA-94/04): A device that uses centrifugal force to separate waste particles.

infectious agent (40CFR259.10.b) means any organism (such as a virus or a bacteria) that is capable of being communicated by invasion and multiplication in body tissues and capable of causing disease or adverse health impacts in humans (under RCRA).

infectious agent (EPA-94/04): Any organism, such as a virus or bacterium, that is pathogenic and capable of being communicated by invasion and multiplication in body tissues.

infectious waste (40CFR240.101-p) means:
(1) Equipment, instruments, utensils, and fomites of a disposable nature from the rooms of patients who are suspected to have or have been diagnosed as having a communicable disease and must, therefore, be isolated as required by public health agencies;
(2) laboratory wastes such as pathological specimens (e.g., all tissues, specimens of blood elements, excreta, and secretions obtained from patients or laboratory animals) and disposable fomites (any substance that may harbor or transmit pathogenic organisms) attendant thereto;
(3) surgical operating room pathologic specimens and disposable fomites attendant thereto and similar disposable materials from outpatient areas and emergency rooms.
(Other identical or similar definitions are provided in 40CFR241.101-h; 243.101-p.)

infectious waste (40CFR245.101-d) means:
(1) Equipment, instruments, utensils, and fomites (any substance that may harbor or transmit pathogenic organisms) of a disposable nature from the rooms of patients who are suspected to have or have been diagnosed as having a communicable disease and must, therefore, be isolated as required by public health agencies;
(2) laboratory wastes, such as pathological specimens (e.g., all tissues, specimens of blood elements, excreta, and secretions obtained from

patients or laboratory animals) and disposable fomites attendant thereto;

(3) surgical operating room pathologic specimens and disposable fomites attendant thereto and similar disposable materials from outpatient areas and emergency rooms.

(Other identical or similar definitions are provided in 40CFR246.101-q.)

infectious waste (EPA-94/04): Hazardous waste with infectious characteristics, including: contaminated animal waste; human blood and blood products; isolation waste, pathological waste; and discarded sharps (needles, scalpels or broken medical instruments.)

infiltration (40CFR35.2005-20) means water other than wastewater that enters a sewer system (including sewer service connections and foundation drains) from the ground through such means as defective pipes, pipe joints, connections, or manholes. Infiltration does not include, and is distinguished from, inflow.

infiltration (40CFR35.905) means water other than wastewater that enters a sewerage system (including sewer service connections) from the ground through such means as defective pipes, pipe joints, connections, or manholes. Infiltration does not include, and is distinguished tom, inflow (under CWA).

infiltration (EPA-94/04):

1. The penetration of water through the ground surface into sub-surface soil or the penetration of water from the soil into sewer or other pipes through defective joints, connections, or manhole walls.
2. The technique of applying large volumes of waste water to land to penetrate the surface and percolate through the underlying soil (cf. percolation).

infiltration gallery (EPA-94/04): A subsurface groundwater collection system, typically shallow in depth, constructed with open-jointed or perforated pipes that discharge collected water into a water-tight chamber from which the water is pumped to treatment facilities and into the distribution system. Usually located close to streams or ponds.

infiltration/inflow (40CFR35.905) means the total quantity of water from both infiltration and inflow without distinguishing the source.

infiltration rate (EPA-94/04): The quantity of water than can enter the soil in a specified time interval.

infiltration water (40CFR440.141-7) means that water which permeates through the earth into the plant site.

inflow (40CFR35.2005-21) means water other than wastewater that enters a sewer system (including sewer service connections) from sources such as, but not limited to, roof leaders, cellar drains, yard drains, area drains, drains from springs and swampy areas, manhole covers, cross connections between storm sewers and sanitary sewers, catch basins, cooling towers, storm waters, surface runoff, street wash waters, or drainage. Inflow does not include, and is distinguished from, infiltration.

inflow (40CFR35.905) means water other than wastewater that enters a sewerage system (including sewer service collections) from sources such as roof leaders, cellar drains, yard drains, area drains, foundation drains, drains from springs and swampy areas, manhole covers, cross connections between storm sewers and sanitary sewers, catch basins, cooling towers, storm waters, surface runoff, street wash waters, or drainage. Inflow does not include, and is distinguished from, infiltration (under the Clean Water Act).

inflow (EPA-94/04): Entry of extraneous rain water into a sewer system from sources other than infiltration, such as basement drains, manholes, storm drains, and street washing.

influencing or attempting to influence (40CFR34.105-h) means making, with the intent to influence, any communication to or appearance before an officer or employee or any agency, a member of Congress, an officer or employee of Congress, or an employee of a member of Congress in connection with any covered Federal action.

influent (EPA-94/04): Water, wastewater, or other liquid flowing into a reservoir, basin, or treatment plant.

information (40CFR2.201-c): See business information.

information file (EPA-94/04): In the Superfund program, a file that contains accurate, up-to-date documents on a Superfund site. The file is usually located in a public building (school, library, or city hall) convenient for local residents.

information which is available to the public (40CFR2.201-g) is information in EPA's possession which EPA will furnish to any member of the public upon request and which EPA may make public, release or otherwise make available to any person whether or not its disclosure has been requested.

infrared incinerator (40CFR260.10) means any enclosed device that uses electric powered resistance heaters as a source of radiant heat and which is not listed as an industrial furnace.

ingredient statement (FIFRA2-7USC136) means a statement which contains:
(1) the name and percentage of each active ingredient, and the total percentage of all inert ingredients, in the pesticide, and
(2) if the pesticide contains arsenic in any form, a statement of the percentages of total and water soluble arsenic, calculated as elementary arsenic.

inground tank (40CFR260.10) means a device meeting the definition of "tank" in 40CFR260.10 whereby a portion of the tank wall is situated to any degree within the ground, thereby preventing visual inspection of that external surface area of the tank that is in the ground.

inhalable diameter (40CFR798.1150-4) refers to that aerodynamic diameter of a particle which is considered to be inhalable for the organism. It is used to refer to particles which are capable of being inhaled and may be deposited anywhere within the respiratory tract from the trachea to the alveoli. For man, the inhalable diameter is considered as 15 micrometers or less (other identical or similar definitions are provided in 40CFR798.2450-4).

inhalable diameter (40CFR798.4350-4) refers to that aerodynamic diameter of a particle which is considered to be inhalable for the organism. It is used to refer to particles which are capable of being inhaled and may be deposited anywhere within the respiratory tract from the trachea to the deep lung (the alveoli). For man, the inhalable diameter is considered here as 15 micrometers or less.

inherently low-emission vehicle (40CFR88.302.93) means any LDV or LDT conforming to the applicable Inherently Low-Emission Vehicle standard, or any HDV with an engine conforming to the applicable Inherently Low-Emission Vehicle standard. No dual-fuel or flexible-fuel vehicles shall be considered Inherently Low-Emission Vehicles unless they are certified to the applicable standard(s) on all fuel types for which they are designed to operate.

inhibition (40CFR797.1060-3) means any decrease in the growth rate of the test algae compared to the control algae.

inhibition (40CFR797.1075-3) means any decrease in the growth rate of the population of test algae compared to the controls.

initial compliance period (40CFR141.2) means the first full three-year compliance period which begins at least 18 months after promulgation, except for contaminants listed at 40CFR141.61(a) (19)-(21), (c)(19)-(33), and 40CFR141.62(b) (11)-(15), initial compliance period means the first full three-year compliance period after promulgation for systems with 150 or more service connections (January 1993-December 1995), and first full three-year compliance period after the effective date of the regulation (January 1996-December 1998) for systems having fewer than 150 service connections.

initial compliance period (water) (EPA-94/04): The first full three-year compliance period which begins at least 18 months after promulgation.

initial crusher (40CFR60.671) means any crusher into which nonmetallic minerals can be fed without prior crushing in the plant.

initial decision (40CFR164.2-k) means the decision of the Administrative Law Judge supported by findings of fact and conclusions regarding all material issues of law, fact, or discretion, as well as reasons therefor. Such decision shall become the final decision and order of the Administrator without further proceedings unless an appeal therefrom is taken or the Administrator orders review thereof as herein provided.

initial decision (40CFR22.03) means the decision issued by the Presiding Officer based upon the record of the proceedings out of which it arises (under CAA).

initial decision (40CFR27.2) means the written decision of the presiding officer required by 40CFR27.2 or 27.37, and includes revised initial decision issued following a remand or a motion for reconsideration.

initial failure rate (40CFR51-AN) means the percentage of vehicles rejected because of excessive emissions of a single pollutant during the first inspection cycle of an inspection/maintenance program. (If inspection is conducted for more than one pollutant, the total failure rate may be higher than the failure rates for each single pollutant.)

initial mixing (40CFR227.29) is defined to be that dispersion or diffusion of liquid, suspended particulate, and solid phases of a waste which occurs within four hours after dumping. The limiting permissible concentration shall not be exceeded beyond the boundaries of the disposal site during initial mixing, and shall not be exceeded at any point in the marine environment after initial mixing. The maximum concentration of the liquid, suspended particulate, and solid phases of a dumped material after initial mixing shall be estimated by one of these methods, in order of preference:

(1) When field data on the proposed dumping are adequate to predict initial dispersion and diffusion of the waste, these shall be used, if necessary, in conjunction with an appropriate mathematical model acceptable to EPA or the District Engineer, as appropriate.
(2) When field data on the dispersion and diffusion of a waste of characteristics similar to that proposed for discharge are available, these shall be used in conjunction with an appropriate mathematical model acceptable to EPA or the District Engineer, as appropriate.
(3) When no field data are available, theoretical oceanic turbulent diffusion relationships may be

applied to known characteristics of the waste and the disposal site.

initial pressure (P_i) (40CFR60-AA (method 27-2.6)) means the pressure applied to the delivery tank at the beginning of the static pressure test, as specified in the appropriate regulation, in mm H_2O.

initial start-up (40CFR63.101) means the first time a new or reconstructed source begins production, or, for equipment added or changed as described in 40CFR63.100 (l) or (m) of this subpart, the first time the equipment is put into operation. Initial start-up does not include operation solely for testing equipment. For purposes of subpart G of this part, initial start-up does not include subsequent start-ups (as defined in this section) of chemical manufacturing process units following malfunctions or shutdowns or following changes in product for flexible operation units or following recharging of equipment in batch operation. For purposes of subpart H of this part, initial start-up does not include subsequent start-ups (as defined in 40CFR63.161 of subpart H of this part) of process units (as defined in 40CFR63.161 of subpart H of this part) following malfunctions or process unit shutdowns.

initial start-up (40CFR63.161) means the first time a new or reconstructed source begins production. Initial start-up does not include operation solely for testing equipment. Initial start-up does not include subsequent start-ups (as defined in this section) of process units following malfunctions or process unit shutdowns.

initial start-up (40CFR63.191) means the first time a new or reconstructed source begins production. Initial start-up does not include operation solely for testing equipment. For purposes of subpart H of this part, initial start-up does not include subsequent start-ups (as defined in 40CFR63.161 of subpart H of this part) of process units (as defined in 40CFR63.161 of subpart H of this part) following malfunctions or process unit shutdowns (under CAA).

initial vacuum (V_i) (40CFR60-AA (method 27-2.7)) means the vacuum applied to the delivery tank at the beginning of the static vacuum test, as specified in the appropriate regulation, in mm H_2O.

initiation of operation (40CFR35.2005-22): The date specified by the grantee on which use of the project begins for the purpose for which it was planned, designed, and built.

injection interval (40CFR146.61) means that part of the injection zone in which the well is screened, or in which the waste is otherwise directly emplaced (other identical or similar definitions are provided in 40CFR148.2).

injection well (40CFR144.3) means a well into which fluids are being injected (other identical or similar definitions are provided in 40CFR146.3; 147.2902; 270.2; also see 144.6^{aaa} for classification of wells and 146.5^{aaa} for classification of injection wells).

injection well (40CFR260.10) means a well into which fluids are injected (cf. underground injection).

injection well (EPA-94/04): A well into which fluids are injected for purposes such as waste disposal, improving the recovery of crude oil, or solution mining.

injection zone (40CFR144.3) means a geological formation, group of formations, or part of a formation receiving fluids through a well (other identical or similar definitions are provided in 40CFR146.3; 147.2902).

injection zone (EPA-94/04): A geological formation receiving fluids through a well.

injury (40CFR112.2) means a measurable adverse change, either long- or short-term, in the chemical or physical quality or the viability of a natural resource resulting either directly or indirectly from exposure to a discharge of oil, or exposure to a product of reactions resulting from a discharge of oil.

ink (40CFR52.741) means a coating used in printing, impressing, or transferring an image onto a substrate (under CAA).

ink (40CFR60.581) means any mixture of ink, coating solids, organic solvents including dilution solvent, and water that is applied to the web of flexible vinyl or urethane on a rotogravure printing line.

ink solids (40CFR60.581) means the solids content of an ink as determined by Reference Method 24, ink manufacturer's formulation data, or plant blending records.

inland oil barge (CWA311-33USC1321) means a non-self-propelled vessel carrying oil in bulk as cargo and certificated to operate only in the inland waters of the United States, while operating in such waters.

inland waters (40CFR300.5), for the purposes of classifying the size of discharges, means those waters of the United States in the inland zone, waters of the Great Lakes, and specified ports and harbors on inland rivers.

inland waters of the United States (CWA311-33USC1321) means those waters of the United States lying inside the baseline from which the territorial sea is measured and those waters outside such baseline which are a part of the Gulf Intracoastal Waterway.

inland zone (40CFR300.5) means the environment inland of the coastal zone excluding the Great Lakes and specified ports and harbors of inland rivers. The term inland zone delineates the area of Federal responsibility for response action. Precise boundaries are determined by EPA/USCG agreement and identified in federal regional contingency plans (under CERCLA).

inner liner (40CFR260.10) means a continuous layer of material placed inside a tank or container which protects the construction materials of the tank or container from the contained waste or reagents used to treat the waste.

innovative control technology (40CFR51.166-19) means any system of air pollution control that has not been adequately demonstrated in practice, but would have a substantial likelihood of achieving greater continuous emissions reduction than any control system in current practice or of achieving at least comparable reductions at lower cost in terms of energy, economics, or nonair quality environmental impacts.

innovative control technology (40CFR52.21-19) means any system of air pollution control that has not been adequately demonstrated in practice, but would have a substantial likelihood of achieving greater continuous emissions reduction than any control system in current practice or of achieving at least comparable reductions at lower cost in terms of energy, economics, or nonair quality environmental impacts.

innovative technologies (EPA-94/04): New or inventive methods to treat effectively hazardous waste and reduce risks to human health and the environment.

innovative technology (40CFR125.22-a) means a production process, a pollution control technique, or a combination of the two which satisfies one of the criteria in 40CFR125.23 and which has not been commercially demonstrated in the industry of which the requesting discharger is a part.

innovative technology (40CFR35.2005-23) means developed wastewater treatment processes and techniques which have not been fully proven under the circumstances of their contemplated use and which represent a significant advancement over the state of the art in terms of significant reduction in life cycle cost or significant environmental benefits through the reclaiming and reuse of water, otherwise eliminating the discharge of pollutants, utilizing recycling techniques such as land treatment, more efficient use of energy and resources, improved or new methods of waste treatment management for combined municipal and industrial systems, or the confined disposal of pollutants so that they will not migrate to cause water or other environmental pollution.

inoculum (EPA-94/04):
1. Bacterium placed in compost to start biological action.
2. A medium containing organisms that is introduced into cultures or living organisms.

inorganic arsenic (40CFR61.161) means the oxides and other noncarbon compounds of the element arsenic included in particulate matter, vapors, and aerosols (other identical or similar definitions are provided in 40CFR61.171; 61.181).

inorganic chemicals (EPA-94/04): Chemical substances of mineral origin, not of basically carbon structure.

inorganic pesticides (40CFR165.1-p): See pesticide.

inprocess wastewater (40CFR61.61-j) means any water which during manufacturing or processing, comes into direct contact with vinyl chloride or polyvinyl chloride or results from the production or use of any raw material, intermediate product, finished product, byproduct, or waste product containing vinyl chloride or polyvinyl chloride but which has not been discharged to a wastewater treatment process or discharged untreated as wastewater. Gasholder seal water is not inprocess wastewater until it is removed from the gasholder (under CAA).

insect (FIFRA2-7USC136) means any of the numerous small invertebrate animals generally having the body more or less obviously segmented, for the most part belonging to the class Insecta, comprising six-legged, usually winged forms, as for example, beetles, bugs, bees, flies, and to other allied classes of arthropods whose members are wingless and usually have more than six legs, as for example, spiders, mites, ticks, centipedes, and wood lice.

insecticide (EPA-94/04): A pesticide compound specifically used to kill or prevent the growth of insects.

inside spray coating operation (40CFR60.491-3) means the system on each beverage can surface coating line used to apply a coating to the interior of a two-piece beverage can body. This coating provides a protective film between the contents of the beverage can and the metal can body. The inside spray coating operation consists of the coating application station, flashoff area, and curing oven. Multiple applications of an inside spray coating are considered to be a single coating operation.

inspection and maintenance (I/M) (EPA-94/04):
1. Activities to assure that vehicles' emissions-controls work properly.
2. Also applies to wastewater treatment plants and other anti-pollution facilities and processes.

inspection and maintenance program

(40CFR52.2485-1) means a program for reducing emissions from in-use gasoline powered vehicles through identifying vehicles that need emission-control-related maintenance and requiring that such maintenance be performed.

inspection and maintenance program (40CFR52.731-1) means a program to reduce emissions from in-use vehicles through identifying vehicles that need emission control related maintenance and requiring that such maintenance be performed (other identical or similar definitions are provided in 40CFR52.1878; 52.2038-1).

inspection criteria (40CFR204.51-h) means the rejection and acceptance numbers associated with a particular sampling plan (other identical or similar definitions are provided in 40CFR205.51-16).

inspection criteria (40CFR86.602.84-5) means the pass and fail numbers associated with a particular sampling (other identical or similar definitions are provided in 40CFR86.1002.84; 86.1002.2001-5; 89.502.96).

inspection/maintenance (40CFR51-AN) means a program to reduce emissions from in-use vehicles through identifying vehicles that need emissions control related maintenance and requiring that maintenance be performed.

installation (40CFR51.165-ii; 51.166; 52.21; 52.24-2): See building, structure, facility, or installation.

installation (40CFR51.301-l) means an identifiable piece of process equipment.

installation (40CFR61.141) means any building or structure or any group of buildings or structures at a single demolition or renovation site that are under the control of the same owner or operator (or owner or operator under common control) (under the Clean Air Act).

installation inspector (40CFR260.10) means a person who, by reason of his knowledge of the physical sciences and the principles of engineering, acquired by a professional education and related practical experience, is qualified to supervise the installation of tank systems.

instant photographic film article (40CFR723.175-8) means a self-developing photographic film article designed so that all the chemical substances contained in the article, including the chemical substances required to process the film, remain sealed during distribution and use.

Institute (OSHA3-29USC652) means the National Institute for Occupational Safety and Health established under this Act.

institution (40CFR26.102-b) means any public or private entity or agency (including federal, state, and other agencies).

institution of higher education (CWA112-33USC1262) means an education institution described in the first sentence of section 1201 of the Higher Education Act of 1965 (other than an institution of any agency of the United States) which is accredited by a nationally recognized accrediting agency of association approved by the Administrator for this purpose. For purposes of this subsection, the Administrator shall publish a list of nationally recognized accrediting agencies or associations which he determines to be reliable authority as to the quality of training offered.

institution of higher education (SMCRA701-30USC1291) as used in subchapters VIII and IX of this chapter, means any such institution as defined by section 1141(a) of title 20.

institutional review board (40CFR26.102-g): See IRB.

institutional solid waste (40CFR243.101-q) means solid wastes generated by educational, health care, correctional, and other institutional facilities.

institutional solid waste (40CFR245.101-e) means solid wastes originating from educational, health care, correctional, and other institutional facilities.

institutional solid waste (40CFR246.101-r) means solid wastes generated by educational, health care, correctional and other institutional facilities (under RCRA).

institutional use (40CFR152.3-n) means any

application of a pesticide in or round any property or facility that functions to provide service to the general public or to public or private organizations, including but not limited to:

(1) Hospitals and nursing homes.
(2) Schools other than preschool and day care. facilities.
(3) Museums and libraries.
(4) Sports facilities.
(5) Office buildings.

instream use (EPA-94/04): Water use taking place within a stream channel, e.g., hydroelectric power generation, navigation, water quality improvement, fish propagation, recreation.

instructions (40CFR205.2-24): See maintenance instructions.

instrument check standard (40CFR136-AC-7): A multielement standard of known concentrations prepared by the analyst to monitor and verify instrument performance on a daily basis.

instrumental detection limit (40CFR136-AC-5) means the concentration equivalent to a signal, due to the analyte, which is equal to three times the standard deviation of a series of ten replicate measurements of a reagent blank signal at the same wavelength.

instrumentation system (40CFR63.161) means a group of equipment components used to condition and convey a sample of the process fluid to analyzers and instruments for the purpose of determining process operating conditions (e.g., composition, pressure, flow, etc.). Valves and connectors are the predominant type of equipment used in instrumentation systems; however, other types of equipment may also be included in these systems. Only valves nominally 0.5 inches and smaller, and connectors nominally 0.75 inches and smaller in diameter are considered instrumentation systems for the purposes of this subpart. Valves greater than nominally 0.5 inches and connectors greater than nominally 0.75 inches associated with instrumentation systems are not considered part of instrumentation systems and must be monitored individually.

insulation board (40CFR429.11-f) means a panel manufactured from interfelted ligno-cellulosic fibers consolidated to a density of less than 0.5 g/cu cm (less than 31 lb/cu ft).

insurance (SF401-42USC9671) means primary insurance, excess insurance, reinsurance, surplus lines insurance, and any other arrangement for shifting and disturbing risk which is determined to be insurance under applicable State or Federal law.

integral vista (40CFR51.301-n) means a view perceived from within the mandatory Class I Federal area of a specific landmark or panorama located outside the boundary of the mandatory Class I Federal area.

integrated exposure assessment (EPA-94/04): Cumulative summation (over time) of the magnitude of exposure to a toxic chemical in all media.

integrated facility (40CFR413.02-h) is defined as a facility that performs electroplating as only one of several operations necessary for manufacture of a product at a single physical location and has significant quantities of process wastewater from non-electroplating manufacturing operations. In addition, to qualify as an "integrated facility" one or more plant electroplating process wastewater lines must be combined prior to or at the point of treatment (or proposed treatment) with one or more plant sewers carrying process wastewater from non-electroplating manufacturing operations.

integrated pest management (IPM) (EPA-94/04): A mixture of chemical and other, non-pesticide, methods to control pests.

integrated refueling emission control system (40CFR86.098.2) means a system where vapors resulting from refueling are stored in a common vapor storage unit(s) with other evaporative emissions of the vehicle and are purged through a common purge system.

integrated steel producer (40CFR63.301) means a company or corporation that produces coke, uses the coke in a blast furnace to make iron, and uses the iron to produce steel. These operations may be performed at different plant sites within the corporation.

integrated system (40CFR158.153-g) means a process for producing a pesticide product that:

(1) Contains any active ingredient derived from a source that is not an EPA-registered product; or

(2) Contains any active ingredient that was produced or acquired in a manner that does not permit its inspection by the Agency under FIFRA section 9(a) prior to its use in the process.

integrated waste management (EPA-94/04): Using a variety of practices to handle municipal solid waste; can include source reduction, recycling, incineration, and landfilling.

interceptor sewer (40CFR21.2-s) means a sewer whose primary purpose is to transport wastewaters from collector sewers to a treatment facility (other identical or similar definitions are provided in 40CFR35.905).

interceptor sewer (40CFR35.2005-24) means a sewer which is designed for one or more of the following purposes:

(i) To intercept wastewater from a final point in a collector sewer and convey such wastes directly to a treatment facility or another interceptor.

(ii) To replace an existing wastewater treatment facility and transport the wastes to an adjoining collector sewer or interceptor sewer for conveyance to a treatment plant.

(iii) To transport wastewater from one or more municipal collector sewers to another municipality or to a regional plant for treatment.

(iv) To intercept an existing major discharge of raw or inadequately treated wastewater for transport directly to another interceptor or to a treatment plant.

interceptor sewers (EPA-94/04): Large sewer lines that, in a combined system, control the flow of sewage to the treatment plant. In a storm, they allow some of the sewage to flow directly into a receiving stream, thus keeping it from overflowing onto the streets. Also used in separate systems to collect the flows from main and trunk sewers and carry them to treatment points.

interconnected (40CFR60.41a-9) means that two or more electric generating units are electrically tied together by a network of power transmission lines, and other power transmission equipment (under CAA).

interested person (40CFR304.12-g) means the Administrator, any EPA employee, any party to the proceeding, any potentially responsible party associated with the facility concerned, any person who filed written comments in the proceeding, any participant or intervenor in the proceeding, all officers, directors, employees, consultants, and agents of any party, and any attorney of record for any of the foregoing persons.

interested person (40CFR72.2) means any person who submitted written comments or testified at a public hearing on the draft permit or other matter subject to notice and comment under the Acid Rain Program or any person who submitted his or her name to the Administrator or the permitting authority, as appropriate, to be placed on a list of persons interested in such matter. The Administrator or the permitting authority may update the list of interested persons from time to time by requesting additional written indication of continued interest from the persons listed and may delete from the list the name of any person failing to respond as requested.

interested person outside the Agency (40CFR124.78-4) includes the permit applicant, any person who filed written comments in the proceeding, any person who requested the hearing, any person who requested to participate or intervene in the hearing, any participant in the hearing and any other interested person not employed by the Agency at the time of the communications, and any attorney of record for those persons.

interested person outside the Agency (40CFR57.809-4) includes the smelter owner, any person who filed written comments in the proceeding, any person who requested the hearing, any person who requested to participate or intervene in the hearing, any participant or party in the hearing and any other interested person not employed by the Agency at the time of the communications, and the attorney of record for such persons.

interface (EPA-94/04): The common boundary

between two substances such as water and a solid, water and a gas, or two liquids such as water and oil.

interference (40CFR403.3-i) means a discharge which, alone or in conjunction with a discharge or discharges from other sources, both:
(1) Inhibits or disrupts the POTW, its treatment processes or operations, or its sludge processes or operations, use or disposal; and
(2) Therefore is a cause of a violation of any requirement of the POTW's NPDES permit (including an increase in the magnitude or duration of a violation) or of the prevention of sewage sludge use or disposal in compliance with the following statutory provisions and regulations or permits issued thereunder (or more stringent State or local regulations): section 405 of the Clean Water Act, the Solid Waste Disposal Act (SWDA) (including Title II, more commonly referred to as the Resource Conservation and Recovery Act (RCRA), and including State regulations contained in any State sludge management plan prepared pursuant to Subtitle D of the SWDA), the Clean Air Act, the Toxic Substances Control Act, and the Marine Protection, Research and Sanctuaries Act.

interference (40CFR425.02-j) means the discharge of sulfides in quantities which can result in human health hazards and/or risks to human life, and an inhibition or disruption of POTW as defined in 40CFR403.3(i).

interference check (40CFR60-AA (method 6C-3.9)) means a method for detecting analytical interferences and excessive biases through direct comparison of gas concentrations provided by the measurement system and by a modified Method 6 procedure. For this check, the modified Method 6 samples are acquired at the sample bypass discharge vent.

interference check sample (40CFR136-AC-8) means a solution containing both interfering and analyte elements of known concentration that can be used to verify background and interelement correction factors.

interference equivalent (40CFR53.23-d): Positive or negative response caused by a substance other than the one being measured.

interference response (40CFR60-AA (method 7E-3.3)) means the output response of the measurement system to a component in the sample gas, other than the gas component being measured.

intergovernmental agreement (40CFR33.005) means any written agreement between units of government under which one public agency performs duties for or in concert with another public agency using EPA assistance. This includes substate and interagency agreements.

intergovernmental agreement (40CFR35.6015-24) means any written agreement between units of government under which one public agency performs duties for or in concert with another public agency using EPA assistance. This includes substate and intergency agreements.

interim (permit) status (EPA-94/04): Period during which treatment, storage and disposal facilities coming under RCRA in 1980 are temporarily permitted to operate while awaiting a permanent permit. Permits issued under these circumstances are usually called "Part A" or "Part B" permits.

interim approval (40CFR281.12) means the approval received by a state program that meets the requirements in 40CFR281.11(c) (1) and (2) for the time period defined in 40CFR281.11(c)(3).

interim authorization (40CFR270.2) means approval by EPA of a State hazardous waste program which has met the requirements of section 3006(c) of RCRA and applicable requirements of part 271, subpart B.

interior body spray coat (40CFR52.741) means a coating applied by spray to the interior of a can body.

intermediate (40CFR704.3) means any chemical substance that is consumed, in whole or in part, in chemical reactions used for the intentional manufacture of other chemical substances or mixtures, or that is intentionally present for the purpose of altering the rates of such chemical reactions.

intermediate (40CFR710.2-n) means any chemical substance:

(1) Which is intentionally removed from the equipment in which it is manufactured, and

(2) which either is consumed in whole or in part in chemical reaction(s) used for the intentional manufacture of other chemical substance(s) or mixture(s), or is intentionally present for the purpose of altering the rate of such chemical reaction(s).

intermediate (40CFR712.3-f) means any chemical substance that is consumed, in whole or in part, in chemical reactions used for the intentional manufacture of other chemical substances or mixtures, or that is intentionally present for the purpose of altering the rates of such chemical reactions (see also paragraph (j) of this section).

intermediate (40CFR720.3-n) means any chemical substance that is consumed, in whole or in part, in chemical reactions used for the intentional manufacture of another chemical substance(s) or mixture(s), or that is intentionally present for the purpose of altering the rates of such chemical reactions.

intermediate (40CFR723.175-9) means any chemical substance which is consumed in whole or in part in a chemical reaction(s) used for the intentional manufacture of another chemical substance.

intermediate cover (40CFR241.101-i) means cover material that serves the same functions as daily cover, but must resist erosion for a longer period of time, because it is applied on areas where additional cells are not to be constructed for extended periods of time.

intermediate handler (40CFR259.10.b) is a facility that either treats regulated medical waste or destroys regulated medical waste but does not do both. The term, as used in this part, does not include transporters.

intermediate PMN (40CFR700.43): See intermediate premanufacture notice.

intermediate premanufacture notice or intermediate PMN (40CFR700.43) means any PMN submitted to EPA for a chemical substance which is an intermediate (as intermediate is defined in 40CFR720.3 of this chapter) in the production of a final product, provided that the PMN for the intermediate is submitted to EPA at the same time as, and together with, the PMN for the final product and that the PMN for the intermediate identifies the final product and describes the chemical reactions leading from the intermediate to the final product. If PMNs are submitted to EPA at the same time for several intermediates used in the production of a final product, each of those is an intermediate PMN if they all identify the final product and every other associated intermediate PMN and are submitted to EPA at the same time as, and together with, the PMN for the final product.

intermediate speed (40CFR86.082.2) means peak torque speed if peak torque speed occurs between 60 and 75 percent of rated speed. If the peak torque speed is less than 60 percent of rated speed, intermediate speed means 60 percent of rated speed. If the peak torque speed is greater than 75 percent of rated speed, intermediate speed means 75 percent of rated speed.

intermediate temperature cold testing (40CFR86.094.2) means testing done pursuant to the driving cycle and testing conditions contained in 40CFR part 86, subpart C, at temperatures between 25°F (-4°C) and 68°F (20°C).

intermittent control system (ICS) (40CFR51.100-nn) means a dispersion technique which varies the rate at which pollutants are emitted to the atmosphere according to meteorological conditions and/or ambient concentrations of the pollutant, in order to prevent ground-level concentrations in excess of applicable ambient air quality standards. Such a dispersion technique is an ICS whether used alone, used with other dispersion techniques, or used as a supplement to continuous emission controls (i.e., used as a supplemental control system).

intermittent emissions (40CFR60.561) means those gas streams containing VOC that are generated at intervals during process line operation and includes both planned and emergency releases.

intermittent operations (40CFR471.02-uu) means the industrial users does not have a continuous operation.

intermittent vapor processing system (40CFR60.501) means a vapor processing system that employs an intermediate vapor holder to accumulate total organic compounds vapors collected from gasoline tank trucks, and treats the accumulated vapors only during automatically controlled cycles.

intermittent vapor processing system (40CFR63.111) means a vapor processing system that employs an intermediate vapor holder to accumulate total organic compound vapors collected from tank trucks or railcars, and treats the accumulated vapors only during automatically controlled cycles.

intermunicipal agency (40CFR40.115-2):
(a) Under the Clean Air Act, an agency of two or more municipalities located in the same State or in different States and having substantial powers or duties pertaining to the prevention and control of air pollution.
(b) Under the Resource Conservation and Recovery Act, an agency established by two or more municipalities with responsibility for planning or administration of solid waste.
(c) In all other cases, an agency of two or more municipalities having substantial powers or duties pertaining to the control of pollution.

intermunicipal agency (RCRA1004-42USC6903) means an agency established by two or more municipalities with responsibility for planning or administration of solid waste.

internal floating roof (40CFR52.741) means a cover or roof in a fixed-roof tank which rests upon and is supported by the volatile organic liquid being contained and is equipped with a closure seal or seals to close the space between the roof edge and tank shell.

internal floating roof (40CFR61.341) means a cover that rests or floats on the liquid surface inside a waste management unit that has a fixed roof (under CAA).

internal floating roof (40CFR63.111) means a cover that rests or floats on the liquid surface (but not necessarily in complete contact with it) inside a storage vessel or waste management unit that has a permanently affixed roof.

internal subunit (40CFR704.25-3) means a subunit that is covalently linked to at least two other subunits. "Internal subunits" of polymer molecules are chemically derived from monomer molecules that have formed covalent links between two or more other molecules (other identical or similar definitions are provided in 40CFR721.350).

internal subunit (40CFR723.250-7) means a monomer unit that is covalently bonded to at least two other molecules. Internal monomer units of polymer molecules are chemically derived from monomer molecules that have formed covalent bonds between two or more other monomer molecules or other reactants.

international shipment (40CFR260.10) means the transportation of hazardous waste into or out of the jurisdiction of the United States.

international system of units (40CFR191.12) is the version of the metric system which has been established by the International Bureau of Weights and Measures and is administered in the United States by the National Institute of Standards and Technology. The abbreviation for this system is "SI."

intersection (40CFR763-AA-11) means nonparallel touching or crossing of fibers, with the projection having an aspect ratio of 5:1 or greater.

interstate agency (40CFR35.905) means an agency of two or more States established under an agreement or compact approved by the Congress, or any other agency of two or more States, having substantial powers or duties pertaining to the control of water (other identical or similar definitions are provided in 40CFR35.2005-25; 40.115-3; 122.2; 124.2; 142.2; 144.3; 233.2; 401.11-m; CWA502-33USC1362; RCRA1004-42USC6903).

interstate air pollution control agency (CAA302-42USC7602) means:
(1) an air pollution control agency established by two or more States, or
(2) an air pollution control agency of two or more municipalities located in different States (cf. air pollution control agency).

interstate carrier water supply (EPA-94/04): A source

of water for drinking and sanitary use on planes, buses, trains, and ships operating in more than one state. These sources are federally regulated.

interstate commerce (40CFR201.1-n) means the commerce between any place in a State and any place in another State, or between places in the same State through another State, whether such commerce moves wholly by rail or partly by rail and partly by motor vehicle, express, or water. This definition of interstate commerce for purposes of this regulation is similar to the definition of interstate commerce in section 203(a) of the Interstate Commerce Act (4USC303(a)).

interstate commerce (40CFR202.10-k) means the commerce between any place in a State and any place in another State or between places in the same State through another State, whether such commerce moves wholly by motor vehicle or partly by motor vehicle and partly by rail, express, water or air. This definition of interstate commerce for purposes of these regulations is the same as the definition of interstate commerce in section 203(a) of the Interstate Commerce Act (49USC303(a)).

interstate commerce (40CFR82.104-l) means the distribution or transportation of any product between one state, territory, possession or the District of Columbia, and another state, territory, possession or the District of Columbia, or the sale, use or manufacture of any product in more than one state, territory, possession or District of Columbia. The entry points for which a product is introduced into interstate commerce are the release of a product from the facility in which the product was manufactured, the entry into a warehouse from which the domestic manufacturer releases the product for sale or distribution, and at the site of United States Customs clearance.

interstate commerce clause (EPA-94/04): A clause of the U.S. Constitution which reserves to the federal government the right to regulate the conduct of business across state lines. Under this clause, for example, the U.S. Supreme Court has ruled that states may not inequitably restrict the disposal out-of-state wastes in their jurisdictions.

interstate waters (EPA-94/04): Waters that flow across or form part of state or international boundaries, e.g., the Great Lakes, the Mississippi River, or coastal waters.

interstitial monitoring (EPA-94/04): The continuous surveillance of the space between the walls of an underground storage tank.

intervener (40CFR209.3-g) means person who files a motion to be made party under 40CFR209.15 or 40CFR209.16, and whose motion is approved (under RCRA).

intervener (40CFR85.1807-2) shall mean a person who files a petition to be made an intervener pursuant to paragraph (g) of this section and whose petition is approved.

inventory (40CFR720.3-o) means the list of chemical substances manufactured or processed in the United States that EPA compiled and keeps current under section 8(b) of the Act (other identical or similar definitions are provided in 40CFR723.250-6; 747.115-1; 747.195-1; 747.200-1).

inventory (EPA-94/04): Inventory of chemicals produced pursuant to section 8 (b) of the Toxic Substances Control Act.

inventory form (40CFR370.2) means the Tier I and Tier II emergency and hazardous chemical inventory forms set forth in subpart D of this part "Material Safety Data Sheet" or "MSDS" means the sheet required to be developed under 40CFR900.1200(g) of Title 29 of the Code of Federal Regulations (under CERCLA).

inventory of open dumps (40CFR256.06) means the inventory required under section 4005(b) and is defined as the list published by EPA of those disposal facilities which do not meet the criteria (under RCRA).

inventory system (40CFR60.581) means a method of physically accounting for the quantity of ink, solvent, and solids used at one or more affected facilities during a time period. The system is based on plant purchase or inventory records.

inversion (EPA-94/04): A layer of warm air

preventing the rise of cooling air and pollutants trapped beneath it. Can cause an air pollution episode.

investigating official (40CFR27.2) means the Inspector General of the United States Environmental Protection Agency or an officer or employee of the Office of Inspector General designated by the Inspector General and serving in a position for which the rate of basic pay is not less than the minimum rate of basic pay for grade GS-16 under the General Schedule.

investor owned utility (40CFR72.2) means a utility that is organized as a tax-paying for-profit business.

invitation for bids (40CFR248.4-t) is the solicitation for prospective suppliers by a purchaser requesting their competitive price quotations.

ion (EPA-94/04): An electrically charged atom that can be drawn from waste water during electrodialysis.

ion exchange treatment (EPA-94/04): A common water-softening method often found on a large scale at water purification plants that remove some organics and radium by adding calcium oxide or calcium hydroxide to increase the pH to a level where the metals will precipitate out.

ionization chamber (EPA-94/04): A device that measures the intensity of ionizing radiation.

ionizing radiation (EPA-94/04): Radiation that can strip electrons from atoms, i.e., alpha, beta, and gamma radiation.

IRB (40CFR26.102-g) means an institutional review board established in accord with and for the purposes expressed in this policy.

IRB approval (40CFR26.102-h) means the determination of the IRB that the research has been reviewed and may be conducted at an institution within the constraints set forth by the IRB and by other institutional and federal requirements.

iron and steel (40CFR420.11-d) means those byproduct cokemaking operations other than merchant cokemaking operations.

iron blast furnace (40CFR420.31-b) means all blast furnaces except ferromanganese blast furnaces.

irradiated food (EPA-94/04): Food subject to brief radioactivity, usually gamma rays, to kill insects, bacteria, and mold, and to permit storage without refrigeration.

irradiation (EPA-94/04): Exposure to radiation of wavelengths shorter than those of visible light (gamma, x-ray, or ultraviolet), for medical purposes, to sterilize milk or other foodstuffs, or to induce polymerization of monomers or vulcanization of rubber.

irreparable harm (40CFR125.121-a) means significant undesirable effects occurring after the date of permit issuance which will not be reversed after cessation or modification of the discharge (under FWPCA).

irreversible effect (EPA-94/04): Effect characterized by the inability of the body to partially or fully repair injury caused by a toxic agent.

irrigation (EPA-94/04): Applying water or wastewater to land areas to supply the water and nutrient needs of plants.

irrigation efficiency (EPA-94/04): The amount of water stored in the crop root zone compared to the amount of irrigation water applied.

irrigation return flow (EPA-94/04): Surface and subsurface water which leaves the field following application of irrigation water.

irritant (EPA-94/04): A substance that can cause irritation of the skin, eyes, or respiratory system. Effects may be acute from a single high level exposure, or chronic from repeated low-level exposures to such compounds as chlorine, nitrogen dioxide, and nitric acid.

is regarded as having an impairment (40CFR7.25): See handicapped person.

ISO standard day conditions (40CFR60.331-g) means 288 degrees Kelvin, 60 percent relative humidity and 101.3 kilopascals pressure.

isodrin (40CFR704.102-3) means the pesticide 1,4:5,8-Dimethanonaphthalene,1,2,3,4,10,10-hexacholoro-1,4,4a,5,8,8a-hexahydro-, (1alpha, 4alpha, 4abeta, 5beta, 8beta, 8abeta)-, CAS Number 465-736.

isokinetic sampling (40CFR60.2) means sampling in which the linear velocity of the gas entering the sampling nozzle is equal to that of the undisturbed gas stream at the sample point.

isolated source (40CFR52.1475-iii): A source that will assume legal responsibility for all violations of the applicable national standards in its designated liability area, as specified in paragraph (e)(8) of this section (under CAA).

isomeric ratio (40CFR704.43-3) means the relative amounts of each isomeric chlorinated naphthalene that composes the chemical substance; and for each isomer the relative amounts of each chlorinated naphthalene designated by the position of the chlorine atom(s) on the naphthalene.

isomeric ratio (40CFR704.45-3) means the ratios of ortho-, meta-, and parachlorinated terphenyls (under TSCA).

isotope (EPA-94/04): A variation of an element that has the same atomic number of protons but a different weight because of the number of neutrons. Various isotopes of the same element may have different radioactive behaviors, and some are highly unstable.

issuance of a part 70 permit (40CFR60.2) will occur, if the State is the permitting authority, in accordance with the requirements of part 70 of this chapter and the applicable, approved State permit program. When the EPA is the permitting authority, issuance of a title V permit occurs immediately after the EPA takes final action on the final permit (other identical or similar definitions are provided in 40CFR63.2).

issuance of a part 70 permit (40CFR61.01) will occur, if the State is the permitting authority, in accordance with the requirements of part 70 of this chapter and the applicable, approved State permit program. When the EPA is the permitting authority, issuance of a title V permit occurs immediately after the EPA takes final action on the final permit (under CAA).

issuance of an NSO (40CFR57.103-n) means the final transmittal of the NSO pursuant to 40CFR57.107(a) by an issuing agency (other than EPA) to EPA for approval, or the publication of an NSO issued by EPA in the Federal Register.

issuing agency, unless otherwise specifically indicated (40CFR57.103-o), means the state or local air pollution control agency to which a smelter's owner has applied for an NSO, or which has issued the NSO, or EPA, when the NSO application has been made to EPA. Any showings or demonstrations required to be made under this part to the issuing agency, when not EPA, are subject to independent determinations by EPA.

ISTEA (40CFR51.392) means the Intermodal Surface Transportation Efficiency Act of 1991 (other identical or similar definitions are provided in 40CFR93.101).

*********** **JJJJJ** ***********

jams and jellies (40CFR407.81-h) shall include the production of jams, jellies and preserves defined as follows: The combination of fruit and fruit concentrate, sugar, pectin, and other additives in an acidic medium resulting in a gelatinized and thickened finished product (under the Clean Water Act).

jar test (EPA-94/04): A laboratory procedure that simulates a water treatment plant's coagulation--flocculation units with differing chemical doses, mix speeds, and settling times to estimate the minimum or ideal coagulant dose required to achieve certain water quality goals.

job shop (40CFR433.11-c) shall mean a facility which

owns not more than 50% (annual area basis) of the materials undergoing metal finishing.

joint sponsor (40CFR790.3) means a person who sponsors testing pursuant to section 4(b)(3)(A) of the Act.

joint sponsorship (40CFR790.3) means the sponsorship of testing by two or more persons in accordance with section 4(b)(3)(A) of the Act.

joint submitters (40CFR700.43) means two or more persons who submit a section 5 notice together.

judicial officer (40CFR164.2-I) shall mean an officer or employee of the Agency appointed as a Judicial Officer by the Administrator pursuant to this section who shall meet the qualifications and perform functions as follows:

(i) Officer. There may be designated for the purposes of this section one or more Judicial Officers. As work requires, there may be a Judicial Officer designated to act for the purposes of a particular case.

(ii) Qualifications. A Judicial Officer my be a permanent or temporary employee of the Agency who performs other duties for the Agency. Such Judicial Officer shall not be employed by the Office of Enforcement or have any connection with the preparation or presentation of evidence for a hearing held pursuant to this subpart.

(iii) Functions. The Administrator may consult with a Judicial Officer or delegate all or part of his authority to act in a given case under this section to a Judicial Officer: Provided, that this delegation shall not preclude the Judicial Officer from referring any motion or case to the Administrator when the Judicial Officer determines such referral to be appropriate.

(Other identical or similar definitions are provided in 40CFR179.3.)

junction box (40CFR60.691) means a manhole or access point to a wastewater sewer system line (under CAA).

junction box (40CFR63.111) means a manhole or access point to a wastewater sewer system line or a lift station.

jurisdiction by law (40CFR1508.15) means agency authority to approve, veto, or finance all or part of the proposal.

*********** **KKKKK** ***********

K_1 (40CFR797.1560): See uptake rate constant.

K_2 (40CFR797.1560): See depuration rate constant.

k_d (40CFR796.2750): See adsorption ratio.

karst (40CFR300-AA) means the terrain with characteristics of relief and drainage arising from a high degree of rock solubility in natural waters. The majority of karst occurs in limestones, but karst may also form in dolomite, gypsum, and salt deposits. Features associated with karst terrains typically include irregular topography, sinkholes, vertical shafts, abrupt ridges, caverns, abundant springs, and/or disappearing streams. Karst aquifers are associated with karst terrain.

karst (EPA-94/04): A geologic formation of irregular limestone deposits with sinks, underground streams, and caverns.

kilowatthour saved or savings (40CFR72.2) means the net savings in electricity use (expressed in Kwh) that result directly from a utility's energy conservation measures or programs.

kinetic energy (EPA-94/04): Energy possessed by a moving body of water as a result of its motion).

kinetic rate coefficient (EPA-94/04): A number that describes the rate at which a water constituent such as a biochemical oxygen demand or dissolved oxygen rises or falls.

known human effects (40CFR717.3-c)
(1) means a commonly recognized human health

effect of a particular substances or mixture as described either in:
(i) Scientific articles or publication abstract in standard reference sources.
(ii) The firm's product labeling or material safety data sheets (MSDS).
(2) However, an effect is not a "known human effect" if it:
(i) Was a significantly more severe toxic effect than previously described.
(ii) Was a manifestation of a toxic effect after a significantly shorter exposure period or lower exposure level than described.
(iii) Was a manifestation of a toxic effect by an exposure route different from that described.

known to or reasonably ascertainable by (40CFR704.3) means all information in a person's possession or control, plus all information that a reasonable person similarly situated might be expected to possess, control, or know.

known to or reasonably ascertainable by (40CFR712.3-g) means all information in a person's possession or control, plus all information that a reasonable person similarly situated might be expected to possess, control, or know, or could obtain without unreasonable burden.

known to or reasonably ascertainable by (40CFR720.3-p) means all information in a person's possession or control, plus all information that a reasonable person similarly situated might be expected to possess, control, or know, or could obtain without unreasonable burden (other identical or similar definitions are provided in 40CFR723.50-2; 723.175-10; 723.250-6).

known to or reasonably ascertainable by (40CFR763.63-f) means all information in a person's possession or control, plus all information that a reasonable person might be expected to possess, control, or know, or could obtain without unreasonable burden or cost.

knows or has reason to know (40CFR27.2) means that a person, with respect to a claim or statement:
(a) Has actual knowledge that the claim or statement is false, fictitious, or fraudulent;

(b) Acts in deliberate ignorance of the truth or falsity of the claim or statement; or
(c) Acts in reckless disregard of the truth or falsity of the claim or statement.

kraft pulp mill (40CFR60.281-a) means any stationary source which produces pulp from wood by cooking (digesting) wood chips in a water solution of sodium hydroxide and sodium sulfide (white liquor) at high temperature and pressure. Regeneration of the cooking chemicals through a recovery process is also considered part of the kraft pulp mill.

*********** LLLLL ***********

LA (40CFR130.2-g): Load allocation.

label (40CFR211.203-o) means that item, as described in this regulation, which is inscribed on, affixed to or appended to a product, its packaging, or both for the purpose of giving noise reduction effectiveness information appropriate to the product.

label (40CFR600.002.85-17) means a sticker that contains fuel economy information and is affixed to new automobiles in accordance with subpart D of this part.

label (40CFR749.68-13) means any written, printed, or graphic material displayed on or affixed to containers of hexavalent chromium-based water treatment chemicals that are to be used in cooling systems.

label and labeling (FIFRA2-7USC136):
(1) Label means the written, printed, or graphic matter on, or attached to, the pesticide or device or any of its containers or wrappers.
(2) Labeling means all labels and all other written, printed, or graphic matter:
(A) accompanying the pesticide or device at any time; or

(B) to which reference is made on the label or in literature accompanying the pesticide or device, except to current official publications of the Environmental Protection Agency, the United States Department of Agriculture and Interior, the Department of Health and Human Services, State experiment stations, State agricultural colleges, and other similar Federal or State institutions or agencies authorized by law to conduct research in the field of pesticides.

laboratory (40CFR259.10.b) means any research, analytical, or clinical facility that performs health care related analysis or service. This includes medical, pathological, pharmaceutical, and other research, commercial, or industrial laboratories.

laboratory (40CFR761.3) means a facility that analyzes samples for PCBs and is unaffiliated with any entity whose activities involve PCBs.

laboratory sample coordinator (40CFR763-AA-12) means that person responsible for the conduct of sample handling and the certification of the testing procedures.

lacquers (40CFR52.741) means any clear wood finishes formulated with nitrocellulose or synthetic resins to dry by evaporation without chemical reaction, including clear lacquer sanding sealers.

LAER (40CFR51): See lowest achievable emission rate.

lag time (40CFR53.23-e): The time interval between a step change in input concentration and the first observable corresponding change in response.

lagoon (EPA-94/04):
1. A shallow pond where sunlight, bacterial action, and oxygen work to purify wastewater; also used for storage of wastewater or spent nuclear fuel rods.
2. Shallow body of water, often separated from the sea by coral reefs or sandbars.

lakewide management plan (CWA118-33USC1268) means a written document which embodies a systematic and comprehensive ecosystem approach to restoring and protecting the beneficial uses of the open waters of each of the Great lakes in accordance with article VI and Annex 2 of the Great lakes Water Quality Agreement.

land (40CFR192.11-b) means any surface or subsurface land that is not part of a disposal site and is not covered by an occupiable building.

land application (40CFR503.11-h) is the spraying or spreading of sewage sludge onto the land surface; the injection of sewage sludge below the land surface; or the incorporation of sewage sludge into the soil so that the sewage sludge can either condition the soil or fertilize crops or vegetation grown in the soil.

land application (EPA-94/04): Discharge of wastewater onto the ground for treatment or reuse.

land ban (EPA-94/04): Phasing out of land disposal of most untreated hazardous wastes, as mandated by the 1984 RCRA amendments.

land disposal (40CFR268.2-c) means placement in or on the land, except in a corrective action management unit, and includes, but is not limited to, placement in a landfill, surface impoundment, waste pile, injection well, land treatment facility, salt dome formation, salt bed formation, underground mine or cave, or placement in a concrete vault, or bunker intended for disposal purposes.

land disposal (RCRA3004.k-42USC6924), when used with respect to a specified hazardous waste, shall be deemed to include, but not be limited to, any placement of such hazardous waste in a landfill, surface impoundment, waste pile, injection well, land treatment facility, salt dome formation, salt bed formation, or under ground mine or cave.

land farming (of waste) (EPA-94/04): A disposal process in which hazardous waste deposited on or in the soil is degraded naturally by microbes.

land treatment facility (40CFR260.10) means a facility or part of a facility at which hazardous waste is applied onto or incorporated into the soil surface; such facilities are disposal facilities if the waste will remain after closure.

land use (CZMA304-16USC1453) means activities which are conducted in, or on the shorelands within, the coastal zone, subject to the requirements outlined in section 1456(g) of this title.

land with a high potential for public exposure (40CFR503.31-d) is land that the public uses frequently. This includes, but is not limited to, a public contact site and a reclamation site located in a populated area (e.g, a construction site located in a city).

land with a low potential for public exposure (40CFR503.31-e) is land that the public uses infrequently. This includes, but is not limited to, agricultural land, forest, and a reclamation site located in an unpopulated area (e.g., a strip mine located in a rural area).

landfill (40CFR259.10.a) means a disposal facility or part of a facility where regulated medical waste is placed in or on the land and which is not a land treatment facility, a surface impoundment, or an injection well.

landfill (40CFR260.10) means a disposal facility or part of a facility where hazardous waste is placed in or on land and which is not a pile, a land treatment facility, a surface impoundment, an underground injection well, a salt dome formation, a salt bed formation, an underground mine or a cave.

landfill cell (40CFR260.10) means a discrete volume of a hazardous waste landfill which uses a liner to provide isolation of wastes from adjacent cells or wastes. Examples of landfill cells are trenches and pits.

landfills (EPA-94/04):
1. Sanitary landfills are disposal sites for non-hazardous solid wastes spread in layers, compacted to the smallest practical volume, and covered by material applied at the end of each operating day.
2. Secure chemical landfills are disposal sites for hazardous waste, selected and designed to minimize the chance of release of hazardous substances into the environment.

lands within any State or lands within such State (SMCRA701) means all lands within a State other than Federal lands and Indian lands.

landscape (EPA-94/04): The traits, patterns, and structure of a specific geographic area, including its biological composition, its physical environment, and its anthropogenic or social patterns. An area where interacting ecosystems are grouped and repeated in similar form.

landscape characterization (EPA-94/04): Documentation of the traits and patterns of the essential elements of the landscape.

landscape ecology (EPA-94/04): The study of the distribution patterns of communities and ecosystems, the ecological processes that affect those patterns, and changes in pattern and process over time.

landscape indicator (EPA-94/04): A measurement of the landscape, calculated from mapped or remotely sensed data, used to describe spatial patterns of land use and land cover across a geographic area. Landscape indicators may be useful as measures of certain kinds of environmental degradation such as forest fragmentation.

langelier index (LI) (EPA-94/04): An index reflecting the equilibrium pH of a water with respect to calcium and alkalinity; used in stabilizing water to control both corrosion and scale deposition.

large (40CFR407.61-w) shall mean a point source that processes a total annual raw material production of fruits, vegetables, specialties and other products that exceeds 9,080 kkg (10,000 tons) per year (other identical or similar definitions are provided in 40CFR407.71-u; 407.81-m).

large appliance (40CFR52.741) means any residential and commercial washers, dryers, ranges, refrigerators, freezers, water heaters, dish washers, trash compactors, air conditioners, and other similar products.

large appliance coating (40CFR52.741) means any coating applied to the component metal parts (including, but not limited to, doors, cases, lids, panels, and interior support parts) of residential and commercial washers, dryers, ranges, refrigerators, freezers, water heaters, dish washers, trash compactors, air conditioners, and other similar products.

large appliance coating facility (40CFR52.741) means a facility that includes one or more large appliance coating line(s).

large appliance coating line (40CFR52.741) means a coating line in which any protective, decorative, or functional coating is applied onto the surface of large appliances.

large appliance part (40CFR60.451) means any organic surface-coated metal lid, door, casing, panel, or other interior or exterior metal part or accessory that is assembled to form a large appliance product. parts subject to in-use temperatures in excess of 250°F are not included in this definition (under CAA).

large appliance product (40CFR60.451) means any organic surface-coated metal range, oven, microwave oven, refrigerator, freezer, washer, dryer, dishwasher, water heater, or trash compactor manufactured for household, commercial, or recreational use (under CAA).

large appliance surface coating line (40CFR60.451) means that portion of a large appliance assembly plant engaged in the application and curing of organic surface coatings on large appliance parts or products.

large municipal separate storm sewer system (40CFR122.26-4) means all municipal separate storm sewers that are either:
(i)   Located in an incorporated place with a population of 250,000 or more as determined by the latest Decennial Census by the Bureau of Census (appendix F); or
(ii) Located in the counties listed in appendix H, except municipal separate storm sewers that are located in the incorporated places, townships or towns within such counties; or
(iii) Owned or operated by a municipality other than those described in paragraph (b)(4)(i) or (ii) of this section and that are designated by the Director as part of the large or medium municipal separate storm sewer system due to

the interrelationship between the discharges of the designated storm sewer and the discharges from municipal separate storm sewers described under paragraph (b)(4)(i) or (ii) of this section. In making this determination the Director may consider the following factors:
(A) Physical interconnections between the municipal separate storm sewers;
(B) The location of discharges from the designated municipal separate storm sewer relative to discharges from municipal separate storm sewers described in paragraph (b)(4)(i) of this section;
(C) The quantity and nature of pollutants discharged to waters of the United States;
(D) The nature of the receiving waters; and
(E) Other relevant factors; or
(iv) The Director may, upon petition, designate as a large municipal separate storm sewer system, municipal separate storm sewers located within the boundaries of a region defined by a storm water management regional authority based on a jurisdictional, watershed, or other appropriate basis that includes one or more of the systems described in paragraph (b)(4)(i), (ii), (iii) of this section.

large MWC plant (40CFR60.31a) means an MWC plant with an MWC plant capacity greater than 225 megagrams per day (250 tons per day) but less than or equal to 1,000 megagrams per day (1,100 tons per day) of MSW (cf. MWC plant).

large MWC plant (40CFR60.51a) means an MWC plant with an MWC plant capacity greater than 25 megagrams per day (250 tons per day) of MSW.

large quantity generator (EPA-94/04): Person or facility generating more than 2200 pounds of hazardous waste per month. Such generators produce about 90 percent of the nation's hazardous waste, and are subject to all RCRA requirements.

large sized plants (40CFR428.71) shall mean plants which process more than 10,430 kg/day (23,000 lbs/day) raw materials.

large water system (40CFR141.2), for the purpose of subpart I of this part only, means a water system that serves more than 50,000 persons.

large water system (EPA-94/04): A water system that serves more than 50,000 persons.

latency (EPA-94/04): Time from the first exposure of a chemical until the appearance of a toxic effect.

lateral expansion (40CFR258.2) means a horizontal expansion of the waste boundaries of an existing MSWLF unit. Leachate means a liquid that has passed through or emerged from solid waste and contains soluble, suspended, or miscible materials removed from such waste.

lateral sewer (40CFR21.2-r) means a sewer which connects the collector sewer to the interceptor sewer (under FWPCA).

lateral sewers (EPA-94/04): Pipes that run under city streets and receive the sewage from homes and businesses, as opposed to domestic feeders and main trunk lines.

latex resin (40CFR61.61-h) means a resin which is produced by a polymerization process which initiates from free radical catalyst sites and is sold undried.

laundering weir (EPA-94/04): Sedimentation basin overflow weir.

law enforcement vehicle (40CFR88.302.94) means any vehicle which is primarily operated by a civilian or military police officer or sheriff, or by personnel of the Federal Bureau of Investigation, the Drug Enforcement Administration, or other agencies of the federal government, or by state highway patrols, municipal law enforcement, or other similar law enforcement agencies, and which is used for the purpose of law enforcement activities including, but not limited to, chase, apprehension, surveillance, or patrol of people engaged in or potentially engaged in unlawful activities. For federal law enforcement vehicles, the definition contained in Executive Order 12759, section 11: Alternative Fueled Vehicle for the Federal Fleet, Guidance Document for Federal Agencies, shall apply.

LC50 (40CFR116.3) means that concentration of material which is lethal to one-half of the test population of aquatic animals upon continuous exposure for 96 hours or less.

LC50 (40CFR797.1350-3) is the median lethal concentration, i.e., that concentration of a chemical in air or water killing 50 percent of a test batch of organisms within a particular period of exposure (which shall be stated) (other identical or similar definitions are provided in 40CFR797.1440-4).

LC50 (40CFR797.1400-8) means that test substance concentration, calculated from experimentally derived mortality data, that is lethal to 50 percent of a test population during continuous exposure over a specified period of time.

LC50 (40CFR797.1930-3) means that experimentally derived concentration of test substance that is calculated to kill 50 percent of a test population during continuous exposure over a specified period of time (other identical or similar definitions are provided in 40CFR797.1950-5).

LC50 (40CFR797.1970-3) means that experimentally derived concentration of test substance that is calculated to have killed 50 percent of a test population during continuous exposure over a specified period of time.

LC50 (40CFR797.2050-2) is the empirically derived concentration of the test substance in the diet that is expected to result in mortality of 50 percent of a population of birds which is exposed exclusively to the treated diet under the conditions of the test (under TSCA).

LC50 (lethal concentration, 50 percent) (40CFR300-AA): Concentration of a substance in air [typically micrograms per cubic meter (ug/m^3) or water [typically micrograms per liter (ug/L)] that kills 50 percent of a group of exposed organisms. The LC(50) is used in the HRS in assessing acute toxicity (under CERCLA).

LC50/lethal concentration (EPA-94/04): Median level concentration, a standard measure of toxicity. It tells how much of a substance is needed to kill half of a group of experimental organisms in a given time (cf. LD50).

LC50 (40CFR795.120) means the median lethal concentration, i.e., that concentration of a chemical in air or water killing 50 percent of the test batch of organisms within a particular period of exposure (which shall be stated).

LD: Lethal dose.

LD50 (40CFR797.2175-2) is the empirically derived dose of the test substance that is expected to result ir mortality of 5 percent of a population of birds which is treated with a single oral dose under the conditions of the test.

LD50 (lethal dose, 50 percent) (40CFR300-AA) means the dose of a substance that kills 50 percent of a group of exposed organisms. The LD(50) is used in the HRS in assessing acute toxicity [milligrams toxicant per kilogram body weight (mg/kg)].

LD50/lethal dose (EPA-94/04): The dose of a toxicant that will kill 50 percent of the test organisms within a designated period. The lower the LD50, the more toxic the compound.

LDT (CAA216): Light-duty truck. See vehicle curb weight.

leachate (40CFR241.101-j) means liquid that has percolated through solid waste and has extracted dissolved or suspended materials from it (under RCRA).

leachate (40CFR257.2) means liquid that has passed through or emerged from solid waste and contains soluble, suspended or miscible materials removed from such wastes.

leachate (40CFR260.10) means any liquid, including any suspended components in the liquid, that has percolated through or drained from hazardous waste.

leachate (EPA-94/04): Water that collects contaminants as it trickles through wastes, pesticides or fertilizers. Leaching may occur in farming areas, feedlots, and landfills, and may result in hazardous substances entering surface water, ground water, or soil.

leachate collection system (40CFR503.21-i) is a system or device installed immediately above a liner that is designed, constructed, maintained, and operated to collect and remove leachate from a sewage sludge unit.

leachate collection system (EPA-94/04): A system that gathers leachate and pumps it to the surface for treatment.

leaching (EPA-94/04): The process by which soluble constituents are dissolved and filtered through the soil by a percolating fluid (cf. leachate).

lead (40CFR415.61-d) shall mean total lead.

lead (40CFR415.671-e) shall mean the total lead present in the process wastewater stream exiting the wastewater treatment system.

lead (40CFR420.02-j) means total lead and is determined by the method specified in 40CFR136.3.

lead (40CFR60.121-c) means elemental lead or alloys in which the predominant component is lead (cf. heavy metals).

lead (Pb) (EPA-94/04): A heavy metal that is hazardous to health if breathed or swallowed. Its use in gasoline, paints, and plumbing compounds has been sharply restricted or eliminated by federal laws and regulations.

lead acid battery manufacturing plant (40CFR60.371-b) means any plant that produces a storage battery using lead and lead compounds for the plates and sulfuric acid for the electrolyte.

lead additive (40CFR80.2-e) means any substance containing lead or lead compounds.

lead additive manufacturer (40CFR80.2-m) means any person who produces a lead additive or sells a lead additive under his own name (under the Clean Air Act).

lead agency (40CFR1508.16) means the agency or agencies preparing or having taken primary responsibility for preparing the environmental impact statement.

lead agency (40CFR300.5) means the agency that has provides the OSC/RPM to plan and implement response action under the NCP, EPA, the USCG, another federal agency, or a state (or political subdivision of a state) operating pursuant to a contract or cooperative agreement executed pursuant to section 104(d)(1) of CERCLA, or designated pursuant to a Superfund Memorandum of Agreement (SMOA) entered into pursuant to subpart F of the NCP or other agreements may be the lead agency for a response action. In the case of a release of a hazardous substance, pollutant, or contaminant, where the release is on, or the sole source of the release is from, any facility or vessel under the jurisdiction, custody, or control of Department of Defense (DOD) or Department of Energy (DOE), then DOD or DOE will be the lead agency. Where the release is on, or the sole source of the release is from, any facility or vessel under the jurisdiction, custody, or control of a federal agency other than EPA, the USCG, DOD, or DOE, then that agency will be the lead agency for remedial actions and removal actions other than emergencies. The federal agency maintains its lead agency responsibilities whether the remedy is selected by the federal agency for non-NPL sites or by EPA and the federal agency or by EPA alone under CERCLA section 120. The lead agency will consult with the support agency, if one exists, throughout the response process.

lead agency (40CFR35.6015-25) means the Federal agency, State agency, political subdivision, or Indian Tribe that has primary responsibility for planning and implementing a response action under CERCLA.

lead free (SDWA1417-42USC300g.6):
(1) when used with respect to solders and flux refers to solders and flux containing not more than 0.2 percent lead, and
(2) when used with respect to pipes and pipe fittings refers to pipes and pipe fittings containing not more than 8.0 percent lead.
(Other identical or similar definitions are provided in 40CFR141.43.d.)

lead matte (40CFR61.171) means any molten solution of copper and other metal sulfides produced by reduction of sinter product from the oxidation of lead sulfide ore concentrates.

lead oxide manufacturing facility (40CFR60.371-c)

means a facility that produces lead oxide from lead, including product recovery.

lead recipe (40CFR60.291) means glass product composition of the following ranges of weight proportions: 50 to 60 percent silicon dioxide, 18 to 35 percent lead oxides, 5 to 20 percent total R_2O (e.g., Na_2M and K_2O), 0 to 8 percent total R_2O_3 (e.g., Al_2O_3), 0 to 15 percent total RO (e.g., CaO, MgO), other than lead oxide, and 5 to 10 percent other oxides.

lead reclamation facility (40CFR60.371-d) means the facility that remelts lead scrap and casts it into lead ingots for use in the battery manufacturing process, and which is not a furnace affected under subpart L of this part.

lead service line (40CFR141.2) means a service line made of lead which connects the water main to the building inlet and any lead pigtail, gooseneck or other fitting which is connected to such lead line.

lead service line (EPA-94/04): A service line made of lead which connects the water main to the building inlet and any lead fitting connected to it.

leaded gasoline (40CFR80.2-f) means gasoline which is produced with the use of any lead additive or which contains more than 0.05 gram of lead per gallon or more than 0.005 gram of phosphorus per gallon.

leak (40CFR61.301) means any instrument reading of 10,000 ppmv or greater using method 21 of 40CFR60, appendix A.

leak (40CFR61.61) means any of several events that indicate interruption of confinement of vinyl chloride within process equipment. Leaks include events regulated under subpart V of this part such as:
(1) An instrument reading of 10,000 ppm or greater measured according to Method 21 (see Appendix A to 40CFR60);
(2) A sensor detection of failure of a seal system, failure of a barrier fluid system, or both;
(3) Detectable emissions as indicated by an instrument reading of greater than 500 ppm above background for equipment designated for no detectable emissions measured according to

Test Method 21 (see appendix A to 40CFR60); and

(4) In the case of pump seals regulated under 40CFR61.242-2, indications of liquid dripping constituting a leak under 40CFR61.242-2. Leaks also include events regulated under 40CFR61.65(b)(8)(i) for detection of ambient concentrations in excess of background concentration. A relief valve discharge is not a leak.

leak definition concentration (40CFR60-AA (method 21-2.1)) means the local VOC concentration at the surface of a leak source that indicates that a VOC emission (leak) is present. The leak definition is an instrument meter reading based on a reference compound.

leak detection system (40CFR260.10) means a system capable of detecting the failure of either the primary or secondary containment structure or the presence of a release of hazardous waste or accumulated liquid in the secondary containment structure. Such a system must employ operational controls (e.g., daily visual inspections for releases into the secondary containment system of aboveground tanks) or consist of an interstitial monitoring device designed to detect continuously and automatically the failure of the primary or secondary containment structure or the presence of a release of hazardous waste into the secondary containment structure (under RCRA).

leak or leaking (40CFR761.3) means any instance in which a PCB Article, PCB Container, or PCB Equipment has any PCBs on any portion of its external surface.

leak tight (40CFR61.141) means that solids or liquids cannot escape or spill out. It also means dust-tight.

least-cost plan or least-cost planning process (40CFR72.2) means an energy conservation and electric power planning methodology meeting the requirements of 40CFR73.82(a)(4) of this chapter.

ledger paper (40CFR250.4-t) means a type of paper generally used in a broad variety of recordkeeping type applications such as in accounting machines (under RCRA).

legal defense cost (40CFR280.92) is any expense that an owner or operator or provided of financial assurance incurs in defending against claims or actions brought:

(1) By EPA or a state to require corrective action or to recover the costs of corrective action;

(2) By or on behalf of a third party for bodily injury or property damage caused by an accidental release; or

(3) By any person to enforce the terms of a financial assurance mechanism.

legal defense costs (40CFR264.141-g) means any expenses that an insurer incurs in defending against claims of third parties brought under the terms and conditions of an insurance policy (under the Resource Conservation Recovery Act).

legal proceedings (40CFR32.105-j) means any criminal proceeding or any civil judicial proceeding to which the Federal Government or a State of local government or quasi-governmental authority is a party. The term includes appeals from such proceedings.

legally authorized representative (40CFR26.102-c) means an individual or judicial or other body authorized under applicable law to consent on behalf of a prospective subject to the subject's participation in the procedure(s) involved in the research.

legionella (40CFR141.2) means a genus of bacteria, some species of which have caused a type of pneumonia called Legionnaires Disease.

legionella (EPA-94/04): A genus of bacteria, some species of which have caused a type of pneumonia called Legionnaires Disease.

legislation (40CFR1508.17) includes a bill or legislative proposal to Congress developed by or with the significant cooperation and support of a Federal agency, but does not include requests for appropriations. The test for significant cooperation is whether the proposal is in fact predominantly that of the agency rather than another source. Drafting does not by itself constitute significant cooperation. Proposals for legislation include requests for ratification of treaties. Only the agency which has primary responsibility for the subject matter involved will prepare a legislative environmental impact statement.

lessee (OPA1001) means a person holding a leasehold interest in an oil or gas lease on lands beneath navigable waters (as that term is defined in section 2(a) of the Submerged Lands Act (43USC1301(a)) or on submerged lands of the outer Continental Shelf, granted or maintained under applicable state law or the Outer Continental Self Lands Act (43USC1331 et seq.).

lesser quantity (40CFR63.2) means a quantity of a hazardous air pollutant that is or may be emitted by a stationary source that the Administrator establishes in order to define a major source under an applicable subpart of this part (other identical or similar definitions are provided in 40CFR63.101; 63.191).

lethal mutation (40CFR798.5275-1) is a change in the genome which, when expressed, causes death to the carrier.

level of concern (LOC) (EPA-94/04): The concentration in air of an extremely hazardous substance above which there may be serious immediate health effects to anyone exposed to it for short periods.

level of detection (40CFR761.3): See quantifiable level.

level of quantitation or LOQ (40CFR766.3) means the lowest concentration at which HDDs/HDFs can be reproducibly measured in a specific chemical substance within specified confidence limits, as described in this part.

liabilities (40CFR144.61-d) means probable future sacrifices of economic benefits arising from present obligations to transfer assets or provide services to other entities in the future as a result of past transactions or events (other identical or similar definitions are provided in 40CFR264.141-f; 265.141-f).

liability (SF101): See liable.

liable or liability (SF101-42USC9601) shall be construed to be the standard of liability which obtains under section 311 of the Federal Water Pollution Control Act (33USC1321) (other identical or similar definitions are provided in OPA1001).

license (10CFR20.3) means a license issued under the regulations in parts 30 through 35, 39, 40, 60, 61, 70, or part 72 of this chapter. Licensee means the holder of such license.

license (10CFR30.4), except where otherwise specified, means a license for byproduct material issued pursuant to the regulations in this part and parts 31 through 35 and 39 of this chapter.

license (10CFR40.4), except where otherwise specified, means a license issued pursuant to the regulations in this part (other identical or similar definitions are provided in 40CFR70.4).

license or permit (40CFR121.1-a) means any license or permit granted by an agency of the Federal Government to conduct any activity which may result in any discharge into the navigable waters of the United States.

licensed material (10CFR20.3) means source material, special nuclear material, or byproduct material received, possessed, used, or transferred under a general or specific License issued by the Commission pursuant to the regulations in this chapter.

licensed site (40CFR192.31-d) means the area contained within the boundary of a location under the control of persons generating or storing uranium byproduct materials under a license issued pursuant to section 84 of the Act. For purposes of this subpart, "licensed site" is equivalent to "regulated unit" in subpart F of part 264 of this chapter (under AEA).

licensee (10CFR20.3): See license.

licensing or permitting agency (40CFR121.1-b) means any agency of the Federal Government to which application is made for a license or permit.

lidar (40CFR60-AA (alt. method 1)): See light detection and ranging.

lidar range (40CFR60-AA(alt. method 1)) means the range or distance from the lidar to a point of interest along the lidar line-of-sight.

life of the unit, firm power contractual arrangement (40CFR72.2) means a unit participation power sales agreement under which a utility or industrial customer reserves, or is entitled to receive, a specified amount or percentage of nameplate capacity and associated energy generated by any specified generating unit and pays its proportional amount of such unit's total costs, pursuant to a contract:

(1) For the life of the unit;

(2) For a cumulative term of no less than 30 years, including contracts that permit an election for early termination; or

(3) For a period equal to or greater than 25 years or 70 percent of the economic useful life of the unit determined as of the time the unit was built, with option rights to purchase or release some portion of the nameplate capacity and associated energy generated by the unit at the end of the period.

life of the unit, firm power contractual arrangement (CAA402-42USC7651a) means a unit participation power sales agreement under which a utility or industrial customer reserves, or is entitled to receive, a specified amount or percentage of capacity and associated energy generated by a specified generating unit (or units) and pays its proportional amount of such unit's total costs, pursuant to a contract either:

(A) for the life of the unit;

(B) for a cumulative term of no less than 30 years, including contracts that permit an election for early termination; or

(C) for a period equal to or greater than 25 years or 70 percent of the economic useful life of the unit determined as of the time the unit was built, with option rights to purchase or re-lease some portion of the capacity and associated energy generated by the unit (or units) at the end of the period.

lifetime exposure (EPA-94/04): Total amount of exposure to a substance that a human would receive in a lifetime (usually assumed to be 70 years).

lift (EPA-94/04): In a sanitary landfill, a compacted layer of solid waste and the top layer of cover material.

lifting station (EPA-94/04): See pumping station.

light duty truck (40CFR52.741) means any motor vehicle rated at 3,850 kg gross vehicle weight or less, designed mainly to transport property.

light duty truck (40CFR60.391) means any motor vehicle rated at 3,850 kilograms gross vehicle weight or less, designed mainly to transport property (under RCRA).

light duty truck (40CFR86.082.2) means any motor vehicle rated at 8,500 pounds GVWR or less which has a vehicle curb weight of 6,000 pounds or less and which has a basic vehicle frontal area of 45 square feet or less, which is:

(1) Designed primarily for purposes of transportation of property or is a derivation of such a vehicle, or

(2) Designed primarily for transportation of persons and has a capacity of more than 12 persons, or

(3) Available with special features enabling off-street or off-highway operation and use.

light duty truck 1 (40CFR86.094.2) means any light light-duty truck up through 3750 lbs loaded vehicle weight.

light duty truck 2 (40CFR86.094.2) means any light light-duty truck greater than 3750 lbs loaded vehicle weight.

light duty truck 3 (40CFR86.094.2) means any heavy light-duty truck up through 5750 lbs adjusted loaded vehicle weight.

light duty truck 4 (40CFR86.094.2) means any heavy light-duty truck greater than 5750 lbs adjusted loaded vehicle weight.

light duty vehicle (40CFR52.731-4) means a gasoline-powered motor vehicle rated at 6,000 lb gross vehicle weight (GVW) or less (other identical or similar definitions are provided in 40CFR52.2038-2; 52.2039-2; 52.2485-2; 52.2490-2; 52.2491-2).

light duty vehicle (40CFR86.082.2) means a

passenger car or passenger car derivative capable of seating 12 passengers or less.

light duty vehicles and engines (CAA202-42USC7521) means new light duty motor vehicles and new light duty motor vehicle engines, as determined under regulations of the Administrator (cf. heavy duty vehicle).

light light duty truck (40CFR86.094.2) means any light-duty truck rated up through 6000 lbs GVWR (under CAA).

light light duty truck (40CFR86.094-2) means any light-duty truck rated up through 6000 lbs GVWR.

light light duty truck (40CFR88.101.94) means any light-duty truck rated through 6000 lbs GVWR.

light liquid (40CFR52.741) means VOM in the liquid state which is not defined as heavy liquid.

light off time or LOT (40CFR85.2122(a)(15)(ii)(C)) means the time required for a catalytic converter (at ambient temperature 68-86 degrees F) to warm-up sufficiently to convert 50% of the incoming HC and CO to CO_2 and H_2O.

light oil condenser (40CFR61.131) means any unit in the light-oil recovery operation that functions to condense benzene-containing vapors.

light oil decanter (40CFR61.131) means any vessel, tank, or other type of device in the light-oil recovery operation that functions to separate light oil from water downstream of the light-oil condenser. A light-oil decanter also may be known as a light-oil separator.

light oil storage tank (40CFR61.131) means any tank, reservoir, or container used to collect or store crude or refined light-oil.

light oil sump (40CFR61.131) means any tank, pit, enclosure, or slop tank in light-oil recovery operations that functions as a wastewater separation device for hydrocarbon liquids on the surface of the water.

lignite (40CFR60.41a-5) means coal that is classified as lignite A or B according to the American Society of Testing and Materials' (ASTM) Standard Specification for Classification of Coals by Rank D388-77 (incorporated by reference, see 40CFR60.17).

lignite (40CFR60.41b) means a type of coal classified as lignite A or lignite B by the American Society of Testing and Materials in ASTM D388-77, Standard Specification for Classification of Coals by Rank (IBR, 40CFR60.17).

lignite coal (SMCRA701-30USC1291) means consolidated lignitic coal having less than 8,300 British thermal units per pound, moist and mineral matter free.

lima beans (40CFR180.1-j8) means the beans and the pod.

lima beans (40CFR407.71-j) shall mean the processing of lima beans into the following product styles: Canned and frozen, green and white, all varieties and sizes.

lime kiln (40CFR60.281-n) means a unit used to calcine lime mud, which consists primarily of calcium carbonate, into quicklime, which is calcium oxide (cf. fluidized bed incinerator).

lime manufacturing plant (40CFR60.341-a) means any plant which uses a rotary lime kiln to produce lime product from limestone by calcination.

lime product (40CFR60.341-b) means the product of the calcination process including, but not limited to, calcitic lime, dolomitic lime, and dead-burned dolomite.

limestone scrubbing (EPA-94/04): Use of a limestone and water solution to remove gaseous stack-pipe sulfur before it reaches the atmosphere (cf. air pollution control device).

limitation (40CFR2.301-3; 2.302-3): See standard.

limited degradation (EPA-94/04): An environmental policy permitting some degradation of natural systems but terminating at a level well beneath an established health standard.

limited water-soluble substances (40CFR797.1060-4) means chemicals which are soluble in water at less than 1,000 mg/l (cf. readily water-soluble substances).

limited water-soluble substances (40CFR797.1075-4) means chemicals which are soluble in water at less than 1,000 mg/l.

limiting factor (EPA-94/04): A condition whose absence or excessive concentration is incompatible with the needs or tolerance of a species or population and which may have a negative influence on their ability to thrive or survive.

limiting permissible concentration (LPC) of the liquid phase of a material (40CFR227.27-a) is:
(1) That concentration of a constituent which, after allowance for initial mixing as provided in 40CFR227.29, does not exceed applicable marine water quality criteria; or, when there are no applicable marine water quality criteria,
(2) That concentration of waste or dredged material in the receiving water which, after allowance for initial mixing, as specified in 40CFR227.29, will not exceed a toxicity threshold defined as 0.01 of a concentration shown to be acutely toxic to appropriate sensitive marine organisms in a bioassay carried out in accordance with approved EPA procedures.
(3) When there is reasonable scientific evidence on a specific waste material to justify the use of an application factor other than 0.01 as specified in paragraph (a)(2) of this section, such alternative application factor shall be used in calculating the LPC.

limiting permissible concentration of the suspended particulate and solid phases of a material (40CFR227.27-b) means that concentration which will not cause unreasonable acute or chronic toxicity or other sublethal adverse effects based on bioassay results using appropriate sensitive marine organisms in the case of the suspended particulate phase, or appropriate sensitive benthic marine organisms in the case of the solid phase; and which will not cause accumulation of toxic materials in the human food chain. These bioassays are to be conducted in accordance with procedures approved by EPA, or, in the case of dredged material, approved by EPA and the Corps of Engineers.

limnology (EPA-94/04): The study of the physical, chemical, hydrological, and biological aspects of fresh water bodies.

lindane (EPA-94/04): A pesticide that causes adverse health effects in domestic water supplies and is toxic to freshwater fish and aquatic life.

linear dynamic range (40CFR136-AC-11) means the concentration range over which the analytical curve remains linear.

liner (40CFR260.10) means a continuous layer of natural or man-made materials, beneath or on the sides of a surface impoundment, landfill, or landfill cell, which restricts the downward or lateral escape of hazardous waste, hazardous waste constituents, or leachate.

liner (40CFR503.21-j) is soil or synthetic material that has a hydraulic conductivity of 1×10^{-7} centimeters per second or less.

liner (EPA-94/04):
1. A relatively impermeable barrier designed to keep leachate inside a landfill. Liner materials include plastic and dense clay.
2. An insert or sleeve for sewer pipes to prevent leakage or infiltration.

lipid solubility (EPA-94/04): The maximum concentration of a chemical that will dissolve in fatty substances. Lipid soluble substances are insoluble in water. They will very selectively disperse through the environment via uptake in living tissue.

liquefaction (EPA-94/04): Changing a solid into a liquid.

liquefied petroleum gas (40CFR80.2-oo) means a liquid hydrocarbon fuel that is stored under pressure and is composed primarily of species that are gases at atmospheric conditions (temperature = 25°C and pressure = 1 atm), excluding natural gas.

liquid/gas method (40CFR52.741) means either of two methods for determining capture which require both gas phase and liquid phase measurements and analysis. The first method requires construction of a TTE. The second method uses the building or room which houses the facility as an enclosure. The second method requires that all other VOM sources within the room be shut down while the test is performed, but all fans and blowers within the room must be operated according to normal procedures.

liquid injection incinerator (EPA-94/04): Commonly used system that relies on high pressure to prepare liquid wastes for incineration breaking them up into tiny droplets to allow easier combustion.

liquid mounted seal (40CFR60.111a-h) means a foam or liquid-filled primary seal mounted in contact with the liquid between the tank wall and the floating roof continuously around the circumference of the tank.

liquid mounted seal (40CFR61.341) means a foam or liquid-filled primary seal mounted in contact with the liquid between the waste management unit wall and the floating roof continuously around the circumference.

liquid mounted seal (40CFR63.111) means a foam- or liquid-filled seal mounted in contact with the liquid between the wall of the storage vessel or waste management unit and the floating roof. The seal is mounted continuously around the circumference of the vessel or unit.

liquid phase of a material (40CFR227.32), subject to the exclusions of paragraph (b) of this section, is the supernatant remaining after one hour undisturbed settling, after centrifugation and filtration through a 0.45 micron filter. The suspended particulate phase is the supernatant as obtained above prior to centrifugation and filtration. The solid phase includes all material settling to the bottom in one hour. Settling shall be conducted according to procedures approved by EPA.

liquid phase process (40CFR60.561) means a polymerization process in which the polymerization reaction is carried out in the liquid phase; i.e., the monomer(s) and any catalyst are dissolved, or suspended in a liquid solvent.

liquid phase slurry process (40CFR60.561) means a liquid phase polymerization process in which the monomer(s) are in solution (completely dissolved) in

a liquid solvent, but the polymer is in the form of solid particles suspended in the liquid reaction mixture during the polymerization reaction; sometimes called a particle form process.

liquid phase solution process (40CFR60.561) means a liquid phase polymerization process in which both the monomer(s) and polymer are in solution (completely dissolved) in the liquid reaction mixture.

liquid service (40CFR52.741) means that the equipment or component contains process fluid that is in a liquid state at operating conditions.

liquid trap (40CFR280.12) means sumps, well cellars, and other traps used in association with oil and gas production, gathering, and extraction operations (including gas production plants), for the purpose of collecting oil, water, and other liquids. These liquid traps may temporarily collect liquids for subsequent disposition or reinjection into a production or pipeline stream, or may collect and separate liquids from a gas stream.

liquids dripping (40CFR60.481) means liquids dripping means any visible leakage from the seal including spraying, misting, clouding, and ice formation.

liquids dripping (40CFR63.161) means any visible leakage from the seal including dripping, spraying, misting, clouding, and ice formation. Indications of liquid dripping include puddling or new stains that are indicative of an existing evaporated drip (under CAA).

list (EPA-94/04): Shorthand term for EPA list of violating facilities or firms debarred from obtaining government contracts because they violated certain sections of the Clean Air or Clean Water Acts. The list is maintained by The Office of Enforcement and Compliance Monitoring.

list of violating facilities (40CFR15.4) means a list of facilities which are ineligible for any agency contract, grant or loan.

list official (40CFR15.4) means an EPA official designated by the Assistant Administrator to maintain the List of Violating Facilities.

listed mixture (40CFR716.3) means any mixture listed in 40CFR716.120.

listed participant (40CFR195.2) in an individual or organization who has met all the requirements for listing in the RMP and RCP programs.

listed waste (EPA-94/04): Wastes listed as hazardous under RCRA but which have not been subjected to the Toxic Characteristics Listing Process because the dangers they present are considered self-evident (cf. hazardous waste characteristics).

listing proceeding (40CFR15.4) means an informal hearing, conducted by the Case Examiner, held to determine whether a facility should be placed on the List of Violating Facilities.

lithographic printing line (40CFR52.741) means a printing line, except that the substrate is not necessarily fed from an unwinding roll, in which each roll printer uses a roll where both the image and non-image areas are essentially in the same plane (planographic).

lithology (40CFR146.3) means the description of rocks on the basis of their physical and chemical characteristics (other identical or similar definitions are provided in 40CFR147.2902).

lithosphere (40CFR191.12) means the solid part of the Earth below the surface, including any ground water contained within it.

litter (EPA-94/04): The highly visible portion of solid waste carelessly discarded outside the regular garbage and trash collection and disposal system.

littoral zone (EPA-94/04):
1. That portion of a body of fresh water extending from the shoreline lakeward to the limit of occupancy of rooted plants.
2. The strip of land along the shoreline between the high and low water levels.

live 18-day embryos (40CFR797.2130-v) are embryos that are developing normally after 18 days of incubation. This is determined by candling the eggs. Values are expressed as a percentage of viable embryos (fertile eggs) (under TSCA).

live 21-day embryos (40CFR797.2150-v) are embryos that are developing normally after 21 days of incubation. This is determined by candling the eggs. Values are expressed as a percentage of viable embryos (fertile eggs).

live weight killed (40CFR432.11-d; 432.21-d; 432.31-d; 432.41-d): See LWK.

load allocation (LA) (40CFR130.2-g) means the portion of a receiving water's loading capacity that is attributed either to one of its existing or future nonpoint sources of pollution or to natural background sources. Load allocations are best estimates of the loading, which may range from reasonably accurate estimates to gross allotments, depending on the availability of data and appropriate techniques for predicting the loading. Wherever possible, natural and nonpoint source loads should be distinguished.

load cell (40CFR201.1-o) means a device external to the locomotive, of high electrical resistance, used in locomotive testing to simulate engine loading while the locomotive is stationary. (Electrical energy produced by the diesel generator is dissipated in the load cell resistors instead of the traction motors.)

load or loading (40CFR130.2-e) means an amount of matter or thermal energy that is introduced into a receiving water to introduce matter or thermal energy into a receiving water. Loading may be either man-caused (pollutant loading) or natural (natural background loading).

loaded emissions test (40CFR51-AN) means a sampling procedure for exhaust emissions which requires exercising the engine under stress (i.e., loading) by use of a chassis dynamometer to stimulate actual driving conditions. As a minimum requirement, the loaded emission test must include running the vehicle and measuring exhaust emissions at two speeds and loads other than idle (under CAA).

loaded vehicle mass (40CFR86.402.78) means curb mass plus 80 kg (176 lb.), average driver mass.

loaded vehicle weight (40CFR86.082.2) means the vehicle curb weight plus 300 pounds.

loaded vehicle weight (40CFR88.101.94) is defined as the curb weight plus 300 lbs.

loading (40CFR61.341) means the introduction of waste into a waste management unit but not necessarily to complete capacity (also referred to as filling).

loading (40CFR795.120) means the ratio of the biomass of gammarids (grams, wet weight) to the volume (liters) of test solution in either a test chamber or passing through it in a 24-hour period.

loading (40CFR797.1300-6) means the ratio of daphnid biomass (grams, wet weight) to the volume (liters) of test solution in a test chamber at a point in time, or passing through the test chamber during a specific interval (other identical or similar definitions are provided in 40CFR797.1330-7).

loading (40CFR797.1400-9) means the ratio of fish biomass (grams, wet weight) to the volume (liters) of test solution in a test chamber or passing through it in a 24-hour period.

loading (40CFR797.1520-8) is the ratio of fish biomass (grams, wet weight) to the volume (liters) of test solution passing through the test chamber during a 24-hour period.

loading (40CFR797.1600-10) is the ratio of biomass (grams of fish, wet weight) to the volume (liters) of test solution passing through the test chamber during a specific interval (normally a 24-hour period).

loading (40CFR797.1830-7) is the ratio of the number of oysters to the volume (liters) of test solution passing through the test chamber per hour.

loading (40CFR797.1930-4) means the ratio of test organism biomass (grams, wet weight) to the volume (liters) of test solution in a test chamber (other identical or similar definitions are provided in 40CFR797.1950-6; 797.1970-4).

loading capacity (40CFR130.2-f) means the greatest amount of loading that a water can receive without violating water quality standards.

loading cycle (40CFR61.301) means the time period

from the beginning of filling a tank truck, railcar, or marine vessel until flow to the control device ceases, as measured by the flow indicator (under the Clean Air Act).

loading cycle (40CFR63.111) means the time period from the beginning of filling a tank truck or railcar until flow to the control device ceases, as measured by the flow indicator.

loading rack (40CFR60.501) means the loading arms, pumps, meters, shutoff valves, relief valves, and other piping and valves necessary to fill delivery tank trucks.

loading rack (40CFR61.301) means the loading arms, pumps, meters, shutoff valves, relief valves, and other piping and valves necessary to fill tank trucks, railcars, or marine vessels.

loading rack (40CFR63.101) means a single system used to fill tank trucks and railcars at a single geographic site. Loading equipment and operations that are physically separate (i.e, do not share common piping, valves, and other equipment) are considered to be separate loading racks (other identical or similar definitions are provided in 40CFR63.111).

loan (40CFR15.4) means an agreement or other arrangement under which any portion of a business, activity, or program is assisted under a loan issued by an agency and includes any subloan issued under a loan issued by an agency.

loan agreement (40CFR39.105-d) means a written agreement between the Bank and the guaranteed borrower stating the terms of the loan.

loan guarantee agreement (40CFR39.105-e) means a written agreement between EPA and the guaranteed borrower stating the terms of the loan guarantee.

loan guarantee and loan insurance (40CFR34.105-i) means an agency's guarantee or insurance of a loan made by a person.

local agency (40CFR51.100-g) means any local government agency other than the State agency, which is charged with responsibility for carrying out a portion of the plan.

local agency (40CFR58.1-m) means any local government agency, other than the State agency, which is charged with the responsibility for carrying out a portion of the plan.

local agency (40CFR60.21-j) means any local governmental agency.

local air quality modeling analysis (40CFR51.852) means an assessment of localized impacts on a scale smaller than the entire nonattainment or maintenance area, including, for example, congested roadway intersections and highways or transit terminals, which uses an air quality dispersion model to determine the effects of emissions on air quality (other identical or similar definitions are provided in 40CFR93.152).

local comprehensive emergency response plan (40CFR310.11-e) means the emergency plan prepared by the local emergency planning committee as required by section 303 of the Emergency Planning and Community Right-To-Know Act of 1986 (SARA Title III).

local education agency (40CFR47.105-c) means any education agency as defined in section 198 of the Elementary and Secondary Education Act of 1965 (20 U.S.C. 3381) and shall include any tribal education agency, as defined in 40CFR47.105(f).

local education agency (40CFR763.103) means:
(1) Any local education agency as defined in section 198(a)(10) of the Elementary and Secondary Education Act of 1965 (20USC2854).
(2) The governing authority of any nonprofit elementary or secondary school, where the term "nonprofit" means owned and operated by one of more nonprofit corporations or associations no part of the net earnings of which inures, or may lawfully inure, to the benefit of any private shareholder or individual.

local education agency (40CFR763.83) means:
(1) Any local education agency as defined in section 198 of the Elementary and Secondary Education Act of 1965 (20USC3381).

(2) The owner of any nonpublic, nonprofit elementary, or secondary school building.
(3) The governing authority of any school operated under the defense dependents' education system provided for under the Defense Dependents' Education Act of 1978 (20USC921, et seq.).

local education agency (LEA) (EPA-94/04): In the asbestos program, an educational agency at the local level that exists primarily to operate schools or to contract for educational services, including primary and secondary public and private schools. A single, unaffiliated school can be considered an LEA for AHERA purposes.

local education agency (TSCA202-15USC2642) means:
(A) any local education agency as defined in section 198 of the Elementary and Secondary Education Act of 1965 (20USC3381),
(B) the owner of any private, nonprofit elementary, or secondary school building, and
(C) the governing authority of any school operated under the defense dependents' education system provided for under the Defense Dependents' Education Act of 1978 (20USC921, et seq.).

local emergency planning committee (40CFR350.1) or committee means the local emergency planning committee appointed by the emergency response commission.

local emergency planning committee (LEPC) (EPA-94/04): A committee appointed by the state emergency response commission, as required by SARA Title III, to formulate a comprehensive emergency plan for its jurisdiction.

local government (40CFR31.3) means a county, municipality, city, town, township, local public authority (including any public and Indian housing agency under the United States Housing Act of 1937) school district, special district, intrastate district, council of governments (whether or not incorporated as a nonprofit corporation under state law), any other regional or interstate government entity, or any agency or instrumentality of a local government (other identical or similar definitions are provided in 40CFR34.105-j; 280.92; CZMA304-16USC1453-90).

local share (40CFR39.105-f) means the amount of the total grant eligible and allowable project costs which a public body is obligated to pay under the grant.

locomotive (40CFR201.1-p) means for the purpose of this regulation, a self-propelled vehicle designed for and used on railroad tracks in the transport of rail cars, including self-propelled rail passenger vehicles.

locomotive (40CFR89.2) means a self-propelled piece of on-track equipment (other than equipment designed for operation both on highways and rails, specialized maintenance equipment, and other similar equipment) designed for moving other equipment, freight or passenger traffic.

locomotive load cell test stand (40CFR201.1-q) means the load cell 40CFR201.1(o) and associated structure, equipment, trackage and locomotive being tested.

log sorting and log storage facilities (40CFR122.27-3) means facilities whose discharges result from the holding of unprocessed wood, for example, logs or roundwood with bark or after removal of bark held in self-contained bodies of water (mill ponds or log ponds) or stored on land where water is applied intentionally on the logs (wet decking) (see 40CFR429, subpart I, including the effluent limitations guidelines).

London Amendments (40CFR82.3-w) means the Montreal Protocol, as amended at the Second Meeting of the parties to the Montreal Protocol in London in 1990 (the Amendments discuss the control of ozone worldwide).

long term contract (RCRA1004-42USC6903) means, when used in relation to solid waste supply, a contract of sufficient duration to assure the viability of a resource recovery facility (to the extent that such viability depends upon solid waste supply).

long term stabilization (40CFR61.221-a) means the addition of material on a uranium mill tailings pile for purpose of ensuring compliance with the requirements of 40CFR192.02(a) or 192.32(b)(i). These actions shall be considered complete when the Nuclear Regulatory Commission determines that the requirements of 40CFR192.02(a) or 192.32(b)(i) have been met.

loose fill insulation (40CFR248.4-u) means insulation in granular, nodular, fibrous, powdery, or similar form, designed to be installed by pouring, blowing or hand placement.

low altitude (40CFR86.082.2) means any elevation less than or equal to 1,219 meters (4,000 feet) (other identical or similar definitions are provided in 40CFR86.1602).

low altitude conditions (40CFR86.082.2) means a test altitude less than 549 meters (1,800 feet).

low concentration PCBs (40CFR761.123) means PCBs that are tested and found to contain less than 500 ppm PCBs, or those PCB-containing materials which EPA requires to be assumed to be at concentrations below 500 ppm (i.e., untested mineral oil dielectric fluid).

low density polyethylene (LDPE) (40CFR60.561) means a thermoplastic polymer or copolymer comprised of at least 50 percent ethylene by weight and having a density of 0.940 g/cm^3 or less.

low dose (40CFR795.232-4) should correspond to 1/10 of the high dose (cf. high dose).

low emission vehicle (40CFR88.101.94) means any light-duty vehicle or light-duty truck conforming to the applicable Low-Emission Vehicle standard, or any heavy-duty vehicle with an engine conforming to the applicable Low-Emission Vehicle standard.

low hazard contents (40CFR1910.35) shall be classified as those of such low combustibility that no self-propagating fire therein can occur and that consequently the only probable danger requiring the use of emergency exits will be from panic, fumes, or smoke, or fire from some external source.

low heat release rate (40CFR60.41b) means a heat release rate of 730,000 J/sec-m^3 (70,000 Btu/hour-ft^3) or less.

low level radioactive waste (LLRW) (EPA-94/04):

Wastes less hazardous than most of those associated with nuclear reactor; generated by hospitals, research laboratories, and certain industries. The Department of Energy, Nuclear Regulatory Commission, and EPA share responsibilities for managing them (cf. high level radioactive wastes).

low loss fitting (40CFR82.152-h) means any device that is intended to establish a connection between hoses, appliances, or recovery or recycling machines and that is designed to close automatically or to be closed manually when disconnected, minimizing the release of refrigerant from hoses, appliances, and recovery or recycling machines.

low noise emission product (NCA15-42USC4914) means any product which emits noise in amounts significantly below the levels specified in noise emission standards under regulations applicable under section 6 at the time of procurement to that type of product.

low noise emission product determination (40CFR203.1-5) means the Administrator's determination whether or not a product, for which a properly filed application has been received, meets the low-noise-emission product criterion.

low NO_x burners and low NO_x burner technology (40CFR76.2) means commercially available combustion modification NO_x controls that minimize NO_x formation by introducing coal and its associated combustion air into a boiler such that initial combustion occurs in a manner that promotes rapid coal devolatilization in a fuel-rich (i.e., oxygen deficient) environment and introduces additional air to achieve a final fuel-lean (i.e., oxygen rich) environment to complete the combustion process. This definition shall include the staging of any portion of the combustion air using air nozzles or registers located inside any waterwall hole that includes a burner. This definition shall exclude the staging of any portion of the combustion air using air nozzles or ports located outside any waterwall hole that includes a burner (commonly referred to as NO_x ports or separated overfire air ports).

low NO_x burners (EPA-94/04): One of several combustion technologies used to reduce emissions of nitrogen oxides (NO_x.)

low pressure appliance (40CFR82.152-i) means an appliance that uses a refrigerant with a boiling point above 10 degrees Centigrade at atmospheric pressure (29.9 inches of mercury). This definition includes but is not limited to equipment utilizing refrigerants -11, -113, and -123.

low pressure process (40CFR60.561) means a production process for the manufacture of low density polyethylene in which a reaction pressure markedly below that used in a high pressure process is used. Reaction pressure of current low pressure processes typically go up to about 300 psig.

low processing packinghouse (40CFR432.31-c) shall mean a packinghouse that processes no more than the total animals killed at that plant, normally processing less than the total kill (cf. high processing packinghouse).

low terrain (40CFR51.166-26) means any area other than high terrain.

low terrain (40CFR52.21-26) means any area other than high terrain.

low viscosity poly(ethylene terephthalate) (40CFR60.561) means a poly(ethylene terephthalate) that has an intrinsic viscosity of less than 0.75 and is used in such applications as clothing, bottle, and film production.

low volume waste sources (40CFR423.11-b) means, taken collectively as if from one source, wastewater from all sources except those for which specific limitations are otherwise established in this part. Low volume wastes sources include, but are not limited to: wastewaters from wet scrubber air pollution control systems, ion exchange water treatment system, water treatment evaporator blowdown, laboratory and sampling streams, boiler blowdown, floor drains, cooling tower basin cleaning wastes, and recirculating house service water systems. Sanitary and air conditioning wastes are not included.

lower detectable limit (40CFR53.23-c): The minimum pollutant concentration which produces a signal of twice the noise level.

lower explosive limit (LEL) (EPA-94/04): The

concentration of a compound in air below which the mixture will not catch on fire.

lower explosive limit for methane gas (40CFR503.21-k) is the lowest percentage of methane gas in air, by volume, that propagates a flame at 25 degrees Celsius and atmospheric pressure (under the Clean Water Act).

lowest achievable emission rate (40CFR51.165-xiii) means, for any source, the more stringent rate of emissions based on the following:
(A) The most stringent emissions limitation which is contained in the implementation plan of any State for such class or category of stationary source, unless the owner or operator of the proposed stationary source demonstrates that such limitations are not achievable; or
(B) The most stringent emissions limitation which is achieved in practice by such class or category of stationary sources. This limitation, when applied to a modification, means the lowest achievable emissions rate for the new or modified emissions units within the stationary source. In no event shall the application of the term permit a proposed new or modified stationary source to emit any pollutant in excess of the amount allowable under an applicable new source standards of performance.

lowest achievable emission rate (40CFR51-AS-18) means, for any source, the more stringent rate of emissions based on the following:
(A) The most stringent emissions limitation which is contained in the implementation plan of any State for such class or category of stationary source, unless the owner or operator of the proposed stationary source demonstrates that such limitations are not achievable; or
(B) The most stringent emissions limitation which is achieved in practice by such class or category of stationary source. This limitation, when applied to a modification, means the lowest achievable emissions rate for the new or modified emissions units within the stationary source. In no event shall the application of this term permit a proposed new or modified stationary source to emit any pollutant in excess of the amount allowable under applicable new source standards of performance.

lowest achievable emission rate (EPA-94/04): Under the Clean Air Act, the rate of emissions that reflects:
(a) the most stringent emission limitation in the implementation plan of any state for such source unless the owner or operator demonstrates such limitations are not achievable; or
(b) the most stringent emissions limitation achieved in practice, whichever is more stringent. A proposed new or modified source may not emit pollutants in excess of existing new source standards.

lowest achievable emission rate (LAER) (CAA171-42USC7501) means for any source, that rate of emissions which reflects:
(A) the most stringent emission limitation which is contained in the implementation plan of any State for such class or category of source, unless the owner or operator of the proposed source demonstrates that such limitations are not achievable, or
(B) the most stringent emission limitation which is achieved in practice by such class or category of source, whichever is more stringent. In no event shall the application of this term permit a proposed new or modified source to emit any pollutant in excess of the amount allowable under applicable new source standards of performance. The best available control technology (BACT), as defined in CAA169, shall be substituted for the lowest achievable emission rate [CAAA182(c)(7)] (cf. best available control technology).

lowest observed adverse effect level (EPA-94/04): The lowest dose in an experiment which produced an observable adverse effect.

lubricating oil (RCRA1004-42USC6903) means the fraction of crude oil which is sold for purposes of reducing friction in any industrial or mechanical device. Such term includes re-refined oil.

luminescent materials (40CFR469.41-a) shall mean materials that emit light upon excitation by such energy sources as photons, electrons, applied voltage, chemical reactions or mechanical energy and which are specifically used as coatings in fluorescent lamps and cathode ray tubes. Luminescent materials include, but are not limited to, calcium halophosphate, yttrium oxide, zinc sulfide, and zinc-cadmium sulfide.

LVW (loaded vehicle weight) (CAA216): See vehicle curb weight.

LWK (live weight killed) (40CFR432.11-d) shall mean the total weight of the total number of animals slaughtered during the time to which the effluent limitations apply; i.e., during any one day or any period of thirty consecutive days (other identical or similar definitions are provided in 40CFR432.21-d; 432.31-d; 432.41-d).

********** **MMMMM** **********

M10 (40CFR435.11) shall mean those offshore facilities continuously manned by ten (10) or more persons (other identical or similar definitions are provided in 40CFR435.41-b).

M9IM (40CFR435.11-n) shall mean those offshore facilities continuously manned by nine (9) or fewer persons or only intermittently manned by any number of persons.

M9IM (40CFR435.41-c) shall mean those coastal facilities continuously manned by nine (9) or fewer persons or intermittently manned by any number of persons.

machine shop (40CFR61.31-d) means a facility performing cutting, grinding, turning, honing, milling, deburring, lapping, electrochemical machining, etching, or other similar operations (under the Clean Air Act).

magic carpet project (40CFR52.2493-7) means the 1-year experimental program supported by the City of Seattle and METRO of free fare transit service in the area bounded by Western Avenue, Stewart Street, Sixth Avenue, and Jackson Street as funded and authorized by City Council Resolution 102452 and METRO Council Resolution 1926.

magnet wire (40CFR52.741) means aluminum or copper wire formed into an electromagnetic coil.

magnet wire coating (40CFR52.741) means any coating or electrically insulating varnish or enamel applied to magnet wire.

magnet wire coating facility (40CFR52.741) means a facility that includes one or more magnet wire coating line(s).

magnet wire coating line (40CFR52.741) means a coating line in which any protective, decorative, or functional coating is applied onto the surface of a magnet wire.

magnetic separation (EPA-94/04): Use of magnets to separate ferrous materials from mixed municipal waste stream.

magnetic tape (40CFR60.711-13) means any flexible substrate that is covered on one or both sides with a coating containing magnetic particles and that is used for audio or video recording or information storage.

mail or serve by mail (40CFR72.2) means to submit or serve by means other than personal service.

maintain (40CFR1516.2) means maintain, collect, use or disseminate.

maintain (40CFR16.2-a): See individual, maintain, record, system of records, and routine use.

maintenance (40CFR280.12) means the normal operational upkeep to prevent an underground storage tank system from releasing product (under RCRA).

maintenance area (40CFR51.392) means any geographic region of the United States previously designated nonattainment pursuant to the CAA Amendments of 1990 and subsequently redesignated to attainment subject to the requirement to develop a maintenance plan under section 175A of the CAA, as amended (other identical or similar definitions are provided in 40CFR93.101).

maintenance area (40CFR51.852) means an area with a maintenance plan approved under section 175A of the Act (other identical or similar definitions are provided in 40CFR93.152).

maintenance event (40CFR85.1402) means a single maintenance activity for which the engine is removed from service. Once the engine is returned to service, the maintenance event is considered done.

maintenance instructions (40CFR204.2-10) means those instructions for maintenance, use, and repair, which the Administrator is authorized to require pursuant to section 6(c)(1) of the Act.

maintenance instructions or instructions (40CFR205.2-24) means those instructions for maintenance, use, and repair, which the Administrator is authorized to require Pursuant to section 6(c)(1) of the Act.

maintenance period (40CFR51.392) with respect to a pollutant or pollutant precursor means that period of time beginning when a State submits and EPA approves a request under section 107(d) of the CAA for redesignation to an attainment area, and lasting for 20 years, unless the applicable implementation plan specifies that the maintenance period shall last for more than 20 years (other identical or similar definitions are provided in 40CFR93.101).

maintenance plan (40CFR51.491) means an implementation plan for an area for which the State is currently seeking designation or has previously sought redesignation to attainment, under section 107(d) of the Act, which provides for the continued attainment of the NAAQS.

maintenance plan (40CFR51.852) means a revision to the applicable SIP, meeting the requirements of section 175A of the Act (other identical or similar definitions are provided in 40CFR93.152).

maintenance wastewater (40CFR63.101) means wastewater generated by the draining of process fluid from components in the chemical manufacturing process unit into an individual drain system prior to or during maintenance activities. Maintenance wastewater can be generated during planned and unplanned shutdowns and during periods not associated with a shutdown. Examples of activities that can generate maintenance wastewaters include descaling of heat exchanger tubing bundles, cleaning of distillation column traps, draining of low legs and high point bleeds, draining of pumps into an individual drain system, and draining of portions of the chemical manufacturing process unit for repair.

major cylinder component (40CFR85.1402) means piston assembly, cylinder liner, connecting rod, or piston ring set.

major disaster (40CFR109.2-d) means any hurricane, tornado, storm, flood, high water, wind-driven water, tidal wave, earthquake, drought, fire, or other catastrophe in any part of the United States which, in the determination of the President, is or threatens to become of sufficient severity and magnitude to warrant disaster assistance by the Federal Government to supplement the efforts and available resources of States and local governments and relief organizations in alleviating the damage, loss, hardship, or suffering caused thereby.

major emitting facility (CAA169-42USC7479) means any of the following stationary sources of air pollutants which emit, or have the potential to emit, one hundred tons per year or more of any air pollutant from the following types of stationary sources: fossil-fuel fired steam electric plants of more than two hundred and fifty million British thermal units per hour heat input, coal cleaning plants (thermal dryers), kraft pulp mills, Portland Cement plants, primary zinc smelters, iron and steel mill plants, primary aluminum ore reduction plants, primary copper smelters, municipal incinerators capable of charging more than two hundred and fifty tons of refuse per day, hydrofluoric, sulfuric, and nitric acid plants, petroleum refineries, lime plants, phosphate rock processing plants, coke oven batteries, sulfur recovery plants, carbon black plants (furnace process) primary lead smelters, fuel conversion plants, sintering plants, secondary metal production facilities, chemical process plants, fossil-fuel boilers of more than two hundred and fifty million British thermal units per hour heat input, petroleum storage and transfer facilities with a capacity exceeding three hundred thousand barrels, taconite ore processing facilities, glass fiber processing plants, charcoal production facilities. Such term also includes any other source with the potential to emit two hundred and fifty tons per year or more of any air pollutant. This term shall not include new or modified facilities which are nonprofit health or education institutions which have been exempted by the State.

major emitting facility (CAA302): See major stationary source.

major employment facility (40CFR52.2296-2) means any single employer location having 250 or more employees.

major facility (40CFR122.2) means any NPDES "facility or activity" classified as such by the Regional Administrator, or, in the case of "approved State programs," the Regional Administrator in conjunction with the State Director.

major facility (40CFR124.2) means any RCRA, UIC, NPDES, or 404 "facility or activity" classified as such by the Regional Administrator, or, in the case of "approved State programs," the Regional Administrator in conjunction with the State Director.

major facility (40CFR144.3) means any UIC "facility or activity" classified as such by the Regional Administrator, or, in the case of approved State programs, the Regional Administrator in conjunction with the State Director.

major facility (40CFR270.2) means any facility or activity classified as such by the Regional Administrator, or, in the case of approved State programs, the Regional Administrator in conjunction with the State Director.

major federal action (40CFR1508.18) includes actions with effects that may be major and which are potentially subject to Federal control and responsibility. Major reinforces but does not have a meaning independent of significantly (40CFR1508.27). Actions include the circumstance where the responsible officials fail to act and that failure to act is reviewable by courts or administrative tribunals under the Administrative Procedure Act or other applicable law as agency action:
(a) Actions include new and continuing activities, including projects and programs entirely or partly financed, assisted, conducted, regulated, or approved by federal agencies; new or revised agency rules, regulations, plans, policies, or procedures; and legislative proposals (40CFR1506.8, 1508.17). Actions do not include funding assistance solely in the form of general revenue sharing funds, distributed under the State and Local Fiscal Assistance Act of 1972, 31 U.S.C. 1221 et seq., with no Federal agency control over the subsequent use of such funds. Actions do not include bringing judicial or administrative civil or criminal enforcement actions.
(b) Federal actions tend to fall within one of the following categories:
  (1) Adoption of official policy, such as rules, regulations, and interpretations adopted pursuant to the Administrative Procedure Act, 5 U.S.C. 551 et seq.; treaties and international conventions or agreements; formal documents establishing an agency's policies which will result in or substantially alter agency programs.
  (2) Adoption of formal plans, such as official documents prepared or approved by federal agencies which guide or prescribe alternative uses of Federal resources, upon which future agency actions will be based.
  (3) Adoption of programs, such as a group of concerted actions to implement a specific policy or plan; systematic and connected agency decisions allocating agency resources to implement a specific statutory program or executive directive.
  (4) Approval of specific projects, such as construction or management activities located in a defined geographic area. Projects include actions approved by permit or other regulatory decision as well as federal and federally assisted activities.

major industrial use sector or sector (40CFR82.172) means an industrial category which EPA has reviewed under the SNAP program with historically high consumption patterns of ozone-depleting substances, including: Refrigeration and air conditioning; foam-blowing; fire suppression and explosion protection; solvents cleaning; aerosols; sterilants; tobacco expansion; pesticides; and adhesives, coatings and inks sectors.

major life activities (40CFR7.25): See handicapped person.

major maintenance, service, or repair (40CFR82.152-j) means any maintenance, service, or repair involving the removal of any or all of the following appliance components: compressor, condenser, evaporator, or auxiliary heat exchanger coil.

major modification (40CFR51.165-v) means:
(A) any physical change in or change in the method of operation of a major stationary source that would result in a significant net emissions increase of any pollutant subject to regulation under the Act.
(B) Any net emissions increase that is considered significant for volatile organic compounds shall be considered significant for ozone.
(C) A physical change or change in the method of operation shall not include:
  (1) Routine maintenance, repair and replacement;
  (2) Use of an alternative fuel or raw material by reason of an order under sections 2 (a) and (b) of the Energy Supply and Environmental Coordination Act of 1974 (or any superseding legislation) or by reason of a natural gas curtailment plan pursuant to the Federal Power Act;
  (3) Use of an alternative fuel by reason of an order or rule section 125 of the Act;
  (4) Use of an alternative fuel at a steam generating unit to the extent that the fuel is generated from municipal solid waste;
  (5) Use of an alternative fuel or raw material by a stationary source which:
    (i) The source was capable of accommodating before December 21, 1976, unless such change would be prohibited under any federally enforceable permit condition which was established after December 12, 1976 pursuant to 40CFR52.21 or under regulations approved pursuant to 40CFR subpart I or 40CFR51.166, or
    (ii) The source is approved to use under any permit issued under regulations approved pursuant to this section;
  (6) An increase in the hours of operation or in the production rate, unless such change is prohibited under any federally enforceable permit condition which was established after December 21, 1976 pursuant to 40CFR52.21 or regulations approved pursuant to 40CFR part 51 subpart I or 40CFR51.166.
  (7) Any change in ownership at a stationary source.
  (8) The addition, replacement or use of a pollution control project at an existing electric utility steam generating unit, unless the reviewing authority determines that such addition, replacement, or use renders the unit less environmentally beneficial, or except:
    (i) When the reviewing authority has reason to believe that the pollution control project would result in a significant net increase in representative actual annual emissions of any criteria pollutant over levels used for that source in the most recent air quality impact analysis in the area conducted for the purpose of title I, if any, and
    (ii) The reviewing authority determines that the increase will cause or contribute to a violation of any national ambient air quality standard or PSD increment, or visibility limitation.
  (9) The installation, operation, cessation, or removal of a temporary clean coal technology demonstration project, provided that the project complies with:
    (i) The State Implementation Plan for the State in which the project is located, and
    (ii) Other requirements necessary to attain and maintain the national ambient air quality standard during the project and after it is terminated.

(Other identical or similar definitions are provided in 40CFR51.165-v; 51.166-2; 51-AS-5; 52.21-2; 52.24-5.)

major modification (EPA-94/04): This term is used to define modifications of major stationary sources of emissions with respect to Prevention of Significant

Deterioration and New Source Review under the Clean Air Act.

major municipal separate storm sewer outfall (or major outfall) (40CFR122.26-5) means a municipal separate storm sewer outfall that discharges from a single pipe with an inside diameter of 36 inches or more or its equivalent (discharge from a single conveyance other than circular pipe which is associated with a drainage area of more than 50 acres); or for municipal separate storm sewers that receive storm water from lands zoned for industrial activity (based on comprehensive zoning plans or the equivalent), an outfall that discharges from a single pipe with an inside diameter of 12 inches or more or from its equivalent (discharge from other than a circular pipe associated with a drainage area of 2 acres or more).

major national holiday (40CFR52.274.d.2-v) means a holiday such as Christmas, New Year's Day, or Independence Day (other identical or similar definitions are provided in 40CFR52.274.h.2-ii; 52.274.p.2-iii; 52.274.r.2-iv).

major outfall (40CFR122.26-6) means a major municipal separate storm sewer outfall.

major PSD modification (40CFR124.41) means a "major modification" as defined in 40CFR52.21.

major PSD stationary source (40CFR124.41) means a "major stationary source" as defined in 40CFR52.21(b)(1).

major source (40CFR63.2) means any stationary source or group of stationary sources located within a contiguous area and under common control that emits or has the potential to emit considering controls, in the aggregate, 10 tons per year or more of any hazardous air pollutant or 25 tons per year or more of any combination of hazardous air pollutants, unless the Administrator establishes a lesser quantity, or in the case of radionuclides, different criteria from those specified in this sentence (other identical or similar definitions are provided in 40CFR63.101; 63.191).

major source (40CFR63.321) means any dry cleaning facility that meets the conditions of 40CFR63.320(g).

major source (40CFR70.2) means any stationary source (or any group of stationary sources that are located on one or more contiguous or adjacent properties, and are under common control of the same person (or persons under common control) belonging to a single major industrial grouping and that are described in paragraph (1), (2), or (3) of this definition. For the purposes of defining "major source," a stationary source or group of stationary sources shall be considered part of a single industrial grouping if all of the pollutant emitting activities at such source or group of sources on contiguous or adjacent properties belong to the same Major Group (i.e., all have the same two-digit code) as described in the Standard Industrial Classification Manual, 1987 (40CFR70.2):

(1) Major source under section 112 of the Act, which is defined as:

  (i) For pollutants other than radionuclides, any stationary source or group of stationary sources located within a contiguous area and under common control that emits or has the potential to emit, in the aggregate, 10 tons per year (tpy) or more of any hazardous air pollutant which has been listed pursuant to section 112(b) of the Act, 25 tpy or more of any combination of such hazardous air pollutants, or such lesser quantity as the Administrator may establish by rule. Notwithstanding the preceding sentence, emissions from any oil or gas exploration or production well (with its associated equipment) and emissions from any pipeline compressor or pump station shall not be aggregated with emissions from other similar units, whether or not such units are in a contiguous area or under common control, to determine whether such units or stations are major sources; or

  (ii) For radionuclides, "major source" shall have the meaning specified by the Administrator by rule.

(2) Major stationary source of air pollutants, as defined in section 302 of the Act, that directly emits or has the potential to emit, 100 tpy or more of any air pollutant (including any major source of fugitive emissions of any such pollutant, as determined by rule by the Administrator). The fugitive emissions of a

stationary source shall not be considered in determining whether it is a major stationary source for the purposes of section 302(j) of the Act, unless the source belongs to one of the following categories of stationary source:

i   Coal cleaning plants (with thermal dryers);
ii   Kraft pulp mills;
iii Portland cement plants;
iv   Primary zinc smelters;
v   Iron and steel mills;
vi   Primary aluminum ore reduction plants;
vii Primary copper smelters;
viii Municipal incinerators capable of charging more than 250 tons of refuse per day;
ix   Hydrofluoric, sulfuric, or nitric acid plants;
x   Petroleum refineries;
xi   Lime plants;
xii Phosphate rock processing plants;
xiii Coke oven batteries;
xiv Sulfur recovery plants;
xv   Carbon black plants (furnace process);
xvi Primary lead smelters;
xvii Fuel conversion plants;
xviii Sintering plants;
xix Secondary metal production plants;
xx   Chemical process plants;
xxi Fossil-fuel boilers (or combination thereof) totaling more than 250 million British thermal units per hour heat input;
xxii Petroleum storage and transfer units with a total storage capacity exceeding 300,000 barrels;
xxiii Taconite ore processing plants;
xxiv Glass fiber processing plants;
xxv Charcoal production plants;
xxvi Fossil-fuel-fired steam electric plants of more than 250 million British thermal units per hour heat input; or
xxvii All other stationary source categories regulated by a standard promulgated under section 111 or 112 of the Act, but only with respect to those air pollutants that have been regulated for that category.

(3) Major stationary source as defined in part D of title I of the Act, including:
   (i) For ozone nonattainment areas, sources with the potential to emit 100 tpy or more of volatile organic compounds or oxides of nitrogen in areas classified as "marginal" or "moderate," 50 tpy or more in areas classified as "serious," 25 tpy or more in areas classified as "severe," and 10 tpy or more in areas classified as "extreme"; except that the references in this paragraph to 100, 50, 25 and 10 tpy of nitrogen oxides shall not apply with respect to any source for which the Administrator has made a finding, under section 182(f) (1) or (2) of the Act, that requirements under section 182(f) of the Act do not apply;
   (ii) For ozone transport regions established pursuant to section 184 of the Act, sources with the potential to emit 50 tpy or more of volatile organic compounds;
   (iii) For carbon monoxide nonattainment areas:
      (A) That are classified as "serious," and
      (B) in which stationary sources contribute significantly to carbon monoxide levels as determined under rules issued by the Administrator, sources with the potential to emit 50 tpy or more of carbon monoxide; and
   (iv) For particulate matter (PM-10) nonattainment areas classified as "serious," sources with the potential to emit 70 tpy or more of PM-10.

major source (CAA112-42USC7412) means any stationary source or group of stationary sources located within a contiguous area and under common control that emits or has the potential to emit considering controls, in the aggregate, 10 tons per year or more of any hazardous air pollutant or 25 tons per year or more of any combination of hazardous air pollutants. The Administrator may establish a lesser quantity, or in the case of radionuclides different criteria, for a major source than that specified in the previous sentence, on the basis of the potency of the air pollutant, persistence, potential for bioaccumulation, other characteristics of the air pollutant, or other relevant factors (cf. major modification).

major source (CAA501-42USC7661) means any stationary source (or any group of stationary sources located within a contiguous area and under common control) that is either of the following:
(A) A major source as defined in section 112.
(B) A major stationary source as defined in section 302 or part D of title I.

major source baseline date (40CFR51.166-14) means:

(i) (a) In the case of particulate matter and sulfur dioxide, January 6, 1975, and

(b) In the case of nitrogen dioxide, February 8, 1988.

(ii) Minor source baseline date means the earliest date after the trigger date on which a major stationary source or a major modification subject to 40CFR52.21 or to regulations approved pursuant to 40CFR51.166 submits a complete application under the relevant regulations. The trigger date is:

(a) In the case of particulate matter and sulfur dioxide, August 7, 1977, and

(b) In the case of nitrogen dioxide, February 8, 1988.

(iii) The baseline date is established for each pollutant for which increments or other equivalent measures have been established if:

(a) The area in which the proposed source or modification would construct is designated as attainment or unclassifiable under section 107(d)(i) (D) or (E) of the Act for the pollutant on the date of its complete application under 40CFR52.21 or under regulations approved pursuant to 40CFR51.166; and

(b) In the case of a major stationary source, the pollutant would be emitted in significant amounts, or, in the case of a major modification, there would be a significant net emissions increase of the pollutant.

(iv) Any minor source baseline date established originally for the TSP increments shall remain in effect and shall apply for purposes of determining the amount of available PM-10 increments, except that the reviewing authority may rescind any such minor source baseline date where it can be shown, to the satisfaction of the reviewing authority, that the emissions increase from the major stationary source, or the net emissions increase from the major modification, responsible for triggering that date did not result in a significant amount of PM-10 emissions.

(Other identical or similar definitions are provided in 40CFR52.21-14.)

major stationary source (40CFR51.165-iv) means:

(A) (1) Any stationary source of air pollutants which emits or has the potential to emit 100 tons per year or more of any pollutant subject to regulation under the Act, or

(2) Any physical change that would occur at a stationary source not qualifying under paragraph (a)(1)(iv)(A)(1) as a major stationary source, if the change would constitute a major stationary source by itself.

(B) A major stationary source that is major for volatile organic compounds shall be considered major for ozone.

(C) The fugitive emissions of a stationary source shall not be included in determining for any of the purposes of this paragraph whether it is a major stationary source, unless the source belongs to one of the following categories of stationary sources:

(1) Coal cleaning plants (with thermal dryers);

(2) Kraft pulp mills;

(3) Portland cement plants;

(4) Primary zinc smelters;

(5) Iron and steel mills;

(6) Primary aluminum ore reduction plants;

(7) Primary copper smelters;

(8) Municipal incinerators capable of charging more than 250 tons of refuse per day;

(9) Hydrofluoric, sulfuric, or nitric acid plants;

(10) Petroleum refineries;

(11) Lime plants;

(12) Phosphate rock processing plants;

(13) Coke oven batteries;

(14) Sulfur recovery plants;

(15) Carbon black plants (furnace process);

(16) Primary lead smelters;

(17) Fuel conversion plants;

(18) Sintering plants;

(19) Secondary metal production plants;

(20) Chemical process plants;

(21) Fossil-fuel boilers (or combination thereof) totaling more than 250 million British thermal units per hour heat input;

(22) Petroleum storage and transfer units with a total storage capacity exceeding 300,000 barrels;

(23) Taconite ore processing plants;

(24) Glass fiber processing plants;

(25) Charcoal production plants;

(26) Fossil fuel-fired steam electric plants of

more than 250 million British thermal units per hour heat input; and

(27) Any other stationary source category which, as of August 7, 1980, is being regulated under section 111 or 112 of the Act.

(Other identical or similar definitions are provided in 40CFR; 51.166-1; 51-AS-4; 52.21-1; 52.24-4; 70.2.)

major stationary source (40CFR65.01-d) shall mean any stationary source which directly emits, or has the potential to emit, 100 tons per year or more of any air pollutant for which a national ambient air quality standard under section 109 of the Act is in effect (including any major stationary source of fugitive emissions of any such pollutant, as determined by rule by the Administrator).

major stationary source (40CFR66.3-h) means any stationary facility or source of air pollutants which directly emits, or has the potential to emit, one hundred tons per year or more of any air pollutant regulated by EPA under the Clean Air Act.

major stationary source (CAA169A-42USC7491) means the following types of stationary sources with the potential to emit 250 tons or more of any pollutant; fossil-fuel fired steam electric plants of more than 250 million British thermal units per hour heat input, coal cleaning plants (thermal dryers), kraft pulp mills, Portland Cement plants, primary zinc smelters, iron and steel mill plants, primary aluminum ore reduction plants, primary copper smelters, municipal incinerators capable of charging more than 250 tons of refuse per day, hydrofluoric, sulfuric, and nitric acid plants, petroleum refineries, lime plants, phosphate rock processing plants, coke oven batteries, sulfur recovery plants, carbon black plants (furnace process), primary lead smelters, fuel conversion plants, sintering plants, secondary metal production facilities, chemical process plants, fossil-fuel boilers of more than 250 million British thermal units per hour heat input, petroleum storage and transfer facilities with a capacity exceeding 300,000 barrels, taconite ore processing facilities, glass fiber processing plants, charcoal production facilities.

major stationary source and major emitting facility (CAA302-42USC7602): Except as otherwise expressly provided, the terms "major stationary source" and "major emitting facility" mean any stationary facility or source of air pollutants which directly emits, or has the potential to emit, one hundred tons per year or more of any air pollutant (including any major emitting facility or source of fugitive emissions of any such pollutant, as determined by rule by the Administrator).

major stationary source and major modification (40CFR51.301-p) mean "major stationary source" and "major modification," respectively, as defined in 40CFR51.24 (40CFR51.24 does not exist. It may mean 40CFR52.24 or 51.165 or 51.166).

major stationary sources (EPA-94/04): Term used to determine the applicability of Prevention of Significant Deterioration and new source regulations. In a nonattainment area, any stationary pollutant source with potential to emit more than 100 tons per year is considered a major stationary source. In PSD areas the cutoff level may be either 100 or 250 tons, depending upon the source.

majors (EPA-94/04): Larger publicly owned treatment works (POTWs) with flows equal to at least one million gallons per day (mgd) or servicing population equivalent to 10,000 persons; certain other POTWs having significant water quality impacts (cf. minors).

make available for use (40CFR171.2) means to distribute, sell, ship, deliver for shipment, or receive and (having so received) deliver, to any person. However, the term excludes transactions solely between persons who are pesticide producers, registrants, wholesalers, or retail sellers, acting only in those capacities.

makes, wherever it appears (40CFR27.2), shall include the terms presents, submits, and causes to be made, presented. or submitted. As the context requires, making or made shall likewise include the corresponding forms of such terms.

makeup solvent (40CFR60.601) means the solvent introduced into the affected facility that compensates for solvent lost from the affected facility during the manufacturing process.

malfunction (40CFR264.1031) means any sudden

failure of a control device or a hazardous waste management unit or failure of a hazardous waste management unit to operate in a normal or usual manner, so that organic emissions are increased.

malfunction (40CFR52.741) means any sudden and unavoidable failure of air pollution control equipment, process equipment, or a process to operate in a normal or usual manner. Failures that are caused entirely or in part by poor maintenance, careless operation, or any other preventable upset condition or preventable equipment breakdown shall not be considered malfunctions.

malfunction (40CFR57.103-p) means any unanticipated and unavoidable failure of air pollution control equipment or process equipment or of a process to operate in a normal or usual manner. Failures that are caused entirely or in part by poor design, poor maintenance, careless operation, or any other preventable upset condition or preventable equipment breakdown shall not be considered malfunctions. A malfunction exists only for the minimum time necessary to implement corrective measures.

malfunction (40CFR60.2) means any sudden and unavoidable failure of air pollution control equipment or process equipment or of a process to operate in a normal or usual manner. Failures that are caused entirely or in part by poor maintenance, careless operation, or any other preventable upset condition or preventable equipment breakdown shall not be considered malfunctions.

malfunction (40CFR61.141) means any sudden and unavoidable failure of air pollution control equipment or process equipment or of a process to operate in a normal or usual manner so that emissions of asbestos are increased. Failures of equipment shall not be considered malfunctions if they are caused in any way by poor maintenance, careless operation, or any other preventable upset conditions, equipment breakdown, or process failure (under CAA).

malfunction (40CFR61.161) means any sudden failure of air pollution control equipment or process equipment or of a process to operate in a normal or usual manner so that emissions of inorganic arsenic are increased (other identical or similar definitions are provided in 40CFR61.171; 61.181).

malfunction (40CFR63.2) means any sudden, infrequent, and not reasonably preventable failure of air pollution control equipment, process equipment, or a process to operate in a normal or usual manner. Failures that are caused in part by poor maintenance or careless operation are not malfunctions (other identical or similar definitions are provided in 40CFR63.101; 63.191).

malfunction (40CFR63.301) means any sudden, infrequent, and not reasonably preventable failure of air pollution control equipment, process equipment, or a process to operate in a normal or usual manner. Failures caused in part by poor maintenance or careless operation are not malfunctions.

malfunction (40CFR63.51) means any sudden failure of air pollution control equipment or process equipment or of a process to operate in a normal or usual manner. Failures that are caused entirely or in part by poor maintenance, careless operation, or any other preventable upset condition or preventable equipment breakdown shall not be considered malfunctions.

malfunction (40CFR86.082.2) means not operating according to specifications (e.g., those specifications listed in the application for certification).

malleable iron (40CFR464.31-e) means a cast iron made by a prolonged anneal of white cast iron in which decarburization or graphitization, or both, take place to eliminate some or all of the cementite. Graphite is present in the form of temper carbon.

man made beta particle and photon emitters (40CFR141.2) means all radionuclides emitting beta particles and/or photons listed in Maximum Permissible Body Burdens and Maximum Permissible Concentration of Radionuclides in Air or Water for Occupational Exposure, NBS Handbook 69, except the daughter products of thorium-232, uranium-235 and uranium-238.

man made beta particle and photon emitters (EPA-94/04): All radionuclides emitting beta particles and/or photos listed in Maximum Permissible Body

Burdens and Maximum Permissible Concentrations of Radionuclides in Air and Water for Occupational Exposure.

management (40CFR191.02-m) means any activity, operation, or process (except for transportation) conducted to prepare spent nuclear fuel or radioactive waste for storage or disposal, or the activities associated with placing such fuel or waste in a disposal system.

management authority (40CFR228.2-e) means the EPA organizational entity assigned responsibility for implementing the management functions identified in 40CFR228.3.

management conference (40CFR35.9010) means a management conference convened by the Administrator under section 320 of the CWA for an estuary in the NEP.

management of migration (40CFR300.5) means actions that are taken to minimize and mitigate the migration of hazardous substances or pollutants or contaminants and the effects of such migration. Measures may include, but are not limited to, management of a plume of contamination, restoration of a drinking water aquifer, or surface water restoration.

management or hazardous waste management (40CFR260.10) means the systematic control of the collection, source separation, storage, transportation, processing, treatment, recovery, and disposal of hazardous waste.

management plan (EPA-94/04): Under the Asbestos Hazard Emergency Response Act (AHERA), a document that each Local Education Agency is required to prepare, describing all activities planned and undertaken by a school to comply with AHERA regulations, including building inspections to identify asbestos-containing materials, response actions, and operations and maintenance programs to minimize the risk of exposure.

management program (CZMA304-16USC1453) includes, but is not limited to, a comprehensive statement in words, maps, illustrations, or other media of communication, prepared and adopted by

the state in accordance with the provisions of this chapter, setting forth objectives, policies, and standards to guide public and private uses of lands and waters in the coastal zone.

managerial controls (EPA-94/04): Methods of nonpoint source pollution control based on decisions about managing agricultural wastes or application times or rates for agrochemicals.

mandatory class I federal area (40CFR51.301-o) means any area identified in part 81, subpart D of this title.

mandatory class I Federal areas (CAA169A-42USC7491) means Federal areas which may not be designated as other than class I under this part.

mandatory recycling (EPA-94/04): Programs which by law require consumers to separate trash so that some or all recyclable materials are recovered for recycling rather than going to landfills.

manifest (40CFR144.3) means the shipping document originated and signed by the "generator" which contains the information required by subpart B of 40CFR part 262.

manifest (40CFR260.10) means the shipping document EPA form 8700-22 and, if necessary, EPA form 8700-22A, originated and signed by the generator in accordance with the instructions included in the Appendix to part 262.

manifest (40CFR270.2) means the shipping document originated and signed by the generator which contains the information required by subpart B of 40CFR262.

manifest (40CFR761.3) means the shipping document EPA form 8700-22 and any continuation sheet attached to EPA form 8700-22, originated and signed by the generator of PCB waste in accordance with the instructions included with the form and subpart K of this part.

manifest (RCRA1004-42USC6903) means the form used for identifying the quantity, composition, and the origin, routing, and destination of hazardous waste during its transportation from the point of generation to the point of disposal, treatment, or storage.

manifest document number (40CFR260.10) means the U.S. EPA twelve digit identification number assigned to the generator plus a unique five digit document number assigned to the Manifest by the generator for recording and reporting purposes (under RCRA).

manifest system (EPA-94/04): Tracking of hazardous waste from "cradle to grave" (generation through disposal) with accompanying documents known as manifests (cf. cradle to grave).

manifold business forms (40CFR250.4-u) means a type of product manufactured by business forms manufacturers that is commonly produced as marginally punched continuous forms in small rolls or fan folded sets with or without carbon paper interleaving. It has a wide variety of uses such as invoices, purchase orders, office memoranda, shipping orders, and computer printout.

manmade air pollution (CAA169A-42USC7491) means air pollution which results directly or indirectly from human activities.

manned control center (40CFR761.3) means an electrical power distribution control room where the operating conditions of a PCB transformer are continuously monitored during the normal hours of operation (of the facility), and, where the duty engineers, electricians, or other trained personnel have the capability to deenergize a PCB transformer completely within 1 minute of the receipt of a signal indicating abnormal operating conditions such as an overtemperature condition or overpressure condition in a PCB transformer.

manual (40CFR66.3-i) means the "Noncompliance Penalties Instruction Manual" which accompanies these regulations. This Manual appears as Appendix B to these regulations.

manual method (40CFR53.1-h) means a method for measuring concentrations of an ambient air pollutant in which sample collection, analysis, or measurement, or some combination thereof, is performed manually (under CAA).

manual separation (EPA-94/04): Hand separation of compostable or recyclable material from waste.

manual separation (EPA-94/04): Hand sorting of reyclable or compostable materials in waste.

manufacture (40CFR125.101) means to produce as an intermediate or final product, or byproduct.

manufacture (40CFR372.3) means to produce, prepare, import, or compound a toxic chemical. Manufacture also applies to a toxic chemical that is produced coincidentally during the manufacture, processing, use, or disposal of another chemical or mixture of chemicals, including a toxic chemical that is separated from that other chemical or mixture of chemicals as a byproduct, and a toxic chemical that remains in that other chemical or mixture of chemicals as an impurity.

manufacture (40CFR704.3) means to manufacture for commercial purposes (other identical or similar definitions are provided in 40CFR716.3).

manufacture (40CFR710.2-o) means to produce or manufacture in the United States or import into the customs territory of the United States.

manufacture (40CFR720.3-q) means to produce or manufacture in the United States or import into the customs territory of the United States (other identical or similar definitions are provided in 40CFR723.50-2; 723.250-6).

manufacture (40CFR761.3) means to produce, manufacture, or import into the customs territory of the United States.

manufacture (40CFR763.163) means to produce or manufacture in the United States.

manufacture (CWA312-33USC1322) means any person engaged in the manufacturing, assembling, or importation of marine sanitation devices or of vessels subject to standards and regulations promulgated under this section.

manufacture (TSCA3-15USC2602) means to import into the customs territory of the United States (as defined in general headnote 2 of the Tariff Schedules of the United States), produce, or manufacture (other identical or similar definitions are provided in 40CFR723.175-5; 762.3-d).

manufacture for commercial purposes (40CFR704.3) means:
(1) To import, produce, or manufacture with the purpose of obtaining an immediate or eventual commercial advantage for the manufacture and includes among other things, such "manufacture" of any amount of a chemical substance or mixture:
  (i) For commercial distribution, including for test marketing.
  (ii) For use by the manufacturer, including use for product research and development, or as an intermediate.
(2) Manufacture for commercial purposes also applies to substances that are produced coincidentally during the manufacture, processing, use, or disposal of another substance or mixture, including both byproducts that are separated from that other substance or mixture and impurities that remain in that substance or mixture. Such byproducts and impurities may, or may not, in themselves have commercial value. They are nonetheless produced for the purpose of obtaining a commercial advantage since they are part of the manufacture of a chemical product for a commercial purpose.
(Other identical or similar definitions are provided in 40CFR712.3-h; 716.3; 717.3; 763.63-g.)

manufacture of electronic crystals (40CFR469.22-c) means the growing of crystals and/or the production of crystals wafers for use in the manufacture of electronic devices.

manufacture of semiconductors (40CFR469.12-c) means those processes, beginning with the use of crystal wafers, which lead to or are associated with the manufacture of semiconductor devices (under CWA).

manufacture or import for commercial purposes (40CFR710.2-p) means to manufacture or import:
(1) For distribution in commerce, including for test marketing purposes, or
(2) For use by the manufacturer, including for use as an intermediate.

manufacture or import for commercial purposes (40CFR720.3-r) means:
(1) To import, produce, or manufacture with the purpose of obtaining an immediate or eventual commercial advantage for the manufacturer or importer, and includes, among other things, "manufacture" of any amount of a chemical substance or mixture:
  (i) For commercial distribution, including for test marketing.
  (ii) For use by the manufacturer, including use for product research and development or as an intermediate.
(2) The term also applies to substances that are produced coincidentally during the manufacture, processing, use, or disposal of another substance or mixture, including byproducts that are separated from that other substance or mixture and impurities without separate commercial value are nonetheless produced for the purpose of obtaining a commercial advantage, since they are part of the manufacture of a chemical substance for commercial purposes.
(Other identical or similar definitions are provided in 40CFR747.200-1.)

manufacture or process (40CFR717.3-d) means to manufacture or process for commercial purposes.

manufacture solely for export (40CFR720.3-s) means: to manufacture for a commercial purposes solely for export from the United States under the following restrictions on domestically activity:
(1) Processing is limited solely to sites under the control of the manufacturer.
(2) Distribution in commerce is limited to purposes of export.
(3) The manufacturer may not use the substance except in small quantities solely for research and development.
(Other identical or similar definitions are provided in 40CFR747.200-1.)

manufactured (40CFR60.531) means completed and ready for shipment (whether or not packaged).

manufactured with a controlled substance (40CFR82.104-m) means that the manufacturer of the product itself used a controlled substance directly in the product's manufacturing, but the product itself does not contain more than trace quantities of the controlled substance at the point of introduction into interstate commerce. The following situations are excluded from the meaning of the phrase "manufactured with" a controlled substance:

(1) Where a product has not had physical contact with the controlled substance;
(2) Where the manufacturing equipment or the product has had physical contact with a controlled substance in an intermittent manner, not as a routine part of the direct manufacturing process;
(3) Where the controlled substance has been transformed, except for trace quantities; or
(4) Where the controlled substance has been completely destroyed.

manufacturer (40CFR129.2-m) means any establishment engaged in the mechanical or chemical transformation of materials or substances into new products including but not limited to the blending of materials such as pesticidal products, resins, or liquors.

manufacturer (40CFR2.301-5) has the meaning given it in section 216(1) of the Act, 42USC550(1).

manufacturer (40CFR2.303-2) has the meaning given it in 42USC4902(6).

manufacturer (40CFR2.311-5) has the meaning given it in section 501(9) of the Act, 15USC2001(9).

manufacturer (40CFR205.2-16) means any person engaged in the manufacturing or assembling of new products, or the importing of new products for resale, or who acts for and is controlled by any such person in connection with the distribution of such products.

manufacturer (40CFR211.203-p) as stated in the Act means any person engaged in the manufacturing or assembling of new products, or the importing of new products for resale, or who acts for, and is controlled by, any such person in connection with the distribution of such products.

manufacturer (40CFR60.531) means any person who constructs or imports a wood heater (under the Clean Air Act).

manufacturer (40CFR610.11-4) means a person or company which is engaged in the business of producing or assembling, and which has primary control over the design specifications, of a retrofit device for which a fuel economy improvement claim is made.

manufacturer (40CFR704.3) means a person who imports, produces, or manufactures a chemical substance. A person who extracts a component chemical substance from a previously existing chemical substance or a complex combination of substances is a manufacturer of that component chemical substance.

manufacturer (40CFR716.3) means a person who produces or manufactures a chemical substance. A person who extracts a component chemical substance from a previously existing chemical substance or a complex combination of substances is a manufacturer of that component chemical substance.

manufacturer (40CFR720.3-t) means a person who imports, produces, or manufactures a chemical substance. A person who extracts a component chemical substance from a previously existing chemical substance or a complex combination of substances is a manufacturer of that component chemical substance. A person who contracts with a manufacturer to manufacture or produce a chemical substance is also a manufacturer if:

(1) the manufacturer manufactures or produces the substance exclusively for that person, and
(2) that person specifies the identity of the substance and controls the total amount produced and the basic technology for the plant process.

(Other identical or similar definitions are provided in 40CFR723.250-6; 747.115-1; 747.195-1; 747.200-1.)

manufacturer (40CFR763.163) means a person who produces or manufactures in the United States.

manufacturer (40CFR82.172) means any person engaged in the direct manufacture of a substitute (under CAA).

manufacturer (40CFR85.1807-3) refers to a manufacturer contesting a recall order directed at that manufacturer.

manufacturer (40CFR85.1902-f) shall be given the meaning ascribed to it by section 214 of the Act (under CAA).

manufacturer (40CFR86.1115.87) means a manufacturer contesting a compliance level or penalty determination sent to the manufacturer.

manufacturer (40CFR86.614.84-2) refers to a manufacturer contesting a suspension or revocation order directed at the manufacturer (other identical or similar definitions are provided in 40CFR86.1014.84-2; 86.1115.87-2).

manufacturer (CAA211.o-42USC7545) includes an importer and the term "manufacture" includes importation.

manufacturer (CAA216-42USC7550) as used in sections 202, 203, 206, 207, and 208 means any person engaged in the manufacturing or assembling of new motor vehicles or new motor vehicle engines, or importing such vehicles or engines for resale, or who acts for and is under the control of any such person in connection with the distribution of new motor vehicles or new motor vehicle engines, but shall not include any dealer with respect to new motor vehicles or new motor vehicle engines received by him in commerce.

manufacturer (NCA3-42USC4902) means any person engaged in the manufacturing or assembling of new products, or the importing of new products for resale, or who acts for, and is controlled by, any such person in connection with the distribution of such products.

manufacturer owned nonroad engine (40CFR89.902.96) means an uncertified nonroad engine owned and controlled by a nonroad engine manufacturer and used in a manner not involving lease or sale by itself or in a vehicle or piece of equipment employed from year to year in the ordinary course of business for product development, production method assessment, and market promotion purposes.

manufacturer parts (40CFR86.1602) are parts produced or sold by the manufacturer of the motor vehicle or motor vehicle engine.

manufacturer's formulation (EPA-94/04): A list of substances or component parts as described by the maker of a coating, pesticide, or other product containing chemicals or other substances.

manufacturer's rated dryer capacity (40CFR60.621) means the dryer's rated capacity of articles, in pounds or kilograms of clothing articles per load, dry basis, that is typically found on each dryer on the manufacturer's name-plate or in the manufacturer's equipment specifications.

manufacturing (40CFR61.141) means the combining of commercial asbestos--or, in the case of woven friction products, the combining of textiles containing commercial asbestos--with any other material(s), including commercial asbestos, and the processing of this combination into a product. Chlorine production is considered a part of manufacturing.

manufacturing activities (40CFR704.203) means all those activities at one site which are necessary to produce a substance identified in subpart D of this part and make it ready for sale or use as the listed substance, including purifying or importing the substance.

manufacturing line (40CFR60.681) means the manufacturing equipment comprising the forming section, where molten glass is fiberized and a fiberglass mat is formed; the curing section, where the binder resin in the mat is thermally "set;" and the cooling section, where the mat is cooled (under CAA).

manufacturing process (40CFR52.741) means a method whereby a process emission source or series of process emission sources is used to convert raw materials, feed stocks, subassemblies, or other components into a product, either for sale or for use as a component in a subsequent manufacturing process.

manufacturing process (40CFR761.3) means all of a series of unit operations operating at a site, resulting in the production of a product.

manufacturing stream (40CFR721.3) means all reasonably anticipated transfer, flow, or disposal of a chemical substance, regardless of physical state or concentration, through all intended operations of manufacture, including the cleaning of equipment.

manufacturing use product (40CFR152.3-o) means any pesticide product that is not an end-use product.

manufacturing use product (40CFR158.153-h) means any pesticide product other than an end use product. A product may consist of the technical grade of active ingredient only, or may contain inert ingredients, such as stabilizers or solvents.

manufacturing-use product (40CFR162.151-b) means any pesticide product other than a product to be labeled with directions for end use. This term includes any product intended for use as a pesticide after re-formulation or repackaging.

margin of safety (EPA-94/04): Maximum amount of exposure producing no measurable effect in animals (or studied humans) divided by the actual amount of human exposure in a population.

marine bays and estuaries (40CFR35.2005-26) are semi-enclosed coastal waters which have a free connection to the territorial sea.

marine environment (40CFR125.121-b) means that territorial seas, the contiguous zone and the oceans.

marine mammal (MMPA3-16USC1362) means any mammal which:
(A) is morphologically adapted to the marine environment (including sea otters and members of the orders Sirenia, Pinnipedia and Cetacea), or
(B) primarily inhabits the marine environment (such as the polar bear); and, for the purposes of this chapter, includes any part of any such marine mammal, including its raw, dressed, or dyed fur or skin.

marine mammal product (MMPA3-16USC1362) means any item of merchandise which consists, or is composed in whole or in part, of any marine mammal.

marine sanitation device (40CFR140.1-c) includes any equipment for installation onboard a vessel and which is designed to receive, retain, treat, or discharge sewage and any process to treat such sewage.

marine sanitation device (CWA312-33USC1322) includes any equipment for installation on board a vessel which is designed to receive, retain, treat, or discharge sewage, and any process to treat such sewage.

marine sanitation device (EPA-94/04): Any equipment or process installed on board a vessel to receive, retain, treat, or discharge sewage.

marine vessel (40CFR61.301) means any tank ship or tank barge which transports liquid product such as benzene.

mark (40CFR761.3) means the descriptive name, instructions, cautions, or other information applied to PCBs and PCB items, or other objects subject to these regulations.

marked (40CFR761.3) means the marking of PCB items and PCB storage areas and transport vehicles by means of applying a legible mark by painting, fixation of an adhesive label, or by any other method that meets the requirements of these regulations.

market/marketers (40CFR761.3) means the processing or distributing in commerce, or the person who processes or distributes in commerce, used oil fuels to burners or other marketers, and may include the generator of the fuel if it markets the fuel directly to the burner.

market response strategies (40CFR51.491) are strategies that create one more incentives for affected sources to reduce emissions, without directly specifying limits on emissions or emission-related parameters that individual sources or even all sources in the aggregate are required to meet.

marking (40CFR11.4-c) means the act of physically indicating the classification assignment on classified material.

MARPOL 73/78 (40CFR110.1) means the International Convention for the Prevention of Pollution from Ships, 1973, as modified by the Protocol of 1978 relating thereto, Annex 1, which regulates pollution from oil and which entered into force on October 2, 1983.

marsh (EPA-94/04): A type of wetland that does not accumulate appreciable peat deposits and is dominated by herbaceous vegetation. Marshes may be either fresh or saltwater, tidal or non-tidal (cf. wetlands).

mass balance (40CFR797.2850-2) means a quantitative accounting of the distributions of chemical in plant components, support medium, and test solutions. It also means a quantitative determination of uptake as the difference between the quantity of gas entering an exposure chamber, the quantity leaving the chamber, and the quantity adsorbed to the chamber walls.

mass balance (SF311.l) means an accumulation of the annual quantities of chemicals transported to a facility, produced at a facility, consumed at a facility, used at a facility, accumulated at a facility, released from a facility, and transported from a facility as a waste or as a commercial product or byproduct or component of a commercial product or byproduct.

mass burn refractory MWC (40CFR60.51a) means a combustor that combusts MSW in a refractory wall furnace. This does not include rotary combustors without waterwalls.

mass burn rotary waterwall MWC (40CFR60.51a) means a combustor that combusts MSW in a cylindrical rotary waterwall furnace. This does not include rotary combustors without waterwalls.

mass burn waterwall MWC (40CFR60.51a) means a combustor that combusts MSW in a conventional waterwall furnace.

mass feed stoker steam generating unit (40CFR60.41b) means a steam generating unit where solid fuel is introduced directly into a retort or is fed directly onto a grate where it is combusted.

mass flow rate (40CFR63.111), as used in the wastewater provisions, means the mass of a constituent in a wastewater stream, determined by multiplying the average concentration of that constituent in the wastewater stream by the annual volumetric flow rate and density of the wastewater stream.

mass of pollutant that can be discharged (40CFR463.2-h) is the pollutant mass calculated by multiplying the pollutant concentration times the average process water usage flow rate.

master inventory file (40CFR710.23-a) means EPA's comprehensive list of chemical substances which constitute the Chemical Substances Inventory compiled under section 8(b) of the Act. It includes substances reported under subpart A of this part and substances reported under part 720 of this chapter for which a Notice of Commencement of Manufacture or Import has been received under 40CFR720.120 of this chapter.

MATC (40CFR797.1330-8): See maximum acceptable toxicant concentration (other identical or similar definitions are provided in 40CFR797.1950-7).

matching funds (40CFR35.4010) means the portion of allowable project costs that a recipient contributes toward completing the technical assistance grant project using non-Federal funds or Federal funds if expressly authorized by statute. The match may include in-kind as well as cash contributions.

material (40CFR220.2-d) means matter of any kind or description, including, but not limited to, dredged material, solid waste, incinerator residue, garbage, sewage, sewage sludge, munitions, radiological, chemical, and biological warfare agents, radioactive materials, chemicals, biological and laboratory waste, wreck or discarded equipment, rock, sand, excavation debris, industrial, municipal, agricultural, and other waste, but such term does not mean sewage from vessels within the meaning of section 312 of the FWPCA. Oil within the meaning of section 311 of the FWPCA shall constitute "material" for purposes of this Subchapter H only to the extent that it is taken on board a vessel or aircraft for the primary purpose of dumping.

material category (EPA-94/04): In the asbestos program, broad classification of materials into thermal surfacing insulation, surfacing material, and miscellaneous material.

material recovery section (40CFR60.561) means the equipment that recovers unreacted or byproduct materials from any process section for return to the process line, off-site purification or treatment, or sale. Equipment designed to separate unreacted or byproduct material from the polymer product are to be included in this process section, provided at least some of the material is recovered for reuse in the process, off-site purification or treatment, or sale, at the time the process section becomes an affected facility. Otherwise, such equipment are to be assigned to one of the other process sections, as appropriate. Equipment that treats recovered materials are to be included in this process section, but equipment that also treats raw materials are not to be included in this process section. The latter equipment are to be included in the raw materials preparation section. If equipment is used to return unreacted or byproduct material directly to the same piece of process equipment from which it was emitted, then that equipment is considered part of the process section that contains the process equipment. If equipment is used to recover unreacted or byproduct material from a process section and return it to another process section or a different piece of process equipment in the same process section or sends it off-site for purification, treatment, or sale, then such equipment are considered part of a material recovery section. Equipment used for the on-site recovery of ethylene glycol from poly(ethylene terephthalate) plants, however, are not included in the material recovery section, but are covered under the standards applicable to the polymerization reaction section (40CFR60.562-1(c)(1)(ii)(A) or (2)(ii)(A)).

material safety data sheet (MSDS) (40CFR370.2) means the sheet required to be developed under 29CFR1910.1200(g) of the Code of Federal Regulations.

material safety data sheet (MSDS) (40CFR721.3): See MSDS.

material safety data sheet (MSDS) (EPA-94/04): A compilation of information required under the OSHA Communication Standard on the identity of hazardous chemicals, health, and physical hazards, exposure limits, and precautions. Section 311 of SARA (Superfund Amendments and Reauthorization Act) requires facilities to submit MSDSs under certain circumstances.

material safety data sheet (SF329) means the sheet required to be developed under section 1901.1200(g) of title 29 of the Code of Federal Regulations, as that section may be amended from time to time.

material specification (40CFR247.101-b) means a specification that stipulates the use of certain materials to meet the necessary performance requirements (other identical or similar definitions are provided in 40CFR249.04-h).

materials recovery facility (EPA-94/04): A facility that processes residentially collected mixed recyclables into new products available for market.

materials recovery facility (MRF) (EPA-94/04)): Facility that processes residentially collected mixed recyclables into new products.

matrix (40CFR763-AA-14) means a fiber of fibers with one end free and the other end embedded in or hidden by a particulate. The exposed fiber must meet the fiber definition.

matter (40CFR1508.19) includes for purposes of part 1504:
(a) With respect to the Environmental Protection Agency, any proposed legislation, project, action or regulation as those terms are used in section 30(a) of the Clean Air Act (42USC7609).
(b) With respect to all other agencies, any proposed major federal action to which section 102(2)(C) of NEPA applies.

maximum acceptable toxicant concentration (4CFR797.1330-8) means the maximum concentration at which a chemical can be present and not be toxic to the test organism (other identical or similar definitions are provided in 40CFR797.1950-7).

maximum achievable control technology (MACT) emission limitation for existing sources (40CFR63.51) means the emission limitation reflecting the maximum degree of reduction in emissions of hazardous air pollutants (including a prohibition on such emissions, where achievable) that the

Administrator, taking into consideration the cost of achieving such emission reductions, and any non-air quality health and environmental impacts and energy requirements, determines is achievable by sources in the category or subcategory to which such emission standard applies. This limitation shall not be less stringent than the MACT floor.

maximum achievable control technology (MACT) emission limitation for new sources (40CFR63.51) means the emission limitation which is not less stringent than the emission limitation achieved in practice by the best controlled similar source, and which reflects the maximum degree of reduction in emissions of hazardous air pollutants (including a prohibition on such emissions, where achievable) that the Administrator, taking into consideration the cost of achieving such emission reduction, and any non-air quality health and environmental impacts and energy requirements, determines is achievable by sources in the category or subcategory to which such emission standard applies.

maximum Achievable Control Technology (MACT) floor (40CFR63.51) means:
(1) For existing sources:
 (i) The average emission limitation achieved by the best performing 12 percent of the existing sources in the United States (for which the Administrator has emissions information), excluding those sources that have, within 18 months before the emission standard is proposed or within 30 months before such standard is promulgated, whichever is later, first achieved a level of emission rate or emission reduction which complies, or would comply if the source is not subject to such standard, with the lowest achievable emission rate (as defined in section 171 of the Act) applicable to the source category and prevailing at the time, in the category or subcategory, for categories and subcategories of stationary sources with 30 or more sources; or
 (ii) The average emission limitation achieved by the best performing five sources in the United States (for which the Administrator has or could reasonably obtain emissions information) in the category or subcategory, for a category or subcategory

of stationary sources with fewer than 30 sources;
(2) For new sources, the emission limitation achieved in practice by the best controlled similar source.

maximum as applied to BAT effluent limitations and NSPS for drilling fluids and drill cuttings (40CFR435.11-k) shall mean the maximum concentration allowed as measured in any single sample of the barite.

maximum contaminant level (40CFR141.2) means the maximum permissable level of a contaminant in water which is delivered to any user of a public water system.

maximum contaminant level (40CFR142.2) means the maximum permissible level of a contaminant in water which is delivered to the free flowing outlet of the ultimate user of a public water system; except in the case of turbidity where the maximum permissible level is measured at the point of entry to the distribution system. Contaminants added to the water under circumstances controlled by the user, except for those resulting from corrosion of piping and plumbing caused by water quality, are excluded from this definition.

maximum contaminant level (40CFR300-AA), under section 1412 of the Safe Drinking Water Act, as amended, means the maximum permissible concentration of a substance in water that is delivered to any user of a public water supply (under CERCLA).

maximum contaminant level (EPA-94/04): The maximum permissible level of a contaminant in water delivered to any user of a public system. MCLs are enforceable standards.

maximum contaminant level (SDWA1401-42USC300f) means the maximum permissible level of a contaminant in water which is delivered to any user of a public water system.

maximum contaminant level goal (40CFR300-AA), Under section 1412 of the Safe Drinking Water Act, as amended, means a nonenforceable concentration for a substance in drinking water that is protective of adverse human health effects and allows an adequate margin of safety.

maximum contaminant level goal (EPA-94/04): Under the Safe Drinking Water Act, a non-enforceable concentration of a drinking water contaminant, set at the level at which no known or anticipated adverse effects on human health occur and which allows an adequate safety margin. The MCLG is usually the starting point for determining the regulated maximum contaminant level (cf. maximum contaminant level).

maximum contaminant level goal or MCLG (40CFR141.2) means the maximum level of a contaminant in drinking water at which no known or anticipated adverse effect on the health of persons would occur, and which allows an adequate margin of safety. Maximum contaminant level goals are nonenforceable health goals.

maximum daily discharge limitation (40CFR122.2) means the highest allowable "daily discharge."

maximum demonstrated MWC unit load (40CFR60.51a) means the maximum 4-hour block average MWC unit load achieved during the most recent dioxin/furan test demonstrating compliance with the applicable standard for MWC organics specified under 40CFR60.53a.

maximum demonstrated particulate matter control device temperature (40CFR60.51a) means the maximum 4-hour block average temperature measured at the final particulate matter control device inlet during the most recent dioxin/furan test demonstrating compliance with the applicable standard for MWC organics specified under 40CFR60.53a. If more than one particulate matter control device is used in series at the affected facility, the maximum 4-hour block average temperature is measured at the final particulate matter control device.

maximum design heat input capacity (40CFR60.41c) means the ability of a steam generating unit to combust a stated maximum amount of fuel (or combination of fuels) on a steady state basis as determined by the physical design and characteristics of the steam generating unit.

maximum extent practicable (40CFR112.2) means the limitations used to determine oil spill planning resources and response times for on-water recovery, shoreline protection, and cleanup for worst case discharges from onshore non-transportation-related facilities in adverse weather. It considers the planned capability to respond to a worst case discharge in adverse weather, as contained in a response plan that meets the requirements in 40CFR112.20 or in a specific plan approved by the Regional Administrator.

maximum for any one day and average of daily values for thirty consecutive days (40CFR407.71-w) shall be based on the daily average mass of raw material processed during the peak thirty consecutive day production period (other identical or similar definitions are provided in 40CFR407.81-o).

maximum for any one day and average of daily values for thirty consecutive days (40CFR407.61-y) shall be based on the daily average mass of material processed during the peak thirty consecutive day production period.

maximum for any one day as applied to BPT, BCT and BAT effluent limitations and NSPS for oil and grease in produced water (40CFR435.11-l) shall mean the maximum concentration allowed as measured by the average of four grab samples collected over a 24-hour period that are analyzed separately. Alternatively, for BAT and NSPS the maximum concentration allowed may be determined on the basis of physical composition of the four grab samples prior to a single analysis.

maximum heat input capacity (40CFR60.41b) means the ability of a steam generating unit to combust a stated maximum amount of fuel on a steady state basis, as determined by the physical design and characteristics of the steam generating unit.

maximum organic vapor pressure (40CFR61.341) means the equilibrium partial pressure exerted by the waste at the temperature equal to the highest calendar-month average of the waste storage temperature for waste stored above or below the ambient temperature or at the local maximum monthly average temperature as reported by the National Weather Service for waste stored at the

ambient temperature, as determined:
(1) In accordance with 40CFR60.17(c); or
(2) As obtained from standard reference texts; or
(3) In accordance with 40CFR60.17(a)(37); or
(4) Any other method approved by the Administrator.

maximum potential NO_x emission rate (40CFR72.2) means the emission rate of nitrogen oxides (in lb/mmBtu) calculated in accordance with section 3 of appendix F of part 75 of this chapter, using the maximum potential nitrogen oxides concentration and either the minimum oxygen concentration (in % O_2) or the maximum carbon dioxide concentration (in % CO_2) as defined in section 2 of appendix A of part 75 of this chapter.

maximum production capacity (40CFR57.103-q) means either the maximum demonstrated rate at which a smelter has produced its principal metallic final product under the process equipment configuration and operating procedures prevailing on or before August 7, 1977, or a rate which the smelter is able to demonstrate by calculation is attainable with process equipment existing on August 7, 1977. The rate may be expressed as a concentrate feed rate to the smelter.

maximum rated capacity (40CFR204.51-b) means that the portable air compressor, operating at the design full speed with the compressor on load, delivers its rated cfm output and pressure, as defined by the manufacturer.

maximum rated horsepower (40CFR86.082.2) means the maximum brake horsepower output of an engine as stated by the manufacturer in his sales and service literature and his application for certification under 40CFR86.082.21.

maximum rated RPM (40CFR205.151-17) leans the engine speed measured in revolutions per minute (RPM) at which peak net brake power (SAE J-245) is developed for motorcycles of a given configuration.

maximum rated torque (40CFR86.082.2) means the maximum torque produced by an engine as stated by the manufacturer in his sales and service literature and his application for certification under 40CFR86.082.21.

maximum sound level (40CFR201.1-r) means the greatest A-weighted sound level in decibels measured during the designated time interval or during the event, with either fast meter response 40CFR201.1(1) or slow meter response 40CFR201.l(ii) as specified. It is abbreviated as L_{max}.

maximum theoretical emissions (40CFR52.741) means the quantity of volatile organic material emissions that theoretically could be emitted by a stationary source before add-on controls based on the design capacity or maximum production capacity of the source and 8760 hours per year. The design capacity or maximum production capacity includes use of coating(s) or ink(s) with the highest volatile organic material content actually used in practice by the source.

maximum thirty (30) day average (40CFR439.1-a) shall mean the maximum average of daily values for 30 consecutive days.

maximum tolerated dose (EPA-94/04): The maximum doses that an animal species can tolerate for a major portion of its lifetime without significant impairment or toxic effect other than carcinogenicity.

maximum total trihalomethane potential (MTP) (40CFR141.2) means the maximum concentration of total trihalomethanes produced in a given water containing a disinfectant residual after 7 days at a temperature of 25°C or above.

maximum true vapor pressure (40CFR60.111b-f) means the equilibrium partial pressure exerted by the stored VOL at the temperature equal to the highest calendar-month average of the VOL storage temperature for VOL's stored above or below the ambient temperature or at the local maximum monthly average temperaure as reported by the National Weather Service for VOL's stored at the ambient temperature, as determined:
(1) In accordance with methods described in American Petroleum Institute Bulletin 2517, Evaporation Loss From External Floating Roof Tanks (incorporated by reference, 40CFR0.17); or
(2) As obtained from standard reference texts; or
(3) As determined by ASTM Method D2879-83 (incorporated by reference, 40CFR0.17);

(4) Any other method approved by the Administrator.

maximum true vapor pressure (40CFR63.111) means the equilibrium partial pressure exerted by the total organic HAP's in the stored or transferred liquid at the temperature equal to the highest calendar-month average of the liquid storage or transfer temperature for liquids stored or transferred above or below the ambient temperature or at the local maximum monthly average temperature as reported by the National Weather Service for liquids stored or transferred at the ambient temperature, as determined:
(1) In accordance with methods described in American Petroleum Institute Publication 2517, Evaporative Loss From External Floating-Roof Tanks (incorporated by reference as specified in 40CFR63.14 of subpart A of this part); or
(2) As obtained from standard reference texts; or
(3) As determined by the American Society for Testing and Materials Method D2879-83 (incorporated by reference as specified in 40CFR63.14 of subpart A of this part); or
(4) Any other method approved by the Administrator.

mayonnaise and salad dressings (40CFR407.81-i) shall be defined as the emulsified and non-emulsified semisolid food prepared from the combining of edible vegetable oil with acidifying, and egg yolk containing ingredients, or gum and starch combinations to which certain colorings, spices, and flavorings have been added.

mbbl (40CFR419.11-f) means one thousand barrels (one barrel is equivalent to 42 gallons).

MBTA (40CFR52.1162-9): Massachusetts Bay Transportation Authority.

mean retention time (40CFR131.35.d-7) means the time obtained by dividing a reservoir's mean annual minimum total storage by the non-zero 30-day, ten-year low-flow from the reservoir.

mean spectral response (40CFR60-AB-2.7) means the wavelength that is the arithmetic mean value of the wavelength distribution for the effective spectral response curve of the transmissometer.

means of emission limitation (CAA302-42USC7602) means a system of continuous emission reduction (including the use of specific technology or fuels with specified pollution characteristics).

measurement (40CFR403.7(b)(1)) refers to the ability of the analytical method or protocol to quantify as well as identify the presence of the substance in question.

measurement method (40CFR195.2) is a means of measuring radon gas or radon decay products encompassing similar measurement devices, sampling techniques, or analysis procedures.

measurement period (40CFR201.1-s) means a continuous period of time during which noise of railroad yard operations is assessed, the beginning and finishing times of which may be selected after completion of the measurements.

measurement system (40CFR60-AA (method 25A-2.1)) means the total equipment required for the determination of the gas concentration. The system consists of the following major subsystems:
(1) Sample interface: That portion of the system that is used for one or more of the following: sample acquisition, sample transportation, sample conditioning, or protection of the analyzer from the effects of the stack effluent.
(2) Organic analyzer: That portion of the system that senses organic concentration and generates an output proportional to the gas concentration.

measurement system (40CFR60-AA (method 6C-3.1)) means the total equipment required for the determination of gas concentration. The measurement system consists of the following major sub-systems:
(1) Sample interface: That portion of a system used for one or more of the following: sample acquisition, sample transport, sample conditioning, or protection of the analyzers from the effects of the stack effluent.
(2) Gas analyzer: That portion of the system that senses the gas to be measured and generates an output proportional to its concentration.
(3) Data recorder: A strip chart recorder, analog computer, or digital recorder for recording measurement data from the analyzer output.

measurement system (40CFR60-AA (method 7E-3.1)) means the total equipment required for the determination of NO_x concentration. The measurement system consists of the following major subsystems:

(1) <u>Sample interface, gas analyzer, and data recorder:</u> Same as Method 6C, sections 3.1.1, 3.1.2, and 3.1.3.

(2) <u>NO_2 to NO converter:</u> A device that converts the nitrogen dioxide (NO_2) in the sample gas to nitrogen oxide (NO).

meat cutter (40CFR432.61-b) shall mean an operation which fabricates, cuts, or otherwise produces fresh meat cuts and related finished products from livestock carcasses, at rates greater than 2730 kg (6000 lb) per day.

mechanical aeration (EPA-94/04): Use of mechanical energy to inject air into water to cause a waste stream to absorb oxygen.

mechanical and thermal integrity (40CFR85.2122(a)-(15)(ii)(G)) means the ability of a converter to continue to operate at its previously determined efficiency and light-off time and be free from exhaust leaks when subject to thermal and mechanical stresses representative of the intended application (under CAA).

mechanical separation (EPA-94/04): Using mechanical means to separate waste into various components.

mechanical torque rate (40CFR85.2122(a)(2)(iii)(F)) means a term applied to a thermostatic coil, defined as the torque accumulation per angular degree of deflection of a thermostatic coil (under the Clean Air Act).

mechanical turbulence (EPA-94/04): Random irregularities of fluid motion in air caused by buildings or other non-thermal, processes.

mechanism (40CFR56.1) means an administrative procedure, guideline, manual, or written statement.

media (EPA-94/04): Specific environments--air, water, soil--which are the subject of regulatory concern and activities.

median diameter (40CFR798.1150-3; 798.2450-3; 798.4350-3): See geometric mean diameter.

medical device (CAA601) means any device (as defined in the Federal Food, Drug, and Cosmetic Act (21USC321)), diagnostic product, drug (as defined in the Federal Food, Drug, and Cosmetic Act), and drug delivery system:

(A) if such device, product, drug, or drug delivery system utilizes a class I or class II substance for which no safe and effective alternative has been developed, and where necessary, approved by the Commissioner; and

(B) if such device, product, drug, or drug delivery system, has, after notice and opportunity for public comment, been approved and determined to be essential by the Commissioner in consultation with the Administrator.

medical emergency (40CFR350.40) means any unforeseen condition which a health professional would judge to require urgent and unscheduled medical attention. Such a condition is one which results in sudden and/or serious symptom(s) constituting a threat to a person's physical or psychological well-being and which requires immediate medical attention to prevent possible deterioration, disability, or death.

medical surveillance (EPA-94/04): A periodic comprehensive review of a worker's health status; acceptable elements of such surveillance program are listed in the Occupational Safety and Health Administration standards for asbestos.

medical use (10CFR30.4) means the intentional internal or external administration of byproduct material, or the radiation therefrom, to human beings in the practice of medicine in accordance with a license issued by a State or Territory of the United States, the District of Columbia, or the Commonwealth of Puerto Rico.

medical waste (40CFR259.10.b) means any solid waste which is generated in the diagnosis, treatment (e.g., provision of medical services), or immunization of human beings or animals, in research pertaining thereto, or in the production or testing of biologicals. The term does not include any hazardous waste identified or listed under part 261 of this chapter or any household waste as defined in 40CFR261.4-(b)(I) of this chapter. NOTE to this definition: Mixtures of hazardous waste and medical waste are subject to this part except as provided in 40CFR259.31.

medical waste (40CFR60.51a) means any solid waste which is generated in the diagnosis, treatment, or immunization of human beings or animals, in research pertaining thereto, or in production or testing of biologicals. Medical waste does not include any hazardous waste identified under subtitle C of the Resource Conservation and Recovery Act or any household waste as defined in regulations under subtitle C of the Resource Conservation and Recovery Act.

medical waste (CWA502-33USC1362) means isolation wastes; infectious agents; human blood and blood products; pathological wastes; sharps; body parts; contaminated bedding; surgical wastes and potentially contaminated laboratory wastes; dialysis wastes; and such additional medical items as the Administrator shall prescribe by regulation.

medical waste (EPA-94/04): Any solid waste generated in the diagnosis, treatment, or immunization of human beings or animals, in research pertaining thereto, or in the production or testing of biologicals, excluding hazardous waste identified or listed under 40CFR part 261 or any household waste as defined in 40CFRSub-section 261.4 (b)(1).

medical waste (RCRA1004-42USC6903) Except as otherwise provided in this paragraph, the term "medical waste" means any solid waste which is generated in the diagnosis, treatment, or immunization of human beings or animals, in research pertaining thereto, or in the production or testing of biologicals. Such term does not include any hazardous waste identified or listed under subtitle C or any household waste as defined in regulations under subtitle C.

Medical Waste Tracking Act (MWTA) of 1988: See act or MWTA.

medium (40CFR407.61-v) shall mean a point source that processes a total annual raw material production of fruits, vegetables, specialties and other products that is between 1,816 kkg (2,000 tons) per year and 9,080 kkg (10,000 tons) per year (other identical or similar definitions are provided in 40CFR407.71-t; 407.81-l).

medium duty vehicle (40CFR52.2038-3) means a gasoline powered motor vehicle rated at more than 6,000 lb GVW and less than 10,000 lb GVW (other identical or similar definitions are provided in 40CFR52.2485-3; 52.2490-3).

medium municipal separate storm sewer system (40CFR122.26-7) means all municipal separate storm sewers that are either:

(i) Located in an incorporated place with a population of 100,000 or more but less than 250,000, as determined by the latest Decennial Census by the Bureau of Census (appendix G); or

(ii) Located in the counties listed in appendix I, except municipal separate storm sewers that are located in the incorporated places, townships or towns within such counties; or

(iii) Owned or operated by a municipality other than those described in paragraph (b)(4)(i) or (ii) of this section and that are designated by the Director as part of the large or medium municipal separate storm sewer system due to the interrelationship between the discharges of the designated storm sewer and the discharges from municipal separate storm sewers described under paragraph (b)(4)(i) or (ii) of this section. In making this determination the Director may consider the following factors:

(A) Physical interconnections between the municipal separate storm sewers;

(B) The location of discharges from the designated municipal separate storm sewer relative to discharges from municipal separate storm sewers described in paragraph (b)(7)(i) of this section;

(C) The quantity and nature of pollutants discharged to waters of the United States;

(D) The nature of the receiving waters; or

(E) Other relevant factors; or

(iv) The Director may, upon petition, designate as a medium municipal separate storm sewer system, municipal separate storm sewers located within the boundaries of a region defined by a storm water management regional authority based on a jurisdictional, watershed, or other appropriate basis that includes one or more of the systems described in paragraphs (b)(7)(i), (ii), (iii) of this section.

medium size water system (40CFR141.2), for the purpose of subpart I of this part only, means a water system that serves greater than 3,300 and less than or equal to 50,000 persons.

medium size water system (EPA-94/04): A water system that serves from 3,300 to 50,000 persons.

medium sized plants (40CFR428.61) shall mean plants which process between 3,720 kg/day (8,200 lbs/day) and 10,430 kg/day (23,000 lbs/day) of raw materials.

meeting (40CFR1517.2) means the deliberations of at least two Council members where such deliberations determine or result in the joint conduct or disposition of official collegial Council business, but does not include deliberations to take actions to open or close a meeting under 40CFR1517.4 and 1517.5 or to release or withhold information under 40CFR1517.4 and 1517.7. Meeting shall not be construed to prevent Council members from considering individually Council business that is circulated to them sequentially in writing (under NEPA).

melt (40CFR409.21) shall mean that amount of raw material (raw sugar) contained within aqueous solution at the beginning of the process for production of refined cane sugar (other identical or similar definitions are provided in 40CFR409.31-c).

meltdown and refining (40CFR60.271-i) means that phase of the steel production cycle when charge material is melted and undesirable elements are removed from the metal.

meltdown and refining period (40CFR60.271-j) means the time period commencing at the termination of the initial charging period and ending at the initiation of the tapping period, excluding any intermediate charging periods.

melting (40CFR60.271a) means that phase of steel production cycle during which the iron and steel scrap is heated to the molten state.

member of the public (40CFR190.02-k) means any individual that can receive a radiation dose in the general environment, whether he may or may not also be exposed to radiation in an occupation associated with a nuclear fuel cycle. However, an individual is not considered a member of the public during any period in which he is engaged in carrying out any operation which is part of a nuclear fuel cycle.

member of the public (40CFR191.02-p) means any individual except during the time when that individual is a worker engaged in any activity, operation, or process that is covered by the Atomic Energy Act of 1954, as amended.

meniscus (EPA-94/04): The curved top of a column of liquid in a small tube.

merchant (40CFR420.11-c) means those byproduct cokemaking operations which provide more than fifty percent of the coke produced to operations, industries, or processes other than iron making blast furnaces associated with steel production.

mercury (40CFR415.61-c) shall mean the total mercury present in the process wastewater stream exiting the mercury treatment system.

mercury (40CFR61.51-a) means the element mercury, excluding any associated elements, and includes mercury in particulates, vapors, aerosols, and compounds.

mercury (EPA-94/04): A heavy metal that can accumulate in the environment and is highly toxic if breathed or swallowed (cf. heavy metals).

mercury chlor-alkali cell (40CFR61.51-e) means a device which is basically composed of an electrolyzer section and a denuder (decomposer) section and utilizes mercury to produce chlorine gas, hydrogen gas, and alkali metal hydroxide.

mercury chlor-alkali electrolyzer (40CFR61.51-f) means an electrolytic device which is part of a mercury chlor-alkali cell and utilizes a flowing mercury cathode to produce chlorine gas and alkali metal amalgam.

mercury ore (40CFR61.51-b) means a mineral mined specifically for its mercury content.

mercury ore processing facility (40CFR61.51-c) means a facility processing mercury ore to obtain mercury.

mesotrophic (EPA-94/04): Reservoirs and lakes which contain moderate quantities of nutrients and are moderately productive in terms of aquatic animal and plant life.

metabolism (40CFR795.228-2) means the study of the sum of the processes by which a particular substance is handled in the body and includes absorption, tissue distribution, biotransformation, and excretion (other identical or similar definitions are provided in 40CFR795.231-2).

metabolism (40CFR795.232-2) means the sum of the enzymatic and nonenzymatic processes by which a particular substance is handled in the body.

metabolite (40CFR723.50-8) means a chemical entity produced by one or more enzymatic or nonenzymatic reactions as a result of exposure of an organism to a chemical substance.

metabolites (EPA-94/04): Any substances produced by biological processes, such as those from pesticides.

metal cladding (40CFR471.02-g): See cladding.

metal cleaning waste (40CFR423.11-d) means any wastewater resulting from cleaning [with or without chemical cleaning compounds] any metal process equipment including, but not limited to, boiler tube cleaning, boiler fireside cleaning, and air preheater cleaning.

metal coil surface coating operation (40CFR60.461) means the application system used to apply an organic coating to the surface of any continuous metal strip with thickness of 0.15 millimeter (mm) (0.006 in.) or more that is packaged in a roll or coil.

metal furniture (40CFR52.741) means a furniture piece including, but not limited to, tables, chairs, waste baskets, beds, desks, lockers, benches, shelving, file cabinets, lamps, and room dividers.

metal furniture coating (40CFR52.741) means any non-adhesive coating applied to any furniture piece made of metal or any metal part which is or will be assembled with other metal, wood, fabric, plastic or glass parts to form a furniture piece including, but not limited to, tables, chairs, waste baskets, beds, desks, lockers, benches, shelving, file cabinets, lamps, and room dividers. This definition shall not apply to any coating line coating miscellaneous metal parts or products.

metal furniture coating facility (40CFR52.741) means a facility that includes one or more metal furniture coating line(s).

metal furniture coating line (40CFR52.741) means a coating line in which any protective, decorative, or functional coating is applied onto the surface of metal furniture.

metal powder production (40CFR471.02-x) operations are mechanical process operations which convert metal to a finely divided form.

metal preparation (40CFR466.02-f) means any and all of the metal processing steps preparatory to applying the enamel slip. Usually this includes cleaning, picking and applying a nickel flash or chemical coating.

metalimnion (EPA-94/04): The middle layer of a thermally stratified lake or reservoir. In this layer there is a rapid decrease in temperature with depth. Also called thermocline.

metallic mineral concentrate (40CFR60.381) means a material containing metallic compounds in concentrations higher than naturally occurring in ore but requiring additional processing if pure metal is to be isolated. A metallic mineral concentrate contains at least one of the following metals in any of its oxidation states and at a concentration that contributes to the concentrate's commercial value: Aluminum, copper, gold, iron, lead, molybdenum, silver, titanium, tungsten, uranium, zinc, and zirconium. This definition shall not be construed as requiring that material containing metallic compounds be refined to a pure metal in order for the material to be considered a metallic mineral concentrate to be covered by the standards.

metallic mineral processing plant (40CFR60.381) means any combination of equipment that produces metallic mineral concentrates from ore. Metallic mineral processing commences with the mining of ore and includes all operations either up to and including the loading of wet or dry concentrates or solutions of metallic minerals for transfer to facilities at non-adjacent locations that will subsequently process metallic concentrates into purified metals (or other products), or up to and including all material transfer and storage operations that precede the operations that produce refined metals (or other products) from metallic mineral concentrates at facilities adjacent to the metallic mineral processing plant. This definition shall not be construed as requiring that mining of ore be conducted in order for the combination of equipment to be considered a metallic mineral processing plant (cf. metallic mineral concentrate).

metallic shoe seal (40CFR60.111a-i) includes but is not limited to a metal sheet held vertically against the tank wall by springs or weighted levers and is connected by braces to the floating roof. A flexible coated fabric (envelope) spans the annular space between the metal sheet and the floating roof.

metallic shoe seal or mechanical shoe seal (40CFR63.111) means a metal sheet that is held vertically against the wall of the storage vessel by springs, weighted levers, or other mechanisms and is connected to the floating roof by braces or other means. A flexible coated fabric (envelope) spans the annular space between the metal sheet and the floating roof.

metallic shoe type seal (40CFR52.741) means a primary or secondary seal constructed of metal sheets (shoes) which are joined together to form a ring, springs, or levels which attach the shoes to the floating roof and hold the shoes against the tank wall, and a coated fabric which is suspended from the shoes to the floating roof.

metallo-organic active ingredients (40CFR455.31-a) means carbon containing active ingredients containing one or more metallic atoms in the structure.

metallo-organic pesticides (40CFR165.1-p): See pesticide.

metalworking fluid (40CFR721.3) means a liquid of any viscosity or color containing intentionally added water used in metal machining operations for the purpose of cooling, lubricating, or rust inhibition (other identical or similar definitions are provided in 40CFR747.115-2; 747.195-2; 747.200-2).

meteorological measurements (40CFR58.1-z) means measurements of wind speed, wind direction, barometric pressure, temperature, relative humidity, and solar radiation.

methane (EPA-94/04): A colorless, nonpoisonous, flammable gas created by anaerobic decomposition of organic compounds.

methanol (40CFR80.2-uu) means any fuel sold for use in motor vehicles and commonly known or commercially sold as methanol or MXX, where XX is the percent methanol (CH_3OH) by volume (under CAA).

methanol (EPA-94/04): An alcohol that can be used as an alternative fuel or as a gasoline additive. It is less volatile than gasoline; when blended with gasoline it lowers the carbon monoxide emissions but increases hydrocarbon emissions. Used as pure fuel, its emissions are less ozone-forming that those from gasoline.

methanol fueled (40CFR86.090.2) means any motor vehicle or motor vehicle engine that is engineered and designed to be operated using methanol fuel (i.e., a fuel that contains at least 50 percent methanol (CH_3OH) by volume) as fuel. Flexible fuel vehicles are methanol-fueled vehicles.

method (40CFR60; 61): See test method.

method 18 (EPA-94/04): An EPA test method which uses gas chromatographic techniques to measure the concentration of volatile organic compounds in a gas stream.

method 24 (EPA-94/04): An EPA reference method to determine density, water content and total volatile content (water and VOC) of coatings.

method 25 (EPA-94/04): An EPA reference method to determine the VOC concentration in a gas stream.

method detection limit (MDL) (40CFR135-AB) is defined as the minimum concentration of a substance that can be measured and reported with 99% confidence that the analyte concentration is greater than zero and is determined from analysis of a sample in a given matrix containing the analyte (under FWPCA).

method detection limit (MDL) (40CFR300-AA) means the lowest concentration of analyte that a method can detect reliably in either a sample or blank.

method of standard addition (40CFR136-AC-14) means the standard addition technique involves the use of the unknown and the unknown plus a known amount of standard.

methods of operation (40CFR21.2-m) means the installation, emplacement, or introduction of materials, including those involved in construction, to achieve a process or procedure to control: Surface water pollution from non-point sources--that is, agricultural, forest practices, mining, construction; ground or surface water pollution from well, subsurface, or surface disposal operations; activities resulting in salt water intrusion; or changes in the movement, flow, or circulation of navigable or ground waters.

methoxychlor (EPA-94/04): Pesticide that causes adverse health effects in domestic water supplies and is toxic to freshwater and marine aquatic life.

methyl methacrylate-acrylonitrile-butadiene-styrene (MABS) resins (40CFR63.191) means styrenic polymers containing methyl methacrylate, acrylonitrile, 1,3-butadiene, and styrene. The MABS copolymers are prepared by dissolving or dispersing polybutadiene rubber in a mixture of methyl methacrylate-acrylonitrile-styrene and butadiene monomer. The graft polymerization is carried out by a bulk or a suspension process.

methyl methacrylate-butadiene-styrene (MBS) resins (40CFR63.191) means styrenic polymers containing methyl methacrylate, 1,3-butadiene, and styrene. Production of MBS terpolymers is achieved using an emulsion process in which methyl methacrylate and styrene are grafted onto a styrene-butadiene rubber (under CAA).

methyl orange alkalinity (EPA-94/04): A measure of the total alkalinity in a water sample in which to color of methyl orange reflects the change in level.

metropolitan planning organization (MPO) (40CFR51.392) is that organization designated as being responsible, together with the State, for conducting the continuing, cooperative, and comprehensive planning process under 23 U.S.C. 134 and 49 U.S.C. 1607. It is the forum for cooperative transportation decision-making (other identical or similar definitions are provided in 40CFR93.101).

metropolitan planning organization (MPO) (40CFR51.852) is that organization designated as being responsible, together with the State, for conducting the continuing, cooperative, and comprehensive planning process under 23 U.S.C. 134 and 49 U.S.C. 1607 (other identical or similar definitions are provided in 40CFR93.152).

metropolitan planning organization (MPO) (40CFR52.138-5) means the organization designated under 23 U.S.C. 134 and 23CFR part 450.106. For the specific purposes of this regulation, MPO means either the Maricopa Association of Governments or the Pima Association of Governments.

metropolitan statistical area or MSA (40CFR60.331-m) as defined by the Department of Commerce.

mg/L (40CFR133.101-h) means milligrams per liter.

mgal (40CFR419.11-f) means one thousand gallons.

michelin-A operation (40CFR60.541) means the operation identified as Michelin-A in the Emission Standards and Engineering Division confidential file as referenced in Docket A-80-9, Entry II-B-12.

michelin-B operation (40CFR60.541) means the operation identified as Michelin-B in the Emission

Standards and Engineering Division confidential file as referenced in Docket A-80-9, Entry II-B-12.

michelin-C automatic operation (40CFR60.541) means the operation identified as Michelin-C-automatic in the Emission Standards and Engineering Division confidential file as referenced in Docket A-80-9, Entry II-B-12.

microbial growth (EPA-94/04): The activity and growth of microorganisms such as bacteria, algae, diatoms, plankton, and fungi.

microbial pesticide (EPA-94/04): A microorganism that is used to control a pest, but of minimum toxicity to man.

microclimate (EPA-94/04): The localized climate conditions with in an urban area or neighborhood.

microcurie (10CFR30.4) means that amount of radioactive material which disintegrates at the rate of 37 thousand atoms per second.

micronuclei (40CFR798.5395) are small particles consisting of acentric fragments of chromosomes or entire chromosomes, which lag behind at anaphase of cell division. After telophase, these fragments may not be included in the nuclei of daughter cells and form single or multiple micronuclei in the cytoplasm.

milestone (40CFR192.31-q) means an enforceable date by which action, or the occurrence of an event, is required for purposes of achieving compliance with the 20 pCi/m^2-s flux standard.

milestone (40CFR51.392) has the meaning given in section 182(g)(1) and section 189(c) of the CAA. A milestone consists of an emissions level and the date on which it is required to be achieved (other identical or similar definitions are provided in 40CFR93.101).

milestone (40CFR51.852) has the meaning given in sections 182(g)(1) and 189(c)(1) of the Act (other identical or similar definitions are provided in 40CFR93.152).

milestones (40CFR51.491) means the reductions in emissions required to be achieved pursuant to section 182(b)(1) and the corresponding

requirements in section 182(c)(2) (B) and (C), 182(d), and 182(e) of the Act for O_3 nonattainment areas, as well as the reduction in emissions of CO equivalent to the total of the specified annual emissions reductions required by December 31, 1995, pursuant to section 187(d)(1).

military engine (40CFR86.082.2) means any engine manufactured solely for the Department of Defense to meet military specifications.

milking center (40CFR412.11-k) shall mean a separate milking area with storage and cooling facilities adjacent to a free stall barn or cowyard dairy operation.

milkroom (40CFR412.11-j) shall mean milk storage and cooling rooms normally used for stall barn dairies.

mill (40CFR440.132-f) is a preparation facility within which the metal ore is cleaned, concentrated, or otherwise processed before it is shipped to the customer, refiner, smelter, or manufacturer. A mill includes all ancillary operations and structures necessary to clean, concentrate, or otherwise process metal ore, such as ore and gangue storage areas and loading facilities.

mill broke (40CFR250.4-v) means any paper waste generated in a paper mill prior to completion of the papermaking process. It is usually returned directly to the pulping process. Mill broke is excluded from the definition of "recovered materials."

milling (40CFR471.02-y) is the mechanical treatment of a nonferrous metal to produce powder, or to coat one component of a powder mixture with another.

million gallons per day (MGD) (EPA-94/04): A measure of water flow.

mimeo paper (40CFR250.4-w) means a grade of writing paper used for making copies on stencil duplicating machines.

mine (40CFR436.181-d) shall mean an area of land, surface or underground, actively used for or resulting from the extraction of a mineral from natural deposits.

mine (40CFR436.21-d) shall mean an area of land, surface or underground, actively mined from the production of crushed and broken tone from natural deposits.

mine (40CFR436.31-d) shall mean an area of land, surface or underground, actively mined for the production of sand and gravel from natural deposits (other identical or similar definitions are provided in 40CFR436.41-d).

mine (40CFR440.132-g) is an active mining area, including all land and property placed under, or above the surface of such land, used in or resulting from the work of extracting metal ore or minerals from their natural deposits by any means or method, including secondary recovery of metal ore from refuse or other storage piles, wastes, or rock dumps and mill tailings derived from the mining, cleaning, or concentration of metal ores.

mine (40CFR440.141-d) means a place where work or other activity related to the extraction or recovery of ore is performed.

mine area (40CFR440.141-9) means the land area from which overburden is stripped and ore is removed prior to moving the ore to the beneficiation area.

mine dewatering (40CFR436.181-b) shall mean any water that is impounded or that collects in the mine and is pumped, drained or otherwise removed from the mine through the efforts of the mine operator (under CWA).

mine dewatering (40CFR436.21-b) shall mean any water that is impounded or that collects in the mine and is pumped, drained or otherwise removed from the mine through the efforts of the mine operator. However, if a mine is also used for treatment of process generated waste water, discharges of commingled water from the facilities shall be deemed discharges of process generated waste water.

mine dewatering (40CFR436.31-b) shall mean any water that is impounded or that collects in the mine and is pumped, drained, or otherwise removed from the mine through the efforts of the mine operator. This term shall also include wet pit overflows caused solely by direct rainfall and ground water seepage. However, if a mine is also used for the treatment of process generated waste water, discharges of commingled water from the mine shall be deemed discharges of process generated waste water (other identical or similar definitions are provided in 40CFR436.41-b).

mine drainage (40CFR434.11-h) means any drainage, and any water pumped or siphoned, from an active mining area or a post-mining area.

mine drainage (40CFR436.381-b) means any water drained, pumped or siphoned from a mine (other identical or similar definitions are provided in 40CFR440.132-h; 440.141-10).

mine drainage alkaline (40CFR434.11-c) means mine drainage which, before any treatment, has a pH equal to or greater than 6.0 and total iron concentration of less than 10 mg/l.

miner of asbestos (40CFR763.63-h) is a person who produces asbestos by mining or extracting asbestos-containing ore so that it may be further milled to produce bulk asbestos for distribution in commerce, and includes persons who conduct milling operations to produce bulk asbestos by processing asbestos-containing ore. Milling involves the separation of the fibers from the ore, grading and sorting the fibers, or fiberizing crude asbestos ore. To mine or mill is to "manufacture" for commercial purposes under TSCA.

mineral fiber insulation (40CFR248.4-v) means insulation (rock wool or fiberglass) which is composed principally of fibers manufactured from rock, slag or glass, with or without binders.

mineral handling and storage facility (40CFR60.471) means the areas in asphalt roofing plants in which minerals are unloaded from a carrier, the conveyor transfer points between the carrier and the storage silos, and the storage silos.

mineral oil PCB transformer (40CFR761.3) means any transformer originally designed to contain mineral oil as the dielectric fluid and which has been tested and found to contain 500 ppm or greater PCBs.

mineralisation (40CFR796.3400-B), (in this context) means extensive degradation of a molecule during which a labelled carbon atom is oxidized quantitatively with release of the appropriate amount of $^{14}CO_2$.

minimal risk (40CFR26.102-i) means that the probability and magnitude of harm or discomfort anticipated in the research are not greater in and of themselves than those ordinarily encountered in daily life or during the performance of routine physical or psychological examinations or tests.

minimization (EPA-94/04): A comprehensive program to minimize or eliminate wastes, usually applied to wastes at their point of origin (cf. waste minimization).

minimize (40CFR6-AA-f) means to reduce to the smallest possible amount or degree.

minimum as applied to BPT and BCT effluent limitations and NSPS for sanitary wastes (40CFR435.11-m) shall mean the minimum concentration value allowed as measured in any single sample of the discharged waste stream (under CWA).

mining of an aquifer (EPA-94/04): Withdrawal of ground water over a period of time that exceeds the rate of recharge of the aquifer.

mining overburden returned to the mine site (40CFR260.10) means any material overlying an economic mineral deposit which is removed to gain access to that deposit and is then used for reclamation of a surface mine.

mining wastes (40CFR243.101-r) means residues which result from the extraction of raw materials from the earth (other identical or similar definitions are provided in 40CFR246.101-s).

minor source baseline date (40CFR51.166-14): See major source baseline date.

minor source baseline date (40CFR52.21-14): See major source baseline date.

minority business enterprise (40CFR33.005) is a business which is:
(1) Certified as socially and economically disadvantaged by the Small Business Administration,
(2) certified as a minority business enterprise by a State or Federal agency, or
(3) an independent business concern which is at least 51 percent owned and controlled by minority group member(s). A minority group member is an individual who is a citizen of the United States and one of the following:
   (i) Black American;
   (ii) Hispanic American (with origins from Puerto Rico, Mexico, Cuba, South or Central America);
   (iii) Native American (American Indian, Eskimo, Aleut, native Hawaiian), or
   (iv) Asian-Pacific American (with origins from Japan, China, the Philippines, Vietnam, Korea, Samoa, Guam, the U.S. Trust Territories of the Pacific, Northern Marianas, Laos, Cambodia, Taiwan or the Indian subcontinent).

minority business enterprise (MBE) (40CFR35.6015-26) means a business which is:
(i) Certified as socially and economically disadvantaged by the Small Business Administration,
(ii) certified as a minority business enterprise by a State or Federal agency, or
(iii) an independent business concern which is at least 51 percent owned and controlled by minority group member(s). A minority group member is an individual who is a citizen of the United States and one of the following:
   (A) Black American;
   (B) Hispanic American (with origins from Puerto Rico, Mexico, Cuba, South or Central America);
   (C) Native American (American Indian, Eskimo, Aleut, native Hawaiian), or
   (D) Asian-Pacific American (with origins from Japan, China, the Philippines, Vietnam, Korea, Samoa, Guam, the U.S. Trust Territories of the Pacific, Northern Marianas, Laos, Cambodia, Taiwan or the Indian subcontinent).

minority group (40CFR8.2-z) as used herein shall

include, where appropriate, female employees and prospective female employees.

minors (EPA-94/04): Publicly owned treatment works with flows less than 1 million gallons per day (cf. majors).

misbranded (FIFRA2-7USC136):
(1) A pesticide is misbranded if:
  (A) its labeling bears any statement, design, or graphic representation relative thereto or to its ingredients which is false or misleading in any particular;
  (B) it is contained in a package or other container or wrapping which does not conform to the standards established by the Administrator pursuant to section 25(c)(3);
  (C) it is an imitation of, or is offered for sale under the name of, another pesticide;
  (D) its label does not bear the registration number assigned under section 7 to each establishment in which it was produced;
  (E) any work, statement, or other information required by or under authority of this Act to appear on the label or labeling is not prominently placed thereon with such conspicuousness (as compared with other words, statements, designs, or graphic matter in the labeling) and in such terms as to render it likely to be read and understood by the ordinary individual under customary conditions of purchase and use;
  (F) the labeling accompanying it does not contain directions for use which are necessary for effecting the purpose for which the product is intended and if complied with, together with any requirements imposed under section 3(d) of this Act, are adequate to protect health and the environment;
  (G) the label does not contain a warning or caution statement which may be necessary and if complied with, together with any requirements imposed under section 3(d) of this Act, is adequate to protect health and the environment; or
  (H) in the case of a pesticide not registered in accordance with section 3 of this Act and

intended for export, the label does not contain, in words prominently placed thereon with such conspicuousness (as compared with other words, statements, designs, or graphic matter in the labeling) as to render it likely to be noted by the ordinary individual under customary conditions of purchase and use, the following: "Not Registered for Use in the United States of America."
(2) A pesticide is misbranded if:
  (A) the label does not bear an ingredient statement on that part of the immediate container (and on the outside container or wrapper of the retail package, if there be one, through which the ingredient statement on the immediate container cannot be clearly read) which is presented or displayed under customary conditions of purchase, except that a pesticide is not misbranded under this subparagraph if:
    (i) the size or form of the immediate container, or the outside container or wrapper of the retail package, makes it impracticable to place the ingredient statement on the part which is presented or displayed under customary conditions of purchase: and
    (ii) the ingredient statement appears prominently on another part of the immediate container, or outside container or wrapper, permitted by the Administrator;
  (B) the labeling does not contain a statement of the use classification under which the product is registered;
  (C) there is not affixed to its container, and to the outside container or wrapper of the retail package, if there be one, through which the required information on the immediate container cannot be clearly read, a label bearing:
    (i) the name and address of the producer, registrant, or person for whom produced;
    (ii) the name, brand, or trademark under which the pesticide is sold;
    (iii) the net weight or measure of the content, except that the Administrator may permit reasonable variations; and

(iv) when required by regulation of the Administrator to effectuate the purposes of the Act, the registration number assigned to the pesticide under this Act, and the use classification; and

(D) the pesticide contains any substances in quantities highly toxic to man, unless the label shall bear, in addition to any other matter required by this Act:

(i) the skull and crossbones;

(ii) the word "poison" prominently in red on a background of distinctly contrasting color; and

(iii) a statement of a practical treatment (first aid or otherwise) in case of poisoning by the pesticide.

miscellaneous ACM (40CFR763.83) means miscellaneous material that is ACM in a school building.

miscellaneous ACM (EPA-94/04): Interior asbestos-containing building material or structural components, members or fixtures, such as floor and ceiling tiles; does not include surfacing materials or thermal system insulation.

miscellaneous fabric product manufacturing process (40CFR52.741) means:

(A) A manufacturing process involving one or more of the following applications, including any drying and curing of formulations, and capable of emitting VOM:

(1) Adhesives to fabricate or assemble components or products.

(2) Asphalt solutions to paper or fiberboard.

(3) Asphalt to paper or felt.

(4) Coatings or dye to leather.

(5) Coatings to plastic.

(6) Coatings to rubber or glass.

(7) Disinfectant material to manufactured items.

(8) Plastic foam scrap or "fluff" from the manufacture of foam containers and packaging material to form resin pallets.

(9) Resin solutions to fiber substances.

(10) Viscose solutions for food casings.

(B) The storage and handling of formulations associated with the process described above, and the use and handling of organic liquids and other substances for clean-up operations associated with the process described in this definition.

miscellaneous formulation manufacturing process (40CFR52.741) means:

(A) A manufacturing process which compounds one or more of the following and is capable of emitting VOM:

(1) Adhesives.

(2) Asphalt solutions.

(3) Caulks, sealants, or waterproofing agents.

(4) Coatings, other than paint and ink.

(5) Concrete curing compounds.

(6) Dyes.

(7) Friction materials and compounds.

(8) Resin solutions.

(9) Rubber solutions.

(10) Viscose solutions.

(B) The storage and handling of formulations associated with the process described above, and the use and handling of organic liquids and other substances for clean-up operations associated with the process described in this definition.

miscellaneous material (40CFR763.83) means interior building material on structural components, structural members or fixtures, such as floor and ceiling tiles, and does not include surfacing material or thermal system insulation.

miscellaneous material (TSCA-AIA1) means building material on structural components, structural members or fixtures, such as floor and ceiling tiles. The term does not include surfacing material or thermal system insulation.

miscellaneous materials (EPA-94/04): Interior building materials on structural components, such as floor or ceiling tiles.

miscellaneous metal parts and products coating (40CFR52.741) means any coating applied to any metal part or metal product, even if attached to or combined with a nonmetal part or product, except cans, coils, metal furniture, large appliances, and magnet wire. Prime coat, prime surfacer coat, topcoat, and final repair coat for automobiles and light-duty trucks are not miscellaneous metal parts and products coatings. However, underbody anti-chip (e.g., underbody plastisol) automobile, and light-duty truck coatings are miscellaneous metal parts and products coatings. Also, automobile or light-duty truck refinishing coatings, coatings applied to the exterior of marine vessels, coatings applied to the exterior of airplanes, and the customized topcoating of automobiles and trucks if production is less than 35 vehicles per day are not miscellaneous metal parts and products coatings.

miscellaneous metal parts or products (40CFR52.741) means any metal part or metal product, even if attached to or combined with a nonmetal part or product, except cans, coils, metal furniture, large appliances, magnet wire, automobiles, ships, and airplane bodies.

miscellaneous metal parts or products coating facility (40CFR52.741) means a facility that includes one or more miscellaneous metal parts or products coating lines.

miscellaneous metal parts or products coating line (40CFR52.741) means a coating line in which any protective, decorative, or functional coating is applied onto the surface of miscellaneous metal parts of products.

miscellaneous oil spill control agent (40CFR300.5) is any product, other than a dispersant, sinking agent, surface collecting agent, biological additive, or burning agent, that can be used to enhance oil sill cleanup, removal, treatment, or mitigation.

miscellaneous organic chemical manufacturing process (40CFR52.741) means:
(A) A manufacturing process which produces by chemical reaction, one or more of the following organic compounds or mixtures of organic compounds and which is capable of emitting VOM:
  (1) Chemicals listed in appendix A of this section
  (2) Chlorinated and sulfonated compounds
  (3) Cosmetic, detergent, soap, or surfactant intermediaries or specialties and products
  (4) Disinfectants
  (5) Food additives

(6) Oil and petroleum product additives
(7) Plasticizers
(8) Resins or polymers
(9) Rubber additives
(10) Sweeteners
(11) Varnishes
(B) The storage and handling of formulations associated with the process described above and the use and handling of organic liquids and other substances for clean-up operations associated with the process described in this definition.

miscellaneous unit (40CFR260.10) means a hazardous waste management unit where hazardous waste is treated, stored, or disposed of and that is not a container, tank, surface impoundment, pile, land treatment unit, landfill, incinerator, boiler, industrial furnace, underground injection well with appropriate technical standards under 40CFR146, or unit eligible for a research, development, and demonstration permit under 40CFR270.65 (under RCRA).

miscellaneous waste stream (40CFR468.02-w) shall mean the following additional waste streams related to forming copper: hydrotesting, sawing, surface milling, and maintenance.

miscellaneous wastewater streams (40CFR461.2-f) shall mean the combined wastewater streams from the process operations listed below for each subcategory. If a plant has one of these streams, then the plant receives the entire miscellaneous waste stream allowance:
(1) Cadmium subcategory. Cell wash, electrolyte preparation, floor and equipment wash, and employee wash.
(2) Lead subcategory. Floor wash, wet air pollution control, battery repair, laboratory, hand wash, and respirator wash.
(3) Lithium subcategory. Floor and equipment wash, cell testing, and lithium scrap disposal.
(4) Zinc subcategory. Cell wash, electrolyte preparation, employee wash, reject cell handling, floor and equipment wash.

mischmetal (40CFR421.271-c) refers to a rare earth metal alloy comprised of the natural mixture of rare earths to about 94-99 percent. The balance of the alloy includes traces of other elements and one to two percent iron.

miscible liquids (EPA-94/04): Two or more liquids that can be mixed and will remain mixed under normal conditions.

missed detection (EPA-94/04): The situation that occurs when a test indicates that a tank is "tight" when in fact it is leaking.

missing data period (40CFR72.2) means the total number of consecutive hours during which any component part of a certified CEMS or approved alternative monitoring system is not providing quality-assured data, regardless of the reason.

mist (EPA-94/04): Liquid particles measuring 40 to 500 microns, are formed by condensation of vapor. By comparison, fog particles are smaller than 40 microns.

mist coating (40CFR60.721): See fog coating.

mitigation (40CFR1508.20) includes:
(a) Avoiding the impact altogether by not taking certain action or parts of an action.
(b) Minimizing impacts by limiting the degree or magnitude of the action and its implementation.
(c) Rectifying the impact by repairing, rehabilitating, or restoring the affected environment.
(d) Reducing or eliminating the impact over time by preservation and maintenance operations during the life of the action.
(e) Compensating for the impact by replacing or providing substitute resources or environments.

mitigation (EPA-94/04): Measures taken to reduce adverse impacts on the environment.

mitotic gene conversion (40CFR798.5575-1) is detected by the change of inactive alleles of the same gene to wild-type alleles through intragenic recombination in mitotic cells.

mixed fertilizer (40CFR418.71-b) shall mean a mixture of wet and/or dry straight fertilizer materials, mixed fertilizer materials, fillers and additives prepared through chemical reaction to a given formulation (cf. blend fertilizer).

mixed funding (EPA-94/04): Settlements in which potentially responsible parties and EPA share the cost of a response action.

mixed liquor (EPA-94/04): A mixture of activated sludge and water containing organic matter undergoing activated sludge treatment in an aeration tank.

mixed radioactive and other hazardous substances (40CFR300-AA) means material containing both radioactive hazardous substances and nonradioactive hazardous substances, regardless of whether these types of substances are physically separated, combined chemically, or simply mixed together.

mixing zone (40CFR125.121-c) means the zone extending from the sea's surface to seabed and extending laterally to a distance of 100 meters in all directions from the discharge point(s) or to the boundary of the zone of initial dilution as calculated by a plume model approved by the director, whichever is greater, unless the director determines that the more restrictive mixing zone or another definition of the mixing zone is more appropriate for a specific discharge.

mixing zone (40CFR230.3-m) means a limited volume of water serving as a zone of initial dilution in the immediate vicinity of a discharge point where receiving water quality may not meet quality standards or other requirements otherwise applicable to the receiving water. The mixing zone should be considered as a place where wastes and water mix and not as a place where effluents are treated (under the Marine Protection, Research, and Sanctuaries Act).

mixing zone or dilution zone (40CFR131.35.d-8) means a limited area or volume of water where initial dilution of a discharge takes place; and where numeric water quality criteria can be exceeded but acutely toxic conditions are prevented from occurring.

mixture (40CFR116.3) means any combination of two or more elements and/or compounds in solid, liquid, or gaseous form except where such substances have undergone a chemical reaction so as to become inseparable by physical means.

mixture (40CFR2.306-5) has the meaning given it in section 3(8) of the Act, 15USC2602(8).

mixture (40CFR355.20) means a heterogenous association of substances where the various individual substances retain their identities and can usually be separated by mechanical means. Includes solutions or compounds but does not include alloys or amalgams.

mixture (40CFR372.3) means any combination of two or more chemicals, if the combination is not, in whole or in part, the result of a chemical reaction. However, if the combination as produced by a chemical reaction but could have been produced without a chemical reaction, it is also treated as a mixture. A mixture also includes any combination which consists of a chemical and associated impurities.

mixture (40CFR710.2-q) means any combination of two or more chemical substances if the combination does not occur in nature and is not, in whole or in part, the result of a chemical reaction; except that "mixture" does include:

(1) Any combination which occurs, in whole or in part, as a result of a chemical reaction if the combination could have been manufactured for commercial purposes without a chemical reaction at the time the chemical substances comprising the combination were combined and if, after the effective date or premanufacture notification requirements, none of the chemical substances comprising the combination is a new chemical substance, and

(2) Hydrates of a chemical substance or hydrated ions formed by association of a chemical substance with water.

(Other identical or similar definitions are provided in 40CFR720.3-u; 723.250-6.)

mixture (40CFR712.3-i) means any combination of two or more chemical substances if the combination does not occur in nature and is not, in whole or in part, the result of a chemical reaction; except that "mixture" does include:

(1) any combination which occurs, in whole or in part, as a result of a chemical reaction if the combination could have been manufactured for commercial purposes without a chemical reaction at the time the chemical substances comprising the combination were combined, and if all of the chemical substances comprising the combination are included in the EPA, TSCA Chemical Substance Inventory after the effective date of the premanufacture notification requirement under 40CFR720, and

(2) hydrates of a chemical substance or hydrated ions formed by association of a chemical substance with water. The term mixture includes alloys, inorganic glasses, ceramics, frits, and cements, including Portland cement.

mixture (40CFR720.3-u) means any combination of two or more chemical substances if the combination does not occur in nature and is not, in whole or in part, the result of a chemical reaction; except "mixture" does include:

(1) any combination which occurs, in whole or in part, as a result of a chemical reaction if the combination could have been manufactured for commercial purposes without a chemical reaction at the time the chemical substances comprising the combination were combined and if all of the chemical substances comprising the combination are not new chemical substances, and

(2) hydrates of a chemical substance or hydrated ions formed by association of a chemical substance with water, so long as the nonhydrated form is itself not a new chemical substance.

(Other identical or similar definitions are provided in 40CFR723.250-6.)

mixture (40CFR761.3) means any combination of two or more chemical substances if the combination does not occur in nature and is not, in whole or in part, the result of a chemical reaction; except that such term does include any combination which occurs, in whole or in part, as a result of a chemical reaction if none of the chemical substances comprising the combination is a new chemical substance and if the combination could have been manufactured for commercial purposes without a chemical reaction at the time the chemical substances comprising the combination were combined (under TSCA).

mixture (40CFR82.172) means any mixture or blend of two or more compounds.

mixture (TSCA3-15USC2602) means any combination of two or more chemical substances if the combination does not occur in nature and is not, in whole or in part, the result of a chemical reaction; except that such term does include any combination which occurs, in whole or in part, as a result of a chemical reaction if none of the chemical substances comprising the combination is a new chemical substance and if the combination could have been manufactured for commercial purposes without a chemical reaction at the time the chemical substances comprising the combination were combined.

ml/l (40CFR434.11-i) means milliliters per liter.

mobile incinerator systems (EPA-94/04): Hazardous waste incinerators that can be transported from one site to another.

mobile offshore drilling unit (OPA1001) means a vessel (other than a self-elevating lift vessel) capable of use as an offshore facility.

mobile source (40CFR117.1) means any vehicle, rolling stock, or other means of transportation which contains or carries a reportable quantity of a hazardous substance.

mobile source (40CFR51.491): On-road (highway) vehicles (e.g., automobiles, trucks and motorcycles) and nonroad vehicles (e.g., trains, airplanes, agricultural equipment, industrial equipment, construction vehicles, off-road motorcycles, and marine vessels).

mobile source (EPA-94/04): Any non-stationary source of air pollution such as cars, trucks, motorcycles, buses, airplanes, locomotives.

MOD director (40CFR85.1402) means Director of Manufacturers Operations Division, Office of Mobile Sources-Office of Air and Radiation of the Environmental Protection Agency.

MOD Director (40CFR85.2102-18) means Director of Manufacturers Operations Division, Office of Mobile Sources--Office of Air and Radiation of the Environmental Protection Agency.

model (40CFR86.082.2) means a specific combination of car line, body style, and drive-train configuration.

model line (40CFR60.531) means all wood heaters offered for sale by a single manufacturer that are similar in all material respects.

model plant (EPA-94/04): A hypothetical plant design used for developing economic, environmental, and energy impact analyses as support for regulations or regulatory guidelines; first step in exploring the economic impact of a potential NSPS.

model specific code (40CFR205.151-18) means the designation used for labeling purposes in sections 205.15 and 205.16 for identifying the motorcycle manufacturer, class, and advertised engine displacement, respectively.

model type (40CFR2.311-6) has the meaning given it in section 501(11) of the Act, 15USC2001(11).

model type (40CFR86.082-2) means a unique combination of car line, basic engine, and transmission class (other identical or similar definitions are provided in 40CFR600.002.85-19).

model year (40CFR204.51-c) means the manufacturers annual production period which includes January 1 of such calendar year; Provided, that if the manufacturer has no annul production period, the term model year shall mean the calendar (other identical or similar definitions are provided in 40CFR205.51-17).

model year (40CFR205.151-19) means the manufacturers annual production period, which includes January 1 of any calendar year, or if the manufacturer has no annual production period, the term model year shall mean the calendar year.

model year (40CFR600.002.85-6) means the manufacturer's annual production period (as determined by the Administrator) which includes January 1 of such calendar year. If a manufacturer has no annual production period, the term "model year" means the calendar year.

model year (40CFR85.1502-7) means the manufacturers annual production period (as determined by the Administrator) which includes January 1 of such calendar year; provided, that if the manufacturer has no annual production period, the term model year shall mean the calendar year in which a vehicle is modified. A certificate holder shall be deemed to have produced a vehicle or engine when the certificate holder has modified the nonconforming vehicle or engine.

model year (40CFR85.2102-6) means the manufacturer's annual production period (as determined by the Administrator) which includes January 1 of such calendar year; however, if the manufacturer has no annual production period, the term model year shall mean the calendar year.

model year (40CFR86.082.2) means the manufacturer's annual production period (as determined by the Administrator) which includes January 1 of such calendar year: Provided, that if the manufacturer has no annual production period, the term model year shall mean the calendar year.

model year (40CFR86.402.78) means the manufacturer's annual production period (as determined by the Administrator) which includes January first of such calendar year. If the manufacturer has no annual production period, the term model year shall mean the calendar year.

model year (40CFR88.302.94), as it applies to the clean fuel vehicle fleet purchase requirements, means September 1 through August 31.

model year (CAA202-42USC7521):
(i) with reference to any specific calendar year means the manufacturer's annual production period (as determined by the Administrator) which includes January 1 of such calendar year. If the manufacturer has no annual production period, the term "model year" shall mean the calendar year.
(ii) For the purpose of assuring that vehicles and engines manufactured before the beginning of a model year were not manufactured for purposes of circumventing the effective date of a standard required to be prescribed by subsection (b), the Administrator may prescribe regulations defining "model year" otherwise than as provided in clause (i).

model year (MY) (40CFR89.2) means the manufacturer's annual new model production period which includes January 1 of the calendar year, ends no later than December 31 of the calendar year, and does not begin earlier than January 2 of the previous calendar year. Where a manufacturer has no annual new model production period, model year means calendar year.

model year for imported engines (40CFR89.602.96): The manufacturer's annual production period (as determined by the Administrator) which includes January 1 of the calendar year; provided, that if the manufacturer has no annual production period, the term "model year" means the calendar year in which a nonroad engine is modified. An independent commercial importer (ICI) is deemed to have produced a nonroad engine when the ICI has modified (including labeling) the nonconforming nonroad engine to meet applicable emission requirements.

modification (40CFR52.2486-4) means any change to a parking facility that increases or may increase the motor vehicle capacity of, or the motor vehicle activity associated with, such parking facility.

modification (40CFR55.2) shall have the meaning given in the applicable requirements incorporated into 40CFR55.13 and 55.14 of this part, except that for two years following the date of promulgation of this part the definition given in section 111(a) of the Act shall apply for the purpose of determining the required date of compliance with this part, as set forth in 40CFR55.3 of this part.

modification (40CFR60.2) means any physical change in, or change in the method of operation of, an existing facility which increases the amount of any air pollutant (to which a standard applies) emitted into the atmosphere by that facility or which results in the emission of any air pollutant (to which a standard applies) into the atmosphere not previously emitted.

modification (40CFR8.2-p) means any alteration in the terms and conditions of a contract, including supplemental agreements, amendments and extensions.

modification (CAA111-42USC7411) means any

physical change in, or change in the method of operation of, a stationary source which increases the amount of any air pollutant emitted by such source or which results in the emission of any air pollutant not previously emitted.

modification (CAA112-42USC7412) means any physical change in, or change in the method of operation of, a major source which increases the actual emissions of any hazardous air pollutant emitted by such source by more than a de minimis amount or which results in the emission of any hazardous air pollutant not previously emitted by more than a de minimis amount.

modification or modified source (40CFR52.01-d) mean any physical change in, or change in the method of operation of, a stationary source which increases the emission rate of any pollutant for which a national standard has been promulgated under part 50 of this chapter or which results in the emission of any such pollutant not previously emitted, except that:

(1) Routine maintenance, repair, and replacement shall not be considered a physical change, and

(2) The following shall not be considered a change in the method of operation:

    (i) An increase in the production rate, if such increase does not exceed the operating design capacity of the source;

    (ii) An increase in the hours of operation;

    (iii) Use of an alternative fuel or raw material, if prior to the effective date of a paragraph in this part which imposes conditions on or limits modifications, the source is designed to accommodate such alternative use.

modified discharge (40CFR125.58-j) means the volume, composition and location of the discharge proposed by the applicant for which a modification under section 301(h) of the Act is requested. A modified discharge may be a current discharge, improved discharge, or altered discharge.

modified solid waste incineration unit (CAA129.g-42USC7429) means a solid waste incineration unit at which modifications have occurred after the effective date of a standard under subsection (a) if:

(A) the cumulative cost of the modifications, over the life of the unit, exceed 50 per centum of the original cost of construction and installation of the unit (not including the cost of any land purchased in connection with such construction or installation) updated to current costs, or

(B) the modification is a physical change in or change in the method of operation of the unit which increases the amount of any air pollutant emitted by the unit for which standards have been established under this section or section 111.

modified source (EPA-94/04): The enlargement of a major stationery source is often referred to as modification, implying that more emissions will occur.

modular excess air MWC (40CFR60.51a) means a combustor that combusts MSW and that is not field-erected and has multiple combustion chambers, all of which are designed to operate at conditions with combustion air amounts in excess of theoretical air requirements.

modular starved air MWC (40CFR60.51a) means a combustor that combusts MSW and that is not field-erected and has multiple combustion chambers in which the primary combustion chamber is designed to operate at substoichiometric conditions.

modulated stem (40CFR85.2122(a)(1)(ii)(G)) means stem attached to the vacuum break diaphragm in such a manner as to allow stem displacement independent of diaphragm displacement.

modulated stem displacement (40CFR85.2122(a)(1)-(ii)(C)) means the distance through which the modulated stem may move when actuated independent of diaphragm displacement.

modulated stem displacement force (40CFR85.2122-(a)(1)(ii)(D)) means the amount of force required to start and finish a modulated stem displacement.

molar absorptivity (e_r) (40CFR796.3700-iv) is defined as the proportionality constant in the Beer-Lambert law when the concentration is given in terms of moles per liter (i.e., molar concentration). Thus, $A = e_r Cl$, where A and e_r represent the absorbance and molar absorptivity at wavelength r and l and C are defined in (3). The units of e_r are $molar^{-1}$ cm^{-1}. Numerical values of molar absorptivity depend upon the nature of the absorbing species (other identical or similar definitions are provided in 40CFR796.3780-iv).

molecule (EPA-94/04): The smallest division of a compound that still retains or exhibits all the properties of the substance.

molten salt reactor (EPA-94/04): A thermal treatment unit that rapidly heats waste in a heat-conducting fluid bath of carbonate salt.

monitor (40CFR52.741) means to measure and record.

monitor accuracy (40CFR72.2) means the closeness of the measurement made by a CEMS or by one of its component parts to the reference value of the emissions or volumetric flow being measured, expressed as the difference between the measurement and the reference value.

monitor operating hour (40CFR72.2) means any hour or portion thereof over which a CEMS, COMS, or other monitoring system approved by the Administrator under part 75 of this chapter is operating regardless of the number of measurements (i.e., data points) collected during the hour.

monitoring (EPA-94/04): Periodic or continuous surveillance or testing to determine the level of compliance with statutory requirements and/or pollutant levels in various media or in humans, plants, and animals.

monitoring device (40CFR60.2) means the total equipment, required under the monitoring of operations sections in applicable subparts, used to measure and record (if applicable) process parameters.

monitoring system (40CFR61.02) means any system, required under the monitoring sections in applicable subparts, used to sample and condition (if applicable), to analyze, and to provide a record of emissions or process parameters.

monitoring well (EPA-94/04): 1. A well used to obtain water quality samples or measure groundwater levels. 2. Well drilled at a hazardous waste management facility or Superfund site to collect ground-water samples for the purpose of physical, chemical, or biological analysis to determine the amounts, types, and distribution of contaminants in the ground water beneath the site.

monoclonal antibodies (EPA-94/04): (Also called MABs and MCAs):
1. Man-made clones of a molecule, produced in quantity for medical or research purposes.
2. Molecules of living organisms that selectively find and attach to other molecules to which their structure conforms exactly. This could also apply to equivalent activity by chemical molecules.

monomer (40CFR704.25-4) means a chemical substance that has the capacity to form links between two or more other molecules (cf. polymer) (other identical or similar definitions are provided in 40CFR723.250-8; 721.350).

monomictic (EPA-94/04): Lakes and reservoirs which are relatively deep, do not freeze over during the winter months, and undergo a single stratification and mixing cycle during the year (usually in the fall).

monovent (40CFR60.61-d) means an exhaust configuration of a building or emission control device (e.g., positive-pressure fabric filter) that extends the length of the structure and has a width very small in relation to its length (i.e., length to width ratio is typically greater than 5:1). The exhaust may be an open vent with or without a roof, louvered vents, or a combination of such features.

month (40CFR60.541) means a calendar month or a prespecified period of 28 days or 35 days (utilizing a 4-4-5-week recordkeeping and reporting schedule).

monthly average (40CFR425.02-i) means the arithmetic average of eight (8) individual data points from effluent sampling and analysis during any calendar month.

monthly average (40CFR503.11-i) is the arithmetic mean of all measurements taken during the month.

monthly average (40CFR503.41-h) is the arithmetic mean of the hourly averages for the hours a sewage sludge incinerator operates during the month.

monthly average regulatory values (40CFR461.3) shall be the basis for the monthly average discharge in direct discharge permits and for pretreatment standards. Compliance with the monthly discharge limit is required regardless of the number of samples analyzed and averaged.

Montreal Protocol (40CFR82.3) means the Montreal Protocol on Substances that Deplete the Ozone Layer, a protocol to the Vienna Convention for the Protection of the Ozone Layer, including adjustments adopted by parties thereto and amendments that have entered into force.

Montreal Protocol (40CFR82.3-x) means the Montreal Protocol on Substances that Deplete the Ozone Layer, a protocol to the Vienna Convention for the Protection of the Ozone Layer, including adjustments adopted by the parties thereto and amendments that have entered into force.

Montreal Protocol and the Protocol (CAA601-42USC7671) mean the Montreal Protocol on Substances that Deplete the Ozone Layer, a protocol to the Vienna Convention for the Protection of the Ozone Layer, including adjustments adopted by parties thereto and amendments that have entered into force.

Montreal protocol parties (40CFR82-AC-92): Countries which are party to the Montreal protocol. They are: Argentina, Australia, Austria, Bahrain, Bangladesh, Belgium, Botswana (3/3/92), Brazil, Bulgaria, Burkina Faso, Byelorussian Soviet Socialist Republic, Cameroon, Canada, Chile, China, Costa Rica, Cyprus (8/26/92), Czechoslovakia, Denmark, Ecuador, Egypt, European Economic Community, Fiji, Finland, France, Gambia, Germany, Ghana, Greece, Guatemala, Guinea (9/23/92), Hungary, Iceland, India (9/17/92), Indonesia (9/24/92), Iran, Ireland, Israel (9/28/92), Italy, Japan, Jordan, Kenya, Libyan Arab Jamahiriya, Liechtenstein, Luxembourg, Malawi, Malaysia, Maldives, Malta, Mexico, Netherlands, New Zealand, Nigeria, Norway, Panama, Philippines, Poland, Portugal, Republic of Korea (5/27/92), Russian Federation, Singapore, South Africa, Spain, Sri Lanka, Sweden, Switzerland, Syrian Arab Republic, Thailand, Togo, Trinidad and Tobago, Tunisia, Turkey, Uganda, Ukrainian Soviet Socialist Republic, United Arab Emirates, United

Kingdom, United States of America, Uruguay, Venezuela, Yugoslavia, Zambia.

moratorium (EPA-94/04): During the negotiation process, a period of 60 to 90 days during which EPA and potentially responsible parties may reach settlement but no site response activities can be conducted.

moratorium (MMPA3-16USC1362) means a complete cessation of the taking of marine mammals and a complete ban on the importation into the United States of marine mammals and marine mammal products, except as provided in this chapter.

morbidity (EPA-94/04): Rate of disease incidence.

morphologic transformation (40CFR795.285-1) is the acquisition of certain phenotypic, characteristics most notably loss of contact inhibition and loss of anchorage dependence which are often but not always associated with the ability to induce tumors in appropriate hosts (under the Toxic Substances Control Act).

most current guidance document (TSCA202-15USC2642) means the Environmental Protection Agency's "Guidance for Controlling Asbestos-Containing Material in Buildings" as modified by the Environmental Protection Agency after march 31, 1986.

most probable number (EPA-94/04): The most probable number of coliform-group organisms per unit volume of sample water.

most stringent federally enforceable emissions limitation (40CFR72.2) means the most stringent emissions limitation for a given pollutant applicable to the unit, which has been approved by the Administrator under the Act, whether in a State implementation plan approved pursuant to title I of the Act, a new source performance standard, or otherwise. To determine the most stringent emissions limitation for sulfur dioxide, each limitation shall be converted to lbs/mmBtu, using the appropriate conversion factors in appendix B of this part; provided that for determining the most stringent emissions limitation for sulfur dioxide for 1985, each limitation shall also be annualized, using the appropriate annualization factors in appendix A of this part.

motor activity (40CFR798.6200-2) is any movement of the experimental animal.

motor carrier (40CFR202.10-l) means a common carrier by motor vehicle, a contract carrier by motor vehicle, or a private carrier of property by motor vehicle as those terms are defined by paragraphs (14), (15), and (1) of section 203(a) of the Interstate Commerce Act (49USC303(a)).

motor controller (40CFR600.002.85-47) means an electronic or electro-mechanical device to convert energy stored in an energy storage device into a form suitable to power the traction motor.

motor fuel (40CFR280.12) means petroleum or a petroleum-based substance that is motor gasoline, aviation gasoline, No. 1 or No. 2 diesel fuel, or any grade of gasohol, and is typically used in the operation of a motor engine.

motor vehicle (40CFR202.10-m) means any vehicle, machine, tractor, trailer, or semitrailer propelled or drawn by mechanical power and used upon the highways in the transportation of passengers or property, or any combination thereof, but does not include any vehicle, locomotive, or car operated exclusively on a rail or rails.

motor vehicle (CAA216-42USC7550) means any self-propelled vehicle designed for transporting persons or property on a street or highway.

motor vehicle air conditioner (MVAC) (40CFR82.152-k) means any appliance that is a motor vehicle air conditioner as defined in 40CFR part 82, subpart B.

motor vehicle air conditioners (40CFR82.32-d) means mechanical vapor compression refrigeration equipment used to cool the driver's or passenger's compartment of any motor vehicle. This definition is not intended to encompass the hermetically sealed refrigeration systems used on motor vehicles for refrigerated cargo and the air conditioning systems on passenger buses using HCFC-22 refrigerant (under CAA).

motor vehicle as used in this subpart (40CFR82.32-c) means any vehicle which is self-propelled and designed for transporting persons or property on a street or highway, including but not limited to passenger cars, light duty vehicles, and heavy duty vehicles. This definition does not include a vehicle where final assembly of the vehicle has not been completed by the original equipment manufacturer.

motor vehicle emissions budget (40CFR51.392) is that portion of the total allowable emissions defined in a revision to the applicable implementation plan (or in an implementation plan revision which was endorsed by the Governor or his or her designee, subject to a public hearing, and submitted to EPA, but not yet approved by EPA) for a certain date for the purpose of meeting reasonable further progress milestones or attainment or maintenance demonstrations, for any criteria pollutant or its precursors, allocated by the applicable implementation plan to highway and transit vehicles. The applicable implementation plan for an ozone nonattainment area may also designate a motor vehicle emissions budget for oxides of nitrogen (NO_x) for a reasonable further progress milestone year if the applicable implementation plan demonstrates that this NO_x budget will be achieved with measures in the implementation plan (as an implementation plan must do for VOC milestone requirements). The applicable implementation plan for an ozone nonattainment area includes a NO_x budget if NO_x reductions are being substituted for reductions in volatile organic compounds in milestone years required for reasonable further progress (other identical or similar definitions are provided in 40CFR93.101).

motor vehicle or engine part manufacturer (CAA216-42USC7550) as used in sections 207 and 208 means any person engaged in the manufacturing, assembling or rebuilding of any device, system, part, component or element of design which is installed in or on motor vehicles or motor vehicle engines.

motor vehicles held for lease or rental to the general public (40CFR88.302.94) means a vehicle that is owned or controlled primarily for the purpose of short-term rental or extended-term leasing (with or without maintenance), without a driver, pursuant to a contract.

motorcycle (40CFR205.151-1) means any motor vehicle, other than a tractor, that:

(i) Has two or three wheels;

(ii) Has a curb mass less than or equal to 680 kg (1499 lb); and

(iii) Is capable, with an 80 kg (176 lb) driver, of achieving a maximum speed of at least 24 km/h (15 mph) over a level paved surface.

motorcycle (40CFR86.402.78) means any motor vehicle with a headlight, taillight, and stoplight and having: two wheels, or three wheels and a curb mass less than or equal to 680 kilograms (1499 pounds).

motorcycle noise level (40CFR205.151-20) means the A-weighted noise level of a motorcycle as measured by the acceleration test procedure.

movement (40CFR260.10) means that hazardous waste transported to a facility in an individual vehicle.

mpc (40CFR427.71-b) shall mean 1000 pieces of floor tile.

MPRSA: See act or MPRSA.

MSBu (40CFR406.11-d) shall mean 1000 standard bushels (other identical or similar definitions are provided in 40CFR406.21-d; 406.41-d).

MSDS (40CFR721.3) means material safety data sheet, the written listing of data for the chemical substance as required under 40CFR721.72(c).

muck cooker (40CFR63.321) means a device for heating perchloroethylene-laden waste material to volatilize and recover perchloroethylene.

muck soils (EPA-94/04): Earth made from decaying plant materials.

mudballs (EPA-94/04): Round material that forms in filters and gradually increases in size when not removed by backwashing.

muffler (40CFR202.10-n) means a device for abating the sound of escaping gases of an internal combustion engine.

mulch (EPA-94/04): A layer of material (wood chips, straw, leaves, etc.) placed around plants to hold moisture, prevent weed growth, and enrich or sterilize the soil.

multi-header generator (40CFR72.2) means a generator served by ductwork from more than one unit.

multi-header unit (40CFR72.2) means a unit with ductwork serving more than one generator.

multi-media (PPA6603) means water, air, and land.

multiple effect evaporator system (40CFR60.281-f) means the multiple-effect evaporators and associated condenser(s) and hotwell(s) used to concentrate the spent cooking liquid that is separated from the pulp (black liquor).

multiple ferrous melting furnace scrubber configuration (40CFR464.31-h) means a multiple ferrous melting furnace scrubber configuration is a configuration where two or more discrete wet scrubbing devices are employed in series in a single melting furnace exhaust gas stream. The ferrous melting furnace scrubber mass allowance shall be given to each discrete wet scrubbing device that has an associated wastewater discharge in a multiple ferrous melting furnace scrubber configuration. The mass allowance for each discrete wet scrubber shall be identical and based on the air flow of the exhaust gas stream that passes through the multiple scrubber configuration.

multiple package coating (40CFR52.741) means a coating made from more than one different ingredient which must be mixed prior to using and has a limited pot life due to the chemical reaction which occurs upon mixing.

multiple stands (40CFR420.101-e) means those recirculation or direct application cold rolling mills which include more than one stand of work rolls (cf. single stand).

multiple use (EPA-94/04): Use of land for more than one purpose; i.e., grazing of livestock, watershed and wildlife protection, recreation, and timber production. Also applies to use of bodies of water for recreational purposes, fishing, and water supply.

multiple use (FLPMA103-43USC1702) means the management of the public lands and their various resource values so that they are utilized in the combination that will best meet the present and future needs of the American people; making the most judicious use of the land for some or all of these resources or related services over areas large enough to provide sufficient latitude for periodic adjustments in use to conform to changing needs and conditions; the use of some land for less than all of the resources; a combination of balanced and diverse resource uses that takes into account the long-term needs of future generations for renewable and nonrenewable resources, including. but not limited to, recreation, range, timber, minerals, watershed, wildlife and fish, and natural scenic, scientific and historical values; and harmonious and coordinated management of the various resources without permanent impairment of the productivity of the land and the quality of the environment with consideration being given to the relative values of the resources and not necessarily to the combination of uses that will give the greatest economic return or the greatest unit output.

multistage remote sensing (EPA-94/04): A strategy for landscape characterization that involves gathering and analyzing information at several geographic scales, ranging from generalized levels of detail at the national level through high levels of detail at the local scale.

municipal discharge (EPA-94/04): Discharge of effluent from waste water treatment plants which receive waste water from households, commercial establishments, and industries in the coastal drainage basin. Combined sewer/separate storm overflows are included in this category.

municipal separate storm sewer (40CFR122.26-8) means a conveyance or system of conveyances (including roads with drainage systems, municipal streets, catch basins, curbs, gutters, ditches, man-made channels, or storm drains):
(i) Owned or operated by a State, city, town, borough, county, parish, district, association, or other public body (created by or pursuant to State law) having jurisdiction over disposal of sewage, industrial wastes, storm water, or other wastes, including special districts under State law

such as a sewer district, flood control district or drainage district, or similar entity, or an Indian tribe or an authorized Indian tribal organization, or a designated and approved management agency under section 208 of the CWA that discharges to waters of the United States;
(ii) Designed or used for collecting or conveying storm water;
(iii) Which is not a combined sewer; and
(iv) Which is not part of a Publicly Owned Treatment Works (POTW) as defined at 40CFR122.2.

municipal sewage (EPA-94/04): Wastes (mostly liquid) originating from a community; may be composed of domestic wastewaters and/or industrial discharges.

municipal solid waste landfill unit (40CFR258.2) means a discrete area of land or an excavation that receives household waste, and that is not a land application unit, surface impoundment, injection well, or waste pile, as those terms are defined under 40CFR257.2. A MSWLF unit also may receive other types of RCRA subtitle D wastes, such as commercial solid waste, nonhazardous sludge, conditionally exempt small quantity generator waste and industrial solid waste. Such a landfill may be publicly or privately owned. A MSWLF unit may be a new MSWLF unit, an existing MSWLF unit or a lateral expansion (under the Resource Conservation Recovery Act).

municipal solid wastes (40CFR761.3) means garbage, refuse, sludges, wastes, and other discarded materials resulting from residential and non-industrial operations and activities, such as household activities, office functions, and commercial housekeeping wastes.

municipal solid wastes (MSW) (40CFR240.101-q) means, normally, residential and commercial solid wastes generated within a community (other identical or similar definitions are provided in 40CFR241.101-k).

municipal type solid waste (40CFR60.41b) means refuse, more than 50 percent of which is waste consisting of a mixture of paper, wood, yard wastes, food wastes, plastics, leather, rubber, and other combustible materials, and noncombustible materials such as glass and rock.

municipal type solid waste or MSW (40CFR60.51a) means household, commercial/retail, and/or institutional waste. Household waste includes material discarded by single and multiple residential dwellings, hotels, motels, and other similar permanent or temporary housing establishments or facilities. Commercial/retail waste includes material discarded by stores, offices, restaurants, warehouses, nonmanufacturing activities at industrial facilities, and other similar establishments or facilities. Institutional waste includes material discarded by schools, hospitals, nonmanufacturing activities at prisons and government facilities and other similar establishments or facilities. Household, commercial/retail, and institutional waste do not include sewage, wood pallets, construction and demolition wastes, industrial process or manufacturing wastes, or motor vehicles (including motor vehicle parts or vehicle fluff). Municipal-type solid waste does not include wastes that are solely segregated medical wastes. However, any mixture of segregated medical wastes and other wastes which contains more than 30 percent waste medical waste discards, is considered to be municipal-type solid waste.

municipal waste (CAA129.g-42USC7429) means refuse (and refuse-derived fuel) collected from the general public and from residential, commercial, institutional, and industrial sources consisting of paper, wood, yard wastes, food wastes, plastics, leather, rubber, and other combustible materials and non-combustible materials such as metal, glass and rock, provided that:

(A) the term does not include industrial process wastes or medical wastes that are segregated from such other wastes; and

(B) an incineration unit shall not be considered to be combusting municipal waste for purposes of section 111 or this section if it combusts a fuel feed stream, 30 percent or less of the weight of which is comprised, in aggregate, of municipal waste.

municipal waste combustor or MWC or MWC unit (40CFR60.51a) means any device that combusts, solid, liquid, or gasified MSW including, but not limited to, field-erected incinerators (with or without heat recovery), modular incinerators (starved air or excess air), boilers (i.e., steam generating units), furnaces (whether suspension-fired, grate-fired, mass-fired, or fluidized bed-fired) and gasification-combustion units. This does not include combustion units, engines, or other devices that combust landfill gases collected by landfill gas collection systems.

municipality (40CFR122.2) means a city, town, borough, county, parish, district, association, or other public body created by or under State law and having jurisdiction over disposal of sewage, industrial wastes, or other wastes, or an Indian tribe or an authorized Indian tribal organization, or a designated and approved management agency under section 208 of CWA.

municipality (40CFR142.2) means a city, town or other public body created by or pursuant to State law, or an Indian tribe which does not meet the requirements of subpart H of this part (under SDWA).

municipality (40CFR35.2005-27) is a city, town, borough, county, parish, district, association, or other public body (including an intermunicipal agency of two or more of the foregoing entities) created under State law, or an Indian tribe or an authorized Indian tribal organization, having jurisdiction over disposal of sewage, industrial wastes, or other waste, or designated and approved management agency under section 208 of the Act:

(i) This definition includes a special district created under State law such as a water district, sewer district, sanitary district, utility district, drainage district or similar entity or an integrated waste management facility, as defined in section 201(e) of the Act, which has as one of its principal responsibilities the treatment, transport, or disposal of domestic wastewater in a particular geographic area.

(ii) This definition excludes the following:

(A) Any revenue producing entity which has as its principal responsibility an activity other than providing wastewater treatment services to the general public, such as an airport, turnpike, port facility or other municipal utility.

(B) Any special district (such as school district or a park district) which has the responsibility to provide wastewater treatment services in support of its principal activity at specific facilities, unless the special district has the responsibility under State law to provide wastewater treatment services to the community surrounding the special districts facility and no other municipality, with concurrent jurisdiction to serve the community, serves or intends to serve the special districts facility or the surrounding community.

municipality (40CFR35.905) means a city, town, borough, county, parish, district, association, or other public body (including an inter-municipal agency of two or more of the foregoing entities) created under State law, or an Indian tribe or an authorized Indian tribal organization, having jurisdiction over disposal of sewage, industrial wastes, or other waste, or designated and approved sewage agency under section 20 of the Act:

(a) This definition includes a special district created under State law such as a water district, sewer district, sanitary district, utility district, drainage district, or similar entity or an integrated waste management facility, as defined in section 201(e) of the Act, which has as one of its principal responsibilities the treatment, transport, or disposal of liquid wastes of the general public in a particular geographic area.

(b) This definition excludes the following:

(1) Any revenue producing entity which has as its principal responsibility an activity other than providing wastewater treatment services to the general public, such as an airport, turnpike, port facility, or other municipal utility.

(2) Any special district (such as school district or a park district) which has the responsibility to provide wastewater treatment services in support of its principal activity at specific facilities, unless the special district has the responsibility under State law to provide waste water treatment services to the community surrounding the special districts facility and no other municipality, with concurrent jurisdiction to serve the community, serves or intends to serve the special districts facility or the surrounding community.

municipality (40CFR40.115-4):

(a) Under the Federal Water Pollution Control Act, a city, town, borough, county, parish, district, association, or other public body created by or pursuant to State law, or an Indian tribe or an authorized Indian tribal organization, with jurisdiction over disposal of sewage, industrial wastes, or other wastes; or a designated and approved management agency under section 208 of the act.

(b) Under the Resource Conservation and Recovery Act, a city, town, borough, county, parish, district, or other public body created by or pursuant to State law, with responsibility for the planning or administration of solid waste management, or an Indian tribe or authorized tribal organization or Alaska Native village or organization, and any rural community or unincorporated town or village or any other public entity for which an application for assistance is made by a State or political subdivision thereof.

(c) In all other cases, a city, town, borough, county, parish, district, or other public body created by or pursuant to State law, or an Indian tribe or an authorized Indian tribal organization, having substantial powers or duties pertaining to the control of pollution.

municipality (40CFR401.11-m) shall be defined in accordance with section 502 of the Act unless the context otherwise requires.

municipality (40CFR501.2) means a city, town, borough, county, parish, district, association, or other public body (including an intermunicipal agency of two or more of the foregoing entities) created under State law (or an Indian tribe or an authorized Indian tribal organization), or a designated and approved management agency under section 208 of the Clean Water Act. This definition includes a special district created under State law such as a water district, sewer district, sanitary district, utility district, drainage district, or similar entity, or an integrated waste management facility as defined in section 201(e) of the CWA, as amended, that has as one of its principal responsibilities the treatment, transport, or disposal of sewage sludge.

municipality (40CFR503.9-o) means a city, town,

borough, county, parish, district, association, or other public body (including an intermunicipal Agency of two or more of the foregoing entities) created by or under State law; an Indian tribe or an authorized Indian tribal organization having jurisdiction over sewage sludge management; or a designated and approved management Agency under section 208 of the CWA, as amended. The definition includes a special district created under State law, such as a water district, sewer district, sanitary district, utility district, drainage district, or similar entity, or an integrated waste management facility as defined in section 201(e) of the CWA, as amended, that has as one of its principal responsibilities the treatment, transport, use, or disposal of sewage sludge.

municipality (CAA302-42USC7602) means a city, town, borough, county, parish, district, or other public body created by or pursuant to State law.

municipality (CWA502-33USC1362) means a city, town, borough, county, parish, district, association, or other public body created by or pursuant to State law and having jurisdiction over disposal of sewage, industrial wastes, or other wastes, or an Indian tribe or an authorized Indian tribal organization, or a designated and approved management agency under section 208 of this Act.

municipality (RCRA1004-42USC6903):
(A) means a city, town, borough, county, parish, district, or other public body created by or pursuant to State law, with responsibility for the planning or administration of solid waste management, or an Indian tribe or authorized tribal organization or Alaska Native village or organization, and
(B) includes any rural community or unincorporated town or village or any other public entity for which an application for assistance is made by a State or political subdivision thereof.

municipality (SDWA1401-42USC300f) means a city, town, or other public body created by or pursuant to State law, or an Indian tribe.

mushrooms (40CFR407.71-k) shall mean the processing of mushrooms into the following product styles: Canned, frozen, dehydrated, all varieties, shapes and sizes.

mutagen/mutagenicity (EPA-94/04): An agent that causes a permanent genetic change in a cell other than that which occurs during normal genetic recombination. Mutagenicity is the capacity of a chemical or physical agent to cause such permanent alternation.

MVAC-like appliance (40CFR82.152-l) means mechanical vapor compression, open-drive compressor appliances used to cool the driver's or passenger's compartment of an non-road motor vehicle. This includes the air-conditioning equipment found on agricultural or construction vehicles. This definition is not intended to cover appliances using HCFC-22 refrigerant.

MWC acid gases (40CFR60.51a) means all acid gases emitted in the exhaust gases from MWC units including, but not limited to, sulfur dioxide and hydrogen chloride gases.

MWC metals (40CFR60.51a) means metals and metal compounds emitted in the exhaust gases from MWC units.

MWC organics (40CFR60.51a) means organic compounds emitted in the exhaust gases from MWC units and includes total tetra- through octa-chlorinated dibenzo-p-dioxins and dibenzofurans.

MWC plant (40CFR60.31a) means one or more MWC units at the same location for which construction, modification, or reconstruction is commenced on or before December 20, 1989 (cf. large MWC plant or very large MWC plant).

MWC plant (40CFR60.51a) means one or more MWC units at the same location for which construction, modification, or reconstruction is commenced after December 20, 1989.

MWC plant capacity (40CFR60.31a) means the aggregate MWC unit capacity of all MWC units at an MWC plant for which construction, modification, or reconstruction is commenced on or before December 20, 1989.

MWC plant capacity (40CFR60.51a) means the aggregate MWC unit capacity of all MWC units at an MWC plant for which construction, modification, or reconstruction commenced after December 20, 1989. Any MWC units for which construction, modification, or reconstruction is commenced on or before December 20, 1989, are not included for determining applicability under this subpart.

MWC unit capacity (40CFR60.51a) means the maximum design charging rate of an MWC unit expressed in megagrams per day (tons per day) of MSW combusted, calculated according to the procedures under 40CFR60.58a, paragraph (j). Municipal waste combustor unit capacity is calculated using a design heating value of 10,500 kilojoules per kilogram (4,500 British thermal units per pound) for MSW and 19,800 kilojoules per kilogram (8,500 British thermal units per pound) for medical waste. The calculational procedures under 40CFR60.58a(j) include procedures for determining MWC unit capacity for batch MWC's and cofired combustors and combustors firing mixtures of medical waste and other MSW.

mwh (40CFR424.11-b) shall mean megawatt hour(s) of electrical energy consumed in the smelting process (furnace power consumption) (other identical or similar definitions are provided in 40CFR424.21-b).

************ NNNNN ************

NAAQS (CAA302-42USC7602) menas the National Ambient Air Quality Standard.

NAAQS and National Ambient Air Quality Standards, unless otherwise specified (40CFR57.103-r), refer only to the National Primary and Secondary Ambient Air Quality Standards for sulfur dioxide.

nameplate capacity (40CFR72.2) means the maximum electrical generating output (expressed in MWe) that a generator can sustain over a specified period of time when not restricted by seasonal or other deratings, as listed in the NADB under the

data field "NAMECAP" if the generator is listed in the NADB or as measured in accordance with the United States Department of Energy standards if the generator is not listed in the NADB.

NAMS (National Air Monitoring Stations) (40CFR58.1-c): Collectively the NAMS are a subset of the SLAMS ambient air quality monitoring network.

naphthalene (or priority pollutant No. 55) (40CFR420.02-o) means the value obtained by the standard method Number 610 specified in 44 FR: 69464, 69571 (December 3, 1979).

naphthalene processing (40CFR61.131) means any operations required to recover naphthalene including the separation, refining, and drying of crude or refined naphthalene.

national air monitoring stations (40CFR58.1-c): See NAMS.

national allowance data base or NADB (40CFR72.2) means the data base established by the Administrator under section 402(4)(C) of the Act.

national ambient air quality standard (NAAQS) (40CFR51.491) means a standard set by the EPA at 40CFR part 50 under section 109 of the Act.

national ambient air quality standards (NAAQS) (40CFR300-AA) means primary standards for air quality established under section 108 and 109 of the Clean Air Act, as amended.

national ambient air quality standards (NAAQS) (40CFR51.392) are those standards established pursuant to section 109 of the CAA (other identical or similar definitions are provided in 40CFR93.101).

national ambient air quality standards (NAAQS) (40CFR51.852) are those standards established pursuant to section 109 of the Act and include standards for carbon monoxide (CO), lead (Pb), nitrogen dioxide (NO_2), ozone, particulate matter (PM-10), and sulfur dioxide (SO_2) (other identical or similar definitions are provided in 40CFR93.152).

national ambient air quality standards (NAAQS)

(EPA-94/04): Standards established by EPA that apply for outside air throughout the country (cf. criteria pollutants, state implementation plans or emissions trading).

national consensus standard (29CFR1910.2) means any standard or modification thereof which:
(1) has been adopted and promulgated by a nationally recognized standards-producing organization under procedures whereby it can be determined by the Secretary of Labor or by the Assistant Secretary of Labor that persons interested and affected by the scope or provisions of the standard have reached substantial agreement on its adoption,
(2) was formulated in a manner which afforded an opportunity for diverse views to be considered, and
(3) has been designated as such a standard by the Secretary or the Assistant Secretary, after consultation with other appropriate Federal agencies.

national consensus standard (OSHA3-29USC652) means any occupational safety and health standard or modification thereof which:
(1) has been adopted and promulgated by a nationally recognized standards-producing organization under procedures whereby it can be determined by the Secretary that persons interested and affected by the scope or provisions of the standard have reached substantial agreement on its adoption,
(2) was formulated in a manner which afforded an opportunity for diverse views to be considered, and
(3) has been designated as such a standard by the Secretary, after consultation with other appropriate Federal agencies.

national contingency plan (40CFR310.11-f) means the National Oil and Hazardous Substances Pollution Contingency Plan (40CFR300).

national contingency plan (CWA311-33USC1321) means the National Contingency Plan prepared and published under subsection (d).

national contingency plan (OPA1001) means the national contingency plan prepared and published under section 311(d) of the Federal Water Pollution Control Act, as amended by this Act, or revised under section 105 of the Comprehensive Environmental Response, Compensation, and Liability Act (42USC9605).

national contingency plan (SF101-42USC9601) means the national contingency plan published under section 311(c) of the Federal Water Pollution Control Act or revised pursuant to section 105 of the Act.

national contingency plan or NCP (40CFR304.12-h) means the National Oil and Hazardous Substances Pollution Contingency Plan, developed under section 311(c)(2) of the Federal Water Pollution Control Act, 33USC1251, et seq., as amended, revised periodically pursuant to section 105 of CERCLA, 42USC9605, and published at 40CFR300.

national emission standards for hazardous air pollutants (NESHAPs) (40CFR300-AA) means standards established for substances listed under section 112 of the Clean Air Act, as amended. Only those NESHAPs promulgated in ambient concentration units apply in the HRS.

national emissions standards for hazardous air pollutants (NESHAPS) (EPA-94/04): Emissions standards set by EPA for an air pollutant not covered by NAAQS that may cause an increase in fatalities or in serious, irreversible, or incapacitating illness. Primary standards are designed to protect human health, secondary standards to protect public welfare (e.g., building facades, visibility, crops, and domestic animals).

National Environmental Policy Act (NEPA) of 1969: See act or NEPA.

national estuary program (EPA-94/04): A program established under the Clean Water Act Amendments of 1987 to develop and implement conservation and management plans for protecting estuaries and restoring and maintaining their chemical, physical, and biological integrity, as well as controlling point and nonpoint pollution sources.

national interim primary drinking water regulations (EPA-94/04): Commonly referred to as NIPDWRs.

national municipal plan (EPA-94/04): A policy created in 1984 by EPA and the states in 1984 to bring all publicly owned treatment works (POTWs) into compliance with Clean Water Act requirements.

national oil and hazardous substances contingency plan (NOHSCP/NCP) (EPA-94/04): The federal regulation that guides determination of the sites to be corrected under both the Superfund program and the program to prevent or control spills into surface waters or elsewhere.

national panel of environmental arbitrators or panel (40CFR304.12-i) means a panel of environmental arbitrators selected and maintained by the Association to arbitrate cost recovery claims under this part.

national pollutant discharge elimination system (40CFR270.2) means the national program for issuing, modifying, revoking and reissuing, terminating, monitoring and enforcing permits, and imposing and enforcing pretreatment requirements, under sections 307, 402, 318, and 405 of the CWA. The term includes an approved program.

national pollutant discharge elimination system (NPDES) (40CFR122.2) means the national program for issuing, modifying, revoking and reissuing, terminating, monitoring and enforcing permits, and imposing and enforcing pretreatment requirements, under sections 307, 402, 318, and 405 of CWA. The term includes an "approved program."

national pollutant discharge elimination system (NPDES) (40CFR136.2-e) means the national system for the issuance of permits under section 402 of the Act and includes any State or interstate program which has been approved by the Administrator, in whole or in part, pursuant to section 402 of the Act.

national pollutant discharge elimination system (NPDES) (EPA-94/04): A provision of the Clean Water Act which prohibits discharge of pollutants into waters of the United States unless a special permit is issued by EPA, a state, or, where delegated, a tribal government on an Indian reservation.

national pretreatment standard or pretreatment standard (40CFR117.1) means any regulation containing pollutant discharge limits promulgated by the EPA in accordance with section 307 (b) and (c) of the Act, which applies to industrial users of a publicly owned treatment works. It further means any State or local pretreatment requirement applicable to a discharge and which is incorporated into a permit issued to a publicly owned treatment works under section 402 of the Act.

national pretreatment standard, pretreatment standard, or standard (40CFR403.3-j) means any regulation containing pollutant discharge limits promulgated by the EPA in accordance with section 307 (b) and (c) of the Act, which applies to industrial users. This term includes prohibitive discharge limits established pursuant to 40CFR03.5.

national primary drinking water regulation (40CFR142.2) means any primary drinking water regulation contained in part 141 of this chapter (under SDWA).

national priorities list (NPL) (40CFR300.5) means the list, compiled by EPA pursuant to CERCLA section 105, of uncontrolled hazardous substance releases in the United States that are priorities for long-term remedial evaluation and response.

national priorities list (NPL) (40CFR35.6015-27) means EPA's list of the most serious uncontrolled or abandoned hazardous waste sites identified for possible long-term remedial action under Superfund. A site must be on NPL to receive money from the Trust Fund for remedial action. The list is based primarily on the score a site receives from the Hazard Ranking System.

national priorities list (NPL) (EPA-94/04): EPA's list of the most serious uncontrolled or abandoned hazardous waste sites identified for possible long-term remedial action under Superfund. The list is based primarily on the score a site receives from the Hazard Ranking System. EPA is required to update the NPL at least once a year. A site must be on the NPL to receive money from the Trust Fund for remedial action.

national program assistance agreements (40CFR35.9010) means assistance agreements approved by the EPA Assistant Administrator for Water for work undertaken to accomplish broad NEP goals and objectives.

national response center (40CFR310.11-g) means the national communications center located in Washington, DC, that receives and relays notice of oil discharge or releases of hazardous substances to appropriate Federal officials.

national response center (EPA-94/04): The federal operations center that receives notifications of all releases of oil and hazardous substances into the environment; open 24 hours a day, is operated by the U.S. Coast Guard, which evaluates all reports and notifies the appropriate agency.

national response team (NRT) (EPA-94/04): Representatives of 13 federal agencies that, as a team, coordinate federal responses to nationally significant incidents of pollution--an oil spill, a major chemical release, or a Superfund response action-- and provide advice and technical assistance to the responding agency(ies) before and during a response action.

national response unit (CWA311-33USC1321) means the National Response Unit established under subsection (j).

national secondary drinking water regulations (EPA-94/04): Commonly referred to as NSDWRs.

national security exemption (40CFR204.2-5) means an exemption from the prohibitions of section 10(a) (1), (2), (3), and (5) of the Act, which may be granted under section 10(b)(l) of the Act for the purpose of national (other identical or similar definitions are provided in 40CFR205.2-5).

national security exemption (40CFR211.102-g) means an exemption from the prohibitions of section 10(a)(3) and (5) of the Act, which may be granted under section 10(b)(l) of the Act in cases involving national security.

national security exemption (40CFR85.1702-2) means an exemption which may be granted under 40CFR203(b)(1) of the Act for the purpose of national security.

national security exemption (40CFR89.902.96) means an exemption which may be granted under 40CFR89.1004(b) for the purpose of national security.

national security information (40CFR11.4-d): As used in this order this term is synonymous with "classified information." It is any information which must be protected against unauthorized disclosure in the interest of the national defense or foreign relations of the United States.

national standard (40CFR51.100-e) means either a primary or secondary standard.

nations complying with, but not joining, the Protocol (40CFR82.3) means any nation listed in Appendix C to this part.

nations complying with, but not joining, the Protocol (40CFR82.3-z) means any nation listed in appendix C, annex 2, to this subpart.

natural barrier (40CFR61.141) means a natural object that effectively precludes or deters access. Natural barriers include physical obstacles such as cliffs, lakes or other large bodies of water, deep and wide ravines, and mountains. Remoteness by itself is not a natural barrier.

natural conditions (40CFR51.301-q) includes naturally occurring phenomena that reduce visibility as measured in terms of visual range, contrast, or coloration.

natural draft opening (40CFR60.711-14) means any opening in a room, building, or total enclosure that remains open during operation of the facility and that is not connected to a duct in which a fan is installed. The rate and direction of the natural draft across such an opening is a consequence of the difference in pressures on either side of the wall containing the opening.

natural draft opening (40CFR60.741) means any opening in a room, building, or total enclosure that remains open during operation of the facility and that is not connected to a duct in which a fan is installed. The rate and direction of the natural draft across such an opening are a consequence of the difference in pressures on either side of the wall or barrier containing the opening.

natural gas (40CFR60.41b) means:

(1) a naturally occurring mixture of hydrocarbon and nonhydrocarbon gases found in geologic formations beneath the earth's surface, of which the principal constituent is methane; or

(2) liquid petroleum gas, as defined by the American Society for Testing and Materials in ASTM D1835-82, "Standard Specification for Liquid Petroleum Gases" (IBR, see 40CFR60.17).

natural gas (40CFR60.41c) means:

(1) a naturally occurring mixture of hydrocarbon and nonhydrocarbon gases found in geologic formations beneath the earth's surface, of which the principal constituent is methane, or

(2) liquefied petroleum (LP) gas, as defined by the American Society for Testing and Materials in ASTM D1835-86, "Standard Specification for Liquefied Petroleum Gases" (incorporated by reference, 40CFR60.17).

natural gas (40CFR60.641) means a naturally occurring mixture of hydrocarbon and nonhydrocarbon gases found in geologic formations beneath the earth's surface. The principal hydrocarbon constituent is methane.

natural gas (40CFR72.2) means a naturally occurring fluid mixture of hydrocarbons containing little or no sulfur (e.g., methane, ethane, or propane), produced in geological formations beneath the Earth's surface, and maintaining a gaseous state at standard atmospheric temperature and pressure conditions under ordinary conditions.

natural gas (40CFR80.2-tt) means a fuel whose primary constituent is methane.

natural gas liquids (40CFR60.631) means the hydrocarbons, such as ethane, propane, butane, and pentane, that are extracted from field gas (under CAA).

natural gas processing plant (gas plant) (40CFR60.631) means any processing site engaged in the extraction of natural gas liquids from field gas, fractionation of mixed natural gas liquids to natural gas products, or both.

natural resources (SF101-42USC9601) means land, fish, wildlife, biota, air, water, ground water, drinking water supplies, and other such resources belonging to, managed by, held in trust by, appertaining to, or otherwise controlled by the United States (including the resources of the fishery conservation zone established by the Fishery Conservation and Management Act of 1976), any State, local government, or any foreign government, and Indian tribe, or, if such resources are subject to a trust restriction or alienation, any member of an Indian tribe (other identical or similar definitions are provided in 40CFR300.5).

navigable waters (40CFR110.1) means the waters of the United States, including the territorial seas. The term includes:

(a) All waters that are currently used, were used in the past, or may be susceptible to use in interstate or foreign commerce, including all waters that are subject to the ebb and flow of the tide;

(b) Interstate waters, including interstate wetlands;

(c) All other waters such as intrastate lakes, rivers, streams (including intermittent streams), mudflats, sandflats, and wetlands, the use, degradation, or destruction of which would affect or could affect interstate or foreign commerce including any such waters:

(1) That are or could be used by interstate or foreign travelers for recreational or other purposes;

(2) From which fish or shellfish are or could be taken and sold in interstate or foreign commerce;

(3) That are used or could be used for industrial purposes by industries in interstate commerce;

(d) All impoundments of waters otherwise defined as navigable waters under this section;

(e) Tributaries of waters identified in paragraphs (a) through (d) of this section, including adjacent wetlands; and

(f) Wetlands adjacent to waters identified in paragraphs (a) through (e) of this section: Provided, that waste treatment systems (other than cooling ponds meeting the criteria of this paragraph) are not waters of the United States. (Other identical or similar definitions are provided in 40CFR117.1; 300.5.)

navigable waters (40CFR116.3) is defined in section 502(7) of the Act to mean "waters of the United States, including the territorial seas," and includes, but is not limited to:

(1) All waters which are presently used, or were used in the past, or may be susceptible to use as a means to transport interstate or foreign commerce, including all waters which are subject to the ebb and flow of the tide, and including adjacent wetlands; the term wetlands as used in this regulation shall include those areas that are inundated or saturated by surface or ground water at a frequency and duration sufficient to support, and that under normal circumstances do support, a prevelance of vegetation typically adapted for life in saturated soil conditions. Wetlands generally include swamps, marshes, bogs and similar areas; the term adjacent means bordering, contiguous or neighboring;

(2) Tributaries of navigable waters of the United States, including adjacent wetlands;

(3) Interstate waters, including wetlands; and

(4) All other waters of the United States such as intrastate lakes, rivers, streams, mudflats, sandflats and wetlands, the use, degradation or destruction of which affect interstate commerce including, but not limited to:

(i) Intrastate lakes, rivers, streams, and wetlands which are utilized by interstate travelers for recreational or other purposes; and

(ii) Intrastate lakes, rivers, streams, and wetlands from which fish or shellfish are or could be taken and sold in interstate commerce; and

(iii) Intrastate lakes, rivers, streams, and wetlands which are utilized for industrial purposes by industries in interstate commerce. Navigable waters do not include prior converted cropland. Notwithstanding the determination of an area's status as prior converted cropland by any other federal agency, for the purposes of the Clean Water Act, the final authority regarding Clean Water Act jurisdiction remains with EPA.

navigable waters (40CFR401.11-l) includes: all navigable waters of the United States; tributaries of navigable waters of the United States; interstate waters; intrastate lakes, rivers, and streams which are utilized by interstate travelers for recreational or other purposes; intrastate lakes, rivers, and streams from which fish or shellfish are taken and sold in interstate commerce; and intrastate lakes, rivers, and streams which are utilized for industrial purposes by industries in interstate commerce (under the Clean Water Act).

navigable waters (CWA502-33USC1362) means the waters of the United States, including the territorial seas (other identical or similar definitions are provided in OPA1001).

navigable waters (EPA-94/04): Traditionally, waters sufficiently deep and wide for navigation by all, or specified vessels; such waters in the United States come under federal jurisdiction and are protected by certain provisions of the Clean Water Act.

navigable waters (SF101) or navigable waters of the United States" means the waters of the United States, including the territorial seas.

navigable waters of the United States (40CFR112.2) means navigable waters as defined in section 502(7) of the FWPCA, and includes:

(1) All navigable waters of the United States, as defined in judicial decisions prior to passage of the 1972 Amendments to the FWPCA (Pub. L. 92-500), and tributaries of such waters;

(2) Interstate waters;

(3) Intrastate lakes, rivers, and streams which are utilized by interstate travelers for recreational or other purposes; and

(4) Intrastate lakes, rivers, and streams from which fish or shellfish are taken and sold in interstate commerce. Navigable waters do not include prior converted cropland. Notwithstanding the determination of an area's status as prior converted cropland by any other federal agency, for the purposes of the Clean Water Act, the final authority regarding Clean Water Act jurisdiction remains with EPA.

navigable waters or navigable waters of the United States (40CFR302.3) means waters of the United

States, including the territorial seas (under CERCLA).

NCA: See act or NCA.

NCN (40CFR457.31): Nitrocarbonitrate.

NCP (40CFR86.1102.87): Nonconformance penalty as described in section 206(g) of the Clean Air Act and in this subpart.

near region (40CFR60-AA(alt. method 1)) means the region of the atmospheric path along the lidar line-of-sight between the lidar's convergence distance and the plume being measured.

near the first service connection (40CFR141.2) means at one of the 20 percent of all service connections in the entire system that are nearest the water supply treatment facility, as measured by water transport time within the distribution system.

nearby (40CFR51.100-jj) as used in 40CFR51.100(ii) of this part is defined for a specific structure or terrain feature and:

(1) For purposes of applying the formulae provided in 40CFR51.100(ii)(2) means that distance up to five times the lesser of the height or the width dimension of a structure, but not greater than 0.8 km (1/2 mile), and

(2) For conducting demonstrations under 40CFR51.100(ii)(3) means not greater than 0.8 km (1/2 mile), except that the portion of a terrain feature may be considered to be nearby which falls within a distance of up to 10 times the maximum height (Ht) of the feature, not to exceed 2 miles if such feature achieves a height (Ht) 0.8 km from the stack that is at least 40 percent of the GEP stack height determined by the formulae provided in 40CFR51.100(ii)(2)(ii) of this part or 26 meters, whichever is greater, as measured from the ground-level elevation at the base of the stack. The height of the structure or terrain feature is measured from the ground-level elevation at the base of the stack.

nearest onshore area (NOA) (40CFR55.2) means, with respect to any existing or proposed OCS source located within 25 miles of a state's seaward boundary, the onshore area that is geographically closest to that source.

neat oil (40CFR467.02-o) is a pure oil with no or few impurities added. In aluminum forming its use is mostly as a lubricant.

neat oil (40CFR471.02-z) is a pure oil with no or few impurities added. In nonferrous metals forming, its use is mostly as a lubricant.

neat soap (40CFR417.11-c) shall mean the solution of completely saponified and purified soap containing about 20-30 percent water which is ready for final formulation into a finished product (other identical or similar definitions are provided in 40CFR417.31-c; 417.61-c; 417.71-c).

necessary and adequate (40CFR21.2-b), For purposes of paragraph (g)(2) of the Small Business Act, necessary and adequate refers to additions, alterations, or methods of operation in the absence of which a small business concern could not comply with one or more applicable standards. This can be determined with reference to design specifications provided by manufacturers, suppliers, or consulting engineers; including, without limitations, additions, alterations, or methods of operation the design specifications of which will provide a measure of treatment or abatement of pollution in excess of that required by the applicable standard.

necessary preconstruction approvals or permits (40CFR51.165-xvii) means those Federal air quality control laws and regulations and those air quality control laws and regulations which are part of the applicable State Implementation Plan (under the Clean Air Act).

necessary preconstruction approvals or permits (40CFR51.166-10) means those permits or approvals required under federal air quality control laws and regulations and those air quality control laws and regulations which are part of the applicable State Implementation Plan.

necessary preconstruction approvals or permits (40CFR51-AS-16) means those permits or approvals required under federal air quality control laws and regulations and those air quality control laws and regulations which are part of the applicable State Implementation Plan.

necessary preconstruction approvals or permits (40CFR52.21-10) means those permits or approvals required under federal air quality control laws and regulations and those air quality control laws and regulations which are part of the applicable State Implementation Plan.

necessary preconstruction approvals or permits (40CFR52.24-16) means those permits or approvals required under federal air quality control laws and regulations and those air quality control laws and regulations which are part of the applicable State Implementation Plan.

necessary preconstruction approvals or permits (CAA169-42USC7479) means those permits or approvals required by the permitting authority as a precondition to undertaking any activity under clauses (i) or (ii) of subparagraph (A) of this paragraph.

necrosis (EPA-94/04): Death of plant or animal cells or tissues. In plants, necrosis can discolor stems or leaves or kill a plant entirely.

negative pressure fabric filter (40CFR60.271a) means a fabric filter with the fans on the downstream side of the filter bags.

negligible residue (40CFR180.1-l) means any amount of a pesticide chemical remaining in or on a raw agricultural commodity or group of raw agricultural commodities that would result in a daily intake regarded as toxicologically insignificant on the basis of scientific judgment of adequate safety data. Ordinarily this will add to the diet an amount which will be less than 1/2,000th of the amount that has been demonstrated to have no effect from feeding studies on the most sensitive animal species tested. Such toxicity studies shall usually include at least 90-day feeding studies in two species of mammals (under FIFRA).

negotiations (EPA-94/04): (Under Superfund) After potentially responsible parties are identified for a site, EPA coordinates with them to reach a settlement that will result in the PRP paying for or conducting the cleanup under EPA supervision. If negotiations fail, EPA can order the PRP to conduct the cleanup or EPA can pay for the cleanup using Superfund monies and then sue to recover the costs.

neighboring company (40CFR60.41a-12) means any one of those electric utility companies with one or more electric power interconnections to the principal company and which have geographically adjoining service areas.

nematocide (EPA-94/04): A chemical agent which is destructive to nematodes.

nematode (FIFRA2-7USC136) means invertebrate animals of the phylum nemathelminthes and class nematoda, that is, unsegmented round worms with elongated, fusiform, or saclike bodies covered with cuticle, and inhabiting soil, water, plants, or plant parts; may also nemas or eelworms.

NEP (40CFR35.9000): See national estuary program.

NEPA (40CFR51.852): See act or NEPA.

NEPA process (40CFR1508.21) means all measures necessary for compliance with the requirements of section 2 and Title I of NEPA.

NEPA process completion (40CFR51.392), for the purposes of this subpart, with respect to FHWA or FTA, means the point at which there is a specific action to make a determination that a project is categorically excluded, to make a Finding of No Significant Impact, or to issue a record of decision on a Final Environmental Impact Statement under NEPA (other identical or similar definitions are provided in 40CFR93.101).

nephelometric (EPA-94/04): A means of measuring turbidity in a sample by passing light through a sample and measuring the amount of light deflected.

NERC region (40CFR72.2) means the North American Electric Reliability Council region or, if any, subregion.

net (40CFR409.21-b) shall mean the addition of pollutants (other identical or similar definitions are provided in 40CFR409.31-b).

net cane (40CFR409.61-c) shall mean that amount of "gross cane" less the weight of extraneous material.

net emissions increase (40CFR51.165-vi) means
(A) the amount by which the sum of the following exceeds zero:
  (1) Any increase in actual emissions from a particular physical change or change in the method of operation at a stationary source; and
  (2) Any other increases and decreases in actual emissions at the source that are contemporaneous with the particular change and are otherwise creditable.
(B) An increase or decrease in actual emissions is contemporaneous with the increase from the particular change only if it occurs before the date that the increase from the particular change occurs.
(C) An increase or decrease in actual emissions is creditable only if:
  (1) It occurs within a reasonable period to be specified by the reviewing authority; and
  (2) The reviewing authority has not relied on it in issuing a permit for the source under regulations approved pursuant to this section which permit is in effect when the increase in actual emissions from the particular change occurs.
(D) An increase in actual emissions is creditable only to the extent that the new level of actual emissions exceeds the old level.
(E) A decrease in actual emissions is creditable only to the extent that:
  (1) The old level of actual emission or the old level of allowable emissions whichever is lower, exceeds the new level of actual emissions;
  (2) It is federally enforceable at and after the time that actual construction on the particular change begins; and
  (3) The reviewing authority has not relied on it in issuing any permit under regulations approved pursuant to 40CFR part 51 subpart I or the state has not relied on it in demonstrating attainment or reasonable further progress;
  (4) It has approximately the same qualitative significance for public health and welfare as

that attributed to the increase from the particular change.
(F) An increase that results from a physical change at a source occurs when the emissions unit on which construction occurred becomes operational and begins to emit a particular pollutant. Any replacement unit that requires shakedown becomes operational only after a reasonable shakedown period, not to exceed 180 days.
(Other identical or similar definitions are provided in 40CFR51.166-3; 51-AS-6; 52.21-3; 52.24-6.)

net income neutrality (40CFR72.2) means, in the case of energy conservation measures undertaken by an investor-owned utility whose rates are regulated by a State utility regulatory authority, rates and charges established by the State utility regulatory authority that ensure that the net income earned by the utility on its State-jurisdictional equity investment will be no lower as a consequence of its expenditures on cost-effective qualified energy conservation measures and any associated lost sales than it would have been had the utility not made such expenditures, or that the State utility regulatory authority has implemented a ratemaking approach designed to meet this objective.

net system capacity (40CFR60.41a-13) means the sum of the net electric generating capability (not necessarily equal to rated capacity) of all electric generating equipment owned by an electric utility company (including steam generating units, internal combustion engines, gas turbines, nuclear units, hydroelectric units, and all other electric generating equipment) plus firm contractual purchases that are interconnected to the affected facility that has the malfunctioning flue gas desulfurization system. The electric generating capability of equipment under multiple ownership is prorated based on ownership unless the proportional entitlement to electric output is otherwise established by contractual arrangement.

net working capital (40CFR144.61-d) means current assets minus current liabilities (other identical or similar definitions are provided in 40CFR264.141-f; 265.141-f).

net worth (40CFR144.61-d) means total assets minus total liabilities and is equivalent to owner's equity

(Other identical or similar definitions are provided in 40CFR264.141-f; 265.141-f.)

netting (EPA-94/04): A concept in which all emissions sources in the same area that are owned or controlled by single company are treated as one large source, thereby allowing flexibility in controlling individual sources in order to meet a single emissions standard (cf. bubble).

neurotoxic effect (40CFR798.6400): See neurotoxicity.

neurotoxic target esterase (NTE) (40CFR798.6450-2) is a membrane-bound neural protein that hydrolyze phenyl valerate and is highly correlated with the initiation of OPIDN. NTE activity is operationally defined as the phenyl valerate hydrolytic activity resistant to paraoxon but sensitive to mipafox or neuropathic O-P ester inhibition.

neurotoxicity (40CFR798.6050-1) is any adverse effect on the structure or function of the central and/or peripheral nervous system related to exposure to a chemical substance (other identical or similar definitions are provided in 40CFR798.6200-1) (under TSCA).

neurotoxicity or a neurotoxic effect (40CFR798.6400) is any adverse change in the structure or function of the nervous system following exposure to a chemical substance.

neurotoxicity or a neurotoxic effect (40CFR798.6500-1) is an adverse change in the structure or function of the nervous system following exposure to a chemical agent. Behavioral toxicity is an adverse change in the functioning of the organism with respect to its environment following exposure to a chemical agent.

neurotoxicity or a neurotoxic effect (40CFR798.6850-1) is an adverse change in the structure or function of the nervous system following exposure to a chemical agent.

neutral sulfite semichemical pulping operation (40CFR60.281-b) means any operation in which pulp is produced from wood by cooking (digesting) wood chips in a solution of sodium sulfite and sodium bicarbonate, followed by mechanical defibrating (grinding).

neutralization (40CFR420.91-i) means those acid pickling operations that do not include acid recovery or acid regeneration processes.

neutralization (EPA-94/04): Decreasing the acidity or alkalinity of a substance by adding alkaline or acidic materials, respectively.

new (40CFR63.321) means commenced construction or reconstruction on or after December 9, 1991.

new (40CFR89.2) for the purposes of this part, means a domestic or imported nonroad engine, nonroad vehicle, or nonroad equipment the equitable or legal title to which has never been transferred to an ultimate purchaser. Where the equitable or legal title to the engine, vehicle, or equipment is not transferred to an ultimate purchaser until after the engine, vehicle or equipment is placed into service, then the engine, vehicle, or equipment will no longer be new after it is placed into service. A nonroad engine, vehicle, or equipment is placed into service when it is used for its functional purposes (under CAA).

new aircraft turbine engine (40CFR87.1) means an aircraft gas turbine engine which has never been in service.

new biochemical and microbial registration review (40CFR152.403-b) means review of an application or registration of a biochemical or microbial pesticide product containing a biochemical or microbial active ingredient not contained in any other pesticide product that is registered under FIFRA at the time the application is made. For purposes of this subpart, the definitions of biochemical and microbial pesticides contained in section 15.6 and (b) of this chapter shall apply.

new chemical (40CFR166.3-g) means an active ingredient not contained in any currently registered pesticide.

new chemical registration review (40CFR152.403-a) means review of an application for registration of a pesticide product containing a chemical active ingredient which is not contained as an active ingredient in any other pesticide product that is registered under FIFRA at the time the application is made.

new chemical substance (40CFR710.2-r) means any chemical substance which is not included in the inventory compiled and published under subsection 8(b) of the Act.

new chemical substance (40CFR720.3-v) means any chemical substance which is not included on the Inventory (other identical or similar definitions are provided in 40CFR723.50-2; 723.250-6; 747.200-1).

new chemical substance (TSCA3-15USC2602) means any chemical substance which is not included in the chemical substance list compiled and published under section 8(b) (other identical or similar definitions are provided in 40CFR723.175-5).

new class II wells (40CFR147.2902) means wells constructed or converted after the effective date of this program, or which are under construction on the effective date of this program.

new covered fleet vehicle (40CFR88.302.94) means a vehicle that has not been previously controlled by the current purchaser, regardless of the model year, except as follows: Vehicles that were manufactured before the start of the fleet program for such vehicle's weight class, vehicles transferred due to the purchase of a company not previously controlled by the purchaser or due to a consolidation of business operations, vehicles transferred as part of an employee transfer, or vehicles transferred for seasonal requirements (i.e., for less than 120 days) are not considered new. States are permitted to discontinue the use of the fourth exception for fleet operators who abuse the discretion afforded them. This definition of new covered fleet vehicle is distinct from the definition of new vehicle as it applies to manufacturer certification, including the certification of vehicles to the clean fuel standards (under the Clean Air Act).

new discharger (40CFR122.2) means any building, structure, facility, or installation:
(a) From which there is or may be a "discharge of pollutants";

(b) That did not commence the "discharge of pollutants" at a particular "site" prior to August 13, 1979;
(c) Which is not a "new source"; and
(d) Which has never received a finally effective NPDES permit for discharges at that "site."
This definition includes an "indirect discharger" which commences discharging into "waters of the United States" after August 13, 1979. It also includes any existing mobile point source (other than an offshore or coastal oil and gas exploratory drilling rig or a coastal oil and gas developmental drilling rig) such as a seafood processing rig, seafood processing vessel, or aggregate plant, that begins discharging at a "site" for which it does not have a permit; and any offshore or coastal mobile oil and gas exploratory drilling rig or coastal mobile oil and gas developmental drilling rig that commences the discharge of pollutants after August 13, 1979, at a "site" under EPA's permitting jurisdiction for which it is not covered by an individual or general permit and which is located in an area determined by the Regional Administrator in the issuance of a final permit to be an area or biological concern. In determining whether an area is an area of biological concern, the Regional Administrator shall consider the factors specified in 40CFR125.122(a) (1) through (10). An offshore or coastal mobile exploratory drilling rig or coastal mobile developmental drilling rig will be considered a "new discharger" only for the duration of its discharge in an area of biological concern.

new emission unit (40CFR63.51) means an emission unit for which construction or reconstruction is commenced after the section 112(j) deadline, or after proposal of a relevant standard under section 112(d) or section 112(h) of the Clean Air Act (as amended in 1990), whichever comes first, except that, as provided by 40CFR63.52(f)(1), an emission unit, at a major source, for which construction or reconstruction is commenced before the date upon which the area source becomes a major source, shall not be considered a new emission unit if, after the addition of such emission unit, the source is still an area source.

new facility (40CFR260.10): See new hazardous waste management facility.

new hazardous waste management facility or new

facility (40CFR260.10) means a facility which began operation, or for which construction commenced after October 21, 1976 (cf. existing hazardous waste management facility).

new HWM facility (40CFR270.2) means a Hazardous Waste Management facility which began operation or for which construction commenced after November 19, 1980.

new independent power production facility or new IPP (40CFR72.2) means a unit that:
(1) Commences commercial operation on or after November 15, 1990;
(2) Is nonrecourse project-financed, as defined in 10CFR part 715;
(3) Sells 80% of electricity generated at wholesale; and
(4) Does not sell electricity to any affiliate or, if it does, demonstrates it cannot obtain the required allowances from such an affiliate.

new injection wells (40CFR144.3) means an "injection well" which began injection after a UIC program for the State applicable to the well is approved or prescribed.

new major source (40CFR63.51) means a major source for which construction or reconstruction is commenced after the section 112(j) deadline, or after proposal of a relevant standard under section 112(d) or section 112(h) of the Clean Air Act (as amended in 1990), whichever comes first (under the Clean Air Act).

new motor vehicle (CAA216-42USC7550): Except with respect to vehicles or engines imported or offered for importation, the term "new motor vehicle" means a motor vehicle the equitable or legal title to which has never been transferred to an ultimate purchaser; and the term "new motor vehicle engine" means an engine in a new motor vehicle or a motor vehicle engine the equitable or legal title to which has never been transferred to the ultimate purchaser; and with respect to imported vehicles or engines, such terms mean a motor vehicle and engine, respectively, manufactured after the effective date of a regulation issued under section 202 which is applicable to such vehicle or engine (or which would be applicable to such vehicle or engine had it

been manufactured for importation into the United States).

new MSWLF unit (40CFR258.2) means any municipal solid waste landfill unit that has not received waste prior to October 9, 1993, or prior to October 9, 1995 if the MSWLF unit meets the conditions of 40CFR258.1(f)(1).

new OCS source (CAA328-42USC7627) means an OCS source which is a new source within the meaning of section 111(a).

new product (40CFR162.151-c) means a pesticide product which is not a federally registered product (under FIFRA).

new product (40CFR205.2-15) means:
(i) a product the equitable or legal title of which has never been transferred to an ultimate purchaser, or
(ii) a product which is imported or offered for importation into the United States and which is manufactured after the effective date of a regulation under section 6 or 8 which would have been applicable to such product had it been manufactured in the United States.

new product (NCA3-42USC4902) means:
(A) a product the equitable or legal title of which has never been transferred to an ultimate purchaser, or
(B) a product which is imported or offered for importation into the United States and which is manufactured after the effective date of a regulation under section 6 or section 8 which would have been applicable to such product had it been manufactured in the United States.

new shed (40CFR63.301) means a shed for which construction commenced after September 15, 1992. The shed at Bethlehem Steel Corporation's Bethlehem plant on Battery A is deemed not to be a new shed.

new solid waste incineration unit (CAA129.g-42USC7429) means a solid waste incineration unit the construction of which is commenced after the Administrator proposes requirements under this section establishing emissions standards or other requirements which would be applicable to such unit or a modified solid waste incineration unit.

new source (40CFR122.2) means any building, structure, facility, or installation from which there is or may be a "discharge of pollutants," the construction of which commenced:

(a) After promulgation of standards of performance under section 306 of CWA which are applicable to such source, or

(b) After proposal of standards of performance in accordance with section 306 of CWA which are applicable to such source, but only if the standards are promulgated in accordance with section 306 within 120 days of their proposal.

(Other identical or similar definitions are provided in 40CFR122.29.)

new source (40CFR129.2-h) means any source discharging a toxic pollutant, the construction of which is commenced after proposal of an effluent standard or prohibition applicable to such source if such effluent standard or prohibition is thereafter promulgated in accordance with section 307.

new source (40CFR401.11-e) means any building, structure, facility or installation from which there is or may be the discharge of pollutants, the construction of which is commenced after the publication of proposed regulations prescribing a standard of performance under section 306 of the Act which will be applicable to such source if such standard is thereafter promulgated in accordance with section 306 of the Act.

new source (40CFR403.3-k) means:

(1) building, structure, facility, or installation from which there is or may be a discharge of pollutants, the construction of which commenced after the publication of proposed Pretreatment Standards under section 307(c) of the Act which will be applicable to such source if such Standards are thereafter promulgated in accordance with that section, provided that:

  (i) The building, structure, facility or installation is constructed at a site at which no other source is located; or

  (ii) The building, structure, facility or installation totally replaces the process or production equipment that causes the

discharge of pollutants at an existing source; or

  (iii) The production or wastewater generating processes of the building, structure, facility or installation are substantially independent of an existing source at the same site. In determining whether these are substantially independent, factors such as the extent to which the new facility is integrated with the existing plant, and the extent to which the new facility is engaged in the same general type of activity as the existing source should be considered.

(2) Construction on a site at which an existing source is located results in a modification rather than a new source if the construction does not create a new building, structure, facility or installation meeting the criteria of paragraphs (k)(1)(ii), or (k)(1)(iii) of this section but otherwise alters, replaces, or adds to existing process or production equipment.

(3) Construction of a new source as defined under this paragraph has commenced if the owner or operator has:

  (i) Begun, or caused to begin as part of a continuous onsite construction program:

    (A) Any placement, assembly, or installation of facilities or equipment; or

    (B) Significant site preparation work including clearing, excavation, or removal of existing buildings, structures, or facilities which is necessary for the placement, assembly, or installation of new source facilities or equipment; or

  (ii) Entered into a binding contractual obligation for the purchase of facilities or equipment which are intended to be used in its operation within a reasonable time. Options to purchase or contracts which can be terminated or modified without substantial loss, and contracts for feasibility, engineering, and design studies do not constitute a contractual obligation under this paragraph.

new source (40CFR435.11-p) means:

(1) any facility or activity of this subcategory that meets the definition of "new source" under

40CFR122.2 and meets the criteria for determination of new sources under 40CFR122.29(b) applied consistently with all of the following definitions:

(i) The term water area as used in the term "site" in 40CFR122.29 and 122.2 shall mean the water area and ocean floor beneath any exploratory, development, or production facility where such facility is conducting its exploratory, development or production activities.

(ii) The term significant site preparation work as used in 40CFR122.29 shall mean the process of surveying, clearing or preparing an area of the ocean floor for the purpose of constructing or placing a development or production facility on or over the site.

(2) "New Source" does not include facilities covered by an existing NPDES permit immediately prior to the effective date of these guidelines pending EPA issuance of a new source NPDES permit.

new source (40CFR61.02) New source means any stationary source, the construction or modification of which is commenced after the publication in the FEDERAL REGISTER of proposed national emission standards for hazardous air pollutants which will be applicable to such source (under the Clean Air Act).

new source (40CFR63.2) means any affected source the construction or reconstruction of which is commenced after the Administrator first proposes a relevant emission standard under this part (other identical or similar definitions are provided in 40CFR63.101; 63.191).

new source (CAA111-42USC7411) New source means any stationary source, the construction or modification of which is commended after the publication of regulations (or, if earlier, proposed regulations) prescribing a standard of performance under this section which will be applicable to such source.

new source (CAA112-42USC7412) means a stationary source the construction or reconstruction of which is commenced after the Administrator first proposes regulations under this section establishing an emission standard applicable to such source.

new source (CWA306-33USC1316) means any source, the construction of which is commenced after the publication of proposed regulations prescribing a standard of performance under this section which will be applicable to such source, if such standard is thereafter promulgated in accordance with this section.

new source (EPA-94/04): Any stationary source built or modified after publication of final or proposed regulations that prescribe a given standard of performance.

new source and new discharger (40CFR122.29-1) are defined in 40CFR122.2 [see note 2.].

new source coal mine (40CFR434.11-j) means:

(1) a coal mine (excluding coal preparation plants and coal preparation plant associated areas) including an abandoned mine which is being re-mined:

(i) The construction of which is commenced after May 4, 1984; or

(ii) Which is determined by the EPA Regional Administrator to constitute a "major alteration." In making this determination, the Regional Administrator shall take into account whether one or more of the following events resulting in a new, altered or increased discharge or pollutants has occurred after May 4, 1984 in connection with the mine for which the NPDES permit is being considered:

(A) Extraction of a coal seam not previously extracted by that mine;

(B) Discharge into a drainage area not previously affected by wastewater discharge from the mine;

(C) Extensive new surface disruption at the mining operation;

(D) A construction of a new shaft, slope, or drift; and

(E) Such other factors as the Regional Administrator deems relevant.

(2) No provision in this part shall be deemed to affect the classification as a new source of a facility which was classified as a new source coal mine under previous EPA regulations, but would not be classified as a new source under this section, as modified. Nor shall any provision

in this part be deemed to affect the standards applicable to such facilities, except as provided in 40CFR434.65 of this chapter.

new source or new OCS source (40CFR55.2) shall have the meaning given in the applicable requirements of 40CFR55.13 and 55.14 of this part, except that for two years following the date of promulgation of this part, the definition given in 40CFR55.3 of this part shall apply for the purpose of determining the required date of compliance with this part.

new source performance standards (NSPS) (EPA-94/04): Uniform national EPA air emission and water effluent standards which limit the amount of pollution allowed from new sources or from modified existing sources.

new source review (NSR) (EPA-94/04): Clean Air Act requirement that requires State Implementation Plans include a permit review that applies to the construction and operation of new and modified major stationary sources in nonattainment areas to assure attainment of the national ambient air quality standards.

new tank component (40CFR260.10): See new tank system or new tank component.

new tank system (40CFR280.12) means a tank system that will be used to contain an accumulation of regulated substances and for which installation has commenced after December 22, 1988 (cf. existing tank system).

new tank system or new tank component (40CFR260.10) means a tank system or component that will be used for the storage or treatment of hazardous waste and for which installation has commenced after July 14, 1986; except, however, for purposes of 40CFR264.193(g)(2) and 265.193(g)(2), a new tank system is one for which construction commences after July 14, 1986.

new underground injection well (SDWA1424-42USC300h-3) means an underground injection well whose operation was not approved by appropriate State and Federal agencies before the date of the enactment of this title.

new unit (40CFR72.2) means a unit that commences commercial operation on or after November 15, 1990, including any such unit that serves a generator with a nameplate capacity of 25 MWe or less or that is a simple combustion turbine.

new unit (CAA402-42USC7651a) means a unit that commences commercial operation on or after the date of enactment of the Clean Air Act Amendments of 1990.

new use (40CFR152.3-p) when used with respect to a product containing a particular active ingredient, means:
(1) Any proposed use pattern that would require the establishment of, the increase in, or the exemption from the requirement of, a tolerance or food additive regulation under section 408 or 409 of the Federal Food, Drug and Cosmetic Act;
(2) Any aquatic, terrestrial, outdoor, or forestry use pattern, if no product containing the active ingredient is currently registered for that use pattern; or
(3) Any additional use pattern that would result in a significant increase in the level of exposure, or a change in the route of exposure, to the active ingredient of man or other organisms.

new use pattern registration review (40CFR152.403-c) means review of an application for registration, or for amendment of a registration entailing major change to the use pattern of an active ingredient contained in a product registered under FIFRA or pending Agency decision on a prior application at the time of application. For purposes of this paragraph, examples of major changes include but are not limited to, changes from non-food to food use, outdoor to indoor use, ground to aerial application, terrestrial to aquatic use, and nonresidential to residential use.

new uses of asbestos (40CFR763.163) means commercial uses of asbestos not identified in 40CFR763.165 the manufacture, importation or processing of which would be initiated for the first time after August 25, 1989 (under the Toxic Substances Control Act).

new vessel (40CFR140.1-e) refers to any vessel on

which construction was initiated on or after January 30, 1975.

new vessel (CWA312-33USC1322) includes every description of watercraft or other artificial contrivance used, or capable of being used, as a means of transportation on the navigable waters, the construction of which is initiated after promulgation of standards and regulation under this section.

new water (40CFR440.141-11) means water from any discrete source such as a river, creek, lake or well which is deliberately allowed or brought into the plant site.

new well (40CFR146.61) means any Class I hazardous waste injection well which is not an existing well.

newsprint (40CFR250.4-x) means paper of the type generally used in the publication of newspapers or special publications like the Congressional Record. It is made primarily from mechanical wood pulps combined with some chemical wood pulp (under RCRA).

nickel (Ni) (40CFR415.361-e) shall mean the total nickel present in the process wastewater stream exiting the wastewater treatment system (other identical or similar definitions are provided in 40CFR415.471-c; 415.651-e; 420.02-k).

ninetieth (90th) percentile (40CFR72.2) means a value that would divide an ordered set of increasing values so that at least 90 percent are less than or equal to the value and at least 10 percent are greater than or equal to the value.

ninety fifth (95th) percentile (40CFR72.2) means a value that would divide an ordered set of increasing values so that at least 95 percent of the set are less than or equal to the value and at least 5 percent are greater than or equal to the value.

ninety six (96)-hour LC50 (40CFR435.11-aa) shall refer to the concentration (parts per million) or percent of the suspended particulate phase (SPP) from a sample that is lethal to 50 percent of the test organisms exposed to that concentration of the SPP after 96 hours of constant exposure.

NIOSH (40CFR721.3) means the National Institute for Occupational Safety and Health of the U.S. Department of Health and Human Services.

NIST/EPA-approved certified reference material or NIST/EPA-approved CRM (40CFR72.2) means a calibration gas mixture that has been approved by EPA and the National Institutes of Standards and Technologies (NIST) as having specific known chemical or physical property values certified by a technically valid procedure as evidenced by a certificate or other documentation issued by a certifying standard-setting body.

nitrate (EPA-94/04): Plant nutrient and inorganic fertilizer that enters water supply sources from septic systems, animal feed lots, agricultural fertilizers, manure, industrial waste waters, sanitary landfills and garbage dumps.

nitric acid plant (40CFR51.100-cc) means any facility producing nitric acid 30 to 70 percent in strength by either the pressure or atmospheric pressure process.

nitric acid production unit (40CFR60.71-a) means any facility producing weak nitric acid by either the pressure or atmospheric pressure process (cf. weak nitric acid).

nitric oxide (NO) (EPA-94/04): A gas formed by combustion under high temperature and high pressure in an internal combustion engine; changes into nitrogen dioxide in the ambient air and contributes to photochemical smog.

nitrification (EPA-94/04): The process whereby ammonia in wastewater is oxidized to nitrite and then to nitrate by bacterial or chemical reactions.

nitrilotriacetic acid (NTA) (EPA-94/04): A compound now replacing phosphates in detergents.

nitrite (EPA-94/04):
1. An intermediate in the process of nitrification.
2. Nitrous oxide salts used in food preservation .

nitrogen dioxide (NO_2) (EPA-94/04): The result of nitric oxide combining with oxygen in the atmosphere; major component of photochemical smog.

Nitrogen oxide (NO$_x$) (EPA-94/04): Product of combustion from transportation and stationary sources and a major contributor to the formation of ozone in the troposphere and to acid deposition.

nitrogen oxides (40CFR60.2) means all oxides of nitrogen except nitrous oxide, as measured by test methods set forth in this part.

nitrogenous wastes (EPA-94/04): Animal or vegetable residues that contain significant amounts of nitrogen.

nitrophenols(EPA-94/04): Synthetic organopesticides containing carbon, hydrogen, nitrogen, and oxygen.

nitrosating agent (40CFR747.115-3) means any substance that has the potential to transfer a nitrosyl group (-NO) to a primary, secondary, or tertiary amine to form the corresponding nitrosamine (other identical or similar definitions are provided in 40CFR747.195-3; 747.200-3).

NMFS (40CFR233.2): National Marine Fisheries Service.

NMOG (CAA241-42USC7581): Nonmethane organic gas.

no adverse effect level: See no-effect level.

no detectable emission (40CFR60-AA (method 21-2.4)) means any VOC concentration at a potential leak source (adjusted for local VOC ambient concentration) that is less than a value corresponding to the instrument readability specification of section 3.1.1(c) indicates that a leak is not present (under CAA).

no detectable emissions (40CFR60.691) means less than 500 ppm above background levels, as measured by a detection instrument in accordance with Method 21 in Appendix A of 40CFR60.

no detectable emissions (40CFR61.341) means less than 500 parts per million by volume (ppmv) above background levels, as measured by a detection instrument reading in accordance with the procedures specified in 40CFR1.355(h) of this subpart.

no discharge of free oil (40CFR435.11-q) shall mean that waste streams may not be discharged when they would cause a film or sheen upon or a discoloration of the surface of the receiving water or fail the static sheen test defined in Appendix 1 to 40CFR435, subpart A.

no discharge of free oil (40CFR435.41-d) shall mean that a discharge does not cause a film or sheen upon or a discoloration on the surface of the water or adjoining shorelines or cause a sludge or emulsion to be deposited beneath the surface of the water or upon adjoining shorelines (under the Clean Water Act).

no effect level/no toxic effect level/no adverse effect level/no observed effect level (40CFR795.260) is the maximum dose used in a test which produces no observed adverse effects. A no-observed-effect level is expressed in terms of the weight of a substance given daily per unit weight of test animal (mg/kg). When administered to animals in food or drinking water, the no-observed-effect level is expressed as mg/kg of food of mg/mL of water.

no effect level/no toxic effect level/no adverse effect level/no observed effect level (40CFR798.2450-6) is the maximum dose used in a test which produces no observed adverse effects. A no-observed effect level is expressed in terms of weight or volume of test substance given daily per unit volume of air (mg/L or ppm).

no effect level/no toxic effect level/no adverse effect level/no observed effect level (40CFR798.2250-3) is the maximum dose used in a test which produces no observed adverse effects. A no-observed effect level is expressed in terms of the weight of a test substance given daily per unit weight of test animal (mg/kg).

no effect level/no toxic effect level/no adverse effect level/no observed effect level (40CFR798.2650-3) is the maximum dose used in a test which produces no-observed adverse effects. A no-observed effect level is expressed in terms of the weight of a substance given daily per unit weight of test animal (mg/kg). When administered to animals in food or drinking water the no-observed-effect level is expressed as mg/kg of food or mg/mL of water (other identical or similar definitions are provided in 40CFR798.2675-3).

no further remedial action planned (EPA-94/04): Determination made by EPA following a preliminary assessment that a site does not pose a significant risk and so requires no further activity under CERCLA.

no observable adverse effect level (NOAEL) (EPA-94/04): From long-term toxicological studies of agriculture chemical active ingredients, levels at which indicate a safe, lifetime exposure level for a given chemical. Used to establish tolerance for human diets. Also written NOEL.

no observed effect concentration (NOEC) (40CFR797.1600-11) is the highest tested concentration in an acceptable early life stage test:
(i) which did not cause the occurrence of any specified adverse effect (statistically different from the control at the 5 percent level); and
(ii) below which no tested concentration caused such an occurrence (under the Toxic Substances Control Act).

no observed effect level (40CFR798.4350-6) is the maximum concentration in a test which produces no observed adverse effects. A no-observed-effect level is expressed in terms of weight or volume of test substance given daily per unit volume of air (cf. no-effect level) (other identical or similar definitions are provided in 40CFR798.4900-3).

no reasonable alternatives (40CFR125.121-d) means:
(1) No land-based disposal sites, discharge point(s) within internal waters, or approved ocean dumping sites within a reasonable distance of the site of the proposed discharge the use of which would not cause unwarranted economic impacts on the discharger, or, notwithstanding the availability of such sites,
(2) On-site disposal is environmentally preferable to other alternative means of disposal after consideration of:
(i) The relative environmental harm of disposal on-site, in disposal sites located on land, from discharge point(s) within internal waters, or in approved ocean dumping sites, and
(ii) The risk to the environment and human

safety posed by the transportation of the pollutants.

no till (EPA-94/04): Planting crops without prior seedbed preparation, into an existing cover crop, sod, or crop residues, and eliminating subsequent tillage operations.

no toxic effect level: See no-effect level.

NO_2 (40CFR58.1-f) means nitrogen dioxide.

noble metal (EPA-94/04): Chemically inactive metal such as gold; does not corrode easily.

noise (40CFR53.23-b): Spontaneous, short duration deviations in output, about the mean output, which are not caused by input concentration changes. Noise is determined as the standard deviation about the mean and is expressed in concentration units (under CAA).

noise (EPA-94/04): Product-level or product-volume changes occurring during a test that are not related to a leak but may be mistaken for one.

Noise Control Act (NCA): See act or NCA.

noise control system (40CFR205.151-21) means any vehicle part, component or system, the purpose of which includes control or the reduction of noise emitted from a vehicle, including all exhaust system components.

noise control system (40CFR205.51-18) includes any vehicle part, component or system the primary purpose of which is to control or cause the reduction of noise emitted from a vehicle.

noise emission standard (40CFR205.151-22) means the noise levels in 40CFR205.152 or 40CFR205.166.

noise emission test (40CFR204.51-g) means a test conducted pursuant to the measurement methodology specified in 40CFR204.54 (other identical or similar definitions are provided in 40CFR205.51-19; 205.151-23).

noise reduction rating (NRR) (40CFR211.203-q) means a single number noise reduction factor in decibels, determined by an empirically derived technique which takes into account performance variation of protectors in noise reducing effectiveness due to differing noise spectra, fit variability and the mean attenuation of test stimuli at the one-third octave band test frequencies.

nominal 1-month period (40CFR60.711-15) means a calendar month or, if established prior to the performance test in a statement submitted with notification of anticipated startup pursuant to 40CFR60.7(a)(2), a similar monthly time period (e.g., 30-day month or accounting month).

nominal 1-month period (40CFR60.721) means either a calendar month, 30-day month, accounting month, or similar monthly time period that is established prior to the performance test (i.e., in a statement submitted with notification of anticipated actual startup pursuant to 40CFR60.7(2)).

nominal 1-month period (40CFR60.741) means a calendar month or, if established prior to the performance test in a statement submitted with notification of anticipated startup pursuant to 40CFR60.7(a)(2), a similar monthly time period (e.g., 30-day month or accounting month).

nominal concentration (40CFR158.153-i) means the amount of an ingredient which is expected to be present in a typical sample of a pesticide product at the time the product is produced, expressed as a percentage by weight.

nominal fuel tank capacity (40CFR86.082.2) means the volume of the fuel tank(s), specified by the manufacturer to the nearest tenth of a U.S. gallon, which may be filled with fuel from the fuel tank filler inlet.

non-agreement state (10CFR70.4): See agreement State.

non-attainment area (40CFR51.392) means any geographic region of the United States which has been designated as nonattainment under 40CFR107 of the CAA for any pollutant for which a national ambient air quality standard exists (other identical or similar definitions are provided in 40CFR93.101) (under CAA).

non-attainment area (40CFR51.491) means any area of the country designated by the EPA at 40CFR part 81 in accordance with section 107(d) of the Act as nonattainment for one or more criteria pollutants. An area could be a nonattainment area for some pollutants and an attainment area for other pollutants.

non-attainment area (40CFR52.138-6) means for the specific purpose of this regulation either the Pima County carbon monoxide nonattainment area as described in 40CFR81.303 or the Maricopa County carbon monoxide nonattainment area as described in 40CFR81.303 (i.e., the MAG urban planning area).

non-attainment area (CAA171-42USC7501) means, for any air pollutant, an area which is designated nonattainment with respect to that pollutant within the meaning of section 107(d).

non-attainment area (EPA-94/04): Area that does not meet one or more of the National Ambient Air Quality Standards for the criteria pollutants designated in the Clean Air Act.

non-attainment area (NAA) (40CFR51.852) means an area designated as nonattainment under section 107 of the Act and described in 40CFR part 81 (other identical or similar definitions are provided in 40CFR93.152).

non-automated monitoring and recording system (40CFR63.111) means manual reading of values measured by monitoring instruments and manual transcription of those values to create a record. Non-automated systems do not include strip charts.

non-binding allocations of responsibility (NBAR) (EPA-94/04): Process for EPA to propose a way for potentially responsible parties to allocate costs among themselves.

non-commercial education broadcasting entities (40CFR47.105-e) means any noncommercial educational broadcasting station (and/or its legal nonprofit affiliates) as defined and licensed by the Federal Communications Commission.

non-commercial purposes (40CFR280.12) with respect to motor fuel means not for resale.

non-commercial scientific institution (40CFR2.100-f) refers to an institution that is not operated on a commercial basis as that term is referenced in paragraph (e) of this section, and which is operated solely for the purpose of conducting scientific research the results of which are not intended to promote any particular product or industry.

non-community water system (40CFR141.2) means a public water system that is not a community water system.

non-community water system (EPA-94/04): A public water system that is not a community water system, e.g., the water supply at a camp site or national park.

non-conforming engine (40CFR85.1502-8): See non-conforming vehicle.

non-conforming nonroad engine (40CFR89.602.96): A nonroad engine which is not covered by a certificate of conformity prior to final orconditional admission (or for which such coverage has not been adequately demonstrated to EPA) and which has not been finally admitted into the United States under the provisions of 40CFR89.605-96 or 40CFR89.609-96.

non-conforming vehicle or engine (40CFR85.1502-8) is a motor vehicle or motor vehicle engine which is not covered by a certificate of conformity prior to final or conditional importation and which has not been finally admitted into the United States under the provisions of 40CFR85.1505, 40CFR85.1509, or the applicable Provisions of 40CFR85.1512. Excluded from this definition are vehicles admitted under provisions of 40CFR85.1512 covering EPA approved manufacturer and U.S. Government Agency catalyst and O_2 sensor control programs.

non-consumer article (40CFR762.3-c) means any article subject to TSCA which is not a "consumer product" within the meaning of the Consumer Product Safety Act (CPSA), 15USC2052.

non-contact cooling water (40CFR401.11-n) means water used for cooling which does not come into direct contact with any raw material, intermediate product, waste product or finished product (under CWA).

non-contact cooling water (40CFR418.21-e) shall mean water which is used in a cooling system designed so as to maintain constant separation of the cooling medium from all contact with process chemicals but which may on the occasion of corrosion, cooling system leakage or similar cooling system failures contain small amounts of process chemicals: provided, that all reasonable measures have been taken to prevent, reduce, eliminate and control to the maximum extent feasible such contamination: And provided further, that all reasonable measures have been taken that will mitigate the effects of such contamination once it has occurred (other identical or similar definitions are provided in 40CFR418.51-e).

non-contact cooling water (EPA-94/04): Water used for cooling which does not come into direct contact with any raw material, product, byproduct, or waste.

non-contact cooling water pollutants (40CFR401.11-o) means pollutants present in noncontact cooling waters.

non-contact cooling water system (40CFR60.691) means a once-through drain, collection and treatment system designed and operated for collecting cooling water which does not come into contact with hydrocarbons or oily wastewater and which is not recirculated through a cooling tower.

non-continental area (40CFR60.41a-27) means the State of Hawaii, the Virgin Islands, Guam, American Samoa, the Commonwealth of Puerto Rico, or the Northern Mariana Islands (other identical or similar definitions are provided in 40CFR60.41b; 60.41c).

non-continuous discharger (40CFR430.01-c) is a mill which is prohibited by the NPDES authority from discharging pollutants during specific periods of time for reasons other than treatment plant upset control, such periods being at least 24 hours in duration. A mill shall not be deemed a non-continuous discharger unless its permit, in addition to setting forth the prohibition described above, requires compliance with the effluent limitations established for non-continuous dischargers and also requires compliance with maximum day and average of 30 consecutive days effluent limitations. Such maximum day and average of 30 consecutive days effluent limitations for non-continuous dischargers shall be established by the NPDES authority in the form of concentrations which reflect wastewater treatment levels that are representative of the application of the best practicable control technology currently available, the best conventional pollutant control technology, or new source performance standards in lieu of the maximum day and average of 30 consecutive days effluent limitations for conventional pollutants set forth in each subpart.

non-continuous discharger (40CFR431.11-c) is a mill which is prohibited by the NPDES authority from discharging pollutants during specific periods of time for reasons other than treatment plant upset control, such periods being at least 24 hours in duration. A mill shall not be deemed a non-continuous discharger unless its permit, in addition to setting forth the prohibition described above, requires compliance with the effluent limitations established by this subpart for non-continuous discharges and also requires compliance with maximum day and average of 30 consecutive days effluent limitations. Such maximum day and average of 30 consecutive days effluent limitations for non-continuous dischargers shall be established by the NPDES authority in the form of concentrations which reflect wastewater treatment levels that are representative of the application of the best practicable control technology currently available, the best conventional pollutant control technology, or new source performance standards in lieu of the maximum day and average of 30 consecutive days effluent limitations for conventional pollutants set forth in this subpart.

non-continuous discharger (40CFR464.02-f) is a plant which does not discharge pollutants during specific periods of time for reasons other than treatment plant upset, such periods being at least 24 hours in duration. A typical example of a non-continuous discharger is a plant where wastewaters are routinely stored for periods in excess of 24 hours to be treated on a batch basis. For non-continuous discharging direct discharging plants, NPDES permit authorities shall apply the mass-based annual average effluent limitations or standards and the concentration-based maximum day and maximum for monthly average effluent limitations or standards established in the regulations. POTWs may elect to establish concentration-based standards for non-continuous discharges to POTWs. They may do so by establishing concentration-based pretreatment standards equivalent to the mass-based standards provided in 4CFR464.15, 464.16, 464.25, 464.26, 464.35, 464.36, 464.45, and 464.46 of the regulations. Equivalent concentration standards may be established by following the procedures outlined in 40CFR464.03(b).

non-conventional pollutant (EPA-94/04): Any pollutant not statutorily listed or which is poorly understood by the scientific community.

non-corrugating medium furnish subdivision mills (40CFR430.51) are mills where recycled corrugating medium is not used in the production of paperboard (cf. corrugating medium furnish subdivision mills).

non-degradation (EPA-94/04): An environmental policy which disallows any lowering of naturally occurring quality regardless of preestablished health standards.

non-discriminatory (40CFR51.491) means that a program in one State does not result in discriminatory effects on other States or sources outside the State with regard to interstate commerce.

non-domestic construction material (40CFR35.936.13-iv) means a construction material other than a domestic construction material.

non-emission related maintenance (40CFR86.084.2) means that maintenance which does not substantially affect emissions and which does not have a lasting effect on the deterioration of the vehicle or engine with respect to emissions once the maintenance is performed at any particular date.

non-emission-related maintenance (40CFR86.088.2) means that maintenance which does not substantially affect emissions and which does not have a lasting effect on the emissions deterioration of the vehicle or engine during normal in-use operation once the maintenance is performed.

non-enclosed process (40CFR721.3) means any equipment system (such as an open-top reactor, storage tank, or mixing vessel) in which a chemical substance is manufactured, processed, or otherwise used where significant direct contact of the bulk chemical substance and the workplace air may occur.

non-excessive infiltration (40CFR35.2005-28) means the quantity of flow which is less than 120 gallons per capita per day (domestic base flow and infiltration) or the quantity of infiltration which cannot be economically and effectively eliminated from a sewer system as determined in a cost-effectiveness analysis (see also 40CFR35.2005 (b)(16) and 35.2120).

non-excessive inflow (40CFR35.2005-29) means the maximum total flow rate during storm events which does not result in chronic operational problems related to hydraulic overloading of the treatment works or which does not result in a total flow of more than 25 gallons per capita per day (domestic base flow plus infiltration plus inflow). Chronic operational problems may include surcharging, backups, bypasses, and overflows (see 40CFR35.2005(b)(16) and 35.2120).

non-expendable personal property (40CFR30.200) means personal property with a useful life of at lest two years and an acquisition cost of $500 or more.

non-ferrous metal (40CFR471.02-a) is any pure metal other than iron or any metal alloy for which a metal other than iron is its major constituent in percent by weight.

non-fractionating plant (40CFR60.631) means any gas plant that does not fractionate mixed natural gas liquids into natural gas products.

non-friable (40CFR763.83) means material in a school building which when dry may not be crumbled, pulverized , or reduced to powder by hand pressure.

non-friable asbestos-containing material (40CFR61.141) means any material containing more than 1 percent asbestos as determined using the method specified in appendix A, subpart F, 40CFR763, section 1, Polarized Light Microscopy, that, when dry, cannot be crumbled, pulverized, or reduced to powder by hand pressure.

non-gaseous losses (40CFR60.601) means the solvent that is not volatilized during fiber production, and

that escapes the process and is unavailable for recovery, or is in a form or concentration unsuitable for economical recovery.

non-impervious solid surfaces (40CFR761.123) means solid surfaces which are porous and are more likely to absorb spilled PCBs prior to completion of the cleanup requirements prescribed in this policy. Non-impervious solid surfaces include, but are not limited to, wood, concrete, asphalt, and plasterboard.

non-industrial source (40CFR125.58-k) means any source of pollutants which is not an industrial source.

non-industrial use (40CFR721.3) means use other than at a facility where chemical substances or mixtures are manufactured, imported, or processed.

non-integrated refueling emission control system (40CFR86.098.2) means a system where fuel vapors from refueling are stored in a vapor storage unit assigned solely to the function of storing refueling vapors.

non-ionizing electromagnetic radiation (EPA-94/04):
1. Radiation that does not change the structure of atoms but does heat tissue and may cause harmful biological effects.
2. Microwaves, radio waves, and low-frequency electromagnetic fields from high-voltage transmission lines.

non-isolated intermediate (40CFR704.3) means any intermediate that is not intentionally removed from the equipment in which it is manufactured, including the reaction vessel in which it is manufactured, equipment which is ancillary to the reaction vessel, and any equipment through which the substance passes during a continuous flow process, but not including tanks or other vessels in which the substance is stored after its manufacture. Mechanical or gravity transfer through a closed system is not considered to be intentional removal, but storage or transfer to shipping containers "isolates" the substance by removing it from process equipment in which it is manufactured.

non-isolated intermediate (40CFR710.23-b) means any intermediate that is not intentionally removed from the equipment in which it is manufactured, including the reaction vessel in which it is manufactured, equipment which is ancillary to the reaction vessel, and any equipment through which the substance passes during a continuous flow process, but not including tanks or other vessels in which the substance is stored after its manufacture.

non-isolated intermediate (40CFR712.3-j) means any intermediate that is not intentionally removed from the equipment in which it is manufactured, including the reaction vessel in which it is manufactured, equipment which is ancillary to the reaction vessel, and any equipment through which the substance passes during a continuous flow process, but not including tanks or other vessels in which the substance is stored after its manufacture (see also paragraph (f) of this section).

non-isolated intermediate (40CFR720.3-w) means any intermediate that is not intentionally removed from the equipment in which it is manufactured, including the reaction vessel in which it is manufactured, equipment which is ancillary to the reaction vessel, and any equipment through which the chemical substance passes during a continuous flow process, but not including tanks or other vessels in which the substance is stored after its manufacture.

non-metallic mineral (40FR60.671) means any of the following minerals or any mixture of which the majority is any of the following minerals:
(a) crushed and broken stone, including limestone, dolomite, granite, traprock, sandstone, quartz, quartzite, marl, marble, slate, shale, oil shale, and shell.
(b) sand and gravel.
(c) clay including kaolin, fireclay, bentonite, fuller's earth, ball clay, and common clay.
(d) rock salt.
(e) gypsum.
(f) sodium compounds, including sodium carbonate, sodium chloride, and sodium sulfate.
(g) pumice.
(h) gilsonite.
(i) talc and pyrophyllite.
(j) boron, including borax, kernite, and colemanite.
(k) barite.
(l) fluorospar.
(m) feldspar.
(n) diatomite.
(o) perlite.
(p) vermiculite.
(q) mica.
(r) kyanite, including andalusite, sillimanite, topaz, and dumortierite.

non-metallic mineral processing plant (40CFR60.671) means any combination of equipment that is used to crush or grind any nonmetallic mineral wherever located, including lime plants, power plants, steel mills, asphalt concrete plants, portland cement plants, or any other facility processing nonmetallic minerals except as provided in 40CFR60.670 (b) and (c) (under CAA).

non-methane hydrocarbon equivalent (40CFR88.101.94) means the sum of the carbon mass emissions of non-oxygenated non-methane hydrocarbons plus the carbon mass emissions of alcohols, aldehydes, or other organic compounds which are separately measured in accordance with the applicable test procedures of 40CFR part 86, expressed as gasoline-fueled vehicle non-methane hydrocarbons. In the case of exhaust emissions, the hydrogen-to-carbon ratio of the equivalent hydrocarbon is 1.85:1. In the case of diurnal and hot soak emissions, the hydrogen-to-carbon ratios of the equivalent hydrocarbons are 2.33:1 and 2.2:1 respectively.

non-methane organic gas (40CFR88.101.94) is defined as in section 241(3) Clean Air Act as amended (42 U.S.C. 7581(3)).

non-methane organic gas (NMOG) (CAA241-42USC7581) means the sum of nonoxygenated and oxygenated hydrocarbons contained in a gas sample, including, at a minimum, all oxygenated organic gases containing 5 or fewer carbon atoms (i.e., aldehydes, ketones, alcohols, ethers, etc.), and all known alkanes, alkenes, alkynes, and aromatics containing 12 or fewer carbon atoms. To demonstrate compliance with a NMOG standard, NMOG emissions shall be measured in accordance with the "California Non-Methane Organic Gas Test Procedures." In the case of vehicles using fuels other than base gasoline, the level of NMOG emissions shall be adjusted based on the reactivity of the emissions relative to vehicles using base gasoline.

non-operational storage tank (RCRA9001-42USC6991) means any underground storage tank in which regulated substances will not be deposited or from which regulated substances will not be dispensed after the date of the enactment of the Hazardous and Solid Waste Amendments of 1984.

non-oxygenated hydrocarbon (40CFR86.090.2) means organic emissions measured by a flame ionization detector, excluding methanol (under CAA).

non-passenger automobile (40CFR600.002.85-41) means an automobile that is not a passenger automobile, as defined by the Secretary of Transportation at 49CFR523.5.

non-PCB transformer (40CFR761.3) means any transformer that contains less than 50 ppm PCB; except that any transformer that has been converted from a PCB transformer or a PCB-contaminated transformer cannot be classified as a non-PCB transformer until reclassification has occurred, in accordance with the requirements of 40CFR761.30(a)(2)(v).

non-perishable raw agricultural commodity (40CFR180.1-m) means any raw agricultural commodity not subject to rapid decay or deterioration that would render it unfit for consumption. Examples are cocoa beans, coffee beans, field-dried beans, field-dried peas, grains, and nuts. Not included are eggs, milk meat, poultry, fresh fruits, and vegetables such as onions, parsnips, potatoes, and carrots.

non-point source (40CFR35.1605-4) means pollution sources which generally are not controlled by establishing effluent limitations under sections 301, 302, and 402 of the Act. Nonpoint source pollutants are not traceable to a discrete identifiable origin, but generally result from land runoff, precipitation, drainage. or seepage.

non-point source (EPA-94/04): Diffuse pollution sources (i.e., without a single point of origin or not introduced into a receiving stream from a specific outlet). The pollutants are generally carried off the land by storm water. Common nonpoint sources are agriculture, forestry, urban, mining, construction, dams, channels, land disposal, saltwater intrusion, and city streets.

non-potable (EPA-94/04): Water that is unsafe or unpalatable to drink because in contains objectionable pollution, contamination, minerals, or infective agents.

non-procurement list (40CFR32.105-k) means the portion of the list of parties excluded from Federal Procurement or Non-procurement Programs complied, maintained and distributed by the General Services Administration (GSA) containing the names and other information about persons who have been debarred, suspended, or voluntarily excluded under Executive Order 12549 and these regulations, and those who have been determined to be ineligible.

non-profit elementary or secondary school (TSCA202-15USC2642) means any elementary or secondary school (as defined in section 198 of the Elementary and Secondary Education Act of 1965 (20USC2854)) owned and operated by one or more nonprofit corporations or associations not part of the net earnings of which inures, or may lawfully inure, to the benefit of any private shareholder or individual.

non-profit elementary or secondary school (TSCA302-15USC2662) has the meaning given such term by section 202(8).

non-recovery coke oven battery (40CFR63.301) means a source consisting of a group of ovens connected by common walls and operated as a unit, where coal undergoes destructive distillation under negative pressure to produce coke, and which is designed for the combustion of the coke oven gas from which byproducts are not recovered (under CAA).

non-regenerative carbon adsorber (40CFR61.131) means a series, over time, of non-regenerative carbon beds applied to a single source or group of sources, where non-regenerative carbon beds are carbon beds that are either never regenerated or are moved from their location for regeneration.

non-repairable (40CFR63.161) means that it is technically infeasible to repair a piece of equipment

443

from which a leak has been detected without a process unit shutdown.

non-restricted access areas (40CFR761.123) means any area other than restricted access, outdoor electrical substations, and other restricted access locations, as defined in this section. In addition to residential/commercial areas, these areas include unrestricted access rural areas (areas of low density development and population where access is uncontrolled by either man-made barriers or naturally occurring barriers, such as rough terrain, mountains, or cliffs).

non-road compression-ignition engine (40CFR89.2) means a nonroad engine which utilizes the compression-ignition combustion cycle.

non-road emissions (EPA-94/04): Pollutants emitted by combustion engines on farm and construction equipment, gasoline-powered lawn and garden equipment, and power boats and outboard motors (under CAA).

non-road engine (40CFR89.2) means:
(1) Except as discussed in paragraph (2) of this definition, a nonroad engine is any internal combustion engine:
   (i)   in or on a piece of equipment that is self-propelled or serves a dual purpose by both propelling itself and performing another function (such as garden tractors, off-highway mobile cranes and bulldozers); or
   (ii) in or on a piece of equipment that is intended to be propelled while performing its function (such as lawnmowers and string trimmers); or
   (iii) that, by itself or in or on a piece of equipment, is portable or transportable, meaning designed to be and capable of being carried or moved from one location to another. Indicia of transportability include, but are not limited to, wheels, skids, carrying handles, dolly, trailer, or platform.
(2) An internal combustion engine is not a nonroad engine if:
   (i)   the engine is used to propel a motor vehicle or a vehicle used solely for competition, or is subject to standards promulgated under section 202 of the Act; or
   (ii) the engine is regulated by a federal New Source Performance Standard promulgated under section 111 of the Act; or
   (iii) the engine otherwise included in paragraph (1)(iii) of this definition remains or will remain at a location for more than 12 consecutive months or a shorter period of time for an engine located at a seasonal source. A location is any single site at a building, structure, facility, or installation. Any engine (or engines) that replaces an engine at a location and that is intended to perform the same or similar function as the engine replaced will be included in calculating the consecutive time period. An engine located at a seasonal source is an engine that remains at a seasonal source during the full annual operating period of the seasonal source. A seasonal source is a stationary source that remains in a single location on a permanent basis (i.e., at least two years) and that operates at that single location approximately three months (or more) each year. This paragraph does not apply to an engine after the engine is removed from the location.

non-road engine (CAA216-42USC7550) means an internal combustion engine (including the fuel system) that is not used in a motor vehicle or a vehicle used solely for competition, or that is not subject to standards promulgated under section 111 or section 202.

non-road equipment (40CFR89.2) means equipment that is powered by nonroad engines.

non-road vehicle (40CFR89.2) means a vehicle that is powered by a nonroad engine as defined in this section and that is not a motor vehicle or a vehicle used solely for competition.

non-road vehicle (CAA216-42USC7550) means a vehicle that is powered by a nonroad engine and that is not a motor vehicle or a vehicle used solely for competition.

non-road vehicle or nonroad equipment

manufacturer (40CFR89.2) means any person engaged in the manufacturing or assembling of new nonroad vehicles or equipment or importing such vehicles or equipment for resale, or who acts for and is under the control of any such person in connection with the distribution of such vehicles or equipment. A nonroad vehicle or equipment manufacturer does not include any dealer with respect to new nonroad vehicles or equipment received by such person in commerce.

non-scheduled renovation operation (40CFR61.141) means a renovation operation necessitated by the routine failure of equipment, which is expected to occur within a given period based on past operating experience, but for which an exact date cannot be predicted.

non-sudden accidental occurrence (40CFR264.141-g) means an occurrence which takes place over time and involves continuous or repeated exposure (other identical or similar definitions are provided in 40CFR265.141-g).

non-target organism (40CFR171.2-17) means a plant or animal other than the one against which the pesticide is applied.

non-transient non-community water system (EPA-94/04): A public water system that regularly serves at least 25 of the same nonresident persons per day for more than six months per year.

non-transient non-community water system or NTNCWS (40CFR141.2) means a public water system that is not a community water system and that regularly serves at least 25 of the same persons over 6 months per year (under the Safe Drinking Water Act).

non-transportation-related onshore and offshore facilities (40CFR112-AA) means:
(A) Mixed onshore and offshore oil well drilling facilities including all equipment and appurtenances related thereto used in drilling operations for exploratory or development wells, but excluding any terminal facility, unit or process integrally associated with the handling or transferring of oil in bulk to or from a vessel.
(B) Mobile onshore and offshore oil well drilling platforms, barges, trucks, or other mobile facilities including all equipment and appurtenances related thereto when such mobile facilities are fixed in position for the purpose of drilling operations for exploratory or development wells, but excluding any terminal facility, unit or process integrally associated with the handling or transferring of oil in bulk to or from a vessel.
(C) Fixed onshore and offshore oil production structures, platforms, derricks, and rigs including all equipment and appurtenances related thereto, as well as completed wells and the wellhead separators, oil separators, and storage facilities used in the production of oil, but excluding any terminal facility, unit or process integrally associated with the handling or transferring of oil in bulk to or from a vessel.
(D) Mobile onshore and offshore oil production facilities including all equipment and appurtenances related thereto as well as completed wells and wellhead equipment, piping from wellheads to oil separators, oil separators, and storage facilities used in the production of oil when such mobile facilities are fixed in position for the purpose of oil production operations, but excluding any terminal facility, unit or process integrally associated with the handling or transferring of oil in bulk to or from a vessel.
(E) Oil refining facilities including all equipment and appurtenances related thereto as well as in-plant processing units, storage units, piping, drainage systems and waste treatment units used in the refining of oil, but excluding any terminal facility, unit or process integrally associated with the handling or transferring of oil in bulk to or from a vessel.
(F) Oil storage facilities including all equipment and appurtenances related thereto as well as fixed bulk plant storage, terminal oil storage facilities, consumer storage, pumps and drainage systems used in the storage of oil, but excluding inline or breakout storage tanks needed for the continuous operation of a pipeline system and any terminal facility, unit or process integrally associated with the handling or transferring of oil in bulk to or from a vessel.
(G) Industrial, commercial, agricultural or public facilities which use and store oil, but excluding

any terminal facility, unit or process integrally associated with the handling or transferring of oil in bulk to or from a vessel.

(H) Waste treatment facilities including in-plant pipelines, effluent discharge lines, and storage tanks, but excluding waste treatment facilities located on vessels and terminal storage tanks and appurtenances for the reception of oily ballast water or tank washings from vessels and associated systems used for off-loading vessels.

(I) Loading racks, transfer hoses, loading arms and other equipment which are appurtenant to a nontransportation-related facility or terminal facility and which are used to transfer oil in bulk to or from highway vehicles or railroad cars.

(J) Highway vehicles and railroad cars which are used for the transport of oil exclusively within the confines of a nontransportation-related facility and which are not intended to transport oil in interstate or intrastate commerce.

(K) Pipeline systems which are used for the transport of oil exclusively within the confines of a nontransportation-related facility or terminal facility and which are not intended to transport oil in interstate or intrastate commerce, but excluding pipeline systems used to transfer oil in bulk to or from a vessel.

non-utility unit (CAA402-42USC7651a) means a unit other than a utility unit.

non-vapor tight (40CFR61.301) means any tank truck, railcar, or marine vessel that does not pass the required vapor-tightness test.

non-wastewaters (40CFR268.2-d) are wastes that do not meet the criteria for wastewaters in paragraph (f) of this section.

noncompliance coal (EPA-94/04): Any coal, when burned, that emits greater than 3.0 pounds of sulfur dioxide per million Btu. Also known as high sulfur coal.

nondischarging treatment plant (EPA-94/04): A treatment plant that does not discharge treated wastewater into any stream or river. Most are pond systems that dispose of the total flow they receive by means of evaporation or percolation to groundwater, or facilities that dispose of their effluent by recycling

or reuse (e.g., spray irrigation or groundwater discharge).

nonfriable asbestos-containing materials (EPA-94/04): Any material containing more than one percent asbestos (as determined by Polarized Light Microscopy) that, when dry, cannot be crumbled, pulverized, or reduced to powder by hand pressure.

normal ambient value (40CFR228.2-h) means that concentration of a chemical species reasonably anticipated to be present in the water column, sediments, or biota in the absence of disposal activities at the disposal site in question (under the Marine Protection, Research, and Sanctuaries Act).

normal liquid detergent operations (40CFR417.161-d) shall mean all such operations except those defined as fast turnaround operation of automated fill lines.

normal operation (40CFR417.151-d) of a spray tower shall mean operation utilizing formulations that present limited air quality problems from stack gases and associated need for extensive wet scrubbing, and without more than 6 turnarounds in a 30 consecutive day period, thus permitting essentially complete recycle of waste water.

normal range of a release (40CFR302.8) is all releases (in pounds or kilograms) of a hazardous substance reported or occurring over any 24-hour period under normal operating conditions during the preceding year. Only releases that are both continuous and stable in quantity and rate may be included in the normal range.

normally containing a quantity of refrigerant (40CFR82.152-m) means containing the quantity of refrigerant within the appliance or appliance component when the appliance is operating with a full charge of refrigerant.

NOS (40CFR261/App8): Not otherwise specified.

not classified area (40CFR51.392) means any carbon monoxide nonattainment area which EPA has not classified as either moderate or serious (other identical or similar definitions are provided in 40CFR93.101).

not feasible to prescribe or enforce a numerical emission limitation (40CFR63.51) means a situation in which the Administrator or a State determines that a pollutant (or stream of pollutants) listed pursuant to section 112(b) of the Act cannot be emitted through a conveyance designed and constructed to emit or capture such pollutant, or that any requirement for, or use of, such a conveyance would be inconsistent with any Federal law; or the application of measurement technology to a particular source is not practicable due to technological or economic limitations (under the Clean Air Act).

not-for-profit organization (40CFR47.105-d) means an organization, association, or institution described in section 501(c)(3) of the Internal Revenue Code of 1986, which is exempt from taxation pursuant to the provisions of section 501(a) of such Code.

note (40CFR39.105-g) means an evidence of the debt, including a bond, obligation to pay, or other evidence of indebtedness where appropriate.

notice (40CFR32.105-l) means a written communication served in person or sent by certified mail, return receipt requested, or its equivalent, to the last known address of a party, its identified counsel, its agent for service of process, or any partner, officer, director, owner, or joint venturer of the party. Notice, if undeliverable, shall be considered to have been received by the addressee five days after being properly sent to the last address known by the agency.

notice (40CFR8.33-g) means a notice of hearing.

notice of deficiency (EPA-94/04): An EPA request to a facility owner or operator requesting additional information before a preliminary decision on a permit application can be made.

notice of intent (40CFR1508.22) means a notice that an environmental impact statement will be prepared and considered. The notice shall briefly:

(a) Describe the proposed action and possible alternatives.
(b) Describe the agency's proposed scoping process including whether, when, and where any scoping meeting will be held.

(c) State the name and address of person within the agency who can answer questions about the proposed action and the environmental impact statement.

notice of intent to deny (EPA-94/04): Notification by EPA of its preliminary intent to deny a permit application.

notice of intent to rescind (40CFR173.2-b) means a notice to a State issued under 40CFR173.3 which initiates a proceeding to rescind the State's primary enforcement responsibility for pesticide use violations.

NO_x (CAA302) means oxides of nitrogen.

NPDES (40CFR110.1; 122.2; 124.2; 133.101-i; 272.2) means National Pollutant Discharge Elimination System.

NPDES permit or permit (40CFR403.3-l) means a permit issued to a POTW pursuant to section 402 of the Act.

NPDES State (40CFR403.3-m) means a State (as defined in 40CFR122.2) or Interstate water pollution control agency with an NPDES permit program approved pursuant to section 402(b) of the Act (CWA).

NRC-licensed facility (40CFR61.101-e) means any facility licensed by the Nuclear Regulatory Commission or any Agreement State to receive title to, receive, possess, use, transfer, or deliver any source, byproduct, or special nuclear material (under CAA).

NSD (40CFR763-AA-15) means no structure detected.

NSO (40CFR57): Non-ferrous smelter order.

NSPS (40CFR467.02-w) means new source performance standards under section 306 of the Act (CWA).

NTNCWS (40CFR141.2): Non-transient non-community water system (under the Safe Drinking Water Act).

NTU (40CFR131.35.d-19): Nephelometric turbidity units.

nuclear fuel cycle (40CFR190.02-a) means the operations defined to be associated with the production of electrical power for public use by any fuel cycle through utilization of nuclear energy (under AEA).

nuclear reactors and support facilities (EPA-94/04): Uranium mills, commercial power reactors, fuel reprocessing plants, and uranium enrichment facilities.

nuclear winter (EPA-94/04): Prediction by some scientists that smoke and debris rising from massive fires of a nuclear war could block sunlight for weeks or months, cooling the earth's surface and producing climate changes that could, for example, negatively effect world agricultural and weather patterns.

nuclide (EPA-94/04): A species of atom characterized by the number of protons, neutrons, and energy in the nucleus.

number average molecular weight (40CFR723.250-9) means the arithmetic average (mean) of the molecular weight of all molecules in a polymer (under TSCA).

nursery (40CFR170.3) means any operation engaged in the outdoor production of any agricultural plant to produce cut flowers and ferns or plants that will be used in their entirety in another location. Such plants include, but are not limited to, flowering and foliage plants or trees; tree seedlings; live Christmas trees; vegetable, fruit, and ornamental transplants; and turfgrass produced for sod (under the Federal Insecticide, Fungicide, and Rodenticide Act).

nutrient (EPA-94/04): Any substance assimilated by living things that promotes growth. The term is generally applied to nitrogen and phosphorus in wastewater, but is also applied to other essential and trace elements.

nutrient pollution (EPA-94/04): Contamination of water resources by excessive inputs of nutrients. In surface waters, excess algal production is a major concern.

NWPA (40CFR191.02-e): Nuclear Waste Policy Act of 1982 (Pub. L. 97-425).

*********** OOOOO ***********

O_3 (40CFR58.1-h) means ozone.

O&G (40CFR420.02-b): See oil and grease.

objective evidence of an emission related repair (40CFR85.2102-15) means all diagnostic information and data, the actual parts replaced during repair, and any other information directly used to support a warranty claim, or to support denial of such a claim.

obligations (40CFR31.3) means the amounts of orders placed, contracts and subgrants awarded, goods and services received, and similar transactions during a given period that will require payment by the grantee during the same or a future period.

observation period (40CFR60-AA (method 22-3.5)) means accumulated time period during which observations are conducted, not to be less than the period specified in the applicable regulation.

observation period (40CFR797.2175-4) is the portion of the test that begins after the test birds have been dosed and extends at least 14 days.

observed effect concentration (OEC) (40CFR797.1600-12) is the lowest tested concentration in an acceptable early life stage test:
(i) which caused the occurrence of any specified adverse effect (statistically different from the control at the 5 percent level); and
(ii) above which all tested concentrations caused such an occurrence.

occupational dose (10CFR20.3) includes exposure of an individual to radiation:
(i) in a restricted area; or

(ii) in the course of employment in which the individual's duties involve exposure to radiation, provided, that "occupational dose" shall not be deemed to include any exposure of an individual to radiation for the purpose of medical diagnosis or medical therapy of such individual.

occurrence (40CFR280.92) means an accident, including continuous or repeated exposure to conditions, which results in a release from an underground storage tank.

ocean discharge waiver (EPA-94/04): A variance from Clean Water Act requirements for discharges into marine waters.

ocean dumping (40CFR165.1-m) means the disposal of pesticides in or on the oceans and seas, as defined in Pub. L. 92-532.

ocean waters (40CFR125.58-l, see also 40CFR220.2-c) means those coastal waters landward of the baseline of the territorial seas, the deep waters of the territorial seas, or the waters of the contiguous zone.

OCS source (40CFR55.2) means any equipment, activity, or facility which:
(1) Emits or has the potential to emit any air pollutant;
(2) Is regulated or authorized under the Outer Continental Shelf Lands Act (OCSLA) (43 U.S.C. 40CFR1331 et seq.); and
(3) Is located on the OCS or in or on waters above the OCS.
This definition shall include vessels only when they are:
(1) Permanently or temporarily attached to the seabed and erected thereon and used for the purpose of exploring, developing or producing resources therefrom, within the meaning of section 4(a)(1) of OCSLA (43 U.S.C. 40CFR1331 et seq.); or
(2) Physically attached to an OCS facility, in which case only the stationary sources aspects of the vessels will be regulated.

octanol/water partition coefficient (K_{ow}) (40CFR796.1550) is defined as the equilibrium ratio of the molar concentrations of a chemical in n-octanol and water, in dilute solution. K_{ow} is a constant for a given chemical at a given temperature. Since K_{ow} is the ratio of two molar concentrations, it is a dimensionless quantity. Sometimes K_{ow} is reported as log 10 K(ow). The mathematical statement of K_{ow} is:
- Equation 1: $K_{ow} = (C_{octanol})/(C_{water})$, where C is the molar concentration of the solute in n-octanol and water at equilibrium at a given temperature.

(Other identical or similar definitions are provided in 40CFR796.1570; 796.1720-i.)

octanol water partition coefficient $(K_{ow}$ **[or P])** (40CFR300-AA) means the measure of the extent of partitioning of a substance between water and octanol at equilibrium. The K_{ow} is determined by the ratio between the concentration in octanol divided by the concentration in water at equilibrium [unitless].

octave band attenuation (40CFR211.203-r) means the amount of sound reduction determined according to the measurement procedure of 40CFR21.20 for one-third octave bands of noise.

odor threshold (EPA-94/04): The minimum odor of a water sample that can just be detected after successive dilutions with odorless water. Also called threshold odor.

off-kg (off-lb) (40CFR471.02-vv) means the mass of metal or metal alloy removed from a forming operation at the end of a process cycle for transfer to a different machine or process.

off-kilogram (off-pound) (40CFR467.02-aa) shall mean the mass of aluminum or aluminum alloy removed from a forming or ancillary operation at the end of a process cycle for transfer to a different machine or process.

off-kilogram (off-pound) (40CFR468.02-n) shall mean the mass of copper or copper alloy removed from a forming or ancillary operation at the end of a process cycle for transfer to a different machine or process.

off-lb (40CFR471.02-vv): See off-kg.

off-pound (40CFR467.02-aa; 468.02-n): See off-kilogram.

off-road motorcycle (40CFR205.151-4) means any motorcycle that is not a street motorcycle or competition motorcycle.

off-site (40CFR270.2) means any site which is not on-site.

off-site facility (EPA-94/04): A hazardous waste treatment, storage or disposal area that is located away from the generating site.

off-take system (40CFR63.301) means any individual oven apparatus that is stationary and provides a passage for gases from an oven to a coke oven battery collecting main or to another oven. Off-take system components include the standpipe and standpipe caps, goosenecks, stationary jumper pipes, mini-standpipes, and standpipe and gooseneck connections.

office (SMCRA701-30USC1291) means the Office of Surface Mining Reclamation and Enforcement established pursuant to subchapter II of this chapter.

Office Director (40CFR85.1402) means the Director for the Office of Mobile Sources-Office of Air and Radiation of the Environmental Protection Agency or an authorized representative of the Office Director (other identical or similar definitions are provided in 40CFR85.2102-2).

Office Director–approved emissions test or emissions short test (40CFR85.2102-5) means any test prescribed under 40CFR85.2201 et seq., and meeting all of the requirements thereunder.

office of civil rights (40CFR8.33-h) means the Office of Civil Rights and Urban Affairs in the Agency.

office of civil rights (OCR) (40CFR7.25) means the Director of the Office of Civil Rights, EPA Headquarters or his/her designated representative.

office of federal contract compliance (40CFR8.33-i) means the Office of Federal Contract Compliance, U.S. Department of Labor.

Office of the Administrator (40CFR179.3) means the Agency's Administrator and Deputy Administrator and their immediate staff, including the judicial officer (under the Federal Insecticide, Fungicide, and Rodenticide Act).

Office of the Assistant Administrator for Enforcement and General Counsel (40CFR8.33-j) means the Office of the Assistant Administrator for Enforcement and General Counsel in the Agency.

office papers (40CFR250.4-y) means note pads, loose-leaf fillers, tablets, and other papers commonly used in offices, but not defined elsewhere (under RCRA).

officer or employee of an agency (40CFR34.105-k) includes the following individuals who are employed by an agency:
(1) An individual who is appointed to a position in the government under title 5, U.S. Code, including a position under a temporary appointment;
(2) A member of the uniformed services as defined in section 101(3), title 37, U.S. Code;
(3) A special government employee as defined in section 202, title 18 U.S. Code; and,
(4) an individual who is a member of a Federal advisory committee, as defined by the Federal Advisory committee Act, title 5, U.S. Code appendix 2.

offset (40CFR52.741) means, with respect to printing and publishing operations, use of a blanket cylinder to transfer ink from the plate cylinder to the surface to be printed.

offset plan (40CFR72.2) means a plan pursuant to part 77 of this chapter for offsetting excess emissions of sulfur dioxide that have occurred at an affected unit in any calendar year.

offset printing paper (40CFR250.4-z) means an uncoated or coated paper designed for offset lithography.

offset ruling: See 40CFR52.24(c)[aaa]

offsets (EPA-94/04): A concept whereby emissions from a proposed new or modified stationary source are balanced by reductions from existing sources to stabilize total emissions (cf. bubble, emissions trading or netting).

offshore facility (40CFR110.1) means any facility of any kind located in, on, or under any of the navigable waters of the United States, and any facility of any kind that is subject to the jurisdiction of the United States and is located in, on, or under any other waters, other than a vessel or a public vessel (under FWPCA).

offshore facility (40CFR112.2) means any facility of any kind located in, on, or under any of the navigable waters of the United States, which is not a transportation-related facility.

offshore facility (40CFR116.3) means any facility of any kind located in, on, or under, any of the navigable waters of the United States, and any facility of any kind which is subject to the jurisdiction of the United States and is located in, on, or under any other waters, other than a vessel or a public vessel.

offshore facility (40CFR300.5), as defined by section 101(17) of CERCLA and section 311(a)(11) of the CWA, means any facility of any kind located in, on, or under any of the navigable waters of the United States and any facility of any kind which is subject to the jurisdiction of the United States and is located in, on, or under any other waters, other than a vessel or a public vessel.

offshore facility (40CFR302.3) means any facility of any kind located in, on, or under, any of the navigable waters of the United States, and any facility of any kind which is subject to the jurisdiction of the United States and is located in, on, or under any other waters, other than a vessel or a public vessel.

offshore facility (CWA311-33USC1321) means any facility of any kind located in, on, or under, any of the navigable waters of the United States, and any facility of any kind which is subject to the jurisdiction of the United States and is located in, on, or under any other waters, other than a vessel or a public vessel (other identical or similar definitions are provided in OPA1001).

offshore facility (SF101-42USC9601) means any facility of any kind located in, on, or under, any of the navigable water of the United States, and any facility of any kind which is subject to the jurisdiction of the United Sates and is located in, on, or under

any other waters, other than a vessel or a public vessel.

offshore platform gas turbines (40CFR60.331-n) means any stationary gas turbine located on a platform in an ocean.

offstream uses (EPA-94/04): Water withdrawn from surface or groundwater sources for use at another place.

oil (40CFR109.2) means oil of any kind or in any form, including, but not limited to, petroleum, fuel oil, sludge, oil refuse, and oil mixed with wastes other than dredged spoil.

oil (40CFR110.1), when used in relation to section 311 of the Act, means oil of any kind or in any form, including, but not limited to, petroleum, fuel oil, sludge, oil refuse, and oil mixed with wastes other than dredged spoil. "Oil," when used in relation to section 18(m)(3) of the Deepwater Port Act of 1974, has the meaning provided in section 3(14) of the Deepwater Port Act of 1974.

oil (40CFR112.2) means oil of any kind or in any form, including, but not limited to, petroleum, fuel oil, sludge, oil refuse and oil mixed with wastes other than dredged spoil (other identical or similar definitions are provided in 40CFR113.3-h).

oil (40CFR300.5), as defined by section 311(a)(1) of the CWA, means oil of any kind or in any form, including, but not limited to, petroleum, fuel oil, sludge, oil refuse, and oil mixed with wastes other than dredged spoil.

oil (40CFR426.81-c) shall mean those components of a waste water amenable to measurement by the technique or techniques described in the most recent addition of "Standard Methods" for the analysis of grease in polluted waters, waste waters, and effluents, such as "Standard Methods," 13th Edition, 2nd Printing, page 407 (other identical or similar definitions are provided in 40CFR426.111-c; 426.121-c).

oil (40CFR60.41b) means crude oil or petroleum or a liquid fuel derived from crude oil or petroleum, including distillate and residual oil (other identical or similar definitions are provided in 40CFR60.41c) (under CAA).

oil (CWA311-33USC1321) means oil of any kind or in any form, including, but not limited to, petroleum, fuel oil, sludge, oil refuse, and oil mixed with wastes other than dredged spoil.

oil (OPA1001) means oil of any kind or in any form, including, but not limited to, petroleum, fuel oil, sludge, oil refuse, and oil mixed with wastes other than dredged spoil, but does not include petroleum, including crude oil or any fraction thereof, which is specifically listed or designated as a hazardous substance under subparagraphs (A) through (F) of section 101(14) of the Comprehensive Environmental Response, Compensation, and Liability Act (42USC9601) and which is subject to the provisions of that Act.

oil and grease (40CFR408.11) shall mean those components of a waste water amenable to measurement by the method described in Methods for Chemical Analysis of Water and Wastes, 1971, Environmental Protection Agency, Analytical Quality Control Laboratory, page 217 (other identical or similar definitions are provided in 40CFR408.21; 408.31; 408.41; 408.51; 408.61; 408.71; 408.81; 408.91; 408.101; 408.111; 408.121; 408.131; 408.141).

oil and grease (40CFR410.11-b) shall mean total recoverable oil and grease as measured by the procedure listed in 40CFR136.

oil and grease (40CFR432.11-f) shall mean those components of process waste water amenable to measurement by the method described in "Methods for Chemical Analysis of Water and Wastes," 1971, EPA, Analytical Quality Control Laboratory, page 217 (other identical or similar definitions are provided in 40CFR432.21-f; 432.31-f; 432.41-f) (under CWA).

oil and grease (or O&G) (40CFR420.02-b) means the value obtained by the method specified in 40CFR136.3.

oil desulfurization (EPA-94/04): Widely used precombustion method for reducing sulfur dioxide emissions from oil-burning power plants. The oil is

treated with hydrogen. which removes some of the sulfur by forming hydrogen sulfide gas.

oil fingerprinting (EPA-94/04): A method that identifies sources of oil and allows spills to be traced to their source.

oil fired (40CFR72.2) means the combustion of: fuel oil for more than 10 percent of the average annual heat input during the previous three calendar years or for more than 15 percent of the annual heat input in any one of those calendar years; and any solid, liquid, or gaseous fuel, other than coal or any other coal-derived fuel (except a coal-derived gaseous fuel with a sulfur content no greater than natural gas), for the remaining heat input, if any; provided that for purposes of the monitoring exceptions of part 75 of this chapter, the supplemental fuel used in addition to fuel oil, if any, shall be limited to gaseous fuels, other than a coal-derived fuel.

oil pollution fund (40CFR300.5) means the fund established by section 311(k) of the CWA.

oil spill (EPA-94/04): An accidental or intentional discharge of oil which reaches bodies of water. Can be controlled by chemical dispersion, combustion, mechanical containment, and/or adsorption. Spills from tanks and pipelines can also occur away from water bodies, contaminating the soil, getting into sewer systems and threatening underground water sources.

oil spill removal organization (40CFR112.2) means an entity that provides oil spill response resources, and includes any for-profit or not-for-profit contractor, cooperative, or in-house response resources that have been established in a geographic area to provide required response resources (under FWPCA).

oil water separator (40CFR60.691) means wastewater treatment equipment used to separate oil from water consisting of a separation tank, which also includes the forebay and other separator basins, skimmers, weirs, grit chambers, and sludge hoppers. Slop oil facilities, including tanks, are included in this term along with storage vessels and auxiliary equipment located between individual drain systems and the oil-water separator. This term does not include storage vessels or auxiliary equipment which do not come in contact with or store oily wastewater (under CAA).

oil water separator (40CFR61.341) means a waste management unit, generally a tank or surface impoundment, used to separate oil from water. An oil-water separator consists of not only the separation unit but also the forebay and other separator basins, skimmers, weirs, grit chambers, sludge hoppers, and bar screens that are located directly after the individual drain system and prior to additional treatment units such as an air flotation unit, clarifier, or biological treatment unit. Examples of an oil-water separator include an API separator, parallel-plate interceptor, and corrugated-plate interceptor with the associated ancillary equipment.

oil water separator or organic-water separator (40CFR63.111) means a waste management unit, generally a tank used to separate oil or organics from water. An oil-water or organic-water separator consists of not only the separation unit but also the forebay and other separator basins, skimmers, weirs, grit chambers, sludge hoppers, and bar screens that are located directly after the individual drain system and prior to additional treatment units such as an air flotation unit, clarifier, or biological treatment unit. Examples of an oil-water or organic-water separator include, but are not limited to, an American Petroleum Institute separator, parallel-plate interceptor, and corrugated-plate interceptor with the associated ancillary equipment.

oily wastewater (40CFR60.691) means wastewater generated during the refinery process which contains oil, emulsified oil, or other hydrocarbons. Oily wastewater originates from a variety of refinery processes including cooling water, condensed stripping steam, tank draw-off, and contact process water.

old chemical registration review (40CFR152.403-d) means review of an application for registration of a new product containing active ingredients and uses which are substantially similar or identical to those currently registered or for which an application is pending Agency decision.

oligotrophic lakes (EPA-94/04): Deep clear lakes with few nutrients, little organic matter and a high dissolved-oxygen level.

olives (40CFR407.61-k) shall mean the processing of olives into the following product styles: Canned, all varieties, fresh and stored, green ripe, black ripe, spanish, sicilian, and any other styles to which spices, acids, and flavorings may have been added.

on premise sales (40CFR244.101-i) means sales transactions in which beverages are purchased by a consumer for immediate consumption within the area under control of the dealer.

on scene coordinator (OSC) (40CFR113.3-g) is the single Federal representative designated pursuant to the National Oil and Hazardous Substances Pollution Contingency Plan and identified in approved Regional Oil and Hazardous Substances Pollution Contingency Plans.

on scene coordinator (OSC) (40CFR300.5) means the federal official predesignated by EPA or the USCG to coordinate and direct federal responses under subpart D, or the official designated by the lead agency to coordinate and direct removal actions under subpart E of the NCP.

on scene coordinator (OSC) (EPA-94/04): The predesignated EPA, Coast Guard, or Department of Defense official who coordinates and directs Superfund removal actions or Clean Water Act oil-or hazardous-spill response actions.

on site (40CFR260.10) means the same or geographically contiguous property which may be divided by public or private right-of-way, provided the entrance and exit between the properties is at a cross-roads intersection, and access is by crossing as opposed to going along, the right-of-way. Non-contiguous properties owned by the same person but connected by a right-of-way which he controls and to which the public does not have access, is also considered on-site property.

on site (40CFR270.2) means on the same or geographically contiguous property which may be divided by public or private right(s)-of-way, provided the entrance and exit between the properties is at a cross-roads intersection, and access is by crossing as opposed to going along, the right(s)-of-way. Non-contiguous properties owned by the same person but connected by a right-of-way which the person controls and to which the public does not have access, is also considered on-site property.

on site (40CFR300.5) means the areal extent of contamination and all suitable areas in very close proximity to the contamination necessary for implementation of the response action.

on site (40CFR761.3) means within the boundaries of a contiguous property unit.

on site facility (EPA-94/04): A hazardous waste treatment, storage or disposal area that is located on the generating site.

on street parking (40CFR52.2051-a) means stopping a motor vehicle on any street, highway, or roadway (except for legal stops at or before intersections and as caution and safety requie) whether or not a person remains in the vehicle. Stopping for the purpose of effecting any pickup or delivery of goods shall be permitted as provided in paragraphs (d) and (e) of this section.

on street parking (40CFR52.733-1) means stopping a motor vehicle on any street, highway, or roadway (except for legal stops at or before intersections and as caution and safety require), whether or not a person remains in the vehicle.

on the premises where stored (40CFR280.12) with respect to heating oil means UST systems located on the same property where the stored heating oil is used.

onboard controls (EPA-94/04): Devices placed on vehicles to capture gasoline vapor during refueling and route it to the engines when the vehicle is starting so that it can be efficiently burned.

once through cooling water (40CFR419.11-e) shall mean those waters discharged that are used for the purpose of heat removal and that do not come into direct contact with any raw material, intermediate, or finished product.

once through cooling water (40CFR423.11-g) means

water passed through the main cooling condensers in one or two passes for the purpose of removing waste heat (cf. recirculated cooling water).

one (1)-year, 2-year, and 10-year, 24-hour precipitation events (40CFR434.11-n) means the maximum 24-hour precipitation event with a probable recurrence interval of once in one, two, and ten years respectively as defined by the National Weather Service and Technical Paper No. 40, "Rainfall Frequency Atlas of the U.S.," May 1961, or equivalent regional or rainfall probability information developed therefrom.

one (1)-year, 24-hour precipitation event (40CFR434.11-s) means the maximum 24-hour precipitation event with a probably recurrence interval or once in one year as defined by the National Weather Service and Technical Paper No. 40, "Rainfall Frequency Atlas of the U.S.," May 1961, or equivalent regional or rainfall probability information developed therefrom.

one hit model (EPA-94/04): Mathematical model based on the biological theory that a single "hit" of some minimum critical amount of a carcinogen at a cellular target such as DNA can initiate an irreversible series of events, eventually leading to a tumor.

one hour period (40CFR60.2) means any 60-minute period commencing on the hour.

one hour period (40CFR63.2), unless otherwise defined in an applicable subpart, means any 60-minute period commencing on the hour (other identical or similar definitions are provided in 40CFR63.101).

one hundred year (100-year) flood (40CFR264.18-(b)(2)(iii)) means a flood that has a one percent chance of being equalled or exceeded in any given year.

one hundred year (100-year) floodplain (40CFR264.18-(b)(2)(i)) means any land area which is subject to a one percent or greater chance of flooding in any given year from any source.

onground tank (40CFR260.10) means a device

meeting the definition of "tank" in 40CFR260.10 and that is situated in such a way that the bottom of the tank is on the same level as the adjacent surrounding surface so that the external tank bottom cannot be visually inspected.

onions, canned (40CFR407.71-l) shall mean the processing of onions into the following product styles: Canned, frozen, and fried (canned), peeled, whole, sliced, and any other piece size but not including frozen, battered onion rings or dehydrated onions.

onshore (40CFR435.51-b) shall mean all land areas landward of the territorial seas as defined in 40CFR125.1(gg).

onshore (40CFR435.61-b) shall mean all land areas landward of the inner boundary of the territorial seas as defined in 40CFR126.1(gg).

onshore (40CFR60.631) means all facilities except those that are located in the territorial seas or on the outer continental shelf (other identical or similar definitions are provided in 40CFR60.641).

onshore area (40CFR55.2) means a coastal area designated as an attainment, nonattainment, or unclassifiable area by EPA in accordance with section 107 of the Act. If the boundaries of an area designated pursuant to section 107 of the Act do not coincide with the boundaries of a single onshore air pollution control agency, then onshore area shall mean a coastal area defined by the jurisdictional boundaries of an air pollution control agency (under CAA).

onshore facility (40CFR110.1) means any facility (including, but not limited to, motor vehicles and rolling stock) of any kind located in, on, or under any land within the United States, other than submerged land.

onshore facility (40CFR112.2) means any facility of any kind located in, on, or under any land within the United States, other than submerged lands, which is not a transportation-related facility.

onshore facility (40CFR116.3) means any facility (including, but not limited to, motor vehicles and rolling stock) of any kind located in, on, or under,

any land within the United States other than submerged land.

onshore facility (40CFR300.5), as defined by section 101(18) of CERCLA, means any facility (including, but not limited to, motor vehicles and rolling stock) of any kind located in, on, or under any land or non-navigable waters within the United States; and as defined by section 311(a)(10) of the CWA, means any facility (including, but not limited to, motor vehicles and rolling stock) of any kind located in, on, or under any land within the United States other than submerged land.

onshore facility (40CFR302.3) means any facility (including, but not limited to, motor vehicles and rolling stock) of any kind located in, on, or under, any land or non-navigable waters within the United States.

onshore facility (CWA311-33USC1321) means any facility (including, but not limited to, motor vehicles and rolling stock) of any kind located in, on, or under, any land within the United States other than submerged land.

onshore facility (SF101-42USC9601) means any facility (including, but not limited to, motor vehicles and rolling stock) of any kind located in, on, or under, any land or nonnavigable waters within the United States.

onshore oil storage facility (40CFR113.3-f) means any facility (excluding motor vehicles and rolling stock) of any kind located in, on, or under, any land within the United States, other than submerged land.

onsite coating mix preparation equipment (40CFR60.741) means those pieces of coating mix preparation equipment located at the same plant as the coating operation they serve.

opacity (40CFR60.2) means the degree to which emissions reduce the transmission of light and obscure the view of an object in the background (under CAA).

opacity (40CFR60-AA(alt. method 1)) means one minus the optical transmittance of a smoke plume, screen target, etc.

opacity (40CFR60-AB-2.4) means the fraction of incident light that is attenuated by an optical medium. Opacity (Op) and transmittance (Tr) are related by: Op = 1 - Tr.

opacity (40CFR61.171) means the degree to which emissions reduce the transmission of light (other identical or similar definitions are provided in 40CFR61.181).

opacity (40CFR63.2) means the degree to which emissions reduce the transmission of light and obscure the view of an object in the background. For continuous opacity monitoring systems, opacity means the fraction of incident light that is attenuated by an optical medium.

opacity (40CFR72.2) means the degree to which emissions reduce the transmission of light and obscure the view of an object in the background.

opacity (40CFR86.082.2) means the fraction of a beam of light, expressed in percent, which fails to penetrate a plume of smoke.

opacity (40CFR89.2) means the fraction of a beam of light, expressed in percent, which fails to penetrate a plume of smoke.

opacity (EPA-94/04): The amount of light obscured by particulate pollution in the air; clear window glass has zero opacity, a brick wall is 100 percent opaque. Opacity is an indicator of changes in performance of particulate control systems.

opaque stains (40CFR52.741) means all stains that are not semi-transparent stains.

ope (40CFR420.41-f) means those basic oxygen furnace steelmaking wet air cleaning systems which are designed to allow excess air enter the air pollution control system for the purpose of combusting the carbon monoxide in furnace gases.

open burning (40CFR165.1-n) means the combustion of a pesticide or pesticide container in any fashion other than incineration.

open burning (40CFR240.101-r) means burning of solid wastes in the open, such as in an open dump

(other identical or similar definitions are provided in 40CFR241.101-l).

open burning (40CFR257.3.7) means the combustion of solid waste without:
(1) control of combustion air to maintain adequate temperature for efficient combustion,
(2) containment of the combustion reaction in an enclosed device to provide sufficient residence time and mixing for complete combustion, and
(3) control of the emission of the combustion products.

open burning (40CFR258.2) means the combustion of solid waste without:
(1) Control of combustion air to maintain adequate temperature for efficient combustion,
(2) Containment of the combustion reaction in an enclosed device to provide sufficient residence time and mixing for complete combustion, and
(3) Control of the emission of the combustion products.

open burning (40CFR260.10) means the combustion of any material without the following characteristics:
(1) Control of combustion air to maintain adequate temperature for efficient combustion,
(2) Containment of the combustion-reaction in an enclosed device to provide sufficient residence time and mixing for complete combustion, and
(3) Control of emission of the gaseous combustion products (cf. incineration or thermal treatment).

open burning (EPA-94/04): Uncontrolled fires in an open dump.

open cut mine (40CFR440.141-12) means any form of recovery of ore from the earth except by a dredge.

Open dump (40CFR257.2) means a facility for the disposal of solid waste which does not comply with this part.

open dump (40CFR240.101-s) means a land disposal site at which solid wastes are disposed of in a manner that does not protect the environment, are susceptible to open burning, and are exposed to the elements, vectors, and scavengers (other identical or similar definitions are provided in 40CFR241.101-m) (under RCRA).

open dump (EPA-94/04): An uncovered site used for disposal of waste without environmental controls (cf. dump).

open dump (RCRA1004-42USC6903) means any facility or site where solid waste is disposed of which is not a sanitary landfill which meets the criteria promulgated under section 4004 and which is not a facility for disposal of hazardous waste.

open dumping (40CFR165.1-o) means the placing of pesticides or containers in a land site in a manner which does not protect the environment and is exposed to the elements, vectors, and scavengers.

open ended valve (40CFR52.741) means any valve, except pressure relief devices, having one side of the valve in contact with process fluid and one side open to the atmosphere, either directly or through open piping.

open ended valve or line (40CFR60.481) means any valve, except safety relief valves, having one side of the valve seat in contact with process fluid and one side open to the atmosphere, either directly or through open piping.

open ended valve or line (40CFR61.241) means any valve, except pressure relief valves, having one side of the valve seat in contact with process fluid and one side open to atmosphere, either directly or through open piping (other identical or similar definitions are provided in 40CFR264.1031).

open ended valve or line (40CFR63.161) means any valve, except pressure relief valves, having one side of the valve seat in contact with process fluid and one side open to atmosphere, either directly or through open piping.

open hearth furnace steelmaking (40CFR420.41-b) means the production of steel from molten iron, steel scrap, fluxes, and various combinations thereof, in refractory lined fuel-fired furnaces equipped with regenerative chambers to recover heat from the flue and combustion gases.

open lot (40CFR412.11-f) shall mean pens or similar confinement areas with dirt, or concrete (or paved or hard) surfaces wherein animals or poultry are substantially or entirely exposed to the outside environment except for possible small portions affording some protection by windbreaks, small shedtype shade areas. For the purposes hereof the term "open lot" is synonymous with the terms "cowyard" (dairy cattle), "pasture lot" (swine), and "dirt lot" (swine, sheep or turkeys), "dry lot" (swine, cattle, sheep, or turkeys) which are terms widely used in the industry.

open site (40CFR202.10-o) means an area that is essentially free of large sound-reflecting objects, such as barriers, walls, board fences, signboards, parked vehicles, bridges, or buildings.

open top vapor depressing (40CFR52.741) means the batch process of cleaning and removing soils from surfaces by condensing hot solvent vapor on the colder metal parts.

opening an appliance (40CFR82.152-n) means any service, maintenance, or repair on an appliance that could be reasonably expected to release refrigerant from the appliance to the atmosphere unless the refrigerant were recovered previously recovered from the appliance.

operable treatment works (40CFR35.905) is a treatment works that:
(a) Upon completion of construction will treat waste water, transport waste water to or from treatment, or transport and dispose of waste water in a manner which will significantly improve an objectionable water quality situation or health hazard, and
(b) Is a component part of a complete waste treatment system which, upon completion of construction for the complete waste treatment system (or completion of construction of other treatment works in the system in accordance with a schedule approved by the Regional Administrator), will comply with all applicable statutory and regulatory requirements.

operable unit (40CFR300.5) means a discrete action that comprises an incremental step toward comprehensively addressing site problems. This discrete portion of a remedial response manages migration, or eliminates or mitigates a release, threat of a release, or pathway of exposure. The cleanup of a site can be divided into a number of operable units, depending on the complexity of the problems associated with the site. Operable units may address geographical portions of a site, specific site problems, or initial phases of an action, or may consist of any set of actions performed over time or any actions that are concurrent but located in different parts of a site.

operable unit (40CFR35.4010) means a response action taken as one part of an overall site response. A number of operable units may occur in the course of a site response.

operable unit (40CFR35.6015-28): A discrete action, as described in the Cooperative Agreement or SSC, that comprises an incremental step toward comprehensively addressing site problems. The cleanup of a site can be divided into a number of operable units, depending on the complexity of the problems associated with the site. Operable units may address geographical portions of a site, specific site problems, or initial phases of an action, or may consist of any set of actions performed over time or any actions that are concurrent but located in different parts of a site.

operable unit (EPA-94/04): Term for each of a number of separate activities undertaken as part of a Superfund site cleanup. A typical operable unit would be removal of drums and tanks from the surface of a site.

operant, operant behavior, operant conditioning (40CFR798.6500-2): An operant is a class of behavioral responses which change or operates on the environment in the same way. Operant behavior is further distinguished as behavior which is modified by its consequences. Operant conditioning is the experimental procedure used to modify some class of behavior by reinforcement or punishment.

operated by the same producer (40CFR152.3-q) when used with respect to two establishments, means that each such establishment is either owned by, or leased for operation by and under the control of, the same person. The term does not include establishments owned or operated by different persons, regardless of contractual agreement between such persons.

operating (40CFR72.2) when referring to a combustion or process source seeking entry into the Opt-in Program, means that the source had documented consumption of fuel input for more than 876 hours in the 6 months immediately preceding the submission of a combustion source's opt-in application under 40CFR74.16(a) of this chapter.

operating conditions (EPA-94/04): Conditions specified in a RCRA permit that dictate how an incinerator must operate as it burns different waste types. A trial burn is used to identify operating conditions needed to meet specified performance standards.

operating day (40CFR60.561) means, for the purposes of these standards, any calendar day during which equipment used in the manufacture of polymer was operating for at least 8 hours or one labor shift, whichever is shorter. Only operating days shall be used in determining compliance with the standards specified in 40CFR60562-1(c)(1)(ii)(B), (1)(ii)(C), (2)(ii)(B), and (2)(ii)(C). Any calendar day in which equipment is used for less than 8 hours or one labor shift, whichever is less, is not an "operating day" and shall not be used as part of the rolling 14-day period for determining compliance with the standards specified in 40CFR60.562-1(c)(1)(ii)(B), (1)(ii)(C), (2)(ii)(B), and (2)(ii)(C).

operating hours (40CFR86.078.7), where vehicle, component, or engine storage areas or facilities are concerned, shall mean all times during which personnel other than custodial personnel are at work in the vicinity of the area or facility and have access to it. Where facilities or areas other than those covered by paragraph (c)(7)(ii) of this section are concerned, operating hours shall mean all times during which an assembly line is in operation or all times during which testing, maintenance, mileage (or service) accumulation, production or compilation of records, or any other procedure or activity related to certification testing, to translation of designs from the test stage to the production stage, or to vehicle (or engine) manufacture (or assembly is being carried out in a facility.

operating hours (40CFR89.2) means:
(1) For engine storage areas or facilities, all times

during which personnel other than custodial personnel are at work in the vicinity of the storage area or facility and have access to it.

(2) For all other areas or facilities, all times during which an assembly line is in operation or all times during which testing, maintenance, service accumulation, production or compilation of records, or any other procedure or activity related to certification testing, to translation of designs from the test stage to the production stage, or to engine manufacture or assembly is being carried out in a facility.

operating period (40CFR76.2) means a period of time of not less than three consecutive months and that occurs not more than one month prior to applying for an alternative emission limitation demonstration period under 40CFR76.10, during which the owner or operator of an affected unit that cannot meet the applicable emission limitation:

(1) Operates the installed NO_x emission controls in accordance with primary vendor specifications and procedures, with the unit operating under normal conditions; and

(2) records and reports quality-assured continuous emission monitoring (CEM) and unit operating data according to the methods and procedures in part 75 of this chapter.

operating permit (40CFR63.101) means a permit required by 40CFR part 70 or 71.

operating permit (40CFR63.111) means a permit required by 40CFR part 70 or part 71.

operating permit (40CFR72.2) means a permit issued under part 70 of this chapter and any other regulations implementing title V of the Act.

operation (40CFR413.11-b) shall mean any step in the electroplating process in which a metal is electrodeposited on a basis material and which is followed by a rinse; this includes the related operations of alkaline cleaning, acid pickle, stripping, and coloring when each operation is followed by a rinse (other identical or similar definitions are provided in 40CFR413.21-b).

operation (40CFR413.41-b) shall mean any step in the anodizing process in which a metal is cleaned,

anodized, or colored when each such step is followed by a rinse.

operation (40CFR413.51-b) shall mean any step in the coating process in which a basis material surface is acted upon by a process solution and which is followed by a rinse; plus the related operations of alkaline cleaning, acid pickle, and sealing, when each operation is followed by a rinse.

operation (40CFR413.71-c) shall mean any step in the electroless plating process in which a metal is deposited on a basis material and which if followed by a rinse; this includes the related operations of alkaline cleaning, acid pickle, and stripping, when each operation is followed by a rinse.

operation (40CFR413.81-b) shall mean any step in the printed circuit board manufacturing process in which the board is immersed in an aqueous process bath which is followed by a rinse.

operation (40CFR413.61-b) shall mean any step in the chemical milling or etching processes in which metal is chemically or electrochemically removed from the work piece and which is followed by a rinse; this includes related metal cleaning operations which preceded chemical milling or etching, when each operation is followed by a rinse.

operation (40CFR61.251-e) means that an impoundment is being used for the continued placement of new tailings or is in standby status for such placement. An impoundment is in operation from the day that tailings are first placed in the impoundment until the day that final closure begins (under CAA).

operation and maintenance (40CFR35.2005-30) are activities required to assure the dependable and economical function of treatment works:

(i) Maintenance: Preservation of functional integrity and efficiency of equipment and structures. This includes preventive maintenance, corrective maintenance and replacement of equipment (see 40CFR35.2005(b)(36)) as needed.

(ii) Operation: Control of the unit processes and equipment which make up the treatment works. This includes financial and personnel management; records, laboratory control, process control, safety and emergency operation planning.

operation and maintenance (O&M) (40CFR35.6015-29) means measures required to maintain the effectiveness of response actions (other identical or similar definitions are provided in 40CFR300.5).

operation and maintenance (EPA-94/04):
1. Activities conducted after a Superfund site action is completed to ensure that the action is effective.
2. Actions taken after construction to assure that facilities constructed to treat waste water will be properly operated and maintained to achieve normative efficiency levels and prescribed effluent limitations in an optimum manner.
3. On-going asbestos management plan in a school or other public building, including regular inspections, various methods of maintaining asbestos in place, and removal when necessary.

operational (40CFR192.31-p) means that a uranium mill tailings pile or impoundment is being used for the continued placement of uranium byproduct material or is in standby status for such placement. A tailings pile or impoundment is operational from the day that uranium byproduct material is first placed in the pile or impoundment until the day final closure begins.

operational (40CFR61.221-b) means a uranium mill tailings pile that is licensed to accept additional tailings, and those tailings can be added without violating subpart W or any other Federal, state or local rule or law. A pile cannot be considered operational if it is filled to capacity or the mill it accepts tailings from has been dismantled or otherwise decommissioned.

operational life (40CFR280.12) refers to the period beginning when installation of the tank system has commenced until the time the tank system is properly closed under subpart G.

operational test period (40CFR60-AB-2.17) means a period of time (168 hours) during which the CEMS is expected to operate within the established performance specifications without any unscheduled maintenance, repair, or adjustment.

operations and maintenance program (40CFR763.83) means means a program of work practices to maintain friable ACBM in good condition, ensure clean up of asbestos fibers previously released, and prevent further release by minimizing and controlling friable ACBM disturbance or damage.

operator (40CFR256.06) includes facility owners and operators.

operator (40CFR258.2) means the person(s) responsible for the overall operation of a facility or part of a facility.

operator (40CFR260.10) means the person responsible for the overall operation of a facility.

operator (40CFR280.12) means any person in control of, or having responsibility for, the daily operation of the UST system.

operator (40CFR52.2285-4) means the person who is directly responsible for the operation of the gasoline storage container(s), whether the person be a lessee or an agent of the owner (other identical or similar definitions are provided in 40CFR52.2286-4).

operator (40CFR610.11-8) means any person who installs, services or maintains a retrofit device in an automobile or who operates an automobile with a retrofit device installed.

operator (40CFR763-AA-16) means a person responsible for the TEM instrumental analysis of the sample.

operator (40CFR85.1402) means transit authority, state, city department, or private or public entity controlling the use of one or more urban buses.

operator (RCRA9001-42USC6991) means any person in control of, or having responsibility for, the daily operation of the underground storage tank.

operator (SMCRA701-30USC1291) means any person, partnership, or corporation engaged in coal mining who removes or intends to remove more than two hundred and fifty tons of coal from the earth by coal mining within twelve consecutive calendar months in any one location.

OPIDN (40CFR798.6450-1): See organophosphorus induced delayed neurotoxicity.

OPPTS (40CFR179.3) means the Agency's Office of Prevention, Pesticides and Toxic Substances.

opt in or opt into (40CFR72.2) means to elect to become an affected unit under the Acid Rain Program through the issuance of the final effective opt-in permit under 40CFR74.14 of this chapter.

opt-in permit (40CFR72.2) means the legally binding written document that is contained within the Acid Rain permit and sets forth the requirements under part 74 of this chapter for a combustion source or a process source that opts into the Acid Rain Program.

opt-in source (40CFR72.2) means a combustion source or process source that has elected to become an affected unit under the Acid Rain Program and whose opt-in permit has been issued and is in effect.

optical density (40CFR60-AB-2.5) means A logarithmic measure of the amount of incident light attenuated. Optical density (D) is related to the transmittance (Tr) and opacity (Op) as follows: D = -log(10) Tr = -log(10)(1 - Op).

optimal corrosion control treatment (EPA-94/04): The corrosion control treatment that minimizes the lead and copper concentration at users' taps while also insuring that the treatment does not cause the water system to violate any national primary drinking water regulations.

optimal corrosion control treatment, for the purpose of subpart I of this part only (40CFR141.2), means the corrosion control treatment that minimizes the lead and copper concentrations at users' taps while insuring that the treatment does not cause the water system to violate any national primary drinking water regulations.

optimum sustainable population (MMPA3-16USC1362) means, with respect to any population stock, the number of animals which will result in the maximum productivity of the population or the species, keeping in mind the carrying capacity of the habitat and the health of the ecosystem of which they form a constituent element.

option (40CFR86.082.2) means any available equipment or feature not standard equipment on a model.

oral toxicity (EPA-94/04): Ability of a pesticide to cause injury when ingested.

order (40CFR108.2-c) means any order issued by the Administrator under section 309 of the Act; any order issued by a State to secure compliance with a permit, or condition thereof, issued under a program approved pursuant to section 402 of the Act; or any order issued by a court in an action brought pursuant to section 309 or section 505 of the Act.

order (40CFR29.2) means Executive Order 12372, issued July 14, 1982, and amended April 8, 1983, and titled "Intergovernmental Review of Federal Programs."

order (40CFR8.2-q) means parts II, III, and IV of Executive Order 11246, dated September 24, 1965 (30 FR 12319), and any Executive Order amending or superseding such orders.

ore (40CFR440.141-13) means gold placer deposit consisting of metallic gold-bearing gravels, which may be: residual, from weathering of rocks in-situ; river gravels in active streams; river gravels in abandoned and often buried channels; alluvial fans; sea-beaches; and sea-beaches now elevated and inland. Ore is the raw "bank run" material measured in place, before being moved by mechanical or hydraulic means to a beneficiation process.

organ (40CFR190.02-i) means any human organ exclusive of the dermis, the epidermis, or the cornea.

organic (EPA-94/04):
1. Referring to or derived from living organisms.
2. In chemistry, any compound containing carbon.

organic active ingredients (40CFR455.21-a) means carbon-containing active ingredients used in pesticides, excluding metalloorganic active ingredients.

organic carbon partition coefficient (K_{oc}) (40CFR300-AA) means the measure of the extent of partitioning of a substance, at equilibrium, between organic carbon in geologic materials and water. The higher the K_{oc}, the more likely a substance is to bind to geologic materials than to remain in water [ml/g].

organic chemicals/compounds (EPA-94/04): Animal or plant-produced substances containing mainly carbon, hydrogen, nitrogen, and oxygen.

organic chlorine (40CFR797.1520-9) is the chlorine associated with all chlorine-containing compounds that elute just before lindane to just after mirex during gas chromatographic analysis using a halogen detector.

organic coating (40CFR60.311) means any coating used in a surface coating operation, including dilution solvents, from which volatile organic compound emissions occur during the application or the curing process. For the purpose of this regulation, powder coatings are not included in this definition (under CAA).

organic coating (40CFR60.451) means any coating used in a surface coating operation, including dilution solvents, from which VOC emissions occur during the application or the curing process. For the purpose of this regulation, powder coatings are not included in this definition.

organic compound (40CFR52.741) means any compound of carbon, excluding carbon monoxide, carbon dioxide, carbonic acid, metallic carbides or carbonates, and ammonium carbonate.

organic hazardous air pollutant or organic HAP (40CFR63.101) means one of the chemicals listed in table 2 of this subpart.

organic hazardous air pollutant or organic HAP (40CFR63.111) means any of the chemicals listed in table 2 of subpart F of this part.

organic material (40CFR52.741) means any chemical compound of carbon including diluents and thinners which are liquids at standard conditions and which are used as dissolvers, viscosity reducers, or cleaning agents, but excluding methane, carbon monoxide, carbon dioxide, carbonic acid, metallic carbonic acid, metallic carbide, metallic carbonates, and ammonium carbonate.

organic material hydrocarbon equivalent (40CFR86.090.2) means the sum of the carbon mass contributions of non-oxygenated hydrocarbons, methanol and formaldehyde as contained in a gas sample, expressed as gasoline fueled vehicle hydrocarbons. In the case of exhaust emissions, the hydrogen-to-carbon ratio of the equivalent hydrocarbon is 1.85:1. In the case of diurnal and hot soak emissions, the hydrogen-to-carbon ratios of the equivalent hydrocarbons are 2.33:1 and 2.2:1, respectively.

organic materials (40CFR52.1145-3) are chemical carbon monoxide, carbon dioxide, carbonic acid, metallic carbides, metallic carbonates, and ammonium carbonate.

organic matter (40CFR796.2750-iii) is the organic fraction of the sediment or soil; it includes plant and animal residues at various stages of decomposition, cells and tissues of soil organisms, and substances synthesized by the microbial population.

organic matter: Carbonaceous waste contained in plant or animal matter and originating from domestic or industrial sources.

organic monitoring device (40CFR63.111) means a unit of equipment used to indicate the concentration level of organic compounds exiting a recovery device based on a detection principle such as infra-red, photoionization, or thermal conductivity.

organic pesticide chemicals (40CFR455.21-c) means the sum of all organic active ingredients listed in 40CFR455.20(b) which are manufactured at a facility subject to this subpart.

organic pesticides (40CFR165.1-p): See pesticide.

organic solvent-based green tire spray (40CFR60.541) means any mold release agent and lubricant applied to the inside or outside of green tires that contains more than 12 percent, by weight, of VOC as sprayed.

organic solvents (40CFR52.1088) means organic materials, including diluents and thinners, which are liquids at standard conditions and which are used as dissolvers, viscosity reducers, or cleaning agents (other identical or similar definitions are provided in 40CFR52.1107; 52.2440-2).

organic solvents (40CFR52.1145-1) include diluents and thinners and are defined as organic materials which are liquids at standard conditions and which are used as dissolvers, viscosity reducers, or cleaning agents, except that such materials which exhibit a boiling point higher than 220°F at 0.5 millimeters of mercury absolute pressure or having an equivalent vapor pressure shall not be considered to be solvents unless exposed to temperatures exceeding 220°F.

organic vapor (40CFR52.741) means the gaseous phase of an organic material or a mixture of organic materials present in the atmosphere.

Organism (EPA-94/04): Any form of animal or plant life.

organization (40CFR195.2) is any individual, sole proprietorship, partnership, business, company, corporation, college or university, government agency (includes Federal, State and local government entities), laboratory, or institution.

organochlorine pesticides (40CFR797.1520-10) are those pesticides which contain carbon and chlorine such as aldrin, DDD, DDE, DDT, dieldrin, endrin, and heptachlor.

organophosphates (EPA-94/04): Pesticides that contain phosphorus; short-lived, but some can be toxic when first applied.

organophosphorus induced delayed neurotoxicity (OPIDN) (40CFR798.6450-1) is a neurological syndrome in which limb weakness and upper motor neuron spasticity are the predominant clinical signs and distal axonopathy of peripheral nerve and spinal cord are the correlative pathological signs. Clinical signs and pathology first appear between 1 and 2 weeks following exposure which normally inhibits greater than 80 percent of NTE (for O-Ps that age) (under TSCA).

organotins (EPA-94/04): Chemical compounds used in anti-foulant paints to protect the hulls of boats and ships, buoys, and pilings from marine organisms such as barnacles.

original AHERA inspection (original inspection; inspection) (EPA-94/04): Examination of school buildings arranged by Local Education Agencies to identify asbestos containing materials, evaluate their condition, take samples of materials suspected to contain asbestos; performed by EPA-accredited inspectors.

original data submitter (40CFR152.83-d) means the person who possesses all rights to exclusive use or compensation under FIFRA section 3(c)(l)(D) in a study originally submitted in support of an application for registration amended registration, reregistration, or experimental use permit, or to maintain an existing registration in effect. The term includes the person who originally submitted the study, any person to whom the rights under FIFRA section 3(c)(l)(D) have been transferred, or the authorized representative of a group of joint data developers.

original engine configuration (40CFR85.1402) means the engine configuration at time of initial sale (under CAA).

original engine manufacturer (OEM) (40CFR89.602.96): The entity which originally manufactured the nonroad engine.

original equipment manufacturer (OEM) (40CFR85.1502-9) is the entity which originally manufactured the motor vehicle or motor vehicle engine prior to conditional importation.

original equipment part (40CFR85.1402) means a part present in or on an engine at the time an urban bus is originally sold to the ultimate purchaser.

original equipment part (40CFR85.2102-7) means a part present in or on a vehicle at the time the vehicle is sold to the ultimate purchaser, except for components installed by a dealer which are not manufactured by the vehicle manufacturer or are not installed at the direction of the vehicle manufacturer.

original generation point (40CFR259.10.b) means the location where regulated medical waste is generated. Waste may be taken from original generation points to a central collection point prior to off-site transport or on-site treatment.

original generation point (EPA-94/04): Where regulated medical or other material first becomes waste.

original production (OP) year (40CFR85.1502-10) is the calendar year in which the motor vehicle or motor vehicle engine was originally produced by the OEM.

original production (OP) year (40CFR89.602.96): The calendar year in which the nonroad engine was originally produced by the OEM.

original production (OP) years old (40CFR85.1502-11) means the age of a vehicle as determined by subtracting the original production year of the vehicle from the calendar year of importation.

original production (OP) years old (40CFR89.602.96): The age of a nonroad engine as determined by subtracting the original production year of the nonroad engine from the calendar year of importation.

ornamental (40CFR171.2-18) means trees, shrubs, and other plantings in and around habitations generally, but not necessarily located in urban and suburban areas, including residences, parks, streets, retail outlets, and industrial and institutional buildings.

osmosis (EPA-94/04): The passage of a liquid from a weak solution to a more concentrate solution across a semipermeable membrane that allows passage of the solvent (water) but not the dissolved solids.

other coatings (40CFR420.121-c) means coating steel products with metals other than zinc or terne metal by the hot dip process including the immersion of the steel product in a molten bath of metal, and the related operations preceding the subsequent to the immersion phase.

other container (40CFR503.11-j) is either an open or closed receptacle. This includes, but is not limited to, a bucket, a box, a carton, and a vehicle or trailer with a load capacity of one metric ton or less.

other lead-emitting operation (40CFR60.371-e)

means any lead-acid battery manufacturing plant operation from which lead emissions are collected and ducted to the atmosphere and which is not part of a grid casting, lead oxide manufacturing, lead reclamation, paste mixing, or three-process operation facility, or a furnace affected under subpart L of this part.

other minerals (SMCRA701-30USC1291) means clay, stone, sand, gravel, metalliferous and nonmetalliferous ores, and any other solid material or substances of commercial value excavated in solid form from natural deposits on or in the earth, exclusive of coal and those minerals which occur naturally in liquid or gaseous form.

other restricted access (nonsubstation) locations (40CFR761.123) means areas other than electrical substations that are at least 0.1 kilometer (km) from a residential/commercial area and limited by man-made barriers (e.g., fences and walls) to substantially limited by naturally occurring barriers such as mountains, cliffs, or rough terrain. These areas generally include industrial facilities and extremely remote rural locations. (Areas where access is restricted but are less than 0.1 km from a residential/commercial area are considered to be residential/commercial areas.)

other significant evidence (40CFR154.3-e) means factually significant information that relates to the uses of the pesticide and their adverse risk to man or to the environment but does not include evidence based only on misuse of the pesticide unless such misuse is widespread and commonly recognized practice.

otherwise subject to the jurisdiction of the United States (CWA311-33USC1321) means subject to the jurisdiction of the United States by virtue of United States citizenship, United States vessel documentation or numbering, or as provided for by international agreement to which the United States is a party (other identical or similar definitions are provided in 40CFR116.3; SF101-42USC9601).

otherwise use or otherwise used (40CFR372.3) means any use of a toxic chemical that is not covered by the terms "manufacture" or "process" and includes use of a toxic chemical contained in a mixture or trade name product. Relabeling or redistributing a container of a toxic chemical where no repackaging of the toxic chemical occurs does not constitute use or processing of the toxic chemicals.

otto cycle (40CFR86.090.2) means type of engine with operating characteristics significantly similar to the theoretical Otto combustion cycle. The use of a throttle during normal operation is indicative of an Otto-cycle engine.

out of control period (40CFR72.2) means any period:
(1) Beginning with the hour corresponding to the completion of a daily calibration error, linearity check, or quality assurance audit that indicates that the instrument is not measuring and recording within the applicable performance specifications; and
(2) Ending with the hour corresponding to the completion of an additional calibration error, linearity check, or quality assurance audit following corrective action that demonstrates that the instrument is measuring and recording within the applicable performance specifications.

outdoor electrical substations (40CFR761.123) means outdoor, fenced-off, and restricted access areas used in the transmission and/or distribution of electrical power. Outdoor electrical substations restrict public access by being fenced or walled off as defined under 40CFR61.30(l)(1)(ii). For purposes of this TSCA policy, outdoor electrical substations are defined as being located at least 0.1 km from a residential/commercial area. Outdoor fenced-off and restricted access areas used in the transmission and/or distribution of electrical power which are located less than 0.1 km from a residential/commercial areas.

outdoor use (40CFR152.161-b) means any pesticide application that occurs outside enclosed manmade structures or the consequences of which extend beyond enclosed manmade structures, including, but not limited to, pulp and paper mill water treatments and industrial cooling water treatments.

outer continental shelf (40CFR55.2) shall have the meaning provided by section 2 of the OCSLA (43 U.S.C. 40CFR1331 et seq.).

outer continental shelf (CAA328-42USC7627) has the meaning provided by section 2 of the Outer Continental Shelf Lands Act (43USC1331).

outer continental shelf (OPA1001) means an offshore facility which is located, in whole or in part, on the outer continental shelf and is or was used for one or more of the following purposes: exploring for, drilling for, producing, storing, handling, transferring, processing, or transporting oil produced from the Outer Continental Shelf.

outer continental shelf energy activity (CZMA304-16USC1453) means any exploration for, or any development or production of, oil or natural gas from the Outer Continental Shelf (as defined in section 1331(a) of title 43) or the siting, construction, expansion, or operation of any new or expanded energy facilities directly required by such exploration, development, or production.

outer continental shelf source and OCS source (CAA328-42USC7627) include any equipment, activity, or facility which:
(i) emits or has the potential to emit any air pollutant,
(ii) is regulated or authorized under the Outer Continental Shelf Lands Act, and
(iii) is located on the Outer Continental Shelf or in or on waters above the Outer Continental Shelf. Such activities include, but are not limited to, platform and drill ship exploration, construction, development, production, processing, and transportation. For purposes of this subsection, emissions from any vessel servicing or associated with an OCS source, including emissions while at the OCS source or en route to or from the OCS source within 25 miles of the OCS source, shall be considered direct emissions from the OCS source.

outfall (40CFR122.26-9) means a point source as defined by 40CFR122.2 at the point where a municipal separate storm sewer discharges to waters of the United States and does not include open conveyances connecting two municipal separate storm sewers, or pipes, tunnels or other conveyances which connect segments of the same stream or other waters of the United States and are used to convey waters of the United States.

outfall (EPA-94/04): The place where effluent is discharged into receiving waters.

outlays (expenditures) (40CFR31.3) mean charges made to the project or program. They may be reported on a cash or accrual basis. For reports prepared on a cash basis, outlays are the sum of actual cash disbursement for direct charges for goods and services, the amount of indirect expense incurred, the value of in-kind contributions applied, and the amount of cash advances and payments made to contractors and subgrantees. For reports prepared on an accrued expenditure basis, outlays are the sum of actual cash disbursements, the amount of indirect expense incurred, the value of in-kind contributions applied, and the new increase (or decrease) in the amounts owed by the grantee for goods and other property received, for services performed by employees, contractors, subgrantees, subcontractors, and other payees, and other amounts becoming owed under programs for which no current services or performance are required, such as annuities, insurance claims, and other benefit payments.

output (40CFR35.105) means an activity or product which the applicant agrees to complete during the budget period.

outside air (40CFR61.141) means the air outside buildings and structures, including, but not limited to, the air under a bridge or in an open air ferry dock.

outside employment or other outside activity (40CFR3.500) is any work or service performed by an employee other than the performance of official duties. It includes such activities as writing and editing, publishing, teaching, lecturing, consulting, self-employment and other work or services. Employees must ensure that their outside activities may not reasonably be construed as implying official EPA endorsement of any statement, activity, product or service.

outstanding data requirement (FIFRA2-7USC136) means:
(1) In general--The term "outstanding data requirement" means a requirement for any study, information, or data that is necessary to make a determination under section 3(c)(5) and

which study, information, or data:
(A) has not been submitted to the Administrator, or
(B) if submitted to the Administrator, the Administrator has determined must be resubmitted because it is not valid, complete, or adequate to make a determination under 3(c)(5) and the regulations and guidelines issued under such section.
(2) Factors--In making a determination under paragraph (1)(B) respecting a study, the Administrator shall examine, at a minimum, relevant protocols, documentation of the conduct and analysis of the study, and the results of the study fulfill the data requirement for which the study was submitted to the Administrator.

oven (40CFR52.741) means a chamber within which heat is used for one or more of the following purposes: Dry, bake, cure, or polymerize a coating or ink.

oven (40CFR60.441) means a chamber which uses heat or irradiation to bake, cure, polymerize, or dry a surface coating.

oven (40CFR63.301) means a chamber in the coke oven battery in which coal undergoes destructive distillation to produce coke.

over-the-head position (40CFR211.203-s) means the mode of use of a device with a headband, in which the headband is worn such that it passes over the users head. This is contrast to the behind-the-head and under-the-chin positions.

overall control (40CFR52.741) means the product of the capture efficiency and the control device efficiency.

overburden (40CFR122.26-10) means any material of any nature, consolidated or unconsolidated, that overlies a mineral deposit, excluding topsoil or similar naturally occurring surface materials that are not disturbed by mining operations.

overburden (EPA-94/04): Rock and soil cleared away before mining.

overdraft (EPA-94/04): The pumping of water from a groundwater basin or aquifer in excess of the supply flowing into the basin; results in a depletion or "mining" of the groundwater in the basin (cf. groundwater mining).

overfill release (40CFR280.12) is a release that occurs when a tank is filled beyond its capacity, resulting in a discharge of the regulated substance to the environment.

overfire air (40CFR240.101-a) means air, under control as to quantity and direction, introduced above or beyond a fuel bed by induced or forced draft.

overfire air (EPA-94/04): Air forced into the top of an incinerator or boiler to fan the flames.

overflow (40CFR403.7(h)) means the intentional or unintentional diversion of flow from the POTW before the POTW Treatment Plant. POTWs which at least once annually overflow untreated wastewater to receiving waters may claim Consistent Removal of a pollutant only by complying with either paragraph (h)(1) or (h)(2) of this section. However, this subsection shall not apply where Industrial User(s) can demonstrate that Overflow does not occur between the Industrial User(s) and the POTW Treatment Plant.

overflow rate (EPA-94/04): One of the guidelines for the design of settling tanks and clarifiers in treatment plants; used by plant operators to determine if tanks and clarifiers are over- or under-loaded.

overland flow (EPA-94/04): A land application technique that cleanses waste water by allowing it to flow over a sloped surface. As the water flows over the surface, contaminants are absorbed and the water is collected at the bottom of the slope for reuse.

oversaturated (supersaturated) solution (40CFR796.1840-v) is a solution that contains a greater concentration of a solute than is possible at equilibrium under fixed conditions of temperature and pressure.

oversized regulated medical waste (40CFR259.10.b) means medical waste that is too large to be placed in a plastic bag or standard container.

oversized regulated medical waste (EPA-94/04): Medical waste that is too large for plastic bags or standard containers.

oversubscription payment deadline (40CFR72.2) means 30 calendar days prior to the allowance transfer deadline.

overturn (EPA-94/04): One complete cycle of top to bottom mixing of previously stratified water masses. This phenomenon may occur in spring or fall, or after storms, and results in uniformity of chemical and physical properties of water at all depths.

overvarnish (40CFR52.741) means a transparent coating applied directly over ink or coating (under CAA).

overvarnish coating operation (40CFR60.491-4) means the system on each beverage can surface coating line used to apply a coating over ink which reduces friction for automated beverage can filling equipment, provides gloss, and protects the finished beverage can body from abrasion and corrosion. The overvarnish coating is applied to two-piece beverage can bodies. The overvarnish coating operation consists of the coating application station, flashoff area, and curing oven.

own or control (40CFR704.3) means ownership of 50 percent or more a company's voting stock or other equity rights, or the poser to control the management and policies of that company. A company may own or control one or more sites. A company may be owned or controlled by a foreign or domestic parent company.

owned or controlled by the parent company (40CFR712.3-k) means the parent owns or controls 50 percent or more of the other company's voting stock or other equity rights, or has the power to control the management and policies of the other company.

owned or operated, leased or otherwise controlled by such person (40CFR88.302.94) means either of the following:

(1) Such person holds the beneficial title to such vehicle; or

(2) Such person uses the vehicle for transportation purposes pursuant to a contract or similar arrangement, the term of such contract or similar arrangement is for a period of 120 days or more, and such person has control over the vehicle pursuant to the definition of control of this section.

owner (40CFR170.3) means any person who has a present possessory interest (fee, leasehold, rental, or other) in an agricultural establishment covered by this part. A person who has both leased such agricultural establishment to another person and granted that same person the right and full authority to manage and govern the use of such agricultural establishment is not an owner for purposes of this part.

owner (40CFR258.2) means the person(s) who owns a facility or part of a facility.

owner (40CFR260.10) means the person who owns a facility or part of a facility.

owner (40CFR280.12) means:

(a) In the case of an UST system in use on November 8, 1984, or brought into use after that date, any person who owns an UST system used for storage, use, or dispensing of regulated substances; and

(b) In the case of any UST system in use before November 8, 1984, but no longer in use on that date, any person who owned such UST immediately before the discontinuation of its use.

owner (40CFR52.2285-3) means the owner of the gasoline storage container(s) (other identical or similar definitions are provided in 40CFR52.2286-3).

owner (40CFR72.2) means any of the following persons:

(1) Any holder of any portion of the legal or equitable title in an affected unit or in a combustion source or process source; or

(2) Any holder of a leasehold interest in an affected unit or in a combustion source or process source; or

(3) Any purchaser of power from an affected unit or from a combustion source or process source under a life-of-the-unit, firm power contractual arrangement as the term is defined herein and used in section 408(i) of the Act. However, unless expressly provided for in a leasehold agreement, owner shall not include a passive lessor, or a person who has an equitable interest through such lessor, whose rental payments are not based, either directly or indirectly, upon the revenues or income from the affected unit; or

(4) With respect to any Allowance Tracking System general account, any person identified in the submission required by 40CFR73.31(c) of this chapter that is subject to the binding agreement for the authorized account representative to represent that person's ownership interest with respect to allowances.

owner (40CFR85.2102-6) means the original purchaser or any subsequent purchaser of a vehicle.

owner (RCRA9001-42USC6991) means:

(A) in the case of an underground storage tank in use on the date of enactment of the Hazardous and Solid Waste Amendments of 1984, or brought into use after that date, any person who owns an underground storage tank used for the storage, use, or dispensing of regulated substances, and

(B) in the case of any underground storage tank in use before the date of enactment of the Hazardous and Solid Waste Amendments of 1984, but no longer in use on the date of enactment of such Amendments, any person who owned such tank immediately before the discontinuation of its use.

owner (RCRA9003) does not include any person who, without participating in the management of an underground storage tank and otherwise not engaged in petroleum production, refining, and marketing, holds indicia of ownership primarily to protect the owner's security interest in he tank.

owner/operator (40CFR147.2902) means the owner or operator of any facility or activity subject to regulation under the Osage UIC program.

owner or operator (40CFR51.100-f) means any

person who owns, leases, operates, controls, or supervises a facility, building, structure, or installation which directly or indirectly result or may result in emissions of any air pollutant for which a national standard is in effect (other identical or similar definitions are provided in 40CFR52.145; 52.741; 52.1881-vi; 60.2; 61.02; 63.2; 63.101; 63.191; 66.3-j; 72.2; 112.2; 122.2; 124.2; 124.41; 129.2-k; 144.3; 146.3; 232.2; 270.2; 280.92; CAA111-42USC7411; CAA112-42USC7412; CWA306-33USC1316; CWA311-33USC1321; SF101-42USC9601).

owner or operator of a demolition or renovation activity (40CFR61.141) means any person who owns, leases, operates, controls, or supervises the facility being demolished or renovated or any person who owns, leases, operates, controls, or supervises the demolition or renovation operation, or both (under CAA).

owner's manual (40CFR85.2102-9) means the instruction booklet normally provided to the purchaser of a vehicle.

oxidant (EPA-94/04): A substance containing oxygen that reacts chemically in air to produce a new substance; the primary ingredient of photochemical smog.

oxidation (EPA-94/04): The addition of oxygen that breaks down organic waste or chemicals such as cyanides, phenols, and organic sulfur compounds in sewage by bacterial and chemical means.

oxidation control system (40CFR60.101-j) means an emission control system which reduces emissions from sulfur recovery plants by converting these emissions to sulfur dioxide.

oxidation pond (EPA-94/04): A man-made body of water in which waste is consumed by bacteria, used most frequently with other waste-treatment processes; a sewage lagoon.

oxidation reduction potential (EPA-94/04): The electric potential required to transfer electrons from one compound or element (the oxidant) to another compound or element (the reductant); used as a qualitative measure of the state of oxidation in water treatment systems.

oxides of nitrogen (40CFR86.082.2) means the sum of the nitric oxide and nitrogen dioxide contained in a gas sample as if the nitric oxide were in the form of nitrogen dioxide (other identical or similar definitions are provided in 40CFR86.402.78).

oxybisphenoxarsine (OBPA)/1,3-diisocyanate (40CFR63.191) means the chemical with CAS number 58-366. The chemical is primarily used for fungicidal and bactericidal protection of plastics. The process uses chloroform as a solvent.

oxygenate (40CFR80.2-jj) means any substance which, when added to gasoline, increases the oxygen content of that gasoline. Lawful use of any of the substances or any combination of these substances requires that they be "substantially similar" under section 211(f)(1) of the Clean Air Act, or be permitted under a waiver granted by the Administrator under the authority of section 211(f)(4) of the Clean Air Act.

oxygenate blender (40CFR80.2-mm) means any person who owns, leases, operates, controls, or supervises an oxygenate blending facility, or who owns or controls the blendstock or gasoline used or the gasoline produced at an oxygenate blending facility.

oxygenate blender records (40CFR80.126-d) shall include laboratory analysis reports; refiner, importer and oxygenate blender contracts; quality assurance program records; product transfer documents; oxygenate purchasing, inventory, and usage records; and daily tank inventory gauging reports, meter tickets, and product transfer documents.

oxygenate blending facility (40CFR80.2-ll) means any facility (including a truck) at which oxygenate is added to gasoline or blendstock, and at which the quality or quantity of gasoline is not altered in any other manner except for the addition of deposit control additives.

oxygenated fuels (EPA-94/04): Gasoline which has been blended with alcohols or ethers that contain oxygen in order to reduce carbon monoxide and other emissions.

oxygenated fuels program reformulated gasoline, or

OPRG (40CFR80.2-nn) means reformulated gasoline which is intended for use in an oxygenated fuels program control area, as defined at paragraph (pp) of this section, during an oxygenated fuels program control period, as defined at paragraph (qq) of this section.

oxygenated gasoline (40CFR80.2-rr) means gasoline which contains a measurable amount of oxygenate.

oxygenated solvent (EPA-94/04): An organic solvent containing oxygen as part of the molecular structure. Alcohols and ketones are oxygenated compounds often used as paint solvents.

ozonation/ozonator (EPA-94/04): Application of ozone to water for disinfection or for taste and odor control. The ozonator is the device that does this.

ozone (O_3) (EPA-94/04): Found in two layers of the atmosphere, the stratosphere and the troposphere. In the stratosphere (the atmospheric layer 7 to 10 miles or more above the earth's surface) ozone is a natural form of oxygen that provides a protective layer shielding the earth from ultraviolet radiation. In the troposphere (the layer extending up 7 to 10 miles from the earth's surface), ozone is a chemical oxidant and major component of photochemical smog. It can seriously impair the respiratory system and is one of the most widespread of all the criteria pollutants for which the Clean Air Act required EPA to set standards. Ozone in the troposphere is produced through complex chemical reactions of nitrogen oxides, which are among the primary pollutants emitted by combustion sources; hydrocarbons, released into the atmosphere through the combustion, handling and processing of petroleum products; and sunlight.

ozone depletion (EPA-94/04): Destruction of the stratospheric ozone layer which shields the earth from ultraviolet radiation harmful to life. This destruction of ozone is caused by the breakdown of certain chlorine and/or bromine containing compounds (chlorofluorocarbons or halons), which break down when they reach the stratosphere and then catalytically destroy ozone molecules.

ozone hole (EPA-94/04): Thinning break in the stratospheric ozone layer. Designation of amount of such depletion as a "ozone hole" is made when detected amount of depletion exceeds fifty percent. Seasonal ozone holes have been observed over both the Antarctic region and the Arctic region and part of Canada and the extreme northeastern United States.

ozone layer (EPA-94/04): The protective layer in the atmosphere, about 15 miles above the ground, that absorbs some of the sun's ultraviolet rays, thereby reducing the amount of potentially harmful radiation reaching the earth's surface.

ozone depletion potential (CAA601-42USC7671) means a factor established by the Administrator to reflect the ozone-depletion potential of a substance, on a mass per kilogram basis, as compared to chlorofluorocarbon-11 (CFC-11). Such factor shall be based upon the substance's atmospheric lifetime, the molecular weight of bromine and chlorine, and the substance's ability to be photolytically disassociated, and upon other factors determined to be an accurate measure of relative ozone-depletion potential.

*********** PPPPP ***********

P-TBB (40CFR704.33-3) means the substance p-tert-butylbenzaldehyde, also identified as 4-(1,1-dimethylethyl)benzaldehyde, CAS No. 939-979.

P-TBBA (40CFR704.33-1) means the substance p-tert-butylbenzoic acid, also identified as 4-(1,1-dimethylethyl)benzoic acid, CAS No. 98-737.

P-TBT (40CFR704.33-2) means the substance p-tert-butyltoluene, also identified as 1-(1,1-dimethylethyl)-4-methylbenzene, CAS No. 98-511.

package (40CFR211.203-t) means the container in which a hearing protective device is presented for purchase or use. The package in some cases may be the same as the carrying case.

package (40CFR259.10.b) means the packaging-containers and its contents.

package or packaging (40CFR152.3-r) means the immediate container or wrapping, including any attached closure(s), in which the pesticide is contained for distribution, sale, consumption, use, or store. The term does not include any shipping or bulk container used for transporting or delivering the pesticide unless it is the only such package (other identical or similar definitions are provided in 40CFR157.21-c).

packaging (40CFR152.3-r; 157.21-c): See package.

packaging (40CFR259.10.b) means the assembly of one or more containers and any other components necessary to assure minimum compliance with 40CFR259.41 of this part.

packaging (EPA-94/04): The assembly of one or more containers and any other components necessary to assure minimum compliance with a program's storage and shipment packaging requirements. Also, the containers, etc., involved.

packaging rotogravure printing (40CFR52.741) means rotogravure printing upon paper, paper board, metal foil, plastic film, and other substrates, which are, in subsequent operations, formed into packaging products or labels for articles to be sold (under CAA).

packaging rotogravure printing line (40CFR52.741) means a rotogravure printing line is which surface coatings are applied to paper, paperboard, foil, film, or other substrates which are to be used to produce containers, packaging products, or labels for articles (under CAA).

packed bed scrubber (EPA-94/04): An air pollution control device in which emissions pass through alkaline water to neutralize hydrogen chloride gas.

packed tower (EPA-94/04): A pollution control device that forces dirty air through a tower packed with crushed rock or wood chips while liquid is sprayed over the packing material. The pollutants in the air stream either dissolve or chemically react with the liquid.

packer (40CFR146.3) means a device lowered into a well to produce a fluid-tight seal.

packer (40CFR147.2902) means a device lowered into a well to produce a fluid-tight seal within the casing.

packinghouse (40CFR432.31-b) shall mean a plant that both slaughters animals and subsequently processes carcasses into cured, smoked, canned or other prepared meat products (other identical or similar definitions are provided in 40CFR432.41-b).

padup rebuild (40CFR63.301) means a coke oven battery that is a complete reconstruction of an existing coke oven battery on the same site and pad without an increase in the design capacity of the coke plant as of November 15, 1990 (including any capacity qualifying under 40CFR63.304(b)(6)), and the capacity of any coke oven battery subject to a construction permit on November 15, 1990, which commenced operation before October 27, 1993. The Administrator may determine that a project is a padup rebuild if it effectively constitutes a replacement of the battery above the pad, even if some portion of the brickwork above the pad is retained.

pail (40CFR52.741) means any cylindrical metal shipping container of 1- to 12-gallon capacity and constructed of 29-gauge and heavier material (under CAA).

paint manufacturing plant (40CFR52.741) means a plant that mixes, blends, or compounds enamels, lacquers, sealers, shellacs, stains, varnishes, or pigmented surface coatings.

palatable water (EPA-94/04): Water at a desirable temperature that is free from objectionable tastes, odors, colors, and turbidity.

PAMS (40CFR58.1-x) means Photochemical Assessment Monitoring Stations.

pandemic (EPA-94/04): A widespread throughout an area, nation or the world.

panel (40CFR304.12-i): See national panel of environmental arbitrators.

paper (40CFR250.4-aa) means one of two broad subdivisions of paper products, the other being paperboard. Paper is generally lighter in basis weight, thinner, and more flexible than paperboard. Sheets 0.012 inch or less in thickness are generally classified as paper. Its primary uses are for printing, writing, wrapping, and sanitary purposes. However, in this guideline, the term paper is also used as a generic term that includes both paper and paperboard. It includes the following types of papers: bleached paper, bond paper, book paper, brown paper, coarse paper, computer paper, cotton fiber content paper, cover stock or cover paper, duplicator paper, form bond, ledger paper, manifold business forms, mimeo paper, newsprint, office papers, offset printing paper, printing paper, stationery, tabulating paper, unbleached papers, writing paper, and xerographic/copy paper.

paper coating (40CFR52.741) means any coating applied on paper, plastic film, or metallic foil to make certain products, including (but not limited to) adhesive tapes and labels, book covers, post cards, office copier paper, drafting paper, or pressure sensitive tapes. Paper coating includes the application of coatings by impregnation and/or saturation.

paper coating facility (40CFR52.741) means a facility that includes one or more paper coating lines.

paper coating line (40CFR52.741) means a coating line in which any protective, decorative, or functional coating is applied on, saturated into, or impregnated into paper, plastic film, or metallic foil to make certain products, including (but not limited to) adhesive tapes and labels, book covers, post cards, office copier paper, drafting paper, and pressure sensitive tapes.

paper napkins (40CFR250.4-bb) means special tissues, white or colored, plain or printed, usually folded, and made in a variety of sizes for use during meals or with beverages.

paper product (40CFR250.4-cc) means any item manufactured from paper or paperboard. The term "paper product" is used in this guideline to distinguish such items as boxes, doilies, and paper towels from printing and writing papers. It includes the following types of products: corrugated boxes, doilies, envelopes, facial tissue, fiber of fiberboard boxes, folding boxboard, industrial wipers, paper napkins, paper towels, tabulating cards, and toilet tissue.

paper towels (40CFR250.4-dd) means paper toweling in folded sheets, or in raw form, for use in drying or cleaning, or where quick absorption is required (under RCRA).

paperboard (40CFR250.4-ee) means one of the two broad subdivisions of paper, the other being paper itself. Paperboard is usually heavier in basis weight and thicker than paper. Sheets 0.012 inch or more in thickness are generally classified as paperboard. The broad classes of paperboard are containerboard, which is used for corrugated boxes, boxboard, which is principally used to make cartons, and all other paperboard.

parameter (EPA-94/04): A variable, measurable property whose value is a determinant of the characteristics of a system; e.g., temperature, pressure, and density are parameters of the atmosphere.

paraquat (EPA-94/04): A standard herbicide used to kill various types of crops, including marijuana.

parent company (40CFR704.3) is a company that owns or controls another company.

parent corporation (40CFR144.61-c) means a corporation which directly owns at least 50 percent of the voting stock of the corporation which is the facility owner or operator; the latter corporation is deemed a "subsidiary" of the parent corporation (other identical or similar definitions are provided in 40CFR264.141-d; 265.141-d).

park and ride lot (40CFR52.2493-2) means a parking facility for the parking of private automobiles and bicycles of mass transit users.

parking facility (40CFR52.1162-7) means a lot, garage, building, or portion thereof in or of which motor vehicles are temporarily parked.

parking facility (also called "facility") (40CFR52.2486-1) means a lot, garage, building or structure, or combination or portion thereof, in or on which motor vehicles are temporarily based.

parking space (40CFR52.1162-8) means the area allocated by a parking facility for the temporary storage of one automobile.

parking space (40CFR52.2486-6) means any area or space below, above, or at ground level, open or enclosed, on-street or off-street, that is used for parking one motor vehicle at any time.

parshall flume (EPA-94/04): Device used to measure the flow of water in an open channel.

part 70 permit (40CFR60.2) means any permit issued, renewed, or revised pursuant to part 70 of this chapter (other identical or similar definitions are provided in 40CFR63.2).

part 70 permit or permit (40CFR70.2) (unless the context suggests otherwise) means any permit or group of permits covering a part 70 source that is issued, renewed, amended, or revised pursuant to this part.

part 70 program or State program (40CFR70.2) means a program approved by the Administrator under this part.

part 70 source (40CFR70.2) means any source subject to the permitting requirements of this part, as provided in 40CFR70.3(a) and 70.3(b) of this part.

part A permit, part B permit (EPA-94/04): See Interim Permit Status.

part time fellow (40CFR46.120): An individual enrolled in an academic educational program directly related to pollution abatement and control and taking at least 6 credit hours but less than 30 credit hours per school year, or an academic workload otherwise defined by the institution as less than a full-time curriculum. The fellow need not be pursuing a degree.

partial closure (40CFR260.10) means the closure of a hazardous waste management unit in accordance with the applicable closure requirements of parts 264 and 265 of this Chapter at a facility that contains other active hazardous waste management units. For example, partial closure may include the closure of a tank (including its associated piping and underlying containment systems), landfill cell, surface impoundment, waste pile, or other hazardous waste management unit, while other units of the same facility continue to operate.

partially covered fleet (40CFR88.302.93) pertains to a vehicle fleet in a covered area which contains both covered fleet vehicles and non-covered fleet vehicles, i.e., exempt from covered fleet purchase requirements.

participant (40CFR172.1-e) means any person acting as a representative of the permittee and responsible for making available for use, or supervising the use or evaluation of, an experimental use pesticide to be applied at a specific application site.

participant (40CFR195.2) is an individual or organization engaged in radon measurement and/or mitigation activities or in offering radon measurement and/or mitigation services to consumers and others, whose proficiency program application EPA has accepted.

participant (40CFR32.105-m) means any person who submits a proposal for, enters into, or reasonably may be expected to enter into a covered transaction. This term also includes any person who acts on behalf of or is authorized to commit a participant in a covered transaction as an agent or representative of another participant.

participating PRP (40CFR304.12-j) is any potentially responsible party who has agreed, pursuant to 40CFR304.21 of this part, to submit one or more issues arising in an EPA claim for resolution pursuant to the procedures established by this part.

participation rate (EPA-94/04): Portion of population participating in a recycling program.

particle count (EPA-94/04): Results of a microscopic examination of treated water with a special "particle counter" that classifies suspended particles by number and size.

particle size analysis (40CFR796.2700-ii) is the

determination of the various amounts of the different particle sizes in a soil sample (i.e., sand, silt, clay) usually by sedimentation, sieving, micrometry, or combinations of these methods. The names and size limits of these particles as widely used in the United States are:

Name	Diameter range
Very coarse sand	2.0 to 1.0 mm
Coarse sand	1.0 to 0.5 mm
Medium sand	0.5 to 0.25 mm
Fine sand	0.25 to 0.125 mm
Very fine sand	0.125 to 0.062 mm
Silt	0.062 to 0.002 mm
Clay	<0.002 mm

(Other identical or similar definitions are provided in 40CFR796.2750-iv.)

particle size distribution (effective hydrodynamic radius): See method A in section 40CFR796.1520.

particulate asbestos material (40CFR61.141) means finely divided particles of asbestos or material containing asbestos.

particulate filter respirator (29CFR1910.94a) means an air purifying respirator, commonly referred to as a dust or a fume respirator, which removes most of the dust or fume from the air passing through the device.

particulate matter (40CFR51.100-oo) means any airborne finely divided solid or liquid material with an aerodynamic diameter smaller than 100 micrometers.

particulate matter (40CFR60.2) means any finely divided solid or liquid material, other than uncombined water, as measured by the reference methods specified under each applicable subpart, or an equivalent or alternative method.

particulate matter (40CFR60.51a) means total particulate matter emitted from MWC units as measured by Method 5 (see also 40CFR60.58a).

particulate matter (40CFR61.171) means any finely divided solid or liquid material, other than uncombined water, as measured by the specified reference method.

particulate matter emissions (40CFR51.100-pp)

means all finely divided solid or liquid material, other than uncombined water, emitted to the ambient air as measured by applicable reference methods, or an equivalent or alternative method, specified in this chapter, or by a test method specified in an approved State implementation plan.

particulates (EPA-94/04):
1. Fine liquid or solid particles such as dust, smoke, mist, fumes, or smog found in air or emissions.
2. Very small solids suspended in water. They vary in size, shape, density, and electrical charge, and can be gathered together by coagulation and flocculation.

partition coefficient (EPA-94/04): Measure of the sorption phenomenon, whereby a pesticide is divided between the soil and water phase; also referred to as adsorption partition coefficient.

parts per billion (ppb)/parts per million (ppm) (EPA-94/04): Units commonly used to express contamination ratios, as in establishing the maximum permissible amount of a contaminant in water, land, or air.

parts per million (volume) (40CFR52.741) means a volume/volume ratio which expresses the volumetric concentration of gaseous air contaminant in a million unit volume of gas.

party (40CFR104.2-e) means the Environmental Protection Agency as the proponent of an effluent standard or standards, and any person who files an objection pursuant to 40CFR104.3 hereof (under FWCPA).

party (40CFR108.2-d) means an employee filing a request under 40CFR108.3, any employee similarly situated, the employer of any such employee, and the Regional Administrator or his designee.

party (40CFR123.64-i) means the petitioner, the State, the Agency, and any other person whose request to participate as a party is granted.

party (40CFR124.72) means the EPA trial staff under 40CFR124.78 and any person whose request for a hearing under 40CFR124.74 or whose request to be admitted as a party or to intervene under 40CFR124.79 or 40CFR124.117 has been granted.

party (40CFR142.202-b) means any "person" or "supplier of water" as defined in section 1401 of the SDWA, 42USC300f, alleged to have violated any regulation implementation section 1412 of the SDWA, 42USC300g-4 and 300g-5, or section 1445 of the SDWA, 42USC300j-4, or any regulation implementing section 1445.

party (40CFR164.2-m) means any person, group, organization, or Federal agency or department that participates in a hearing.

party (40CFR209.3-h) means the Environmental Protection Agency, the respondents and any interveners.

party (40CFR22.03) means any person that participates in a hearing as complainant, respondent, or intervenor.

party (40CFR304.12-k) means EPA and any person who has agreed, pursuant to 40CFR304.21 of this part, to submit one or more issues arising in an EPA claim for resolution pursuant to the procedures established by this part, and any person who has been granted leave to intervene pursuant to 40CFR304.24(a) of this part.

party (40CFR791.3-g) refers to a person subject to a section 4 test rule, who:
(1) Seeks reimbursement from another person under these rules, or from
(2) whom reimbursement is sought under these rules.

party (40CFR8.33-k) means a respondent; the Director; and any person or organization participating in a proceeding pursuant to section 8.

party (40CFR82.3-aa) means any foreign state that is listed in appendix C to this subpart (pursuant to instruments of ratification, acceptance, or approval deposited with the Depositary of the United Nations Secretariat), as having ratified the specified control measure in effect under the Montreal Protocol. Thus, for purposes of the trade bans specified in 40CFR 82.4(k)(2) pursuant to the London Amendments, only those foreign states that are listed in appendix C to this subpart as having ratified both the 1987 Montreal Protocol and the London Amendments shall be deemed to be parties.

party (40CFR85.1807-4) shall include the Environmental Protection Agency, the manufacturer, and any interveners.

party (40CFR86.614.84-3) shall include the Agency and the manufacturer (other identical or similar definitions are provided in 40CFR86.1014.84-3; 86.1115.87).

party to the proceeding (40CFR173.2-d) shall mean the State or the Agency's Office of Enforcement (under FIFRA).

pascal (Pa) (40CFR796.1950-ii): is the standard international unit of vapor pressure and is defined as newtons per square meter (N/m^2). A newton is the force necessary to give acceleration of one meter per second squared to one kilogram of mass.

pass through (40CFR403.3-n) means a discharge which exits the POTW into waters of the United States in quantities or concentrations which, alone or in conjunction with a discharge or discharges from other sources, is a cause of a violation of any requirement of the POTW's NPDES permit (including an increase in the magnitude or duration of a violation).

passenger automobile (40CFR600.002.85-5) means any automobile which the Secretary determines is manufactured primarily for use in the transportation of no more than 10 individuals.

passive institutional control (40CFR191.12) means:
(1) Permanent markers placed at a disposal site,
(2) public records and archives,
(3) government ownership and regulations regarding land or resource use, and
(4) other methods of preserving knowledge about the location, design, and contents of a disposal system.

past year (40CFR167.3) means the calendar year immediately prior to that in which the report is submitted.

paste mixing facility (40CFR60.371-f) means the facility including lead oxide storage, conveying, weighing, metering, and charging operations; paste blending, handling, and cooling operations; and plate pasting, takeoff, cooling, and drying operations (under CAA).

pasture (40CFR503.11-k) is land on which animals feed directly on feed crops such as legumes, grasses, grain stubble, or stover.

pastures crops (40CFR257.3.5) means crops such as legumes, grasses, grain stubble and stover which are consumed by animals while grazing (under the Resource Conservation Recovery Act).

path length (40CFR60-AB-2.18) means the depth of effluent in the light beam between the receiver and the transmitter of a single-pass transmissometer, or the depth of effluent between the transceiver and reflector of a double-pass transmissometer. Two path lengths are referenced by this specification as follows:
(1) Monitor path length: The path length (depth of effluent) at the installed location of the CEMS.
(2) Emission outlet path length: The path length (depth of effluent) at the location where emissions are released to the atmosphere. For noncircular outlets, $D(e) = (2LW) \div (L + W)$, where L is the length out the outlet and W is the width of the outlet. Note that this definition does not apply to pressure baghouse outlets with multiple stacks, side discharge vents, ridge roof monitors, etc.

pathogenic organisms (40CFR503.31-f) are disease-causing organisms. These include, but are not limited to, certain bacteria, protozoa, viruses, and viable helminth ova.

pathogens (EPA-94/04): Microorganisms that can cause disease in other organisms or in humans, animals and plants (e.g., bacteria, viruses, or parasites) found in sewage, in runoff from farms or rural areas populated with domestic and wild animals, and in water used for swimming. Fish and shellfish contaminated by pathogens, or the contaminated water itself, can cause serious illness.

payment (40CFR35.3105-h): An action by the EPA to increase the amount of capitalization grant funds available for cash draw from an LOC (letter of credit).

Pb (40CFR58.1-v) means lead.

PBDD (40CFR766.3): See polybrominated dibenzo-p-dioxin.

PCA (40CFR86.1102.87): Production Compliance Audit as described in 40CFR86.1106.87 of this subpart.

PCB and PCBs (40CFR761.3) means any chemical substance that is limited to the biphenyl molecule that has been chlorinated to varying degrees or any combination of substances which contains such substance. Refer to 40CFR761.1(b) for applicable concentrations of PCBs. PCB and PCBs as contained in PCB items are defined in 40CFR761.3. For any purposes under this part, inadvertently generated non-Aroclor PCBs are defined as the total PCBs calculated following division of the quantity of monochlorinated biphenyls by 50 and dichlorinated biphenyls by 5.

PCB article (40CFR761.3) means any manufactured article, other than a PCB container, that contains PCBs and whose surface(s) has been in direct contact with PCBs. "PCB article" includes capacitors, transformers, electric motors, pumps, pipes and any other manufactured item:
(1) which is formed to a specific shape or design during manufacture,
(2) which has end use function(s) dependent in whole or in part upon its shape or design during end use, and
(3) which has either no change of chemical composition during its end use or only those changes of composition which have no commercial purpose separate from that of the PCB Article.

PCB article container (40CFR761.3) means any package, can, bottle, bag, barrel, drum, tank, or other device used to contain PCB articles or PCB equipment, and whose surface(s) has not been in direct contact with PCBs.

PCB container (40CFR761.3) means any package, can, bottle, bag, barrel, drum, tank, or other device that contains PCBs or PCB articles and whose surface(s) has been in direct contact with PCBs.

PCB contaminated electrical equipment (40CFR761.3) means any electrical equipment, including but not limited to transformers (including those used in railway locomotives and self-propelled cars), capacitors, circuit breakers, reclosers, voltage regulators, switches (including sectionalizers and motor starters), electromagnets, and cable, that contain 50 ppm or greater PCB, but less than 500 ppm PCB. Oil-filled electrical equipment other than circuit breakers, reclosers, and cable whose PCB concentration is unknown must be assumed to be PCB-contaminated electrical equipment (see 40CFR761.30(a) and (h) for provisions permitting reclassification of electrical equipment containing 500 ppm or greater PCBs to PCB-contaminated electrical equipment).

PCB equipment (40CFR761.3) means any manufactured item, other than a PCB container or a PCB article container, which contains a PCB article or other PCB equipment, and includes microwave ovens, electronic equipment, and fluorescent light ballasts and fixtures.

PCB item (40CFR761.3) is defined as any PCB article, PCB article container, PCB container, or PCB equipment, that deliberately or unintentionally contains or has a part of it any PCB or PCBs (under TSCA).

PCB manufacturer (40CFR129.105-1) means a manufacturer who produces polychlorinated biphenyls.

PCB transformer (40CFR761.3) means any transformer that contains 500 ppm PCB or greater (cf. non-PCB transformer).

PCB wastes (40CFR761.3) means those PCBs and PCB items that are subject to the disposal requirements of subpart D of this part.

PCBs (40CFR761.123) means polychlorinated biphenyls as defined under 40CFR61.3 as specified under 40CFR61.1(b), no requirements may be avoided through dilution of the PCB concentration (under TSCA).

PCDD (40CFR766.3): See polychlorinated dibenzo-p-dioxin.

PCM (40CFR763-AA-17) means phase contrast microscopy.

PCV (positive crankcase ventilation) valve (40CFR85.2122(a)(4)(ii)) means a device to control the flow of blow-by gasses and fresh air from the crankcase to the fuel induction system of the engine.

peaches (40CFR407.61-l) shall mean the processing of peaches into the following product styles: Canned or frozen, all varieties, peeled, pitted and unpitted, whole, halves, sliced, diced, and any other cuts, nectar, and concentrate but not dehydrated.

peak air flow (40CFR85.2122(a)(15)(ii)(D)) means the maximum engine intake mass air flow rate measure during the 195 second to 202 second time interval of the Federal Test Procedure.

peak electricity demand (EPA-94/04): The maximum electricity used to meet the cooling load of a building or buildings in a given area.

peak levels (EPA-94/04): Levels of airborne pollutant contaminants much higher than average or occurring for short periods of time in response to sudden releases.

peak load (40CFR60.331-i) means 100 percent of the manufacturer's design capacity of the gas turbine at ISO standard day conditions.

peak spectral response (40CFR60-AB-2.6) means the wavelength of maximum sensitivity of the transmissometer.

peak torque speed (40CFR86.082.2) means the speed at which an engine develops maximum torque.

peaking unit (40CFR72.2) means a unit that has:
(1) an average capacity factor of no more than 10 percent during the previous 3 calendar years and
(2) a capacity factor of no more than 20 percent in each of those calendar years.

peanuts (40CFR180.1-j9) means the peanut meat after removal of the hulls.

pears (40CFR407.61-m) shall mean the processing of pears into the following product styles: Canned, peeled, halved, sliced, diced, and any other cuts, nectar and concentrate but not dehydrated.

peas (40CFR407.71-m) shall mean the processing of peas into the following product styles: Canned and frozen, all varieties and sizes, whole.

peel-apart film article (40CFR723.175-11) means a self-developing photographic film article consisting of a positive image receiving sheet, a light sensitive negative sheet, and a sealed reagent pod containing a developer reagent and designed so that all the chemical substances required to develop or process the film will not remain sealed within the article during and after the development of the film.

PEL (40CFR763.121): Permissible exposure limit.

pellet burning wood heater (40CFR60-AA (method 28-2.13)) means a wood heater which meets the following criteria:
(1) The manufacturer makes no reference to burning cord wood in advertising or other literature,
(2) the unit is safety listed for pellet fuel only,
(3) the unit operating and instruction manual must state that the use of cordwood is prohibited by law, and
(4) the unit must be manufactured and sold including the hopper and auger combination as integral parts.

percent absorption (40CFR795.228-3) means 100 times the ratio between total excretion of radioactivity following oral or dermal administration and total excretion following intravenous administration of test substance.

percent load (40CFR86.082.2) means the fraction of the maximum available torque at a specified engine speed.

percent removal (40CFR133.101-j) means a percentage expression of the removal efficiency across a treatment plant for a given pollutant parameter, as determined from the 30-day average values of the raw wastewater influent pollutant concentrations to the facility and the 30-day average values of the effluent pollutant concentrations for a given time period.

percent saturation (EPA-94/04): The amount of a substance that is dissolved in a solution compared to the amount that could be dissolved in it.

percentage of completion method (40CFR31.3) refers to a system under which payments are made for construction work according to the percentage of completion of the work, rather than to the grantee's cost incurred.

perceptible (40CFR63.321) leaks mean any perchloroethylene vapor or liquid leaks that are obvious from:
(1) The odor of perchloroethylene;
(2) Visual observation, such as pools or droplets of liquid; or
(3) The detection of gas flow by passing the fingers over the surface of equipment.

perceptible leaks (40CFR60.621) means any petroleum solvent vapor or liquid leaks that are conspicuous from visual observation or that bubble after application of a soap solution, such as pools or droplets of liquid, open containers or solvent, or solvent laden waste standing open to the atmosphere.

perched water (EPA-94/04): Zone of unpressurized water held above the water table by impermeable rock or sediment.

perchloroethylene consumption (40CFR63.321) means the total volume of perchloroethylene purchased based upon purchase receipts or other reliable measures.

percolating water (EPA-94/04): Water that passes through rocks or soil under the force of gravity.

percolation (EPA-94/04):
1. The movement of water downward and radially through subsurface soil layers, continuing downward to groundwater. Can also involve upward movement of the water.
2. Slow seepage of water through a filter.

performance assessment (40CFR191.12) means an analysis that:

(1) Identifies the processes and events that might affect the disposal system;

(2) examines the effects of these processes and events on the performance of the disposal system; and

(3) estimates the cumulative releases of radionuclides, considering the associated uncertainties, caused by all significant processes and events. These estimates shall be incorporated into an overall probability distribution of cumulative release to the extent practicable.

performance audit (40CFR63.2) means a procedure to analyze blind samples, the content of which is known by the Administrator, simultaneously with the analysis of performance test samples in order to provide a measure of test data quality.

performance averaging period (40CFR60.431) means 30 calendar days, one calendar month, or four consecutive weeks as specified in sections of this subpart.

performance data (for incinerators) (EPA-94/04): Information collected, during a trial burn, on concentrations of designated organic compounds and pollutants found in incinerator emissions. Data analysis must show that the incinerator meets performance standards under operating conditions specified in the RCRA permit (see trial burn or performance standards).

performance evaluation (40CFR63.2) means the conduct of relative accuracy testing, calibration error testing, and other measurements used in validating the continuous monitoring system data (other identical or similar definitions are provided in 40CFR63.101; 63.191).

performance evaluation sample (40CFR141.2) means a reference sample provided to a laboratory for the purpose of demonstrating that the laboratory can successfully analyze the sample within limits of performance specified by the Agency. The true value of the concentration of the reference material is unknown to the laboratory at the time of the analysis.

performance specification (40CFR247.101-c) means a specification that states the desired operation or function of a product but does not specify the materials from which the product must be constructed.

performance standards (EPA-94/04):
(1) Regulatory requirements limiting the concentrations of designated organic compounds, particulate matter, and hydrogen chloride in emissions from incinerators.
(2) Operating standards established by EPA for various permitted pollution control systems, asbestos inspections, and various program operations and maintenance requirements.

performance test (40CFR63.2) means the collection of data resulting from the execution of a test method (usually three emission test runs) used to demonstrate compliance with a relevant emission standard as specified in the performance test section of the relevant standard (other identical or similar definitions are provided in 40CFR63.101; 63.191).

periodic application of cover material (40CFR257.3.6-4) means the application and compaction of soil and other suitable material over disposed solid waste at the end of each operating day or at such frequencies and in such a manner as to reduce the risk of fire and to impede vectors access to the waste (other identical or similar definitions are provided in 40CFR257.3.8-6).

periphyton (EPA-94/04): Microscopic underwater plants and animals that are firmly attached to solid surfaces such as rocks, logs, pilings, and other structures.

perlite composite board (40CFR248.4-w) means insulation board composed of expanded perlite and fibers formed into rigid, flat, rectangular units with a suitable sizing material incorporated in the product. It may have on one or both surfaces a facing or coating to prevent excessive hot bitumen strike-in during roofing installation.

permanent opening (40CFR60.541) means an opening designed into an enclosure to allow tire components to pass through the enclosure by conveyor or other mechanical means, to provide access for permanent mechanical or electrica' equipment, or to direct air flow into the enclosure. A permanent opening is not equipped with a door or other means of obstruction of air flow.

permanent radon barrier (40CFR192.31-l) means the final radon barrier constructed to achieve compliance with, including attainment of, the limit on releases of radon-222 in 40CFR192.32(b)(1)(ii).

permanent storage capacity (40CFR60.301-d) means grain storage capacity which is inside a building, bin, or silo.

permeability (EPA-94/04): The rate at which liquids pass through soil or other materials in a specified direction.

permissible dose (EPA-94/04): The dose of a chemical that may be received by an individual without the expectation of a significantly harmful result.

permissible exposure limit (PEL) (40CFR763.121): The employer shall ensure that no employee is exposed to an airborne concentration in excess of 0.2 fiber per cubic centimeter of air as an 8-hour time-weighted average (TWA), as determined by the method described in Appendix A of this section or by an equivalent method.

permit (40CFR256.06) is an entitlement to commence and continue operation of a facility as long as both procedural and performance standards are met. The term "permit" includes any functional equivalent such as a registration or license.

permit (40CFR122.2) means an authorization, license, or equivalent control document issued by EPA or an "approved State" to implement the requirements of this part and parts 123 and 124. "Permit" includes an NPDES "general permit" (40CFR122.28). Permit does not include any permit which has not yet been the subject of final agency action, such as a "draft permit" or a "proposed permit."

permit (40CFR124.2) means an authorization, license, or equivalent control document issued by EPA or an "approved State" to implement the requirements of this part and parts 122, 123, 144,

145, 233, 270, and 271. "Permit" includes RCRA "permit by rule" (40CFR270.60), UIC area permit (40CFR144.33), NPDES or 404 "general permit" (40CFR270.61, 144.34, and 233.38). Permit does not include RCRA interim status (40CFR270.70), UIC authorization by rule (40CFR144.21), or any permit which has not yet been the subject of final agency action, such as a "draft permit" or a "proposed permit."

permit (40CFR129.2-e) means a permit for the discharge of pollutants into navigable waters under the National Pollutant Discharge Elimination System established by section 402 of the Act and implemented in regulations in 40CFR124 and 125.

permit (40CFR144.3) means an authorization, license, or equivalent control document issued by EPA or an approved State to implement the requirements of this part, parts 145, 146 and 124. "Permit" includes an area permit (40CFR144.33) and an emergency permit (40CFR144.34). Permit does not include UIC authorization by rule (40CFR144.21), or any permit which has not yet been the subject of final agency action, such as a "draft permit."

permit (40CFR146.3) means an authorization, license, or equivalent control document issued by EPA or an "approved State" to implement the requirements of this part and parts 124, 144, and 145. Permit does not include RCRA interim status (122.23), UIC authorization by rule (40CFR144.21 to 144.26 and 144.15), or any permit which has not yet been the subject of final agency action, such as a "draft permit" or a "proposed permit."

permit (40CFR147.2902) means an authorization issued by EPA to implement UIC program requirements. Permit does not include the UIC authorization by rule or any permit which has not yet been the subject of final Agency action.

permit (40CFR2.309-2) means any permit applied for or granted under the Act.

permit (40CFR21.2-g) means any permit issued by either EPA or a State under the authority of section 402 of the Act; or by the Corps of Engineers under section 404 of the Act.

permit (40CFR22.03) means a permit issued under section 102 of the Marine, Protection, Research, and Sanctuaries Act.

permit (40CFR232.2) means a written authorization issued by an approved State to implement the requirements of part 233, or by the Corps under 33CFR320-330. When used in these regulations, "permit" includes "general regulation permit" as well as individual permit.

permit (40CFR270.2) means an authorization, license, or equivalent control document issued by EPA or an approved State to implement the requirements of this part and parts 271 and 124. Permit includes permit by rule (40CFR270.60), and emergency permit (40CFR270.61). Permit does not include RCRA interim status (subpart G of this part), or any permit which has not yet been the subject of final agency action, such as a draft permit or a proposed permit.

permit (40CFR403.3-l): See NPDES permit.

permit (40CFR501.2) means an authorization, license, or equivalent control document issued by EPA or an "approved State program" to implement the requirements of this part.

permit (EPA-94/04): An authorization, license, or equivalent control document issued by EPA or an approved state agency to implement the requirements of an environmental regulation; e.g., a permit to operate a wastewater treatment plant or to operate a facility that may generate harmful emissions.

permit (SMCRA701-30USC1291) means a permit to conduct surface coal mining and reclamation operations issued by the State regulatory authority pursuant to a State program or by the Secretary pursuant to a Federal program.

permit applicant or applicant (SMCRA701-30USC1291) means a person applying for a permit.

permit area (40CFR440.141-14) means the area of land specified or referred to in an NPDES permit in which active mining and related activities may occur that result in the discharge regulated under the terms of the permit. Usually this is specifically delineated in an NPDES permit or permit application, but in other cases may be ascertainable from an Alaska Tri-agency permit application or similar document specifying the mine location, mining plan and similar data.

permit area (SMCRA701) means the area of land indicated on the approved map submitted by the operator with his application, which area of land shall be covered by the operator's bond as required by section 1259 of this title and shall be readily identifiable by appropriate markers on the site.

permit modification (40CFR63.2) means a change to a title V permit as defined in regulations codified in this chapter to implement title V of the Act (42 U.S.C. 7661).

permit modification (40CFR70.2) means a revision to a part 70 permit that meets the requirements of 40CFR70.7(e) of this part.

permit or license applicant (ESA3-16USC1531) means, when used with respect to an action of a Federal agency for which exemption is sought under section 1536 of this title, any person whose application to such agency for a permit or license has been denied primarily because of the application of section 1536(a) of this title to such agency action.

permit or PSD permit (40CFR124.41) means a permit issued under 40CFR52.21 or by an approved State.

permit program (40CFR61.02) means a comprehensive State operating permit system established pursuant to title V of the Act (42 U.S.C. 7661) and regulations codified in part 70 of this chapter and applicable State regulations, or a comprehensive Federal operating permit system established pursuant to title V of the Act and regulations codified in this chapter.

permit program (40CFR63.2) means a comprehensive State operating permit system established pursuant to title V of the Act (42 U.S.C. 7661) and regulations codified in part 70 of this chapter and applicable State regulations, or a comprehensive Federal operating permit system established pursuant to title V of the Act and regulations codified in this chapter (other identical or similar definitions are provided in 40CFR63.101; 63.191).

permit program costs (40CFR70.2) means all reasonable (direct and indirect) costs required to develop and administer a permit program, as set forth in 40CFR70.9(b) of this part (whether such costs are incurred by the permitting authority or other State or local agencies that do not issue permits directly, but that support permit issuance or administration).

permit revision (40CFR63.2) means any permit modification or administrative permit amendment to a title V permit as defined in regulations codified in this chapter to implement title V of the Act (42 U.S.C. 7661).

permit revision (40CFR70.2) means any permit modification or administrative permit amendment.

permit revision (40CFR72.2) means a permit modification, fast track modification, administrative permit amendment, or automatic permit amendment, as provided in subpart H of this part.

permit-by-rule (40CFR270.2) means a provision of these regulations stating that a facility or activity is deemed to have a RCRA permit if it meets the requirements of the provision.

permittee (40CFR172.1-g) means any applicant to whom an experimental use permit has been granted.

permittee (OPA1001) means a person holding an authorization, license, or permit for geological exploration issued under section 11 of the Outer Continental Shelf Lands Act (43USC1340) or applicable State law.

permittee (SMCRA701-30USC1291) means a person holding a permit.

permitting authority (40CFR230.3-n) means the District Engineer of the U.S. Army Corps of Engineers or such other individual as may be designated by the Secretary of the Army to issue or deny permits under section 404 of the Act; or the State Director of a permit program approved by EPA under section 404(g) and section 404(h) or his delegated representative.

permitting authority (40CFR503.9-p) is either EPA or a State with an EPA-approved sludge management program.

permitting authority (40CFR61.02) means:
(1) The State air pollution control agency, local agency, other State agency, or other agency authorized by the Administrator to carry out a permit program under part 70 of this chapter; or
(2) The Administrator, in the case of EPA-implemented permit programs under title V of the Act (42 U.S.C. 7661).

permitting authority (40CFR63.2) means:
(1) The State air pollution control agency, local agency, other State agency, or other agency authorized by the Administrator to carry out a permit program under part 70 of this chapter; or
(2) The Administrator, in the case of EPA-implemented permit programs under title V of the Act (42 U.S.C. 7661).
(Other identical or similar definitions are provided in 40CFR63.101; 63.191.)

permitting authority (40CFR63.51) means either a State agency with an approved permitting program under Title V of the Act or the Administrator in cases where the State does not have an approved permitting program.

permitting authority (40CFR63.51) means the permitting authority as defined in part 70 of this chapter.

permitting authority (40CFR70.2) means either of the following:
(1) The Administrator, in the case of EPA-implemented programs; or
(2) The State air pollution control agency, local agency, other State agency, or other agency authorized by the Administrator to carry out a permit program under this part.

permitting authority (40CFR72.2) means either:

(1) the Administrator in the case of issuance and administration of Acid Rain permits or

(2) the State air pollution control agency, local agency, other State agency, or other agency authorized by the Administrator to issue proposed Acid Rain permits under subpart G of this part and the other regulations promulgated pursuant to titles IV and V of the Act.

permitting authority (CAA402-42USC7651a) means the Administrator, or the State or local air pollution control agency, with an approved permitting program under part B of title III of the Act.

permitting authority (CAA501-42USC7661) means the Administrator or the air pollution control agency authorized by the Administrator to carry out a permit program under this title.

persistence (EPA-94/04): Refers to the length of time a compound stays in the environment, once introduced. A compound may persist for less than a second or indefinitely.

persistent pesticides (EPA-94/04): Pesticides that do not break down chemically or break down very slowly and remain in the environment after a growing season.

person (40CFR2.201-a) means an individual, partnership, corporation, association, or other public or private organization or legal entity, including Federal, State or local governmental bodies and agencies and their employees (other identical or similar definitions are provided in 40CFR2.305-2; 2.310-2; 8.2-r; 8.33-l; 13.2-j; 15.4; 22.03; 22.42; 27.2; 32.105-n; 34.105-l; 40.115-5; 52.741; 72.2; 82.3-bb; 82.172; 82.152-o; 104.2-f; 110.1; 112.2; 122.2; 123.64-ii; 124.2; 124.41; 141.2; 142.2; 144.3; 149.101-e; 154.3-f; 160.3; 163.2-f; 164.2-n; 205.2-12; 209.3-i; 231.2-d; 232.2; 248.4-x; 249.04-i; 250.4-ff; 252.4-f; 253.4; 259.10.b; 260.10; 270.2; 280.12; 300.5; 302.3; 304.12-l; 355.20; 370.2; 501.2; 700.43; 704.3; 710.2-s; 712.3-l; 716.3; 720.3-x; 723.50-2; 723.175-3; 723.250-6; 747.115-1; 747.195-1; 747.200-1; 717.3-f; 761.3; 762.3-b; 763.63-i; 763.163; 790.3; 792.3; CAA302-42USC7602; CWA311-33USC1321; CWA312-33USC1322; CZMA304-16USC1453; ESA3-16USC1531; FIFRA2-7USC136; MMPA3-16USC1362; NCA3-42USC4902; OPA1001; OSHA3-29USC652; RCRA1004-42USC6903; RCRA11006.b-42USC6992e; SDWA1401-42USC300f; SMCRA701-30USC1291; SF101-42USC9601; SF329).

person (40CFR503.9-q) is an individual, association, partnership, corporation, municipality, State or Federal agency, or an agent or employee thereof.

person (40CFR749.68-14) means any natural person, firm, company, corporation, joint venture, partnership, sole proprietorship, association, or any other business entity; any State or political subdivision thereof; any municipality; any interstate body; and any department, agency, or instrumentality of the Federal Government.

person (40CFR88.302.94) includes an individual, corporation, partnership, association, State, municipality, political subdivision of a State, and any agency, department, or instrumentality of the United States and any officer, agent, or employee thereof.

person who prepares sewage sludge (40CFR503.9-r) is either the person who generates sewage sludge during the treatment of domestic sewage in a treatment works or the person who derives a material from sewage sludge.

personal air samples (EPA-94/04): Air samples taken with a pump is directly attached to the worker with the collecting filter and cassette placed in the worker's breathing zone (required under OSHA asbestos standards and EPA worker protection rule).

personal property (40CFR30.200) means property other than real property. It may be tangible (having physical existence), such as equipment and supplies, or intangible (having no physical existence), such as patents, inventions, and copyrights (other identical or similar definitions are provided in 40CFR35.6015-30; 247.101-d).

personal protective equipment (40CFR721.3) means any chemical protective clothing or device placed on the body to prevent contact with, and exposure to, an identified chemical substance or substances in the work area. Examples include, but are not limited to, chemical protective clothing, aprons, hoods, chemical goggles, face splash shields, or equivalent eye protection, and various types of respirators. Barrier creams are not included in this definition.

personnel or facility personnel (40CFR260.10) means all persons who work, at, or oversee the operations of, a hazardous waste facility, and whose actions or failure to act may result in noncompliance with the requirements of parts 264 or 265 of this chapter.

pest (40CFR455.10-d) means:
(1) Any insect, rodent, nematode, fungus, weed, or
(2) any other form of terrestrial or aquatic plant or animal life or virus, bacteria, or other micro-organism (except viruses, bacteria, or other micro-organisms on or in living man or other living animals) which the Administrator declares to be a pest under section 25(c)(1) of Pub. L. 94-140, Federal Insecticide, Fungicide and Rodenticide Act.
(Other identical or similar definitions are provided in 40CFR152.5.)

pest (EPA-94/04): An insect, rodent, nematode, fungus, weed or other form of terrestrial or aquatic plant or animal life that is injurious to health or the environment.

pest (FIFRA2-7USC136) means:
(1) any insect, rodent, nematode, fungus, weed or
(2) any other form of terrestrial or aquatic plant or animal life or virus, bacterial or other micro-organism (except viruses, bacteria, or other micro-organisms on or in living man or other living animals) which the administrator declares to be a pest under section 25(C)(1).

pest problem (40CFR162.151-d) means:
(1) a pest infestation and its consequences, or
(2) any condition for which the use of plant regulators, defoliants, or desiccants would be appropriate.

pesticidal product report (40CFR167.3) means information showing the types and amounts of pesticidal products which were:
(1) Produced in the past calendar year;
(2) produced in the current calendar year; and
(3) sold or distributed in the past calendar year.
For active ingredients, the pesticidal product report must include information on the types and amounts of an active ingredient for which there is actual or constructive knowledge of its use or intended use as a pesticide. This pesticidal product report also pertains to those products produced for export only which must also be reported. A positive or a negative annual report is required in order to maintain registration for the establishment.

pesticidal report (40CFR167.3) means a pesticide, active ingredient, or device.

pesticide (40CFR125.58-m) means any substance or mixture of substances intended for preventing, destroying, repelling, or mitigating any pest, or any substance or mixture of substances intended for use as a plant regulator, defoliant, or desiccant (other identical or similar definitions are provided in 40CFR710.2-b; 720.3-a).

pesticide (40CFR152.3-s) means any substance or mixture of substances intended for preventing, destroying, repelling, or mitigating any pest, or intended for use as a plant regulator, defoliant, or desiccant, other than any article that:
(1) Is a new animal drug under FFDCA sec. 20(w), or
(2) Is an animal drug that has been determined by regulation of the Sectary of Health and Human Services not be a new animal drug, or
(3) Is an animal feed under FFDCA sec. 201(x) that bears or contains any substances described by paragraph(s) (1) or (2) of this section.

pesticide (40CFR165.1-p) means any substance or mixture of substances intended for preventing, destroying, repelling, or mitigating any pest, or any substance or mixture of substances intended for use as a plant regulator, defoliant, or desiccant. Pesticides include four types:
(1) Excess pesticides means all pesticides which cannot be legally sold pursuant to the Act or which are to be discarded;
(2) Organic pesticides means carbon-containing substances used as pesticides, excluding metallo-organic compounds;
(3) Inorganic pesticides means noncarbon-containing substances used as pesticides;
(4) Metallo-organic pesticides means a class of organic pesticides containing one or more metal or metalloid atoms in the structure.

pesticide (40CFR455.10-a) means any substance or mixture of substances intended for preventing, destroying, repelling, or mitigating any pest.

pesticide (40CFR710.2-b) shall have the meaning contained in the Federal Insecticide, Fungicide, and Rodenticide Act, 7USC136 et seq., and the regulations issued thereunder. The meaning of the term, pesticide, defined under the Act was provided below:

Pesticide (7USC136(u)) means:
(1) any substance or mixture of substances intended for preventing, destroying, repelling, or mitigating any pest, and
(2) any substance or mixture of substances intended for use as a plant regulator, defoliant, or desiccant, except that the term "pesticide" shall not include any article that is a "new animal drug" within the meaning of section 201(w) of the Federal Food, Drug, and Cosmetic Act (21 U.S.C. 321(w)) [21 USCS 321(w)], that has been determined by the Secretary of Health and Human Services not to be a new animal drug by a regulation establishing conditions of use for the article, or that is an animal feed within the meaning of section 201(x) of such Act (21 U.S.C. 321(x)) bearing or containing a new animal drug.

(Other identical or similar definitions are provided in 40CFR720.3-a.)

pesticide (40CFR82.172) has the meaning contained in the Federal Insecticide, Fungicide, and Rodenticide Act, 7 U.S.C. 136 et seq. and the regulations issued under it.

pesticide (EPA-94/04): Substances or mixture there of intended for preventing, destroying, repelling, or mitigating any pest. Also, any substance or mixture intended for use as a plant regulator, defoliant, or desiccant.

pesticide chemical (40CFR163.2-d) shall have the same meaning as it has in paragraph (q) and (r), respectively, of section 201 of the Act.

pesticide chemical (40CFR177.3) means any substance which alone, or in chemical combination with or in formulation with one or more other substances, is a "pesticide" within the meaning of FIFRA and which is used in the production, storage, or transportation of any raw agricultural commodity or processed food. The term includes any substance that is an active ingredient, intentionally added inert ingredient, or impurity of such a "pesticide."

pesticide chemical (40CFR180.1-k) as defined in section 201(q) of the act, means any substance which, alone, in chemical combination, or in formulation with one or more other substances, is an economic poison within the meaning of the Federal Insecticide, Fungicide, and Rodenticide Act (7USC136(u)) and as defined in 40CFR152.3 of regulations for its enforcement (40CFR152.3), as now in force or as hereafter amended, and which is used in the production, storage, or transportation of raw agricultural commodities (see below for the listing of pesticide chemicals regulated under FIFRA. The chemicals provided herewith in an alphabetical order were excerpted from 40CFR180.102-1147, 1995 edition which used the source: 36 FR 22540, Nov 25, 1971).

40CFR #	pesticide chemicals (alphabetical order)
180.108	acephate
180.1031	acetaldehyde
180.1029	acetic acid
180.249	alachlor
180.269	aldicarb
180.113	allethrin (allyl homolog of cinerin i)
180.1002	
180.1091	aluminum isopropoxide and aluminum secondary butoxide
180.225	aluminum phosphide
180.415	aluminum tris (o,ethylphosphonate)
180.258	ametryn

180.1065	2-amino-4,5-dihydro-6-methyl-4-propyl-s-triazolo (1,5-alpha) pyrimidin-5-one
180.332	4-amino-6-(1,1-dimethyl-ethyl)-3-(methylthio) 1,2,4-triazin-5 (4h)-one
180.312	4-aminopyridine
180.287	amitraz
180.1003	ammonia
180.217	ammoniates of [ethylenebis (dithiocarbamato)] zinc and ethylenebis (dithiocarbamic acid) bimolecular and trimolecular cyclic anhydrosulfides and disulfides
180.319	ammoniates of [ethylenebis (dithiocarbamato)] zinc and ethylenebis [dithiocarbamic acid] bimolecular and trimolecular cyclic anhydrosulfides and disulfides; interim tolerance
180.1018	ammonium nitrate
180.188	ammonium sulfamate
180.1042	aqueous extract of seaweed meal
180.360	asulam
180.220	atrazine
180.449	avermectin b1 and its delta-8,9-isomer
180.1076	bacillus popilliae, viable spores
180.1111	bacillus subtillis gb03
180.1011	bacillus thuringiensis berliner, viable spores
180.268	barban
180.294	benomyl
180.445	bensulfuron methyl ester
180.355	bentazon
180.376	6-benzyladenine
180.457	beta-([1,1'-biphenyl]-4-yloxy)-alpha-(1,1-dimethylethyl)-1h-1,2,4-triazole-1-ethanol
180.351	bifenox
180.442	bifenthrin
180.141	biphenyl
180.163	1,1-bis(p-chlorophenyl)-2,2,2-trichloroethanol
180.271	boron
180.210	bromacil
180.324	bromoxynil
180.255	bufencarb
180.1034	butanoic anhydride
180.358	butralin
180.321	sec-butylamine
180.1062	butyl benzyl phthalate
180.208	n-butyl-n-ethyl-a,a,a-trifluoro-2,6-dinitro-p-toluidine
180.311	cacodylic acid
180.125	calcium cyanide
180.1054	calcium hypochlorite
180.267	captafol
180.103	captan
180.1110	3-carbamyl-2,4,5-trichlorbenzoic acid
180.169	carbaryl
180.319	carbaryl (1-naphthyl n-methylcarbamate and its metabolite 1-naphthol, calculated as carbaryl; interim tolerance
180.254	carbofuran
180.1049	carbon dioxide
180.1005	carbon tetrachloride
180.156	carbophenothion

180.301	carboxin
180.266	chloramben
180.285	chlordimeform
180.429	chlorimuron ethyl
180.1095	chlorine gas
180.247	2-chloroallyl diethyldithiocarbamate
180.282	2-chloro-n,n-diallylacetamide
180.322	2-chloro-1-(2,4-dichlorophenyl)vinyl diethyl phosphate
180.307	2-((4-chloro-6-(ethylamino)-s-triazin-2-yl) amino)-2-methylpropionitrile
180.211	2-chloro-n-isopropylacetanilide
180.257	chloroneb
180.202	p-chlorophenoxyacetic acid
180.450	beta-(4-chlorophenoxy)-alpha-(1,1-dimethylethyl)-1h-1,2,4-triazole-1-1-ethanol
180.410	1-(4-chlorophenoxy)-3,3-dimethyl- (1h-1,2,4-triazol-1-yl)-2-butanone
180.325	2-(m-chlorophenoxy)propionic acid
180.425	2-(2-chlorophenyl)methyl-4,4-dimethyl 3-isoxazolidinone
180.1008	chloropicrin
180.201	chlorosulfamic acid
180.275	chlorothalonil
180.1045	chlorotoluene
180.252	2-chloro-1-(2,4,5-trichlorophenyl)vinyl dimethyl phosphate
180.216	chloroxuron
180.342	chlorpyrifos
180.419	chlorpyrifos-methyl
180.405	chlorsulfuron
180.398	chlorthiophos
180.181	cipc
180.458	clethodim ((e)-(±)-2-[1-[[(3-chloro-2-propenyl)oxy]imino] propyl]-5-[2-(ethylthio)propyl]-3-hydroxy-2-cyclohexen-1-one)
180.446	clofentezine
180.431	clopyralid
180.1075	colletotrichum gloeosporioides fsp aeschynomene
180.1051	combustion product gas
180.176	coordination product of zinc ion and maneb
180.319	coordination product of zinc ion and maneb; interim tolerance
180.1021	copper
180.136	copper carbonate, basic
180.189	coumaphos
180.1028	cross-linked nylon-type encapsulating polymer
180.1039	cross-linked polyurea-type encapsulating polymer
180.1082	cross-linked polyurea-type encapsulating polymer (alachlor)
180.295	crufomate
180.379	cyano(3-phenoxyphenyl)methyl-4-chloro-a-(1-methylethyl) benzenacetate
180.438	[1 alpha-(s),3 alpha(z)]-(±)-cyano(3-phenoxyphenyl) methyl 3-(2-chloro-3,3,3-trifluoro-1-propenyl)-2,2-dimethylcyclopropane-carboxylate
180.336	cycloheximide
180.436	cyfluthrin
180.144	cyhexatin
180.418	cypermethrin
180.306	cyprazine

180.414	cyromazine
180.142	2,4-d
180.150	dalapon
180.246	daminozide
180.1097	(z)-9-dedecenyl acetate and (z)-11-tetradecenyl acetate (gbm-rope)
180.1107	delta endotoxin of bacillus thuringiensis variety kurstaki encapsulated into killed pseudomonas fluorescens
180.1108	delta endotoxin of bacillus thuringiensis variety San Diego encapsulated into killed pseudomonas fluorescens
180.435	deltamethrin
180.105	demeton
180.353	desmedipham
180.326	dialifor
180.1026	n,n-diallyl dichloroacetamide
180.1017	diatomaceous earth
180.153	diazinon
180.227	dicamba
180.231	dichlobenil
180.118	dichlone
180.460	4-(dichloroacetyl)-3,4-dihydro-3-methyl-2h-1,4-benzoxazine
180.277	s-2,3-dichloroallyl diisopropylthiocarbamate
180.139	1,1-dichloro-2,2-bis(p-ethylphenyl) ethane
180.158	2,4-dichloro-6-o-chloroanilino-s-triazine
180.317	3,5-dichloro-n-(1,1-dimethyl-2-propynyl) benzamine
180.1077	2,2-dichloro-n-(1,3-dioxolan-2-ylmethyl)-n-2-propenylacetamide
180.200	2,6-dichloro-4-nitroaniline
180.331	4-(2,4-dichlorophenoxy) butyric acid
180.380	3-(3,5-dichlorophenyl)-5-ethenyl-5-methyl-2,4-oxazolidinedione
180.434	1-[[2-(2,4-dichlorophenyl)-4-propyl-1,3-dioxolan-2-yl] methyl]-1h-1,2,4-triazole
180.424	2-(3,5-dichlorophenyl)-2-(2,2,2-trichloroethyl)oxirane
180.235	2,2-dichlorovinyl dimethyl phosphate
180.385	diclofop-methyl
180.402	diethatyl-ethyl
180.183	o,o-diethyl s-(2-(ethylthio)ethyl phosphorodithioate
180.234	o,o-diethyl o-(p-(methylsulfinyl) phenyl) phosphorothioate
180.328	n,n-diethyl-2-(1-naphthalenyloxy)propionamide
180.1066	o,o-diethyl-o-phenylphosphorothioate
180.369	difenzoquat
180.377	diflubenzuron
180.426	2-[4,5-dihydro-4-methyl-4(1-methylethyl)-5-oxo-1h-imidazol-2-yl]-3-quinoline carboxylic acid
180.241	s-(o,o-diisopropyl phosphorodithioate) of n-(2-mercaptoethyl) benzenesulfonamide
180.406	dimethipin
180.204	dimethoate including its oxygen analog
180.233	o,o-dimethyl o-p-(dimethylsulfamoyl) phenyl phosphorothioate including its oxygen analog
180.320	3,5-dimethyl-4-(methylthio)phenyl methylcarbamate
180.154	o,o-dimethyl s-[(4-oxo-1,2,3-benzotriazin-3(4h)-yl)methyl] phosphorodithioate
180.154a	o,o-dimethyl s-[(4-oxo-1,2,3-benzotriazin-3(4h)-yl)methyl] phosphorodithioate residues and/or its metabolites in milk
180.299	dimethyl phosphate of 3-hydroxy-n,n-dimethyl cis-crotonamide
180.296	dimethyl phosphate of 3-hydroxy-n-methyl-cis-crotonamide

180.280	dimethyl phosphate of a-methylbenzyl 3-hydroxy-cis-crotonate
180.185	dimethyl tetrachloroterephthalate
180.1046	dimethylformamide
180.384	n n-dimethylpiperidium chloride
180.1083	dimethyl sulfoxide
180.198	dimethyl (2,2,2-trichloro-1-hydroxyethyl) phosphonate
180.372	2,6-dimethyl-4-tridecylmorpholine
180.327	dinitramine
180.344	4,6-dinitro-o-cresol and its sodium salt
180.341	2,4-dinitro-6-octylphenyl crotonate and 2,6-dinitro-4-octylphenyl crotonate
180.281	dinoseb
180.171	dioxathion
180.230	diphenamid
180.190	diphenylamine
180.329	dipropetryn
180.143	dipropyl isocinchomeronate
180.226	diquat
180.106	diuron
180.172	dodine
180.1071	egg solids (whole)
180.182	endosulfan
180.293	endothall
180.319	endothall (7-oxabicyclo-(2,2,1) heptane 2,3-dicarboxylic acid); interim tolerance
180.131	endrin
180.119	epn
180.416	ethalfluralin
180.300	ethephon
180.173	ethion
180.262	ethoprop
180.345	ethofumesate
180.178	ethoxyquin
180.370	5-ethoxy-3-trichloromethyl-1,2,4-thiadiazole
180.412	2-[1-(ethoxyimino(butyl)-5-[2-(ethylthio)propyl]-3-hydroxy-2-cyclohexane-1-one
180.212	s-ethyl cyclohexylethylthiocarbamate
180.109	ethyl 4,4'-dichlorobenzilate
180.232	s-ethyl diisobutylthiocarbamate
180.117	s-ethyl dipropylthiocarbamate
180.228	s-ethyl hexahydro-1h-azepine-1-carbothioate
180.349	ethyl 3-methyl-4-(methylthio) phenyl (1-methylethyl) phosphoramidate
180.374	o-ethyl o-[4-(methylthio)phenyl s-propyl phosphorodithioate
180.221	o-ethyl s-phenyl ethylphosphonodithioate
180.1016	ethylene
180.397	ethylene dibromide
180.1040	ethylene glycol
180.151	ethylene oxide
180.330	s-(2-(ethylsulfinyl)ethyl) o,o-dimethyl phosphorothioate
180.1074	f d & c blue no 1
180.421	fenarimol
180.430	fenoxaprop-ethyl
180.423	fenridazon

180.214	fenthion
180.114	ferbam
180.411	fluazifop-butyl
180.363	fluchloralin
180.400	flucythrinate
180.229	fluometuron
180.420	fluridone
180.145	fluorine compounds
180.427	(alpha rs,2r)-fluvalinate[(rs)-alpha-cyano-3-phenoxybenzyl (r)-2-[2-chloro-4-(trifluoromethyl)anilino] 3-methylbutanoate]
180.191	folpet
180.1032	formaldehyde
180.276	formetanate hydrochloride
180.224	gibberellins
180.1098	gibberellins (ga3)
180.1100	gliocladium virens gl-21
180.124	glyodin
180.364	glyphosate
180.1043	gossyplure
180.319	heptachlor; interim tolerance
180.302	hexachlorophene
180.1069	(z)-11-hexadecenal
180.362	hexakis(2-methyl-2-phenylpropyl) distannoxane
180.396	hexazinone
180.448	hexythiazox
180.1061	hirsutella thompsonii
180.130	hydrogen cyanide
180.1036	hydrogenated castor oil
180.413	imazalil
180.447	imazethapyr, ammonium salt
180.1099	indole butyric acid
180.126	inorganic bromides resulting from soil treatment with ethylene dibromide
180.126a	inorganic bromide residues in peanut hay and peanut hulls
180.123	inorganic bromides resulting from fumigation with methyl bromide
180.199	inorganic bromides resulting from soil treatment with combinations of chloropicrin, methyl bromide, and propargyl bromide
180.319	interim tolerances
180.1022	iodine-detergent complex
180.399	iprodione
180.1030	isobutyric acid
180.1103	isomate-c
180.1073	isomate-m
180.319	isopropyl m-chlorocarbanilate (cipc); interim tolerance
180.319	isopropyl carbanilate (ipc); interim tolerance
180.1063	kontrol hv
180.1090	lactic acid
180.432	lactofen
180.133	lindane
180.184	linuron
180.1056	linseed oil, boiled

180.375	magnesium phosphide
180.111	malathion
180.175	maleic hydrazide
180.110	maneb
180.161	manganous dimethyldithiocarbamate
180.386	mefluidide
180.1092	menthol
180.261	n - (mercaptomethyl) phthalimide s - (o,o - dimethyl phosphorodithioate) and its oxygen analog
180.408	metalaxyl
180.315	methamidophos
180.289	methanearsonic acid
180.357	methazole
180.298	methidathion
180.253	methomyl
180.359, 180.1033	methoprene
180.1079	1-(8-methoxy-4,8-dimethylnonyl)-4-(1-methylethyl)benzene
180.120	methoxychlor
180.1059	methyl a-eleostearate
180.1010	methylene chloride
180.339	2-methyl-4-chlorophenoxyacetic acid
180.387	1-methyl 2-[ethoxy[(1-methylethyl)amino] phosphinothioyl] benzoate
180.318	4-(2-methyl-4-chlorophenoxy) butyric acid
180.157	methyl 3-((dimethoxyphosphinyl)oxy) butenoate, a and b isomers
180.338	6-methyl-1,3-dithiolo (4,5-b) quinoxalin-2-one
180.1067	methyl eugenol/malathion combination
180.437	methyl 2-(4-isopropyl-4-methyl-5-oxo-2-imidazolin-2-yl)-p-toluate and methyl 6-(4-isopropyl-4-methyl-5-oxo-2-imidazolin-2-yl)-m-toluate
180.439	methyl-3-[[[[(4-methoxy-6-methyl-1,3,5-triazin-2-yl) amino] carbonyl]amino]sulfonyl]-2-thiophenecarboxylate
180.451	methyl 2-[[[[n-(4-methoxy-6-methyl-1,3,5-triazin 2-yl) methylamino]carbonyl]amino]sulfonyl]benzoate
180.250	metobromuron
180.368	metolachlor
180.428	metsulfuron methyl
180.149	mineral oil
180.1084	monocarbamide dihydrogen sulfate
180.443	myclobutanil
180.215	naled
180.309	a-naphthaleneacetamide
180.155	1-naphthaleneacetic acid
180.148	b-naphthoxyacetic acid
180.297	n-1-naphthyl phthalamic acid
180.454	nicosulfuron [3-pyridinecarboxamide, 2-((((4,6-dimethoxypyrimidin-2-yl)aminocarbonyl)aminosulfonyl))-n,n-dimethyl]
180.167	nicotine-containing compounds
180.167a	nicotine
180.350	nitrapyrin
180.1050	nitrogen

180.260	norea
180.356	norflurazon
180.1041	nosema locustae
180.1027	nuclear polyhedrosis virus of heliothis zea
180.367	n-octyl bicycloheptenedicarboximide
180.366	2-n-octyl-4-isothiazolin-3-one
180.1055	(e,z)-3, 13-octadecadien-1-ol-acetate and (z,z)-3,13-octadecadien-1-ol acetate
180.180	orthoarsenic acid
180.304	oryzalin
180.346	oxadiazon
180.456	oxadixyl
180.303	oxamyl
180.381	oxyfluorfen
180.337	oxytetracycline
180.1024	paraformaldehyde
180.205	paraquat
180.1101	parasitic (parasitoid) and predatory insects
180.121	parathion or its methyl homolog
180.319	parathion (o,o,diethyl-o-p-nitrophenylthiophosphate) or its methyl homolog; interim tolerance
180.361	pendimethalin
180.291	pentachloronitrobenzene
180.319	pentachloronitrobenzene; interim tolerance
180.1014	pentane
180.165	perfluidone
180.378	permethrin
180.278	phenmedipham
180.319	phenothiazine; interim tolerance
180.129	o-phenylphenol and its sodium salt
180.206	phorate
180.263	phosalone
180.239	phosphamidon
180.1057	phytophthora palmivora
180.292	picloram
180.1035	pine oil
180.127	piperonyl butoxide
180.308	pirimiphos-ethyl
180.409	pirimiphos-methyl
180.1080	plant volatiles/pheromone
180.1089	poly-n-acetyl-d-glucosamine
180.1053	polyamide polymer derived from sebacic acid
180.1037	polybutenes
180.1072	poly-d-glucosamine
180.1078	poly (oxy-1,2-ethanediyl), alpha-isooctadyl-omega-hydroxy
180.1038	polyoxymethylene copolymer
180.1060	polyvinyl chloride
180.1104	poly(vinylpyrrolidone/1-eicosene)
180.1105	poly(vinylpyrrolidone/1-hexadecene)
180.1068	potassium oleate and related c12-c18 fatty acid potassium salts
180.1085	potassium ricinoleate and related c12-c18 fatty acid potassium salts

180.452	primisulfuron-methyl
180.455	procymidone
180.404	profenofos
180.348	profluralin
180.222	prometryn
180.274	propanil
180.259	propargite
180.243	propazine
180.1023	propionic acid
180.238	s-propyl butylethylthiocarbamate
180.240	s-propyl dipropylthiocarbamate
180.1088	pseudomonas fluorescens eg-1053
180.316	pyrazon
180.128	pyrethrins
180.441	quizalofop-ethyl
180.177	ronnel
180.102	sesone
180.340	silvex
180.319	silvex (2-(2,4,5-trichlorophenoxy)-propionic acid); interim tolerance
180.213	simazine (2-chloro-4,6-bis(ethylamino)-s-triazine)
180.213a	simazine (2-chloro-4,6-bis(ethylamino)-s-triazine) and metabolites
180.335	sodium arsenite
180.319	sodium arsenite (calculated as As_2O_3): interim tolerance
180.1020	sodium chlorate
180.1070	sodium chlorite
180.1058	sodium diacetate
180.159	sodium dehydroacetate
180.152	sodium dimethyldithiocarbamate
180.1015	sodium propionate
180.383	sodium salt of acifluorfen
180.433	sodium salt of fomesafen
180.310	sodium trichloroacetate
180.245	streptomycin
180.444	sulfur dioxide
180.1013	
180.1019	sulfuric acid
180.179	tartar emetic
180.390	tebuthiuron
180.440	tefluthrin
180.170	temephos
180.209	terbacil
180.352	terbufos
180.333	terbuthylazine
180.265	terbutryn
180.203	1,2,4,5-tetrachloro-3-nitrobenzene
180.174	tetradifon
180.347	tetraethyl pyrophosphate
180.395	tetrahydro-5,5-dimethyl-2(1h)-pyrimidinone (3-(4-trifluoromethyl)phenyl)-1-(2-4-(trifluoromethyl)phenyl) ethenyl)2-propenylidene)hydrazone
180.162	tetraiodoethylene

180.242	thiabendazole
180.403	thidiazuron
180.401	thiobencarb
180.288	2-(thiocyanomethylthio) benzothiazole
180.407	thiodicarb
180.371	thiophanate-methyl
180.132	thiram
180.138	toxaphene
180.319	toxaphene (chlorinate camphene containing 67-69% chlorine); interim tolerance
180.422	tralomethrin
180.1081	1-triacontanol
180.272	s,s,s-tributyl phosphorotrithioate
180.186	tributylphosphorotrithioite
180.314	s-2,3,3-trichloroallyl diisopropylthiocarbamate
180.1102	trichoderma harzianum, rifai strain krl-ag2
180.417	triclopyr
180.1012	1,1,1-trichloroethane
180.283	2,3,6-trichlorophenylacetic acid
180.1064	(e)-4-tridecen-1-yl-acetate and (z)-4-tridecen-1-yl acetate
180.207	trifluralin
180.382	triforine
180.459	trisulfuron
180.219	2,3,5-triiodobenzoic acid
180.1052	2,2,5-trimethyl-3-dichloroacetyl-1,3-oxazolidine
180.1086	3,7,11-trimethyl-1,6,10-dodecatriene-1-ol and 3,7,11-trimethyl-2,6,10-dodecatriene-3-ol
180.305	3,4,5,-trimethylphenyl methylcarbamate and 2,3,5-trimethylphenyl methylcarbamate
180.236	triphenyltin hydroxide
180.1106	vinylpyrrolidone-vinyl acetate copolymer
180.1025	xylene
180.284	zinc phosphide
180.244	zinc sulfate, basic
180.115	zineb
180.319	zineb (zinc ethylenebisdithiocarbamate); interim tolerance
180.116	ziram

==

pesticide chemicals (40CFR455.10-c) means the sum of all active ingredients manufactured at each facility covered by this part.

pesticide incinerator (40CFR165.1-r) means any installation capable of the controlled combustion of pesticides, at a temperature of 1000 degrees C (1832 degrees F) for two second dwell time in the combustion zone, or lower temperatures and related dwell times that will assure complete conversion of the specific pesticide to inorganic gases and solid ash residues. Such installation complies with the Agency Guidelines for the Thermal Processing of Solid Wastes as prescribed in 40CFR240.

pesticide product (40CFR152.3-t) means a pesticide in the particular form (including composition, packaging, and labeling) in which the pesticide is, or is intended to be, distributed or sold. The term includes any physical apparatus used to deliver or apply the pesticide if distributed or sold with the pesticide.

pesticide product (40CFR162.151-e): See product.

pesticide related wastes (40CFR165.1-q) means all pesticide-containing wastes or byproducts which are produced in the manufacturing or processing of a pesticide and which are to be discarded, but which, pursuant to acceptable pesticide manufacturing or processing operations, are not ordinarily a part of or contained within an industrial waste stream discharged into a sewer or the waters of a state.

pesticide residue (40CFR177.3) means a residue of a pesticide chemical or of any metabolite or degradation product of a pesticide chemical.

pesticide tolerance (EPA-94/04): The amount of pesticide residue allowed by law to remain in or on a harvested crop. EPA sets these levels well below the point where the compounds might be harmful to consumers.

pesticide use (40CFR154.3-g) means a use of a pesticide (described in terms of the application site and other applicable identifying factors) that is included in the labeling of a pesticide product which is registered, or for which an application for registration is pending, and the terms and conditions (or proposed terms and conditions) of registration for the use.

petition (40CFR163.2-h) means a petition filed with the Administrator, Environmental Protection Agency pursuant to section 408(d)(1) of the Act.

petition (40CFR2.308-2) means a petition for the issuance of a regulation establishing a tolerance for a pesticide chemical or exempting the pesticide chemical from the necessity of a tolerance, pursuant to section 408(d) of the Act, 21USC346a(d).

petitioner (40CFR2.308-3) means a person who has submitted a petition to EPA (or to a predecessor agency) (other identical or similar definitions are provided in 40CFR123.64-iii; 164.2-o; 350.1).

petrochemical operations (40CFR419.31-b) shall mean the production of second-generation petrochemicals (i.e., alcohols, ketones, cumene, styrene, etc.) or first generation petrochemicals and isomerization products (i.e. BTX, olefins, cyclohexane, etc.) when 15 percent or more of refinery production is as first-generation petrochemicals and isomerization products (under CWA).

petroleum (40CFR52.741) means the crude oil removed from the earth and the oils derived from tar sands, shale, and coal (other identical or similar definitions are provided in 40CFR60.101-b; 60.111-d; 60.111a-d; 60.111b-h; 60.591; 60.691; 61.341) (under CAA).

petroleum (RCRA9001-42USC6991) means petroleum, including crude oil or any fraction thereof which is liquid at standard conditions of temperature and pressure (60 degrees Fahrenheit and 14.7 pounds per square inch absolute).

petroleum derivatives (EPA-94/04): Chemicals formed when gasoline breaks down in contact with ground water.

petroleum dry cleaner (40CFR60.621) means a dry cleaning facility that uses petroleum solvent in a combination of washers, dryers, filters, stills, and settling tanks.

petroleum liquids (40CFR60.111-b) means petroleum, condensate, and any finished or intermediate products manufactured in a petroleum refinery but does not mean Nos. 2 through 6 fuel oils as specified in ASTM D39678, gas turbine fuel oils Nos. 2GT through 4GT as specified in ASTM D288078, or diesel fuel oils Nos. 2D and 4D as specified in ASTM D97578. (These three methods are incorporated by reference, see 40CFR60.17.) (Other identical or similar definitions are provided in 40CFR60.111a-b.)

petroleum liquids (40CFR60.111b-i) means petroleum, condensate, and any finished or intermediate products manufactured in a petroleum refinery.

petroleum marketing facilities (40CFR280.92) include all facilities at which petroleum is produced or refined and all facilities from which petroleum is sold or transferred to other petroleum marketers or to the public.

petroleum marketing firms (40CFR280.92) are all firms owning petroleum marketing facilities. Firms owning other types of facilities with USTs as well as petroleum marketing facilities are considered to be petroleum marketing firms (under the Resource Conservation Recovery Act).

petroleum refinery (40CFR52.741) means any facility engaged in producing gasoline, kerosene, distillate fuel oils, residual fuel oils, lubricants, or other products through distillation of petroleum, or through redistillation, cracking, or reforming of unfinished petroleum derivatives (other identical or similar definitions are provided in 40CFR60.101-a; 60.111-c; 60.111a-c; 60.591; 60.691; 61.341).

petroleum refinery (40CFR60.41b) means industrial plants as classified by the Department of Commerce under Standard Industrial Classification (SIC) Code 29.

petroleum refining process (40CFR63.101), also referred to as a petroleum refining process unit, means a process that for the purpose of producing transportation fuels (such as gasoline and diesel fuels), heating fuels (such as fuel gas, distillate, and residual fuel oils), or lubricants separates petroleum or separates, cracks, or reforms unfinished derivatives. Examples of such units include, but are not limited to, alkylation units, catalytic hydrotreating, catalytic hydrorefining, catalytic hydrocracking, catalytic reforming, catalytic cracking, crude distillation, and thermal processes (under the Clean Air Act).

petroleum UST system (40CFR280.12) means an underground storage tank system that contains petroleum or a mixture or petroleum with de minimis quantities of other regulated substances. Such systems include those containing motor fuels, jet fuels, distillate fuel oils, residual fuel oils, lubricants, petroleum solvents, and used oils (under RCRA).

pH (EPA-94/04): An expression of the intensity of the basic or acid condition of a liquid. The pH may range from 0 to 14, where 0 is the most acid, 7 is neutral. Natural waters usually have a pH between 6.5 and 8.5.

pH (40CFR131.35.d-9) means the negative logarithm of the hydrogen ion concentration (under the Federal Water Pollution Control Act).

pH (40CFR257.3.5) means the logarithm of the reciprocal of hydrogen ion concentration (under RCRA).

pH (40CFR420.02-q) means the value obtained by the standard method specified in 40CFR136.3 (under CWA).

pH (40CFR796.2750-v) of a sediment or soil is the negative logarithm to the base ten of the hydrogen ion activity of the sediment or soil suspension. It is usually measured by a suitable sensing electrode coupled with a suitable reference electrode at a l/l solid/solution ratio by weight (under the Toxic Substances Control Act).

pharmaceutical (40CFR52.741) means any compound or mixture, other than food, used in the prevention, diagnosis, alleviation, treatment, or cure of disease in man and animal (under the Clean Air Act).

pharmaceutical coating operation (40CFR52.741) means a device in which a coating is applied to a pharmaceutical, including air drying or curing of the coating.

pharmaceutical production process (40CFR63.191) means a process that synthesizes pharmaceutical intermediate or final products using carbon tetrachloride or methylene chloride as a reactant or process solvent. Pharmaceutical production process does not mean process operations involving formulation activities such as tablet coating or spray coating of drug particles.

pharmacist (10CFR40.4) means an individual registered by a state or territory of the United States, the District of Columbia or the Commonwealth of Puerto Rico to compound and dispense drugs, prescriptions and poisons.

pharmacokinetics (40CFR795.228-4) means the study of the rates of absorption, tissue distribution, biotransformation, and excretion (other identical or similar definitions are provided in 40CFR795.231-3; 795.232-3).

pharmakinetics (EPA-94/04): The dynamic behavior of chemicals inside biological systems, including uptake, distribution, metabolism, and excretion.

phase I (40CFR270.2) means that phase of the Federal hazardous waste management program commencing on the effective date of the last of the following to be initially promulgated: 40CFR260, 261, 262, 263, 265, 270 and 271. Promulgation of Phase I refers to promulgation of the regulations necessary for Phase I to begin.

phase I (40CFR72.2) means the Acid Rain Program period beginning January 1, 1995 and ending December 31, 1999.

phase I unit (40CFR72.2) means any affected unit, except an affected unit under part 74 of this chapter, that is subject to an Acid Rain emissions reduction requirement or Acid Rain emissions limitations beginning in Phase I.

phase II (40CFR270.2) means that phase of Federal hazardous waste management program commencing on the effective date of the first subpart of 40CFR264, subparts F through R to be initially promulgated. Promulgation of Phase II refers to promulgation of the regulations necessary for Phase II to begin.

phase II (40CFR72.2) means the Acid Rain Program period beginning January 1, 2000, and continuing into the future thereafter.

phase II bonus allowance allocations (CAA402-42USC7651a) means, for calendar year 2000 through 2009, inclusive, and only for such years, allocations made by the Administrator pursuant to section 403, subsections (a)(2), (b)(2), (c)(4), (d)(3) (except as otherwise provided therein), and (h)(2) of section 405, and section 406.

phase II of the interim period (40CFR51.392) with respect to a pollutant or pollutant precursor means that period of time after the effective date of this rule, lasting until the earlier of the following:
(1) Submission to EPA of the relevant control strategy implementation plan revisions which have been endorsed by the Governor (or his or her designee) and have been subject to a public hearing, or
(2) The date that the Clean Air Act requires relevant control strategy implementation plans to be submitted to EPA, provided EPA has notified the State, MPO, and DOT of the State's failure to submit any such plans. The precise end of Phase II of the interim period is defined in 40CFR51.448.
(Other identical or similar definitions are provided in 40CFR 93.101.)

phase II unit (40CFR72.2) means any affected unit, except an affected unit under part 74 of this chapter, that is subject to an Acid Rain emissions reduction requirement or Acid Rain emissions limitation during Phase II only.

phased disposal (40CFR61.251-f) means a method of tailings management and disposal which uses lined impoundments which are filled and then immediately dried and covered to meet all applicable Federal standards.

PHDD (40CFR766.3): See polyhalogenated dibenzo-p-dioxin.

PHDF (40CFR766.3): See polyhalogenated dibenzofuran.

phenolic compounds (40CFR420.02-e): See phenols 4AAP.

phenolic insulation (40CFR248.4-y) means insulation made with phenolic plastics which are plastics based on resins made by the condensation of phenols, such as phenol or cresol, with aldehydes (under the Resource Conservation Recovery Act).

phenolphthalein alkalinity (EPA-94/04): The alkalinity in a water sample measured by the amount of standard acid needed to lower the pH to a level of 8.3 as indicated by the change of color of the phenolphthalein from pink to clear.

phenols (40CFR410.01-b) shall mean total phenols as measured by the procedure listed in 40CFR136.

phenols (EPA-94/04): Organic compounds that are byproducts of petroleum refining, tanning, and textile, dye, and resin manufacturing. Low concentrations cause taste and odor problems in water; higher concentrations can kill aquatic life and humans. Phosphates: Certain chemical compounds containing phosphorus.

phenols 4AAP (or phenolic compounds)

(40CFR420.02-e) means the value obtained by the method specified in 40CFR136.3 (item 48 in Table 1B). (4AAP means 4-aminoantipyrine.) Because of complexity, section (a) of 40CFR136.3 is provided below:

(a) Parameters or pollutants, for which methods are approved, are listed together with test procedure descriptions and references in Tables IA, IB, IC, ID, and IE. The full text of the referenced test procedures are incorporated by reference into Tables IA, IB, IC, ID, and IE. The references and the sources from which they are available are given in paragraph (b) of this section. These test procedures are incorporated as they exist on the day of approval and a notice of any change in these test procedures will be published in the Federal Register. The discharge parameter values for which reports are required must be determined by one of the standard analytical test procedures incorporated by reference and described in Tables IA, IB, IC, ID, and IE, or by any alternate test procedure which has been approved by the Administrator under the provisions of paragraph (d) of this section and 40CFR136.4 and 136.5 of this part 136. Under certain circumstances (40CFR136.3 (b) or (c) or 40CFR401.13) other test procedures may be used that may be more advantageous when such other test procedures have been previously approved by the Regional Administrator of the Region in which the discharge will occur, and providing the Director of the State in which such discharge will occur does not object to the use of such alternate test procedure.

phenotypic expression time (40CFR798.5300-4) is a period during which unaltered gene products are depleted from newly mutated cells.

Philadelphia CBD (40CFR52.2041-1) is defined as the area within the City of Philadelphia, Pennsylvania, bounded by, but not including Vine Street, South Street, the Schuylkill River, and the Delaware River.

phosphate rock feed (40CFR60.401-f) means all material entering the process unit including, moisture and extraneous material as well as the following ore minerals: Fluorapatite, hydroxylapatite, chlorapatite, and carbonateapatite.

phosphate rock plant (40CFR60.401-a) means any plant which produces or prepares phosphate rock product by any or all of the following processes: Mining, beneficiation, crushing, screening, cleaning, drying, calcining, and grinding.

phosphogypsum (40CFR61.201-b) is the waste or other form of byproduct which results from the process of wet acid phosphorus production and which contains greater than [up to 10] pCi/g radium.

phosphogypsum piles (stacks) (EPA-94/04): Principal byproduct generated in production of phosphoric acid from phosphate rock. These piles may generate radioactive radon gas.

phosphogypsum stacks or stacks (40CFR61.201-c) are:
(1) piles of waste from phosphorus fertilizer production containing phosphogypsum. Stacks shall also include phosphate mines that are used for the disposal of phosphogypsum; or
(2) piles of waste or other form of byproduct which results from wet acid phosphorus production containing phosphogypsum. Stacks shall also include phosphate mines that are used for the disposal of phosphogypsum.

phosphorous plants (EPA-94/04): Facilities using electric furnaces to produce elemental phosphorous for commercial use, such as high grade phosphoric acid, phosphate-based detergent, and organic chemicals use.

phosphorus (EPA-94/04): An essential chemical food element that can contribute to the eutrophication of lakes and other water bodies. Increased phosphorus levels result from discharge of phosphorus-containing materials into surface waters.

photochemical oxidants (EPA-94/04): Air pollutants formed by the action of sunlight on oxides of nitrogen and hydrocarbons.

photochemical smog (EPA-94/04): Air pollution caused by chemical reactions of various pollutants emitted from different sources.

photochemically reactive solvent (40CFR52.1088) means any solvent with an aggregate of more than 20 percent of its total volume composed of the chemical compounds classified below or which exceeds any of the following individual percentage composition limitations, as applied to the total volume of solvent:

(i) A combination of hydrocarbons, alcohols, aldehydes, esters, ethers, or ketones having an olefinic or cycloolefinic type of unsaturation: 5 percent;

(ii) A combination of aromatic compounds with 8 or more carbon atoms to the molecule except ethylbenyene: 8 percent;

(iii) A combination of ethylbenyene or ketones having branched hydrocarbon structures, trichloroethylene or toluene: 20 percent.

(Other identical or similar definitions are provided in 40CFR52.1107; 52.2440-3.)

photographic article (40CFR723.175-12) means any article which will become a component of an instant photographic or peel-apart film article.

photolysis (40CFR300-AA) means chemical reaction of a substance caused by direct absorption of solar energy (direct photolysis) or caused by other substances that absorb solar energy (indirect photolysis).

photoperiod (40CFR797.2130-iii) means the light and dark periods in a 24 hour day. This is usually expressed in a form such as 17 hours light/7 hours dark or 17L/7D (other identical or similar definitions are provided in 40CFR797.2150-iii).

photosynthesis (EPA-94/04): The manufacture by plants of carbohydrates and oxygen from carbon dioxide mediated by chlorophyll in the presence of sunlight.

physical and chemical treatment (EPA-94/04): Processes generally used in large-scale wastewater treatment facilities. Physical processes may include air-stripping or filtration. Chemical treatment includes coagulation, chlorination, or ozonation. The term can also refer to treatment of toxic materials in surface and ground waters, oil spills, and some methods of dealing with hazardous materials on or in the ground.

physical and thermal integrity (40CFR85.2122(a)(7)-(ii)(C)) means the ability of the material of the cap and/or rotor to resist physical and thermal breakdown.

physical chemical treatment system (40CFR420.11-g) means those full scale coke plant wastewater treatment systems incorporating full scale granular activated carbon adsorption units which were in operation prior to January 7, 1981, the date of proposal of this regulation.

physical construction (40CFR270.2) means excavation, movement of earth, erection of forms or structures, or similar activity to prepare an HWM facility to accept hazardous waste.

physical or mental impairment (40CFR7.25): See handicapped person.

physician (10CFR30.4) means a medical doctor or doctor of osteopathy licensed by a State or Territory of the United States, the District of Columbia, or the Commonwealth of Puerto Rico to prescribe drugs in the practice of medicine (other identical or similar definitions are provided in 40CFR40.4).

phytoplankton (EPA-94/04): That portion of the plankton community comprised of tiny plants, e.g., algae, diatoms.

phytotoxic (EPA-94/04): Harmful to plants.

pick interval (40CFR60-AA(alt. method 1)) means the time or range intervals in the lidar backscatter signal whose minimum average amplitude is used to calculate opacity. Two pick intervals are required, one in the near region and one in the far region.

pickles, fresh (40CFR407.61-n) shall mean the processing of fresh cucumbers and other vegetables, all varieties, all sizes from whole to relish, all styles, cured after packing.

pickles processed (40CFR407.61-o) shall mean the processing of pickles, cucumbers and other vegetables, all varieties, sizes and types, made after fermentation and storage.

pickles, salt stations (40CFR407.61-p) shall mean the handling and subsequent preserving of cucumbers and other vegetables at salting stations or tankyards, by salt and other chemical additions necessary to achieve proper fermentation for the packing of processed pickle products. Limitations include allowances for the discharge of spent brine, tank wash, tank soak, and cucumber wash waters. At locations where both salt station and process pack operations (40CFR407.61(o)) occur, additive allowances shall be made for both of these sources in formulation of effluent limitations. The effluent limitations are to be calculated based upon the total annual weight (1000 lb, kkg) of raw product processed at each of the salt station and process pack operations. Allowances for contaminated stormwater runoff should be considered in NPDES permit formulation on a case-by-case basis.

pickling bath (40CFR468.02-k) shall mean any chemical bath (other than alkaline cleaning) through which a workpiece is processed.

pickling fume scrubber (40CFR468.02-l) shall mean the process of using an air pollution control device to remove particulates and fumes from air above a pickling bath by entraining the pollutants in water.

pickling rinse (40CFR468.02-m) shall mean a rinse, other than an alkaline cleaning rinse, through which a workpiece is processed. A rinse consisting of a series of rinse tanks is considered as a single rinse.

pickling rinse for forged parts (40CFR468.02-t) shall mean a rinse, other than an alkaline cleaning rinse, through which forged parts are processed. A rinse consisting of a series of rinse tanks is considered as a single rinse.

pickup truck (40CFR600.002.85-35) means a nonpassenger automobile which has a passenger compartment and an open cargo bed.

picocurie (pCi) (40CFR141.2) means the quantity of radioactive material producing 2.22 nuclear transformations per minute.

picocuries per liter pci/l) (EPA-94/04): A unit of measure for levels of radon gas.

pieces (40CFR427.71-c) shall mean floor tile measured in the standard size of 12" x 12" x 3/32" (under CWA).

pigmented coatings (40CFR52.741) means opaque coatings containing binders and colored pigments which are formulated to conceal the wood surface either as an undercoat or topcoat.

pile (40CFR260.10) means any non-containerized accumulation of solid, nonflowing hazardous waste that is used for treatment or storage.

pilot program (40CFR52.2043-4) means a program that is initiated on a limited basis for the purpose of facilitating a future full scale regional program.

pilot tests (EPA-94/04): Testing a cleanup technology under actual site conditions to identify potential problems prior to full-scale implementation.

pineapples (40CFR407.61-q) shall mean the processing of pineapple into the following product styles: Canned, peeled, sliced, chunk, tidbit, diced, crushed, and any other related piece size, juice and concentrate. It also specifically includes the on-site production of byproducts such as alcohol, sugar or animal feed.

pipe and tube batch (40CFR420.81-e) means those descaling operations that remove surface scale from pipe tube products in batch processes.

pipe and tube mill (40CFR420.71-e) means those steel hot forming operations that produce butt welded or seamless tubular steel products.

pipe or piping (40CFR280.12) means a hollow cylinder or tubular conduit that is constructed of non-earthen materials.

pipe, tube and other (40CFR420.91-n) means those acid pickling operations that pickle pipes, tubes or any steel product other than those included in paragraphs (k), (l) and (m) of this section.

pipeline facilities (including gathering lines) (40CFR280.12) are new and existing pipe rights-of-way and any associated equipment, facilities, or buildings.

piping (40CFR280.12): See pipe or piping.

Pittsburgh CBD (40CFR52.2041-2) is defined as the

area enclosed by the Allegheny River, the Monongahela River, and I-876. I-876 is not included.

Pittsburgh CBD (40CFR52.2050-3) means the area enclosed by the Allegheny River, the Monongahela River, and I-876. I-876 is not included.

place sewage sludge or sewage sludge placed (40CFR503.9-s) means disposal of sewage sludge on a surface disposal site.

plan (40CFR51.100-j) means an implementation plan approved or promulgated under section 110 of 172 of the Act.

plan (40CFR58.1-i) means an implementation plan, approved or promulgated pursuant to section 110 of the Clean Air Act.

plan (40CFR60.21-c) means a plan under section 111(d) of the Act which establishes emission standards for designated pollutants from designated facilities and provides for the implementation and enforcement of such emission standards.

plan or plan item (40CFR52.30-4) mean an implementation plan or portion of an implementation plan or action needed to prepare such plan required by the Clean Air Act, as amended in 1990, or in response to a SIP call issued pursuant to section 110(k)(5) of the Act.

plankton (EPA-94/04): Tiny plants and animals that live in water.

planned renovation operations (40CFR61.141) means a renovation operation, or a number of such operations, in which some RACM will be removed or stripped within a given period of time and that can be predicted. Individual nonscheduled operations are included if a number of such operations can be predicted to occur during a given period of time based on operating experience.

planning (40CFR256.06) includes identifying problems, defining objectives, collecting information, analyzing alternatives and determining necessary activities and courses of action.

planning target (40CFR35.105) means the amount of

Federal financial assistance which the Regional Administrator suggests that an applicant consider in developing its application and work program.

plans (40CFR240.101-t) means reports and drawings, including a narrative operating description, prepared to describe the facility and its proposed operation.

plans (40CFR241.101-n) means reports and drawings, including a narrative operating description, prepared to describe the land disposal site and its proposed operation.

plant (40CFR52.741) means all of the pollutant-emitting activities which belong to the same industrial grouping, are located on one or more contiguous or adjacent properties, and are under the control of the same person (or persons under common control), except the activities of any marine vessel. Pollutant-emitting activities shall be considered as part of the same industrial grouping if they belong to the same "Major Group" (i.e., which have the same two-digit code) as described in the "Standard Industrial Classification Manual, 1987" (incorporated by reference as specified in 40CFR52.742). Plasticizers means a substance added to a polymer composition to soften and add flexibility to the product.

plant (40CFR82.3-cc) means one or more facilities at the same location owned by or under common control of the same person.

plant (ESA3-16USC1531) means any member of the plant kingdom, including seeds, roots and other parts thereof.

plant blending records (40CFR60.581) means those records which document the weight fraction of organic solvents and solids used in the formulation or preparation of inks at the vinyl or urethane printing plant where they are used.

plant regulator (FIFRA2-7USC136) means any substance or mixture of substances intended, through physiological action, for accelerating or retarding the rate of growth or rate of maturation, or for otherwise altering the behavior of plants or the produce thereof, but shall not include substances to the extent that they are intended as plant nutrients, trace elements, nutritional chemicals, plant inoculants, and soil amendments. Also, the term, plant regulator shall not be required to include any of such of those nutrient mixtures or soil amendments as are commonly known as vitamin-hormone horticultural products, intended for improvement, maintenance, survival, health, and propagation of plants, and as are not for pest destruction and are non-toxic, nonpoisonous in the undiluted packaged concentration.

plant site (40CFR440.141-15) means the area occupied by the mine, necessary haulage ways from the mine to the beneficiation process, the beneficiation area, the area occupied by the wastewater treatment facilities and the storage areas for waste materials and solids removed from the wastewaters during treatment.

plant site (40CFR63.101) means all contiguous or adjoining property that is under common control, including properties that are separated only by a road or other public right-of-way. Common control includes properties that are owned, leased, or operated by the same entity, parent entity, subsidiary, or any combination thereof.

plant site (40CFR63.161) means all contiguous or adjoining property that is under common control, including properties that are separated only by a road or other public right-of-way. Common control includes properties that are owned, leased, or operated by the same entity, parent entity, subsidiary, or any combination thereof.

plasma arc incinerator (40CFR260.10) means any enclosed device using a high intensity electrical discharge or arc as a source of heat and which is not listed as an industrial furnace.

plasma arc reactor (EPA-94/04): An incinerator that operates at extremely high temperatures; treats highly toxic wastes that do not burn easily (cf. incinerator).

plasmid (EPA-94/04): A circular piece of DNA that exists apart from the chromosome and replicates independently of it. Bacterial plasmids carry information that renders the bacteria resistant to antibiotics. Plasmids are often used in genetic engineering to carry desired genes into organisms.

plastic body (40CFR60.391) means an automobile or light-duty truck body constructed of synthetic organic material.

plastic body component (40CFR60.391) means any component of an automobile or light-duty truck exterior surface constructed of synthetic organic material.

plastic material (40CFR463.2-f) is a synthetic organic polymer (i.e., a thermoset polymer, a thermoplastic polymer, or a combination of a natural polymer and a thermoset or thermoplastic polymer) that is solid in its final form and that was shaped by flow. The material can either be a homogeneous polymer and a polymer combined with fillers, plasticizers, pigments, stabilizers, or other additives (under CWA).

plastic parts (40CFR60.721) means panels, housings, bases, covers, and other business machine components formed of synthetic polymers.

plastic rigid foam (40CFR248.4-z) means cellular polyurethane insulation, cellular polyisocyanurate insulation, glass fiber reinforced polyisocyanurate-polyurethane foam insulation, cellular polystyrene insulation, phenolic foam insulation, spray-in-place foam and foam-in-place insulation.

plasticizers (40CFR52.741) means a substance added to a polymer composition to soften and add flexibility to the product.

plastics (EPA-94/04): Non-metallic chemoreactive compounds molded into rigid or pliable construction materials, fabrics, etc.

plastics molding and forming (40CFR463.2-a) is a manufacturing process in which materials are blended, molded, formed or otherwise processed into intermediate or final products.

plate mill (40CFR420.71-g) means those steel hot forming operations that produce flat hot-rolled products which are:
(1) between 8 and 48 inches wide and over 0.23 inches thick; or
(2) greater than 48 inches wide and over 0.18 inches thick.

plate soak (40CFR461.2-d) shall mean the process operation of soaking or reacting lead subcategory battery plates, that are more than 2.5 mm (0.100 in) thick, in sulfuric acid.

plate tower scrubber (EPA-94/04): An air pollution control device that neutralizes hydrogen chloride gas by bubbling alkaline water through holes in a series of metal plates.

plug flow (EPA-94/04): Type of flow that occurs in tanks, basins, or reactors when a slug of water moves through without ever dispersing or mixing with the rest of the water flowing through.

plugging (40CFR144.3) means the act or process of stopping the flow of water, oil or gas into or out of a formation through a borehole or well penetrating that formation (other identical or similar definitions are provided in 40CFR146.3; 147.2902).

plugging (EPA-94/04): Act or process of stopping the flow of water, oil, or gas into or out of a formation through a borehole or well penetrating that formation.

plugging and abandonment plan (40CFR144.61-a) means the plan for plugging and abandonment prepared in accordance with the requirements of 40CFR144.28 and 40CFR144.51.

plugging record (40CFR146.3) means a systematic listing of permanent or temporary abandonment of water, oil, gas, test, exploration and waste injection wells, and may contain a well log, description of amounts and types of plugging material used, the method employed for plugging, a description of formations which are sealed and a graphic log of the well showing formation location, formation thickness, and location of plugging structures.

plume (40CFR60-AA(alt. method 1)): The plume being measured by lidar.

plume (EPA-94/04):
1. A visible or measurable discharge of a contaminant from a given point of origin. Can be visible or thermal in water as it extends downstream from the pollution source, or visible in air as, for example, a plume of smoke.

2. The area of radiation leaking from a damaged reactor.
3. Area downwind within which a release could be dangerous for those exposed to leaking fumes.

plume signal (40CFR60-AA(alt. method 1)) means the backscatter signal resulting from the laser light pulse passing through a plume.

plums (40CFR407.61-r) shall mean the processing of plums into the following product styles: Canned and frozen, pitted and unpitted, peeled and unpeeled, blanched and unblanched, whole, halved, and other piece size.

plunger (40CFR85.2122(a)(3)(ii)): See accelerator pump.

plutonium (EPA-94/04): A radioactive metallic element chemically similar to uranium.

plutonium processing and fuel fabrication plant (10CFR70.4) means a plant in which the following operations or activities are conducted:
(1) Operations for manufacture of reactor fuel containing plutonium including any of the following:
  (i) Preparation of fuel material;
  (ii) formation of fuel material into desired shapes;
  (iii) application of protective cladding;
  (iv) recovery of scrap material; and
  (v) storage associated with such operations; or
(2) research and development activities involving any of the operations described in paragraph (r)(1) of this section, except for research and development activities utilizing unsubstantial amount of plutonium.

PM-10 (EPA-94/04): A new standard for measuring the amount of solid or liquid matter suspended in the atmosphere, i.e., the amount of particulate matter over 10 micrometers in diameter; smaller PM-10 particles penetrate to the deeper portions of the lung, affecting sensitive population groups such as children and individuals with respiratory ailments.

PM10 (40CFR51.100-qq) means particulate matter with an aerodynamic diameter less than or equal to a nominal 10 micrometers as measured by a reference method based on Appendix J of part 50 of this chapter and designated in accordance with part 53 of this chapter or by an equivalent method designated in accordance with part 53 of this chapter.

PM10 (CAA302-42USC7602) means particulate matter with an aerodynamic diameter less than or equal to a nominal ten micrometers, as measured by such method as the Administrator may determine (other identical or similar definitions are provided in 40CFR58.1-u).

PM10 emissions (40CFR51.100-rr) means finely divided solid or liquid material, with an aerodynamic diameter less than or equal to a nominal 10 micrometers emitted to the ambient air as measured by an applicable reference method, or an equivalent or alternative method, specified in this chapter or by a test method specified in an approved State implementation plan.

PM10 sampler (40CFR53.1-m) means a device, associated with a manual method for measuring PM10, designed to collect PM10 from an ambient air sample, but lacking the ability to automatically analyze or measure the collected sample to determine the mass concentration of PM10 in the sampled air.

PMN (40CFR700.43): See premanufacture notice.

pneumatic coal-cleaning equipment (40CFR60.251-f) means any facility which classifies bituminous coal by size or separates bituminous coal from refuse by application of air stream(s).

podiatrist (10CFR30.4) means an individual licensed by a State or Territory of the United States, the District of Columbia, or the Commonwealth of Puerto Rico to practice podiatry.

point of disinfectant application (40CFR141.2) is the point where the disinfectant is applied and water downstream of that point is not subject to recontamination by surface water runoff.

point of disinfectant application (EPA-94/04): The point where disinfectant is applied and water downstream of that point is not subject to recontamination by surface water runoff.

point of entry treatment device (40CFR141.2) is a treatment device applied to the drinking water entering a house or building for the purpose of reducing contaminants in the drinking water distributed throughout the house or building.

point of entry treatment device (EPA-94/04): A treatment device applied to the drinking water entering a house or building to reduce the contaminants in the water distributed throughout the house or building.

point of generation (40CFR63.111) means the location where process wastewater exits the process unit equipment (40CFR63.111). [Note: The regulation allows determination of the characteristics of a wastewater stream:
(1) at the point of generation or
(2) downstream of the point of generation if corrections are made for changes in flow rate and VOHAP concentration. Such changes include losses by air emissions; reduction of VOHAP concentration or changes in flow rate by mixing with other water or wastewater streams; and reduction in flow rate or VOHAP concentration by treating or otherwise handling the wastewater stream to remove or destroy HAP's.]

point of use treatment device (40CFR141.2) is a treatment device applied to a single tap used for the purpose of reducing contaminants in drinking water at that one tap.

point of use treatment device (EPA-94/04): Treatment device applied to a single tap to reduce contaminants in the drinking water at that one tap.

point of waste generation (40CFR61.341) means the location where samples of a waste stream are collected for the purpose of determining the waste flow rate, water content, or benzene concentration in accordance with procedures specified in 40CFR1.355 of this subpart. For a chemical manufacturing plant or petroleum refinery, the point of waste generation is a location after the waste stream exits the process unit component, product tank, or waste management unit generating the waste, and before the waste is exposed to the atmosphere or mixed with other wastes. For a coke-byproduct recovery plant subject to and complying with the control requirements of 40CFR61.132, 61.133, or 61.134 of this part, the point of waste generation is a location after the waste stream exits the process unit component or waste management unit controlled by that subpart, and before the waste is exposed to the atmosphere. For other facilities subject to this subpart, the point of waste generation is a location after the waste enters the facility, and before the waste is exposed to the atmosphere or placed in a facility waste management unit.

point of waste generation (40CFR61.341) means the location where the waste stream exits the process unit component or storage tank prior to handling or treatment in an operation that is not an integral part of the production process, or in the case of waste management units that generate new wastes after treatment, the location where the waste stream exits the waste management unit component.

point source (40CFR122.2) Point source means any discernible, confined, and discrete conveyance, including but not limited to any pipe, ditch, channel, tunnel, conduit, well, discrete fissure, container, rolling stock, concentrated animal feeding operation, landfill leachate collection system, vessel, or other floating craft from which pollutants are or may be discharged. This term does not include return flows from irrigated agriculture or agricultural storm water runoff (see also 40CFR122.3).

point source (40CFR257.3.3) can be found (is defined) in the Clean Water Act, as amended, 33USC1251 et seq., and implementing regulations, specifically 33CFR323 (42FR37122, July 19, 1977) (under RCRA).

point source (40CFR260.10) means any discernible, confined, and discrete conveyance, including, but not limited to any pipe, ditch, channel, tunnel, conduit, well, discrete fissure, container, rolling stock, concentrated animal feeding operation, or vessel or other floating craft, from which pollutants are or may be discharged. This term does not include return flows from irrigated agriculture.

point source (40CFR401.11-d) means any discernible, confined and discrete conveyance, including but not limited to any pipe, ditch, channel, tunnel, conduit, well, discrete fissure, container, rolling stock, concentrated animal feeding operation, or vessel or other floating craft, from which pollutants are or may be discharged.

point source (40CFR51.100-k) Point source means the following:

(1) For particulate matter, sulfur oxides, carbon monoxide, volatile organic compounds (VOC) and nitrogen oxide:
   (i)   Any stationary source the actual emissions of which are in excess of 90.7 metric tons (100 tons) per year of the pollutant in a region containing an area whose 1980 "urban place" population, as defined by the U.S. Bureau of the Census, was equal to or greater than 1 million.
   (ii) Any stationary source the actual emissions of which are in excess of 22.7 metric tons (25 tons) per year of the pollutant in a region containing an area whose 1980 "urban place" population, as defined by the U.S. Bureau of the Census, was less than 1 million; or
(2) For lead or lead compounds measured as elemental lead, any stationary source that actually emits a total of 4.5 metric tons (5 tons) per year or more.

point source (CWA502-33USC1362) means any discernible, confined and discrete conveyance, including but not limited to any pipe, ditch, channel, tunnel, conduit, well, discrete fissure, container, rolling stock, concentrated animal feeding operation, or vessel or other floating craft, from which pollutants are or may be discharged. This term does not include agricultural stormwater discharges and return flows from irrigated agriculture.

point source (EPA-94/04): A stationary location or fixed facility from which pollutants are discharged; any single identifiable source of pollution, e.g., a pipe, ditch, ship, ore pit, factory smokestack.

polishing and buffing wheels (29CFR1910.94b) means all power-driven rotatable wheels composed all or in part of textile fabrics, wood, felt, leather, paper, and may be coated with abrasives on the periphery of the wheel for purposes of polishing, buffing, and light grinding.

political subdivision (40CFR35.6015-31) means the unit of government that the State determines to have met the States legislative definition of a political subdivision.

political subdivision (40CFR52.30-1) refers to the representative body that is responsible for adopting and/or implementing air pollution controls for one, or any combination of one or more of the following: city, town, borough, county, parish, district, or any other geographical subdivision created by, or pursuant to, Federal or State law. This will include any agency designated under section 174, 42 U.S.C. 7504, by the State to carry out the air planning responsibilities under part D.

pollen (EPA-94/04): The fertilizing element of flowering plants; background air pollutant.

pollutant (40CFR122.2) means dredged spoil, solid waste, incinerator residue, filter backwash, sewage, garbage, sewage sludge, munitions, chemical wastes, biological materials, radioactive materials (except those regulated under the Atomic Energy Act of 1954, as amended (42USC2011 et seq.)), heat, wrecked or discarded equipment, rock, sand, cellar dirt and industrial, municipal, and agricultural waste discharged into water. It does not mean:
(a) Sewage from vessels; or
(b) Water, gas, or other material which is injected into a well to facilitate production of oil or gas, or water derived in association with oil and gas production and disposed of in a well, if the well used either to facilitate production or for disposal purposes is approved by authority of the State in which the well is located, and if the State determines that the injection or disposal will not result in the degradation of ground or surface water resources.
Note: Radioactive materials covered by the Atomic Energy Act are those encompassed in its definition of source, byproduct, or special nuclear materials. Examples of materials not covered include radium and accelerator-produced isotopes. See Train v. Colorado Public Interest Research Group, Inc., 426 U.S. 1 (1976).

pollutant (40CFR21.2) means dredged spoil, solid waste, incinerator residue, sewage, garbage, sewage sludge, munitions, chemical wastes, biological materials, radioactive materials, heat, wrecked or discarded equipment, rock, sand, cellar dirt and industrial, municipal, and agricultural waste discharged into water. For the purposes of this section, the term also means sewage from vessels within the meaning of section 312 of the Act (under FWPCA).

pollutant (40CFR230.3-o) means dredged spoil, solid waste, incinerator residue, sewage, garbage, sewage sludge, munitions, chemical wastes, biological materials, radioactive materials not covered by the Atomic Energy Act, heat, wrecked or discarded equipment, rock, sand, cellar dirt, and industrial, municipal, and agricultural waste discharged into water. The legislative history of the Act reflects that "radioactive materials" as included within the definition of "pollutant" in section 502 of the Act means only radioactive materials which are not encompassed in the definition of source, byproduct, or special nuclear materials as defined by the Atomic Energy Act of 1954, as amended, and regulated under the Atomic Energy Act. Examples of radioactive materials not covered by the Atomic Energy Act and, therefore, included within the term "pollutant," are radium and accelerator produced isotopes. See Train v. Colorado Public Interest Research Group, Inc. 426 U.S. 1 (1976).

pollutant (40CFR257.3.3) can be found (is defined) in the Clean Water Act, as amended, 33USC1251 et seq., and implementing regulations, specifically 33CFR323 (42FR37122, July 19, 1977).

pollutant (40CFR401.11-f) means dredged spoil, solid waste, incinerator residue, sewage, garbage, sewage sludge, munitions, chemical wastes, biological materials, radioactive materials, heat, wrecked or discarded equipment, rock, sand, cellar dirt and industrial, municipal and agricultural waste discharged into water. It does not mean:
(1) sewage from vessels or
(2) water, gas or other material which is injected into a well to facilitate production of oil or gas, or water derived in association with oil or gas production and disposed of in a well, if the well, used either to facilitate production or for disposal purposes, is approved by authority of the State in which the well is located, and if such State determines that such injection or disposal

will not result in degradation of ground or surface water resources.

pollutant (40CFR503.9-t) is an organic substance, an inorganic substance, a combination of organic and inorganic substances, or a pathogenic organism that, after discharge and upon exposure, ingestion, inhalation, or assimilation into an organism either directly from the environment or indirectly by ingestion through the food chain, could, on the basis of information available to the Administrator of EPA, cause death, disease, behavioral abnormalities, cancer, genetic mutations, physiological malfunctions (including malfunction in reproduction), or physical deformations in either organisms or offspring of the organisms.

pollutant (CWA502-33USC1362) means dredged spoil, solid waste, incinerator residue, sewage, garbage, sewage sludge, munitions, chemical wastes, biological materials, radioactive materials, heat, wrecked or discarded equipment, rock, sand, cellar dirt and industrial, municipal, and agricultural waste discharged into water. This term does not mean
(A) "sewage from vessels" within the meaning of section 312 of this Act; or
(B) water, gas, or other material which is injected into a well to facilitate production of oil or gas, or water derived in association with oil or gas production and disposed of in a well, if the well used either to facilitate production or for disposal purposes is approved by authority of the State in which the well is located, and if such State determines that such injection or disposal will not result in the degradation of ground or surface water resources.

pollutant (EPA-94/04): Generally, any substance introduced into the environment that adversely affects the usefulness of a resource.

pollutant concentration monitor (40CFR72.2) means that component of the continuous emission monitoring system that measures the concentration of a pollutant in a unit's flue gas.

pollutant limit (40CFR503.9-u) is a numerical value that describes the amount of a pollutant allowed per unit amount of sewage sludge (e.g., milligrams per kilogram of total solids); the amount of a pollutant

that can be applied to a unit area of land (e.g., kilograms per hectare); or the volume of a material that can be applied to a unit area of land (e.g., gallons per acre).

pollutant or contaminant (40CFR300.5), as defined by section 101(33) of CERCLA, shall include, but not be limited to, any element, substance, compound, or mixture, including disease causing agents, which after release into the environment and upon exposure, ingestion, inhalation, or assimilation into any organism, either directly from the environment or indirectly by ingesting through food chains, will or may reasonably be anticipated to cause death, disease, behavioral abnormalities, cancer, genetic mutation, physiological malfunctions (including malfunctions in reproduction), or physical deformation in such organisms or their offspring. The term does not include petroleum, including crude oil and any fraction thereof which is not otherwise specifically listed or designated as a hazardous substance under section 101(14) (A) through (F) of CERCLA, nor does it include natural gas, liquified natural gas, or synthetic gas of pipeline quality (or mixtures of natural gas and synthetic gas). For purposes of subpart F of the NCP, the term pollutant or contaminant means any pollutant or contaminant which may present an imminent and substantial danger to public health or welfare (under CERCLA).

pollutant or contaminant (40CFR310.11-h), as defined by section 104(a)(2) of CERCLA, includes, but is not limited to, any element, substance, compound, or mixture, including disease-causing agents, which after release into the environment and upon exposure, ingestion, inhalation, or assimilation into any organism, either directly from the environment or indirectly by ingestion through food chains, will or may reasonably be anticipated to cause death, disease, behavioral abnormalities, cancer, genetic mutation, physiological malfunctions (including malfunctions in reproduction) or physical deformations, in such organisms or their offspring. The term does not include petroleum, including crude oil and any fraction thereof that is not otherwise specifically listed or designated as a hazardous substance under section 101(14)(A) through (F) of CERCLA, nor does it include natural gas, liquefied natural gas, or synthetic gas of pipeline quality (or mixtures of natural gas and such synthetic gas).

pollutant or contaminant (SF101-42USC9601) shall include, but not be limited to, any element, substance, compound, or mixture, including disease-causing agents, which after release into the environment and upon exposure, ingestion, inhalation, or assimilation into any organism, either directly from the environment or indirectly by ingestion through food chains, will or may reasonably be anticipated to cause death, disease, behavioral abnormalities, cancer, genetic mutation, physiological malfunctions (including malfunctions in reproduction) or physical deformations, in such organisms or their offspring; except that the term "pollutant or contaminant" shall not include petroleum, including crude oil or any fraction thereof which is not otherwise specifically listed or designated as a hazardous substance under subparagraphs (A) through (F) of paragraph (14) and shall not include natural gas, liquefied natural gas, or synthetic gas of pipeline quality (or mixtures of natural gas and such synthetic gas).

pollutant standard index (psi) (EPA-94/04): Measure of adverse health effects of air pollution levels in major cities.

pollution (40CFR230.3-p) means the man-made or man-induced alteration of the chemical, physical, biological or radiological integrity of an aquatic ecosystem.

pollution (CWA502-33USC1362) means the man-made or man-induced alteration of the chemical, physical, biological and radiological integrity of water (other identical or similar definitions are provided in 40CFR130.2-c; 401.11-g).

pollution (EPA-94/04): Generally, the presence of matter or energy whose nature, location, or quantity produces undesired environmental effects. Under the Clean Water Act, for example, the term is defined as the manmade or man-induced alteration of the physical, biological, chemical, and radiological integrity of water.

pollution control project (40CFR51.165-xxv) means any activity or project at an existing electric utility steam generating unit for purposes of reducing emissions from such unit. Such activities or projects are limited to:

(A) The installation of conventional or innovative pollution control technology, including but not limited to advanced flue gas desulfurization, sorbent injection for sulfur dioxide and nitrogen oxides controls and electrostatic precipitators:

(B) An activity or project to accommodate switching to a fuel which is less polluting than the fuel used prior to the activity or project, including, but not limited to natural gas or coal reburning, or the cofiring of natural gas and other fuels for the purpose of controlling emissions;

(C) A permanent clean coal technology demonstration project conducted under title II, sec. 101(d) of the Further Continuing Appropriations Act of 1985 (sec. 5903(d) of title 42 of the United States Code), or subsequent appropriations, up to a total amount of $2,500,000,000 for commercial demonstration of clean coal technology, or similar projects funded through appropriations for the Environmental Protection Agency; or

(D) A permanent clean coal technology demonstration project that constitutes a repowering project.

pollution control project (40CFR51.166-31) means any activity or project undertaken at an existing electric utility steam generating unit for purposes of reducing emissions from such unit. Such activities or projects are limited to:

(i) The installation of conventional or innovative pollution control technology, including but not limited to advanced flue gas desulfurization, sorbent injection for sulfur dioxide and nitrogen oxides controls and electrostatic precipitators;

(ii) An activity or project to accommodate switching to a fuel which is less polluting than the fuel used prior to the activity or project, including but not limited to natural gas or coal re-burning, or the co-firing of natural gas and other fuels for the purpose of controlling emissions;

(iii) A permanent clean coal technology demonstration project conducted under title II, section 101(d) of the Further Continuing Appropriations Act of 1985 (section 5903(d) of title 42 of the United States Code), or subsequent appropriations, up to a total amount of $2,500,000,000 for commercial demonstration of clean coal technology, or similar projects funded through appropriations for the Environmental Protection Agency, or

(iv) A permanent clean coal technology demonstration project that constitutes a repowering project.

pollution control project (40CFR52.21-32) means any activity or project undertaken at an existing electric utility steam generating unit for purposes of reducing emissions from such unit. Such activities or projects are limited to:

(i) The installation of conventional or innovative pollution control technology, including but not limited to advanced flue gas desulfurization, sorbent injection for sulfur dioxide and nitrogen oxides controls and electrostatic precipitators;

(ii) An activity or project to accommodate switching to a fuel which is less polluting than the fuel in use prior to the activity or project, including, but not limited to natural gas or coal re-burning, or the co-firing of natural gas and other fuels for the purpose of controlling emissions;

(iii) A permanent clean coal technology demonstration project conducted under title II, section 101(d) of the Further Continuing Appropriations Act of 1985 (sec. 5903(d) of title 42 of the United States Code), or subsequent appropriations, up to a total amount of $2,500,000,000 for commercial demonstration of clean coal technology, or similar projects funded through appropriations for the Environmental Protection Agency; or

(iv) A permanent clean coal technology demonstration project that constitutes a repowering project.

pollution control project (40CFR52.24-24) means any activity or project undertaken at an existing electric utility steam generating unit for purposes of reducing emissions from such unit. Such activities or projects are limited to:

(i) The installation of conventional or innovative pollution control technology, including but not limited to advanced flue gas desulfurization, sorbent injection for sulfur dioxide and nitrogen oxides controls and electrostatic precipitators;

(ii) An activity or project to accommodate switching to a fuel which is less polluting than the fuel in use prior to the activity or project including, but not limited to natural gas or coal re-burning, co-firing of natural gas and other fuels for the purpose of controlling emissions;

(iii) A permanent clean coal technology demonstration project conducted under title II, section 101(d) of the Further Continuing Appropriations Act of 1985 (section 5903(d) of title 42 of the United States Code), or subsequent appropriations, up to a total amount of $2,500,000,000 for commercial demonstration of clean coal technology, or similar projects funded through appropriations for the Environmental Protection Agency; or

(iv) A permanent clean coal technology demonstration project that constitutes a repowering project.

pollution liability (SF401-42USC9671) means liability for injuries arising from the release of hazardous substances or pollutants or contaminants.

pollution prevention (EPA-94/04): The active process of identifying areas, processes, and activities which create excessive waste byproducts for the purpose of substitution, alteration, or elimination of the process to prevent waste generation.

polonium (EPA-94/04): A radioactive element that occurs in pitchblende and other uranium containing ores.

polybrominated dibenzo-p-dioxin or PBDD (40CFR766.3) means to any member of a class of dibenzo-p-dioxins with two to eight bromine substituents.

polybrominated dibenzofurans (40CFR766.3) refers to any member of a class of dibenzofurans with two to eight bromine substituents.

polybutadiene production (40CFR63.191) means a process that produces polybutadiene through the polymerization of 1,3-butadiene.

polycarbonates (40CFR63.191) means a special class of polyester formed from any dihydroxy compound and any carbonate diester or by ester interchange. Polycarbonates may be produced by solution or emulsion polymerization, although other methods may be used. A typical method for the manufacture of polycarbonates includes the reaction of bisphenol-A with phosgene in the presence of pyridine to form a polycarbonate. Methylene chloride is used as a solvent in this polymerization reaction.

polychlorinated biphenyl (40CFR704.43-4) means any chemical substance that is limited to the biphenyl molecule and that has been chlorinated to varying degrees.

polychlorinated biphenyl (40CFR704.45-4) means any chemical substance that is limited to the biphenyl molecule that has been chlorinated to varying degrees.

polychlorinated biphenyls (PCBs) (40CFR129.4-f) means a mixture of compounds composed of the biphenyl molecule which has been chlorinated to varying degrees.

polychlorinated biphenyls or PCBs (40CFR268.2-e) are halogenated organic compounds defined in accordance with 40CFR761.3.

polychlorinated dibenzo-p-dioxin or PCDD (40CFR766.3) means any member of a class of dibenzo-p-dioxins with two to eight chlorine substituents.

polychlorinated dibenzofuran (40CFR766.3) means any member of a class of dibenzofurans with two to eight chlorine substituents.

polyelectrolytes (EPA-94/04): Synthetic chemicals that help solids to clump during sewage treatment.

polyester (40CFR723.250-10) means a chemical substance that meets the definition of polymer and whose polymer molecules contain at least two carboxylic acid ester linkages, at least one of which links internal monomer units together.

polyethylene (40CFR60.561) means a thermoplastic polymer or copolymer comprised of at least 50 percent ethylene by weight; see low density polyethylene and high density polyethylene.

poly(ethylene terephthalate) (PET) (40CFR60.561) means a polymer or copolymer comprised of at least 50 percent bis-(2- hydroxyethyl)-terephthalate (BHET) by weight.

poly(ethylene terephthalate) (PET) manufacture using dimethyl terephthalate (40CFR60.561) means the manufacturing of poly(ethylene terephthalate) based on the esterification of dimethyl terephthalate (DMT) with ethylene glycol to form the intermediate monomer bis-(2-hydroxyethyl)-terephthalate polymerized to form PET.

poly(ethylene terephthalate) (PET) manufacture using terephthalic acid (40CFR60.561) means the manufacturing of poly(ethylene terephthalate) based on the esterification reaction of terephthalic acid (TPA) with ethylene glycol to form the intermediate monomer bis-(2-hydroxyethyl)-terephthalate (BHET) that is subsequently polymerized to form PET.

polyhalogenated dibenzo-p-dioxin or PHDD (40CFR766.3) means any member of a class of dibenzo-p-dioxins containing two to eight chlorine substituents or two to eight bromine substituents.

polyhalogenated dibenzofuran or PHDF (40CFR766.3) means any member of a class of dibenzofurans containing two to eight chlorine, bromine, or a combination of chlorine and bromine substituents.

polymer (40CFR60.601) means any of the natural or synthetic compounds of usually high molecular weight that consist of many repeated links, each link being a relatively light and simple molecule.

polymer (40CFR704.25-5) means a chemical substance that consists of at least a simple weight majority of polymer molecules but consists of less than a simple weight majority of molecules with the same molecular weight. Collectively, such polymer molecules must be distributed over a range of molecular weights wherein differences in molecular weight are primarily attributable to differences in the number of internal subunits (monomer) (other identical or similar definitions are provided in 40CFR721.350).

polymer (40CFR723.250-11) means a chemical substance that consists of at least a simple weight majority of polymer molecules but consists of less than a simple weight majority of molecules of the same molecular weight. Collectively, such polymer molecules must be distributed over a range of molecular weights wherein differences in the molecular weight are primarily attributable to differences in the number of internal subunits.

polymer (EPA-94/04): Basic molecular ingredients in plastic.

polymer molecule (40CFR704.25-6) means a molecule which includes at least four covalently linked subunits, at least two of which are internal subunits (other identical or similar definitions are provided in 40CFR721.350).

polymer molecule (40CFR723.250-12) means a molecule which includes at least four covalently linked subunits, at least two of which are internal subunits.

polymeric coating of supporting substrates (40CFR60.741) means a web coating process that applies elastomers, polymers, or prepolymers to a supporting web other than paper, plastic film, metallic foil, or metal coil.

polymerization reaction section (40CFR60.561) means the equipment designed to cause monomer(s) to react to form polymers, including equipment designed primarily to cause the formation of short polymer chains (oligomers or low polymers), but not including equipment designed to prepare raw materials for polymerization, e.g., esterification vessels. For the purposes of these standards, the polymerization reaction section begins with the equipment used to transfer the materials from the raw materials preparation section and ends with the last vessel in which polymerization occurs. Equipment used for the on-site recovery of ethylene glycol from poly(ethylene terephthalate) plants, however, are included in this process section, rather than in the material recovery process section.

polymerizing monomer (40CFR63.161) means a molecule or compound usually containing carbon and of relatively low molecular weight and simple structure (e.g., hydrogen cyanide, acrylonitrile, styrene), which is capable of conversion to polymers, synthetic resins, or elastomers by combination with

itself due to heat generation caused by a pump mechanical seal surface, contamination by a seal fluid (e.g., organic peroxides or chemicals that will form organic peroxides), or a combination of both with the resultant polymer buildup causing rapid mechanical seal failure.

polypropylene (PP) (40CFR60.561) means a thermoplastic polymer or copolymer comprised of at least 50 percent propylene by weight.

polystyrene (PS) (40CFR60.561) means a thermoplastic polymer or copolymer comprised of at least 80 percent styrene or para-methylstyrene by weight.

polysulfide rubber (40CFR63.191) means a synthetic rubber produced by reaction of sodium sulfide and p-dichlorobenzene at an elevated temperature in a polar solvent. This rubber is resilient and has low temperature flexibility.

polyvinyl chloride (PVC) (EPA-94/04): A tough, environmentally indestructible plastic that releases hydrochloric acid when burned.

polyvinyl chloride plant (40CFR61.61-c) includes any plant where vinyl chloride alone or in combination with other materials is polymerized.

pond water surface area (40CFR421.11-e) for the purpose of calculating the volume of waste water shall mean the area within the impoundment for rainfall and the actual water surface area for evaporation.

pond water surface area (40CFR421.61-d) when used for the purpose of calculating the volume of waste water which may be discharged shall mean the water surface area of the pond created by the impoundment for storage of process wastewater at normal operating level. This surface shall in no case be less than one-third of the surface area of the maximum amount of water which could be contained by the impoundment. The normal operating level shall be the average level of the pond during the preceding calendar month.

population (EPA-94/04): A group of interbreeding organisms occupying a particular space; the number of humans or other living creatures in a designated area.

population at risk (EPA-94/04): A population subgroup that is more likely to be exposed to a chemical, or is more sensitive to the chemical, than is the general population.

population stock or stock (MMPA3-16USC1362) means a group of marine mammals of the same species or smaller taxa in a common spatial arrangement, that interbreed when mature.

porcelain enameling (40CFR466.02-a) means the entire process of applying a fused vitreous enamel coating to a metal basis material. Usually this includes metal preparation and coating operations.

porosity (EPA-94/04): Degree to which soil, gravel, sediment or rock is permeated with pores or cavities through which water or air can move.

portable air compressor or compressor (40CFR204.51-a) means any wheel, skid, truck, or railroad car mounted, but not self-propelled, equipment designed to activate pneumatic tools. This consists of an air compressor (air end), and a reciprocating rotary or turbine engine rigidly connected in permanent alignment and mounted on a common frame. Also included are all cooling, lubricating, regulating, starting, and fuel systems, and all equipment necessary to constitute a complete, self-contained unit with a rated capacity of 75 cfm or greater which delivers air at pressures greater than 50 psig, but does not include any pneumatic tools themselves.

portable grinder (29CFR1910.94b) means any power-driven rotatable grinding, polishing, or buffing wheel mounted in such manner that it may be manually manipulated.

portable plant (40CFR60.671) means any nonmetallic mineral processing plant that is mounted on any chassis or skids and may be moved by the application of a lifting or pulling force. In addition, there shall be no cable, chain, turn buckle, bolt or other means (except electrical connections) by which any piece of equipment is attached or clamped to any anchor, slab, or structure, including bedrock that must be removed prior to the application of a lifting or pulling force for the purpose of transporting the unit.

portland cement plant (40CFR60.61-a) means any facility manufacturing portland cement by either the wet or dry process.

posing an exposure risk to food or feed (40CFR761.3) means being in any location where human food or animal feed products could be exposed to PCBs released from a PCB item. A PCB item poses an exposure risk to food or feed if PCBs released in any way from the PCB item have a potential pathway to human food or animal feed. EPA considers human food or animal feed to include items regulated by the U.S. Department of Agriculture or the Food and Drug Administration as human food or animal feed; this includes direct additives. Food or feed is excluded from this definition if it is used or stored in private homes (under TSCA).

positive crankcase ventilation valve (40CFR85.2122-(a)(4)(ii)): See PCV valve.

positive pressure fabric filter (40CFR60.271a) means a fabric filter with the fans on the upstream side of the filter bags (other identical or similar definitions are provided in 40CFR60.341-c).

positive temperature coefficient type choke heaters (40CFR85.2122(a)(2)(iii)(I)): See PTC type choke heaters.

positive test result (40CFR766.3) means:
(1) Any resolvable gas chromatographic peak for any 2,3,7,8-HDD or HDF which exceeds the LOQ listed under 40CFR766.27 for that congener, or
(2) exceeds LOQs approved by EPA under 40CFR766.28.

possession or control (40CFR704.3) means in the possession or control of any person, or of any subsidiary, partnership in which the person is a general partner, parent company, or any company or partnership which the parent company owns or controls, if the subsidiary, parent company, or other company or partnership is associated with the person in the research, development, test marketing, or commercial marketing or the substance in question. Information is in the possession or control of a person if it is:

(1) In the person's own files including files maintained by employees of the person in the course of their employment.

(2) In commercially available data bases to which the person has purchased access.

(3) Maintained in the files in the course of employment by other agents of the person who are associated with research, development, test marketing or commercial marketing of the chemical substance in question.

(Other identical or similar definitions are provided in 40CFR720.3-y; 723.50-2; 723.250-6.)

post closure (EPA-94/04): The time period following the shutdown of a waste management or manufacturing facility; for monitoring purposes, often considered to be 30 years.

post closure plan (40CFR265.141-e) means the plan for post-closure care prepared in accordance with the requirements of 40CFR265.117 through 40CFR265.120.

post consumer recovered paper (40CFR248.4-aa) means:

(1) Paper, paperboard and fibrous wastes from retail stores, office buildings, homes and so forth, after they have passed through their end-usage as a consumer item including: used corrugated boxes; old newspapers; old magazines; mixed waste paper; tabulating cards and used cordage; and

(2) All paper, paperboard and fibrous wastes that enter and are collected from municipal solid waste.

post consumer recycling (EPA-94/04): Reuse of materials generated from residential and consumer waste, e.g. converting wastepaper from offices into corrugated boxes or newsprint.

post consumer waste (PCW) (40CFR246.101-t) means a material or product that has served its intended use and has been discarded for disposal or recovery after passing through the hands of a final consumer.

post consumer waste (PCW) (40CFR247.101-e) means a material or product that has served its intended use and has been discarded for disposal after passing through the hands of a final user. "Post consumer waste" is a part of the broader category, "recycled material."

post exposure period (40CFR797.2050-5) is the portion of the test that begins with the test birds being returned from a treated diet to the basal diet. This period is typically 3 days in duration, but may be extended if birds continue to die or demonstrate other toxic effects.

post impregnation suspension process (40CFR60.561) means a manufacturing process in which polystyrene beads are first formed in a suspension process, washed, dried, or otherwise finished and then added with a blowing agent to another reactor in which the beads and blowing agent are reacted to produce expandable polystyrene (under CAA).

post mining area (40CFR434.11-k) means:

(1) A reclamation area or

(2) the underground workings of an underground coal mine after the extraction, removal, or recovery of coal from its natural deposit has ceased and prior to bond release (under the Clean Water Act).

post removal site control (40CFR300.5) means those activates that are necessary to sustain the integrity of a Fund-financed removal action following its conclusion. Post-removal site control may be a removal or remedial action under CERCLA. The term includes, without being limited to, activities such as relighting gas flares, replacing filters, and collecting leachate.

postchlorination (EPA-94/04): Addition of chlorine to plant effluent for disinfectant purposes after the effluent has been treated.

pot furnace (40CFR61.161) means a glass melting furnace that contains one or more refractory vessels in which glass is melted by indirect heating. The openings of the vessels are in the outside wall of the furnace and are covered with refractory stoppers during melting.

potable water (EPA-94/04): Water that is safe for drinking and cooking.

potatoes (40CFR407.71-s) shall mean the processing of sweet potatoes into the following product styles: Canned, peeled, solid, syrup, and vacuum packed. The following white potato product styles are also included: Canned, peeled, white, all varieties, whole and sliced.

potential combustion concentration(40CFR60.41a-7) means the theoretical emissions (ng/J, lb/million Btu heat input) that would result from combustion of a fuel in an uncleaned state (without emission control systems) and:
(a) For particulate matter is:
    (1) 3,000 ng/J (7.0 lb/million Btu) heat input for solid fuel; and
    (2) 75 ng/J (0.17 lb/million Btu) heat input for liquid fuels.
(b) For sulfur dioxide is determined under 40CFR60.48a(b).
(c) For nitrogen oxides is:
    (1) 290 ng/J (0.67 lb/million Btu) heat input for gaseous fuels;
    (2) 310 ng/J (0.72 lb/million Btu) heat input for liquid fuels; and
    (3) 990 ng/J (2.30 lb/million Btu) heat input for solid fuels.

potential damage (40CFR763.83) means circumstances in which:
(1) Friable ACBM is in an area regularly used by building occupants, including maintenance personnel, in the course of their normal activities.
(2) There are indications that there is a reasonable likelihood that the material or its covering will become damaged, deteriorated, or delaminated due to factors such as changes in building use, changes in operations and maintenance practices, changes in occupancy, or recurrent damage.

potential electrical output capacity (40CFR60.41a-22) is defined as 33 percent of the maximum design heat input capacity of the steam generating unit (e.g., a steam generating unit with a 100 MW (340 million Btu/hr) fossil-fuel heat input capacity would have a 33MW potential electrical output capacity). For electric utility combined cycle gas turbines the potential electrical output capacity is determined on the basis of the fossil-fuel firing capacity of the steam generator exclusive of the heat input and electrical power contribution by the gas turbine.

potential electrical output capacity (40CFR72.2) means the MWe capacity rating for the units which shall be equal to 33 percent of the maximum design heat input capacity of the steam generating unit, as calculated according to appendix D of part 72 (under CAA).

potential emissions (40CFR55.2) means the maximum emissions of a pollutant from an OCS source operating at its design capacity. Any physical or operational limitation on the capacity of a source to emit a pollutant, including air pollution control equipment and restrictions on hours of operation or on the type or amount of material combusted, stored, or processed, shall be treated as a limit on the design capacity of the source if the limitation is federally enforceable. Pursuant to section 328 of the Act, emissions from vessels servicing or associated with an OCS source shall be considered direct emissions from such a source while at the source, and while enroute to or from the source when within 25 miles of the source, and shall be included in the "potential to emit" for an OCS source. This definition does not alter or affect the use of this term for any other purposes under 40CFR55.13 or 55.14 of this part, except that vessel emissions must be included in the "potential to emit" as used in 40CFR55.13 and 55.14 of this part.

potential for industry-wide application (40CFR125.22-b) means that an innovative technology can be applied in two or more facilities which are in one or more industrial categories (under FWPCA).

potential hydrogen chloride emission rate (40CFR60.51a) means the hydrogen chloride emission rate that would occur from combustion of MSW in the absence of any hydrogen chloride emissions control.

potential production allowances (40CFR82.3-dd) means the production allowances obtained under 40CFR82.9.

potential significant damage (40CFR763.83) means circumstances in which:

(1) Friable ACBM is in an area regularly used by building occupants, including maintenance personnel, in the course of their normal activities.

(2) There are indications that there is a reasonable likelihood that the material or its covering will become damaged, deteriorated, or delaminated due to factors such as changes in building use, changes in operations and maintenance practices, changes in occupancy, or recurrent damage.

(3) The material is subject to major or continuing disturbance, due to factors including, but not limited to, accessibility or, under certain circumstances, vibration or air erosion.

potential sulfur dioxide emission rate (40CFR60.41b) means the theoretical sulfur dioxide emissions (ng/J, lb/million Btu heat input) that would result from combusting fuel in an uncleaned state and without using emission control systems.

potential sulfur dioxide emission rate (40CFR60.41c) means the theoretical SO_2 emissions (nanograms per joule [ng/J], or pounds per million Btu [lb/million Btu] heat input) that would result from combusting fuel in an uncleaned state and without using emission control systems.

potential sulfur dioxide emission rate (40CFR60.51a) means the sulfur dioxide emission rate that would occur from combustion of MSW in the absence of any sulfur dioxide emissions control.

potential to emit (40CFR51.165-iii) means the maximum capacity of a stationary source to emit a pollutant under its physical and operational design. Any physical or operational limitation on the capacity of the source to emit a pollutant, including air pollution control equipment and restrictions on hours of operation or on the type or amount of material combusted, stored, or processed, shall be treated as part of its design only if the limitation or the effect it would have on emissions is federally enforceable. Secondary emissions do not count in determining the potential to emit of a stationary source.

potential to emit (40CFR51.166-4) means the maximum capacity of a stationary source to emit a pollutant under its physical and operational design. Any physical or operational limitation on the capacity of the source to emit a pollutant, including air pollution control equipment and restrictions on hours of operation or on the type or amount of material combusted, stored, or processed, shall be treated as part of its design if the limitation or the effect it would have on emissions is federally enforceable. Secondary emissions do not count in determining the potential to emit of a stationary source.

potential to emit (40CFR51.301-r) means the maximum capacity of a stationary source to emit a pollutant under its physical and operational design. Any physical or operational limitation on the capacity of the source to emit a pollutant including air pollution control equipment and restrictions on hours of operation or on the type or amount of material combusted, stored, or processed, shall be treated as part of its design if the limitation or the effect it would have on emissions is federally enforceable. Secondary emissions do not count in determining the potential to emit of a stationary source.

potential to emit (40CFR51-AS-3) means the maximum capacity of a stationary source to emit a pollutant under its physical and operational design. Any physical or operational limitation on the capacity of the source to emit a pollutant, including air pollution control equipment and restrictions on hours of operation or on the type or amount of material combusted, stored, or processed, shall be treated as part of its design only if the limitation or the effect it would have on emissions is federally enforceable. Secondary emissions do not count in determining the potential to emit of a stationary source.

potential to emit (40CFR52.21-4) means the maximum capacity of a stationary source to emit a pollutant under its physical and operational design. Any physical or operational limitation on the capacity of the source to emit a pollutant, including air pollution control equipment and restrictions on hours of operation or on the type or amount of material combusted, stored, or processed, shall be treated as part of its design if the limitation or the effect it would have on emissions is federally enforceable. Secondary emissions do not count in determining the potential to emit of a stationary source.

potential to emit (40CFR52.24-3) means the maximum capacity of a stationary source to emit a pollutant under its physical and operational design. Any physical or operational limitation on the capacity of the source to emit a pollutant, including air pollution control equipment and restrictions on hours of operation or on amount of material combusted, stored, or processed, shall be treated as part of its design only if the limitation or the effect it would have on emissions is federally enforceable. Secondary emissions do not count in determining the potential to emit of a stationary source.

potential to emit (40CFR63.2) means the maximum capacity of a stationary source to emit a pollutant under its physical and operational design. Any physical or operational limitation on the capacity of the stationary source to emit a pollutant, including air pollution control equipment and restrictions on hours of operation or on the type or amount of material combusted, stored, or processed, shall be treated as part of its design if the limitation or the effect it would have on emissions is federally enforceable.

potential to emit (40CFR66.3-k) means the capability at maximum design capacity to emit a pollutant after the application of air pollution control equipment. Annual potential shall be based on the larger of the maximum annual rated capacity of the stationary source assuming continuous operation, or on a projection of actual annual emissions. Enforceable permit conditions on the type of materials combusted or processed may be used in determining the annual potential. Fugitive emissions, to the extent quantifiable, will be considered in determining annual potential for those stationary sources whose fugitive emissions are regulated by the applicable state implementation plan.

potential to emit (40CFR70.2) means the maximum capacity of a stationary source to emit any air pollutant under its physical and operational design. Any physical or operational limitation on the capacity of a source to emit an air pollutant, including air pollution control equipment and restrictions on hours of operation or on the type or amount of material combusted, stored, or processed, shall be treated as part of its design if the limitation is enforceable by the Administrator. This term does not alter or affect the use of this term for any other purposes under the Act, or the term "capacity factor" as used in title IV of the Act or the regulations promulgated thereunder.

potentially available (40CFR82.104-n) means that adequate information exists to make a determination that the substitute is technologically feasible, environmentally acceptable and economically viable.

potentially available (40CFR82.172) is defined as any alternative for which adequate health, safety, and environmental data, as required for the SNAP notification process, exist to make a determination of acceptability, and which the Agency reasonably believes to be technically feasible, even if not all testing has yet been completed and the alternative is not yet produced or sold.

potentially responsible party (PRP) (EPA-94/04): Any individual or company--including owners, operators, transporters or generators--potentially responsible for, or contributing to a spill or other contamination at a Superfund site. Whenever possible, through administrative and legal actions, GPA requires PRPs to clean up hazardous sites they have contaminated.

potentially responsible party (PRP) (40CFR35.4010) means any individual(s) or company(ies) (such as owner, operators, transporters, or generators) potentially responsible under sections 106 or 107 of CERCLA for the contamination problems at a Superfund site.

potentially responsible party (PRP) (40CFR35.6015-32) means any individual(s), or company(ies) identified as potentially liable under CERCLA for cleanup or payment for costs of cleanup of Hazardous Substance sites. PRPs may include individual(s) or company(ies) identified as having owned, operated, or in some other matter contributed wastes to Hazardous Substance sites.

potentially responsible party or PRP (40CFR304.12-m) means any person who may be liable pursuant to section 107(a) of CERCLA, 42USC9607(a), for response costs incurred and to be incurred by the United States not inconsistent with NCP (under CERCLA).

potentiation (EPA-94/04): The effect of one chemical to increase the effect of another chemical.

potentiometric surface (EPA-94/04): The level to which water will rise in cased wells or other cased excavations into aquifers.

potroom (40CFR60.191) means a building unit which houses a group of electrolytic cells in which aluminum is produced.

potroom group (40CFR60.191) means an uncontrolled potroom, a potroom which is controlled individually, or a group of potrooms or potroom segments ducted to a common control system.

POTW (40CFR122.2; 464.02; 501.2) means publicly owned treatment works.

POTW pretreatment program (40CFR403.3-d): See approved POTW pretreatment program.

POTW treatment plant (40CFR403.3-p) means that portion of the POTW which is designed to provide treatment (including recycling and reclamation) of municipal sewage and industrial waste.

pouring (40CFR61.171) means the removal of blister copper from the copper converter bath.

powder coating (40CFR60.311) means any surface coating which is applied as a dry powder and is fused into a continuous coating film through the use of heat (other identical or similar definitions are provided in 40CFR60.451).

powder forming (40CFR471.02-aa) includes forming and compressing powder into a fully dense finished shape, and is usually done within closed dies.

powder or dry solid form (40CFR721.3) means a state where all or part of the substance would have the potential to become fine, loose, solid particles.

power distribution system (40CFR72.2) means the portion of an electricity grid owned or operated by a utility that is dedicated to delivering electric energy to customers.

power purchase commitment (40CFR72.2) means a commitment or obligation of a utility to purchase electric power from a facility pursuant to:
(1) A power sales agreement;
(2) A state regulatory authority order requiring a utility to:
   (i) Enter into a power sales agreement with the facility;
   (ii) Purchase from the facility; or
   (iii) Enter into arbitration concerning the facility for the purpose of establishing terms and conditions of the utility's purchase of power;
(3) A letter of intent or similar instrument committing to purchase power (actual electrical output or generator output capacity) from the source at a previously offered or lower price and a power sales agreement applicable to the source is executed within the time frame established by the terms of the letter of intent but no later than November 15, 1992 or, where the letter of intent does not specify a timeframe, a power sales agreement applicable to the source is executed on or before November 15, 1992; or
(4) A utility competitive bid solicitation that has resulted in the selection of the qualifying facility or independent power production facility as the winning bidder.

power sales agreement (40CFR72.2) is a legally binding agreement between a QF, IPP, new IPP, or firm associated with such facility and a regulated electric utility that establishes the terms and conditions for the sale of power from the facility to the utility.

power setting (40CFR87.1) means the power or thrust output of an engine in terms of kilonewtons thrust for turbojet and turbofan engines and shaft power in terms of kilowatts for turboprop engines.

ppm (40CFR52.274.d.2-i; f.2-ii; r.2-ii) means parts per million by volume.

practicable (40CFR157.21-d) when used with respect to child-resistant packaging, means that the packaging can be mass produced and can he used in assembly line production.

practicable (40CFR230.3-q) means available and

capable of being done after taking into consideration cost, existing technology, and logistics in light of overall project purposes.

practicable (40CFR248.4-bb) means capable of being used consistent with: performance in accordance with applicable specifications, availability at a reasonable price, availability within a reasonable period of time, and maintenance of a satisfactory level of competition (other identical or similar definitions are provided in 40CFR250.4-gg).

practicable (40CFR252.4-g) means capable of being used consistent with: Performance in accordance with applicable specifications, availability at a reasonable price, availability within a reasonable period of time, and maintenance of a satisfactory level of competition (other identical or similar definitions are provided in 40CFR253.4).

practicable (40CFR6-AA-g) means capable of being done within existing constraints. The test of what is practicable depends upon the situation and includes consideration of the pertinent factors such as environment, community welfare, cost, or technology.

practical knowledge (40CFR171.2-19) means the possession of pertinent facts and comprehension together with the ability to use them in dealing with specific problems and situations.

practice (40CFR257.2) means the act of disposal waste.

pre-certification vehicle (40CFR85.1702-3) means an uncertified vehicle which a manufacturer employs in fleets from year to year in the ordinary course of business for product development, production method assessment, and market promotion purposes, but in a manner not involving lease or sale.

pre-certification vehicle engine (40CFR85.1702-4) means an uncertified heavy-duty engine owned by a manufacturer and used in a manner not involving lease or sale in a vehicle employed from year to year in the ordinary course of business for product development, production method assessment, and market promotion purposes.

pre-existing discharge (CWA301.p-33USC1311)

means any discharge at the time of permit application under this subsection.

prechlorination (EPA-94/04): The addition of chlorine at the headworks of a plant prior to other treatment processes. Done mainly for disinfection and control of tastes, odors, and aquatic growths, and to aid in coagulation and settling.

precious metals (40CFR421.261-b) shall mean gold, platinum, palladium, rhodium, iridium, osmium, and ruthenium.

precious metals (40CFR466.02-h) means gold, silver, or platinum group metals and the principal alloys of those metals.

precious metals (40CFR468.02-x) shall mean gold, platinum, palladium and silver and their alloys. Any alloy containing 30 or greater percent by weight of precious metals is considered a precious metal.

precious metals (40CFR471.02-bb) include gold, platinum, palladium, and silver and their alloys. Any alloy containing 30 or greater percent by weight of precious metals is considered a precious metal alloy.

precipitate (EPA-94/04): A solid that separates from a solution.

precipitation (EPA-94/04): Removal of hazardous solids from liquid waste to permit safe disposal; removal of particles from airborne emissions.

precipitation bath (40CFR60.601) means the water, solvent, or other chemical bath into which the polymer or prepolymer (partially reacted material) solution is extruded, and that causes physical or chemical changes to occur in the extruded solution to result in a semihardened polymeric fiber.

precipitator (EPA-94/04): Pollution control device that collects particles from an air stream.

precision (40CFR53.23-e): Variation about the mean of repeated measurements of the same pollutant concentration, expressed as one standard deviation about the mean.

precision (40CFR53.43-c): The variation in the

measured particle concentration among identical samplers under typical sampling conditions.

precision (40CFR86.082.2) means the standard deviation of replicated measurements.

precoat (40CFR60.441) means a coating operation in which a coating other than an adhesive or release is applied to a surface during the production of a pressure sensitive tape or label product.

preconditioning (40CFR610.11-12) means the operation of an automobile through one (1) EPA Urban Dynamometer Driving Schedule, described in 40CFR86.

precontrolled vehicles (40CFR51-AN) means light duty vehicles sold nationally (except in California) prior to the 1968 model year and light-duty vehicles sold in California prior to the 1966 model year.

precursor (40CFR766.3) means a chemical substance which is not contaminated due to the process conditions under which it is manufactured, but because of its molecular structure, and under favorable process conditions, it may cause or aid the formation of HDDs/HDFs in other chemicals in which it is used as a feedstock or intermediate.

precursor (EPA-94/04): In photochemistry, a compound antecedent to a volatile organic compound (VOC). Precursors react in sunlight to form ozone or other photochemical oxidants.

precursors of a criteria pollutant (40CFR51.852) are:
(1) For ozone, nitrogen oxides (NO_x), unless an area is exempted from NO_x requirements under section 182(f) of the Act, and volatile organic compounds (VOC); and
(2) For PM-10, those pollutants described in the PM-10 nonattainment area applicable SIP as significant contributors to the PM-10 levels.
(Other identical or similar definitions are provided in 40CFR93.152.)

preliminary analysis (40CFR610.11-22) means the engineering analysis performed by EPA prior to testing prescribed by the Administrator based on data and information submitted by a manufacturer or available from other sources.

preliminary assessment (EPA-94/04): The process of collecting and reviewing available information about a known or suspected waste site or release.

preliminary assessment (PA) (40CFR300.5) means review of existing information and an off-site reconnaissance, if appropriate, to determine if a release may require additional investigation or action. A PA may include an on-site reconnaissance, if appropriate.

premanufacture notice (PMN) program (40CFR82.172) has the meaning described in 40CFR part 720, subpart A promulgated under the Toxic Substances Control Act, 15 U.S.C. 2601 et seq.

premanufacture notice or PMN (40CFR700.43) means any notice submitted to EPA pursuant to 40CFR720 of this chapter or 40CFR723.250 of this chapter.

preponderance of evidence (40CFR32.105-o) Proof by information that, compared with that opposing it, leads to the conclusion that the fact at issue is more probably true than not.

prescriptive (EPA-94/04): Water rights which are acquired by diverting water and putting it to use in accordance with specified procedures, e.g., filing a request with a state agency to use unused water in a stream, river, or lake.

present in the same form and concentration as a product packaged for distribution and use by the general public (40CFR370.2) means a substance packaged in a similar manner and present in the same concentration as the substance when packaged for use by the general public, whether or not it is intended for distribution to the general public or used for the same purpose as when it is packaged for use by the general public.

presentation of credentials (40CFR86.078.7) shall mean display of the document designating a person as an EPA Enforcement Officer.

presentation of credentials (40CFR89.2) means the display of the document designating a person as an EPA enforcement officer or EPA authorized representative.

preserve (40CFR6-AA-h) means to prevent modification to the natural floodplain environment or to maintain it as closely as possible to its natural state.

presiding officer (40CFR164.2-p) means any person designated by the Administrator to conduct an expedited hearing (other identical or similar definitions are provided in 40CFR17.2-e; 27.2; 22.03; 72.2; 85.1807-5; 86.614.84-4; 86.1014.84-4; 86.1115.87-4; 104.2-h; 123.64; 124.72; 173.2-e).

pressed and blown glass (40CFR60.291) means glass which is pressed, blown, or both, including textile fiberglass, noncontinuous flat glass, noncontainer glass, and other products listed in SIC 3229. It is separated into:
(1) Glass of borosilicate recipe.
(2) Glass of soda-lime and lead recipes.
(3) Glass of opal, fluoride, and other recipes.

pressure (40CFR146.3) means the total load or force per unit area acting on a surface (other identical or similar definitions are provided in 40CFR147.2902).

pressure drop (40CFR85.2122(a)(16)(ii)(B)) means a measure, in kilopascals, of the difference in static pressure measured immediately upstream and downstream of the air filter element.

pressure release (40CFR60.481) means the emission of materials resulting from system pressure being greater than set pressure of the pressure relief device.

pressure release (40CFR61.241) means the emission of materials resulting from the system pressure being greater than the set pressure of the pressure relief device (other identical or similar definitions are provided in 40CFR264.1031).

pressure release (40CFR63.161) means the emission of materials resulting from the system pressure being greater than the set pressure of the pressure relief device. This release can be one release or a series of releases over a short time period due to a malfunction in the process.

pressure relief device or valve (40CFR63.161) means a safety device used to prevent operating pressures from exceeding the maximum allowable working pressure of the process equipment. A common pressure relief device is a spring-loaded pressure relief valve. Devices that are actuated either by a pressure of less than or equal to 2.5 psig or by a vacuum are not pressure relief devices.

pressure sewers (EPA-94/04): A system of pipes in which water, wastewater, or other liquid is pumped to a higher elevation.

pretreatment (40CFR403.3-q) means the reduction of the amount of pollutants, the elimination of pollutants, or the alteration of the nature of pollutant properties in wastewater prior to or in lieu of discharging or otherwise introducing such pollutants into a POTW. The reduction or alteration may be obtained by physical, chemical or biological processes, process changes or by other means, except as prohibited by 40CFR403.6(d). Appropriate pretreatment technology includes control equipment, such as equalization tanks or facilities, for protection against surges or slug loadings that might interfere with or otherwise be incompatible with the POTW. However, where wastewater from a regulated process is mixed in an equalization facility with unregulated wastewater or with wastewater from another regulated process, the effluent from the equalization facility must meet an adjusted pretreatment limit calculated in accordance with 40CFR403.6(e).

pretreatment (EPA-94/04): Processes used to reduce, eliminate, or alter the nature of wastewater pollutants from non-domestic sources before they are discharged into publicly owned treatment works (POTWs).

pretreatment control authority (40CFR414.10-b) means:
(1) The POTW if the POTW's submission for its pretreatment program has been approved in accordance with the requirements of 40CFR403.11, or
(2) The Approval Authority if the submission has not been approved.

pretreatment requirements (40CFR403.3-r) means any substantive or procedural requirement related to Pretreatment, other than a National Pretreatment Standard, imposed on an Industrial User.

pretreatment standard (40CFR117.1; 403.3-j): See national pretreatment standard.

prevalent level samples (EPA-94/04): Air samples taken under normal conditions (also known as ambient background samples).

prevalent levels (EPA-94/04): Levels of airborne contaminant occurring under normal conditions.

prevention of air pollution emergency episodes— examples regulations: See appendix L to 40CFR51.

prevention of significant deterioration (PSD) (EPA-94/04): EPA program in which state and/or federal permits are required in order to restrict emissions from new or modified sources in places where air quality already meets or exceeds primary and secondary ambient air quality standards.

prevention of significant deterioration of air quality: See 40CFR52.21.

preventive measures (40CFR763.83) means actions taken to reduce disturbance of ACBM or otherwise eliminate the reasonable likelihood of the material's becoming damaged or significantly damaged (under TSCA).

price analysis (40CFR33.005) means the process of evaluating a prospective price without regard to the contractor's separate cost elements and proposed profit. Price analysis determines the reasonableness of the proposed subagreement price based on adequate price competition, previous experience with similar work, established catalog or market price, law, or regulation.

price analysis (40CFR35.6015-33) means the process of evaluating prospective price without regard to the contractor's separate cost elements and proposed profit. Price analysis determines the reasonableness of the proposed contract price based on adequate price competition, previous experience with similar work, established catalog or market price, law, or regulation.

primacy (EPA-94/04): Having the primary responsibility for by administrating and enforcing regulations.

primary aluminum reduction plant (40CFR60.191) means any facility manufacturing aluminum by electrolytic reduction.

primary contact recreation (40CFR131.35.d-10) means activities where a person would have direct contact with water to the point of complete submergence, including but not limited to skin diving, swimming, and water skiing (cf. secondary contact recreation).

primary control system (40CFR60.191) means an air pollution control system designed to remove gaseous and particulate flourides from exhaust gases which are captured at the cell.

primary copper smelter (40CFR60.161-a) means any installation or any intermediate process engaged in the production of copper from copper sulfide ore concentrates through the use of pyrometallurgical techniques.

primary copper smelter (40CFR61.171) means any installation or intermediate process engaged in the production of copper from copper-bearing materials through the use of pyrometallurgical techniques.

primary drinking water regulation (EPA-94/04): Applies to public water systems and specifies a contaminant level, which, in the judgment of the EPA Administrator, will not adversely affect human health.

primary drinking water regulation (SDWA1401-42USC300f) means a regulation which:
(A) applies to public water systems;
(B) specifies contaminants which, in the judgment of the Administrator, may have any adverse effect on the health of persons;
(C) specifies for each such contaminant either:
   (i) a maximum contaminant level, if, in the judgment of the Administrator, it is economically and technologically feasible to ascertain the level of such contaminant in water in public water systems, or
   (ii) if, in the judgment of the Administrator, it is not economically or technologically feasible to so ascertain the level of such contaminant, each treatment technique known to the Administrator which leads to

a reduction in the level of such contaminant sufficient to satisfy the requirements of section 1412; and

(D) contains criteria and procedures to assure a supply of drinking water which dependably complies with such maximum contaminant levels; including quality control and testing procedures to insure compliance with such levels and to insure proper operation and maintenance of the system, and requirements as to:

(i) the minimum quality of water which may be taken into the system and

(ii) siting for new facilities for public water systems (cf. secondary drinking water regulation).

primary emission control system (40CFR60.141a) means the combination of equipment used for the capture and collection of primary emissions (e.g., an open hood capture system used in conjunction with a particulate matter cleaning device such as an electrostatic precipitator or a closed hood capture used in conjunction with matter cleaning device such as a scrubber).

primary emission control system (40CFR61.171) means the hoods, ducts, and control devices used to capture, convey, and collect process emissions (under CAA).

primary emission control system (40CFR61.181) means the hoods, enclosures, ducts, and control devices used to capture, convey, and remove particulate matter from exhaust gases which are captured directly at the source of generation.

primary emissions (40CFR60.141-b) means particulate matter emissions from the BOPF generated during the steel production cycle and captured by the BOPF primary control system.

primary emissions (40CFR60.141a) means particulate matter emissions from the BOPF generated during the steel production cycle which are captured by, and do not thereafter escape from, the BOPF primary control system.

primary enforcement responsibility (40CFR142.2) means the primary responsibility for administration and enforcement of primary drinking water regulations and related requirements applicable to public water systems within a State.

primary exporter (40CFR262.51) means any person who is required to originate the manifest for a shipment of hazardous waste in accordance with 40CFR262, subpart B, or equivalent State provision, which specifies a treatment, storage, or disposal facility in a receiving country as the facility to which the hazardous waste will be sent and any intermediary arranging for the export.

primary fuel (40CFR60.701) means the fuel fired through a burner or a number of similar burners. The primary fuel provides the principal heat input to the device, and the amount of fuel is sufficient to sustain operation without the addition of other fuels.

primary fuel (40CFR63.111) means the fuel that provides the principal heat input to the device. To be considered primary, the fuel must be able to sustain operation without the addition of other fuels (under CAA).

primary fuel or primary fuel supply (40CFR72.2) means the main fuel type (expressed in mmBtu) consumed by an affected unit for the applicable calendar year.

primary industry category (40CFR122.2) means any industry category listed in the NRDC settlement agreement (Natural Resources Defense Council et al. v. Train, 8 E.R.C. 2120 (D.D.C. 1976), modified 12 E.R.C. 1833 (D.D.C. 1979)); also listed in Appendix A of part 122 (cf. secondary industry category).

primary intended service class (40CFR86.085.2) means:

(a) The primary service application group for which a heavy-duty diesel engine is designed and marketed, as determined by the manufacturer. The primary intended service classes are designated as light, medium, and heavy heavy-duty diesel engines. The determination is based on factors such as vehicle GVW, vehicle usage and operating patterns, other vehicle design characteristics, engine horsepower, and other engine design and operating characteristics:

(1) Light heavy-duty diesel engines usually are

non-sleeved and not designed for rebuild; their rated horsepower generally ranges from 70 to 170. Vehicle body types in this group might include any heavy-duty vehicle built for a light-duty truck chassis, van trucks, multi-stop vans, recreational vehicles, and some single axle straight trucks. Typical applications would include personal transportation, light-load commercial hauling and delivery, passenger service, agriculture, and construction. The GVWR of these vehicles is normally less than 19,500 lbs.

(2) Medium heavy-duty diesel engines may be sleeved or non-sleeved and may be designed for rebuild. Rated horsepower generally ranges from 170 to 250. Vehicle body types in this group would typically include school buses, tandem axle straight trucks, city tractors, and a variety of special purpose vehicles such as small dump trucks, and trash compactor trucks. Typical applications would include commercial short haul and intra-city delivery and pickup. Engines in this group are normally used in vehicles whose GVWR varies from 19,500-33,000 lbs.

(3) Heavy heavy-duty diesel engines are sleeved and designed for multiple rebuilds. Their rated horsepower generally exceeds 250. Vehicles in this group are normally tractors, trucks, and buses used in inter-city, long-haul applications. These vehicles normally exceed 33,000 lbs. GVWR.

(Other identical or similar definitions are provided in 40CFR86.090.2.)

primary lead smelter (40CFR60.181-a) means any installation or any intermediate process engaged in the production of lead from lead sulfide ore concentrates through the use of pyrometallurgical techniques.

primary measurement services (primary) (40CFR195.2) refers to radon measurement services using a specific device which services include the capability to read and/or analyze the results generated from the device.

primary metal cast (40CFR464.31-g) shall mean the metal that is poured in the greatest quantity at an individual plant.

primary mill (40CFR420.71-b) means whose steel hot forming operations at reduce ingots to blooms or slabs by passing the ingots between rotating steel rolls. The first hot forming operation performed on solidified steel after it is removed from the ingot molds is carried out on a "primary mill."

primary oxygen blow (40CFR60.141-c) means the period in the steel production cycle of a BOPF during which a high volume of oxygen-rich gas is introduced to the bath of molten iron by means of a lance inserted from the top of the vessel or though tuyeres in the bottom or through the bottom and sides of the vessel. This definition does not include any additional or secondary oxygen blows made after the primary blow or the introduction of nitrogen or other inert gas through tuyeres in the bottom or bottom and the sides of the vessel.

primary oxygen blow (40CFR60.141a) means the period in the steel production cycle of a BOPF during which a high volume of oxygen-rich gas is introduced to the bath of molten iron by means of a lance inserted from the top of the vessel. This definition does not include any additional, or secondary, oxygen blows made after the primary blow.

primary panel (40CFR211.203-u) means the surface that is considered to be the front surface or that surface which is intended for initial viewing at the point of ultimate sale or the point of distribution for use.

primary processor of asbestos (40CFR763.63-j) is a person who processes for commercial purposes bulk asbestos.

primary resistor (40CFR85.2122(a)(10)(ii)) means a device used in the primary circuit of an inductive ignition system to limit the flow of current.

primary standard (40CFR51.100-c) means a national primary ambient air quality standard promulgated pursuant to section 109 of the Act.

primary standard attainment date (CAA302-

42USC7602) means the date specified in the applicable implementation plan for the attainment of a national primary ambient air quality standard for any air pollutant.

primary standards (EPA-94/04): National ambient air quality standards designed to protect human health with "an adequate margin for safety" (cf. national ambient air quality standards or secondary standards).

primary vendor (40CFR76.2) means the vendor of the NO_x emission control system who has primary responsibility for providing the equipment, service, and technical expertise necessary for detailed design, installation, and operation of the controls, including process data, mechanical drawings, operating manuals, or any combination thereof.

primary waste treatment (EPA-94/04): First steps in wastewater treatment; screens and sedimentation tanks are used to remove most materials that float or will settle. Primary treatment removes about 30 percent of carbonaceous biochemical oxygen demand from domestic sewage.

primary zinc smelter (40CFR52.1881-vii) means any installation engaged in the production, or any intermediate process in the production, of zinc or zinc oxide from the zinc sulfide ore concentrates through the use of pyrometallurgical techniques.

primary zinc smelter (40CFR60.171-a) means any installation engaged in the production, or any intermediate process in the production, of zinc or zinc oxide from zinc sulfide ore concentrates through the use of pyrometallurgical techniques.

prime coat (40CFR52.741) means the first of two or more coatings applied to a surface.

prime coat (40CFR60.721) means the initial cost applied to a part when more than one coating is applied, not including conductive sensitizers or electromagnetic interference/radio frequency interference shielding coatings.

prime coat operation (40CFR60.391) means the prime coat spray booth or dip tank, flash-off area, and bake oven(s) which are used to apply and dry or cure the initial coating on components of automobile or light-duty truck bodies.

prime coat operation (40CFR60.461) means the coating application station, curing oven, and quench station used to apply and dry or cure the initial coating(s) on the surface of the metal coil.

prime contractor (40CFR8.2-s) means any person holding a contract, and for the purposes of subpart B (General Enforcement, Compliance Review, and Complaint Procedure) of the rules, regulations, and relevant orders of the Secretary of Labor, any person who has held a contract subject to the order.

prime farmland (SMCRA701-30USC1291) shall have the same meaning as that previously prescribed by the Secretary of Agriculture on the basis of such factors as moisture availability, temperature regime, chemical balance, permeability, surface layer composition, susceptibility to flooding, and erosion characteristics, and which historically have been used for intensive agricultural purposes, and as published in the Federal Register.

prime surfacer coat (40CFR52.741) means a coating used to touch up areas on the surface of automobile or light-duty truck bodies not adequately covered by the prime coat before application of the top coat. The prime surfacer coat is applied between the prime coat and topcoat. An anti-chip coating applied to main body parts (e.g., rocker panels, bottom of doors and fenders, and leading edge of roof) is a prime surfacer coat.

primers (40CFR52.741) means any coatings formulated and applied to substrates to provide a firm bond between the substrate and subsequent coats.

principal (40CFR32.105-p): Officer, director, owner, partner, key employee, or other person within a participant with primary management or supervisory responsibilities; or a person who has a critical influence on or substantive control over a covered transaction, whether or not employed by the participant. Persons who have critical influence on or substantive control over a covered transaction are:
(1) Principal investigators.
(2) Bid and proposal estimators and preparers.

principal company (40CFR60.41a-11) means the electric utility company or companies which own the affected facility.

principal display panel (PDP) (40CFR82.104-o) means the entire portion of the surface of a product, container or its outer packaging that is most likely to be displayed, shown, presented, or examined under customary conditions of retail sale. The area of the PDP is not limited to the portion of the surface covered with existing labeling; rather it includes the entire surface, excluding flanges, shoulders, handles, or necks.

principal importer (40CFR720.3-z) means the first importer who, knowing that a new chemical substance will be imported rather than manufactured domestically, specifies the identity of the chemical substance and the total amount to be imported. Only persons who are incorporated, licensed, or doing business in the United States may be principal importers.

principal importer (40CFR721.3) means the first importer who, knowing that a chemical substance will be imported for a significant new use rather than manufactured in the United States, specifies the chemical substance and the amount to be imported. Only persons who are incorporated, licensed, or doing business in the United States may be principal importers.

principal or major uses (FLPMA103-43USC1702) includes. and is limited to, domestic livestock grazing, fish and wildlife development and utilization, mineral exploration and production, rights-of-way, outdoor recreation, and timber production.

principal organic hazardous constituents (POHCs) (EPA-94/04): Hazardous compounds monitored during an incinerator's trial burn, selected for high concentration in the waste feed and difficulty of combustion.

principal place of business (40CFR171.2) means the principal location, either residence or office, in the State in which an individual, partnership, or corporation applies pesticides.

principal residence (40CFR35.2005-31) means, for the purposes of 40CFR35.2034, the habitation of a family or household for at least 51 percent of the year. Second homes, vacation or recreation residences are not included in this definition.

principal source aquifer (40CFR146.3; 149.2): See sole source aquifer.

principal sponsor (40CFR790.3) means an individual sponsor or the joint sponsor who assumes primary responsibility for the direction of a study and for oral and written communication with EPA.

printing (40CFR52.741) means the application of words, designs, and pictures to a substrate using ink.

printing line (40CFR52.741) means an operation consisting of a series of one or more roll printers and any associated roll coaters, drying areas, and ovens wherein one or more coatings are applied, dried, and/or cured.

printing paper (40CFR250.4-hh) means paper designed for printing, other than newsprint, such as offset and book paper.

prior appropriation (EPA-94/04): A doctrine of water law that allocates the rights to use water on a first-come first-serve basis.

prior approval (40CFR31.3) means documentation evidencing consent prior to incurred specific cost.

priority pollutant No. 3 (40CFR420.02): See benzo(a)pyrene.

priority pollutant No. 4 (40CFR420.02): See benzene.

priority pollutant No. 55 (40CFR420.02): See naphthalene.

priority pollutant No. 85 (40CFR420.02): See tetrachloroethylene.

priority pollutants (40CFR-AA-92) means also the 126 pollutants listed in 40CFR423 Appendix A.

priority pollutants (40CFR414.10) means the toxic pollutants listed in 40CFR401.15.

priority pollutants (40CFR455.10-f) means the toxic pollutants listed in 40CFR part 423, appendix A.

priority water quality areas (40CFR35.2005-34) means, for the purposes of section 35.2015, specific stream segments or bodies of water, as determined by the State, where municipal discharges has resulted in the impairment of a designated use or significant public health risks, and where the reduction of pollution from such discharges will substantially restore surface or groundwater uses.

private applicator (40CFR171.2-20) means a certified applicator who uses or supervises the use of any pesticide which is classified for restricted use for purposes of producing any agricultural commodity on property owned or rented by him or his employer or (if applied without compensation other than trading of personal services between producers of agricultural commodities) on the property of another person.

private carrier of property by motor vehicle (40CFR202.10-p) means any person not included in terms common carrier by motor vehicle or contract carrier by motor vehicle, who or which transports in interstate or foreign commerce by motor vehicle property of which such person is the owner, lessee, or bailee, when such transportation is for sale, lease, rent or bailment, or in furtherance of any commercial enterprise.

privately owned treatment works (40CFR122.2) means any device or system which is:
(a) used to treat wastes from any facility whose operator is not the operator of the treatment works and
(b) not a "POTW."

probability of detection (EPA-94/04): The likelihood, expressed as a percentage, that a test method will correctly identify a leaking tank.

proceeding (40CFR17.2-f) means an adversary adjudication as defined in section 40CFR17.2(b).

proceeding (40CFR2.301-4) means any rulemaking, adjudication, or licensing conducted by EPA under the Act or under regulations which implement the Act, except for determinations under this subpart

(other identical or similar definitions are provided in 40CFR2.302-4; 2.303-4; 2.306-6).

proceeding (40CFR2.304-3) means any rulemaking, adjudication, or licensing process conducted by EPA under the Act or under regulations which implement the Act, except for any determination under this part.

proceeding (40CFR2.305-4) means any rulemaking, adjudication, or licensing conducted by EPA under the Act or under regulations which implement the Act including the issuance of administrative orders and the approval or disapproval of plans (e.g., closure plans) submitted by persons subject to regulation under the Act, but not including determinations under this subpart.

proceeding (40CFR2.310-6) means any rulemaking or adjudication conducted by EPA under the Act or under regulations which implement the Act (including the issuance of administrative orders under section 106 of the Act and cost recovery pre-litigation settlement negotiations under section 107 or 122 of the Act), any cost recovery litigation under section 107 of the Act, or any administrative determination made under section 104 of the Act, but not including determinations under this subpart.

process (40CFR372.3) means the preparation of a toxic chemical, after its manufacture, for distribution in commerce:
(1) In the same form or physical state as, or in a different form or physical state from, that in which it was received by the person so preparing such substance, or
(2) As part of an article containing the toxic chemical. Process also applies to the processing of a toxic chemical contained in a mixture or trade name product.

process (40CFR52.1881-viii) means any source operation including any equipment, devices, or contrivances and all appurtenances thereto, for changing any material whatever or for storage or handling of any materials, the use of which may cause the discharge of an air contaminant into the open air, but not including that equipment defined as fossil fuel fired steam generating units in these regulations. Duplicate or similar parallel operations within a structure, building, or shop shall be considered as a single process for purposes of this regulation.

process (40CFR52.741) means any stationary emission source other than a fuel combustion emission source or an incinerator.

process (40CFR704.3; 716.3) means to process for commercial purposes.

process (40CFR710.2-t) means the preparation of a chemical substance or mixture, after its manufacture, for distribution in commerce:
(1) in the same form or physical state as, or in a different form or physical state from, that in which it was received by the person so preparing such substance or mixture, or
(2) as part of a mixture or article containing the chemical substance or mixture.
(Other identical or similar definitions are provided in 40CFR720.3-aa; 747.115-1; 747.195-1; 747.200-1; 723.250-6.)

process (40CFR761.3) means the preparation of a chemical substance or mixture, after its manufacture, for distribution in commerce:
(1) In the same form or physical state as, or in a different form or physical state from, that in which it was received by the person so preparing such substance or mixture, or
(2) As part of an article containing the chemical substance or mixture.

process (TSCA3-15USC2602) means the preparation of a chemical substance or mixture, after its manufacture, for distribution in commerce:
(A) in the same form or physical state as, or in a different form or physical state from, that in which it was received by the person so preparing such substance or mixture, or
(B) as part of an article containing the chemical substance or mixture.
(Other identical or similar definitions are provided in 40CFR723.175-5; 762.3-d; 763.163.)

process emission (40CFR60.301-g) means the particulate matter which is collected by a capture system.

process emissions (40CFR61.171) means inorganic arsenic emissions from copper converters that are captured directly at the source of generation.

process emissions (40CFR61.181) means inorganic arsenic emissions that are captured and collected in a primary emission control system.

process for commercial purposes (40CFR704.3) means the preparation of a chemical substance or mixture after its manufacture for distribution in commerce with the purpose of obtaining an immediate or eventual commercial advantage for the processor. Processing of any amount of a chemical substance or mixture is included in this definition. If a chemical substance or mixture containing impurities is processed for commercial purposes, then the impurities also are processed for commercial purposes.

process for commercial purposes (40CFR710.2-u) means to process:
(1) for distribution in commerce, including for test marketing purposes, or
(2) for use as an intermediate.

process for commercial purposes (40CFR712.3-m) means the preparation of a chemical substance or mixture, after its manufacture, for distribution in commerce with the purpose of obtaining an immediate or eventual commercial advantage for the processor. Processing of any amount of a chemical substance or mixture is included. If a chemical or mixture containing impurities is processed for commercial purposes, then those impurities are also processed for commercial purposes.

process for commercial purposes (40CFR716.3) means the preparation of a chemical substance or mixture, after its manufacture, for distribution in commerce with the purpose of obtaining an immediate or eventual commercial advantage for the processor. Processing of any amount of a chemical substance or mixture containing impurities is processed for commercial purposes, then those impurities are also processed for commercial purposes.

process for commercial purposes (40CFR717.3-g) means the preparation of a chemical substance or mixture, after its manufacture, for distribution in commerce with the purpose of obtaining advantage for the processor. Processing any amount of a chemical substance or mixture containing impurities is included. If a chemical substance or mixture containing impurities is processed for commercial purposes, then those impurities are also processed for commercial purposes.

process for commercial purposes (40CFR721.3) means the preparation of a chemical substance or mixture containing the chemical substance, after manufacture of the substance, for distribution in commerce with the purpose of obtaining an immediate or eventual commercial advantage for the processor. Processing of any amount of a chemical substance or mixture containing the chemical substance is included in this definition. If a chemical substance or mixture containing impurities is processed for commercial purposes, the impurities also are processed for commercial purposes.

process for commercial purposes (40CFR763.63-k) means the preparation of a chemical substance or mixture, after its manufacture, for distribution in commerce with the purpose of obtaining an immediate or eventual commercial advantage for the processor. Processing of any amount of a chemical substance or mixture is included. If a chemical or mixture containing impurities is processed for commercial purposes, then those impurities are also processed for commercial purposes.

process fugitive emissions (40CFR60.381) means particulate matter emissions from an affected facility that are not collected by a capture system (under CAA).

process gas (40CFR60.101-c) means any gas generated by a petroleum refinery process unit, except fuel gas and process upset gas as defined in this section.

process generated waste water (40CFR412.11-d) shall mean water directly or indirectly used in the operation of a feedlot for any or all of the following: Spillage or overflow from animal or poultry watering systems; washing, cleaning or flushing pens, barns, manure pits or other feedlot facilities; direct contact swimming, washing or spray cooling of animals; and dust control.

process generated waste water (40CFR412.21-d) shall mean water directly or indirectly used in the operation of a feedlot for any or all of the following: Spillage or overflow from animal or poultry watering systems; washing cleaning or flushing pens, barns, manure pits or other feedlot facilities; direct contact swimming, washing or spray cooling of animals; and dust control.

process generated waste water (40CFR436.181-e) shall mean any waste water used in the slurry transport of mined material, air emissions control, or processing exclusive of mining. The term shall also include any other water which becomes commingled with such waste water in a pit, pond lagoon, mine, or other facility used for settling or treatment of such waste water.

process generated waste water (40CFR436.21-e) shall mean any waste water used in the slurry transport of mined material, air emissions control, or processing exclusive of mining. The term shall also include any other water which becomes commingled with such waste water in a pit pond, lagoon, mine, or other facility used for treatment of waste water.

process generated waste water (40CFR436.31-e) shall mean any waste water used in the slurry transport of mined material, air emissions control, or processing exclusive of mining. The term shall also include any other water which becomes commingled with such waste water in a pit, pond, lagoon, mine or other facility used for treatment of such waste water. The term does not include waste water used for the suction dredging of deposits in a body of water and returned directly to the body of waste without being used for other purposes or combined with other waste water (other identical or similar definitions are provided in 40CFR436.41-e).

process heater (40CFR264.1031) means a device that transfers heat liberated by burning fuel to fluids contained in tubes, including all fluids except water that are heated to produce steam.

process heater (40CFR60.41b) means a device that is primarily used to heat a material to initiate or promote a chemical reaction in which the material participates as a reactant or catalyst (other identical or similar definitions are provided in 40CFR60.41c).

process heater (40CFR60.561) means a device that transfers heat liberated by burning fuel to fluids contained in tubular coils, including all fluids except water that is heated to produce steam.

process heater (40CFR60.661) means a device that transfers heat liberated by burning fuel to fluids contained in tubes, including all fluids except water that is heated to produce steam.

process heater (40CFR60.701) means a device that transfers heat liberated by burning fuel directly to process streams or to heat transfer liquids other than water.

process heater (40CFR61.301) means a device that transfers heat liberated by burning fuel to fluids contained in tubes, except water that is heated to produce steam.

process heater (40CFR63.111) means a device that transfers heat liberated by burning fuel directly to process streams or to heat transfer liquids other than water.

process improvement (40CFR60.481) means routine changes made for safety and occupational health requirements, for energy savings, for better utility, for ease of maintenance and operation, for correction of design deficiencies, for bottleneck removal, for changing product requirements, or for environmental control.

process line (40CFR60.561) means a group of equipment assembled that can operate independently if supplied with sufficient raw materials to produce polypropylene, polyethylene, polystyrene (general purpose, crystal, or expandable) or poly(ethylene terephthalate) or one of their copolymers. A process line consists of the equipment in the following process sections (to the extent that these process sections are present at a plant): raw materials preparation, polymerization reaction, product finishing, product storage, and material recovery (under CAA).

process or distribute in commerce solely for export (40CFR747.115-4) means to process or distribute in commerce solely for export from the United States under the following restrictions on domestic activity:

(i) Processing must be performed at sites under the control of the processor.

(ii) Distribution in commerce is limited to purposes of export.

(iii) The processor or distributor may not use the substance except in small quantities solely for research and development.

(Other identical or similar definitions are provided in 40CFR747.195-4.)

process section (40CFR60.561) means the equipment designed to accomplish a general but well-defined task in polymer production. Process sections include raw materials preparation, polymerization reaction, material recovery, product finishing, and product storage and may be dedicated to a single process line or common to more than one process line.

process solely for export (40CFR721.3) means to process for commercial purposes solely for export from the United States under the following restrictions on activity in the United States: Processing must be performed at sites under the control of the processor; distribution in commerce is limited to purposes of export; and the processor may not use the chemical substance except in small quantities solely for research and development.

process stream (40CFR721.3) means all reasonably anticipated transfer, flow, or disposal of a chemical substance, regardless of physical state or concentration, through all intended operations of processing, including the cleaning of equipment.

process stub (40CFR82.152-p) means a length of tubing that provides access to the refrigerant inside a small appliance or room air conditioner and that can be resealed at the conclusion of repair or service.

process unit (40CFR60.481) means components assembled to produce, as intermediate or final products, one or more of the chemicals listed in 40CFR60.489 of this part. A process unit can operate independently if supplied with sufficient feed or raw materials and sufficient storage facilities for the product.

process unit (40CFR60.561) means equipment assembled to perform any of the physical and chemical operations in the production of polypropylene, polyethylene, polystyrene (general purpose, crystal, or expandable), or poly(ethylene terephthalate) or one of their copolymers. A process unit can operate independently if supplied with sufficient feed or raw materials and sufficient storage facilities for the product. Examples of process units are raw materials handling and monomer recovery.

process unit (40CFR60.591) means components assembled to produce intermediate or final products from petroleum, unfinished petroleum derivatives, or other intermediates; a process unit can operate independently if supplied with sufficient feed or raw materials and sufficient storage facilities for the product.

process unit (40CFR60.611) means equipment assembled and connected by pipes or ducts to produce, as intermediates or final products, one or more of the chemicals in 40CFR60.617. A process unit can operate independently if supplied with sufficient fuel or raw materials and sufficient product storage facilities.

process unit (40CFR60.631) means equipment assembled for the extraction of natural gas liquids from field gas, the fractionation of the liquids into natural gas products, or other operations associated with the processing of natural gas products. A process unit can operate independently if supplied with sufficient feed or raw materials and sufficient storage facilities for the products.

process unit (40CFR60.661) means equipment assembled and connected by pipes or ducts to produce, as intermediates or final products, one or more of the chemicals in 40CFR60.667. A process unit can operate independently if supplied with sufficient fuel or raw materials and sufficient product storage facilities.

process unit (40CFR60.701) means equipment assembled and connected by pipes or ducts to produce, as intermediates or final products, one or more of the chemicals in 40CFR60.707. A process unit can operate independently if supplied with sufficient feed or raw materials and sufficient product storage facilities.

process unit (40CFR61.241) means equipment

assembled to produce a VHAP or its derivatives as intermediates or final products, or equipment assembled to use a VHAP in the production of a product. A process unit can operate independently if supplied with sufficient feed or raw materials and sufficient product storage facilities.

process unit (40CFR61.341) means equipment assembled and connected by pipes or ducts to produce intermediate or final products. A process unit can be operated independently if supplied with sufficient fuel or raw materials and sufficient product storage facilities.

process unit (40CFR63.111) has the same meaning as chemical manufacturing process unit as defined in 40CFR63.101 of subpart F of this part and means the equipment assembled and connected by pipes or ducts to process raw materials and to manufacture an intended product. For the purpose of this subpart, process unit or chemical manufacturing process unit includes air oxidation reactors and their associated product separators and recovery devices; reactors and their associated product separators and recovery devices; distillation units and their associated distillate receivers and recovery devices; associated unit operations (as defined in this section); and any feed, intermediate and product storage vessels, product transfer racks, and connected ducts and piping. A chemical manufacturing process unit includes pumps, compressors, agitators, pressure relief devices, sampling connection systems, open-ended valves or lines, valves, connectors, instrumentation systems, and control devices or systems. A chemical manufacturing process unit is identified by its primary product.

process unit (40CFR63.161) means a chemical manufacturing process unit as defined in subpart F of this part, a process subject to the provisions of subpart I of this part, or a process subject to another subpart in 40CFR part 63 that references this subpart.

process unit shutdown (40CFR60.481) means a work practice or operational procedure that stops production from a process unit or part of a process unit. An unscheduled work practice or operational procedure that stops production from a process unit or part of a process unit for less than 24 hours is not a process unit shutdown. The use of spare equipment and technically feasible bypassing of equipment without stopping production are not process unit shutdowns (other identical or similar definitions are provided in 40CFR61.241).

process unit shutdown (40CFR63.161) means a work practice or operational procedure that stops production from a process unit or part of a process unit during which it is technically feasible to clear process material from a process unit or part of a process unit consistent with safety constraints and during which repairs can be effected. An unscheduled work practice or operational procedure that stops production from a process unit or part of a process unit for less than 24 hours is not a process unit shutdown. An unscheduled work practice or operational procedure that would stop production from a process unit or part of a process unit for a shorter period of time than would be required to clear the process unit or part of the process unit of materials and start up the unit, and would result in greater emissions than delay of repair of leaking components until the next scheduled process unit shutdown, is not a process unit shutdown. The use of spare equipment and technically feasible bypassing of equipment without stopping production are not process unit shutdowns.

process unit turnaround (40CFR61.341) means the shutting down of the operations of a process unit, the purging of the contents of the process unit, the maintenance or repair work, followed by restarting of the process.

process unit turnaround waste (40CFR61.341) means a waste that is generated as a result of a process unit turnaround.

process upset gas (40CFR60.101-e) means any gas generated by a petroleum refinery process unit as a result of start-up, shut-down, upset or malfunction.

process variable (EPA-94/04): A physical or chemical quantity which is usually measured and controlled in the operation of a water treatment plant or industrial plant.

process vent (40CFR264.1031) means any open-ended pipe or stack that is vented to the atmosphere either directly, through a vacuum-producing system, or through a tank (e.g., distillate receiver, condenser, bottoms receiver, surge control tank, separator tank, or hot well) associated with hazardous waste distillation, fractionation, thin-film evaporation, solvent extraction, or air or steam stripping operations.

process vent (40CFR63.101) means a gas stream containing greater than 0.005 weight percent total organic hazardous air pollutants that is continuously discharged during operation of the unit from an air oxidation reactor, other reactor, or distillation unit (as defined in this section) within a chemical manufacturing process unit that meets all applicability criteria specified in 40CFR63.100(b)(1) through (b)(3) of this subpart. Process vents include vents from distillate receivers, product separators, and ejector-condensers. Process vents include gas streams that are either discharged directly to the atmosphere or are discharged to the atmosphere after diversion through a product recovery device. Process vents exclude relief valve discharges and leaks from equipment regulated under subpart H of this part.

process verification (EPA-94/04): Verifying that process raw materials, water usage, waste treatment processes, production rate and other facts relative to quantity and quality of pollutants contained in discharges are substantially described in the permit application and the issued permit.

process vessel (40CFR61.131) means each tar decanter, flushing-liquor circulation tank, light-oil condenser, light-oil decanter, wash-oil decanter, or wash-oil circulation tank.

process wastes (40CFR129.2-n) means any designated toxic pollutant, whether in wastewater or otherwise present, which is inherent to or unavoidably resulting from any manufacturing process, including that which comes into direct contact with or results from the production or use of any raw material, intermediate product, finished product, byproduct or waste product and is discharged into the navigable waters.

process wastewater (40CFR117.1) means any water which, during manufacturing or processing, comes into direct contact with or results from the production or use of any raw material, intermediate product, finished product, byproduct, or waste product (other identical or similar definitions are provided in 40CFR401.11-r).

process wastewater (40CFR122.2) means any water which, during manufacturing or processing, comes into direct contact with or results from the production or use of any raw material, intermediate product, finished product, byproduct, or waste product.

process wastewater (40CFR412.11-c) shall mean any process generated waste water and any precipitation (rain or snow) which comes into contact with any manure, litter or bedding, or any other raw material or intermediate or final material or product used in or resulting from the production of animals or poultry or direct products (e.g. milk, eggs) (other identical or similar definitions are provided in 40CFR412.21-c).

process wastewater (40CFR415.241-c) means any water which, during manufacturing or processing, comes into direct contact with or results from the production or use of any raw material, intermediate product, finished product, byproduct, or waste product. The term "process wastewater" does not include contaminated nonprocess wastewater, as defined below (other identical or similar definitions are provided in 40CFR415.311-b; 415.331-c; 415.381-b; 415.401-b; 415.431-b; 415.441-b; 415.531-c; 415.551-b; 415.601-b; 415.631-b).

process wastewater (40CFR415.411-b) means any water which, during manufacturing or processing, comes into direct contact with or results from the production or use of any raw material, intermediate product, finished product, byproduct, or waste product. The term "process wastewater" does not include contaminated nonprocess wastewater, as defined below.

process wastewater (40CFR415.91-d) means any water which, during manufacturing or processing, comes into direct contact with or results from the production or use of any raw material, intermediate product, finished product, byproduct, or waste product. The term "process wastewater" does not

include contaminated non-process wastewater, as defined below.

process wastewater (40CFR418.11-b) shall mean any water which, during manufacturing or processing, comes into direct contact with or results from the production or use of any raw material, intermediate product, finished product, byproduct, or waste product. The term "process wastewater" does not include non-contact cooling water, as defined below (other identical or similar definitions are provided in 40CFR418.21-d).

process wastewater (40CFR422.41-b) means any water which, during manufacturing or processing, comes into direct contact with or results from the production or use of any raw material, intermediate product, finished product or waste product. The term "process waste water" does not include contaminated nonprocess waste water, as defined below.

process wastewater (40CFR422.51-b) means any water which, during manufacturing or processing, comes into direct contact with or results from the production or use of any raw material, intermediate product, finished product, byproduct, or waste product. The term "process waste water" does not include contaminated non-process waste water, as defined below.

process wastewater (40CFR428.11-d) shall mean, in the case of tire and inner tube plants constructed before 1959, discharges from the following: Soapstone solution applications; steam cleaning operations; air pollution control equipment; unroofed process oil unloading areas; mold cleaning operations; latex applications: and air compressor receivers. Discharges from other areas of such plants shall not be classified as process waste water for the purposes of this section. Except as provided in paragraphs (c) and (e) of this section, shall have the meaning set forth in 40CFR401.11(q) of this chapter. Water used only for tread cooling shall be classified as "nonprocess waste water."

process wastewater (40CFR429.11-c) specifically excludes noncontact cooling water, material storage yard runoff (either raw material or processed wood storage), and boiler blowdown. For the dry process hardboard, veneer, finishing, particleboard, and sawmills and planing mills subcategories, fire control water is excluded from the definition.

process wastewater (40CFR440.141-16) means all water used in and resulting from the beneficiation process, including but not limited to the water used to move the ore to and through the beneficiation process, the water used to aid in classification, and the water used in gravity separation, mine drainage, and infiltration and drainage waters which commingle with mine drainage or waters resulting from the beneficiation process.

process wastewater (40CFR443.21-b) shall mean any water which, during the manufacturing process, comes into direct contact with any raw material, intermediate product, byproduct, or product used in or resulting from the production of paving asphalt concrete.

process wastewater (40CFR443.31-b) shall mean any water which, during the manufacturing process, comes into direct contact with any raw material, intermediate product, byproduct, or product used in or resulting from the production of asphalt roofing materials.

process wastewater (40CFR443.41-b) shall mean any water which, during the manufacturing process, comes into direct contact with any raw material, intermediate product, byproduct, or product used in or resulting from the production of linoleum and printed asphalt felt floor coverings.

process wastewater (40CFR458.11-c) shall mean waters which result from baghouse operations or thermal quench operations (other identical or similar definitions are provided in 40CFR458.21-c).

process wastewater (40CFR61.341) means water which comes in contact with benzene during manufacturing or processing operations conducted within a process unit. Process wastewater is not organic wastes, process fluids, product tank drawdown, cooling tower blowdown, steam trap condensate, or landfill leachate.

process wastewater (40CFR63.101) means wastewater which, during manufacturing or processing, comes into direct contact with or results from the production or use of any raw material, intermediate product, finished product, byproduct, or waste product. Examples are product tank drawdown or feed tank drawdown; water formed during a chemical reaction or used as a reactant; water used to wash impurities from organic products or reactants; water used to cool or quench organic vapor streams through direct contact; and condensed steam from jet ejector systems pulling vacuum on vessels containing organics.

process wastewater (EPA-94/04): Any water that comes into contact with any raw material, product, byproduct, or waste.

process wastewater flow (40CFR455.21-d) means the sum of the average daily flows from the following wastewater streams: Process stream and product washes, equipment and floor washes, water used as solvent for raw materials, water used as reaction medium, spent acids, spent bases, contact cooling water, water of reaction, air pollution control blowdown, steam jet blowdown, vacuum pump water, pump seal water, safety equipment cleaning water, shipping container cleanout, safety shower water, contaminated storm water, and product/process laboratory quality control wastewater. Notwithstanding any other regulation, process wastewater flow for the purposes of this subpart does not include wastewaters from the production of intermediate chemicals.

process wastewater pollutants (40CFR401.11-r) means pollutants present in process wastewater (other identical or similar definitions are provided in 40CFR415.91-e; 415.241-d; 415.311-c; 415.331-d; 415.381-c; 415.401-c; 415.411-c; 415.431-c; 415.441-c; 415.531-d; 415.551-c; 415.601-c; 415.631-c).

process wastewater pollutants (40CFR443.11-c) shall mean any pollutants present in the process wastewaters and runoff.

process wastewater pollutants (40CFR443.21-c) shall mean any pollutants present in the process water.

process wastewater pollutants (40CFR443.31-c) shall mean any pollutants present in the process wastewater (other identical or similar definitions are provided in 40CFR443.41-c).

process wastewater pollutants (40CFR455.21-e) means those pollutants present in process wastewater flow.

process wastewater stream (40CFR61.341) means a waste stream that contains only process wastewater (under CAA).

process wastewater stream (40CFR63.111) means a stream that contains process wastewater as defined in 40CFR63.101 of subpart F of this part.

process water (40CFR463.2-b) is any raw, service, recycled, or reused water that contracts the plastic product or contacts the shaping equipment surfaces such as molds and mandrels that are, or have been, in contract with the plastic product.

process weight (40CFR52.1881-ix) means the total weight of all materials and solid fuels introduced into any specific process. Liquid and gaseous fuels and combustion air will not be considered as part of the process weight unless they become part of the product. For a cyclical or batch operation, the process weight per hour will be derived by dividing the total process weight by the number of hours from the beginning of any given process to the completion thereof, excluding any time during which the equipment is idle. For a continuous operation, the process weight per hour will be derived by dividing the process weight for the number of hours in a given period of time by the number of hours in that period. For fluid catalytic cracking units, process weight shall mean the total weight of material introduced as fresh feed to the cracking unit. For sulfuric acid production units, the nitrogen in the air feed shall not be included in the calculation of process weight.

process weight (EPA-94/04): Total weight of all materials, including fuel, used in a manufacturing process; used to calculate the allowable particulate emission rate.

processing activities (40CFR704.203) means all those activities which include:
(1) preparation of a substance identified in subpart D of this part after its manufacture to make another substance for sale or use,
(2) repackaging of the identified substance, or

(3) purchasing and preparing the identified substance for use or distribution in commerce.

processing site (40CFR192.10) means:
(a) Any site, including the mill, containing residual radioactive materials at which all or substantially all of the uranium was produced for sale to any Federal agency prior to January 1, 1971, under a contract with any Federal agency, except in the case of a site at or near Slick Rock, Colorado, unless:
  (1) Such site was owned or controlled as of January 1, 1978, or is thereafter owned or controlled, by any Federal agency, or
  (2) A license (issued by the (Nuclear Regulatory) Commission or its predecessor agency under the Atomic Energy Act of 1954 or by a State as permitted under section 274 of such Act) for the production at site of any uranium or thorium product derived from ores is in effect on January 1, 1978, or is issued or renewed after such date; and
(b) Any other real property or improvement thereon which:
  (1) Is in the vicinity of such site, and
  (2) Is determined by the Secretary, in consultation with the Commission, to be contaminated with residual radioactive materials derived from such site (under AEA).

processor (TSCA3-15USC2602) means any person who processes a chemical substance or mixture (other identical or similar definitions are provided in 40CFR704.3; 710.2-v; 720.3-bb; 747.115-1; 747.195; 747.200-1; 762.3-d; 763.163).

procurement item (40CFR248.4-cc) means any device, good, substance, material, product, or other item, whether real or personal property, which is the subject of any purchase, barter, or other exchange made to procure such item.

procurement item (40CFR249.04-j) means any device, goods, substance, material, product, or other item whether real or personal property which is the subject of any purchase, barter, or other exchange made to procure such item (Pub. L. 94-580, 90 Stat. 2800, 42USC6903).

procurement item (40CFR250.4-ii) means any device, good, substance, material, product, or other item, whether real or personal property, that is the subject of any purchase, barter, or other exchange made to procure such item (other identical or similar definitions are provided in 40CFR252.4-h; 253.4).

procurement item (RCRA1004-42USC6903) means any device, good, substances, material, product, or other item whether real or personal property which is the subject of any purchase, barter, or other exchange made to procure such item.

procuring agency (40CFR249.04-k) means any Federal agency, or any State agency or agency of a political subdivision of a State which is using appropriated Federal funds for such procurement, or any person contracting with any such agency with respect to work performed under such contract (Pub. L. 94-580, 90 Stat. 2800, 42USC6903).

procuring agency (40CFR250.4-jj) means any Federal agency, or any State agency or agency of a political subdivision of a State that is using appropriated Federal funds for such procurement, or any person contracting with any such agency with respect to work performed under such contract (other identical or similar definitions are provided in 40CFR252.4-i; 253.4).

procuring agency (RCRA1004-42USC6903) means any Federal agency, or any State agency or agency of a political subdivision of a State which is using appropriated Federal funds for such procurement, or any person contracting with any such agency with respect to work performed under such contract (other identical or similar definitions are provided in 40CFR248.4-dd).

produce (10CFR70.4), when used in relation to special nuclear material, means:
(1) to manufacture, make, produce, or refine special nuclear material;
(2) to separate special nuclear material from other substances in which such material may be contained; or
(3) to make or to produce new special nuclear material.

produce (40CFR167.3) means to manufacture, prepare, propagate, compound, or process any pesticide, including any pesticide produced pursuant to section 5 of the Act, ny active ingredient or device, or to package, repackage, label, relabel, or otherwise change the container of any pesticide or device (under FIFRA).

produce, produced, and production (CAA601-42USC7671), refer to the manufacture of a substance from any raw material or feedstock chemical, but such terms do not include:
(A) the manufacture of a substance that is used and entirely consumed (except for trace quantities) in the manufacture of other chemicals, or
(B) the reuse or recycling of a substance.

produced sand (40CFR435.11-r) shall refer to slurried particles used in hydraulic fracturing, the accumulated formation sands and scales particles generated during production. Produced sand also includes desander discharge from the produced water waste stream, and blowdown of the water phase from the produced water treating system.

produced water (40CFR435.11-s) shall refer to the water (brine) brought up from the hydrocarbon-bearing strata during the extraction of oil and gas, and can include formation water, injection water, and any chemicals added downhole or during the oil/water separation process.

producer (40CFR167.3) means any person, as defined by the Act, who produces any pesticide, active ingredient, or device (including packaging, repackaging, labeling and relabeling).

producer (40CFR169.1-e) means the person, as defined by the Act, who produces or imports any pesticide or device or active ingredient used in producing a pesticide.

producer (40CFR82.172) means any person who manufactures, formulates or otherwise creates a substitute in its final form for distribution or use in interstate commerce.

producer and produce (FIFRA2-7USC136): The term "producer" means the person who manufacturers, prepares, compounds, propagates, or processes any pesticide or device or active ingredient used in producing a pesticide. The term "produce" means to manufacture, prepare, compound, propagate, or process any pesticide or device, or active ingredient used in producing a pesticide. The dilution by individuals of formulated pesticides for their own use and according to the directions on registered labels shall not of itself result in such individuals being included in the definition of this Act (Federal Insecticide, Fungicide, and Rodenticide Act).

product (40CFR203.1-4) means any manufactured article or goods or component thereof; except that such term does not include:
(i) Any aircraft, aircraft engine, propeller or appliance, as such terms are defined in section 101 of the Federal Aviation Act of 1958; or
(ii) (a) Any military weapons or equipment which are designed for combat use;
(b) any rockets or equipment which are designed for research, experimental or developmental work to be performed by the National Aeronautics and Space Administration; or
(c) to the extent provided by regulations of the Administrator, any other machinery or equipment designed for use in experimental work done by or for the Federal Government.

product (40CFR204.2-17) means any construction equipment for which regulations have been promulgated under this part and includes test product.

product (40CFR205.2-27) means any transportation equipment for which regulations have been promulgated under this part and includes test product.

product (40CFR211.102-h) means any noise-producing or noise-reducing product for which regulations have been promulgated under part 211; the term includes test product.

product (40CFR408.251) shall mean the weight of the oyster meat after shucking.

product (40CFR408.291) shall mean the weight of the scallop meat after processing.

product (40CFR409.11-c) shall mean crystallized refined sugar.

product (40CFR410.01-e), except where a specialized definition is included in the subpart, shall mean the final material produced or processed at the mill.

product (40CFR410.61-a) shall mean the final carpet produced or processed including the primary backing but excluding the secondary backing.

product (40CFR415.161-b) shall mean sodium chloride.

product (40CFR415.171-b) shall mean sodium dichromate.

product (40CFR415.201-b) shall mean sodium sulfite.

product (40CFR415.221-b) shall mean titanium dioxide.

product (40CFR415.231-b) means aluminum fluoride produced by the dry process in which partially dehydrated alumina hydrate is reacted with hydrofluoric acid gas.

product (40CFR415.241-b) shall mean ammonium chloride.

product (40CFR415.281-b) shall mean boric acid.

product (40CFR415.301-b) shall mean calcium carbonate.

product (40CFR415.331-b) shall mean carbon monoxide plus hydrogen.

product (40CFR415.341-c) means chrome pigments.

product (40CFR415.361-b) shall mean copper salts.

product (40CFR415.41-b) shall mean calcium chloride.

product (40CFR415.421-b) means hydrogen cyanide.

product (40CFR415.451) shall mean lithium carbonate.

product (40CFR415.451-b) shall mean cadmium pigment or cadmium salt.

product (40CFR415.471-b) shall mean nickel salts.

product (40CFR415.511-b) shall mean potassium iodide.

product (40CFR415.531-b) shall mean silver nitrate.

product (40CFR415.541-b) means sodium bisulfite.

product (40CFR415.61-b) shall mean chlorine.

product (40CFR415.631) shall mean cadmium pigment or cadmium salt.

product (40CFR415.641-b) shall mean cadmium pigment or cadmium salt.

product (40CFR415.651-b) shall mean cobalt salts.

product (40CFR415.661-b) shall mean sodium chlorate.

product (40CFR415.671-b) shall mean zinc chloride.

product (40CFR415.91-b) shall mean hydrogen peroxide as a one hundred percent hydrogen peroxide solution.

product (40CFR418.21) shall mean the anhydrous ammonia content of the compound manufactured (under CWA).

product (40CFR418.21-b) shall mean the anhydrous ammonia content of the compound manufactured (under CWA).

product (40CFR418.31) shall mean the 100 percent urea content of the material manufactured.

product (40CFR418.41) shall mean the 100 percent ammonium nitrate content of the material manufactured.

product (40CFR418.51-b) shall mean nitric acid on the basis of 100 percent HNO_3.

product (40CFR421.11-c) shall mean alumina.

product (40CFR421.21-b; 421.31-b) shall mean hot aluminum metal.

product (40CFR421.51-b) means electrolytically refined copper.

product (40CFR421.81-b) shall mean zinc metal.

product (40CFR421.91-a) means 100 percent equivalent sulfuric acid, H_2SO_4 capacity.

product (40CFR439.11-b) shall mean pharmaceutical products derived from fermentation processes.

product (40CFR439.21-b) shall mean biological and natural extraction products. This subcategory shall include blood fractions, vaccines, serums, animal bile derivatives, endocrine products, and isolation of medicinal products, such as alkaloids, from botanical drugs and herbs.

product (40CFR439.31-b) shall mean pharmaceutical products derived from chemical synthesis processes.

product (40CFR439.41-b) shall mean products from plants which blend, mix, compound, and formulate pharmaceutical ingredients. Pharmaceutical preparations for human and veterinary use such as ampules, tablets, capsules, vials, ointments, medicinal powders, solutions, and suspensions are included.

product (40CFR439.51-b) shall mean products or services resulting from pharmaceutical research, which includes microbiological, biological, and chemical operations.

product (40CFR454.11-b) shall mean char and charcoal briquets.

product (40CFR454.21-b) shall mean gum rosin and turpentine.

product (40CFR454.31-b) shall mean products from wood rosin, turpentine and pine oil.

product (40CFR454.41-b) shall mean tall oil rosin, pitch and fatty acids.

product (40CFR454.51) shall mean essential oils (under CWA).

product (40CFR454.61-61) shall mean rosin-based derivatives.

product (40CFR457.11-b) shall mean dynamite, nitroglycerin, cyclotrimethylene trinitramine (RDX), cyclotetramethylene tetranitramine (HMX), and trinitrotoluene (TNT).

product (40CFR457.31-b) shall mean products from plants which blend explosives and market a final product, and plants that fill shells and blasting caps. Examples of such installations would be plants manufacturing ammonium nitrate and fuel oil (ANFO), nitrocarbonitrate (NCN), slurries, water gels, and shells.

product (40CFR458.11-b) shall mean carbon black manufactured by the furnace process (other identical or similar definitions are provided in 40CFR458.21-b; 458.31-b).

product (40CFR458.41-b) shall mean carbon black manufactured by the lamp process.

product (40CFR459.11-b) shall mean articles developed or printed by photographic processes, such as paper prints, slides, negatives, enlargements, movie film and other sensitized materials.

product (40CFR460.11-b) shall mean service resulting from the hospital activity in terms of 1,000 occupied beds.

product (40CFR60.661) means any compound or chemical listed in 40CFR60.667 that is produced for sale as a final product as that chemical, or for use in the production of other chemicals or compounds. Byproducts, co-product and intermediates are considered to be products.

product (40CFR60.701) means any compound or chemical listed in 40CFR60.707 which is produced for sale as a final product as that chemical, or for use in the production of other chemicals or compounds. Byproducts, co-products, and intermediates are considered to be products.

product (40CFR63.101) means a compound or chemical which is manufactured as the intended product of the chemical manufacturing process unit.

Byproducts, isolated intermediates, impurities, wastes, and trace contaminants are not considered products.

product (40CFR82.104-p) means an item or category of items manufactured from raw or recycled materials, or other products, which is used to perform a function or task.

product (NCA3-42USC4902) means any manufactured article or goods or component thereof; except that such term does not include:
(A) any aircraft, aircraft engine, propeller, or appliance, as such terms are defined in section 101 of the Federal Aviation Act of 1958; or
(B) (i) any military weapons or equipment which are designed for combat use;
(ii) any rockets or equipment which are designed for research, experimental, or developmental work to be performed by the National Aeronautics and Space Administration; or
(iii) to the extent provided by regulations of the Administrator, any other machinery or equipment designed for use in experimental work done by or for the Federal Government.

product accumulator vessel (40CFR61.241) means any distillate receiver, bottoms receiver, surge control vessel, or product separator in VHAP service that is vented to atmosphere either directly or through a vacuum-producing system. A product accumulator vessel is in VHAP service if the liquid or the vapor in the vessel is at least 10 percent by weight VHAP.

product change (40CFR60.261-c) means any change in the composition of the furnace charge that would cause the electric submerged arc furnace to become subject to a different mass standard applicable under this subpart.

product containing (40CFR82.104-q) means a product including, but not limited to, containers, vessels, or pieces of equipment, that physically holds a controlled substance at the point of sale to the ultimate consumer which remains within the product.

product finishing section (40CFR60.561) means the equipment that treats, shapes, or modifies the polymer or resin to produce the finished end product of the particular facility, including equipment that prepares the product for product finishing. For the purposes of these standards, the product finishing section begins with the equipment used to transfer the polymerized product from the polymerization reaction section and ends with the last piece of equipment that modifies the characteristics of the polymer. Product finishing equipment may accomplish product separation, extruding and pelletizing, cooling and drying, blending, additives introduction, curing, or annealing. Equipment used to separate unreacted or byproduct material from the product are to be included in this process section, provided the material separated from the polymer product is not recovered at the time the process section becomes an affected facility. If the material is being recovered, then the separation equipment are to be included in the material recovery section. Product finishing does not include polymerization, the physical mixing of the pellets to obtain a homogenous mixture of the polymer (except as noted below), or the shaping (such as fiber spinning, molding, or fabricating) or modification (such as fiber stretching and crimping) of the finished end product. If physical mixing occurs in equipment located between product finishing equipment (i.e., before all the chemical and physical characteristics have been "set" by virtue of having passed through the last piece of equipment in the product finishing section), then such equipment are to be included in this process section. Equipment used to physically mix the finished product that are located after the last piece of equipment in the product finishing section are part of the product storage section.

product frosted (40CFR426.121-d) shall mean that portion of the "furnace pull" associated with the fraction of finished incandescent lamp envelopes which is frosted; this quantity shall be calculated by multiplying "furnace pull" by the fraction of finished incandescent lamp envelopes which is frosted (under CWA).

product level (EPA-94/04): The level of a product in a storage tank.

product or pesticide product (40CFR162.151-e) means a pesticide offered for distribution and use, and includes any labeled container and any supplemental labeling.

product packaging station (40CFR60.381) means the equipment used to fill containers with metallic compounds or metallic mineral concentrates.

product separator (40CFR63.101) means phase separators, flash drums, knock-out drums, decanters, degassers, and condenser(s) including ejector-condenser(s) associated with a reactor or an air oxidation reactor.

product separator (40CFR63.111) means phase separators, flash drums, knock-out drums, decanters, degassers, and condenser(s) including ejector-condenser(s) associated with a reactor or an air oxidation reactor.

product storage section (40CFR60.561) means the equipment that is designed to store the finished polymer or resin end product of the particular facility. For the purposes of these standards, the product storage section begins with the equipment used to transfer the finished product out of the product finishing section and ends with the containers used to store the final product. Any equipment used after the product finishing section to recover unreacted or byproduct material are to be considered part of a material recovery section. Product storage does not include any intentional modification of the characteristics of any polymer or resin product, but does include equipment that provide a uniform mixture of product, provided such equipment are used after the last product finishing piece of equipment. This process section also does not include the shipment of a finished polymer or resin product to another facility for further finishing or fabrication.

product tank (40CFR61.341) means a stationary unit that is designed to contain an accumulation of materials that are fed to or produced by a process unit, and is constructed primarily of non-earthen materials (e.g., wood, concrete, steel, plastic) which provide structural support.

product tank (40CFR63.111), as used in the wastewater provisions, means a stationary unit that is designed to contain an accumulation of materials that are fed to or produced by a process unit, and is constructed primarily of non-earthen materials (e.g., wood, concrete, steel, plastic) which provide structural support. This term has the same meaning as a product storage vessel.

product tank drawdown (40CFR61.341) means any material or mixture of materials discharged from a product tank for the purpose of removing water or other contaminants from the product tank.

product tank drawdown (40CFR63.111) means any material or mixture of materials discharged from a product tank for the purpose of removing water or other contaminants from the product tank (under CAA).

product testing (40CFR471.02-cc) includes operations such as dye penetrant testing, hydrotesting, and ultrasonic testing.

product transfer documents (40CFR80.126-e) shall include documents that reflect the transfer of ownership or physical custody of gasoline or blendstock, including invoices, receipts, bills of lading, manifests, and pipeline tickets.

product water (EPA-94/04): Water that has passed through a water treatment plant and is ready to be delivered to the consumers.

production (40CFR430.01-a) shall be defined as the annual off-the-machine production (including off-the-machine coating where applicable) divided by the number of operating days during that year. Paper and paperboard production shall be measured at the off-the-machine moisture content, except for subparts A, B, D, and E where paper and paperboard production shall be measured in air-dry-tons (10% moisture content). Market pulp shall be measured in air-dry-tons (10% moisture). Production shall be determined for each mill based upon past production practices, present trends, or committed growth.

production (40CFR431.11-b) shall be defined as the annual off-the-machine production (including off-the-machine coating where applicable) divided by the number of operating days during that year. Production shall be measured at the off-the-machine moisture content. Production shall be determined for each mill based upon past production practices, present trends, or committed growth.

production (40CFR82.3-ee) means the manufacture of a controlled substance from any raw material or feedstock chemical, but does not include:
(1) The manufacture of a controlled substance that is subsequently transformed;
(2) The reuse or recycling of a controlled substance;
(3) Amounts that are destroyed by the approved technologies; or
(4) Amounts that are spilled or vented unintentionally.

production allowances (40CFR82.3) means the privileges granted by this part to produce calculated levels of controlled substances; however, production allowances may be used to produce controlled substances only in conjunction with consumption allowances. A person's production allowances are the total of the allowances he obtains under 40CFR82.7 (1991 allowances for Group I, Group II and Group III controlled substances), and 40CFR82.8 (1991 allowances for Group IV and Group V controlled substances), and 40CFR82.9 (b), (c) and (d) (additional production allowances).

production allowances (40CFR82.3-ff) means the privileges granted by this subpart to produce controlled substances; however, production allowances may be used to produce controlled substances only in conjunction with consumption allowances. A person's production allowances are the total of the allowances obtained under 40CFR82.7; 82.5 and 82.9, and as may be modified under 40CFR82.12 (transfer of allowances).

production area size (40CFR443.11-b) shall mean that area in which the oxidation towers, loading facilities, and all buildings that house product processes are located.

production changes (40CFR89.602.96): Those changes in nonroad engine configuration, equipment, or calibration which are made by an OEM or ICI in the course of nonroad engine production and required to be reported under 40CFR89.123-96.

production compliance audit (40CFR86.1102.87): See PCA.

production equipment exhaust system (40CFR52.741) means a system for collecting and directing into the atmosphere emissions of volatile organic material from reactors, centrifuges, and other process emission sources.

production facility (10CFR30.4) means production facility as defined in the regulations contained in part 50 of this chapter.

production facility (40CFR435.11-t) shall mean any fixed or mobile structure subject to this subpart that is either engaged in well completion or used for active recovery of hydrocarbons from producing formations.

production line (40CFR60.671) means all affected facilities (crushers, grinding mills, screening operations, bucket elevators, belt conveyors, bagging operations, storage bins, and enclosed truck and railcar loading stations) which are directly connected or are connected together by a conveying system.

production normalizing mass (/kkg) (40CFR467.02-z) for each core or ancillary operation is the mass (off-kkg or off-lb) processed through that operation.

production volume (40CFR600.002.85-32) means, for a domestic manufacturer, the number of vehicle units domestically produced in a particular model year but not exported, and for a foreign manufacturer, means the number of vehicle units of a particular model imported into the United States.

production volume (40CFR704.3) means the quantity of a substance which is produced by a manufacturer, as measured in kilograms or pounds.

production weighted average (40CFR86.085.2) means the manufacturer's production-weighted average particulate emission level, for certification purposes, of all of its diesel engine families included in the particulate averaging program. It is calculated at the end of the model year by multiplying each family particulate emission limit by its respective production, summing these terms, and dividing the sum by the total production of the effected families. Those vehicles produced for sale in California or at high altitude shall each be averaged separately from those produced for sale in any other area.

production weighted average (40CFR86.087.2)

means the manufacturer's production-weighted average particulate emission level, for certification purposes, of all of its diesel engine families included in the particulate averaging program. It is calculated at the end of the model year by multiplying each family particulate emission limit by its respective production, summing these terms, and dividing the sum by the total production of the affected families. Those vehicles produced for sale in California or at high altitude shall each be averaged separately from those produced for sale in any other area. Diesel light-duty trucks with a loaded vehicle weight equal to or greater than 3,751 lbs (LDDT2s) shall only be averaged with other diesel light-duty trucks with a loaded vehicle weight equal to or greater than 3,751 lbs produced by that manufacturer.

production weighted NO$_x$ average (40CFR86.088.2) means the manufacturer's production-weighted average NO$_x$ emission level, for certification purposes, of all of its light-duty truck engine families included in the NO$_x$ averaging program. It is calculated at the end of the model year by multiplying each family NO$_x$ emission limit by its respective production, summing those terms, and dividing the sum by the total production of the effected families. Those vehicles produced for sale in California or at high altitude shall each be averaged separately from those produced for sale in any other area.

production weighted particulate average (40CFR86.090.2) means the manufacturer's production-weighted average particulate emission level, for certification purposes, of all of its diesel engine families included in the light-duty particulate averaging program. It is calculated at the end of the model year by multiplying each family particulate emission limit by its respective production, summing those terms, and dividing the sum by the total production of the effected families. Those vehicles produced for sale in California or at high altitude shall each be averaged separately from those produced for sale in any other area.

production weighted particulate average (40CFR86.088.2) means the manufacturer's production-weighted average particulate emission level, for certification purposes, of all of its diesel engine families included in the particulate averaging program. It is calculated at the end of the model year by multiplying each family particulate emission limit by its respective production, summing those terms, and dividing the sum by the total production of the effected families. Those vehicles produced for sale in California or at high altitude shall each be averaged separately from those produced for sale in any other area.

products of incomplete combustion (PICs) (EPA-94/04): Organic compounds formed by combustion. Usually generated in small amounts and sometimes toxic, PICs are heat-altered versions of the original material fed into the incinerator (e.g., charcoal is a P.C. from burning wood).

profit (40CFR33.005) means the net proceeds obtained by deducting all allowable costs (direct and indirect) from the price. (Because this definition of profit is based on applicable Federal cost principles, it may vary from many firms' definition of profit and may correspond to those firms' definition of fee.) (Other identical or similar definitions are provided in 40CFR35.6015-34.)

program (40CFR403.3-d): See approved POTW pretreatment program.

program (40CFR610.11-21): See evaluation program.

program baseline (40CFR51.491) means the level of emissions, or emission-related parameter(s), for each affected source or group of affected sources, from which program results (e.g., quantifiable emissions reductions) shall be determined.

program directive (40CFR56.1) means any formal written statement by the Administrator, the Deputy Administrator, the Assistant Administrator, a Staff Office Director, the General Counsel, a Deputy Assistant Administrator, an Associate General Counsel, or a division Director of an Operational Office that is intended to guide or direct Regional Offices in the implementation or enforcement of the provisions of the Act.

program element (40CFR35.105) means one of the major groupings of outputs of a continuing environmental program (e.g., administration, enforcement, monitoring).

program income (40CFR30.200) means gross income the recipient earns during its project period from charges for the project. This may include income from service fees, sale of commodities, trade-in allowances, or usage or rental fees. Fees from royalties are program income only if the assistance agreement so states. Revenue generated under the governing powers of a State or local government which could have been generated without an award is not considered program income. Such revenues include fines or penalties levied under judicial or penal power and used as a means to enforce laws. (Revenue from wastewater treatment construction grant projects under Title of the Clean Water Act, as amended, is not program income. It must be used for operation and maintenance costs of the recipients wastewater facilities.)

program of requirements (40CFR6.901-b) means a comprehensive document (booklet) describing program activities to be accomplished in the new special purpose facility or improvement. It includes architectural, mechanical, structural, and space requirements.

program office (CWA118-33USC1268) means the Great Lakes National Program Office established by this section.

program uncertainty factor (40CFR51.491) means a factor applied to discount the amount of emissions reductions credited in an implementation plan demonstration to account for any strategy-specific uncertainties in an EIP (economic incentive program).

prohibit specification (40CFR231.2-b) means to prevent the designation of an area as a present or future disposal site.

prohibited (40CFR129.2-d) means that the constituent shall be absent in any discharge subject to these standards, as determined by any analytical method.

project (40CFR144.3) means a group of wells in a single operation (other identical or similar definitions are provided in 40CFR146.3).

project (40CFR149.101-f) means a program or

action for which an application for Federal financial assistance has been made.

project (40CFR30.200) means the activities or tasks EPA identifies in the assistance agreement.

project (40CFR35.2005-32) means the activities or tasks the Regional Administrator identifies in the grant agreement for which the grantee may expend, obligate or commit funds.

project (40CFR35.6015-35) means the activities or tasks EPA identifies in the cooperative agreement and/or Superfund State Contract.

project (40CFR35.905) means the scope of work for which a grant or grant amendment is awarded under this subpart. The scope of work is defined as step 1, step 2, or step 3 of treatment works construction or segments (see definition of treatment works segment and 40CFR35.930-4).

project (40CFR51.392) means a highway project or transit project (other identical or similar definitions are provided in 40CFR93.101).

project costs (40CFR30.200) means all costs the recipient incurs in carrying out the project. EPA considers all allowable project costs to include the Federal share.

project manager (40CFR35.6015-36): The recipient official designated in the Cooperative Agreement or SSC as the program contact with EPA.

project officer (40CFR30.200) means the EPA official designated in the assistance agreement as EPA's program contact with the recipient. Project officers are responsible for monitoring the project.

project officer (40CFR35.6015-37) means the EPA official designated in the cooperative agreement as EPA's program contact with the recipient. Project officers are responsible for monitoring the project.

project officer (40CFR7.25) means the EPA official designated in the assistance agreement (as defined in EPA assistance) as EPA's program contact with the recipient; project Officers are responsible for monitoring the project.

project performance standards (40CFR35.2005-33) means the performance and operations requirements applicable to a project including the enforceable requirements of the Act and the specifications, including the quantity of excessive infiltration and inflow proposed to be eliminated, which the project is planned and designed to meet.

project period (40CFR30.200) means the length of time EPA specifies in the assistance agreement for completion of all project work. It may be composed of more than one budget period.

project period (40CFR35.6015-38) means the length of time EPA specifies in the cooperative agreement and/or Superfund State Contract for completion of all project work. It may be composed of more than one budget period.

project schedule (40CFR35.2005-35) means a timetable specifying the dates of key project events including public notices of proposed procurement actions, subagreement awards, issuance of notice to proceed with building, key milestones in the building schedule, completion of building, initiation of operation and certification of the project.

promotional printed material (40CFR82.104-r) means any informational or advertising material (including, but not limited to, written advertisements, brochures, circulars, desk references and fact sheets) that is prepared by the manufacturer for display or promotion concerning a product or container, and that does not accompany the product to the consumer.

proof press (40CFR60.431) means any device used only to check the quality of the image formation of newly engraved or etched gravure cylinders and prints only non-saleable items.

propellant (40CFR61.31-i) means a fuel and oxidizer physically or chemically combined which undergoes combustion to provide rocket propulsion.

propellant (EPA-94/04): Liquid in a self-pressurized pesticide product that expels the active ingredient from its container.

propellant plant (40CFR61.31-k) means any facility

engaged in the mixing, casting, or machining of propellant.

properly using (40CFR82.32-e) means using equipment in conformity with Recommended Service Procedures and Recommended Practices for the Containment of R-12 (CFC-12) set forth in appendix A to this subpart. In addition, this term includes operating the equipment in accordance with the manufacturer's guide to operation and maintenance and using the equipment only for the controlled substance for which the machine is designed. For equipment that extracts and recycles refrigerant, properly using also means to recycle refrigerant before it is returned to a motor vehicle air conditioner. For equipment that only recovers refrigerant, properly using includes the requirement to recycle the refrigerant on-site or send the refrigerant off-site for reclamation. Refrigerant from reclamation facilities that is used for the purpose of recharging motor vehicle air conditioners must be at or above the standard of purity developed by the Air-conditioning and Refrigeration Institute (ARI 700-93) (available at 4301 North Fairfax Drive, Suite 425, Arlington, Virginia 22203). Refrigerant may be recycled off-site only if the refrigerant is extracted using recover only equipment, and is subsequently recycled off-site by equipment owned by the person that owns both the recover only equipment and owns or operates the establishment at which the refrigerant was extracted. In any event, approved equipment must be used to extract refrigerant prior to performing any service during which discharge of refrigerant from the motor vehicle air conditioner can reasonably be expected. Intentionally venting or disposing of refrigerant to the atmosphere is an improper use of equipment.

property damage (40CFR280.92) shall have the meaning given this term by applicable state law. This term shall not include those liabilities which, consistent with standard insurance industry practices, are excluded from coverage in liability insurance policies for property damage. However, such exclusions for property damage shall not include corrective action associated with releases from tanks which are covered by the policy (under the Resource Conservation Recovery Act).

proportional sampling (40CFR60.2) means sampling at a rate that produces a constant ration of sampling rate to stack gas flow rate.

proposal (40CFR1508.23) exists at that stage in the development of an action when an agency subject to the Act has goal and is actively preparing to make a decision on one or more alternative means of accomplishing that goal and the effects can be meaningfully evaluated. Preparation of an environmental impact statement on a proposal should be timed (40CFR1502.5) so that the final statement may be completed in time for the statement to be included in any recommendation or report on the Proposal. A proposal may exist in fact as well as by agency declaration that one exists.

proposal (40CFR32.105-q) means a solicited or unsolicited bid, application, request, invitation to consider or similar communication by or on behalf of a person seeking to participate or to receive a benefit, directly or indirectly, in or under a covered transaction.

propose to manufacture, import, or process (40CFR704.3) means that a person has made a firm management decision to commit financial resources for the manufacture, import, or processing of a specified chemical substance or mixture.

propose to manufacture, import, or process (40CFR716.3) means that a person has made a management decision to commit financial resources toward the manufacture, importation, or processing a substance or mixture.

proposed acid rain permit or proposed permit (40CFR72.2) means, in the case of a State operating permit program, the version of an Acid Rain permit that the permitting authority submits to the Administrator after the public comment period, but prior to completion of the EPA permit review period, as provided for in part 70 of this chapter.

proposed permit (40CFR122.2) means a State NPDES "permit" prepared after the close of the public comment period (and, when applicable, any public hearing and administrative appeals) which is sent to EPA for review before final issuance by the State. A "proposed permit" is not a "draft permit" (under FWPCA).

proposed permit (40CFR70.2) means the version of a permit that the permitting authority proposes to issue and forwards to the Administrator for review in compliance with 40CFR70.8.

proposed plan (EPA-94/04): A plan for a site cleanup that is available to the public for comment.

protect health and the environment (FIFRA2-7USC136) means protection against any unreasonable adverse effects on the environment.

protective equipment (40CFR171.2-21) means clothing or any other materials or devices that shield against unintended exposure to pesticides.

proteins (EPA-94/04): Complex nitrogenous organic compounds of high molecular weight made of amino acids; essential for growth and repair of animal tissue. Many, but not all, proteins are enzymes.

protocol (40CFR790.3) means the plan and procedures which are to be followed in conducting a test.

protocol (EPA-94/04): A series of formal steps for conducting a test.

protocol (TSCA-AIA1) means any procedure for taking, handling, and preserving samples of asbestos and asbestos-containing material and for testing and analyzing such samples for the purpose of determining the person who manufactured such samples and the identifying characteristics of such samples.

protocol 1 gas (40CFR72.2) means a calibration gas mixture prepared and analyzed according to the "Procedure for NBS-Traceable Certification of Compressed Gas Working Standards Used for Calibration and Audit of Continuous Emission Monitors ("Revised Traceability Protocol No. 1"), Quality Assurance Handbook for Air Pollution Measurement Systems, Volume III, Stationary Source Specific Methods, section 3.04, EPA-600/4-77-027b, June 1987 (set forth in Appendix H of part 75 of this chapter) or such revised procedure as approved by the Administrator.

protoplast (EPA-94/04): A membrane-bound cell from which the outer wall has been partially or completely removed. The term often is applied to plant cells.

protozoa (EPA-94/04): One-celled animals that are larger and more complex than bacteria. May cause disease.

proven emission control systems (40CFR86.092.2-a) are emission control components or systems (and fuel metering systems) that have completed full durability testing evaluation over a vehicle's useful life in some other certified engine family, or have completed bench or road testing demonstrated to be equal or more severe than certification mileage accumulation requirements. Alternatively, proven components or systems are those that are determined by EPA to be of comparable functional quality and manufactured using comparable materials and production techniques as components or systems which have been durability demonstrated in some other certified engine family. In addition, the components or systems must be employed in an operating environment (e.g.. temperature, exhaust flow, etc.,) similar to that experienced by the original or comparable components or systems in the original certified engine family.

provide for (40CFR256.06) in the phrase "the plan shall (should) provide for" means explain, establish or set forth steps or courses of action.

provider of financial assurance (40CFR280.92) means an entity that provides financial assurance to an owner or operator of an underground storage tank through one of the mechanisms listed in 4CFR280.95-280.103, including a guarantor, insurer, risk retention group, surety, issuer of a letter of credit, issuer of a state-required mechanism, or a state.

PSD permit (40CFR124.41): See permit.

PSD station (40CFR58.1-d) means any station operated for the purpose of establishing the effect on air quality of the emissions from a proposed source for purposes of prevention of significant deterioration as required by 51.24(n) of part 51 of this chapter.

PSES (40CFR467.02-x) means pretreatment

standards for existing sources, under section 307(b) of the Act.

PSI (pollutant standards index) (40CFR58-AG-a): See uniform air quality required for the daily reporting of air quality.

PSNS (40CFR467.02-y) means pretreatment standards for new sources, under section 307(c) of the Act.

PTC (positive temperature coefficient) type choke heaters (40CFR85.2122(a)(2)(iii)(I)) means a positive temperature coefficient resistant ceramic disc capable of providing heat to the thermostatic coil when electrically energized (under the Clean Air Act).

public and commercial building (TSCA202-15TSCA202) means any building which is not a school building, except that the term does not include any residential apartment building of fewer than 10 units.

public body (40CFR39.105-h) means a State, interstate agency, a municipality, or an inter-municipal agency, defined as follows:
(1) State. A State, the District of Columbia, the Commonwealth of Puerto Rico, the Virgin Islands, Guam, American Samoa, and the Trust Territory of the Pacific Islands.
(2) Interstate agency. An agency of two or more States established by or pursuant to an agreement or compact approved by the Congress, or any other agency of two or more States, having substantial powers or duties pertaining to the control of water pollution.
(3) Municipality. A city, town, borough, county, parish, district, association, or other public body (including an intermunicipal agency of two or more of the foregoing entities) created by or pursuant to State law, or an Indian Tribe or an authorized Indian tribal organization, having jurisdiction over disposal of sewage, industrial wastes, or other wastes, or a designated and approved management agency under section 208 of the Act. This definition excludes a special district, such as a school district, which does not have as one of its principal responsibilities the treatment, transport, or disposal of liquid wastes.

public comment period (EPA-94/04): The time allowed for the public to express its views and concerns regarding an action by EPA (e.g., a Federal Register Notice of proposed rule-making, a public notice of a draft permit, or a Notice of Intent to Deny).

public contact site (40CFR503.11-l) is land with a high potential for contact by the public. This includes, but is not limited to, public parks, ball fields, cemeteries, plant nurseries, turf farms, and golf courses.

public facilities and public services (CZMA304-16USC1453) means facilities or services which are financed, in whole or in part, by any state or political subdivision thereof, including, but not limited to, highways and secondary roads, parking, mass transit, docks, navigation aids, fire and police protection, water supply, waste collection and treatment (including drainage), schools and education, and hospitals and health care. Such term may also include any other facility or service so financed which the Secretary finds will support increased population.

public hearing (EPA-94/04): A formal meeting wherein EPA officials hear the public's views and concerns about an EPA action or proposal. EPA is required to consider such comments when evaluating its actions. Public hearings must be held upon request during the public comment period (cf. public notice).

public involvement (FLPMA103-43USC1702) means the opportunity for participation by affected citizens in rulemaking, decisionmaking, and planning with respect to the public lands. including public meetings or hearings held at locations near the affected lands, or advisory mechanisms, or such other procedures as may be necessary to provide pubic comment in a particular instance.

public lands (FLPMA103-43USC1702) means any land and interest in land owned by the United States within the several States and administered by the Secretary of the Interior through the Bureau of Land Management, without regard to how the United States acquired ownership, except:
(1) lands located on the Outer Continental Shelf; and

(2) lands held for the benefit of Indians, Aleuts, and Eskimos.

public notice (EPA-94/04):
1. Notification by EPA informing the public of Agency actions such as the issuance of a draft permit or scheduling of a hearing. EPA is required to ensure proper public notice, including publication in newspapers and broadcast over radio stations.
2. In the safe drinking water program, water suppliers are required to publish and broadcast notices when pollution problems are discovered.

public or private agricultural research agency or educational institution (40CFR172.21) means any organization engaged in research pertaining to the agricultural use of pesticides, or any educational institution engaged in pesticides research. Any research agency or educational institution whose principal function is to promote, or whose principal source of income is directly derived from the sale or distribution of pesticides (or their active ingredients) does not come within the meaning of this term.

public participation (40CFR300.5): See community relations.

public record (40CFR117.1) means the NPDES permit application or the NPDES permit itself and the "record for final permit" as defined in 40CFR124.122.

public vessel (40CFR300.5), as defined by section 311(a)(4) of the CWA, means a vessel owned or bareboat-chartered and operated by the United States, or by a state or political subdivision thereof, or by a foreign nation, except when such vessel is engaged in commerce.

public vessel (CWA311) means a vessel owned or bare-boat-chartered and operated by the United States, or by a State or political subdivision thereof, or by a foreign nation, except when such vessel is engaged in commerce (other identical or similar definitions are provided in CWA312).

public vessel (CWA312-33USC1322) means a vessel owned or bareboat chartered and operated by the United States, or by a State or political subdivision

thereof, or by a foreign nation, except when such vessel is engaged in commerce (other identical or similar definitions are provided in 40CFR110.1; 116.3).

public vessel (OPA1001) means a vessel owned or bare boat chartered and operated by the United States, or by a State or political subdivision thereof, or by a foreign nation, except when such vessel is engaged in commerce.

public water supplies (40CFR125.58-n) means water distributed from a public water system (under SDWA).

public water system (40CFR125.58-o) means a system for the provision to the public of piped water for human consumption, if such system has at least fifteen (15) service connections or regularly serves at least twenty-five (25) individuals. This term includes:
(1) any collection, treatment, storage and distribution facilities under the control of the operator of the system and used primarily in connection with the system, and
(2) any collection or pretreatment storage facilities not under the control of the operator of the system which are used primarily in connection with the system.

public water system (40CFR141.2) means a system for the provision to the public of piped water for human consumption, if such system has at least fifteen service connections or regularly serves an average of at least twenty-five individuals daily at least 60 days out of the year. Such term includes:
(1) any collection, treatment, storage, and distribution facilities under control of the operator of such system and used primarily in connection with such system, and
(2) any collection or pretreatment storage facilities not under such control which are used primarily in connection with such system. A public water system is either a "community water system" or a "noncommunity water system."

public water system (40CFR142.2) means a system for the provision to the public of piped water for human consumption, if such system has at least fifteen service connections or regularly serves an average of at least twenty-five individuals daily at least 60 days out of the year. Such term includes:

(1) any collection, treatment, storage, and distribution facilities under control of the operator of such system and used primarily in connection with such system, and

(2) any collection or pretreatment storage facilities not under such control which are used primarily in connection with such system.

public water system (40CFR143.2-c) means a system for the provision to the public of piped water for human consumption, if such a system has at least fifteen service connections or regularly serves an average of at least twenty-five individuals daily at least 60 days out of the year. Such term includes:

(1) any collection, treatment, storage, and distribution facilities under control of the operator of such system and used primarily in connection with such system, and

(2) any collection or pretreatment storage facilities not under such control which are used primarily in connection with such system. A public water system is either a "community water system" or a "non-community water system."

public water system (EPA-94/04): A system that provides piped water for human consumption to at least 15 service connections or regularly serves 25 individuals.

public water system (SDWA1401-42USC300f) means a system for the provision to the public of piped water for human consumption, if such system has at least fifteen service connections or regularly serves at least twenty-five individuals. Such term includes:

(A) any collection, treatment, storage, and distribution facilities under control of the operator of such system and used primarily in connection with such system, and

(B) any collection or pretreatment storage facilities not under such control which are used primarily in connection with such system.

public water system supervision program (SDWA1443-42USC300j.2) means a program for the adoption and enforcement of drinking water regulations (with such variances and exemptions from such regulations under conditions and in a manner which is not less stringent than the conditions under, and the manner in, which variances and exemptions

may be granted under sections 1415 and 1416) which are no less stringent than the national primary drinking water regulations under section 1412, and for keeping records and making reports required by section 1413(a)(3).

publication rotogravure printing line (40CFR52.741) means a rotogravure printing line in which coatings are applied to paper which is subsequently formed into books, magazines, catalogues, brochures, directories, newspaper supplements, or other types of printed material.

publication rotogravure printing press (40CFR60.431) means any number of rotogravure printing units capable of printing simultaneously on the same continuous web or substrate and includes any associated device for continuously cutting and folding the printed web, where the following saleable paper products are printed:

- Catalogues, including mail order and premium,
- Direct mail advertisements, including circulars, letters, pamphlets, cards, and printed envelopes,
- Display advertisements, including general posters, outdoor advertisements, car cards, window posters; counter and floor displays; point-of-purchase, and other printed display material,
- Magazines,
- Miscellaneous advertisements, including brochures, pamphlets, catalogue sheets, circular folders, announcements, package inserts, book jackets, market circulars, magazine inserts, and shopping news,
- Newspapers, magazine and comic supplements for newspapers, and preprinted newspaper inserts, including hi-fi and spectacolor rolls and sections, periodicals, and
- Telephone and other directories, including business reference services.

publicly owned freshwater lake (40CFR35.1605-3): A freshwater lake that offers public access to the lake through publicly owned contiguous land so that any person has the same opportunity to enjoy non-consumptive privileges and benefits of the lake as any other person. If user fees are charged for public use and access through State or substate operated facilities, the fees must be used for maintaining the public access and recreational facilities of this lake or other publicly owned freshwater lakes in the State, or for improving the quality of these lakes.

publicly owned treatment works (EPA-94/04): A waste-treatment works owned by a state, unit of local government, or Indian tribe, usually designed to treat domestic wastewaters.

publicly owned treatment works (POTW) (40CFR122.2) means any device or system used in the treatment (including recycling and reclamation) of municipal sewage or industrial wastes of a liquid nature which is owned by a "State" or "municipality." This definition includes sewers, pipes, or other conveyances only if they convey wastewater to a POTW providing treatment (cf. privately owned treatment works).

publicly owned treatment works (POTW) (40CFR125.58-p) means a treatment works, as defined in section 212(2) of the Act, which is owned by a State, municipality or intermunicipal or interstate agency.

publicly owned treatment works (POTW) (40CFR270.2) means any device or system used in the treatment (including recycling and reclamation) of municipal sewage or industrial wastes of a liquid nature which is owned by a State or municipality. This definition includes sewers, pipes, or other conveyances only if they convey wastewater to a POTW providing treatment.

publicly owned treatment works or POTW (40CFR117.1) means a treatment works as defined by section 212 of the Act, which is owned by a State or municipality (as defined by section 502(4) of the Act). This definition includes any sewers or other conveyances not connected to a facility providing treatment. The term also means the municipality as defined in section 502(4) of the Act, which has jurisdiction over the indirect discharges to and the discharges from such a treatment works.

publicly owned treatment works or POTW (40CFR260.10) means any device or system used in the treatment (including recycling and reclamation) of municipal sewage or industrial wastes of a liquid nature which is owned by a "State" or "municipality" (as defined by section 502(4) of the CWA). This definition includes sewers, pipes, or other conveyances only if they convey wastewater to a POTW providing treatment.

publicly owned treatment works or POTW (40CFR403.3-o) means a treatment works as defined by section 212 of the Act, which is owned by a State or municipality (as defined by section 502(4) of the Act). This definition includes any devices and systems used in the storage, treatment, recycling and reclamation of municipal sewage or industrial wastes of a liquid nature. It also includes sewers, pipes and other conveyances only if they convey wastewater to a POTW Treatment Plant. The term also means the municipality as defined in section 502(4) of the Act, which has jurisdiction over the indirect discharges to and the discharges from such a treatment works.

publicly owned treatment works (40CFR501.2) means a treatment works treating domestic sewage that is owned by a municipality or State.

pulverized coal-fired steam generating unit (40CFR60.41b) means a steam generating unit in which pulverized coal is introduced into an air stream that carries the coal to the combustion chamber of the steam generating unit where it is fired in suspension. This includes both conventional pulverized coal-fired and micropulverized coal-fired steam generating units.

pumping station (EPA-94/04): Mechanical devices installed in sewer or water systems or other liquid-carrying pipelines that move the liquids to a higher level.

purchasing (40CFR248.4-ee) means the act of and the function of responsibility for the acquisition of equipment, materials, supplies, and services, including: buying, determining the need, selecting the supplier, arriving at a fair and reasonable price and terms and conditions, preparing the contract or purchase order, and follow up.

purchasing activities (40CFR248.4-ff) means all activities included in the purchasing function (under RCRA).

purchasing group (SF401-42USC9671) means any group of persons which has as one of its purposes the purchase of pollution liability insurance on a group basis.

purge or line purge (40CFR60.391) means the coating material expelled from the spray system when clearing it.

pushing (40CFR63.301), for the purposes of 40CFR63.305, means that coke oven operation that commences when the pushing ram starts into the oven to push out coke that has completed the coking cycle and ends when the quench car is clear of the coke side shed.

putrefaction (EPA-94/04): Biological decomposition of organic matter; associated with anaerobic conditions.

putrescible (EPA-94/04): Able to rot quickly enough to cause odors and attract flies.

putrescible wastes (40CFR257.3.8-7) means solid waste which contains organic matter capable of being decomposed by microorganisms and of such a character and proportion as to be capable of attracting or providing food for birds (under the Resource Conservation Recovery Act).

pyrolysis (EPA-94/04): Decomposition of a chemical by extreme heat.

pyrolytic gas and oil (40CFR245.101-f) means gas or liquid products that possess usable heating value that is recovered from the heating of organic material (such as that found in solid waste), usually in an essentially oxygen-free atmosphere.

*********** QQQQQ ***********

QA (40CFR766.3) means quality assurance.

QC (40CFR766.3) means quality control.

qualified ground-water scientist (40CFR260.10) means a scientist or engineer who has received a baccalaureate or post-graduate degree in the natural sciences or engineering, and has sufficient training and experience in ground-water hydrology and related fields as may be demonstrated by state registration, professional certifications, or completion of accredited university courses that enable that individual to make sound professional judgements regarding ground-water monitoring and contaminant fate and transport.

qualified ground-water scientist (40CFR503.21-l) is an individual with a baccalaureate or post-graduate degree in the natural sciences or engineering who has sufficient training and experience in ground-water hydrology and related fields, as may be demonstrated by State registration, professional certification, or completion of accredited university programs, to make sound professional judgments regarding ground-water monitoring, pollutant fate and transport, and corrective action.

qualified handicapped person (40CFR7.25) means:
(a) With respect to employment: A handicapped person who, with reasonable accommodation, can perform the essential functions of the job in question.
(b) With respect to services: A handicapped person who meets the essential eligibility requirements for the receipt of such services.

qualified incinerator (40CFR761.3) means one of the following:
(1) An incinerator approved under the provisions of 40CFR761.70. Any level of PCB concentration can be destroyed in an incinerator approved under 40CFR761.70.
(2) A high efficiency boiler which complies with the criteria of 40CFR761.60(a)(2)(iii)(A), and for which the operator has given written notice to the appropriate EPA Regional Administrator in accordance with the notification requirements for the burning of mineral oil dielectric fluid under 40CFR761.60(a)(2)(iii)(B).
(3) An incinerator approved under section 3005(c) of the Resource Conservation and Recovery Act (42 U.S.C. 6925(c)) (RCRA).
(4) Industrial furnaces and boilers which are identified in 40CFR260.10 and

40CFR279.61(a)(1) and (2) when operating at their normal operating temperatures (this prohibits feeding fluids, above the level of detection, during either startup or shutdown operations).

qualified individual with handicaps (40CFR12.103) means:

(1) With respect to any agency program or activity under which a person is required to perform services or to achieve a level of accomplishment, an individual with handicaps who meets the essential eligibility requirements and who can achieve the purpose of the program or activity, without modifications in the program or activity that the agency can demonstrate would result in a fundamental alteration in its nature; or

(2) With respect to any other program or activity an individual with handicaps who meets the essential eligibility requirements for participation in, or receipt of benefits from, that program or activity.

(3) Qualified handicapped person as that term is defined for purposes of employment in 29CFR1613.702(f), which is made applicable to this part by part 12.140.

qualifying facility (QF) (40CFR72.2) means a "qualifying small power production facility" within the meaning of section 3(17)(C) of the Federal Power Act or a "qualifying cogeneration facility" within the meaning of section 3(18)(B) of the Federal Power Act.

qualifying phase I technology (40CFR72.2) means a technological system of continuous emission reduction that is demonstrated to achieve a ninety (90) percent (or greater) reduction in emissions of sulfur dioxide from the emissions that would have resulted from the use of fossil fuels that were not subject to treatment prior to combustion, as provided in 40CFR72.42.

qualifying phase I technology (CAA402-42USC7651a) means a technological system of continuous emission reduction which achieves a 90 percent reduction in emissions of sulfur dioxide from the emissions that would have resulted from the use of fuels which were not subject to treatment prior to combustion.

qualifying power purchase commitment (40CFR72.2) means a power purchase commitment in effect as of November 15, 1990 without regard to changes to that commitment so long as:

(1) The identity of the electric output purchaser; or

(2) The identity of the steam purchaser and the location of the facility, remain unchanged as of the date the facility commences commercial operation; and

(3) The terms and conditions of the power purchase commitment are not changed in such a way as to allow the costs of compliance with the Acid Rain Program to be shifted to the purchaser (under CAA).

qualifying repowering technology (40CFR72.2) means:

(1) Replacement of an existing coal-fired boiler with one of the following clean coal technologies: atmospheric or pressurized fluidized bed combustion, integrated gasification combined cycle, magnetohydrodynamics, direct and indirect coal-fired turbines, integrated gasification fuel cells, or as determined by the Administrator, in consultation with the Secretary of Energy, a derivative of one or more of these technologies, and any other technology capable of controlling multiple combustion emissions simultaneously with improved boiler or generation efficiency and with significantly greater waste reduction relative to the performance of technology in widespread commercial use as of the date of enactment of the Clean Air Act Amendments of 1990; or

(2) Any oil- or gas-fired unit that has been awarded clean coal technology demonstration funding as of January 1, 1991, by the Department of Energy.

quality assurance narrative statement (40CFR30.200) means a description of how precision, accuracy, representativeness, completeness, and compatibility will be assessed, and which is sufficiently detailed to allow an unambiguous determination of the quality assurance practices to be followed throughout a research project.

quality assurance program plan (40CFR30.200) means a formal document which describes an orderly assembly of management policies, objectives,

principles, organizational responsibilities, and procedures by which an agency or laboratory specifies how it intends to:
(a) Produce data of documented quality, and
(b) Provide for the preparation of quality assurance project plans and standard operating procedures.

quality assurance project plan (40CFR30.200) means an organization's written procedures which delineate how it produces quality data for a specific project or measurement method.

quality assurance project plan (40CFR35.6015-39) means a written document, associated with remedial site sampling, which presents in specific terms the organization (where applicable), objectives, functional activities, and specific quality assurance and quality control activities and procedures designed to achieve the data quality objectives of a specific project(s) or continuing operation(s).

quality assurance project plan (QAPP) (40CFR300.5) is a written document, associated with all remedial site sampling activities, which presents in specific terms the organization (where applicable), objectives, functional activities, and specific quality assurance (QA) and quality control (QC) activities designed to achieve the data quality objectives of a specific project(s) or continuing operation(s). The QAPP is prepared for each specific project or continuing operation (or group of similar projects or continuing operations). The QAPP will be prepared by the responsible program office, regional office, laboratory, contractor, recipient of an assistance agreement, or other organization. For an enforcement action, potentially responsible parties may prepare a QAPP subject to lead agency approval.

quality assurance/quality control (EPA-94/04): A system of procedures, checks, audits, and corrective actions to ensure that all EPA research design and performance, environmental monitoring and sampling, and other technical and reporting activities are of the highest achievable quality.

quality assurance unit (40CFR160.3) means any person or organizational element, except the study director, designated by testing facility management to perform the duties relating to quality assurance of the studies (other identical or similar definitions are provided in 40CFR792.3).

quality assured monitor operating hour (40CFR72.2) means any hour over which a certified CEMS, COMS, or other monitoring system approved by the Administrator under part 75 of this chapter, is operating: in accordance with 40CFR75.10(e) and (f); and within the performance specifications set forth in part 75, Appendix A of this chapter and the quality control/quality assurance procedures set forth in part 75, appendix B of this chapter, without unscheduled maintenance, repair, or adjustment.

quality control sample (40CFR136-AC-9) means a solution obtained from an outside source having known, concentration values to be used to verify the calibration standards..

quantifiable level/level of detection (40CFR761.3) means 2 micrograms per gram from any resolvable gas chromatographic peak, i.e., 2 ppm.

quarter (40CFR60.481) means a 3-month period; the first quarter concludes on the last day of the last full month during the 180 days following initial startup.

quench station (40CFR60.461) means that portion of the metal coil surface coating operation where the coated metal coil is cooled, usually by a water spray, after baking or curing.

quench tank (EPA-94/04): A water-filled tank used to cool incinerator residues or hot materials during industrial processes.

*********** **RRRRR** ***********

racial classifications (40CFR7.25):
(a) American Indian or Alaskan native. A person having origins in any of the original peoples of North America, and who maintains cultural

identification through tribal affiliation or community recognition.

(b) Asian or Pacific Islander. A person having origins in any of the original peoples of the Far East, Southeast Asia, the Indian subcontinent, or the Pacific Islands. This area includes, for example, China, Japan, Korea, the Philippine Islands, and Samoa.

(c) Black and not of Hispanic origin. A person having origins in any of the black racial groups of Africa.

(d) Hispanic. A person of Mexican, Puerto Rican, Cuban, Central or South American or other Spanish culture or origin, regardless of race.

(e) White, not of Hispanic origin. A person having origins in any of the original peoples of Europe, North Africa, or the Middle East.

rack dryer (40CFR60.301-n) means any equipment used to reduce the moisture content of grain in which the grain flows from the top to the bottom in a cascading flow around rows of baffles (racks).

rack weighted average partial pressure (40CFR63.111) means the throughput weighted average of the average maximum true vapor pressure of liquids containing organic HAPs transferred at a transfer rack. The rack-weighted average partial pressure shall be calculated using the equation below:

- $P = (\Sigma\ P_i G_i)/(\Sigma\ G_i)$; where:
- P = Rack-weighted average partial pressure, kilopascals.
- P_i = Individual HAP maximum true vapor pressure, kilopascals, $P_i = X_i{}^*P$, where X_i is the mole fraction of compound i in the liquid.
- G_i = Yearly volume of each liquid that contains organic HAP that is transferred at the rack, liters.
- i = Each liquid that contains HAP that is transferred at the rack.

radiant energy or radiation (40CFR796.3700-i) is defined as the energy traveling as a wave unaccompanied by transfer of matter. Examples include x-rays, visible light, ultraviolet light, radio waves, etc. (other identical or similar definitions are provided in 40CFR796.3780-i; 796.3800-i).

radiation (10CFR20.3) means any or all of the following: alpha rays, beta rays, gamma rays, X-rays, neutrons, high-speed electrons, high-speed protons; and other atomic particles; but not sound or radio waves, or visible, infra red, or ultraviolet light.

radiation (40CFR190.02-e) means any or all of the following: alpha, beta, gamma, or X-rays; neutrons; and high-energy electrons, protons, or other atomic particles; but not sound or radio waves, nor visible, infrared, or ultraviolet light.

radiation (40CFR300-AA) means particles (alpha, beta, neutrons) or photons (x- and gamma-rays) emitted by radionuclides.

radiation (EPA-94/04): Transmission of energy through space or any medium. Also known as radiant energy.

radiation standards (EPA-94/04): Regulations that set maximum exposure limits for protection of the public from radioactive materials.

radiation (40CFR796.3700-i; 796.3780-i; 796.3800-i): See radiant energy.

radicle (40CFR797.2750-5) means that portion of the plant embryo which develops into the primary root.

radio frequency radiation (EPA-94/04): See non-ionizing electromagnetic radiation.

radioactive decay (40CFR300-AA) means the process of spontaneous nuclear transformation, whereby an isotope of one element is transformed into an isotope of another element, releasing excess energy in the form of radiation.

radioactive decay (EPA-94/04): Spontaneous change in an atom by emission of charged particles and/or gamma rays; also known as radioactive disintegration and radioactivity.

radioactive half-life (40CFR300-AA) means the time required for one-half the atoms in a given quantity of a specific radionuclide to undergo radioactive decay.

radioactive material (10CFR20.3) includes any such material whether or not subject to licensing control of the commission.

radioactive material (40CFR190.02-f) means any material which spontaneously emits radiation.

radioactive material (40CFR191.12) means matter composed of or containing radionuclides, with radiological half-lives greater than 20 years, subject to the Atomic Energy Act of 1954, as amended.

radioactive substance (40CFR300-AA) means solid, liquid, or gas containing atoms of a single radionuclide or multiple radionuclides.

radioactive substances (EPA-94/04): Substances that emit ionizing radiation.

radioactive waste (40CFR144.3) means any waste which contains radioactive material in concentrations which exceed those listed in 10CFR20, Appendix B, Table II, Column 2 (other identical or similar definitions are provided in 40CFR146.3).

radioactive waste (40CFR191.02-j), as used in this part, means the high-level and transuranic radioactive waste covered by this part.

radioactivity (40CFR300-AA) means the property of those isotopes of elements that exhibit radioactive decay and emit radiation.

radiographer (10CFR30.4) means any individual who performs or who, in attendance at the site where the sealed source or sources are being used, personally supervises radiographic operations and who is responsible to the licensee for assuring compliance with the requirements of the Commission's regulations and the conditions of the license.

radiographer assistant (10CFR30.4) means any individual who, under the personal supervision of a radiographer, uses radiographic exposure devices, sealed sources or related handling tools, or radiation survey instruments in radiography.

radiography (10CFR30.4) means the examination of the structure of materials by nondestructive methods, utilizing sealed sources of byproduct materials.

radioisotopes (EPA-94/04): Chemical variants of an element with potentially oncogenic, teratogenic, and mutagenic effects on the human body.

radionuclide (40CFR61.91-c) means a type of atom which spontaneously undergoes radioactive decay (other identical or similar definitions are provided in 40CFR61.101-f).

radionuclide (EPA-94/04): Radioactive particle, man-made or natural, with a distinct atomic weight number. Can have a long life as soil or water pollutants.

radionuclide/radioisotope (40CFR300-AA) means isotope of an element exhibiting radioactivity. For HRS purposes, "radionuclide" and "radioisotope" are used synonymously. (Radionuclides in the alphabetical order from 40CFR302.4 Appendix B, 1995 edition, are provided below).

atomic number	radionuclides
89	Actinium-224
89	Actinium-225
89	Actinium-226
89	Actinium-227
89	Actinium-228
13	Aluminum-26
95	Americium-237
95	Americium-238
95	Americium-239
95	Americium-240
95	Americium-241
95	Americium-242m
95	Americium-242
95	Americium-243
95	Americium-244m
95	Americium-244
95	Americium-245
95	Americium-246m
95	Americium-246
51	Antimony-115
51	Antimony-116m
51	Antimony-116
51	Antimony-117
51	Antimony-118m
51	Antimony-119
51	Antimony-120 (16 min)
51	Antimony-120 (576 day)
51	Antimony-122
51	Antimony-124m
51	Antimony-124
51	Antimony-125
51	Antimony-126m

51	Antimony-126		35	Bromine-74m
51	Antimony-127		35	Bromine-74
51	Antimony-128 (104 min)		35	Bromine-75
51	Antimony-128 (901 hr)		35	Bromine-76
51	Antimony-129		35	Bromine-77
51	Antimony-130		35	Bromine-80m
51	Antimony-131		35	Bromine-80
18	Argon-39		35	Bromine-82
18	Argon-41		35	Bromine-83
33	Arsenic-69		35	Bromine-84
33	Arsenic-70		48	Cadmium-104
33	Arsenic-71		48	Cadmium-107
33	Arsenic-72		48	Cadmium-109
33	Arsenic-73		48	Cadmium-113m
33	Arsenic-74		48	Cadmium-113
33	Arsenic-76		48	Cadmium-115m
33	Arsenic-77		48	Cadmium-115
33	Arsenic-78		48	Cadmium-117m
85	Astatine-207		48	Cadmium-117
85	Astatine-211		20	Calcium-41
56	Barium-126		20	Calcium-45
56	Barium-128		20	Calcium-47
56	Barium-131m		98	Californium-244
56	Barium-131		98	Californium-246
56	Barium-133m		98	Californium-248
56	Barium-133		98	Californium-249
56	Barium-135m		98	Californium-250
56	Barium-139		98	Californium-251
56	Barium-140		98	Californium-252
56	Barium-141		98	Californium-253
56	Barium-142		98	Californium-254
97	Berkelium-245		6	Carbon-11
97	Berkelium-246		6	Carbon-14
97	Berkelium-247		58	Cerium-134
97	Berkelium-249		58	Cerium-135
97	Berkelium-250		58	Cerium-137m
4	Beryllium-7		58	Cerium-137
4	Beryllium-10		58	Cerium-139
83	Bismuth-200		58	Cerium-141
83	Bismuth-201		58	Cerium-143
83	Bismuth-202		58	Cerium-144
83	Bismuth-203		55	Cesium-125
83	Bismuth-205		55	Cesium-127
83	Bismuth-206		55	Cesium-129
83	Bismuth-207		55	Cesium-130
83	Bismuth-210m		55	Cesium-131
83	Bismuth-210		55	Cesium-132
83	Bismuth-212		55	Cesium-134m
83	Bismuth-213		55	Cesium-134
83	Bismuth-214		55	Cesium-135m

55 Cesium-135	63 Europium-146
55 Cesium-136	63 Europium-147
55 Cesium-137	63 Europium-148
55 Cesium-138	63 Europium-149
17 Chlorine-36	63 Europium-150 (126 hr)
17 Chlorine-38	63 Europium-150 (342 yr)
17 Chlorine-39	63 Europium-152m
24 Chromium-48	63 Europium-152
24 Chromium-49	63 Europium-154
24 Chromium-51	63 Europium-155
27 Cobalt-55	63 Europium-156
27 Cobalt-56	63 Europium-157
27 Cobalt-57	63 Europium-158
27 Cobalt-58m	100 Fermium-252
27 Cobalt-58	100 Fermium-253
27 Cobalt-60m	100 Fermium-254
27 Cobalt-60	100 Fermium-255
27 Cobalt-61	100 Fermium-257
27 Cobalt-62m	9 Fluorine-18
29 Copper-60	87 Francium-222
29 Copper-61	87 Francium-223
29 Copper-64	64 Gadolinium-145
29 Copper-67	64 Gadolinium-146
96 Curium-238	64 Gadolinium-147
96 Curium-240	64 Gadolinium-148
96 Curium-241	64 Gadolinium-149
96 Curium-242	64 Gadolinium-151
96 Curium-243	64 Gadolinium-152
96 Curium-244	64 Gadolinium-153
96 Curium-245	64 Gadolinium-159
96 Curium-246	31 Gallium-65
96 Curium-247	31 Gallium-66
96 Curium-248	31 Gallium-67
96 Curium-249	31 Gallium-68
66 Dysprosium-155	31 Gallium-70
66 Dysprosium-157	31 Gallium-72
66 Dysprosium-159	31 Gallium-73
66 Dysprosium-165	32 Germanium-66
66 Dysprosium-166	32 Germanium-67
99 Einsteinium-250	32 Germanium-68
99 Einsteinium-251	32 Germanium-69
99 Einsteinium-253	32 Germanium-71
99 Einsteinium-254m	32 Germanium-75
99 Einsteinium-254	32 Germanium-77
68 Erbium-161	32 Germanium-78
68 Erbium-165	79 Gold-193
68 Erbium-169	79 Gold-194
68 Erbium-171	79 Gold-195
68 Erbium-172	79 Gold-198m
63 Europium-145	79 Gold-198

79 Gold-199	53 Iodine-129
79 Gold-200m	53 Iodine-130
79 Gold-200	53 Iodine-131
79 Gold-201	53 Iodine-132m
72 Hafnium-170	53 Iodine-132
72 Hafnium-172	53 Iodine-133
72 Hafnium-173	53 Iodine-134
72 Hafnium-175	53 Iodine-135
72 Hafnium-177m	77 Iridium-182
72 Hafnium-178m	77 Iridium-184
72 Hafnium-179m	77 Iridium-185
72 Hafnium-180m	77 Iridium-186
72 Hafnium-181	77 Iridium-187
72 Hafnium-182m	77 Iridium-188
72 Hafnium-182	77 Iridium-189
72 Hafnium-183	77 Iridium-190m
72 Hafnium-184	77 Iridium-190
67 Holmium-155	77 Iridium-192m
67 Holmium-157	77 Iridium-192
67 Holmium-159	77 Iridium-194m
67 Holmium-161	77 Iridium-194
67 Holmium-162m	77 Iridium-195m
67 Holmium-162	77 Iridium-195
67 Holmium-164m	26 Iron-52
67 Holmium-164	26 Iron-55
67 Holmium-166m	26 Iron-59
67 Holmium-166	26 Iron-60
67 Holmium-167	36 Krypton-74
1 Hydrogen-3	36 Krypton-76
49 Indium-109	36 Krypton-77
49 Indium-110 (691 min)	36 Krypton-79
49 Indium-110 (49 hr)	36 Krypton-81
49 Indium-111	36 Krypton-83m
49 Indium-112	36 Krypton-85m
49 Indium-113m	36 Krypton-85
49 Indium-114m	36 Krypton-87
49 Indium-115m	36 Krypton-88
49 Indium-115	57 Lanthanum-131
49 Indium-116m	57 Lanthanum-132
49 Indium-117m	57 Lanthanum-135
49 Indium-117	57 Lanthanum-137
49 Indium-119m	57 Lanthanum-138
53 Iodine-120m	57 Lanthanum-140
53 Iodine-120	57 Lanthanum-141
53 Iodine-121	57 Lanthanum-142
53 Iodine-123	57 Lanthanum-143
53 Iodine-124	82 Lead-195m
53 Iodine-125	82 Lead-198
53 Iodine-126	82 Lead-199
53 Iodine-128	82 Lead-200

82 Lead-201	60 Neodymium-139
82 Lead-202m	60 Neodymium-141
82 Lead-202	60 Neodymium-147
82 Lead-203	60 Neodymium-149
82 Lead-205	60 Neodymium-151
82 Lead-209	93 Neptunium-232
82 Lead-210	93 Neptunium-233
82 Lead-211	93 Neptunium-234
82 Lead-212	93 Neptunium-235
82 Lead-214	93 Neptunium-236 (12 E 5 yr)
71 Lutetium-169	93 Neptunium-236 (225 hr)
71 Lutetium-170	93 Neptunium-237
71 Lutetium-171	93 Neptunium-238
71 Lutetium-172	93 Neptunium-239
71 Lutetium-173	93 Neptunium-240
71 Lutetium-174m	28 Nickel-56
71 Lutetium-174	28 Nickel-57
71 Lutetium-176m	28 Nickel-59
71 Lutetium-176	28 Nickel-63
71 Lutetium-177m	28 Nickel-65
71 Lutetium-177	28 Nickel-66
71 Lutetium-178m	41 Niobium-88
71 Lutetium-178	41 Niobium-89 (66 min)
71 Lutetium-179	41 Niobium-89 (122 min)
12 Magnesium-28	41 Niobium-90
25 Manganese-51	41 Niobium-93m
25 Manganese-52m	41 Niobium-94
25 Manganese-52	41 Niobium-95m
25 Manganese-53	41 Niobium-95
25 Manganese-54	41 Niobium-96
25 Manganese-56	41 Niobium-97
101 Mendelevium-257	41 Niobium-98
101 Mendelevium-258	76 Osmium-180
80 Mercury-193m	76 Osmium-181
80 Mercury-193	76 Osmium-182
80 Mercury-194	76 Osmium-185
80 Mercury-195m	76 Osmium-189m
80 Mercury-195	76 Osmium-191m
80 Mercury-197m	76 Osmium-191
80 Mercury-197	76 Osmium-193
80 Mercury-199m	76 Osmium-194
80 Mercury-203	46 Palladium-100
42 Molybdenum-90	46 Palladium-101
42 Molybdenum-93m	46 Palladium-103
42 Molybdenum-93	46 Palladium-107
42 Molybdenum-99	46 Palladium-109
42 Molybdenum-101	15 Phosphorus-32
60 Neodymium-136	15 Phosphorus-33
60 Neodymium-138	78 Platinum-186
60 Neodymium-139m	78 Platinum-188

78 Platinum-189	61 Promethium-151
78 Platinum-191	91 Protactinium-227
78 Platinum-193m	91 Protactinium-228
78 Platinum-193	91 Protactinium-230
78 Platinum-195m	91 Protactinium-231
78 Platinum-197m	91 Protactinium-232
78 Platinum-197	91 Protactinium-233
78 Platinum-199	91 Protactinium-234
78 Platinum-200	88 Radium-223
94 Plutonium-234	88 Radium-224
94 Plutonium-235	88 Radium-225
94 Plutonium-236	88 Radium-226φ
94 Plutonium-237	88 Radium-227
94 Plutonium-238	88 Radium-228
94 Plutonium-239	86 Radon-220
94 Plutonium-240	86 Radon-222
94 Plutonium-241	75 Rhenium-177
94 Plutonium-242	75 Rhenium-178
94 Plutonium-243	75 Rhenium-181
94 Plutonium-244	75 Rhenium-182 (127 hr)
94 Plutonium-245	75 Rhenium-182 (640 hr)
84 Polonium-203	75 Rhenium-184m
84 Polonium-205	75 Rhenium-184
84 Polonium-207	75 Rhenium-186m
84 Polonium-210	75 Rhenium-186
19 Potassium-40	75 Rhenium-187
19 Potassium-42	75 Rhenium-188m
19 Potassium-43	75 Rhenium-188
19 Potassium-44	75 Rhenium-189
19 Potassium-45	45 Rhodium-99m
59 Praseodymium-136	45 Rhodium-99
59 Praseodymium-137	45 Rhodium-100
59 Praseodymium-138m	45 Rhodium-101m
59 Praseodymium-139	45 Rhodium-101
59 Praseodymium-142m	45 Rhodium-102m
59 Praseodymium-142	45 Rhodium-102
59 Praseodymium-143	45 Rhodium-103m
59 Praseodymium-144	45 Rhodium-105
59 Praseodymium-145	45 Rhodium-106m
59 Praseodymium-147	45 Rhodium-107
61 Promethium-141	37 Rubidium-79
61 Promethium-143	37 Rubidium-81m
61 Promethium-144	37 Rubidium-81
61 Promethium-145	37 Rubidium-82m
61 Promethium-146	37 Rubidium-83
61 Promethium-147	37 Rubidium-84
61 Promethium-148m	37 Rubidium-86
61 Promethium-148	37 Rubidium-88
61 Promethium-149	37 Rubidium-89
61 Promethium-150	37 Rubidium-87

44 Ruthenium-94	38 Strontium-85
44 Ruthenium-97	38 Strontium-87m
44 Ruthenium-103	38 Strontium-89
44 Ruthenium-105	38 Strontium-90
44 Ruthenium-106	38 Strontium-91
62 Samarium-141m	38 Strontium-92
62 Samarium-141	16 Sulfur-35
62 Samarium-142	73 Tantalum-172
62 Samarium-145	73 Tantalum-173
62 Samarium-146	73 Tantalum-174
62 Samarium-147	73 Tantalum-175
62 Samarium-151	73 Tantalum-176
62 Samarium-153	73 Tantalum-177
62 Samarium-155	73 Tantalum-178
62 Samarium-156	73 Tantalum-179
21 Scandium-43	73 Tantalum-180m
21 Scandium-44m	73 Tantalum-180
21 Scandium-44	73 Tantalum-182m
21 Scandium-46	73 Tantalum-182
21 Scandium-47	73 Tantalum-183
21 Scandium-48	73 Tantalum-184
21 Scandium-49	73 Tantalum-185
34 Selenium-70	73 Tantalum-186
34 Selenium-73m	43 Technetium-93m
34 Selenium-73	43 Technetium-93
34 Selenium-75	43 Technetium-94m
34 Selenium-79	43 Technetium-94
34 Selenium-81m	43 Technetium-96m
34 Selenium-81	43 Technetium-96
34 Selenium-83	43 Technetium-97m
14 Silicon-31	43 Technetium-97
14 Silicon-32	43 Technetium-98
47 Silver-102	43 Technetium-99m
47 Silver-103	43 Technetium-99
47 Silver-104m	43 Technetium-101
47 Silver-104	43 Technetium-104
47 Silver-105	52 Tellurium-116
47 Silver-106m	52 Tellurium-121m
47 Silver-106	52 Tellurium-121
47 Silver-108m	52 Tellurium-123m
47 Silver-110m	52 Tellurium-123
47 Silver-111	52 Tellurium-125m
47 Silver-112	52 Tellurium-127m
47 Silver-115	52 Tellurium-127
11 Sodium-22	52 Tellurium-129m
11 Sodium-24	52 Tellurium-129
38 Strontium-80	52 Tellurium-131m
38 Strontium-81	52 Tellurium-131
38 Strontium-83	52 Tellurium-132
38 Strontium-85m	52 Tellurium-133m

52 Tellurium-133	50 Tin-123m
52 Tellurium-134	50 Tin-123
65 Terbium-147	50 Tin-125
65 Terbium-149	50 Tin-126
65 Terbium-150	50 Tin-127
65 Terbium-151	50 Tin-128
65 Terbium-153	22 Titanium-44
65 Terbium-154	22 Titanium-45
65 Terbium-155	74 Tungsten-176
65 Terbium-156m (50 hr)	74 Tungsten-177
65 Terbium-156m (244 hr)	74 Tungsten-178
65 Terbium-156	74 Tungsten-179
65 Terbium-157	74 Tungsten-181
65 Terbium-158	74 Tungsten-185
65 Terbium-160	74 Tungsten-187
65 Terbium-161	74 Tungsten-188
81 Thallium-194m	92 Uranium-230
81 Thallium-194	92 Uranium-231
81 Thallium-195	92 Uranium-232
81 Thallium-197	92 Uranium-233
81 Thallium-198m	92 Uranium-234ϕ
81 Thallium-198	92 Uranium-235ϕ
81 Thallium-199	92 Uranium-236
81 Thallium-200	92 Uranium-237
81 Thallium-201	92 Uranium-238ϕ
81 Thallium-202	92 Uranium-239
81 Thallium-204	92 Uranium-240
90 Thorium-226	23 Vanadium-47
90 Thorium-227	23 Vanadium-48
90 Thorium-228	23 Vanadium-49
90 Thorium-229	54 Xenon-120
90 Thorium-230	54 Xenon-121
90 Thorium-231	54 Xenon-122
90 Thorium-232ϕ	54 Xenon-123
90 Thorium-234	54 Xenon-125
69 Thulium-162	54 Xenon-127
69 Thulium-166	54 Xenon-129m
69 Thulium-167	54 Xenon-131m
69 Thulium-170	54 Xenon-133m
69 Thulium-171	54 Xenon-133
69 Thulium-172	54 Xenon-135m
69 Thulium-173	54 Xenon-135
69 Thulium-175	54 Xenon-138
50 Tin-110	70 Ytterbium-162
50 Tin-111	70 Ytterbium-166
50 Tin-113	70 Ytterbium-167
50 Tin-117m	70 Ytterbium-169
50 Tin-119m	70 Ytterbium-175
50 Tin-121m	70 Ytterbium-177
50 Tin-121	70 Ytterbium-178

39 Yttrium-86m
39 Yttrium-86
39 Yttrium-87
39 Yttrium-88
39 Yttrium-90m
39 Yttrium-90
39 Yttrium-91m
39 Yttrium-91
39 Yttrium-92
39 Yttrium-93
39 Yttrium-94
39 Yttrium-95
30 Zinc-62
30 Zinc-63
30 Zinc-65
30 Zinc-69m
30 Zinc-69
30 Zinc-71m
30 Zinc-72
40 Zirconium-86
40 Zirconium-88
40 Zirconium-89
40 Zirconium-93
40 Zirconium-95
40 Zirconium-97

(Notes):

- Ci-Curie. The curie represents a rate of radioactive decay. One curie is the quantity of any radioactive nuclide which undergoes 3.7E 10 disintegrations per second.
- Bq-Becquerel. The becquerel represents a rate of radioactive decay. One becquerel is the quantity of any radioactive nuclide which undergoes one disintegration per second. One curie is equal to 3.7E 10 becquerel.
- @-Final RQs for all radionuclides apply to chemical compounds containing the radionuclides and elemental forms regardless of the diameter of pieces of solid material.
- &-The adjusted RQ of one curie applies to all radionuclides not otherwise listed. Whenever the RQs in Table 302.4 and this appendix to the table are in conflict, the lowest RQ shall apply. For example, uranyl acetate and uranyl nitrate have adjusted RQs shown in Table 302.4 of 100 pounds, equivalent to about one-tenth the RQ level for uranium-238 listed in this appendix.
- E-Exponent to the base 10. For example, 1.3E 2 is equal to 130 while 1.3E 3 is equal to 1300.
- m-Signifies a nuclear isomer which is a

radionuclide in a higher energy metastable state relative to the parent isotope.

- ϕ-Notification requirements for releases of mixtures or solutions of radionuclides can be found in Section 302.6(b) of this rule. Final RQs for the following four common radionuclide mixtures are provided: radium-226 in secular equilibrium with its daughters (0.053 curie); natural uranium (0.1 curie); natural uranium in secular equilibrium with its daughters (0.052 curie); and natural thorium in secular equilibrium with its daughters (0.011 curie).

radius of vulnerability zone (EPA-94/04): The maximum distance from the point of release of a hazardous substance in which the airborne concentration could reach the level of concern under specified weather conditions.

radon (EPA-94/04): A colorless naturally occurring, radioactive, inert gas formed by radioactive decay of radium atoms in soil or rocks.

radon (TSCA302-15USC2662) means the radioactive gaseous element and its short-lived decay products produced by the disintegration of the element radium occurring in the air, water, soil, or other media.

radon contractor proficiency (RCP) program (40CFR195.2) refers to EPA's program to evaluate radon mitigation contractors and the contractor's ability to communicate information to the public.

radon daughters/radon progeny (EPA-94/04): Short-lived radioactive decay products of radon that decay into longer-lived lead isotopes, The daughter isotopes can attach themselves to airborne dust and other particles and, if inhaled, damage to lining of the lung. Also known as radon decay products.

radon decay products (EPA-94/04): A term used to refer collectively to the immediate products of the radon decay chain. These include Po-218, Pb-214, Bi-214, and Po-214, which have an average combined half-life of about 30 minutes.

radon measurement proficiency (RMP) program (40CFR195.2) refers to EPA's program to evaluate organizations and individuals offering measurement services to consumers. It provides a means for organizations to demonstrate their proficiency in measuring radon and its decay products in indoor air.

radon mitigation contractor (40CFR195.2) means a contractor who provides radon mitigation services to the public.

rail car (40CFR201.1-t) means a non-self-propelled vehicle designed for and used on railroad tracks (under the Noise Control Act).

railcar (40CFR60.301-e) means railroad hopper car or boxcar.

railcar loading station (40CFR60.381) means that portion of a metallic mineral processing plant where metallic minerals or metallic mineral concentrates are loaded by a conveying system into railcars.

railcar unloading station (40CFR60.381) means that portion of a metallic mineral processing plant where metallic ore is unloaded from a railcar into a hopper, screen, or crusher.

railroad (40CFR201.1-u) means all the roads in use by any common carrier operating a railroad, whether owned or operated under a contract, agreement, or lease.

raisins (40CFR407.61-s) shall mean the production of raisins from the following products: Dried grapes, all varieties, bleached and unbleached, which have been cleaned and washed prior to packaging.

random incident field (40CFR211.203-aa) means a sound field in which the angle of arrival of sound at a given point in space is random in time.

range (40CFR53.23-a): Nominal minimum and maximum concentration which a method is capable of measuring.

range land (40CFR503.11-m) is open land with indigenous vegetation.

range of concentration (40CFR79.2-g) means the highest concentration, the lowest concentration, and the average concentration of an additive in a fuel.

rare earth metals (40CFR421.271) refers to the elements scandium, yttrium, and lanthanum to lutetium, inclusive.

rasp (EPA-94/04): A machine that grinds waste into a manageable material and helps prevent odor.

rated output (RO) (40CFR87.1) means the maximum power/thrust available for takeoff at standard day conditions as approved for the engine by the Federal Aviation Administration, including reheat contribution where applicable, but excluding any contribution due to water injection.

rated pressure ratio (RPR) (40CFR87.1) means the ratio between the combustor inlet pressure and the engine inlet pressure achieved by an engine operating at rated output.

rated speed (40CFR86.082.2) means the speed at which the manufacturer specifies the maximum rated horsepower of an engine.

raw agricultural commodities (40CFR180.1) include, among other things, fresh fruits, whether or not they have been washed and colored or otherwise treated in their unpeeled natural form; vegetables in their raw or natural state, whether or not they have been stripped of their outer leaves, waxed, prepared into fresh green salads, etc.; grains, nuts, eggs, raw milk, meats, and similar agricultural produce. It does not include foods that have been processed, fabricated, or manufactured by cooking, freezing, dehydrating, or milling.

raw data (40CFR160.3) means any laboratory worksheets, records, memoranda notes, or exact copies thereof, that are the result of original observations and activities of a study and are necessary for the reconstruction and evaluation of the report of that study. In the event that exact transcripts of raw data have been prepared (e.g., tapes which have been transcribed verbatim, dated, and verified accurate by signature), the exact copy or exact transcript may be substituted for the original source as raw data. Raw data may include photographs, microfilm or microfiche copies, computer printouts, magnetic media, including dictated observations, and recorded data from automated instruments (other identical or similar definitions are provided in 40CFR792.3).

raw ink (40CFR60.431) means all purchased ink (under CAA).

raw material (40CFR425.02-h) means the hides received by the tannery except for facilities covered by subpart D and subpart I where "raw material" means the hide or split in the condition in which it is first placed into a wet process.

raw material (40CFR428.101-b) shall mean all latex solids used in the manufacture of latex-dipped, latex-extruded, and latex-molded products.

raw material (40CFR428.11-b) shall mean all natural and synthetic rubber, carbon black, oils, chemical compounds, fabric and wire used in the manufacture of pneumatic tires and inner tubes or components thereof.

raw material (40CFR428.111-b) shall mean all latex solids used in the manufacture of latex foam.

raw material (40CFR428.51-b) shall mean all natural and synthetic rubber, carbon black, oils, chemical compounds, and fabric used in the manufacture of general molded, extruded, and fabricated rubber products (other identical or similar definitions are provided in 40CFR428.61-b; 428.71-b).

raw material (40CFR432.101-e) or as abbreviated herein, "RM," shall mean the basic input materials to a renderer composed of animal and poultry trimmings, bones, meat scraps, dead animals, feathers and related usable byproducts.

raw material equivalent (40CFR428.51-c) shall be equal to the raw material usage multiplied by the volume of air scrubbed via wet scrubbers divided by the total volume of air scrubbed (other identical or similar definitions are provided in 40CFR428.61-c; 428.71-c).

raw materials preparation section (40CFR60.561) means the equipment located at a polymer manufacturing plant designed to prepare raw materials, such as monomers and solvents, for polymerization. For the purposes of these standards, this process section begins with the equipment used to transfer raw materials from storage and recovered material from material recovery process sections, and ends with the last piece of equipment that prepares the material for polymerization. The raw materials preparation section may include equipment that accomplishes purification, drying, or other treatment of raw materials or of raw and recovered materials together, activation of catalysts, and esterification including the formation of some short polymer chains (oligomers), but does not include equipment that is designed primarily to accomplish the formation of oligomers, the treatment of recovered materials alone, or the storage of raw materials.

raw sewage (EPA-94/04): Untreated wastewater and its contents.

raw water (EPA-94/04): Intake water prior to any treatment or use.

rayon fiber (40CFR60.601) means a manufactured fiber composed of regenerated cellulose, as well as manufactured fibers composed of regenerated cellulose in which substituents have replaced not more than 15 percent of the hydrogens of the hydroxyl groups.

RCRA: See act or RCRA.

RDF stoker (40CFR60.51a) means a steam generating unit that combusts RDF in a semi-suspension firing mode using air-fed distributors.

re-refined oil (RCRA1004-42USC6903) means used oil from which the physical and chemical contaminants acquired through previous use have been removed through a refining process (other identical or similar definitions are provided in 40CFR252.4-j).

reactant (40CFR723.250-14) means a chemical substance that is used intentionally in the manufacture of a polymer to become chemically a part of the polymer composition.

reaction quantum yield for an excited-state process (40CFR796.3700-xi) is defined as the fraction of absorbed light that results in photoreaction at a fixed wavelength. It is the ratio of the number of molecules that photoreact to the number of quanta of light absorbed or the ratio of the number of moles that photoreact to the number of einsteins of light absorbed at a fixed wavelength (other identical or similar definitions are provided in 40CFR796.3780-xii; 796.3800-xi).

reaction spinning process (40CFR60.601) means the fiber-forming process where a prepolymer is extruded into a fluid medium and solidification takes place by chemical reaction to form the final polymeric material.

reactivation of a very clean coal-fired electric utility steam generating unit (40CFR51.166-37) means any physical change or change in the method of operation associated with the commencement of commercial operations by a coal-fired utility unit after a period of discontinued operation where the unit:

(i) Has not been in operation for the two-year period prior to the enactment of the Clean Air Act Amendments of 1990, and the emissions from such unit continue to be carried in the permitting authority's emissions inventory at the time of enactment;

(ii) Was equipped prior to shutdown with a continuous system of emissions control that achieves a removal efficiency for sulfur dioxide of no less than 85 percent and a removal efficiency for particulates of no less than 98 percent;

(iii) Is equipped with low-NO_x burners prior to the time of commencement of operations following reactivation; and

(iv) Is otherwise in compliance with the requirements of the Clean Air Act.

reactivation of a very clean coal-fired electric utility steam generating unit (40CFR60.2) means any physical change or change in the method of operation associated with the commencement of commercial operations by a coal-fired utility unit after a period of discontinued operation where the unit:

(1) Has not been in operation for the two-year period prior to the enactment of the Clean Air Act Amendments of 1990, and the emissions from such unit continue to be carried in the permitting authority's emissions inventory at the time of enactment;

(2) Was equipped prior to shut-down with a continuous system of emissions control that

achieves a removal efficiency for sulfur dioxide of no less than 85 percent and a removal efficiency for particulates of no less than 98 percent;

(3) Is equipped with low-NO_x burners prior to the time of commencement of operations following reactivation; and

(4) Is otherwise in compliance with the requirements of the Clean Air Act.

reactivation of a very clean coal-fired electric utility steam generating unit (40CFR52.21-38) means any physical change or change in the method of operation associated with the commencement of commercial operations by a coal-fired utility unit after a period of discontinued operation where the unit:

(i) Has not been in operation for the two-year period prior to the enactment of the Clean Air Act Amendments of 1990, and the emissions from such unit continue to be carried in the permitting authority's emissions inventory at the time of enactment;

(ii) Was equipped prior to shut-down with a continuous system of emissions control that achieves a removal efficiency for sulfur dioxide of no less than 85 percent and a removal efficiency for particulates of no less than 98 percent;

(iii) Is equipped with low-NO_x burners prior to the time of commencement of operations following reactivation; and

(iv) Is otherwise in compliance with the requirements of the Clean Air Act.

reactive functional group (40CFR723.250-13) means an atom or associated group of atoms in a chemical substance that is intended or can reasonably be anticipated to undergo facile chemical reaction.

reactor (40CFR52.741) means a vat, vessel, or other device in which chemical reactions take place.

reactor (40CFR61.61-q) includes any vessel in which vinyl chloride is partially or totally polymerized into polyvinyl chloride.

reactor (40CFR63.101) means a device or vessel in which one or more chemicals or reactants, other than air, are combined or decomposed in such a way that their molecular structures are altered and one or more new organic compounds are formed. Reactor includes the product separator and any associated vacuum pump or steam jet (other identical or similar definitions are provided in 40CFR63.191; 63.111).

reactor opening loss (40CFR61.61-r) means the emissions of vinyl chloride occurring when a reactor is vented to the atmosphere for any purpose other than an emergency relief discharge as defined in 40CFR61.65(a).

reactor processes (40CFR60.701) are unit operations in which one or more chemicals, or reactants other than air, are combined or decomposed in such a way that their molecular structures are altered and one or more new organic compounds are formed.

readily water-soluble substances (40CFR797.1060-5) means chemicals which are soluble in water at a concentration equal to or greater than 1,000 mg/l (cf. limited water-soluble substances).

readily water-soluble substances (40CFR797.1075) means chemicals which are soluble in water at a concentration equal to or greater than 1,000 mg/l.

ready biodegradability (40CFR796.3100-iii) is an expression used to describe those substances which, in certain biodegradation test procedures, produce positive results that are unequivocal and which lead to the reasonable assumption that the substance will undergo rapid and ultimate biodegradation in aerobic aquatic environments.

reaeration (EPA-94/04): Introduction of air into the lower layers of a reservoir. As the air bubbles form and rise through the water, the oxygen from the air dissolves into the water and replenishes the dissolved oxygen. The rising bubbles also cause the lower waters to rise to the surface where they take on oxygen from the atmosphere.

reagent blank (40CFR136-AC-12) means a volume of deionized, distilled water containing the same acid matrix as the calibration standards carried through the entire analytical scheme.

real ear protection at threshold (40CFR211.203-bb) is the mean value in decibels of the occluded threshold of audibility (hearing protector in place) minus the open threshold of audibility (ears open and uncovered) for all listeners on all trials under otherwise identical test conditions.

real property (40CFR247.101-f) means any property that is immovable and attached to the land (under RCRA).

real property (40CFR30.200) means land, including land improvements, and structures and appurtenances, excluding movable machinery and equipment (other identical or similar definitions are provided in 40CFR31.3; 35.6015-40).

reasonable assistance (40CFR204.2-13) means providing timely and unobstructed access to test products or products and records required by this part and opportunity for copying such records or testing such test (other identical or similar definitions are provided in 40CFR205.2-13).

reasonable assistance (40CFR86.078.7) includes, but is not limited to, clerical, copying, interpretation and translation services, the making available on request of personnel of the facility being inspected during their working hours to inform the EPA Enforcement Officer of how the facility operates and to answer his questions, and the performance on request of emissions tests on any vehicle (or engine) which is being, has been, or will be used for certification testing. Such tests shall be nondestructive, but may require appropriate mileage (or service) accumulation. A manufacturer may be compelled to cause the personal appearance of any employee at such a facility before an EPA Enforcement Officer by written request for his appearance, signed by the Assistant Administrator for Enforcement, served on the manufacturer. Any such employee who has been instructed by the manufacturer to appear will be entitled to be accompanied, represented, and advised by counsel.

reasonable compensation (40CFR34.105-m) means, with respect to a regularly employed officer or employee of any person, compensation that is consistent with the normal compensation for such officer or employee for work that is not furnished to, not funded by, or not furnished in cooperation with the Federal Government.

reasonable expense (40CFR85.2102-17) means any expense incurred due to repair of a warranty failure caused by a non-original equipment certified part, including, but not limited to, all charges in any expense categories that would be considered payable by the involved vehicle manufacturer to its authorized dealer under a similar warranty situation where an original equipment part was the cause of the failure. Included in "reasonable expense" are any additional costs incurred specifically due to the processing of a claim involving a certified aftermarket part or parts as covered in these regulations. The direct parts and labor expenses of carrying out repairs is immediately chargeable to the part manufacturer. All charges beyond the actual parts and labor repair expenses must be amortized over the number of claims and/or over a number of years in a manner that would be considered consistent with generally accepted accounting principles. These expense categories shall include but are not limited to the cost of labor, materials, record keeping, special handling, and billing as a result of replacement of a certified aftermarket part.

reasonable further progress (CAA171-42USC7501) means such annual incremental reductions in emissions of the relevant air pollutant as are required by this part or may reasonably be required by the Administrator for the purpose of ensuring attainment of the applicable national ambient air quality standard by the applicable date.

reasonable further progress (EPA-94/04): Annual incremental reductions in air pollution emissions as reflected in a State Implementation Plan, that EPA deems sufficient to provide for attainment of the applicable national ambient air quality standard by the statutory deadline.

reasonable further progress (RFP) plan (40CFR51.491) means any incremental emissions reductions required by the CAA (e.g., section 182(b)) and approved by the EPA as meeting these requirements.

reasonable maximum exposure (EPA-94/04): The maximum exposure reasonably expected to occur in a population.

reasonable payment (40CFR34.105-n) means, with respect to professional and other technical services, a payment in an amount that is consistent with the amount normally paid for such services in the private sector.

reasonable terms (40CFR39.105-i) means rates determined by the Secretary of the Treasury with relationship to the current average average on outstanding marketable obligations of municipalities of comparable maturity.

reasonable time (40CFR51.110) is defined in two ways as follows:
(i) Reasonable time for attainment of a secondary standard must not be more than three years from plan submission unless the State shows that good cause exists for postponing application of the control technology. This definition applies only in a region where the degree of emission reduction necessary for attainment of the secondary standard can be achieved through the application of reasonably available control technology.
(ii) Reasonable time will depend on the degree of emission reduction needed for attainment of the secondary standard and on the social, economic, and technological problems involved in carrying out a control strategy adequate for attainment of the secondary standard. This definition applies only in a region where application of reasonably available control technology will not be sufficient for attainment of the secondary standard in three years.

reasonably anticipated (40CFR723.250-15) means that a knowledgeable person would expect a given physical or chemical composition or characteristic to occur based on such factors as the nature of the precursors used to manufacture the polymer, the type of reaction, the type of manufacturing process, the products produced in polymerization, the intended uses of the substance, or associated use conditions.

reasonably attributable (40CFR51.301-s) means attributable by visual observation or any other technique the State deems appropriate.

reasonably available control measures (RACM) (EPA-94/04): A broadly defined term referring to technological and other measures for pollution control.

reasonably available control technology (RACT) (40CFR51.100-o) means devices, systems process modifications, or other apparatus or techniques that are reasonably available taking into account:

(1) the necessity of imposing such controls in order to attain and maintain a national ambient air quality standard,

(2) the social, environmental and economic impact of such controls, and

(3) alternative means of providing for attainment and maintenance of such standard. (This provision defines RACT for the purposes of 40CFR51.110(c)(2) and 51.341(b) only.)

reasonably available control technology (RACT) (EPA-94/04): Control technology that is both reasonably available, and both technologically and economically feasible. Usually applied to existing sources in nonattainment areas; in most cases is less stringent than new source performance standards.

reasonably foreseeable emissions (40CFR51.852) are projected future indirect emissions that are identified at the time the conformity determination is made; the location of such emissions is known and the emissions are quantifiable, as described and documented by the Federal agency based on its own information and after reviewing any information presented to the Federal agency (other identical or similar definitions are provided in 40CFR93.152) (under CAA).

reasons of business confidentiality (40CFR2.201-e) include the concept of trade secrecy and other related legal concepts which give (or may give) a business the right to preserve the confidentiality of business information and to limit its use or disclosure by others in order that the business may obtain or retain business advantages it derives from its rights in the information. The definition is meant to encompass any concept which authorizes a Federal agency to withhold business information under 5USC552(b)(4), as well as any concept which requires EPA to withhold information from the public for the benefit of a business under 18USC1905 or any of the various statutes cited in 40CFR2.301 through 40CFR2.309.

rebricking (40CFR60.291) means cold replacement of damaged or worn refractory parts of the glass melting furnace. Rebricking includes replacement of the refractories comprising the bottom, sidewalls, or roof of the melting vessel; replacement of refractory work in the heat exchanger; replacement of refractory portions of the glass conditioning and distribution system (other identical or similar definitions are provided in 40CFR61.161).

reburning (40CFR76.2) means reducing the coal and combustion air to the main burners and injecting a reburn fuel (such as gas or oil) to create a fuel-rich secondary combustion zone above the main burner zone and final combustion air to create a fuel-lean burnout zone. The formation of NO_x is inhibited in the main burner zone due to the reduced combustion intensity, and NO_x is destroyed in the fuel-rich secondary combustion zone by conversion to molecular nitrogen.

recarbonization (EPA-94/04): Process in which carbon dioxide is bubbled into water being treated to lower the pH.

receive or receipt of (40CFR72.2) means the date the Administrator or a permitting authority comes into possession of information or correspondence (whether sent in writing or by authorized electronic transmission), as indicated in an official correspondence log, or by a notation made on the information or correspondence, by the Administrator or the permitting authority in the regular course of business.

receiving country (40CFR262.51) means a foreign country to which a hazardous waste is sent for the purpose of treatment, storage or disposal (except short-term storage incidental to transportation).

receiving property (40CFR201.1-w) means any residential or commercial property that receives the sound from railroad facility operations, but that is not owned or operated by a railroad; except that occupied residences located on property owned or controlled by the railroad are included in the definition of receiving property. For purposes of this definition railroad crew sleeping quarters located on property owned or controlled by the railroad are not considered as residences. If, subsequent to the publication date of these regulations, the use of any property that is currently not applicable to this regulation changes, and it is newly classified as either residential or commercial, it is not receiving property until four years have elapsed from the date of the actual change in use.

receiving property measurement location (40CFR201.1-v) means a location on receiving property that is on or beyond the railroad facility boundary and that meets the receiving property measurement location criteria of subpart C.

receiving waters (EPA-94/04): A river, lake, ocean, stream or other watercourse into which wastewater or treated effluent is discharged.

recessive mutation (40CFR798.5275-2) is a change in the genome which is expressed in the homozygous or homozygous condition.

recharge (40CFR149.2-c) means a process, natural or artificial, by which water is added to the saturated zone of an aquifer.

recharge (EPA-94/04): The process by which water is added to a zone of saturation, usually by percolation from the soil surface, e.g., the recharge of an aquifer.

recharge area (40CFR149.2-b) means an area in which water reaches the zone of saturation (ground water) by surface infiltration; in addition, a "major recharge area" is an area where a major part of the recharge to an aquifer occurs.

recharge area (EPA-94/04): A land area in which water reaches the zone of saturation from surface infiltration, e.g., where rainwater soaks through the earth to reach an aquifer.

recharge rate (EPA-94/04): The quantity of water per unit of time that replenishes or refills an aquifer.

recharge zone (40CFR149.101-c) means the area through which water enters the Edwards Underground Reservoir as defined in the December 16, 1975, Notice of Determination.

recipient (40CFR30.200) means any entity which has been awarded and accepted an EPA assistance agreement.

recipient (40CFR34.105-o) includes all contractors, subcontractors at any tier, and subgrantees at any tier of the recipient of funds received in connection with a Federal contract, grant, loan, or cooperative agreement. The term excludes an Indian tribe, tribal organization, or any other Indian organization with respect to expenditures specifically permitted by other Federal law.

recipient (40CFR35.4010) means any group of individuals that hss been awarded a technical assistance grant.

recipient (40CFR35.6015-41) means any State, political subdivision thereof, or federally recognized Indian Tribe which has been awarded and has accepted an EPA cooperative agreement.

recipient (40CFR7.25) means, for the purposes of this regulation, any state or its political subdivision, any instrumentality of a state or its political subdivision, any public or private agency, institution, organization, or other entity, or any person to which Federal financial assistance is extended directly or through another recipient, including any successor, assignee, or transferee of a recipient, but excluding the ultimate beneficiary of the assistance.

recipient (40CFR721.3) means any person who purchases or otherwise obtains a chemical substance directly from a person who manufacturers, imports, or processes the substance (under the Toxic Substances Control Act).

recipient of funds (40CFR51.392) designated under title 23 U.S.C. or the Federal Transit Act means any agency at any level of State, county, city, or regional government that routinely receives title 23 U.S.C. or Federal Transit Act funds to construct FHWA/FTA projects, operate FHWA/FTA projects or equipment, purchase equipment, or undertake other services or operations via contracts or agreements. This definition does not include private landowners or developers, or contractors or entities that are only paid for services or products created by their own employees (other identical or similar definitions are provided in 40CFR93.101).

recipient's project manager (40CFR35.4010) means the person legally authorized to obligate the organization to the terms and conditions of EPA's regulations and the grant agreement, and designated by the recipient to serve as its principal contact with EPA.

reciprocal translocations (40CFR798.5955-2) are chromosomal translocations resulting from reciprocal exchanges between two or more chromosomes.

reciprocating compressor (40CFR60.631) means a piece of equipment that increases the pressure of a process gas by positive displacement, employing linear movement of the driveshaft.

recirculated cooling water (40CFR423.11-h) means water which is passed through the main condensers for the purpose of removing waste heat, passed through a cooling device for the purpose of removing such heat from the water and then passed again, except for blowdown, through the main condenser (cf. once through cooling water).

recirculation (40CFR420.101-a) means those cold rolling operations which include recirculation of rolling solutions at all mill stands.

reclaim refrigerant (40CFR82.152-q) means to reprocess refrigerant to at least the purity specified in appendix A to 40CFR part 82, subpart F (based on ARI Standard 700-1993, Specifications for Fluorocarbon and Other Refrigerants) and to verify this purity using the analytical methodology prescribed in appendix A. In general, reclamation involves the use of processes or procedures available only at a reprocessing or manufacturing facility.

reclaimed (40CFR261.1-4): A material is "reclaimed" if it is processed to recover a usable product, or if it is regenerated. Examples are recovery of lead values from spent batteries and regeneration of spent solvents.

reclaimer (40CFR63.321) means a machine used to remove perchloroethylene from articles by tumbling them in a heated air stream (cf. dryer).

reclamation (EPA-94/04): (In recycling) Restoration of materials found in the waste stream to a beneficial use which may be for purposes other than the original use.

reclamation area (40CFR434.11-l) means surface area of a coal mine which has been returned to required contour and on which revegetation (specifically, seeding or planting) work has commenced.

reclamation plan (SMCRA701-30USC1291) means a plan submitted by an applicant for a permit under a State program or Federal program which sets forth a plan for reclamation of the proposed surface coal mining operations pursuant to section 1258 of this title.

reclamation site (40CFR503.11-n) is drastically disturbed land that is reclaimed using sewage sludge. This includes, but is not limited to, strip mines and construction sites.

recombinant bacteria (EPA-94/04): A microorganism whose genetic makeup has been altered by deliberate introduction of new genetic elements. The offspring of these altered bacteria also contain these new genetic elements, i.e. they "breed true."

recombinant DNA (EPA-94/04): The new DNA that is formed by combining pieces of DNA from different organisms or cells.

recommencing discharger (40CFR122.2) means a source which recommences discharge after terminating operations.

recommendation to list (40CFR15.4) means a written request which has been signed and sent by a recommending person to the Listing Official asking that EPA place a facility on the List of Violating Facilities.

recommended decision (40CFR164.2-q) means the recommended findings and conclusions of the Presiding Officer in an expedited hearing.

recommended maximum contaminant level (RMCL) (EPA-94/04): The maximum level of a contaminant in drinking water at which no known or anticipated adverse affect on human health would occur, and that includes an adequate margin of safety.

Recommended levels are nonenforceable health goals (cf. maximum contaminant level).

recommending person (40CFR15.4) means a Regional Administrator, the Associate Enforcement Counsel for Air or the Associate Enforcement Counsel for Water or their successors, the Assistant Administrator for Air and Radiation or the Assistant Administrator for Water or their successors, a Governor, or a member of the public.

reconfigured emission-data vehicle (40CFR86.082.2) means an emission-data vehicle obtained by modifying a previously used emission-data vehicle to represent another emission-data vehicle.

reconstructed source (EPA-94/04): Facility in which components are replaced to such an extent that the fixed capital cost of the new components exceed 50 percent of the capital cost of constructing a comparable brand-new facility. New-source performance standards may be applied to sources reconstructed after the proposal of the standard if it is technologically and economically feasible to meet the standard.

reconstruction (40CFR51.301-t) will be presumed to have taken place where the fixed capital cost of the new component exceeds 50 percent of the fixed capital cost of a comparable entirely new source. Any final decision as to whether reconstruction has occurred must be made in accordance with the provisions of 60.15 (f) (1) through (3) of this title.

reconstruction (40CFR60.15-b) means the replacement of components of an existing facility to such an extent that:
(1) The fixed capital cost of the new components exceeds 50 percent of the fixed capital cost that would be required to construct a comparable entirely new facility, and
(2) It is technologically and economically feasible to meet the applicable standards set forth in this part.

reconstruction (40CFR63.2) means the replacement of components of an affected or a previously unaffected stationary source to such an extent that:
(1) The fixed capital cost of the new components exceeds 50 percent of the fixed capital cost that

would be required to construct a comparable new source; and
(2) It is technologically and economically feasible for the reconstructed source to meet the relevant standard(s) established by the Administrator (or a State) pursuant to section 112 of the Act. Upon reconstruction, an affected source, or a stationary source that becomes an affected source, is subject to relevant standards for new sources, including compliance dates, irrespective of any change in emissions of hazardous air pollutants from that source.
(Other identical or similar definitions are provided in 40CFR63.101; 63.191.)

reconstruction (40CFR63.321), for purposes of this subpart, means replacement of a washer, dryer, or reclaimer; or replacement of any components of a dry cleaning system to such an extent that the fixed capital cost of the new components exceeds 50 percent of the fixed capital cost that would be required to construct a comparable new source.

record (29CFR1910.20) means any item, collection, or grouping of information regardless of the form or process by which it is maintained (e.g., paper, document, microfiche, microfilm, X-ray film, or automated data processing).

record (40CFR1516.2) means any item or collection or grouping of information about an individual that is maintained by the Council (including, but not limited to, his or her employment history, payroll information, and financial transactions), and that contains his or her name, or an identifying number, symbol, or other identifying particular assigned to the individual such as a social security number.

record (40CFR16.2-a): See individual, maintain, record, system of records, and routine use.

record (40CFR2.100-b): See EPA record.

record of decision (ROD) (EPA-94/04): A public document that explains which cleanup alternative(s) will be used at National Priorities List sites where, under CERCLA, Trust Funds pay for the cleanup.

recordation, record, or recorded (40CFR72.2) means, with regard to allowances, the transfer of allowances by the Administrator from one Allowance Tracking System account or subaccount to another (under CAA).

recorded (40CFR2.201-j) means written or otherwise registered in some form for preserving information, including such forms as drawings, photographs, videotape, sound recordings, punched cards, and computer tape or disk.

recover refrigerant (40CFR82.152-r) means to remove refrigerant in any condition from an appliance and to store it in an external container without necessarily testing or processing it in any way.

recoverable (RCRA1004-42USC6903) refers to the capability and likelihood of being recovered from solid waste for a commercial or industrial use.

recoverable resources (40CFR245.101-g) means materials that still have useful physical, chemical, or biological properties after serving their original purpose and can, therefore, be reused or recycled for the same or other purposes (other identical or similar definitions are provided in 40CFR246.101-u).

recovered material (40CFR249.04-l) means waste material and byproducts which have been recovered or diverted from solid waste, but such term does not include those materials and byproducts generated from, and commonly reused within, an original manufacturing process (Pub. L. 94-580, 90 Stat. 2800, 42USC6903, as amended by Pub. L. 96-482) (under RCRA).

recovered material (RCRA1004-42USC6903) means waste material and byproducts which have been recovered or diverted from solid waste, but such term does not include those materials and byproducts generated from, and commonly reused within, an original manufacturing process.

recovered materials (40CFR248.4-gg) means waste material and byproducts which have been recovered or diverted from solid waste, but such term does not include those materials and byproducts generated from, and commonly reused within, an original manufacturing process.

recovered materials (40CFR250.4-kk) means waste material and byproducts that have been recovered or diverted from solid waste, but such term does not include those materials and byproducts generated from, and commonly reused within, an original manufacturing process. In the case of paper and paper products, the term "recovered materials" includes:

(1) Postconsumer materials such as:
  (i) Paper, paperboard, and fibrous wastes from retail stores, office buildings, homes, and so forth, after they have passed through their end usage as a consumer item, including: Used corrugated boxes, old newspapers, old magazines, mixed waste paper, tabulating cards, and used cordage, and
  (ii) All paper, paperboard, and fibrous wastes that enter and are collected from municipal solid waste; and
(2) Manufacturing, forest residues, and other wastes such as:
  (i) Dry paper and paperboard waste generated after completion of the papermaking process (that is, those manufacturing operations up to and including the cutting and trimming of the paper machine reel into smaller rolls or rough sheets) including: envelope cuttings, bindery trimmings, and other paper and paperboard waste, resulting from printing, cutting, forming, and other converting operations, bag, box and carton manufacturing wastes, and butt rolls, mill wrappers, and rejected unused stock, and
  (ii) Finished paper and paperboard from obsolete inventories of paper and paperboard manufacturers, merchants, wholesalers, dealers, printers, converters, or others,
  (iii) Fibrous byproducts of harvesting, manufacturing, extractive, or woodcutting processes, flax, straw, linters, bagasse, slash, and other forest residues,
  (iv) Wastes generated by the conversion of goods made from fibrous material (e.g., waste rope from cordage manufacture, textile mill waste, and cuttings);
  (v) Fibers recovered from waste water that otherwise would enter the waste stream (under RCRA).

recovered resources (RCRA1004-42USC6903) means material or energy recovered from solid waste.

recovered solvent (40CFR60.601) means the solvent captured from liquid and gaseous process streams that is concentrated in a control device and that may be purified for reuse.

recovery (40CFR245.101-h) means the process of obtaining materials or energy resources from solid waste (other identical or similar definitions are provided in 40CFR246.101-v).

recovery device (40CFR60.611) means an individual unit of equipment, such as an absorber, condenser, and carbon adsorber, capable of and used to recover chemicals for use, reuse or sale.

recovery device (40CFR60.661) means an individual unit of equipment such as an absorber, carbon adsorber, or condenser, capable of and used for the purpose of recovering chemicals for use, reuse, or sale (other identical or similar definitions are provided in 40CFR60.701).

recovery device (40CFR63.101) means an individual unit of equipment capable of and used for the purpose of recovering chemicals for use, reuse, or sale. Recovery devices include, but are not limited to, absorbers, carbon adsorbers, and condensers (other identical or similar definitions are provided in 40CFR63.111).

recovery efficiency (40CFR82.152-s) means the percentage of refrigerant in an appliance that is recovered by a piece of recycling or recovery equipment.

recovery furnace (40CFR60.281-h) means either a straight kraft recovery furnace or a cross recovery furnace, and includes the direct-contact evaporator for a direct-contact furnace.

recovery rate (EPA-94/04): Percentage of usable recycled materials that have been removed from the total amount of municipal solid waste generated in a specific area or by a specific business.

recovery system (40CFR60.561) means an individual unit or series of material recovery units, such as absorbers, condensers, and carbon adsorbers, used for recovering volatile organic compounds (under CAA).

recovery system (40CFR60.611) means an individual recovery device or series of such devices applied to the same process stream.

recovery system (40CFR60.661) means an individual recovery device or series of such devices applied to the same vent stream (other identical or similar definitions are provided in 40CFR60.701).

recruiting and training agency (40CFR8.2-t) means any person who refers workers to any contractor or subcontractor, or who provides or supervises apprenticeship or training for employment by any contractor or subcontractor.

recurrent expenditures (40CFR35.105) means those expenses associated with the activities of a continuing environmental program. All expenditures, except those for equipment purchases with a unit acquisition cost of $5.000 or more, are considered recurrent unless justified by the applicant as unique and approved as such by the Regional Administrator in the assistance award.

recyclable paper (40CFR250.4-ll) means any paper separated at its point of discard or from the solid waste stream for utilization as a raw material in the manufacture of a new product. It is often called "waste paper" or "paper stock." Not all paper in the waste stream is recyclable. It may be heavily contaminated or otherwise unusable.

recycle refrigerant (40CFR82.152-t) means to extract refrigerant from an appliance and clean refrigerant for reuse without meeting all of the requirements for reclamation. In general, recycled refrigerant is refrigerant that is cleaned using oil separation and single or multiple passes through devices, such as replaceable core filter-driers, which reduce moisture, acidity, and particulate matter. These procedures are usually implemented at the field job site.

recycle/reuse (EPA-94/04): Minimizing waste generation by recovering and reprocessing usable products that might otherwise become waste (i.e., recycling of aluminum cans, paper, and bottles, etc.).

recycled (40CFR261.1-7): A material is "recycled" if it is used, reused, or reclaimed.

recycled material (40CFR245.101-j) means a material that is utilized in place of a primary, raw, or virgin material in manufacturing a product.

recycled material (40CFR246.101-w) means a material that is used in place of a primary, raw or virgin material in manufacturing a product.

recycled material (40CFR247.101-g) means a material that can be utilized in place of a raw or virgin material in manufacturing a product and consists of materials derived from post consumer waste, industrial scrap, material derived from agricultural wastes and other items, all of which can be used in the manufacture of new products (under RCRA).

recycled oil (RCRA1004-42USC6903) means any used oil which is reused, following its original use, for any purpose (including the purpose for which the oil was originally used). Such term includes oil which is re-refined, reclaimed, burned, or reprocessed.

recycled PCBs (40CFR761.3) means those PCBs which appear in the processing of paper products or asphalt roofing materials from PCB-contaminated raw materials. Processes which recycle PCBs must meet the following requirements:

(1) There are no detectable concentrations of PCBs in asphalt roofing material products leaving the processing site.
(2) The concentration of PCBs in paper products leaving any manufacturing site processing paper products, or in paper products imported into the United States, must have an annual average of less than 25 ppm with a 50 ppm maximum.
(3) The release of PCBs at the point at which emissions are vented to ambient air must be less than 10 ppm.
(4) The amount of Aroclor PCBs added to water discharged from an asphalt roofing processing site must at all times be less than 3 micrograms per liter (μg/L) for total Aroclors (roughly 3 parts per billion (3 ppb)). Water discharges from the processing of paper products must at all times be less than 3 micrograms per liter (μg/1) for total Aroclors (roughly 3 ppb), or comply

with the equivalent mass-based limitation.
(5) Disposal of any other process wastes at concentrations of 50 ppm or greater must be in accordance with subpart D of this part.

recycling (40CFR244.101-j) means the process by which recovered materials are transformed into new products.

recycling (40CFR245.101-j) means the process by which recovered materials are transformed into new products (other identical or similar definitions are provided in 40CFR246.101-x).

red bag waste (EPA-94/04): See infectious waste.

red border (EPA-94/04): An EPA document undergoing review before being submitted for final management decision-making.

red tide (EPA-94/04): A proliferation of a marine plankton toxic and often fatal to fish, perhaps stimulated by the addition of nutrients. A tide can be red, green, or brown, depending on the coloration of the plankton.

reduced sulfur compounds (40CFR60.101-l) means hydrogen sulfide (H_2S), carbonyl sulfide (COS) and carbon disulfide (CS_2).

reduced sulfur compounds (40CFR60.641) means H_2S, carbonyl sulfide (COS), and carbon disulfide (CS_2).

reduced utilization (40CFR72.2) means a reduction, during any calendar year in Phase I, in the heat input (expressed in mmBtu for the calendar year) at a Phase I unit below the unit's baseline, where such reduction subjects the unit to the requirement to submit a reduced utilization plan under 40CFR72.43; or, in the case of an opt-in source, means a reduction in the average utilization, as specified in 40CFR74.44 of this chapter, of an opt-in source below the opt-in source's baseline.

reduction (EPA-94/04): The addition of hydrogen, removal of oxygen, or addition of electrons to an element or compound.

reduction control system (40CFR60.101-k) means an

emission control system which reduces emissions from sulfur recovery plants by converting these emissions to hydrogen sulfide.

reentry interval (EPA-94/04): The period of time immediately following the application of a pesticide during which unprotected workers should not enter a field.

reference compound (40CFR60-AA (method 21-2.2)) means the VOC species selected as an instrument calibration basis for specification of the leak definition concentration. (For example: If a leak definition concentration is 10,000 ppmv as methane, then any source emission that results in a local concentration that yields a meter reading of 10,000 on an instrument calibrated with methane would be classified as a leak. In this example, the leak definition is 10,000 ppmv, and the reference compound is methane.)

reference conditions (40CFR50.3): All measurements of air quality are corrected to a reference temperature of 25 C. and to a reference pressure of 760 millimeters of mercury (1,013.2 millibars) (under CAA).

reference control technology for process vents (40CFR63.111) means a combustion device used to reduce organic HAP emissions by 98 percent, or to an outlet concentration of 20 parts per million by volume.

reference control technology for storage vessels (40CFR63.111) means an internal floating roof meeting the specifications of 40CFR63.119(b) of this subpart, an external floating roof meeting the specifications of 40CFR63.119(c) of this subpart, an external floating roof converted to an internal floating roof meeting the specifications of 40CFR63.119(d) of this subpart, or a closed-vent system to a control device achieving 95-percent reduction in organic HAP emissions. For purposes of emissions averaging, these four technologies are considered equivalent.

reference control technology for transfer racks (40CFR63.111) means a combustion device or recovery device used to reduce organic HAP emissions by 98 percent, or to an outlet

concentration of 20 parts per million by volume; or a vapor balancing system.

reference control technology for wastewater (40CFR63.111) means the use of:
(1) Controls specified in 40CFR63.133 through 40CFR63.137;
(2) A steam stripper meeting the specifications of 40CFR63.138(g) of this subpart or any of the other alternative control measures specified in 40CFR63.138 (b), (c), (d), and (e) of this subpart; and
(3) A control device to reduce by 95 percent (or to an outlet concentration of 20 parts per million by volume for combustion devices) the organic HAP emissions in the vapor streams vented from wastewater tanks, oil-water separators, containers, surface impoundments, individual drain systems, and treatment processes (including the design steam stripper) managing wastewater.

reference dose (RfD) (40CFR300-AA) means the estimate of a daily exposure level of a substance to a human population below which adverse noncancer health effects are not anticipated [milligrams toxicant per kilogram body weight per day (mg/kg-day)].

reference dose (RfD) (EPA-94/04): The concentration of a chemical known to cause health problems; also be referred to as the ADI, or acceptable daily intake.

reference method (40CFR50.1-f) means a method of sampling and analyzing the ambient air for an air pollutant that is specified as a reference method in an appendix to this part, or a method that has been designated as a reference method in accordance with part 53 of this chapter; it does not include a method for which a reference method designation has been cancelled in accordance with 40CFR53.11 or 40CFR53.16 of this chapter.

reference method (40CFR53.1-e) means a method of sampling and analyzing the ambient air for an air pollutant that is specified as a reference method in an appendix to part 50 of this chapter, or a method that has been designated as a reference method in accordance with this part; it does not include a method for which a reference method designation has been cancelled in accordance with 40CFR53.11 or 40CFR53.16.

reference method (40CFR60.2) means any method of sampling and analyzing for an air pollutant as specified in the applicable subpart.

reference method (40CFR61.02) means any method of sampling and analyzing for an air pollutant, as described in Appendix B to this part.

reference method (40CFR72.2) means any direct test method of sampling and analyzing for an air pollutant as specified in part 60, Appendix A of this chapter.

reference signal (40CFR60-AA(alt. method 1)) means the backscatter signal resulting from the laser light pulse passing through ambient air.

reference substance (40CFR160.3) means any chemical substance or mixture or analytical standard, or material other than a test substance, feed, or water, that is administered to or used in analyzing the test system in the course of a study for the purposes of establishing a basis for comparison with the test substance for known chemical or biological measurements (other identical or similar definitions are provided in 40CFR792.3).

reference substance (40CFR797.1350-4) is a chemical used to access the constancy of response of a given species of test organisms to that chemical, usually by use of the acute LC_{50}. (It is assumed that any change in sensitivity to the reference substance will indicate the existence of some similar change in degree of sensitivity to other chemicals whose toxicity is to be determined.)

reference value or reference signal (40CFR72.2) means the known concentration of a calibration gas, the known value of an electronic calibration signal, or the known value of any other measurement standard approved by the Administrator, assumed to be the true value for the pollutant or diluent concentration or volumetric flow being measured.

referring agency (40CFR1508.24) means the federal agency which has referred any matter to the Council after a determination that the matter is unsatisfactory from the standpoint of public health or welfare or environmental quality.

refillable beverage container (40CFR244.101-k) means a beverage container that when returned to a distributor or bottler is refilled with a beverage and reused.

refiner (40CFR52.741) means any person who owns, leases, operates, controls, or supervises a refinery (other identical or similar definitions are provided in 40CFR80.2-i).

refinery (40CFR80.2-h) means a plant at which gasoline or diesel fuel is produced.

refinery process unit (40CFR60.101-f) means any segment of the petroleum refinery in which a specific processing operation is conducted.

refinery unit, process unit or unit (40CFR52.741) means a set of components which are a part of a basic process operation such as distillation, hydrotreating, cracking, or reforming of hydrocarbons.

refining (40CFR60.271a) means that phase of the steel production cycle during which undesirable elements are removed from the molten steel and alloys are added to reach the final metal chemistry (under CAA).

reformulated gasoline (40CFR80.2-ee) means any gasoline whose formulation has been certified under 40CFR80.40, which meets each of the standards and requirements prescribed under 40CFR80.41, and which contains less than the maximum concentration of the marker specified in 40CFR80.82 that is allowed for reformulated gasoline under 40CFR80.82.

reformulated gasoline (CAA211.k-42USC7545) means any gasoline which is certified by the Administrator under this section as complying with this subsection (cf. conventional gasoline).

reformulated gasoline (EPA-94/04): Gasoline with a different composition from conventional gasoline (e.g., lower aromatics content) that cuts air pollutants.

reformulated gasoline blendstock for oxygenate blending, or RBOB (40CFR80.2-kk) means a petroleum product which, when blended with a specified type and percentage of oxygenate, meets the definition of reformulated gasoline, and to which the specified type and percentage of oxygenate is added other than by the refiner or importer of the RBOB at the refinery or import facility where the RBOB is produced or imported.

reformulated gasoline credit (40CFR80.2-ii) means the unit of measure for the paper transfer of oxygen or benzene content resulting from reformulated gasoline which contains more than 2.1 weight percent of oxygen or less than 0.95 volume percent benzene.

refractory metals (40CFR471.02-dd) includes the metals of columbium, tantalum, molybdenum, rhenium, tungsten and vanadium and their alloys.

refrigerant (40CFR82.32-f) means any class I or class II substance used (40CFR82.32-) in a motor vehicle air conditioner. Class I and class II substances are listed in part 82, subpart A, appendix A. Effective November 15, 1995, refrigerant shall also include any substitute substance.

refrigerated condenser (40CFR52.741) means a surface condenser in which the coolant supplied to the condenser has been cooled by a mechanical device, other than by a cooling tower or evaporative spray cooling, such as refrigeration unit or steam chiller unit.

refrigerated condenser (40CFR63.321) means a vapor recovery system into which an air-perchloroethylene gas-vapor stream is routed and the perchloroethylene is condensed by cooling the gas-vapor stream.

refrigerated condenser coil (40CFR63.321) means the coil containing the chilled liquid used to cool and condense the perchloroethylene.

refueling emissions (40CFR86.098.2) means evaporative emissions that emanate from a motor vehicle fuel tank(s) during a refueling operation.

refueling emissions (EPA-94/04): Emissions released during vehicle refueling.

refueling emissions canisters (40CFR86.098.2) means any vapor storage unit(s) that is exposed to the vapors generated during refueling.

refund (40CFR244.101-l) means the sum, equal to the deposit, that is given to the consumer or the dealer or both in exchange for empty returnable beverage containers.

refurbishment (40CFR60.501) means, with reference to a vapor processing system, replacement of components of, or addition of components to, the system within any 2-year period such that the fixed capital cost of the new components required for such component replacement or addition exceeds 50 percent of the cost of a comparable entirely new system.

refuse (EPA-94/04): See solid waste.

refuse reclamation (EPA-94/04): Conversion of solid waste into useful products, e.g., composting organic wastes to make soil conditioners or separating aluminum and other metals for recycling.

refuse-derived fuel or RDF (40CFR60.51a) means a type of MSW through shredding and size classification. This includes all classes of RDF including low density fluff RDF through densified RDF and RDF fuel pellets.

regeneration (EPA-94/04): Manipulation of cells to cause them to develop into whole plants.

regenerative carbon adsorber (40CFR61.131) means a carbon adsorber applied to a single source or group of sources, in which the carbon beds are regenerated without being moved from their location (under CAA).

regenerative cycle gas turbine (40CFR60.331-c) means any stationary gas turbine that recovers thermal energy from the exhaust gases and utilizes the thermal energy to preheat air prior to entering the combustor (cf. simple cycle gas turbine).

regenerative cycle gas turbine (40CFR60.331-s) means any stationary gas turbine which recovers heat from the gas turbine exhaust gases to preheat the inlet combustion air to the gas turbine.

region (40CFR51.100-m) means an area designated as an air quality control region (AQCR) under section 107(c) of the Act.

region (40CFR60.21-i) means an air quality control region designated under section 107 of the Act and described in part 81 of this chapter.

Regional Administrator (40CFR20.2-e) means the Regional designee appointed by the Administrator to certify facilities under this part (other identical or similar definitions are provided in 40CFR21.2-d; 22.03; 52.1128-6; 58.1-k; 65.01-c; 108.2; 112.2; 121.1-d; 122.2; 124.2; 131.35.d-11; 136.2-c; 144.3; 147.2902; 165.1-c; 232.2; 270.2; 260.10; 270.2; 403.3-s).

regional authority (RCRA1004-42USC6903) means the authority established or designated under section 4006.

regional hearing clerk (40CFR124.72) means an employee of the Agency designated by a Regional Administrator to establish a repository for all books, records, documents, and other materials relating to hearings under this subpart.

regional hearing clerk (40CFR22.03) means an individual duly authorized by the Regional Administrator to serve as hearing clerk for a given region. Correspondence may be addressed to the Regional Hearing Clerk, U.S. Environmental Protection Agency (address of Regional Office, Appendix). In a case where the complainant is the Assistant Administrator for Enforcement or his delegate, the term Regional Hearing Clerk as used in these rules shall mean the Hearing Clerk.

regional judicial officer (40CFR22.03) means a person designated by the Regional Administrator under 40CFR22.04(b) to serve as Regional Judicial Officer.

regional office (40CFR51.100-h) means one of the ten (10) EPA Regional Offices.

regional response team (RRT) (EPA-94/04): Representatives of federal, local, and state agencies who may assist in coordination of activities at the request of the On-Scene Coordinator before and during a significant pollution incident such as an oil spill, major chemical release, or a Superfund response.

regional water and/or wastewater projects (40CFR51.852) include construction, operation, and maintenance of water or wastewater conveyances, water or wastewater treatment facilities, and water storage reservoirs which affect a large portion of a nonattainment or maintenance area (other identical or similar definitions are provided in 40CFR93.152).

regionally significant action (40CFR51.852) means a Federal action for which the direct and indirect emissions of any pollutant represent 10 percent or more of a nonattainment or maintenance area's emissions inventory for that pollutant (other identical or similar definitions are provided in 40CFR93.152).

regionally significant project (40CFR51.392) means a transportation project (other than an exempt project) that is on a facility which serves regional transportation needs (such as access to and from the area outside of the region, major activity centers in the region, major planned developments such as new retail malls, sports complexes, etc., or transportation terminals as well as most terminals themselves) and would normally be included in the modeling of a metropolitan area's transportation network, including at a minimum all principal arterial highways and all fixed guideway transit facilities that offer an alternative to regional highway travel (other identical or similar definitions are provided in 40CFR93.101) (under CAA).

register (40CFR52.1128-1) as applied to a motor vehicle, means the licensing of such motor vehicle for general operation on public roads or highways by the appropriate agency of the Federal Government or by the Commonwealth.

registrant (40CFR153.62-b) includes any person who holds a registration for a pesticide product issued under FIFRA section 3 or 24(c).

registrant (40CFR164.2-r) means any person who has registered a pesticide pursuant to the provisions of the Act.

registrant (40CFR2.307-3) means any person who

has obtained registration under the Act of a pesticide or of an establishment.

registrant (EPA-94/04): Any manufacturer or formulator who obtains registration for a pesticide active ingredient or product.

registrant (FIFRA2-7USC136) means a person who has registered any pesticide pursuant to the provisions of this Act.

registration (EPA-94/04): Formal listing with EPA of a new pesticide before it can be sold or distributed. Under the Federal Insecticide, Fungicide, and Rodenticide Act, EPA is responsible for registration (pre-market licensing) of pesticides on the basis of data demonstrating no unreasonable adverse effects on human health or the environment when applied according to approved label directions.

registration (FIFRA2-7USC136): This term includes re-registration.

registration division (40CFR180.1-d) means the unit established within the Environmental Protection Agency charged with administration of the Pesticide Residue amendment to the federal Food, Drug, ad cosmetic Act (section 408).

registration office (40CFR5.2) means any of the several offices in EPA which have been designated to receive applications for attendance at direct training courses (see also 40CFR5.4 for a listing of such courses).

registration standards (EPA-94/04): Published documents which include summary reviews of the data available on a pesticide's active ingredient, data gaps, and the Agency's existing regulatory position on the pesticide.

regularly employed (40CFR34.105-p) means, with respect to an officer or employee of a person requesting or receiving a Federal contract, grant, loan, or cooperative agreement or a commitment providing for the United States to insure or guarantee a loan, an officer or employee who is employed by such person for at least 130 working days within one year immediately preceding the date of the submission that initiates agency consideration of such person for receipt of such contract, grant, loan, cooperative agreement, loan insurance commitment, or loan guarantee commitment. An officer or employee who is employed by such person for less than 130 working days within one year immediately preceding the date of the submission that initiates agency consideration of such person shall be considered to be regularly employed as soon as he or she is employed by such person for 130 working days.

regulated activity or activity subject to regulation (40CFR124.41) means a major PSD stationary source or major PSD (prevention of significant deterioration) modification.

regulated air pollutant (40CFR70.2) means the following:
(1) Nitrogen oxides or any volatile organic compounds;
(2) Any pollutant for which a national ambient air quality standard has been promulgated;
(3) Any pollutant that is subject to any standard promulgated under section 111 of the Act;
(4) Any Class I or II substance subject to a standard promulgated under or established by title VI of the Act; or
(5) Any pollutant subject to a standard promulgated under section 112 or other requirements established under section 112 of the Act, including sections 112(g), (j), and (r) of the Act, including the following:
   (i) Any pollutant subject to requirements under section 112(j) of the Act. If the Administrator fails to promulgate a standard by the date established pursuant to section 112(e) of the Act, any pollutant for which a subject source would be major shall be considered to be regulated on the date 18 months after the applicable date established pursuant to section 112(e) of the Act; and
   (ii) Any pollutant for which the requirements of section 112(g)(2) of the Act have been met, but only with respect to the individual source subject to section 112(g)(2) requirement.

regulated area (40CFR763.121) means an area established by the employer to demarcate areas where airborne concentrations of asbestos exceed or can reasonably be expected to exceed the permissible exposure limit. The regulated area may take the form of:

(1) A temporary enclosure, as required by paragraph (e)(6) of this section, or

(2) An area demarcated in any manner that minimizes the number of employees exposed to asbestos.

regulated asbestos-containing material (RACM) (40CFR61.141) means:

(a) Friable asbestos material,

(b) Category I nonfriable ACM that has become friable,

(c) Category I nonfriable ACM that will be or has been subjected to sanding, grinding, cutting, or abrading, or

(d) Category II nonfriable ACM that has a high probability of becoming or has become crumbled, pulverized, or reduced to powder by the forces expected to act on the material in the course of demolition or renovation operations regulated by this subpart.

regulated asbestos-containing material (RACM) (EPA-94/04): Friable asbestos material or nonfriable ACM that will be or has been subjected to sanding, grinding, cutting, or abrading or has crumbled, or been pulverized or reduced to powder in the course of demolition or renovation operations.

regulated chemical (40CFR707.63-c) means any chemical substance or mixture for which export notice is required under 40CFR707.60.

regulated entities (CAA183.e-42USC7511b) means:

(i) manufacturers, processors, wholesale distributors, or importers of consumer or commercial products for sale or distribution in interstate commerce in the United States; or

(ii) manufacturers, processors, wholesale distributors, or importers that supply the entities listed under clause (i) with such products for sale or distribution in interstate commerce in the United States.

regulated medical waste (40CFR259.10.b) means those medical wastes that have been listed in 40CFR259.30(a) of this part and that must be managed in accordance with the requirements of this part.

regulated medical waste (EPA-94/04): Under the Medical Waste Tracking Act of 1988, any solid waste generated in the diagnosis, treatment, or immunization of human beings or animals, in research pertaining thereto, or in the production or testing of biologicals. Included are cultures and stocks of infectious agents; human blood and blood products; human pathological body wastes from surgery and autopsy; contaminated animal carcasses from medical research; waste from patients with communicable diseases; and all used sharp implements, such as needles and scalpels, etc., and certain unused sharps.

regulated pest (40CFR171.2-22) means a specific organism considered by a State or Federal agency to be a pest requiring regulatory restrictions, regulations, or control procedures in order to protect the host, man and/or his environment.

regulated pollutant (40CFR70.2) (for presumptive fee calculation), which is used only for purposes of 40CFR70.9(b)(2), means any "regulated air pollutant" except the following:

(1) Carbon monoxide;

(2) Any pollutant that is a regulated air pollutant solely because it is a Class I or II substance to a standard promulgated under or established by title VI of the Act; or

(3) Any pollutant that is a regulated air pollutant solely because it is subject to a standard or regulation under section 112(r) of the Act.

regulated substance (40CFR280.12) means:

(a) Any substance defined in section 101(14) of the Comprehensive Environmental Response, Compensation and Liability Act (CERCLA) of 1980 (but not including any substance regulated as a hazardous waste under subtitle C), and

(b) Petroleum, including crude oil or any fraction thereof that is liquid at standard conditions of temperature and pressure (60 degrees Fahrenheit and 14.7 pounds per square inch absolute). The term "regulated substance" includes but is not limited to petroleum and petroleum-based substances comprised of a complex blend of hydrocarbons derived from crude oil though processes of separation, conversion, upgrading, and finishing, such as motor fuels, jet fuels, distillate fuel oils, residual fuel oils, lubricants, petroleum solvents, and used oils.

regulated substance (RCRA9001-42USC6991) means:
(A) any substance defined in section 101(14) of the Comprehensive Environmental Response, Compensation, and Liability Act of 1980 (but not including any substance regulated as a hazardous waste under subtitle C), and
(B) petroleum.

regulation VII (40CFR52.274.h.2-iii) in this paragraph means Regulation VII, "Emergencies," of the MBUAPCD, adopted May 25, 1977, and submitted to the Environmental Protection Agency as a revision to the California State Implementation Plan by the California Air Resources Board on November 4, 1977.

regulation promulgation schedule (40CFR63.2) means the schedule for the promulgation of emission standards under this part, established by the Administrator pursuant to section 112(e) of the Act and published in the Federal Register.

regulations published under this part (40CFR211.102-i) means all subparts to 40CFR211.

regulatory agency (40CFR190.02-l) means the government agency responsible for issuing regulations governing the use of sources of radiation or radioactive materials or emissions therefrom and carrying out inspection and enforcement activities to assure compliance with such regulations.

regulatory agency (40CFR192.31-g) means the U.S. Nuclear Regulatory Commission.

regulatory authority (SMCRA701-30USC1291) means the State regulatory authority where the State is administering this chapter under an approved State program or the Secretary where the Secretary is administering this chapter under a Federal program.

reid vapor (40CFR60.111-l) pressure is the absolute vapor pressure of volatile crude oil and volatile nonviscous petroleum liquids, except liquified petroleum gases, as determined by ASTM D323-82 (incorporated by reference, see 40CFR60.17) (other identical or similar definitions are provided in 40CFR60.111a-g; 60.111b-g).

reimbursement period (40CFR766.3) means the period that begins when the data from the last test to be completed under this part for a specific chemical substance listed in 40CFR766.25 is submitted to EPA, and ends after an amount of time equal to that which had been required to develop that data or 5 years, whichever is later.

reimbursement period (40CFR790.3) refers to a period that begins when the data from the last non-duplicative test to be completed under a test rule are submitted to EPA and ends after an amount of time equal to that which had been required to develop data or after five years, whichever is later.

reimbursement period (40CFR791.3-h) refers to a period that begins when the data from the last non-duplicative test to be completed under a test rule is submitted to EPA and ends after an amount of time equal to that which had been required to develop that data or after 5 years, whichever is later (under TSCA).

rejection of a batch (40CFR204.51-q) means that the number of non-complying compressors in the batch sample is greater than or equal to the rejection number as determined by the appropriate sampling plan.

rejection of a batch (40CFR205.51-21) means the number of noncomplying vehicles in the batch sample is greater than or equal to the rejection number as determined by the appropriate sampling plan.

rejection of a batch sequence (40CFR204.51-s) means that the number of rejected batches in a sequence is greater than or equal to the sequence rejection number as determined by the appropriate sampling plan.

rejection of a batch sequence (40CFR205.51-11) means that the number of rejected batches in a sequence is equal to or greater than the rejection number as determined by the appropriate sampling plan.

related coatings (40CFR60.431) means all non-ink purchased liquids and liquid-solid mixtures containing VOC solvent, usually referred to as extenders or varnishes, that are used at publication rotogravure printing presses.

relative accuracy (40CFR72.2) means a statistic designed to provide a measure of the systematic and random errors associated with data from continuous emission monitoring systems, and is expressed as the absolute mean difference between the pollutant concentration or volumetric flow measured by the pollutant concentration or flow monitor and the value determined by the applicable reference method(s) plus the 2.5 percent error confidence coefficient of a series of tests divided by the mean of the reference method tests in accordance with part 75 of this chapter.

relative accuracy (RA) (40CFR60-AF-2.6) means the absolute mean difference between the gas concentration or emission rate determined by the CEMS and the value determined by the RM's plus the 2.5 percent error confidence coefficient of a series of tests divided by the mean of the RM tests or the applicable emission limit.

relative percent of percutaneous absorption (40CFR795.223-2) is defined as 100 times the ratio between total urinary excretion of compound following topical administration and total urinary excretion of compound following intravenous injection.

release (40CFR2.310) has the meaning given it in section 101(22) of the Act, 42USC9601(22).

release (40CFR280.12) means any spilling, leaking, emitting, discharging, escaping, leaching or disposing from an UST into ground water, surface water or subsurface soils.

release (40CFR300.5), as defined by section 101(22) of CERCLA, means any spilling, leaking, pumping, pouring, emitting, emptying, discharging, injection, escaping, leaching, dumping, or disposing into the environment (including the abandonment or discarding of barrels, containers, and other closed receptacles containing any hazardous substance or pollutant or contaminant), but excludes: Any release which results in exposure to persons solely within a workplace, with respect to a claim which such persons may assert against the employer of such persons; emissions from the engine exhaust of a motor vehicle, rolling stock, aircraft, vessel, or pipeline pumping station engine; release of source, byproduct or special nuclear material from a nuclear incident, as those terms are defined in the Atomic Energy Act of 1954, if such release is subject to requirements with respect to financial protection established by the Nuclear Regulatory Commission under section 170 of such Act, or, for the purpose of section 104 of CERCLA or any other response action, any release of source, byproduct, or special nuclear material from any processing site designated under section 102(a)(1) or 302(a) of the Uranium Mill Tailings Radiation Control Act of 1978; and the normal application of fertilizer. For the purpose of the NCP, release also means substantial threat of release (other identical or similar definitions are provided in 40CFR310.11-i).

release (40CFR302.3) means any spilling, leaking, pumping, pouring, emitting, emptying, discharging, injecting, escaping, leaching, dumping, or disposing into the environment, but excludes:

(1) any release which results in exposure to persons solely within a workplace, with respect to a claim which such persons may assert against the employer of such persons,

(2) emissions from the engine exhaust of a motor vehicle, rolling stock, aircraft, vessel, or pipeline pumping station engine,

(3) release of source, byproduct, or special nuclear material from a nuclear incident, as those terms are defined in the Atomic Energy Act of 1954, if such release is subject to requirements with respect to financial protection established by the Nuclear Regulatory Commission under section 170 of such Act, or for the purposes of section 104 of the Comprehensive Environmental Response, Compensation, and Liability Act or any other response action, any release of source, byproduct, or special nuclear material from any processing site designated under section 102(a)(1) or 302(a) of the Uranium Mill Tailings Radiation Control Act of 1978, and

(4) the normal application of fertilizer (under CERCLA).

release (40CFR355.20) means any spilling, leaking, pumping, pouring, emitting, emptying, discharging, injecting, escaping, leaching, dumping, or disposing into the environment (including the abandonment or discarding of barrels, containers, and other closed receptacles) of any hazardous chemical, extremely hazardous substance, or CERCLA hazardous substance.

release (40CFR372.3) means any spilling, leaking, pumping, pouring, emitting, emptying, discharging, injecting, escaping, leaching, dumping, or disposing into the environment (including the abandonment or discarding of barrels, containers, and other closed receptacles) of any toxic chemical.

release (40CFR373.4-c) is defined as specified by CERCLA 101(22).

release (EPA-94/04): Any spilling, leaking, pumping, pouring, emitting, emptying, discharging, injecting, escaping, leaching, dumping, or disposing into the environment of a hazardous or toxic chemical or extremely hazardous substance.

release (PPA6603) has the same meaning as provided by section 329(8) of the Superfund Amendments and Reauthorization Act of 1986.

release (RCRA9001-42USC6991) means any spilling, leaking, emitting, discharging, escaping, leaching, or disposing from an underground storage tank into ground water, surface water or subsurface soils.

release (SP101-42USC9601) means any spilling, leaking, pumping, pouring, emitting, emptying, discharging, injecting, escaping, leaching, dumping, or disposing into the environment (including the abandonment or discarding of barrels, containers, or other closed receptacles containing any hazardous substance or pollutant or contaminant), but excludes:
(A) any release which results in exposure to persons solely within a workplace, with respect to a claim which such persons may assert against the employer of such persons,
(B) emissions from the engine exhaust of a motor vehicle, rolling stock, aircraft, vessel, or pipeline

pumping station engine,
(C) release of source, byproduct, or special nuclear material from a nuclear incident, as those terms are defined in the Atomic Energy Act of 1954, if such release is subject to requirements with respect to financial protection established by the Nuclear Regulatory Commission under section 170 of such Act, or, for the purposes of section 104 of this title or any other response action, any release of source byproduct, or special nuclear material from any processing site designated under section 102(a)(1) or 302(a) of the Uranium Mill Tailings Radiation Control Act of 1978, and
(D) the normal application of fertilizer.

release (SP329) means any spilling, leaking, pumping, pouring, emitting, emptying, discharging, injecting, escaping, leaching, dumping, or disposing into the environment (including the abandonment or discarding of barrels, containers, and other closed receptacles) of any hazardous chemical, extremely hazardous substance, or toxic chemical.

release detection (40CFR280.12) means determining whether a release of a regulated substance has occurred form the UST system into the environment or into the interstitial space between the UST system and its secondary barrier or secondary containment around it.

release zone (40CFR227.28) is the area swept out by the locus of points constantly 100 meters from the perimeter of the conveyance engaged in dumping activities, beginning at the first moment in which dumping is scheduled to occur and ending at the last moment in which dumping is scheduled to occur. No release zone shall exceed the total surface area of the dumpsite.

relevant and appropriate requirements (40CFR300.5) means those cleanup standards, standards of control, and other substantive requirements, criteria, or limitations promulgated under federal environmental or state environmental or facility siting laws that, while not "applicable" to a hazardous substance, pollutant, contaminant, remedial action, location, or other circumstance at a CERCLA site, address problems or situations sufficiently similar to those encountered at the CERCLA site that their use is well suited to the particular site. Only those state standards that are identified in a timely manner and are more stringent than federal requirements may be relevant and appropriate.

relevant standard (40CFR63.2) means:
(1) An emission standard;
(2) An alternative emission standard;
(3) An alternative emission limitation; or
(4) An equivalent emission limitation established pursuant to section 112 of the Act that applies to the stationary source, the group of stationary sources, or the portion of a stationary source regulated by such standard or limitation. A relevant standard may include or consist of a design, equipment, work practice, or operational requirement, or other measure, process, method, system, or technique (including prohibition of emissions) that the Administrator (or a State) establishes for new or existing sources to which such standard or limitation applies. Every relevant standard established pursuant to section 112 of the Act includes subpart A of this part and all applicable appendices of this part or of other parts of this chapter that are referenced in that standard.
(Other identical or similar definitions are provided in 40CFR63.101; 63.191.)

relief (40CFR60.701) valve means a valve used only to release an unplanned, nonroutine discharge. A relief valve discharge results from an operator error, a malfunction such as a power failure or equipment failure, or other unexpected cause that requires immediate venting of gas from process equipment in order to avoid safety hazards or equipment damage (under CAA).

relief valve (40CFR61.61-v) means each pressure relief device including pressure relief valves, rupture disks and other pressure relief systems used to protect process components from overpressure conditions. "Relief valve" does not include polymerization shortstop systems, refrigerated water systems or control valves or other devices used to control flow to an incinerator or other air pollution control device.

relief valve (40CFR63.111) means a valve used only to release an unplanned, nonroutine discharge. A

relief valve discharge can result from an operator error, a malfunction such as a power failure or equipment failure, or other unexpected cause that requires immediate venting of gas from process equipment in order to avoid safety hazards or equipment damage.

relief valve discharge (40CFR61.61-y) means any nonleak discharge through a relief valve.

rem (40CFR141.2) means the unit of dose equivalent from ionizing radiation to the total body or any internal organ or organ system. A "millirem (mrem)" is 1/1000 of a rem.

remedial action (40CFR192.01-b) means any action performed under section 108 of the Act.

remedial action (RA) (EPA-94/04): The actual construction or implementation phase of a Superfund site cleanup that follows remedial design.

remedial action plan (CWA118-33USC1268) means a written document which embodies a systematic and comprehensive ecosystem approach to restoring and protecting the beneficial uses of areas of concern, in accordance with article VI and Annex 2 of the Great Lakes Water Quality Agreement.

remedial design (EPA-94/04): A phase of remedial action that follows the remedial investigation/feasibility study and includes development of engineering drawings and specifications for a site cleanup.

remedial design (RD) (40CFR300.5) means the technical analysis and procedures which follow the selection of remedy for a site and result in a detailed set of plans and specifications for implementation of the remedial action.

remedial investigation (EPA-94/04): An in-depth study designed to gather data needed to determine the nature and extent of contamination at a Superfund site; establish site cleanup criteria; identify preliminary alternatives for remedial action; and support technical and cost analyses of alternatives. The remedial investigation is usually done with the feasibility study. Together they are usually referred to as the "RI/FS."

remedial investigation (RI) (40CFR300.5) is a process undertaken by the lead agency to determine the nature and extent of the problem presented by the release. The RI emphasizes data collection and site characterization, and is generally performed concurrently and in an interactive fashion with the feasibility study. The RI includes sampling and monitoring, as necessary, and includes the gathering of sufficient information to determine the necessity for remedial action and to support the evaluation of remedial alternatives.

remedial project manager (RPM) (40CFR300.5) means the official designated by the lead agency to coordinate, monitor, or direct remedial or other response actions under subpart E of the NCP.

remedial project manager (RPM) (EPA-94/04): The EPA or state official responsible for overseeing on-site remedial action.

remedial response (EPA-94/04): Long-term action that stops or substantially reduces a release or threat of a release of hazardous substances that is serious but not an immediate threat to public health.

remediation (EPA-94/04):
1. Cleanup or other methods used to remove or contain a toxic spill or hazardous materials from a Superfund site;
2. for the Asbestos Hazard Emergency Response program, abatement methods including evaluation, repair, enclosure, encapsulation, or removal of greater than 3 linear feet or square feet of asbestos-containing materials from a building.

remedy or remedial action (40CFR300.5) means those actions consistent with permanent remedy taken instead of, or in addition to, removal action in the event of a release or threatened release of a hazardous substance into the environment, to prevent or minimize the release of hazardous substances so that they do not migrate to cause substantial danger to present or future public health or welfare or the environment. The term includes, but is not limited to, such actions at the location of the release as storage, confinement, perimeter protection using dikes, trenches, or ditches, clay cover, neutralization, cleanup of released hazardous substances and associated contaminated materials, recycling or reuse, diversion, destruction, segregation of reactive wastes, dredging or excavations, repair or replacement of leaking containers, collection of leachate and runoff, on-site treatment or incineration, provision of alternative water supplies, any monitoring reasonably required to assure that such actions protect the public health and welfare and the environment and, where appropriate, post-removal site control activities. The term includes the costs of permanent relocation of residents and businesses and community facilities (including the cost of providing "alternative land of equivalent value" to an Indian tribe pursuant to CERCLA section 126(b)) where EPA determines that, alone or in combination with other measures, such relocation is more cost-effective than, and environmentally preferable to, the transportation, storage, treatment, destruction, or secure disposition off-site of such hazardous substances, or may otherwise be necessary to protect the public health or welfare; the term includes off-site transport and off-site storage, treatment, destruction, or secure disposition of hazardous substances and associated contaminated materials. For the purpose of the NCP, the term also includes enforcement activities related thereto.

remedy or remedial action (SP101SP101-42USC9601) means those actions consistent with permanent remedy taken instead of or in addition to removal actions in the event of a release or threatened release of a hazardous substance into the environment, to prevent or minimize the release of hazardous substances so that they do not migrate to cause substantial danger to present or future public health or welfare or the environment. The term includes, but is not limited to, such actions at the location of the release as storage, confinement, perimeter protection using dikes, trenches, or ditches, clay cover, neutralization, cleanup of released hazardous substances and associated contaminated materials, recycling or reuse, diversion, destruction, segregation of reactive wastes, dredging or excavations, repair or replacement of leaking containers, collection of leachate and runoff, onsite treatment or incineration, provision of alternative water supplies, and any monitoring reasonably required to assure that such actions protect the public health and welfare and the environment. The term includes the costs of permanent relocation of residents and businesses and community facilities where the President determines that, alone or in combination with other measures, such relocation is more cost-effective than and environmentally preferable to the transportation, storage, treatment, destruction, or secure disposition offsite of hazardous substances, or may otherwise be necessary to protect the public health or welfare; the term includes offsite transport and offsite storage, treatment, destruction, or secure disposition of hazardous substances and associated contaminated materials.

remined area (CWA301.p-33USC1311) means only that area of any coal remining operation on which coal mining was conducted before the effective date of the Surface Mining Control and Reclamation Act of 1977.

remote sensing (EPA-94/04): The collection and interpretation of information about an object without physical contact with the object; e.g., satellite imaging and aerial photograph.

removal (40CFR117.1): See remove.

removal (40CFR403.7(a)(1)-i) means a reduction in the amount of a pollutant in the POTW's effluent or alteration of the nature of a pollutant during treatment at the POTW. The reduction or alteration can be obtained by physical, chemical or biological means and may be the result of specifically designed POTW capabilities or may be incidental to the operation of the treatment system. Removal as used in this subpart shall not mean dilution of a pollutant in the POTW.

removal (40CFR763.121) means the taking out or stripping of asbestos or materials containing asbestos.

removal (40CFR763.83) means the taking out or the stripping of substantially all ACBM from a damaged area, a functional space, or a homogeneous area in a school building.

removal action (40CFR300-AA) means an action that removes hazardous substances from the site for proper disposal or destruction in a facility permitted under the Resource Conservation and Recovery Act or the Toxic Substances Control Act or by the Nuclear Regulatory Commission.

removal action (EPA-94/04): Short-term immediate actions taken to address releases of hazardous substances that require expedited response (cf. cleanup).

removal costs (OPA1001) means the costs of removal that are incurred after a discharge of oil has occurred or, in any case in which there is a substantial threat of a discharge of oil, the costs to prevent, minimize, or mitigate oil pollution from such an incident.

remove (40CFR61.141) means to take out RACM or facility components that contain or are covered with RACM from any facility.

remove or removal (40CFR109.2-c) refers to the removal of the oil from the water and shorelines or the taking of such other actions as may be necessary to minimize or mitigate damage to the public health or welfare, including, but not limited to, fish, shellfish, wildlife, and public and private property, shorelines, and beaches.

remove or removal (40CFR113.3-i) means the removal of the oil from the water and shorelines or the taking of such other actions as the Federal On-scene Coordinator may determine to be necessary to minimize or mitigate damage to the public health or welfare, including but not limited to, fish, shellfish, wildlife, and public and private property, shorelines, and beaches. Additionally, the terms not otherwise defined herein shall have the meanings assigned them by section 311(a) of the Act.

remove or removal (40CFR117.1-g) refers to removal of the oil or hazardous substances from the water and shoreline or the taking of such other actions as may be necessary to minimize or mitigate damage to the public health or welfare, including, but not limited to, fish, shellfish, wildlife, and public and private property, shorelines, and beaches.

remove or removal (40CFR300.5), as defined by section 311(a)(8) of the CWA, refers to removal of oil or hazardous substances from the water and shorelines or the taking of such other actions as may be necessary to minimize or mitigate damage to the public health or welfare or to the environment. As defined by section 101(23) of CERCLA, remove or removal means the cleanup or removal of released hazardous substances from the environment; such actions as may be necessary taken in the event of threat of release of hazardous substances into the environment; such actions as may be necessary to monitor, assess, and evaluate the release or threat of release of hazardous substances; the disposal of removed material; or the taking of such other actions as may be necessary to prevent, minimize, or mitigate damage to the public health or welfare or to the environment, which may otherwise result from a release or threat of release. The term includes, in addition, without being limited to, security fencing or other measures to limit access, provision of alternative water supplies, temporary evacuation and housing of threatened individuals not otherwise provided for, action taken under section 104(b) of CERCLA, post-removal site control, where appropriate, and any emergency assistance which may be provided under the Disaster Relief Act of 1974. For the purpose of the NCP, the term also includes enforcement activities related thereto.

remove or removal (CWA311-33USC1321) refers to containment and removal of the oil or hazardous substances from the water and shorelines or the taking of such other actions as may be necessary to minimize or mitigate damage to the public health or welfare, including, but not limited to, fish, shellfish, wildlife, and public and private property, shorelines, and beaches.

remove or removal (OPA1001) means containment and removal of oil or a hazardous substance from water and shorelines or the taking of other actions as may be necessary to minimize or mitigate damage to the public health or welfare, including, but not limited to, fish, shellfish, wildlife, and public and private property, shorelines, and beaches.

remove or removal (SP101-42USC9601) means the cleanup or removal of released hazardous substances from the environment, such actions as may be necessary taken in the event of the treat of release of hazardous substances into the environment, such actions as may be necessary to monitor, assess, and evaluate the release or threat of release of hazardous substances, the disposal of removed material, or the taking of such other actions as may be necessary to prevent, minimize, or mitigate damage to the public

health or welfare or to the environment, which may otherwise result from a release or threat of release. The term includes, in addition, without being limited to, security fencing or other measures to limit access, provision of alternative water supplies, temporary evacuation and housing of threatened individuals not otherwise provided for, action taken under section 104(b) of this Act, and any emergency assistance which may be provided under the Disaster Relief Act of 1974.

renderer (40CFR432.101-b) shall mean an independent or off-site rendering operation, conducted separate from a slaughterhouse, packinghouse or poultry dressing or processing plant, which manufactures at rates greater than 75,000 pounds of raw material per day of meat meal, tankage, animal fats or oils, grease, and tallow, and may cure cattle hides, but excluding marine oils, fish meal, and fish oils.

renewable energy (CAA808-42USC7171) means energy from photovoltaic, solar thermal, wind, geothermal, and biomass energy production technologies.

renewal (40CFR70.2) means the process by which a permit is reissued at the end of its term.

renewal system (40CFR797.1330-9) means the technique in which test organisms are periodically transferred to fresh test solution of the same composition.

renewal test (40CFR797.1350-6) is a test without continuous flow of solution, but with occasional renewal of test solutions after prolonged periods, e.g., 24 hours.

renovation (40CFR61.141) means altering a facility or one or more facility components in any way, including the stripping or removal of RACM from a facility component. Operations in which load-supporting structural members are wrecked or taken out are demolitions.

Resilient floor covering (40CFR61.141) means asbestos-containing floor tile, including asphalt and vinyl floor tile, and sheet vinyl floor covering containing more than 1 percent asbestos as determined using polarized light microscopy according to the method specified in appendix A, subpart F, 40CFR763, section 1, Polarized Light Microscopy.

renovation (40CFR763.121) means the modifying of any existing structure, or portion thereof, where exposure to airborne asbestos may result.

repackager (40CFR704.203) means a person who buys a substance identified in subpart D of this part or mixture, removes the substance or mixture from the container in which it was bought, and transfers this substance, as is, to another container for sale.

repair (40CFR280.12) means to restore a tank or UST system component that has caused a release of product from the UST system.

repair (40CFR763.121) means overhauling, rebuilding, reconstructing, or reconditioning of structures or substrates where asbestos is present.

repair (40CFR763.83) means returning damaged ACBM to an undamaged condition or to an intact state so as to prevent fiber release.

repair coatings (40CFR52.741) means coatings used to correct imperfections or damage to furniture surface.

repaired (40CFR52.741) means, for the purpose of paragraph (i) of this section, that equipment component has been adjusted, or otherwise altered, to eliminate a leak.

repaired (40CFR60.481) means that equipment is adjusted, or otherwise altered, in order to eliminate a leak as indicated by one of the following: an instrument reading or 10,000 ppm or greater, indication of liquids dripping, or indication by a sensor that a seal or barrier fluid system has failed (under CAA).

repaired (40CFR61.241) means that equipment is adjusted, or otherwise altered, to eliminate a leak (other identical or similar definitions are provided in 40CFR264.1031).

repaired (40CFR63.161) means that equipment is adjusted, or otherwise altered, to eliminate a leak as defined in the applicable sections of this subpart (under CAA).

repeat compliance period (40CFR141.2) means any subsequent compliance period after the initial compliance period.

repeat compliance period (EPA-94/04): Any subsequent compliance period after the initial one.

replacement (40CFR35.2005-36) means obtaining and installing equipment, accessories, or appurtenances which are necessary during the design or useful life, whichever is longer, of the treatment works to maintain the capacity and performance for which such works were designed and constructed.

replacement (40CFR35.905) means expenditures for obtaining and installing equipment, accessories, or appurtenances which are necessary during the useful life of the treatment works to maintain the capacity and performance for which such works were designed and constructed. The term operation and maintenance includes replacement.

replacement (CWA212-33USC1292) as used in this title means those expenditures for obtaining and installing equipment, accessories, or appurtenances during the useful life of the treatment works necessary to maintain the capacity and performance for which such works are designed and constructed.

replacement cost (40CFR60.481) means the capital needed to purchase all the depreciable components in a facility.

replacement unit (40CFR260.10) means a landfill, surface impoundment, or waste pile unit:
(1) from which all or substantially all of the waste is removed, and
(2) that is subsequently reused to treat, store, or dispose of hazardous waste. "Replacement unit" does not apply to a unit from which waste is removed during closure, if the subsequent reuse solely involves the disposal of waste from that unit and other closing units or corrective action areas at the facility, in accordance with an approved closure plan or EPA or State approved corrective action.

replacement unit (40CFR72.2) means an affected unit replacing the thermal energy provided by an opt-in source, where both the affected unit and the opt-in source are governed by a thermal energy plan.

replicable (40CFR51.491) refers to methods which are sufficiently unambiguous such that the same or equivalent results would be obtained by the application of the methods by different users.

replicate (40CFR797.1600-13) is two or more duplicate tests, samples, organisms, concentrations, or exposure chambers.

reportable quantities (40CFR117.1) means quantities that may be harmful as set forth in 40CFR117.3, the discharge of which is a violation of section 311(b)(3) and requires notice as set forth in 40CFR117.21 (under FWPCA).

reportable quantity (40CFR302.3) means that quantity, as set forth in this part, the release of which requires notification pursuant to this part.

reportable quantity (40CFR355.20) means, for any CERCLA hazardous substance, the reportable quantity established in Table 302.4 of 40CFR302, for such substance, for any other substance, the reportable quantity is one pound.

reportable quantity (RQ) (EPA-94/04): Quantity of a hazardous substance that triggers reports under CERCLA. If a substance exceeds its RQ, the release must be reported to the National Response Center, the SERC, and community emergency coordinators for areas likely to be affected.

reporting agency (40CFR58-AG-b) means the applicable state agency or, in metropolitan areas, a local air pollution control agency designated by the State to carry out the provisions of 40CFR58.40.

reporting area (40CFR58-AG-c) means the geographical area for which the daily index is representative for the reporting period. This area(s) may be the total urban area (or subpart thereof) or each of any number of distinct geographical subregions of the urban area deemed necessary by the reporting agency for adequate presentation of local air quality conditions.

reporting day (40CFR58-AG-d) means the calendar day during which the daily report is given.

reporting period (40CFR58-AG-e) means the time interval for which the daily report is representative. Normally, the reporting period is the 24-hour period immediately preceding the time of the report and should coincide to the extent practicable with the reporting day. In cases where the index will be forecasted the reporting period will include portions of the reporting day for which no monitoring data are available at the time of the report.

reporting period (40CFR704.203) means the time period during which CAIR reporting forms are to be submitted to EPA.

reporting year (40CFR704.203) means the most recent complete corporate fiscal year during which a person manufactures, imports, or processes the listed substance, and which falls within a coverage period identified with a substance in subpart D of this part (under TSCA).

repowering (40CFR51.166-36) means:
(i) replacement of an existing coal-fired boiler with one of the following clean coal technologies: atmospheric or pressurized fluidized bed combustion, integrated gasification combined cycle, magnetohydrodynamics, direct and indirect coal-fired turbines, integrated gasification fuel cells, or as determined by the Administrator, in consultation with the Secretary of Energy, a derivative of one or more of these technologies, and any other technology capable of controlling multiple combustion emissions simultaneously with improved boiler or generation efficiency and with significantly greater waste reduction relative to the performance of technology in widespread commercial use as of November 15, 1990.
(ii) Repowering shall also include any oil and/or gas-fired unit which has been awarded clean coal technology demonstration funding as of January 1, 1991, by the Department of Energy.
(iii) The reviewing authority shall give expedited consideration to permit applications for any source that satisfies the requirements of this subsection and is granted an extension under section 409 of the Clean Air Act.

repowering (40CFR52.21-37) means:
(i) replacement of an existing coal-fired boiler with one of the following clean coal technologies: atmospheric or pressurized fluidized bed combustion, integrated gasification combined cycle, magnetohydrodynamics, direct and indirect coal-fired turbines, integrated gasification fuel cells, or as determined by the Administrator, in consultation with the Secretary of Energy, a derivative of one or more of these technologies, and any other technology capable of controlling multiple combustion emissions simultaneously with improved boiler or generation efficiency and with significantly greater waste reduction relative to the performance of technology in widespread commercial use as of November 15, 1990.
(ii) Repowering shall also include any oil and/or gas-fired unit which has been awarded clean coal technology demonstration funding as of January 1, 1991, by the Department of Energy.
(iii) The Administrator shall give expedited consideration to permit applications for any source that satisfies the requirements of this subsection and is granted an extension under section 409 of the Clean Air Act.

repowering (40CFR60.2) means replacement of an existing coal-fired boiler with one of the following clean coal technologies: atmospheric or pressurized fluidized bed combustion, integrated gasification combined cycle, magnetohydrodynamics, direct and indirect coal-fired turbines, integrated gasification fuel cells, or as determined by the Administrator, in consultation with the Secretary of Energy, a derivative of one or more of these technologies, and any other technology capable of controlling multiple combustion emissions simultaneously with improved boiler or generation efficiency and with significantly greater waste reduction relative to the performance of technology in widespread commercial use as of November 15, 1990. Repowering shall also include any oil and/or gas-fired unit which has been awarded clean coal technology demonstration funding as of January 1, 1991, by the Department of Energy (under CAA).

repowering (CAA402-42USC7651a) means replacement of an existing coal-fired boiler with one of the following clean coal technologies: atmospheric or pressurized fluidized bed combustion, integrated gasification combined cycle, magnetohydrodynamics, direct and indirect coal-fired turbines, integrated gasification fuel cells, or as determined by the Administrator, in consultation with the Secretary of Energy, a derivative of one or more of these technologies, and any other technology capable of controlling multiple combustion emissions simultaneously with improved boiler or generation efficiency and with significantly greater waste reduction relative to the performance of technology in widespread commercial use as of the date of enactment of the Clean Air Act Amendments of 1990. Notwithstanding the provisions of section 409(a), for the purpose of this title, the term "repowering" shall also include any oil and/or gas-fired unit which has been awarded clean coal technology demonstration funding as of January 1, 1991, by the Department of Energy.

repowering (EPA-94/04): Rebuilding and replacing major components of a power plant instead of building a new one.

representative (40CFR27.2) means an attorney who is a member in good standing of the bar of any State, Territory, or possession of the United States or of the District of Columbia or the Commonwealth of Puerto Rico, or other representative who must conform to the standards of conduct and ethics required of practitioners before the courts of the United States.

representative actual annual emissions (40CFR51.165-xxi) means the average rate, in tons per year, at which the source is projected to emit a pollutant for the two-year period after a physical change or change in the method of operation of a unit (or a different consecutive two-year period within 10 years after that change, where the reviewing authority determines that such period is more representative of source operations), considering the effect any such change will have on increasing or decreasing the hourly emissions rate and on projected capacity utilization. In projecting future emissions the reviewing authority shall:
(A) Consider all relevant information, including but not limited to, historical operational data, the company's own representations, filings with the State or Federal regulatory authorities, and

compliance plans under title IV of the Clean Air Act; and

(B) Exclude, in calculating any increase in emissions that results from the particular physical change or change in the method of operation at an electric utility steam generating unit, that portion of the unit's emissions following the change that could have been accommodated during the representative baseline period and is attributable to an increase in projected capacity utilization at the unit that is unrelated to the particular change, including any increased utilization due to the rate of electricity demand growth for the utility system as a whole.

representative actual annual emissions (40CFR51.166-32) means the average rate, in tons per year, at which the source is projected to emit a pollutant for the two-year period after a physical change or change in the method of operation of a unit (or a different consecutive two-year period within 10 years after that change, where the reviewing authority determines that such period is more representative of normal source operations), considering the effect any such change will have on increasing or decreasing the hourly emissions rate and on projected capacity utilization. In projecting future emissions the reviewing authority shall:

(i) Consider all relevant information, including but not limited to, historical operational data, the company's own representations, filings with the State or Federal regulatory authorities, and compliance plans under title IV of the Clean Air Act; and

(ii) Exclude, in calculating any increase in emissions that results from the particular physical change or change in the method of operation at an electric utility steam generating unit, that portion of the unit's emissions following the change that could have been accommodated during the representative baseline period and is attributable to an increase in projected capacity utilization at the unit that is unrelated to the particular change, including any increased utilization due to the rate of electricity demand growth for the utility system as a whole.

representative actual annual emissions (40CFR52.21-33) means the average rate, in tons per year, at which the source is projected to emit a pollutant for the two-year period after a physical change or change in the method of operation of a unit (or a different consecutive two-year period within 10 years after that change, where the Administrator determines that such period is more representative of normal source operations), considering the effect any such change will have on increasing or decreasing the hourly emissions rate and on projected capacity utilization. In projecting future emissions the Administrator shall:

(i) Consider all relevant information, including but not limited to, historical operational data, the company's own representations, filings with the State or Federal regulatory authorities, and compliance plans under title IV of the Clean Air Act; and

(ii) Exclude, in calculating any increase in emissions that results from the particular physical change or change in the method of operation at an electric utility steam generating unit, that portion of the unit's emissions following the change that could have been accommodated during the representative baseline period and is attributable to an increase in projected capacity utilization at the unit that is unrelated to the particular change, including any increased utilization due to the rate of electricity demand growth for the utility system as a whole.

representative actual annual emissions (40CFR52.24-20) means the average rate, in tons per year, at which the source is projected to emit a pollutant for the two-year period after a physical change or change in the method of operation of a unit (or a different consecutive two-year period within 10 years after that change, where the Administrator determines that such period is more representative of normal source operations), considering the effect any such change will have on increasing or decreasing the hourly emissions rate and on projected capacity utilization. In projecting future emissions the Administrator shall:

(i) Consider all relevant information, including but not limited to, historical operational data, the company's own representations, filings with the State or Federal regulatory authorities, and compliance plans under title IV of the Clean Air Act; and

(ii) Exclude, in calculating any increase in emissions that results from the particular physical change or change in the method of operation at an electric utility steam generating unit, that portion of the unit's emissions following the change that could have been accommodated during the representative baseline period and is attributable to an increase in projected capacity utilization at the unit that is unrelated to the particular change, including any increased utilization due to the rate of electricity demand growth for the utility system as a whole.

representative affected facility (40CFR60.531) means an individual wood heater that is similar in all material respects to other wood heaters within the model line it represents.

representative important species (40CFR125.71-b) means species which are representative, in terms of their biological needs, of a balanced, indigenous community of shellfish, fish and wildlife in the body of water into which a discharge of heat is made.

representative of the news media (40CFR2.100-h) refers to any person actively gathering news for an entity that is organized and operated to publish or broadcast news to the public. The term news means information that is about current events or that would be of current interest to the public. Examples of news media entities include television or radio stations broadcasting to the public at large, and publishers of periodicals (but only in those instances when they can qualify as disseminators of news) who make their products available for purchase or subscription by the general public. These examples are not intended to be all-inclusive. Moreover, as traditional methods of news delivery evolve (e.g., electronic dissemination of newspapers through telecommunications services), such alternative media would be included in this category. In the case of freelance journalists, they may be regarded as working for a news organization if they can demonstrate a solid basis for expecting publication through that organization, even though not actually employed by it. A publication contract would be the clearest proof, but EPA may also look to the past publication record of a requestor in making this determination.

representative sample (40CFR260.10) means a sample of a universe or whole (e.g., waste pile, lagoon, ground water) which can be expected to exhibit the average properties of the universe or whole.

representative sample (EPA-94/04): A portion of material or water that is as nearly identical in content and consistency as possible to that in the larger body of material or water being sampled.

request (40CFR2.100-c) means a request to inspect or obtain a copy of one or more records.

request for proposal (40CFR248.4-hh) is a request for an offer by one party to another of terms and conditions with references to some work or undertaking; the initial overture or preliminary statement for consideration by the other party to a proposed agreement.

requester (40CFR403.13-a) means an industrial user or a POTW or other interested person seeking a variance from the limits specified in a categorical Pretreatment Standard.

requestor (40CFR2.100-d) means any person who has submitted a request to EPA.

required activity (40CFR52.30-2) means the submission of a plan or plan item, or the implementation of a plan or plan item.

requirements and standards (40CFR761.123) means:
(1) "Requirements" as used in this policy refers to both the procedural responses and numerical decontamination levels set forth in this policy as constituting adequate cleanup of PCBs.
(2) "Standards" refers to the numerical decontamination levels set forth in this policy.

reregistration (EPA-94/04): The reevaluation and relicensing of existing pesticides originally registered prior to current scientific and regulatory standards. EPA reregisters pesticides through its Registration Standards Program.

research (40CFR26.102-d) means a systematic investigation, including research development, testing and evaluation, designed to develop or contribute to generalizable knowledge. Activities which meet this definition constitute research for purposes of this policy, whether or not they are conducted or supported under a program which is considered research for other purposes. For example, some demonstration and service programs may include research activities.

research and development (10CFR30.4) means:
(1) Theoretical analysis, exploration, or experimentation; or
(2) the extension of investigative findings and theories of a scientific or technical nature into practical application for experimental and demonstration purposes, including the experimental production and testing of models, devices, equipment, materials and processes. "Research and development" as used in this part and parts 31 through 35 does not include the internal or external administration of byproduct material, or the radiation therefrom, to human beings.

research and development (10CFR70.4) means:
(1) theoretical analysis, exploration, or experimentation; or
(2) the extension of investigative findings and theories of a scientific or technical nature into practical application for experimental and demonstration purposes, including the experimental production and testing of models, devices, equipment, materials, and processes (under CAA).

research and development (40CFR82.172) means quantities of a substitute manufactured, imported, or processed or proposed to be manufactured, imported, or processed solely for research and development.

research and development facility (40CFR63.101) means laboratory and pilot plant operations whose primary purpose is to conduct research and development into new processes and products, where the operations are under the close supervision of technically trained personnel, and is not engaged in the manufacture of products for commercial sale, except in a de minimis manner.

research office (CWA118-33USC1268) means the Great Lakes Research Office established by subsection (d).

research subject to regulation (40CFR26.102-e), and similar terms are intended to encompass those research activities for which a federal department or agency has specific responsibility for regulating as a research activity (for example, Investigational New Drug requirements administered by the Food and Drug Administration). It does not include research activities which are incidentally regulated by a federal department or agency solely as part of the department's or agency's broader responsibility to regulate certain types of activities whether research or non-research in nature (for example, Wage and Hour requirements administered by the Department of Labor).

reseller (40CFR80.2-n) means any person who purchases gasoline or diesel fuel identified by the corporate, trade, or brand name of a refiner from such refiner or a distributor and resells or transfers it to retailers or wholesale purchaser-consumers displaying the refiner's brand, and whose assets or facilities are not substantially owned, leased, or controlled by such refiner.

reservation (40CFR131.35.d-12) means all land within the limits of the Colville Indian Reservation, established on July 2, 1872 by Executive Order, presently containing 1,389,000 acres more or less, and under the jurisdiction of the United States government, notwithstanding the issuance of any patent, and including rights-of-way running through the reservation.

reserve (40CFR35.105) means a portion of the State's construction grant allotment which the State proposes to set aside to use for construction or permit program management or water quality management planning activities.

reserve (CAA402-42USC7651a) means any bank of allowances established by the Administrator under this title.

reserve capacity (EPA-94/04): Extra treatment capacity built into solid waste and wastewater treatment plants and interceptor sewers to accommodate flow increases due to future population growth.

reservoir (EPA-94/04): Any natural or artificial

holding area used to store, regulate, or control water (under SDWA).

residence (40CFR61.91-d) means any home, house, apartment building, or other place of dwelling which is occupied during any portion of the relevant year.

residential/commercial areas (40CFR761.123) means those areas where people live or reside, or where people work in other than manufacturing or farming industries. Residential areas include housing and the property on which housing is located, as well as playgrounds, roadways, sidewalks, parks, and other similar areas within a residential community. Commercial areas are typically accessible to both members of the general public and employees and include public assembly properties, institutional properties, stores, office buildings, and transportation centers.

residential parking facility (40CFR52.2486-7) means a parking facility the use of which is limited exclusively to residents (and guests) of a residential building or group of buildings under common control and in which no commercial parking is permitted.

residential property (40CFR201.1-x) means any property that is used for any of the purposes described in the following standard land use codes (ref. Standard Land Use Coding Manual, U.S. DOT/FHWA Washington D.C., reprinted March 1977): 1, Residential; 651, Medical and other Health Services; 68, Educational Services; 691, Religious Activities; and 711, Cultural Activities.

residential solid waste (40CFR243.101-s) means the wastes generated by the normal activities of households, including, but not limited to, food wastes, rubbish, ashes, and bulky wastes.

residential solid waste (40CFR245.101-k) means the garbage, rubbish, trash, and other solid waste resulting from the normal activities of households.

residential solid waste (40CFR246.101-y) means the wastes generated by the normal activities of households, including but not limited to, food wastes, rubbish, ashes, and bulky wastes.

residential tank (40CFR280.12) is a tank located on property used primarily for dwelling purposes (under RCRA).

residential use (40CFR152.3-u) means use of a pesticide directly:
(1) On humans or pets,
(2) In, on, or around any structure, vehicle, article, surface, or area associated with the household, including but not limited to areas such as nonagricultural outbuildings, non-commercial greenhouses, pleasure boats and recreational vehicles, or
(3) In any preschool or day care facility.

residential use (40CFR157.21-e) means use of a pesticide or device:
(1) Directly on humans or pets;
(2) In, on, or around any structure, vehicle, article, surface or area associated with the household, including but not limited to areas such as non-agricultural outbuildings, non-commercial greenhouses, pleasure boats and recreational vehicles; or
(3) In or around any preschool or day care facility (under FIFRA).

residential use (40CFR82.172) means use by a private individual of a chemical substance or any product containing the chemical substance in or around a permanent or temporary household, during recreation, or for any personal use or enjoyment. Use within a household for commercial or medical applications is not included in this definition, nor is use in automobiles, watercraft, or aircraft.

residual (40CFR63.111) means any HAP-containing water or organic that is removed from a wastewater stream by a waste management unit or treatment process that does not destroy organics (nondestructive unit). Examples of residuals from nondestructive wastewater management units are: The organic layer and bottom residue removed by a decanter or organic-water separator and the overheads from a steam stripper or air stripper. Examples of materials which are not residuals are: Silt; mud; leaves; bottoms from a steam stripper or air stripper; and sludges, ash, or other materials removed from wastewater being treated by destructive devices such as biological treatment units and incinerators.

residual (EPA-94/04): Amount of a pollutant remaining in the environment after a natural or technological process has taken place, e.g., the sludge remaining after initial wastewater treatment, or particulates remaining in air after it passes through a scrubbing or other process.

residual risk (EPA-94/04): The extent of health risk from air pollutants remaining after application of the Maximum Achievable Control Technology (MACT). Residue; The dry solids remaining after the evaporation of a sample of water or sludge.

residual disinfectant concentration (C in CT calculations) (40CFR141.2) means the concentration of disinfectant measured in mg/l in a representative sample of water.

residual emissions (40CFR55.2) means the difference in emissions from an OCS source if it applies the control requirements(s) imposed pursuant to 40CFR55.13 or 40CFR55,14 of this part and emissions from that source if it applies a substitute control requirement pursuant to an exemption granted under 40CFR55.7 of this part.

residual oil (40CFR60.41b) means crude oil, fuel oil numbers 1 and 2 that have a nitrogen content greater than 0.05 weight percent, and all fuel oil numbers 4, 5 and 6, as defined by the American Society of Testing and Materials in ASTM D396-78, Standard Specifications for Fuel Oils IBR (see 40CFR60.17).

residual oil (40CFR60.41c) means crude oil that does not comply with the specifications under the definition of distillate oil, and all fuel oil numbers 4, 5, and 6, as defined by the American Society for Testing and Materials in ASTM D396-78, "Standard Specification for Fuel Oils" (incorporated by reference, 40CFR60.17).

residual radioactive material (10CFR40.4) means:
(1) Waste (which the Secretary of Energy determines to be radioactive) in the form of tailings resulting from the processing of ores for the extraction of uranium and other valuable constituents of the ores; and
(2) other waste (which the Secretary of Energy determines to be radioactive) at a processing site

which relates to such processing, including any residual stock of unprocessed ores or low-grade materials. This term is used only with respect to materials at sites subject to remediation under Title I of the Uranium Mill Tailings Radiation Control Act of 1978, as amended.

residue (40CFR240.101-u) means all the solids that remain after completion of thermal processing, including bottom ash, fly ash, and grate siftings (other identical or similar definitions are provided in 40CFR241.101-o).

resilient floor covering (40CFR61.141) means asbestos-containing floor tile, including asphalt and vinyl floor tile, and sheet vinyl floor covering containing more than 1 percent asbestos as determined using polarized light microscopy according to the method specified in appendix E, subpart E, 40CFR part 763, section 1, Polarized Light Microscopy.

resistance (EPA-94/04): For plants and animals, the ability to withstand poor environmental conditions or attacks by chemicals or disease. May be inborn or acquired.

resource conservation (RCRA1004-42USC6903) means reduction of the amounts of solid waste that are generated, reduction of overall resource consumption, and utilization of recovered resources.

Resource Conservation and Recovery Act: See act or RCRA.

resource recovery (EPA-94/04): The process of obtaining matter or energy from materials formerly discarded.

resource recovery (RCRA1004-42USC6903) means the recovery of material or energy from solid waste.

resource recovery facility (40CFR245.101-l) means any physical plant that processes residential, commercial, or institutional solid wastes biologically, chemically, or physically, and recovers useful products, such as shredded fuel, combustible oil or gas, steam, metal, glass, etc. for recycling.

resource recovery facility (40CFR51-AS-19) means

any facility at which solid waste is processed for the purpose of extracting, converting to energy, or otherwise separating and preparing solid waste for reuse. Energy conversion facilities must utilize solid waste to provide more than 50 percent of the heat input to be considered a resource recovery facility under this ruling.

resource recovery facility (RCRA1004-42USC6903) means any facility at which solid waste is processed for the purpose of extracting, converting to energy, or otherwise separating and preparing solid waste for reuse.

resource recovery system (RCRA1004-42USC6903) means a solid waste management system which provides for collection, separation, recycling, and recovery of solid wastes, including disposal of nonrecoverable waste residues.

resource recovery unit (40CFR60.41a-26) means a facility that combusts more than 75 percent non-fossil fuel on a quarterly (calendar) heat input basis (under CAA).

respirable dust (29CFR1910.94a) means airborne dust in sizes capable of passing through the upper respiratory system to reach the lower lung passages.

respond or response (40CFR300.5), as defined by section 101(25) of CERCLA, means remove, removal, remedy, or remedial action, including enforcement activities related thereto.

respond or response (SP101-42USC9601) means remove, removal, remedy, and remedial action, all such terms (including the terms "removal" and "remedial action") include enforcement activities related thereto.

respondent (40CFR164.2-s) means the Assistant Administrator of the Office of Hazardous Materials Control of the Agency.

respondent (40CFR209.3-j) means any person against whom a complaint has been issued under this subpart.

respondent (40CFR22.03) means any person proceeded against in the complaint.

respondent (40CFR32.105-r) means a person against whom a debarment or suspension action has been initiated.

respondent (40CFR8.33-m) means a person against whom sanctions are proposed because of alleged violations of the Executive Order and rules, regulations, and orders thereunder.

response (40CFR300.5): See respond.

response action (40CFR304.12-n) means remove, removal, remedy and remedial action, as those terms are defined by section 101 of CERCLA, 42USC9601, including enforcement activities related thereto.

response action (40CFR35.4010) means all activities undertaken to address the problems created by hazardous substances at a National Priorities List site.

response action (40CFR763.83) means a method, including removal, encapsulation, enclosure, repair, operations and maintenance, that protects human health and the environment from friable ACBM.

response action (EPA-94/04):
1. Generic term for actions taken in response to actual or potential health-threatening environmental events such as spills, sudden releases, and asbestos abatement/management problems;
2. A CERCLA-authorized action involving either a short-term removal action or a long-term removal response. This may include but is not limited to: removing hazardous materials from a site to an EPA-approved hazardous waste facility for treatment, containment or treating the waste on-site, identifying and removing the sources of ground-water contamination and halting further migration of contaminants;
3. Any of the following actions taken in school buildings in response to AHERA to reduce the risk of exposure to asbestos: removal, encapsulation, enclosure, repair, and operations and maintenance (cf. cleanup).

response action (TSCA202-15USC2642) means methods that protect humans health and the environment from asbestos-containing material. Such methods include methods described in chapters 3 and 5 of the Environmental Protection Agency's "Guidance for Controlling Asbestos-Containing Materials in Buildings."

response costs (40CFR304.12-o) means all costs of removal or remedial action incurred and to be incurred by the United States at a facility pursuant to section 104 of CERCLA, 42USC9604, including, but not limited to, all costs of investigation and information gathering, planning and implementing a response action, administration, enforcement, litigation, interest and indirect costs.

response factor (40CFR60-AA (method 21-2.5)) means the ratio of the known concentration of a VOC compound to the observed meter reading when measured using an instrument calibrated with the reference compound specified in the application regulation.

response factor (RF) (40CFR796.1720-v) is the solute concentration required to give a one unit area chromatographic peak or one unit output from the HPLC recording integrator at a particular recorder and detector attenuation. The factor is required to convert from units of area to units of concentration. The determination of the response factor is given in paragraph (b)(3)(ii)(C)(2) of this section.

response factor (RF) (40CFR796.1860-viii) is the solute concentration required to give a one unit area chromatographic peak or one unit output from the HPLC recording integrator at a particular recorder attenuation. The factor is required to convert from units of area to units of concentration. The determination of the response factor is given in paragraph (b)(3)(i)(B)(2) of this section.

response time (40CFR60-AA (method 21-2.7)) means the time interval from a step change in VOC concentration at the input of the sampling system to the time at which 90 percent of the corresponding final value is reached as displayed on the instrument readout meter.

response time (40CFR60-AA (method 25A-2.6)) means the time interval from a step change in pollutant concentration at the inlet to the emission measurement system to the time at which 95 percent

of the corresponding final value is reached as displayed on the recorder.

response time (40CFR60-AA (method 6C-3.8)) means the amount of time required for the measurement system to display 95 percent of a step change in gas concentration on the data recorder (other identical or similar definitions are provided in 40CFR60-AA (method 7E-3.2)).

response time (40CFR60-AB-2.15) means the amount of time it takes the CEMS to display on the data recorder 95 percent of a step change in opacity.

responsible agency (40CFR240.101-v) means the organizational element that has the legal duty to ensure that owners, operators, or users of facilities comply with these guidelines.

responsible agency (40CFR241.101-p) means the organizational element that has the legal duty to ensure that owners, operators or users of land disposal sites comply with these guidelines.

responsible agency (40CFR243.101-t) means the organizational element that has the legal duty to ensure compliance with these guidelines.

responsible agency (40CFR247.101-h) means a department, agency, establishment or instrumentality of the executive branch of the Federal Government or the organizational element within such responsible agency that has the primary responsibility for procurement of materials or products or the preparation of specifications for the procurement of materials or products.

responsible official (40CFR56.1) means the EPA Administrator or any EPA employee who is accountable to the Administrator for carrying out a power or duty delegated under section 301(a)(1) of the Act, or is accountable in accordance with EPA's formal organization for a particular program or function as described in part 1 of this Title.

responsible official (40CFR6.1003) is either the EPA Assistant Administrator or Regional Administrator as appropriate for the particular EPA program. Also, an action "significantly" affects the environment if it does significant harm to the environment even though on balance the action may be beneficial to the environment. To the extent applicable, the responsible official shall address the considerations under 40CFR1508.27 in determining significant effect.

responsible official (40CFR6.501-g) means a Federal or State official authorized to fulfill the requirements of this subpart. The responsible federal official is the EPA Regional Administrator and the responsible State official is as defined in a delegation agreement under 205(g) of the Clean Water Act. The responsibilities of the State official are subject to the limitations in 40CFR6.514 of this subpart.

responsible official (40CFR63.2) means one of the following:

(1) For a corporation: A president, secretary, treasurer, or vice president of the corporation in charge of a principal business function, or any other person who performs similar policy- or decision-making functions for the corporation, or a duly authorized representative of such person if the representative is responsible for the overall operation of one or more manufacturing, production, or operating facilities and either:
  (i) The facilities employ more than 250 persons or have gross annual sales or expenditures exceeding $25 million (in second quarter 1980 dollars); or
  (ii) The delegation of authority to such representative is approved in advance by the Administrator.
(2) For a partnership or sole proprietorship: a general partner or the proprietor, respectively.
(3) For a municipality, State, Federal, or other public agency: either a principal executive officer or ranking elected official. For the purposes of this part, a principal executive officer of a Federal agency includes the chief executive officer having responsibility for the overall operations of a principal geographic unit of the agency (e.g., a Regional Administrator of the EPA).
(4) For affected sources (as defined in this part) applying for or subject to a title V permit: "responsible official" shall have the same meaning as defined in part 70 or Federal title V regulations in this chapter (42 U.S.C. 7661), whichever is applicable.

(Other identical or similar definitions are provided in 40CFR63.101; 63.191; 63.321; 70.2.)

responsible official (40CFR63.51) means one of the following:

(1) For a corporation, a president, secretary, treasurer, or vice-president of the corporation in charge of a principal business function, or any other person who performs similar policy- or decision-making functions for the corporation; or a duly authorized representative of such person if the representative is responsible for the overall operation of one or more manufacturing, production, or operating facilities applying for or subject to a permit and either:
  (i) The facilities employ more than 250 persons or have gross annual sales or expenditures exceeding $25 million (in second quarter 1980 dollars); or
  (ii) The delegation of authority to such representative is approved in advance by the permitting authority.
(2) For a partnership or sole proprietorship, a general partner or the proprietor, respectively.
(3) For a municipality, State, Federal, or other public agency, either a principal executive officer or ranking elected official. For the purposes of this part, a principal executive officer of a Federal agency includes the chief executive officer having responsibility for the overall operations of a principal geographic unit of the agency (e.g., Regional Administrators of EPA).

responsible party (40CFR761.123) means the owner of the PCB equipment, facility, or other source of PCBs or his/her designated agent (e.g., a facility manager or foreman).

responsiveness summary (EPA-94/04): A summary of oral and/or written public comments received by EPA during a comment period on key EPA documents, and EPA's response to those comments.

resting losses (40CFR86.096.2) means evaporative emissions that may occurcontinuously, that are not diurnal emissions, hot soak emissions, running losses, or spitback emissions.

resting losses (40CFR86.098.2) means evaporative emissions that may occur continuously, that are not diurnal emissions, hot soak emissions, refueling emissions, running losses, or spitback emissions.

restoration (EPA-94/04): Measures taken to return a site to pre-violation conditions.

restore (40CFR6-AA-i) means to re-establish a setting or environment in which the natural functions of the floodplain can again operate.

restricted area (10CFR20.3) means any area access to which is controlled by the licensee for purposes of protection of individuals from exposure to radiation and radioactive materials. "Restricted area" shall not include any areas used as residential quarters, although a separate room or rooms in a residential building may be set apart as a restricted area.

restricted data (10CFR70.4) means all data concerning:
(1) design, manufacture or utilization of atomic weapons;
(2) the production of special nuclear material; or
(3) the use of special nuclear material in the production of energy, but shall not include data declassified or removed from the Restricted Data category pursuant to section 142 of the Act.

restricted entry interval (40CFR170.3) means the time after the end of a pesticide application during which entry into the treated area is restricted.

restricted use (EPA-94/04): A pesticide may be classified (under FIFRA regulations) for restricted use if the it requires special handling because of its toxicity, and, if so, it may be applied only by trained, certified applicators or those under their direct supervision.

restricted use pesticide (40CFR171.2-23) means a pesticide that is classified for restricted use under the provisions of section 3(d)(1)(C) of the Act.

restricted use pesticide retail dealer (40CFR171.2) means any person who makes available for use any restricted use pesticide, or who offers to make available for use any such pesticide.

restriction enzymes (EPA-94/04): Enzymes that recognize specific regions of a long DNA molecule and cut it at those points.

retail outlet (40CFR80.2-j) means any establishment at which gasoline, diesel fuel, methanol, natural gas or liquefied petroleum gas is sold or offered for sale for use in motor vehicles.

retail price (NCA15-42USC4914) means:
(A) the maximum statutory price applicable to any type of product; or
(B) in any case where there is no applicable maximum statutory price, the most recent procurement price paid for any type of product.

retailer (40CFR717.3-h) means a person who distributes in commerce a chemical substance, mixture, or article to the ultimate purchasers who are not commercial entities.

retailer (40CFR80.2-k) means any person who owns, leases, operates, controls, or supervises a retail outlet (under CAA).

retailer (40CFR82.104-s) means a person to whom a product is delivered or sold, if such delivery or sale is for purposes of sale or distribution in commerce to consumers who buy such product for purposes other than resale.

retan-wet finish (40CFR425.02-c) means the final processing steps performed on a tanned hide including, but not limited to, the following wet processes: retan, bleach, color, and fat liquor.

retarder (active) (40CFR201.1-y) means a device or system for decelerating rolling rail cars and controlling the degree of deceleration on a car by car basis.

retarder sound (40CFR201.1-z) means a sound which is heard and identified by the observer as that of a retarder, and that causes a sound level meter indicator at fast meter response 40CFR201.1(1) to register an increase of at least ten decibels above the level observed immediately before hearing the sound.

retention chamber (40CFR797.1930-5) means a structure within a flow-through test chamber which confines the test organisms, facilitating observation of test organisms and eliminating loss of organisms in outflow water.

retention chamber (40CFR797.1950-8) means a structure within a flow-through test chamber which confines the test organisms, facilitating observation of test organisms and eliminating washout from test chambers.

retention time, t_R, (40CFR796.1570) is the time in minutes elapsed between sample injection into the chromatograph and the peak maximum (concentration) as recorded on a chromatogram. The retention time is characteristic of the substance, the liquid phase flow rate, and the stationary phase, at a given temperature. With proper flow and temperature control, it can be reproduced to within one percent and used to identify multiple peaks. Although several substances can have nearly identical retention times, each substance has only one retention time. This retention time is not influenced by the presence of other components. Retention times for this method vary between several minutes for substances with a lower K_{ow} to thirty minutes or greater for substances with higher K_{ow}'s (under TSCA).

retread tire (40CFR253.4) means a worn automobile, truck, or other motor vehicle tire whose tread has been replaced.

retrofill (40CFR761.3) means to remove PCB or PCB-contaminated dielectric fluid and to replace it with either PCB, PCB-contaminated, or non-PCB dielectric fluid.

retrofit (40CFR51-AN) means the addition or removal of an item of equipment, or a required adjustment, connection, or disconnection of an existing item of equipment, for the purpose of reducing emissions.

retrofit (40CFR610.11-5) means the addition of a new item, modification or removal of an existing item of equipment beyond that of regular maintenance, on an automobile after its initial manufacture.

retrofit (EPA-94/04): Addition of a pollution control device on an existing facility without making major changes to the generating plant.

retrofit device or device (40CFR610.11-1) means:
(i) Any component, equipment, or other device (except a flow measuring instrument or other driving aid, or lubricant or lubricant additive) which is designed to be installed in or on an automobile as an addition to, as a replacement for, or through alteration or modification of, any original component, or other devices; or
(ii) Any fuel additive which is to be added to the fuel supply of an automobile by means other than fuel dispenser pumps; and
(iii) Which any manufacturer, dealer, or distributor of such device represents will provide higher fuel economy than would have resulted with the automobile as originally equipped, as determined under rules of the Administrator.

retrofitted configuration (40CFR610.11-16) means the test configuration after adjustment of engine calibrations to the retrofit specifications and after all retrofit hardware has been installed.

returnable beverage container (40CFR244.101-m) means a beverage container for which a deposit is paid upon purchase and for which a refund of equal value is payable upon return.

reuse (EPA-94/04): Using a product or component of municipal solid waste in its original form more than once, e.g., refilling a glass bottle that has been returned or using a coffee can to hold nuts and bolts.

reused or used (40CFR261.1-5): See used or reused.

reverberation time (40CFR211.203-cc) is the time that would be required for the mean-square sound pressure level, originally in a steady state, to fall 60 dB after the source is stopped.

reverberatory furnace (40CFR60.121-a) includes the following types of reverberatory furnaces: stationary, rotating, rocking, and tilting (other identical or similar definitions are provided in 40CFR60.131-b) (under CAA).

reverberatory smelting furnace (40CFR60.161-j) means any vessel in which the smelting of copper sulfide ore concentrates or calcines is performed and in which the heat necessary for smelting is provided primarily by combustion of a fossil fuel.

reverse mutation assay in E. coli (40CFR798.5100) detects mutation in a gene of tryptophan requiring strain to produce a tryptophan independent strain of this organism (cf. forward mutation).

reverse mutation assay in salmonella typhimurium (40CFR798.5265-1) detects mutation in a gene of histidine requiring strain to produce a histidine independent strain of this organism.

reverse osmosis (EPA-94/04): A treatment process used in water systems by adding pressure to force water through a semi-permeable membrane. Reverse osmosis removes most drinking water contaminants. Also used in wastewater treatment. Large scale reverse osmosis plants are being developed.

reversible effect (EPA-94/04): An effect which is not permanent; especially adverse effects which diminish when exposure to a toxic chemical is ceased.

review (40CFR2.100-j) refers to the process of examining documents located in response to a request that is for a commercial use (see paragraph (e) of this section) to determine whether any portion of any document located is permitted to be withheld. It also includes processing any documents for disclosure, e.g., doing all that is necessary to excise them and otherwise prepare them for release. Review does not include time spent resolving legal or policy issues regarding the application of exemptions. (Documents must be reviewed in responding to all requests; however, review time may only be charged to commercial use requesters.)

reviewing agency (40CFR63.51) means a State agency with an approved permitting program under Title V of the Act. An EPA Regional Office is the reviewing agency where the State does not have such an approved permitting program.

reviewing official (40CFR27.2) means the General Counsel of the Authority or his designee who is:
(a) Not subject to supervision by, or required to report to, the investigating official;
(b) Not employed in the organizational unit of the Authority in which the investigating official is employed; and
(c) Serving in a position for which the rate of basic pay is not less than the minimum rate of basic pay for grade GS-16 under the General Schedule.

r_f (40CFR796.2700-iii) is the furthest distance traveled by a test material on a thin-layer chromatography plate divided by the distance traveled by a solvent front (arbitrarily set at 10.0 cm in soil TLC studies).

RF (40CFR796.1720-v; 796.1860-viii): See response factor.

RFP baseline (40CFR51.491) means the total of actual volatile organic compounds or nitrogen oxides emissions from all anthropogenic sources in an O_3 nonattainment area during the calendar year 1990 (net of growth and adjusted pursuant to section 182(b)(1)(B) of the Act), expressed as typical O_3 season, weekday emissions (under the Clean Air Act).

ribonucleic acid (RNA) (EPA-94/04): A molecule that carries the genetic message from DNA to a cellular protein-producing mechanisms.

rice (40CFR406.61-b) shall mean rice delivered to a plant before processing.

right-of-way (FLPMA103-43USC1702) includes an easement, lease, permit, or license to occupy, use, or traverse public lands granted for the purpose listed in subchapter V of this chapter.

rill (EPA-94/04): A small channel eroded into the soil surface by runoff; can be easily smoothed out obliterated by normal tillage.

Ringlemann chart (EPA-94/04): A series of shaded illustrations used to measure the opacity of air pollution emissions, ranging from light grey through black; used to set and enforce emissions standards.

riparian habitat (EPA-94/04): Areas adjacent to rivers and streams with a high density, diversity, and productivity of plant and animal species relative to nearby uplands.

riparian rights (EPA-94/04): Entitlement of a land owner to certain uses of water on or bordering his property, including the right to prevent diversion or misuse of upstream waters. Generally a matter of state law.

rise time (40CFR53.23-e): Rise time means interval between initial response and 95 percent of final response after a step decrease in input concentration.

rise time (40CFR85.2122(a)(9)(ii)(D)) means the time required for the spark voltage to increase from 10% to 90% of its maximum value.

risk (EPA-94/04): A measure of the probability that damage to life, health, property, and/or the environment will occur as a result of a given hazard.

risk assessment (EPA-94/04): Qualitative and quantitative evaluation of the risk posed to human health and/or the environment by the actual or potential presence and/or use of specific pollutants.

risk based targeting (EPA-94/04): The direction of resources to those areas that have been identified as having the highest potential or actual adverse effects on human health and/or the environment.

risk characterization (EPA-94/04): This last step in the risk assessment process characterizes the potential for adverse health effects and evaluates the uncertainty involved.

risk communication (EPA-94/04): The exchange of information about health or environmental risks among risk assessors and managers, the general public, news media, interest groups, etc.

risk estimate (EPA-94/04): A description of the probability that organisms exposed to a specific dose of a chemical or other pollutant will develop an adverse response (e.g., cancer)

risk factor (EPA-94/04): Characteristic (e.g., race, sex, age, obesity) or variable (e.g., smoking, occupational exposure level) associated with increased probability of a toxic effect.

risk management (EPA-94/04): The process of evaluating and selecting alternative regulatory and non-regulatory responses to risk. The selection process necessarily requires the consideration of legal, economic, and behavioral factors.

risk retention group (SP401-42USC9671) means any corporation or other limited liability association taxable as a corporation, or as an insurance company, formed under the laws of any State:
(A) whose primary activity consists of assuming and spreading all, or any portion, of the pollution liability of its group members;
(B) which is organized for the primary purpose of conducting the activity described under subparagraph (A);
(C) which is chartered or licensed as an insurance company and authorized to engage in the business of insurance under the laws of any State; and
(D) which does not exclude any person from membership in the group solely to provide for members of such a group a competitive advantage over such a person.

risk specific concentration (40CFR503.41-i) is the allowable increase in the average daily ground level ambient air concentration for a pollutant from the incineration of sewage sludge at or beyond the property line of the site where the sewage sludge incinerator is located.

risk specific dose (EPA-94/04): The dose associated with a specified risklevel.

river basin (EPA-94/04): The land area drained by a river and its tributaries.

rO (40CFR87.1): See rated output.

roadways (40CFR61.141) means surfaces on which vehicles travel. This term includes public and private highways, roads, streets, parking areas, and driveways (under CAA).

roaster (40CFR60.161-c) means any facility in which a copper sulfide ore concentrate charge is heated in the presence of air to eliminate a significant portion (5 percent or more) of the sulfur contained in the charge.

roaster (40CFR60.171-b) means any facility in which a zinc sulfide ore concentrate charge is heated in the presence of air to eliminate a significant portion (more than 10 percent) of the sulfur contained in the charge.

roasting (40CFR61.181) means the use of a furnace to heat arsenic plant feed material for the purpose of eliminating a significant portion of the volatile materials contained in the feed.

rock crushing and gravel washing facilities (40CFR122.27-2) means facilities which process crushed and broken stone, gravel, and riprap (see 40CFR436, subpart B, including the effluent limitations guidelines).

rock wool insulation (40CFR248.4-ii) means insulation which is composed principally from fibers manufactured from slag or natural rock, with or without binders.

rocket motor test site (40CFR61.41-a) means any building, structure, facility, or installation where the static test firing of a beryllium rocket motor and/or the disposal of beryllium propellant is conducted (under CAA).

rod, wire and coil (40CFR420.91-k) means those acid pickling operations that pickle rod, wire or coiled rod and wire products.

rodenticide (EPA-94/04): A chemical or agent used to destroy rats or other rodent pests, or to prevent them from damaging food, crops, etc.

roentgen (R) (40CFR300-AA) means a measure of external exposures to ionizing radiation. One roentgen equals that amount of x-ray or gamma radiation required to produce ions carrying a charge of 1 electrostatic unit (esu) in 1 cubic centimeter of dry air under standard conditions. One microroentgen (μR) equals 10^{-6}R.

roll bonding (40CFR471.02-ff) is the process by which a permanent bond is created between two metals by rolling under high pressure in a bonding mill (co-rolling).

roll coater (40CFR52.741) means an apparatus in which a uniform layer of coating is applied by means of one or more rolls across the entire width of a moving substrate.

roll printer (40CFR52.741) means an apparatus used in the application of words, designs, or pictures to a substrate, usually by means of one or more rolls each with only partial coverage.

roll printing (40CFR52.741) means the application of words, designs, and pictures to a substrate usually by means of a series of hard rubber or metal rolls each with only partial coverage.

rollboard (40CFR763.163) means an asbestos-containing product made of paper that is produced in a continuous sheet, is flexible, and is rolled to achieve a desired thickness. Asbestos rollboard consists of two sheets of asbestos paper laminated together. Major applications of this product include: office partitioning; garage paneling; linings for stoves and electric switch boxes; and fire-proofing agent for security boxes, safes, and files.

roller coating (40CFR52.741) means a method of applying a coating to a sheet or strip in which the coating is transferred by a roller or series of rollers.

rolling (40CFR467.02-p) is the reduction in thickness or diameter of a workpiece by passing it between lubricated steel rollers. There are two subcategories based on the rolling process. In the rolling with neat oils subcategory, pure or neat oils are used as lubricants for the rolling process. In the rolling with emulsions subcategory, emulsions are used as lubricants for the rolling process.

rolling (40CFR468.02-o) shall mean the reduction in the thickness or diameter of a workpiece by passing it between rollers.

rolling (40CFR471.02-ee) is the reduction in thickness or diameter of a workpiece by passing it between lubricated steel rollers.

roof monitor (40CFR60.191) means that portion of the roof of a potroom where gases not captured at the cell exit from the potroom.

room enclosure (40CFR63.321) means a stationary structure that encloses a transfer machine system, and is vented to a carbon adsorber or an equivalent control device during operation of the transfer machine system.

root crops (40CFR257.3.5-8) means plants whose

edible parts are grown below the surface of the soil (under RCRA).

rotary blast cleaning table (29CFR1910.94a) means an enclosure where the pieces to be cleaned are positioned on a rotating table and are passed automatically through a series of blast sprays.

rotary kiln incinerator (EPA-94/04): An incinerator with a rotating combustion chamber that keeps waste moving, thereby allowing it to vaporize for easier burning.

rotary lime kiln (40CFR60.341-d) means a unit with an inclined rotating drum that is used to produce a lime product from limestone by calcination.

rotary spin (40CFR60.681) means a process used to produce wool fiberglass insulation by forcing molten glass through numerous small orifices in the side wall of a spinner to form continuous glass fibers that are then broken into discrete lengths by high velocity air flow.

rotogravure print station (40CFR60.581) means any device designed to print or coat inks on one side of a continuous web or substrate using the intaglio printing process with a gravure cylinder.

rotogravure printing (40CFR52.741) means the application of words, designs, and pictures to a substrate by means of a roll printing technique in which the pattern to be applied is recessed relative to the non-image area.

rotogravure printing line (40CFR52.741) means a printing line in which each roll printer uses a roll with recessed areas for applying an image to a substrate (under CAA).

rotogravure printing line (40CFR60.581) means any number of rotogravure print stations and associated dryers capable of printing or coating simultaneously on the same continuous vinyl or urethane web or substrate, which is fed from a continuous roll.

rotogravure printing unit (40CFR60.431) means any device designed to print one color ink on one side of a continuous web or substrate using a gravure cylinder.

rough fish (EPA-94/04): Fish not prized for eating, such as gar and suckers. Most are more tolerant of changing environmental conditions than game species.

rounded (40CFR600.002.85-30) means a number shortened to the specific number of decimal places in accordance with the "Round Off Method" specified in ASTM E 29-67.

route of exposure (EPA-94/04): The avenue by which a chemical comes into contact with an organism (e.g., inhalation, ingestion, dermal contact, injection.)

routine maintenance area (40CFR763.83) means an area, such as a boiler room or mechanical room, that is not normally frequented by students and in which maintenance employees or contract workers regularly conduct maintenance activities.

routine release (40CFR302.8) is a release that occurs during normal operating procedures or processes.

routine use (40CFR1516.2) means with respect to the disclosure of a record, the use of such record for a purpose which is compatible with the purpose for which it was collected.

routine use (40CFR16.2-a): See individual, maintain, record, system of records, and routine use.

RPR (40CFR87.1): See rated pressure ratio.

rubbish (40CFR243.101-u) means a general term for solid waste, excluding food wastes and ashes, taken from residences, commercial establishments, and institutions.

rubbish (EPA-94/04): Solid waste, excluding food waste and ashes, from homes, institutions, and work-places.

run off (EPA-94/04): That part of precipitation, snow melt, or irrigation water that runs off the land into streams or other surface-water. It can carry pollutants from the air and land into receiving waters.

running losses (EPA-94/04): Evaporation of motor

vehicle fuels from the fuel tank while the vehicle is in use.

rule compliance factor (40CFR51.491) means a factor applied to discount the amount of emissions reductions credited in an implementation plan demonstration to account for less-than-complete compliance by the affected sources in an EIP.

rules, regulations, and relevant orders of the Secretary of Labor used in both paragraph (4) of the equal opportunity clause and elsewhere herein (40CFR8.2-u) means rules, regulations, and relevant orders of the Secretary of Labor or his designee issued pursuant to the Order.

run (40CFR52.1881-x) means the net period of time during which an emission sample is collected. Unless otherwise specified, a run may be either intermittent or continuous within the limits of good engineering practice as determined by the Administrator.

run (40CFR60.2) means the net period of time during which an emission sample is collected. Unless otherwise specified, a run may be either intermittent or continuous within the limits of good engineering practice (other identical or similar definitions are provided in 40CFR61.02).

run (40CFR61.61-n) means the net period of time during which an emission sample is collected.

run (40CFR63.2) means one of a series of emission or other measurements needed to determine emissions for a representative operating period or cycle as specified in this part (other identical or similar definitions are provided in 40CFR63.101; 63.191).

run (40CFR63.301) means the observation of visible emissions from topside port lids, offtake systems, coke oven doors, or the charging of a coke oven that is made in accordance with and is valid under Methods 303 or 303A in appendix A to this part.

run-of-pile triple superphosphate (40CFR60.231-b) means any triple superphosphate that has not been processed in a granulator and is composed of particles at least 25 percent by weight of which (when not caked) will pass through a 16 mesh screen.

run-off (40CFR241.101-q) means the portion of precipitation that drains from an area as surface flow.

run-off (40CFR258.2) means any rainwater, leachate, or other liquid that drains over land from any part of a facility.

run-off (40CFR260.10) means any rainwater, leachate, or other liquid that drains over land from any part of a facility.

run-off coefficient (40CFR122.26-11) means the fraction of total rainfall that will appear at a conveyance as runoff.

run-on (40CFR258.2) means any rainwater, leachate, or other liquid that drains over land onto any part of a facility.

run-on (40CFR260.10) means any rainwater, leachate, or other liquid that drains over land onto any part of a facility.

running changes (40CFR85.1502-12) are those changes in vehicle or engine configuration, equipment or calibration which are made by an OEM or ICI in the course of motor vehicle or motor vehicle engine production.

running loss (40CFR86.082.2) means fuel evaporative emissions resulting from an average trip in an urban area or the simulation of such a trip.

running losses (40CFR86.096.2) means evaporative emissions that occur during vehicle operation.

runoff (40CFR419.11-b) shall mean the flow of storm water resulting from precipitation coming into contact with petroleum refinery property.

runoff (40CFR503.9-v) is rainwater, leachate, or other liquid that drains overland on any part of a land surface and runs off of the land surface (under CWA).

rupture of a PCB transformer (40CFR761.3) means a violent or non-violent break in the integrity of a PCB transformer caused by an overtemperature and/or overpressure condition that results in the release of PCBs.

rural transport ozone nonattainment area (40CFR51.392) means an ozone nonattainment area that does not include, and is not adjacent to, any part of a Metropolitan Statistical Area or, where one exists, a Consolidated Metropolitan Statistical Area (as defined by the United States Bureau of the Census) and is classified under Clean Air Act section 182(h) as a rural transport area (other identical or similar definitions are provided in 40CFR93.101) (under CAA).

*********** SSSSS ***********

S/cm^3 (40CFR763-AA-22) means structures per cubic centimeter.

S/mm^2 (40CFR763-AA-23) means structures per square millimeter.

S1S (40CFR429.11-g): See smooth-one-side.

S2S (40CFR429.11-h): See smooth-two-sides.

sacrificial anode (EPA-94/04): An easily corroded material deliberately installed in a pipe or take to give it up (sacrifice) to corrosion while the rest of the water supply facility remains relatively corrosion free.

safe (EPA-94/04): Condition of exposure under which there is a practical certainty that no harm will result to exposed individuals.

safe disposal (40CFR165.1-s) means discarding pesticides or containers in a permanent manner so as to comply with these proposed procedures and so as to avoid unreasonable adverse effects on the environment.

Safe Drinking Water Act: See act or SDWA.

safe water (EPA-94/04): Water that does not contain harmful bacteria, toxic materials, or chemicals and is considered safe for drinking even though it may have taste, and odor, color and certain mineral problems.

safe yield (EPA-94/04): The annual amount of water that can be taken from a source or supply over a period of years without depleting that source beyond its ability to be replenished naturally in "wet years."

safener (EPA-94/04): A chemical added to a pesticide to keep it from injuring plants.

safety relief valve (40CFR52.741) means a valve which is normally closed and which is designed to open in order to relieve excessive pressures within a vessel or pipe.

salad dressings (40CFR407.81-i): See mayonnaise and salad dressings.

sale (40CFR60.531) means the transfer of ownership or control, except that transfer of control shall not constitute a sale for purposes of section 60.530(f) (under CAA).

sale for purposes other than resale (40CFR761.3) means sale of PCBs for purposes of disposal and for purposes of use, except where use involves sale for distribution in commerce. PCB equipment which is first leased for purposes of use any time before July 1, 1979, will be considered sold for purposes other than resale.

sales (40CFR88.202.94) means vehicles that are produced, sold, and distributed (in accordance with normal business practices and applicable franchise agreements) in the State of California, including owners of covered fleets under subpart C of part 86 of this chapter. The manufacturer can choose at their option from one of the following three methods for determining sales:
(i) Sales is defined as sales to the ultimate purchaser.
(ii) Sales is defined as vehicle sales by a manufacturer to a dealer, distributer, fleet operator, broker, or any other entity which comprises the first point of sale.
(iii) Sales is defined as equivalent to the production of vehicles for the state of California. This option can be revoked if it is determined that the production and actual sales numbers do not

exhibit a functional equivalence per the language of 40CFR86.708-94(b)(1) of this chapter.

saline estuarine waters (40CFR125.58-q) means those semi-enclosed coastal waters which have a free connection to the territorial sea, undergo net seaward exchange with ocean waters, and have salinities comparable to those of the ocean. Generally, these waters are near the mouth of estuaries and have cross-sectional annual mean salinities greater than twenty-five (25) parts per thousand.

salt bath descaling, oxidizing (40CFR420.81-a) means the removal of scale from semi-finished steel products by the action of molten salt baths other than those containing sodium hydride.

salt bath descaling, reducing (40CFR420.81-b) means the removal of scale from semi-finished steel products by the action of molten salt baths containing sodium hydride.

salt water intrusion (EPA-94/04): The invasion of fresh surface or ground water by salt water. If it comes from the ocean it may be called sea water intrusion.

salts (EPA-94/04): Minerals that water picks up as it passes through the air, over and under the ground, or from households and industry.

salvage (EPA-94/04): The utilization of waste materials.

salvaging (40CFR241.101-r) means the controlled removal of waste materials for utilization.

same location (40CFR60.51a) means the same or contiguous property that is under common ownership or control, including properties that are separated only by a street, road, highway, or other public right-of-way. Common ownership or control includes properties that are owned, leased, or operated by the same entity, parent entity, subsidiary, subdivision, or any combination thereof, including any municipality or other governmental unit, or any quasi-governmental authority (e.g., a public utility district or regional waste disposal authority).

sample interval (40CFR60-AA(alt. method 1)) means

the time period between successive samples for a digital signal or between successive measurements for an analog signal.

sample loop (40CFR796.1720-iv) is a 1/16 in. O.D. (1.6 mm) stainless steel tube with an internal volume between 20 and 50 uL. The loop is attached to the sample injection valve of the HPLC and is used to inject standard solutions into the mobile phase of the HPLC when determining the response factor for the recording integrator. The exact volume of the loop must be determined as described in paragraph (b)(3)(ii)(C)(1) of this section when the HPLC method is used.

sample loop (40CFR796.1860-vii) is a 1/16 in. O.D. (1.6 mm) stainless steel tube with an internal volume between 20 and 50 uL (micro liter). The loop is attached to the sample injection valve of the HPLC and is used to inject standard solutions into the mobile phase of the HPLC when determining the response factor for the recording integrator. The exact volume of the loop must be determined as described in paragraph (b)(3)(i)(B)(1) of this section when the HPLC method is used.

sample quantitation limit (SQL) (40CFR300-AA) means the quantity of a substance that can be reasonably quantified given the limits of detection for the methods of analysis and sample characteristic that may affect quantitation (for example, dilution, concentration).

sample system (40CFR87.1) means the system which provides for the transportation of the gaseous emission sample from the sample probe to the inlet of the instrumentation system.

sampling area (40CFR763.103-f) means any area, whether contiguous or not, within a school building which contains friable material that is homogeneous in texture and appearance.

sampling effectiveness (40CFR53.43-a): The ratio (expressed as a percentage) of the mass concentration of particles of a given size reaching the sampler filter or filters to the mass concentration of particles of the same size approaching the sampler.

sampling system bias (40CFR60-AA (method 6C-

3.5)) means the difference between the gas concentrations exhibited by the measurement system when a known concentration gas is introduced at the outlet of the sampling probe and when the same gas is introduced directly to the analyzer (other identical or similar definitions are provided in 40CFR60-AA (method 7E-3.2)).

sanctions (EPA-94/04): Actions taken by the federal government for failure to plan or implement a State Improvement Plan (SIP). Such action may include withholding of highway funds and a ban on construction of new sources of potential pollution.

sand filters (EPA-94/04): Devices that remove some suspended solids from sewage. Air and bacteria decompose additional wastes filtering through the sand so that cleaner water drains from the bed.

sanding sealers (40CFR52.741) means any coatings formulated for and applied to bare wood for sanding and to seal the wood for subsequent application of varnish. To be considered a sanding sealer a coating must be clearly labelled as such.

sanitary landfill (40CFR165.1-t) means a disposal facility employing an engineered method of disposing of solid wastes on land in a manner which minimizes environmental hazards by spreading the solid wastes in thin layers, compacting the solid wastes to the smallest practical volume, and applying cover material at the end of each working day. Such facility complies with the Agency Guidelines for the Land Disposal of Solid Wastes as prescribed in 40CFR241.

sanitary landfill (40CFR240.101-w) means a land disposal site employing an engineered method of disposing of solid wastes on land in a manner that minimizes environmental hazards by spreading the solid wastes in thin layers, compacting the solid wastes to the smallest practical volume, and applying and compacting cover material at the end of each operating day (other identical or similar definitions are provided in 40CFR242.101-s).

sanitary landfill (40CFR257.2) means a facility for the disposal of solid waste which complies with this part.

sanitary landfill (EPA-94/04): See landfills.

sanitary landfill (RCRA1004-42USC6903) means a facility for the disposal of solid waste which meets the criteria published under section 4004.

sanitary sewer (40CFR35.2005-37) means a conduit intended to carry liquid and water-carried wastes from residences, commercial buildings, industrial plants and institutions together with minor quantities of ground, storm and surface waters that are not admitted intentionally.

sanitary sewer (40CFR35.905) means a sewer intended to carry only sanitary or sanitary and industrial waste waters from residences, commercial buildings, industrial plants, and institutions.

sanitary sewers (EPA-94/04): Underground pipes that carry off only domestic or industrial waste, not storm water.

sanitary survey (EPA-94/04): An on-site review of the water sources, facilities, equipment, operation and maintenance of a public water system to evaluate the adequacy of those elements for producing and distributing safe drinking water.

sanitary water (also known as gray water) (EPA-94/04): Water discharged from sinks, showers, kitchens, or other nonindustrial operations, but not from commodes.

sanitation (EPA-94/04): Control of physical factors in the human environment that could harm development, health, or survival.

sanitary survey (40CFR141.2) means an onsite review of the water source, facilities, equipment, operation and maintenance of a public water system for the purpose of evaluating the adequacy of such source, facilities, equipment, operation and maintenance for producing and distributing safe drinking water (other identical or similar definitions are provided in 40CFR142.2).

sanitary waste (40CFR435.11-u) shall refer to human body waste discharged from toilets and urinals located within facilities subject to this subpart.

sanitized (40CFR350.1) means a version of a document from which information claimed as trade secret or confidential has been omitted or withheld (cf. unsanitized).

saprophytes (EPA-94/04): Organisms living on dead or decaying organic matter that help natural decomposition of organic matter in water.

SARA: See act or SARA.

SAROAD site identification form (40CFR58.1-q) is one of the several forms in the SAROAD system. It is the form which provides a complete description of the site (and its surroundings) of an ambient air quality monitoring station.

satellite vehicle (40CFR243.101-v) means a small collection vehicle that transfers its load into a larger vehicle operating in conjunction with it.

saturated solution (40CFR796.1840-vi) is a solution in which the dissolved solute is in equilibrium with an excess of undissolved solute; or a solution in equilibrium such that at a fixed temperature and pressure, the concentration of the solution is at its maximum value and will not change even in the presence of an excess of solute.

saturated solution (40CFR796.1860-iii) is a solution in which the dissolved solute is in equilibrium with an excess of undissolved solute; or a solution in equilibrium such that at a fixed temperature and pressure, the concentration of the solute in the solution is at its maximum value and will not change even in the presence of an excess of solute.

saturated zone (40CFR258.2) means that part of the earth's crust in which all voids are filled with water.

saturated zone (EPA-94/04): The area below the water table where all open spaces are filled with water.

saturated zone or zone of saturation (40CFR260.10) means that part of the earth's crust in which all voids are filled with water.

saturation (EPA-94/04): The condition of a liquid (water) when it has taken into solution the maximum possible quantity of a given substance at a given temperature and pressure.

saturator (40CFR60.471) means the equipment in which asphalt is applied to felt to make asphalt roofing products. The term saturator includes the saturator, wet looper, and coater.

sauerkraut canning (40CFR407.71-p) shall mean the draining and subsequent filling and canning of fermented cabbage and juice.

sauerkraut cutting (40CFR407.71-o) shall mean the trimming, cutting, and subsequent preparatory handling of cabbage necessary for and including brining and fermentation, and subsequent tank soaking.

sausage and luncheon meat processor (40CFR432.71-b) shall mean an operation which cuts fresh meats, grinds, mixes, seasons, smokes or otherwise produces finished products such as sausage, bologna and luncheon meats at rates greater than 2730 kg (6000 lb) per day.

sawing (40CFR471.02-gg) is cutting a workpiece with a band, blade, or circular disc having teeth.

scarfing (40CFR420.71-f) means those steel surface conditioning operations in which flames generated by the combustion of oxygen and fuel are used to remove surface metal imperfections from slabs, billets, or blooms.

scavenging (40CFR241.101-t) means uncontrolled removal of solid waste materials.

scavenging (40CFR243.101-w) means the uncontrolled and unauthorized removal of materials at any point in the solid waste management system.

SCF (40CFR464.02-i): Standard cubic feet.

schedule and timetable of compliance (CAA302-42USC7602) means a schedule of required measures including an enforceable sequence of actions or operations leading to compliance with an emission limitation, other limitation, prohibition, or standard.

schedule of compliance (40CFR122.2) means a schedule of remedial measures included in a "permit," including an enforceable sequence of interim requirements (for example, actions, operations, or milestone events) leading to compliance with the CWA and regulations.

schedule of compliance (40CFR124.2) means a schedule of remedial measures included in a "permit," including an enforceable sequence of interim requirements (for example, actions, operations, or milestone events) leading to compliance with the "appropriate Act and regulations" (other identical or similar definitions are provided in 40CFR144.3).

schedule of compliance (40CFR270.2) means a schedule of remedial measures included in a permit, including an enforceable sequence of interim requirements (for example, actions, operations, or milestone events) leading to compliance with the Act and regulations.

schedule of compliance (40CFR401.11-m) shall be defined in accordance with section 502 of the Act unless the context otherwise requires.

schedule of compliance (40CFR72.2) means an enforceable sequence of actions, measures, or operations designed to achieve or maintain compliance, or correct non-compliance, with an applicable requirement of the Acid Rain Program, including any applicable Acid Rain permit requirement.

schedule of compliance (CAA501-42USC7661) means a schedule of remedial measures, including an enforceable sequence of actions or operations, leading to compliance with an applicable implementation plan, emission standard, emission limitation, or emission prohibition.

schedule of compliance (CWA502-33USC1362) means a schedule of remedial measures including an enforceable sequence of actions or operations leading to compliance with an effluent limitation, other limitation, prohibition, or standard.

schedule of reinforcement (40CFR798.6500-3) specifies the relation between behavioral responses and the delivery of reinforcers, such as food or water.

scheduled maintenance (40CFR57.103-s) means any periodic procedure, necessary to maintain the integrity or reliability of emissions control performance, which can be anticipated and scheduled in advance. In sulfuric acid plants, it includes among other items the screening or replacement of catalyst, the re-tubing of heat exchangers, and the routine repair and cleaning of gas handling/cleaning equipment.

scheduled maintenance (40CFR85.1402) means those maintenance events required by the equipment certifier in order to ensure that the retrofitted engine will maintain its emissions performance over the in-use compliance period.

scheduled maintenance (40CFR86.082.2) means any adjustment, repair, removal, disassembly, cleaning, or replacement of vehicle components or systems which is performed on a periodic basis to prevent part failure or vehicle (if the engine were installed in a vehicle) malfunction.

scheduled maintenance (40CFR86.084.2) means any adjustment, repair, removal, disassembly, cleaning, or replacement of vehicle components or systems which is performed on a periodic basis to prevent part failure or vehicle (if the engine were installed in a vehicle) malfunction, or anticipated as necessary to correct an overt indication of vehicle malfunction or failure for which periodic maintenance is not appropriate. This definition applies beginning with the 14 model year.

scheduled maintenance (40CFR86.402.78) means any adjustment, repair, removal, disassembly, cleaning, or replacement of vehicle components or systems which is performed on a periodic basis to prevent part failure or vehicle malfunction, or anticipated as necessary to correct an overt indication of vehicle malfunction or failure for which periodic maintenance is not appropriate.

school (40CFR763.103-g) means any public or private day or residential school that provides elementary or secondary education for grade 12 or under as determined under State law, or any school of any Agency of the United States.

school (TSCA202-15USC2642) means any elementary or secondary school as defined in section 198 of the Elementary and Secondary Education Act of 1965 (20USC2854) (other identical or similar definitions are provided in 40CFR763.83).

school building (40CFR763.83) means:
(1) Any structure suitable for use as a classroom, including a school facility such as a laboratory, library, school eating facility, or facility used for the preparation of food.
(2) Any gymnasium or other facility which is specially designed for athletic or recreational activities for an academic course in physical education.
(3) Any other facility used for the instruction or housing of students or for the administration of educational or research programs.
(4) Any maintenance, storage, or utility facility, including any hallway, essential to the operation of any facility described in this definition of "school building" under paragraphs (1), (2), or (3).
(5) Any portico or covered exterior hallway or walkway.
(6) Any exterior portion of a mechanical system used to condition interior space (under the Toxic Substances Control Act).

school building (TSCA202-15USC2642) means:
(1) Any structure suitable for use as a classroom, including a school facility such as a laboratory, library, school eating facility, or facility used for the preparation of food.
(2) Any gymnasium or other facility which is specially designed for athletic or recreational activities for an academic course in physical education.
(3) Any other facility used for the instruction or housing of students or for the administration of educational or research programs.
(4) Any maintenance, storage, or utility facility, including any hallway, essential to the operation of any facility described in this definition of "school building" under paragraphs (1), (2), or (3).

school building (TSCA302-15USC2662) has the meaning given such term by section 202(13).

school buildings (40CFR763.103-h) means:
(1) Structures used for the instruction of school children, including classrooms, laboratories, libraries, research facilities and administrative facilities.
(2) School eating facilities, and school kitchens.
(3) Gymnasiums or other facilities used for athletic or recreational activities, or for courses in physical education.
(4) Dormitories or other living areas of residential schools.
(5) Maintenance, storage, or utility facilities essential to the operation of the facilities described in paragraphs (h)(1) through (4) of this section.

science advisory board (SAB) (EPA-94/04): A group of external scientists who advise EPA on science and policy.

scope (environmental impact statement) (40CFR1508.25) consists of the range of actions, alternatives, and impacts to be considered in an environmental impact statement. The scope of an individual statement may depend on its relationship to other statements (40CFR1502.20 and 1508.28). To determine the scope of environmental impact statements, agencies shall consider 3 types of actions, 3 of types of alternatives, and 3 types of impacts. They include:
(a) Actions (other than unconnected single actions) which may be:
  (1) Connected actions, which means that they are closely related and therefore should be discussed in the same impact statement. Actions are connected if they:
    (i) Automatically trigger other actions which may require environmental impact statements.
    (ii) Cannot or will not proceed unless other actions are taken previously or simultaneously.
    (iii) Are interdependent parts of a larger action and depend on the larger action for their justification.
  (2) Cumulative actions, which when viewed with other proposed actions have cumulatively significant impacts and should therefore be discussed in the same impact statement.
  (3) Similar actions, which when viewed with other reasonably foreseeable or proposed agency actions, have similarities that provide a basis for evaluating their

environmental consequences together, such as common timing or geography. An agency may wish to analyze these actions in the same impact statement. It should do so when the best way to assess adequately the combined impacts of similar actions or reasonable alternatives to such actions is to treat them in a single impact statement.

(b) Alternatives, which include:
  (1) No action alternative.
  (2) Other reasonable courses of actions.
  (3) Mitigation measures (not in the proposed action).

(c) Impacts, which my be:
  (1) Direct;
  (2) Indirect;
  (3) Cumulative.

scope of work (40CFR6.901-c) means a document similar in content to the program of requirements but substantially abbreviated. It is usually prepared for small-scale projects.

scrap (EPA-94/04): Materials discarded from manufacturing operations that may be suitable for reprocessing.

scrap metal (40CFR261.1-6) is bits and pieces of metal parts (e.g., bars, turnings, rods, sheets, wire) or metal pieces that may be combined together with bolts or soldering (e.g., radiators, scrap automobiles, railroad box cars), which when worn or superfluous can be recycled.

scratch brush wheels (29CFR1910.94b) means all power-driven rotatable wheels made from wire or bristles, and used for scratch cleaning and brushing purposes.

screen (40CFR60.381) means a device for separating material according to size by passing undersize material through one or more mesh surfaces (screens) in series and retaining oversize material on the mesh surfaces (screens).

screening (EPA-94/04): Use of screens to remove coarse floating and suspended solids from sewage.

screening concentration (40CFR300-AA) means media-specific benchmark concentration for a hazardous substance that is used in the HRS for comparison with the concentration of that hazardous substance in a sample from that media. The screening concentration for a specific hazardous substance corresponds to its reference dose of inhalation exposures or for oral exposures, as appropriate, and, if the substance is a human carcinogen with a weight-of-evidence classification of A, B, or C, to that concentration that corresponds to its 10^{-6} individual lifetime excess cancer risk for inhalation exposures or for oral exposures, as appropriate.

screening operation (40CFR60.671) means a device for separating material according to size by passing undersize material through one or more mesh surfaces (screens) in series, and retaining oversize material on the mesh surfaces (screens).

screwed connector (40CFR63.161) means a threaded pipe fitting where the threads are cut on the pipe wall and the fitting requires only two pieces to make the connection (i.e., the pipe and the fitting).

scrubber (EPA-94/04): An air pollution device that uses a spray of water or reactant or a dry process to trap pollutants in emissions.

scrubbing (40CFR165.1-u) means the washing of impurities from any process gas stream.

SDWA: See act or SDWA.

seafood (40CFR408.11) shall mean the raw material, including freshwater and saltwater fish and shellfish, to be processed, in the form in which it is received at the processing plant (other identical or similar definitions are provided in 40CFR408.21; 408.31; 408.41; 408.51; 408.61; 408.71; 408.81; 408.91; 408.101; 408.111; 408.121; 408.131; 408.141; 408.151; 408.161; 408.171; 408.181; 408.191; 408.201; 408.211; 408.221; 408.231; 408.241; 408.281; 408.311; 408.321; 408.331).

sealed source (10CFR30.4) means any byproduct material that is encased in a capsule designed to prevent leakage or escape of the byproduct material.

sealed source (10CFR70.4) means any special nuclear material that is encased in a capsule designed to prevent leakage or escape of the special nuclear material.

sealer (40CFR52.741) means a coating containing binders which seals wood prior to the application of the subsequent coatings.

search (40CFR2.100-i) includes all time spent looking for material that is responsive to a request, including page-by-page or line-by-line identification of material within documents. Searching for material must be done in the most efficient and least expensive manner so as to minimize costs for both the EPA and the requestor. For example, EPA will not engage in line-by-line search when merely duplicating an entire document would prove the less expensive and quicker method of complying with a request. Search will be distinguished, moreover, from review of material in order to determine whether the material is exempt from disclosure (see paragraph (j) of this section). Searches may be done manually or by computer using existing programming.

Seattle central business district (40CFR52.2486-8) means the area enclosed by Yesler Way, the I-5 freeway, Eighth Avenue, Virginia Street, and the Alaska Way Viaduct. Streets forming boundaries (excluding the I-5 freeway and the Alaska Way Viaduct) shall be part of the central business district (CBD) (other identical or similar definitions are provided in 40CFR52.2489-1; 52.2493-3).

secondary air supply (40CFR60-AA (method 28-2.4)) means an air supply that introduces air to the wood heater such that the bum rate is not altered by more than 25 percent when the secondary air supply is adjusted during the test run. The wood heater manufacturer can document this through design drawings that show the secondary air is introduced only into a mixing chamber or secondary chamber outside the firebox.

secondary contact recreation (40CFR131.35.d-13) means activities where a person's water contact would be limited to the extent that bacterial infections of eyes, ears, respiratory, or digestive systems or urogenital areas would normally be avoided (such as wading or fishing).

secondary drinking water regulation (SDWA1401-42USC300f) means a regulation which applies to public water systems and which specifies the maximum contaminant level which, in the judgment of the Administrator, are requisite to protect the public welfare. Such regulations may apply to any contaminant in drinking water:

(A) which may adversely affect the odor or appearance of such water and consequently may cause a substantial number of the persons served by the public water system providing such water to discontinue its use, or

(B) which may otherwise adversely affect the public welfare. Such regulations may vary according to geographic and other circumstances (cf. primary drinking water regulation).

secondary drinking water regulations (EPA-94/04): Non-enforceable regulations applying to public water systems and specifying the maximum contamination levels that, in the judgment of EPA, are required to protect the public welfare. These regulations apply to any contaminants that may adversely affect the odor or appearance of such water and consequently may cause people served by the system to discontinue its use.

secondary emission control system (40CFR60.141a) means the combination of equipment used for the capture and collection of secondary emissions, e.g.,:

(1) An open hood system for the capture and collection of primary and secondary emissions from the BOPF, with local hooding ducted to a secondary emission collection device such as a baghouse for the capture and collection of emissions from the hot metal transfer and skimming station; or

(2) An open hood system for the capture and collection of primary and secondary emissions from the furnace, plus a furnace enclosure with local hooding ducted to a secondary emission collection device, such as a baghouse, for additional capture and collection of secondary emissions from the furnace, with local hooding ducted to a secondary emission collection device, such as a baghouse, for the capture and collection of emissions from hot metal transfer and skimming station; or

(3) A furnace enclosure with local hooding ducted to a secondary emission collection device such as a baghouse for the capture and collection of

secondary emissions from a BOPF controlled by a closed hood primary emission control system, with local hooding ducted to a secondary emission collection device, such as a baghouse, for the capture and collection of emissions from hot metal transfer and skimming stations.

secondary emissions (40CFR51.165) means emissions which would occur as a result of the construction or operation of a major stationary source or major modification, but do not come from the major stationary source or major modification itself. For the purpose of this section, secondary emissions must be specific, well defined, quantifiable, and impact the same general area as the stationary source or modification which causes the secondary emissions. Secondary emissions include emissions from any offsite support facility which would not be constructed or increase its emissions except as a result of the construction of operation of the major stationary source of major modification. Secondary emissions do not include any emissions which come directly from a mobile source such as emissions from the tailpipe of a motor vehicle, from a train, or from a vessel (other identical or similar definitions are provided in 40CFR51-AS-8).

secondary emissions (40CFR51.166-18) means emissions which occur as a result of the construction or operation of a major stationary source or major modification, but do not come from the major stationary source or major modification itself. For the purposes of this section, secondary emissions must be specific, well defined, quantifiable, and impact the same general areas the stationary source modification which causes the secondary emissions. Secondary emissions include emissions from any offsite support facility which would not be constructed or increase its emissions except as a result of the construction or operation of the major stationary source or major modification. Secondary emissions do not include any emissions which come directly from a mobile source, such as emissions from the tailpipe of a motor vehicle, from a train, or from a vessel.

secondary emissions (40CFR51.301-u) means emissions which occur as a result of the construction or operation of an existing stationary facility but do not come from the existing stationary facility. Secondary emissions may include, but are not limited to, emissions from ships or trains coming to or from the existing stationary facility.

secondary emissions (40CFR52.21-18) means emissions which would occur as a result of the construction or operation of a major stationary source or major modification, but do not come from the major stationary source or major modification itself. Secondary emissions include emissions from any offsite support facility which would not be constructed or increase its emissions except as a result of the construction or operation of the major stationary source or major modification. Secondary emissions do not include any emissions which come directly from a mobile source, such as emissions from the tailpipe of a motor vehicle, from a train, or from a vessel:

(i) Emissions from ships or trains coming to or from the new or modified stationary source; and

(ii) Emissions from any offsite support facility which would not otherwise be constructed or increase its emissions as a result of the construction or operation of the major stationary source or major modification.

secondary emissions (40CFR52.24-8) means emissions which would occur as a result of the construction or operation of a major stationary source or major modification, but do not come from the major stationary source or major modification itself. For the purpose of this section, secondary emissions must be specific, well defined, quantifiable, and impact the same general area as the stationary source or modification which causes the secondary emissions. Secondary emissions include emissions from any offsite support facility which would otherwise not be constructed or increase its emissions except as a result of the construction or operation of the major stationary source or major modification. Secondary emissions do not include any emissions which come directly from a mobile source, such as emissions from the tailpipe of a motor vehicle, from a train, or from a vessel.

secondary emissions (40CFR60.141a) means particulate matter emissions that are not captured by the BOPF primary control system, including emissions from hot metal transfer and skimming stations. This definition also includes particulate matter emissions that escape from openings in the primary emission control system, such as from lance hole openings, gaps or tears in the ductwork of the primary emission control system, or leaks in hoods.

secondary emissions (40CFR61.171) means inorganic arsenic emissions that escape capture by a primary emission control system (other identical or similar definitions are provided in 40CFR61.181).

secondary emissons (40CFR51.165-viii) means emissions which would occur as a result of the construction or operation of a major stationary source or major modification, but do not come from the major stationary source or major modification itself. For the purpose of this section, secondary emissions must be specific, well defined, quantifiable, and impact the same general area as the stationary source or modification which causes the secondary emissions. Secondary emissions include emissions from any offsite support facility which would not be constructed or increase its emissions except as a result of the construction or operation of the major stationary source of major modification. Secondary emissions do not include any emissions which come directly from a mobile source such as emissions from the tailpipe of a motor vehicle, from a train, or from a vessel.

secondary fuel (40CFR60.701) means a fuel fired through a burner other than a primary fuel burner. The secondary fuel may provide supplementary heat in addition to the heat provided by the primary fuel.

secondary fuel (40CFR63.111) means a fuel fired through a burner other than the primary fuel burner that provides supplementary heat in addition to the heat provided by the primary fuel.

secondary hood system (40CFR61.171) means the equipment (including hoods, ducts, fans, and dampers) used to capture and transport secondary inorganic arsenic emissions.

secondary industry category (40CFR122.2) means any industry category which is not a "primary industry category" (cf. primary industry category).

secondary lead smelter (40CFR60.121-b) means any facility producing lead from a leadbearing scrap material by smelting to the metallic form.

secondary materials (EPA-94/04): Materials that have been manufactured and used at least once and are to be used again.

secondary maximum contaminant levels (40CFR143.2-f) means SMCLs which apply to public water systems and which, in the judgement of the Administrator, are requisite to protect the public welfare. The SMCL means the maximum permissible level of a contaminant in water which is delivered to the free flowing outlet of the ultimate user of public water system. Contaminants added to the water under circumstances controlled by the user, except those resulting from corrosion of piping and plumbing caused by water quality, are excluded from this definition.

secondary processor of asbestos (40CFR763.63-l) is a person who processes for commercial purposes an asbestos mixture.

secondary radon measurement services (secondary) (40CFR195.2) refers to radon measurement services that do not include the reading or the ability to analyze the results of the measurement devices used. These services may include placement and retrieval of devices, reporting results, and/or consultation with consumers.

secondary standard (40CFR51.100-d) means a national secondary ambient air quality standard promulgated pursuant to section 109 of the Act.

secondary standards (EPA-94/04): National ambient air quality standards designed to protect welfare, including effects on soils, water, crops, vegetation, manmade materials, animals, wildlife, weather, visibility, and climate, damage to property, transportation hazards, effects on economic values, and on personal comfort and well-being.

secondary treatment (40CFR125.58-r) means the term as defined in 40CFR133.

secondary treatment (EPA-94/04): The second step in most publicly owned waste treatment systems in which bacteria consume the organic parts of the waste. It is accomplished by bringing together waste, bacteria, and oxygen in trickling filters or in the activated sludge process. This treatment removes floating and settleable solids and about 90 percent of the oxygen-demanding substances and suspended solids. Disinfection is the final stage of secondary treatment (cf. primary treatment or tertiary treatment).

secrecy (40CFR350-AA): The subject matter of a trade secret must be secret. Matters of public knowledge or of general knowledge in an industry cannot be appropriated by one as his secret. Matters which are completely disclosed by the goods which one markets cannot be his secret. Substantially, a trade secret is known only in the particular business in which it is used. It is not requisite that only the proprietor of the business know it. He may, without losing his protection, communicate it to employees involved in its use. He may likewise communicate it to others pledged to secrecy. Others may also know of it independently, as, for example, when they have discovered the process or formula by independent invention and are keeping it secret. Nevertheless, a substantial element of secrecy must exist, so that, except by the use of improper means, there would be difficulty in acquiring the information. An exact definition of a trade secret is not possible. Some factors to be considered in determining whether given information is one's trade secret are:
(1) The extent to which the information is known outside of his business;
(2) the extent to which it is known by employees and others involved in his business;
(3) the extent of measures taken by him to guard the secrecy of the information;
(4) the value of the information to him and to his competitors;
(5) the amount of effort or money expended by him in developing the information;
(6) the ease or difficulty with which the information could be properly acquired or duplicated by others.

secret (40CFR11.4-f): See security classification category.

secretary (40CFR122.2) means the Secretary of the Army, acting through the Chief of Engineers.

secretary (40CFR232.2) means the Secretary of the Army acting through the Chief of Engineers (under RCRA).

secretary (40CFR52.1161-9) means the Secretary of Transportation and Construction of the Commonwealth of Massachusetts.

secretary (40CFR600.002.85-3) means the Secretary of Transportation or his authorized representative.

secretary (40CFR87.1) means the Secretary of Transportation and any other officer or employee of the Department of Transportation to whom the authority involved may be delegated.

secretary (CZMA304-16USC1453) means the Secretary of Commerce.

secretary (ESA3-16USC1531) means, except as otherwise herein provided, the Secretary of the Interior or the Secretary of Commerce as program responsibilities are vested pursuant to the provisions of Reorganization Plan Numbered 4 of 1970; except that with respect to the enforcement of the provisions of this chapter and the Convention which pertain to the importation or exportation of terrestrial plants, the term also means the Secretary of Agriculture.

secretary (FLPMA103-43USC1702), unless specifically designated otherwise, means the Secretary of the Interior.

secretary (MMPA3-16USC1362) means:
(A) the Secretary of the department in which the National Oceanic and Atmospheric Administration is operating, as to all responsibility, authority, funding, and duties under this chapter with respect to members of the order Cetacea and members, other than walruses, of the order Pinnipedia, and
(B) the Secretary of the Interior as to all responsibility, authority, funding, and duties under this chapter with respect to all other marine mammals covered by this chapter.

secretary (OPA1001) means the Secretary of the department in which the Coast Guard is operating.

secretary (OSHA3-29USC652) means the Secretary of Labor.

secretary (SMCRA701-30USC1291) means the

Secretary of the Interior, except where otherwise described.

secretary of energy (40CFR600.002.85-44) means the Secretary of Energy or his authorized representative.

secretary of energy (40CFR72.2) means the Secretary of the United States Department of Energy or the Secretary's duly authorized representative.

section 112(j) deadline (40CFR63.51) means the date 18 months after the date for which a relevant standard is scheduled to be promulgated under this part. The applicable date for categories of major sources is contained in the source category schedule for standards.

section 13 (40CFR7.25) refers to section 13 of the Federal Water Pollution Control Act Amendments of 12.

section 304(a) criteria (40CFR131.3-c) are developed by EPA under authority of section 304(a) of the Act based on the latest scientific information on the relationship that the effect of a constituent concentration has on particular aquatic species and/or human health. This information is issued periodically to the States as guidance for use in developing criteria.

section 404 program or State 404 program or 404 (40CFR124.2) means an "approved State program" to regulate the "discharge of dredged material" and the "discharge of fill material" under section 404 of the Clean Water Act in "State regulated waters."

section 5 notice (40CFR700.43) means any PMN, consolidated PMN, intermediate PMN, significant new use notice, exemption notice, or exemption application.

section 502(b)(10) changes (40CFR70.2) are changes that contravene an express permit term. Such changes do not include changes that would violate applicable requirements or contravene federally enforceable permit terms and conditions that are monitoring (including test methods), recordkeeping, reporting, or compliance certification requirements.

section 504 (40CFR12.103) means section 504 of the Rehabilitation Act of 1973 (Pub. L. 93-112, 87 Stat. 394 (29USC794)), as amended by the Rehabilitation Act Amendments of 1974 (Pub. L. 93-516, Stat. 1617); and the Rehabilitation, Comprehensive Services, and Developmental Disabilities Amendments of 1978 (Pub. L. 95-602, 92 Stat. 2955); and the Rehabilitation Act Amendments of 1986 (Pub. L. 99-506, 100 Stat. 1810). As used in this part, section 504 applies only to programs or activities conducted by Executive agencies and not to federally assisted programs.

section mill (40CFR420.71-c) means those steel hot forming operations that produce a variety of finished and semi-finished steel products other than the products of those mills specified below in paragraphs (d), (e), (g) and (h) of this section.

secure chemical landfill (EPA-94/04): See landfills.

secure maximum contaminant level (EPA-94/04): Maximum permissible level of a contaminant in water delivered to the free flowing outlet of the ultimate user, or of contamination resulting from corrosion of piping and plumbing caused by water quality.

security classification assignment (40CFR11.4-e) means the prescription of a specific security classification for a particular area or item of information. The information involved constitutes the sole basis for determining the degree of classification assigned.

security classification category (40CFR11.4-f) means the specific degree of classification (Top Secret, Secret or Confidential) assigned to classified information to indicate the degree of protection required:

(1) Top Secret. Top Secret refers to national security information or material which requires the highest degree of protection. The test for assigning Top Secret classification shall be whether its unauthorized disclosure could reasonably be expected to cause exceptionally grave damage to the national security. Examples of "exceptionally grave damage" include armed hostilities against the United States or its allies; disruption of foreign relations vitally affecting the national security; the compromise of vital

national defense plans or complex cryptologic and communications intelligence systems; the revelation of sensitive intelligence operations; and the disclosure of scientific or technological developments vital to national security. This classification shall be used with the utmost restraint.

(2) Secret. Secret refers to that national security information or material which requires a substantial degree of protection. The test for assigning Secret classification shall be whether its unauthorized disclosure could reasonably be expected to cause serious damage to the national security. Examples of "serious damage" include disruption of foreign relations significantly affecting the national security; significant impairment of a program or policy directly related to the national security; revelation of significant military plans or intelligence operations; and compromise of scientific or technological developments relating to national security. The classification Secret shall be sparingly used.

(3) Confidential. Confidential refers to that national security information or material which requires protection. The test for assigning Confidential classification shall be whether its unauthorized disclosure could reasonably be expected to cause damage to the national security.

sediment (40CFR796.2750-vii) is the unconsolidated inorganic and organic material that is suspended in and being transported by surface water, or has settled out and has deposited into beds.

sediment yield (EPA-94/04): The quantity of sediment arriving at a specific location.

sedimentation (40CFR141.2) means a process for removal of solids before filtration by gravity or separation.

sedimentation (EPA-94/04): Letting solids settle out of wastewater by gravity during treatment.

sedimentation tanks (EPA-94/04): Wastewater tanks in which floating wastes are skimmed off and settled solids are removed for disposal.

sediments (EPA-94/04): Soil, sand, and minerals washed from land into water, usually after rain. They pile up in reservoirs, rivers and harbors, destroying fish and wildlife habitat, and clouding the water so that sunlight cannot reach aquatic plants. Careless farming, mining, and building activities will expose sediment materials, allowing them to wash off the land after rainfall.

seed protectant (EPA-94/04): A chemical applied before planting to protect seeds and seedlings from disease or insects.

seepage (EPA-94/04): Percolation of water through the soil from unlined canals, ditches, laterals, watercourses, or water storage facilities.

segregated stormwater sewer system (40CFR61.341) means a drain and collection system designed and operated for the sole purpose of collecting rainfall runoff at a facility, and which is segregated from all other individual drain systems.

seismic impact zone (40CFR503.21-m) is an area that has a 10 percent or greater probability that the horizontal ground level acceleration of the rock in the area exceeds 0.10 gravity once in 250 years (under CWA).

selective catalytic reduction (40CFR76.2) means a noncombustion control technology that destroys NO_x by injecting a reducing agent (e.g., ammonia) into the flue gas that, in the presence of a catalyst (e.g., vanadium, titanium, or zeolite), converts NO_x into molecular nitrogen and water.

selective noncatalytic reduction (40CFR76.2) means a noncombustion control technology that destroys NO_x by injecting a reducing agent (e.g., ammonia, urea, or cyanuric acid) into the flue gas, downstream of the combustion zone that converts NO_x to molecular nitrogen, water, and when urea or cyanuric acid are used, to carbon dioxide (CO_2).

selective pesticide (EPA-94/04): A chemical designed to affect only certain types of pests, leaving other plants and animals unharmed.

selenium (40CFR415.361-d) shall mean the total selenium present in the process wastewater stream exiting the wastewater treatment system (other identical or similar definitions are provided in 40CFR415.631).

selenium (40CFR415.451) shall mean the total selenium present in the process wastewater stream exiting the wastewater treatment system.

selenium (40CFR415.641-d) shall mean the total selenium present in the process wastewater stream exiting the wastewater treatment system.

self-contained recovery equipment (40CFR82.152-u) means refrigerant recovery or recycling equipment that is capable of removing the refrigerant from an appliance without the assistance of components contained in the appliance.

SEM (40CFR763-AA-19) means scanning electron microscope.

semi-confined aquifer (EPA-94/04): An aquifer partially confined by soil layers of low permeability through which recharge and discharge can still occur.

semi-static test (40CFR797.1440-6) is a test without flow of solution, but with occasional batchwise renewal of test solutions after prolonged periods (e.g., 24 hours).

semi-transparent stains (40CFR52.741) means stains containing dyes or semi-transparent pigments which are formulated to enhance wood grain and change the color of the surface but not to conceal the surface, including, but not limited to, sap stain, toner, non-grain raising stains, pad stain, or spatter stain.

semi-wet (40CFR420.41-e) means those steelmaking air cleaning systems that use water for the sole purpose of conditioning the temperature and humidity of furnace gases such that the gases may be cleaned in dry air pollution control systems (under CWA).

semiannual (40CFR61.111) means a 6-month period; the first semiannual period concludes on the last day of the last month during the 180 days following initial startup for new sources; and the first semiannual period concludes on the last day of the last full month during the 180 days after June 6, 1984 for existing sources.

semiannual (40CFR61.131) means a 6-month period; the first semiannual period concludes on the last day of the last full month during the 180 days following initial startup for new sources; the first semiannual period concludes on the last day of the last full month during the 180 days after the effective date of the regulation for existing sources (other identical or similar definitions are provided in 40CFR61.241).

semiconductors (40CFR469.12-b) means solid state electrical devices which perform functions such as information processing and display, power handling, and interconversion between light energy and electrical energy.

senescence (EPA-94/04): The aging process. Sometimes used to describe lakes or other bodies of water in advanced stages of eutrophication.

senior management official (40CFR350.1) means an official with management responsibility for the person or persons completing the report, or the manager of environmental programs for the facility or establishments, or for the corporation owning or operating the facility or establishments responsible for certifying similar reports under other environmental regulatory requirements (other identical or similar definitions are provided in 40CFR372.3).

sensitivity (40CFR136-AC-6) means the slope of the analytical curve, i.e. functional relationship between emission intensity and concentration.

sensor (40CFR60.481) means a device that measures a physical quantity or the change in a physical quantity such as temperature, pressure, flow rate, pH, or liquid level (other identical or similar definitions are provided in 40CFR61.241; 264.1031).

sensor (40CFR63.161) means a device that measures a physical quantity or the change in a physical quantity, such as temperature, pressure, flow rate, pH, or liquid level.

separate collection (40CFR246.101-z) means collecting recyclable materials which have been separated at the point of generation and keeping those materials separate from other collected solid waste in separate compartments of a single collection vehicle or through the use of separate collection vehicles.

separator tank (40CFR264.1031) means a device used for separation of two immiscible liquids.

septage (40CFR122.2) means the liquid and solid material pumped from a septic tank, cesspool, or similar domestic sewage treatment system, or a holding tank when the system is cleaned or maintained (other identical or similar definitions are provided in 40CFR501.2).

septic system (EPA-94/04): An onsite system designed to treat and dispose of domestic sewage. A typical septic system consists of a tank that receives waste from a residence or business and a system of tile lines or a pit for disposal of the liquid effluent (sludge) that remains after decomposition of the solids by bacteria in the tank and must be pumped out periodically.

septic tank (40CFR280.12) is a water-tight covered receptacle designed to receive or process, through liquid separation or biological digestion, the sewage discharged from a building sewer. The effluent from such receptacle is distributed for disposal through the soil and settled solids and scum from the tank are pumped out periodically and hauled to a treatment facility.

septic tank (EPA-94/04): An underground storage tank for wastes from homes not connected to a sewer line. Waste goes directly from the home to the tank (cf. septic system).

serial number (40CFR205.151-25) means the identification number assigned by the manufacturer to a specific production unit.

serial number (40CFR72.2) means, when referring to allowances, the unique identification number assigned to each allowance by the Administrator, pursuant to 40CFR73.34(d) of this chapter.

series resistance (40CFR85.2122(a)(6)(ii)(B)) means the sum of resistances from the condenser plates to the condensers external connections.

serious acute effects (40CFR721.3) means human injury or human disease processes that have a short latency period for development, result from short-term exposure to a chemical substance, or are a combination of these factors and which are likely to result in death or severe or prolonged incapacitation (under TSCA).

serious acute effects (40CFR723.50-9) means human disease processes or other adverse effects that have short latency periods for development, result from short-term exposure, or are a combination of these factors and that are likely to result in death, severe or prolonged incapacitation, disfigurement, or severe or prolonged loss of the ability to use a normal bodily or intellectual function with a consequent impairment of normal activities.

serious chronic effects (40CFR721.3) means human injury or human disease processes that have a long latency period for development, result from long-term exposure to a chemical substance, or are a combination of these factors and which are likely to result in death or severe or prolonged incapacitation.

serious chronic effects (40CFR723.50-10) means human disease processes or other adverse effects that have long latency periods for development, result from long-term exposure, are long-term illnesses, or are a combination of these factors and that are likely to result in death, severe or prolonged incapacitation, disfigurement, or severe or prolonged loss of the ability to use a normal bodily or intellectual function with a consequent impairment of normal activities.

service connector (EPA-94/04): The pipe that carries tap water from a public water main to a building.

service for consideration (40CFR82.32-g) means being paid to perform service, whether it is in cash, credit, goods, or services. This includes all service except that done for free.

service involving refrigerant (40CFR82.32-h) means any service during which discharge or release of refrigerant from the motor vehicle air conditioner to the atmosphere can reasonably be expected to occur.

service line sample (40CFR141.2) means a one-liter sample of water collected in accordance with

40CFR141.86(b)(3), that has been standing for at least 6 hours in a service line.

service line sample (EPA-94/04): A one-liter sample of water collected according to federal regulations that has been standing for at least 6 hours in a service pipeline.

service pipe (EPA-94/04): The pipeline extending from the water main to the building served or to the consumer's system.

service station dealer (SF101-42USC9601) means any person:
(A) (i) who owns or operates a motor vehicle service station, filling station, garage, or similar retail establishment engaged in the business of selling, repairing, or servicing motor vehicles, where a significant percentage of the gross revenue of the establishment is derived from the fueling, repairing, or servicing of motor vehicles, and
(ii) who accepts for collection, accumulation, and delivery to an oil recycling facility, recycled oil that (I) has been removed from the engine of a light duty motor vehicle or household appliances by the owner of such vehicle or appliances, and (II) is presented, by such owner, to such person for collection, accumulation, and delivery to an oil recycling facility.
(B) For purposes of section 114(c), the term "service station dealer" shall, notwithstanding the provisions of subparagraph (A), include any government agency that establishes a facility solely for the purpose of accepting recycled oil that satisfies the criteria set forth in subclauses (I) and (II) of subparagraph (A)(ii), and, with respect to recycled oil that satisfies the criteria set forth in subclauses (I) and (II), owners or operators of refuse collection services who are compelled by State law to collect, accumulate, and deliver such oil to an oil recycling facility.
(C) The President shall promulgate regulations regarding the determination of what constitutes a significant percentage of the gross revenues of an establishment for purposes of this paragraph.

services (40CFR35.2005-38) means a contractor's labor, time, or efforts which do not involve the delivery of a specific end item, other than documents, (e.g., reports, design drawing, specifications). This term does not include employment agreements or collective bargaining agreements (other identical or similar definitions are provided in 40CFR33.005).

services (40CFR35.6015-42) means a recipient's in-kind or a contractors labor, time, or efforts which do not involve the delivery of a specific end item, other than documents (e.g., reports, design drawing, specifications). This term does not include employment agreements or collective bargaining agreements.

set of safety relief valves (40CFR52.741) means one or more safety relief valves designed to open in order to relieve excessive pressures in the same vessel or pipe.

set pressure (40CFR63.161) means the pressure at which a properly operating pressure relief device begins to open to relieve atypical process system operating pressure.

settle (40CFR14.2-d) means the act of considering, ascertaining, adjusting, determining or otherwise resolving a claim.

settleable solids (40CFR434.11-m) is that matter measure by the volumetric method specified in 40CFR434.64.

settleable solids (40CFR440.141-17) means the particulate material (both organic or inorganic) which will settle in one hour expressed in milliliters per liter (ml/l) as determined using an Imhoff cone and the method described for residue settleable in 40CFR136.

settleable solids (EPA-94/04): Material heavy enough to sink to the bottom of a wastewater treatment tank.

settling chamber (EPA-94/04): A series of screens placed in the way of flue gases to slow the stream of air, thus helping gravity to pull particles into a collection device.

settling tank (40CFR60.621) means a container that

gravimetrically separates oils, grease, and dirt from petroleum solvent, together with the piping and ductwork used in the installation of this device.

settling tank (EPA-94/04): A holding area for wastewater, where heavier particles sink to the bottom for removal and disposal.

seven (7)-day average (40CFR133.101-a) means the arithmetic mean of pollutant parameter values for samples collected in a period of 7 consecutive days.

severe property damage (40CFR122.41.m-ii) means substantial physical damage to property, damage to the treatment facilities which causes them to become inoperable, or substantial and permanent loss of natural resources which can reasonably be expected to occur in the absence of a bypass. Severe property damage does not mean economic loss caused by delays in production (other identical or similar definitions are provided in 40CFR403.17-2).

sewage (40CFR140.1-a) means human body wastes and the wastes from toilets and other receptacles intended to receive or retain body wastes.

sewage (CWA312-33USC1322) means human body wastes and the wastes from toilets and other receptacles intended to receive or retain body wastes except that, with respect to commercial vessels on the Great Lakes such term shall include graywater.

sewage (EPA-94/04): The waste and wastewater produced by residential and commercial sources and discharged into sewers.

sewage collection system (40CFR35.905) means, for the purpose of section 35.925-13, each, and all, of the common lateral sewers, within a publicly owned treatment system, which are primarily installed to receive waste waters directly from facilities which convey waste water from individual structures or from private property, and which include service connection Y fittings designed for connection with those facilities. The facilities which convey waste water from individual structures, from private property to the public lateral sewer, or its equivalent, are specifically excluded from the definition, with the exception of pumping units, and pressurized lines, for individual structures or groups of structures when

such units are cost effective and are owned and maintained by the grantee.

sewage from vessels (40CFR122.2) means human body wastes and the wastes from toilets and other receptacles intended to receive or retain body wastes that are discharged from vessels and regulated under section 312 of CWA, except that with respect to commercial vessels on the Great Lakes this term includes graywater. For the purposes of this definition, "graywater" means galley, bath, and shower water.

sewage lagoon (EPA-94/04): See lagoon.

sewage sludge (40CFR122.2) means any solid, semi-solid, or liquid residue removed during the treatment of municipal waste water or domestic sewage. Sewage sludge includes, but is not limited to, solids removed during primary, secondary, or advanced waste water treatment, scum, septage, portable toilet pumpings, type III marine sanitation device pumpings (33CFR159), and sewage sludge products. Sewage sludge does not include grit or screenings, or ash generated during the incineration of sewage sludge (other identical or similar definitions are provided in 40CFR501.2).

sewage sludge (40CFR503.9-w) is solid, semi-solid, or liquid residue generated during the treatment of domestic sewage in a treatment works. Sewage sludge includes, but is not limited to, domestic septage; scum or solids removed in primary, secondary, or advanced wastewater treatment processes; and a material derived from sewage sludge. Sewage sludge does not include ash generated during the firing of sewage sludge in a sewage sludge incinerator or grit and screenings generated during preliminary treatment of domestic sewage in a treatment works.

sewage sludge (EPA-94/04): Sludge produced at a Publicly Owned Treatment Works, the disposal of which is regulated under the Clean Water Act.

sewage sludge feed rate (40CFR503.41-j) is either the average daily amount of sewage sludge fired in all sewage sludge incinerators within the property line of the site where the sewage sludge incinerators are located for the number of days in a 365 day period that each sewage sludge incinerator operates, or the average daily design capacity for all sewage sludge incinerators within the property line of the site where the sewage sludge incinerators are located.

sewage sludge incinerator (40CFR503.41-k) is an enclosed device in which only sewage sludge and auxiliary fuel are fired.

sewage sludge unit (40CFR503.21-n) is land on which only sewage sludge is placed for final disposal. This does not include land on which sewage sludge is either stored or treated. Land does not include waters of the United States, as defined in 40CFR122.2.

sewage sludge unit boundary (40CFR503.21-o) is the outermost perimeter of an active sewage sludge unit (under CWA).

sewage sludge use or disposal practice (40CFR122.2) means the collection, storage, treatment, transportation, processing, monitoring, use, or disposal of sewage sludge.

sewage treatment works (40CFR220.2-f) means municipal or domestic waste treatment facilities of any type which are publicly owned or regulated to the extent that feasible compliance schedules are determined by the availability of funding provided by Federal, State, or local governments.

sewer (EPA-94/04): A channel or conduit that carries wastewater and storm-water runoff from the source to a treatment plant or receiving stream. "Sanitary" sewers carry household, industrial, and commercial waste. "Storm" sewers carry runoff from rain or snow. "Combined" sewers handle both.

sewer line (40CFR60.691) means a lateral, trunk line, branch line, ditch, channel, or other conduit used to convey refinery wastewater to downstream components of a refinery wastewater treatment system. This term does not include buried, below-grade sewer lines.

sewer line (40CFR61.341) means a lateral, trunk line, branch line, or other enclosed conduit used to convey waste to a downstream waste management unit.

sewer line (40CFR63.111) means a lateral, trunk line, branch line, or other conduit including, but not limited to, grates, trenches, etc., used to convey wastewater streams or residuals to a downstream waste management unit.

sewerage (EPA-94/04): The entire system of sewage collection, treatment, and disposal.

sex-linked genes (40CFR798.5275-3) are present on the sex (X or Y) chromosomes. Sex-linked genes in the context of this guideline refer only to those located on the X-chromosome.

shaft power (40CFR87.1) means only the measured shaft power output of turboprop engine.

shall (40CFR256.06) denotes requirements for the development and implementation of the State plan.

share (40CFR31.3), when referring to the awarding agency's portion of real property, equipment or supplies, means the percentage as the awarding agency's portion of the acquiring party's total costs under the grant to which the acquisition costs under the grant to which the acquisition cost of the property was charged. Only costs are to be counted-- not the value of third party in-kind contributions.

sharps (EPA-94/04): Hypodermic needles, syringes (with or without the attached needle) pasteur pipettes, scalpel blades, blood vials, needles with attached tubing, and culture dishes used in animal or human patient care or treatment, or in medical, research or industrial laboratories. Also included are other types of broken or unbroken glassware that were in contact with infectious agents, such as used slides and cover slips, and unused hypodermic and suture needles, syringes, and scalpel blades.

shed (40CFR63.301) means a structure for capturing coke oven emissions on the coke side or pusher side of the coke oven battery, which routes the emissions to a control device or system.

sheen (40CFR110.1) means an iridescent appearance on the surface of the water.

sheet basecoat (40CFR52.741) means a coating applied to metal when the metal is in sheet form to serve as either the exterior or interior of a can for either two-piece or three-piece cans.

shell deposition (40CFR797.1800-3) is the measured length of shell growth that occurs between the time the shell is ground at test initiation and test termination 96 hours later.

shellfish, fish and wildlife (40CFR125.58-s) means any biological population or community that might be adversely affected by the applicant's modified discharge.

shift (40CFR204.51-t) means the regular production work period for one group of workers (other identical or similar definitions are provided in 40CFR205.51-22).

shift supervisor (40CFR60.51a) means the person in direct charge and control of the operation of an MWC and who is responsible for on-site supervision, technical direction, management, and overall performance of the facility during an assigned shift.

shipment (40CFR749.68-15) means the act or process of shipping goods by any form of conveyance.

shipped liquid ammonia (40CFR418.51-d) shall mean liquid ammonia commercially shipped for which the Department of Transportation requires 0.2 percent minimum water content.

shipping losses (40CFR418.21-c) shall mean: Discharges resulting from loading tank cars or tank trucks; discharges resulting from cleaning tank cars or tank trucks; and discharges from air pollution control scrubbers designed to control emissions from loading or cleaning tank cars or tank trucks (other identical or similar definitions are provided in 40CFR418.51-c).

shock load (EPA-94/04): The arrival at a water treatment plant of raw water containing unusual amounts of algae, colloidal mater, color, suspended solids, turbidity, ore other pollutants.

shop (40CFR60.271-m) means the building which houses one or more EAFs.

shop (40CFR60.271a) means the building which houses one or more EAFs or AOD vessels (under CAA).

shop opacity (40CFR60.271-k) means the arithmetic average of 24 or more opacity observations of emissions from the shop taken in accordance with Method 9 of Appendix A of this part for the applicable time periods.

shop opacity (40CFR60.271a) means the arithmetic average of 24 observations of the opacity of emissions from the shop taken in accordance with Method 9 of Appendix A of this part.

short circuiting (EPA-94/04): When some of the water in tanks or basins flows faster than the rest; usually undesirable since it may result in shorter contact, reaction, or settling times in comparison with the calculated or presumed detention times.

short coke oven battery (40CFR63.301) means a coke oven battery with ovens less than 6 meters in height.

short term test indicative of carcinogenic potential (40CFR721.3) means either any limited bioassay that measures tumor or preneoplastic induction, or any test indicative of interaction of a chemical substance with DNA (i.e., positive response in assays for gene mutation, chromosomal aberrations, DNA damage and repair, or cellular transformation).

short term test indicative of the potential to cause a developmentally toxic effect (40CFR721.3) means either any in vivo preliminary development toxicity screen conducted in a mammalian species, or any in vitro developmental toxicity screen, including any test system other than the intact pregnant mammal, that has been extensively evaluated and judged reliable for its ability to predict the potential to cause developmentally toxic effects in intact systems across a broad range of chemicals or within a class of chemicals that includes the substance of concern.

shortfall (40CFR51.491) means the difference between the amount of emissions reductions credited in an implementation plan for a particular EIP and those that are actually achieved by that EIP, as determined through an approved reconciliation process.

shot casting (40CFR471.02-hh) is the production of shot by pouring molten metal in finely divided streams to form spherical particles.

should (40CFR256.06) denotes recommendations for the development and implementation of the State plan.

shutdown (40CFR60.2) means the cessation of operation of an affected facility for any purpose.

shutdown (40CFR61.161) means the cessation of operation of an affected source for any purpose.

shutdown (40CFR61.171) means the cessation of operation of a stationary source for any reason (other identical or similar definitions are provided in 40CFR61.181).

shutdown (40CFR63.101) means the cessation of operation of a chemical manufacturing process unit or a reactor, air oxidation reactor, distillation unit, or the emptying and degassing of a storage vessel for purposes including, but not limited to, periodic maintenance, replacement of equipment, or repair. Shutdown does not include the routine rinsing or washing of equipment in batch operation between batches.

shutdown (40CFR63.2) means the cessation of operation of an affected source for any purpose.

shutdown (40CFR63.301) means the operation that commences when pushing has occurred on the first oven with the intent of pushing the coke out of all of the ovens in a coke oven battery without adding coal, and ends when all of the ovens of a coke oven battery are empty of coal or coke.

SI unit (40CFR191.12) means a unit of measure in the International System of Units.

SIC codes (PPA6603) refers to the 2-digit code numbers used for classification of economic activity in the Standard Industrial Classification Manual.

side seam spray coat (40CFR52.741) means a coating applied to the seam of a three-piece can.

sidewall cementing operation (40CFR60.541) means

the system used to apply cement to a continuous strip of sidewall component or any other continuous strip component (except combined tread/sidewall component) that is incorporated into the sidewall of a finished tire. A sidewall cementing operation consists of a cement application station and all other equipment, such as the cement supply system and feed and takeaway conveyors, necessary to apply cement to sidewall strips or other continuous strip component (except combined tread/sidewall component) and to allow evaporation of solvent from the cemented rubber.

sievert (40CFR191.12) is the SI unit of effective dose and is equal to 100 rem or one joule per kilogram. The abbreviation is Sv.

signal (EPA-94/04): The volume or product-level change produced by a leak in a tank.

signal spike (40CFR60-AA(alt. method 1)) means an abrupt, momentary increase and decrease in signal amplitude.

signal words (EPA-94/04): The words used on a pesticide label--Danger, Warning, Caution--to indicate level of toxicity.

signed (40CFR86.902.93) means a certification request which results in a signed Certificate of Conformity.

significant (40CFR51.165-x) means, in reference to a net emissions increase or the potential of a source to emit any of the following pollutions, as rate of emissions that would equal or exceed any of the following rates:
- Carbon monoxide: 100 tons per year (tpy)
- Nitrogen oxides: 40 tpy
- Sulfur dioxide: 40 tpy
- Ozone: 40 tpy of volatile organic compounds
- Lead: 0.6 tpy.

(Other identical or similar definitions are provided in 40CFR51.166-23; 51-AS-10; 52.21-23; 52.24-10.)

significant adverse environmental effects (40CFR721.3) means injury to the environment by a chemical substance which reduces or adversely affects the productivity, utility, value, or function of biological, commercial, or agricultural resources, or which may adversely affect a threatened or endangered species. A substance will be considered to have the potential for significant adverse environmental effects if it has one of the following:

(1) An acute aquatic EC50 of 1 mg/L or less.

(2) An acute aquatic EC50 of 20 mg/L or less where the ratio of aquatic vertebrate 24-hour to 48-hour EC50 is greater than or equal to 2.0.

(3) A Maximum Acceptable Toxicant Concentration (MATC) of less than or equal to 100 parts per billion (100 ppb).

(4) An acute aquatic EC50 of 20 mg/L or less coupled with either a measured bioconcentration factor (BCF) equal to or greater than 1,000x or in the absence of bioconcentration data a log P value equal to or greater than 4.3.

significant adverse reactions (40CFR717.3-i) are reactions that may indicate a substantial impairment of normal activities, or long-lasting or irreversible damage to health or the environment.

significant biological treatment (40CFR133.101-k) means the use of an aerobic or anaerobic biological treatment process in a treatment works to consistently achieve a 30-day average of a least 65 percent removal of BOD(5).

significant deterioration (EPA-94/04): Pollution resulting from a new source in previously "clean" areas (cf. prevention of significant deterioration).

significant economic loss (40CFR166.3-h) means that, under the emergency conditions: for a productive activity, the profitability would be substantially below the expected profitability for that activity; or, for other types of activities, where profits cannot be calculated, the value of public or private fixed assets would be substantially below the expected value for those assets. Only losses caused by the emergency conditions, specific to the impacted site, and specific to the geographic area affected by the emergency conditions are included. The contribution of obvious mismanagement to the loss will not be considered in determining loss. In evaluating the significance of an economic loss for productive activities, the Agency will consider whether the expected reduction in profitability exceeds what would be expected as a result of normal fluctuations over a number of years, and whether the loss would affect the long-term financial viability expected from the productive activity. In evaluating the significance of an economic loss for situations other than productive activities, the Agency will consider reasonable measures of expected loss.

significant environmental effects (40CFR723.50-11) means:

(i) Any irreversible damage to biological, commercial, or agricultural resources of importance to society;

(ii) Any reversible damage to biological, commercial, or agricultural resources of importance to society if the damage persists beyond a single generation of the damaged resource or beyond a single year; or

(iii) Any known or reasonably anticipated loss of members of an endangered or threatened species. Endangered or threatened species are those species identified as such by the Secretary of the Interior in accordance with the Endangered Species Act, as amended (16 U.S.C. 1531).

significant hazard to public health (40CFR149.101-j) means any level of contaminant which causes or may cause the aquifer to exceed any maximum, contaminant level set forth in any promulgated National Primary Drinking Water Standard at any point where the water may be used for drinking purposes or which may otherwise adversely affect the health of persons, or which may require a public water system to install additional treatment to prevent such adverse effect.

significant impairment (40CFR51.301-v) means, for purposes of section 303, visibility impairment which, in the judgment of the Administrator, interferes with the management, protection, preservation, or enjoyment of the visitor's visual experience of the mandatory Class I Federal area. This determination must be made on a case-by-case basis taking into account the geographic extent, intensity, duration, frequency and time of the visibility impairment, and how these factors correlate with (1) times of visitor use of the mandatory Class I Federal area, and (2) the frequency and timing of natural conditions that reduce visibility.

significant industrial user (40CFR403.3-t):

(1) Except as provided in paragraph (t)(2) of this section, the term Significant Industrial User means:

   (i) All industrial users subject to Categorical Pretreatment Standards under 40CFR403.6 and 40CFR Chapter I, Subchapter N; and

   (ii) Any other industrial that: discharges an average of 25,000 gallons per day or more of process wastewater to the POTW (excluding sanitary, noncontact cooling and boiler blowdown wastewater); contributes a process wastestream which makes up 5 percent or more of the average dry weather hydraulic or organic capacity of the POTW treatment plant; or is designated as such by the Control Authority as defined in 40CFR403.12(a) on the basis that the industrial user has a reasonable potential for adversely affecting the POTW's operation or for violating any pretreatment standard or requirement (in accordance with 40CFR403.8(f)(6)).

(2) Upon a finding that an industrial user meeting the criteria in paragraph (t)(1)(ii) of this section has no reasonable potential for adversely affecting the POTW's operation or for violating any pretreatment standard or requirement, the Control Authority (as defined in 40CFR403.12(a)) may at any time, on its own initiative or in response to a petition received from an industrial user or POTW, and in accordance with 40CFR403.8(f)(6), determine that such industrial user is not a significant industrial user.

significant materials (40CFR122.26-12) includes, but is not limited to: raw materials; fuels; materials such as solvents, detergents, and plastic pellets; finished materials such as metallic products; raw materials used in food processing or production; hazard substances designated under section 101(14) of CERCLA; any chemical the facility is required to report pursuant to section 313 of title III of SARA; fertilizers; pesticides; and waste products such as ashes, slag and sludge that have the potential to be released with storm water discharges.

significant municipal facilities (EPA-94/04): Those publicly owned sewage treatment plants that discharge a million gallons per day or more and are therefore considered by states to have the potential to substantially effect the quality of receiving waters.

significant new use (40CFR721.1175) is any use other than as an intermediate in the manufacture of fluorinated substances in an enclosed process.

significant new use (40CFR721.1750) is: Use in a consumer product at concentrations greater than five percent by weight.

significant new use (40CFR82.172) means use of a new or existing substitute in a major industrial use sector as a result of the phaseout of ozone-depleting compounds.

significant new use notice (40CFR700.43) means any notice submitted to EPA pursuant to 40CFR5(a)(l)(B) of the Act (the Toxic Substances Control Act) in accordance with 40CFR721 of this chapter (see below for the listing of significant new use chemicals regulated under the Toxic Substances Control Act. The chemicals provided herewith in an alphabetical order were excerpted from 40CFR721, 1995 edition).

==

40CFR #	significant new use chemicals
721.225	2-Chloro-N-methyl-N-substituted acetamide (generic name).
721.275	Halogenated-N-(2-propenyl)-N-(substituted phenyl) acetamide.
721.285	Acetamide, N-[4-(pentyloxy)phenyl]-, acetamide, N-[2-nitro-4-(pentyloxy)phenyl]-, and acetamide, N-[2-amino-4-(pentyloxy)phenyl]-.
721.320	Acrylamide-substituted epoxy.
721.323	Substituted acrylamide.
721.325	Certain acrylates.
721.370	Substituted diacrylate.
721.390	Monoacrylate .

721.400	Polyalkylpolysilazane, bis(substituted acrylate).
721.415	Aliphatic diurethane acrylate ester.
721.430	Oxo-substituted aminoalkanoic acid derivative.
721.445	Substituted ethyl alkenamide.
721.460	Amino acrylate monomer.
721.470	Aliphatic difunctional acrylic acid ester.
721.490	Modified acrylic ester (generic name).
721.505	Halogenated acrylonitrile.
721.520	Alanine, N-(2-carboxyethyl)-N-alkyl-, salt.
721.530	Substituted aliphatic acid halide (generic name).
721.536	Halogenated phenyl alkane.
721.540	Alkylphenoxypolyalkoxyamine (generic name).
721.550	Alkyl alkenoate, azobis-.
721.562	Substituted alkylamine salt.
721.575	Substituted alkyl halide.
721.625	Alkylated diarylamine, sulfurized (generic name).
721.639	Amine aldehyde condensate.
721.642	Amines, N-(C14-18 and C16-18 unsaturated alkyl)] dipropylene-tri-, tripropylenetetra-, and tetrapropylenepenta-.
721.650	11-Aminoundecanoic acid.
721.700	Methylenebistrisubstituted aniline (generic name).
721.715	Trisubstituted anthracene.
721.750	Aromatic amine compound .
721.757	Polyoxyalkylene substituted aromatic azo colorant.
721.775	Brominated aromatic compound (generic name).
721.805	Benzenamine, 4,4'-[1,3-phenylenebis(1-methylethylidene)]bis[2,6-dimethyl-.
721.825	Certain aromatic ether diamines.
721.840	Alkyl substituted diaromatic hydrocarbons.
721.875	Aromatic nitro compound .
721.925	Substituted aromatic (generic).
721.950	Sodium salt of an alkylated, sulfonated aromatic (generic name).
721.982	Calcium, bis(2,4-pentanedionato-O,O').
721.1000	Benzenamine, 3-chloro-2,6-dinitro-N,N-dipropyl-4-(trifluoromethyl)-.
721.1025	Benzenamine, 4-chloro-2-methyl-; benzenamine, 4-chloro-2-methyl-, hydrochloride; and benzenamine, 2-chloro-6-methyl-.
721.1050	Benzenamine, 2,5-dibutoxy-4-(4-morpholinyl)-, sulfate
721.1068	Benzenamine, 4-isocyanato-N,N-bis(4-isocyanatophenyl)-2,5-dimethoxy-.
721.1075	Benzenamine, 4-(1-methylbutoxy)-, hydrochloride.
721.1120	Benzenamine, 4,4'-[1,4-phenylenebis (1-methylethylidene)]bis[2,6-dimethyl-.
721.1150	Substituted polyglycidyl benzeneamine.
721.1175	Benzene, substituted, alkyl acrylate derivative (generic name).
721.1187	Bis(imidoethylene) benzene.
721.1193	Benzene, 2-bromo-1,4-dimethoxy-.
721.1210	Benzene, (2-chloroethoxy)-.
721.1225	Benzene, 1,2-dimethyl-, polypropene derivatives, sulfonated, potassium salts.
721.1300	[(Dinitrophenyl)azo]-[2,4-diamino-5-methoxybenzene] derivatives.
721.1325	Benzene, 1-(1-methylbutoxy)-4-nitro-.
721.1350	Benzene, (1-methylethyl)(2-phenylethyl)-.
721.1372	Substituted nitrobenzene.
721.1375	Disubstituted nitrobenzene (generic name).

721.1425	Pentabromoethylbenzene.
721.1430	Pentachlorobenzene.
721.1435	1,2,4,5-Tetrachlorobenzene.
721.1440	1,3,5-Trinitrobenzene.
721.1450	1,3-Benzenediamine, 4-(1,1-dimethylethyl)-ar-methyl.
721.1500	1,2-Benzenediamine, 4-ethoxy, sulfate.
721.1525	Mixture of: 1,3-benzenediamine, 2-methyl-4,6-bis(methylthio)-(CAS NO. 104983-85-9) and 1,3-benzenediamine, 4-methyl-2,6-bis(methylthio)- (CAS NO. 102093-68-5).
721.1550	Benzenediazonium, 4-(dimethylamino)-, salt with 2-hydroxy-5-sulfobenzoic acid (1:1).
721.1555	Substituted phenyl azo substituted benzenediazonium salt.
721.1568	Substituted benzenediazonium.
721.1575	Substituted benzenedicarboxylic acid, poly(alkyl acrylate) derivative .
721.1612	Substituted 2-nitro- and 2-aminobenzesulfonamide.
721.1625	Alkylbenzene sulfonate, amine salt.
721.1630	1,2-Ethanediol bis(4-methylbenzenesulfonate); 2,2-oxybis-ethane bis(4-methylbenzenesulfonate); ethanol, 2,2'-[oxybis(2,1-ethanediyl oxy)]bis-, bis(4-methylbenzenesulfonate); ethanol, 2,2'-[oxybis (2,1-ethane diyloxy)] bis-, bis(4-methylbenzenesulfonate); ethanol, 2,2'-[[1-[(2-propenyloxy) methyl]-1,2-ethanediyl] bis(oxy)]bis-, bis(4-methylbenzene sulfonate); and ethanol, 2-[1-[[2-[2-[[(4-methylphenyl)sulfonyl] oxy]ethoxy]ethoxy]methyl]-2-(2-propenyloxy)ethoxy]-, 4-methylbenzenesulfonate.
721.1637	1,2-Propanediol, 3-(2-propenyloxy)-, bis(4-methylbenzene sulfonate); 2-propanol, 1-[2-[[(4-methylphenyl)sulfonyl] oxy]ethoxy]-3-(2-propenyloxy)-4-methylbenzenesulfonate; and 2-propanol, 1-[2-[2-[[(4-methylphenyl)sulfonyl]oxy] ethoxy]ethoxy]-3-(2-propenyloxy)-, 4-methylbenzenesulfonate.
721.1640	3,6,9,12,-Tetraoxatetradecane-1,14-diol, bis(4-methylbenzenesulfonate; 3,6,9,13-tetraoxahexadec-15-ene-1,11-diol, bis(4-methylbenzenesulfonate); 3,6,9,12,16-pentaoxanonadec-18-ene-1,14-diol, bis(4-methyl benzenesulfonate); and 3,6,9,12-tetraoxatetradecane-1,14-diol, 7-[(2-propenyloxy)methyl]-, bis(4-methylbenzenesulfonate).
721.1643	Benzenesulfonic acid, amino substituted phenylazo-.
721.1645	Benzenesulfonic acid, 4-methyl-, reaction products with oxirane mono[(C10-16-alkyloxy)methyl] derivatives and 2,2,4(or 2,4,4)-trimethyl-1,6-hexanediamine.
721.1650	Alkylbenzenesulfonic acid and sodium salts.
721.1675	Disulfonic acid rosin amine salt of a benzidine derivative (generic name).
721.1700	Halonitrobenzoic acid, substituted (generic name).
721.1725	Benzoic acid, 3,3'-methylenebis [6 amino-, di-2-propenyl ester.
721.1728	Benzoic acid, 2-(3-phenylbutylidene)amino-, methyl ester.
721.1732	Nitrobenzoic acid octyl ester.
721.1735	Alkylbisoxyalkyl (substituted-1,1-dimethylethylphenyl) benzotriazole (generic name).
721.1740	Substituted dichlorobenzothiazoles.
721.1745	Ethoxybenzothiazole disulfide.
721.1750	1H-Benzotriazole, 5-(pentyloxy)- and 1H-benzotriazole, 5-(pentyloxy)-, sodium and potassium salts.
721.1755	Methylenebisbenzotriazole.
721.1760	Substituted benzotriazole derivatives.
721.1765	2-Substituted benzotriazole.
721.1775	6-Nitro-2(3H)-benzoxazolone.
721.1790	Polybrominated biphenyls.
721.1820	Bisphenol derivative.
721.1825	Bisphenol A, epichlorohydrin, polyalkylenepolyol and polyisocyanato derivative.

721.1850	Toluene sulfonamide bisphenol A epoxy adduct.
721.1875	Boric acid, alkyl and substituted alkyl esters (generic name).
721.1900	Substituted bromothiophene.
721.1907	Butanamide, 2,2'-[3,3'-dichloro [1,1'-biphenyl]-4,4'-diyl)bisazobis[N-2,3-dihydro-2-oxo-1H-benzimidazol-5-yl)-3-oxo-.
721.1920	1,4-Bis(3-hydroxy-4-benzoylphenoxy)butane.
721.1950	2-Butenedioic acid (Z), mono(2-((1-oxopropenyloxy)ethyl) ester .
721.2025	Substituted phenylimino carbamate derivative.
721.2050	Carbamic acid, (trialkyloxy silyalkyl)-substituted acrylate ester.
721.2075	Carbamodithioic acid, methyl-, compound with methanamine (1:1).
721.2084	Carbon oxyfluoride (Carbonic difluoride).
721.2085	Hydroxyalkylquinoline dioxoindandialkylcarboxamide.
721.2086	Coco acid triamine condensate, polycarboxylic acid salts.
721.2088	Carboxylic acids, (C6-C9) branched and linear.
721.2089	Tetrasubstituted aminocarboxylic acid.
721.2092	3-Methylcholanthrene.
721.2094	Substituted carboheterocyclic butane tetracarboxylate.
721.2100	Derivative of tetrachloroethylene.
721.2120	Cyclic amide.
721.2132	Tetraglycidalamines (generic name).
721.2140	Carbopolycyclicol azoalkylaminoalkylcarbomonocyclic ester, halogen acid salt.
721.2150	N,N,N',N'-Tetrakis(oxiranylmethyl)-1,3-cyclohexane dimethanamine.
721.2155	3,3',5,5'-Tetramethylbiphenyl-4,4'-diol.
721.2170	Cyclic phosphazene, methacrylate derivative.
721.2175	Salt of cyclodiamine and mineral acid.
721.2180	Substituted thiazino hydrazine salt (generic name).
721.2184	Titanate [Ti6O13 (2-)], dipotassium.
721.2188	1,3,5-Triazine-2,4,6-triamine, hydrobromide.
721.2192	Disubstituted alkyl triazines (generic name).
721.2194	Substituted triazine isocyanurate (generic name).
721.2196	Poly(substituted triazinyl) piperazine (generic name).
721.2198	Substituted triphenylmethane.
721.2200	Tris (2,3-dibromopropyl) phosphate.
721.2225	Cyclohexanecarbonitrile, 1,3,3-trimethyl-5-oxo-.
721.2250	1,4-Cyclohexanediamine, cls- and trans-.
721.2260	1,2-Cyclohexanedicarboxylic acid, 2,2-bis[[[[2-[(oxiranylmethoxy) carbonyl]cyclohexy]carbonyl]oxy]methyl]-1,3-propanediyl bis(oxiranylmethyl) ester.
721.2270	Aliphatic dicarboxylic acid salt.
721.2287	DDT (Dichlorodiphenyltrichloroethane).
721.2340	Dialkenylamide (generic name).
721.2355	Diethylstilbestrol.
721.2380	Disubstituted diamino anisole.
721.2410	Alkoxylated alkyldiethylenetriamine, alkyl sulfate salts.
721.2420	Alkoxylated dialkyldiethylenetriamine, alkyl sulfate salt.
721.2475	Dimetridazole.
721.2480	Urea, condensate with poly[oxy(methyl-1,2ethanediyl)]-α- (2-aminomethylethyl)-μ-(2-aminoethylethoxy) (generic name).
721.2490	Urea, (hexahydro-6-methyl-2-oxopyrimidinyl)-.
721.2500	Polyamine ureaformaldehyde condensate (specific name).
721.2520	Alkylated diphenyls.

721.2540	Diphenylmethane diisocyanate (MDI) modified.
721.2550	Urethane.
721.2555	Urethane acrylate.
721.2560	Alkylated diphenyl oxide (generic name).
721.2565	Alkylated sulfonated diphenyl oxide, alkali and amine salts.
721.2575	Disubstituted diphenylsulfone.
721.2585	3-Alkyl-2-(2-anilino)vinyl thiazolinium salt (generic name).
721.2600	Epibromohydrin.
721.2625	Reaction product of alkanediol and epichlorohydrin.
721.2650	Acid modified acrylated epoxide.
721.2675	Perfluoroalkyl epoxide (generic name).
721.2725	Trichlorobutylene oxide.
721.2750	Epoxy resin .
721.2800	Erionite fiber.
721.2815	Aliphatic ester.
721.2825	Alkyl ester (generic name).
721.2840	Alkylcarbamic acid, alkynyl ester.
721.2860	Unsaturated amino ester salt (generic name).
721.2880	Unsaturated amino alkyl ester salt(generic name).
721.2900	Substituted aminobenzoic acid ester (generic name).
721.2920	tert-Amyl peroxy alkylene ester (generic name).
721.2930	Substituted benzenedicarboxylic acid ester.
721.2940	Benzoate ester.
721.2950	Carboxylic acid glycidyl esters.
721.2980	Substituted cyclohexyldiamino ethyl esters.
721.3000	Dicarboxylic acid monoester.
721.3020	1,1-Dimethylpropyl peroxyester (generic name).
721.3028	Methacrylic ester.
721.3034	Methylamine esters.
721.3040	Alkenoic acid, trisubstituted-benzyl-disubstituted-phenyl ester.
721.3060	Alkenoic acid, trisubstituted-phenylalkyl-disubstituted-phenyl ester.
721.3080	Substituted phosphate ester (generic).
721.3100	Oligomeric silicic acid ester compound with a hydroxylalkylamine.
721.3120	Propenoate-terminated alkyl substituted silyl ester.
721.3140	Vinyl epoxy ester .
721.3152	Ethanaminium, N-ethyl-2-hydroxy-N,N-bis(2-hydroxyethyl)-, diester with C12-18 fatty acids, ethyl sulfates (salts).
721.3160	1-Chloro-2-bromoethane.
721.3180	Ethane, 2-chloro-1,1,1,2-tetrafluoro-.
721.3200	Ethane, 1,1-dichloro-1-fluoro-.
721.3220	Pentachloroethane.
721.3240	Ethane, 1,1,1,2,2-pentafluoro-.
721.3248	Ethane, 1,2,2-trichlorodifluoro-.
721.3254	Ethane, 1,1,1 trifluoro-.
721.3260	Ethanediimidic acids.
721.3320	Ethanol, 2-amino-, compound with N-hydroxy-N-nitrosobenzenamine (1:1).
721.3350	N-Nitrosodiethanolamine.
721.3360	Substituted ethanolamine.
721.3364	Aliphatic ether.
721.3374	Alkylenediolalkyl ether.

721.3380	Anilino ether.
721.3420	Brominated arylalkyl ether.
721.3430	4-Bromophenyl phenyl ether.
721.3437	Dialkyl ether.
721.3440	Haloalkyl substituted cyclic ethers.
721.3435	Butoxy-substituted ether alkane.
721.3460	Diglycidyl ether of disubstituted carbopolycyle (generic name).
721.3480	Halogenated biphenyl glycidyl ethers.
721.3486	Polyglycerin mono(4-nonylphenyl) ether.
721.3500	Perhalo alkoxy ether.
721.3520	Aliphatic polyglycidyl ether .
721.3620	Fatty acid amine condensate, polycarboxylic acid salts.
721.3625	Fatty acid amine salt (generic name).
721.3627	Branched synthetic fatty acid.
721.3629	Triethanolamine salts of fatty acids.
721.3640	Trimethylolpropane fatty acid diacrylate.
721.3680	Ethylene oxide adduct of fatty acid ester with pentaerythritol.
721.3700	Fatty acid, ester with styrenated phenol, ethylene oxide adduct.
721.3720	Fatty amide.
721.3740	Bisalkylated fatty alkyl amine oxide
721.3760	Fluorene-containing diaromatic amines.
721.3764	Fluorene substituted aromatic amine.
721.3790	Polyfluorocarboxylates.
721.3800	Formaldehyde, condensated polyoxyethylene fatty acid, ester with styrenated phenol, ethylene oxide adduct.
721.3815	Furan, 2-(ethoxymethyl)- tetrahydro-.
721.3860	Glycol monobenzoate.
721.3870	Monomethoxy neopentyl glycol propoxylate monoacrylate.
721.3880	Polyalkylene glycol substituted acetate.
721.3900	Alkyl polyethylene glycol phosphate, potassium salt.
721.4020	Polyalkylene glycol alkyl ether acrylate.
721.4040	Glycols, polyethylene-, 3-sulfo-2-hydroxypropyl-p-(1,1,3,3-tetramethylbutyl)phenyl ether, sodium salt.
721.4060	Alkylene glycol terephthalate and substituted benzoate esters (generic name).
721.4080	MNNG (N-methyl-N'-nitro-N-nitrosoguanidine).
721.4100	Tris(disubstituted alkyl) heterocycle.
721.4110	Allyloxysubstituted heterocycle.
721.4128	Dimethyl-3-substituted heteromonocycle.
721.4133	Dimethyl-3-substituted heteromonocyclic amine.
721.4140	Hexachloronorbornadiene.
721.4155	Hexachloropropene.
721.4160	Benzene, substituted, alkyl acrylate derivative (generic name).
721.4180	Hexamethylphosphoramide.
721.4200	Substituted alkyl peroxyhexane carboxylate (mixed isomers) (generic name).
721.4215	Hexanedioic acid, diethenyl ester.
721.4220	Hexanedioic acid, polymer with 1,2-ethanediol and 1,6-diisocyanato-2,2,4(or 2,4,4)-trimethylhexane, 2-hydroxyethyl-acrylate-blocked.
721.4240	Alkyl peroxy-2-ethyl hexanoate.
721.4250	Hexanoic acid, 2-ethyl-, ethenyl ester.
721.4255	1,4,7,10,13,16-Hexaoxacyclooctadecane, 2-[(2-propenyl oxy)methyl]-.

721.4260	Hydrazine, [4-(1-methylbutoxy)phenyl]-, monohydrochloride.
721.4270	Nitrophenoxylalkanoic acid substituted thiazino hydrazide (generic name).
721.4280	Substituted hydrazine.
721.4300	Hydrazinecarboxamide, N,N'-1,6-hexanediylbis [2,2-dimethyl-.
721.4320	Hydrazinecarboxamide, N,N'-(methylenedi-4,1-phenylene)bis[2,2-dimethyl-.
721.4360	Certain hydrogen containing chlorofluorocarbons.
721.4380	Modified hydrocarbon resin.
721.4390	Trisubstituted hydroquinone diester.
721.4400	Substituted hydroxyalkyl alkenoate, [(1-oxo-2-propenyl)oxy]alkoxy] carbonylamino] substituted] aminocarbonyl]oxy-.
721.4420	Substituted hydroxylamine.
721.4460	Amidinothiopropionic acid hydrochloride.
721.4463	Hydrochlorofluorocarbon.
721.4466	3-Hydroxy-1,1-dimethylbutyl derivative.
721.4470	2,4-Imidazolidinedione, bromochloro-5,5-dimethyl-.
721.4473	Dialkylamidoimidazoline.
721.4480	2-Imino-1,3-thiazin-4-one-5,6-dihydromonohydrochloride.
721.4490	Capped aliphatic isocyanate.
721.4500	Isopropylamine distillation residues and ethylamine distillation residues.
721.4520	Isopropylidene, bis(1,1-dimethylpropyl) derivative.
721.4550	Diperoxy ketal.
721.4568	Methylpolychloro aliphatic ketone.
721.4585	Lecithins, phospholipase A2-hydrolyzed.
721.4590	Mannich-based adduct.
721.4594	Substituted azo metal complex dye.
721.4596	Diazo substituted carbomonocyclic metal complex.
721.4600	Recovered metal hydroxide.
721.4620	Dialkylamino alkanoate metal salt.
721.4640	Substituted benzenesulfonic acid, alkali metal salt.
721.4660	Alcohol, alkali metal salt.
721.4680	Metal salts of complex inorganic oxyacids (generic name).
721.4700	Metalated alkylphenol copolymer (generic name).
721.4720	Disubstituted phenoxazine, chlorometalate salt.
721.4740	Alkali metal nitrites.
721.4780	Hydroxyalkyl methacrylate, alkyl ester.
721.4790	2-(2-Hydroxy-3-tert-butyl-5-methylbenzyl)-4-methyl-6-tert-butylphenyl methacrylate.
721.4794	Polypiperidinol-acrylate methacrylate.
721.4800	Methacrylic ester .
721.4820	Methane, bromodifluoro- .
721.4880	Methanol, trichloro-, carbonate (2:1).
721.4925	Methyl n-butyl ketone.
721.5050	2,2'-[(1-Methylethylidene)bis[4,1-phenyloxy[1-(butoxymethyl)-(2,1-ethanediyl]oxymethylene]]bisoxirane, reaction product with a diamine.
721.5075	Mixed methyltin mercaptoester sulfides.
721.5175	Mitomycin C.
721.5192	Substituted 1,6-dihydroxy naphthalene.
721.5200	Disubstituted phenylazo trisubstituted naphthalene.
721.5225	Naphthalene,1,2,3,4-tetrahydro(1-phenylethyl) (specific name).
721.5275	2-Napthalenecarboxamide-N-aryl-3-hydroxy-4-arylazo (generic name).
721.5278	Substituted naphthalenesulfonic acid, alkali salt.

721.5282	Trisodium chloro [(trisubstituted heteromonocycle amino) propylamino]triazinylamino hydroxyazo naphthalenetrisulfonate.
721.5285	Ethoxylated substituted naphthol.
721.5300	Neodecaneperoxoic acid, 1,1,3,3-tetramethylbutyl ester.
721.5310	Neononanoic acid, ethenyl ester.
721.5325	Nickel acrylate complex .
721.5330	Nickel salt of an organo compound containing nitrogen.
721.5350	Substituted nitrile (generic name).
721.5375	Nitrothiophenecarboxylic acid, ethyl ester, bis[[[[(substituted)]amino]alkylphenyl]azo] (generic name).
721.5385	Octanoic acid, hydrazide.
721.5400	3,6,9,12,15,18,21-Heptaoxatetratriaoctanoic acid, sodium salt.
721.5425	α-Olefin sulfonate, potassium salts.
721.5450	α-Olefin sulfonate, sodium salt.
721.5475	1-Oxa-4-azaspiro[4.5]decane, 4-dichloroacetyl-.
721.5500	7-Oxabicyclo[4.1.0]heptane, 3-ethenyl, homopolymer, ether with 2-ethyl-2-(hydroxymethyl)-1,3-propanediol (3:1), epoxidized.
721.5525	Substituted spiro oxazine.
721.5540	1H,3H,5H-oxazolo [3,4-c] oxazole, dihydro-7a-methyl-.
721.5550	Substituted dialkyl oxazolone (generic name).
721.5575	Oxirane, 2,2'-(1,6-hexanediylbis (oxymethylene)) bis-.
721.5600	Substituted oxirane.
721.5700	Pentanenitrile, 3-amino-.
721.5705	2,5,8,10,13-Pentaoxahexadec-15-enoic acid, 9,14-dioxo-2-[(1-oxo-2-propenyl)oxy]ethyl ester.
721.5710	Phenacetin.
721.5740	Phenol, 4,4'-methylenebis(2,6-dimethyl-.
721.5760	Phenol, 4,4'-[methylenebis(oxy-2,1-ethanediylthio)]bis-.
721.5763	Methylenebisbenzotriazolyl phenols.
721.5769	Mixture of nitrated alkylated phenols.
721.5780	Phenol, 4,4'-(oxybis(2,1-ethanediylthio)bis-.
721.5800	Sulfurized alkylphenol.
721.5820	Aminophenol.
721.5840	Ethylated aminophenol.
721.5860	Methylphenol, bis(substituted)alkyl.
721.5867	Substituted phenol.
721.5880	Sulfur bridged substituted phenols (generic name).
721.5900	Trisubstituted phenol (generic name).
721.5910	Acrylated epoxy phenolic resin.
721.5915	Polysubstituted phenylazopolysubstitutedphenyl dye.
721.5920	Phenyl(disubstitutedpolycyclic).
721.5960	N,N'-Bis(2-(2-(3-alkyl)thiazoline) vinyl)-1,4-phenylenediamine methyl sulfate double salt (generic name).
721.5970	Phosphated polyarylphenol ethoxylate, potassium salt.
721.5980	Dialkyl phosphorodithioate phosphate compounds.
721.5990	Halogenated phosphate ester.
721.6020	Phosphine, dialkylyphenyl.
721.6060	Alkylaryl substituted phosphite.
721.6070	Alkyl phosphonate ammonium salts.
721.6080	Phosphonium salt (generic name).
721.6085	Phosphonocarboxylate salts.

721.6090	Phosphoramide.
721.6100	Phosphoric acid, C6-12-alkyl esters, compounds with 2-(dibutylamino) ethanol.
721.6110	Alkyldi(alkyloxyhydroxypropyl) derivative, phosphoric acid esters, potassium salts.
721.6120	Phosphoric acid, 1,2-ethanediyl tetrakis(2-chloro-1-methylethyl) ester.
721.6140	Dialkyldithiophosphoric acid, aliphatic amine salt.
721.6160	Piperazinone, 1,1',1"-[1,3,5-triazine-2,4,6-triyltris[(cyclohexylimino)-2,1-ethanedlyl]]tris-[3,3,4,5,5-pentamethyl].
721.6186	Polyamine dithiocarbamate.
721.6200	Fatty acid polyamine condensate, phosphoric acid ester salts.
721.6220	Aryl sulfonate of a fatty acid mixture, polyamine condensate.
721.6470	Polyaminopolyacid.
721.6500	Polymer.
721.6520	Acrylamide, polymer with substituted alkylacrylamide salt (generic name).
721.6540	Acrylamide, polymers with tetraalkyl ammonium salt and polyalkyl, aminoalkyl methacrylamide salt.
721.6560	Acrylic acid, polymer with substituted ethene.
721.6580	Polymer of adipic acid, alkanepolyol, alkyldiisocyanatocarbomonocycle, hydroxyalkyl acrylate ester.
721.6620	Alkanaminium, polyalkyl-[(2-methyl-1-oxo-2-propenyl)oxy] salt, polymer with acrylamide and substituted alkyl methacrylate.
721.6625	Oxiranemethanamine, N,N'-[methylenebis(2-ethyl-4,1-phenylene)]bis[N-(oxiranylmethyl)]-.
721.6640	Polymer of alkanedioic acid, methylenebiscarbomonocyclic diisocyanate, and alkylene glycols, hydroxyalkyl acrylate ester.
721.6660	Polymer of alkanepolyol and polyalkylpolyiso-cyanatocarbomonocycle, acetone oxime-blocked (generic name).
721.6680	Alkanoic acid, butanediol and cyclohexanealkanol polymer (generic name).
721.6700	Polymer of alkenoic acid, substituted alkylacrylate sodium salt (generic name).
721.6720	Alkyldicarboxylic acids, polymers with alkanepolyol and TDI, alkanol blocked, acrylate.
721.6740	Polymer of alkyl carbomonocycle diisocyanate with alkanepolyol polyacrylate.
721.6760	Alkylenebis (substituted carbomonocycle), epichlorohydrin, disubstituted heteromonocycle, acrylate polymer.
721.6780	Polymer of substituted alkylphenol formaldehyde and phthalic anhydride, acrylate (generic name).
721.6820	Polymer of substituted aryl olefin.
721.6840	Substituted bis(hydroxyalkane) polymer with epichlorohydrin, acrylate.
721.6880	Bisphenol A, epichlorohydrin, methylenebis (substituted carbomonocycle), polyalkylene glycol, alkanol, methacrylate polymer.
721.6900	Polymer of bisphenol A diglycidal ether, substituted alkenes, and butadiene.
721.6920	Butyl acrylate, polymer with substituted methyl styrene, methyl methacrylate, and substituted silane.
721.6940	Caprolactone, polymer with hexamethylene diisocyanate, hydroxyalkyl acrylate ester, reaction products with substituted alkanoic acid and metal heteromonocycle.
721.6960	E-Caprolactone modified 2-hydroxyethyl acrylate monomer.
721.6980	Dimer acids, polymer with polyalkylene glycol, bisphenol A-diglycidyl ether, and alkylenepolyols polyglycidyl ethers (generic name).
721.7000	Polymer of disodium maleate, allyl ether, and ethylene oxide.
721.7020	Distillates (petroleum), C(3-6), polymers with styrene and mixed terpenes (generic name).
721.7040	Formaldehyde, polymer with (chloromethyl)oxirane, 4,4'-(1-methyl ethylidene)bis[2,6-dibromophenol] and phenol, 2-methyl-2-propenoate.
721.7046	Formaldehyde, polymer with substituted phenols, glycidyl ether.

721.7080	Polymer of hydroxyethyl acrylate and polyisocyanate.
721.7100	Polymer of isophorone diisocyanate, trimethylolpropane, polyalkylenepolyol, disubstituted alkanes and hydroxyethyl acrylate.
721.7140	Methylenebis(4-isocyanato benzene), polymer with polycaprolactone triol and alkoxylated alkanepolyol, hydroxyalkyl methacrylate ester.
721.7160	2-Oxepanone, polymer with 4,4'-(1-methylethylidene)bisphenol and 2,2-[(1-methylethylidene)bis(4,1-phenyleneoxymethylene)]bisoxirane, graft.
721.7180	Substituted oxide-alkylene polymer, methacrylate.
721.7200	Perfluoroalkyl aromatic carbamate modified alkyl methacrylate copolymer.
721.7210	Epoxidized copolymer of phenol and substituted phenol.
721.7220	Polymer of substituted phenol, formaldehyde, epichlorohydrin, and disubstituted benzene.
721.7240	Polymer of disubstituted phthalate, dioxoheteropolycycle, and methacrylic acid.
721.7260	Polymer of polyethylenepolyamine and alkanediol diglycidyl ether.
721.7280	1,3-Propanediamine, N,N'-1,2-ethanediylbis-, polymer with 2,4,6-trichloro-1,3,5-triazine, reaction products with N-butyl-2,2,6,6-tetramethyl-4-piperidinamine.
721.7300	2-Propenenitrile, polymer with 1,3-butadiene, 3-carboxy-1-cyano-1-methylpropyl-terminated, polymers with bisphenol A, epichlorohydrin, and 4,4'-(1methylethylidene)bis[2,6-dibromophenol], dimethacrylate.
721.7320	2-Propenenitrile, polymer with 1,3-butadiene, 3-carboxy-1-cyano-1-methylpropyl-terminated, polymers with epichlorohydrin, formaldehyde, 4,4'-(1-methylethylidene)bis[2,6-dibromophenol], and phenol, 2-methyl-2-propenoate.
721.7340	Polymer of styrene, substituted alkyl methacrylates, 2-ethylhexyl acrylate, methacrylic acid and substituted bis(benzene).
721.7360	Carbamodithioic acid, methyl-, compound with methanamine (1:1).
721.7370	Acrylates of aliphatic polyol.
721.7400	Di(alkanepolyol) ether, polyacrylate.
721.7420	Oxyalkanepolyol polyacrylate.
721.7440	Polyalkylenepolyol alkylamine. (generic name).
721.7450	Aromatic amine polyols.
721.7460	Polyol carboxylate ester.
721.7480	Isocyanate terminated polyols.
721.7500	Nitrate polyether polyol (generic name).
721.7540	Polysubstituted polyol.
721.7560	Alkoxylated alkane polyol, polyacrylate ester.
721.7580	Substituted acrylated alkoxylated aliphatic polyol.
721.7600	Alkyl(heterocyclicyl) phenylazohetero monocyclic polyone (generic name).
721.7620	Alkyl(heterocyclicyl) phenylazohetero monocyclic polyone, ((alkylimidazolyl) methyl) derivative (generic name).
721.7655	Alkylsulfonium salt.
721.7660	Poly(oxy-1,4-butanediyl), α-(1-oxo-2-propenyl)-w-[(1-oxo-2-propenyl)oxy].
721.7680	Poly(oxy-1,2-ethanediyl), .α.-hydro-.w.hydroxy-, ether with 2-ethyl-2-(hydroxymethyl)-1,3-propanediol (3:1) di-2-propenoate, methyl ether
721.7700	Poly(oxy-1,2-ethanediyl), α-hydro-w-(oxiranylmethoxy)-, ether with 2-ethyl-2-(hydroxymethyl)-1,3-propanediol (3:1).
721.7710	Polyepoxy polyol.
721.7720	Poly(oxy-1,2-ethanediyl), α,α'-[(1-methylethylidene) di-4,1-phenylene] bis [w-(oxiranylmethoxy)-.
721.7740	Poly(oxy-1,2-ethanediyl), α-(2-methyl-1-oxo-2-propenyl)-w-hydroxy-, C10-16-alkyl ethers.
721.7760	Poly(oxy-1,2-ethanediyl), α-(1-oxo-2-propenyl)-w-hydroxy-, C10-16-alkyl ethers.

721.7770	Alkylphenoxypoly(oxyethylene) sulfuric acid ester, substituted amine salt.
721.7780	Poly[oxy(methyl-1,2-ethanediyl)], α,α'-(2,2-dimethyl-1,3-propanediyl)bis[w-(oxiranymethoxy)-.
721.8075	Polyurethane.
721.8082	Polyester polyurethane acrylate.
721.8090	Polyurethane polymer.
721.8100	Potassium N,N-bis (hydroxyethyl) cocoamine oxide phosphate, and potassium N,N-bis (hydroxyethyl) tallowamine oxide phosphate.
721.8125	Propane, 1,1,1,2,3,3,3-heptafluoro-.
721.8155	Propanenitrile, 3-[amino, N-tallowalkyl] dipropylenetri-and tripropylenetri- and propanenitrile, 3-[amino, (C14-18 and C16-18 unsaturated alkyl)] trimethylenedi-, dipropylenetri-, and tripropylenetetra-.
721.8160	Propanoic acid, 2,2-dimethyl-, ethenyl ester.
721.8170	Propanol, [2-(1,1-dimethylethoxy)methylethoxy]-.
721.8225	2-Propenamide, N-[3-dimethylamino)propyl]-.
721.8250	1-Propanol, 3,3'-oxybis[2,2-bis(bromomethyl)
721.8265	2-Propenoic acid, C18-26 and C>20 alkyl esters.
721.8275	2-Propenoic acid, 3-(dimethylamino)-2,2-dimethylpropyl ester.
721.8290	2-Propenoic acid, docosyl ester.
721.8300	2-Propenoic acid, 2-hydroxybutyl ester.
721.8325	2-Propenoic acid, 1-(hydroxymethyl) propyl ester.
721.8335	2-Propenoic acid, 2-[[(1-methylethoxy)carbonyl]amino]ethyl ester.
721.8350	2-Propenoic acid, 7-oxabicyclo[4.1.0]hept-3-ylmethyl ester.
721.8375	2-Propenoic acid, 2-(2-oxo-3-oxazolidinyl)ethyl ester.
721.8400	2-Propenoic acid, 3,3,5-trimethylcyclohexyl ester.
721.8425	2-Propenoic acid, 2-[[[[[1,3,3-trimethyl-5- [[[2-[(1-oxo-2-propenyl)oxy] ethoxy]carbonyl]amino]cyclohexyl]methyl] amino]carbonyl]oxy]ethyl ester.
721.8450	2-Propenoic acid, 2-methyl-, 2-[3-(2H-benzotriazol-2-yl)-4-hydroxyphenyl]ethyl ester.
721.8475	2-Propenoic acid, 2-methyl-, 1,1-dimethylethyl ester.
721.8500	2-Propenoic acid, 2-methyl-, 7-oxabicyclo [4.1.0]hept-3-ylmethyl ester.
721.8525	2-Propenoic acid, 2-methyl-, 3,3,5-trimethylcyclohexyl ester.
721.8550	2-Propenoic acid, 2-methyl-, 7,7,9-trimethyl-4,13-dioxo-3,14-dioxo-5,12-diazahexadecane, 1,16-diyl ester.
721.8575	2-Propenoic acid [octahydro-4,7-methano-1H-indene-1, 5(1,6 or 2,5)-diyl]bis(methylene) ester.
721.8600	2-Propenoic acid, octahydro-4, 7-methano-1H-indenyl ester.
721.8650	2-Propenoic acid, reaction product with 2-oxepanone and alkyltriol.
721.8654	2-Propenoic acid 3-(trimethoxy silyl)propyl ester.
721.8670	Alkylcyano substituted pyridazo benzoate.
721.8675	Halogenated pyridines.
721.8700	Halogenated alkyl pyridine.
721.8750	Halogenated substituted pyridine.
721.8775	Substituted pyridines.
721.8825	Substituted methylpyridine and substituted 2-phenoxypyridine.
721.8850	Disubstituted halogenated pyridinol.
721.8875	Substituted halogenated pyridinol.
721.8900	Substituted halogenated pyridinol, alkali salt.
721.8965	1H-Pyrole-2, 5-dione, 1-(2,4,6-tribromophenyl)-.
721.9000	N-Nitrosopyrrolidine.
721.9075	Quaternary ammonium salt of fluorinated alkylaryl amide.

721.9100	Substituted quinoline.
721.9220	Reaction products of secondary alkyl amines with a substituted benzenesulfonic acid and sulfuric acid (generic name).
721.9240	Reaction product of alkyl carboxylic acids, alkane polyols, alkyl acrylate, and isophorone diisocyanate.
721.9260	Reaction product of alkylphenol, tetraalkyl titanate and tin complex.
721.9280	Reaction product of ethoxylated fatty acid oils and a phenolic pentaerythritol tetraester.
721.9300	Reaction products of substituted hydroxyalkanes and polyalkylpolyisocyanatocarbomonocycle.
721.9320	Reaction product of hydroxyethyl acrylate and methyl oxirane.
721.9360	Reaction product of a monoalkyl succinic anhydride with an w-hydroxy methacrylate.
721.9400	Reaction product of phenolic pentaerythritol tetraesters with fatty acid esters and oils, and glyceride triesters.
721.9420	Polymethylcarbomonocycle, reaction product with 2-hydroxyethyl acrylate.
721.9460	Tall oil fatty acids, reaction products with polyamines, alkyl substituted.
721.9470	Reserpine.
721.9480	Resorcinol, formaldehyde substituted carbomonocycle resin.
721.9500	Silane, (1,1-dimethylethoxy)dimethoxy(2-methyl propyl)-.
721.9505	Silanes substituted macrocycle polyethyl.
721.9510	Silicone ester polyacrylate.
721.9525	Acrylate substituted siloxanes and silicones.
721.9526	Sodium perthiocarbonate.
721.9527	Bis(1,2,2,6,6-pentamethyl-4-piperidin-4-ol) ester of cycloaliphatic spiroketal.
721.9530	Bis(2,2,6,6-tetramethylpiperidinyl) ester of cycloalkyl spiroketal.
721.9540	Polysulfide mixture.
721.9550	Sulfonamide.
721.9570	Halophenyl sulfonamide salt.
721.9580	Ethyl methanesulfonate.
721.9620	Aromatic sulfonic acid compound with amine.
721.9630	Polyfluorosulfonic acid salt.
721.9650	Tetramethylammonium salts of alkylbenzenesulfonic acid.
721.9656	Thiaalkanethiol.
721.9658	Thiadiazole derivative.
721.9660	Methylthiouracil.
721.9665	Organotin catalysts.
721.9700	Monosubstituted alkoxyaminotrazines (generic name).
721.9730	1,3,5-Triazin-2-amine, 4-dimethylamino-6-substituted-.
721.9740	Brominated triazine derivative.
721.9750	2-Chloro-4,6-bis(substituted)-1,3,5-triazine, dihydrochloride.
721.9820	Substituted triazole.
721.9850	2,4,8,10-Tetraoxa-3,9-diphosphaspiro[5.5]undecane, 3,9-bis[2,4,6-tris(1,1-dimethylethyl)phenoxy]-.
721.9870	Unsaturated organic compound.
721.9892	Alkylated urea.
721.9925	Aminoethylethylene urea methacrylamide.
721.9957	N-Nitroso-N-methylurethane.
721.9962	Trifunctional aliphatic blocked urethane cross-linker.
721.9975	Zirconium(IV), [2,2-bis[(2-propenyloxy)methyl]-1-butanolato-01,02]tris(2-propenoato-O-)-.

===

significant non-compliance (EPA-94/04): See significant violations.

significant source of ground water (40CFR191.12), as used in this part, means:
(1) An aquifer that:
   (i) Is saturated with water having less than 10,000 milligrams per liter of total dissolved solids;
   (ii) is within 2,500 feet of the land surface;
   (iii) has a transmissivity greater than 200 gallons per day per foot, provided that any formation or part of a formation included within the source of ground water has a hydraulic conductivity greater than 2 gallons per day per square foot; and
   (iv) is capable of continuously yielding at least 10,000 gallons per day to a pumped or flowing well for a period of at least a year; or
(2) an aquifer that provides the primary source of water for a community water system as of the effective date of this subpart.

significant violations (EPA-94/04): Violations by point source dischargers of sufficient magnitude or duration to be a regulatory priority.

significantly (40CFR1508.27) as used in NEPA requires considerations of both context and intensity:
(a) Context. This means that the significance of an action must be analyzed in several contexts such as society as a whole (human, national), the affected region, the affected interests, and the locality. Significance varies with the setting of the proposed action. For instance, in the case of a site-specific action, significance would usually depend upon the effects in the locale rather than in the world as a whole. Both short- and long-term effects are relevant.
(b) Intensity. This refers to the severity of impact. Responsible officials must bear in mind that more than one agency may make decisions about partial aspects of a major action. (Items 1 through 10 in 40CFR1508.27 should be considered in evaluating intensity.)

significantly damaged friable miscellaneous ACM (40CFR763.83) means damaged friable miscellaneous ACM where the damage is extensive and severe.

significantly damaged friable surfacing ACM (40CFR763.83) means damaged friable surfacing ACM in a functional space where the damage is extensive and severe.

significantly greater effluent reduction than BAT (40CFR125.22-c) means that the effluent reduction over BAT produced by an innovative technology is significant when compared to the effluent reduction over best practicable control technology currently available (BPT) produced by BAT.

significantly lower cost (40CFR125.22-d) means that an innovative technology must produce a significant cost advantage when compared to the technology used to achieve BAT limitations in terms of annual capital costs and annual operation and maintenance expenses over the useful life of the technology.

significantly more stringent limitation (40CFR133.101-m) means BOD5 and SS limitations necessary to meet the percent removal requirements of at least 5 mg/L more stringent than the otherwise applicable concentration-based limitations (e.g., less than 25 mg/L in the case of the secondary treatment limits for BOD5 and SS), or the percent removal limitations in 4CFR133.102 and 133.105, if such limits would, by themselves, force significant construction or other significant capital expenditure.

silicomanganese (40CFR60.261-o) means that alloy as defined by ASTM Designation A48364 (Reapproved 1974) (incorporated by reference, 40CFR60.17).

silicomanganese zirconium (40CFR60.261-u) means that alloy containing 60 to 65 percent by weight silicon, 1.5 to 2.5 percent by weight calcium, 5 to 7 percent by weight zirconium, 0.75 to 1.25 percent by weight aluminum, 5 to 7 percent by weight manganese, and 2 to 3 percent by weight barium.

silicon metal (40CFR60.261-x) means any silicon alloy containing more than 96 percent silicon by weight.

silt (EPA-94/04): Sedimentary materials composed of fine or intermediate sized mineral particles.

silvery iron (40CFR60.261-s) means any ferrosilicon,

as defined by ASTM Designation A100-69 (Reapproved 1974) (incorporated by reference, 40CFR60.17), which contains less than 30 percent silicon.

silvicultural point source (40CFR122.27-1) means any discernible, confined and discrete conveyance related to rock crushing, gravel washing, log sorting, or log storage facilities which are operated in connection with silvicultural activities and from which pollutants are discharged into waters of the United States. The term does not include non-point source silvicultural activities such as nursery operations, site preparation, reforestation and subsequent cultural treatment, thinning, prescribed burning, pest and fire control, harvesting operations, surface drainage, or road construction and maintenance from which there is natural runoff. However, some of these activities (such as stream crossing for roads) may involve point source discharges of dredged or fill material which may require a CWA section 404 permit (see 33CFR209.120 and part 233).

silviculture (EPA-94/04): Management of forest land for timber.

similar composition (40CFR162.151-f) refers to a pesticide product which contains only the same active ingredient(s), or combination of active ingredients, and which is in the same category of toxicity, as a federally registered pesticide product.

similar in all material respects (40CFR60.531) means that the construction materials, exhaust and inlet air system, and other design features are within the allowed tolerances for components identified in 40CFR60.533(k).

similar product (40CFR162.151-g) means a pesticide product which, when compared to a federally registered product, has a similar composition and a similar use pattern.

similar source (40CFR63.51) means an emission unit that has comparable emissions and is structurally similar in design and capacity to other emission units such that the emission units could be controlled using the same control technology.

similar systems (40CFR86.092.2) are engine, fuel metering and emission control system combinations which use the same fuel (e.g., gasoline, diesel, etc.), combustion cycle (i.e., two or four stroke), general type of fuel system (i.e., carburetor or fuel injection), catalyst system (e.g., none, oxidization, three-way plus oxidization, three-way only, etc.), fuel control system (i.e., feedback or non-feedback), secondary air system (i.e., equipped or not equipped) and EGR (i.e., equipped or not equipped) (EGR means exhaust gase recirculation).

similar use pattern (40CFR162.151-h) refers to a use of a pesticide product which, when compared to a federally registered use of a product with a similar composition, does not require a change in precautionary labeling under 40CFR156.10(h), and which is substantially the same as the federally registered use. Registrations involving changed use patterns are not included in this term.

simple combustion turbine (40CFR72.2) means a unit that is a rotary engine driven by a gas under pressure that is created by the combustion of any fuel. This term includes combined cycle units without auxiliary firing. This term excludes combined cycle units with auxiliary firing, unless the unit did not use the auxiliary firing from 1985 through 1987 and does not use auxiliary firing at any time after November 15, 1990.

simple cycle gas turbine (40CFR60.331-b) means any stationary gas turbine which does not recover heat from the gas turbine exhaust gases to preheat the inlet combustion air to the gas turbine, or which does not recover heat from the gas turbine exhaust gases to heat water or generate steam (cf. regenerative cycle gas turbine).

simple manufacturing operation (40CFR410.41-a) shall mean all the following unit processes: Desizing, fiber preparation and dyeing (other identical or similar definitions are provided in 40CFR410.51-a) (cf. complex manufacturing operation).

simple manufacturing operation (40CFR410.61-b) shall mean the following unit processes: fiber preparation and dyeing with or without carpet backing (cf. complex manufacturing operation).

simple slaughterhouse (40CFR432.11-c) shall mean

a slaughterhouse which accomplishes very limited byproduct processing, if any, usually no more than two of such operations as rendering, paunch and viscera handling, blood processing, hide processing, or hair processing.

simplify (40CFR29.12-1) means that a State may develop its own format, choose its own submission date, and select the planning period for a State plan.

simultaneous loading (40CFR63.111) means, for a shared control device, loading of organic HAP materials from more than one transfer arm at the same time such that the beginning and ending times of loading cycles coincide or overlap and there is no interruption in vapor flow to the shared control device.

single coat (40CFR52.741) means one coating application applied to a metal surface.

single family structure (40CFR141.2), for the purpose of subpart I of this part only, means a building constructed as a single-family residence that is currently used as either a residence or a place of business.

single passenger commuter vehicle (40CFR52.1161-7) means a motor-driven vehicle with four or more wheels with capacity for a driver plus one or more passengers which is used by a commuter traveling alone to work or classes and is not customarily required to be used in the course of his employment or studies.

single passenger commuter vehicle (40CFR52.2297-10) means a private motor vehicle with four or more wheels with capacity for a driver plus one or more passengers which is used by a commuter traveling alone to work or classes, and is not customarily required to be used in the course of his employment or studies.

single response (40CFR310.11-j) means all of the concerted activities conducted in response to a single episode, incident or threat causing or contributing to a release or threatened release of hazardous substances or pollutants or contaminants.

single seal system (40CFR63.111) means a floating roof having one continuous seal that completely covers the space between the wall of the storage vessel and the edge of the floating roof. This seal may be a vapor-mounted, liquid-mounted, or metallic shoe seal.

single stand (40CFR420.101-d) means those recirculation or direct application cold rolling mills which include only one stand of work rolls (cf. multiple stand).

single unit heavy duty vehicle (40CFR88.302.93) means a self-propelled motor vehicle with a GVWR greater than 8,500 pounds (3,900 kilograms) built on one chassis which encompasses the engine, passenger compartment, and cargo carrying function, and not coupled to trailered equipment. All buses, whether or not they are articulated, are considered single-unit vehicles.

sink (EPA-94/04): Place in the environment where a compound or material collects.

sinking (EPA-94/04): Controlling oil spills by using an agent to trap the oil and sink it to the bottom of the body of water where the agent and the oil are biodegraded.

sinking agents (40CFR300.5) means those additives applied to oil discharges to sink floating pollutants below the water surface.

sinter bed (40CFR60.181-c) means the lead sulfide ore concentrate charge within a sintering machine.

sintering machine (40CFR60.171-c) means any furnace in which calcines are heated in the presence of air to agglomerate the calcines into a hard porous mass called "sinter."

sintering machine (40CFR60.181-b) means any furnace in which a lead sulfide ore concentrate charge is heated in the presence of air to eliminate sulfur contained in the charge and to agglomerate the charge into a hard porous mass called "sinter."

sintering machine discharge end (40CFR60.181-d) means any apparatus which receives sinter as it is discharged from the conveying grate of a sintering machine.

SIP (40CFR57.102-87): See State implementation plan.

SIP call (EPA-94/04): EPA action requiring a state to resubmit all or part of its State Implementation Plan to demonstrate attainment of the required national air quality standards within the statutory deadline. A SIP Revision is a portion of an SIP altered at the request of EPA or on a state's initiative.

sister chromatid exchanges (40CFR798.5900) represent reciprocal interchanges of the two chromatid arms within a single chromosome. These exchanges are visualized during the metaphase portion of the cell cycle and presumably require enzymatic incision, translocation and ligation of at least two DNA helices (other identical or similar definitions are provided in 40CFR798.5915) (under TSCA).

site (40CFR122.2) means the land or water area where any "facility or activity" is physically located or conducted, including adjacent land used in connection with the facility or activity (other identical or similar definitions are provided in 40CFR122.29-4; 124.2).

site (40CFR124.41) means the land or water area upon which a "major PSD stationary source" or "major PSD modification" is physically located or conducted, including but not limited to adjacent land used for utility systems; as repair, storage, shipping or processing areas; or otherwise in connection with the "major PSD stationary source" or "major PSD modification."

site (40CFR144.3) means the land or water area where any "facility or activity" is physically located or conducted, including adjacent land used in connection with the facility or activity.

site (40CFR146.3) means the land or water area where any facility or activity is physically located or conducted, including adjacent land used in connection with the facility or activity.

site (40CFR190.02-d) means the area contained within the boundary of a location under the control of persons possessing or using radioactive material on which is conducted one or more operations covered by this part.

site (40CFR191.02-n) means an area contained within the boundary of a location under the effective control of persons possessing or using spent nuclear fuel or radioactive waste that are involved in any activity, operation, or process covered by this subpart.

site (40CFR270.2) means the land or water area where any facility or activity is physically located or conducted, including adjacent land used in connection with the facility or activity.

site (40CFR300-AA) means area(s) where a hazardous substance has been deposited, stored, disposed, or placed, or has otherwise come to be located. Such areas may include multiple sources and may include the area between sources.

site (40CFR704.3) means a contiguous property unit. Property divided only by a public right-of-way shall be considered one site. There may be more than one plant on a single site. The site for a person who imports a substance is the site of the operating unit within the person's organization which is directly responsible for importing the substance and which controls the import transaction and may in some cases be the organization's headquarters office in the United States.

site (40CFR710.2-w) means a contiguous property unit. Property divided only by a public right-of-way shall be considered one site. There may be more than one manufacturing plant on a single site. For the purposes of imported chemical substances, the site shall be the business address of the importer.

site (40CFR712.3-n) means a contiguous property unit. Property divided only by a public right-of-way shall be considered one site. There may be more than one manufacturing plant on a single site (other identical or similar definitions are provided in 40CFR717.3-j).

site (40CFR721.3) means a contiguous property unit. Property divided only by a public right-of-way is one site. There may be more than one manufacturing plant on a single site.

site (40CFR723.175-3) has the same meaning as in 40CFR710.2 of this chapter.

site (40CFR723.50-12) means a contiguous property unit. Property divided only by a public right-of-way is one site. There may be more than one manufacturing plant on a single site.

site (40CFR763.63-m) means a contiguous property unit. Property divided only by a public right-of-way shall be considered one site. There may be more than one manufacturing plant on a single site.

site (EPA-94/04): An area or place within the jurisdiction of the EPA and/or a state.

site area emergency (10CFR30.4) means events may occur, are in progress, or have occurred that could lead to a significant release of radioactive material and that could require a response by offsite response organizations to protect persons offsite.

site area emergency (10CFR40.4) means events may occur, are in progress, or have occurred that could lead to a significant release of radioactive material and that could require a response by offsite response organizations to protect persons offsite (other identical or similar definitions are provided in 40CFR70.4).

site assessment program (EPA-94/04): A means of evaluating hazardous waste sites through preliminary assessments and site inspections to develop a Hazard Ranking System score.

site inspection (EPA-94/04): The collection of information from a Superfund site to determine the extent and severity of hazards posed by the site. It follows and is more extensive than a preliminary assessment. The purpose is to gather information necessary to score the site, using the Hazard Ranking System, and to determine if it presents an immediate threat requiring prompt removal.

site inspection (SI) (40CFR300.5) means an on-site investigation to determine whether there is a release or potential release and the nature of the associated threats. The purpose is to augment the data collected in the preliminary assessment and to generate, if necessary, sampling and other field data to determine

if further action or investigation is appropriate (under CERCLA).

site lease (40CFR72.2), as used in part 73, subpart E of this chapter, means a legally binding agreement signed between a new IPP (independent power production) or a firm associated with a new IPP and a site owner that establishes the terms and conditions under which the new IPP or the firm associated with the new IPP has the binding right to utilize a specific site for the purposes of operating or constructing the new IPP.

site limited (40CFR710.23-c) means a chemical substance is manufactured and processed only within a site and is not distributed for commercial purposes as a substance or as part of a mixture or article outside the site. Imported substances are never site-limited.

site limited intermediate (40CFR721.3) means an intermediate manufactured, processed, and used only within a site and not distributed in commerce other than as an impurity or for disposal. Imported intermediates cannot be "site-limited."

site of construction (40CFR8.2-v) means the general physical location of any building, highway, or other change or improvement to real property which is undergoing construction, rehabilitation, alteration, conversion, extension, demolition, and repair and any temporary location or facility at which a contractor, subcontractor, or other participating party meets a demand or performs a function relating to the contract or subcontract.

site safety plan (EPA-94/04): A crucial element in all removal actions, it includes information on equipment being used, precautions to be taken, and steps to take in the event of an on-site emergency.

siting (EPA-94/04): The process of choosing a location for a facility.

six minute period (40CFR60.2) means any one of the 10 equal parts of a one-hour period.

six minute period (40CFR63.2) means, with respect to opacity determinations, any one of the 10 equal parts of a 1-hour period.

size (40CFR60.671) means the rated capacity in tons per hour of a crusher, grinding mill, bucket elevator, bagging operation, or enclosed truck or railcar loading station; the total surface area of the top screen of a screening operation; the width of a conveyor belt; and the rated capacity in tons of a storage bin.

size classes of discharges (40CFR300.5) refers to the following size classes of oil discharges which are provided as guidance to the OSC and serve as the criteria for the actions delineated in subpart D. They are not meant to imply associated degrees of hazard to public health or welfare, nor are they a measure of environmental damage. Any oil discharge that poses a substantial threat to public health or welfare or the environment or results in significant public concern shall be classified as a major discharge regardless of the following quantitative measures:

(a) Minor discharge means a discharge to the inland waters of less than 1,000 gallons of oil or a discharge to the coastal waters of less than 10,000 gallons of oil.

(b) Medium discharge means a discharge of 1,000 to 10,000 gallons of oil to the inland waters or a discharge of 10,000 to 100,000 gallons of oil to the coastal waters.

(c) Major discharge means a discharge of more than 10,000 gallons of oil to the inland waters or more than 100,000 gallons of oil to the coastal waters.

size classes of releases (40CFR300.5) refers to the following size classifications which are provided as guidance to the OSC for meeting pollution reporting requirements in subpart B. The final determination of the appropriate classification of a release will be made by the OSC based on consideration of the particular release (e.g., size, location, impact, etc.):

(a) Minor release means a release of a quantity of hazardous substance(s), pollutant(s), or contaminant(s) that poses minimal threat to public health or welfare or the environment.

(b) Medium release means all releases not meeting the criteria for classification as a minor or major release.

(c) Major release means a release of any quantity of hazardous substance(s), pollutant(s), or contaminant(s) that poses a substantial threat to public health or welfare or the environment or results in significant public concern (under CERCLA).

skimming (40CFR61.171) means the removal of slag from the molten converter bath.

skimming (EPA-94/04): Using a machine to remove oil or scum from the surface of the water.

skimming station (40CFR60.141a) means the facility where slag is mechanically raked from the top of the bath of molten iron.

skin sensitization (allergic contact dermatitis) (40CFR798.4100) is an immunologically mediated cutaneous reaction to a substance. In the human, the responses may be characterized by pruritis, erythema, edema, papules, vesicles, bullae, or a combination of these. In other species the reactions may differ and only erythema and edema may be seen.

slag (40CFR60.261-d) means the more or less completely fused and vitrified matter separated during the reduction of a metal from its ore.

SLAMS (40CFR58.1-b): State or Local Air Monitoring Station(s). The SLAMS make up the ambient air quality monitoring network which is required by 40CFR58.20 to be provided for in the State's implementation plan. This definition places no restrictions on the use of the physical structure or facility housing the SLAMS. Any combination of SLAMS and any other monitors (Special Purpose, NAMS, PSD) may occupy the same facility or structure without affecting the respective definitions of those monitoring station.

slaughterhouse (40CFR432.11-b) shall mean a plant that slaughters animals and has as its main product fresh meat as whole, half or quarter carcasses or smaller meat cuts (other identical or similar definitions are provided in 40CFR432.21-b).

slip gauge (40CFR61.61-d) means a gauge which has a probe that moves through the gas/liquid interface in a storage or transfer vessel and indicates the level of vinyl chloride in the vessel by the physical state of the material the gauge discharges.

slop oil (40CFR60.691) means the floating oil and

solids that accumulate on the surface of an oil-water separator (other identical or similar definitions are provided in 40CFR61.341).

slope factor (also referred to as cancer potency factor) (40CFR300-AA) means estimate of the probability of response (for example, cancer) per unit intake of a substance over a lifetime. The slope factor is typically used to estimate upper-bound probability of an individual developing cancer as a result of exposure to a particular level of a human carcinogen with a weight-of-evidence classification of A, B, or C. [$(mg/kg\text{-}day)^{-1}$ for non-radioactive substances and $(pC(i))^{-1}$ for radioactive substances].

slow meter response (40CFR201.1-ii) means that the slow response of the sound level meter shall be used. The slow dynamic response shall comply with the meter dynamic characteristics in paragraph 5.4 of the American National Standard Specification for Sound Level Meters, ANSI S1.4-1971. This publication is available from the American National Standards Institute Inc., 1430 Broadway, New York, New York 10018.

slow meter response (40CFR204.2-14) means the meter ballistics of meter dynamic characteristics as specified by American National Standard S1.4-1971 or subsequent approved revisions.

slow sand filtration (40CFR141.2) means a process involving passage of raw water through a bed of sand at low velocity (generally less than 0.4 m/h) resulting in substantial particulate removal by physical and biological mechanisms.

slow sand filtration (EPA-94/04): Passage of raw water through a bed of sand at low velocity, resulting in substantial removal of chemical and biological contaminants.

sludge (40CFR110.1) means an aggregate of oil or oil and other matter of any kind in any form other than dredged spoil having a combined specific gravity equivalent to or greater than water.

sludge (40CFR240.101-x) means the accumulated semiliquid suspension of settled solids deposited from wastewaters or other fluids in tanks or basins. It does not include solids or dissolved material in domestic sewage or other significant pollutants in water resources, such as silt, dissolved or suspended solids in industrial wastewater effluents, dissolved materials in irrigation return flows or other common water pollutants.

sludge (40CFR241.101-u) means the accumulated semiliquid suspension of settled solids deposited from wastewaters or other fluids in tanks or basins. It does not include solids or dissolved material in domestic sewage or other significant pollutants in water resources, such as silt, dissolved or suspended solids in industrial wastewater effluents, dissolved materials in irrigation return flows or other common water pollutants.

sludge (40CFR243.101-x) means the accumulated semiliquid suspension of settled solids deposited from wastewaters or other fluids in tanks or basins. It does not include solids or dissolved material in domestic sewage or other significant pollutants in water resources, such as silt, dissolved materials in irrigation return flows or other common water pollutants.

sludge (40CFR246.101-aa) means the accumulated semiliquid suspension of settled solids deposited from wastewaters or other fluids in tanks or basins. It does not include solid or dissolved material in domestic sewage or other significant pollutants in water resources, such as silt, dissolved material in irrigation return flows or other common water pollutants.

sludge (40CFR257.2) means any solid, semisolid, or liquid waste generated from a municipal, commercial, or industrial wastewater treatment plant, water supply treatment plant, or air pollution control facility or any other such waste having similar characteristics and effect.

sludge (40CFR258.2) means any solid, semi-solid, or liquid waste generated from a municipal, commercial, or industrial wastewater treatment plant, water supply treatment plant, or air pollution control facility exclusive of the treated effluent from a wastewater treatment plant.

sludge (40CFR260.10) means any solid, semi-solid, or liquid waste generated from a municipal, commercial, or industrial wastewater treatment plant, water supply treatment plant, or air pollution control facility exclusive of the treated effluent from a wastewater treatment plant.

sludge (40CFR261.1-3) has the same meaning used in 40CFR260.10 of this chapter.

sludge (40CFR61.51-l) means sludge produced by a treatment plant that processes municipal or industrial waste waters.

sludge (EPA-94/04): A semi-solid residue from any of a number of air or water treatment processes; can be a hazardous waste.

sludge (RCRA1004-42USC6903) means any solid, semisolid or liquid waste generated from a municipal, commercial, or industrial wastewater treatment plant, water supply treatment plant, or air pollution control facility or any other such waste having similar characteristics and effects.

sludge digester (EPA-94/04): Tank in which complex organic substances like sewage sludges are biologically dredged. During these reactions, energy is released and much of the sewage is converted to methane, carbon dioxide, and water.

sludge dryer (40CFR260.10) means any enclosed thermal treatment device that is used to dehydrate sludge and that has a maximum total thermal input, excluding the heating value of the sludge itself, of 2,500 Btu/lb of sludge treated on a wet-weight basis.

sludge dryer (40CFR61.51-m) means a device used to reduce the moisture content of sludge by heating to temperatures above 65°C (ca. 150°F) directly with combustion gases.

sludge only facility (40CFR122.2) means any "treatment works treating domestic sewage" whose methods of sewage sludge use or disposal are subject to regulations promulgated pursuant to section 405(d) of the CWA, and is required to obtain a permit under 40CFR122.1(b)(3) of this part.

sludge requirements (40CFR403.7(a)(1)-ii) shall mean the following statutory provisions and regulations or permits issued thereunder (or more stringent State or local regulations): section 405 of the Clean Water Act; the Solid Waste Disposal Act (SWDA) (including Title II more commonly referred to as the Resource Conservation Recovery Act (RCRA) and State regulations contained in any State sludge management plan prepared pursuant to Subtitle D of SWDA); the Clean Air Act; the Toxic Substances Control Act; and the Marine Protection, Research and Sanctuaries Act.

slurry (EPA-94/04): A watery mixture of insoluble matter resulting from some pollution control techniques.

sm^3 (40CFR464.02-h) shall mean standard cubic meters.

small appliance (40CFR82.152-v) means any of the following products that are fully manufactured, charged, and hermetically sealed in a factory with five (5) pounds or less of refrigerant: refrigerators and freezers designed for home use, room air conditioners (including window air conditioners and packaged terminal air conditioners), packaged terminal heat pumps, dehumidifiers, under-the-counter ice makers, vending machines, and drinking water coolers.

small business (40CFR33.005) means a business as defined in section 3 of the Small Business Act, as amended (15USC632) (other identical or similar definitions are provided in 40CFR35.6015-43).

small business (40CFR704.102-4) means any manufacturer, importer, or processor who meets either paragraph (a)(4)(i) or (ii) of this section:
(i) A business is small if its total annual sales, when combined with those of its parent (if any), are less than $40 million. However, if the annual manufacture, importation, or processing volume of a particular chemical substance at any individual site owned or controlled by the business is greater than 45,400 kilograms (100,000 pounds), the business shall not qualify as small for purposes of reporting on the manufacture, importation, or processing of that chemical substance at that site, unless the business qualifies as small under paragraph (a)(4)(ii) of this section.
(ii) A business is small if its total annual sales, when combined with those of its parent company (if

any), are less than $4 million, regardless of the quantity of the particular chemical substance manufactured, imported, or processed by that business.

(iii) For imported and processed mixtures containing HEX-BCH, the 45,400 kilograms (100,000 pounds) standard in paragraph (a)(4)(i) of this section applies only to the amount of HEX-BCH in a mixture and not the other components of the mixture.

small business (40CFR791.3-i) refers to a manufacturer or importer whose annual sales, when combined with those of its parent company (if any), are less than $30 million (other identical or similar definitions are provided in 40CFR704.102).

small business concern (40CFR21.2-a) means a concern defined by section 2[3] of the Small Business Act, 15USC632, 13CFR121, and regulations of the Small Business Administration promulgated thereunder.

small business concern (40CFR700.43) means any person whose total annual sales in the person's fiscal year preceding the date of the submission of the applicable section 5 notice, when combined with those of the parent company (if any), are less than $40 million.

small commercial establishments (40CFR35.2005-39) means, for purposes of section 35.2034, private establishments such as restaurants, hotels, stores, filling stations, or recreational facilities and private, nonprofit entities such as churches, schools, hospitals, or charitable organizations with dry weather wastewater flows less than 25,000 gallons per day.

small community (40CFR35.2005-40) means, for purposes of 40CFR35.2020(b) and 35.2032, any municipality with a population of 3,500 or less or highly dispersed sections of larger municipalities, as determined by the Regional Administrator.

small diesel refinery (40CFR72.2) means a domestic motor diesel fuel refinery or portion of a refinery that, as an annual average of calendar years 1988 through 1990 and as reported to the Department of Energy on Form 810, had bona fide crude oil throughput less than 18,250,000 barrels per year, and

the refinery or portion of a refinery is owned or controlled by a refiner with a total combined bona fide crude oil throughput of less than 50,187,500 barrels per year.

small manufacturer (40CFR704.43-5) means a manufacturer (including importers) who meets either paragraph (a)(5)(i) or (ii) of this section:

(i) A manufacture of a chemical substance is small if its total annual sales, when combined with those of its parent company (if any), are less than $40 million. However, if the annual production volume of a particular chemical substance at any individual site owned or controlled by the manufacturer is greater than 45,400 kilograms (100,000 pounds), the manufacturer shall not qualify as small for purposes of reporting on the production of that chemical substance at that site, unless the manufacturer qualifies as small under paragraph (1)(5)(ii) of this section.

(ii) A manufacturer of a chemical substance is small if its total annual sales, when combined with those of its parent company (if any), are less than $4 million, regardless of the quantity of the particular chemical substance produced by that manufacturer.

(iii) For imported mixtures containing a chemical substance identified in paragraph (b) of this section, the 45,400 kilograms (100,000 pounds) standard in paragraph (a)(5)(i) of this section applied only to the amount of the chemical substance in a mixture and not the other components of the mixture.

small manufacturer (40CFR704.45-5) means a manufacturer (importers are defined as manufacturers under TSCA) who meets either of the following standards under this rule:

(i) First standard. A manufacturer of an existing chemical substance is small if its total annual sales, when combined with those of its parent company (if any), are less than $40 million. However, if the annual production volume of a particular chemical substance at any individual site owned or controlled by the manufacturer is greater than 45,400 kilograms (100,000 pounds), the manufacturer shall not qualify as small for purposes of reporting on the production of that chemical substance at the site, unless the manufacturer qualified as small under paragraph (a)(5)(ii) of this section.

(ii) Second standard. A manufacturer of an existing chemical substance is small if its total annual sales, when combined with those of its parent company (if any), are less than $4 million, regardless of the quantity of chemicals produced by that manufacturer.

(Other identical or similar definitions are provided in 40CFR704.3; 710.2-x.)

small manufacturer, processor, or importer (40CFR763.63-n) means a manufacturer or processor who employed no more than 10 full-time employees at any one time in 1981.

small processor (40CFR432.51-b) shall mean an operation that produces up to 2730 kg (6000 lb) per day of any type or combination of finished products (under CWA).

small processor (40CFR704.25-7) means a processor that meets either the standard in paragraph (a)(7)(i) of this section or the standard in paragraph (a)(7)(ii) of this section:

(i) First standard. A processor of a chemical substance is small if its total annual sales, when combined with those of its parent company, if any, are less than $40 million. However, if the annual processing volume of a particular chemical substance at any individual site owned or controlled by the processor is greater than 45,400 kilograms (100,00 pounds), the processor shall not qualify as small for purposes of reporting on the processing of that chemical substance at that site, unless the processor qualifies as small under paragraph (a)(7)(ii) of this section.

(ii) Second standard. A processor of a chemical substance is small if its total annual sales, when combined with those of its parent company (if any), are less than $4 million, regardless of the quantity of the particular chemical substance processed by that company.

(iii) Inflation index. EPA will use the Inflation Index described in the definition of "small manufacturer" set forth in 40CFR704.3, for purposes of adjusting the total annual sales values of this small processor definition. EPA will provide notice in the Federal Register when

changing the total annual sales values of this definition.

(Other identical or similar definitions are provided in 40CFR704.33-4; 704.104-3; 704.203.)

small quantities (40CFR710.2-y), for purposes of scientific experimentation or analysis or chemical research on, or analysis of, such substance or another substance, including any such research or analysis for the development of a product (hereinafter sometimes shortened to "small quantities for research and development"), means quantities of a chemical substance manufactured, imported, or processed or proposed to be manufactured, imported, or processed that:

(1) are no greater than reasonably necessary for such purposes and

(2) after the publication of the revised inventory, are used by, or directly under the supervision of, a technically qualified individual(s).

small quantities for research and development (40CFR710.2-y): See small quantities for purposes of scientific experimentation or analysis or chemical research on, or analysis of, such substance or another substance, including any such research or analysis for the development of a product.

small quantities for research and development (40CFR761.3) means any quantity of PCBs:

(1) that is originally packaged in one or more hermetically sealed containers of a volume of no more than five (5.0) milliliters, and

(2) that is used only for purposes of scientific experimentation or analysis, or chemical research on, or analysis of, PCBs, but not for research or analysis for the development of a PCB product (under TSCA).

small quantities solely for purposes of scientific experimentation or analysis or chemical research on, or analysis of, such substance or another substance, including such research or analysis for the development of a product: See small quantities solely for research and development (40CFR704.3; 720.3-cc; 747.115-1; 747.195-1; 747.200-1).

small quantities solely for research and development (or "small quantities solely for purposes of scientific experimentation or analysis or chemical research on, or analysis of, such substance or another substance, including such research or analysis for the development of a product") (40CFR704.3) means quantities of a chemical substance manufactured, imported, or processed or proposed to be manufactured, imported, or processed solely for research and development that are not greater than reasonably necessary for such purposes (other identical or similar definitions are provided in 40CFR20.3-cc; 747.115-1; 747.195-1; 747.200-1).

small quantity generator (40CFR260.10) means a generator who generates less than 1000 kg of hazardous waste in a calendar month.

small quantity generator (SQG-sometimes referred to as "squeegee") (EPA94/04): Persons or enterprises that produce 220-2200 pounds per month of hazardous waste; are required to keep more records than conditionally exempt generators. The largest category of hazardous waste generators, SQGs, include automotive shops, dry cleaners, photographic developers, and a host of other small businesses (cf. conditionally exempt generators).

small refinery (40CFR80.2-aa) means a domestic diesel fuel refinery:
(1) Which has a crude oil or bonafide feedstock capacity of 50,000 barrels per day or less, and
(2) Which is not owned or controlled by any refiner with a total combined crude oil or bonafide feedstock capacity greater than 137,500 barrels per day.
The above capacities shall be measured in terms of the average of the actual daily utilization rates of the affected refiners or refineries during the period January 1, 1988 to December 31, 1990. These averages will be calculated as barrels per calendar day.

small sized plants (40CFR428.51) shall mean plants which process less than 3,720 kg/day (8,200 lbs/day) of raw materials.

small source (CAA302-42USC7602) means a source that emits less than 100 tons of regulated pollutants per year, or any class of persons that the Administrator determines, through regulation, generally lack technical ability or knowledge regarding control of air pollution.

small uses (40CFR82.172) means any use of a substitute in a sector other than a major industrial use sector, or production by any producer for use of a substitute in a major industrial sector of 10,000 lbs. or less per year.

small water system (40CFR141.2), for the purpose of subpart I of this part only, means a water system that serves 3,300 persons or fewer (under the Safe Drinking Water Act).

SMCL (40CFR143.2-f): See secondary maximum contaminant levels.

smelt dissolving tank (40CFR60.281-m) means a vessel used for dissolving the smelt collected from the recovery furnace.

smelter (EPA-94/04): A facility that melts or fuses ore, often with an accompanying chemical change, to separate its metal content. Emissions cause pollution. "Smelting" is the process involved.

smelter owner and operator (40CFR57.103-t) means the owner or operator of the smelter, without distinction.

smelting (40CFR60.161-e) means processing techniques for the melting of a copper sulfide ore concentrate or calcine charge leading to the formation of separate layers of molten slag, molten copper, and/or copper matte (under the Clean Air Act).

smelting furnace (40CFR60.161-f) means any vessel in which the smelting of copper sulfide ore concentrates or calcines is performed and in which the heat necessary for smelting is provided by an electric current, rapid oxidation of a portion of the sulfur contained in the concentrate as it passes through an oxidizing atmosphere, or the combustion of a fossil fuel.

smog (EPA-94/04): Air pollution associated with oxidants (cf. photochemical smog).

smoke (40CFR86.082.2) means the matter in the exhaust emission which obscures the transmission of light (other identical or similar definitions are provided in 40CFR87.1).

smoke (EPA-94/04): particles suspended in air after incomplete combustion.

smoke emissions (40CFR60-AA (method 22-3.4)) means pollutant generated by combustion in a flare and occurring immediately downstream of the flame. Smoke occurring within the flame, but not downstream of the flame, is not considered a smoke emission.

smoke number (SN) (40CFR87.1) means the dimensionless term quantifying smoke emissions.

smooth-one-side (S1S) hardboard (40CFR429.11-g) means hardboard which is produced by the wet-matting, wet-pressing process.

smooth-two-sides (S2S) hardboard (40CFR429.11-h) means hardboard which is produced by the wet-matting, dry-pressing process.

SN (40CFR87.1): See smoke number.

snap beans (40CFR407.71-q) shall mean the processing of snap beans into the following product styles: Canned and frozen green, Italian, wax, string, bush, and other related varieties, whole, French, fancy, Extra Standard, Standard, and other cuts.

SO_2 (40CFR58.1-e) means sulfur dioxide.

soda lime recipe (40CFR60.291) means glass product composition of the following ranges of weight proportions: 60 to 75 percent silicon dioxide, 10 to 17 percent total R_2O (e.g., Na_2O and K_2O), 8 to 20 percent total RO but not to include any PbO (e.g., CaO, and MgO), 0 to 8 percent total R_2O_3 (e.g., Al_2O_3), and 1 to 5 percent other oxides.

soft detergents (EPA-94/04): Cleaning agents that break down in nature.

soft water (EPA-94/04): Any water that does not contain a significant amount of dissolved minerals such as salts of calcium or magnesium.

soil (40CFR192.11-d) means all unconsolidated materials normally found on or near the surface of the earth including, but not limited to, silts, clays, sands, gravel, and small rocks.

soil (40CFR761.123) means all vegetation, soils and other ground media, including but not limited to, sand, grass, gravel, and oyster shells. It does not include concrete and asphalt.

soil (40CFR796.2700-iv) is the unconsolidated mineral material on the immediate surface of the earth that serves as a natural medium for the growth of land plants; its formation and properties are determined by various factors such as parent material, climate, macro- and micro-organisms, topography, and time (other identical or similar definitions are provided in 40CFR796.2750-viii).

soil (40CFR796.3400-A) is a mixture of mineral and organic chemical constituents, the latter containing compounds of high carbon and nitrogen content and of high molecular weights, animated by small (mostly micro-) organisms. Soil may be handled in two states:
(1) Undisturbed, as it has grown with time, in characteristic layers of a variety of soil types,
(2) Disturbed, as it is usually sampled by digging and used in the test described here.

soil adsorption field (EPA-94/04): A sub-surface area containing a trench or bed with clean stones and a system of piping through which treated sewage may seep into the surrounding soil for further treatment and disposal.

soil aggregate (40CFR796.2700-v) is the combination or arrangement of soil separates (sand, silt, clay) into secondary units. These units may be arranged in the profile in a distinctive characteristic pattern that can be classified on the basis of size, shape, and degree of distinctness into classes, type, and grades (under TSCA).

soil aggregate (40CFR796.2750-ix) is the combination or arrangement of soil separates (sand, silt, clay) into secondary units. These units may be arranged in the oil profile in a distinctive characteristic pattern that can be classified according to size, shape, and degree of distinctness into classes, types, and grades.

soil and water conservation practices (EPA-94/04): control measures consisting of managerial, vegetative, and structural practices to reduce the loss of soil and water.

soil classification (40CFR796.2700-vi) is the systematic arrangement of soils into groups or categories. Broad groupings are made on the basis of general characteristics, subdivisions, on the basis of more detailed differences in specific properties. The soil classification system used today in the United States is the 7th Approximation Comprehensive System. The ranking of subdivisions under the system is: order, suborder, great group, family and series.

soil classification (40CFR796.2750-x) is the systematic arrangement of soils into groups or categories. Broad groupings are based on general soil characteristics while subdivisions are based on more detailed differences in specific properties. The soil classification system used in this standard and the one used today in the United States is the 7th Approximation Comprehensive System. The ranking of subdivisions under this system is: order, suborder, great group, family, and series.

soil conditioner (EPA-94/04): An organic material like humus or compost that helps soil absorb water, build a bacterial community, and take up mineral nutrients.

soil erodibility (EPA-94/04): A measure of the soil's susceptibility to raindrop impact, runoff, and other erosional processes.

soil gas (EPA-94/04): Gaseous elements and compounds in the small spaces between particles of the earth and soil. Such gases can be moved or driven out under pressure.

soil horizon (40CFR796.2700-vii) is a layer of soil approximately parallel to the land surface. Adjacent layers differ in physical, chemical, and biological properties or characteristics such as color, structure, texture, consistency, kinds, and numbers of organisms present, and degree of acidity or alkalinity.

soil horizon (40CFR796.2750-xi) is a layer of soil approximately parallel to the land surface. Adjacent layers differ in physical, chemical, and biological properties such as color, structure, texture, consistency, kinds and numbers of organisms present, and degree of acidity or alkalinity.

soil injection (40CFR165.1-v) means the emplacement of pesticides by ordinary tillage practices within the plow layer of soil.

soil order (40CFR796.2700-viii) is the broadest category of soil classification and is based on general similarities of physical/chemical properties. The formation by similar genetic processes causes these similarities. The soil orders found in the United States are: Alfisol, Aridisol, Entisol, Histosol, Inceptisol, Mollisol, Oxisol, Spodosol, Ultisol, and Vertisol.

soil order (40CFR796.2750-xii) is the broadest category of soil classification and is based on the general similarities of soil physical/chemical properties. The formation of soil by similar general genetic processes causes these similarities. The Soil Orders found in the United States are: Alfisol, Aridisol, Entisol, Histosol, Inceptisol, Mollisol, Oxisol, Spodosol, Ultisol, and Vertisol.

soil organic matter (40CFR796.2700-ix) is the organic fraction of the soil; it includes plant and animal residues at various stages of decomposition, cells and tissues of soil organisms, and substances synthesized by the microbial population.

soil pH (40CFR257.3.5-9) is the value obtained by sampling the soil to the depth of cultivation or solid waste placement, whichever is greater, and analyzing by the electrometric method. ("Methods of Soil Analysis, Agronomy Monograph No. 9," C.A. Black, ed., American Society of Agronomy, Madison, Wisconsin, pp. 914-926, 1965.)

soil pH (40CFR796.2700-x) is the negative logarithm to the base 10 of the hydrogen ion activity of a soil as determined by means of a suitable sensing electrode coupled with a suitable reference electrode at a 1:1 soil:water ratio.

soil series (40CFR796.2700-xi) is the basic unit of soil classification and is a subdivision of a family. A series consists of soils that were developed under comparable climatic and vegetational conditions. The soils comprising a series are essentially alike in all major profile characteristics except for the texture of the "A" horizon (i.e., the surface layer of soil) (other identical or similar definitions are provided in 40CFR796.2750-xiii).

soil sterilant (EPA-94/04): A chemical that temporarily or permanently prevents the growth of all plants and animals. depending on the chemical.

soil texture (40CFR796.2700-xii) refers to the classification of soils based on the relative proportions of the various soil separates present. The soil textural classes are: clay, sandy clay, silty clay, clay loam, silty clay loam, sandy clay loam, loam, silt loam, silt, sandy loam loamy sand, and sand.

soil texture (40CFR796.2750-xiv) is a classification of soils that is based on the relative proportions of the various soil separates present. The soil textural classes are: clay, sandy clay, silty clay, clay loam, silty clay loam, sandy clay loam, loam, silt loam, silt, sandy loam, loamy sand, and sand.

solar irradiance in water (40CFR796.3700-viii) is related to the sunlight intensity in water and is proportional to the average light flux (in the units of 10^{-3} einsteins cm^{-2}- day^{-1} that is available to cause photoreaction in a wavelength interval centered at (wavelength) over a 24-hour day at a specific latitude and season date (other identical or similar definitions are provided in 40CFR796.3780-ix).

sold or distributed (40CFR167.3) means the aggregate amount of a pesticidal product released for shipment by the establishment in which the pesticidal product was produced.

solder (EPA-94/04): Metallic compound used to seal joints between pipes. Until recently, most solder contained 50 percent lead. Use of lead solder containing more than 0.2 percent lead is now prohibited for pipes carrying drinking water.

sole or principal source aquifer (40CFR146.3) means an aquifer which has been designated by the Administrator pursuant to section 1424 (a) or (e) of the SDWA.

sole or principal source aquifer (40CFR149.2-d) means an aquifer which is designated as an SSA under section 1424(e) of the SDWA.

sole source aquifer (EPA-94/04): An aquifer that supplies 50-percent or more of the drinking water of an area.

solid derived fuel (40CFR60.41a-24) means any solid, liquid, or gaseous fuel derived from solid fuel for the purpose of creating useful heat and includes, but is not limited to, solvent refined coal, liquified coal, and gasified coal.

solid waste (40CFR243.101-y) means garbage, refuse, sludges, and other discarded solid materials, including solid waste materials resulting from industrial, commercial, and agricultural operations, and from community activities, but does not include solid or dissolved materials in domestic sewage or other significant pollutants in water resources, such as silt, dissolved or suspended solids in industrial wastewater effluents, dissolved materials in irrigation return flows or other common water pollutants. Unless specifically noted otherwise, the term "solid 40CFR Chapter 1 (7-1-86 Edition) waste" as used in these guidelines shall not include mining, agricultural, and industrial solid wastes; hazardous wastes; sludges; construction and demolition wastes; and infectious wastes.

solid waste (40CFR246.101-bb) means garbage, refuse, sludge, and other discarded solid materials, including solid waste materials resulting from industrial, commercial, and agricultural operations, and from community activities, but does not include solids or dissolved materials in domestic sewage or other significant pollutants in water resources, such as silt, dissolved or suspended solids in industrial wastewater effluents, dissolved materials in irrigation return flows or other common water pollutants. Unless specifically noted otherwise, the term "solid waste" as used in these guidelines shall not include mining, agricultural, and industrial solid wastes; hazardous wastes; sludges; construction and demolition wastes; and infectious wastes.

solid waste (40CFR247.101-i) means garbage, refuse, sludges, and other discarded solid materials, including solid waste materials resulting from industrial, commercial, and agricultural operations, and from community activities, but does not include solids or dissolved materials in domestic sewage or other significant pollutants in water resources, such as silt, dissolved or suspended solids in industrial waste water effluents, dissolved materials in irrigation return flow, or other common water pollutants.

solid waste (40CFR257.2) means any garbage, refuse,

sludge from a waste treatment plant, water supply treatment plant, or air pollution control facility and other discarded material, including solid, liquid, semisolid, or contained gaseous material resulting from industrial, commercial, mining, and agricultural operations, and from community activities, but does not include solid or dissolved materials in domestic sewage, or solid or dissolved material in irrigation return flows or industrial discharges which are point sources subject to permits under section 402 of the Federal Water Pollution Control Act, as amended (86 Stat. 880), or source, special nuclear, or byproduct material as defined by the Atomic Energy Act of 1954, as amended (68 Stat. 923).

solid waste (40CFR258.2) means any garbage, or refuse, sludge from a wastewater treatment plant, water supply treatment plant, or air pollution control facility and other discarded material, including solid, liquid, semi-solid, or contained gaseous material resulting from industrial, commercial, mining, and agricultural operations, and from community activities, but does not include solid or dissolved materials in domestic sewage, or solid or dissolved materials in irrigation return flows or industrial discharges that are point sources subject to permit under 33 U.S.C. 1342, or source, special nuclear, or byproduct material as defined by the Atomic Energy Act of 1954, as amended (68 Stat. 923).

solid waste (40CFR259.10.a) means a solid waste defined in section 1004 (27) of RCRA.

solid waste (40CFR260.10) means a solid waste as defined in 261.2 of this chapter.

solid waste (40CFR261.2) is any discarded material that is not excluded by 40CFR261.4(a) or that is not excluded by variance granted under 40CFR260.30 and 40CFR260.31.

solid waste (40CFR60.51) means refuse, more than 50 percent of which is municipal type waste consisting of a mixture of paper, wood, yard wastes, food wastes, plastics, leather, rubber, and other combustibles, and noncombustible materials such as glass and rock.

solid waste (EPA-94/04): Non-liquid, non-soluble materials ranging from municipal garbage to industrial wastes that contain complex and sometimes hazardous substances. Solid wastes also include sewage sludge, agricultural refuse, demolition wastes, and mining residues. Technically, solid waste also refers to liquids and gases in containers.

solid waste (RCRA1004-42USC6903) means any garbage, reuse, sludge from a waste treatment plan, water supply treatment plant, or air pollution control facility and other discarded material, including solid, liquid, semisolid, or contained gaseous material resulting from industrial, commercial, mining, and agricultural operations, and from community activities, but does not include solid or dissolved material in domestic sewage, or solid or dissolved materials in irrigation return flows or industrial discharges which are point sources subject to permits under section 402 of the Federal Water Pollution Control Act, as amended (86 Stat. 880), or source, special nuclear, or byproduct material as defined by the Atomic Energy Act of 1954, as amended (68 Stat. 923).

solid waste and medical waste (CAA129.g-42USC7429) shall have the meanings established by the Administrator pursuant to the Solid Waste Disposal Act.

solid waste boundary (40CFR257.3.4-5) means the outermost perimeter of the solid waste (projected in the horizontal plane) as it would exist at completion of the disposal activity.

solid waste derived fuel (40CFR247.101-j) means a fuel that is produced from solid waste that can be used as a primary or supplementary fuel in conjunction with or in place of fossil fuels. The solid-waste-derived fuel can be in the form of raw (unprocessed) solid waste, shredded (or pulped) and classified solid waste, gas or oil derived from pyrolyzed solid waste, or gas derived from the biodegradation of solid waste.

solid waste disposal (EPA-94/04): The final placement of refuse that is not salvaged or recycled.

solid waste incineration unit (CAA129.g-42USC7429) means a distinct operating unit of any facility which combusts any solid waste material from commercial or industrial establishments or the general public

(including single and multiple residences, hotels, and motels). Such term does not include incinerators or other units required to have a permit under section 3005 of the Solid Waste Disposal Act. The term "solid waste incineration unit" does not include:
(A) materials recovery facilities (including primary or secondary smelters) which combust waste for the primary purpose of recovering metals,
(B) qualifying small power production facilities, as defined in section 3(17)(C) of the Federal Power Act (16USC769(17)(C)), or qualifying cogeneration facilities, as defined in section 3(18)(B) of the Federal Power Act (16USC796(18)(B)), which burn homogeneous waste (such as units which burn tires or used oil, but not including refuse-derived fuel) for the production of electric energy or in the case of qualifying cogeneration facilities which burn homogeneous waste for the production of electric energy and steam or forms of useful energy (such as heat) which are used for industrial, commercial, heating or cooling purposes, or
(C) air curtain incinerators provided that such incinerators only burn wood wastes, yard wastes and clean lumber and that such air curtain incinerators comply with opacity limitations to be established by the Administrator by rule.

solid waste incinerator (40CFR72.2) means a source as defined in section 129(g)(1) of the Act.

solid waste management (EPA-94/04): Supervised handling of waste materials from their source through recovery processes to disposal.

solid waste management (RCRA1004-42USC6903) means the systematic administration of activities which provide for the collection, source separation, storage, transportation, transfer, processing, treatment, and disposal of solid waste.

solid waste management facility (RCRA1004-42USC6903) includes:
(A) any resource recovery system or component thereof,
(B) any system, program, or facility for resource conservation, and
(C) any facility for the collection, source separation, storage, transportation, transfer, processing,

treatment or disposal of solid wastes, including hazardous wastes, whether such facility is associated with facilities generating such wastes or otherwise.

solid waste planning, solid waste management, and comprehensive planning (RCRA1004-42USC6903) include planning or management respecting resource recovery and resource conservation.

solid waste storage container (40CFR243.101-bb) means a receptacle used for the temporary storage of solid waste while awaiting collection.

Solid Waste Disposal Act: See act or SWDA.

solid wastes (40CFR240.101-y) means garbage, refuse, sludges, and other discarded solid materials resulting from industrial and commercial operations and from community activities. It does not include solids or dissolved material in domestic sewage or other significant pollutants in water resources, such as silt, dissolved or suspended solids in industrial wastewater effluents, dissolved materials in irrigation return flows or other common water pollutants (other identical or similar definitions are provided in 40CFR241.101-v).

solidification and stabilization (EPA-94/04): Removal of wastewater from a waste or changing it chemically to make it less permeable and susceptible to transport by water.

solution (40CFR796.1840-vii) is a homogeneous mixture of two or more substances constituting a single phase (other identical or similar definitions are provided in 40CFR796.1860-iv).

solution heat treatment (40CFR468.02-p) shall mean the process introducing a workpiece into a quench bath for the purpose of heat treatment following rolling, drawing or extrusion.

solvent (40CFR52.741) means a liquid substance that is used to dissolve or dilute another substance.

solvent (40CFR795.120) means a substance (e.g., acetone) which is combined with the test substance to facilitate introduction of the test substance into the dilution water.

solvent applied in the coating (40CFR60.441) means all organic solvent contained in the adhesive, release, and precoat formulations that is metered into the coating applicator from the formulation area.

solvent borne (40CFR60.391) means a coating which contains five percent or less water by weight in its volatile fraction.

solvent borne ink systems (40CFR60.431) means ink and related coating mixtures whose volatile portion consists essentially of VOC solvent with not more than five weight percent water, as applied to the gravure cylinder.

solvent cleaning (40CFR52.741) means the process of cleaning soils from surfaces by cold cleaning, open top vapor degreasing, or conveyorized degreasing.

solvent extraction operation (40CFR264.1031) means an operation or method of separation in which a solid or solution is contacted with a liquid solvent (the two being mutually insoluble) to preferentially dissolve and transfer one or more components into the solvent.

solvent feed (40CFR60.601) means the solvent introduced into the spinning solution preparation system or precipitation bath. This feed stream includes the combination of recovered solvent and makeup solvent.

solvent filter (40CFR60.621) means a discrete solvent filter unit containing a porous medium that traps and removes contaminants from petroleum solvent, together with the piping and ductwork used in the installation of this device.

solvent inventory variation (40CFR60.601) means the normal changes in the total amount of solvent contained in the affected facility.

solvent of high photochemical reactivity (40CFR52.1145-2) means any solvent with an aggregate of more than 20 percent of its total volume composed of the chemical compounds classified below or which exceeds any of the following individual percentage composition limitations in reference to the total volume of solvent:
(i) A combination of hydrocarbons, alcohols,

aldehydes, esters, ethers, or ketones having an olefinic or cycloolefinic type of unsaturation: 5 percent;
(ii) A combination of aromatic compounds with eight or more carbon atoms to the molecule except ethylbenzene: 8 percent;
(iii) A combination of ethylbenzene, ketones having branched hydrocarbon structures, trichloroethylene or toluene: 20 percent. Whenever any organic solvent or any constituent of an organic solvent may be classified from its chemical structure into more than one of the above groups of organic compounds, it shall be considered as a member of the most reactive chemical group, that is, that group having the least allowable percentage of total volume of solvents.

solvent recovery dryer (40CFR60.621) means a class of dry cleaning dryers that employs a condenser to condense and recover solvent vapors evaporated in a closed-loop stream of heated air, together with the piping and ductwork used in the installation of this device.

solvent recovery system (40CFR60.431) means an air pollution control system by which VOC solvent vapors in air or other gases are captured and directed through a condenser(s) or a vessel(s) containing beds of activated carbon or other adsorbents. For the condensation method, the solvent is recovered directly from the condenser. For the adsorption method, the vapors are adsorbed, then desorbed by steam or other media, and finally condensed and recovered.

solvent recovery system (40CFR60.601) means the equipment associated with capture, transportation, collection, concentration, and purification of organic solvents. It may include enclosures, hoods, ducting, piping, scrubbers, condensers, carbon absorbers, distillation equipment, and associated storage vessels.

solvent spun synthetic fiber (40CFR60.601) means any synthetic fiber produced by a process that uses an organic solvent in the spinning solution, the precipitation bath, or processing of the sun fiber (under CAA).

solvent spun synthetic fiber process (40CFR60.601)

means the total of all equipment having a common spinning solution preparation system or a common solvent recovery system, and that is used in the manufacture of solvent-spun synthetic fiber. It includes spinning solution preparation, spinning, fiber processing and solvent recovery, but does not include the polymer production equipment.

soot (EPA-94/04): Carbon dust formed by incomplete combustion.

sorbent (40CFR260.10) means a material that is used to soak up free liquids by either adsorption or absorption, or both. Sorb means to either adsorb or absorb, or both.

sorption (EPA-94/04): The action of soaking up or attracting substances; process used in many pollution control systems.

sound exposure level (40CFR201.1-bb) means the level in decibels calculated as ten times the common logarithm of time integral of squared A-weighted sound pressure over a given time period or event divided by the square of the standard reference sound pressure of 20 micropascals and a reference duration of one second.

sound level (40CFR201.1-aa) means the level, in decibels, measured by instrumentation which satisfies the requirements of American National Standard Specification for Sound Level Meters S1.4-1971 Type 1 (or SIA) or Type 2 if adjusted as shown in Table 1. This publication is available from the American National Standards Institute, Inc., 1430 Broadway, New York, New York 10018. For the purpose of these procedures the sound level is to be measured using the A-weighting of spectrum and either the FAST or SLOW dynamic averaging characteristics, as designated. It is abbreviated as L_A (under the Noise Control Act).

sound level (40CFR202.10-q) means the quantity in decibels measured by a sound level meter satisfying the requirements of American National Standards Specification for Sound Level Meters S1.4-1971. This publication is available from the American National Standards Institute, Inc., 1430 Broadway, New York, New York 10018. Sound level is the frequency-weighted sound pressure level obtained with the

standardized dynamic characteristic fast or slow and weighting B, or C; unless indicated otherwise, the A-weighting is understood.

sound level (40CFR204.2-15) means the weighted sound pressure level measured by the use of a metering characteristic and weighting B, or C as specified in American National Standard Specification for Sound Level Meters SI.4-11 or subsequent approved revision. The weighting employed must be specified, otherwise A-weighting is understood.

sound level (40CFR205.2-7) means 20 times the logarithm to base 10 of the ratio of pressure of a sound to the reference pressure. The reference pressure is 20 micropascals (20 micronewtons per square meter). Note: Unless otherwise explicitly stated, it is to be understood that the sound pressure is the effective (rms) sound pressure, per American National Standards Institute, Inc., 1430 Broadway, New York, New York 10018.

sound pressure level (40CFR201.1-cc) (in stated frequency band) means the level, in decibels, calculated as 20 times the common logarithm of the ratio of a sound pressure to the reference sound pressure of 20 micropascals.

sound pressure level (40CFR204.2-16) means, in decibels, 20 times the logarithm to the base ten of the ratio of a sound pressure to the reference sound pressure of 20 micropascals (20 micronewtons per square meter). In the absence of any modifier, the level is understood to be that of a root-mean-square pressure.

sound pressure level (40CFR205.2-8) means in decibels, 20 times the logarithm to the base 10 of the ratio of a sound pressure to the reference sound pressure of 20 micropascals (20 micronewtons per square meter). In the absence of any modifier, the level is understood to be that of a root-mean-square pressure. The unit of any sound level is the decibel, having the unit symbol dB.

soups (40CFR407.81-j) shall mean the combination of various fresh and pre-processed meats, fish, diary products, eggs, flours, starches, vegetables, spices, and other similar raw ingredients into a variety of finished mixes and styles but not including dehydrated soups.

sour water stream (40CFR61.341) means a stream that:
(1) Contains ammonia or sulfur compounds (usually hydrogen sulfide) at concentrations of 10 ppm by weight or more;
(2) is generated from separation of water from a feed stock, intermediate, or product that contained ammonia or sulfur compounds; and
(3) requires treatment to remove the ammonia or sulfur compounds.

sour water stripper (40CFR61.341) means a unit that:
(1) Is designed and operated to remove ammonia or sulfur compounds (usually hydrogen sulfide) from sour water streams;
(2) has the sour water streams transferred to the stripper through hard piping or other enclosed system; and
(3) is operated in such a manner that the offgases are sent to a sulfur recovery unit, processing unit, incinerator, flare, or other combustion device.

source (40CFR129.2-j) means any building, structure, facility, or installation from which there is or may be the discharge of toxic pollutants designated as such by the Administration under section 307(a)(1) of the Act.

source (40CFR300-AA) means any area where a hazardous substance has been deposited, stored, disposed, or placed, plus those soils that have become contaminated from migration of a hazardous substance. Sources do not include those volumes of air, ground water, surface water, or surface water sediments that have become contaminated by migration, except: in the case of either a ground water plume with no identified source or contaminated surface water sediments with no identified source, the plume or contaminated sediments may be considered a source (under CERCLA).

source (40CFR52.2285-6) means both storage containers and delivery vessels (other identical or similar definitions are provided in 40CFR52.2286-6).

source (40CFR60-AA(alt. method 1)) means the source being tested by lidar.

source (40CFR61.191-b) means any building structure, pile, impoundment or area used for interim storage or disposal that is or contains waste material containing radium in sufficient concentration to emit radon-222 in excess of this standard prior to remedial action.

source (40CFR63.101) means the collection of emission points to which this subpart applies as determined by the criteria in 40CFR63.100 of this subpart. For purposes of subparts F, G, and H of this part, the term affected source as used in subpart A of this part has the same meaning as the term source defined here.

source (40CFR63.321), for purposes of this subpart, means each dry cleaning system.

source (40CFR66.3-l) means any source of air pollution subject to applicable legal requirements as defined in paragraph (c).

source (40CFR72.2) means any governmental, institutional, commercial, or industrial structure, installation, plant, building, or facility that emits or has the potential to emit any regulated air pollutant under the Act. For purposes of section 502(c) of the Act, a "source," including a "source" with multiple units, shall be considered a single "facility."

source (CWA306-33USC1316) means any building, structure, facility, or installation from which there is or may be the discharge of pollutants (other identical or similar definitions are provided in 40CFR129.29).

source category schedule for standards (40CFR63.51) means the schedule for promulgating MACT standards issued pursuant to section 112(e) of the Act.

source control action (40CFR300.5) is the construction or installation and start-up of those actions necessary to prevent the continued release of hazardous substances or pollutants or contaminants (primarily from a source on top of or within the ground, or in buildings or other structures) into the environment.

source control maintenance measures (40CFR300.5) are those measures intended to maintain the effectiveness of source control actions once such actions are operating and functioning properly, such as the maintenance of landfill caps and leachate collection systems.

source material (10CFR20.3) means:

(1) uranium or thorium, or any combination thereof in any physical or chemical form or

(2) ores which contain by weight one-twentieth of one percent (0.05%) or more of:

   (i) Uranium,

   (ii) thorium or

   (iii) any combination thereof. Source material does not include special nuclear material.

(Other identical or similar definitions are provided in 40CFR40.4.)

source material (10CFR30.4) means source material as defined in the regulations contained in part 40 of this chapter.

source material (10CFR70.4) means source material as defined in section 11(z). of the Act and in the regulations contained in part 40 of this chapter.

source material (40CFR710.2-c) shall have the meaning contained in the Atomic Energy Act of 1954, 42USC2014 et seq., and the regulations issued thereunder (other identical or similar definitions are provided in 40CFR720.3-a).

source operation (40CFR52.1881-xi) means the last operation preceding the emission of an air contaminant, which operation:

(a) results in the separation of the air contaminant from process materials or in the conversion of the process materials into air contaminants, as in the case of combustion of fuel; and

(b) is not primarily an air pollution abatement operation.

source/receptor areas (40CFR52.274.d.2-vi) are defined for each episode occurrence based on air monitoring, geographical, and meteorological factors: Source area is that area in which contaminants are discharged and a receptor area is that area in which the contaminants accumulate and are measured (under CAA).

source reduction (EPA-94/04): Reducing the amount of materials entering the waste stream by redesigning products or patterns of production or consumption (e.g., using returnable beverage containers). Synonymous with waste reduction.

source reduction (PPA6603):

(A) means any practice which

   (i) reduces the amount of any hazardous substance, pollutant, or contaminant entering any waste stream or otherwise released into the environment (including fugitive emissions) prior to recycling, treatment, or disposal; and

   (ii) reduces the hazards to public health and the environment associated with the release of such substances, pollutants, or contaminants. The term includes equipment or technology modifications, process or procedure modifications, reformulation or redesign of products, substitution of raw materials, and improvements in housekeeping, maintenance, training, or inventory control.

(B) The term "source reduction" does not include any practice which alters the physical, chemical, or biological characteristics or the volume of a hazardous substance, pollutant, or contaminant through a process or activity which itself is not integral to and necessary for the production of a product or the providing of a service.

source separation (40CFR246.101-cc) means the setting aside of recyclable materials at their point of generation by the generator.

source separation (EPA-94/04): Segregating various wastes at the point of generation (e.g., separation of paper, metal and glass from other wastes to make recycling simpler and more efficient).

span (40CFR60-AA (method 6C-3.2)) means the upper limit of the gas concentration measurement range displayed on the data recorder (other identical or similar definitions are provided in 40CFR60-AA (method 7E-3.2)).

span (40CFR72.2) means the range of values that a monitor component is required to be capable of measuring under part 75 of this chapter.

span drift (40CFR53.23-e): The percent change in response to an up-scale pollutant concentration over a 24-hour period of continuous unadjusted operation (cf. zero drift).

span gas (40CFR86.082.2) means a gas of known concentration which is used routinely to set the output level of an analyzer.

span gas (40CFR86.402.78) means a gas of known concentration which is used routinely to set the output level of any analyzer.

span value (40CFR60-AA (method 25A-2.2)) means the upper limit of a gas concentration measurement range that is specified for affected source categories in the applicable part of the regulations. The span value is established in the applicable regulation and is usually 1.5 to 2.5 times the applicable emission limit. If no span value is provided, use a span value equivalent to 1.5 to 2.5 times the expected concentration. For convenience, the span value should correspond to 100 percent of the recorder scale.

span value (40CFR60-AB-2.10) means the opacity value at which the CEMS is set to produce the maximum data display output as specified in the applicable subpart.

span value (40CFR60-AF-2.3) means the upper limit of a gas concentration measurement range that is specified for affected source categories in the applicable subpart of the regulation.

spandex fiber (40CFR60.601) means a manufactured fiber in which the fiber-forming substance is a long chain synthetic polymer comprised of at least 85 percent of a segmented polyurethane.

spare flue gas desulfurization system module (40CFR60.41a-19) means a separate system of sulfur dioxide emission control equipment capable of treating an amount of flue gas equal to the total amount of flue gas generated by an affected facility when operated at maximum capacity divided by the total number of nonspare flue gas desulfurization modules in the system.

spark ignition powered motor vehicle (40CFR52.731-

3) means a self-propelled over-the-road vehicle that is powered by a spark ignition type of internal combustion engine, including but not limited to engines fueled by gasoline, propane, butane, and methane compounds.

spark plug (40CFR85.2122(a)(7)(ii)(A)) means a device to suitably deliver high tension electrical ignition voltage to the spark gap in the engine combustion chamber.

special aquatic sites (40CFR230.3-q1) means those sites identified in subpart E. They are geographic areas, large or small, possessing special ecological characteristics of productivity, habitat, wildlife protection, or other important and easily disrupted ecological values. These areas are generally recognized as significantly influencing or positively contributing to the general overall environmental health or vitality of the entire ecosystem of a region (see 40CFR230.10(a)(3)).

special area management plan (CZMA304-16USC1453) means a comprehensive plan providing for natural resource protection and reasonable coastal-dependent economic growth containing a detailed and comprehensive statement of policies; standards and criteria to guide public and private uses of lands and waters; and mechanisms for timely implementation in specific geographic areas within the coastal zone.

special expertise (40CFR1508.26) means statutory responsibility, agency mission, or related program experience.

special features enabling off-street or off-highway operation and use (40CFR86.084.2) means:
(1) That has 4-wheel drive; and
(2) That has at least four of the following characteristics calculated when the automobile is at curb weight, on a level surface, with the front wheels parallel to the vehicle's longitudinal centerline, and the tires inflated to the manufacturer's recommended pressure;
  (i)   Approach angle of not less than 2 degrees.
  (ii) Breakover angle of not less than 14 degrees.
  (iii) Departure angle of not less than 20 degrees.

(iv) Running clearance of not less than inches.

(v) Front and rear axle clearances of not less than 7 inches each.

special fellow (40CFR46.120): An individual enrolled in an educational program relating to environmental sciences, engineering, professional schools, and allied sciences.

special government employee (40CFR3.102-b) means an officer or employee of the Environmental Protection Agency who is retained, designated, appointed or employed to perform, with or without compensation, temporary duties either on a full-time or intermittent basis, for not to exceed 30 days during any period of 365 consecutive days.

special local need (40CFR162.151-i) means an existing or imminent pest problem within a State for which the State lead agency, based upon satisfactory supporting information, has determined that an appropriate federally registered pesticide product is not sufficiently available.

special nuclear material (10CFR20.3) means:

(i) Plutonium, uranium-233, uranium enriched in the isotope 233 or in the isotope 235, and any other material which the Commission, pursuant to the provisions of section 51 of the act, determines to be special nuclear material, but does not include source material; or

(ii) any material artificially enriched by any of the foregoing but does not include source material.

special nuclear material (10CFR30.4) means special nuclear material as defined in the regulations contained in part 70 of this chapter.

special nuclear material (10CFR40.4) means:

(1) Plutonium, uranium 233, uranium enriched in the isotope 233 or in the isotope 235, and any other material which the Commission, pursuant to the provisions of section 51 of the Act, determines to be special nuclear material; or

(2) any material artificially enriched by any of the foregoing.

special nuclear material (10CFR70.4) means:

(1) plutonium, uranium 233, uranium enriched in the isotope 233 or in the isotope 235, and any

other material which the Commission, pursuant to the provisions of section 51 of the act, determines to be special nuclear material, but does not include source material; or

(2) any material artificially enriched by any of the foregoing but does not include source material.

special nuclear material (40CFR710.2-c) shall have the meaning contained in the Atomic Energy Act of 1954, 42USC2014 et seq., and the regulations issued thereunder (other identical or similar definitions are provided in 40CFR720.3-a).

special nuclear material of low strategic significance (10CFR70.4) means:

(1) Less than amount of special nuclear material of moderate strategic significance, as defined in 40CFR70.4(z)(1), but more than 15 grams of uranium-235 (contained in uranium enriched to 20 percent or more in the U^{235} isotope) or 15 grams of uranium-233 or 15 grams of plutonium or the combination of 15 grams when computed by the equation, grams = (grams contained U^{235}) + (grams plutonium) + (grams U^{233}); or

(2) Less than 10,000 grams but more than 1000 grams of uranium-235 (contained in uranium enriched to 10 percent or more but less than 20 percent in the U^{235} isotope); or

(3) 10,000 grams or more of uranium-235 (contained in uranium enriched above natural but less than 10 percent in the U^{235} isotope).

special nuclear material of moderate strategic significance (10CFR70.4) means:

(1) Less than a formula quantity of strategic special nuclear material but more than 1000 grams of uranium-235 (contained in uranium enriched to 20 percent or more in the U^{235} isotope) or more than 500 grams of uranium-233 or plutonium or in a combined quantity of more than 1000 grams when computed by the equation, grams = (grams contained U^{235}) + 2 (grams U^{233} + grams plutonium); or

(2) 10,000 grams or more of uranium-235 (contained in uranium enriched to 10 percent or more but less than 20 percent in the U^{235} isotope).

special nuclear material scrap (10CFR70.4) means the various forms of special nuclear material generated during chemical and mechanical processing, other than recycle material and normal process intermediates, which are unsuitable for use in their present form, but all or part of which will be used after further processing.

special production area (40CFR723.175-13) means a demarcated area within which all manufacturing, processing, and use of a new chemical substance takes place, except as provided in paragraph (f) of this section, in accordance with the requirements of paragraph (e) of this section.

special purpose equipment (40CFR201.1-dd) means maintenance-of-way equipment which may be located on or operated from rail cars including: Ballast cribbing machines, ballast regulators, conditioners and scarifiers, bolt machines, brush cutters, compactors, concrete mixers, cranes and derricks, earth boring machines, electric welding machines, grinders, grouters, pile drivers, rail heaters, rail layers, sandblasters, snow plows, spike drivers, sprayers and other types of such maintenance-of-way equipment.

special purpose facility (40CFR6.901-a) means a building or space, including land incidental to its use, which is wholly or predominantly utilized for the special purpose of an agency and not generally suitable for other uses, as determined by the General Services Administration.

special review (40CFR166.3-i) refers to any interim administrative review of the risks and benefits of the use of a pesticide conducted pursuant to the provisions of EPA's Rebuttable Presumption Against Registration rules, 40CFR162.4ll(a), or any subsequent version of those rules.

special review (EPA-94/04): Formerly known as Rebuttable Presumption Against Registration (RPAR), this is the regulatory process through which existing pesticides suspected of posing unreasonable risks to human health, non-target organisms, or the environment are referred for review by EPA. Such review requires an intensive risk/benefit analysis with opportunity for public comment. If risk is found to outweigh social and economic benefits, regulatory actions ranging from label revisions and use-restriction to cancellation or suspended registration can be initiated.

special source of ground water (40CFR191.12), as used in this part, means those Class I ground waters identified in accordance with the Agency's Ground-Water Protection Strategy published in August 1984 that:
(1) Are within the controlled area encompassing a disposal system or are less than five kilometers beyond the controlled area;
(2) are supplying drinking water for thousands of persons as of the date that the Department chooses a location within that area for detailed characterization as a potential site for a disposal system (e.g., in accordance with section 112(b)(1)(B) of the NWPA); and
(3) are irreplaceable in that no reasonable alternative source of drinking water is available to that population.

special track work (40CFR201.1-ee) means track other than normal tie and ballast bolted or welded rail or containing devices such as retarders or switching mechanisms.

special waste (EPA-94/04): Items such as household hazardous waste, bulky wastes (refrigerators, pieces of furniture, etc.) tires, and used oil.

special wastes (40CFR240.101-z) means nonhazardous solid wastes requiring handling other than that normally used for municipal solid waste (under RCRA).

specially designated landfill (40CFR165.1-w) means a landfill at which complete long term protection is provided for the quality of surface and subsurface waters from pesticides, pesticide containers, and pesticide-related wastes deposited therein, and against hazard to public health and the environment. Such sites should be located and engineered to avoid direct hydraulic continuity with surface and subsurface waters, and any leachate or subsurface flow into the disposal area should be contained within the site unless treatment is provided. Monitoring wells should be established and a sampling and analysis program conducted. The location of the disposal site should be permanently recorded in the appropriate local office of legal jurisdiction. Such facility complies with the Agency Guidelines for the Disposal of Solid Wastes as prescribed in 40CFR241.

specialty (40CFR420.71-l): See specialty hot forming operation.

specialty hot forming operation (or specialty) (40CFR420.71-l) applies to all hot forming operations other than "carbon hot forming operations."

specialty paper (40CFR763.163) means an asbestos-containing product that is made of paper intended for use as filters for beverages or other fluids or as paper fill for cooling towers. Cooling tower fill consists of asbestos paper that is used as a cooling agent for liquids from industrial processes and air conditioning systems.

specialty steel (40CFR420.71-i) means those steel products containing allowing elements which are added to enhance the properties of the steel product when individual alloying elements (e.g., aluminum, chromium, cobalt, columbium, molybdenum, nickel, titanium, tungsten, vanadium, zirconium) exceed 3% or the total of all alloying elements exceed 5% (under CWA).

species (EPA-94/04): A reproductively isolated aggregate of interbreeding organisms.

species (ESA3-16USC1531) includes any subspecies of fish or wildlife or plants, and any distinct population segment of any species of vertebrate fish or wildlife which interbreeds when mature.

specific chemical identity (29CFR1910.20) means the chemical name, Chemical Abstracts Service (CAS) Registry Number, or any other information that reveals the precise chemical designation of the substance.

specific chemical identity (40CFR350.1) means the chemical name, Chemical Abstracts Service (CAS) Registry Number, or any other information that reveals the precise chemical designation of the substance. Where the trade name is reported in lieu of the specific chemical identity, the trade name will be treated as the specific chemical identity for purposes of this part.

specific conductance (EPA-94/04): Rapid method of estimating the dissolved-solid content of a water supply by testing the capacity of the water to carry an electrical current.

specific emissions (40CFR89.302.96), g/kW-hr, is expressed on the basis of observed gross brake power. When it is not possible to test the engine in the gross conditions, for example, if the engine and transmission form a single integral unit, the engine may be tested in the net condition. Power corrections from net to gross conditions will be allowed with prior approval of the Administrator.

specific emissions (40CFR89.402.96), (g/kW-hr), shall be expressed on the basis of observed gross power. When it is not possible to test the engine in the gross conditions, for example, if the engine and transmission form a single integral unit, the engine may be tested in the net condition. Power corrections from net to gross conditions will be allowed with prior approval of the Administrator.

specific gravity monitoring device (40CFR63.111) means a unit of equipment used to monitor specific gravity and having an accuracy of ± 0.02 specific gravity units.

specific oxygen uptake rate (SOUR) (40CFR503.31-h) is the mass of oxygen consumed per unit time per unit mass of total solids (dry weight basis) in the sewage sludge.

specific written consent (29CFR1910.20):
(i) means a written authorization containing the following:
  (A) The name and signature of the employee authorizing the release of medical information,
  (B) The date of the written authorization,
  (C) The name of the individual or organization that is authorized to release the medical information,
  (D) The name of the designated representative (individual or organization) that is authorized to receive the released information,
  (E) A general description of the medical information that is authorized to be released,
  (F) A general description of the purpose for the release of the medical information, and

(G) A date or condition upon which the written authorization will expire (if less than one year).

(ii) A written authorization does not operate to authorize the release of medical information not in existence on the date of written authorization, unless the release of future information is expressly authorized, and does not operate for more than one year from the date of written authorization.

(iii) A written authorization may be revoked in writing prospectively at any time.

specific yield (EPA-94/04): The amount of water that a unit volume of saturated permeable rock will yield when drained by gravity.

specification (40CFR246.101-dd) means a clear and accurate description of the technical requirements for materials, products or services, identifying the minimum requirements for quality and construction of materials and equipment necessary for an acceptable product. In general, specifications are in the form of written descriptions, drawings, prints, commercial designations, industry standards, and other descriptive references (other identical or similar definitions are provided in 40CFR247.101-k).

specification (40CFR248.4-jj) means a description of the technical requirements for a material, product, or service that includes the criteria for determining whether these requirements are met. In general, specifications are in the form of written commercial designations, industry standards, and other descriptive references.

specification (40CFR249.04-m) means a clear and accurate description of the technical requirement for materials, products, or services, which specifies the minimum requirement for quality and construction of materials and equipment necessary for an acceptable product. In general, specifications are in the form of written descriptions, drawings, prints, commercial designations, industry standards, and other descriptive references.

specification (40CFR250.4-mm) means a detailed description of the technical requirements for materials, products, or services that specifies the minimum requirement for quality and construction of

materials and equipment necessary for an acceptable product. Specifications are generally in the form of a written description, drawings, prints, commercial designations, industry standards, and other descriptive references.

specification (40CFR252.4-k) means a description of the technical requirements for a material, product, or service that includes the criteria for determining whether these requirements are met. In general, specifications are in the form of written commercial designations, industry standards, and other descriptive references (other identical or similar definitions are provided in 40CFR253.4).

specified air contaminant (40CFR52.741) means any air contaminant as to which this section contains emission standards or other specific limitations.

specified ports and harbors (40CFR300.5) means those port and harbor areas on inland rivers, and land areas immediately adjacent to those waters, where the USCG acts as predesignated on-scene coordinator. Precise locations are determined by EPA/USCG regional agreements and identified in Federal regional contingency plans.

specimen (40CFR160.3) means any material derived from a test system for examination or analysis (other identical or similar definitions are provided in 40CFR792.3).

spectral uncertainty (40CFR211.203-v) means possible variation in exposure to the noise spectra in the workplace. (To avoid the underprotection that would result from these variations relative to the assumed Pink Noise used to determine the NRR, an extra three decibel reduction is included when computing the NRR.)

spent acid solution (or spent pickle liquor) (40CFR420.91-j) means those solutions of steel pickling acids which have been used in the pickling process and are discharged or removed therefrom (under CWA).

spent lubricant (40CFR468.02-q) shall mean water or an oil-water mixture which is used in forming operations to reduce friction, heat and wear and ultimately discharged.

spent material (40CFR261.1-1) is any material that has been used and as a result of contamination can no longer serve the purpose for which it was produced without processing.

spent nuclear fuel (40CFR191.02-g) means fuel that has been withdrawn from a nuclear reactor following irradiation, the constituent elements of which have not been separated by reprocessing.

spent pickle liquor (40CFR420.91-j): See spent acid solution.

spill (40CFR761.123) means both intentional and unintentional spills, leaks, and other uncontrolled discharges where the release results in any quantity of PCBs running off or about to run off the external surface of the equipment or other PCB source, as well as the contamination resulting from those releases. This policy applies to spills of 50 ppm or greater PCBs. The concentration of PCBs spilled is determined by the PCB concentration in the material onto which the PCBs were spilled. Where a spill of untested mineral oil occurs, the oil is presumed to contain greater than 50 ppm, but less than 500 ppm, PCBs and is subject to the relevant requirements of this policy.

spill area (40CFR761.123) means the area of soil on which visible traces of the spill can be observed plus a buffer zone of 1 foot beyond the visible traces. Any surface or object (e.g., concrete sidewalk or automobile) within the visible traces area or on which visible traces of the spilled material are observed is included in the spill area. This area represents the minimum area assumed to be contaminated by PCBs in the absence of precleanup sampling data and is thus the minimum area which must be cleaned (under TSCA).

spill boundaries (40CFR761.123) means the actual area of contamination as determined by postcleanup verification sampling or by precleanup sampling to determine actual spill boundaries. EPA can require additional cleanup when necessary to decontaminate all areas within the spill boundaries to the levels required in this policy (e.g., additional cleanup will be required if postcleanup sampling indicates that the area decontaminated by the responsible party, such as the spill area as defined in this section, did not

encompass that actual boundaries of PCB contamination).

spill event (40CFR112.2) means a discharge of oil into or upon the navigable waters of the United States or adjoining shorelines in harmful quantities, as defined at 40CFR part 110.

spill prevention control and countermeasures plan (SPCP) (EPA-94/04): Plan covering the release of hazardous substances as defined in the Clean Water Act.

spinach (40CFR407.71-r) shall mean the processing of spinach and leafy greens into the following product styles: Canned or frozen, whole leaf, chopped, and other related cuts.

spinning reserve (40CFR60.41a-17) means the sum of the unutilized net generating capability of all units of the electric utility company that are synchronized to the power distribution system and that are capable of immediately accepting additional load. The electric generating capability of equipment under multiple ownership is prorated based on ownership unless the proportional entitlement to electric output is otherwise established by contractual arrangement (under CAA).

spinning solution (40CFR60.601) means the mixture of polymer, prepolymer, or copolymer and additives dissolved in solvent. The solution is prepared at a viscosity and solvent-to-polymer ratio that is suitable for extrusion into fibers.

spinning solution preparation system (40CFR60.601) means the equipment used to prepare spinning solutions; the system includes equipment for mixing, filtering, blending, and storage of the spinning solutions.

spitback emissions (40CFR86.096.2) means evaporative emissions resulting from the loss of liquid fuel that is emitted from a vehicle during a fueling operation.

splash loading (40CFR52.741) means a method of loading a tank, railroad tank car, tank truck, or trailer by use of other than a submerged loading pipe.

spoil (EPA-94/04): Dirt or rock removed from its original location--destroying the composition of the soil in the process--as in strip-mining, dredging, or construction.

Spokane central business district (40CFR52.2486-9) means the area enclosed by Trent Avenue, Monroe Street, Third Avenue, and Division Street. Streets forming boundaries shall be part of the central business district (other identical or similar definitions are provided in 40CFR52.2489-2; 52.2493-4; 52.2494-1).

sponsor (40CFR160.3) means:
(1) A person who initiates and supports, by provision of financial or other resources, a study;
(2) A person who submits a study to the EPA in support of an application for a research or marketing permit; or
(3) A testing facility, if it both initiates and actually conducts the study.

sponsor (40CFR790.3) means the person or persons who design, direct and finance the testing of a substance or mixture.

sponsor (40CFR792.3) means:
(1) A person who initiates and supports, by provision of financial or other resources, a study;
(2) A person who submits a study to the EPA in response to a section 4(a) test rule and/or a person who submits a study under a negotiated testing agreement or section 5 rule/order to the extent the agreement or rule/order references this part; or
(3) A testing facility, if it both initiates and actually conducts the study (under the Toxic Substances Control Act).

spot allowance (40CFR72.2) means an allowance that may be used for purposes of compliance with a unit's Acid Rain sulfur dioxide emissions limitation requirements beginning in the year in which the allowance is offered for sale.

spot auction (40CFR72.2) means an auction of a spot allowance.

spot sale (40CFR72.2) means a sale of a spot allowance.

sprawl (EPA-94/04): Unplanned development of open land.

spray application (40CFR60.311) means a method of applying coatings by atomizing and directing the atomized spray toward the part to be coated.

spray application (40CFR60.391) means a method of applying coatings by atomizing the coating material and directing the atomized material toward the part to be coated. Spray applications can be used for prime coat, guide coat, and topcoat operations.

spray application (40CFR721.3) means any method of projecting a jet of vapor of finely divided liquid onto a surface to be coated; whether by compressed air, hydraulic pressure, electrostatic forces, or other methods of generating a spray.

spray booth (40CFR60.391) means a structure housing automatic or manual spray application equipment where prime coat, guide coat, or topcoat is applied to components of automobile or light-duty truck bodies.

spray booth (40CFR60.451) means the structure housing automatic or manual spray application equipment where a coating is applied to large appliance parts or products.

spray booth (40CFR60.721) means the structure housing automatic or manual spray application equipment where a coating is applied to plastic parts for business machines.

spray finishing operations (29CFR1910.94c): Spray finishing operations are employment of methods wherein organic or inorganic materials are utilized in dispersed form for deposit on surfaces to be coated, treated, or cleaned. Such methods of deposit may involve either automatic, manual, or electrostatic deposition but do not include metal spraying or metallizing, dipping, flow coating, roller coating, tumbling, centrifuging, or spray washing and degreasing as conducted in self-contained washing and degreasing machines or systems.

spray in place foam (40CFR248.4-ll) is rigid cellular polyurethane or polyisocyanurate foam produced by catalyzed chemical reactions that hardens at the site of the work. The term includes spray-applied and injected applications.

spray in place insulation (40CFR248.4-kk) means insulation material that is sprayed onto a surface or into cavities and includes cellulose fiber spray-on as well as plastic rigid foam products.

spray room (29CFR1910.94c): A spray room is a room in which spray-finishing operations not conducted in a spray booth are performed separately from other areas.

spray tower scrubber (EPA-94/04): A device that sprays alkaline water into a chamber where acid gases are present to aid in the neutralizing of the gas.

spreader stoker steam generating unit (40CFR60.41b) means a steam generating unit in which solid fuel is introduced to the combustion zone by a mechanism that throws the fuel onto a grate from above. Combustion takes place both in suspension and on the grate.

spring (EPA-94/04): Ground water seeping out of the earth where the water table intersects the ground surface.

spring melt/thaw (EPA-94/04): The process by which warm temperatures melt winter snow and ice. Because various forms of acid deposition may have been stored in the frozen water, the melt can result in abnormally large amounts of acidity entering streams and rivers, sometimes causing fish kills.

sq ft (sq m) (40CFR413.81-a) shall mean the area of the printed circuit board immersed in an aqueous process bath.

sq m (sq ft) (40CFR413.11-a) shall mean the area plated expressed in square meters [square feet] (other identical or similar definitions are provided in 40CFR413.21-a; 413.41-a; 413.51-a; 413.71-a).

sq m (sq ft) (40CFR413.51-a) shall mean the area processed expressed in square meters [square feet].

sq m (sq ft) (40CFR413.61-a) shall mean the area exposed to process chemicals expressed in square meters (square feet).

squash (40CFR407.71-n) shall include the processing of pumpkin and squash into canned and frozen styles.

SRF (40CFR35.3105-i): State water pollution control revolving fund.

SS (40CFR133.101-l): The pollutant parameter total suspended solids. See also suspended solids.

SSA (40CFR149.2-d): See sole source aquifer.

stabilization (EPA-94/04): Conversion of the active organic matter in sludge into inert, harmless material.

stabilization ponds (EPA-94/04): See lagoon.

stable air (EPA-94/04): A motionless mass of air that holds instead of dispersing pollutants.

stable in quantity and rate release (40CFR302.8) is stable in quantity and rate is a release that is predictable and regular in amount and rate of emission.

stack (40CFR51.100-ff) means any point in a source designed to emit solids, liquids, or gases into the air, including a pipe or duct but not including flares (under CAA).

stack (40CFR72.2) means a structure that includes one or more flues and the housing for the flues.

stack (40CFR129.2-q) means any chimney, flue, conduit, or duct arranged to conduct emissions to the ambient air.

stack (40CFR52.1881-xii) means any chimney, flue, vent, roof monitor, conduit or duct arranged to vent emissions to the ambient air.

stack (EPA-94/04): A chimney, smokestack, or vertical pipe that discharges used air.

stack effect (EPA-94/04): Air, as in a chimney, that moves upward because it is warmer than the ambient atmosphere.

stack emission (40CFR60.671) means the particulate

matter that is released to the atmosphere from a capture system.

stack emissions (40CFR60.381): Stack emissions means the particulate matter captured and released to the atmosphere through a stack, chimney, or flue.

stack gas (EPA-94/04): See flue gas.

stack height (40CFR503.41-l) is the difference between the elevation of the top of a sewage sludge incinerator stack and the elevation of the ground at the base of the stack when the difference is equal to or less than 65 meters. When the difference is greater than 65 meters, stack height is the creditable stack height determined in accordance with 40CFR51.100 (ii) (see also part 40CFR52.21(h)).

stack height procedures: See 40CFR51.164.

stack height provisions: See 40CFR51.118.

stack in existence (40CFR51.100-gg) means that the owner or operator had:
(1) begun, or caused to begin, a continuous program of physical on-site construction of the stack or
(2) entered into binding agreements or contractual obligations, which could not be cancelled or modified without substantial loss to the owner or operator, to undertake a program of construction of the stack to be completed within a reasonable time.

stage II controls (EPA-94/04): Systems placed on service station gasoline pumps to control and capture gasoline vapors during refuelling.

stagnation (EPA-94/04): Lack of motion in a mass of air or water that holds pollutants in place.

stall barn (40CFR412.11-h) shall mean specialized facilities wherein producing cows and replacement cows are milked and fed in a fixed location (under CWA).

standard (29CFR1910.2) means a standard which requires conditions, or the adoption or use of one or more practices, means, methods, operations, or processes, reasonably necessary or appropriate to provide safe or healthful employment and places of employment.

standard (40CFR171.2-24) means the measure of knowledge and ability which must be demonstrated as a requirement for certification.

standard (40CFR403.3-j): See national pretreatment standard.

standard (40CFR51.392) means a national ambient air quality standard (other identical or similar definitions are provided in 40CFR93.101).

standard (40CFR60.2) means a standard of performance proposed or promulgated under this part.

standard (40CFR61.02) means a national emission standard including a design, equipment, work practice or operational standard for a hazardous air pollutant proposed or promulgated under this part (under CAA).

standard boiling point (40CFR796.1220) is described as the temperature at which the pressure of the saturated vapor of a liquid is the same as the standard pressure. The measured boiling point is dependent on the atmospheric pressure.

standard bushel (40CFR406.11-c) shall mean a bushel of shelled corn weighing 56 pounds (other identical or similar definitions are provided in 40CFR406.21-c; 406.41-c).

standard conditions (40CFR52.741) means a temperature of 70°F and a pressure of 14.7 psia.

standard conditions (40CFR60.2) means a temperature of 293°K (68°F) and a pressure of 101.3 kilopascals (29.92 in Hg).

standard conditions (40CFR60.51a) means a temperature of 293° Kelvin (68° Fahrenheit) and a pressure of 101.3 kilopascals (29.92 inches of mercury).

standard conditions (40CFR63.2) means a temperature of 293°K (68°F) and a pressure of 101.3 kilopascals (29.92 in. Hg) (other identical or similar definitions are provided in 40CFR63.101; 63.191).

standard conditions (40CFR72.2) means 68°F at 1 atm (29.92 in. of mercury).

standard cubic foot (scf) (40CFR52.741) means the volume of one cubic foot of gas at standard conditions.

standard day conditions (40CFR87.1) means standard ambient conditions as described in the United States Standard Atmosphere, 1976 (i.e., Temperature = 15 degrees C, specific humidity = 0.00 kg/H$_2$O/kg dry air, and pressure = 101325 Pa).

standard equipment (40CFR86.082.2) means those features or equipment which are marketed on a vehicle over which the purchaser can exercise no choice.

standard ferromanganese (40CFR60.261-n) means that alloy as defined by ASTM Designation A9976 (incorporated by reference, 40CFR60.17).

standard industrial classification manual (40CFR52.741) means the Standard Industrial Classification Manual (1987), Superintendent of Documents, U.S. Government Printing Office, Washington, DC 20402 (incorporated by reference as specified in 40CFR52.742).

standard of performance (40CFR401.11-k) means any restriction established by the Administrator pursuant to section 306 of the Act on quantities, rates, and concentrations of chemical, physical, biological, and other constituents which are or may be discharged from new sources into navigable waters, the waters of the contiguous zone or the ocean.

standard of performance (CAA111-42USC7411) means:
(A) with respect to any air pollutant emitted from a category of fossil fuel fired stationary sources to which subsection (b) applies, a standard:
  (i) establishing allowable emission limitations for such category of sources, and
  (ii) requiring the achievement of a percentage reduction in the emissions from such category of sources from the emissions which would have resulted from the use of fuels which are not subject to treatment prior to combustion,

(B) with respect to any air pollutant emitted from a category of stationary sources (other than fossil fuel fired sources) to which subsection (b) applies, a standard such as that referred to in subparagraph (A)(i); and

(C) with respect to any air pollutant emitted from a particular source to which subsection (d) applies, a standard which the State (or the Administrator under the conditions specified in subsection (d)(2)) determines is applicable to that source and which reflects the degree of emission reduction achievable through the application of the best system of continuous emission reduction which (taking into consideration the cost of achieving such emission reduction, and any nonair quality health and environmental impact and energy requirements) the Administrator determines has been adequately demonstrated for that category of source. For the purpose of subparagraphs (A) (i) and (ii) and (B), a standard of performance shall reflect the degree of emission limitation and the percentage reduction achievable through application of the best technological system of continuous emission reduction which (taking into consideration the cost of achieving such emission reduction, any nonair quality health and environmental impact and energy requirements) the Administrator determines has been adequately demonstrated. for the purpose of subparagraph (1)(A)(ii), any cleaning of the fuel or reduction in the pollution characteristics of the fuel after extraction and prior to combustion may be credited, as determined under regulations promulgated by the Administrator, to a source which burns such fuel.

standard of performance (CWA306-33USC1316) means a standard for the control of the discharge of pollutants which reflects the greatest degree of effluent reduction which the Administrator determines to be achievable through application of the best available demonstrated control technology, processes, operating methods, or other alternatives, including, where practicable, a standard permitting no discharge of pollutants.

standard operating procedure (40CFR30.200) means a document which describes in detail an operation analysis, or action which is commonly accepted as the preferred method for performing certain routine or repetitive tasks.

standard operating procedure (40CFR61.61-m) means a formal written procedure officially adopted by the plant owner or operator and available on a routine basis to those persons responsible for carrying out the procedure.

standard or limitation (40CFR2.301-3) means any emission standard or limitation established or publicly proposed pursuant to the Act or pursuant to any regulation under the Act.

standard or limitation (40CFR2.302-3) means any prohibition, any effluent limitation, or any toxic, pre-treatment or new source performance standard established or publicly proposed pursuant to the Act or pursuant to regulations under the Act, including limitations or prohibitions in a permit issued or proposed by EPA or by a State under section 402 of the Act, 33USC1342.

standard pressure (40CFR61.61-u) means a pressure of 760 mm of Hg (29.92 in. of Hg).

standard reference material or SRM (40CFR72.2) means a calibration gas mixture issued and certified by NIST as having specific known chemical or physical property values.

standard sample (40CFR141.2) means the aliquot of finished drinking water that is examined for the presence of coliform bacteria.

standard sample (EPA-94/04): The part of finished drinking water that is examined for the presence of coliform bacteria.

standard temperature (40CFR61.61-t) means a temperature of 20°C (69°F).

standard wipe test (40CFR761.123) means, for spills of high-concentration PCBs on solid surfaces, a cleanup to numerical surface standards and sampling by a standard wipe test to verify that the numerical standards have been met. This definition constitutes the minimum requirements for an appropriate wipe testing protocol. A standard-size template (10 centimeters (cm) x 10 cm) will be used to delineate the area of cleanup; the wiping medium will be a gauze pad or glass wool of known size which has been saturated with hexane. It is important that the wipe be performed very quickly after the hexane is exposed to air. EPA strongly recommends that the gauze (or glass wool) be prepared with the hexane in the laboratory and that the wiping medium be stored in sealed glass vials until it is used for the wipe test. Further, EPA requires the collection and testing field blanks and replicates.

standards (EPA-94/04): Norms that impose limits on the amount of pollutants or emissions produced. EPA establishes minimum standards, but states are allowed to be stricter.

standards for sewage sludge use or disposal (40CFR122.2): Standards for sewage sludge use or disposal means the regulations promulgated pursuant to section 405(d) of the CWA which govern minimum requirements for sludge quality, management practices, and monitoring and reporting applicable to sewage sludge or the use or disposal of sewage sludge by any person.

standards for sewage sludge use or disposal (40CFR501.2) means the regulations promulgated at 40CFR part 503 pursuant to section 405(d) of the CWA which govern minimum requirements for sludge quality, management practices, and monitoring and reporting applicable to the generation or treatment of sewage sludge from a treatment works treating domestic sewage or use or disposal of that sewage sludge by any person.

standards for the development of test data (TSCA3-15USC2602) means a prescription of:
(A) (i)   the health and environmental effects, and
    (ii) information relating to toxicity, persistence, and other characteristics which affect health and the environment, for which test data for a chemical substance or mixture are to be developed and any analysis that is to be performed on such data, and
(B) to the extent necessary to assure that data respecting such effects and characteristics are reliable and adequate:

(i) the manner in which such data are to be developed,

(ii) the specification of any test protocol or methodology to be employed in the development of such data, and

(iii) such other requirements as are necessary to provide such assurance.

standpipe cap (40CFR63.301) means an apparatus used to cover the opening in the gooseneck of an offtake system.

Stark Einstein law (40CFR796.3700-x): The second law of photochemistry, states that only one molecule is activated to an excited state per photon or quantum of light absorbed (other identical or similar definitions are provided in 40CFR796.3780-xi; 796.3800-x).

start of a response action (EPA-94/04): The point in time when there is a guarantee or set-aside of funding either by EPA, other federal agencies, states or Principal Responsible parties in order to begin response actions at a Superfund site.

start of response action (40CFR35.4010) means the point in time when there is a guarantee or set-aide of funding either by EPA, other Federal agencies, States, or PRPs in order to begin response activities at a site. The document which reflects the set-aside of, or formally guarantees, funding during the coming fiscal year, is EPA's annual Superfund Comprehensive Accomplishments Plan (SCAP).

starting material (40CFR158.153-j) means a substance used to synthesize or purify a technical grade of active ingredient (or the practical equivalent of the technical grade ingredient if the technical grade cannot be isolated) by chemical reaction.

startup (40CFR264.1031) means the setting in operation of a hazardous waste management unit or control device for any purpose.

startup (40CFR52.01-e) means the setting in operation of a source for any purpose.

startup (40CFR52.741) means the setting in operation of an emission source for any purpose (under CAA).

startup (40CFR60.2) means the setting in operation of an affected facility for any purpose.

startup (40CFR61.02) means the setting in operation of a stationary source for any purpose (other identical or similar definitions are provided in 40CFR63.2).

startup (40CFR63.161) means the setting in operation of a piece of equipment or a control device that is subject to this subpart.

startup (40CFR63.101) means the setting into operation of a chemical manufacturing process unit for the purpose of production. Startup does not include operation solely for testing equipment. Startup does not include the recharging of equipment in batch operation. Startup does not include changes in product for flexible operation units.

startup (40CFR63.301) means that operation that commences when the coal begins to be added to the first oven of a coke oven battery that either is being started for the first time or that is being restarted and ends when the doors have been adjusted for maximum leak reduction and the collecting main pressure control has been stabilized. Except for the first startup of a coke oven battery, a startup cannot occur unless a shutdown has occurred.

startup, shutdown, and malfunction plan (40CFR63.101) means the plan required under 40CFR63.6(e)(3) of subpart A of this part. This plan details the procedures for operation and maintenance of the source during periods of startup, shutdown, and malfunction.

state (40CFR15.4) means a State, the District of Columbia, the Commonwealth of Puerto Rico, the Virgin Islands, Guam, American Samoa, the Commonwealth of the Northern Mariana Islands, or the Trust Territories of the Pacific Islands (other identical or similar definitions are provided in 40CFR20.2-g; 21.2-h; 29.2; 31.3; 32.105-s; 32.605-10; 34.105-q; 35.105; 35.905; 35.2005-41; 35.6015-44; 40.115-6; 51.491; 55.2; 61.02; 63.51; 63.2; 63.101; 63.191; 70.2; 72.2; 122.2; 124.2; 124.41; 131.3-j; 141.2; 142.2; 143.2-d; 144.3; 171.2-25; 173.2-c; 205.2-19; 231.2-f; 232.2; 250.4-nn; 257.2; 258.2; 260.10;

270.2; 300.5; 350.1; 355.20; 370.2; 372.3; 501.2; 710.2-z; 720.3-dd; 762.3-d; 763.83; 763.163; CAA302-42USC7602; CAA402-42USC7651a; CWA502-33USC1362; ESA3-16USC1531; FIFRA2-7USC136; NCA3-42USC4902; RCRA1004-42USC6903; SDWA1401-42USC300f; SMCRA701-30USC1291; SF329; SF401-42USC9671; TSCA3-15USC2602; TSCA202-15USC2642).

state (40CFR503.9-x) is one of the United States of America, the District of Columbia, the Commonwealth of Puerto Rico, the Virgin Islands, Guam, American Samoa, the Trust Territory of the Pacific Islands, the Commonwealth of the Northern Mariana Islands, and an Indian Tribe eligible for treatment as a State pursuant to regulations promulgated under the authority of section 518(e) of the CWA.

state 404 program (40CFR124.2): See section 404 program.

state 404 program or State program (40CFR233.2) means a State program which has been approved by EPA under section 404 of the Act to regulate the discharge of dredged or fill material into certain waters as defined in 40CFR232.2(p).

state agency (40CFR35.905) means the State water pollution control agency designated by the Governor having responsibility for enforcing State laws relating to the abatement of pollution (other identical or similar definitions are provided in 40CFR35.2005-42; 51.100-i; 58.1-l; ESA3-16USC1531).

state authority (RCRA1004-42USC6903) means the agency established or designated under section 4007.

state certifying authority (40CFR20.2-b) means:
(1) For water pollution control facilities, the State pollution control agency as defined in section 502 of the Act;
(2) For air pollution control facilities, the air pollution control agency designated pursuant to 40CFR302(b)(1) of the Act; or
(3) For both air and water pollution control facilities, any interstate agency authorized to act in place of the certifying agency of a State.

state delayed compliance order (40CFR65.01-g) shall mean a delayed compliance order issued by a State or by a political subdivision of a State (under CAA).

State Director (40CFR122.2) means the chief administrative officer of any State or interstate agency operating an "approved program," or the delegated representative of the State Director. If responsibility is divided among two or more State or interstate agencies, "State Director" means the chief administrative officer of the State or interstate agency authorized to perform the particular procedure or function to which reference is made (other identical or similar definitions are provided in 40CFR124.2; 129.2-s; 133.101-n; 144.3; 146.3; 233.2; 270.2).

state emergency response commission (SERC) (EPA-94/04): Commission appointed by each state governor according to the requirements of SARA Title III. The SERCs designate emergency planning districts, appoint local emergency planning committees, and supervise and coordinate their activities.

State/EPA agreement (40CFR122.2) means an agreement between the Regional Administrator and the State which coordinates EPA and State activities, responsibilities and programs including those under the CWA programs.

State/EPA agreement (40CFR144.3) means an agreement between the Regional Administrator and the State which coordinates EPA and State activities, responsibilities and programs.

State/EPA Agreement (40CFR270.2) means an agreement between the Regional Administrator and the State which coordinates EPA and State activities, responsibilities and programs.

state implementation plan (40CFR65.01-h) shall mean the plan, including the most recent revision thereof, which has been approved or promulgated by the Administrator under section 110 of the Act, and which implements the requirements of section 110 (under CAA).

state implementation plan (SIP) (40CFR51.491) means a plan developed by an authorized governing body, including States, local governments, and

Indian-governing bodies, in a nonattainment area, as required under titles I; II of the Clean Air Act, and approved by the EPA as meeting these same requirements.

state implementation plans (SIP) (EPA-94/04): EPA approved state plans for the establishment, regulation, and enforcement of air pollution standards. Static Water Depth: The vertical distance from the centerline of the pump discharge down to the surface level of the free pool while no water is being drawn rom the pool or water table.

state lead agency (40CFR162.151-j): See State or State lead agency.

state operating permit program (40CFR72.2) means an operating permit program that the Administrator has approved as meeting the requirements of titles IV and V of the Act, part 70 of this chapter, and this part, including subpart G of this part.

state or local air monitoring stations (40CFR58.1-b): See SLAMS.

state or State lead agency (40CFR162.151-j) as used in this subpart means the State agency designated by the State to be responsible for registering pesticides to meet special local needs under sec. 24(c) of the Act.

state primary drinking water regulation (40CFR142.2) means a drinking water regulation of a State which is comparable to a national primary drinking water regulation (under the Safe Drinking Water Act).

state program (SMCRA701-30USC1291) means a program established by a State pursuant to section 1253 of this title to regulate surface coal mining and reclamation operations, on lands within such State in accord with the requirements of this chapter and regulations issued by the Secretary pursuant to this chapter.

state program revision (40CFR142.2) means a change in an approved State primacy program (under SDWA).

state program director or director (40CFR501.2)

means the chief executive officer of the State sewage sludge management agency.

state program (40CFR233.2): See state 404 program.

state regulated waters (40CFR232.2) means those waters of the United States in which the Corps suspends the issuance of section 404 permits upon approval of a State's section 404 permit program by the Administrator under section 404(h). The program cannot be transferred for those waters which are presently used, or are susceptible to use in their natural condition or by reasonable improvement as a means to transport interstate or foreign commerce shoreward to their ordinary high water mark, including all waters which are subject to the ebb and flow of the tide shoreward to the high tide line, including wetlands adjacent thereto. All other waters of the United States in a State with an approved program shall be under jurisdiction of the State program, and shall be identified in the program description as required by part 233.

state regulatory authority (SMCRA701-30USC1291) means the department or agency in each State which has primary responsibility at the State level for administering this chapter.

state sewage sludge management agency (40CFR501.2) means the agency designated by the Governor as having the lead responsibility for managing or coordinating the approved State program under this part.

state water pollution control agency (40CFR401.11-m) shall be defined in accordance with section 502 of the Act unless the context otherwise requires.

state water pollution control agency (CWA502-33USC1362) means the State agency designated by the Governor having responsibility for enforcing State laws relating to the abatement of pollution.

statement (40CFR21.2-i) means a written approval by EPA, or if appropriate, a State, of the application.

statement (40CFR27.2) means any representation, certification, affirmation, document, record. or accounting or bookkeeping entry made:
(a) With respect to a claim or to obtain the

approval or payment of a claim (including relating to eligibility to make a claim); or

(b) With respect to (including relating to eligibility for):

(1) A contract with, or a bid or proposal for a contract with; or

(2) A grant, loan, or benefit from, the Authority, or any State, political subdivision of a State, or other party, if the United States Government provides any portion of the money or property under such contract or for such grant, loan, or benefit, or if the Government will reimburse such State, political subdivision, or party for any portion of the money or property under such contract or for such grant, loan, or benefit.

statement of work (SOW) (40CFR35.6015-45) means the portion of the cooperative agreement application and/or Superfund State Contract that describes the purpose and scope of activities and tasks to be carried out as a part of the proposed project.

static (40CFR797.1400-10) means the test solution is not renewed during the period of the test.

static loaded radius arc (40CFR86.084.2) means a portion of a circle whose center is the center of standard tire-rim combination of an automobile and whose radius is the distance from that center to the level surface on which the automobile is standing, measured with the automobile at curb weight, the wheel parallel to the vehicles longitudinal centerline, and the tire inflated to the manufacturer's recommended pressure.

static replacement test (40CFR797.1160-6) means a test method in which the test solution is periodically replaced at specific intervals during the test (under TSCA).

static sheen test (40CFR435.11-v) shall refer to the standard test procedure that has been developed for this industrial subcategory for the purpose of demonstrating compliance with the requirement of no discharge of free oil. The methodology for performing the static sheen test is presented in Appendix 1 to 40CFR435, subpart A.

static system (40CFR795.120) means a test chamber in which the test solution is not renewed during the period of the test.

static system (40CFR797.1050-5) means a test container in which the test solution is not renewed during the period of the test.

static system (40CFR797.1300-7) means a test system in which the test solution and test organisms are placed in the test chamber and kept there for the duration of the test without renewal of the test solution.

static system (40CFR797.1930-6) means a test chamber in which the test solution is not renewed during the period of the test.

static test (40CFR797.1350-5) is a toxicity test with aquatic organisms in which no flow of test solution occurs. Solutions may remain uncharged throughout the duration of the test (other identical or similar definitions are provided in 40CFR797.1440-5).

static water level (EPA-94/04):
1. Elevation or level of the water table in a well when the pump is not operating.
2. The level or elevation to which water would rise in a tube connected to an artesian aquifer, or basin, in a conduit under pressure.

station wagon (40CFR600.002.85-36) means a passenger automobile with an extended roof line to increase cargo or passenger capacity, cargo compartment open to the passenger compartment, a tailgate, and one or more rear seats readily removed or folded to facilitate cargo carrying.

stationary casting (40CFR467.02-r) is the pouring of molten aluminum into molds and allowing the metal to air cool.

stationary casting (40CFR471.02-ii) is the pouring of molten metal into molds and allowing the metal to cool.

stationary compactor (40CFR243.101-z) means a powered machine which is designed to compact solid waste or recyclable materials, and which remains stationary when in operation (other identical or similar definitions are provided in 40CFR246.101-ee) (under RCRA).

stationary emission source and stationary source (40CFR52.741) mean an emission source which is not self-propelled.

stationary gas turbine (40CFR60.331-a) means any simple cycle gas turbine, regenerative cycle gas turbine or any gas turbine portion of a combined cycle steam/electric generating system that is not self propelled. It may, however, be mounted on a vehicle for portability.

stationary gas turbine (40CFR72.2) means a turbine that combusts natural gas, coal-derived gaseous fuel with a sulfur content no greater than natural gas, or fuel oil in order to heat inlet combustion air and thereby turn a turbine, in addition to or instead of producing steam or heating water, and that is not self propelled.

stationary source (40CFR51.165-i): Any building, structure, facility, or installation which emits or may emit any air pollutant subject to regulation under the Act.

stationary source (40CFR51.166-5): Any building, structure, facility, or installation which emits or may emit any air pollutant subject to regulation under the Act.

stationary source (40CFR51.301-w) means any building, structure, facility, or installation which emits or may emit any air pollutant (under the Clean Air Act).

stationary source (40CFR51.491): Any building, structure, facility or installation, other than an area or mobile source, which emits or may emit any criteria air pollutant or precursor subject to regulation under the Act.

stationary source (40CFR51-AS-1) means any building, structure, facility, or installation which emits or may emit any air pollutant subject to regulation under the Act.

stationary source (40CFR52.01-a) means any building, structure, facility, or installation which emits

or may emit an air pollutant for which a national standard is in effect.

stationary source (40CFR52.21-5): Any building, structure, facility, or installation which emits or may emit any air pollutant subject to regulation under the Act (other identical or similar definitions are provided in 40CFR52.24-1).

stationary source (40CFR61.02) means any building, structure, facility, or installation which emits or may emit any air pollutant which has been designated as hazardous by the Administrator.

stationary source (40CFR63.2) means any building, structure, facility, or installation which emits or may emit any air pollutant (other identical or similar definitions are provided in 40CFR63.101; 63.191).

stationary source (40CFR65.01) Stationary source shall mean any stationary building, facility, equipment, installation, or operation (or combination thereof) which is located on one or more contiguous or adjacent properties and which is owned or operated by the same person (or by persons under common control), and which emits an air pollutant for which a national ambient air quality standard promulgated under section 109 of the Act is in effect.

stationary source (40CFR70.2) means any building, structure, facility, or installation that emits or may emit any regulated air pollutant or any pollutant listed under section 112(b) of the Act (under the Clean Air Act).

stationary source (CAA111-42USC7411) means any building, structure, facility, or installation which emits or may emit any air pollutant.

stationary source (CAA112-42USC7412) shall have the same meaning as such term has under section 111(a).

stationary source (CAA302-42USC7602) means generally any source of an air pollutant except those emissions resulting directly from an internal combustion engine for transportation purposes or from a nonroad engine or nonroad vehicle as defined in section 216.

stationary source (EPA-94/04): A fixed-site producer of pollution, mainly power plants and other facilities using industrial combustion processes.

stationery (40CFR250.4-oo) means writing paper suitable for pen and ink, pencil, or typing. Matching envelopes are included in this definition.

statistical significance (40CFR228.2-f) shall mean the statistical significance determined by using appropriate standard techniques of multivariate analysis with results interpreted at the 95 percent confidence level and based on data relating species which are present in sufficient numbers at control areas to permit a valid statistical comparison with the areas being tested.

statistical sound level (40CFR201.1-ff) means the level in decibels that is exceeded in a stated percentage (x) of the duration of the measurement period. It is abbreviated as L_x.

statistically significant increase in a release (40CFR302.8) is an increase in the quantity of the hazardous substance released above the upper bound of the reported normal range of the release.

statutory economic incentive program (40CFR51.491) means any EIP submitted to the EPA as an implementation plan revision to comply with sections 182(g)(3), 182(g)(5), 187(d)(3), or 187(g) of the Act.

steady-state (40CFR797.1520-11) is the time period during which the amounts of test substance being taken up and depurated by the test organisms are equal, i.e., equilibrium.

steady-state (40CFR797.1830-8) is the time period during which the amounts of test chemical being taken up and depurated by the test oysters are equal, i.e., equilibrium.

steady-state bioconcentration factor (40CFR797.1520-12) is the mean concentration of the test substance in test organisms during steady-state divided by the mean concentration in the test solution during the same period.

steady-state bioconcentration factor (40CFR797.1830-9) is the mean concentration of the test chemical in test organisms during steady-state divided by the mean concentration of the test chemical in the test solution during the same period.

steady-state or apparent plateau (40CFR797.1560-5) is a condition in which the amount of test material being taken up and depurated is equal at a given water concentration.

steam generating unit (40CFR60.41a-1) means any furnace, boiler, or other device used for combusting fuel for the purpose of producing steam (including fossil-fuel-fired steam generators associated with combined cycle gas turbines; nuclear steam generators are not included).

steam generating unit (40CFR60.41b) means a device that combusts any fuel or byproduct/waste to produce steam or to heat water or any other heat transfer medium. This term includes any municipal-type solid waste incinerator with a heat recovery steam generating unit or any steam generating unit that combusts fuel and is part of a cogeneration system or a combined cycle system. This term does not include process heaters as they are defined in this subpart.

steam generating unit (40CFR60.41c) means a device that combusts any fuel and produces steam or heats water or any other heat transfer medium. This term includes any duct burner that combusts fuel and is part of a combined cycle system. This term does not include process heaters as defined in this subpart.

steam generating unit (40CFR61.301) means any enclosed combustion device that uses fuel energy in the form of steam.

steam generating unit operating day (40CFR60.41b) means a 24-hour period between 12:00 midnight and the following midnight during which any fuel is combusted at any time in the steam generating unit. It is not necessary for fuel to be combusted continuously for the entire 24-hour period (other identical or similar definitions are provided in 40CFR60.41c).

steam jet ejector (40CFR63.111) means a steam nozzle which discharges a high-velocity jet across a suction chamber that is connected to the equipment to be evacuated.

steam sales agreement (40CFR72.2) is a legally binding agreement between a QF, IPP, new IPP, or firm associated with such facility and an industrial or commercial establishment requiring steam that establishes the terms and conditions under which the facility will supply steam to the establishment.

steam stripping operation (40CFR264.1031) means a distillation operation in which vaporization of the volatile constituents of a liquid mixture takes place by the introduction of steam directly into the charge.

steel (40CFR464.31-f) means an iron-base alloy containing carbon, manganese, and often other alloying elements. Steel is defined here to include only those iron-carbon alloys containing less than 1.2 percent carbon by weight.

steel basis material (40CFR465.02-e) means cold rolled steel, hot rolled steel, and chrome, nickel and tin coated steel which are processed in coil coating.

steel production cycle (40CFR60.141-d) means the operations conducted within the BOPF steelmaking facility that are required to produce each batch of steel and includes the following operations: scrap charging, preheating (when used), hot metal charging, primary oxygen blowing, sampling (vessel turndown and turnup), additional oxygen blowing (when used), tapping, and deslagging. This definition applies to an affected facility constructed, modified, or reconstructed after January 20, 1983. For an affected facility constructed, modified, or reconstructed after June 11, 1973, but on or before January 20, 1983, "steel production cycle" means the operations conducted within the BOPF steelmaking facility that are required to produce each batch of steel and includes the following operations: scrap charging, preheating (when used), hot metal charging, primary oxygen blowing, sampling (vessel turndown and turnup), additional oxygen blowing (when used), and tapping.

steel production cycle (40CFR60.141a) means the operations conducted within the BOPF steelmaking facility that are required to produce each batch of steel, including the following operations: scrap charging, preheating (when used), hot metal charging, primary oxygen blowing, sampling (vessel turndown and turnup), additional oxygen blowing (when used), tapping, and deslagging. Hot metal transfer and skimming operations for the next steel production cycle are also included when the hot metal transfer station or skimming station is an affected facility.

STEM (40CFR763-AA-20) means scanning transmission electron microscope.

step 1. facilities planning (40CFR35.2005-43): See 40CFR6.501-a for the definition.
- step 2 (40CFR35.2005-44): Preparation of design drawings and specifications.
- step 3 (40CFR35.2005-45): Building of a treatment works and related services and supplies.
- step 2 + 3 (40CFR35.2005-46): Design and building of a treatment works and building related services and supplies.
- step 7 (40CFR35.2005-47): Design/building of treatment works wherein a grantee awards a single contract for designing and building certain treatment works.

step 1 facilities planning (40CFR6.501-a) means preparation of a plan for facilities as described in 40CFR35, subpart E or I.
- Step 2 (40CFR6.501-b) means a project to prepare design drawings and specifications as described in 40CFR35, subpart E or I.
- Step 3 (40CFR6.501-c) means a project to build a publicly owned treatment works as described in 40CFR35, subpart E or I.
- Step 2+3 (40CFR6.501-d) means a project which combines preparation of design drawings and specifications as described in 40CFR6.501(b) and building as described in 40CFR6.501(c).

sterilization (EPA-94/04): The removal or destruction of all microorganisms, including pathogenic and other bacteria, vegetative forms and spores.

still (40CFR60.621) means a device used to volatilize, separate, and recover petroleum solvent from contaminated solvent, together with the piping and ductwork used in the installation of this device.

still (40CFR63.321) means any device used to volatilize and recover perchloroethylene from contaminated perchloroethylene.

stipend (40CFR45.115) means supplemental financial assistance, other than tuition and fees, paid directly to the trainee by the recipient organization.

stipend (40CFR46.120): Supplemental financial assistance other than tuition fees, and book allowance, paid directly to the fellow.

stock configuration (40CFR205.165-7) means that no modifications have been made to the original equipment motorcycle that would affect the noise emissions of the vehicle when measured according to the acceleration test procedure.

stock on hand (40CFR763.163) means the products which are in the possession, direction, or control of a person and are intended for distribution in commerce.

stock solution (40CFR797.1520-13) is the concentrated solution of the test substance which is dissolved and introduced into the dilution water.

stock solution (40CFR797.1600-14) is the source of the test solution prepared by dissolving the test substance in dilution water or a carrier which is then added to dilution water at a specified, selected concentration by means of the test substance delivery system.

stoker boiler (40CFR76.2) means a boiler that burns solid fuel in a bed, on a stationary or moving grate, that is located at the bottom of the furnace.

stone feed (40CFR60.341-e) means limestone feedstock and millscale or other iron oxide additives that become part of the product.

storage (40CFR191.02-k) means retention of spent nuclear fuel or radioactive wastes with the intent and capability to readily retrieve such fuel or waste for subsequent use, processing, or disposal.

storage (40CFR243.101-aa) means the interim containment of solid waste after generation and prior to collection for ultimate recovery or disposal (other

identical or similar definitions are provided in 40CFR246.101-ff).

storage (40CFR259.10.a) means the temporary holding of regulated medical wastes at a designated accumulation area before treatment, disposal, or transport to another location.

storage (40CFR260.10) means the holding of hazardous waste for a temporary period, at the end of which the hazardous waste is treated, disposed of, or stored elsewhere.

storage (40CFR270.2) means the holding of hazardous waste for a temporary period, at the end of which the hazardous waste is treated, disposed, or stored elsewhere.

storage (40CFR373.4-b) means the holding of hazardous substances for a temporary period, at the end of which the hazardous substance is either used, neutralized, disposed of, or stored elsewhere.

storage (EPA-94/04): Temporary holding of waste pending treatment or disposal, as in containers, tanks, waste piles, and surface impoundments.

storage (RCRA1004-42USC6903), when used in connection with hazardous waste, means the containment of hazardous waste, either on a temporary basis or for a period of years, in such a manner as not to constitute disposal of such hazardous waste.

storage and retrieval of aerometric data (SAROAD) system (40CFR58.1-p) is a computerized system which stores and reports information relating to ambient air quality.

storage bin (40CFR60.381) means a facility for storage (including surge bins and hoppers) or metallic minerals prior to further processing or loading.

storage bin (40CFR60.671) means a facility for storage (including surge bins) or nonmetallic minerals prior to further processing or loading.

storage container (40CFR52.2285-2) means any stationary vessel of more than 1,000 gallons (3,785 liters) nominal capacity. Stationary vessels include portable vessels placed temporarily at the location; e.g., tanks on skids (other identical or similar definitions are provided in 40CFR52.2286-2) (under CAA).

storage for disposal (40CFR761.3) means temporary storage of PCBs that have been designated for disposal.

storage tank or storage vessel (40CFR52.741) means any stationary tank, reservoir or container used for the storage of VOL's.

storage vessel (40CFR60.111-a) means any tank, reservoir, or container used for the storage of petroleum liquids, but does not include:
(1) Pressure vessels which are designed to operate in excess of 15 pounds per square inch gauge without emissions to the atmosphere except under emergency conditions,
(2) Subsurface caverns or porous rock reservoirs, or
(3) Underground tanks if the total volume of petroleum liquids added to and taken from a tank annually does not exceed twice the volume of the tank.

storage vessel (40CFR60.111a-a) means each tank, reservoir, or container used for the storage of petroleum liquids, but does not include:
(1) Pressure vessels which are designed to operate in excess of 204.9 kPa (15 psig) without emissions to the atmosphere except under emergency conditions;
(2) Subsurface caverns or porous rock reservoirs; or
(3) Underground tanks if the total volume of petroleum liquids added to and taken from a tank annually does not exceed twice the volume of the tank.

storage vessel (40CFR60.111b-j) means each tank, reservoir, or container used for the storage of volatile organic liquids but not include:
(1) Frames, housing, auxiliary supports, or other components that are not directly involved in the containment of liquids or vapors; or
(2) Subsurface caverns or porous rock reservoirs (under CAA).

storage vessel (40CFR60.691) means any tank, reservoir, or container used for the storage of petroleum liquids, including oily wastewater.

storage vessel (40CFR63.101) means a tank or other vessel that is used to store organic liquids that contain one or more of the organic HAP's listed in table 2 of this subpart and that has been assigned, according to the procedures in 40CFR63.100(g) of this subpart, to a chemical manufacturing process unit that is subject to this subpart. Storage vessel does not include:
(1) Vessels permanently attached to motor vehicles such as trucks, railcars, barges, or ships;
(2) Pressure vessels designed to operate in excess of 204.9 kilopascals and without emissions to the atmosphere;
(3) Vessels with capacities smaller than 38 cubic meters;
(4) Vessels storing organic liquids that contain organic hazardous air pollutants only as impurities;
(5) Bottoms receiver tanks;
(6) Surge control vessels; or
(7) Wastewater storage tanks. Wastewater storage tanks are covered under the wastewater provisions.

store or storage of sewage sludge (40CFR503.9-y) is the placement of sewage sludge on land on which the sewage sludge remains for two years or less. This does not include the placement of sewage sludge on land for treatment.

storm sewer (40CFR35.905) means a sewer intended to carry only storm waters, surface runoff, street wash waters, and drainage (other identical or similar definitions are provided in 40CFR35.2005-48).

storm sewer (EPA-94/04): A system of pipes (separate from sanitary sewers) that carries only water runoff from buildings and land surfaces.

storm water (40CFR122.26-13) means storm water runoff, snow melt runoff, and surface runoff and drainage.

storm water discharge associated with industrial activity (40CFR122.26-14) means the discharge from any conveyance which is used for collecting and conveying storm water and which is directly related to manufacturing, processing or raw materials storage areas at an industrial plant. The term does not include discharges from facilities or activities excluded from the NPDES program under 40CFR part 122. For the categories of industries identified in paragraphs (b)(14) (i) through (x) of this section, the term includes, but is not limited to, storm water discharges from industrial plant yards; immediate access roads and rail lines used or traveled by carriers of raw materials, manufactured products, waste material, or by-products used or created by the facility; material handling sites; refuse sites; sites used for the application or disposal of process waste waters (as defined at 40CFR part 401); sites used for the storage and maintenance of material handling equipment; sites used for residual treatment, storage, or disposal; shipping and receiving areas; manufacturing buildings; storage areas (including tank farms) for raw materials, and intermediate and finished products; and areas where industrial activity has taken place in the past and significant materials remain and are exposed to storm water. For the categories of industries identified in paragraph (b)(14)(xi) of this section, the term includes only storm water discharges from all the areas (except access roads and rail lines) that are listed in the previous sentence where material handling equipment or activities, raw materials, intermediate products, final products, waste materials, by-products, or industrial machinery are exposed to storm water. For the purposes of this paragraph, material handling activities include the storage, loading and unloading, transportation, or conveyance of any raw material, intermediate product, finished product, by-product or waste product. The term excludes areas located on plant lands separate from the plant's industrial activities, such as office buildings and accompanying parking lots as long as the drainage from the excluded areas is not mixed with storm water drained from the above described areas. Industrial facilities (including industrial facilities that are Federally, State, or municipally owned or operated that meet the description of the facilities listed in this paragraph (b)(14)(i)-(xi) of this section) include those facilities designated under the provisions of paragraph (a)(1)(v) of this section. The following categories of facilities are considered to be engaging in "industrial activity" for purposes of this subsection:

i. Facilities subject to storm water effluent limitations guidelines, new source performance standards, or toxic pollutant effluent standards under 40CFR subchapter N (except facilities with toxic pollutant effluent standards which are exempted under category (xi) in paragraph (b)(14) of this section);

ii. Facilities classified as Standard Industrial Classifications 24 (except 2434), 26 (except 265 and 267), 28 (except 283), 29, 311, 32 (except 323), 33, 3441, 373;

iii. Facilities classified as Standard Industrial Classifications 10 through 14 (mineral industry) including active or inactive mining operations (except for areas of coal mining operations no longer meeting the definition of a reclamation area under 40CFR434.11(1) because the performance bond issued to the facility by the appropriate SMCRA authority has been released, or except for areas of non-coal mining operations which have been released from applicable State or Federal reclamation requirements after December 17, 1990) and oil and gas exploration, production, processing, or treatment operations, or transmission facilities that discharge storm water contaminated by contact with or that has come into contact with, any overburden, raw material, intermediate products, finished products, byproducts or waste products located on the site of such operations; (inactive mining operations are mining sites that are not being actively mined, but which have an identifiable owner/operator; inactive mining sites do not include sites where mining claims are being maintained prior to disturbances associated with the extraction, beneficiation, or processing of mined materials, nor sites where minimal activities are undertaken for the sole purpose of maintaining a mining claim);

iv. Hazardous waste treatment, storage, or disposal facilities, including those that are operating under interim status or a permit under subtitle C of RCRA;

v. Landfills, land application sites, and open dumps that receive or have received any industrial wastes (waste that is received from any of the facilities described under this subsection) including those that are subject to regulation under subtitle D of RCRA;

vi. Facilities involved in the recycling of materials, including metal scrapyards, battery reclaimers, salvage yards, and automobile junkyards, including but limited to those classified as Standard Industrial Classification 5015 and 5093;

vii. Steam electric power generating facilities, including coal handling sites;

viii. Transportation facilities classified as Standard Industrial Classifications 40, 41, 42 (except 4221-25), 43, 44, 45, and 5171 which have vehicle maintenance shops, equipment cleaning operations, or airport deicing operations. Only those portions of the facility that are either involved in vehicle maintenance (including vehicle rehabilitation, mechanical repairs, painting, fueling, and lubrication), equipment cleaning operations, airport deicing operations, or which are otherwise identified under paragraphs (b)(14) (i)-(vii) or (ix)-(xi) of this section are associated with industrial activity;

ix. Treatment works treating domestic sewage or any other sewage sludge or wastewater treatment device or system, used in the storage treatment, recycling, and reclamation of municipal or domestic sewage, including land dedicated to the disposal of sewage sludge that are located within the confines of the facility, with a design flow of 1.0 mgd or more, or required to have an approved pretreatment program under 40CFR part 403. Not included are farm lands, domestic gardens or lands used for sludge management where sludge is beneficially reused and which are not physically located in the confines of the facility, or areas that are in compliance with section 405 of the CWA;

x. Construction activity including clearing, grading and excavation activities except: operations that result in the disturbance of less than five acres of total land area which are not part of a larger common plan of development or sale; (xi) Facilities under Standard Industrial Classifications 20, 21, 22, 23, 2434, 25, 265, 267, 27, 283, 285, 30, 31 (except 311), 323, 34 (except 3441), 35, 36, 37 (except 373), 38, 39, 4221-25, (and which are not otherwise included within categories (ii)-(x)).

storm water or wastewater collection system (40CFR280.12) means piping, pumps, conduits, and any other equipment necessary to collect and transport the flow of surface water run-off resulting from precipitation, or domestic, commercial or industrial wastewater to and from retention areas or any areas where treatment is designated to occur. The collection of storm water and wastewater does not include treatment except where incidental to conveyance.

stormwater sewer system (40CFR60.691) means a drain and collection system designed and operated for the sole purpose of collecting stormwater and which is segregated from the process wastewater collection system.

straight kraft recovery furnace (40CFR60.281-i) means a furnace used to recover chemicals consisting primarily of sodium and sulfur compounds by burning black liquor which on a quarterly basis contains 7 weight percent or less of the total pulp solids from the neutral sulfite semichemical process or has green liquor sulfidity of 28 percent or less.

strategic special nuclear material (10CFR70.4) means uranium-235 (contained in uranium enriched to 20 percent or more in the U^{235} isotope), uranium-233, or plutonium.

stratification (EPA-94/04): Separating into layers.

stratosphere (CAA111-42USC7411) means that part of the atmosphere above the tropopause.

stratosphere (EPA-94/04): The portion of the atmosphere 10-to-25 miles above the earth's surface.

stratum (plural strata) (40CFR144.3) means a single sedimentary bed or layer, regardless of thickness, that consists of generally the same kind of rock material (other identical or similar definitions are provided in 40CFR146.3).

strawberries (40CFR407.61-t) shall mean the processing of strawberries into the following product styles: Canned and frozen, whole, sliced, and pureed.

streamflow source zone (40CFR149.101-i) means the upstream headwaters area which drains into the recharge zone as defined in the December 16, 1975, Notice of Determination.

street motorcycle (40CFR205.151-2) means:
(i) Any motorcycle that:

(A) With an 80 kg (176 lb) driver, is capable of achieving a maximum speed of at least 40 km/h (25 mph) over a level paved surface; and

(B) Is equipped with features customarily associated with practical street or highway use, such features including but not limited to any of the following: stoplight horn rear view mirror, turn signals; or

(ii) Any motorcycle that:

(A) Has an engine displacement less than 50 cubic centimeters;

(B) Produces no more than two brake horse power;

(C) With a 80 kg (176 lb) driver, cannot exceed 48 km/h (30 mph) over a level paved surface.

street wastes (40CFR243.101-cc) means materials picked up by manual or mechanical sweepings of alleys, streets, and sidewalks; wastes from public waste receptacles; and material removed from catch basins.

stressed waters (40CFR125.58-t) means those receiving environments in which an applicant can demonstrate to the satisfaction of the Administrator, that the absence of a balanced, indigenous population is caused solely by human perturbations other than the applicant's modified discharge.

strip (40CFR61.141) means to take off RACM from any part of a facility or facility components (under CAA).

strip cropping (EPA-94/04): Growing crops in a systematic arrangement of strips or bands that serve as barriers to wind and water erosion.

strip mining (EPA-94/04): A process that uses machines to scrape soil or rock away from mineral deposits just under the earth's surface.

strip, sheet and plate (40CFR420.91-m) means those acid pickling operations that pickle strip, sheet or plate products.

strip, sheet, and miscellaneous products (40CFR420.121-e) means steel products other than wire products and fasteners.

stripper (40CFR61.61-s) includes any vessel in which residual vinyl chloride is removed from polyvinyl chloride resin, except bulk resin, in the slurry form by the use of heat and/or vacuum. In the case of bulk resin, stripper includes any vessel which is used to remove residual vinyl chloride from polyvinyl chloride resin immediately following the polymerization step in the plant process flow.

strong chelating agents (40CFR413.02-f) is defined as all compounds which, by virtue of their chemical structure and amount present, form soluble metal complexes which are not removed by subsequent metals control techniques such as pH adjustment followed by clarification or filtration.

structural deformation (EPA-94/04): Distortion in walls of a tank after liquid has been added or removed.

structural member (40CFR61.141) means any load-supporting member of a facility, such as beams and load supporting walls; or any nonload-supporting member, such as ceilings and nonload-supporting walls.

structure (40CFR51.165; 51.166; 52.21; 52.24): See building, structure, facility, or installation.

structure (40CFR763-AA-21) means a microscopic bundle, cluster, fiber, or matrix which may contain asbestos.

student (40CFR52.1161-4) means any full-time day student who does not live at the educational institution and who travels to and from classes by any mode of travel.

student (40CFR52.2297-2) means any full-time day student who does not live at the educational facility and who travels to and from classes by any mode of travel.

study (40CFR160.3) means any experiment at one or more test sites, in which a test substance is studied in a test system under conditions or in the environment to determine or help predict its effects, metabolism, product performance (efficacy studies only as required by 40CFR158.640), environmental and chemical fate, persistence, or other characteristics in humans, other living organisms, or media. The term does not include basic exploratory studies carried out to determine whether a test substance has any potential utility.

study (40CFR716.3; 720.3-k; 723.50): See health and safety study.

study (40CFR792.3) means any experiment at one or more test sites, in which a test substance is studied in a test system under conditions or in the environment to determine or help predict its effects, metabolism, environmental and chemical fate, persistence, or other characteristics in humans, other living organisms, or media. The term does not include basic exploratory studies carried out to determine whether a test substance has any potential utility (under TSCA).

study completion date (40CFR160.3) means the date the final report is signed by the study director (other identical or similar definitions are provided in 40CFR792.3).

study director (40CFR160.3) means the individual responsible for the overall conduct of a study (other identical or similar definitions are provided in 40CFR792.3).

study initiation date (40CFR160.3) means the date the protocol is signed by the study director (other identical or similar definitions are provided in 40CFR792.3).

stuffing box pressure (40CFR61.241) means the fluid (liquid or gas) pressure inside the casing or housing of a piece of equipment, on the process side of the inboard seal.

styrene-butadiene rubber production (40CFR63.191) means a process that produces styrene-butadiene copolymers, whether in solid (elastomer) or emulsion (latex) form.

subagreement (40CFR30.200) means a written agreement between an EPA recipient and another party (other than another public agency) and any lower tier agreement for services, supplies, or construction necessary to complete the project. Subagreements include contracts and subcontracts for personal and professional services, agreements with consultants, and purchase orders.

subagreement (40CFR33.005) means a written agreement between an EPA recipient and another party (other than another public agency) and any lower tier agreement for services, supplies, or construction necessary to complete the project. Subagreements include contracts and subcontracts for personal and professional services, agreements with consultants, and purchase orders.

subagreement (40CFR35.4010) means a written agreement between the technical assistance grant recipient and another party (a contractor other than a public agency) for services or supplies necessary to complete the technical assistance grant project. Subagreements include contracts and subcontracts for personal and professional services or supplies necessary to complete the technical assistance grant project, and agreements with consultant, and purchase orders.

subagreement (40CFR35.936.1-b) means a written agreement between ar EPA grantee and another party (other than another public agency) and any tier of agreement thereunder for the furnishing of services, supplies, or equipment necessary to complete the project for which a grant was awarded, including contracts and subcontracts for personal and professional services, agreements with consultants and purchase orders, but excluding employment agreements subject to State or local personnel systems. (See also sections 40CFR35.937-12 and 40CFR35.938-9 regarding subcontracts of any tier under prime contracts for architectural or engineering services or construction awarded by the grantee--generally applicable only to subcontracts in excess of $10,000.)

subbituminous coal (40CFR60.41a-4) means coal that is classified as subbituminous A, B, or C according to the American Society of Testing and Materials (ASTM) Standard Specification for Classification of Coals by Rank D388-77 (incorporated by reference, see 40CFR60.17).

subchronic (EPA-94/04): Of intermediate duration, usually used to describe studies or levels of exposure between 5 and 90 days.

subchronic delayed neurotoxicity (40CFR798.6560) is a prolonged, delayed onset locomotor ataxia resulting from repeated daily administration of the test substance.

subchronic dermal toxicity (40CFR798.2250-1) is the adverse effects occurring as a result of the repeated daily exposure of experimental animals to a chemical by dermal application for part (approximately 10 percent) of a life span.

subchronic inhalation toxicity (40CFR798.2450-1) is the adverse effects occurring as a result of the repeated daily exposure of experimental animals to a chemical by inhalation for part (approximately 10 percent) of a life span.

subchronic oral toxicity (40CFR795.260-1) is the adverse effects occurring as a result of the repeated daily exposure of experimental animals to a chemical for a part (approximately 10 percent for rats) of a life span.

subchronic oral toxicity (40CFR798.2650-1) is the adverse effects occurring as a result of the repeated daily exposure of experimental animals to a chemical by the oral route for a part (approximately 10 percent for rats) of a life span (other identical or similar definitions are provided in 40CFR798.2675-1).

subclass (40CFR86.1102.87) means a classification of heavy-duty engines or heavy-duty vehicles based on such factors as gross vehicle weight rating, fuel usage, vehicle usage, engine horsepower or additional criteria that the Administrator shall apply. Subclasses include, but are not limited to:
i. Light-duty gasoline trucks (6,001-8,500 lb. GVW)
ii. Light-duty diesel trucks (6,001-8,500 lb. GVW)
iii. Light-duty petroleum-fueled diesel trucks (6,001-8,500 lb. GVW)
iv. Light-duty methanol-fueled diesel trucks (6,001-8,500 lb. GVW)
v. Light heavy-duty gasoline-fueled Otto cycle engines (for use in vehicles of 8,501-14,000 lb. GVW)
vi. Light heavy-duty methanol-fueled Otto cycle engines (for use in vehicles of 8,501-14,000 lb. GVW)
vii. Heavy heavy-duty gasoline-fueled Otto cycle engines (for use in vehicles of 14,001 lb and above GVW)
viii. Heavy heavy-duty methanol-fueled Otto cycle engines (for use in vehicles of 14,001 lb. and above GVW)
ix. Light heavy-duty petroleum-fueled diesel engines (see 40CFR86.085-2(a)(1))
x. Light heavy-duty methanol-fueled diesel engines (see 40CFR86.085-2(a)(1))
xi. Medium heavy-duty petroleum-fueled diesel engines (see 40CFR86.085-2(a)(2))
xii. Medium heavy-duty methanol-fueled diesel engines (see 40CFR86.085-2(a)(2))
xiii. Heavy heavy-duty petroleum-fueled diesel engines (see 40CFR86.085-2(a)(3))
xiv. Heavy heavy-duty methanol-fueled diesel engines (see 40CFR86.085-2(a)(3))
xv. Petroleum-fueled Urban Bus engines (see 40CFR86.091-2)
xvi. Methanol-fueled Urban Bus engines (see 40CFR86.091-2).

For NCP purposes, all optionally certified engines and/or vehicles (engines certified in accordance with 40CFR86.087-10(a)(3) and vehicles certified in accordance with 40CFR86.085-1(b)) shall be considered part of, and included in the FRAC calculation of, the subclass for which they are optionally certified.

subconfiguration (40CFR600.002.85-51) means a unique combination, within a vehicle configuration of equivalent test weight, road-load horsepower, and any other operational characteristics or parameters which the Administrator determines may significantly affect fuel economy within a vehicle configuration (under MVICSA).

subcontract (40CFR8.2-w) means any agreement or arrangement between a contractor and any person (in which the parties do not stand in the relationship of any employer and an employee):
(1) For the furnishing of supplies or services or for the use of real or personal property, including lease arrangements, which, in whole or in part, is necessary to the performance of any one or more contracts; or
(2) Under which any portion of the contractor's obligation under any one or more contracts is performed, undertaken or assumed.

subcontractor (40CFR35.6015-46) means any first tier party that has a contract with the recipient's prime contractor.

subcontractor (40CFR8.2-x) means any person holding a subcontract and, for the purposes of subpart B (General Enforcement; Compliance Review; and Complaint Procedure) of the rules, regulations, and relevant orders of the Secretary of Labor any person who had held a subcontract subject to the order.

subgrant (40CFR31.3) means an award of financial assistance in the form of money, or property in lieu of money, made under a grant by a grantee to an eligible subgrantee. The term includes financial assistance when provided by contractual legal agreement, but does not include procurement purchases, nor does it include any form of assistance which is excluded from the definition of grant in this part.

subgrantee (40CFR31.3) means the government or other legal entity to which a subgrant is awarded and which is accountable to the grantee for the use of the funds provided.

subindex (40CFR58-AG-g) means the calculated index value for a single pollutant as described in section 7.

submarginal area (40CFR51.392) means any ozone nonattainment area which EPA has classified as submarginal in 40CFR part 81 (other identical or similar definitions are provided in 40CFR93.101).

submerged aquatic vegetation (EPA-94/04): Vegetation such as sea grasses that cannot withstand excessive drying and therefore live with their leaves at or below the water surface; an important habitat for young fish and other aquatic organisms.

submerged loading pipe (40CFR52.741) means any discharge pipe or nozzle which meets either of the following conditions:
(A) Where the tank is filled from the top, the end of the discharge pipe or nozzle must be totally submerged when the liquid level is 15 cm (6 in.) above the bottom of the tank.
(B) Where the tank is filled from the side, the discharge pipe or nozzle must be totally submerged when the liquid level is 46 cm (18 in.) above the bottom of the tank.

submission (40CFR403.3-u) means:
(1) A request by a POTW for approval of a Pretreatment Program to the EPA or a Director;
(2) a request by a POTW to the EPA or a Director for authority to revise the discharge limits in categorical Pretreatment Standards to reflect POTW pollutant removals; or
(3) a request to the EPA by an NPDES State for approval of its State pretreatment program.

submit or serve (40CFR72.2) means to send or transmit a document, information, or correspondence to the person specified in accordance with the applicable regulation:
(1) In person;
(2) By United States Postal Service certified mail with the official postmark or, if service is by the Administrator or the permitting authority, by any other mail service by the United States Postal Service; or
(3) By other means with an equivalent time and date mark used in the regular course of business to indicate the date of dispatch or transmission and a record of prompt delivery. Compliance with any "submission," "service," or "mailing" deadline shall be determined by the date of dispatch, transmission, or mailing and not the date of receipt.

submitter (40CFR350.1) means a person filing a required report or making a claim of trade secrecy to EPA under sections 303 (d)(2) and (d)(3), 311, 312, and 313 of Title III.

subsidence (40CFR146.3) means the lowering of the natural land surface in response to: Earth movements; lowering of fluid pressure; removal of underlying supporting material by mining or solution of solids, either artificially or from natural causes; compaction due to wetting (hydrocompaction); oxidation of organic matter in soils; or added load on the land (other identical or similar definitions are provided in 40CFR147.2902).

substance (40CFR704.3) means either a chemical

substance or mixture unless otherwise indicated (under TSCA).

substance (40CFR716.3) means "chemical substance" as defined at section 3(2)(A) of TSCA, 15USC2602(2)(A).

substances (40CFR717.3-k) means a chemical substance or mixture unless otherwise indicated.

substantial business relationship (40CFR264.141-h) means the extent of a business relationship necessary under applicable State law to make a guarantee contract issued incident to that relationship valid and enforceable. A "substantial business relationship" must arise from a pattern of recent or ongoing business transactions, in addition to the guarantee itself, such that a currently existing business relationship between the guarantor and the owner or operator is demonstrated to the satisfaction of the applicable EPA Regional Administrator (other identical or similar definitions are provided in 40CFR265.141-h).

substantial business relationship (40CFR280.92) means the extent of a business relationship necessary under applicable state law to make a guarantee contract issued incident to that relationship valid and enforceable. A guarantee contract is issued "incident to that relationship" if it arises from and depends on existing economic transactions between the guarantor and the owner or operator.

substantial governmental relationship (40CFR280.92) means the extent of a governmental relationship necessary under applicable state law to make an added guarantee contract issued incident to that relationship valid and enforceable. A guarantee contract is issued "incident to that relationship" if it arises from a clear commonality of interest in the event of an UST release such as coterminous boundaries, overlapping constituencies, common ground-water aquifer, or other relationship other than monetary compensation that provides a motivation for the guarantor to provide a guarantee (under RCRA).

substantiation (40CFR350.1) means the written answers submitted to EPA by a submitter to the specific questions set forth in this regulation in support of a claim in chemical identity is a trade secret.

substate (40CFR256.06) refers to any public regional, local, county, municipal, or intermunicipal agency, or regional or local public (including interstate) solid or hazardous waste management authority, or other public agency below the State level.

substitute (40CFR29.12-3) means that a State may use a plan or other document that it has developed for its own purposes to meet Federal requirements.

substitute data (40CFR72.2) means emissions or volumetric flow data provided to assure 100 percent recording and reporting of emissions when all or part of the continuous emission monitoring system is not functional or is operating outside applicable performance specifications.

substitute or alternative (40CFR82.172) means any chemical, product substitute, or alternative manufacturing process, whether existing or new, intended for use as a replacement for a class I or II compound.

substitution unit (40CFR72.2) means an affected unit, other than a unit under section 410 of the Act, that is designated as a Phase I unit in a substitution plan under 40CFR72.41.

substrate (40CFR52.741) means the surface onto which a coating is applied or into which a coating is impregnated.

substrate (40CFR60.741) means the surface to which a coating is applied.

subunit (40CFR704.25-8) means an atom or group of associated atoms chemically derived from corresponding reactants (other identical or similar definitions are provided in 40CFR721.350).

subunit (40CFR723.250-16) means an atom or group of associated atoms chemically derived from corresponding reactants.

sudden accidental occurrence (40CFR264.141-g) means an occurrence which is not continuous or repeated in nature (other identical or similar definitions are provided in 40CFR265.141-g).

suitable substitute decision (40CFR203.1-6) means the Administrator's decision whether a product which the Administrator has determined to be a low noise-emission product is a suitable substitute for a product or products presently being purchased by the Federal Government.

sulfide (40CFR410.01-a) shall mean total sulfide (dissolved and acid soluble) as measured by the procedures listed in 40CFR136.

sulfide (40CFR425.02-a) shall mean total sulfide as measured by the potassium ferricyanide titration method described in Appendix A or the modified Monier-Williams method described in Appendix B.

sulfite cooking liquor (40CFR430.101-b) shall be defined as bisulfite cooking liquor when the pH of the liquor is between 3.0 and 6.0 and as acid sulfite cooking liquor when the pH is less than 3.0 (other identical or similar definitions are provided in 40CFR430.211-b).

sulfur (40CFR52.1881-xiii) recovery plant means any plant that recovers elemental sulfur from any gas stream.

sulfur dioxide (SO$_2$) (EPA-94/04): A pungent, colorless, gaseous pollutant formed primarily by the combustion of fossil fuels.

sulfur free generation (40CFR72.2) means the generation of electricity by a process that does not have any emissions of sulfur dioxide, including hydroelectric, nuclear, solar, or wind generation. A "sulfur-free generator" is a generator that is located in one of the 48 contiguous States or the District of Columbia and produces "sulfur-free generation."

sulfur percentage (40CFR80.2-y) is the percentage of sulfur as determined by ASTM standard test method D 2622-87, entitled "Standard Test Method for Sulfur in Petroleum Products by X-Ray Spectrometry." ASTM test method D 2622-87 is incorporated by reference. This incorporation by reference was approved by the Director of the Federal Register in accordance with 5 U.S.C. 552(a)

and 1CFR part 51. A copy may be obtained from the American Society for Testing and Materials, 1916 Race Street, Philadelphia, PA 19103. A copy may be inspected at the Air Docket section (A-130), room M-1500, U.S. Environmental Protection Agency, Docket No. A-86-03, 401 M Street SW., Washington DC 20460 or at the Office of the Federal Register, 1100 L Street NW, room 8401, Washington, DC 20005.

sulfur production rate (40CFR60.641) means the rate of liquid sulfur accumulation from the sulfur recovery unit.

sulfur recovery unit (40CFR60.641) means a process device that recovers elemental sulfur from acid gas.

sulfuric acid pickling (40CFR420.91-a) means those operations in which steel products are immersed in sulfuric acid solutions to chemically remove oxides and scale, and those rinsing operations associated with such immersions.

sulfuric acid plant (40CFR51.100-dd) means any facility producing sulfuric acid by the contact process by burning elemental sulfur, alkylation acid, hydrogen sulfide, or acid sludge, but not including facilities where conversion to sulfuric acid is utilized primarily as a means of preventing emissions to the atmosphere of sulfur dioxide or other sulfur compounds.

sulfuric acid plant (40CFR60.161-h) means any facility producing sulfuric acid by the contact process (other identical or similar definitions are provided in 40CFR60.171-d; 60.181-i).

sulfuric acid production unit (40CFR52.1881-xiv) means any facility producing sulfuric acid by the contact process by burning elemental sulfur, alkylation acid, hydrogen sulfide, organic sulfides and mercaptans, or acid sludge.

sulfuric acid production unit (40CFR60.81-a) means any facility producing sulfuric acid by the contact process by burning elemental sulfur, alkylation acid, hydrogen sulfide, organic sulfides and mercaptans, or acid sludge, but does not include facilities where conversion to sulfuric acid is utilized primarily as a means of preventing emissions to the atmosphere of sulfur dioxide or other sulfur compounds (under CAA).

sump (40CFR260.10) means any pit or reservoir that meets the definition of tank and those troughs/trenches connected to it that serves to collect hazardous waste for transport to hazardous waste storage, treatment, or disposal facilities.

sump (EPA-94/04): A pit or tank that catches liquid runoff for drainage or disposal.

sunlight direct aqueous photolysis rate constant (40CFR796.3700-vii) is the first-order rate constant in the units of day^{-1} and is a measure of the rate of disappearance of a chemical dissolved in a water body in sunlight (other identical or similar definitions are provided in 40CFR796.3780-viii; 796.3800-vii).

superchlorination (EPA-94/04): Chlorination with doses that are deliberately selected to produce free or combined residuals so large as to require dechlorination.

supercritical water (EPA-94/04): A type of thermal treatment using moderate temperatures and high pressures to enhance the ability of water to break down large organic molecules into smaller, less toxic ones. Oxygen injected during this process combines with simple organic compounds to form carbon dioxide and water.

superfund (EPA-94/04): The program operated under the legislative authority of CERCLA and SARA that funds and carries out EPA solid waste emergency and long-term removal and remedial activities. These activities include establishing the National Priorities List, investigating sites for inclusion on the list, determining their priority, and conducting and/or supervising the cleanup and other remedial actions.

superfund innovative technology evaluation (EPA-94/04): EPA program to promote development and use of innovative treatment technologies in Superfund site cleanups.

superfund memorandum of agreement (SMOA) (40CFR300.5) means a nonbinding, written document executed by an EPA Regional Administrator and the head of a state agency that may establish the nature and extent of EPA and state interaction during the removal, pre-remedial, remedial, and/or enforcement response process. The SMOA is not a site-specific document although attachments may address specific sites. The SMOA generally defines the role and responsibilities of both the lead and the support agencies.

superfund state contract (40CFR300.5) is a joint, legally binding agreement between EPA and a state to obtain the necessary assurances before a federal-lead remedial action can begin at a site. In the case of a political subdivision-lead remedial response, a three-party Superfund state contract among EPA, the state, and political subdivision thereof, is required before a political subdivision takes the lead for any phrase of remedial response to ensure state involvement pursuant to section 121(f)(1) of CERCLA. The Superfund state contract may be amended to provide the state's CERCLA section 104 assurances before a political subdivision can take the lead for remedial action.

superfund state contract (SSC) (40CFR35.6015-47) means a joint, legally binding agreement between EPA and another party(s) to obtain the necessary assurances before a EPA-lead remedial action or any political subdivision-lead activities can begin at a site, and to ensure State or Indian Tribe involvement as required under CERCLA section 121(f).

superphosphoric acid plant (40CFR60.211-a) means any facility which concentrates wet-process phosphoric acid to 66 percent or greater P_2O_5 content by weight for eventual consumption as a fertilizer.

supersaturated solution (40CFR796.1840-v): See oversaturated solution.

supplemental printed material (40CFR82.104-t) means any informational material (including, but not limited to, package inserts, fact sheets, invoices, material safety data sheets, procurement and specification sheets, or other material) which accompanies a product or container to the consumer at the time of purchase.

supplementary control system (40CFR52.1475-i)

means any system which limits the amount of pollutant emissions during periods when meteorological conditions conducive to ground-level concentrations in excess of national standards exist or are anticipated.

supplementary control system (SCS) (40CFR57.103-u) means any technique for limiting the concentration of a pollutant in the ambient air by varying the emissions of that pollutant according to atmospheric conditions. For the purposes of this part, the term supplementary control system does not include any dispersion technique based solely on the use of a stack the height of which exceeds good engineering practice (as determined under regulations implementing section 123 of the Act).

supplier of water (40CFR141.2): Any person who owns or operates a public water system (other identical or similar definitions are provided in 40CFR142.2; 143.2; EPA-94/04).

supplier of water (SDWA1401-42USC300f) means any person who owns or operates a public water system (other identical or similar definitions are provided in 40CFR141.2; 142.2; 143.2-e).

supplies (40CFR31.3) means all tangible personal property other than equipment as defined in this part.

supplies (40CFR33.005) means all property, including equipment, materials, printing, insurance, and leases of real property, but excluding land or a permanent interest in land.

supplies (40CFR35.6015-48) means all tangible personal property other than equipment as defined in this subpart.

supply-side measure (40CFR72.2) means a measure to improve the efficiency of the generation, transmission, or distribution of electricity, implemented by a utility in connection with its operations or facilities to provide electricity to its customers, and includes the measures set forth in part 73, Appendix A, section 2 of this chapter.

support agency (40CFR300.5) means the agency or agencies that provide the support agency coordinator to furnish necessary data to the lead agency, review response data and documents, and provide other assistance as requested by the OSC or RPM, EPA, the USCG, another federal agency, or a state may be support agencies for a response action if operating pursuant to a contract executed under section 104(d)(1) of CERCLA, or designated pursuant to a Superfund Memorandum of Agreement entered into pursuant to subpart F of the NCP or other agreement. The support agency may also concur on decision documents.

support agency (40CFR35.6015-49) means the agency that furnishes necessary data to the lead agency, reviews response data and documents, and provides other assistance as required by the lead agency.

support agency coordinator (SAC) (40CFR300.5) means the official designated by the support agency, as appropriate, to interact and coordinate with the lead agency in response actions under subpart E of this part.

support media (40CFR797.2800) means the quartz sand or glass beads used to support the plant.

support media (40CFR797.2850-3) means the sand or glass beads used to support the plant.

suppressed combustion (40CFR420.41-g) means those basic oxygen furnace steelmaking wet air cleaning systems which are designed to limit or suppress the combustion of carbon monoxide in furnace gases by restricting the amount of excess air entering the air pollution control system.

surface casing (40CFR146.3) means the first string of well casing to be installed in the well.

surface coal mining and reclamation operations (SMCRA701-30USC1291) means surface mining operations and all activities necessary and incident to the reclamation of such operations after August 3, 1977.

surface coal mining operations (SMCRA701) means:
(A) activities conducted on the surface of lands in connection with a surface coal mine or subject to the requirement of section 1266 of this title

surface operations and surface impacts incident to an underground coal mine, the products of which enter commerce or the operations of which directly or indirectly affect interstate commerce. Such activities include excavation for the purpose of obtaining coal including such common methods as contour, strip, auger, mountain-top removal, box cut, open pit, and area mining, the uses of explosives and blasting, and in situ distillation or retorting, leaching or other chemical or physical processing, and the cleaning, concentrating, or other processing or preparation, loading of coal for interstate commerce at or near the mine site: provided, however, that such activities do not include the extraction of coal incidental to the extraction of other minerals where coal does not exceed 16 2/3 per centum of the tonnage of minerals removed for purposes of commercial use or sale or coal explorations subject to section 1262 of this title; and

(B) the areas upon which such activities occur or where such activities disturb the natural land surface. Such areas shall also include any adjacent land the use of which is incidental to any such activities, all lands affected by the construction of new roads or the improvement or use of existing roads to gain access to the site of such activities and for haulage, and excavations, workings, impoundments, dams, ventilation shafts, entryways, refuse banks, dumps, stockpiles, overburden piles, spoil banks, culm banks, tailings, holes or depressions, repair areas, storage areas, processing areas, shipping areas and other areas upon which are sited structures, facilities, or other property or materials on the surface, resulting from or incident to such activities.

surface coating (40CFR468.02-v) shall mean the process of coating a copper workpiece as well as the associated surface finishing and flattening.

surface coating operation (40CFR60.311) means the system on a metal furniture surface coating line used to apply and dry or cure an organic coating on the surface of the metal furniture part or product. The surface coating operation may be a prime coat or a top coat operation and includes the coating application station(s), flash-off area, and curing oven.

surface coating operation (40CFR60.391) means any prime coat, guide coat, or topcoat operation on an automobile or light-duty truck surface coating line.

surface coating operation (40CFR60.451) means the system on a large appliance surface coating line used to apply and dry or cure an organic coating on the surface of large appliance parts or products. The surface coating operation may be a prime coat or a topcoat operation and includes the coating application station(s), flashoff area, and curing oven.

surface collecting agents (40CFR300.5) means those chemical agents that form a surface film to control the layer thickness of oil.

surface condenser (40CFR52.741) means a device which removes a substance from a gas stream by reducing the temperature of the stream, without direct contact between the coolant and the stream.

surface disposal site (40CFR503.21-p) is an area of land that contains one or more active sewage sludge units.

surface impoundment (40CFR280.12) is a natural topographic depression, man-made excavation, or diked area formed primarily of earthen materials (although it may be lined with man-made materials) that is not an injection well.

surface impoundment (40CFR61.341) means a waste management unit which is a natural topographic depression, man-made excavation, or diked area formed primarily of earthen materials (although it may be lined with man-made materials), which is designed to hold an accumulation of liquid wastes or waste containing free liquids, and which is not an injection well. Examples of surface impoundments are holding, storage, settling, and aeration pits, ponds, and lagoons.

surface impoundment (40CFR63.111) means a waste management unit which is a natural topographic depression, manmade excavation, or diked area formed primarily of earthen materials (although it may be lined with manmade materials), which is designed to hold an accumulation of liquid wastes or waste containing free liquids. A surface impoundment is used for the purpose of treating, storing, or disposing of wastewater or residuals, and is not an injection well. Examples of surface impoundments are equalization, settling, and aeration pits, ponds, and lagoons.

surface impoundment (EPA-94/04): Treatment, storage, or disposal of liquid hazardous wastes in ponds.

surface impoundment or impoundment (40CFR260.10) means a facility or part of a facility which is a natural topographic depression, man-made excavation, or diked area formed primarily of earthen materials (although it may be lined with man-made materials), which is designed to hold an accumulation of liquid wastes or wastes containing free liquids, and which is not an injection well. Examples of surface impoundments are holding, storage, settling, and aeration pits, ponds, and lagoons.

surface material (TSCA-AIA1) means material in a building that is sprayed on surfaces, troweled on surfaces, or otherwise applied to surfaces for acoustical, fireproofing, or other purposes, such as acoustical plaster on ceilings and fireproofing material on structural members.

surface moisture (40CFR60.381) means water that is not chemically bound to a metallic mineral or metallic mineral concentrate.

surface runoff (EPA-94/04): Precipitation, snow melt, or irrigation in excess of what can infiltrate the soil surface and be stored in small surface depressions; a major transporter of nonpoint source pollutants.

surface treatment (40CFR471.02-jj) is a chemical or electrochemical treatment applied to the surface of a metal. Such treatments include pickling, etching, conversion coating, phosphating, and chromating. Surface treatment baths are usually followed by a water rinse. The rinse may consist of single or multiple stage rinsing. For the purposes of this part, a surface treatment operation is defined as a bath followed by a rinse, regardless of the number of stages. Each surface treatment bath, rinse combination is entitled to discharge allowance.

surface uranium mines (EPA-94/04): Strip mining operations for removal of uranium-bearing ore.

surface water (40CFR131.35.d-14) means all water above the surface of the ground within the exterior boundaries of the Colville Indian Reservation including but not limited to lakes, ponds, reservoirs, artificial impoundments, streams, rivers, springs, seeps and wetlands.

surface water (40CFR141.2) means all water which is open to the atmosphere and subject to surface runoff.

surface water (EPA-94/04): All water naturally open to the atmosphere (rivers, lakes, reservoirs, ponds, streams, impoundments, seas, estuaries, etc.) and all springs, wells, or other collectors directly influenced by surface water.

surfacing ACM (40CFR763.83) means surfacing material that is ACM.

surfacing ACM (EPA-94/04): Asbestos-containing material that is sprayed or troweled on or otherwise applied to surfaces, such as acoustical plaster on ceilings and fireproofing materials on structural members.

surfacing material (40CFR763.83) means material in a school building that is sprayed-on, troweled-on, or otherwise applied to surfaces, such as acoustical plaster on ceilings and fireproofing materials on structural members, or other materials on surfaces for acoustical, fireproofing, or other purposes (under TSCA).

surfacing material (EPA-94/04): Material sprayed or troweled onto structural members (beams, columns, or decking) for fire protection; or on ceilings or walls for fireproofing, acoustical or decorative purposes. Includes textured plaster, and other textured wall and ceiling surfaces.

surfactant (40CFR417.91-c) shall mean those methylene blue active substances amendable to measurement by the method described in "Methods for Chemical Analysis of Water and Wastes," 1971, Environmental Protection Agency, Analytical Quality Control Laboratory, page 131 (other identical or similar definitions are provided in 40CFR417.101-c; 417.111-c; 417.121-c; 417.131-c; 417.141-c; 417.151-c; 417.161-c; 417.171-c; 417.181-c; 417.191-c).

surfactant (EPA-94/04): A detergent compound that promotes lathering.

surge control tank (40CFR264.1031) means a large-sized pipe or storage reservoir sufficient to contain the surging liquid discharge of the process tank to which it is connected.

surge control vessel (40CFR63.101) means feed drums, recycle drums, and intermediate vessels. Surge control vessels are used within a chemical manufacturing process unit when in-process storage, mixing, or management of flow rates or volumes is needed to assist in production of a product.

surge control vessel (40CFR63.111) means feed drums, recycle drums, and intermediate vessels. Surge control vessels are used within a chemical manufacturing process unit when in-process storage, mixing, or management of flow rates or volumes is needed to assist in production of a product (under CAA).

surge control vessel (40CFR63.161) means feed drums, recycle drums, and intermediate vessels. Surge control vessels are used within a process unit (as defined in the specific subpart that references this subpart) when in-process storage, mixing, or management of flow rates or volumes is needed to assist in production of a product.

surge control vessel (40CFR63.191) means feed drums, recycle drums, and intermediate vessels. Surge control vessels are used within a process unit when in-process storage, mixing, or management of flow rates or volumes is needed to assist in production of a product.

surplus (40CFR51.491) means, at a minimum, emissions reductions in excess of an established program baseline which are not required by SIP requirements or State regulations, relied upon in any applicable attainment plan or demonstration, or credited in any RFP or milestone demonstration, so as to prevent the double-counting of emissions reductions.

surveillance system (EPA-94/04): A series of monitoring devices designed to check on environmental conditions.

susceptibility (40CFR171.2-26) means the degree to which an organism is affected by a pesticide at a particular level of exposure.

suspect material (EPA-94/04): Building material suspected of containing asbestos, e.g., surfacing material, floor tile, ceiling tile, thermal system insulation, and other miscellaneous materials.

suspended (40CFR136-AC-2) means those elements which are retained by a 0.45 μm membrane filter (under FWPCA).

suspended loads (EPA-94/04): Sediment particles maintained in the water column by turbulence and carried with the flow of water.

suspended solids (EPA-94/04): Small particles of solid pollutants that float on the surface of, or are suspended in, sewage or other liquids. They resist removal by conventional means.

suspending official (40CFR32.105-t) means an official authorized to impose suspension. The suspending official is either:
(1) The agency head, or
(2) An official designated by the agency head,
(3) The Director, Grants Administration Division, is the authorized suspending official.

suspension (40CFR31.3) means, depending on the context, either:
(1) temporary withdrawal of the authority to obligate grant funds pending corrective action by the grantee or subgrantee or a decision to terminate the grant, or
(2) an action taken by a suspending official in accordance with agency regulations implementing E.O. 12549 to immediately exclude a person from participating in grant transactions for a period, pending completion of an investigation and such legal or debarment proceedings as may ensue.

suspension (40CFR32.105-u) means an action taken by a suspending official in accordance with these regulations that immediately excludes a person from participating in covered transactions for a temporary period, pending completion of an investigation and such legal, debarment, or Program Fraud Civil

Remedies Act proceedings as may ensue. A person so excluded is "suspended."

suspension (EPA-94/04): Suspending the use of a pesticide when EPA deems it necessary to prevent an imminent hazard resulting from its continued use. An emergency suspension takes effect immediately; under an ordinary suspension a registrant can request a hearing before the suspension goes into effect. Such a hearing process might take six months.

suspension culture (EPA-94/04): Cells growing in a liquid nutrient medium.

sustained yield (FLPMA103-43USC1702) means the achievement and maintenance in perpetuity of a high-level annual or regular periodic output of the various renewable resources of the public lands consistent with multiple use.

swaging (40CFR471.02-kk) is a process in which a solid point is formed at the end of a tube, rod, or bar by the repeated blows of one or more pairs of opposing dies.

swamp (EPA-94/04): A type of wetland dominated by woody vegetation but without appreciable peat deposits. Swamps may be fresh or salt water and tidal or non-tidal (cf. wetlands).

SWDA: See act or SWDA.

sweet water (40CFR417.41-c) shall mean the solution of 8-10 percent crude glycerine and 90-22 percent water that is a byproduct of saponification or fat splitting.

sweetening unit (40CFR60.641) means a process device that separates the H_2S and CO_2 contents from the sour natural gas stream.

switcher locomotive (40CFR201.1-gg) means any locomotive designated as a switcher by the builder or reported to the ICC as a switcher by the operator-owning-railroad and including, but not limited to, all locomotives of the builder/model designations listed in Appendix A to this subpart.

symmetrical tetrachloropyridine (40CFR63.191) means the chemical with CAS number 2402-791.

synergism (EPA-94/04): An interaction of two or more chemicals which results in an effect that is greater than the sum of their effects taken independently.

synthetic ammonium sulfate manufacturing plant (40CFR60.421) means any plant which produces ammonium sulfate by direct combination of ammonia and sulfuric acid.

synthetic fiber (40CFR60.601) means any fiber composed partially or entirely of materials made by chemical synthesis, or made partially or entirely from chemically-modified naturally-occurring materials.

synthetic organic chemicals (SOCS) (EPA-94/04): Man-made organic chemicals. Some SOCs are volatile, others tend to stay dissolved in water instead of evaporating.

synthetic organic chemicals manufacturing industry (40CFR60.481) means the industry that produces, as intermediates or final products, one or more of the chemicals listed in 40CFR60.489.

system (40CFR86.082.2) includes any motor vehicle engine modification which controls or causes the reduction of substances emitted from motor vehicles.

system (40CFR86.402.78) includes any motor vehicle modification which controls or causes the reduction of substances emitted from motor vehicles.

system dependent recovery equipment (40CFR82.152-w) means refrigerant recovery equipment that requires the assistance of components contained in an appliance to remove the refrigerant from the appliance.

system emergency reserves (40CFR60.41a-15) means an amount of electric generating capacity equivalent to the rated capacity of the single largest electric generating unit in the electric utility company (including steam generating units, internal combustion engines, gas turbines, nuclear units, hydroelectric units, and all other electric generating equipment) which is interconnected with the affected facility that has the malfunctioning flue gas desulfurization system. The electric generating capability of equipment under multiple ownership is prorated based on ownership unless the proportional entitlement to electric output is otherwise established by contractual arrangement.

system load (40CFR60.41a-15) means the entire electric demand of an electric utility company's service area interconnected with the affected facility that has the malfunctioning flue gas desulfurization system plus firm contractual sales to other electric utility companies. Sales to other electric utility companies (e.g., emergency power) not on a firm contractual basis may also be included in the system load when no available system capacity exists in the electric utility company to which the power is supplied for sale.

system of records (40CFR1516.2) means a group of any records under the control of the Council from which information is retrieved by the name of the individual or by some identifying number, symbol, or other identifying particular assigned to the individual.

system of records (40CFR16.2-a): See individual, maintain, record, system of records, and routine use.

system with a single service connection (40CFR141.2) means a system which supplies drinking water to consumers via a single service line (under SDWA).

system with a single service connection (EPA-94/04): A system that supplies drinking water to consumers via a single service line.

systemic pesticide (EPA-94/04): A chemical absorbed by an organism that makes the organism toxic to pests.

*********** TTTTT ***********

T, as in "cyanide, T" (40CFR433.11-a) shall mean total.

tablet coating operation (40CFR52.741) means a pharmaceutical coating operation in which tablets are coated.

tabulating cards (40CFR250.4-pp) means cards used in automatic tabulating machines.

tabulating paper (40CFR250.4-pp) means paper used in tabulating forms for use on automatic data processing equipment.

tag (40CFR211.203-w) means stiff paper, metal or other hard material that is tied or otherwise affixed to the packaging of a protector.

tail water (EPA-94/04): The runoff of irrigation water from the lower end of an irrigated field.

tailings (40CFR61.251-g): See uranium byproduct material.

tailings (EPA-94/04): Residue of raw material or waste separated out during the processing of crops or mineral ores.

tailings closure plan (radon) (40CFR192.31-n) means the Nuclear Regulatory Commission or Agreement State approved plan detailing activities to accomplish timely emplacement of a permanent radon barrier. A tailings closure plan shall include a schedule for key radon closure milestone activities such as wind blown tailings retrieval and placement on the pile, interim stabilization (including dewatering or the removal of freestanding liquids and recontouring), and emplacement of a permanent radon barrier constructed to achieve compliance with the 20 pCi/m^2-s flux standard as expeditiously as practicable considering technological feasibility (including factors beyond the control of the licensee).

tailpipe standards (EPA-94/04): Emissions limitations applicable to engine exhausts from mobile sources.

take (ESA3-16USC1531) means to harass, harm, pursue, hunt, shoot, wound, kill, trap, capture, or collect, or to attempt to engage in any such conduct.

take (MMPA3-16USC1362) means to harass, hunt, capture, or kill, or attempt to harass, hunt, capture, or kill any marine mammal.

taking (40CFR257.3.2-3) means harassing, harming, pursuing, hunting, wounding, killing, trapping, capturing, or collecting or attempting to engage in such conduct.

tall coke oven battery (40CFR63.301) means a coke oven battery with ovens 6 meters or more in height.

tallow (40CFR432.101-d) shall mean a product made from beef cattle or sheep fat that has a melting point of 40°C or greater.

tamper (SDWA1432-42USC300i.1) means:
(1) to introduce a contaminant into a public water system with the intention of harming persons; or
(2) to otherwise interfere with the operation of a public water system with the intention of harming persons.

tampering (40CFR204.2-9) means those acts prohibited by section 10(a)(2) of the Act (other identical or similar definitions are provided in 40CFR204.51-x; 205.2-23; 205.51-26).

tampering (40CFR205.151-26) mens the removal or rendering inoperative by any person, other than for purposes of maintenance, repair, or replacement, of any device or element of design incorporated into any product in compliance with regulations under section 6, prior to its sale or delivery to the ultimate purchaser or while it is in use; or the use of a product after such deice or element of design has been removed or rendered inoperative by any person.

tampering (EPA-94/04): Adjusting, negating, or removing pollution control equipment on a motor vehicle.

tangentially fired boiler (40CFR76.2) means a boiler that has coal and air nozzles mounted in each corner of the furnace where the vertical furnace walls meet. Both pulverized coal and air are directed from the furnace corners along a line tangential to a circle lying in a horizontal plane of the furnace.

tangible net worth (40CFR144.61-d) means the tangible assets that remain after deducting liabilities; such assets would not include intangibles such as goodwill and rights to patents or royalties.

tangible net worth (40CFR264.141-f) means the tangible assets that remain after deducting liabilities; such assets would not include intangibles such as goodwill and rights to patents or royalties (other identical or similar definitions are provided in 40CFR265.141-f; 280.92).

tank (40CFR260.10) means a stationary device, designed to contain an accumulation of hazardous waste which is constructed primarily of non-earthen materials (e.g., wood, concrete, steel, plastic) which provide structural support.

tank (40CFR280.12) is a stationary device designed to contain an accumulation of regulated substances and constructed of non-earthen materials (e.g., concrete, steel, plastic) that provide structural support.

tank (40CFR61.341) means a stationary waste management unit that is designed to contain an accumulation of waste and is constructed primarily of nonearthen materials (e.g., wood, concrete, steel, plastic) which provide structural support.

tank fuel volume (40CFR86.082.2) means the volume of fuel in the fuel tank(s), which is determined by taking the manufacturers nominal fuel tank(s) capacity and multiplying by 0.40, the result being rounded using ASTM E 29-67 to the nearest tenth of a U.S. gallon.

tank system (40CFR260.10) means a hazardous waste storage or treatment tank and its associated ancillary equipment and containment system (under RCRA).

tank system (40CFR280.12): See UST system.

tank vessel (OPA1001) means a vessel that is constructed or adapted to carry, or that carries, oil or hazardous material in bulk as cargo or cargo residue, and that:
(A) is a vessel of the United States;
(B) operates on the navigable waters; or
(C) transfers oil or hazardous material in a place subject to the jurisdiction of the United States.

tankage (40CFR432.101-c) shall mean dried animal byproduct residues used in feedstuffs.

tap (40CFR60.271-g) means the pouring of molten steel from an EAF.

tap (40CFR60.271a) means the pouring of molten steel from an EAF or AOD vessel.

tapping (40CFR60.261-e) means the removal of slag or product from the electric submerged arc furnace under normal operating conditions such as removal of metal under normal pressure and movement by gravity down the spout into the ladle.

tapping period (40CFR60.261-f) means the time duration from initiation of the process of opening the tap hole until plugging of the tap hole is complete.

tapping period (40CFR60.271) means the time period commencing at the moment an EAF begins to tilt to pour and ending either three minutes after an EAF returns to an upright position or six minutes after commencing to tilt, whichever is longer.

tapping station (40CFR60.261-h) means that general area where molten product or slag is removed from the electric submerged arc furnace.

tar decanter (40CFR61.131) means any vessel, tank, or container that functions to separate heavy tar and sludge from flushing liquor by means of gravity, heat, or chemical emulsion breakers. A tar decanter also may be known as a flushing-liquor decanter (under CAA).

tar intercepting sump (40CFR61.131) means any tank, pit, or enclosure that serves to receive or separate tars and aqueous condensate discharged from the primary cooler. A tar-intercepting sump also may be known as a primary-cooler decanter.

tar storage tank (40CFR61.131) means any vessel, tank, reservoir, or other type of container used to collect or store crude tar or tar-entrained naphthalene, except for tar products obtained distillation, such as coal tar pitch, creosotes, or carbolic oil. This definition also includes any vessel, tank, reservoir, or container used to reduce the water content of the tar by means of heat, residence time, chemical emulsion breakers, or centrifugal separation. A tar storage tank also may be known as a tar-dewatering tank.

target distance limit (40CFR300-AA) means the maximum distance over which targets for the site are evaluated. The target distance limit varies by HRS pathway.

task (40CFR35.6015-50) means an element of a Superfund response activity identified in the Statement of work of a Superfund Cooperative Agreement or a Superfund State Contract.

taxi/idle (in) (40CFR87.1) means those aircraft operations involving taxi and idle between the time of landing roll-out and final shutdown of all propulsion engines.

taxi/idle (out) (40CFR87.1) means those aircraft operations involving taxi and idle between the time of initial starting of the propulsion engine(s) used for the taxi and turn on to duty runway.

technical assistance grant (EPA-94/04): As part of the Superfund program, Technical Assistance Grants of up to $50,000 are provided to citizens' groups to obtain assistance in interpreting information related to cleanups at Superfund sites or those proposed for the National Priorities List. Grants are used by such groups to hire technical advisors to help them understand the site-related technical information for the duration of response activities.

technical grade of active ingredient (40CFR158.153-k) means a material containing an active ingredient:
(1) Which contains no inert ingredient, other than one used for purification of the active ingredient; and
(2) Which is produced on a commercial or pilot-plant production scale (whether or not it is ever held for sale).

technical support document (40CFR66.3-m) means the "Noncompliance Penalties Technical Support Document" which accompanies these regulations. The Technical Support Document appears as Appendix A to these regulations (under the Clean Air Act).

technically feasible (40CFR157.21-f) when applied to child-resistant packaging, means that the technology exists to produce the child-resistant packaging for a particular pesticide.

technically qualified individual (40CFR710.2-aa) means a person:
(1) Who because of his education, training, or experience, or a combination of these factors, is capable of appreciating the health and environmental risks associated with the chemical substance which is used under his supervision,
(2) who is responsible for enforcing appropriated methods of conducting scientific experimentation, analysis, or chemical research in order to minimize such risks, and
(3) who is responsible for the safety assessments and clearances related to the procurement, storage, use, and disposal of the chemical substance as may be appropriate or required within the scope of conducting the research and development activity. The responsibilities in paragraph (aa)(3) of this section may be delegated to another individual, or other individuals, as long as each meets the criteria in paragraph (aa)(1) of this section.

technically qualified individual (40CFR720.3-ee) means a person or persons:
(1) who, because of education, training, or experience, or a combination of these factors, is capable of understanding the health and environmental risks associated with the chemical substance which is used under his or her supervision,
(2) who is responsible for enforcing appropriate methods of conducting scientific experimentation, analysis, or chemical research to minimize such risks, and
(3) who is responsible for the safety assessments and clearances related to the procurement, storage, use, and disposal of the chemical substance as may be appropriate or required within the scope of conducting a research and development activity.

technician (40CFR82.152-x) means any person who performs maintenance, service, or repair that could be reasonably expected to release class I or class II refrigerants from appliances into the atmosphere, including but not limited to installers, contractor employees, in-house service personnel, and in some cases, owners. Technician also means any person disposing of appliances except for small appliances (under CAA).

technological system of continuous emission reduction (CAA111-42USC7411) means:
(A) a technological process for production or operation by any source which is inherently low-polluting or nonpolluting, or
(B) a technological system for continuous reduction of the pollution generated by a source before such pollution is emitted into the ambient air, including precombustion cleaning or treatment of fuels.

technology-based limitations (EPA-94/04): Industry-specific effluent limitations applied to a discharge when it will not cause a violation of water quality standards at low stream flows. Usually applied to discharges into large rivers.

technology-based standards (EPA-94/04): Effluent limitations applicable to direct and indirect sources which are developed on a category-by-category basis using statutory factors, not including water-quality effects.

TEM (40CFR763-AA-24) means transmission electron microscope.

temperature (40CFR131.35.d-15) means water temperature expressed in centigrade degrees (C).

temperature monitoring device (40CFR63.111) means a unit of equipment used to monitor temperature and having an accuracy of ± 1 percent of the temperature being monitored expressed in degrees Celsius or ± 0.5 degrees Celsius (°C), whichever is greater.

temperature sensor (40CFR63.321) means a thermometer or thermocouple used to measure temperature.

tempering (40CFR426.61-b) shall mean the process whereby glass is heated near the melting point and then rapidly cooled to increase its mechanical and thermal endurance.

temporary clean coal technology demonstration project (40CFR51.165-xxii) means a clean coal technology demonstration project that is operated for a period of 5 years or less, and which complies with the State Implementation Plan for the State in which the project is located and other requirements necessary to attain and maintain the national ambient air quality standards during the project and after it is terminated.

temporary clean coal technology demonstration project (40CFR51.166-35) means a clean coal technology demonstration project that is operated for a period of 5 years or less, and which complies with the State implementation plan for the State in which the project is located and other requirements necessary to attain and maintain the national ambient air quality standards during and after the project is terminated.

temporary clean coal technology demonstration project (40CFR52.21-36) means a clean coal technology demonstration project that is operated for a period of 5 years or less, and which complies with the State implementation plans for the State in which the project is located and other requirements necessary to attain and maintain the national ambient air quality standards during the project and after it is terminated.

temporary clean coal technology demonstration project (40CFR52.24-21) means a clean coal technology demonstration project that is operated for a period of 5 years or less, and which complies with the State implementation plans for the State in which the project is located and other requirements necessary to attain and maintain the national ambient air quality standards during the project and after it is terminated.

temporary enclosure (40CFR60.711-16) means a total enclosure that is constructed for the sole purpose of measuring the fugitive emissions from an affected facility. A temporary enclosure must be constructed and ventilated (through stacks suitable for testing) so that it has minimal impact on the performance of the permanent capture system. A temporary enclosure will be assumed to achieve total capture of fugitive VOC emissions if it conforms to the requirements found in 40CFR60.713(b)(5)(i) and if all natural draft openings are at least four duct or hood equivalent diameters away from each exhaust duct or hood. Alternatively, the owner or operator may apply to the Administrator for approval of a temporary enclosure on a case-by-case basis.

temporary enclosure (40CFR60.741) means a total enclosure that is constructed for the sole purpose of measuring the fugitive VOC emissions from an affected facility.

temporary opening (40CFR60.541) means an opening into an enclosure that is equipped with a means of obstruction, such as a door, window, or port, that is normally closed.

temporary seal (40CFR63.301) means any measure, including but not limited to, application of luting or packing material, to stop a collecting main leak until the leak is repaired.

ten (10)-year, 24-hour fall event (40CFR411.31-b) shall mean a rainfall event with a probable recurrence interval of once in ten years as defined by the National Weather Service in Technical Paper No. 40. "Rainfall Frequency Atlas of the United States," May 1961, and subsequent amendments, or equivalent regional or state rainfall probability information developed therefrom.

ten (10)-year, 24-hour precipitation event (40CFR436.21-c) shall mean the maximum 24-hour precipitation even with a probable reoccurrence interval of once in 10 years. This information is available in "Weather Bureau Technical Paper No. 40," May 1961 and "NOAA Atlas 2," 1973 for the 11 Western Sates, and may be obtained from the National Climatic Center of Environmental Data Service, National Oceanic and Atmospheric Administration, U.S. Department of Commerce (other identical or similar definitions are provided in 40CFR436.31-c; 436.41-c; 436.181-c).

ten (10)-year, 24-hour precipitation event (40CFR440.132-i) is the maximum 24-hour precipitation event with a probable recurrence interval of once in 10 years as established by the Department of Commerce, National Oceanic and Atmospheric Administration, National Weather Service, or equivalent regional or rainfall probability information.

ten (10)-year, 24-hour rainfall event (40CFR129.2-r) means the maximum precipitation event with a probable recurrence interval of once in 10 years as defined by the National Weather Service in Technical paper No. 40, "Rainfall Frequency Atlas of the United States," May 1961, and subsequent amendments or equivalent regional or State rainfall probability information developed therefrom (other identical or similar definitions are provided in 40CFR418.11-d; 422.41-d; 422.51-d).

ten (10)-year, 24-hour rainfall event (40CFR423.11-i) means a rainfall event with a probable recurrence interval of once in ten years as defined by the National Weather Service in Technical Paper No. 40. "Rainfall Frequency Atlas of the United States," May 1961 or equivalent regional rainfall probability information developed therefrom.

ten (10)-year, 24-hour rainfall event and 25-year, 24-hour rainfall event (40CFR412.11-e) shall mean a rainfall event with a probable recurrence interval of once in ten years or twenty-five years, respectively, as defined by the National Weather Service in Technical Paper Number 40, "Rainfall Frequency Atlas of the United States", May 1961, and subsequent amendments, or equivalent regional or state rainfall probability information developed therefrom (other identical or similar definitions are provided in 40CFR412.21-e).

tender (40CFR80.126-f) means the physical transfer of custody of a volume of gasoline or other petroleum product all of which has the same identification (reformulated gasoline, conventional gasoline, RBOB, and other non-finished gasoline petroleum products), and characteristics (time and place of use restrictions for reformulated gasoline) (under CAA).

terminal velocity of a small sphere falling under the influence of gravity in a viscous fluid: See 40CFR796.1520.

termination (10CFR20.3) means the end of employment with the licensee or, in the case of individuals not employed by the licensee, the end of a work assignment in the licensee's restricted areas in a given calendar quarter, without expectation or specific scheduling of reentry into the licensee's restricted areas during the remainder of that calendar quarter.

termination (40CFR31.3) means permanent withdrawal of the authority to obligate previously-awarded grant funds before that authority would otherwise expire. It also means the voluntary relinquishment of that authority by the grantee or subgrantee. Termination does not include:

(1) Withdrawal of funds awarded on the basis of the grantee's underestimate of the unobligated balance in a prior period;

(2) Withdrawal of the unobligated balance as of the expiration of a grant;

(3) Refusal to extend a grant or award additional funds, to make a competing or noncompeting continuation, renewal, extension, or supplemental award; or

(4) Voiding of a grant upon determination that the award was obtained fraudulently, or was otherwise illegal or invalid from inception.

termination under 40CFR280.97(b)(1) and 40CFR280.97(b)(2) (40CFR280.92) means only those changes that could result in a gap in coverage as where the insured has not obtained substitute coverage or has obtained substitute coverage with a different retroactive date than the retroactive date of the original policy.

terminology (40CFR6.101-a): All terminology used in this part will be consistent with the terms as defined in 40CFR1508 (the CEQ Regulations). Any qualifications will be provided in the definitions set forth in each subpart of this regulation.

terms and conditions of registration (40CFR154.3-h) means the terms and conditions governing lawful sale, distribution, and use approved in conjunction with registration, including labeling, use classification, composition, and packaging.

terms of a grant or subgrant (40CFR31.3) mean all requirements of the grant or subgrant, whether in statute, regulations, or the award document.

terne coating (40CFR420.121-b) means coating steel products with terne metal by the hot dip process including the immersion of the steel product in a molten bath of lead and tin metals, and the related operations preceding and subsequent to the immersion phase.

terracing (EPA-94/04): Dikes built along the contour

of sloping farm land that hold runoff and sediment to reduce erosion.

territorial sea (40CFR116.3) means the belt of the seas measured from the line of ordinary low water along that portion of the coast which is in direct contact with the open sea and the line marking the seaward limit of inland waters, and extending seaward a distance of 3 miles.

territorial sea (40CFR230.3-r) means the belt of the sea measured from the baseline as determined in accordance with the Convention on the Territorial Sea and the Contiguous Zone and extending seaward a distance of three miles.

territorial sea (40CFR401.11-m) shall be defined in accordance with section 502 of the Act unless the context otherwise requires.

territorial sea (CWA502-33USC1362) means the belt of the seas measured from the line of ordinary low water along that portion of the coast which is in direct contact with the open sea and the line marking the seaward limit of inland waters, and extending seaward a distance of three miles (other identical or similar definitions are provided in 40OPA1001).

territorial sea and contiguous zone (SF101-42USC9601) shall have the meaning provided in section 502 of the Federal Water Pollution Control Act.

tertiary treatment (EPA-94/04): Advanced cleaning of wastewater that goes beyond the secondary or biological stage, removing nutrients such as phosphorus, nitrogen, and most BOD and suspended solids.

test analyzer (40CFR53.1-j) means an analyzer subjected to testing as a candidate method in accordance with subparts B, C, and/or D of this part, as applicable.

test chamber (40CFR797.1520-14) is the container in which the test organisms are maintained during the test period.

test chamber (40CFR797.1600-15) is defined as the individual containers in which test organisms are

maintained during exposure to test solution (under TSCA).

test compressor (40CFR204.51-w) means a compressor used to demonstrate compliance with the applicable noise emissions standard.

test data (40CFR610.11-10) means any information which is a quantitative measure of any aspect of the behavior of a retrofit device.

test data (40CFR720.3-ff) means data from a formal or informal test or experiment, including information concerning the objectives, experimental methods and materials, protocols, results, data analyses, recorded observations, monitoring data, measurements, and conclusions from a test or experiment (other identical or similar definitions are provided in 40CFR723.50-2; 723.250-6).

test data (40CFR723.175-14) means:
(i) Data from a formal or informal study, test, experiment, recorded observation, monitoring, or measurement.
(ii) Information concerning the objectives, experimental methods and materials, protocols, results, data analyses (including risk assessments), and conclusions from a study, test, experiment, recorded observation, monitoring, or measurement.

test data (40CFR723.250-6) have the same meanings as in 40CFR720.3 of this chapter.

test engine (40CFR86.1002.2001-6) means an engine in a test sample.

test engine (40CFR86.1002.84) means an engine in a test sample.

test engine (40CFR89.502.96) means an engine in a test sample.

test exhaust system (40CFR205.165-8) means an exhaust system in Selective Enforcement Audit test sample.

test facility (40CFR211.203-x) for this subpart, a laboratory that has been set up and calibrated to conduct ANSI Std. S3.19-1974 tests on hearing protective devices. It must meet the applicable requirements of these regulations.

test facility (40CFR60-AA (method 28-2.5)) means the area in which the wood heater is installed, operated, and sampled for emissions.

test fleet (40CFR89.2) means the engine or group of engines that a manufacturer uses during certification to determine compliance with emission standards.

test for differential growth inhibition of repair proficient and repair deficient bacteria (40CFR798.5500) measure differences in chemically induced cell killing between wild-type strains with full repair capacity and mutant strains deficient in one or more of the enzymes which govern repair of damaged DNA.

test fuel charge (40CFR60-AA (method 28-2.6)) means the collection of test fuel pieces placed in the wood heater at the start of the emission test run.

test fuel crib (40CFR60-AA (method 28-2.7)) means the arrangement of the test fuel charge with the proper spacing requirements between adjacent fuel pieces.

test fuel loading density (40CFR60-AA (method 28-2.8)) means the weight of the as-fired test fuel charge per unit volume of usable firebox.

test fuel piece (40CFR60-AA (method 28-2.9)) means the 2 x 4 or 4 x 4 wood piece cut to the length required for the test fuel charge and used to construct the test fuel crib.

test hearing protector (40CFR211.203-y) means a hearing protector that has been selected for testing to verify the value to be put on the label, or which has been designated for testing to determine compliance of the protector with the labeled value.

test marketing (40CFR704.3) means the distribution in commerce of no more than a predetermined amount of a chemical substance, mixture, article containing that chemical substance or mixture, or a mixture containing that substance, by a manufacturer or processor, to no more than a defined number of potential customers to explore market capability in a competitive situation during a predetermined testing period prior to the broader distribution of that chemical substance, mixture, or article in commerce (under TSCA).

test marketing (40CFR710.2-bb) means the distribution in commerce of no more than a predetermined amount of a chemical substance, mixture, or article containing that chemical substance or mixture, by a manufacturer or processor to no more than a defined number of potential customers to explore market capability in a competitive situation during a predetermined testing period prior to the broader distribution of that chemical substance, mixture or article in commerce (other identical or similar definitions are provided in 40CFR720.3-gg).

test marketing (40CFR712.3-o) means distributing in commerce a limited amount of a chemical substance or mixture, or article containing such substance or mixture, to a defined number of potential customers, during a predetermined testing period, to explore market capability prior to broader distribution in commerce.

test marketing (40CFR82.172) means the distribution in interstate commerce of a substitute to no more than a limited, defined number of potential customers to explore market viability in a competitive situation. Testing must be restricted to a defined testing period before the broader distribution of that substitute in interstate commerce.

test method (40CFR63.2) means the validated procedure for sampling, preparing, and analyzing for an air pollutant specified in a relevant standard as the performance test procedure. The test method may include methods described in an appendix of this chapter, test methods incorporated by reference in this part, or methods validated for an application through procedures in Method 301 of Appendix A of this part.

test period (40CFR797.2050-6) is the combination of the exposure period and the post-exposure period; or, the entire duration of the test.

test product (40CFR204.2-18) means any product that is required to be tested pursuant to this part (other identical or similar definitions are provided in 40CFR205.2-28).

test product (40CFR211.102-k) means any product that must be tested according to regulations published under part 211.

test request (40CFR211.203-z) means a request submitted to the manufacturer by the Administrator that will specify the hearing protector category, and test sample size to be tested according to 40CFR211.212.1, and other information regarding the audit.

test rule (40CFR791.3-j) refers to a regulation ordering the development of data on health or environmental effects or chemical fate for a chemical substance or mixture pursuant to TSCA sec. 4(a).

test run (40CFR60-AA (method 28-2.10)) means an individual emission test which encompasses the time required to consume the mass of the test fuel charge.

test sample (40CFR204.51-m) means the collection of compressors from the same category or configuration which is randomly drawn from the batch sample and which will receive emissions tests.

test sample (40CFR205.51-23) means the collection of vehicles from the same category, configuration or subgroup thereof which is drawn from the batch sample and which will receive noise emissions tests.

test sample (40CFR86.1002.2001-4) means the collection of vehicles or engines of the same configuration which have been drawn from the population of vehicles or engines of that configuration and which will receive emission testing (under CAA).

test sample (40CFR86.1002.84) means the collection of vehicles or engines of the same configuration which have been drawn from the population of engines or vehicles of that configuration and which will receive exhaust emission testing.

test sample (40CFR86.1102.87) means a group of heavy-duty engines or heavy-duty vehicles of the same configuration which have been selected to receive emission testing.

test sample (40CFR86.602.84-4) means the collection of vehicles of the same configuration which have been drawn from the population of vehicles of that configuration and which will receive exhaust emission testing.

test sample (40CFR86.602.98-(b)(4)) means the collection of vehicles of the same configuration which have been drawn from the population of vehicles of that configuration and which will receive emission testing.

test sample (40CFR89.502.96) means the collection of engines selected from the population of an engine family for emission testing.

test sample size (40CFR204.51-o) means the number of compressors of the same configuration in a test sample.

test sample size (40CFR205.51-27) means the number of vehicles of the same category or configuration in a test sample.

test sampler (40CFR53.1-n) means a sampler subjected to testing as part of a candidate method in accordance with subpart C or D of this part.

test solution (40CFR797.1400-11) means the test substance and the dilution water in which the test substance is dissolved or suspended.

test solution (40CFR797.1520-15) is dilution water containing the dissolved test substance to which test organisms are exposed.

test solution (40CFR797.1600-16) is dilution water with a test substance dissolved or suspended in it.

test solution (40CFR797.2750-6) means the test chemical and the dilution water in which the test chemical is dissolved or suspended.

test substance (40CFR160.3) means a substance or mixture administered or added to a test system in a study, which substance or mixture:
(1) Is the subject of an application for a research or marketing permit supported by the study, or is the contemplated subject of such an application; or

(2) Is an ingredient impurity, degradation product, metabolite, or radioactive isotope of a substance described by paragraph (1) of this definition or some other substance related to a substance described by that paragraph, which is used in the study to assist in characterizing the toxicity, metabolism, or other characteristics of a substance described by that paragraph.

test substance (40CFR790.3) means the form of chemical substance or mixture that is specified for use in testing.

test substance (40CFR792.3) means a substance or mixture administered or added to a test system in a study, which substance or mixture is used to develop data to meet the requirements of a TSCA section 4(a) test rule and/or is developed under a negotiated testing agreement or section 5 rule or order to the extent the agreement or rule or order references this part.

test substance (40CFR795.232) refers to the unlabeled and both radiolabeled mixtures (^{14}C-n-hexane and ^{14}C-methylcyclopentane) of commercial hexane used in the testing.

test substance (40CFR797.1600-17) is the specific form of a chemical substance or mixture that is used to develop data.

test substance (40CFR797.2050-3) is the specific form of a chemical or mixture of chemicals that is used to develop the data (other identical or similar definitions are provided in 40CFR797.2130-ii; 797.2150-ii; 797.2175-3).

test system (40CFR160.3) means any animal, plant, microorganism, or chemical or physical matrix, including but not limited to, soil or water, or components thereof, to which the test, control or reference substance is administered or added for study. Test system also includes appropriate groups or components of the system not treated with the test, control, or reference substances (other identical or similar definitions are provided in 40CFR792.3) (under FIFRA).

test vehicle (40CFR205.151-27) means a vehicle in a Selective Enforcement Audit test sample.

test vehicle (40CFR205.51-28) means a vehicle selected and used to demonstrate compliance with the applicable noise emission standards.

test vehicle (40CFR86.1002.2001-7) means a vehicle in a test sample.

test vehicle (40CFR86.602.84-7) means a vehicle in a test (other identical or similar definitions are provided in 40CFR86.1002.84).

test weight (40CFR600.002.85-43) means the weight within an inertia weight class which is used in the dynamometer testing of a vehicle, and which is based on its loaded vehicle weight in accordance with the provisions of part 86.

test weight (40CFR88.101.94) is defined as the average of the curb weight and the GVWR.

test weight and the abbreviation (TW) (CAA216-42USC7550) mean the vehicle curb weight added to the gross vehicle weight rating (GVWR) and divided by 2.

test weight basis (40CFR86.094.2) means the basis on which equivalent test weight is determined in accordance with 40CFR86.129-94 of subpart B of this part.

testing agent (40CFR610.11-11) means any person who develops test data on a retrofit device.

testing exemption (40CFR204.2-7) means an exemption from the prohibitions of section 10(a) (1), (2), (3), and (5) of the Act, which may be granted under section 10(b)(l) of the Act for the purpose of research, investigations, studies, demonstrations, or training, but not including national security where lease or sale of the exempted product is involved (other identical or similar definitions are provided in 40CFR205.2-26; 211.102-j).

testing exemption (40CFR85.1702-5) means an exemption which may be granted under 40CFR203(b)(1) for the purpose of research investigations, studies, demonstrations or training, but not including national security.

testing exemption (40CFR89.902.96) means an

exemption which may be granted under 40CFR89.1004(b) for the purpose of research investigations, studies, demonstrations or training, but not including national security.

testing facility (40CFR160.3) means person who actually conducts a study, i.e., actually uses the test substance in a test system. Testing facility encompasses only those operational units that are being or have been used to conduct studies (other identical or similar definitions are provided in 40CFR792.3).

tetrachloroethylene (or priority pollutant No. 85) (40CFR420.02-p) means the value obtained by the standard method Number 610 specified in 44 FR 69464, 69571 (December 3, 1979).

tetratogenesis (EPA-94/04): The introduction of nonhereditary birth defects in a developing fetus by exogenous factors such as physical or chemical agents acting in the womb to interfere with normal embryonic development.

textile fiberglass (40CFR60.291) means fibrous glass in the form of continuous strands having uniform thickness.

textiles (40CFR763.163) means an asbestos-containing product such as: yarn, thread, wick; cord; braided and twisted rope; braided and woven tubing; mat; roving; cloth; slit and woven tape; lap; felt; and other bonded or non-woven fabrics.

texture coat (40CFR60.721) means the rough coat that is characterized by discrete, raised spots on the exterior surface of the part. This definition does not include conductive sensitizers or EMI/RFI shielding coatings.

the 33/50 program (40CFR63.111) means a voluntary pollution prevention initiative established and administered by the EPA to encourage emissions reductions of 17 chemicals emitted in large volumes by industrial facilities. The EPA Document Number 741-K-92-001 provides more information about the 33/50 program.

theoretical arsenic emissions factor (40CFR61.161) means the amount of inorganic arsenic, expressed in grams per kilogram of glass produced, as determined based on a material balance.

therapeutic index (EPA-94/04): The ratio of the dose required to produce toxic or lethal effects to dose required to produce nonadverse or therapeutic response.

thermal deflection rate (40CFR85.2122(a)(2)(iii)(G)) means the angular degrees of rotation per degree of temperature change of the thermostatic coil.

thermal dryer (40CFR60.251-e) means any facility in which the moisture content of bituminous coal is reduced by contact with a heated gas stream which is exhausted to the atmosphere.

thermal dryer (40CFR60.381) means a unit in which the surface moisture content of a metallic mineral or a metallic mineral concentrate is reduced by direct or indirect contact with a heated gas stream.

thermal energy (40CFR72.2) means the thermal output produced by a combustion source used directly as part of a manufacturing process but not used to produce electricity.

thermal pollution (EPA-94/04): Discharge of heated water from industrial processes that can kill or injure aquatic organisms.

thermal processing (40CFR240.101-aa) means processing of waste material by means of heat (under RCRA).

thermal stratification (EPA-94/04): The formation of layers of different temperatures in a lake or reservoir.

thermal system insulation (40CFR763.83) means material in a school building applied to pipes, fittings, boilers, breeching, tanks, ducts, or other interior structural components to prevent heat loss or gain, or water condensation, or for other purposes.

thermal system insulation (TSCA-AIA1) means material in a building applied to pipes, fittings, boilers, breeching, tanks, ducts, or other interior structural components to prevent heat loss or gain, or water condensation, or for other purposes.

thermal system insulation (TSI) (EPA-94/04): Asbestos-containing material applied to pipes, fittings, boilers, breeching, tanks, ducts, or other interior structural components to prevent heat loss or gain or water condensation.

thermal system insulation ACM (40CFR763.83) means thermal system insulation that is ACM.

thermal treatment (40CFR260.10) means the treatment of hazardous waste in a device which uses elevated temperatures as the primary means to change the chemical, physical, or biological character or composition of the hazardous waste. Examples of thermal treatment processes are incineration, molten salt, pyrolysis, calcination, wet air oxidation, and microwave discharge (see also incinerator; open burning).

thermal treatment (EPA-94/04): Use of elevated temperatures to treat hazardous wastes (cf. incineration or pyrolysis).

thermocline (EPA-94/04): The middle layer of a thermally stratified lake or reservoir. In this layer there is a rapid decrease in temperature with depth. Also called the metalimnion.

thermostatic coil (40CFR85.2122(a)(2)(iii)(D)) means a spiral-wound coil of thermally-sensitive material which provides rotary force (torque) and/or displacement as a function of applied temperature.

thermostatic switch (40CFR85.2122(a)(2)(iii)(E)) means an element of thermally-sensitive material which acts to open or close an electrical circuit as a function of temperature.

thin-film evaporation operation (40CFR264.1031) means a distillation operation that employs a heating surface consisting of a large diameter tube that may be either straight or tapered, horizontal or vertical. Liquid is spread on the tube wall by a rotating assembly of blades that maintain a close clearance from the wall or actually ride on the film of liquid on the wall.

third party in-kind contribution (40CFR31.3) means property or services which benefit a federally assisted project or program and which are contributed by in-Federal third parties without charge to the grantee, or a cost-type contractor under the grant agreement.

thirty-day (30-day) average (40CFR133.101-b) means the arithmetic mean of pollutant parameter values of samples collected in a period of 30 consecutive days.

thirty-day (30-day) limitation (40CFR429.11-j) is a value that should not be exceeded by the average of daily measurements taken during any 30-day period.

thirty-day rolling average (40CFR52.741) means any value arithmetically averaged over any consecutive thirty-days.

THM (40CFR141.2): Trihalomethane.

threat of discharge or release (40CFR300.5): See discharge; release.

threat of release (40CFR300.5): See of release.

threatened species (40CFR723.50): See endangered species.

threatened species (ESA3-16USC1531) means any species which is likely to become an endangered species within the foreseeable future throughout all or a significant portion of its range.

three (3)-hour period (40CFR61.61-z) means any three consecutive 1-hour periods (each hour commencing on the hour), provided that the number of 3-hour periods during which the vinyl chloride concentration exceeds 10 ppm does not exceed the number of 1-hour periods during which the vinyl chloride concentration exceeds 10 ppm.

three-piece can (40CFR52.741) means a can which is made from a rectangular sheet and two circular ends.

three-process operation facility (40CFR60.371-g) means the facility including those processes involved with plate stacking, burning or strap casting, and assembly of elements into the battery case.

threshold (EPA-94/04): The lowest dose of a chemical at which a specified measurable effect is observed and below which it is not observed.

threshold level (EPA-94/04): Time-weighted average pollutant concentration values, exposure beyond which is likely to adversely affect human health (cf. environmental exposure).

threshold limit value (TLV) (EPA-94/04): The concentration of an airborne substance that an average person can be repeatedly exposed to without adverse effects. TLVs may be expressed in three ways: TLV-TWA-Time weighted average, based on an allowable exposure averaged over a normal 8-hour workday or 40-hour workweek; TLV-STEL-Short-term exposure limit or maximum concentration for a brief specified period of time, depending on a specific chemical (TWA must still be met); and TLV-C Ceiling Exposure Limit or maximum exposure concentration not to be exceeded under any circumstances. (TWA must still be met.)

threshold odor (EPA-94/04): See odor threshold.

threshold planning quantity (EPA-94/04): A quantity designated for each chemical on the list of extremely hazardous substances that triggers notification by facilities to the State Emergency Response Commission that such facilities are subject to emergency planning requirements under SARA Title III.

threshold planning quantity (TPQ) (40CFR355.20) means, for a substance listed in Appendices A and B, the quantity listed in the column "threshold planning quantity" for that substance (other identical or similar definitions are provided in 40CFR370.2).

throttle (40CFR86.082.2) means the mechanical linkage which either directly or indirectly controls the fuel flow to the engine.

throttle (40CFR86.090-2) means a device used to control an engine's power output by limiting the amount of air entering the combustion chamber.

tidal marsh (EPA-94/04): Low, flat marshlands traversed by channels and tidal hollows, subject to tidal inundation; normally, the only vegetation present is salt tolerant bushes and grasses (cf. wetlands).

tiering (40CFR1508.28) refers to the coverage of general matters in broader environmental impact statements (such as national program or policy statements) with subsequent narrower statements or environmental analyses (such as regional or basinwide program statements or ultimately site-specific statements) incorporating by reference the general discussions and concentrating solely on the issues specific to the statement subsequently prepared. Tiering is appropriate when the sequence of statements or analyses is:

(a) From a program, plan, or policy environmental impact statement to a program plan, or policy statement or analysis of lesser scope or to a site-specific statement or analysis.

(b) From an environmental impact statement on a specific action at an early stage (such as need and site selection) to a supplement (which is preferred) or a subsequent statement or analysis at a later stage (such as environmental mitigation). Tiering in such cases is appropriate when it helps the lead agency to focus on the issues which are ripe for decision and exclude from consideration issues already decided or not yet ripe.

tillage (EPA-94/04): Plowing, seedbed preparation, and cultivation practices.

time origin destination (TOD) information (40CFR52.2043-3) means information that identifies a commuters' work schedule, home and work location, or the location of other desired origins and destinations of trips (such as shopping or recreational trips).

time origin destination (TOD) information (40CFR52.2296-4) means specifications of a driver or rider's work schedule, home and work locations.

time period (40CFR51.100-x) means any period of time designated by hour, month, season, calendar year, averaging time, or other suitable characteristics, for which ambient air quality is estimated (under CAA).

time period of the pressure or vacuum test (t) (40CFR60-AA (method 27-2.5)) means the time period of the test, as specified in the appropriate regulation, during which the change in pressure or vacuum is monitored, in minutes.

time reference (40CFR60-AA(alt. method 1)) means the time (t_o) when the laser pulse emerges from the laser, used as the reference in all lidar time or range measurements.

time response curve (40CFR797.1350-8) is the curve relating cumulative percentage response of a test batch of organisms, exposed to a single dose or single concentration of a chemical, to a period of exposure (other identical or similar definitions are provided in 40CFR797.1440-8).

time weighted average (TWA) (EPA-94/04): In air sampling, the average air concentration of contaminants during a given period.

timed delay (40CFR85.2122(a)(1)(ii)(B)) means a delayed diaphragm displacement controlled to occur within a given time period.

tire (40CFR253.4) means the following types of tires: passenger car tires, light- and heavy-duty truck tires, high speed industrial tires, bus tires, and special service tires (including military, agricultural, off-the-road, and slow speed industrial).

tire (40CFR60.541) means any agricultural, airplane, industrial, mobile home, light-duty truck and/or passenger vehicle tire that has a bead diameter less than or equal to 0.5 meter (m) (19.7 inches) and a cross section dimension less than or equal to 0.325 m (12.8 in.) and that is mass produced in an assembly-line fashion.

title (40CFR35.6015-51) means the valid claim to property which denotes ownership and the rights of ownership, including the rights of possession, control, and disposal of property.

Title III (40CFR350.1) means Title III of the Superfund Amendments and Reauthorization Act of 1986, also titled the Emergency Panning and Community Right-to-Know Act of 1986.

Title III (40CFR372.3) means Title III of the Superfund Amendments and Reauthorization Act of 1986, also titled the Emergency Planning and Community Right-To-Know Act of 1986 (other identical or similar definitions are provided in 40CFR372.3).

Title V permit (40CFR61.02) means any permit issued, renewed, or revised pursuant to Federal or State regulations established to implement Title V of the Act (42 U.S.C. 7661). A Title V permit issued by a State permitting authority is called a part 70 permit in this part.

Title V permit (40CFR63.2) means any permit issued, renewed, or revised pursuant to Federal or State regulations established to implement Title V of the Act (42 U.S.C. 7661). A Title V permit issued by a State permitting authority is called a part 70 permit in this part.

to distribute or sell (FIFRA2-7USC136) means to distribute, sell, offer for sale, hold for distribution, hold for sale, hold for shipment, ship, deliver for shipment, release for shipment, or receive and (having so received) deliver or offer to deliver. The term does not include the holding or application of registered pesticides or use dilutions thereof by any applicator who provides a service of controlling pests without delivering any unapplied pesticide to any person so served.

to use any registered pesticide in a manner inconsistent with its labeling (FIFRA2-7USC136) means to use any registered pesticide in a manner not permitted by the labeling, except that the term shall not include:
(1) applying a pesticide at any dosage, concentration, or frequency less than that specified on the labeling unless the labeling specifically prohibits deviation from the specified dosage, concentration, or frequency,
(2) applying a pesticide against any target pest not specified on the labeling if the application is to the crop, animal, or site specified on the labeling, unless the Administrator has required that the labeling specifically state that the pesticide may be used only for the pests specified on the labeling after the administrator has determined that the use of the pesticide against other pests would cause an unreasonable adverse effect on the environment,
(3) employing any method of application not prohibited by the labeling unless the labeling specifically states that the product may be applied only by the methods specified on the labeling,

(4) mixing a pesticide or pesticides with a fertilizer when such mixture is not prohibited by the labeling,

(5) any use of a pesticide in conformance with section 5, 18, or 24 of this Act, or

(6) any use of a pesticide in a manner that the Administrator determines to be consistent with the purposes of this Act. After March 31, 1979, the term shall not include the use of a pesticide for agricultural or forestry purposes at a dilution less than label dosage unless before or after that date the Administrator issues a regulation or advisory opinion consistent with the study provided for in section 27(b) of the Federal Pesticide Act of 1978, which regulation or advisory opinion specifically requires the use of definite amounts of dilution.

toilet tissue (40CFR250.4-qq) means a sanitary tissue paper. The principal characteristics are softness, absorbency, cleanliness, and adequate strength (considering easy disposability). It is marketed in rolls of varying sizes or in interleaved packages.

tolerance (40CFR177.3) means:
(1) The amount of a pesticide residue that legally may be present in or on a raw agricultural commodity under the terms of a tolerance under FFDCA section 408 or a processed food under the terms of a food additive regulation under FFDCA section 409. Tolerances are usually expressed in terms of parts of the pesticide residue per million parts of the food (ppm), by weight.
(2) [Reserved]

tolerance with regional registration (40CFR180.1-n) means any tolerance which is established for pesticide residues resulting from the use of the pesticide pursuant to a regional registration. Such a tolerance is supported by residue data from specific growing regions for a raw agricultural commodity. Individual tolerances with regional registration are designated in separate subsections in 40CFR180.101 through 180.999, as appropriate. Additional residue data which are representative of the proposed use area are required to expand the geographical area of usage of a pesticide on a raw agricultural commodity having an established tolerance with regional registration. Persons seeking geographically broader registration of a crop having a tolerance with regional registration should contact the appropriate EPA product manager concerning additional residue data required to expand the use area.

tolerances (EPA-94/04): Permissible residue levels for pesticides in raw agricultural produce and processed foods. Whenever a pesticide is registered for use on a food or a feed crop, a tolerance (or exemption from the tolerance requirement) must be established. EPA establishes the tolerance levels, which are enforced by the Food and Drug Administration and the Department of Agriculture.

tomato starch cheese canned specialties (40CFR407.81-k) shall mean canned specialties resulting from a combination of fresh and pre-processed tomatoes, starches, cheeses, spices, and other flavorings necessary to produce a variety of products similar to but not exclusively ravioli, spaghetti, tamales, and enchiladas.

tomatoes (40CFR407.61-u) shall mean the processing of tomatoes into canned, peeled, whole, stewed, and related piece sizes; and processing of tomatoes into the following products and product styles: Canned, peeled and unpeeled paste, concentrate, puree, sauce, juice, catsup and other similar formulated items requiring various other pre-processed food ingredients.

ton or tonnage (40CFR72.2) means any "short ton" (i.e., 2,000 pounds). For the purpose of determining compliance with the Acid Rain emissions limitations and reduction requirements, total tons for a year shall be calculated as the sum of all recorded hourly emissions (or the tonnage equivalent of the recorded hourly emissions rates) in accordance with part 75 of this chapter, with any remaining fraction of a ton equal to or greater than 0.50 ton deemed to equal one ton and any fraction of a ton less than 0.50 ton deemed not to equal any ton.

tonnage (EPA-94/04): The amount of waste that a landfill accepts, usually expressed in tons per month. The rate at which a landfill accepts waste is limited by the landfill's permit.

tons per day (40CFR245.101-m) means annual tonnage divided by 260 days.

too numerous to count (40CFR141.2) means that the total number of bacterial colonies exceeds 200 on a 47-mm diameter membrane filter used for coliform detection.

top blown furnace (40CFR60.141a) means any BOPF in which oxygen is introduced to the bath of molten iron by means of an oxygen lance inserted from the top of the vessel.

top security (40CFR11.4-f): See security classification category.

topcoat (40CFR52.741) means a coating applied in a multiple coat operation other than prime coat, final repair coat, or prime surfacer coat.

topcoat operation (40CFR52.741) means all topcoat spray booths, flash-off areas, and bake ovens at a facility which are used to apply, dry, or cure the final coatings (except final off-line repair) on components of automobile or light-duty truck bodies.

topcoat operation (40CFR60.391) means the topcoat spray booth, flash-off area, and bake oven(s) which are used to apply and dry or cure the final coating(s) on components of automobile and light-duty truck bodies.

topography (EPA-94/04): The physical features of a surface area including relative elevations and the position of natural and man-made features.

topside port lid (40CFR63.301) means a cover, removed during charging or decarbonizing, that is placed over the opening through which coal can be charged into the oven of a byproduct coke oven battery.

Tordon acid(TM) (40CFR63.191) means the synthetic herbicide 4-amino-3,5,6-trichloropicolinic acid, picloram. The category includes, but is not limited to, chlorination processes utilized in Tordon(TM) acid production.

torr (40CFR796.1950-iii) is a unit of pressure which equals 133.3 pascals or 1 mm Hg at $0^{\circ}C$.

total (40CFR136-AC-3) means the concentration determined on an unfiltered sample following vigorous digestion (section 9.3), or the sum of the dissolved plus suspended concentrations (section 9.1 plus 9.2).

total annual sales (40CFR704.3) means the total annual revenue (in dollars) generated by the sale of all products of a company. Total annual sales must include the total annual sales revenue of all sites owned or controlled by that company and the total annual sales revenue of that company's subsidiaries and foreign or domestic parent company, if any (under TSCA).

total chromium (40CFR410.01-c) shall mean hexavalent and trivalent chromium as measured by the procedures listed in 40CFR136.

total dissolved phosphorous (EPA-94/04): The total phosphorous content of all material that will pass through a filter, which is determined as orthophosphate without prior digestion or hydrolysis. Also called soluble P. or ortho P.

total dissolved solids (40CFR122.2) means the total dissolved (filterable) solids as determined by use of the method specified in 40CFR136.

total dissolved solids (40CFR144.3) means the total dissolved (filterable) solids as determined by use of the method specified in 40CFR part 136.

total dissolved solids (TDS) (40CFR131.35.d-16) means the total filterable residue that passes through a standard glass fiber filter disk and remains after evaporation and drying to a constant weight at 180 degrees C; it is considered to be a measure of the dissolved salt content of the water.

total dissolved solids (TDS) (40CFR146.3) means the total dissolved (filterable) solids as determined by use of the method specified in 40CFR136.

total dissolved solids (TDS) (EPA-94/04): All material that passes the standard glass river filter; now called total filtrable reside. Term is used to reflect salinity.

total enclosure (40CFR60.441) means a structure or building around the coating applicator and flashoff area or the entire coating line for the purpose of confining and totally capturing fugitive VOC emissions.

total enclosure (40CFR60.711-17) means a structure that is constructed around a source of emissions so that all VOC emissions are collected and exhausted through a stack or duct. With a total enclosure, there will be no fugitive emissions, only stack emissions. The only openings in a total enclosure are forced makeup air and exhaust ducts and any natural draft openings such as those that allow raw materials to enter and exit the enclosure for processing. All access doors or windows are closed during routine operation of the enclosed source. Brief, occasional openings of such doors or windows to accommodate process equipment adjustments are acceptable, but if such openings are routine or if an access door remains open during the entire operation, the access door must be considered a natural draft opening. The average inward face velocity across the natural draft openings of the enclosure must be calculated including the area of such access doors. The drying oven itself may be part of the total enclosure. A permanent enclosure that meets the requirements found in 40CFR60.713(b)(4)(i) is assumed to be a total enclosure. The owner or operator of a permanent enclosure that does not meet the requirements may apply to the Administrator for approval of the enclosure as a total enclosure on a case-by-case basis. Such approval shall be granted upon a demonstration to the satisfaction of a Administrator that all VOC emissions are contained and vented to the control device.

total enclosure (40CFR60.741) means a structure that is constructed around a source of emissions and operated so that all VOC emissions are collected and exhausted through a stack or duct. With a total enclosure, there will be no fugitive emissions, only stack emissions. The drying oven itself may be part of the total enclosure.

total fluorides (40CFR60.191) means elemental fluorine and all fluoride compounds as measured by reference methods specified in 40CFR60.195 or by equivalent or alternative methods (see 40CFR60.8(b)).

total fluorides (40CFR60.201-b) means elemental fluorine and all fluoride compounds as measured by reference methods specified in 40CFR60.204, or equivalent or alternative methods.

total fluorides (40CFR60.221-b) means elemental fluorine and all fluoride compounds as measured by reference methods specified in 40CFR60.224, or equivalent or alternative methods.

total fluorides (40CFR60.231-c) means elemental fluorine and all fluoride compounds as measured by reference methods specified in 40CFR60.234, or equivalent or alternative methods.

total fluorides (40CFR60.241-b) means elemental fluorine and all fluoride compounds as measured by reference methods specified in 40CFR60.244, or equivalent or alternative methods.

total hydrocarbons (40CFR503.41-m) means the organic compounds in the exit gas from a sewage sludge incinerator stack measured using a flame ionization detection instrument referenced to propane.

total maximum daily load (TMDL) (40CFR130.2-i) means the sum of the individual nonpoint sources and natural background. If a receiving water has only one point source discharger, the TMDL is the sum of that point source WLA plus the LAs for any nonpoint sources of pollution and natural background sources, tributaries, or adjacent segments. TMDLs can be expressed in terms of either mass per time, toxicity, or other appropriate measure. If Best Management Practices (BMPs) or other nonpoint source pollution controls make more stringent load allocations practicable, then wasteload allocations can be made less stringent. Thus, the TMDL process provides for nonpoint source control tradeoffs.

total metal (40CFR413.02-e) is defined as the sum of the concentration or mass of copper (Cu), nickel (Ni), chromium (Cr) (total) and zinc (Zn).

total of direct and indirect emissions (40CFR51.852) means the sum of direct and indirect emissions increases and decreases caused by the Federal action; i.e., the "net" emissions considering all direct and indirect emissions. The portion of emissions which are exempt or presumed to conform under

40CFR51.853, (c), (d), (e), or (f) are not included in the "total of direct and indirect emissions." The "total of direct and indirect emissions" includes emissions of criteria pollutants and emissions of precursors of criteria pollutants (other identical or similar definitions are provided in 40CFR93.152).

total organic active ingredients (40CFR455.20) means the sum of all organic active ingredients covered by 40CFR455.20(a) which are manufactured at a facility subject to this subpart (other identical or similar definitions are provided in 40CFR455.21-b).

total organic compounds (40CFR60.501) means those compounds measured according to the procedures in 40CFR60.503.

total organic compounds (TOC) (40CFR60.561) means those compounds measured according to the procedures specified in 40CFR60.564.

total organic compounds (TOC) (40CFR60.611) means those compounds measured according to the procedures in 40CFR60.614(b)(4). For the purposes of measuring molar composition as required in 40CFR60.614(d)(2)(i), hourly emissions rate as required in 40CFR60.614(d)(5) and 40CFR60.614(e) and TOC concentration as required in 40CFR60.615(b)(4) and 40CFR60.615(g)(4), those compounds which the Administrator has determined do not contribute appreciably to the formation of ozone are to be excluded. The compounds to be excluded are identified in Environmental Protection Agency's statements on ozone abatement policy for SIP revisions (42 FR 35314; 44 FR 32042; 45 FR 32424; 45 FR 48942).

total organic compounds (TOC) (40CFR60.661) means those compounds measured according to the procedures in 40CFR60.664(b)(4). For the purposes of measuring molar composition as required in 40CFR60.664(d)(2)(i); hourly emissions rate as required in 40CFR60.664(d)(5) and 40CFR60.664(e); and TOC concentration as required in 40CFR60.665(b)(4) and 40CFR60.665(g)(4), those compounds which the Administrator has determined do not contribute appreciably to the formation of ozone are to be excluded. The compounds to be excluded are identified in Environmental Protection Agency's

statements on ozone abatement policy for State Implementation Plan (SIP) revisions (42 FR 35314; 44 FR 32042; 45 FR 32424; 45 FR 48942).

total organic compounds or TOC (40CFR60.701) means those compounds measured according to the procedures in 40CFR60.704(b)(4). For the purposes of measuring molar composition as required in 40CFR60.704(d)(2)(i) and 40CFR60.704(d)(2)(ii), hourly emission rate as required in 40CFR60.704(d)(5) and 40CFR60.704(e), and TOC concentration as required in 40CFR60.705(b)(4) and 40CFR60.705(f)(4), those compounds which the Administrator has determined do not contribute appreciably to the formation of ozone are to be excluded.

total organic compounds or TOC (40CFR63.111), as used in the process vents provisions, means those compounds measured according to the procedures of Method 18 of 40CFR part 60, appendix A.

total phenols (40CFR464.02-g) shall mean total phenolic compounds as measured by the procedure listed in 40CFR136 (distillation followed by colorimetric--4AAP).

total planned net output capacity (40CFR72.2) means the planned generator output capacity, excluding that portion of the electrical power which is designed to be used at the power production facility, as specified under one or more qualifying power purchase commitments or contemporaneous documents as of November 15, 1990; "Total installed net output capacity" shall be the generator output capacity, excluding that portion of the electrical power actually used at the power production facility, as installed.

total rated capacity (40CFR52.01-h) means the sum of the rated capacities of all fuel-burning equipment connected to a common stack. The rated capacity shall be the maximum guaranteed by the equipment manufacturer or the maximum normally achieved during use, whichever is greater.

total rated capacity (40CFR52.1881-xv) means the sum of the rated capacities of all fuel-burning equipment connected to a common stack. The rated capacity shall be the maximum guaranteed by the equipment manufacturer or the maximum normally achieved during use as determined by the Administrator, whichever is greater.

total recoverable (40CFR136-AC-4) means the concentration determined on an unfiltered sample following treatment with hot, dilute mineral acid.

total reduced sulfur (TRS) (40CFR60.281-c) means the sum of the sulfur compounds hydrogen sulfide, methyl mercaptan, dimethyl sulfide, and dimethyl disulfide, that are released during the kraft pulping operation and measured by Reference Method 16.

total residual chlorine (or total residual oxidants for intake water with bromides) (40CFR423.11-a) means the value obtained using the amperometric method for total residual chlorine described in 40CFR130.

total residual chlorine (TRC) (40CFR420.02-f) means the value obtained by the iodometric titration with an amperometric endpoint method specified in 40CFR136.3.

total residual oxidants for intake water with bromides (40CFR423.11-a): See total residual chlorine.

total resource effectiveness (TRE) index value (40CFR60.611) means a measure of the supplemental total resource requirement per unit reduction of TOC associated with an individual air oxidation vent stream, based on vent stream flow rate, emission rate of TOC, net heating value, and corrosion properties (whether or not the vent stream is halogenated), as quantified by the equation given under 40CFR60.614(e).

total resource effectiveness index value or TRE index value (40CFR63.111) means a measure of the supplemental total resource requirement per unit reduction of organic HAP associated with a process vent stream, based on vent stream flow rate, emission rate of organic HAP, net heating value, and corrosion properties (whether or not the vent stream contains halogenated compounds), as quantified by the equations given under 40CFR63.115 of this subpart.

total resource effectiveness or TRE (40CFR60.701) index value means a measure of the supplemental

total resource requirement per unit reduction of TOC associated with a vent stream from an affected reactor process facility, based on vent stream flow rate, emission rate of TOC, net heating value, and corrosion properties (whether or not the vent stream contains halogenated compounds), as quantified by the equation given under 40CFR60.704(e).

total smelter charge (40CFR60.161-k) means the weight (dry basis) of all copper sulfide ore concentrates processed at a primary copper smelter, plus the weight of all other solid materials introduced into the roasters and smelting furnaces at a primary copper smelter, except calcine, over a one-month period.

total SO_2 equivalents (40CFR60.641) means the sum of volumetric or mass concentrations of the sulfur compounds obtained by adding the quantity existing as SO_2 to the quantity of SO_2 that would be obtained if all reduced sulfur compounds were converted to SO_2 (ppmv or kg/DSCM).

total solids (40CFR503.31-i) are the materials in sewage sludge that remain as residue when the sewage sludge is dried at 103 to 105 degrees Celsius.

total suspended particles (EPA-94/04): A method of monitoring particulate matter by total weight.

total suspended particulate (40CFR51.100-ss) means particulate matter as measured by the method described in Appendix B of part 50 of this chapter.

total suspended particulate (TSP) (40CFR58.1-t) means particulate matter as measured by the method described in Appendix B of part 50 of this chapter.

total suspended solids (TSS) (EPA-94/04): A measure of the suspended solids in wastewater, effluent, or water bodies, determined by tests for "total suspended nonfilterable solids" (cf. suspended solids).

total test distance (40CFR86.402.78) is defined for each class of motorcycles in 40CFR86.427.78.

total toxic organics (TTO) (400CFR464.31-a) is a regulated parameter under PSES (40CFR464.35) and PSNS (40CFR464.36) for the ferrous subcategory and is comprised of a discrete list of toxic organic pollutants for each process segment where it is regulated, as follows:

(1) Casting Quench (40CFR464.35(b) and 40CFR464.36(b)):
23. chloroform (trichloromethane)
34. 2,4-dimethylphenol

(2) Dust Collection Scrubber (40CFR464.35(c) and 40CFR464.36(b)):
1. acenaphthene
23. chloroform (trichloromethane)
31. 2,4-dichlorophenol
34. 2,4-dimethylphenol
39. fluoranthene
44. methylene chloride (dichloromethane)
55. naphthalene
64. pentachlorophenol
65. phenol
66. bis(2-ethylhexyl)phthalate
67. butyl benzyl phthalate
68. di-n-butyl phthalate
70. diethyl phthalate
71. dimethyl phthalate
72. benzo (a)anthracene (1,2-benzanthracene)
76. chrysene
77. acenaphthylene
78. anthracene
80. fluorene
81. phenanthrene
84. pyrene

(3) Investment Casting (40CFR464.35(e) and 40CFR464.36(e)):
23. chloroform (trichloromethane)
44. methylene chloride (dichloromethane)
66. bis (2-ethylhexyl) phthalate
77. acenaphthylene
84. pyrene

(4) Melting Furnace Scrubber (40CFR464.35(f) and 4CFR464.36(f)):
23. chloroform (trichloromethane)
31. 2,4-dichlorophenol
34. 2,4-dimethylphenol
39. fluoranthene
44. methylene chloride (dichloromethane)
55. naphthalene
65. phenol
66. bis (2-ethylhexyl) phthalate
67. butyl benzyl phthalate
68. di-n-butyl phthalate
72. benzo (a)anthracene (1,2-benzanthracene)

76. chrysene
77. acenaphthylene
78. anthracene
80. fluorene
81. phenanthrene
84. pyrene

(5) Mold Cooling (40CFR464.35(g) and 40CFR464.36(g)):
23. chloroform (trichloromethane)
34. 2,4-dimethylphenol

(6) Slag Quench (40CFR464.35(h) and 40CFR464.36(h)):
34. 2,4-dimethylphenol
71. dimethyl phthalate

(7) Wet Sand Reclamation (40CFR464.35(i) and 40CFR464.36(i)):
1. acenaphthene
34. 2,4-dimethylphenol
39. fluoranthene
44. methylene chloride (dichloromethane)
55. naphthalene
65. phenol
66. bis (2-ethylhexyl) phthalate
68. di-n-butyl phthalate
70. diethyl phthalate
71. dimethyl phthalate
72. benzo(a)anthracene(1,2-benzanthracene)
77. acenaphthylene
84. pyrene

total toxic organics (TTO) (40CFR413.02-i; 433.11-e) shall mean total toxic organics, which is the summation of all quantifiable values greater than 0.01 milligrams per liter for the following toxic organics:
1. Acenaphthene
2. Acrolein
3. Acrylonitrile
4. Benzene
5. Benzidine
6. Carbon tetrachloride = Tetrachloromethane
7. Chlorobenzene
8. 1,2,4-Trichlorobenzene
9. Hexachlorobenzene
10. 1,2-Dichloroethane
11. 1,1,1-Trichloroethane
12. Hexachloroethane
13. 1,1-Dichloroethane
14. 1,1,2-Trichloroethane

15.	1,1,2,2-Tetrachloroethane		65.	Di-n-butyl phthalate
16.	Chloroethane		66.	Di-n-octyl phthalate
17.	Bis(2-chloroethyl) ether		67.	Diethyl phthalate
18.	2-Chloroethyl vinyl ether (mixed)		68.	Dimethyl phthalate
19.	2-Chloronaphthalene		69.	1,2-Benzanthracene = Benzo[a]anthracene
20.	2,4,6-Trichlorophenol		70.	Benzo[a]pyrene = 3,4-benzopyrene
21.	Parachlorometa cresol		71.	3,4-Benzofluoranthene = Benzo[b]fluoranthene
22.	Chloroform = Trichloromethane		72.	11,12-benzofluoranthene = Benzo[k]fluoranthene
23.	2-Chlorophenol		73.	Chrysene
24.	1,2-Dichlorobenzene		74.	Acenaphthylene
25.	1,3-Dichlorobenzene		75.	Anthracene
26.	1,4-Dichlorobenzene		76.	1,12-benzoperylene = Benzo[ghi]perylene
27.	3,3-Dichlorobenzidine		77.	Fluorene
28.	1,1-Dichloroethylene		78.	Phenanthrene
29.	1,2-Trans-dichloroethylene		79.	1,2,5,6-dibenzanthracene = Dibenzo[a,h]anthracene
30.	2,4-Dichlorophenol		80.	Indeno(1,2,3-cd) pyrene = 2,3-o-phenylene pyrene
31.	1,2-Dichloropropane		81.	Pyrene
32.	1,3-Dichloropropene		82.	Tetrachloroethylene
33.	2,4-Dimethylphenol		83.	Toluene
34.	2,4-Dinitrotoluene		84.	Trichloroethylene
35.	2,6-Dinitrotoluene		85.	Vinyl chloride = Chloroethylene
36.	1,2-Diphenylhydrazine		86.	Aldrin
37.	Ethylbenzene		87.	Dieldrin
38.	Fluoranthene		88.	Chlordane (technical mixture and metabolites)
39.	4-Chlorophenyl phenyl ether		89.	4,4-DDT
40.	4-Bromophenyl phenyl ether		90.	4,4-DDE (p,p-DDX)
41.	Bis(2-chloroisopropyl) ether		91.	4,4-DDD (p,p-TDE)
42.	Bis(2-chloroethoxy) methane		92.	Alpha-endosulfan
43.	Methylene chloride = Dichloromethane		93.	Beta-endosulfan
44.	Methyl chloride = Chloromethane		94.	Endosulfan sulfate
45.	Methyl bromide = Bromomethane		95.	Endrin
46.	Bromoform = Tribromomethane		96.	Endrin aldehyde
47.	Dichlorobromomethane		97.	Heptachlor
48.	Chlorodibromomethane		98.	Heptachlor epoxide = BHC-hexachlorocyclohexane
49.	Hexachlorobutadiene		99.	Alpha-BHC
50.	Hexachlorocyclopentadiene		100.	Beta-BHC
51.	Isophorone		101.	Gamma-BHC
52.	Naphthalene		102.	Delta-BHC (PCB-polychlorinated biphenyls)
53.	Nitrobenzene		103.	PCB-1242 (Arochlor 1242)
54.	2-Nitrophenol		104.	PCB-1254 (Arochlor 1254)
55.	4-Nitrophenol		105.	PCB-1221 (Arochlor 1221)
56.	2,4-Dinitrophenol		106.	PCB-1232 (Arochlor 1232)
57.	4,6-Dinitro-o-cresol			
58.	N-Nitrosodimethylamine			
59.	N-Nitrosodiphenylamine			
60.	N-Nitrosodi-n-propylamine			
61.	Pentachlorophenol			
62.	Phenol			
63.	Bis(2-ethylhexyl) phthalate			
64.	Butyl benzyl phthalate			

107. PCB-1248 (Arochlor 1248)
108. PCB-1260 (Arochlor 1260)
109. PCB-1016 (Arochlor 1016)
110. Toxaphene
111. 2,3,7,8-Tetrachlorodibenzo-p-dioxin (TCDD)

total toxic organics (TTO) (40CFR464.02-j) shall mean the sum of the mass of each of the toxic organic compounds which are found at a concentration greater than 0.010 mg/L. The specialized definitions for each subpart contain a discrete list of toxic organic compounds comprising TTO for each process segment in which TTO is regulated.

total toxic organics (TTO) (40CFR464.11-a) is a regulated parameter under PSES (40CFR464.15) and PSNS (40CFR464.16) for the aluminum subcategory and is comprised of a discrete list of toxic organic pollutants for each process segment where it is regulated, as follows:
(1) Casting Quench (40CFR464.15(b) and 40CFR464.16(b)):
4. benzene
21. 2,4,6-trichlorophenol
22. Para-chloro meta-cresol
23. chloroform (trichloromethane)
34. 2,4-dimethylphenol
39. fluoranthene
44. methylene chloride (dichloromethane)
65. phenol
66. bis(2-ethylhexyl)phthalate
67. butyl benzyl phthalate
84. pyrene
85. tetrachloroethylene
87. trichloroethylene
(2) Die Casting (40CFR464.15(c) and 40CFR464.16(c)):
1. acenaphthene
4. benzene
7. chlorobenzene
11. 1,1,1-trichloroethane
21. 2,4,6-trichlorophenol
22. para-chloro meta-cresol
23. chloroform (trichloromethane)
34. 2,4-dimethylphenol
39. fluoranthene
44. methylene chloride (dichloromethane)
55. naphthalene

65. phenol
66. bis(2-ethylhexyl)phthalate
67. butyl benzyl phthalate
68. di-n-butyl phthalate
70. diethyl phthalate
72. benzo (a)anthracene (1,2-benzanthracene)
73. benzo (a)pyrene (3,4-benzopyrene)
76. chrysene
78. anthracene
80. fluorene
81. phenanthrene
84. pyrene
85. tetrachloroethylene
86. toluene
(3) Dust Collection Scrubber (40CFR464.15(d) and 40CFR464.16(d)):
1. acenaphthene
21. 2,4,6-trichlorophenol
23. chloroform (trichloromethane)
34. 2,4-dimethylphenol
39. fluoranthene
44. methylene chloride (dichloromethane)
65. phenol
66. bis (2-ethylhexyl) phthalate
68. di-n-butyl phthalate
70. diethyl phthalate
73. benzo (a)pyrene (3,4-benzopyrene)
84. pyrene
(4) Investment Casting (40CFR464.15(f) and 40CFR464.16(f)):
11. 1,1,1-trichloroethane
23. chloroform (trichloromethane)
44. methylene chloride (dichloromethane)
66. bis (2-ethylhexyl) phthalate
84. pyrene
85. tetrachloroethylene
87. trichloroethylene
(5) Melting Furnace Scrubber (40CFR464.15(g) and 40CFR464.16(g)):
1. acenaphthene
21. 2,4.6-trichlorophenol
23. chloroform (trichloromethane)
34. 2,4-dimethylphenol
39. fluoranthene
44. methylene chloride (dichloromethane)
65. phenol
66. bis (2-ethylhexyl) phthalate
68. di-n-butyl phthalate
70. diethyl phthalate
73. benzo (a)pyrene (3,4-benzopyrene)

84. pyrene

(6) Mold Cooling (40CFR464.15(h) and 40CFR464.16(h)):

4. benzene
21. 2,4,6-trichlorophenol
22. para-chloro meta-cresol
23. chloroform (trichloromethane)
34. 2,4-dimethylphenol
39. fluoranthene
44. methylene chloride
65. phenol
66. bis(2-ethylhexyl) phthalate
67. butyl benzyl phthalate
84. pyrene
85. tetrachloroethylene
87. trichloroethylene

total toxic organics (TTO) (40CFR464.21-a) is a regulated parameter under PSES (40CFR464.25) and PSNS (40CFR464.26) for the copper subcategory and is comprised of a discrete list of toxic organic pollutants for each process segment where it is regulated, as follows:

(1) Casting Quench (40CFR464.25(a) and 40CFR464.26(a)):

23. chloroform (trichloromethane)
64. pentachlorophenol
66. bis(2-ethylhexyl)phthalate
71. dimethyl phthalate

(2) Dust Collection Scrubbers (40CFR464.25(c) and 464.26(c)):

1. acenaphthene
22. para-chloro meta-cresol
23. chloroform (trichloromethane)
34. 2,4-dimethylphenol
55. naphthalene
58. 4-nitrophenol
64. pentachlorophenol
65. phenol
66. bis(2-ethylhexyl)phthalate
67. butyl benzyl phthalate
68. di-n-butyl phthalate
70. diethyl phthalate
71. dimethyl phthalate
72. benzo(a)anthracene (1,2-benzanthracene)
74. 3,4-benzofluoranthene
75. benzo(k) fluoranthene
76. chrysene
77. acenaphthylene
78. anthracene

81. phenanthrene
84. pyrene

(3) Investment Casting (40CFR464.25(e) and 40CFR464.26(e)):

1. acenaphthene
22. para-chloro meta-cresol
23. chloroform (trichloromethane)
34. 2,4-dimethylphenol
55. naphthalene
58. 4-nitrophenol
64. pentachlorophenol
65. phenol
66. bis (2-ethylhexyl)phthalate
67. butyl benzyl phthalate
68. di-n-butyl phthalate
70. diethyl phthalate
71. dimethyl phthalate
72. benzo(a)anthracene (1,2-benzanthracene)
74. 3.4-benzofluoranthene
75. benzo(k) fluoranthene
76. chrysene
77. acenaphthylene
78. anthracene
81. Phenanthrene
84. pyrene

(4) Melting Furnace Scrubber (40CFR464.25(f) and 4CFR464.26(f)):

1. acenaphthene
22. para-chloro meta-cresol
23. chloroform (trichloromethane)
34. 2,4-dimethylphenol
55. naphthalene
58. 4-nitrophenol
64. pentachlorophenol
65. phenol
66. bis (2-ethylhexyl) phthalate
67. butyl benzyl phthalate
68. di-n-butyl phthalate
70. diethyl phthalate
71. dimethyl phthalate
72. benzo(a)anthracene (1,2-benzanthracene)

74. 3,4-benzoflouranthene
75. benzo(k) flouranthene
76. chrysene
77. acenaphthylene
78. anthracene
81. phenanthrene
84. pyrene

(5) Mold Cooling (40CFR464.25(g) and

40CFR464.26(g)):

   23.   chloroform (trichloromethane)
   64.   pentachlorophenol
   66.   bis(2-ethylhexyl)phthalate
   71.   dimethyl phthalate

total toxic organics (TTO) (40CFR464.41-a) is a regulated parameter under PSES (40CFR464.45) and PSNS (40CFR464.46) for the zinc subcategory and is comprised of a discrete list of toxic organic pollutants for each process segment where it is regulated, as follows:

(1) Casting Quench (40CFR464.45(a) and 40CFR464.46(a)):

   21.   2,4,6-trichlorophenol
   22.   para-chloro meta-cresol
   31.   2,4-dichlorophenol
   34.   2,4-dimethylphenol
   39.   fluoranthene
   44.   methylene chloride (dichloromethane)
   65.   phenol
   66.   bis(2-ethylhexyl) phthalate
   68.   di-n-butyl phthalate
   70.   diethyl phthalate
   85.   tetrachloroethylene

(2) Die Casting (40CFR465.45(b) and 40CFR464.46(b)):

   1.   acenaphthene
   21.   2,4,6-trichlorophenol
   22.   para-chloro meta-cresol
   24.   2-chlorophenol
   34.   2,4-dimethylphenol
   44.   methylene chloride (dichloromethane)
   55.   naphthalene
   65.   phenol
   66.   bis (2-ethylhexyl) phthalate
   68.   di-n-butyl phthalate
   70.   diethyl phthalate
   85.   tetrachloroethylene
   86.   toluene
   87.   trichloroethylene

(3) Melting Furnace Scrubber (40CFR464.45(c) and 4CFR464.46(c)):

   31.   2,4-dichlorophenol
   34.   2,4-dimethylphenol
   39.   fluoranthene
   44.   methylene chloride (dichloromethane)
   55.   naphthalene
   65.   phenol
   66.   bis(2-ethylhexyl) phthalate

   68.   di-n-butyl phthalate
   85.   tetrachloroethylene
   86.   toluene
   87.   trichloroethylene

(4) Mold Cooling (40CFR464.45(d) and 40CFR464.46(d)):

   21.   2,4,6-trichlorophenol
   22.   para-chloro meta-cresol
   31.   2,4-dichlorophenol
   34.   2,4-dimethylphenol
   39.   fluoranthene
   44.   methylene chloride (dichloromethane)
   65.   phenol
   66.   bis(2-ethylhexyl) phthalate
   68.   di-n-butyl phthalate
   70.   diethyl phthalate
   85.   tetrachloroethylene

total toxic organics (TTO) (40CFR465.02-j) shall mean the sum of the mass of each of the following toxic organic compounds which are found at a concentration greater than 0.010 mg/L:

- 1,1,1-Trichloroethane
- 1,1-Dichloroethane
- 1,1,2,2-Tetrachloroethane
- Bis (2-chloroethyl) ether
- Chloroform
- 1,1-Dichloroethylene
- Methylene chloride (dichloromethane)
- Pentachlorophenol
- Bis (2-ethylhexyl) phthalate
- Butyl benzyl-phthalate
- Di-N-butyl phthalate
- Phenanthrene
- Tetrachloroethylene
- Toluene

total toxic organics (TTO) (40CFR467.02-q) shall mean the sum of the masses or concentrations of each of the following toxic organic compounds which is found in the discharge at a concentration greater than 0.010 mg/L:

- p-Chloro-m-cresol
- 2-Chlorophenol
- 2,4-Dinitrotoluene
- 1,2-Diphenylhydrazine
- Ethylbenzene
- Fluoranthene
- Isophorone
- Naphthalene

- N-Nitrosodiphenylamine
- Phenol
- Benzo[a]pyrene
- Benzo[ghi]perylene
- Fluorene
- Phenanthrene
- Dibenzo[a,h] anthracene
- Indeno(1,2,3-c,d)pyrene
- Pyrene
- Tetrachloroethylene
- Toluene
- Trichloroethylene
- Endosulfan sulfate
- Bis(2-ethyl hexyl)phthalate
- Diethylphthalate
- 3,4-Benzofluoranthene
- Benzo[k]fluoranthene
- Chrysene
- Acenaphthylene
- Anthracene
- Di-n-butyl phthalate
- Endrin
- Endrin aldehyde
- PCB-1242, 1254, 1221
- PCB-1232, 1248, 1260, 1016
- Acenaphthene

total toxic organics (TTO) (40CFR468.02-r) shall mean the sum of the masses or concentrations of each of the following toxic organic compounds which is found at a concentration greater than 0.010 mg/L:
- Benzene
- 1,1,1-Trichloroethane chloroform
- 2,6-Dinitrotoluene ethylbenzene methylene chloride naphthalene
- N-nitrosodiphenylamine anthracene phenanthrene toluene trichloroethylene

total toxic organics (TTO) (40CFR469.12-a) means the sum of the concentrations for each of the following toxic organic compounds which is found in the discharge at a concentration greater the ten (10) micrograms per liter:
- 1,2,4 Trichlorobenzene chloroform
- 1,2 Dichlorobenzene
- 1,3 Dichlorobenzene
- 1,4 Dichlorobenzene ethylbenzene
- 1,1,1 Trichlorobenzene ethylbenzene chloride naphthalene
- 2 Nitrophenol phenol bis (2-ethylhexyl) phthalate

tetrachloroethylene toluene trichloroethylene
- 2 Chloroethylene
- 2,4 Dichlorophenol
- 4 Nitrophenol pentachlorophenol di-n-butyl phthalate anthracene
- 1,2 Diphenylhydrazine isophorone butyl benzyl phthalate
- 1,2 Dichloroethylene
- 2,4,6 Trichlorophenol carbon tetrachloride
- 1,1 Dichloroethane
- 1,1,2 Trichloroethane dichlorobromomethane

(Other identical or similar definitions are provided in 40CFR469.22-a.)

total toxic organics (TTO) (40CFR469.31-b) means the sum of the concentrations for each of the following toxic organic compounds which is found in the discharge at a concentration greater than ten (10) micrograms per liter:
- 1,1,1 Chloroform
- Trichloroethane
- Methylene chloride
- Bis (2-ethylhexyl) phthalate
- Toluene
- Trichloroethylene

total trihalomethanes (TTHM) (40CFR141.2) means the sum of the concentration in milligrams per liter of the trihalomethane compounds (trichloromethane [chloroform], dibromochloromethane, bromodichloromethane and tribromomethane [bromoform]), rounded to two significant figures.

total volatile organic hazardous air pollutant concentration (40CFR63.111) means the sum of the concentrations of all individually-speciated organic HAP's, as measured by Method 305 in appendix A of this part.

totally enclosed manner (40CFR761.3) means any manner that will ensure no exposure of human beings or the environment to any concentration of PCBs.

totally enclosed treatment facility (40CFR260.10) means a facility for the treatment of hazardous waste which is directly connected to an industrial production process and which is constructed and operated in a manner which prevents the release of any hazardous waste or any constituent thereof into the environment during treatment. An example is a pipe in which waste acid is neutralized.

touch-up coat (40CFR60.721) means the coat applied to correct any imperfections in the finish after color or texture coats have been applied. This definition does not include conductive sensitizers or EMI/RFI shielding coatings.

toxaphene (40CFR129.4-d) means a material consisting of technical grade chlorinated camphene having the approximate formula of $C_{10}H_{10}C_{18}$ and normally containing 67-69 percent chlorine by weight (FWPCA).

toxaphene (EPA-94/04): Chemical that causes adverse health effects in domestic water supplies and is toxic to freshwater and marine aquatic life.

toxaphene formulator (40CFR129.103-2) means a person who produces, prepares or processes a formulated product comprising a mixture of toxaphene and inert materials or other diluents into a product intended for application in any use registered under the Federal Insecticide, Fungicide and Rodenticide Act, as amended (7USC135, et seq.).

toxaphene manufacturer (40CFR129.103-1) means a manufacturer, excluding any source which is exclusively a toxaphene formulator, who produces, prepares or processes toxaphene or who uses toxaphene as a material in the production, preparation or processing of another synthetic organic substance.

toxic air pollutants (CAA211.k-42USC7545) means the aggregate emissions of the following: Benzene; 1,3 Butadiene; Polycyclic organic matter (POM); Acetaldehyde; Formaldehyde.

toxic chemical (40CFR372.3) means a chemical or chemical category listed in 40CFR372.65.

toxic chemical (EPA-94/04): Any chemical listed in EPA rules as "Toxic Chemicals Subject to section 313 of the Emergency Planning and Community Right-to-Know Act of 1986."

toxic chemical (PPA6603) means any substance on the list described in section 313(c) of the Superfund Amendments and Reauthorization Act of 1986.

toxic chemical (SF329) means a substance on the list described in section 313(c).

toxic chemical release form (EPA-94/04): Information form required of facilities that manufacture, process, or use (in quantities above a specific amount) chemicals listed under SARA Title III.

toxic chemical use substitution (EPA-94/04): Replacing toxic chemicals with less harmful chemicals in industrial processes.

toxic cloud (EPA-94/04): Airborne plume of gases, vapors, fumes, or aerosols containing toxic materials.

toxic effect (40CFR798.6050-2) is an adverse change in the structure or function of an experimental animal as a result of exposure to a chemical substance (other identical or similar definitions are provided in 40CFR798.6200-3).

toxic pollutant (40CFR501.2) means any pollutant listed as toxic under section 307(a)(1) or any pollutant identified in regulations implementing section 405(d) of the CWA.

toxic pollutants (40CFR122.2) means any pollutant listed as toxic under section 307(a)(1), or, in the case of "sludge use or disposal practices," any pollutant identified in regulations implementing section 405(d) of the CWA.

toxic pollutants (40CFR125.58-u) means those substances listed in 40CFR401.15 (see below for the listing of toxic pollutants regulated under the Clean Water Act. The listing was excerpted from 40CFR401.15, 1995 edition).
1. Acenaphthene
2. Acrolein
3. Acrylonitrile
4. Aldrin/Dieldrin{1}
5. Antimony and compounds{2}
6. Arsenic and compounds
7. Asbestos
8. Benzene
9. Benzidine{1}

10. Beryllium and compounds
11. Cadmium and compounds
12. Carbon tetrachloride
13. Chlordane (technical mixture and metabolites)
14. Chlorinated benzenes (other than di-chlorobenzenes)
15. Chlorinated ethanes (including 1,2-dichloroethane, 1,1,1-trichloroethane, and hexachloroethane)
16. Chloroalkyl ethers (chloroethyl and mixed ethers)
17. Chlorinated naphthalene
18. Chlorinated phenols (other than those listed elsewhere; includes trichlorophenols and chlorinated cresols)
19. Chloroform
20. 2-chlorophenol
21. Chromium and compounds
22. Copper and compounds
23. Cyanides
24. DDT and metabolites{1}
25. Dichlorobenzenes (1,2-, 1,3-, and 1,4-dichlorobenzenes)
26. Dichlorobenzidine
27. Dichloroethylenes (1,1-, and 1,2-dichloroethylene)
28. 2,4-dichlorophenol
29. Dichloropropane and dichloropropene
30. 2,4-dimethylphenol
31. Dinitrotoluene
32. Diphenylhydrazine
33. Endosulfan and metabolites
34. Endrin and metabolites{1}
35. Ethylbenzene
36. Fluoranthene
37. Haloethers (other than those listed elsewhere; includes chlorophenylphenyl ethers, bromophenylphenyl ether, bis(dichloroisopropyl) ether, bis-(chloroethoxy) methane and polychlorinated diphenyl ethers)
38. Halomethanes (other than those listed elsewhere; includes methylene chloride, methylchloride, methylbromide, bromoform, dichlorobromomethane
39. Heptachlor and metabolites
40. Hexachlorobutadiene
41. Hexachlorocyclohexane
42. Hexachlorocyclopentadiene
43. Isophorone
44. Lead and compounds
45. Mercury and compounds
46. Naphthalene
47. Nickel and compounds
48. Nitrobenzene
49. Nitrophenols (including 2,4-dinitrophenol, dinitrocresol)
50. Nitrosamines
51. Pentachlorophenol
52. Phenol
53. Phthalate esters
54. Polychlorinated biphenyls (PCBs){1}
55. Polynuclear aromatic hydrocarbons (including benzanthracenes, benzopyrenes, benzofluoranthene, chrysenes, dibenz-anthracenes, and indenopyrenes)
56. Selenium and compounds
57. Silver and compounds
58. 2,3,7,8-tetrachlorodibenzo-p-dioxin (TCDD)
59. Tetrachloroethylene
60. Thallium and compounds
61. Toluene
62. Toxaphene{1}
63. Trichloroethylene
64. Vinyl chloride
65. Zinc and compounds

{1} Effluent standard promulgated (40CFR part 129).
{2} The term compounds shall include organic and inorganic compounds.

toxic pollutants (40CFR131.3-d) are those pollutants listed by the Administrator under section 307(a) of the Act (see also 40CFR401.15 and 40CFR129.4 for the listing of toxic pollutants).

toxic pollutants (CWA502-33USC1362) means those pollutants, or combinations of pollutants, including disease-causing agents, which after discharge and upon exposure, ingestion, inhalation or assimilation into any organism either directly from the environment or indirectly by ingestion through food chains, will, on the basis of information available to the Administrator, cause death, disease, behavioral abnormalities, cancer, genetic mutations, physiological malfunctions (including malfunctions in reproduction) or physical deformations in such organisms or their offspring.

toxic pollutants (EPA-94/04): Materials that cause death, disease, or birth defects in organisms that ingest or absorb them. The quantities and exposures necessary to cause these effects can vary widely.

toxic release inventory (EPA-94/04): Database of toxic releases in the United States compiled from SARA Title III section 313 reports.

toxic substance (EPA-94/04): A chemical or mixture that may present an unreasonable risk of injury to health or the environment.

toxic substance or harmful physical agent (29CFR1910.20) means any chemical substance, biological agent (bacteria, virus, fungus, etc.), or physical stress (noise, heat, cold, vibration, repetitive motion, ionizing and non-ionizing radiation, hypo- or hyperbaric pressure, etc.) which:
(i)   Is listed in the least printed edition of the National Institute for Occupational Safety and Health (NIOSH) Registry of Toxic Effects of Chemical Substances (RTECS); or
(ii) Has yielded positive evidence of an acute or chronic health hazard in testing conducted by, or known to, the employer; or
(iii) Is the subject of a material safety data sheet kept by or known to the employer indicating that the material may pose a hazard to human health.

Toxic Substance and Control Act: See act or TSCA.

toxic waste (EPA-94/04): A waste that can produce injury if inhaled, swallowed, or absorbed through the skin.

toxicant (EPA-94/04): A harmful substance or agent that may injure an exposed organism.

toxicity (40CFR131.35.d-17) means acute and/or chronic toxicity.

toxicity (40CFR171.2-27) means the property of a pesticide to cause any adverse physiological effects (under FIFRA).

toxicity as applied to BAT effluent limitations and NSPS for drilling fluids and drill cuttings (40CFR435.11-w) shall refer to the bioassay test procedure presented in Appendix 2 of 40CFR435, subpart A.

toxicity assessment (EPA-94/04): Characterization of the toxicological properties and effects of a chemical, with special emphasis on establishment of dose response characteristics.

toxicity curve (40CFR797.1350-9) is the curve produced from toxicity tests when LC50 values are plotted against duration of exposure. (This term is also used in aquatic toxicology, but in a less precise sense, to describe the curve produced when the median period of survival is plotted against test concentrations) (other identical or similar definitions are provided in 40CFR797.1440-9).

toxicity testing (EPA-94/04): Biological testing (usually with an invertebrate, fish, or small mammal) to determine the adverse effects of a compound or effluent.

toxicological profile (EPA-94/04): An examination, summary, and interpretation of a hazardous substance to determine levels of exposure and associated health effects.

traceable (40CFR50.1-h) means that a local standard has been compared and certified either directly or via not more than one intermediate standard, to a primary standard such as a National Bureau of Standards Standard Reference Material (NBS SRM), or a USEPA/NBS-approved Certified Reference Material (CRM) (other identical or similar definitions are provided in 40CFR58.1-r).

tracking form (40CFR259.10.b) means the Federal Medical Tracking Form that must accompany all applicable shipments of regulated medical wastes generated within one of the covered States (under RCRA).

tractor (40CFR205.151-28) means for the purposes of this subpart, any two or three wheeled vehicle used exclusively for agricultural purposes, or for snow plowing, including self-propelled machines used exclusively in growing, harvesting or handling farm produce.

trade name product (40CFR372.3) means a chemical or mixture of chemicals that is distributed to other persons and that incorporates a toxic chemical component that is not identified by the applicable chemical name or Chemical Abstracts Service Registry number listed in 40CFR372.65.

trade secrecy (40CFR350.1) means any confidential formula, pattern, process, device, information or compilation of information that is used in a submitter's business, and that gives the submitter an opportunity to obtain and advantage over competitors who do not know or use it. EPA intends to be guided by the Restatement of Torts, section 757, comment b.

trade secrecy claim (40CFR350.1) is a submittal under sections 303 (d)(2) or (d)(3), 311, 312 or 313 of Title III in which a chemical identity is claimed as trade secret, and is accompanied by substantiation in support of the claim of trade secrecy for chemical identity.

trade secret (29CFR1910.20) means any confidential formula, pattern, process, device, or information or compilation of information that is used in an employer's business and that gives the employer an opportunity to obtain an advantage over competitors who do not know or use it.

trade secret (40CFR350-AA): A trade secret may consist of any formula, pattern, device or compilation of information which is used in one's business, and which gives him an opportunity to obtain an advantage over competitors who do not know or use it. It may be a formula for a chemical compound, a process of manufacturing, treating of preserving materials, a pattern for a machine or other device, or a list of customers. It differs from other secret information in a business (see section 759) in that it is not simply information as to single or ephemeral events in the conduct of the business, as, for example, the amount or other terms of a secret bid for a contract or the salary of certain employees, or the security investments made or contemplated, or the date fixed for the announcement of a new policy or for bringing out a new model or the like. A trade secret is a process or device for continuous use in the operation of the business. Generally it relates to the production of goods, as, for example, a machine or formula for the production of an article. It may, however, relate to the sale of goods or to other operations in the business, such as a code for determining discounts, rebates or other concessions

in a price list or catalogue, or a list of specialized customers, or a method of bookkeeping or other office management (cf. secrecy).

trading (40CFR88.202.94) means the exchange of credits between manufacturers.

trading (40CFR89.202.96) means the exchange of nonroad engine emission credits between manufacturers.

trading (40CFR86.090.2) means the exchange of heavy-duty engine NO_x or particulate emission credits between manufacturers.

trainee (40CFR45.115) means a student selected by the recipient organization who receives support to meet the objectives in section 45.110.

transboundary pollutants (EPA-94/04): Air pollution that travels from one jurisdiction to another, often crossing state or international boundaries.

transfer and loading system (40CFR60.251-i) means any facility used to transfer and load coal for shipment.

transfer efficiency (40CFR52.741) means the ratio of the amount of coating solids deposited onto a part or product to the total amount of coating solids used.

transfer efficiency (40CFR60.311) means the ratio of the amount of coating solids deposited onto the surface of a part or product to the total amount of coating solids used (other identical or similar definitions are provided in 40CFR60.391; 60.451; 60.721).

transfer facility (40CFR259.10.a) means any transportation-related facility including loading docks, parking areas, storage areas and other similar areas where shipments of regulated medical waste are held (come to rest or are managed) during the course of transportation. For example, a location at which regulated medical waste is transferred directly between two vehicles is considered a transfer facility. A transfer facility is a "transporter."

transfer facility (40CFR260.10) means any transportation related facility including loading docks, parking areas, storage areas and other similar areas where shipments of hazardous waste are held during the normal course of transportation (other identical or similar definitions are provided in 40CFR270.2).

transfer facility (40CFR761.3) means any transportation-related facility including loading docks, parking areas, and other similar areas where shipments of PCB waste are held during the normal course of transportation. Transport vehicles are not transfer facilities under this definition, unless they are used for the storage of PCB waste, rather than for actual transport activities. Storage areas for PCB waste at transfer facilities are subject to the storage facility standards of 40CFR761.65, but such storage areas are exempt from the approval requirements of 40CFR761.65(d) and the recordkeeping requirements of 40CFR761.180, unless the same PCB waste is stored there for a period of more than 10 consecutive days between destinations.

transfer machine system (40CFR63.321) means a multiple-machine dry cleaning operation in which washing and drying are performed in different machines. Examples include, but are not limited to:
(1) A washer and dryer(s);
(2) A washer and reclaimer(s); or
(3) A dry-to-dry machine and reclaimer(s).

transfer operation (40CFR63.101) means the loading, into a tank truck or railcar, of organic liquids that contain one or more of the organic hazardous air pollutants listed in table 2 of this subpart from a transfer rack (as defined in this section). Transfer operations do not include loading at an operating pressure greater than 204.9 kilopascals.

transfer point (40CFR60.671) means a point in a conveying operation where the nonmetallic mineral is transferred to or from a belt conveyor except where the nonmetallic mineral is being transferred to a stockpile.

transfer rack (40CFR63.101) means the collection of loading arms and loading hoses, at a single loading rack, that are assigned to a chemical manufacturing process unit subject to this subpart according to the procedures specified in 40CFR63.100(h) of this subpart and are used to fill tank trucks and railcars with organic liquids that contain one or more of the

organic hazardous air pollutants listed in table 2 of this subpart. Transfer rack includes the associated pumps, meters, shutoff valves, relief valves, and other piping and valves. Transfer rack does not include:
(1) Racks, arms, or hoses that only transfer liquids containing organic hazardous air pollutants as impurities;
(2) Racks, arms, or hoses that vapor balance during all loading operations; or
(3) Racks transferring organic liquids that contain organic hazardous air pollutants only as impurities.

transfer station (40CFR243.101-dd) means a site at which solid wastes are concentrated for transport to a processing facility or land disposal site. A transfer station may be fixed or mobile.

transfer unit (40CFR72.2) means a Phase I unit that transfers all or part of its Phase I emission reduction obligations to a control unit designated pursuant to a Phase I extension plan under 40CFR72.42.

transferee (40CFR144.3) means the owner or operator receiving ownership and/or operational control of the well.

transferor (40CFR144.3) means the owner or operator transferring ownership and/or operational control of the well.

transform (40CFR82.104-u) means to use and entirely consume a class I or class II substance, except for trace quantities, by changing it into one or more substances not subject to this subpart in the manufacturing process of a product or chemical.

transform (40CFR82.3-gg) means to use and entirely consume (except for trace quantities) a controlled substance in the manufacture of other chemicals for commercial purposes.

transhipment (40CFR82.3-hh) means the continuous shipment of a controlled substance from a foreign state of origin through the United States, its territories, to a second foreign state of final destination, as long as the shipment does not enter into United States jurisdiction.

transient non-community water system or TWS

(40CFR141.2) means a non-community water system that does not regularly serve at least 25 of the same persons over six months per year.

transient shipment (10CFR40.4) means a shipment of nuclear material, originating and terminating in foreign countries, on a vessel or aircraft that stops at a United States port (other identical or similar definitions are provided in 40CFR70.4).

transient water system (EPA-94/04): A non-community water system that does not serve 25 of the same nonresidents per day for more than six months per year.

transit (40CFR51.392) is mass transportation by bus, rail, or other conveyance which provides general or special service to the public on a regular and continuing basis. It does not include school buses or charter or sightseeing services (other identical or similar definitions are provided in 40CFR93.101).

transit country (40CFR262.51) means any foreign country, other than a receiving country, through which a hazardous waste is transported.

transit improvement measures (40CFR52.2493-1) means those actions or combinations of actions taken by the Metropolitan Transit Division of the Municipality of Metropolitan Seattle (METRO) and the city of Spokane to promote the attractiveness and increased use of public mass transit systems.

transit project (40CFR51.392) is an undertaking to implement or modify a transit facility or transit-related program; purchase transit vehicles or equipment; or provide financial assistance for transit operations. It does not include actions that are solely within the jurisdiction of local transit agencies, such as changes in routes, schedules, or fares. It may consist of several phases. For analytical purposes, it must be defined inclusively enough to:

(1) Connect logical termini and be of sufficient length to address environmental matters on a broad scope;

(2) Have independent utility or independent significance, i.e., be a reasonable expenditure even if no additional transportation improvements in the area are made; and

(3) Not restrict consideration of alternatives for other reasonably foreseeable transportation improvements.
(Other identical or similar definitions are provided in 40CFR93.101.)

transitional area (40CFR51.392) means any ozone nonattainment area which EPA has classified as transitional in 40CFR part 81 (other identical or similar definitions are provided in 40CFR93.101).

transitional low-emission vehicle (40CFR88.101.94) means any light-duty vehicle or light-duty truck conforming to the applicable Transitional Low-Emission Vehicle standard.

transitional period (40CFR51.392) with respect to a pollutant or pollutant precursor means that period of time which begins after submission to EPA of the relevant control strategy implementation plan which has been endorsed by the Governor (or his or her designee) and has been subject to a public hearing. The transitional period lasts until EPA takes final approval or disapproval action on the control strategy implementation plan submission or finds it to be incomplete. The precise beginning and end of the transitional period is defined in 40CFR51.448 (other identical or similar definitions are provided in 40CFR93.101).

translocation (40CFR797.2850-4) means the transference or transport of chemical from the site of uptake to other plant components.

transmission class (40CFR600.002.85-22) means a group of transmissions having the following common features: Basic transmission type (manual, automatic, or semi-automatic); number of forward gears used in fuel economy testing (e.g., manual four-speed, three-speed automatic, two-speed semi-automatic); drive system (e.g., front wheel drive, rear wheel drive; four wheel drive), type of overdrive, if applicable (e.g., final gear ratio less than 1.00, separate overdrive unit); torque converter type, if applicable (e.g., non-lockup, lockup, variable ratio); and other transmission characteristics that may be determined to be significant by the Administrator.

transmission class (40CFR86.082.2) means the basic type of transmission, e.g., manual, automatic, semiautomatic.

transmission configuration (40CFR600.002.85-27) means the Administrator may further subdivide within a transmission class if the Administrator determines that sufficient fuel economy differences exist. Features such as gear ratios, torque converter multiplication ratio, stall speed, shift calibration, or shift speed may be used to further distinguish characteristics within a transmission class.

transmission configuration (40CFR86.082.2) means a unique combination, within a transmission class, of the number of the forward gears and, if applicable, over-drive. The Administrator may further subdivide a transmission configuration (based on such criteria as gear ratios, torque convertor multiplication ratio, stall speed and shift calibration, etc.), if he determines that significant fuel economy or exhaust emission differences exist within that transmission configuration.

transmission lines (EPA-94/04): Pipelines that transport raw water from its source to a water treatment, then to the distribution grid system.

transmissive fault or fracture (40CFR146.61) is a fault or fracture that has sufficient permeability and vertical extent to allow fluids to move between formations (other identical or similar definitions are provided in 40CFR148.2).

transmissive fracture (40CFR148.2): See transmissive fault.

transmissivity (40CFR191.12) means the hydraulic conductivity integrated over the saturated thickness of an underground formation. The transmissivity of a series of formations is the sum of the individual transmissivities of each formation comprising the series.

transmissivity (EPA-94/04): The ability of an aquifer to transmit water.

transmissometer (40CFR60-AB-2.2) means that portion of the CEMS that includes the sample interface and the analyzer.

transmittance (40CFR60-AB-2.3) means the fraction of incident light that is transmitted through an optical medium.

transpiration (EPA-94/04): The process by which water vapor is lost to the atmosphere from living plants. The term can also be applied to the quantity of water thus dissipated.

transport or transportation (SF101-42USC9601) means the movement of a hazardous substance by any mode, including pipeline (as defined in the Pipeline Safety Act), and in the case of a hazardous substance which has been accepted for transportation by a common or contract carrier, the term "transport" or "transportation" shall include any stoppage in transit which is temporary, incidental to the transportation movement, and at the ordinary operating convenience of a common or contract carrier, and any such stoppage shall be considered as a continuity of movement and not as the storage of a hazardous substance.

transport vehicle (40CFR260.10) means a motor vehicle or rail car used for the transportation of cargo by any mode. Each cargo-carrying body (trailer, railroad freight car, etc.) is a separate transport vehicle.

transport vehicle (40CFR761.3) means a motor vehicle or rail car used for the transportation of cargo by any mode. Each cargo-carrying body (e.g., trailer, railroad freight car) is a separate transport vehicle.

transportation (40CFR259.10.a) means the shipment or conveyance or regulated medical waste by air, rail, highway, or water.

transportation (40CFR260.10) means the movement of hazardous waste by air, rail, highway, or water.

transportation (SF101): See transport.

transportation control measure (40CFR51.100-r) means any measure that is directed toward reducing emissions of air pollutants from transportation sources. Such measures include, but are not limited to, those listed in section 108(f) of the Clean Air Act.

transportation control measure (TCM) (40CFR51.392) is any measure that is specifically identified and committed to in the applicable implementation plan that is either one of the types listed in section 108 of the CAA, or any other measure for the purpose of reducing emissions or concentrations of air pollutants from transportation sources by reducing vehicle use or changing traffic flow or congestion conditions. Notwithstanding the above, vehicle technology-based, fuel-based, and maintenance-based measures which control the emissions from vehicles under fixed traffic conditions are not TCMs for the purposes of this subpart (other identical or similar definitions are provided in 40CFR93.101).

transportation control measure (TCM) (40CFR51.491) is any measure of the types listed in section 108(F) of the Act, or any measure in an applicable implementation plan directed toward reducing emissions of air pollutants from transportation sources by a reduction in vehicle use or changes in traffic conditions.

transportation control measure (TCM) (40CFR52.138-7) means any measure in an applicable implementation plan which is intended to reduce emissions from transportation sources.

transportation control measures (EPA-94/04): Steps taken by a locality to improve air quality by reducing or changing the flow of traffic, e.g., public transit, carpools, HOV lanes, etc.

transportation improvement program (TIP) (40CFR51.392) means a staged, multiyear, intermodal program of transportation projects covering a metropolitan planning area which is consistent with the metropolitan transportation plan, and developed pursuant to 23CFR part 450 (other identical or similar definitions are provided in 40CFR93.101).

transportation improvement program (TIP) (40CFR52.138-8) means the staged multiyear program of transportation improvements including an annual (or biennial) element which is required in 23CFR part 450.

transportation plan (40CFR51.392) means the official intermodal metropolitan transportation plan that is developed through the metropolitan planning process for the metropolitan planning area, developed pursuant to 23CFR part 450 (other identical or

similar definitions are provided in 40CFR93.101) (under CAA).

transportation project (40CFR93.101) is a highway project or a transit project.

transportation related and non-transportation related as applied to an onshore or offshore facility (40CFR112.2), are defined in the Memorandum of Understanding between the Secretary of Transportation and the Administrator of the Environmental Protection Agency, dated November 24, 1971, 36 FR 24080.

transportation related onshore and offshore facilities (40CFR112-AA) means:
(A) Onshore and offshore terminal facilities including transfer hoses, loading arms and other equipment and appurtenances used for the purpose of handling or transferring oil in bulk to or from a vessel as well as storage tanks and appurtenances for the reception of oily ballast water or tank washings from vessels, but excluding terminal waste treatment facilities and terminal oil storage facilities.
(B) Transfer hoses, loading arms and other equipment appurtenant to a non-transportation-related facility which is used to transfer oil in bulk to or from a vessel.
(C) Interstate and intrastate onshore and offshore pipeline systems including pumps and appurtenances related thereto as well as in-line or breakout storage tanks needed for the continuous operation of a pipeline system, and pipelines from onshore and offshore oil production facilities, but excluding onshore and offshore piping from wellheads to oil separators and pipelines which are used for the transport of oil exclusively within the confines of a non-transportation-related facility or terminal facility and which are not intended to transport oil in interstate or intrastate commerce or to transfer oil in bulk to or from a vessel.
(D) Highway vehicles and railroad cars which are used for the transport of oil in interstate or intrastate commerce and the equipment and appurtenances related thereto, and equipment used for the fueling of locomotive units, as well as the rights-of-way on which they operate. Excluded are highway vehicles and railroad cars

and motive power used exclusively within the confines of a non-transportation-related facility or terminal facility and which are not intended for use in interstate or intrastate commerce (cf. non-transportation-related onshore and offshore facilities).

transporter (40CFR259.10.a) means a person engaged in the off-site transportation of regulated medical waste by air, rail, highway, or water.

transporter (40CFR260.10) means a person engaged in the offsite transportation of hazardous waste by air, rail, highway, or water.

transporter (40CFR270.2) means a person engaged in the off-site transportation of hazardous waste by air, rail, highway or water.

transporter of PCB waste (40CFR761.3) means, for the purposes of subpart K of this part, any person engaged in the transportation of regulated PCB waste by air, rail, highway, or water for purposes other than consolidation by a generator.

transuranic radioactive waste (40CFR191.02-i), as used in this part, means waste containing more than 100 nanocuries of alpha-emitting transuranic isotopes, with half-lives greater than twenty years, per gram of waste, except for:
(1) High-level radioactive wastes;
(2) wastes that the Department has determined, with the concurrence of the Administrator, do not need the degree of isolation required by this part; or
(3) wastes that the Commission has approved for disposal on a case-by-case basis in accordance with 10CFR61.

trash (EPA-94/04): Material considered worthless or offensive that is thrown away. Generally defined as dry waste material, but in common usage it is a synonym for garbage, rubbish, or refuse.

trash-to-energy plan (EPA-94/04): Burning trash to produce energy.

TRC (40CFR420.02-f): Total residual chlorine.

TRE index value (40CFR60.661) means a measure of the supplemental total resource requirement per unit reduction of TOC associated with an individual distillation vent stream, based on vent stream flow rate, emission rate of TOC net heating value, and corrosion properties (whether or not the vent stream is halogenated), as quantified by the equation given under 40CFR60.664(e).

tread end cementing operation (40CFR60.541) means the system used to apply cement to one or both ends of the tread or combined tread/sidewall component. A tread end cementing operation consists of a cement application station and all other equipment, such as the cement supply system and feed and takeaway conveyors, necessary to apply cement to tread ends and to allow evaporation of solvent from the cemented tread ends.

treat or treatment of sewage sludge (40CFR503.9-z) is the preparation of sewage sludge for final use or disposal. This includes, but is not limited to, thickening, stabilization, and dewatering of sewage sludge. This does not include storage of sewage sludge.

treatability studies (EPA-94/04): Tests of potential cleanup technologies conducted in a laboratory (cf. bench scale tests).

treatability study (40CFR260.10) means a study in which a hazardous waste is subjected to a treatment process to determine:
(1) Whether the waste is amenable to the treatment process,
(2) what pretreatment (if any) is required,
(3) the optimal process conditions needed to achieve the desired treatment,
(4) the efficiency of a treatment process for a specific waste or wastes, or
(5) the characteristics and volumes of residuals from a particular treatment process. Also included in this definition for the purpose of the 40CFR261.4 (e) and (f) exemptions are liner compatibility, corrosion, and other material compatibility studies and toxicological and health effects studies. A "treatability study" is not a means to commercially treat or dispose of hazardous waste.

treated area (40CFR170.3) means any area to which

a pesticide is being directed or has been directed (under FIFRA).

treated regulated medical waste (40CFR259.10.b) means regulated medical waste that has been treated to substantially reduce or eliminate its potential for causing disease, but has not yet been destroyed.

treated regulated medical waste (EPA-94/04): Medical waste treated to substantially reduce or eliminate its pathogenicity, but that has not yet been destroyed.

treated wastewater (EPA-94/04): Wastewater that has been subjected to one or more physical, chemical, and biological processes to reduce its pollution of health hazards.

treatment (40CFR259.10.a) when used in the context of medical waste management means any method, technique, or process designed to change the biological character or composition of any regulated medical waste so as to reduce or eliminate its potential for causing disease. When used in the context of 40CFR259.30(a) of this part, treatment means either the provision of medical services or the preparation of human or animal remains for interment or cremation.

treatment (40CFR260.10) means any method, technique, or process, including neutralization, designed to change the physical, chemical, or biological character or composition of any hazardous waste so as to neutralize such waste, or so as to recover energy or material resources from the waste, or so as to render such waste non-hazardous, or less hazardous; safer to transport, store, or dispose of; or amenable for recovery, amenable for storage, or reduced in volume (other identical or similar definitions are provided in 40CFR270.2) (under RCRA).

treatment (EPA-94/04):
(1) Any method, technique, or process designed to remove solids and/or pollutants from solid waste, wastestreams, effluents, and air emissions.
(2) Methods used to change the biological character or composition of any regulated medical waste so as to substantially reduce or eliminate its potential for causing disease.

treatment (RCRA1004-42USC6903), when used in connection with hazardous waste, means any method, technique, or process, including neutralization, designed to change the physical, chemical, or biological character or composition of any hazardous waste so as to neutralize such waste or so as to render such waste nonhazardous, safer for transport, amenable for recovery, amenable for storage, or reduced in volume. Such term includes any activity or processing designed to change the physical form or chemical composition of hazardous waste so as to render it nonhazardous.

treatment facility and treatment system (40CFR434.11-o) mean all structures which contain, convey, and as necessary, chemically or physically treat coal mine drainage, coal preparation plant process wastewater, or drainage from coal preparation plant associated areas, which remove pollutants regulated by this part from such waters. This includes all pipes, channels, ponds, basins, tanks and all other equipment serving such structures.

treatment plant (EPA-94/04): A structure built to treat wastewater before discharging it into the environment.

treatment process (40CFR61.341) means a stream stripping unit, thin-film evaporation unit, waste incinerator, or any other process used to comply with 40CFR1.348 of this subpart.

treatment process (40CFR63.111) means a specific technique that removes or destroys the organics in a wastewater or residual stream such as a steam stripping unit, thin-film evaporation unit, waste incinerator, biological treatment unit, or any other process applied to wastewater streams or residuals to comply with 40CFR63.138 of this subpart. Most treatment processes are conducted in tanks. Treatment processes are a subset of waste management units.

treatment, storage, and disposal facility (EPA-94/04): Site where a hazardous substance is treated, stored, or disposed of. TSD facilities are regulated by EPA and states under RCRA.

treatment system (40CFR434.11-o): See treatment faciltiy.

treatment technique requirement (40CFR142.2) means a requirement of the national primary drinking water regulations which specifies for a contaminant a specific treatment technique(s) known to the Administrator which leads to a reduction in the level of such contaminant sufficient to comply with the requirements of part 141 of this chapter.

treatment technology (40CFR300.5) means any unit operation or series of unit operations that alters the composition of a hazardous substance or pollutant or contaminant through chemical, biological, or physical means so as to reduce toxicity, mobility, or volume of the contaminated materials being treated. Treatment technologies are an alternative to land disposal of hazardous wastes without treatment.

treatment works (40CFR35.2005-49) mean any devices and systems for the storage, treatment, recycling, and reclamation of municipal sewage, domestic sewage, or liquid industrial wastes used to implement section 201 of the Act, or necessary to recycle or reuse water at the most economical cost over the design life of the works. These include intercepting sewers, outfall sewers, sewage collection systems, individual systems, pumping, power, and other equipment and their appurtenances; extensions, improvement, remodeling, additions, and alterations thereof; elements essential to provide a reliable recycled supply such as standby treatment units and clear well facilities; and any works, including acquisition of the land that will be an integral part of the treatment process or is used for ultimate disposal of residues resulting from such treatment (including land for composting sludge, temporary storage of such compost and land used for the storage of treated wastewater in land treatment systems before land application); or any other method or system for preventing, abating, reducing, storing, treating, separating, or disposing of municipal waste or industrial waste, including waste in combined storm water and sanitary sewer systems.

treatment works (40CFR35.905) are any devices and systems for the storage, treatment, recycling, and reclamation of municipal sewage, domestic sewage, or liquid industrial wastes used to implement section 201 of the Act, or necessary to recycle or reuse water at the most economical cost over the useful life of the works. These include intercepting sewers, outfall

sewers, sewage collection systems, individual systems, pumping, power, and other equipment and their appurtenances; extensions, improvement, remodeling, additions, and alterations thereof; elements essential to provide a reliable recycled supply such as standby treatment units and clear well facilities; and any works, including site acquisition of the land that will be an integral part of the treatment process or is used for ultimate disposal of residues resulting from such treatment (including land for composting sludge, temporary storage of such compost, and land used for the storage of treated wastewater in land treatment systems before land application); or any other method or system for preventing, abating, reducing, storing, treating, separating, or disposing of municipal waste or industrial waste, including waste in combined storm water and sanitary sewer systems.

treatment works (40CFR503.9-aa) is either a federally owned, publicly owned, or privately owned device or system used to treat (including recycle and reclaim) either domestic sewage or a combination of domestic sewage and industrial waste of a liquid nature.

treatment works (CWA212-33USC1292):
(A) means any devices and systems used in the storage, treatment, recycling, and reclamation of municipal sewage or industrial wastes of a liquid nature to implement section 201 of this act, or necessary to recycle or reuse water at the most economical cost over the estimated life of the works, including intercepting sewers, outfall sewers, sewage collection systems, pumping, power, and other equipment, and their appurtenances; extensions, improvements, remodeling, additions, and alterations thereof; elements essential to provide a reliable recycled supply such as standby treatment units and clear well facilities; and any works, including site acquisition of the land that will be an integral part of the treatment process (including land use for the storage of treated wastewater in land treatment systems prior to land application) or is used for ultimate disposal of residues resulting for such treatment.
(B) In addition to the definition contained in subparagraph (A) of this paragraph, "treatment works" means any other method or system for preventing, abating, reducing, storing, treating,

separating, or disposing of municipal waste, including storm water runoff, or industrial waste, including waste in combined storm water and sanitary sewer systems. Any application for construction grants which includes wholly or in part such methods or systems shall, in accordance with guidelines published by the Administrator pursuant to subparagraph (C) of this paragraph, contain adequate data and analysis demonstrating such proposal to be, over the life of such works, the most efficient alternative to comply with sections 301 or 302 of this act, or the requirements of section 201 of this act.

(C) For the purposes of subparagraph (B) of this paragraph, the Administrator shall, within one hundred and eighty days after the date of enactment of this title, publish and thereafter revise no less than annually, guidelines for the evaluation of methods, including cost effective analysis, described in subparagraph (B) of this paragraph.

treatment works phase or segment (40CFR35.2005-50) may be any substantial portion of a facility and its interceptors described in a facilities plan under 40CFR35.2030, which can be identified as a subagreement or discrete subitem. Multiple subagreements under a project shall not be considered to be segments or phases. Completion of building of a treatment works phase or segment may, but need not in and of itself, result in an operable treatment works.

treatment works segment (40CFR35.905) may be any portion of an operable treatment works described in an approved facilities plan under 40CFR35.917, which can be identified as a contract or discrete subitem or subcontract for steps 1, 2, or 3 work. Completion of construction of a treatment works segment may, but need not, result in an operable treatment works.

treatment works treating domestic sewage (40CFR122.2) means a POTW or any other sewage sludge or waste water treatment devices or systems, regardless of ownership (including federal facilities), used in the storage, treatment, recycling, and reclamation of municipal or domestic sewage, including land dedicated for the disposal of sewage

sludge. This definition does not include septic tanks or similar devices. For purposes of this definition, "domestic sewage" includes waste and waste water from humans or household operations that are discharged to or otherwise enter a treatment works. In States where there is no approved State sludge management program under section 405(f) of the CWA, the Regional Administrator may designate any person subject to the standards for sewage sludge use and disposal in 40CFR503 as a "treatment works treating domestic sewage," where he or she finds that there is a potential for adverse effects on public health and the environment from poor sludge quality or poor sludge handling, use or disposal practices, or where he or she finds that such designation is necessary to ensure that such person is in compliance with 40CFR503.

treatment works treating domestic sewage (40CFR501.2) means a POTW or any other sewage sludge or wastewater treatment devices or systems, regardless of ownership (including Federal facilities), used in the storage, treatment, recycling, and reclamation of municipal or domestic sewage, including land dedicated for the disposal of sewage sludge. This definition does not include septic tanks or similar devices. For purposes of this definition, "domestic sewage includes waste and waste water from humans or household operations that are discharged to or otherwise enter a treatment works.

treatment zone (40CFR260.10) means a soil area of the unsaturated zone of a land treatment unit within which hazardous constituents are degraded, transformed, or immobilized.

tremie (EPA-94/04): Device used to place concrete or grout under water.

trenching or burial operation (40CFR257.3.6-5) means the placement of sewage sludge or septic tank pumpings in a trench or other natural or man-made depression and the covering with soil or other suitable material at the end of each operating day such that the wastes do not migrate to the surface.

trend assessment survey (40CFR228.2-b): See baseline survey.

trial burn (EPA-94/04): An incinerator test in which

emissions are monitored for the presence of specific organic compounds, particulates, and hydrogen chloride.

tribal education agency (40CFR47.105-f) means a school or community college which is controlled by an Indian tribe, band, or nation, including any Alaska Native village, which is recognized as eligible for special programs and services provided by the United States to Indians because of their status as Indians and which is not administered by the Bureau of Indian Affairs.

tribe or tribes (40CFR131.35.d-18) means the Colville Confederated Tribes.

trichloroethylene (TCE) (EPA-94/04): A stable, low boiling-point colorless liquid, toxic if inhaled. Used as a solvent or metal decreasing agent, and in other industrial applications.

trickle irrigation (EPA-94/04): Method in which water drips to the soil from perforated tubes or emitters.

trickling filter (EPA-94/04): A coarse treatment system in which wastewater is trickled over a bed of stones or other material covered with bacteria that break down the organic waste and produce clean water.

trihalomethane (THM) (40CFR141.2) means one of the family of organic compounds, named as derivatives of methane, wherein three of the four hydrogen atoms in methane are each substituted by a halogen atom in the molecular structure.

trihalomethane (THM) (EPA-94/04): One of a family of organic compounds named as derivative of methane. THMs are generally byproducts of chlorination of drinking water that contains organic material.

triple rinse (40CFR165.1-x) means the flushing of containers three times, each time using a volume of the normal diluent equal to approximately ten percent of the container's capacity, and adding the rinse liquid to the spray mixture or disposing of it by a method prescribed for disposing of the pesticide (under FIFRA).

triple superphosphate plant (40CFR60.231-a) means any facility manufacturing triple superphosphate by reacting phosphate rock with phosphoric acid. A run-of-pile triple superphosphate plant includes curing and storing.

tris (40CFR704.205) means tris(2,3-dibromopropyl) phosphate (also commonly named DBPP, TBPP, and Tris-BP).

trophic conditions (40CFR35.1605-6): A relative description of a lake's biological productivity based on the availability of plant nutrients. The range of trophic conditions is characterized by the terms of oligotrophic for the least biologically productive, to eutrophic for the most biologically productive.

troposphere (EPA-94/04): The layer of the atmosphere closest to the earth's surface.

truck dumping (40CFR60.671) means the unloading of nonmetallic minerals from movable vehicles designed to transport nonmetallic minerals from one location to another. Movable vehicles include but are not limited to: trucks, front end loaders, skip hoists, and railcars.

truck loading station (40CFR60.381) means that portion of a metallic mineral processing plant where metallic minerals or metallic mineral concentrates are loaded by a conveying system into trucks.

truck unloading station (40CFR60.381) means that portion of a metallic mineral processing plant where metallic ore is unloaded from a truck into a hopper, screen, or crusher.

trucked batteries (40CFR461.2-g) shall mean batteries moved into or out of the plant by truck when the truck is actually washed in the plant to remove residues left in the truck from the batteries.

true vapor pressure (40CFR52.741) means the equilibrium partial pressure exerted by a volatile organic liquid as determined in accordance with methods described in American Petroleum Institute Bulletin 2517, "Evaporation Loss From Floating Roof Tanks," second edition, February 1980 (incorporated by reference as specified in 40CFR52.742).

true vapor pressure (40CFR60.111-i) means the equilibrium partial pressure exerted by a petroleum liquid as determined in accordance with methods described in American Petroleum Institute Bulletin 2517, Evaporation Loss from External Floating-Roof Tanks, Second Edition, February 1980 (incorporated by reference, see 40CFR60.17).

true vapor pressure (40CFR60.111a-f) means the equilibrium partial pressure exerted by a petroleum liquid such as determined in accordance with methods described in American Petroleum Institute Bulletin 2517, Evaporation Loss from External Floating-Roof Tanks, Second Edition, February 1980 (incorporated by reference, see 40CFR60.17).

trust fund (40CFR300.5): See fund.

trust fund (CERCLA) (EPA-94/04): A fund set up under the Comprehensive Environmental Response, Compensation and Liability Act (CERCLA) to help pay for cleanup of hazardous waste sites and for legal action to force those responsible for the sites to clean them up.

trustee (40CFR144.70-b) means the Trustee who enters into this Agreement and any successor Trustee.

trustee (40CFR300.5) means an official of a federal natural resources management agency designated in subpart G of the NCP or a designated state official Indian tribe who may pursue claims for damages under section 107(f) of CERCLA.

TSCA: See act or TSCA (40CFR707.63).

TSS (or total suspended solids, or total suspended residue) (40CFR420.02-a) means the value obtained by the method specified in 40CFR136.3.

TTHM (40CFR141.2): Total trihalomethanes.

tube reducing (40CFR471.02-ll) is an operation which reduces the diameter and wall thickness of tubing with a mandrel and a pair of rolls with tapered grooves.

tube settler (EPA-94/04): Device using bundles of tubes to let solids in water settle to the bottom for removal by conventional sludge collection means; sometimes used in sedimentation basins and clarifiers to improve particle removal.

tuberculation (EPA-94/04): Development or formation of small mounds of corrosion products on the inside of iron pipe. These tubercules roughen the inside of the pipe, increasing its resistance to water flow.

tumbling or barrel finishing (40CFR471.02-mm) is an operation in which castings, forgings, or parts pressed from metal powder are rotated in a barrel with ceramic or metal slugs or abrasives to remove scale, fins, or burrs. It may be done dry or with an aqueous solution.

tumbling or burnishing (40CFR468.02-u) shall mean the process of polishing, deburring, removing sharp corners, and generally smoothing parts for both cosmetic and functional purposes, as well as the process of washing the finished parts and cleaning the abrasion media.

tundra (EPA-94/04): A type of ecosystem dominated by lichens, mosses, grasses, and woody plants. Tundra is found at high latitudes (arctic tundra) and high altitudes (alpine tundra). Arctic tundra is underlain by permafrost and is usually saturated (cf. wetlands).

turbidimeter (EPA-94/04): A device that measures the density of suspended solids in a liquid.

turbidity (40CFR131.35.d-19) means the clarity of water expressed as nephelometric turbidity units (NTU) and measured with a calibrated turbidimeter (under FWPCA).

turbidity (EPA-94/04):
1. Haziness in air caused by the presence of particles and pollutants.
2. A cloudy condition in water due to suspended silt or organic matter.

turbines employed in oil/gas production or oil/gas transportation (40CFR60.331-l) means any stationary gas turbine used to provide power to extract crude oil/natural gas from the earth or to move crude oil/natural gas, or products refined from these substances through pipelines.

turbo-fired boiler (40CFR76.2) means a pulverized coal, wall-fired boiler with burners arranged on walls so that the individual flames extend down toward the furnace bottom and then turn back up through the center of the furnace.

TW (CAA216): See test weight.

twenty five (25)-hour daily average or 24-hour daily average (40CFR60.51a) means the arithmetic or geometric mean (as specified in 40CFR60.59a (e), (g), or (h) as applicable) of all hourly emission rates when the affected facility is operating and firing MSW measured over a 24-hour period between 12 midnight and the following midnight.

twenty five (25)-year, 24-hour rainfall event (40CFR418.11-e) shall mean the maximum 24-hour precipitation event with a probable recurrence interval of once in 25 years as defined by the National Weather Service in technical paper No. 40, "Rainfall Frequency Atlas of the United States," May 1961, and subsequent amendments in effect, as of the effective date of this regulation (other identical or similar definitions are provided in 40CFR422.41-e; 422.51-e).

twenty five kilotonne party (40CFR82.3) means any nation listed in Appendix D to this part.

twenty four (24)-hour period (40CFR60.41a-25) means the period of time between 12:01 a.m. and 12:00 midnight.

twenty four (24)-hour period (40CFR60-AG) means the period of time between 12:00 midnight and the following midnight.

two (2)-year, 24-hour precipitation event (40CFR434.11-t) means the maximum 24-hour precipitation event with a probable recurrence interval of once in two years as defined by the National Weather Service and Technical Paper No. 40, "Rainfall Frequency Atlas of the U.S.," May 1961, or equivalent regional or rainfall probability information developed therefrom.

two or more animal feeding operations under common ownership (40CFR122.23-2) are considered, for the purposes of these regulations, to be a single animal feeding operation if they adjoin each other or if they use a common area or system for the disposal of wastes.

two-piece can (40CFR52.741) means a can which is drawn from a shallow cup and requires only one end to be attached.

two-piece can (40CFR60.491-5) means any beverage can that consists of a body manufactured from a single piece of steel or aluminum and a top. Coatings for a two-piece can are usually applied after fabrication of the can body.

type I sound level meter (40CFR204.2-11) means a sound level meter which meets the Type I requirements of American National Standard Specification S1.4-1971 for sound level meters. This publication is available from the American National Standards Institute, Inc., 1430 Broadway, New York, New York 10018.

type I sound level meter (40CFR205.2-25) means a sound level meter which meets the type I requirements of ANSI SI.4-1972 specification for sound level meters. This publication is available from the American National Standards Institute, Inc., 1430 Broadway, New York, New York 1001.

type III foci of transformed cell (40CFR795.285-2) are multilayered aggregations of densely staining cells with random orientation and criss-cross arrays at the periphery of the aggregate. They appear as dark stained areas on a light staining background monolayer which is one-cell thick.

type of pesticidal product (40CFR167.3) refers to each individual product as identified by: the product name; EPA Registration Number (or EPA File Symbol, if any, for planned products, or Experimental Permit Number, if the pesticide is produced under an Experimental Use Permit); active ingredients; production type (technical, formulation, repackaging, etc.); and, market for which the product was produced (domestic, foreign, etc.). In cases where a pesticide is not registered, registration is not applied for, or the pesticide is not produced under an Experimental Use Permit, the term shall also include the chemical formulation (under the Federal Insecticide, Fungicide, and Rodenticide Act).

type of resin (40CFR61.61-e) means the broad classification of resin referring to the basic manufacturing process for producing that resin, including, but not limited to, the suspension, dispersion, latex, bulk, and solution processes (under CAA).

type size (40CFR82.104-v) means the actual height of the printed image of each capital letter as it appears on a label.

*********** UUUUU ***********

U (40CFR440.132-j): See uranium.

$\mu g/m^3$ (40CFR52.274.d.2-iii; f.2-iii; p.2-ii; r.2-iii) means micrograms per cubic meter.

UIC (40CFR144.3) means the Underground Injection Control program under part C of the Safe Drinking Water Act, including an "approved State program."

ultimate biodegradability (40CFR796.3100-ii) is the breakdown of an organic compound to CO_2, water, the oxides or mineral salts of other elements and/or to products associated with normal metabolic processes of microorganisms.

ultimate consumer (40CFR600.002.85-38) means the first person who purchases an automobile for purposes other than resale or leases an automobile.

ultimate consumer (40CFR82.104-w) means the first commercial or non-commercial purchaser of a container or product that is not intended for re-introduction into interstate commerce as a final product or as part of another product.

ultimate purchaser (40CFR205.2-14) means the first person who in good faith purchases a product for purposes other than resale.

ultimate purchaser (40CFR53.1-l) means the first person who purchases a reference method or an equivalent method for purposes other than resale.

ultimate purchaser (40CFR85.1902-e) shall be given the meaning ascribed to it by section 24 of the Act.

ultimate purchaser (40CFR89.2) means, with respect to any new nonroad engine, new nonroad vehicle, or new nonroad equipment, the first person who in good faith purchases such new nonroad engine, nonroad vehicle, or nonroad equipment for purposes other than resale.

ultimate purchaser (CAA216-42USC7550) means, with respect to any new motor vehicle or new motor vehicle engine, the first person who in good faith purchases such new motor vehicle or new engine for purposes other than resale.

ultimate purchaser (NCA3-42USC4902) means the first person who in good faith purchases a product for purposes other than resale.

ultra clean coal (UCC) (EPA-94/04): Coal that is washed, ground into fine particles, then chemically treated to remove sulfur, ash, silicone, and other substances; usually briquetted and coated with a sealant made from coal.

ultra low-emission vehicle (40CFR88.101.94) means any light-duty vehicle or light-duty truck conforming to the applicable Ultra Low-Emission Vehicle standard, or any heavy-duty vehicle with an engine conforming to the applicable Ultra Low-Emission Vehicle standard.

ultrasonic testing (40CFR471.02-nn) is a nondestructive test which applies sound, at a frequency above about 20 HJz, to metal, which has been immersed in liquid (usually water) to locate inhomogeneities or structural discontinuities.

ultraviolet rays (EPA-94/04): Radiation from the sun that can be useful or potentially harmful. UV rays from one part of the spectrum (UV-A) enhance plant life and are useful in some medical and dental procedures; UV rays from other parts of the spectrum (UV-B) can cause skin cancer or other tissue damage. The ozone layer in the atmosphere partly shields us from ultraviolet rays reaching the earth's surface.

umbo (40CFR797.1800-4) means the narrow end (apex) of the oyster (other identical or similar definitions are provided in 40CFR797.1830-10).

unacceptable adverse effect (40CFR231.2-e) means impact on an aquatic or wetland ecosystem which is likely to result in significant degradation of municipal water supplies (including surface or ground water) or significant loss of or damage to fisheries, shellfishing, or wildlife habitat or recreation areas. In evaluating the unacceptability of such impacts, consideration should be given to the relevant portions of the section 404(b)(1) guidelines (40CFR230).

unauthorized dispersion technique (40CFR57.103-v) refers to any dispersion technique which, under section 123 of the Act and the regulations promulgated pursuant to that section, may not be used to reduce the degree of emission limitation otherwise required in the applicable SIP.

unbleached papers (40CFR250.4-rr) means papers made of pulp that have not been treated with bleaching agents.

uncertified person (40CFR171.2) means any person who is not holding a currently valid certification document indicating that he is certified under section 4 of FIFRA in the category of the restricted use pesticide made available for use.

unconfined aquifer (EPA-94/04): An aquifer containing water that is not under pressure; the water level in a well is the same as the water table outside the well.

uncontrolled sanitary landfill (40CFR122.26-15) means a landill or open dump, whether in operation or closed, that does not meet the requirements for runon or runoff controls established pursuant to subtitle D of the Solid Waste Disposal Act (under FWPCA).

uncontrolled total arsenic emissions (40CFR61.161) means the total inorganic arsenic in the glass melting furnace exhaust gas preceding any add-on emission control device.

under common control with (40CFR66.3-f): See control.

under normal circumstances garaged at personal residence (40CFR88.302.94) means a vehicle that, when it is not in use, is normally parked at the personal residence of the individual who usually operates it, rather than at a central refueling, maintenance, and/or business location. Such vehicles are not considered to be capable of being central fueled (as defined in this subpart) and are exempt from the program unless they are, in fact, centrally fueled.

under the direct supervision of (40CFR171.2-28) means the act or process whereby the application of a pesticide is made by a competent person acting under the instructions and control of a certified applicator who is responsible for the actions of that person and who is available if and when needed, even though such certified applicator is not physically present at the time and place the pesticide is applied.

undercoaters (40CFR52.741) means any coatings formulated for and applied to substrates to provide a smooth surface for subsequent coats.

underfire air (40CFR240.101-a) means any forced or induced air, under control as to quantity and direction, that is supplied from beneath and which passes through the solid wastes fuel bed.

underground area (40CFR280.12) means an underground room, such as a basement, cellar, shaft or vault, providing enough space for physical inspection of the exterior of the tank situated on or above the surface of the floor.

underground drinking water source (40CFR257.3.4-4) means:
(i) An aquifer supplying drinking water for human consumption, or
(ii) An aquifer in which the ground water contains less than 10,000 mg/L total dissolved solids.

underground injection (40CFR144.3; 146.3; 270.2) means a well injection.

underground injection (40CFR260.10) means the subsurface emplacement of fluids through a bored, drilled or driven well; or through a dug well, where the depth of the dug well is greater than the largest surface dimension (see also injection well) (under RCRA).

underground injection (SDWA1421-42USC300h) means the subsurface emplacement of fluids by well injection. Such term does not include the underground injection of natural gas for purposes of storage.

underground injection control (UIC) (40CFR124.2) means the program under part C of the Safe Drinking Water Act, including an "approved program" (other identical or similar definitions are provided in 40CFR146.3; 270.2).

underground injection control (UIC) (EPA-94/04): The program under the Safe Drinking Water Act that regulates the use of wells to pump fluids into the ground.

underground release (40CFR280.12) means any belowground release.

underground source of drinking water (40CFR144.3) means an aquifer or its portion:
(a) (1) Which supplies any public water system; or
  (2) Which contains a sufficient quantity of ground water to supply a public water system; and:
    (i) Currently supplies drinking water for human consumption; or
    (ii) Contains fewer than 10,000 mg/L total dissolved solids; and
(b) Which is not an exempted aquifer.
(Other identical or similar definitions are provided in 40CFR146.3; 147.2902; 270.2.)

underground sources of drinking water (EPA-94/04): Aquifers currently being used as a source of drinking water or those capable of supplying a public water system. They have a total dissolved solids content of 10,000 milligrams per liter or less, and are not "exempted aquifers" (cf. exempted aquifer).

underground storage tank (EPA-94/04): A tank located at least partially underground and designed to hold gasoline or other petroleum products or chemicals.

underground storage tank (RCRA9001-42USC6991) means any one or combination of tanks (including underground pipes connected thereto) which is used to contain an accumulation of regulated substances, and the volume of which (including the volume of the underground pipes connected thereto) is 10 per centum or more beneath the surface of the ground. Such term does not include any:
(a) farm or residential tank of 1,100 gallons or less capacity used for storing motor fuel for noncommercial purposes,
(b) tank used for storing heating oil for consumptive use on the premises where stored,
(c) septic tank,
(d) pipeline facility (including gathering lines) regulated under:
    (1) the Natural Gas Pipeline Safety Act of 1968 (49USC App. 1671, et seq.),
    (2) the Hazardous Liquid Pipeline Safety Act of 1979 (49USC App. 2001, et seq.), or
    (3) which is an intrastate pipeline facility regulated under State laws comparable to the provisions of law referred to in clause (d)(1) or (d)(2) of this subparagraph,
(e) surface impoundment, pit, pond, or lagoon,
(f) storm water or waste water collection system,
(g) flow-through process tank,
(h) liquid trap or associated gathering lines directly related to oil or gas production and gathering operations, or
(i) storage tank situated in an underground area (such as a basement, cellar, mineworking, drift, shaft, or tunnel) if the storage tank is situated upon or above the surface of the floor. The term "underground storage tank" shall not include any pipes connected to any tank which is described in subparagraphs (a) through (i).
(Other identical or similar definitions are provided in 40CFR280.12; 280.12.)

underground tank (40CFR260.10) means a device meeting the definition of "tank" in 40CFR60.10 whose entire surface area is totally below the surface of and covered by the ground.

underground uranium mine (40CFR61.21-c) means a man-made underground excavation made for the purpose of removing material containing uranium for the principal purpose of recovering uranium (under CAA).

underground water source protection program (SDWA1443-42USC300j.2) means a program for the adoption and enforcement of a program which meets the requirements of regulations under section 1421 and for keeping records and making reports required by section 1422(b)(1)(A)(ii). Such term includes, where applicable, a program which meets the requirements of section 1425.

underlying hazardous constituent (40CFR268.2-i) means any constituent listed in 40CFR 268.48, Table UTS-Universal Treatment Standards, except vanadium and zinc, which can reasonably be expected to be present at the point of generation of the hazardous waste, at a concentration above the constituent-specific UTS treatment standards.

undertread cementing operation (40CFR60.541) means the system used to apply cement to a continuous strip of tread or combined tread/sidewall component. An undertread cementing operation consists of a cement application station and all other equipment, such as the cement supply system and feed and takeaway conveyors, necessary to apply cement to tread or combined tread/sidewall strips and to allow evaporation or solvent from the cemented tread or combined tread/sidewall.

underutilization (40CFR72.2) means a reduction, during any calendar year in Phase I, of the heat input (expressed in mmBtu for the calendar year) at a Phase I unit below the unit's baseline.

undisturbed performance (40CFR191.12) means the predicted behavior of a disposal system, including consideration of the uncertainties in predicted behavior, if the disposal system is not disrupted by human intrusion or the occurrence of unlikely natural events.

unexpended consumption allowances (40CFR82.3-ii) means consumption allowances that have not been used. At any time in any control period a person's unexpended consumption allowances are the total of the level of consumption allowances the person has authorization under this subpart to hold at that time for that control period, minus the level of controlled substances that the person has produced or imported (not including transhipments and used controlled substances) in that control period until that time.

unexpended production allowances (40CFR82.3-jj) means production allowances that have not been used. At any time in any control period a person's unexpended production allowances are the total of the level of production allowances he has authorization under this subpart to hold at that time for that control period, minus the level of controlled substances that the person has produced in that control period until that time.

unfit for use tank system (40CFR260.10) means a tank system that has been determined through an integrity assessment or other inspection to be no longer capable of storing or treating hazardous waste without posing a threat of release of hazardous waste to the environment.

unified planning work program or UPWP (40CFR52.138-9) means the program required by 23CFR 450.108(c) and endorsed by the metropolitan planning organization which describes urban transportation and transportation-related planning activities anticipated in the area during the next 1- to 2-year period including the planning work to be performed with federal planning assistance and with funds available under the Urban Mass Transportation Act (49 U.S.C.) section 9 or 9A. UPWPs are also known as overall work programs or OWPs.

uniform air quality required for the daily reporting of air quality (40CFR58-AG-a) is a modified form of the Pollutant Standards Index (PSI).

uniforming (40CFR60.721): See fog coating.

unit (40CFR72.2) means a fossil fuel-fired combustion device.

unit (CAA402-42USC7651a) means a fossil fuel-fired combustion device.

unit account (40CFR72.2) means an Allowance Tracking System account, established by the Administrator for an affected unit pursuant to 40CFR73.31 (a) or (b) of this chapter.

unit acquisition cost (40CFR35.6015-52) means the net invoice unit price of the property including the cost of modifications, attachments, accessories, or auxiliary apparatus necessary to make the property usable for the purpose for which it was acquired. Other charges, such as the cost of installation, transportation, taxes, duty, or protective in-transit insurance, shall be included or excluded from the unit acquisition cost in accordance with the recipient's regular accounting practices.

unit load (40CFR72.2) means the total (i.e., gross) output of a unit or source in any calendar year (or other specified time period) produced by combusting a given heat input of fuel, expressed in terms of:

(1) The total electrical generation (MWe) for use within the plant and for sale; or

(2) In the case of a unit or source that uses part of its heat input for purposes other than electrical generation, the total steam pressure (psia) produced by the unit or source.

unit operating hours (40CFR72.2) means, when referring to the availability of a monitoring system, the number of hours (or fraction of an hour) that a unit combusts any fuel; and, in the case of a COMS availability, includes the number of hours (or fraction of an hour) after combustion has ceased when fans are operating.

unit operation (40CFR63.101) means one or more pieces of process equipment used to make a single change to the physical or chemical characteristics of one or more process streams. Unit operations include, but are not limited to, reactors, distillation columns, extraction columns, absorbers, decanters, dryers, condensers, and filtration equipment.

unit packaging (40CFR157.21-g) means a package that is labeled with directions to use the entire contents of the package in a single application.

unit-week of maintenance (40CFR52.145) means a period of 7 days during which a fossil fuel-fired steam-generating unit is under repair, and no coal is combusted in the unit.

United States (40CFR7.25) means the States of the United States, the District of Columbia, the Commonwealth of Puerto Rico, the Virgin Islands, American Samoa, Guam, Wake Island, the Canal Zone, and all other territories and possessions of the United States; the term state includes any one of the foregoing (other identical or similar definitions are provided in 40CFR8.2; 30.4; 40.4; 70.4; 85.1502; 89.602.96; 109.2; 112.2; 110.1; 260.10; 300.5; 302.3; 710.2-cc; 720.3-hh; 762.3-d; 763.163; TSCA3-15USC2602).

units (40CFR797.1350-10; 797.1440-10): All concentrations are given in weight per volume (e.g., in mg/liter).

universal biohazard symbol (40CFR259.10.b) means the symbol design that conforms to the design shown in 29CFR1910.145(f)(8)(ii).

unleaded gasoline (40CFR80.2-g) means gasoline which is produced without the use of any lead additive and which contains not more than 0.05 gram of lead per gallon and not more than 0.005 gram of phosphorus per gallon.

unliquidated obligations for reports prepared on a cash basis (40CFR31.3) mean the amount of obligations incurred by the grantee that has not been paid. For reports prepared on an accrued expenditure basis, they represent the amount of obligations incurred by the grantee, for which an outlay has not been recorded.

unloading leg (40CFR60.301-o) means a device which includes a bucket-type elevator which is used to remove grain from a barge or ship.

unobligated balance (40CFR31.3) means the portion of the funds authorized by the Federal agency that has not been obligated by the grantee and is determined by deducting the cumulative obligations from the cumulative funds authorized.

unproven emission control systems (40CFR86.092.2-b) are emission control components or systems (and fuel metering systems) that do not qualify as proven emission control systems.

unreasonable adverse effects on the environment (40CFR166.3-j) means any unreasonable risk to man or the environment, taking into account the economic, social, and environmental costs and benefits of the use of any pesticide.

unreasonable adverse effects on the environment

(FIFRA2-7USC136) means any unreasonable risk to man or the environment, taking into account the economic, social, and environmental costs and benefits of the use of any pesticide.

unreasonable degradation of the marine environment (40CFR125.121-e) means:
(1) Significant adverse changes in ecosystem diversity, productivity and stability of the biological community within the area of discharge and surrounding biological communities,
(2) Threat to human health through direct exposure to pollutants or through consumption of exposed aquatic organisms, or
(3) Loss of esthetic, recreational, scientific or economic values which is unreasonable in relation to the benefit derived from the discharge.

unreasonable risk (EPA-94/04): Under the Federal Insecticide, Fungicide, and Rodenticide Act (FIFRA), "unreasonable adverse effects" means any unreasonable risk to man or the environment, taking into account the medical, economic, social, and environmental costs and benefits of any pesticide.

unreclaimable residues (40CFR165.1-y) means residual materials of little or no value remaining after incineration.

unrefined and unprocessed ore (10CFR40.4) means ore in its natural form prior to any processing, such as grinding, roasting or beneficiating, or refining.

unregulated hazardous substance (SF211/2704) means a hazardous substance:
(A) for which no standard, requirement, criteria, or limitation is in effect under the Toxic Substances Control Act, the Safe Drinking Water Act, the Clean Air Act, or the Clean Water Act; and
(B) for which no water quality criteria are in effect under any provision of the Clean Water Act.

unregulated safety relief valve (40CFR52.741) means a safety relief valve which cannot be actuated by a means other than high pressure in the pipe or vessel which it protects.

unrestricted area (10CFR20.3) means any area access to which is not controlled by the licensee for purposes of protection of individuals from exposure to radiation and radioactive materials, and any area used for residential quarters.

unsanitized (40CFR350.1) means a version of a document from which information claimed as trade secret or confidential has not been withheld or omitted (cf. sanitized).

unsaturated zone (EPA-94/04): The area above the water table where soil pores are not fully saturated, although some water may be present.

unsaturated zone or zone of aeration (40CFR260.10) means the zone between the land surface and the water table.

unscheduled DNA synthesis in mammalian cells in culture (40CFR798.5550) is defined as the incorporation of tritium labelled thymidine (^{3}H-TdR) into the DNA of cells which are not in the S phase of the cell cycle.

unscheduled maintenance (40CFR86.082.2) means any adjustment, repair, removal, disassembly, cleaning, or replacement of vehicle components or systems which is performed to correct a part failure or vehicle (if the engine were installed in a vehicle) malfunction.

unscheduled maintenance (40CFR86.084.2) means any adjustment, repair, removal disassembly, cleaning, or replacement of vehicle components or systems which is performed to correct a part failure or vehicle (if the engine were installed in a vehicle) malfunction which was not anticipated.

unscheduled maintenance (40CFR86.402.78) means any inspection, adjustment repair, removal, disassembly, cleaning, or replacement of vehicle components or systems which is performed to correct or diagnose a part failure or vehicle malfunction which was not anticipated.

unsigned (40CFR86.902.93) means a certification request which does not result in a signed Certificate of Conformity because it is either voluntarily withdrawn by the manufacturer or does not receive approval from the EPA.

unsolicited proposal (40CFR30.200) means an informal written offer to perform EPA funded work for which EPA did not publish a solicitation.

unstabilized solids (40CFR503.31-j) are organic materials in sewage sludge that have not been treated in either an aerobic or anaerobic treatment process.

unstable area (40CFR503.21-q) is land subject to natural or human-induced forces that may damage the structural components of an active sewage sludge unit. This includes, but is not limited to, land on which the soils are subject to mass movement (under CWA).

untreated regulated medical waste (40CFR259.10.b) means regulated medical waste that has not been treated to substantially reduce or eliminate its potential for causing disease.

unwarranted failure to comply (SMCRA701-30USC1291) means the failure of a permittee to prevent the occurrence of any violation of his permit or any requirement of this chapter due to indifference, lack of diligence, or lack of reasonable care, or the failure to abate any violation of such permit or the chapter due to indifference, lack of diligence, or lack of reasonable care.

upgrade (40CFR280.12) means the addition or retrofit of some systems such as cathodic protection, lining, or spill and overfill controls to improve the ability of an underground storage tank system to prevent the release of product.

upper limit (40CFR86.1102.87) means the emission level for a specific pollutant above which a certificate of conformity may not be issued or may be suspended or revoked.

uppermost aquifer (40CFR258.2) means the geologic formation nearest the natural ground surface that is an aquifer, as well as lower aquifers that are hydraulically interconnected with this aquifer within the facility's property boundary (other identical or similar definitions are provided in 40CFR260.10).

upscale calibration value (40CFR60-AB-2.11) means the opacity value at which a calibration check of the CEMS is performed by simulating an upscale opacity condition as viewed by the receiver.

upset (40CFR122.41.n) means an exceptional incident in which there is unintentional and temporary noncompliance with technology based permit effluent limitations because of factors beyond the reasonable control of the permittee. An upset does not include noncompliance to the extent caused by operational error, improperly designed treatment facilities, inadequate treatment facilities, lack of preventive maintenance, or careless or improper operation.

upset (40CFR403.16-a) means an exceptional incident in which there is unintentional and temporary noncompliance with categorical pretreatment standards because of factors beyond the reasonable control of the industrial user. An upset does not include noncompliance to the extent caused by operational error, improperly designed treatment facilities, inadequate treatment facilities, lack of preventive maintenance, or careless or improper operation.

uptake (40CFR797.1520-16) is the sorption of a test substance into and onto aquatic organisms during exposure.

uptake (40CFR797.1830-11) is the sorption of a test chemical into and onto aquatic organisms during exposure.

uptake (u) (40CFR797.1560-6) is the process of sorbing testing material into and/or onto the test organisms.

uptake phase (40CFR797.1520-17) is the initial portion of a bioconcentration test during which the organisms are exposed to the test solution (other identical or similar definitions are provided in 40CFR797.1830-12).

uptake phase (40CFR797.1560-7) is the time during the test when test organisms are being exposed to the test material.

uptake rate constant (k_1) (40CFR797.1560-8) is the mathematically determined value that is used to define the uptake of test material by exposed test organisms, usually reported in units of liters/gram/hour.

uranium byproduct material (40CFR192.31-b) means the tailings or wastes produced by the extraction or concentration of uranium from any ore processed primarily for its source material content. Ore bodies depleted by uranium solution extraction operations and which remain underground do not constitute "byproduct material" for the purpose of this subpart (under AEA).

uranium byproduct material or tailings (40CFR61.221-c) means the waste produced by the extraction or concentration of uranium from any ore processed primarily for its source material content. Ore bodies depleted by uranium solution extraction and which remain underground do not constitute byproduct material for the purposes of this subpart (other identical or similar definitions are provided in 40CFR61.251-g).

uranium fuel cycle (40CFR190.02-b) means the operations of milling of uranium ore, chemical conversion of uranium, isotopic enrichment of uranium, fabrication of uranium fuel, generation of electricity by a light-water-cooled nuclear power plant using uranium fuel, and reprocessing of spent uranium fuel, to the extent that these directly support the production of electrical power for public use utilizing nuclear energy, but excludes mining operations, operations at waste disposal sites, transportation of any radioactive material in support of these operations, and the reuse of recovered non-uranium special nuclear and byproduct materials from the cycle.

uranium mill tailings piles (EPA-94/04): Former uranium ore processing sites that contain leftover radioactive materials (wastes), including radium and unrecovered uranium.

Uranium Mill Tailings Radiation Control Act (UMTRCA) Standards (40CFR300-AA): Standards for radionuclides established under sections 102, 104, and 108 of the Uranium Mill Tailings Radiation Control Act, as amended.

uranium mill tailings waste piles (EPA-94/04): Licensed active mills with tailings piles and evaporation ponds created by acid or alkaline leaching processes.

uranium milling (10CFR40.4) means any activity that results in the production of byproduct material as defined in this part.

urban area population (40CFR58.1-s) means the population defined in the most recent decennial U.S. Census of Population Report.

urban bus (40CFR85.1402) has the meaning set forth in 40CFR86.091-2 of this chapter.

urban bus (40CFR86.091.2) means a heavy heavy-duty diesel-powered passenger-carrying vehicle with a load capacity of fifteen or more passengers and intended primarily for intra-city operation, i.e., within the confines of a city or greater metropolitan area. Urban bus operation is characterized by short rides and frequent stops. To facilitate this type of operation, more than one set of quick-operating entrance and exit doors would normally be installed. Since fares are usually paid in cash or tokens rather than purchased in advance in the form of tickets, urban buses would normally have equipment installed for collection of fares. Urban buses are also typically characterized by the absence of equipment and facilities for long distance travel, e.g., rest rooms, large luggage compartments, and facilities for stowing carry-on luggage. The useful life for urban buses is the same as the useful life for other heavy heavy-duty diesel engines.

urban bus (40CFR86.093.2) means a passenger-carrying vehicle powered by a heavy heavy-duty diesel engine, or of a type normally powered by a heavy heavy-duty diesel engine, with a load capacity of fifteen or more passengers and intended primarily for intracity operation, i.e., within the confines of a city or greater metropolitan area. Urban bus operation is characterized by short rides and frequent stops. To facilitate this type of operation, more than one set of quick-operating entrance and exit doors would normally be installed. Since fares are usually paid in cash or tokens, rather than purchased in advance in the form of tickets, urban buses would normally have equipment installed for collection of fares. Urban buses are also typically characterized by the absence of equipment and facilities for long distance travel, e.g., rest rooms, large luggage compartments, and facilities for stowing carry-on luggage. The useful life for urban buses is the same as the useful life for other heavy heavy-duty diesel engines.

urban runoff (EPA-94/04): Storm water from city streets and adjacent domestic or commercial properties that carries pollutants of various kinds into the sewer systems and receiving waters.

urban waterfront and port (CZMA306a-16USC1455a) means any developed area that is densely populated and is being used for, or has been used for, urban residential recreational, commercial, shipping or industrial purposes.

urea-formaldehyde foam insulation (EPA-94/04): A material once used to conserve energy by sealing crawl spaces, attics, etc.; no longer used because emissions were found to be a health hazard.

usable firebox volume (40CFR60-AA (method 28-2.11)) means the volume of the firebox determined using the following definitions:
(1) Height: The vertical distance extending above the loading door, if fuel could reasonably occupy that space, but not more than 2 inches above the top (peak height) of the loading door, to the floor of the firebox (i.e., below a permanent grate) if the grate allows a 1-inch diameter piece of wood to pass through the grate, or, if not, to the top of the grate. Firebox height is not necessarily uniform but must account for variations caused by internal baffles, air channels, or other permanent obstructions.
(2) Length: The longest horizontal fire chamber dimension that is parallel to a wall of the chamber.
(3) Width: The shortest horizontal fire chamber dimension that is parallel to a wall of the chamber.

USDW (40CFR144.3) means "underground source of drinking water."

use (40CFR82.172) means any use of a substitute for a Class I or Class II ozone-depleting compound, including but not limited to use in a manufacturing process or product, in consumption by the end-user, or in intermediate uses, such as formulation or packaging for other subsequent uses (under the Clean Air Act).

use attainability analysis (40CFR131.3-g) is a structured scientific assessment of the factors affecting the attainment of the use which may include physical, chemical, biological, and economic factors as described in 40CFR131.10(g).

use in agricultural or wildlife propagation (40CFR435.51) means that the produced water is of good enough quality to be used for wildlife or livestock watering or other agricultural uses and that the produced water is actually put to such use during periods of discharge.

use of asbestos (40CFR763.103-i) means the presence of asbestos-containing material in school buildings.

use restrictions (40CFR82.172) means restrictions on the use of a substitute imposing either conditions on how the substitute can be used across a sector end-use or limits on the end-uses or specific applications where it can be used within a sector.

use stream (40CFR721.3) means all reasonably anticipated transfer, flow, or disposal of a chemical substance, regardless of physical state or concentration, through all intended operations of industrial, commercial, or consumer use.

used controlled substances (40CFR82.3-kk) means controlled substances that have been recovered from their intended use systems (may include controlled substances that have been, or may be subsequently, recycled or reclaimed).

used in or for the manufacturing or processing of an instant photographic or peel-apart film article (40CFR723.175-15), when used to describe activities involving a new chemical substance, means the new chemical substance
(i) is included in the article, or
(ii) is an intermediate to a chemical substance included in the article or is one of a series of intermediates used to manufacture a chemical substance included in the article.

used oil (40CFR260.10) means any oil that has been

refined from crude oil, or any synthetic oil, that has been used and as a result of such use is contaminated by physical or chemical impurities.

used oil (RCRA1004-42USC6903) means any oil which has been:
(A) refined from crude oil,
(B) used, and
(C) as a result of such use, contaminated by physical or chemical impurities.

used oil fuel (40CFR266.40) includes any fuel produced from used oil by processing, blending, or other treatment.

used or reused (40CFR261.1-5): For purposes of sections 261.2 and 261.6, a material is used or reused if it is either:
(i) Employed as an ingredient (including use as an intermediate) in an industrial process to make a product (for example, distillation bottoms from one process used as feedstock in another process). However, a material will not satisfy this condition if distinct components of the material are recovered as separate end products (as when metals are recovered from metal-containing secondary materials); or
(ii) Employed in a particular function or application as an effective substitute for a commercial product (for example, spent pickle liquor used as phosphorous precipitant and sludge conditioner in wastewater treatment).

used solely for competition (40CFR89.2) means exhibiting features that are not easily removed and that would render its use other than in competition unsafe, impractical, or highly unlikely.

useful life (40CFR35.2005-51) means the period during which a treatment works operates. (Not "design life" which is the period during which a treatment works is planned and designed to be operated.)

useful life (40CFR35.905) means estimated period during which a treatment works will be operated.

useful life (40CFR85.1902-c) shall be given the meaning ascribed to it by section 202(d) of the Act and regulations promulgated thereunder.

useful life (40CFR85.2102-10) means that period established pursuant to section 202(d) of the Act and regulations promulgated thereunder.

useful life (40CFR86.001-2) means:
(1) For light-duty vehicles, and for light light-duty trucks not subject to the Tier 0 standards of 40CFR86.094-9(a), intermediate useful life and/or full useful life. Intermediate useful life is a period of use of 5 years or 50,000 miles, whichever occurs first. Full useful life is a period of use of 10 years or 100,000 miles, whichever occurs first, except as otherwise noted in 40CFR86.094-9. The useful life of evaporative and/or refueling emission control systems on the portion of these vehicles subject to the evaporative emission test requirements of 40CFR86.130-96, and/or the refueling emission test requirements of 40CFR86.151-2001, is defined as a period of use of 10 years or 100,000 miles, whichever occurs first.
(2) For light light-duty trucks subject to the Tier 0 standards of 40CFR86.094-9(a), and for heavy light-duty truck engine families, intermediate and/or full useful life. Intermediate useful life is a period of use of 5 years or 50,000 miles, whichever occurs first. Full useful life is a period of use of 11 years or 120,000 miles, whichever occurs first. The useful life of evaporative emission and/or refueling control systems on the portion of these vehicles subject to the evaporative emission test requirements of 40CFR86.130-96, and/or the refueling emission test requirements of 40CFR86.151-2001, is also defined as a period of 11 years or 120,000 miles, whichever occurs first.
(3) For an Otto-cycle heavy-duty engine family:
   i.   For hydrocarbon and carbon monoxide standards, a period of use of 8 years or 110,000 miles, whichever first occurs.
   ii.   For the oxides of nitrogen standard, a period of use of 10 years or 110,000 miles, whichever first occurs.
   iii.   For the portion of evaporative emission control systems subject to the evaporative emission test requirements of 40CFR86.1230-96, a period of use of 10 years or 110,000 miles, whichever occurs first.
(4) For a diesel heavy-duty engine family:

i. For light heavy-duty diesel engines, for hydrocarbon, carbon monoxide, and particulate standards, a period of use of 8 years or 110,000 miles, whichever first occurs.

ii. For light heavy-duty diesel engines, for the oxides of nitrogen standard, a period of use of 10 years or 110,000 miles, whichever first occurs.

iii. For medium heavy-duty diesel engines, for hydrocarbon, carbon monoxide, and particulate standards, a period of use of 8 years or 185,000 miles, whichever first occurs.

iv. For medium heavy-duty diesel engines, for the oxides of nitrogen standard, a period of use of 10 years or 185,000 miles, whichever first occurs.

v. For heavy heavy-duty diesel engines, for hydrocarbon, carbon monoxide, and particulate standards, a period of use of 8 years or 290,000 miles, whichever first occurs, except as provided in paragraph (4)(vii) of this definition.

vi. For heavy heavy-duty diesel engines, for the oxides of nitrogen standard, a period of use of 10 years or 290,000 miles, whichever first occurs.

vii. For heavy heavy-duty diesel engines used in urban buses, for the particulate standard, a period of use of 10 years or 290,000 miles, whichever first occurs.

(Other identical or similar definitions are provided in 40CFR85.1502-14; 86.082.2; 86.084.2; 86.085.2; 86.090.2; 86.094.2; 86.098.2; 86.096.2; 86.402.78.)

useful life (40CFR89.602.96): A period of time as specified in subpart B of this part which for a nonconforming nonroad engine begins at the time of resale (for a nonroad engine owned by the ICI at the time of importation) or release to the owner (for a nonroad engine not owned by the ICI at the time of importation) of the nonroad engine by the ICI after modification and/or testing pursuant to 40CFR89.605-96 or 40CFR89.609-96.

user (40CFR403.3-h): See industrial user.

user charge (40CFR35.905) means a charge levied on users of a treatment works, or that portion of the ad valorem taxes paid by a user, for taxes users proportionate share of the cost of operation and maintenance (including replacement) of such works under sections 2(b)(l)(A) and 201(h)(2) of the Act and this (other identical or similar definitions are provided in 40CFR35.2005-52).

user fee (EPA-94/04): Fee collected from only those persons who use a particular service, as compared to one collected from the public in general.

UST system or tank system (40CFR280.12) means an underground storage tank, connected underground piping, underground ancillary equipment, and containment system, if any.

utility (40CFR72.2) means any person that sells electricity.

utility competitive bid solicitation (40CFR72.2) is a public request from a regulated utility for offers to the utility for meeting future generating needs. A qualifying facility, independent power production facility, or new IPP may be regarded as having been "selected" in such solicitation if the utility has named the facility as a project with which the utility intends to negotiate a power sales agreement.

utility load (EPA-94/04): The total electricity demand for a utility district.

utility regulatory authority (40CFR72.2) means an authority, board, commission, or other entity (limited to the local-, State-, or federal-level, whenever so specified) responsible for overseeing the business operations of utilities located within its jurisdiction, including, but not limited to, utility rates and charges to customers.

utility system (40CFR72.2) means all interconnected units and generators operated by the same utility operating company.

utility unit (40CFR72.2) means a unit owned or operated by a utility:
(1) That serves a generator in any State that produces electricity for sale, or
(2) That during 1985, served a generator in any State that produced electricity for sale.
(3) Notwithstanding paragraphs (1) and (2) of this

definition, a unit that was in operation during 1985, but did not serve a generator that produced electricity for sale during 1985, and did not commence commercial operation on or after November 15, 1990 is not a utility unit for purposes of the Acid Rain Program.

(4) Notwithstanding paragraphs (1) and (2) of this definition, a unit that cogenerates steam and electricity is not a utility unit for purposes of the Acid Rain Program, unless the unit is constructed for the purpose of supplying, or commences construction after November 15, 1990 and supplies, more than one-third of its potential electrical output capacity and more than 25 MWe output to any power distribution system for sale.

utility unit (CAA402-42USC7651a):

(A) means:

    (i)   a unit that serves a generator in any State that produces electricity for sale, or

    (ii) a unit that, during 1985, served a generator in any State that produced electricity for sale.

(B) Notwithstanding subparagraph (A), a unit described in subparagraph (A) that:

    (i)   was in commercial operation during 1985, but

    (ii) did not, during 1985, serve a generator in any State that produced electricity for sale shall not be a utility unit for purposes of this title.

(C) A unit that cogenerates steam and electricity is not a "utility unit" for purposes of this title unless the unit is constructed for the purpose of supplying, or commences construction after the date of enactment of this title and supplies, more than one-third of its potential electric output capacity and more than 25 megawatts electrical output to any utility power distribution system for sale.

utilization (40CFR72.2) means the heat input (expressed in mmBtu/time) for a unit.

utilization facility (10CFR30.4) means a utilization facility as defined in the regulations contained in part 50 of this chapter.

utilize (40CFR60.711-18) refers to the use of solvent

that is delivered to coating mix preparation equipment for the purpose of formulating coatings to be applied on an affected coating operation and any other solvent (e.g., dilution solvent) that is added at any point in the manufacturing process.

UV-VIS absorption spectrum of a solution (40CFR796.1050) is a function of the concentration, c_1, expressed in mol/L, of all absorbing species present; the path length, d, of the spectrophotometer cell, expressed in cm; and the molar absorption (extinction) coefficient, e_i, of each species. The absorbance (optical density) A of the solution is then given by:

$$A = d \times \Sigma_i(e_i c_i)$$

For a resolvable absorbance peak, the band width lambda is the wavelength range, expressed in $nm = 10^{-9}m$, of the peak at half the absorbance maximum.

************ VVVVV ************

vacuum break (choke pull-off) (40CFR85.2122(a)(1)(ii)(F)) means a vacuum-operated device to open the carburetor choke plate a predetermined amount on cold start.

vacuum leakage (40CFR85.2122(a)(1)(ii)(E)) means leakage into the vacuum cavity of a vacuum break.

vacuum producing system (40CFR52.741) means any reciprocating, rotary, or centrifugal blower or compressor or any jet ejector or device that creates suction from a pressure below atmospheric and discharges against a greater pressure.

vacuum purge system (40CFR85.2122(a)(1)(ii)(H)) means a vacuum system with a controlled air flow to purge the vacuum system of undesirable manifold vapors.

valid day (40CFR60.101-q) means a 24-hour period

in which at least 18 valid hours of data are obtained. A "valid hour" is one in which at least 2 valid data points are obtained.

valid emission performance warranty claim (40CFR85.2102-16) means a claim in which there is no evidence that the vehicle had not been properly maintained and operated in accordance with manufacturer instructions, the vehicle failed to conform to applicable emission standards as measured by an Office Director-approved type of emission warranty test during its useful life, and the owner is subject to sanction as a result of the test failure.

valid study (40CFR152.83-e) means a study that has been conducted in accordance with the Good Laboratory Practice standards of 40CFR160 or generally accepted scientific methodology and that EPA has not determined to be invalid.

validated test (40CFR154.3-i) means a test determined by the Agency to have been conducted and evaluated in a manner consistent with accepted scientific procedures.

valuable commercial and recreational species (40CFR228.2-g) shall mean those species for which catch statistics are compiled on a routine basis by the Federal or State agency responsible for compiling such statistics for the general geographical area impacted, or which are under current study by such Federal or State agencies for potential development for commercial or recreational use.

value engineering (40CFR35.2005-53) means a specialized cost control technique which uses a systematic and creative approach to identify and to focus on unnecessarily high cost in a project in order to arrive at a cost saving without sacrificing the reliability or efficiency of the project.

value engineering (40CFR35.6015-53) means a systematic and creative analysis of each contract term or task to ensure that its essential function is provided at the overall lowest cost.

value engineering (VE) (40CFR35.905) means a specialized cost control technique which uses a systematic and creative approach to identify and to

focus on unnecessarily high cost in a project in order to arrive at a cost saving without sacrificing the reliability or efficiency of the project.

value for pesticide purposes (40CFR172.1-f) means that characteristic of a substance or mixture of substances which produces an efficacious action on a pest.

valve height (40CFR797.1800-5) means the greatest linear dimension of the oyster as measured from the umbo to the ventral edge of the valves (the farthest distance from the umbo (other identical or similar definitions are provided in 40CFR797.1830-13).

valves not externally regulated (40CFR52.741) means valves that have no external controls, such as in-line check valves.

van (40CFR600.002.85-39) means any light truck having an integral enclosure fully enclosing the driver compartment and load-carrying device, and having no body sections protruding more than 30 inches ahead of the leading edge of the windshield.

van (40CFR86.082.2) means a light-duty truck having an integral enclosure, fully enclosing the driver compartment and load carrying device, and having no body sections protruding more than 30 inches ahead of the leading edge of the windshield.

vapor balance system (40CFR52.741) means any combination of pipes or hoses which creates a closed system between the vapor spaces of an unloading tank and a receiving tank such that vapors displaced from the receiving tank are transferred to the tank being unloaded.

vapor balancing system (40CFR63.101) means a piping system that is designed to collect organic HAP vapors displaced from tank trucks or railcars during loading; and to route the collected organic HAP vapors to the storage vessel from which the liquid being loaded originated, or to compress collected organic HAP vapors and commingle with the raw feed of a chemical manufacturing process unit.

vapor capture system (40CFR60.581) means any device or combination of devices designed to contain, collect, and route organic solvent vapors emitted from the flexible vinyl or urethane rotogravure printing line.

vapor capture system (40CFR60.741) means any device or combination of devices designed to contain, collect, and route solvent vapors released from the coating mix preparation equipment or coating operation.

vapor capture system (EPA-94/04): Any combination of hoods and ventilation system that captures or contains organic vapors so they may be directed to an abatement or recovery device.

vapor collection system (40CFR52.741) means all piping, seals, hoses, connections, pressure-vacuum vents, and other possible sources between the gasoline delivery vessel and the vapor processing unit and/or the storage tanks and vapor holder.

vapor collection system (40CFR60.501) means any equipment used for containing total organic compounds vapors displaced during the loading of gasoline tank trucks.

vapor collection system (40CFR61.301) means any equipment located at the affected facility used for containing benzene vapors displaced during the loading of tank trucks, railcars, or marine vessels. This does not include the vapor collection system that is part of any tank truck, railcar, or marine vessel vapor collection manifold system.

vapor collection system (40CFR63.111), as used in the transfer provisions, means the equipment used to collect and transport organic HAP vapors displaced during the loading of tank trucks or railcars. This does not include the vapor collection system that is part of any tank truck or railcar vapor collection manifold system.

vapor control system (40CFR52.741) means any system that limits or prevents release to the atmosphere of organic material in the vapors displaced from a tank during the transfer of gasoline (under CAA).

vapor dispersion (EPA-94/04): The movement of vapor clouds in air due to wind, thermal action, gravity spreading, and mixing.

vapor incinerator (40CFR264.1031) means any enclosed combustion device that is used for destroying organic compounds and does not extract energy in the form of steam or process heat (under RCRA).

vapor incinerator (40CFR61.131) means any enclosed combustion device that is used for destroying organic compounds and does not necessarily extract energy in the form of steam or process heat.

vapor mounted seal (40CFR60.111a-j) means a foam-filled primary seal mounted continuously around the circumference of the tank so there is an annular vapor space underneath the seal. The annular vapor space is bounded by the bottom of the primary seal, the tank wall, the liquid surface, and the floating roof.

vapor mounted seal (40CFR61.341) means a foam-filled primary seal mounted continuously around the perimeter of a waste management unit so there is an annular vapor space underneath the seal. The annular vapor space is bounded by the bottom of the primary seal, the unit wall, the liquid surface, and the floating roof.

vapor mounted seal (40CFR63.111) means a continuous seal that completely covers the annular space between the wall of the storage vessel or waste management unit and the edge of the floating roof and is mounted such that there is a vapor space between the stored liquid and the bottom of the seal.

vapor plumes (EPA-94/04): Flue gases visible because they contain water droplets.

vapor pressure (40CFR300-AA) means the pressure exerted by the vapor of a substance when it is in equilibrium with its solid or liquid form at a given temperature. For HRS purposes, use the value reported at or near 25°C. [atmosphere or torr].

vapor pressure (40CFR796.1950-iv) is the pressure at which a liquid or solid is in equilibrium with its vapor at a given temperature.

vapor processing system (40CFR60.501) means all equipment used for recovering or oxidizing total organic compounds vapors displaced from the affected facility.

vapor recovery system (40CFR52.741) means a vapor gathering system capable of collecting all VOM vapors and gases discharged from the storage tank and a vapor disposal system capable of processing such VOM vapors and gases so as to prevent their emission to the atmosphere.

vapor recovery system (40CFR60.111-k) means a vapor gathering system capable of collecting all hydrocarbon vapors and gases discharged from the storage vessel and a vapor disposal system capable of processing such hydrocarbon vapors and gases so as to prevent their emission to the atmosphere.

vapor tight gasoline tank truck (40CFR60.501) means a gasoline tank truck which has demonstrated within the 2 preceding months that its product delivery tank will sustain a pressure change of not more than 750 pascals (75 mm of water) within 5 minutes after it is pressurized to 4,500 pascals (450 mm of water). This capability is to be demonstrated using the pressure test procedure specified in Reference Method 27.

vapor tight marine vessel (40CFR61.301) means a marine vessel with a benzene product tank that has been demonstrated within the preceding 12 months to have no leaks. This demonstration shall be made using method 21 of part 60, appendix A, during the last 20 percent of loading and during a period when the vessel is being loaded at its maximum loading rate. A reading of greater than 10,000 ppm as methane shall constitute a leak. As an alternative, a marine vessel owner or operator may use the vapor-tightness test described in 40CFR61.304(f) to demonstrate vapor tightness. A marine vessel operated at negative pressure is assumed to be vapor-tight for the purpose of this standard.

vapor tight tank truck or vapor-tight railcar (40CFR61.301) means a tank truck or railcar for which it has been demonstrated within the preceding 12 months that its product tank will sustain a pressure change of not more than 750 pascals within 5 minutes after it is pressurized to a minimum of 4,500 pascals. This capability is to be demonstrated using the pressure test procedure specified in method

27 of part 60, appendix A, and a pressure measurement device which has a precision of ±2.5 mm water and which is capable of measuring above the pressure at which the tank truck or railcar is to be tested for vapor tightness.

variance (40CFR122.2) means any mechanism or provision under section 301 or 316 of CWA or under 40CFR125, or in the applicable "effluent limitations guidelines" which allows modification to or waiver of the generally applicable effluent limitation requirements or time deadlines of CWA. This includes provisions which allow the establishment of alternative limitations based on fundamentally different factors or on sections 301(c), 301(g), 301(h), 301(i), or 316(a) of CWA.

variance (40CFR51.100-y) means the temporary deferral of a final compliance date for an individual source subject to an approved regulation, or a temporary change to an approved regulation as it applies to an individual source.

variance (EPA-94/04): Government permission for a delay or exception in the application of a given law, ordinance, or regulation.

variance (NPDES) (40CFR124.2) means any mechanism or provision under section 301 or 316 of CWA or under 40CFR part 125, or in the applicable "effluent limitations guidelines" which allows modification to or waiver of the generally applicable effluent limitation requirements or time deadlines of CWA. This includes provisions which allow the establishment of alternative limitations based on fundamentally different factors or on sections 301(c), 301(g), 301(h), 301(i), or 316(a) of CWA.

vector (40CFR240.101-bb) means a carrier, usually an arthropod, that is capable of transmitting a pathogen from one organism to another (other identical or similar definitions are provided in 40CFR241.101-w).

vector (40CFR243.101-ee) means a carrier that is capable of transmitting a pathogen from one organism to another.

vector (EPA-94/04):
1. An organism, often an insect or rodent, that

carries disease.

2. Plasmids, viruses, or bacteria used to transport genes into a host cell. A gene is placed in the vector; the vector then "infects" the bacterium.

vector attraction (40CFR503.31-k) is the characteristic of sewage sludge that attracts rodents, flies, mosquitos, or other organisms capable of transporting infectious agents.

vegetable tan (40CFR425.02) means the process of converting hides into leather using chemicals either derived from vegetable matter or synthesized to produce effects similar to those chemicals.

vegetative controls (EPA-94/04): Nonpoint source pollution control practices that involve vegetative cover to reduce erosion and minimize loss of pollutants.

vehicle (40CFR160.3) means any agent which facilitates the mixture, dispersion, or solubilization of a test substance with a carrier (other identical or similar definitions are provided in 40CFR792.3; 792.226).

vehicle (40CFR205.151-29) means any motorcycle regulated pursuant to this subpart.

vehicle (40CFR205.51-29) means any motor vehicle, machine, or tractor, which is propelled by mechanical power and capable of transportation of property on a street or highway and which has a gross vehicle weight rating in excess of 10,000 pounds and a partially or fully enclosed operator compartment.

vehicle (40CFR52.741) means a device by which any person or property may be propelled, moved, or drawn upon a highway, excepting a device moved exclusively by human power or used exclusively upon stationary rails or tracks.

vehicle (40CFR85.2102-11) means a light duty vehicle or a light duty truck.

vehicle (40CFR86.602.84-6) means any new production light-duty vehicle as defined in subpart A of this part.

vehicle configuration (40CFR600.002.85-24) means a unique combination of basic engine, engine code, inertia weight class, transmission configuration, and axle ratio within a base level.

vehicle configuration (40CFR86.082.2) means a unique combination of basic engine, engine code, inertia weight class, transmission configuration, and axle ratio.

vehicle curb weight (40CFR86.082.2) means the actual or the manufacturer's estimated weight of the vehicle in operational status with all standard equipment, and weight of fuel at nominal tank capacity, and the weight of optional equipment computed in accordance with 40CFR86.02.24; incomplete light-duty trucks shall have the curb weight specified by the manufacturer.

vehicle curb weight, gross vehicle weight rating (GVWR), light-duty truck (LDT), light-duty vehicle, and loaded vehicle weight (LVW) (CAA216-42USC7550) have the meaning provided in regulations promulgated by the Administrator and in effect as of the enactment of the Clean Air Act Amendments of 1990. The abbreviations in parentheses corresponding to any term referred to in this paragraph shall have the same meaning as the corresponding term.

vehicle for hire (40CFR52.731-2) means any chauffeur-driven, spark-ignition-powered motor vehicle used for the purpose of providing transportation for a fee or charge, such as taxicabs and limousine services.

vehicle miles travelled (VMT) (EPA-94/04): A measure of the extent of motor vehicle operation; the total number of vehicle miles travelled within a specific geographic area over a given period of time.

vehicle or engine configuration (40CFR85.2113) means the specific subclassification unit of an engine family as determined by engine displacement, fuel system, engine code, transmission, and inertia weight class as applicable.

vehicle used for motor vehicle manufacturer product evaluations and tests (40CFR88.302.94) means a vehicle that is owned and operated by a motor vehicle manufacturer (as defined in section 216(1) of the Act), or motor vehicle component manufacturer, or owned or held by a university research department, independent testing laboratory, or other such evaluation facility, solely for the purpose of evaluating the performance of such vehicle for engineering, research and development, or quality control reasons.

velocity pressure (vp) (29CFR1910.94b) means the kinetic pressure in the direction of flow necessary to cause a fluid at rest to flow at a given velocity. It is usually expressed in inches of water gauge.

vent (40CFR60.671) means an opening through which there is mechanically induced air flow for the purpose of exhausting from a building air carrying particulate matter emissions from one or more affected facilities.

vent stream (40CFR60.561) means any gas stream released to the atmosphere directly from an emission source or indirectly either through another piece of process equipment or a material recovery device that constitutes part of the normal recovery operations in a polymer process line where potential emissions are recovered for recycle or resale, and any gas stream directed to an air pollution control device. The emissions released from an air pollution control device are not considered a vent stream unless, as noted above, the control device is part of the normal material recovery operations in a polymer process line where potential emissions are recovered for recycle or resale.

vent stream (40CFR60.611) means any gas stream containing nitrogen which was introduced as air to the air oxidation reactor, released the atmosphere directly from any air oxidation reactor recovery train or indirectly, after division through other process equivalent. The vent stream excludes equipment leaks and relief valve discharges including, but not limited to, pumps, compressors, and valves.

vent stream (40CFR60.661) means any gas stream discharged directly from a distillation facility to the atmosphere or indirectly to the atmosphere after diversion through other process equipment. The vent stream excludes relief valve discharges and equipment leaks including, but not limited to, pumps, compressors, and valves.

vent stream (40CFR60.701) means any gas stream discharged directly from a reactor process to the atmosphere or indirectly to the atmosphere after diversion through other process equipment. The vent stream excludes relief valve discharges and equipment leaks.

vent stream (40CFR63.111), as used in the process vent provisions, means a process vent as defined in 40CFR63.101 of subpart F of this part.

vented (40CFR264.1031) means discharged through an opening, typically an open-ended pipe or stack, allowing the passage of a stream of liquids, gases, or fumes into the atmosphere. The passage of liquids, gases, or fumes is caused by mechanical means such as compressors or vacuum-producing systems or by process-related means such as evaporation produced by heating and not caused by tank loading and unloading (working losses) or by natural means such as diurnal temperature changes.

ventilation/suction (EPA-94/04): The act of admitting fresh air into a space in order to replace stale or contaminated air; achieved by blowing air into the space. Similarly, suction represents the admission of fresh air into an interior space by lowering the pressure outside of the space, thereby drawing the contaminated air outward.

venturi scrubbers (EPA-94/04): Air pollution control devices that use water to remove particulate matter from emissions.

vertical spindle disc grinder (29CFR1910.94b) means a grinding machine having a vertical, rotatable, power-driven spindle carrying a horizontal abrasive disc wheel.

very high pressure appliance (40CFR82.152-y) means an appliance that uses a refrigerant with a boiling point below -50 degrees Centigrade at atmospheric pressure (29.9 inches of mercury). This definition includes but is not limited to equipment utilizing refrigerants -13 and -503.

very large MWC plant (40CFR60.31a) means an MWC plant with an MWC plant capacity greater than 1,000 megagrams per day (1,100 tons per day) of MSW (cf. MWC plant).

very low sulfur oil (40CFR60.41b) means an oil that contains no more than 0.5 weight percent sulfur or that, when combusted without sulfur dioxide emission control, has a sulfur dioxide emission rate equal to or less than 215 ng/J (0.5 lb/million Btu) heat input.

vessel (40CFR110.1) means every description of watercraft or other artificial contrivance used, or capable of being used, as a means of transportation on water other than a public vessel (other identical or similar definitions are provided in 40CFR112.2; 116.3).

vessel (40CFR140.1-d) includes every description of watercraft or other artificial contrivance used, or capable of being used, as a means of transportation on waters of the United States.

vessel (40CFR21.2-n) means every description of watercraft or other artificial contrivance used, or capable of being used, as a means of transportation on the navigable waters of the Unite States other than a vessel owned or operated by the United States or a State or a political subdivision thereof, or a foreign nation; and is used for commercial purposes by a small business concern.

vessel (40CFR260.10) includes every description of watercraft, used or capable of being used as a means of transportation on the water.

vessel (40CFR300.5), as defined by section 101(28) of CERCLA, means every description of watercraft or other artificial contrivance used, or capable of being used, as a means of transportation on water; and, as defined by section 311(a)(3) of the CWA, means every description of watercraft or other artificial contrivance used, or capable of being used, as a means of transportation on water other than a public vessel.

vessel (40CFR302.3) means every description of watercraft or other artificial contrivance used, or capable of being used, as a means of transportation on water.

vessel (CWA311-33USC1321) means every description of watercraft or other artificial contrivance used, or capable of being used, as a means of transportation on water other than a public vessel (other identical or similar definitions are provided in OPA1001).

vessel (SF101-42USC9601) means every description of watercraft or other artificial contrivance used, or capable of being used, as a means of transportation on water.

viable embryos (fertility) (40CFR797.2130-iv) are eggs in which fertilization has occurred and embryonic development has begun. This is determined by candling the egg 11 days after incubation has begun. It is difficult to distinguish between the absence of fertilization and early embryonic death. The distinction can be made by breaking out eggs that appear infertile and examining further. This distinction is especially important when a test compound induces early embryo mortality. Values are expressed as a percentage of eggs set.

viable embryos (fertility) (40CFR797.2150-iv) are eggs in which fertilization has occurred and embryonic development has begun. This is determined by candling the eggs 14 days after incubation has begun. It is difficult to distinguish between the absence of fertilization and early embryonic death. The distinction can be made by breaking out eggs that appear infertile and examining further. This distinction is especially important when a test compound induces early embryo mortality. Values are expressed as a percentage of eggs set.

vibration (40CFR763.83) means the periodic motion of friable ACBM which may result in the release of asbestos fibers.

vinyl asbestos floor tile (40CFR763.163) means an asbestos-containing product composed of vinyl resins and used as floor tile.

vinyl chloride (EPA-94/04): A chemical compound, used in producing some plastics, that is believed to be oncogenic.

vinyl chloride plant (40CFR61.61-b) includes any plant which produces vinyl chloride by any process (under CAA).

vinyl chloride purification (40CFR61.61-p) includes

any part of the process of vinyl chloride production which follows vinyl chloride formation.

vinyl coating (40CFR52.741) means any topcoat or printing ink applied to vinyl coated fabric or vinyl sheets. Vinyl coating does not include plastisols.

vinyl coating facility (40CFR52.741) means a facility that includes one or more vinyl coating line(s) (under CAA).

vinyl coating line (40CFR52.741) means a coating line in which any protective, decorative or functional coating is applied onto vinyl coated fabric or vinyl sheets.

violating facility (40CFR30.200) means any facility that is owned, leased, or supervised by an applicant, recipient, contractor, or subcontractor that EPA lists under 40CFR15 as not in compliance with Federal, State, or local requirements under the Clean Air Act or Clean Water Act. A facility includes any building, plant, installation, structure, mine, vessel, or other floating craft.

virgin material (40CFR246.101-gg) means a raw material used in manufacturing that has been mined or harvested and has not as yet become a product (other identical or similar definitions are provided in 40CFR247.101-l).

virgin material (RCRA1004-42USC6903) means a raw material, including previously unused copper, aluminum, lead, zinc, iron, or other metal or metal ore, any undeveloped resource that is, or with new technology will become, a source of raw materials.

virgin materials (EPA-94/04): Resources extracted from nature in their raw form, such as timber or metal ore.

virus (40CFR141.2) means a virus of fecal origin which is infectious to humans by waterborne transmission.

viscose process (40CFR60.601) means the fiber forming process where cellulose and concentrated caustic soda are reacted to form soda or alkali cellulose. This reacts with carbon disulfide to form sodium cellulose xanthate, which is then dissolved in a solution of caustic soda. After ripening, the solution is spun into an acid coagulating bath. This precipitates the cellulose in the form of a regenerated cellulose filament.

visibility impairment (40CFR51.301-x) means any humanly perceptible change in visibility (visual range, contrast, coloration) from that which would have existed under natural conditions.

visibility impairment and impairment of visibility (CAA169A-42USC7491) shall include reduction in visual range and atmospheric discoloration.

visibility in any mandatory class I federal area (40CFR51.301-y) includes any integral vista associated with that area.

visibility protection area (40CFR52.26-1) means any area listed in 40CFR81.401-81.436 (1984) (other identical or similar definitions are provided in 40CFR52.28-1).

visible emission (40CFR63.2) means the observation of an emission of opacity or optical density above the threshold of vision.

visible emissions (40CFR61.141) means any emissions, which are visually detectable without the aid of instruments, coming from RACM or asbestos-containing waste material, or from any asbestos milling, manufacturing, or fabricating operation. This does not include condensed, uncombined water vapor.

visible specific locus mutation (40CFR798.5200-1) is a genetic change that alters factors responsible for coat color and other visible characteristics of certain mouse strains.

VOC (40CFR58.1-x) means volatile organic compound (other identical or similar definitions are provided in 40CFR60.431; 60.441).

VOC (CAA302-42USC7602): Volatile organic compound, as defined by the Administrator.

VOC content (40CFR60.311) means the proportion of a coating that is volatile organic compounds (VOCs), expressed as kilograms of VOC's per liter of coating solids (other identical or similar definitions are provided in 40CFR60.451).

VOC content (40CFR60.391) means all volatile organic compounds that are in a coating expressed as kilograms of VOC per liter of coating solids.

VOC content (40CFR60.461) means the quantity, in kilograms per liter of coating solids, of volatile organic compounds (VOC's) in a coating.

VOC content (40CFR60.491-6) means all volatile organic compounds (VOC) that are in a coating. VOC content is expressed in terms of kilograms of VOC per litre of coating solids.

VOC content of the coating applied (40CFR60.711-19) means the product of Method 24 VOC analyses or formulation data (if the data are demonstrated to be equivalent to Method 24 results) and the total volume of coating fed to the coating applicator. This quantity is intended to include all VOC that actually are emitted from the coating operation in the gaseous phase. Thus, for purposes of the liquid-liquid VOC material balance in 40CFR60.713(b)(1), any VOC (including dilution solvent) added to the coatings must be accounted for, and any VOC contained in waste coatings or retained in the final product may be measured and subtracted from the total. (These adjustments are not necessary for the gaseous emission test compliance provisions of 40CFR60.713(b).)

VOC emission control device (40CFR60.541) means equipment that destroys or recovers VOC.

VOC emission reduction system (40CFR60.541) means a system composed of an enclosure, hood, or other device for containment and capture of VOC emissions and a VOC emission control device.

VOC emissions (40CFR60.311) means the mass of volatile organic compounds (VOCs), expressed as kilograms of VOCs per liter of applied coating solids, emitted from a metal furniture surface coating operation (other identical or similar definitions are provided in 40CFR60.451).

VOC emissions (40CFR60.721) means the mass of VOCs emitted from the surface coating of plastic parts for business machines expressed as kilograms of VOCs per liter of coating solids applied (i.e., deposited on the surface).

VOC in the applied coating (40CFR60.741) means the product of Method 24 VOC analyses or formulation data (if those data are demonstrated to be equivalent to Method 24 results) and the total volume of coating fed to the coating applicator.

VOC solvent (40CFR60.431) means an organic liquid or liquid mixture consisting of VOC components.

VOC used (40CFR60.741) means the amount of VOC delivered to the coating mix preparation equipment of the affected facility (including any contained in premixed coatings or other coating ingredients prepared off the plant site) for the formulation of polymeric coatings to be applied to supporting substrates at the coating operation, plus any solvent added after initial formulation is complete (e.g., dilution solvent added at the coating operation). If premixed coatings that require no mixing at the plant site are used, "VOC used" means the amount of VOC delivered to the coating applicator(s) of the affected facility.

volatile (EPA-94/04): Any substance that evaporates readily.

volatile hazardous air pollutant or VHAP (40CFR61.241) means a substance regulated under this part for which a standard for equipment leaks of the substance has been proposed and promulgated. Benzene is a VHAP. Vinyl chloride is a VHAP (under CAA).

volatile liquids (EPA-94/04): Liquids which easily vaporize or evaporate at room temperature.

volatile organic compound (40CFR60.2) means any organic compound which participates in atmospheric photochemical reactions; or which is measured by a reference method, an equivalent method, an alternative method, or which is determined by procedures specified under any subpart.

volatile organic compound (VOC) (EPA-94/04): Any organic compound that participates in atmospheric photochemical reactions except those designated by EPA as having negligible photochemical reactivity.

volatile organic compounds (VOC) (40CFR51.100-s) means any compound of carbon, excluding carbon monoxide, carbon dioxide, carbonic acid, metallic carbides or carbonates, and ammonium carbonate, which participates in atmospheric photochemical reactions:

(1) This includes any such organic compound other than the following, which have been determined to have negligible photochemical reactivity:
- methane;
- ethane;
- methylene chloride (dichloromethane);
- 1,1,1-trichloroethane (methyl chloroform);
- 1,1,2-trichloro-1,2,2-trifluoroethane (CFC-113);
- trichlorofluoromethane (CFC-11);
- dichlorodifluoromethane (CFC-12);
- chlorodifluoromethane (HCFC-22);
- trifluoromethane (HFC-23);
- 1,2-dichloro 1,1,2,2-tetrafluoroethane (CFC-114);
- chloropentafluoroethane (CFC-115);
- 1,1,1-trifluoro 2,2-dichloroethane (HCFC-123);
- 1,1,1,2-tetrafluoroethane (HFC-134a);
- 1,1-dichloro 1-fluoroethane (HCFC-141b);
- 1-chloro 1,1-difluoroethane (HCFC 142b);
- 2-chloro-1,1,1,2-tetrafluoroethane (HCFC-124);
- pentafluoroethane (HFC-125);
- 1,1,2,2-tetrafluoroethane (HFC-134);
- 1,1,1-trifluoroethane (HFC-143a);
- 1,1-difluoroethane (HFC-152a);
- parachlorobenzotrifluoride (PCBTF);
- cyclic, branched, or linear completely methylated siloxanes;
- acetone; and
- perfluorocarbon compounds which fall into these classes:
  (i) Cyclic, branched, or linear, completely fluorinated alkanes;
  (ii) Cyclic, branched, or linear, completely fluorinated ethers with no unsaturations;
  (iii) Cyclic, branched, or linear, completely fluorinated tertiary amines with no unsaturations; and

  (iv) Sulfur containing perfluorocarbons with no unsaturations and with sulfur bonds only to carbon and fluorine.

(2) For purposes of determining compliance with emissions limits, VOC will be measured by the test methods in the approved State implementation plan (SIP) or 40CFR part 60, appendix A, as applicable. Where such a method also measures compounds with negligible photochemical reactivity, these negligibility-reactive compounds may be excluded as VOC if the amount of such compounds is accurately quantified, and such exclusion is approved by the enforcement authority.

(3) As a precondition to excluding these compounds as VOC or at any time thereafter, the enforcement authority may require an owner or operator to provide monitoring or testing methods and results demonstrating, to the satisfaction of the enforcement authority, the amount of negligibly-reactive compounds in the source's emissions.

(4) For purposes of Federal enforcement for a specific source, the EPA shall use the test methods specified in the applicable EPA-approved SIP, in a permit issued pursuant to a program approved or promulgated under title V of the Act, or under 40CFR part 51, subpart I or appendix S, or under 40CFR parts 52 or 60. The EPA shall not be bound by any State determination as to appropriate methods for testing or monitoring negligibly-reactive compounds if such determination is not reflected in any of the above provisions.

volatile organic compounds (VOC) (40CFR51.165-xix) is defined in 40CFR51.100(s).

volatile organic compounds (VOC) (40CFR51.166-29) is as defined in 40CFR51.100(s) of this part (under CAA).

volatile organic compounds (VOC) (40CFR51-AS-20) is as defined in 40CFR51.100(s) of this part.

volatile organic compounds (VOC) (40CFR52.21-30) is as defined in 40CFR51.100(s) of this chapter.

volatile organic compounds (VOC) (40CFR52.24-18) is as defined in 40CFR51.100(s) of this chapter.

volatile organic compounds (VOC) (40CFR60.561) means, for the purposes of these standards, any reactive organic compounds as defined in 40CFR60.2 definitions.

volatile organic compounds or VOC (40CFR60.481) means, for the purposes of this subpart, any reactive organic compounds as defined in 40CFR60.2 definitions.

volatile organic compounds or VOC (40CFR60.711-20) means any organic compounds that participate in atmospheric photochemical reactions or that are measured by Method 18, 24, 25, or 25A or an equivalent or alternative method as defined in 40CFR60.2.

volatile organic compounds or VOC (40CFR60.741) means any organic compounds that participate in atmospheric photochemical reactions; or that are measured by a reference method, an equivalent method, an alternative method, or that are determined by procedures specified under any subpart.

volatile organic concentration or VO concentration (40CFR63.111) refers to the concentration of organic compounds (including both HAP and non-HAP organic compounds) in a wastewater stream that is measured by Method 25D, as found in 40CFR part 60, appendix A.

volatile organic hazardous air pollutant concentration or VOHAP concentration (40CFR63.111) means the concentration of an individually-speciated organic HAP in a wastewater stream or a residual that is measured by Method 305 in appendix A of this part.

volatile organic liquid (40CFR52.741) means any substance which is liquid at storage conditions and which contains volatile organic compounds.

volatile organic liquid (VOL) (40CFR60.111b-k) means any organic liquid which can emit volatile organic compounds into the atmosphere except those VOL's that emit only those compounds which the Administrator has determined do not contribute appreciably to the formation of ozone. These compounds are identified in EPA statements on ozone abatement policy for SIP revisions (42 FR 35314, 44 FR 32042, 45 FR 32424, and 45 FR 48941).

volatile organic material (VOM) or volatile organic compound (VOC) (40CFR52.741) is as defined in 40CFR51.100(s) of this chapter.

volatile solids (40CFR503.31-l) is the amount of the total solids in sewage sludge lost when the sewage sludge is combusted at 550 degrees Celsius in the presence of excess air.

volatile solids (EPA-94/04): Those solids in water or other liquids that are lost on ignition of the dry solids at 550 degrees Centigrade.

volatile synthetic organic chemicals (EPA-94/04): Chemicals that tend to volatilize or evaporate.

volatilization (40CFR300-AA) means a physical transfer process through which a substance undergoes a change of state from a solid or liquid to a gas.

volatilization (40CFR796.1950-v) is the loss of a substance to the air from a surface or from solution by evaporation.

volume of process water used per year (40CFR463.11-b) is the volume of process water that flows through a contact cooling and heating water process and comes in contact with the plastic product over a period of one year.

volume of process water used per year (40CFR463.21-b) is the volume of the process water that flows through a cleaning process and comes in contact with the plastic product over a period of one year.

volume of process water used per year (40CFR463.31-b) is the volume of process water that flows through a finishing water process and comes in contact with the plastics product over a period of one year.

volume records (40CFR80.126-g) shall include summaries of gasoline produced or imported that account for the volume of each type of gasoline produced or imported. The volumes shall be based on tank gauges or meter reports and temperature adjusted to 60 degrees Fahrenheit.

volume reduction (EPA-94/04): Processing waste materials to decrease the amount of space they occupy, usually by compacting or shredding, incineration, or composting.

volumetric flow (40CFR72.2) means the rate of movement of a specified volume of gas past a cross-sectional area (e.g., cubic feet per hour).

volumetric tank test (EPA-94/04): One of several tests to determine the physical integrity of a storage tank; the volume of fluid in the tank is measured directly or calculated from product-level changes. A marked drop in volume indicates a leak.

voluntarily submitted information (40CFR2.201-i) means business information in EPA's possession:
(1) The submission of which EPA had no statutory or contractual authority to require; and
(2) The submission of which was not prescribed by statute or regulation as a condition of obtaining some benefit (or avoiding some disadvantage) under a regulatory program of general applicability, including such regulatory programs as permit, licensing, registration, or certification programs, but excluding programs concerned solely or primarily with the award or administration by EPA of contracts or grants.

voluntary emissions recall (40CFR85.1902-d) shall mean a repair, adjustment, or modification program voluntarily initiated and conducted by a manufacturer to remedy any emission-related defect for which direct notification of vehicle or engine owners has been provided.

voluntary exclusion or voluntarily excluded (40CFR32.105-v): A status of nonparticipation or limited participation in covered transactions assumed by a person pursuant to the terms of a settlement.

volunteer (40CFR300.5) means any individual accepted to perform services by the lead agency which has authority to accept volunteer services (examples: See 16USC742f(c)). A volunteer is subject to the provisions of the authorizing statute and the NCP.

vulnerability analysis (EPA-94/04): Assessment of elements in the community that are susceptible to damage should a release of hazardous materials occur.

vulnerable zone (EPA-94/04): An area over which the airborne concentration of a chemical accidentally released could reach the level of concern.

********** WWWWW **********

waiver (40CFR13.2-l) means the cancellation, remission, forgiveness, or non-recovery of a debt or debt-related charge as permitted or required by law.

waiver (40CFR35.4010) means excusing recipients from following certain anticipated regulatory or administrative requirements if the authority to issue a waiver is provided in the regulation itself; and the Agency believes sufficient justification exists to approve such action. The Award Official has the authority to issue a waiver. Deviation means an exemption from certain provisions of existing regulations, which may be necessary in some unforeseen instances. The Director, Grants Administration Division, is authorized under 40CFR30.1001(b) to approve deviations from the requirements of regulations (except for those that implement statutory or executive order requirements) when such situations warrant special consideration.

wall fired boiler (40CFR76.2) means a boiler that has pulverized coal burners arranged on the walls of the furnace. The burners have discrete, individual flames that extend perpendicularly into the furnace area.

wall fired boilers (EPA-94/04): Coal-fired furnaces in which burners are installed and fired on opposite walls of the unit.

wall insulation (40CFR248.4-mm) means a material,

primarily designed to resist heat flow, which is installed within or on the walls between conditioned areas of a building and unconditioned areas of a building or the outside, as well as common wall assemblies between separately conditioned units in multiple unit structures.

warning device (40CFR201.1-hh) means a sound emitting device used to alert and warn people of the presence of railroad equipment.

warning label (40CFR82.104-x) means the warning statement required by section 611 of the Act. The term warning statement shall be synonymous with warning label for purposes of this subpart.

warranty (40CFR204.2-8) means the warranty required by section 6(c)(1) of the Act (other identical or similar definitions are provided in 40CFR205.2-22).

warranty (40CFR205.151-30) means the warranty required by section 6(d)(1) of the Act.

warranty booklet (40CFR85.2102-12) means booklet, separate from the owner's manual, containing all warranties provided with the vehicle.

wash coat (40CFR52.741) means a coating containing binders which seals wood surfaces, prevents undesired staining, and controls penetration.

wash oil circulation tank (40CFR61.131) means any vessel that functions to hold the wash oil used in light-oil recovery operations or the wash oil used in the wash-oil final cooler.

wash oil decanter (40CFR61.131) means any vessel that functions to separate, by gravity, the condensed water from the wash oil received from a wash-oil final cooler or from a light-oil scrubber.

washer (40CFR60.621) means a machine which agitates fabric articles in a petroleum solvent bath and spins the articles to remove the solvent, together with the piping and ductwork used in the installation of this device.

washer (40CFR63.321) means a machine used to clean articles by immersing them in perchloroethylene. This includes a dry-to-dry machine when used with a reclaimer.

washout (40CFR257.3.1-3) means the carrying away of solid waste by waters of the base flood.

washout (40CFR264.18-(b)(2)(ii)) means the movement of hazardous waste from the active portion of the facility as a result of flooding.

waste (40CFR191.12), as used in this subpart, means any spent nuclear fuel or radioactive waste isolated in a disposal system.

waste (40CFR60.111b-l) means any liquid resulting from industrial, commercial, mining or agricultural operations, or from community activities that is discarded or is being accumulated, stored, or physically, chemically, or biologically treated prior to being discarded or recycled.

waste (40CFR61.341) means any material resulting from industrial, commercial, mining or agricultural operations, or from community activities that is discarded or is being accumulated, stored, or physically, chemically, thermally, or biologically treated prior to being discarded, recycled, or discharged.

waste (40CFR704.43-6) means any solid liquid, semisolid, or contained gaseous material that results from the production of a chemical substance identified in paragraph (b) of this section and which is to be disposed.

waste (EPA-94/04):
1. Unwanted materials left over from a manufacturing process.
2. Refuse from places of human or animal habitation.

waste category (40CFR259.10.b) means either untreated regulated medical waste or treated regulated medical waste.

waste characterization (EPA-94/04): Identification of chemical and microbiological constituents of a waste material.

waste exchange (EPA-94/04): Arrangement in which

companies exchange their wastes for the benefit of both parties.

waste feed (EPA-94/04): The continuous or intermittent flow of wastes into an incinerator.

waste form (40CFR191.12) means the materials comprising the radioactive components of waste and any encapsulating or stabilizing matrix.

waste generator (40CFR61.141) means any owner or operator of a source covered by this subpart whose act or process produces asbestos-containing waste material.

waste load allocation (EPA-94/04): The maximum load of pollutants each discharger of waste is allowed to release into a particular waterway. Discharge limits are usually required for each specific water quality criterion being, or expected to be, violated. The portion of a stream's total assimilative capacity assigned to an individual discharge.

waste load allocation (WLA) (40CFR130.2-h) means the portion of a receiving water's loading capacity that is allocated to one of its existing or future point sources of pollution. WLAs constitute a type of water quality-based effluent limitation.

waste management unit (40CFR61.341) means a piece of equipment, structure, or transport mechanism used in handling, storage, treatment, or disposal of waste. Examples of a waste management unit include a tank, surface impoundment, container, oil-water separator, individual drain system, steam stripping unit, thin-film evaporation unit, waste incinerator, and landfill.

waste management unit (40CFR63.111) means any component, piece of equipment, structure, or transport mechanism used in conveying, storing, treating, or disposing of wastewater streams or residuals. Examples of waste management units include wastewater tanks, air flotation units, surface impoundments, containers, oil-water or organic-water separators, individual drain systems, biological treatment units, waste incinerators, and organic removal devices such as decanters, steam and air stripper units, and thin-film evaporation units (under CAA).

waste management unit boundary (40CFR258.2) means a vertical surface located at the hydraulically downgradient limit of the unit. This vertical surface extends down into the uppermost aquifer (under RCRA).

waste minimization (EPA-94/04): Measures or techniques that reduce the amount of wastes generated during industrial production processes; term is also applied to recycling and other efforts to reduce the amount of waste going into the waste stream.

waste oil (40CFR761.3) means used products primarily derived from petroleum, which include, but are not limited to, fuel oils, motor oils, gear oils, cutting oils transmission fluids, hydraulic fluids, and dielectric fluids.

waste paper (40CFR250.4-ss) means any of the following "recovered materials":
(1) Postconsumer materials such as:
  (i) Paper, paperboard, and fibrous wastes from retail stores, office buildings, homes, and so forth, after they have passed through their end usage as a consumer item, including: Used corrugated boxes, old newspapers, old magazines, mixed waste paper, tabulating cards, and used cordage, and
  (ii) All paper, paperboard, and fibrous wastes that enter and are collected from municipal solid waste; and
(2) Manufacturing, forest residues, and other wastes such as:
  (i) Dry paper and paperboard waste generated after completion of the papermaking process (that is, those manufacturing operations up to and including the cutting and trimming of the paper machine reel into smaller rolls or rough sheets) including: Envelope cuttings, bindery trimmings, and other paper and paperboard waste, resulting from printing, cutting, forming, and other converting operations, bag, box, and carbon manufacturing wastes, and butt rolls, mill wrappers, and rejected unused stock; and
  (ii) Finished paper and paperboard from obsolete inventories of paper and

paperboard manufacturers, merchants, wholesalers, dealers, printers, converters, or others.

waste reduction (EPA-94/04): Using source reduction, recycling, or composting to prevent or reduce waste generation.

waste shipment record (40CFR61.141) means the shipping document, required to be originated and signed by the waste generator, used to track and substantiate the disposition of asbestos-containing waste material.

waste stream (40CFR61.341) means the waste generated by a particular process unit, product tank, or waste management unit. The characteristics of the waste stream (e.g., flow rate, benzene concentration, water content) are determined at the point of waste generation. Examples of a waste stream include process wastewater, product tank drawdown, sludge and slop oil removed from waste management units, and landfill leachate.

waste stream (EPA-94/04): The total flow of solid waste from homes, businesses, institutions, and manufacturing plants that are recycled, burned, or disposed of in landfills, or segments thereof such as the "residential waste stream" or the "recyclable waste stream."

waste treatment lagoon (EPA-94/04): Impoundment made by excavation or earth fill for biological treatment of wastewater.

waste treatment plant (EPA-94/04): A facility containing a series of tanks, screens, filters and other processes by which pollutants are removed from water.

waste treatment stream (EPA-94/04): The continuous movement of waste from generator to treater and disposer.

waste treatment systems (40CFR122.2), including treatment ponds or lagoons designed to meet the requirements of CWA (other than cooling ponds as defined in 40CFR423.11(m) which also meet the criteria of this definition) are not waters of the United States. This exclusion applies only to manmade bodies of water which neither were originally created in waters of the United States (such as disposal area in wetlands) nor resulted from the impoundment of waters of the United States. [See Note 1 of this section.] Waters of the United States do not include prior converted cropland. Notwithstanding the determination of an area's status as prior converted cropland by any other federal agency, for the purposes of the Clean Water Act, the final authority regarding Clean Water Act jurisdiction remains with EPA.

waste treatment systems (40CFR232.2), including treatment ponds or lagoons designed to meet the requirements of the Act (other than cooling ponds as defined in 40CFR123.11(m) which also meet the criteria of this definition) are not waters of the United States.

wastewater (40CFR63.101) means organic hazardous air pollutant-containing water, raw material, intermediate, product, byproduct, co-product, or waste material that exits equipment in a chemical manufacturing process unit that meets all of the criteria specified in 40CFR63.100(b)(1) through (b)(3) of this subpart; and either:
(1) Contains a total volatile organic hazardous air pollutant concentration of at least 5 parts per million by weight and has a flow rate of 0.02 liter per minute or greater; or
(2) Contains a total volatile organic hazardous air pollutant concentration of at least 10,000 parts per million by weight at any flow rate. Wastewater includes process wastewater and maintenance wastewater.

wastewater (EPA-94/04): The spent or used water from a home, community, farm, or industry that contains dissolved or suspended matter.

wastewater infrastructure (EPA-94/04): The plan or network for the collection, treatment, and disposal of sewage in a community. The level of treatment will depend on the size of the community, the type of discharge, and/or the designated use of the receiving water.

wastewater operations and maintenance (EPA-94/04): Actions taken after construction to assure that facilities constructed to treat wastewater will be operated, maintained, and managed to reach prescribed effluent levels in an optimum manner.

wastewater stream (40CFR63.111) means a stream that contains only wastewater as defined in 40CFR63.101 of subpart F of this part.

wastewater system (40CFR60.691) means any component, piece of equipment, or installation that receives, treats, or processes oily wastewater from petroleum refinery process units.

wastewater tank (40CFR63.111) means a stationary waste management unit that is designed to contain an accumulation of wastewater or residuals and is constructed primarily of non-earthen materials (e.g., wood, concrete, steel, plastic) which provide structural support. Wastewater tanks used for flow equalization are included in this definition (under CAA).

wastewater treatment process (40CFR61.61-k) includes any process which modifies characteristics such as BOD, COD, TSS, and pH, usually for the purpose of meeting effluent guidelines and standards; it does not include any process the purpose of which is to remove vinyl chloride from water to meet requirements of this subpart.

wastewater treatment system (40CFR61.341) means any component, piece of equipment, or installation that receives, manages, or treats process wastewater, product tank drawdown, or landfill leachate prior to direct or indirect discharge in accordance with the National Pollutant Discharge Elimination System permit regulations under 40CFR122. These systems typically include individual drain systems, oil-water separators, air flotation units, equalization tanks, and biological treatment units.

wastewater treatment tank (40CFR280.12) means a tank that is designed to receive and treat an influent wastewater through physical, chemical, or biological methods.

wastewater treatment unit (40CFR260.10) means a device which:
(1) Is part of a wastewater treatment facility that is subject to regulation under either section 402 or 307(b) of the Clean Water Act; and

(2) Receives and treats or stores an influent wastewater that is a hazardous waste as defined in 40CFR261.3 of this chapter, or that generates and accumulates a wastewater treatment sludge that is a hazardous waste as defined in 40CFR261.3 of this chapter, or treats or stores a wastewater treatment sludge which is a hazardous waste as defined in 40CFR261.3 of this Chapter; and
(3) Meets the definition of tank or tank system in 40CFR260.10 of this chapter.

wastewater treatment unit (40CFR270.2) means a device which:
(a) Is part of a wastewater treatment facility which is subject to regulation under either section 402 or section 307(b) of the Clean Water Act; and
(b) Receives and treats or stores an influent wastewater which is a hazardous waste as defined in 40CFR261.3 of this chapter, or generates and accumulates a wastewater treatment sludge which is a hazardous waste as defined in 40CFR261.3 of this chapter, or treats or stores a wastewater treatment sludge which is a hazardous waste as defined in 40CFR261.3 of this chapter; and
(c) Meets the definition of tank or tank system in 40CFR260.10 of this chapter.

wastewaters (40CFR268.2-f) are wastes that contain less than 1% by weight total organic carbon (TOC) and less than 1% by weight total suspended solids (TSS), with the following exceptions:
(1) F001, F002, F003, F004, F005, wastewaters are solvent-water mixtures that contain less than 1% by weight TOC or less than 1% by weight total F001, F002, F003, F004, F005 solvent constituents listed in 40CFR 268.41, Table CCWE.
(2) K011, K013, K014 wastewaters contain less than 5% by weight TOC and less than 1% by weight TSS, as generated.
(3) K103 and K104 wastewaters contain less than 4% by weight TOC and less than 1% by weight TSS.

water (bulk shipment) (40CFR260.10) means the bulk transportation of hazardous waste which is loaded or carried on board a vessel without containers or labels.

water based green tire spray (40CFR60.541) means any mold release agent and lubricant applied to the inside or outside of green tires that contains 12 percent or less, by weight, of VOC as sprayed.

water dumping (40CFR165.1-z) means the disposal of pesticides in or on lakes, ponds, rivers, sewers, or other water systems as defined in Pub. L. 92-500.

water jet weaving (40CFR410.31-b) shall mean the internal subdivision of the low water use processing subcategory for facilities primarily engaged in manufacturing woven greige goods through the water jet weaving process.

Water Management Division Director (40CFR403.3-f) means one of the Directors of the Water Management Divisions within the Regional offices of the Environmental Protection Agency or this person's delegated representative.

water pollution (EPA-94/04): The presence in water of enough harmful or objectionable material to damage the water's quality.

water pollution control agency (40CFR15.4) means any agency which is defined in section 502(1) or section 502(2), 33USC1362(1) or (2), of the CWA.

water purveyor (EPA-94/04): An agency or person that supplies water (usually potable water).

water quality criteria (EPA-94/04): Levels of water quality expected to render a body of water suitable for its designated use. Criteria are based on specific levels of pollutants that would make the water harmful if used for drinking, swimming, farming, fish production, or industrial processes.

water quality limited segment (40CFR130.2-j) means any segment where it is known that water quality does not meet applicable water quality standards, and/or is not expected to meet applicable water quality standards, even after the application of the technology-based effluent limitations required by sections 301(b) and 306 of the Act.

water quality limited segment (40CFR131.3-h) means any segment where it is known that water quality does not meet applicable water quality standards,

and/or is not expected to meet applicable water quality standards, even after the application of the technology-based effluent limitations required by sections 301(b) and 306 of the Act.

water quality management (WQM) plan (40CFR130.2-k) means a State or areawide waste treatment management plan developed and updated in accordance with the provisions of sections 205(j), 208 and 303 of the Act and this regulation.

water quality standards (40CFR121.1-g) means standards established pursuant to section 10(c) of the Act, and State-adopted water quality standards for navigable waters which are not interstate waters.

water quality standards (40CFR125.58-v) means applicable water quality standards which have been approved, left in effect, or promulgated under section 303 of the Clean Water Act.

water quality standards (40CFR131.3-i) are provisions of State or Federal law which consist of a designated use or uses for the waters of the United States and water quality criteria for such waters based upon such uses. Water quality standards are to protect the public health or welfare, enhance the quality of water and serve the purposes of the Act.

water quality standards (EPA-94/04): State-adopted and EPA-approved ambient standards for water bodies. The standards prescribe the use of the water body and establish the water quality criteria that must be met to protect designated uses.

water quality standards (WQS) (40CFR130.2-d) means provisions of State or Federal law which consist of a designated use or uses for the water quality criteria for such waters based upon such uses. Water quality standards are to protect the public health or welfare, enhance the quality of water and serve the purposes of the Act.

water quality-based limitations (EPA-94/04): Effluent limitations applied to dischargers when mere technology-based limitations would cause violations of water quality standards. Usually applied to discharges into small streams.

water quality-based permit (EPA-94/04): A permit

with an effluent limit more stringent than one based on technology performance. Such limits may be necessary to protect the designated use of receiving waters (i.e., recreation, irrigation, industry or water supply).

water reducible (40CFR60.391): See waterborne or water reducible.

water seal controls (40CFR60.691) means a seal pot, p-leg trap, or other type of trap filled with water that has a design capability to create a water barrier between the sewer and the atmosphere.

water seal controls (40CFR61.341) means a seal pot, p-leg trap, or other type of trap filled with water that has a design capability to create a water barrier between the sewer line and the atmosphere (under CAA).

water seal controls (40CFR63.111) means a seal pot, p-leg trap, or other type of trap filled with water (e.g, flooded sewers that maintain water levels adequate to prevent air flow through the system) that creates a water barrier between the sewer line and the atmosphere. The water level of the seal must be maintained in the vertical leg of a drain in order to be considered a water seal.

water separator (40CFR63.321) means any device used to recover perchloroethylene from a water-perchloroethylene mixture.

water solubility (40CFR300-AA) means the maximum concentration of a substance in pure water at a given temperature. For HRS purposes, use the value reported at or near 25°C. [milligrams per liter (mg/L)].

water solubility (EPA-94/04): The maximum possible concentration of a chemical compound dissolved in water. If a substance is water soluble it can very readily disperse through the environment.

water storage pond (EPA-94/04): An impound for liquid wastes designed to accomplish some degree of biochemical treatment.

water supplier (EPA-94/04): One who owns or operates a public water system.

water supply system (EPA-94/04): The collection, treatment, storage, and distribution of potable water from source to consumer.

water table (40CFR241.101-x) means the upper water level of a body of groundwater (under RCRA).

water table (EPA-94/04): The level of groundwater.

water treatment chemicals (40CFR749.68-16) means any combination of chemical substances used to treat water in cooling systems and can include corrosion inhibitors, antiscalants, dispersants, and any other chemical substances except biocides.

water treatment lagoon (EPA-94/04): An impound for liquid wastes designed to accomplish some degree of biochemical treatment.

water use (CZMA304-16USC1453) means activities which are conducted in or on the water; but does not mean or include the establishment of any water quality standard or criteria or the regulation of the discharge or runoff of water pollutants except the standards, criteria, or regulations which are incorporated in any program as required by the provisions of section 1456(f) of this title.

water well (EPA-94/04): An excavation where the intended use is for location, acquisition, development, or artificial recharge of ground water (excluding sandpoint wells).

waterborne coating (40CFR60.741) means a coating which contains more than 5 weight percent water in its volatile fraction.

waterborne disease outbreak (40CFR141.2) means the significant occurrence of acute infectious illness, epidemiologically associated with the ingestion of water from a public water system which is deficient in treatment, as determined by the appropriate local or State agency.

waterborne disease outbreak (EPA-94/04): The significant occurrence of acute infection illness associated with drinking water from a public water system that is deficient in treatment, as determined by the appropriate local or state agency.

waterborne ink systems (40CFR60.431) means ink and related coating mixtures whose volatile portion consists of a mixture of VOC solvent and more than five weight percent water, as applied to the gravure cylinder.

waterborne or water reducible (40CFR60.391) means a coating which contains more than five weight percent water in its volatile fraction.

waters of the United States (40CFR257.3.3) can be found (is defined) in the Clean Water Act, as amended, 33USC1251 et seq., and implementing regulations, specifically 33CFR323 (42FR37122, July 19, 1977).

waters of the United States (40CFR721.3) has the meaning set forth in 40CFR122.2.

waters of the United States or waters of the U.S. (40CFR122.2) means:
(a) All waters which are currently used, were used in the past, or may be susceptible to use in interstate or foreign commerce, including all waters which are subject to the ebb and flow of the tide;
(b) All interstate waters, including interstate "wetlands";
(c) All other waters such as intrastate lakes, rivers, streams (including intermittent streams), mudflats, sandflats, "wetlands," sloughs, prairie potholes, wet meadows, playa lakes, or natural ponds the use, degradation, or destruction of which would affect or could affect interstate or foreign commerce including any such waters:
 (1) Which are or could be used by interstate or foreign travelers for recreational or other purposes;
 (2) From which fish or shellfish are or could be taken and sold in interstate or foreign commerce; or
 (3) Which are used or could be used for industrial purposes by industries in interstate commerce;
(d) All impoundments of waters otherwise defined as waters of the United States under this definition;
(e) Tributaries of waters identified in paragraphs (a) through (d) of this definition;
(f) The territorial sea; and

(g) "Wetlands" adjacent to waters (other than waters that are themselves wetlands) identified in paragraphs (a) through (f) of this definition. (Other identical or similar definitions are provided in 40CFR230.3-s; 232.2.)

waters under the jurisdiction of the United States (MMPA3-16USC1362) means:
(A) the territorial sea of the United States, and
(B) the waters included within a zone, contiguous to the territorial sea of the United States, of which the inner boundary is a line coterminous with the seaward boundary of each coastal State, and the outer boundary is a line drawn in such a manner that each point on it is 200 nautical miles from the baseline from which the territorial sea is measured.

watershed (EPA-94/04): The land area that drains into a stream; the watershed for a major river may encompass a number of smaller watersheds that ultimately combine at a common delivery point.

weak nitric acid (40CFR60.71-b) means acid which is 30 to 70 percent in strength (cf. nitric acid production unit).

web (40CFR52.741) means a substrate which is printed in continuous roll-fed presses.

web coating (40CFR60.741) means the coating of products, such as fabric, paper, plastic film, metallic foil, metal coil, cord, and yarn, that are flexible enough to be unrolled from a large roll; and coated as a continuous substrate by methods including, but not limited to, knife coating, roll coating, dip coating, impregnation, rotogravure, and extrusion.

weed (FIFRA2-7USC136) means any plant which grows where not wanted.

weight of evidence (40CFR300-AA) means EPA classification system for characterizing the evidence supporting the designation of a substance as a human carcinogen. EPA weight of evidence groupings include:
• Group A: Human carcinogen-sufficient evidence of carcinogenicity in humans.
• Group B1: Probable human carcinogen-limited evidence of carcinogenicity in humans.

- Group B2: Probable human carcinogen-limited sufficient evidence of carcinogenicity in humans.
- Group C: Probable human carcinogen-limited evidence of carcinogenicity in animals.
- Group D: Not classifiable as to human carcinogenicity--applicable when there is no animal evidence is inadequate.
- Group E: Evidence of noncarcinogenicity for humans.

weir (EPA-94/04): 1. A wall or plate placed in an open channel to measure the flow of water. 2. A wall or obstruction used to control flow from settling tanks and clarifiers to assure a uniform flow rate and avoid short-circuiting (cf. short circuiting).

welfare (CAA302-42USC7602): All language referring to effects on welfare includes, but is not limited to, effects on soils, water, crops, vegetation, man-made materials, animals, wildlife, weather, visibility, and climate, damage to and deterioration of property, and hazards to transportation, as well as effects on economic values and on personal comfort and well-being.

well (40CFR144.3) means a bored, drilled or driven shaft, or a dug hole, whose depth is greater than the largest surface dimension (other identical or similar definitions are provided in 40CFR146.3; 147.2902) (under SDWA).

well (40CFR260.10) means any shaft or pit dug or bored into the earth, generally of a cylindrical form, and often walled with bricks or tubing to prevent the earth from caving in.

well (40CFR435.61-c) shall means crude oil producing wells and shall not include gas wells or wells injecting water for disposal or for enhanced recovery of oil or gas.

well (EPA-94/04): A bored, drilled, or driven shaft, or a dug hole whose depth is greater than the largest surface dimension and whose purpose is to reach underground water supplies or oil, or to store or bury fluids below ground.

well completion fluids (40CFR435.11-x) shall refer to salt solutions, weighted brines, polymers, and various additives used to prevent damage to the well bore during operations which prepare the drilled well for hydrocarbon production.

well field (EPA-94/04): Area containing one or more wells that produce usable amounts of water (or oil).

well injection (40CFR144.3) means the subsurface emplacement of fluids through a bored, drilled, or driven well; or through a dug well, where the depth of the dug well is greater than the largest surface dimension (cf. injection well) (other identical or similar definitions are provided in 40CFR146.3; 147.2902).

well injection (40CFR165.1-aa) means disposal of liquid wastes through hole or shaft to a subsurface stratum.

well injection (40CFR260.10): See underground injection.

well injection (EPA-94/04): The subsurface emplacement of fluids into a well.

well monitoring (40CFR146.3) means the measurement, by on-site instruments or laboratory methods, of the quality of water in a well.

well monitoring (EPA-94/04): Measurement, by on-site instruments or laboratory methods, of the quality of water in a well.

well plug (40CFR146.3) means a watertight and gastight seal installed in a borehole or well to prevent movement of fluids.

well plug (EPA-94/04): A watertight and gastight seal installed in a bore hole or well to prevent movement of fluids.

well stimulation (40CFR146.3) means several processes used to clean the well bore, enlarge channels, and increase pore space in the interval to be injected thus making it possible for wastewater to move more readily into the formation, and includes:
(1) surging,
(2) jetting,
(3) blasting,
(4) acidizing,
(5) hydraulic fracturing.

well treatment fluids (40CFR435.11-y) shall refer to any fluid used to restore or improve productivity by chemically or physically altering hydrocarbon-bearing strata after a well has been drilled.

well workover (40CFR147.2902) means any reentry of an injection well; including, but not limited to, the pulling of tubular goods, cementing or casing repairs; and excluding any routine maintenance (e.g. re-seating the packer at the same depth, or repairs to surface equipment).

wellhead protection area (EPA-94/04): A protected surface and subsurface zone surrounding a well or wellfield supplying a public water system to keep contaminants from reaching the well water.

wellhead protection area (SDWA1428-42USC300h.7) means the surface and subsurface area surrounding a water well or wellfield, supplying a public water system, through which contaminants are reasonably likely to move toward and reach such water well or wellfield. The extent of a wellhead protection area, within a State, necessary to provide protection from contaminants which may have any adverse effect on the health of persons is to be determined by the State in the program submitted under subsection (a). Not later than one year after the enactment of the Safe Drinking Water Act Amendments of 1986, the Administrator shall issue technical guidance which States may use in making such determinations. Such guidance may reflect such factors as the radius of influence around a well or wellfield, the depth of drawdown of the water table by such well or wellfield at any given point, the time or rate of travel of various contaminants in various hydrologic conditions, distance from the well or wellfield at any given point, the time or rate of travel of various contaminants in various hydrologic conditions, distance from the well or wellfield, or other factors affecting the likelihood of contaminants reaching the well or wellfield, taking into account available engineering pump tests or comparable data, field reconnaissance, topographic information, and the geology of the formation in which the well or wellfield is located.

wet (40CFR420.41-d) means those steelmaking air cleaning systems that primarily use water for furnace gas cleaning (cf. semi-wet).

wet air pollution control scrubbers (40CFR471.02-oo) are air pollution control devices used to remove particulates and fumes from air by entraining the pollutants in a water spray.

wet barking operations (40CFR430.01-b) shall be defined to include hydraulic barking operations and wet drum barking operations which are those drum barking operations that use substantial quantities of water in either water sprays in the barking drums or in a partial submersion of the drums in a "tub" of water.

wet bottom (40CFR76.2) means the boiler has a furnace bottom temperature above the ash melting point and the bottom ash is removed as a liquid.

wet desulfurization system (40CFR420.11-e) means those systems which remove sulfur compounds from coke oven gases and produce a contaminated process wastewater.

wet electrostatic precipitator (40CFR503.41-n) is an air pollution control device that uses both electrical forces and water to remove pollutants in the exit gas from a sewage sludge incinerator stack.

wet flue gas desulfurization technology (40CFR60.41b) means a sulfur dioxide control system that is located downstream of the steam generating unit and removes sulfur oxides from the combustion gases of the steam generating unit by contacting the combustion gas with an alkaline slurry or solution and forming a liquid material. This definition applies to devices where the aqueous liquid material product of this contact is subsequently converted to other forms. Alkaline reagents used in wet flue gas desulfurization technology include, but are not limited to, lime, limestone, and sodium.

wet flue gas desulfurization technology (40CFR60.41c) means an SO_2 control system that is located between the steam generating unit and the exhaust vent or stack, and that removes sulfur oxides from the combustion gases of the steam generating unit by contacting the combustion gases with an alkaline slurry or solution and forming a liquid material. This definition includes devices where the liquid material is subsequently converted to another form. Alkaline reagents used in wet flue gas desulfurization systems include, but are not limited to, lime, limestone, and sodium compounds.

wet lot (40CFR412.21-g) shall mean a confinement facility for raising ducks which is open to the environment with a small portion of shelter area, and with open water runs and swimming areas to which ducks have free access.

wet mixture (40CFR723.175-16) means a water or organic solvent-based suspension, solution, dispersion, or emulsion used in the manufacture of an instant photographic or peel-apart film article.

wet process phosphoric acid plant (40CFR60.201-a) means any facility manufacturing phosphoric acid by reacting phosphate rock and acid.

wet scrubber (40CFR503.41-o) is an air pollution control device that uses water to remove pollutants in the exit gas from a sewage sludge incinerator stack.

wet scrubber system (40CFR60.41b) means any emission control device that mixes an aqueous stream or slurry with the exhaust gases from a steam generating unit to control emissions of particulate matter or sulfur dioxide (other identical or similar definitions are provided in 40CFR60.41c).

wet scrubbers (40CFR467.02-s) are air pollution control devices used to remove particulates and fumes from air by entraining the pollutants in a water spray.

wetlands (40CFR110.1) means those areas that are inundated or saturated by surface or ground water at a frequency or duration sufficient to support, and that under normal circumstances do support, a prevalence of vegetation typically adapted for life in saturated soil conditions. Wetlands generally include playa lakes, swamps, marshes, bogs and similar areas such as sloughs, prairie potholes, wet meadows, prairie river overflows, mudflats, and natural ponds.

wetlands (40CFR122.2) means those areas that are inundated or saturated by surface or groundwater at a frequency and duration sufficient to support, and that under normal circumstances do support, a prevalence of vegetation typically adapted for life in

saturated soil conditions. Wetlands generally include swamps, marshes, bogs, and similar areas.

wetlands (40CFR230.3-t) means those areas that are inundated or saturated by surface or ground water at a frequency and duration sufficient to support, and that under normal circumstances do support, a prevalence of vegetation typically adapted for life in saturated soil conditions. Wetlands generally include swamps, marshes, bogs and similar areas (other identical or similar definitions are provided in 40CFR232.2).

wetlands (40CFR257.3.3) can be found (is defined) in the Clean Water Act, as amended, 33USC1251 et seq., and implementing regulations, specifically 33CFR323 (42FR37122, July 19, 1977).

wetlands (40CFR435.41-f) shall mean those surface areas which are inundated or saturated by surface or ground water at a frequency and duration sufficient to support, and that under normal circumstances do support, a prevalence of vegetation typically adapted for life in saturated soil conditions. Wetlands generally include swamps, marshes, bogs, and similar areas.

wetlands (40CFR503.9-bb) means those areas that are inundated or saturated by surface water or ground water at a frequency and duration to support, and that under normal circumstances do support, a prevalence of vegetation typically adapted for life in saturated soil conditions. Wetlands generally include swamps, marshes, bogs, and similar areas.

wetlands (40CFR6-AA-j) means those areas that are inundated by surface or ground water with a frequency sufficient to support and under normal circumstances does or would support a prevalence of vegetative or aquatic life that requires saturated or seasonally saturated soiled conditions for growth and reproduction. Wetlands generally include swamps, marshes, bogs, and similar areas such as sloughs, potholes, wet meadows, river overflows, mud flats, and natural ponds.

wetlands (EPA-94/04): An area that is saturated by surface or ground water with vegetation adapted for life under those soil conditions, as swamps, bogs, fens, marshes, and estuaries.

wheat (40CFR406.41-b) shall mean wheat delivered to a plant before processing.

whole effluent toxicity (40CFR122.2) means the aggregate toxic effect of an effluent measured directly by a toxicity test.

whole program (40CFR70.2) means a part 70 permit program, or any combination of partial programs, that meet all the requirements of these regulations and cover all the part 70 sources in the entire State. For the purposes of this definition, the term "State" does not include local permitting authorities, but refers only to the entire State, Commonwealth, or Territory.

wholesale purchaser-consumer (40CFR80.2-o) means any organization that is an ultimate consumer of gasoline, diesel fuel, methanol, natural gas or liquefied petroleum gas and which purchases or obtains gasoline, diesel fuel, natural gas or liquefied petroleum gas from a supplier for use in motor vehicles and, in the case of gasoline, diesel fuel, methanol or liquefied petroleum gas, receives delivery of that product into a storage tank of at least 550-gallon capacity substantially under the control of that organization.

wholesaler (40CFR82.104-y) means a person to whom a product is delivered or sold, if such delivery or sale is for purposes of sale or distribution to retailers who buy such product for purposes of resale.

wilderness (FLPMA103-43USC1702) as used in section 1782 of this title shall have the same meaning as it does in section 1131(c) of title 16.

wildlife habitat (40CFR131.35.d-20) means the waters and surrounding land areas of the Reservation used by fish, other aquatic life and wildlife at any stage of their life history or activity (under FWPCA).

wildlife refuge (EPA-94/04): An area designated for the protection of wild animals, within which hunting and fishing are either prohibited or strictly controlled.

wire products and fasteners (40CFR420.121-f) means steel wire, products manufactured from steel wire, and steel fasteners manufactured from steel wire or other steel shapes.

wire-to-wire efficiency (EPA-94/04): The efficiency of a pump and motor together.

with modified-processes (40CFR60.291) means using any technique designed to minimize emissions without the use of add-on pollution controls.

withdraw specification (40CFR231.2-a) means to remove from designation any area already specified as a disposal site by the U.S. Army Corps of Engineers or by a state which has assumed the section 404 program, or any portion of such area.

withdrawal (FLPMA103-43USC1702) means withholding an area of Federal land from settlement, sale, location, or entry, under some or all of the general land laws, for the purpose of limiting activities under those laws in order to maintain other public values in the area or reserving the area for a particular public purpose or program; or transferring jurisdiction over an area of Federal land, other than "property" governed by the Federal Property and Administrative Services Act, as amended (40USC472) from one department, bureau or agency to another department, bureau or agency.

within the impoundment (40CFR421.11-d), For all impoundments the term "within the impoundment" for purposes of calculating the volume of process wastewater which may be discharged, shall mean the surface area within the impoundment at the maximum capacity plus the area of the inside and outside slopes of the impoundment dam and the surface area between the outside edge of the impoundment dam and seepage ditches upon which rain falls and is returned to the impoundment. For the purpose of such calculations, the surface area allowance for external appurtenances to the impoundment shall not be more than 30 percent of the water surface area within the impoundment dam at maximum capacity.

within the impoundment (40CFR421.41-c): For all impoundments constructed prior to the effective date of the interim final regulation (40 FR 8513), when used to calculate the volume of process wastewater which may be discharged, the term "within the impoundment" means the water surface area within the impoundment at maximum capacity plus the surface area of the inside and outside slopes of the impoundment dam and any seepage ditch adjacent to the dam upon which rain falls and is returned to the impoundment. For the purpose of such calculations, the surface area allowances set forth above shall not exceed more than 30 percent of the water surface area within the impoundment dam at maximum capacity.

within the impoundment (40CFR421.41-d): For all impoundments constructed on or after the effective date of the interim final regulation (the interim regulation was effective February 27, 1975; 40 FR 8513, February 27, 1975), the terms "within the impoundment," for purposes of calculating the volume of process wastewater which may be discharged, means the water surface area within the impoundment at maximum capacity.

within the impoundment (40CFR421.61-b): For all impoundments constructed prior to the effective date of this regulation the term "within the impoundment" when used for purposes of calculating the volume of process wastewater which may be discharged shall mean the water surface area within the impoundment at maximum capacity plus the surface area of the inside and outside slopes of the impoundment dam as well as the surface area between the outside edge of the impoundment dam and seepage ditch immediately adjacent to the dam upon which rain falls and is returned to the impoundment. For the purpose of such calculations, the surface area allowances set forth above shall not be more than 30 percent of the water surface area within the impoundment dam at maximum capacity.

within the impoundment (40CFR421.61-c): For all impoundments constructed on or after the effective date of this regulation, the term "within the impoundment" for purposes of calculating the volume of process wastewater which may be discharged shall mean the water surface area within the impoundment at maximum capacity.

WLA (40CFR130.2-h): Wasteload allocation.

women's business enterprise (40CFR33.005) is a business which is certified as such by a State or Federal agency, or which meets the following definition: A women's business enterprise is an independent business concern which is at least 51 percent owned by a woman or women who also control and operate it. Determination of whether a business is at least 51 percent owned by a woman or women shall be made without regard to community property laws. For example, an otherwise qualified WBE which is 51 percent owned by a married woman in a community property state will not be disqualified because her husband has a 50 percent interest in her share. Similarly, a business which is 51 percent owned by a married man and 49 percent owned by an unmarried woman will not become a qualified WBE by virtue of his wife's 50 percent interest in his share of the business.

women's business enterprise (WBE) (40CFR35.6015-54) means a business which is certified as a Women's Business Enterprise by a State or Federal agency, or which meets the following definition. A women's business enterprise is an independent business concern which is at least 51 percent owned by a woman or women who also control and operate it. Determination of whether a business is at least 51 percent owned by a woman or women shall be made without regard to community property laws.

wood (40CFR60.41b) means wood, wood residue, bark, or any derivative fuel or residue thereof, in any form, including, but not limited to, sawdust, sanderdust, wood chips, scraps, slabs, millings, shavings, and processed pellets made from wood or other forest residues (other identical or similar definitions are provided in 40CFR60.41c).

wood burning stove pollution (EPA-94/04): Air pollution caused by emissions of particulate matter, carbon monoxide, total suspended particulates, and polycyclic organic matter from wood-burning stoves.

wood fiber furnish subdivision mills (40CFR430.181-c) are those mills where cotton fibers are not used in the production of fine papers (cf. cotton fiber furnish subdivision mills).

wood furniture (40CFR52.741) means room furnishings including cabinets (kitchen, bath, and vanity), tables, chairs, beds, sofas, shutters, art objects, wood paneling, wood flooring, and any other coated furnishings made of wood, wood composition, or fabricated wood materials.

wood furniture coating facility (40CFR52.741) means a facility that includes one or more wood furniture coating line(s).

wood furniture coating line (40CFR52.741) means a coating line in which any protective, decorative, or functional coating is applied onto wood furniture.

wood heater (40CFR60.531) means an enclosed, woodburning appliance capable of and intended for space heating and domestic water heating that meets all of the following criteria:
(a) An air-to-fuel ratio in the combustion chamber averaging less than 35-to-1 as determined by the test procedure prescribed in 40CFR60.534 performed at an accredited laboratory,
(b) A usable firebox volume of less than 20 cubic feet,
(c) A minimum burn rate less than 5 kg/hr as determined by the test procedure prescribed in 40CFR60.534 performed at an accredited laboratory, and
(d) A maximum weight of 800 kg. In determining the weight of an appliance for these purposes, fixtures and devices that are normally sold separately, such as flue pipe, chimney, and masonry components that are not an integral part of the appliance or heat distribution ducting, shall not be included.

wood heater (40CFR60-AA (method 28-2.12)) means an enclosed, wood-burning appliance capable of and intended for space heating or domestic water heating, as defined in the applicable regulation.

wood residue (40CFR60.41) means bark, sawdust, slabs, chips, shavings, mill trim, and other wood products derived from wood processing and forest management operations.

wood treatment facility (EPA-94/04): An industrial facility that treats lumber and other wood products for outdoor use. The process employs chromated copper arsenate, which is regulated as a hazardous material.

woodworking (40CFR52.741) means the shaping, sawing, grinding, smoothing, polishing, and making into products of any form or shape of wood.

wool (40CFR410.11-a) shall mean the dry raw wool as it is received by the wool scouring mill.

wool fiberglass (40CFR60.291) means fibrous glass of random texture, including fiberglass insulation, and other products listed in SIC 3296.

wool fiberglass insulation (40CFR60.681) means a thermal insulation material composed of glass fibers and made from glass produced or melted at the same facility where the manufacturing line is located.

work area (40CFR721.3) means a room or defined space in a workplace where a chemical substance is manufactured, processed, or used and where employees are present.

work program (40CFR35.105) means the document which identifies how and when the applicant will use program funds to produce specific outputs.

work program (40CFR35.9010) means the scope of work of an assistance application, which identifies how and when the applicant will use funds to produce specific outputs.

worker (40CFR170.3) means any person, including a self-employed person, who is employed for any type of compensation and who is performing activities relating to the production of agricultural plants on an agricultural establishment to which subpart B of this part applies. While persons employed by a commercial pesticide handling establishment are performing tasks as crop advisors, they are not workers covered by the requirements of subpart of this part.

working day (40CFR129.2-f) means the hours during a calendar day in which a facility discharges effluents subject to this part.

working day (40CFR2.201-o) is any day on which Federal government offices are open for normal business. Saturdays, Sundays, and official Federal holidays are not working days; all other days are (other identical or similar definitions are provided in 40CFR350.1).

working day (40CFR61.141) means Monday through Friday and includes holidays that fall on any of the days Monday through Friday.

working day (40CFR85.1502-15) means any day on which Federal government offices are open for normal business. Saturdays, Sundays, and official Federal holidays are not working days.

working day (40CFR89.602.96): Any day on which federal government offices are open for normal business. Saturdays, Sundays, and official federal holidays are not working days.

working days (40CFR16.2-c) means calendar days excluding Saturdays, Sundays, and legal public holidays.

working face (40CFR241.101-y) means that portion of the land disposal site where solid wastes are discharged and are spread and compacted prior to the placement of cover material (under the Resource Conservation Recovery Act).

working level (WL) (40CFR192.11-c) means any combination of short-lived radon decay products in one liter of air that will result in the ultimate emission of alpha particles with a total energy of 130 billion electron volts.

working level (WL) (EPA-94/04): A unit of measure for documenting exposure to radon decay products, the so-called "daughters." One working level is equal to approximately 200 picocuries per liter.

working level month (WLM) (EPA-94/04): A unit of measure used to determine cumulative exposure to radon.

workmen's compensation commission (OSHA3-29USC652) means the National Commission on State Workmen's Compensation laws established under this Act.

workover fluids (40CFR435.11-z) shall refer to salt solutions, weighted brines, polymers, or other specialty additives used in a producing well to allow for maintenance, repair or abandonment procedures.

workplace (40CFR721.3) means an establishment at one geographic location containing one or more work areas.

worst case discharge (CWA311-33USC1321) means:
(A) in the case of a vessel, a discharge in adverse weather conditions of its entire cargo; and
(B) in the case of an offshore facility or onshore facility, the largest foreseeable discharge in adverse weather conditions.

worst case discharge for an onshore non-transportation-related facility (40CFR112.2) means the largest foreseeable discharge in adverse weather conditions as determined using the worksheets in Appendix D to this part (under the Federal Water Pollution Control Act).

WQM (40CFR130.2-k): See water quality management.

WQS (40CFR130.2-d): See water quality standards.

writing paper (40CFR250.4-tt) means a paper suitable for pen and ink, pencil, typewriter or printing.

written instructions for proper maintenance and use (40CFR85.1402) means those maintenance and operation instructions specified in the warranty as being necessary to assure compliance of the retrofit/rebuild equipment with applicable emission standards for the in-use compliance period (under CAA).

written instructions for proper maintenance and use (40CFR85.2102-13) means those maintenance and operation instructions specified in the owners manual as being necessary to assure compliance of a vehicle with applicable emission standards for the useful life of the vehicle that are:
(i) In accordance with the instructions specified for performance on the manufacturers Prototype vehicle used in certification (including those specified for vehicles used under special circumstances), and
(ii) In compliance with the requirements of 40CFR6.XXX-38 (as appropriate for the applicable model year vehicle/engine classification), and
(iii) In compliance with any other regulations

promulgated by EPA governing maintenance and use instructions.

********** XYZ **********

x (40CFR409.61-d) shall mean that fraction of the "net cane" harvested by the advanced harvesting systems.

xenobiote (EPA-94/04): Any biotum displaced from its normal habitat; a chemical foreign to a biological system.

xerographic/copy paper (40CFR250-uu) means any grade of paper suitable for copying by the xerographic process (a dry method of reproduction) (under RCRA).

yard waste (EPA-94/04): The part of solid waste composed of grass clippings, leaves, twigs, branches, and garden refuse.

year (40CFR704.102) means corporate fiscal year.

year or yearly (40CFR63.321) means any consecutive 12-month period of time.

yellow boy (EPA-94/04): Iron oxide flocculent (clumps of solids in waste or water); usually observed as orange-yellow deposits in surface streams with excess iron content (cf. floc or flocculation).

yield (EPA-94/04): The quantity of water (expressed as a rate of flows or total quantity per year) that can be collected for a given use from surface or groundwater sources.

yield point (40CFR1910.66-AD): The stress at which the material exhibits a permanent set of 0.2 percent.

Z-list (EPA-94/04): OSHA's tables of toxic and hazardous air contaminants.

zero ambient air material (40CFR72.2) means a gas that consists of ambient air and that is vendor-certified to contain: traceable concentrations respectively of SO_2, NO_x, and THC below 0.1 ppm; a concentration of CO below 1 ppm; and a concentration of CO_2 below 400 ppm.

zero device miles (40CFR610.11-19) means the period of time between retrofit installation and the accumulation of 100 miles of automobile operation after installation.

zero drift (40CFR53.23-e): The change in response to zero pollutant concentration, over 12- and 24-hour periods of continuous unadjusted operation (cf. span drift).

zero drift (40CFR60-AA (method 25A-2.4)): The difference in the measurement system response to a zero level calibration gas before and after a stated period of operation during which no unscheduled maintenance, repair, or adjustment took place (under CAA).

zero drift (40CFR60-AA (method 6C-3.6)) means the difference in the measurement system output reading from the initial calibration response at the zero concentration level after a stated period of operation during which no unscheduled maintenance, repair, or adjustment took place (other identical or similar definitions are provided in 40CFR60-AA (method 7E-3.2)).

zero drift (40CFR60-AB-2.3) means the difference in the CEMS output readings from the zero calibration value after a stated period of normal continuous operation during which no unscheduled maintenance, repair, or adjustment took place. A calibration value of 10 percent opacity or less may be used in place of the zero calibration value.

zero emission vehicle (40CFR88.101.94) means any light-duty vehicle or light-duty truck conforming to the applicable Zero-Emission Vehicle standard, or any heavy-duty vehicle conforming to the applicable Zero-Emission Vehicle standard.

zero hours (40CFR86.082.2) means that point after normal assembly line operations and adjustments are completed and before ten (10) additional operating hours have been accumulated, including emission testing, if performed.

zero kilometers (40CFR86.402.78) means that point after normal assembly line operations and adjustments, after normal dealer setup and preride inspection operations have been completed, and before 100 kilometers of vehicle operation or three hours of engine operation have been accumulated, including emission testing if performed.

zero, low-level, and high-level values (40CFR60-AF-2.4) means the CEMS response values related to the source specific span value. Determination of zero, low-level, and high-level values is defined in the appropriate PS in Appendix B of this part (under CAA).

zero miles (40CFR86.082.2) means that point after initial engine starting (not to exceed 100 miles of vehicle operation, or three hours of engine operation) at which normal assembly line operations and adjustments are completed, and including emission testing, if performed.

zero order reaction (40CFR796.3780-vi) is defined as a reaction in which the rate of disappearance of a chemical is independent of the concentration of the chemical or the concentration of any other chemical present in the reaction mixture.

ZID (40CFR125.58-w): See zone of initial dilution.

zinc (40CFR415.631; 415.671-d) shall mean the total zinc present in the process wastewater stream exiting the wastewater treatment system.

zinc (40CFR415.641-e) shall mean the total zinc present in the process wastewater stream exiting the wastewater treatment system.

zinc (40CFR420.02-l) means total zinc and is determined by the method specified in 40CFR136.3.

zinc casting (40CFR464.02-d): The remelting of zinc or zinc alloy to form a cast intermediate or final product by pouring or forcing the molten metal into a mold, except for ingots, pigs, or other cast shapes related to nonferrous (primary) metals manufacturing (40CFR421) and nonferrous metals forming (40CFR471). Processing operations following the cooling of castings not covered under nonferrous metals forming are covered under the electroplating and metal finishing point source categories (40CFR413 and 433).

zone of aeration (EPA-94/04): The comparatively dry soil or rock located between the ground surface and the top of the water table.

zone of engineering control (40CFR260.10) means an area under the control of the owner/operator that, upon detection of a hazardous waste release, can be readily cleaned up prior to the release of hazardous waste or hazardous constituents to ground waster or surface water.

zone of initial dilution (ZID) (40CFR125.58-w) means the region of initial mixing surrounding or adjacent to the end of the outfall pipe or diffuser ports, provided that the ZID may not be larger than allowed by mixing zone restrictions in applicable water quality standards.

zone of saturation (40CFR260.10): See saturated zone.

zone of saturation (EPA-94/04): See saturated zone.

zooplankton (EPA-94/04): Tiny aquatic animals eaten by fish.

APPENDIX: ENVIRONMENTAL ACRONYMS

4-AAP: 4-Aminoantipyrine (40CFR420.02)

A/B or AB: Afterburner (40CFR60.471)

A/WPR: Air/Water Pollution Report (EPA-94/04)

a.i.: Active Ingredient (EPA-92/12)

a.n.: Atomic Number

a.w.: Atomic Weight

A: Ampere (40CFR60)

A: Argon

A&C: Abatement and Control (EPA-94/04)

A&I: Alternative and Innovative (Wastewater Treatment System) (EPA-94/04)

A&R: Air and Radiation (EPA-94/04)

A&C: Abatement and Control

A&R or AR: Air and Radiation

AA: Accountable Area (EPA-94/04)

AA: Adverse Action (EPA-94/04)

AA: Advices of Allowance (EPA-94/04)

AA: Associate Administrator (EPA-94/04)

AA: Atomic Absorption (EPA-94/04)

AA: Attainment Area

AAA: American Automobile Association

AAAS: American Association for the Advancement of Science

AADI: Adjusted Acceptable Daily Intake (EPA-92/12)

AADT: Annual Average Daily Traffic (DOE-91/4)

AAEE: American Academy of Environmental Engineers (EPA-94/04)

AALACS: Ambient Aquatic Life Advisory Concentrations (40CFR300)

AAMA: American Academy of Medical Administration

AANWR: Alaskan Arctic National Wildlife Refuge (EPA-94/04)

AAOHN: American Association of Occupational Health Nurses

AAP: Acoustical Assurance Period (40CFR205.151)

AAP: Affirmative Action Plan

AAP: Affirmative Action Program

AAP: Asbestos Action Program (EPA-94/04)

AAPCO: American Association of Pesticide Control Officials (EPA-94/04)

AAQSS: Ambient Air Quality Standards (DOE-91/4)

AAR: Applications Analysis Report

AARC: Alliance for Acid Rain Control (EPA-94/04)

AARP: American Association of Retired Persons

AAS: Atomic Absorption Spectrophotometry (40CFR266-A9)

AATCC: American Association of Textile Chemists and Colorists

ABA: American Bar Association

ABAG: Association of Bay Area Governments

ABEL: EPA's computer model for analyzing a violator's ability to pay a civil penalty (EPA-94/04)

ABES: Alliance for Balanced Environmental Solutions (EPA-94/04)

ABMA: American Boiler Manufacturers Association

ABS: Alkyl Benzene Sulfonate

Ac: Actinium

AC: Actual Commitment (EPA-94/04)

AC: Advisory Circular (EPA-94/04)

AC: Alternating Current

ACA: American Conservation Association (EPA-94/04)

ACBM: Asbestos Containing Building Material (40CFR763.83)

ACE: Alliance for Clean Energy (EPA-94/04)

ACEC: American Consulting Engineers Council

ACEEE: American Council for an Energy Efficient Economy (EPA-94/04)

ACFM: Actual Cubic Feet Per Minute (EPA-94/04)

ACGIH: American Conference of Governmental Industrial Hygienists (29CFR1910.1200)

AChE: Acetylcholinesterase (EPA-92/12)

ACHP: Advisory Council on Historic Preservation

ACI: American Concrete Institute

ACL: Alternate Concentration Limit (EPA-94/04)

ACL: Analytical Chemistry Laboratory (EPA-94/04)

ACM: Asbestos-Containing Material (40CFR763.83)

ACP: Agriculture Control Program (Water Quality Management) (EPA-94/04)

ACP: Air Carcinogen Policy (EPA-94/04)

ACQUIRE: Aquatic Information Retrieval (EPA-94/04)

ACR: Acute-to-Chronic Ratio (EPA-91/3)

ACS: Acid Scrubber

ACS: American Chemical Society (EPA-94/04)

ACSH: American Council on Science and Health

ACT: Action (EPA-94/04)

ACTS: Asbestos Contractor Tracking System (EPA-94/04)

ACUS: Administrative Conference of United States

ACWA: American Clean Water Association (EPA-94/04)

ACWM: Asbestos-Containing Waste Material (EPA-94/04)

ADABA: Acceptable Data Base (EPA-94/04)

ADB: Applications Data Base (EPA-94/04)

ADI: Acceptable Daily Intake (EPA-94/04)

ADP: Adenosine Diphosphate

ADP: AHERA Designated Person (EPA-94/04)

ADP: Automated Data Processing (EPA-94/04)

ADQ: Audits of Data Quality (EPA-94/04)

ADR: Alternate Dispute Resolution (EPA-94/04)

ADSS: Air Data Screening System (EPA-94/04)

ADT: Average Daily Traffic (EPA-94/04)

AEA: Atomic Energy Act (40CFR23.9)

AEC: Associate Enforcement Counsels (EPA-94/04)

AEC: Atomic Energy Commission (DOE-91/4)

AECD: Auxiliary emission control device (40CFR89.3-94)

AEE: Alliance for Environmental Education (EPA-94/04)

AEERL: Air and Energy Engineering Research Laboratory (EPA-94/04)

AEM: Acoustic Emission Monitoring (EPA-94/04)

AER: Advanced Electrical Reactor

AERE: Association of Environmental and Resource Economists (EPA-94/04)

AEROS: Aerometric and Emissions Reporting System (40CFR51.100-l)

AES: Auger Electron Spectrometry (EPA-94/04)

AFA: American Forestry Association (EPA-94/04)

AFCA: Area Fuel Consumption Allocation (EPA-94/04)

AFCEE: Air Force Center for Environmental Excellence (EPA-94/04)

AFRCE: Air Force Regional Civil Engineers

AFS: AIRS Facility Subsystem (EPA-94/04)

AFUG: AIRS Facility Users Group (EPA-94/04)

Ag: Silver

AGC: Associate General Counsels

AGMA: American Gear Manufacturer's Association

AH: Allowance Holders (EPA-94/04)

AHA: American Hospital Association

AHERA: Asbestos Hazard Emergency Response Act of 1986 (40CFR22.41)

AHFS: American Hospital Formulary Service (EPA-92/12)

AI: Artificial Intelligence

AIA: American Institute of Architects

AIA: Asbestos Information Association

AIADA: American International Automobile Dealers Association

AIC: Active to Inert Conversion (EPA-94/04)

AICC: Annual Installed Capital Cost

AICE: American Institute of Civil Engineers

AIChE: American Institute of Chemical Engineers

AICUZ: Air Installation Compatible Use Zones (EPA-94/04)

AID: Agency for International Development (EPA-94/04)

AIDS: Acquired Immune Deficiency Syndrome

AIF: Atomic Industrial Forum

AIG: Assistant Inspector General

AIHA: American Industrial Health Association (EPA-92/12)

AIHC: American Industrial Health Council (EPA-94/04)

AIME: American Institute of Metallurgical, Mining and Petroleum Engineers

AIP: Auto Ignition Point (EPA-94/04)

AIRFA: American Indian Religious Freedom Act (DOE-91/4)

AIRS: Aerometric Information Retrieval System (EPA-94/04)

AISC: American Institute of Steel Construction

AISI: American Iron and Steel Institute

AK: Alaska

Al: Aluminum

AL: Acceptable Level (EPA-94/04)

AL: Administrative Leave

AL: Alabama

AL: Annual Leave

ALA-O: Delta-Aminolevulinic Acid Dehydrates (EPA-94/04)

ALA: American League of Anglers

ALA: American Lung Association

ALA: Delta-Aminolevulinic Acid (EPA-94/04)

ALAPO: Association of Local Air Pollution Control Officers (EPA-94/04)

ALARA: As Low As Reasonably Achievable (EPA-94/04)

ALC: Application Limiting Constituent (EPA-94/04)

ALE: Arid Land Ecology (DOE-91/4)

ALEC: American Legislative Exchange Council

ALJ: Administrative Law Judge (EPA-94/04)

ALMS: Atomic Line Molecular Spectroscopy (EPA-94/04)

ALR: Action Leakage Rate (EPA-94/04)

ALVW: Adjusted Loaded Vehicle Weight (40CFR86.094.3-94)

Am: Americium

AMA: American Medical Association (EPA-92/12)

AMBIENS: Atmospheric Mass Balance of Industrially Emitted and Natural Sulfur (EPA-94/04)

AMC: American Mining Congress

AMCA: Air Moment and Control Association

AML: Average Monthly Limit (EPA-91/3)

AMPS: Automatic Mapping and Planning System

AMOS: Air Management Oversight System (EPA-94/04)

AMPS: Automatic Mapping and Planning System (EPA-94/04)

AMS: Administrative Management Staff

AMS: American Meteorological Society

AMS: Army Map Service

AMSA: Association of Metropolitan Sewer Agencies (EPA-94/04)

AMSD: Administrative and Management Services Division

ANC: Acid Neutralizing Capacity (EPA-94/04)

ANEC: American Nuclear Energy Council

ANFO: Ammonium Nitrate and Fuel Oil (40CFR457.31)

ANL: Argonne National Laboratory

ANOVA: Analysis of Variance (40CFR264.97)

ANPR: Advance Notice of Proposed Rulemaking (EPA-94/04)

ANRHRD: Air, Noise, & Radiation Health Research Division/ORD (EPA-94/04)

ANS: American Nuclear Society

ANSI: American National Standards Institute

ANSS: American Nature Study Society (EPA-94/04)

AO: Administrative Officer

AO: Administrative Order (on consent)

AO: Administrator's Office

AO: Area Office

AO: Awards and Obligations

AOAC: Association of Official Analytical Chemists (40CFR60.17.b)

AOC: Abnormal Operating Conditions (EPA-94/04)

AOD: Argon-Oxygen Decarbonization (40CFR60.271a)

AOML: Atlantic Oceanographic and Meteorological Laboratory (EPA-94/04)

AP: Accounting Point (EPA-94/04)

APA: Administrative Procedures Act (EPA-94/04)

APA: American Pharmaceutical Association (EPA-92/12)

APCA: Air Pollution Control Association (EPA-94/04)

APCD: Air Pollution Control Device

APCD: Air Pollution Control District (EPA-94/04)

APCE: Air Pollution Control Equipment

APCS: Air Pollution Control System

APDS: Automated Procurement Documentation System (EPA-94/04)

APEG: Alkaline Polyethylene Glycol

APHA: American Public Health Association (EPA-94/04)

API: American Paper Institute

API: American Petroleum Industry (40CFR86)

APR: Air Purifying Respirator

APRAC: Urban Diffusion Model for Carbon Monoxide from Motor Vehicle Traffic (EPA-94/04)

APT: Associated Pharmacists and Toxicologists

APTI: Air Pollution Training Institute (EPA-94/04)

APWA: American Public Works Association (EPA-94/04)

AQ-7: Non-reactive Pollutant Modelling (EPA-94/04)

AQA: Air Quality Act

AQC: Air Quality Committee

AQCCT: Air-Quality Criteria and Control Techniques (EPA-94/04)

AQCP: Air Quality Control Program (EPA-94/04)

AQCR: Air Quality Control Region (CAA)

AQCR: Air-Quality Control Region (EPA-94/04)

AQD: Air-Quality Digest (EPA-94/04)

AQDHS: Air-Quality Data Handling System (EPA-94/04)

AQDM: Air-Quality Display Model (EPA-94/04)

AQL: Acceptable Quality Level (40CFR86)

AQMA: Air Quality Maintenance Area (40CFR51.40)

AQMP: Air-Quality Maintenance Plan (EPA-94/04)

AQMP: Air-Quality Management Plan (EPA-94/04)

AQSM: Air-Quality Simulation Model (EPA-94/04)

AQTAD: Air-Quality Technical Assistance Demonstration (EPA-94/04)

Ar: Argon

AR: Administrative Record (EPA-94/04)

AR: Arkansas

ARA: Assistant Regional Administrator (EPA-94/04)

ARA: Associate Regional Administrator (EPA-94/04)

ARAR: Applicable or Relevant and Appropriate Requirements (RCRA) (40CFR300.4)

ARAR: Applicable or Relevant and Appropriate Standards, Limitations, Criteria, and Requirements (EPA-94/04)

ARB: Air Resources Board (EPA-94/04)

ARC: Agency Ranking Committee (EPA-94/04)

ARCC: American Rivers Conservation Council (EPA-94/04)

ARCS: Alternative Remedial Contract Strategy (EPA-94/04)

ARG: American Resources Group (EPA-94/04)

ARIP: Accidental Release Information Program (EPA-94/04)

ARL: Air Resources Laboratory (EPA-94/04)

ARM: Agricultural Runoff Model (EPA-91/3)

ARM: Air Resources Management (EPA-94/04)

ARO: Alternate Regulatory Option (EPA-94/04)

ARPA: Advanced Research Projects Agency

ARPO: Acid Rain Policy Office

ARPS: Atmospheric Research Program Staff

ARRARs: Applicable or Relevant and Appropriate Requirements (EPA-91/12)

ARRP: Acid Rain Research Program (EPA-94/04)

ARRPA: Air Resources Regional Pollution Assessment Model (EPA-94/04)

ARS: Agricultural Research Service (EPA-94/04)

ARTS: Augusta Regional Transportation Study (DOE-91/4)

ARZ: Auto Restricted Zone (EPA-94/04)

As: Arsenic

AS: Absorber

AS: Area Source (EPA-94/04)

ASA: American Standards Association

ASAP: As Soon As Possible

ASC: Area Source Category (EPA-94/04)

ASCE: American Society of Civil Engineers

ASCII: American Standard Code for Information Interchange

ASCP: American Society of Consulting Planners

ASDWA: Association of State Drinking Water Administrators (EPA-94/04)

ASHAA: Asbestos in Schools Hazard Abatement Act (EPA-94/04)

ASHP: American Society of Hospital Pharmacists (EPA-92/12)

ASIFHESE: Alphabet Soup Index For Health and Enviromnental Science and Engineering

ASIWCPA: Association of State and Interstate Water Pollution Control Administrators (EPA-94/04)

ASMDHS: Airshed Model Data Handling System (EPA-94/04)

ASME: American Society of Mechanical Engineers

ASPA: American Society of Public Administration

ASRL: Atmospheric Sciences Research Laboratory (EPA-94/04)

AST: Advanced Secondary (Wastewater) Treatment (EPA-94/04)

ASTHO: Association of State and Territorial Health Officers (EPA-94/04)

ASTM: American Society for Testing and Materials (EPA-94/04)

ASTSWMO: Association of State and Territorial Solid Waste Management Officials (EPA-94/04)

ASV: Anodic Stripping Voltammetry

At: Astatine

AT: Advanced Treatment (EPA-94/04)

AT: Advanced Treatment (water)

AT: Alpha Track Detection (EPA-94/04)

AT: Ash Trap

ATC: Acceptable Tissue Concentration (EPA-91/3)

ATE: Acute Toxicity Endpoint (EPA-91/3)

ATERIS: Air Toxics Exposure and Risk Information System (EPA-94/04)

ATMI: American Textile Manufacturers Institute

ATP: Adenosine Triphosphate

ATS: Action Tracking System (EPA-94/04)

ATS: Administrator's Tracking System

ATSAC: Administrator's Toxic Substances Advisory Committee

ATSDR: Agency for Toxic Substances and Disease Registry (40CFR300.4)

ATTF: Air Toxics Task Force (EPA-94/04)

ATTIC: Alternative Treatment Technology Information Center

Au: Gold

AUR: Air Unit Risk (EPA-92/12)

AUSA: Assistant United States Attorney

AUSM: Advanced Utility Simulation Model (EPA-94/04)

AVS: Acid Volatile Sulfides (EPA-91/3)

AWMA: Air and Waste Management Association

AWQC: Ambient Water Quality Criteria (40CFR300-AA)

AWRA: American Water Resources Association (EPA-94/04)

AWS: American Welding Society

AWT: Advanced Waste Treatment

AWT: Advanced Wastewater Treatment (EPA-94/04)

AWWA: American Water Works Association (EPA-94/04)

AWWARF: American Water Works Association Research Foundation (EPA-94/04)

AZ: Arizona

b.p.: Boiling Point

B.P.: Before Present (DOE-91/4)

B: Boron

b:a lambda(a): Blood-to-air partition coefficient (EPA-92/12)

Ba: Barium

BAA: Board of Assistance Appeals (EPA-94/04)

BAC: Biotechnology Advisory Committee (EPA-94/04)

BACER: Biological and Climatological Effects Research

BACM: Best Available Control Measures (EPA-94/04)

BACT: Best Available Control Technology (40CFR51)

BADCT: Best Available Demonstrated Control Technology

BADT: Best Available Demonstrated Technology (EPA-94/04)

BAF: Bioaccumulation Factor (EPA-91/3)

BaP: Benzo(a)Pyrene (EPA-94/04)

BAP: Benefits Analysis Program (EPA-94/04)

BARF: Best Available Retrofit Facility

BART: Best Available Retrofit Technology (40CFR51.301-c)

BASIS: Battelle's Automated Search Information System (EPA-94/04)

BAT: Best Available Technology (40CFR141)

BAT: Best Available Technology Economically Achievable (40CFR467)

BATEA: Best Available Treatment Economically Achievable (EPA-94/04)

BBS: Bulletin Board System

BBS: OSWER Electronic Bulletin Board System

BCC: Blind Carbon Copy

BCCM: Board for Certified Consulting Meteorologists

BCF: Bioconcentration Factor (40CFR797)

BCPCT: Best Conventional Pollutant Control Technology (EPA-94/04)

BCT: Best Control Technology (EPA-94/04)

BCT: Best Conventional Pollutant Control Technology (40CFR430)

BCT: Best Conventional Technology (EPA-91/3)

BD: Background Document

BDAT: Best Demonstrated Achievable Technology (EPA-94/04)

BDCT: Best Demonstrated Control Technology (EPA-94/04)

BDT: Best Demonstrated Technology (EPA-94/04)

Be: Beryllium

BEA: Bureau of Economic Advisors

BEA: Bureau of Economic Analysis

BEIR: Biological Effects of Ionizing Radiation (DOE-91/4)

BEJ: Best Engineering Judgement

BEJ: Best Expert Judgment (EPA-94/04)

BEP: Black Employment Program

BF: Bonifide Notice of Intent to Manufacture or Import (IMD/OTS) (EPA-94/04)

BG: Billion Gallons

BGD: Billion Gallons per Day

BHET: Bis-(2-hydroxyethyl)-terephthalate (40CFR60.561)

BHNL: Brookhaven National Laboratory

BHP: Biodegradation, Hydrolysis, and Photolysis (EPA-92/12)

BHP: Brake Horsepower (40CFR86)

Bi: Bismuth

BI: Brookings Institution

BIA: Bureau of Indian Affairs (40CFR147.2902)

BIAC: Business Industry Advisory Committee

BID: Background Information Document (EPA-94/04)

BID: Buoyancy Induced Dispersion (EPA-94/04)

BIOPLUME: Model to Predict the Maximum Extent of Existing Plumes (EPA-94/04)

BIOS: Basic Input Output System (computer)

EIP: Economic Incentive Program (40CFR51.490)

Bk: Berkelium

BLM: Bureau of Land Management

BLOB: Biologically Liberated Organo-Beasties

BLP: Buoyant Line and Point Source Model

BLS: Bureau of Labor Statistics

BMP: Best Management Practice(s) (40CFR122.2)

BMR: Baseline Monitoring Report (EPA-94/04)

BNA: Base, Neutral, and Acid (BNA) Compounds

BO: Budget Obligations (EPA-94/04)

BOA: Basic Ordering Agreement (Contracts) (EPA-94/04)

BOD: Biochemical Oxygen Demand (40CFR133; EPA-94/04)

BOD: Biological Oxygen Demand (EPA-94/04)

BOD5: Biochemical Oxygen Demand as Measured in the Standard 5-Day Test (EPA-92/12)

BOF: Basic Oxygen Furnace (EPA-94/04)

BOM: Bureau of Mines or U.S. Bureau of Mines

BOP: Basic Oxygen Process (EPA-94/04)

BOPF: Basic Oxygen Process Furnace (40CFR60.141)

BOYSNC: Beginning of Year Significant Non-Compliers (EPA-94/04)

BP: Boiling Point (EPA-94/04)

BPA: Blanket Purchase Agreement

BPA: Bonneville Power Administration (DOE-91/4)

BPCT: Best Practicable Control Technolgy

BPJ: Best Professional Judgment (EPA-94/04)

BPT: Best Practicable Control Technology Currently Available (40CFR430.222)

BPT: Best Practicable Technology (EPA-94/04)

BPT: Pest Practicable Treatment (EPA-94/04)

BPWTT: Best Practicable Waste Treatment Technoloy (40CFR35.2005-7)

BPWTT: Best Practical Wastewater Treatment Technology (EPA-94/04)

Bq: Becquerel (a radiation unit) (40CFR302.4-AB)

BQU: Bacterial Quantity Unit

Br: Bromine

BRS: Bibliographic Retrieval Service (EPA-94/04)

BSCO: Brake Specific Carbon Monoxide (40CFR86)

BSHC: Brake Specific Hydrocarbons (40CFR86)

BSI: British Standards Institute (EPA-94/04)

$BSNO_x$: Brake Specific Oxides of Nitrogen (40CFR86)

BSO: Benzene Soluble Organics (EPA-94/04)

BST: Benzene Study Team (MCA)

BTG: Benzene Task Group (MCA)

Btu: British Thermal Unit (40CFR60)

BTX: Benzene-Toluene-Xylene (40CFR61.131)

BTZ: Below the Treatment Zone (EPA-94/04)

BU: Bargaining Unit

BUN: Blood Urea Nitrogen (EPA-94/04)

bw: Body Weight (EPA-92/12)

BY: Budget Year

C: Carbon

C: Cyclone

C: Degree Celsius (centigrade) (40CFR60-94)

Ca: Calcium

CA: California

CA: Capacity

CA: Carbon Absorber

CA: Citizen Act (EPA-94/04)

CA: Competition Advocate (EPA-94/04)

CA: Cooperative Agreements (EPA-94/04)

CA: Corrective Action (EPA-94/04)

CAA: Clean Air Act (40CFR89.3-94)

CAA: Compliance Assurance Agreement (EPA-94/04)

CAAA: Clean Air Act Amendments of 1990 (40CFR89.3-94)

CAB: Civil Aeronautics Board

CAD: Computer Assisted Design

CAER: Community Awareness and Emergency Response (EPA-94/04)

CAFE: Corporate Average Fuel Economy (EPA-94/04)

CAFO: Consent Agreement/Final Order (EPA-94/04)

CAG: Carcinogen Assessment Group, U.S. EPA (EPA-92/12)

CAG: Carcinogenic Assessment Group (EPA-94/04)

CAIR: Comprehensive Assessment of Information Rule (40CFR704.200)

cal: Calorie (40CFR60)

CALINE: California Line Source Model (EPA-94/04)

CAMP: Continuous Air Monitoring Program (EPA-94/04)

CAMU: Corrective Action for Solid Waste Management Unit (40CFR264.552)

CAN: Common Account Number (EPA-94/04)

CAO: Corrective Action Order (EPA-94/04)

CAP: Corrective Action Plan (EPA-94/04)

CAP: Cost Allocation Procedure (EPA-94/04)

CAP: Criteria Air Pollutant (EPA-94/04)

CAR: Corrective Action Report (EPA-94/04)

CARB: California Air Resources Board

CAS: Center for Automotive Safety (EPA-94/04)

CAS: Chemical Abstract Service (EPA-94/04)

CASAC: Clean Air Scientific Advisory Committee (EPA-94/04)

CASLP: Conference on Alternative State and Local Practices (EPA-94/04)

CAST: Council on Agricultural Science and Technology

CATS: Corrective Action Tracking System (EPA-94/04)

CAU: Carbon Adsorption Unit (EPA-94/04)

CAU: Command Arithmetic Unit (EPA-94/04)

CB: Carbon Bed

CB: Continuous Bubbler (EPA-94/04)

CBA: Chesapeake Bay Agreement (EPA-94/04)

CBA: Cost Benefit Analysis (EPA-94/04)
CBB: Chesapeake Bay Basin
CBC: Circulating Bed Combustor
CBD: Central Business District (40CFR52.2486)
CBD: Central Business District (EPA-94/04)
CBD: Commerce Business Daily
CBI: Compliance Biomonitoring Inspection (EPA-94/04)
CBI: Confidential Business Information (EPA-94/04)
CBO: Congressional Budget Office
CBOD: Carbonaceous Biochemical Oxygen Demand (40CFR133.101)
CBOD: Carbonaceous Biochemical Oxygen Demand (EPA-94/04)
CBP: Chesapeake Bay Program (EPA-94/04)
CBP: Combustion By-Product
CBP: County Business Patterns (EPA-94/04)
CC/RTS: Chemical Collection/Request Tracking System (EPA-94/04)
cc: Cubic Centimeter (40CFR60-94)
CC: Carbon Copy
CC: Closed Cup (EPA-92/12)
CCA: Competition in Contracting Act
CCA: Competition in Contracting Act (EPA-94/04)
CCA: Council of Chemical Associations
CCAA: Canadian Clean Air Act (EPA-94/04)
CCAP: Center for Clean Air Policy (EPA-94/04)
CCC: Criteria Continuous Concentration (EPA-91/3)
CCDF: Complementary Cumulative Distribution Function (DOE-91/4)
CCE: Carbon Chloroform Extract
CCEA: Conventional Combustion Environmental Assessment (EPA-94/04)
CCERP: Committee to Coordinate Environmental and Related Programs
CCHW: Citizens Clearinghouse for Hazardous Wastes (EPA-94/04)
CCID: Confidential Chemicals Identification System (EPA-94/04)
CCMA: Certified Color Manufacturers Association
CCMS/NATO: Committee on Challenges of a Modern Society/North Atlantic Treaty Organization (EPA-94/04)
CCP: Composite Correction Plan (EPA-94/04)
CCP: Comprehensive Carcinogen Policy (OSHA)
CCS: Counter Current Scrubber
CCTP: Clean Coal Technology Program (EPA-94/04)
CCW: Constituent Concentrations in Waste (40CFR268.43)

CCWE: Constituent Concentrations in Waste Extract (40CFR268.41)
Cd: Cadmium
CD: Climatological Data (EPA-94/04)
CD: Criterion Document
CDB: Consolidated Data Base (EPA-94/04)
CDBA: Central Data Base Administrator (EPA-94/04)
CDBG: Community Development Block Grant (EPA-94/04)
CDC: Centers for Disease Control (EPA-92/12)
CDC: Communicable Disease Center
CDD: Chlorinated dibenzo-p-dioxin (EPA-94/04)
CDF: Chlorinated dibenzofuran (EPA-94/04)
CDHS: Comprehensive Data Handling System (EPA-94/04)
CDI: Case Development Inspection (EPA-94/04)
CDM: Climatological Dispersion Model (EPA-94/04)
CDM: Comprehensive Data Management (EPA-94/04)
CDMQC: Climatological Dispersion Model with Calibration and Source Contribution (EPA-94/04)
CDNS: Climatological Data National Summary (EPA-94/04)
CDP: Census Designated Places (EPA-94/04)
CDS: Compliance Data System (EPA-94/04)
Ce: Cerium
CE: Categorical Exclusion (EPA-94/04)
CE: Conditionally Exempt Generator (EPA-94/04)
CE: Cost Effectiveness
CEA: Cooperative Enforcement Agreement (EPA-94/04)
CEA: Cost and Economic Assessment (EPA-94/04)
CEA: Council of Economic Advisors
CEAM: Center for Exposure Assessment Modeling (EPA-91/3)
CEARC: Canadian Environmental Assessment Research Council (EPA-94/04)
CEAT: Contractor Evidence Audit Team (EPA-94/04)
CEB: Chemical Element Balance (EPA-94/04)
CEC: Cation Exchange Capacity (40CFR257.3.5)
CEC: Clearinghouse on Environmental Carcinogens
CEC: Commission of European Communities
CECATS: CSB Existing Chemicals Assessment Tracking System (EPA-94/04)
CED: Committee for Economic Development
CEDE: Committed Effective Dose Equivalent (DOE-91/4)
CEE: Center for Environmental Education

CEE: Center for Environmental Education (EPA-94/04)

CEEM: Center for Energy and Environmental Management (EPA-94/04)

CEFIC: Counseil Europeen Des Federations De L'Industrie Chimique

CEI: Compliance Evaluation Inspection (CWA)

CEI: Compliance Evaluation Inspection (EPA-94/04)

CELRF: Canadian Environmental Law Research Foundation (EPA-94/04)

CEM: Continuous Emission Monitoring (EPA-94/04)

CEMA: Conveyor Equipment Manufacturer's Association

CEMS: Continuous Emission Monitoring System (40CFR60.51a)

CEO: Chief Executive Officer

CEP: Council on Economic Priorities

CEPP: Chemical Emergency Preparedness Plan (EPA-94/04)

CEQ: Council on Environmental Quality (40CFR6.101

CEQA: California Environmental Quality Act

CERCLA: Comprehensive Environmental Response, Compensation, and Liability Act (1980) (EPA-94/04)

CERCLIS: CERCLA Information System (40CFR300.5)

CERCLIS: Comprehensive Environmental Response, Compensation, and Liability Information System (EPA-94/04)

CERI: Center for Environmental Research Information (EPA)

CERT: Certificate of Eligibility (EPA-94/04)

CETIS: Complex Effluent Toxicity Information System (EPA-91/3)

CETTP: Complex Effluent Toxicity Testing Program (EPA-91/3)

CEU: Continuing Education Units

cf.: Compare

Cf: Californium

CF: Conservation Foundation (EPA-94/04)

CFA: Consumer Federation of American

CFC: Chlorofluorocarbons (EPA-94/04)

CFC: Combined Federal Campaign

cfh: Cubic feet per hour (40CFR86.403.78-94)

cfm: Cubic feet per minute (40CFR86.403.78-94)

CFM: Chlorofluoromethanes (EPA-94/04)

CFR: Code of Federal Regulations (EPA-94/04)

CFS: Cubic Feet per Second

CFU: Colony Forming Unit

CFV: Critical Flow Venturi (40CFR86)

CFV-CVS: Critical Flow Venturi-Constant Volume Sampler (40CFR86)

CGS unit: Centimeter-Gram-Second unit (absolute metric unit)

CHABA: Committee on Hearing and Bio-Acoustics (EPA-94/04)

CHAMP: Community Health Air Monitoring Program (EPA-94/04)

CHC: Chemical of Highest Concern (EPA-91/3)

ChE: Cholinesterase (EPA-92/12)

CHEMNET: A mutual aid network of chemical shippers and contractors (NRT-87/3)

CHEMNET: Chemical Industry Emergency Mutual Aid Network (EPA-94/04)

CHEMTREC: Chemical Transportation Emergency Center (NRT-87/3)

CHESS: Community Health and Environmental Surveillance System (EPA-94/04)

CHIP: Chemical Hazard Information Profiles (EPA-94/04)

CHLOREP: Chlorine Emergency Plan (NRT-87/3)

CHNTRN: Channel Transport Model (EPA-91/3)

CHRIS/HACS: Chemical Hazards Response Information System/Hazard Assessment Computer System (NRT-87/3)

Ci/L: Curies per Liter (DOE-91/4)

Ci/m^3: Curies per Cubic Meter (DOE-91/4)

Ci/yr: Curies per Year (DOE-91/4)

CI: Color index

Ci: Curie (40CFR190.02)

CI: Compression-ignition (40CFR89.3-94)

CI: Compression Ignition (EPA-94/04)

CI: Confidence Interval (EPA-94/04)

CIA: Central Intelligence Agency

CIAQ: Council on Indoor Air Quality (EPA-94/04)

CIBL: Convective Internal Boundary Layer (EPA-94/04)

CIBO: Council of Industrial Boiler Owners

CICA: Competition in Contracting Act (EPA-94/04)

CICIS: Chemicals in Commerce Information System (EPA-94/04)

CIDAC: Cancer Information Dissemination and Analysis Center

CIDRS: Cascade Impactor Data Reduction System (EPA-94/04)

CIIT: Chemical Industry Institute of Toxicology (EPA-92/12)

CIMI: Committee on Integrity and Management Improvement (EPA-94/04)

CIRLG: Chemical Industry Regulatory Liaison Group

CIS: Chemical Information System (EPA-94/04)

CIS: Contracts Information System (EPA-94/04)

CJE: Critical Job Element

CJO: Chief Judicial Officer

CKD: Cement Kiln Dust

CKRC: Cement Kiln Recycling Coalition

CL: Chemiluminescence (40CFR86)

CL: Chlorine

CLC: Capacity Limiting Constituents (EPA-94/04)

CLEANS: Clinical Laboratory for Evaluation and Assessment of Toxic Substances (EPA-94/04)

CLEVER: Clinical Laboratory for Evaluation and Validation of Epidemiologic Research (EPA-94/04)

CLF: Conservation Law Foundation (EPA-94/04)

CLIPS: Chemical List Index and Processing System (EPA-94/04)

CLP: Contract Laboratory Program (40CFR300-AA)

CLS: Chlorine Scrubber

cm: Centimetre(s) (40CFR86.403.78-94)

Cm: Curium

CM: Corrective Measure (EPA-94/04)

CMA: Chemical Manufacturers Association (EPA-94/04)

CMAA: Crane Manufacturer's Association of American

CMB: Chemical Mass Balance (EPA-94/04)

CMC: Criteria Maximum Concentration (EPA-91/3)

CME: Comprehensive (groundwater) Monitoring Evaluation

CME: Comprehensive Monitoring Evaluation (EPA-94/04)

CMEL: Comprehensive Monitoring Evaluation Log (EPA-94/04)

CMEP: Critical Mass Energy Project (EPA-94/04)

CMPU: Chemical Manufacturing Process Unit (40CFR63 Subpart G Appendix)

CN ratio: Carbon Nitrogen Ratio

CNG: Compressed Natural Gas (EPA-94/04)

CNP ratio: Carbon Nitrogen Phosphorus ratio

CNR: Composite Noise Rating (DOE-91/4)

CNS: Central Nervous System (EPA-92/12)

Co: Cobalt

CO: Carbon Monoxide (40CFR88.103-94)

CO: Colorado

CO_2: Carbon dioxide (40CFR86.403.78-94)

COA: Corresponding Onshore Area (40CFR55.2)

COB: Close of Business

COCO: Contractor-Owned/Contractor-Operated (EPA-94/04)

COD: Chemical Oxygen Demand (EPA-94/04)

COE: Corps of Engineers (DOE-91/4)

COE: U.S. Army Corps of Engineers

COFA: Certification of Fund Availability (EPA)

COG: Council of Governments

COH: Coefficient Of Haze (EPA-94/04)

COHb: Carboxyhemoglobin

COLA: Cost of Living Adjustment

COLIS: Computerized On-Line Information Service

COM: Continuous Opacity Monitor

COMPLEX: Complex Terrain Screening Model

COMPTER: Multiple Source Air Quality Model

COMS: Continuous Opacity Monitoring System

CON: Selected Contractor or "Awardee"

Conc: Concentration (40CFR86.403.78-94)

CONG: Congressional Committee

CORMIX 1: Cornell Mixing Zone Expert System (EPA-91/3)

CORPS: Army Corps of Engineers

COS: Conservative Opportunity Society

COWPS: Council on Wage and Price Stability

CP: Construction Permit (DOE-91/4)

CPCCP: Spill Prevention Control and Countermeasure Plan

CPF: Carcinogenic Potency Factor (EPA-94/04)

CPI: Consumer Price Index

CPK: Creatine Phosphokinase (EPA-92/12)

CPL: Chemistry and Physics Laboratory

CPM: Continuous Particle Monitor

CPO: Certified Project Officer (EPA-94/04)

CPP: Compliance Policy and Planning

CPR: Campaign for Pesticide Reform

CPR: Center for Public Resources

CPR: Coalition for Pesticide Reform

CPS: Compliance Program and Schedule

CPSA: Consumer Product Safety Act (40CFR762.3)

CPSC: Consumer Product Safety Commission

CPSDAA: Compliance and Program Staff to the Deputy Assistant Administrator

CPU: Central Processing Unit (computer)

CQA: Construction Quality Assurance (40CFR265.19)

Cr: Chromium

CR: Community Relations

CR: Congressional Record

CR: Continuous Radon Monitoring (EPA-94/04)

CRA: Civil Rights Act

CRA: Classification Review Area

CRAC: Chemical Regulations Advisory Committee (CMA)

CRAVE: Carcinogen Risk Assessment Verification Endeavor (EPA-92/12)

CRC: Community Relations Coordinator (40CFR300.4)

CRDL: Contract-Required Detection Limit (40CFR300-AA)

CRGS: Chemical Regulations and Guidelines System

CRL: Central Regional Laboratory

CRM: Certified Reference Material (40CFR50.1-h).

CROP: Consolidated Rules of Practice (EPA-94/04)

CRP: Community Relations Plan (40CFR300.4)

CRP: Conservation Reserve Program (EPA-94/04)

CRQL: Contract-Required Quantitation Limit (40CFR300-AA)

CRR: Center for Renewable Resources (EPA-94/04)

CRS: Congressional Research Service

CRSTER: Single Source Dispersion Model (EPA-94/04)

CRT: Cathode Ray Tube

CRTK: Community Right-To-Know (40CFR372.1)

CRWI: Coalition for Responsible Waste Incineration

Cs: Cesium

CS: Caustic Scrubber

CS: Compliance Staff

CSDT: California State Department of Transportation (DOE-91/4)

CSI: Clean Sites, Inc.

CSI: Compliance Sampling Inspection (EPA-94/04)

CSIN: Chemical Substances Information Network (EPA-94/04)

CSMA: Chemical Specialties Manufacturers Association

CSO: Combined Sewer Overflow (40CFR35.2024)

CSPA: Council of State Planning Agencies (EPA-94/04)

CSPI: Center for Science in the Public Interest

CSRL: Center for the Study of Responsive Law (EPA-94/04)

CSS: Commodity Stabilization Service

CST: Certification Short Test (40CFR86.096.3-94)

CT: Chimney Tray

CT: Closed Throttle (40CFR86)

CT: Connecticut

CTAP: Chemical Transport and Analysis Program (EPA-91/3)

CTARC: Chemical Testing and Assessment Research Commission (EPA-94/04)

CTE: Chronic Toxicity Endpoint (EPA-91/3)

CTFA: Cosmetics, Toiletries and Frangances Association

CTG: Control Technique Guideline (40CFR52.25)

cu.: Cubic (40CFR86.403.78-94)

cu.m: Cubic Meter (EPA-92/12)

cu.m/day: Cubic meter per day, used in the HEC derivation of an RfC (EPA-92/12)

Cu: Copper

cu-ft: Cubic Feet (40CFR60)

CV: Chemical Vocabulary (EPA-94/04)

CV: Coefficient of Variation (EPA-91/3)

CVAAS: Cold Vapor Atomic Absorption Spectroscopy (40CFR266-A9)

CVS: Constant volume sampler (40CFR86.403.78-94)

CW: Congress Watch

CW: Continuous working-level monitoring (EPA-94/04)

CWA: Clean Water Act (aka FWPCA) (EPA-94/04)

CWAP: Clean Water Action Project (EPA-94/04)

cwt: Hundred Weight (40CFR406.61)

CWTC: Chemical Waste Transportation Council (EPA-94/04)

CZARA: Coastal Zone Management Act Reauthorization Amendments (EPA-94/04)

CZMA: Coast Zone Management Act (16USC1451-1464)

d.: Density

D.P.H: Doctor of Public Health

D.P.Hy.: Doctor of Public Hygiene

d: Day (DOE-91/4)

D&D: Decontamination and Decommissioning (DOE-91/4)

DA: Deputy Administrator

DA: Designated Agent

DA: Dilution Air

DAA: Deputy Assistant Administrator

D_{ae}: Aerodynamic equivalent diameter (EPA-8/90b)

D_{ae}: Aerodynamic resistance diameter (EPA-8/90b)

DAFGDS: Dual Alkali Flue Gas Desulfurization System (40CFR60-AG)

DAIG: Deputy Assistant Inspector General

DAMDF: Durham Air Monitoring Demonstration Facility

DAPSS: Document and Personnel Security System (IMD) (EPA-94/04)

DAR: Defense Acquisition Regulations

DAR: Direct Assistance Request

DASD: Direct Access Storage Drive

DASE: Dutch Association of Safety Experts (EPA-

92/12)
dB: Decibel (40CFR201.1)
dBA: Decibel, A weighted (DOE-91/4)
DBA: Design-Basis Accident (DOE-91/4)
DBCP: Dibromochloropropane
DBE: Design-Basis Earthquake (DOE-91/4)
DCA: Document Control Assistant
DCCLC: Dynamic Coupled Column Liquid Chromatographic (40CFR796.1720)
dcf: Dry Cubic Feet (40CFR60)
DCI: Data Call-In (EPA-94/04)
dcm: Dry Cubic Meter (40CFR60)
DCN: Document Control Number
DCO: Delayed Compliance Order
DCO: Delayed Compliance Order (EPA-94/04)
DCO: Document Control Officer (EPA-94/04)
DCP: Direct Current Plasma
DCP: Discrimination Complaints Program
DD: Deputy Director
DDD: Dichloro-diphenyl dichloroethane
DDO: Dispute Decision Official (40CFR35.3030)
DDT: Dichloro-Diphenyl Trichloroethane (EPA-94/04)
DE: Delaware
DE: Department of Education
DE: Destruction Efficiency
DEC: Department of Environmental Conservation
DEC: Direct-shell Evacuation Control (40CFR60.271a)
DEIS: Draft Environmental Impact Statement (DOE-91/4)
DEMA: Diesel Engine Manufacturers Association
DEPP: Dredge and Fill Permit Program
DERs: Data Evaluation Records (EPA-94/04)
DES: Diethylstilbesterol (EPA-94/04)
DESCON: Computer Program that Estimates Design Condition (EPA-91/3)
DETA: Dyes Environmental and Toxicology Organization
DF: Dilution Factor (EPA-91/3)
DFLOW: Computer Program that Calculates Biologically based design flows (EPA-91/3)
DHEW: U.S. Department of Health, Education, and Welfare (now U.S. Department of Health and Human Services) (EPA-92/12)
DHHS: Department of Health and Human Services
DI: Diagnostic Inspection (EPA-94/04)
DI: Dry Injection
DL: Detection Limit (40CFR300-AA)
DLA: Designated Liability Area (40CFR57.401)

DM: Demister
DMA: Designated Management Agency (40CFR130.2)
DMA: Dimethylaniline (40CFR57.302)
DMR: Discharge Monitoring Report (40CFR122.2)
DMT: Dimethyl Terephthalate (40CFR60.561)
DMT: Dispersion Modelling and Transport (MCA)
DNA: Deoxyribonucleic acid (EPA-94/04)
DNPH: 2,4-dinitrophenylhydrazine (4086.090.3-94)
DNT: Dinitrotoluene
DO: Dissolved Oxygen (EPA-94/04)
DOA: Department of Agriculture
DOC/l: Dissolved Organic Carbon/liter (40CFR796.3180.a)
DOC: Department of Commerce (40CFR300.4)
DOD: Department of Defense (40CFR300.4)
DOE: Department of Energy (40CFR300.4)
DOE-RL: Department of Energy, Richland Operations Office (DOE-91/4)
DOE-SR: Department of Energy, Savannah River Operations Office (DOE-91/4)
DOEd: Department of Education
DOI: Department of Interior (40CFR300.4)
DOJ: Department of Justice (40CFR300.4)
DOL: Department of Labor (40CFR300.4)
DOS: Department of State (40CFR300.4)
DOS: Disk Operating System (computer)
DOT: Department of Transportation (40CFR300.4)
DOTr: Department of Treasury
DOW: Defenders Of Wildlife (EPA-94/04)
D_p: Particle diameter (EPA-8/90b)
DPA: Deepwater Ports Act (EPA-94/04)
DPD: Method of measuring chlorine residual in water (EPA-94/04)
DPD: n,n-Diethyl-Paraphenylene Diamine
DPLH: Direct Productive Labor Hour (DOE)
DPN: Diphosphopyridine Nucleotide
DQO: Data Quality Objective (EPA-94/04)
DRA: Deputy Regional Administrator
DRC: Deputy Regional Counsel
DRE: Destruction and Removal Efficiency (EPA-94/04)
DRES: Dietary Risk Evaluation System (EPA-94/04)
DRMS: Defense Reutilization and Marketing Service (EPA-94/04)
DRR: Data Review Record (EPA-94/04)
DS: Dichotomous Sampler (EPA-94/04)
DS: Dissolved Solids
DS: Dry Scrubber
DSAP: Data Self Auditing Program (EPA-94/04)

dscf: Dry Standard Cubic Feet (40CFR60)

dscm: Dry Standard Cubic Meter (40CFR60)

DSS: Decision Support System (EPA-94/04)

DSS: Domestic Sewage Study (EPA-94/04)

DT: Detectors (radon) damaged or lost (EPA-94/04)

DT: Detention Time (EPA-94/04)

DU: Decision Unit (EPA-94/04)

DU: Ducks Unlimited (EPA-94/04)

DUC: Decision Unit Coordinator (EPA-94/04)

DW: Drinking Water (EPA-92/12)

DWEL: Drinking Water Equivalent Level (EPA-94/04)

DWS: Drinking Water Standard (EPA-94/04)

Dy: Dysprosium

DYNHYD4: Hydrodynamic Model (EPA-91/3)

DYNTOX: Dynamic Toxic Model (EPA-91/3)

E-MAIL: Electronic Mail

E: Exponent (e.g., 1.5E-6 = 1.5 x 10 to the power of -6) (EPA-92/12)

E-PERM: Electret-Passive Environmental Radon Monitor

EA: Endangerment Assessment (EPA-94/04)

EA: Enforcement Agreement (EPA-94/04)

EA: Environmental Action (EPA-94/04)

EA: Environmental Assessment (EPA-94/04)

EA: Environmental Audit (EPA-94/04)

EAB: Exclusion Area Boundary (DOE-91/4)

EADS: Environmetal Assessment Data Systems

EAF: Electric Arc Furnaces (40CFR60)

EAG: Exposure Assessment Group (EPA-94/04)

EAP: Environmental Action Plan (EPA-94/04)

EAR: Environmental Auditing Roundtable (EPA-94/04)

EAS: Economic Analysis Staff

EB: Emissions Balancing (EPA-94/04)

EBCDIC: Extended Binary Coded Decimal Interchange Code

EBR: Experimental Breeder Reactor (DOE-91/4)

EBS: Emergency Broadcasting System (NRT-87/3)

EC: Effect Concentration (EPA-91/3)

EC: Effective Concentration (EPA-94/04)

EC: Emulsifiable Concentrate (EPA-94/04)

EC: Environment Canada (EPA-94/04)

EC: European Communities; or Ecology Subcommittee

EC: European Community (Common Market)

ECA: Economic Community for Africa (EPA-94/04)

ECAO: Environmental Criteria and Assessment Office, Superfund Health Risk (EPA-91/12)

ECAP: Employee Counselling and Assistance Program (EPA-94/04)

ECD: Electron Capture Detector (EPA-94/04)

ECE: Economic Commission for Europe

ECETOC: European Chemical Industry Ecology and Toxicology Centre

ECHH: Electro-Catalytic Hyper-Heaters (EPA-94/04)

ECHO: Each Community Helps Others (EPA)

ECIETC: European Chemical Industry Ecology & Toxicology Center

ECL: Environmental Chemical Laboratory (EPA-94/04)

ECL: Executive Control Language

ECLA: Economic Commission for Latin America

ECR: Enforcement Case Review (EPA-94/04)

ECRA: Economic Cleanup Responsibility Act (EPA-94/04)

ECSL: Enforcement Compliance Schedule Letters

ED: Department of Education

ED: Effective Dose (EPA-94/04)

ED: Electron Diffraction (40CFR763-AA)

ED10: 10 Percent Effective Dose (40CFR300-AA)

EDA: Economic Development Administration

EDA: Emergency Declaration Area (EPA-94/04)

EDB: Ethylene Dibromide (EPA-94/04)

EDC: Ethylene Dichloride (EPA-94/04)

EDD: Enforcement Decision Document (EPA-94/04)

EDE: Effective Dose Equivalent (DOE-91/4)

EDF: Environmental Defense Fund (EPA-94/04)

EDNA: Environmental Designation for Noise Abatement (DOE-91/4)

EDP: Electrodeposition (40CFR60.311)

EDP: Electronic Data Processing

EDRS: Enforcement Document Retrieval System (EPA-94/04)

EDS: Electronic Data System (EPA-94/04)

EDS: Energy Data System (EPA-94/04)

EDT: Edit Data Transmission

EDTA: Ethylene Diamine Triacetic Acid (EPA-94/04)

EDX: Electronic Data Exchange (EPA-94/04)

EDXA: Energy Dispersive X-ray Analysis (40CFR763-AA)

EDZ: Emission Density Zoning (EPA-94/04)

EEA: Energy and Environmental Analysis (EPA-94/04)

EEC: European Economic Commission

EECs: Estimated Environmental Concentrations (EPA-94/04)

EED: Estimated Exposure Dose (EPA-92/12)
EEG: Electroencephalogram (EPA-92/12)
EEI: Edison Electrical Institute (DOE-91/4)
EEOC: Equal Employment Opportunity Commission

EER: Excess Emission Report (EPA-94/04)
EERL: Eastern Environmental Radiation Laboratory (EPA-94/04)
EERU: Environmental Emergency Response Unit (EPA-94/04)
EESI: Environment and Energy Study Institute (EPA-94/04)
EESL: Environmental Ecological and Support Laboratory (EPA-94/04)
EETFC: Environmental Effects, Transport, and Fate Committee (EPA-94/04)
EF: Emission Factor (EPA-94/04)
EF: Exposure Frequency (EPA-91/12)
EFO: Equivalent Field Office (EPA-94/04)
EFTC: European Fluorocarbon Technical Committee (EPA-94/04)
e.g.: For Example
EG&G: EG&G Idaho, Inc. (DOE-91/4)
EGR: Exhaust Gas Recirculation (40CFR86.403.78-94)
EH: Redox Potential (EPA-94/04)
EHC: Environmental Health Committee (EPA-94/04)
EHS: Extremely Hazardous Substance (EPA-94/04)
EI: Emissions Inventory (EPA-94/04)
EIA: Economic Impact Assessment (EPA-94/04)
EIA: Environmental Impact Assessment (EPA-94/04)
EIL: Environmental Impairment Liability (EPA-94/04)
EIR: Endangerment Information Report (EPA-94/04)
EIR: Environmental Impact Report (EPA-94/04)
EIS/AS: Emissions Inventory System/Area Source (EPA-94/04)
EIS/PS: Emissions Inventory System/Point Source (EPA-94/04)
EIS: Environmental Impact Statement (EPA-94/04)
EIS: Environmental Inventory System (EPA-94/04)
EKG: Electrocardiogram (EPA-92/12)
EKMA: Empirical Kinetic Modeling Approach (EPA-94/04)
EL: Exposure Level (EPA-94/04)
ELEVEN-AA: 11-Aminoundecanoic Acid (40CFR704.25)

ELI: Environmental Law Institute (EPA-94/04)
ELISA: Enzyme-Linked Immunosorbent Assay (EPA-92/12)
ELR: Environmental Law Reporter (EPA-94/04)
ELWK: Equivalent Live Weight Killed (40CFR432.11)
EM: Electromagnetic Conductivity (EPA-94/04)
EMAS: Enforcement Management and Accountability System (EPA-94/04)
EMI/RFI: Electromagnetic Interference/Radio Frequency Interference (40CFR60.721)
EMI: Emergency Management Institute (NRT-87/3)
EMR: Environmental Management Report (EPA-94/04)
EMS: Enforcement Management System (EPA-94/04)
EMSL: Environmental Monitoring Support Laboratory (EPA-94/04)
EMSL: Environmental Monitoring Systems Laboratory (EPA-94/04)
EMTD: Estimated Maximum Tolerated Dose (EPA-92/12)
EMTS: Environmental Monitoring Testing Site (EPA-94/04)
EMTS: Exposure Monitoring Test Site (EPA-94/04)
ENM: Environmental Noise Model (DOE-91/4)
EO: Ethylene Oxide (EPA-94/04)
EO: Executive Officer
EO: Executive Order
EOB: Executive Office Building
EOC: Emergency Operating Center (EPA-94/04)
EOD: Entrance on Duty
EOE: Equal Opportunity Employer
EOF: Emergency Operations Facility (RTP) (EPA-94/04)
EOJ: End of Job
EOP: Emergency Operations Plan (NRT-87/3)
EOP: End Of Pipe (EPA-94/04)
EOT: Emergency Operations Team (EPA-94/04)
EOY: End of Year
EP tox: extraction procedure toxicity
EP: Earth Protectors (EPA-94/04)
EP: End point (40CFR86.403.78-94)
EP: End-use Product (EPA-94/04)
EP: Environmental Profile (EPA-94/04)
EP: Equilibrium Partitioning (EPA-91/3)
EP: Experimental Product (EPA-94/04)
EP: Extraction Procedure (EPA-94/04)
EPA: Environmental Protection Agency (40CFR86.403.78-94)

EPAA: Environmental Programs Assistance Act (EPA-94/04)

EPAAR: EPA Acquisition Regulations (EPA-94/04)

EPACASR: EPA Chemical Activities Status Report (EPA-94/04)

EPAYS: EPA Payroll System

EPCA: Energy Policy and Conservation Act (EPA-94/04)

EPCRA: Emergency Planning and Community Right-To-Know Act of 1986 (40CFR22.40)

EPCRA: Emergency Preparedness and Community Right to Know Act (EPA-94/04)

EPD: Emergency Planning District (EPA-94/04)

EPI: Environmental Policy Institute (EPA-94/04)

EPIC: Environmental Photographic Interpretation Center (EPA-94/04)

EPNL: Effective Perceived Noise Level (EPA-94/04)

EPO: Estuarian Programs Office (NOAA)

EPRI: Electric Power Research Institute (EPA-94/04)

EPTC: Extraction Procedure Toxicity Characteristic (EPA-94/04)

eq: Equivalent (40CFR60)

Er: Erbium

ER: Electrical Resistivity (EPA-94/04)

ER: Extrarespiratory (EPA-92/12)

ERA: Economic Regulatory Agency (EPA-94/04)

ERA: Equal Rights Amendment

ERAB: Energy Research Advisory Board (DOE-91/4)

ERAMS: Environmental Radiation Ambient Monitoring System (EPA-94/04)

ERC: Emergency Response Commission (EPA-94/04)

ERC: Emissions Reduction Credit (EPA-94/04)

ERC: Environmental Research Center (EPA-94/04)

ERCS: Emergency Response Cleanup Services (EPA-94/04)

ERD&DAA: Environmental Research, Development and Demonstration Authorization Act (EPA-94/04)

ERISA: Employee Retirement Income Security Act

ERL: Environmental Research Laboratory (EPA-94/04)

ERNS: Emergency Response Notification System (EPA-94/04)

ERP: Enforcement Response Policy (EPA-94/04)

ERPG: Emergency Response Planning Guideline (DOE-91/4)

ERT: Emergency Response Team (EPA-94/04)

ERT: Environmental Response Team (40CFR300.4)

ERTAQ: ERT Air Quality Model (EPA-94/04)

Es: Einsteinium

ES: Enforcement Strategy (EPA-94/04)

ES: Entrainment Separator

ESA: Endangered Species Act (EPA-94/04)

ESA: Environmentally Sensitive Area (EPA-94/04)

ESC: Endangered Species Committee (EPA-94/04)

ESCA: Electron Spectroscopy for Chemical Analysis (EPA-94/04)

ESCAP: Economic and Social Commission for Asia and the Pacific (EPA-94/04)

ESECA: Energy Supply and Environmental Coordination Act (EPA-94/04)

ESH: Environmental Safety and Health (EPA-94/04)

ESP: Electrostatic Precipitator (40CFR60.471)

ESU: Electrostatic Unit

et al.: and elsewhere

et seq.: and the following one(s)

ET: Emissions Trading (EPA-94/04)

ET: Extrathoracic Region of the Respiratory Tract (EPA-92/12)

etc. (et cetera): and so forth

ETP: Emissions Trading Policy (EPA-94/04)

ETS: Emergency Temporary Standard

ETS: Environmental Tobacco Smoke (EPA-94/04)

ETV: Electrothermal Vaporization

Eu: Europium

EUP: End-Use Product (EPA-94/04)

EUP: Environmental Use Permit

EUP: Experimental Use Permit (40CFR168.22)

EVA: Ethylene Vinyl Acetate

EWCC: Environmental Workforce Coordinating Committee (EPA-94/04)

EX: Executive Level Appointment

EXAMS: Exposure Analysis Modeling System (EPA-94/04)

EXAMS-ll: Exposure Analysis Modeling System (EPA-91/3)

ExEx: Expected Exceedance

ExEx: Expected Exceedance (EPA-94/04)

EXIMBANK: Export-Import Bank of the U.S.

F/M: Food to Microorganism Ratio (EPA-94/04)

F: Fahrenheit (40CFR86.403.78-94)

F: Fluorine

F1: First Filial Generation (in experimental animals) (EPA-92/12)

FAA: Federal Aviation Administration, Department of Transportation (40CFR87.2)

FAC: Free Available Chlorine

FACA: Federal Advisory Committee Act (EPA-94/04)

FAME: Framework for Achieving Managerial Excellence

FAN: Fixed Account Number (EPA-94/04)

FAO: Food and Agriculture Organization

FAR: Federal Acquisition Regulation (40CFR34.105)

FASB: Financial Accounting Standards Board

FAST: Fugitive Assessment Sampling Train

FATES: FIFRA and TSCA Enforcement System (EPA-94/04)

FAV: Final Acute Value (EPA-91/3)

FBC: Fluidized Bed Combustion (EPA-94/04)

FC: Fluorocarbon

FCC: Federal Communications Commission

FCC: Fluid Catalytic Converter (EPA-94/04)

FCCU: Fluid Catalytic Cracking Unit (EPA-94/04)

FCM2: WASP Food Chain Model (EPA-91/3)

FCO: Federal Coordinating Officer (40CFR300.4)

FCO: Forms Control Officer (40FR300)

FDA: Food and Drug Administration (40CFR160.3)

FDAAL: Food and Drug Administration Action Level (40CFR300)

FDF: Fundamentally Different Factors (EPA-94/04)

FDIC: Federal Deposit Insurance Corporation

FDL: Final Determination Letter (EPA-94/04)

FDO: Fee Determination Official (EPA-94/04)

Fe: Iron

FE: Fugitive Emissions (EPA-94/04)

FEA: Federal Energy Administration

FEC: Federal Executive Council

FEDS: Federal Energy Data System (EPA-94/04)

FEFx: Forced Expiratory Flow (EPA-94/04)

FEGLI: Federal Employee Group Life Insurance

FEHB: Federal Employees Health Benefits

FEI: Federal Executive Institute

FEIS: Final Environmental Impact Statement (DOE-91/4)

FEIS: Fugitive Emissions Information System (EPA-94/04)

FEL: Family emission limit (40CFR86.090.3-94)

FEL: Frank Effect Level (EPA-94/04)

FEMA: Federal Emergency Management Agency (40CFR300.5)

FEPCA: Federal Environmental Pesticide Control Act; enacted as amendments to FIFRA (EPA-94/04)

FERC: Federal Energy Regulatory Commission (EPA-94/04)

FERSA: Federal Employee Retirement System Act

FES: Factor Evaluation System (EPA-94/04)

FETRA: Finite Element Transport Model (EPA-91/3)

FEV: Forced Expiratory Volume (EPA-94/04)

FEV_1: Forced expiratory volume at one second (EPA-8/90b)

FEV1: Forced Expiratory Volume-one second (EPA-94/04)

FEVI: Front End Volatility Index (EPA-94/04)

FEW: Federally Employed Women

FF: Fabric Filter

FF: Federal Facilities (EPA-94/04)

FFA: Flammable Fabrics Act

FFAR: Fuel and Fuel Additive Registration (EPA-94/04)

FFDCA: Federal Food, Drug, and Cosmetic Act (40CFR160.3)

FFF: Firm Financial Facility (EPA-94/04)

FFFSG: Fossil-Fuel-Fired Steam Generator (EPA-94/04)

FFIS: Federal Facilities Information System (EPA-94/04)

FFP: Firm Fixed Price (EPA-94/04)

FFPA: Farmland Protection Policy Act

FGD: Flue Gas Desulfurization

FGD: Flue-Gas Desulfurization (EPA-94/04)

FGETS: Food and Gill Exchange of Toxic Substances (EPA-91/3)

FHA: Farmers Home Administration

FHA: Federal Housing Administration

FHLBB: Federal Home Loan Bank Board

FHSA: Federal Hazardous Substances Act

FHWA: Federal Highway Administration (40CFR93.101)

FI: Formaldehyde Institute

FIA: Federal Insurance Administration

FIC: Federal Information Center

FICA: Federal Insurance Contributions Act

FID: Flame Ionization Detector (40CFR86)

FIFO: First In/First Out

FIFRA: Federal Insecticide, Fungicide, and Rodenticide Act (EPA-94/04)

FIM: Friable Insulation Material (EPA-94/04)

FINDS: Facility Index System (EPA-94/04)

FIP: Federal Implementation Plan (40CFR52.741.a)

FIP: Federal Information Plan

FIP: Final Implementation Plan (EPA-94/04)

FIPS: Federal Information Procedures System (EPA-94/04)

FIT: Field Investigation Team (EPA-94/04)

FL: Florida

FL: Full Load (40CFR86)

FLETC: Federal Law Enforcement Training Center (EPA-94/04)

FLM: Federal Land Manager (EPA-94/04)

FLOSTAT: U.S. Geological Survey Computer Program that estimates the arithmetic mean flow and 7Q10 of rivers and streams (EPA-91/3)

FLP: Flash Point (EPA-94/04)

FLPMA: Federal Land Policy and Management Act (EPA-94/04)

FLSA: Fair Labor Standards Act

Fm: Fermium

FM: Food Chain Multipliers (EPA-91/3)

FMAP: Financial Management Assistance Project (EPA-94/04)

FMC: Federal Maritime Commission

FMFIA: Federal Managers Financial Integrity Act

FMIA: Federal Meat Inspection Act (21USC60)

FML: Flexible Membrane Liner (40CFR258.40-93)

FMO: Financial Management Officer

FMP: Facility Management Plan (EPA-94/04)

FMP: Financial Management Plan (EPA-94/04)

FMS: Financial Management System (EPA-94/04)

FMVCP: Federal Motor Vehicle Control Program (EPA-94/04)

FN: Fog Nozzle

FNSI: Finding of No Significant Impact (40CFR1508.13)

FOE: Friends Of the Earth (EPA-94/04)

FOI: Freedom of Information (EPA-92/12)

FOIA: Freedom Of Information Act (EPA-94/04)

FOISD: Fiber Optic Isolated Spherical Dipole Antenna (EPA-94/04)

FONSI: Finding Of No Significant Impact (EPA-94/04)

FORAST: Forest Response to Anthropogenic Stress (EPA-94/04)

FORTRAN: Formula Translation

FP: Fine Particulate (EPA-94/04)

FPA: Federal Pesticide Act (EPA-94/04)

FPAS: Foreign Purchase Acknowledgement Statements (EPA-94/04)

FPC: Federal Power Commission

FPD: Flame Photometric Detector (EPA-94/04)

FPEIS: Fine Particulate Emissions Information System (EPA-94/04)

FPM: Federal Personnel Manual (EPA-94/04)

FPPA: Federal Pollution Prevention Act (EPA-94/04)

FPR: Federal Procurement Regulation (EPA-94/04)

FPRS: Federal Program Resources Statement (EPA-94/04)

FPRS: Formal Planning and Supporting System (EPA-94/04)

Fr: Francium

FR: Federal Register (EPA-94/04)

FR: Final Rulemaking (EPA-94/04)

FRA: Federal Register Act (EPA-94/04)

FRB: Federal Reserve Board

FRC: Federal Records Center

FRC: Functional Reserve Capacity (EPA-92/12)

FRDS: Federal Reporting Data System

FREDS: Flexible Regional Emissions Data System (EPA-94/04)

Freon-113: trichlorotrifluoroethane (DOE-91/4)

FRES: Forest Range Environmental Study (EPA-94/04)

FRM: Federal Reference Methods (EPA-94/04)

FRN: Federal Register Notice (EPA-94/04)

FRN: Final Rulemaking Notice (EPA-94/04)

FRS: Formal Reporting System (EPA-94/04)

FRTIB: Federal Retirement Thrift Investment Board

FS: Feasibility Study (40CFR300.4)

FS: Forest Service

FSA: Food Security Act (EPA-94/04)

FSP: Field Sampling Plan

FSS: Facility Status Sheet (EPA-94/04)

FSS: Federal Supply Schedule (EPA-94/04)

ft: Feet (40CFR60)

FT: Full Time

ft^2: Square Feet (40CFR60)

ft^3: Cubic Feet (40CFR60)

FTA: Federal Transit Administration (40CFR93.101)

FTC: Federal Trade Commission

FTE: Full Time Equivalent

FTIR: Fourier Transform Infrared

FTP: Federal Test Procedure (40CFR89.3-94)

FTS: Federal Telecommunications System

FTS: File Transfer Service (EPA-94/04)

FTT: Full Time Temporary

FTTS: FIFRA/TSCA Tracking System (EPA-94/04)

FTU: Formazin Turbidity Unit

FUA: Fuel Use Act (EPA-94/04)

FURS: Federal Underground Injection Control Reporting System (EPA-94/04)

FVC: Forced Vital Capacity (EPA-8/90b)

FVMP: Federal Visibility Monitoring Program

(EPA-94/04)

FWCA: Fish and Wildlife Coordination Act (EPA-94/04)

FWP: Federal Womens Program

FWPCA: Federal Water Pollution and Control Act (40CFR220.2)

FWPCA: Federal Water Pollution and Control Administration (EPA-94/04)

FWQC: Federal Water Quality Criteria (EPA-91/12)

FWS: U.S. Fish and Wildlife Service (40CFR233.2)

FY: Fiscal Year

FYI: For Your Information

G: Gram (40CFR60)

G/MI: Grams per mile (EPA-94/04)

G-EQ: gram-equivalent (40CFR60)

G/DSCM: Grams/(dry standard cubic meter)

G/Ncm: Grams/(normal cubic meter), 1 normal cubic meter = 1 dry standard cubic meter

G/KW-HR: Grams per kilowatt hour (40CFR89.3-94)

G/MI: Grams per mile

G/ML: Grams/milliliter, 1 milliliter = 1 cc = 1 cm^3

G&A: General and Administrative Cost

Ga: Gallium

GA: Georgia

GAAP: Generally Accepted Accounting Principles (EPA-94/04)

GAC: Granular Activated Carbon (40CFR141.61.b)

GAC: Ground-water Activated Carbon

GACT: Granular Activated Carbon Treatment (EPA-94/04)

gal: Gallon (40CFR60)

GAO: General Accounting Office

Gas Research Institute

GATT: General Agreement on Tariffs and Trade

GBL: Government Bill of Lading

GC/MS: Gas Chromatograph/Mass Spectograph (EPA-94/04)

GC: Gas chromatograph (40CFR86.090.3-94)

GC: Gas Cooler

GC: General Counsel

GCP: Good Combustion Practices

GCWR: Gross Combination Weight Rating (40CFR202.10)

Gd: Gadolinium

GD: Guidance Document

GDE: Generic Data Exemption (EPA-94/04)

Ge: Germanium

GEA: Glossary of EPA Acronyms

GEI: Geographic Enforcement Initiative (EPA-94/04)

GEMS: Global Environmental Monitoring System (EPA-94/04)

GEMS: Graphical Exposure Modeling System (EPA-94/04)

GEP: Good Engineering Practice (40FR51)

GF: General Files

GFF: Glass Fiber Filter (EPA-94/04)

GFO: Grant Funding Order (EPA-94/04)

GFP: Government-Furnished Property (EPA-94/04)

GI: Gastrointestinal (EPA-92/12)

GICS: Grant Information and Control System (EPA-94/04)

GIS: Geographic Information Systems (EPA-94/04)

GIS: Global Indexing System (EPA-94/04)

GLBC: Great Lakes Basin Commission

GLC: Gas Liquid Chromatography (EPA-94/04)

GLERL: Great Lakes Environmental Research Laboratory (EPA-94/04)

GLICP: Great Lakes Initiative Contract Program

GLNPO: Great Lakes National Program Office (EPA-94/04)

GLP: Good Laboratory Practices (EPA-94/04)

GLWQA: Great Lakes Water Quality Agreement (EPA-94/04)

GMA: Grocery Manufacturers Association

GMCC: Global Monitoring for Climatic Change (EPA-94/04)

GMT: Greenwich Mean Time

GNP: Gross National Product

GOCM: Goals, Objectives, Commitments, and Measures

GOCO: Government-Owned/Contractor-Operated (EPA-94/04)

GOGO: Government-Owned/Government-Operated (EPA-94/04)

GOP: General Operating Procedures (EPA-94/04)

GOPO: Government-Owned/Privately Operated (EPA-94/04)

GOTCHA: Generalized Overall Toxics Control and Hazards Act

GPAD: Gallons-per-acre per-day (EPA-94/04)

GPD: Gallons Per Day

GPG: Grams-per-Gallon (EPA-94/04)

GPM: Gallons Per Minute

GPO: Government Printing Office

GPR: Ground-Penetrating Radar (EPA-94/04)

GPS: Groundwater Protection Strategy (EPA-94/04)

GPT: Glutamic-Pyruvic Transaminase (EPA-92/12)

GR: Grain

GR: Grab Radon Sampling (EPA-94/04)

GR/DSCF: Grains/(dry standard cubic feet), 1 gr/dscf = 2,300 milligrams/dscm

GRAS: Generally Regarded As Safe

GRCDA: Government Refuse Collection and Disposal Association (EPA-94/04)

GRGL: Groundwater Residue Guidance Level (EPA-94/04)

GS: General Schedule

GSA: General Services Administration

GTN: Global Trend Network (EPA-94/04)

GTR: Government Transportation Request (EPA-94/04)

GVP: Gasoline Vapor Pressure (EPA-94/04)

GVW: Gross Vehicle Weight (40CFR51.731)

GVWR: Gross Vehicle Weight Rating (40CFR88.103-94)

GW: Grab Working-Level Sampling (EPA-94/04)

GW: Groundwater (EPA-94/04)

GWM: Groundwater Monitoring (EPA-94/04)

GWPS: Groundwater Protection Standard (EPA-94/04)

GWPS: Groundwater Protection Strategy (EPA-94/04)

H.R.: House of Representatives

h: hour (40CFR86.403.78-94)

H: Humidifier

H: Hydrogen

H_2O: Water (40CFR86.403.78-94)

ha: Hectare (40CFR401.11)

Ha: Hahnium

HA: Health Advisory (EPA-94/04)

HAC: Heating and Air Conditioning

HAD: Health Assessment Document (EPA-94/04)

HAF: Halogen Acid Furnace (40CFR260.10)

HAP: Hazardous Air Pollutant (EPA-94/04)

HAPEMS: Hazardous Air Pollutant Enforcement Management System (EPA-94/04)

HAPPS: Hazardous Air Pollutant Prioritization System (EPA-94/04)

HAS: Health Assessment Summary (EPA-92/12)

HATREMS: Hazardous and Trace Emissions System (40CFR51.117)

HAZMAT: Hazardous Materials

HAZMAT: Hazardous Materials (EPA-94/04)

HAZOP: Hazard and Operability Study (EPA-94/04)

HB: Health Benefits

HBEP: Hispanic and Black Employment Programs

HBV: Hepatitis B Virus (29CFR1910.1030)

HC: Hazardous Constituents (EPA-94/04)

HC: Hydrocarbon(s) (40CFR86.403.78-94)

HCA: Hydrogen Chloride Absorber

HCC: House Commerce Committee

HCCPD: Hexachlorocyclo-pentadiene (EPA-94/04)

HCHO: Formaldehyde (40CFR88.103-94)

HCP: Hypothermal Coal Process (EPA-94/04)

HCS: Hydrogen Chloride Scrubber

HCT: Hematocrit (EPA-92/12)

HDD: Halogenated Dibenzodioxins (40CFR766.3-93)

HDD: Heavy-Duty Diesel (EPA-94/04)

HDE: Heavy-Duty Engine (EPA-94/04)

HDF: Halogenated Dibenzofurans (40CFR766.3-93)

HDG: Heavy-Duty Gasoline-Powered Vehicle (EPA-94/04)

HDPE: High Density Polyethylene (EPA-94/04)

HDT: Heavy-Duty Truck (EPA-94/04)

HDT: Highest Dose Tested in a study (EPA-94/04)

HDV: Heavy-Duty Vehicle (40CFR88.103-94)

He: Helium

HE: Heat Exchanger

HEAL: Human Exposure Assessment Location (EPA-94/04)

HEAST: Health Effects Assessment Summary Tables (EPA-91/12)

HEC: Human Equivalent Concentration (EPA-92/12)

HECC: House Energy and Commerce Committee (EPA-94/04)

HEEP: Health and Environmental Effects Profile (EPA-92/12)

HEI: Health Effects Institute (EPA-94/04)

HEM: Human Exposure Modeling (EPA-94/04)

HEP: Hispanic Employment Program

HEPA filter: High-Efficiency Particulate Air filter (40CFR763.121)

HEPA: High-Efficiency Particulate Air (EPA-94/04)

HERL: Health Effects Research Laboratory

HERS: Hyperion Energy Recovery System (EPA-94/04)

HES: High Energy Scrubber

HEX-BCH: Hexachloronorbornadiene

Hf: Hafnium

HFID: Heated Flame Ionization Detector (40CFR86)

HFPO: Hexafluoropropylene oxide (40CFR704.104)

Hg: Mercury

Hgb: Hemoglobin (EPA-92/12)

HHC: Human Health Criteria (EPA-91/3)

HHDFLOW: Historic Daily Flow Program (EPA-91/3)

HHE: Human Health and the Environment (EPA-94/04)

HHEM: Human Health Evaluation Manual (EPA-91/12)

HHFA: Housing and Home Finance Agency

HHS: Department of Health and Human Services (40CFR26.101)

HHS: Department of Health and Human Services (Formerly HEW) (40CFR300.4)

HHV: Higher Heating Value (EPA-94/04)

HI-VOL: High-Volume Sampler (EPA-94/04)

HI: Hawaii

HI: Hazard Index (EPA-94/04)

HIT: Hazard information transmission (NRT-87/3)

HIV: Human Immunodeficiency Virus (29CFR1910.1030)

HIWAY: A Line Source Model for Gaseous Pollutants (EPA-94/04)

HLLW: High-Level Liquid Waste (DOE-91/4)

HLRW: High Level Radioactive Waste (40CFR191.02)

HLW: High-Level Waste (DOE-91/4)

HMIS: Hazardous Materials Information System (EPA-94/04)

HMS: Hanford Meteorological Station (DOE-91/4)

HMS: Highway Mobile Source (EPA-94/04)

HMTA: Hazardous Materials Transportation Act (EPA-94/04)

HMTR: Hazardous Materials Transportation Regulations (EPA-94/04)

Ho: Holmium

HO: Headquarters Offices

HOC: Halogenated Organic Carbons (EPA-94/04)

HOC: Halogenated Organic Compound (40CFR268.2)

HON: Hazardous Organic NESHAP (EPA-94/04)

HOV: High-Occupancy Vehicle (EPA-94/04)

HP: Horse Power (40CFR401.11)

HPLC: High-Performance Liquid Chromatography (EPA-94/04)

HPLC: High-pressure liquid chromatography (40CFR86.090.3-94)

HPV: High Priority Violator (EPA-94/04)

HQ: Hazard Quotient (EPA-91/12)

HQ: Headquarters

HQCDO: Headquarters Case Development Officer (EPA-94/04)

hr.: Hour(s) (40CFR87.2-94)

HRA: Hourly Rolling Average

HRC: Human Resources Council

HRGC: High Resolution Gas Chromatography (40CFR766.3-93)

HRMS: High Resolution Mass Spectrometry (40CFR766.3-93)

HRS: Hazard Ranking System (40CFR300.4)

HRS: Hazardous Ranking System (EPA-94/04)

HRUP: High-Risk Urban Problem (EPA-94/04)

HSA: Historic Sites Act of 1935 (40CFR6.301)

HSDB: Hazardous Substance Data Base (EPA-94/04)

HSL: Hazardous Substance List (EPA-94/04)

HSP: Health and Safety Plan

HSPF: Hydrologic Simulation Program (EPA-91/3)

HSWA: Hazardous and Solid Waste Amendments of 1984 (EPA-94/04)

HT: Hypothermally Treated (EPA-94/04)

HTHE: High Temperature Heat Exchanger

HTP: High Temperature and Pressure (EPA-94/04)

HUD: Department of Housing and Urban Development

HVAC: Heating, Ventilating, and Air Conditioning

HVAF: High Velocity Air Filter (40FR60.471)

HVIO: High Volume Industrial Organics (EPA-94/04)

HVL: Highly Volatile Liquid (40CFR195.2)

HW: Hazardous Waste (EPA-94/04)

HWDGR: Hazardous Waste Disposal Guidelines and Regulations

HWDMS: Hazardous Waste Data Management System (EPA-94/04)

HWERL: Hazardous Waste Engineering Research Laboratory (See RREL)

HWGTF: Hazardous Waste Groundwater Task Force (EPA-94/04)

HWGTF: Hazardous Waste Groundwater Test Facility (EPA-94/04)

HWLT: Hazardous Waste Land Treatment (EPA-94/04)

HWM: Hazardous Waste Management (40CFR270.2)

HWPF: Hazardous Waste Processing Facility (DOE-91/4)

HWR: Heavy-Water Reactor (DOE-91/4)

HWRTF: Hazardous Waste Restrictions Task Force (EPA-94/04)

HWTC: Hazardous Waste Treatment Council (EPA-94/04)

Hz: Hertz (40CFR60)

I/A: Innovative/Alternative (EPA-94/04)

I/I: Infiltration/inflow (40CFR35.905)

I/M: Inspection/Maintenance (40CFR51.350)

I/M: Inspection/Maintenance (EPA-94/04)

i.m.: Intramuscular (EPA-92/12)

i.p.: Intraperitoneal (EPA-92/12)

i.v.: Intravenous (EPA-92/12)

I: Interstate (DOE-91/4)

I: Iodine

IA: Interagency Agreement (EPA-94/04)

IA: Iowa

IAAC: Interagency Assessment Advisory Committee (EPA-94/04)

IAEA: International Atomic Energy Agency (DOE-91/4)

IAG: Interagency Agreement (EPA-94/04)

IAP: Incentive Awards Program (EPA-94/04)

IAP: Individual Annoyance Prediction (DOE-91/4)

IAP: Indoor Air Pollution (EPA-94/04)

IARC: International Agency for Research on Cancer (EPA-94/04)

IATA: International Air Transport Association

IATDB: Interim Air Toxics Data Base (EPA-94/04)

IBA: Industrial Biotechnology Association

IBP: Initial Boiling Point (40CFR86)

IBR: Incorporation by Reference (40CFR60.17)

IBRD: International Bank for Reconstruction and Development

IBT: Industrial Biotest Laboratory (EPA-94/04)

IC: Inhibition Concentration (EPA-91/3)

ICAIR: Interdisciplinary Planning and Information Research (EPA-94/04)

ICAO: International Civil Aviation Organization

ICAP: Inductively Coupled Argon Plasma (40CFR266-A9)

ICB: Information Collection Budget (EPA-94/04)

ICBEN: International Commission on the Biological Effects of Noise (EPA-94/04)

ICC: Interstate Commerce Commission

ICE: Industrial Combustion Emissions Model (EPA-94/04)

ICE: Internal Combustion Engine (EPA-94/04)

ICGEC: Interagency Collaborative Group on Environmental Carcinogenesis

ICI: Independent Commercial Importer (40CFR89.3-94)

ICIE: International Center for Industry and Environment

ICP: Inductively Coupled Plasma (EPA-94/04)

ICR: Information Collection Request (EPA-94/04)

ICR: Institute of Cancer Research (EPA-92/12)

ICRDB: International Cancer Research Data Bank

ICRE: Ignitability, Corrosivity, Reactivity, Extraction (EPA-94/04)

ICRP: International Commission for Radiological Protection (EPA-92/12)

ICRP: International Commission on Radiological Protection (EPA-94/04)

ICRU: International Commission on Radiological Units (40CFR141.2)

ICS: Incident Command System (EPA-94/04)

ICS: Institute for Chemical Studies (EPA-94/04)

ICS: Intermittent Control Strategies (EPA-94/04)

ICS: Intermittent Control System (40CFR51.100-nn)

ICTCB: International Congress on Toxic Combustion Byproduct

ICWM: Institute for Chemical Waste Management (EPA-94/04)

ID fan: Induced Draft Fan

ID: Idaho

ID: Inside Diameter (40CFR60)

ID50: Infectious Dose 50

IDHW: Idaho Department of Health and Welfare (DOE-91/4)

IDLH: Immediately Dangerous to Life and Health (EPA-94/04)

IDLH: Immediately Dangerous to Life or Health (29CFR1910.120)

IDOD: Immediate Dissolved Oxygen Demand

i.e.: That Is

IEB: International Environment Bureau (EPA-94/04)

IEEE: Institute of Electrical and Electronics Engineers

IEMP: Integrated Environmental Management Project (EPA-94/04)

IEMS: Integrated Emergency Management System (NRT-87/3)

IES: Institute for Environmental Studies (EPA-94/04)

IFB: Invitation for Bid (EPA-94/04)

IFCAM: Industrial Fuel Choice Analysis Model (EPA-94/04)

IFIS: Industry File Information System (EPA-94/04)

IFMS: Integrated Financial Management System (EPA-94/04)

IFPP: Industrial Fugitive Process Particulate (EPA-94/04)

IG: Inspector General

IGCC: Integrated Gasification Combined Cycle (EPA-94/04)

IGCI: Industrial Gas Cleaning Institute (EPA-94/04)

IIS: Inflationary Impact Statement (EPA-94/04)

IJC: International Joint Commission (on Great Lakes) (EPA-94/04)

IL: Illinois

ILO: International Labor Organization

IMEP: Indicated Mean Effective Pressure (40CFR85.2122)

IMF: International Monetary Fund

IMM: Intersection Midblock Model (EPA-94/04)

IMPACT: Integrated Model of Plumes and Atmosphere in Complex Terrain (EPA-94/04)

IMPROVE: Interagency Monitoring of Protected Visual Environment (EPA-94/04)

in H_2O: Inches of Water (40CFR60)

in Hg: Inches of Mercury (40CFR60)

in.: Inch(es) (40CFR86.403.78-94)

In: Indium

IN: Indiana

INEL: Idaho National Engineering Laboratory (DOE-91/4)

INPUFF: A Gaussian Puff Dispersion Model (EPA-94/04)

INT: Intermittent (EPA-94/04)

IO: Immediate Office

IOAA: Immediate Office of the Assistant Administrator

IOAU: Input/Output Arithmetic Unit

IOB: Iron Ore Beneficiation (EPA-94/04)

IOU: Input/Output Unit (EPA-94/04)

IP: Inhalable Particles (EPA-94/04)

IPA: Intergovernmental Personnel Act

IPA: Intergovernmental Personnel Agreement

IPCS: International Program on Chemical Safety (EPA-94/04)

IPDWS: Interim Primary Drinking Water Standard (SWDA)

IPM: Inhalable Particulate Matter (EPA-94/04)

IPM: Integrated Pest Management (EPA-94/04)

IPP: Implementation Planning Program (EPA-94/04)

IPP: Independent Power Production (40CFR72.2)

IPP: Integrated Plotting Package (EPA-94/04)

IPP: Intermedia Priority Pollutant (document) (EPA-94/04)

Ir: Iridium

IR: Infrared

IRA: Initial Rate of Absorption

IRAC: International Association of Research on Cancer (France)

IRB: Institutional Review Board (40CFR26.102)

IRG: Interagency Review Group (EPA-94/04)

IRI: Industrial Risk Insurance

IRIS: Instructional Resources Information System (EPA-94/04)

IRIS: Integrated Risk Information System (EPA-94/04)

IRLG: Interagency Regulatory Liaison Group (Composed of EPA, CPSC, FDA, and OSHA) (EPA-94/04)

IRM: Intermediate Remedial Measures (EPA-94/04)

IRMC: Inter-Regulatory Risk Management Council (EPA-94/04)

IRP: Installation Restoration Program (EPA-94/04)

IRPTC: International Register of Potentially Toxic Chemicals (EPA-94/04)

IRR: Institute of Resource Recovery (EPA-94/04)

IRS: Internal Revenue Service

IRS: International Referral Systems (EPA-94/04)

IS: Interim Status (EPA-94/04)

ISAC: Industry Sector Advisory Committee

ISA: Instrument Society of American

ISAM: Indexed Sequential File Access Method (EPA-94/04)

ISC: Industrial Source Complex (EPA-94/04)

ISCL: Interim Status Compliance Letter (EPA-94/04)

ISCLT: Industrial Source Complex Long Term Model (EPA-94/04)

ISCST: Industrial Source Complex Short Term Model (EPA-94/04)

ISD: Interim Status Document (EPA-94/04)

ISE: Ion-specific electrode (EPA-94/04)

ISMAP: Indirect Source Model for Air Pollution (EPA-94/04)

ISO: International Organization for Standardization

ISO: International Science Organization

ISO: International Standard Organization

ISPF: (IBM) Interactive System Productivity Facility (EPA-94/04)

ISS: Interim Status Standards (EPA-94/04)

ISTEA: Intermodal Surface Transportation Efficiency Act of 1991 (40CFR93.101)

ISV: In Situ Vitrification

ITC: Interagency Testing Committee (EPA-94/04)

ITC: International Trade Commission

ITDP: Individual Training and Development Plan

ITII: International Technical Information Institute (EPA-92/12)

ITP: Individual Training Plan

ITP: International Travel Plan

ITSDC: Interagency Toxic Substances Data

Committee
IUP: Intended Use Plan (40CFR35.3150)
IUPAC: International Union of Pure and Applied Chemistry (29CFR1910.1200)
IW: Infectious Waste
IUR: Inventory Update Rule (EPA-94/04)
IWC: In-Stream Waste Concentration (EPA-94/04)
IWS: Ionizing Wet Scrubber (EPA-94/04)
J: Joule (40CFR60)
JAPCA: Journal of Air Pollution Control Association (EPA-94/04)
JCL: Job Control Language (EPA-94/04)
JEC: Joint Economic Committee (EPA-94/04)
JECFA: Joint Expert Committee of Food Additives (EPA-94/04)
JLC: Justification for Limited Competition (EPA-94/04)
JMPR: Joint Meeting on Pesticide Residues (EPA-94/04)
JNCP: Justification for Non-Competitive Procurement (EPA-94/04)
JOFOC: Justification for Other Than Full and Open Competition (EPA-94/04)
JPA: Joint Permitting Agreement (EPA-94/04)
JPL: Jet Propulsion Laboratory
JSD: Jackson Structured Design (EPA-94/04)
JSP: Jackson Structured Programming (EPA-94/04)
JTU: Jackson Turbidity Unit (EPA-94/04)
k: 1000
K: Kelvin (40CFR86.403.78-94)
K: Potassium
kg: Kilogram(s) (40CFR86.403.78-94)
kkg: 1000 kilogram(s) (40CFR401.11)
km: Kilometre(s) (40CFR86.403.78-94)
KOV: Knock Out Vessel
kPa: Kilopascals (one thousand newtons per square meter) (40CFR52.741-94)
KPEG: potassium polyethylene glycolate
Kr: Krypton
KS: Kansas
Ku: Kurchatovium
kW: Kilowatt (40CFR89.3-94)
KWH: Kilowatt Hour (40CFR401.11)
KY: Kentucky
L: Liter (40CFR401.11)
La: Lanthanum
LA: Load Allocation (40CFR130.2)
LA: Louisiana
LAA: Lead Agency Attorney (EPA-94/04)
LADD: Lowest Acceptable Daily Dose (EPA-94/04)

LAER: Lowest Achievable Emission Rate (40CFR51)
LAER: Lowest Achievable Emission Rate (EPA-94/04)
LAI: Laboratory Audit Inspection (EPA-94/04)
LAMP: Lake Acidification Mitigation Project (EPA-94/04)
LARC: International Agency for Research on Cancer (29CFR1910.1200)
LAS: Linear Alkylate Sulfonate
LASER: Light Amplification by Stimulated Emission of Radiation
lb: Pound(s) (40CFR86.403.78-94)
LC LO: Lethal Concentration Low
LC: Lethal Concentration (40CFR116)
LC: Liquid Chromatography (EPA-94/04)
LC50: Lethal Concentration 50; concentration lethal to 50% of the animals (EPA-92/12)
LCD: Local Climatological Data (EPA-94/04)
LCL: Lower Control Limit (EPA-94/04)
LCLO: Lethal Concentration Low; the lowest concentration at which death occurred (EPA-92/12)
LCM: Life Cycle Management (EPA-94/04)
LCRS: Leachate Collection and Removal System (EPA-94/04)
LD 0: Lethal Dose Zero
LD L0: The lowest dosage of a toxic substance that kills test organisms (EPA-94/04)
LD LO: Lethal Dose Low
LD: Land Disposal (EPA-94/04)
LD: Lethal Dose
LD: Light Duty (EPA-94/04)
LD50: Lethal Dose 50; dose lethal to 50% of the animals (EPA-92/12)
LDC: London Dumping Convention (EPA-94/04)
LDCRS: Leachate Detection, Collection, and Removal System (EPA-94/04)
LDD: Light-Duty Diesel (EPA-94/04)
LDH: Lactic Acid Dehydrogenase (EPA-92/12)
LDIP: Laboratory Data Integrity Program (EPA-94/04)
LDR: Land Disposal Restrictions (EPA-94/04)
LDRTF: Land Disposal Restrictions Task Force (EPA-94/04)
LDS: Leak Detection System (EPA-94/04)
LDT: Light-Duty Truck (40CFR88.103-94)
LDT: Lowest Dose Tested (EPA-94/04)
LDV: Light-Duty Vehicle (40CFR88.103-94)
LEL: Lower Explosive Limit (40CFR52.741)

LEL: Lowest Effect Level (EPA-94/04)

LEP: Laboratory Evaluation Program (EPA-94/04)

LEPC: Local Emergency Planning Committee (40CFR300.4)

LERC: Local Emergency Response Committee (EPA-94/04)

LEV: Low-Emission Vehicle (40CFR88.103-94)

LFL: Lower Flammability Limit (EPA-94/04)

LGR: Local Governments Reimbursement Program (EPA-94/04)

Li: Lithium

LI: Langelier Index (EPA-94/04)

LIDAR: Light Detection and Ranging (EPA-94/04)

LIFO: Last In/First Out

LIMB: Limestone-Injection Multi-Stage Burner (EPA-94/04)

LLNL: Larwence Livermore National Laboratory

LLRW: Low Level Radioactive Waste (EPA-94/04)

LLW: Low Level Radioactive Waste (EPA-91/12)

LMFBR: Liquid Metal Fast Breeder Reactor (EPA-94/04)

LMR: Labor Management Relations

LNEP: Low Noise Emission Product

LNG: Liquified Natural Gas

LOAEL: Lowest-Observed-Adverse-Effect-Level (EPA-94/04)

LOC: Letter of Credit (40CFR35.3105)

LOC: Level of Concern

LOC: Library of Congress

LOE: Level of Effort

LOEC: Lowest Observed Effect Concentration (EPA-91/3)

LOEL: Lowest Observed Effect Level (EPA-8/90b)

LOIS: Loss of Interim Status (SDWA)

LONGZ: Long Term Terrain Model

LOQ: Level of Quantitation (40CFR766.3-93)

LOT: Light off Time (40CFR85.2122)

LP: Legislative Proposal

LPC: Limiting Permissible Concentration (40CFR227.27)

LPG: Liquefied Petroleum Gas (40CFR86.094.3-94)

Lpm: Liter per minute (40CFR60)

LRC: Lewis Research Center (NASA)

LRMS: Low Resolution Mass Spectroscopy

LRTAP: Long Range Transportation of Air Pollution

LSI: Legal Support Inspection (CWA)

LSL: Lump Sum Leave

LST: Low Solvent Technology

LTA: Lead Trial Attorney

LTA: Long Term Average (EPA-91/3)

LTD: Land Treatment Demonstration

LTHE: Low Temperature Heat Exchanger

LTO: Landing Takeoff (40CFR87.2-94)

LTOP: Lease to Purchase

LTR: Lead Technical Representative

LTU: Land Treatment Unit

Lu: Lutetium

LUIS: Label Use Information System (EPA-94/04)

LUST: Leaking Underground Storage Tank(s) (current usage omits the "L")

LVW: Loaded Vehicle Weight (40CFR88.103-94)

Lw: Lawrencium

LWCF: Land and Water Conservation Fund

LWDF: Liquid Waste Derived Fuel

LWK: Live Weight Killed (40CFR432.11)

LWOP: Lease with Option to Purchase

LWOP: Leave Without Pay

m.p.: Melting Point

m: Metre(s) (40CFR86.403.78-94)

M: Molar (40CFR60)

m^2: Square Meter (40CFR60)

m^3: Cubic Meter (40CFR60)

M5: Method 5

MA: Massachusetts

MAB or MCA: Monoclonal Antibodies

MAB: Man and Biosphere Program

MAC: Management Advisory Committee

MACT: Maximum Achievable Control Technology (CAA-title III)

MADCAP: Model of Advection, Diffusion, and Chemistry for Air Pollution

MAER: Maximum Allowable Emission Rate

MAG: Management Advisory Group

MAOP: Maximum Allowable Operating Pressure (40CFR192.3)

MAPCC: Michigan Air Pollution Control Commission

MAPPER: Maintaining, Preparing, and Producing Executive Reports

MAPS: Multistate Atmospheric Power Production Pollution Study

MAPSIM: Mesoscale Air Pollution Simulation Model (EPA-94/04)

MARC: Mining and Reclamation Council

MARPOL 73/78: International Convention for the Prevention of Oil Pollution from Ships, 1973, as Modified by the Protocol of 1978 (40CFR110.1)

MATC: Maximum Acceptable Toxicant Concentration (40CFR797.1330)

MATC: Maximum Allowable Toxicant Concentration

MAWP: Maximum Allowable Working Pressure

MBAS: Methylene-Blue-Active Substances (EPA-94/04)

Mbbl: 1000 barrels (one barrel is equivalent to 42 gallons) (40CFR419.11)

MBDA: Minority Business Development Agency

MBE: Minority Business Enterprises (40CFR35.6015)

MBTA: Massachusetts Bay Transportation Authority (40CFR52.1161)

MC: Multiple Cyclones

MCA: Manufacturing Chemists Association

MCEF: Mixed Cellulose Ester Fiber

MCL: Maximum Contaminant Level (40CFR142.2)

MCLG: Maximum Contaminent Level Goal (40CFR141.2)

MCP: Methylcyclopentane (40CFR799.2155)

MCP: Municipal Compliance Plan (CWA)

Md: Mendelevium

MD: Mail Drop

MD: Maryland

MDA: Methylenedianilline

MDL: Method Detection Limit (40CFR300-AA)

ME: Maine

MED: Minimum Effective Dose (EPA-92/12)

MEFS: Midterm Energy Forecasting System

MEFV: Maximum Expiratory Flow Volume (EPA-92/12)

MEG: Multimedia Environmental Goal

MEI: Maximum Exposed Individual

MEK: Methyl Ethyl Ketone

MEM: Modal Emission Model

MENS: Mission Element Needs Statement

MeOH: Methanol ($CH3OH$) (40CFR86.090.3-94)

MEP: Multiple Extraction Procedure (EPA-94/04)

meq: Milliequivalent (40CFR60)

MERS: Monticello Ecological Research Station (EPA-91/3)

MESOPAC: Mesoscale Meteorological Reprocesser Program

MESOPLUME: Mesoscale "Bent Plume" Model

MESOPUFF: Mesoscale Puff Model

MESS: Model Evaluation Support System

MEXAMS: Metals Exposure Analysis Modeling System (EPA-91/3)

MF: Membrane Filter (40CFR121.21)

MF: Modifying Factor (EPA-92/12)

MFBI: Major Fuel Burning Installation

MFC: Metal Finishing Category

MFL: Million Fibers per Liter (141.23.a)

mg/kg: Milligrams per Kilogram = (approx.) ppm (40CFR116)

mg/L: Milligrams per liter = (approx.) ppm (40CFR116)

mg: Milligram = 10^{-3} gram (40CFR60)

Mg: Magnesium

Mg: Megagram = 10^6 gram (metric tons or tonnes) (40CFR72.741-94)

Mgal: One Thousand Gallons (40CFR419.11)

MGD: Million Gallons Per Day

MH: Man Hours

MHD: Magnetohydrodynamics

MI: Michigan

MIBK: Methyl Isobutyl Ketone

MIC: Methyl Isocaynate

MICE: Management Information Capability for Enforcement

MICH: Michigan fiver Model (EPA-91/3)

MICROMORT: A One-in-a-Million Chance of Death from an Environmental Hazard

min.: Minute(s) (40CFR87.2-94)

MINTEQA2: Equilibrium Metals Speciation Model (EPA-91/3)

MIPS: Millions of Instructions Per Second

MIS: Management Information System

MIS: Mineral Industry Surveys

MITS: Management Information Tracking System

MJ: Megajoule(s) (1 million joules) (40CFR86.090.3-94)

mL/L: Milliliters per Liter (40CFR434.11)

mL: Milliliter = 10^{-3} Liter (40CFR60)

ML: Meteorology Laboratory

ML: Military Leave

ML: Minimum Level (EPA-91/3)

MLAP: Migrant Legal Action Program

MLSS: Mixed Liquor Suspended Solids

MLVSS: Mixed Liquor Volatile Suspended Solids

mm: Millimeter = 10^{-3} meter (40CFR60)

MM5: Modified Method 5 (sampling method)

MMAD: Mass Median Aerodynamic Diameter (EPA-92/12)

mmHg: Millimeters of Mercury; a measure of pressure (EPA-92/12)

MMPA: Marine Mammal Protection Act

MMS: Minerals Management Service (DOI)

MMT: Million Metric Tons

Mn: Manganese

MN: Minnesota

MNT: Mononitrotoluene

Mo: Molybdenium
MO: Missouri
MOA: Memorandum of Agreement
MOBILE: Mobile Source Emission Model
MOD: Miscellaneous Obligation Document
MOD: Modification
MOE: Margin Of Exposure (EPA-94/04)
MOI: Memorandum of Intent
mol. wt.: Molecular weight (40CFR60)
mol: Mole (40CFR60)
MOS: Margin of Safety (EPA-94/04)
MOS: Metal Oxide Semiconductor
MOU: Memorandum of Understanding
MP: Manufacturing-use Product (EPA-94/04)
MP: Melting Point (EPA-94/04)
mpc: 1000 Pieces (40CFR427.71)
MPC: Max. Permissible Concentration (radiation)
MPG: miles per gallon (40CFR600.003.77-94)
mph: Miles per hour (40CFR86.403.78-94)
MPN: Maximum Probable Number (EPA-94/04)
MPN: Most Probable Number
MPO: Metropolitan Planning Organization (40CFR51.138)
MPP: Merit Promotion Plan
MPRSA: Marine Protection, Research and Sanctuaries Act (40CFR2.309)
MPTDS: MPTER Model with Deposition and Settling of Pollutants
MPTER: Multiple Point Source Model with Terrain
MRA: Minimum Retirement Age
mrem: Millirem = 10^{-3} rem (40CFR60)
MRF: Materials Recovery Facility (EPA-94/04)
MRID: Master Record Identification number (EPA-94/04)
MRL: Maximum-Residue Limit (Pesticide Tolerance) (EPA-94/04)
MRP: Multi-Roller Press (in sludge drying unit)
MS: Mail Stop
MS: Mass Spectrometry
MS: Mississippi
MSA: Management System Audits
MSA: Metropolitan Statistical Area (40CFR60.331)
MSAM: Multi-Keyed Indexed Sequential File Access Method
MSBu: 1000 Standard Bushels (40CFR406.11)
MSDS: Material Safety Data Sheet (40CFR370.2)
MSEE: Major Source Enforcement Effort
MSHA: Mine Safety and Health Administration (DOL)
MSIS: Model State Information System

MSL: Mean Sea Level
MSPB: Merit System Protection Board
MSW: Medical Solid Waste
MSW: Municipal Solid Waste (40CFR240.101)
MSWLF: Municipal Solid Waste Landfill (40CFR258.1)
MT: Montana
MTB: Materials Transportation Bureau
MTBE: Methyl Tertiary Butyl Ether
MTD: Maximum Tolerated Dose (EPA-94/04)
MTDDIS: Mesoscale Transport Diffusion and Deposition Model for Industrial Sources
MTF: Multiple-Tube Fermentation (40CFR141.21)
MTG: Media Task GroupMTF: Multiple-Tube Fermentation
MTL: Median Threshold Limit (EPA-92/12)
MTP: Maximum Total Trihalomethane Potential (40CFR142.2)
MTS: Management Tracking System (OW)
MTSL: Monitoring and Technical Support Laboratory
MTU: Mobile Treatment Unit (40CFR261.4(f))
MTZ: Mass Transfer Zone
MUP: Manufacturing-Use Product (EPA-94/04)
MUTA: Mutagenicity (EPA-94/04)
MVA: Multivariate Analysis
MVAC: Motor Vehicle Air Conditioner (40CFR82.152)
MVAPCA: Motor Vehicle Air Pollution Control Act
MVEL: Motor Vehicle Emissions Laboratory
MVh: Minute ventilatory volume for human (composite value expressed in cu.m/day) used in the HEC derivation of an RfC (EPA-92/12)
MVho: Minute ventilatory volume for human in an occupational environment, assuming 8 hour/day exposure (composite value expressed in MVl/M) Motor Vehicle Inspection/Maintenance
MVIACSA: Motor Vehicle Information and Cost Saving Act (40CFR600.002)
MVICSA: Motor Vehicle Information and Cost Savings Act
MVMA: Motor Vehicle Manufacturers Association
MVRS: Marine Vapor Recovery System
MVTS: Motor Vehicle Tampering Survey
MW: Megawatt(40CFR401.11)
MW: Molecular Weight
MWC: Municipal Waste Combustor (40CFR60.51a)
MWG: Model Work Group
Mwh: Megawatt Hour(s) (40CFR401.11)
MWI: Medical Waste Incinerator

MWL: Municipal Waste Leachate

MWTA: Medical Waste Trucking Act of 1988

MYDP: Multi-Year Development Plans

N/A: Not Applicable

N/A: Not Available

n.a.: Not Available (EPA-92/12)

N: Newton (40CFR60)

N: Nitrogen

N: Normal (40CFR60)

N_2: Nitrogen (40CFR86.403.78-94)

Na: Sodium

NA: National Archives

NA: Nonattainment

NAA: Nonattainment Areas

NAAQS: National Ambient Air Quality Standards Program (CAA) (40CFR57.103)

NAAS: National Air Audit System (OAR)

NAC: National Academy of Sciences

NAC: Nnitric Acid Concentrator

NACA: National Agricultural Chemicals Association

NAD: Nicotinamide Adenine Dinucleotide

NADB: National Atmospheric Data Bank

NADP: National Atmospheric Deposition Program

NAIS: Neutral Administrative Inspection System)

NALD: Nonattainment Areas Lacking Demonstrations

NAM: National Association of Manufacturers

NAMA: National Air Monitoring Audits

NAMS: National Air Monitoring System (40CFR58.1-c)

NANCO: National Association of Noise Control Officials

NAPAP: National Acid Precipitation Assessment Program (CAA402)

NAPBN: National Air Pollution Background Network

NAPBTAC: National Air Pollution Control Technical Advisory Committe

NAPCTAC: National Air Pollution Control Techniques Advisory Committee

NAPIM: National Association of Printing Ink Manufacturers

NAR: National Asbestos Registry

NARA: National Air Resources Act

NARA: National Archives and Records Administration

NARM: Naturally-ccurring or Accelerator-produced Radioactive Material

NARS: National Asbestos-Contractor Registry System

NAS: National Academy of Sciences

NAS: National Audubon Society

NASA: National Aeronautices and Space Administration

NATICH: National Air Toxics Information Clearinghouse

NATO: North Atlantic Treaty Organization

NATS: National Air Toxics Strategy

NAWC: National Association of Water Companies

NAWDEX: National Water Data Exchange

Nb: Niobium

NBAR: Non-Binding Allocation of Authority

NBS: National Bureau of Standards (See NIST)

NC fines: Nitrocellulose fines

NC: North Carolina

NCA: National Coal Association

NCA: Noise Control Act of 1972

NCAC: National Clean Air Coalition

NCAF: National Clean Air Fund

NCAMP: National Coalition Against the Misuse of Pesticides

NCAQ: National Commission on Air Quality

NCAR: National Center for Atmospheric Research

NCASI: National Council of the Paper Industry for Air and Stream Improvements

NCC: National Climatic Center

NCC: National Computer Center

NCF: Network Control Facility

NCHS: National Center for Health Statistics (NIH)

nCi/L: Nanocuries per Liter (DOE-91/4)

NCI: National Cancer Institute

NCIC: National Crime Information Center

NCLP: National Contract Laboratory Program

NCM: National Coal Model

NCM: Notice of Commencement of Manufacture (TSCA)

NCN: Nitrocarbonitrate (40CFR457.31)

NCO: Negotiated Consent Order

NCP: National Contingency Plan (40CFR300.4)

NCP: National Oil and Hazardous Substances Pollution Contingency Plan (EPA-91/12)

NCP: National Oil and Hazardous Substances Pollution Contingency Plan (NRT-87/3)

NCP: Noncompliance Penalties (CAA)

NCP: Nonconformance Penalty (40CFR86.1102.87)

NCR: Noncompliance Report (CWA)

NCR: Nonconformance Report

NCRIC: National Chemical Response and Information Center

NCRP: National Council on Radiation Protection

and Measurements (DOE-91/4)

NCS: National Compliance Strategy:

NCTR: National Center for Toxicological Research

NCV: Nerve Conduction Velocity

NCVECS: National Center for Vehicle Emissions Control and Safety

NCWQ: National Commission on Water Quality

NCWS: Non-Community Water System (EPA-94/04)

Nd: Neodymium

ND: North Dakota

NDD: Negotiation Decision Document

NDDN: National Dry Deposition Network

NDIR: Nondispersive Infrared Analysis (40CFR86)

NDS: National Dioxin Study

NDS: National Disposal Site

NDUV: Nondispersion Ultraviolet

NDWAC: National Drinking Water Adviosry Council

Ne: Neon

NE: Nebraska

NEA: National Energy Act

NEDA: National Environmental Development Association

NEDS: National Emissions Data System

NEEC: National Environmental Enforcement Council

NEEJ: National Environmental Enforcement Journal

NEIC: National Earthquake Information Center (DOE-91/4)

NEIC: National Enforcement Investigations Center (40CFR16.13)

NEMA: National Electrical Manufacturer's Association

NEP: National Energy Plan

NEP: National Estuary Program (40CFR35.9000)

NEPA: National Environmental Policy Act of 1969 (40CFR1508.21)

NER: National Emissions Report

NEROS: Northeast Regional Oxidant Study

NESCAUM: Northeast States for Coordinated Air Use Management

NESHAP: National Emissions Standards for Hazardous Air Pollutants (CAA) (40CFR300)

NETA: National Environmental Training Association (EPA-94/04)

NETTING: Emission Trading Used to Avoid PSD/NSR Permit Review Requirements

NFA: National Fire Academy (NRT-87/3)

NFAN: National Filter Analysis Network

NFFE: National Federation of Federal Employees

NFIP: National Flood Insurance Program (40CFR6-AA)

NFPA: National Fire Protection Association

NFRAP: No Further Remedial Action Planned (EPA-94/04)

NFWF: National Fish and Wildlife Foundation

ng: Nanogram = 10^{-9} gram (40CFR60)

NGA: Natural Gas Association

NGPA: Natural Gas Policy Act

NGWIC: National Ground Water Information Center

NH: New Hampshire

NHANES: National Health and Nutrition Examination Study

NHMIE: National Hazardous Materials Information Exchange (NRT-87/3)

NHPA: National Historic Preservation Act of 1966

NHTSA: National Highway Traffic Safety Act

NHTSA: National Highway Traffic Safety Administration (DOT)

NHWP: Northeast Hazardous Waste Project

Ni: Nickel

NICS: National Institute for Chemical Studies

NICT: National Incident Coordination Team (EPA-94/04)

NIEHS: National Institute of Environmental Health Sciences

NIEI: National Indoor Environmental Institute

NIH: National Institutes of Health (EPA-92/12)

NIM: National Impact Model

NIMBY: Not In My Backyard

NIOSH: National Institute for Occupational Safety and Health of the U.S. Department of Health and Human Services (29CFR1910.20)

NIPDWR: National Interim Primary Drinking Water Regulations (EPA-94/04)

NIS: Noise Information System

NISAC: National Industrial Security Advisory Committee (EPA-94/04)

NIST: National Institute for Standards and Testing (40CFR89.3-94)

NIST: National Institute of Standards and Technology (Formerly NBS)

NITEP: National Incinerator Testing and Evaluation Program

Nitrogenous BOD: Nitrogenous Bacteria Oxygen Demand

NJ: New Jersey

NJIT: New Jersey Institute of Technology

NLA: National Lime Association

NLAP: National Laboratory Audit program

NLETS: National Law Enforcement Teletype Systems

NLGI number: National Lubricating Grease Institute number

NLM: National Library of Medicine (EPA-92/12)

NLT: Not Later Than

nm: Nanometer = 10^{-9} meter

NM: New Mexico

NMC: National Meteorological Center

NMFS: National Marine Fisheries Service (40CFR233.2)

NMHC: Nonmethane Hydrocarbons (40CFR86.094.3-94)

NMHCE: Non-Methane Hydrocarbon Equivalent (40CFR86.094.3-94)

NMOC: Nonmethane Organic Compound

NMOG: Non-Methane Organic Gas (40CFR88.103-94)

NMP: National Municipal Policy

NMR: Nuclear Magnetic Resonance

NNC: Notice of Noncompliance

NNPSPP: National Non-Point Source Pollution Program

No.: Number (40CFR86.403.78-94)

No: Nobelium

NO: Nitric oxide (40CFR89.3-94)

NO_2: Nitrogen Dioxide (40CFR58.1)

NOA: Nearest Onshore Area (40CFR55.2)

NOA: New Obligation Authority

NOA: Notice of Arrival (EPA-94/04)

NOAA: National Oceanic and Atmospheric Administration (DOC) (40FR300.4)

NOAC: Nature of Action Code (EPA-94/04)

NOAEL (NOEL): No Observable Adverse Effect Level (EPA-94/04)

NOAEL: No Observed Adverse Effect Level

NOC: Notice of Commencement

NOC: Notice of Construction (DOE-91/4)

NOD: Notice of Deficiency (RCRA)

NOD: same as Nitrogenous BOD

NOEC: No Observed Effect Concentration (40CFR797.1600)

NOEL: No Observed Effects Level (40CFR798.4350)

NOHSCP: National Oil and Hazardous Substances Contingency Plan (40CFR300)

NOI: Notice of Intent

NON: Notice of Noncompliance (TSCA)

NOPES: Non-Occupational Pesticide Exposure Study

NORA: National Oil Recyclers Association

NOS: National Ocean Survey (NOAA)

NOS: Not Otherwise Specified (40CFR261-AVIII)

NOV/CD: Notice of Violation/Compliance Demand

NOV: Notice of Violation

NO_x: Nitrogen Oxides (CAA302)

Np: Neptunium

NP: Office of New Production Reactors (DOE-91/4)

NPAA: Noise Pollution and Abatement Act

NPCA: National Parks and Conservation Association

NPDES: National Pollutant Discharge Elimination System (40CFR110.1)

NPHAP: National Pesticide Hazard Assessment Program (EPA-94/04)

NPIRS: National Pesticide Information Retrieval System

NPL: National Priority(ies) List (40CFR300.4)

NPM: National Program Manager

NPN: National Particulate Network

NPR: New Production Reactor (DOE-91/4)

NPRM: Notice of Proposed Rulemaking

NPS: National Park Service

NPS: National Permit Strategy

NPS: National Pesticide Survey (OW)

NPS: Non-Point Source

NPS: Nonpoint Source Model for Urban and Rural Areas (EPA-91/3)

NPUG: National Prime User Group

NRA: National Recreation Area

NRC: National Research Council (EPA-92/12)

NRC: National Response Center (40CFR300.4)

NRC: Non-Reusable Containers

NRC: Nuclear Regulatory Commission (40CFR61.101)

NRCA: National Resource Council of America

NRDC: Natural Resources Defense Council

NREL: National Renewable Engineering Laboratory (Formerly SERI)

NRHP: National Register of Historic Places (DOE-91/4)

NRMRL: National Risk Management Research Laboratory (Formerly RREL, before June 1995)

NRR: Noise Reduction Rating (40CFR211.203)

NRT: National Response Team (40CFR300.4)

NRWA: National Rural Water Association

NSC: National Security Council

NSD: No Structure Detected (40CFR763-AA)

NSDWR: National Secondary Drinking Water Regulations (EPA-94/04)

NSEC: National System for Emergency Coordination (EPA-94/04)

NSEP: National System for Emergency Preparedness (EPA-94/04)

NSF: National Sanitation Foundation

NSF: National Science Foundation

NSF: National Strike Force (40CFR300.4)

NSO: Nonferrous Smelter Orders (CAA) (40CFR57.101)

NSPE: National Society of Professional Engieers

NSPS: New Source Performance Standards (CAA) (40CFR467.02)

NSR: New Source Review (EPA-94/04)

NSSC: Neutral Sulfite Semi-Chemical

NSTL: National Space Technology Laboratory

NSWMA: National Solid Waste Management Association

NSWS: National Surface Water Survey

NTA: Negotiated Testing Agreement

NTA: Nitrilotriacetic Acid

NTE: Not to Exceed

NTIS: National Technical Information Service (40CFR89.3-94)

NTN: National Trends Network

NTNCW: Non-Transient Non-Community Water System (40CFR141.2)

NTNCWS: Non-transient Non-community Water System (40CFR141)

NTP: National Toxicology Program (29CFR1910.1200)

NTS: Nevada Test Site (DOE-91/4)

NTSP: National Transportation Safety Board

NTU: Nephelometric Turbidity Units (40CFR131.35)

NURF: National Utility Reference File (CAA402)

NV: Nevada

NVPP: National Vehicle Population Poll

NWA: National Water Alliance

NWF: National Wildlife Federation

NWPA: Nuclear Waste Policy Act of 1982 (40CFR191.02)

NWRC: National Weather Records Center

NWS: National Weather Service (NOAA)

NY: New York

O/O: Owners or Operators (40CFR260-AI-93)

O: Oxygen

O&M: Operations and Maintenance (EPA-94/04)

O&G: Oil and Gas

O&G: Oil and Grease (40CFR420.02)

O&M: Operation and Maintenance (40CFR300.4)

O&M: Operations and Maintenance

O_2: Oxygen (40CFR89.3-94)

OAIAP: Office of Atmospheric and Indoor Air Programs (EPA)

OALJ: Office of Administrative Law Judges (EPA)

OAQPS: Office of Air Quality Planning and Standards (EPA)

OARM: Office of Administration and Resources Management (EPA)

OASDI: Old Age and Survivor Insurance

OC: Object Class

OC: Office of Controller (EPA)

OC: Open Cup (EPA-92/12)

OCAPO: Office of Compliance Analysis and Program Operations (EPA)

OCD: Offshore and Coastal Dispersion (EPA-94/04)

OCE: Office of Civil Enforcement (EPA)

OCE: Office of Criminal Enforcement (EPA)

OCEM: Office of Cooperative Environmental Management (EPA)

OCI: Office of Criminal Investigations (40CFR16.13)

OCI: Organizational Conflicts of Interest

OCM: Office of Compliance Monitoring (EPA)

OCPSF: Organic Chemicals, Plastics, and Synthetic Fibers (40CFR414.11)

OCR: Office of Civil Rights (40CFR7.25)

OCR: Optical Character Reader

OCS: Outer Continental Shelf (40CFR55.2)

OCSLA: Outer Continental Shelf Lands Act (40CFR55.2)

OD: Organizational Development

OD: Outside Diameter (40CFR60)

ODBA: Ocean Dumping Ban Act of 1988

ODBU: Office of Disadvantaged Business Utilization (EPA)

OEC: Observed Effect Concentration (40CFR797.1600)

OEM: Original equipment manufacturer (40CFR89.3-94)

OEM: Original Equipment Manufacturer (40CFR85.1502)

OEP: Office of External Programs

OER: Office of Exploratory Research (EPA)

OERR: Office of Emergency and Remedial Response (EPA) (EPA-91/12)

OERR: Office of Emergency and Remedial Response (Superfund) (EPA)

OF: Optional Form (EPA-94/04)

OFA: Office of Federal Activities (EPA)

OFFE: Office of Federal Facilities Enforcement (EPA)

OFR: Office of the Federal Register

OGC: Office of General Counsel (EPA)

OGWDW: Office of Ground Water and Drinking Water (EPA)

OH: Ohio

OHEA: Office of Health and Environmental Assessment (EPA)

OHM/TADS: Oil and Hazardous Materials Technical Assistance Data Systems (EPA-92/12)

ohm: Electrical Resistance Unit (40CFR60)

OHR: Office of Health Research (EPA)

OHRM: Office of Human Resources Management (EPA)

OHSS: Occupation Health and Safety Staff

OIA: Office of International Activities (EPA)

OIG: Office of the Inspector General (EPA)

OIRM: Office of Information Resources Management (EPA)

OK: Oklahoma

OLTS: On Line Tracking System (EPA-94/04)

OMB: Office of Management and Budget (40CFR31.3)

OMEP: Office of Marine and Esturine Protection (EPA)

OMHCE: Organic Material Hydrocarbon Equivalent (40CFR86)

OMMSQA: Office of Modeling, Monitoring Systems, and Quality Assurance (EPA)

OMNMHCE: Organic Material Non-Methane Hydrocarbon Equivalent (40CFR86)

OMS: Office of Mobile Sources (EPA)

ONRW: Outstanding National Resource Waters (EPA-91/3)

OP Year: Original Production Year (40CFR85.1502)

OP: Office of Prevention (EPA)

OP: Operating Plan

OPA: Office of Policy Analysis (EPA)

OPA: Oil Pollution Act of 1990 (Public Law 101-380)

OPAC: Overall Performance Appraisal Certification

OPAR: Office of Policy Analysis and Review (EPA)

OPF: Official Personnel Folder

OPIDN: Organophosphorus Induced Delayed Neurotoxicity (40CFR798.6450)

OPMO: Office of Program Management Operations (EPA)

OPP: Office of Pesticide Programs (EPA)

OPPE: Office of Policy, Planning and Evaluation (EPA)

OPTS: Office of Pesticides and Toxic Substances (40CFR179.3)

OR: Office of Enforcement (EPA)

OR: Oregon

ORA: Office of Radiation Programs (EPA)

ORD: Research and Development (EPA)

ORM: Other Regulated Material

ORM: Other Regulated Material (EPA-94/04)

ORME: Office of Regulatory Management and Evaluation (EPA)

ORNL: Oak Ridge National Laboratory (DOE-91/4)

ORP recorders: Oxidation Reduction Potential Recorders

ORP: Oxidation-Reduction Potential (EPA-94/04)

ORPM: Office of Research Program Management (EPA)

ORV: Off-road Vehicle

OS/VS: Operationg System/Virtual Storage

Os: Osmium

OS: Orifice Scrubber

OSC: On Scene Coordinator (40CFR300.4)

OSHA: Occupational Health and Safety Administration (DOE-91/4)

OSHA: Occupational Safety and Health Act

OSHA: Occupational Safety and Health Administration (DOL)

OST: Office of Science and Technology (EPA)

OST: Office of Science and Technology, U.S. EPA (EPA-92/12)

OSTP: Office of Science and Technology Policy (White House)

OSW: Office of Solid Waste (EPA)

OSWER: Office of Solid Waste and Emergency Response (EPA-91/12)

OT: Overtime

OTA: Office of Technology Assessment (U.S. Congress)

OTEC: Ocean Thermal Energy Conversion

OTS: Office of Toxic Substances (EPA)

OTTRS: Office of Technology Transfer and Regulatory Support (EPA)

OUST: Office of Underground Storage Tanks (EPA)

OW: Office of Water (EPA)

OWEC: Office of Wastewater Enforcement and Compliance (EPA)

OWF: On the Weight of the Fiber

OWOW: Office of Wetlands, Oceans and Watersheds (EPA)

OWP: Overall Work Programs (40CFR51.138)

OWPE: Office of Waste Programs Enforcement (EPA)

OWRS: Office of Water Regulations and Standards, U.S. EPA (EPA-92/12)

OX: Oxidation (40CFR141.61.b)

OY: Operating Year

OYG: Operating Year Guidance

oz: Ounces (40CFR60)

OZIPP: Ozone Isopleth Plotting Package

OZIPPM: Modified Ozone Isopleth Plotting Package

P-TBBA: p-tert-butylbenzoic acid (40CFR704.33)

P-TBT: p-tert-butyltoluene (40CFR704.33)

P.O.: Per os (by mouth) (EPA-92/12)

p: Probit Dose Extrapolation Model (EPA-92/12)

P: Phosphorus

P-TBB: p-tert-butylbenzaldehyde (40CFR704.33)

P&A: Precision and Accuracy

PA/SI: Preliminary Assessment/Site Inspection (EPA-91/12)

Pa: Pascal(pressure) (40CFR86.403.78-94)

Pa: Protactinium

PA: Pennsylvaina

PA: Policy Analyst

PA: Preliminary Assessment (40CFR300.4)

PAA: Priority Abatement Areas

PAAT: Public Affairs Assist Team (40CFR300.5)

PADRE: Particle Analysis and Data Reduction Program

PAG: Protective Action Guide (DOE-91/4)

PAGM: Permit Applications Guidance Manual

PAH: Polycyclic Aromatic Hydrocarbon

PAH: Polynuclear Aromatic Hydrocarbon

PAHO: Pan Americn Health Organization

PAI: Performance Audit Inspection (CWA) (EPA-94/04)

PAI: Pure Active Ingredient compound (EPA-94/04)

PAIR: Preliminary Assessment Information Rule

PAL: Point, Area, and Line Source Air Quality Mode

PALDS: PAL Model with Deposition and Settling of Pollutants

PAM: Pesticide Analytical Manual (EPA-94/04)

PAMS: Photochemical Assessment Monitoring Station (40CFR58.1)

PAN: Perooxyacetyl Nitrate

PAPR: Powered Air Purifying Respirator

PARS: Precision and Accuracy Reporting System

PASS: Procurement Automated Source System

PAT: Packed Tower Aeration (40CFR141.61.b)

PAT: Permit Assistance Team (RCRA) (EPA-94/04)

PATS: Pesticide Action Tracking System (EPA-94/04)

PATS: Pesticides Analytical Transport Solution (EPA-94/04)

Pb: Lead

PBA: Preliminary Benefit Analysis (BEAD) (EPA-94/04)

PBB: Polybromated Biphenyls (40CFR704.195)

PBC: Pacific Basin Conference

PBC: Packed Bed Condenser

PBDD: Polybrominated Dibenzo-p-Dioxin (40CFR766.3)

PBL: Planetary Boundary Layer

PBLSQ: The Lead Line Source Model

PBPK: Physiologically Based Pharmacokinetic (EPA-92/12)

PBS: Packed Bed Scrubber

PC: Personal Computer

PC: Planned Commitment

PC: Polycarbonate

PC: Position Classification

PC: Pulverized Coal

PC&B: Personnel Compensation and Benefits

PCA: Principle Component Analysis (EPA-94/04)

PCA: Production Compliance Audit (40CFR86.1102.87)

PCB: Polychlorinated Biphenyls (40CFR129.4)

PCC: Primary Combustion Chamber

PCDD: Polychlorinated Dibenzodioxin (40CFR766.3)

PCDF: Polychlorinated Dibendzofuran (40CFR766)

PCE: Pollution Control Equipment

PCE: Pyrometric Cone Equivalent

pCi/L: Picocuries Per Litre

PCi: Picocurie = 10^{-12} Curie (40CFR60)

PCIE: President's Council on Integrity and Efficiency in Government

PCIOS: Processor Common Input/Output System

PCM: Phase Contrast Microscopy (40CFR763-AA)

PCN: Policy Criteria Notice (EPA-94/04)

PCNB: Pentachloronitrobenzene

PCO: Pest Control Operator (EPA-94/04)

PCO: Printing Control Officer

PCON: Potential Contractor

PCP: Pentachlorophenyl

PCRC: Pesticedes Chemical Review Committee

PCS: Permanent Change of Station

PCS: Permit Compliance System (CWA)

PCS: Primary Coolant System (DOE-91/4)

PCS: Probable Cargcinogenic Substances

PCSC: PC Site Coordinator

PCV: Positive Crankcase Ventilation

(40CFR85.2122)

PCW: Post Consumer Waste (40CFR246.101)

Pd: Palladium

PD: Position Description

PD: Position Document (EPA-92/12)

PD: Project Description

PDCI: Product Data Call-In (EPA-94/04)

PDFID: Preconstruction Direct Flame Ionization Detection

PDMS: Pesticide Document Management System (OPP)

PDP: Positive Displacement Pump (40CFR86)

PDR: Particulate Data Reduction

PE: Polyethylene

PE: Population Equivalent

PE: Program Element

PEF: Particulate Emission Factor (EPA-91/12)

PEI: Petroleum Equipment Institute

PEL: Permissible Exposure Level (DOE-91/4)

PEL: Permissible Exposure Limit (40CFR763.121)

PEL: Personal Exposure Limit

PEM: Partial Equilibrium Multimarket Model

PEM: Personal Exposure Model

PEPE: Prolonged Elevated Pollution Episode

PEQI: Perceived Environmental Quality Indices

PERF: Police Executive Research Forum

PESTAN: Pesticides Analytical Transport Solution

PET: Poly(ethylene terephthalate) (40CFR60.561)

PF: Potency Factor

PFCRA: Program Fraud Civil Remedies Act (EPA-94/04)

PFT: Permanent Full Time

PFTE: Permanent Full Time Equivalent

PHC: Principal Hazardous Constituent (EPA-94/04)

PHDD: Polyhalogenated Dibenzo-p-Dioxin (40CFR766.3)

PHDF: Polyhalogenated Dibenzofuran (40CFR766.3)

PHS: U.S. Public Health Service (EPA-92/12)

PHSA: Public Health Service Act (EPA-94/04)

PI: Preliminary Injunction (EPA-94/04)

PI: Program Information (EPA-94/04)

PIAT: Public Information Assist Team (40CFR300.4)

PIC: Products of Incomplete Combustion (EPA-94/04)

PIC: Public Information Center

PID: Photon Ionization Detector

PIGS: Pesticides in Groundwater Strategy (EPA-94/04)

PIMS: Pesticide Incident Monitoring System (EPA-94/04)

PIN: Pesticide Information Network (EPA-94/04)

PIN: Procurement Information Notice (EPA-94/04)

PIP: Public Involvement Program (EPA-94/04)

PIPQUIC: Program Integration Project Queries Used in Interactive Command (EPA-94/04)

PIR: Product of Incomplete Reaction

PIRG: Public Interest Research Group (EPA-94/04)

PIRT: Pretreatment Implementation Review Task Force (EPA-94/04)

PIS: Public Information Specialist

PITS: Project Information Tracking System (EPA-94/04)

PL: Public Law (DOE-91/4)

PLIRRA: Pollution Liability Insurance and Risk Retention Act (EPA-94/04)

PLM: Polarized Light Microscopy (EPA-94/04)

PLO: Public Land Order (DOE-91/4)

PLUVUE: Plume Visibility Model (EPA-94/04)

Pm: Promethium

PM: Particulate Matter (40CFR86.094.3-94)

PM: Program Manager

PM10: Particulate Matter (nominally 10 um (micrometers) diameter) and Less (EPA-94/04)

PM15: Particulate Matter (nominally 15 um (micrometers) diameter) (EPA-94/04)

PMA: Pharmceutical Manufacturers Association

PMEL: Pacific Marine Environmental Laboratory (EPA-94/04)

PMIP: Presidential Management Intern Program

PMIS: Personnel Management Information System (OARM)

PMN: Premanufacture Notification (TSCA) (40CFR700.43)

PMNF: Premanufacture Notification Form (EPA-94/04)

PMR: Pollutant Mass Rate (EPA-94/04)

PMRS: Performance Management and Recognition System (EPA-94/04)

PMS: Personnel Management Specialist

PMS: Program Management System (EPA-94/04)

PNA: Polynuclear Aromatic Hydrocarbons (EPA-94/04)

PNA: Probablistic Noise Audibility (DOE-91/4)

PNL: Pacific Northwest Laboratory (DOE-91/4)

Po: Polonium

PO: Project Officer (EPA-94/04)

PO: Purchase Order

POC: Point Of Compliance (EPA-94/04)

POC: Program Office Contacts

POE: Point Of Exposure (EPA-94/04)

POGO: Privately-Owned/Government-Operated (EPA-94/04)

POHC: Principal Organic Hazardous Constituent (EPA-94/04)

POI: Point Of Interception (EPA-94/04)

POLREP:Pollution Report (EPA-94/04)

POM: Particulate Organic Matter (EPA-94/04)

POM: Polycyclic Organic Matter (EPA-94/04)

POR: Program of Requirements (EPA-94/04)

POTW: Publicly Owned Treatment Works (40CFR117.1)

POV: Privately Owned Vehicle (EPA-94/04)

PP: Pay Period

PP: Pollution Prevention

PP: Polypropylene (40CFR60.561)

PP: Priority Pollutant (CWA)

PP: Program Planning (EPA-94/04)

PPA: Pesticide Producers Association

PPA: Planned Program Accomplishment (EPA-94/04)

PPA: Pollution Prevention Act of 1990

ppb: Parts Per Billion (40CFR60)

PPIA: Poultry Products Inspection Act (21USC453)

PPIC: Pesticide Programs Information Center (EPA-94/04)

PPIS: Pesticide Product Information System (EPA-94/04)

PPM/PPB: Parts per million/parts per billion (EPA-94/04)

ppm: Parts Per Million = (approx.) mg/kg (milligrams per kilogram) = (approx.) mg/L (milligrams per liter) (40CFR116)

PPMAP: Power Planning Modeling Application Procedure (EPA-94/04)

PPPA: Poison Prevention Packaging Act

PPSP: Power Plant Siting Program (EPA-94/04)

PPT: Parts Per Trillion (EPA-94/04)

PPT: Permanent Part Time

PPTH: Parts Per Thousand (EPA-94/04)

PQL: Practical Quantitation Limits (40CFR264-A9)

PQUA: Preliminary Quantitative Usage Analysis (EPA-94/04)

Pr: Praseodymium

PR: Preliminary Review (EPA-94/04)

PR: Procurement Request

PRA: Paperwork Reduction Act (EPA-94/04)

PRA: Planned Regulatory Action (EPA-94/04)

PRA: Probablistic Risk Assessment (DOE-91/4)

PRATS: Pesticides Regulatory Action Tracking System (EPA-94/04)

PRC: Planning Research Corporation (EPA-94/04)

PRG: Preliminary Remediation Goal (EPA-91/12)

PRI: Periodic Reinvestigation (EPA-94/04)

PRM: Prevention Reference Manuals (EPA-94/04)

PRMO: Policy and Resources Management Office (EPA)

PRN: Pesticide Registration Notice (EPA-94/04)

PRP: Potentially Responsible Party (40CFR35.4010)

PRP: Potentially Responsible Party (EPA-94/04)

PRZM: Pesticide Root Zone Model (EPA-94/04)

PS: Point Source (EPA-94/04)

PS: Polystyrene (40CFR60.561)

PSAM: Point Source Ambient Monitoring (EPA-94/04)

PSC: Program Site Coordinator (EPA-94/04)

PSD: Prevention of Significant Deterioration (40CFR52.21)

PSE: Program Subelement

PSES: Pretreatment Standards for Existing Sources (40CFR467.02)

psi: Pounds Per Square Inch (Pressure)

PSI: Pollutant Standards Index (40CFR58-AG-a)

PSI: Pounds Per Square Inch (EPA-94/04)

PSI: Pressure Per Square Inch (EPA-94/04)

psia: Pounds Per Square Inch Absolute (40CFR60)

psig: Pounds Per Square Inch Gauge (40CFR60)

PSIG: Pressure Per Square Inch Gauge (EPA-94/04)

PSM: Point Source Monitoring (EPA-94/04)

PSNS: Pretreatment Standards for New Sources (40CFR467.02)

PSP: Payroll Savings Plan

PSS: Personnel Staffing Specialist

PSTN: Pesticide Safety Team Network (NRT-87/3)

PSU: Primary Sampling Unit (EPA-94/04)

PSY: Steady-State, Two-Dimensional Plume Model (EPA-91/3)

Pt: Platinum

PT: Packed Tower

PT: Part Time

PTA: Part Throttle Acceleration (40CFR86)

PTC: Positive Temperature Coefficient (40CFR85)

PTD: Part Throttle Deceleration (40CFR86)

PTDIS: Single Stack Meteorological Model in EPA UNAMAP Series (EPA-94/04)

PTE: Permanent Total Enclosure (40CFR52.741.a.4)

PTE: Potential to Emit (EPA-94/04)

PTFE: Polytetrafluoroethylene (Teflon) (EPA-94/04)

PTMAX: Single Stack Meteorological Model in EPA

UNAMAP series (EPA-94/04)

PTO: Patent and Trademark Office (DOC)

PTPLU: Point Source Gaussian Diffusion Model (EPA-94/04)

Pu: Plutonium

PU: Pulmonary region of the respiratory tract (EPA-92/12)

PUC: Public Utility Commission (EPA-94/04)

PUREX: Plutonium-Uranium Extraction Plant (DOE-91/4)

PV: Project Verification (EPA-94/04)

PVC: Polyvinyl Chloride (EPA-94/04)

PWS: Public Water Supply/System (EPA-94/04)

PWS: Public Water System (SDWA)

PWSS: Public Water Supply System (EPA-94/04)

PWSS: Public Water Supply System (SDWA)

PY: Prior Year

Q: Quench

QA/QC: Quality Assistance/Quality Control (EPA-94/04)

QA: Quality Assurance (40CFR766.3)

QAC: Quality Assurance Coordinator (EPA-94/04)

QAMIS: Quality Assurance Management and Information System (EPA-94/04)

QAO: Quality Assurance Officer (EPA-94/04)

QAPP: Quality Assurance Program (or Project) Plan (EPA-94/04)

QAT: Quality Action Team (EPA-94/04)

QBTU: Quadrillion British Thermal Units (EPA-94/04)

QC: Quality Control (40CFR766.3)

QC: Quench Column

QCA: Quiet Communities Act (EPA-94/04)

QCI: Quality Control Index (EPA-94/04)

QCP: Quiet Community Program (EPA-94/04)

ql*: Cancer Potency Factor (EPA-91/3)

QNCR: Quarterly Noncompliance Report (EPA-94/04)

QS: Quench Separator

QSAR: Quantitative Structure-Activity Relationships (EPA-91/3)

QSI: Quality Step Increase

QT: Quench Tower

QUA: Qualitative Use Assessment (EPA-94/04)

QUIPE: Quarterly Update for Inspector in Pesticide Enforcement (EPA-94/04)

R: Alkyl (C_nH_{2n+1}) group

R: Degree Rankine (40CFR60)

R: Rankine (40CFR86.403.78-94)

R&D: Research and Development (EPA-94/04)

Ra: Radium

RA: Reasonable Alternative (EPA-94/04)

RA: Regional Administrator

RA: Regulatory Alternatives (EPA-94/04)

RA: Regulatory Analysis (EPA-94/04)

RA: Remedial Action (40CFR300.4)

RA: Resource Allocation (EPA-94/04)

RA: Risk Analysis (EPA-94/04)

RA: Risk Assessment (EPA-94/04)

RAATS: RCRA Administrative Action Tracking System (EPA-94/04)

RAC: Radiation Advisory Committee (EPA-94/04)

RAC: Reference Air Concentration (ug/m^3)

RAC: Reference Ambient Concentration (EPA-91/3)

RAC: Regional Asbestos Coordinator (EPA-94/04)

RAC: Response Action Coordinator (EPA-94/04)

RACM: Reasonably Available Control Measures (EPA-94/04)

RACM: Regulated Asbestos-Containing Material (40CFR61.141)

RACT: Reasonably Available Control Technology (40CFR51.100-o)

RAD: Radiation Adsorbed Dose (unit of measurement of radiation absorbed by humans) (EPA-94/04)

RADM: Random Walk Advection and Dispersion Model (EPA-94/04)

RADM: Regional Acid Deposition Model (EPA-94/04)

RAGS: Risk Assessment Guidance for Superfund (EPA-91/12)

RAM: Random Access Memory (computer)

RAM: Urban Air Quality Model for Point and Area Source in EPA UNAMAP Series (EPA-94/04)

RAMP: Rural Abandoned Mine Program (EPA-94/04)

RAMS: Regional Air Monitoring System (EPA-94/04)

RAP: Radon Action Program (EPA-94/04)

RAP: Remedial Accomplishment Plan (EPA-94/04)

RAP: Reregistration Assessment Panel (EPA-94/04)

RAP: Response Action Plan (EPA-94/04)

RAPS: Regional Air Pollution Study (EPA-94/04)

RARG: Regulatory Analysis Review Group (EPA-94/04)

RAS: Routine Analytical Service (EPA-94/04)

RAT: Radiological Assistance Team (40CFR300.4)

RAT: Relative Accuracy Test (EPA-94/04)

Rb: Rubidium

RB: Red Border

RB: Request for Bid (EPA-94/04)

RBC: Red Blood Cell(s) (EPA-92/12)

RBC: Rotating Biological Contactor (40CFR35.2035)

RC: Regional Counsel

RC: Responsibility Center (EPA-94/04)

RCC: Radiation Coordinating Council (EPA-94/04)

RCDO: Regional Case Development Officer (EPA-94/04)

RCO: Regional Compliance Officer (EPA-94/04)

RCP: Regional Contingency Plan (40CFR300.4)

RCP: Research Centers Program (EPA-94/04)

RCRA: Resource Conservation and Recovery Act of 1976 (40CFR124)

RCRIS: Resource Conservation and Recovery Information System (EPA-94/04)

RD/RA: Remedial Design/ Remedial Action (EPA-94/04)

RD: Remedial Design (40CFR300.4)

RD&D: Research, Development and Demonstration (EPA-94/04)

RDD: Regional Deposited Dose (EPA-8/90b)

RDF: Refuse-Derived Fuel (40CFR60)

rDNA: Recombinant DNA (EPA-94/04)

RDU: Regional Decision Units (EPA-94/04)

RDV: Reference Dose Values (EPA-94/04)

Re: Rhenium

RE: Reasonable Efforts (EPA-94/04)

RE: Reportable Event (EPA-94/04)

REAG: Reproductive Effects Assessment Group

REAP: Regional Enforcement Activities Plan (EPA-94/04)

redox: Oxidation Reduction

REE: Rare Earth Elements (EPA-94/04)

REEP: Review of Environmental Effects of Pollutants (EPA-94/04)

REF: Reference

REM (Roentgen Equivalent Man) (EPA-94/04)

REM/FIT: Remedial/Field Investigation Team (EPA-94/04)

REM: Remedial Engineering Management

REM: Roentgen Equivalent Man (radiation unit)

REMS: RCRA Enforcement Management System (EPA-94/04)

REP: Reasonable Efforts Program (EPA-94/04)

REPS: Regional Emissions Projection System (EPA-94/04)

RESOLVE: Center for Environmental Conflict Resolution (EPA-94/04)

Rf: Rutherfordium

RF: Radio Frequency

RF: Response Factor (40CFR796)

RFA: Regulatory Flexibility Act (EPA-94/04)

RFB: Request for Bid (EPA-94/04)

RFBAT: Reasonably Foreseeable Best Available Technology

RfC: Chronic inhalation reference dose (EPA-8/90b)

RfC: Inhalation Reference Concentration (EPA-92/12)

RfC: Reference Concentration (EPA-91/12)

RfC_s: Subchronic inhalation reference dose (EPA-8/90b)

RfD: Oral Reference Dose (EPA-92/12)

RfD: Reference Dose (EPA-91/12)

RfD: Reference Dose Values (40CFR300)

RFI: Radio Frequency Interference (40CFR60)

RFI: Remedial Field Investigation (EPA-94/04)

RFP: Reasonable Further Programs (EPA-94/04)

RFP: Request for Proposal (EPA-94/04)

RFP: Rocky Flats Plant (DOE-91/4)

RFQ: Request for Quote

RgD: Regulatory Dose (EPA-92/12)

RGS: Research Grants Staff

Rh: Rhodium

RH: Reheat

RHRS: Revised Hazard Ranking System (EPA-94/04)

RI/FS: Remedial Information/Feasibility Study (EPA-94/04)

RI/FS: Remedial Investigation/Feasibility Study (EPA-91/12)

RI/FS: Reports of Investigation/Feasibility Studies (DOE-91/4)

RI: Reconnaissance Inspection (EPA-94/04)

RI: Remedial Investigation (40CFR300.4)

RI: Rhode Island

RIA: Regulatory Impact Analysis (EPA-94/04)

RIA: Regulatory Impact Assessment (EPA-94/04)

RIC: Radon Information Center (EPA-94/04)

RIC: RTP Information Center

RICC: Retirement Information and Counseling Center (EPA-94/04)

RICO: Racketeer Influenced and Corrupt Organizations Act (EPA-94/04)

RID: Regulatory Integration Division

RIF: Reduction in Force

RIM: Regulatory Interpretation Memorandum (EPA-94/04)

RIN: Regulatory Identifier Number (EPA-94/04)

RIP: RCRA Implementation Plan (EPA-94/04)

RISC: Regulatory Information Service Center (EPA-94/04)

RJE: Remote Job Entry (EPA-94/04)

RJS: Reverse Jet Scrubber

RK: Rotary Kiln

RLL: Rapid and Large Leakage (Rate) (EPA-94/04)

RM: Risk Management (EPA-92/12)

RMCL: Recommended Maximum Contaminant Level (this phrase is being discontinued in favor of MCLG) (EPA-94/04)

RMDHS: Regional Model Data Handling System (EPA-94/04)

RME: Reasonable Maximum Exposure (EPA-91/12)

RMIS: Resources Management Information System

RMIS: Resources Management Information System (EPA-94/04)

RMO: Records Management Officer

RMP: Revolutions Per Minute

Rn: Radon

RNA: Ribonucleic Acid (EPA-94/04)

rO: Rated output (40CFR87.2-94)

RO: Regional Office

ROADCHEM: Roadway Version that Includes Chemical Reactions of BI, NO_2, and O_3 (EPA-94/04)

ROADWAY: A Model to Predict Pollutant Concentrations Near a Roadway (EPA-94/04)

ROC: Record Of Communication (EPA-94/04)

ROD: Record of Decision (40CFR300.4)

ROD: Record of Decision (CERCLA)

RODS: Records Of Decision System (EPA-94/04)

ROG: Reactive Organic Gases (EPA-94/04)

ROI: Region of Influence (DOE-91/4)

ROLLBACK: A Proportional Reduction Model (EPA-94/04)

ROM: Read Only Memory (computer)

ROM: Regional Oxidant Model (EPA-94/04)

ROMCOE: Rocky Mountain Center on Environment (EPA-94/04)

ROP: Regional Oversight Policy (EPA-94/04)

ROPA: Record Of Procurement Action (EPA-94/04)

RP: Radon Progeny Integrated Sampling (EPA-94/04)

RP: Respirable Particulates

RP: Respirable Particulates (EPA-94/04)

RP: Responsible Party (EPA-94/04)

RP: Responsible Party

RPAR: Rebuttable Presumption Against Registration (FIFRA) (40CFR166.3)

rpm: Revolutions per minute (40CFR86.403.78-94)

RPM: Reactive Plume Model (EPA-94/04)

RPM: Remedial Project Manager (40CFR300.4)

RPO: Regional Planning Officer

RPO: Regional Program Officer

rPR: Rated pressure ratio (40CFR87.2-94)

RQ: Reportable Quantities (40CFR355-AA)

RRC: Regional Response Center (40CFR300.4)

RREL: Risk Reduction Engineering Laboratory (EPA) (Formerly HWERL)

RRT: Regional Response Team (40CFR300.4)

RRT: Requisite Remedial Technology (EPA-94/04)

RS: Registration Standard (EPA-94/04)

RSCC: Regional Sample Control Center (EPA-94/04)

RSD: Risk-Specific Dose (EPA-94/04)

RSD: Risk Specific Dose (ug/m^3)

RSE: Removal Site Evaluation (EPA-94/04)

RSKERL: Robert S. Kerr Environmental Research Laboratory

RSPA: Research and Special Programs Administration (40CFR300.4)

RT: Regional Total

RTCM: Reasonable Transportation Control Measure (EPA-94/04)

RTD: Return to Duty

RTDM: Rough Terrain Diffusion Model (EPA-94/04)

RTECS: Registry of Toxic Effects of Chemical Substances (29CFR1910.20)

RTF: Reactivity Task Force (MCA)

RTM: Regional Transport Model (EPA-94/04)

RTP: Research Triangle Park

Ru: Ruthenium

RUP: Restricted Use Pesticide (EPA-94/04)

RV: Residual Volume (EPA-92/12)

RVP: Reid Vapor Pressure (40CFR86)

RWC: Receiving Water Concentration (EPA-91/3)

RWC: Residential Wood Combustion (EPA-94/04)

S/cm^3: Structures per Cubic Centimeter (40CFR763-AA)

S/mm^2: Structures per Square Millimeter (40CFR763-AA)

S/TCAC: Scientific/Technical Careers Advisory Committee

s.c.: Subcutaneous (EPA-92/12)

s: Second (40CFR60)

S: Scrubber

S: Sulphur

S&A: Sampling and Analysis (EPA-94/04)

S&A: Surveillance and Analysis (EPA-94/04)

S&E: Salaries and Expensses

S1S: Smooth-One-Side (40CFR429.11)

S2S: Smooth-Two-Sides (40CFR42.119)

SA: Special Assistant

SA: Sunshine Act

SAB: Science Advisory Board (EPA-94/04)

SAC: Secretarial Advisory Board

SAC: Sulfuric Acid Concentrator

SAC: Support Agency Coordinator (40CFR300.4)

SAC: Suspended and Cancelled Pesticides (EPA-94/04)

SACTI: Seasonal/Annual Cooling Tower Impacts (computer code) (DOE-91/4)

SADAA: Science Assistant to the Deputy Administrator

SAE: Society of Automotive Engineers (40CFR89.3-94)

SAED: Selected Area Electron Diffraction (40CFR763-AA)

SAEWG: Standing Air Emissions Work Group (EPA-94/04)

SAIC: Special-Agents-In-Charge (EPA-94/04)

SAIP: Systems Acquisition and Implementation Program (EPA-94/04)

SAMWG: Standing Air Monitoring Work Group (EPA-94/04)

SANE: Sulfur and Nitrogen Emissions (EPA-94/04)

SANSS: Structure and Nomenclature Search System (EPA-94/04)

SAP: Sampling and Analysis Plan

SAP: Scientific Advisory Panel (EPA-94/04)

SAP: Serum Alkaline Phosphatase (EPA-92/12)

SAR: Sodium Absorption Ratio

SAR: Start Action Request (EPA-94/04)

SAR: Structural Activity Relationship (of a qualitative assessment) (EPA-94/04)

SARA: Superfund Amendments and Reauthorization Act of 1986 (40CFR280)

SARAH2: Surface Water Assessment Model for Back Calculating Reductions in Biotic Hazardous Wastes (EPA-91/3)

SAROAD: Storage and Retrieval of Aerometric Data (40CFR51.117)

SAS: Special Analytical Service (EPA-94/04)

SAS: Statistical Analysis System (EPA-94/04)

SASS: Source Assessment Sampling System (EPA-94/04)

SATO: Scheduled Airline Traffic Office

SAV: Submerged Aquatic Vegetation (EPA-94/04)

Sb: Antimony

SBA: Small Business Act

SBA: Small Business Administration (40CFR21.2)

SBO: Small Business Ombudsman

Sc: Scandium

SC: Sierra Club (EPA-94/04)

SC: South Carolina

SC: Steering Committee

SCAC: Support Careers Advisory Committee

SCAP: Superfund Comprehensive Accomplishments Plan (40CFR35.4010)

SCAP: Superfund Consolidated Accomplishments Plan (EPA-94/04)

SCBA: Self-Contained Breathing Apparatus (EPA-94/04)

SCC: Secondary Combustion Chamber

SCC: Source Classification Code (EPA-94/04)

SCD/SWDC: Soil or Soil and Water Conservation District (EPA-94/04)

SCDHEC: South Carolina Department of Health and Environmental Control (DOE-91/4)

SCE&G: South Carolina Electric and Gas Company (DOE-91/4)

scf: Standard Cubic Feet (40CFR60)

SCF: Supercritical Fluid

scfh: Standard Cubic Feet Per Hour (40CFR60)

SCFM: Standard Cubic Feet Per Minute (EPA-94/04)

SCLDF: Sierra Club Legal Defense Fund (EPA-94/04)

scm: Standard Cubic Meter (40CFR60)

SCORPIO: Subject Content-Oriented Retriever for Processing Information On-Line

SCR: Selective Catalytic Reduction (EPA-94/04)

SCRAM: State Consolidated RCRA Authorization Manual (EPA-94/04)

SCRC: Superfund Community Relations Coordinator (EPA-94/04)

SCRP: Superfund Community Relations Program

SCS: Soil Conservation Service, U.S. Department of Agriculture

SCS: Supplementary Control Strategy/System (EPA-94/04)

SCS: Supplementary Control System (40CFR57.103)

SCSA: Soil Conservation Society of America (EPA-94/04)

SCSP: Storm and Combined Sewer Program (EPA-94/04)

SCW: Supercritical Water Oxidation (EPA-94/04)

SD: South Dakota

SD: Spray Dryer
SD: Standard Deviation
SDBE: Small Disadvantaged Business Enterprise
SDC: Systems Decision Plan (EPA-94/04)
SDI: Sludge Density Index
SDWA: Safe Drinking Water Act (40CFR124.2)
SDWC: Secondary Drinking Water Criteria (SDWA)
Se: Selenium
SEA: Selective Enforcement Auditing (40CFR89.3-94)
SEA: State Enforcement Agreement (EPA-94/04)
SEA: State/EPA Agreement (EPA-94/04)
SEA: Supplementary Environmental Analysis (like EIS)
SEAM: Surface, Environment, and Mining (EPA-94/04)
SEAS: Strategic Environmental Assessment System (EPA-94/04)
sec.: Seconds (40CFR87.2-94)
SEC: Securities and Exchange Commission
SEE: Senior Environmental Employee
SEIA: Socioeconomic Impact Analysis (EPA-94/04)
SEIA: Solar Energy Industries Association
SEM: Scanning Electronic Microscope (40CFR763-AA)
SEM: Standard Error of the Means (EPA-94/04)
SEP: Standard Evaluation Procedures (EPA-94/04)
SEP: Structural Integrity Procedure (RCRA)
SEPWC: Senate Environment and Public Works Committee (EPA-94/04)
SERATRA: Sediment Contaminant Transport Model Simplified Lake/Stream Analysis (EPA-91/3)
SERC: State Emergency Response Commission (40CFR300.4)
SERI: Solar Energy Research Institute (See NREL)
SES: Secondary Emissions Standard (EPA-94/04)
SES: Senior Executive Service
SES: Socioeconomic Status
SET: Site Evaluation Team (DOE-91/4)
SETS: Site Enforcement Tracking System (EPA-94/04)
SF: Safety Factor (EPA-92/12)
SF: Slope Factor (EPA-91/12)
SF: Standard Form (EPA-94/04)
SF: Superfund (EPA-94/04)
SFA: Spectral Flame Analyzers (EPA-94/04)
SFDS: Sanitary Facility Data System (EPA-94/04)
SFFAS: Superfund Financial Assessment System (EPA-94/04)

SFIREG: State FIFRA Issues Research and Evaluation Group (EPA-94/04)
SFS: State Funding Study (EPA-94/04)
sft^3: Standard Cubic Feet (DOE-91/4)
SGOT: Serum Glutamic Oxaloacetic Transaminase (EPA-92/12)
SGPT: Serum Glutamic Pyruvic Transaminase (EPA-92/12)
SHORTZ: Short Term Terrain Model (EPA-94/04)
SHPO: State Historic Preservation Officer (DOE-91/4)
SHWL: Seasonal High Water Level (EPA-94/04)
SI unit: System International (Meter-Kg-Second) unit
Si: Silicon
SI: International System of Units (EPA-94/04)
SI: Site Inspection (40CFR300.4)
SI: Spark Ignition (40CFR89.3-94)
SI: Surveillance Index (EPA-94/04)
SIC: Standard Industrial Classification (EPA-94/04)
SIC: Standard Industrial Code (DOE-91/4)
SICEA: Steel Industry Compliance Extension Act (EPA-94/04)
sigma g: Geometric Standard Deviation (EPA-92/12)
SIMS: Secondary Ion-Mass Spectrometry (EPA-94/04)
SIP: State Implementation Plan (40CFR51.350)
SIS: Stay In School
SITE: Superfund Innovative Technology Evaluation (EPA-94/04)
SIWSA: Standard Metropolitan Statistical Area (DOE-91/4)
SL: Sick Leave
SLAEM: Single Layer Analytical Element Model (DOE-91/4)
SLAMS: State/Local Air Monitoring Station (EPA-94/04)
SLAMS: State or Local Air Monitoring Station (40CFR58.1-b)
SLANG: Selected Letter and Abbreviated Name Guide
SLRL: Sex Linked Recessive Lethal (40CFR798.5275)
SLSM: Simple Line Source Model (EPA-94/04)
Sm: Samarium
sm^3: Standard Cubic Meter (40CFR464.02)
SMART: Simple Maintenance of ARTS (EPA-94/04)
SMCL: Secondary Maximum Contaminant Levels (40CFR143)
SMCRA: Surface Mining Control and Reclamation

Act (EPA-94/04)

SME: Subject Matter Expert (EPA-94/04)

SMO: Sample Management Office (EPA-94/04)

SMOA: Superfund Memorandum of Agreement (40CFR300.4)

SMR: Standard Mortality Ratio (EPA-92/12)

SMSA: Standard Metropolitan Statistical Area (EPA-94/04)

SMYS: Specified Minimum Yield Strength (40CFR192.3)

Sn: Tin

SN: Smoke Number (40CFR87.2-94)

SNA: System Network Architecture

SNA: System Network Architecture (EPA-94/04)

SNAAQS: Secondary National Ambient Air Quality Standards (EPA-94/04)

SNAP: Significant Noncompliance Action Program (EPA-94/04)

SNARL: Suggested No Adverse Response Level (EPA-94/04)

SNC: Significant Noncompliers (EPA-94/04)

SNCR: Selective Non-Catalytic Reduction

SNG: Synthetic Natural Gas

SNL: Sandia National Laboratories (DOE-91/4)

SNUR: Significant New Use Rule (EPA-94/04)

SOC: Synthetic Organic Chemicals (EPA-94/04)

SOCMA: Synthetic Organic Chemical Manufacturers Association

SOCMI: Synthetic Organic Chemical Manufacturing Industry (40CFR60.700)

SOEH: Society for Occupational and Environmental Health

SOM: Sensitivity of Method (FDA)

SOP: Standard Operating Procedure (40CFR61.61)

SOT: Society of Toxicology

SOTDAT: Source Test Data (EPA-94/04)

SOW: Scope Of Work (EPA-94/04)

SOW: Statement of Work (40CFR35.6015-45)

SP: Shaft Power (40CFR87.2-94)

SPAR: Status of Permit Application Report (EPA-94/04)

SPCC: Spill Prevention, Containment, and Countermeasure (40CFR112.3)

SPE: Secondary Particulate Emissions (EPA-94/04)

SPECS: Specifications

SPF: Structured Programming Facility (EPA-94/04)

SPI: Society of the Plastics Industry

SPI: Strategic Planning Initiative (EPA-94/04)

SPL: Sound Pressure Level (DOE-91/4)

SPLMD: Soil-pore Liquid Monitoring Device (EPA-94/04)

SPMS: Special Purpose Monitoring Stations (EPA-94/04)

SPMS: Strategic Planning and Management System (EPA-94/04)

SPOC: Single Point Of Contact (EPA-94/04)

SPS: State Permit System (EPA-94/04)

SPSS: Statistical Package for the Social Sciences (EPA-94/04)

SPUR: Software Package for Unique Reports (EPA-94/04)

sq ft: Square Feet

sq m: Square Meter

sq.cm.: Square Centimeters (EPA-92/12)

SQBE: Small Quantity Burner Exemption (EPA-94/04)

SQC: Sediment Quality Criteria (EPA-91/3)

SQG: Small Quantity Generator (40CFR261.5)

Sr: Strontium

SR: State Route (DOE-91/4)

SRAB: Source Receptor Analysis Branch (EPA)

SRAP: Superfund Remedial Accomplishment Plan (EPA-94/04)

SRC: Solvent-Refined Coal (EPA-94/04)

SRF: State Water Pollution Control Revolving Fund (40CFR35.3105-i)

SRM: Standard Reference Method (EPA-94/04)

SRP: Special Review Procedure (EPA-94/04)

SRR: Second Round Review (EPA-94/04)

SRR: Submission Review Record (EPA-94/04)

SRS: Savannah River Site (DOE-91/4)

SRTS: Service Request Tracking System (EPA-94/04)

SS: Settleable Solids (EPA-94/04)

SS: Spray Saturator

SS: Superfund Surcharge (EPA-94/04)

SS: Suspended Solids (40CFR133.101)

SSA: Social Security Administration

SSA: Sole Source Aquifer (40CFR149.2)

SSAC: Soil Site Assimilated Capacity (EPA-94/04)

SSAC: Soil Site Assimulated Capacity

SSC: Scientific Support Coordinator (40CFR300.4)

SSC: Superfund State Contracts (40CFR35.6015-47)

SSD: Standards Support Document (EPA-94/04)

SSE: Safe Shutdown Earthquake (DOE-91/4)

SSEIS: Standard Support and Environmental Impact Statement (EPA-94/04)

SSEIS: Stationary Source Emissions and Inventory System (EPA-94/04)

SSI: Size Selective Inlet (EPA-94/04)

SSMS: Spark Source Mass Spectrometry (EPA-94/04)

SSN: Social Security Number

SSO: Source Selection Official (EPA-94/04)

SST: Supersonic Transport

SSTS: Section Seven Tracking System (EPA-94/04)

SSU: Standard Saybolt Unit (a viscosity unit)

SSURO: Stop Sale, Use and Removal Order (EPA-94/04)

ST: Spray Tower

STALAPCO: State and Local Air-Pollution Control Officials (EPA-94/04)

STAPPA: State and Territorial Air Pollution (EPA-94/04)

STAPPA: State and Territorial Air Pollution Program Administrators

STAR: Stability Wind Rose (EPA-94/04)

STAR: State Acid Rain Projects (EPA-94/04)

START: Superfund Technical Assistance Response Team

std: Standard (40CFR60)

STEL: Short Term Exposure Limit (EPA-94/04)

STEM: Scanning Transmission Electron Microscope (40CFR763-AA)

STN: Scientific and Technical Information Network (EPA-94/04)

STORET: Storage and Retrieval of Water Quality Information (EPA-91/3)

STORET: Storage and Retrieval of Water-Related Data (EPA-94/04)

STP: Sewage Treatment Plant (EPA-94/04)

STP: Standard Temperature and Pressure (EPA-94/04)

SUNFED: Special United Nations Fund for Economic Development

SUP: Standard Unit of Processing (EPA-94/04)

SURE: Sulfate Regional Experiment Program (EPA-94/04)

SUS: Saybolt Universal Seconds (29CFR1910.1200)

SV: Sampling Visit (EPA-94/04)

SVI: Sludge Volume Index

SW: Slow Wave (EPA-94/04)

SWC: Settlement With Conditions (EPA-94/04)

SWDA: Solid Waste Disposal Act (EPA-94/04)

SWDF: Solid Waste Derived Fuel

SWDF: Solvent Waste Dervide Fuel

SWIE: Southern Waste Information Exchange (EPA-94/04)

SWMU: Solid Waste Management Unit (EPA-94/04)

SWTR: Surface Water Treatment Rule (EPA-94/04)

sy.: Synonym

SYSOP: Systems Operator (EPA-94/04)

T-R: Transformer-Rectifier (EPA-94/04)

t: Metric Ton

T: Temperature, degrees Kelvin (40CFR87.2-94)

T&A: Time and Attendance

Ta: Tantalum

TA: Travel Authorization

TACB: Texas Air Control Board

TAG: Technical Assistance Grants (40CFR35.4005)

TALMS: Tunable Atomic Line Molecular Spectroscopy (EPA-94/04)

TAMS: Toxic Air Monitoring System (EPA-94/04)

TAMTAC: Toxic Air Monitoring System Advisory Committee (EPA-94/04)

TAO: TSCA Assistance Office

TAP: Technical Assistance Program (EPA-94/04)

TAPDS: Toxic Air Pollutant Data System (EPA-94/04)

TAPP: Time and Attendance, Payroll, and Personnel

TAPPI: Association of the Pulp and Paper Industry (40CFR60.17.d)

TARC: Toxics Testing and Assessment Research Committee

TAS: Tolerance Assessment System (EPA-94/04)

Tb: Terbium

TB: Tracheobronchial region of the respiratory tract (EPA-92/12)

TB: Trial Burn

TBD: To Be Determined

TBP: Trial Burn Plan

TBT: Tributyltin (EPA-94/04)

Tc: Technetium

TC: Target Concentration (EPA-94/04)

TC: Technical Center (EPA-94/04)

TC: Total Carbon

TC: Toxic Concentration (EPA-94/04)

TC: Toxicity Characteristics (EPA-94/04)

TCB: Toxic Combustion Byproduct

TCC: Tagliabue closed cup, a standard method of determining flash points (EPA-92/12)

TCDD: Dioxin (Tetrachlorodibenzo-p-dioxin) (EPA-94/04)

TCDF: Tetrachlorodi-benzofurans (EPA-94/04)

TCE: Trichloroethylene (EPA-94/04)

TCLP: Total Concentrate Leachate Procedure (EPA-94/04)

TCLP: Toxicity Characteristic Leachate Procedure (EPA-94/04)

TCM: Transportation Control Measure (40CFR51.138)

TCP: Transportation Control Plan (EPA-94/04)

TCP: Trichloroethylene (EPA-94/04)

TCP: Trichloropropane (EPA-94/04)

TCRI: Toxic Chemical Release Inventory (EPA-94/04)

T_D: Dispensed Fuel Temperature (40CFR86.098.3-94)

TD: Toxic Dose (EPA-94/04)

TDB: Toxicology Data Base (EPA-92/12)

TDD: Telecommunication Devices for Deaf (40CFR12.103)

TDE: Tetrachlorodiphenylethane

TDF: Tire Derived Fuel

TDI: Toluene Diisocyanate

TDS: Total Dissolved Solids (40CFR131.35)

TDY: Temporary Duty

Te: Tellurium

TEAM: Total Exposure Assessment Model (EPA-94/04)

TEC: Technical Evaluation Committee (EPA-94/04)

TEG: Tetraethylene Glycol (EPA-94/04)

TEGD: Technical Enforcement Guidance Document (EPA-94/04)

TEL: Tetraethyl Lead (40CFR86)

TEM: Texas Episodic Model (EPA-94/04)

TEM: Transmission Electron Microscope (40CFR763-AA)

TEP: Technical Evaluation Panel (EPA-94/04)

TEP: Typical End-use Product (EPA-94/04)

TERA: TSCA Environmental Release Application (EPA-94/04)

TES: Technical Enforcement Support (EPA-94/04)

TEXIN: Texas Intersection Air Quality Model (EPA-94/04)

TFS: Tin Free Steel

TFT: Temporary Full Time

TFTE Temporary Full Time Equivalent

TGAI: Technical Grade of the Active Ingredient (EPA-94/04)

TGIF: Thank God It's Friday

TGO: Total Gross Output (EPA-94/04)

TGP: Technical Grade Product (EPA-94/04)

Th: Thorium

THC: Total Hydrocarbons (40CFR86.094.3-94)

THCE: Total Hydrocarbon Equivalent (40CFR86.090.3-94)

THM: Trihalomethane (40CFR141.2)

Ti: Titanium

TI: Temporary Intermittent (EPA-94/04)

TI: Therapeutic Index (EPA-94/04)

TIBL: Thermal Internal Boundary Layer (EPA-94/04)

TIC: Technical Information Coordinator (EPA-94/04)

TIC: Tentatively Identified Compounds (EPA-94/04)

TIC: Total Inorganic Carbon

TIE: Toxicity Identification Evaluation (EPA-91/3)

TIM: Technical Information Manager (EPA-94/04)

TIM: Time in Mode (40CFR87.2-94)

TIP: Transportation Improvement Program (40CFR51.392)

TIS: Tolerance Index System (EPA-94/04)

TISE: Take It Somewhere Else (EPA-94/04)

TISE: Take It Somewhere Else (Solid Waste Syndrome. See: NIMBY)

TITC: Toxic Substance Control Act Interagency Testing Committee (EPA-94/04)

TITC: TSCA Interagency Testing Committee

TKN: Total Kjeldahl Nitrogen

Tl: Thallium

TLC: Thin Layer Chromatography (40CFR796.2700.a)

TLEV: Transitional Low-Emission Vehicle (40CFR88.103-94)

TLM: Tolerance Limit Median

TLV-C: TLV-Ceiling (EPA-94/04)

TLV-STEL: TLV-Short Term Exposure Limit (EPA-94/04)

TLV-TWA: TLV-Time Weighted Average (EPA-94/04)

TLV: Threshold Limit Value (EPA-94/04)

Tm: Thulium

TMDC: Total Maximum Daily Load

TMDL: Total Maximum Daily Load (40CFR130.2)

TMI: Three Mile Island

TML: Tetramethyl Lead (40CFR86)

TMRC: Theoretical Maximum Residue Contribution (EPA-94/04)

TN: Tennessee

TNCWS: Transient Non-Community Water System (EPA-94/04)

TNT: Trinitrotoluene (40CFR457.11)

TO: Task Order (EPA-94/04)

TO: Travel Order

TOA: Trace Organic Analysis (EPA-94/04)

TOC: Total Organic Carbon (40CFR268.2)

TOC: Total Organic Carbon/Compound (EPA-94/04)

TOC: Total Organic Compound (40CFR60.561)

TOD: Theoretical Oxygen Demand

TOD: Time-Origin-Destination (40CFR52.2294)

TOD: Total Oxygen Demand

TODAM: Transport One-Dimensional Degradation and Migration Model (EPA-91/3)

TOT: Time-of-Travel

TOTAL: Total Respiratory Tract (EPA-92/12)

TOX: Tetradichloroxylene (EPA-94/04)

TOX: Total Organic Halogen

TOXI4: A Subset of WASP4 (EPA-91/3)

TOXIC: Toxic Organic Transport and Bioaccumulation Model (EPA-91/3)

TOXIWASP: Chemical Transport and Fate Model (EPA-91/3)

TP: Technical Product (EPA-94/04)

TPA: Terephthalic acid (40CFR60.561)

TPC: Testing Priorities Committee (EPA-94/04)

TPD: Tons Per Day

TPES: Toxic and Pretreatment Effluent Standards

TPI: Technical Proposal Instructions (EPA-94/04)

TPQ: Threshold Planning Quantity (40CFR355.10)

TPSIS: Transportation Planning Support Information System (EPA-94/04)

TPTH: Triphenyltinhydroxide (EPA-94/04)

TPW: Tons Per Week

TPY: Tons Per Year (EPA-94/04)

TQM: Total Quality Management (EPA-94/04)

TR: Target Risk (EPA-91/12)

TRC: Technical Review Committee (EPA-94/04)

TRC: Total Residual Chlorine (40CFR420.02)

TRD: Technical Review Document (EPA-94/04)

TRE: Total Resource Effectiveness (40CFR60.611)

TRE: Toxicity Reduction Evaluation (EPA-91/3)

TRI: Toxic Release Inventory (EPA-94/04)

TRIP: Toxic Release Inventory Program (EPA-94/04)

TRIS: Toxic Chemical Release Inventory System (EPA-94/04)

TRLN: Triangle Research Library Network (EPA-94/04)

TRLN: Triangle Research Library Network

TRO: Temporary Restraining Order (EPA-94/04)

TRS: Total Reduced Sulfur (40CFR60.281)

TRU: Transuranic, a Classification of Wastes (DOE-91/4)

TRW: Transuranic Radioactive Waste (40CFR191.02)

TS: Total Solids

TSA: Technical Systems Audit (EPA-94/04)

TSCA: Toxic Substances Control Act (EPA-94/04)

TSCA: Toxic Substances Control Act of 1976 (40CFR704.3)

TSCATS: TSCA Test Submissions Database (EPA-94/04)

TSCATS: TSCA Test Submissions Database (OTS)

TSCC: Toxic Substances Coordinating Committee (EPA-94/04)

TSD: Technical Support Document (EPA-94/04)

TSD: Treatment, Storage, and Disposal (40CFR260-AI-93)

TSDF: Treatment, Storage, and Disposal Facility (EPA-94/04)

TSDG: Toxic Substances Dialogue Group (EPA-94/04)

TSI: Thermal System Insulation (EPA-94/04)

TSM: Transportation System Management (EPA-94/04)

TSO: Time Sharing Option (EPA-94/04)

TSP: Teleprocessing Services Program

TSP: Thrift Savings Plan

TSP: Total Suspended Particulates (40CFR51.100)

TSPC: Toxic Substances Priority Committee

TSS: Terminal Security System

TSS: Total Suspended (non-filterable) Solids (40CFR268.2)

TSSC: Toxic Substances Strategy Committee (16 Fed. Agencies)

TSSG: Toxic Substances Strategy Group

TSSMS: Time Sharing Services Management System

TTE: Temporary Total Enclosure (40CFR52.741.a.4)

TTFA: Target Transformation Factor Analysis (EPA-94/04)

TTHM: Total Trihalomethane (40CFR141.30)

TTO: Total Toxic Organics (40CFR413.02)

TTY: Teletypewriter (EPA-94/04)

TU: Toxic unit (EPA-91/3)

TU: Turbidity unit (40CFR141.13)

TU_a: Acute Toxic Unit (EPA-91/3)

TU_c: Chronic Toxic Unit (EPA-91/3)

TUCC: Triangle University Computer Center

TVA: Tennessee Valley Authority (EPA-94/04)

TVS: Total Volatile Solids

TW: Test Weight (40CFR88.103-94)

TWA: Time Weighted Authority

TWA: Time Weighted Average (40CFR763.121)

TWMD: Toxics and Waste Management Division

TWS: Transient Water System (EPA-94/04)

TX: Texas

TZ: Treatment Zone (EPA-94/04)

U: Uranium

UAC: User Advisory Committee (EPA-94/04)

UAM: Urban Airshed Model (EPA-94/04)

UAO: Unilateral Administrative Order (EPA-94/04)

UAPSP: Utility Acid Precipitation Study Program (EPA-94/04)

UAQI: Uniform Air Quality Index (EPA-94/04)

UARG: Utility Air Regulatory Group (EPA-94/04)

UCC: Ultra Clean Coal (EPA-94/04)

UCCI: Urea-Formaldehyde Foam Insulation (EPA-94/04)

UCL: Upper Confidence Limit (EPA-92/12)

UCL: Upper Control Limit (EPA-94/04)

UDDS: Urban Dynamometer Driving Schedule (40CFR86)

UDKHDEN: Three-Dimensional Model Used for Single or Multiple Port Diffusers (EPA-91/3)

UDMH: Unsymmetrical Dimethyl Hydrazine (EPA-94/04)

UEL: Upper Explosive Limit (EPA-94/04)

UF: Uncertainty Factor (EPA-92/12)

UFL: Upper Flammability Limit (EPA-94/04)

UHWM: Uniform Hazardous Waste Manifest (40CFR259.10)

UIC: Underground Injection Control (40CFR124.2)

UL: Underwriters Laboratories, 333 Pfingsten Road, North Brook, IL 60062 (40CFR60.17.f)

ULEV: Ultra Low-Emission Vehicle (40CFR88.103-94)

ULINE: Uniform Linear Density Flume Model (EPA-91/3)

ULP: Unfair Labor Practices

UMERGE: Two-Dimensional Model Used to Analyze Positively Buoyant Discharge (EPA-91/3)

umol: Micromoles (EPA-92/12)

UMTRCA: Uranium Mill Tailings Radiation Control Act (EPA-94/04)

UMTRCA: Uranium Mill Tailings Radiation Control Act of 1978 (40CFR23.8; 300-AA)

UMW: United Mine Workers Union

UN: United Nations

UNAMAP: Users' Network for Applied Modeling of Air Pollution (EPA-94/04)

UNEP: United Nations Environment Program (EPA-94/04)

UNEP: United Nations Environment Programme

UNESCO: United Nations Educational, Scientific and Cultural Organization

UNIDO: United Nations Industrial Development Organization

UO$_3$: Uranium Oxide (DOE-91/4)

UOD: Ultimate Oxygen Demand

UOUTPLM: Cooling Tower Plume Model Adapted for Marine Discharges (EPA-91/3)

UPLUME: Numerical Model That Produces Flux-Average Dilutions (EPA-91/3)

UPS: United Parcel Service (40CFR86)

UPWP: Unified Planning Work Program (40CFR51.138)

URT: Upper Respiratory Tract (EPA-8/90b)

USAO: United States Attorney's Office

USATHAMA: U.S. Army Toxic and Hazardous Materials Agency (DOD)

USBM: United States Bureau of Mines

USC: Unified Soil Classification (EPA-94/04)

USC: United States Code

USCA: United States Code Annotated

USCG: United States Coast Guard (40CFR300.4)

USDA: United States Department of Agriculture (40CFR300.4)

USDOI: United States Department of the Interior

USDW: Underground Sources of Drinking Water (40CFR144.3)

USEPA: United States Environmental Protection Agency

USFS: United States Forest Service (EPA-94/04)

USFWS: U.S. Fish and Wildlife Service (DOE-91/4)

USGS: United States Geological Survey (DOE-91/4)

USP: U.S. Pharmacopeia (EPA-94/04)

USP: United States Pharmacopoeia

USPHS: United States Public Health Service

USPS: United States Postal Service

USS: United States Senate

UST: Underground Storage Tank (40CFR280)

USTR: United States Trade Representative

UT: Utah

UTM: Universal Transverse Mercator (EPA-94/04)

UTP: Urban Transportation Planning (EPA-94/04)

UV: Ultraviolet (40CFR86.090.3-94)

UV-VIS: Ultraviolet Visible (40CFR796.1050)

UZM: Unsaturated Zone Monitoring (EPA-94/04)

v/v: Volume for Volume (EPA-92/12)

v/v: Volume Per Volume (40CFR60)

V: Vanadium

V: Volt (40CFR60)

VA: Veterans Administration

VA: Virginia

VALLEY: Meteorological Model to Calculate Concentrations on Elevated Terrain (EPA-94/04)

VAT: Value Added Tax

VCM: Vinyl Chloride Monomer (EPA-94/04)

VDT: Video Display Terminal

VE: Value Engineering (40CFR35.905)

VE: Visual Emissions (EPA-94/04)

VEO: Visible Emission Observation (EPA-94/04)

VF: Volatilization Factor (EPA-91/12)

VHAP: Volatile Hazardous Air Pollutant (40CFR61.241)

VHS: Vertical and Horizontal Spread Model (EPA-94/04)

VHT: Vehicle-Hours of Travel (EPA-94/04)

VISTTA: Visibility Impairment from Sulfur Transformation and Transport in the Atmosphere (EPA-94/04)

VKT: Vehicle Kilometers Traveled (EPA-94/04)

VMT: Vehicle Miles Traveled (EPA-94/04)

VOC: Volatile Organic Compounds (40CFR60.431-94)

VOL: Volatile Organic Liquid (40CFR60.111b)

VOM: Volatile Organic Materials (40CFR52.741)

VOS: Vehicle Operating Survey (EPA-94/04)

VOST: Volatile Organic Sampling Train (EPA-94/04)

VP: Vapor Pressure (EPA-94/04)

VQ: Venturi Quench

VS: Venturi Scrubber

VSD: Virtually Safe Dose (EPA-94/04)

VSI: Visual Site Inspection (EPA-94/04)

VSS: Volatile Suspended Solids (EPA-94/04)

V_T: Tidal volume (EPA-90/8b)

VT: Vermont

W: Watt(s) (40CFR87.2-94)

W: Wolfram

WA: Washington

WA: Work Assignment (EPA-94/04)

WAC: Washington Administrative Codes (DOE-91/4)

WADTF: Western Atmospheric Deposition Task Force (EPA-94/04)

WAP: Waste Analysis Plan (EPA-94/04)

WASP4: Water Quality Analysis Program (EPA-91/3)

WASTOX: Estuary and Stream Quality Model (EPA-91/3)

WB: Wet Bulb (EPA-94/04)

WB: World Bank

WBC: White Blood Cell(s) (EPA-92/12)

WBE: Women's Business Enterprise (40CFR33.005)

WCED: World Commission on Environment and Development (EPA-94/04)

WDF: Waste Derived Fuel

WDOE: Washington Department of Ecology (DOE-91/4)

WDROP: Distribution Register of Organic Pollutants in Water (EPA-94/04)

WENDB: Water Enforcement National Data Base (EPA-94/04)

WERL: Water Engineering Research Laboratory (EPA-94/04)

WF: Weighting Factor (40CFR86)

WG: Wage Grade

WG: Work Group

WGI: Within Grade Increase

WHB: Waste Heat Boiler

WHC: Westinghouse Hanford Company (DOE-91/4)

WHO: World Health Organization (EPA-94/04)

WHWT: Water and Hazardous Waste Team (EPA-94/04)

WI: Wisconsin

WIC: Washington Information Center

WICEM: World Industry Conference on Environmental Management (EPA-94/04)

WINCO: Westinghouse Idaho Nuclear Company, Inc. (DOE-91/4)

WIPP: Waste Isolation Pilot Plant (DOE-91/4)

WISE: Women In Science and Engineering

WL: Warning Letter (EPA-94/04)

WL: Working Level (radon measurement) (40CFR192.11)

WLA/TMDL: Wasteload Allocation/Total Maximum Daily Load (EPA-94/04)

WLA: Wasteload Allocation (40CFR130.2)

WLM: Working Level Months (EPA-94/04)

WMO: World Meteorological Organization (EPA-94/04)

WNP-1: Washington Nuclear Plant No. 1 (DOE-91/4)

WNP-2: Washington Nuclear Plant No. 2 (DOE-91/4)

WOT: Wide Open Throttle (40CFR86)

WPCF: Water Pollution Control Federation (EPA-94/04)

WPPSS: Washington Public Power Supply System (DOE-91/4)

WQA: Water Quality Act of 1987

WQAB FLOW: Water Quality Analysis System Flow Data Subroutine (EPA-91/3)

WQC: Water Quality Criteria (EPA-92/12)

WQM: Water Quality Management (40CFR130.2)

WQS: Water Quality Standards (40CFR130.2)
WRC: Water Resources Council (EPA-94/04)
WRDA: Water Resources Development Act (EPA-94/04)
WRI: World Resources Institute (EPA-94/04)
WS: Wet Scrubber
WS: Work Status (EPA-94/04)
WSF: Water Soluble Fraction (EPA-94/04)
WSRA: Wild and Scenic Rivers Act (EPA-94/04)
WSRC: Westinghouse Savannah River Company (DOE-91/4)
WSTB: Water Sciences and Technology Board (EPA-94/04)
WSTP: Wastewater Sewage Treatment Plant (EPA-94/04)
wt: Weight (40CFR86.403.78-94)
WTP: Water Treatment Plant
WV: West Virginia
WWEMA: Waste and Wastewater Equipment Manufacturers Association (EPA-94/04)
WWF: World Wildlife Fund (EPA-94/04)
WWTP: Wastewater Treatment Plant (EPA-94/04)
WWTU: Wastewater Treatment Unit (EPA-94/04)
WY: Wyoming
Xe: Xenon
XRF: X-ray fluorescence

Y: Yttrium
Yb: Ytterbium
yd^2: Square Yard (40CFR60)
yr: Year (40CFR60)
YTD: Year to Date
ZBB: Zero Base Budgeting
ZEV: Zero-Emission Vehicle (40CFR88.103-94)
ZHE: Zero Headspace Extractor (EPA-94/04)
ZID: Zone of Initial Dilution (40CFR125.58)
Zn: Zinc
ZOI: Zone Of Incorporation (EPA-94/04)
Zr: Zirconium
ZRL: Zero Risk Level (EPA-94/04)

===========================

μCi/L: Microcuries per Liter (DOE-91/4)
μCi/mL: Microcuries per Milliliter (DOE-91/4)
$\mu g/m^3$: Micrograms per Cubic Liter (DOE-91/4)
μg: Microgram = 10^{-6} gram (40CFR60)
μL: Microliter = 10^{-6} liter (40CFR60)
μm: Micrometer = 10^{-6} meter

REFERENCES

(10CFRxxx) means 10 Code of Federal Regulations, Parts xxx.

(29CFRxxx) means 29 Code of Federal Regulations, Parts xxx.

(40CFRxxx) means 40 Code of Federal Regulations, Parts xxx.

(40CFRxxx-Ay-zz) means 40CFR parts xxx, Appendix y, item zz.
- For example: The term **"federally enforceable"** has the refernece (40CFR51-AS-12) which means that the source of the term was 40CFR part 51 Appendix S, item 12.

(40CFRxxx)[aaa]: "aaa" is a reference that the information provided was not excerpted from 40CFR definition sections, but from some other sections.
- For example: The term **"conventional pollutants"** has the reference (40CFR401.16)[aaa].

(CAAxxx-42USCyyyy):
- CAAxxx means the Clean Air Act, section xxx.
- 42USCyyyy means Title 42, United States Code (USC), section yyyy.
- For example: The term **"air pollutant"** has the reference (CAA302-42USC7602).

(CWAxxx-33USCyyyy):
- CWAxxx means the Clean Water Act, section xxx.
- 33USCyyyy means Title 33, United States Code (USC), section yyyy.
- For example: The term **"area contingency plan"** has the reference (CAA302-42USC7602).

(CZMAxxx-16USCyyyy):
- CZMAxxx means the Coast Zone

Management Act (16USC1451-1464), section xxx.
- 16USCyyyy: means Title 16, United States Code (USC), section yyyy.
- For example: the term **"coastaal zone"** has the reference (CZMA304-16USC1453).

(EPA-94/04), "Terms of Environment: Glossary, Abbreviations, and Acronyms," EPA175-B-94-015, April 1994.

(ESAx-16USCyyyy):
- ESAx means the Endangered Species Act, section x.
- 16USCyyyy means the Title 16, United States Code (USC), section yyyy.
- For example: the term **"endangered species"** has the (ESA3-16USC1531).

(FIFRAx-7USCyyyy):
- FIFRAx means the Federal Insecticide, fungicide, and Rodenticide Act, section x.
- 7USCyyyy means Title 7, United States Code (USC), section yyyy
- For example: the term **"active ingredient"** has the reference (FIFRA2-7USC136).

(FLPMAxxx-43USCyyyy):
- FLPMAxxx means the Federal Land Policy and Management Act, section xxx.
- 43USCyyyy means Title 43, United States Code (USC), section yyyy.
- For example: the term **"areas of critical environmental concern"** has the refeerence (FLPMA103-43USC1702).

(MMPAx-16USCyyyy):
- MMPAx means Marine Mammal Protection Act, section x.
- 16USCyyyy means Title 16, United States

Code (USC), section yyyy.
- For example: The term **"humane"** has the refernce (MMPA3-16USC1362).

(NCAx-42USCxxxx):

- NCAx means the Noise Control Act, section x.
- 42USCyyyy means Title 42, United States Code (USC), section yyyy.
- For example: The term **"environmental noise"** has the reference (NCA3-42USC4902).

(OPAxxxx):

- OPAxxxx means the Oil Pollution Act of 1990 (Public Law 101-380, August 18, 1990), section xxx.
- For example: The term **"damage"** has the reference (OPA1001).

(OSHAx-29USC651):

- OSHAx means the Occupational Safety and Health Act, section x.
- 29USC651 means Title 29, United States Code (USC), section yyyy.
- For example: The term **"workmen's compensation commission"** has the reference (OSHA3-29USC652).

(PPAxxxx):

- PPAxxxx means the Pollution Prevention Act of 1990, section xxxx.
- For example: The term **"multi-media"** has the reference (PPA6603).

(RCRAxxxx-42USCyyyy):

- RCRAxxxx means the Resource Conservation and Recovery Act, section xxxx.
- 42USCyyyy means Title 42, United States Code (USC), section yyyy.
- For example: The term **"disposal"** has the reference (RCRA1004-42USC6903).

(SDWAxxxx-42USCyyyy):

- SDWAxxxx means the Safe Drinking Water Act, section xxxx.
- 42USCyyyy means Title 42, United States Code (USC), section yyyy.
- For example: The term **"contaminant"** has the reference (SDWA1401-42USC300f).

(SFxxx-42USCyyyy):

- SFxxx means the Superfund, section xxx. Superfund is also known as the Comprehensive Environmental Response, Compensation and Liability Act (CERCLA).
- 42USCyyyy means Title 42, United States Code (USC), section yyyy.
- For example: The term **"damages"** has the reference (SF101-42USC9601).

(SMCRAxxx-30USCyyyy):

- SMCRAxxx means the Surface Mining control and Reclamation Act, section xxx.
- 30USCyyyy means Title 30, United States Code (USC), section yyyy.
- For example: The term **"approximate original contour"** has the reference (SMCRA701-30USC1291).

(TSCAxxx-15USCyyyy):

- TSCAxxx means the Toxic Substances Control Act, section xxx.
- 15USCyyyy means Title 15, United States Code (USC), section yyyy.
- For example: The term **"radon"** has the reference (TSCA302-15USC2662).

(TSCA-AIA1):

- TSCA-AIA1 means the Toxic Substances Control Act, section 1, Asbestos Information Act of 1988 (PL100-577, 10/31/88).
- For example: The term **"asbestos"** has the reference (TSCA-AIA1).

MAJOR ENVIRONMENTAL LAWS AND ENVIRONMENTAL REGULATIONS	
Environmental Laws	Environmental Regulations (CFR Cites)
• EPA's purpose and functions	• Subchapter A--General; 40CFR1-29
• EPA's regulatory authorities	• Subchapter B--Grants and Other Federal Assistance; 40CFR30-47
• Clean Air Act (CAA) of 1970	• Subchapter C--Air Programs; 40CFR50-99
• Federal Water Pollution Control Act (FWPCA) of 1972	• Subchapter D--Water Programs; 40CFR100-140 (oil discharge and prevention related regulations)
• Safe Drinking Water Act (SDWA) of 1974	• Subchapter D--Water Programs; 40CFR141-149 (drinking water regulations)
• Federal Insecticide, Fungicide, and Rodenticide Act (FIFRA) of 1947	• Subchapter E--Pesticide Programs; 40CFR150-189
• Atomic Energy Act (AEA) of 1954	• Subchapter F--Radiation Protection Programs; 40CFR190-192
• Noise Control Act (NCA) 1972	• Subchapter G--Noise Abatement Programs; 40CFR201-211
• Marine Protection, Research, and Sanctuaries Act (MPRSA) of 1972	• Subchapter H--Ocean Dumping; 40CFR220-238
• Resource Conservation Recovery Act (RCRA) of 1976	• Subchapter I--Solid Wastes; 40CFR240-259 (municipal waste, land disposal, and resource recovery regulations)
• Hazardous and Solid Waste Act (HSWA) of 1984	• Subchapter I--Solid Wastes; 40CFR260-299 (hazardous waste)
• Comprehensive Environmental Response, Compensation, and Liability Act of 1980 (CERCLA) • Superfund Amendments and Reauthorization Act (SARA) of 1986	• Subchapter J--Superfund, Emergency Planning, and Community Right-To-Know Programs; 40CFR300-399
• Clean Water Act (CWA) of 1977	• Subchapter N--Effluent Guidelines and Standards; 40CFR400-599
• Motor Vehicle Information and Cost Savings Act (MVICSA) (15USC1901)	• Subchapter P (reserved) • Subchapter Q--Energy Policy; 40CFR600-699
• Toxic Substances Control Act (TSCA) of 1976	• Subchapter R--Toxic Substances Control Act; 40CFR700-799
• National Environmental Policy Act (NEPA) 1969	• Subchapter V--Council on Environmental Quality; 40CFR1500-1517

7 X